Conversion Factors

Mass	1 kg = 1000 g	Power	1 W = 1 J/s
	1 g = 10^{-3} kg		1 hp = 745.7 W
	1 u = 1.66×10^{-27} kg $\leftrightarrow$ 931.5 MeV/c^2	Pressure	1 Pa = 1 N/m^2
	1 slug = 14.6 kg		1 atmosphere = 1 atm = 1.01×10^5 Pa = 14.7 lb/in.^{2}
Length	1 m = 100 cm = 1000 mm		1 millimeter of Hg = 1 mm Hg = 133 Pa
	1 foot = 1 ft = 0.305 m		1 bar = 10^5 Pa
	1 inch = 1 in. = 2.54 cm		1 lb/in.^{2} = 6.89×10^3 Pa
	1 mile = 1 mi = 5280 ft = 1609 m	Speed/velocity	1 m/s = 3.60 km/h
	1 light-year (ly) = 9.46×10^{15} m		1 mi/h = 0.447 m/s
Time	1 min = 60 s		100 mi/h $\approx$ 45 m/s
	1 h = 60 min = 3600 s	Temperature	$T(\text{K}) = T(°\text{C}) + 273.15$
	1 day = 24 h = 8.64×10^4 s		$T(°\text{F}) = \frac{9}{5}T(°\text{C}) + 32$
	1 yr = 365.24 days = 3.16×10^7 s		$T(°\text{C}) = \frac{5}{9}[T(°\text{F}) - 32]$
Force	1 N = 0.225 lb	Angle	180° = π rad
	1 dyne = 10^{-5} N		
	1 lb = 4.45 N = 16 oz		
Energy	1 J = 10^7 erg = 0.239 cal		
	1 cal = 4.186 J		
	1 kW · h = 3.60×10^6 J		
	1 eV = 1.60×10^{-19} J		

Greek Alphabet

Letter	Symbol (uppercase)	Symbol (lowercase)	Letter	Symbol (uppercase)	Symbol (lowercase)
alpha	A	α	nu	N	ν
beta	B	β	xi	Ξ	ξ
gamma	Γ	γ	omicron	O	o
delta	Δ	δ	pi	Π	π
epsilon	E	ε	rho	P	ρ
zeta	Z	ζ	sigma	Σ	σ
eta	E	η	tau	T	τ
theta	Θ	θ	upsilon	Υ	υ
iota	I	ι	phi	Φ	ϕ
kappa	K	κ	chi	X	χ
lambda	Λ	λ	psi	Ψ	ψ
mu	M	μ	omega	Ω	ω

COLLEGE PHYSICS

Reasoning and Relationships

SECOND EDITION

Volume 1

Nicholas J. Giordano

Purdue University

Australia • Brazil • Japan • Korea • Mexico • Singapore • Spain • United Kingdom • United States

College Physics: Reasoning and Relationships, **Second Edition**
Volume 1
Nicholas J. Giordano

Publisher, Physical Sciences: Mary Finch
Publisher, Physics and Astronomy: Charles Hartford
Senior Development Editor: Susan Dust Pashos
Associate Development Editor: Brandi Kirksey
Editorial Assistant: Brendan Killion
Senior Media Editor: Rebecca Berardy Schwartz
Marketing Manager: Jack Cooney
Marketing Coordinator: Julie Stefani
Marketing Communications Manager: Darlene Macanan
Senior Content Project Manager: Cathy Brooks
Senior Art Director: Cate Rickard Barr
Print Buyer: Diane Gibbons
Manufacturing Planner: Douglas Bertke
Rights Acquisition Specialist: Shalice Shah-Caldwell
Production Service: Lachina Publishing Services
Text Designer: Roy Neuhaus
Cover Designer: Roy Neuhaus
Cover Image: Feature—Lights © Stephan Jansen/epa/Corbis
Compositor: Lachina Publishing Services

Library of Congress Control Number: 2011933489
ISBN-13: 978-1-111-57095-8
ISBN-10: 1-111-57095-7

Brooks/Cole
20 Channel Center Street
Boston, MA 02210
USA

Cengage Learning is a leading provider of customized learning solutions with office locations around the globe, including Singapore, the United Kingdom, Australia, Mexico, Brazil, and Japan. Locate your local office at **international.cengage.com/region**

Cengage Learning products are represented in Canada by Nelson Education, Ltd.

For your course and learning solutions, visit **www.cengage.com**

Purchase any of our products at your local college store or at our preferred online store **www.cengagebrain.com**

Printed in the United States of America
1 2 3 4 5 6 7 15 14 13 12 11

Brief Table of Contents

VOLUME 1

Mechanics, Waves, & Thermal Physics

Contents

© Rubberball/Mike Kemp/Getty Images

Contents

© Rubberball/Mike Kemp/Getty Images

CHAPTER 8

CHAPTER 9

CHAPTER 10

The Africa Image Library/Alamy

CHAPTER 11

CHAPTER 12

Waves 383

CHAPTER 13

Sound 412

© John Elk III/Alamy

CHAPTER 14

Temperature and Heat 440

© Travelwide/Alamy

CHAPTER 15

Gases and Kinetic Theory 479

CHAPTER 16

Thermodynamics 506

Preface

Changing the Way Students View Physics

The *relationships* between physics and other areas of science are rapidly becoming stronger and are transforming the way all fields of science are understood and practiced. Examples of this transformation abound, particularly in the life sciences. Many students of college physics are engaged in majors relating to the life sciences, and the manner in which they need and will use physics differs from only a few years ago. For their benefit, and for the benefit of students in virtually all technical and even nontechnical disciplines, textbooks must place a greater emphasis on how to apply the *reasoning* of physics to real-world examples. Such examples come quite naturally from the life sciences, but many everyday objects are filled with good applications of fundamental physics principles as well.

Goals of This Book

Reasoning and Relationships

Students often view physics as merely a collection of loosely related equations. We who teach physics work hard to overcome this perception and help students understand how our subject is part of a broader science context. But what does "understanding" in this context really mean? Many physics textbooks assume understanding will result if a solid problem-solving methodology is introduced early and followed strictly. Students in this model can be viewed as successful if they deal with a representative collection of quantitative problems. However, physics education research has shown that students can succeed in such narrow problem-solving tasks and at the same time have fundamentally flawed notions of the basic principles of physics. For these students, physics *is* simply a collection of equations and facts without a firm connection to the way the world works. Although students do need a solid problem-solving framework, I believe such a framework is only one component to learning physics. For real learning to occur, students must know how to reason and must see the relationships between the ideas of physics and their direct experiences. Until the reasoning is sound and the relationships are clear, fundamental learning will remain elusive.

The central theme of this book is to weave reasoning and relationships into the way we teach introductory physics. Three important results of this approach are the following:

1. Early in the text, common student misconceptions about physics and how to study are addressed.
2. Reasoning and relationships examples and problems coach students on how to reason correctly.
3. A systematic approach to problem solving provides the model for how students should approach quantitative problems.

1. Addressing Student Misconceptions. Many students come to their college physics course with a common set of pre-Newtonian misconceptions about physics. I believe the best way to help students overcome these misconceptions is to address them directly and help students see where and how their pre-Newtonian ideas fail. Accordingly, *College Physics: Reasoning and Relationships* devotes Chapter 2 to a qualitative and conceptual discussion of Newton's laws of motion and what they tell us about the relationship between forces and motion. The goal is to arm students with an understanding of this relationship to address many of their pre-Newtonian misconceptions and prepare for the discussion of the application of those laws in Chapter 3 and beyond. Armed with an understanding of the proper relationship between kinematics and forces, students can then reason about a variety of problems in mechanics, providing instructors with the flexibility to introduce a wider variety of problems much sooner in the course.

This approach also presents students with a model for how to study for a course in the sciences. Too often, students believe that a focus on "formulas" and their application to a narrow set of drill problems results in understanding of the material. The qualitative overview in Chapter 2 models for students the keys to the course: Begin with an understanding of key physics principles, identify the *relationships* between those principles, and apply *reason* to the selection of equations needed to solve quantitative problems, as well as analyze the solution itself.

2.5 Why Did It Take Newton to Discover Newton's Laws?

Newton's second law tells us that the acceleration of an object is given by $\vec{a} = (\sum \vec{F})/m$, where $\sum \vec{F}$ is the total force acting on the object. In the simplest situations, there may only be one or two forces acting on an object, but in some cases there may be a very large number of forces acting on an object. Multiple forces can make things appear to be very complicated, which is perhaps why the correct laws of motion—Newton's laws—were not discovered sooner.

Forces on a Swimming Bacterium

Figure 2.34 shows a photo of the single-celled bacterium *Escherichia coli*, usually referred to as *E. coli*. An individual *E. coli* propels itself by moving thin strands of protein that extend away from its body (rather like a tail) called flagella. Most *E. coli* possess several flagella as in the photo in Figure 2.34A, but to understand their function, we consider a single flagellum as sketched in Figure 2.34B. A flagellum is fairly rigid, and because it has a spiral shape, one can think of it as a small propeller. An *E. coli* bacterium moves about by rotating this propeller, thereby exerting a force $\vec{F}_w$ on the nearby water. According to Newton's third law, the water exerts a force $\vec{F}_E$ of equal magnitude and opposite direction on the *E. coli* as sketched in Figure 2.34B. One might be tempted to apply Newton's second law with the force $\vec{F}_E$ and conclude that the *E. coli* will move with an acceleration that is proportional to this force. However, this conclusion is incorrect because we have not included the forces from the water on the body of the *E. coli*. These forces are also indicated in Figure 2.34B; to properly describe the total force from the water, we must draw in many force vectors, pushing the *E. coli* in virtually all directions. At the molecular level, we can understand these forces as follows. Water is composed of molecules that are in constant motion and bombard the *E. coli* from all sides. Each time a water molecule collides with the *E. coli*, the molecule exerts a force on the bacterium, much like the collision of the baseball and bat in Figure 2.32. As we saw in that case, the two colliding objects both experience a recoil force, another example of action–reaction forces; hence, the *E. coli* and the water molecule exert forces on each other. An individual *E. coli* is not very large, but a water molecule is much smaller than the bacterium, and the force from one such collision will have only a small effect on the

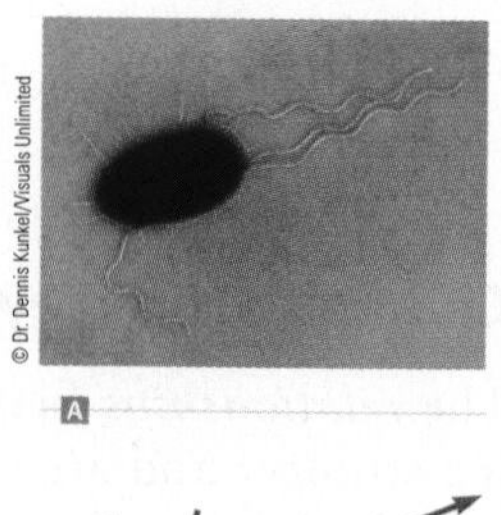

A

$\vec{F}_w$ $\vec{F}_E$ Flagellum

B

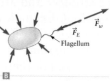

▲ **Figure 2.34** A *E. coli* use the action–reaction principle to propel themselves. An individual *E. coli* bacterium is a few micrometers in diameter (1 μm = 0.000001 m). B The flagellum exerts a force $\vec{F}_w$ on the water, and the water exerts a force $\vec{F}_E = -\vec{F}_w$ on the *E. coli*. There are also forces from the water molecules (indicated by the other red arrows) that act on the cell body.

From Chapter 2, page 45

2. Reasoning and Relationships Problems. Many interesting real-world physics problems cannot be solved exactly with the mathematics appropriate for a college physics course, but they can often be handled in an approximate way using the simple methods (based on algebra and trigonometry) developed in such a course. Professional physicists are familiar with these back of-the-envelope calculations. For instance, we may want to know the approximate force on

PROBLEM SOLVING Dealing with Reasoning and Relationships Problems

1. **RECOGNIZE THE PRINCIPLE.** Determine the key physics ideas that are central to the problem and that connect the quantity you want to calculate with the quantities you know. In the examples found in this section, this physics involves motion with constant acceleration.
2. **SKETCH THE PROBLEM.** Make a drawing that shows all the given information and everything else you know about the problem. For problems in mechanics, your drawing should include all the forces, velocities, and so forth.
3. **IDENTIFY THE RELATIONSHIPS.** Identify the important physics relationships; for problems concerning motion with constant acceleration, they are the relationships between position, velocity, and acceleration in Table 3.1. For many reasoning and relationships problems, values for some of the essential unknown quantities may not be given. You must then use common sense to make reasonable estimates for these quantities. Don't worry or spend time trying to obtain precise values of every quantity (such as the amount that the child's knees flex in Fig. 3.24). Accuracy to within a factor of 3 or even 10 is usually fine because the goal is to calculate the quantity of interest to within an order of magnitude (a factor of 10). Use the Internet, the library, and (especially) your own intuition and experiences.
4. **SOLVE.** Since an exact mathematical solution is not required, cast the problem into one that is easy to solve mathematically. In the examples in this section, we were able to use the results for motion with constant acceleration.
5. Always *consider what your answer means* and check that it makes sense.

As is often the case, practice is a very useful teacher.

From Chapter 3, page 78

a skydiver's knees when she hits the ground. The precise value of the force depends on the details of the landing, but we are often interested in only an approximate (usually order-of-magnitude) answer. We call such problems ***reasoning and relationships*** problems because solving them requires us to identify the key physics relationships and quantities needed for the problem and that we also estimate values of some important quantities (such as the mass of the skydiver and how she flexes her knees) based on experience and common sense.

From Chapter 3, page 78

EXAMPLE 3.7 Ⓡ Cars and Bumpers and Walls

Consider a car of mass 1000 kg colliding with a rigid concrete wall at a speed of 2.5 m/s (about 5 mi/h). This impact is a fairly low-speed collision, and the bumpers on a modern car should be able to handle it without significant damage to the car. Estimate the force exerted by the wall on the car's bumper.

RECOGNIZE THE PRINCIPLE

The motion we want to analyze starts when the car's bumper first touches the wall and ends when the car is stopped. To treat the problem approximately, we *assume* the force on the bumper is constant during the collision period, so the acceleration is also constant. We can then use our expressions from Table 3.1 to analyze the motion. Our strategy is to first find the car's acceleration and then use it to calculate the associated force exerted by the wall on the car from Newton's second law.

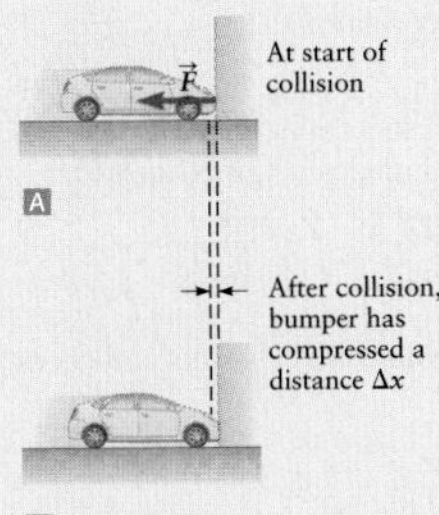

▲ **Figure 3.25** Example 3.7. A When a car collides with a wall, the wall exerts a force F on the bumper. This force provides the acceleration that stops the car. B During the collision and before the car comes to rest, the bumper deforms by an amount Δx. The car travels this distance while it comes to a complete stop.

SKETCH THE PROBLEM

Figure 3.25 shows a sketch of the car along with the force exerted by the wall on the car. There are also two vertical forces—the force of gravity on the car and the normal force exerted by the road on the car—but we have not shown them because we are concerned here with the car's horizontal (x) motion, which we can treat using $\sum F = ma$ for the components of force and acceleration along x.

IDENTIFY THE RELATIONSHIPS

To find the car's acceleration, we need to estimate either the stopping time or the distance Δx traveled during this time. Let's take the latter approach. We are given the initial velocity ($v_0 = 2.5$ m/s) and the final velocity ($v = 0$). Both of these quantities are in Equation 3.4:

$$v^2 = v_0^2 + 2a(x - x_0) \quad (1)$$

From ENHANCED WebAssign *Enhanced WebAssign*

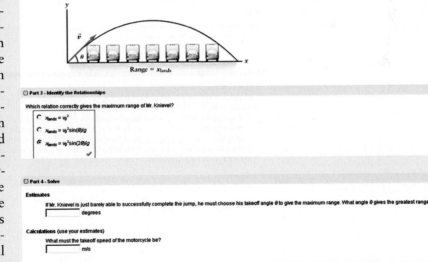

⊟ Reasoning Problem 4.81

Although Evel Knievel never succeeded in jumping over the Grand Canyon, he was famous for jumping (with the help of his motorcycle) over, among other things, 12 large trucks. This jump covered an approximate distance of 30 m. What was Mr. Knievel's minimum initial velocity for this jump? Ignore air drag.

⊟ Part 1 - Recognize the Principles

You need to calculate the *range* of Mr. Knievel's motorcycle. This is the distance between his location at takeoff and landing. Which of the following principles can be used to find the range of a projectile?

- If the total force acting on an object is zero its acceleration is zero.
- The acceleration due to gravity, *g*, varies slightly from place to place on Earth.
- The range of a projectile launched from ground level is determined by its initial velocity.
- At terminal velocity there is no further acceleration.

⊟ Part 2 - Sketch the Problem

This sketch shows the problem. The range (the value of x_{lands}) depends on the initial speed v_0 (the magnitude of $\vec{v}$) and the angle θ.

y
$\vec{v}$
θ
x
Range = x_{lands}

⊟ Part 3 - Identify the Relationships

Which relation correctly gives the maximum range of Mr. Knievel?

- $x_{lands} = v_0^2$
- $x_{lands} = v_0^2\sin(\theta)/g$
- $x_{lands} = v_0^2\sin(2\theta)/g$ ✔

⊟ Part 4 - Solve

Estimates

If Mr. Knievel is just barely able to successfully complete the jump, he must choose his takeoff angle θ to give the maximum range. What angle θ gives the greatest range?

[ ] degrees

Calculations (use your estimates)

What must the takeoff speed of the motorcycle be?

[ ] m/s

⊟ Part 5 - What Does It Mean?

Evel Knievel's actual jump covered about 135 feet (about 41 m). A minimum takeoff speed of 20 m/s would have allowed him to successfully cover the required distance. 20 m/s is equivalent to about 45 mph, which is certainly a realistic value for a motorcycle to achieve. In fact, his motorcycle probably had a much greater initial speed. Our calculation has ignored the effect of air drag - in reality, air drag would have been very important in this case and we would need to include it to get an accurate value of Mr. Knievel's initial speed. One more point—our calculation did not account for the fact that to be successful Mr. Knievel must clear the front corner of the first truck and the back corner of the last truck. This means that his range would have to be a bit larger than simply the distance covered by the trucks.

Reasoning and relationships problems provide physical insight and a chance to practice critical thinking (reasoning), and they can help students see more clearly the fundamental principles associated with a problem. A truly unique feature of this book is the inclusion of these problems in both the worked examples and the end-of-chapter problems, and in Enhanced WebAssign. The reasoning and relationships problems in Enhanced WebAssign are assignable, offer detailed conditional feedback, and provide students with the opportunity to truly internalize the problem-solving process. These detailed tutorials take students step-by-step through the solution. The steps include useful hints to guide student thinking and extensive feedback to assist when student responses show a lack of understanding.

In order to ensure that the structure of the problems and the feedback offered is as specific to my approach as possible, I personally adapted those problems for Enhanced WebAssign. I authored all of the feedback in the tutorials and designed them to have the same "feel" as if I were in the room helping the student reason through the problem.

3. A Systematic Approach to Problem Solving. Although reasoning and relationships problems are used to help students develop a broad understanding of physics, this book also contains a strong component of traditional quantitative problem solving. Quantitative problems are a component of virtually all college physics courses, and students can benefit by developing a systematic approach to such problems. *College Physics: Reasoning and Relationships* therefore places extra emphasis on step-by-step approaches students can use in a wide range of situations. This approach can be seen in the worked examples, which use a five step solution process: (1) recognize the physics principles central to the problem, (2) draw a sketch showing the problem and all the given information, (3) identify the relationships between the known and unknown quantities, (4) solve for the desired quantity, and (5) ask what the answer means and if it makes sense. Explicit problem-solving strategies are also given for major classes of quantitative problems, such as applying the principle of conservation of mechanical energy.

PROBLEM SOLVING Applying the Principle of Conservation of Energy to Problems of Rotational Motion

1. **RECOGNIZE THE PRINCIPLE.** The mechanical energy of an object is conserved only if all the forces that do work on the object are conservative forces. Only then can we apply the conservation of mechanical energy condition in Equation 9.10.
2. **SKETCH THE PROBLEM.** Always use a sketch to collect your information concerning the initial and final states of the system.
3. **IDENTIFY THE RELATIONSHIPS.** Find the initial and final kinetic and potential energies of the object. They will usually involve the initial and final linear and angular velocities, and the initial and final heights, as for the rolling ball in Figure 9.7. You will usually also need to find the moment of inertia.
4. **SOLVE** for the unknown quantities using the conservation of energy condition,
$$KE_i + PE_i = KE_f + PE_f$$
The kinetic energy terms must account for both the translational and the rotational kinetic energies (Eq. 9.5):
$$KE_{\text{total}} = \underbrace{\tfrac{1}{2}mv_{CM}^2}_{\text{translational } KE} + \underbrace{\tfrac{1}{2}I\omega^2}_{\text{rotational } KE}$$
5. Always *consider what your answer means* and check that it makes sense.

From Chapter 9, page 289

EXAMPLE 13.3 The Lowest Note of a Clarinet

A clarinet is a musical instrument we can model as a pipe open at one end and closed at the other (Fig. 13.10A). A standard clarinet is approximately $L = 66$ cm long. What is the frequency of the lowest tone such a clarinet can produce?

RECOGNIZE THE PRINCIPLE

According to our work with pipes, there must be a pressure antinode at the closed end of the clarinet and a pressure node at the open end. Frequency and wavelength are related by $v_{\text{sound}} = f\lambda$, so the standing wave with the lowest frequency will be the one with the longest wavelength.

SKETCH THE PROBLEM

Figure 13.10B shows the clarinet modeled as a pipe open at one end. The allowed standing waves will be just like those found in Figures 13.7 and 13.8.

IDENTIFY THE RELATIONSHIPS

The standing wave with the longest wavelength is sketched in Figure 13.7. Its frequency is given by Equation 13.14, with $n = 1$:

$$f_1 = \frac{v_{\text{sound}}}{4L}$$

This is the fundamental frequency for the clarinet with all tone holes closed.

SOLVE

The speed of sound in air (at room temperature) is $v_{\text{sound}} = 343$ m/s and the length of the tube is $L = 0.66$ m, so we have

$$f_1 = \frac{v_{\text{sound}}}{4L} = \frac{343 \text{ m/s}}{4(0.66 \text{ m})} = \boxed{130 \text{ Hz}}$$

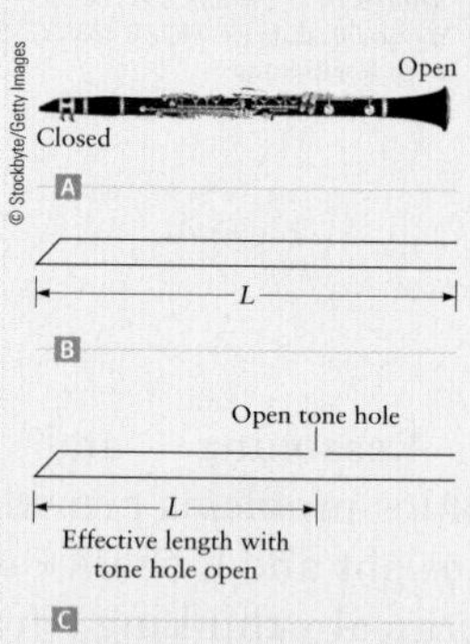

▲ **Figure 13.10** Example 13.3. **A** A clarinet. **B** Physicist's model of a clarinet, as a pipe closed at one end and open at the other. The standing waves will be the same as shown in Figures 13.7 and 13.8. **C** If a tone hole is opened, it will act as the open end of the pipe as far as the standing wave is concerned, changing the frequency of the standing waves.

► ***What does it mean?***

This result is the lowest frequency—that is, the lowest "note"—a clarinet can produce according to our simple model. To play other notes, the musician uncovers one of the many different tone holes along the body of the clarinet. If an open hole is sufficiently large, the pressure node moves from the end of the clarinet to a spot very near the tone hole (Fig. 13.10C), thereby shortening the wavelength of the fundamental standing wave and increasing f_1. In this way, a musician can use the many tone holes to produce many different notes.

From Chapter 13, page 423

Changes in the Second Edition

The entire text was reviewed and many of the derivations rewritten to improve clarity and strengthen the connection to basic principles. As an example, the discussion of electric field lines (Chapter 17) was rewritten to better emphasize how they represent the electric field. Many electric circuit problems were added (Chapter 19) to give students more practice in the *qualitative* analysis of circuits and a better intuitive understanding of Kirchhoff's rules. The section on cameras (Chapter 26) was completely reworked to emphasize digital cameras, how they work, and how they can be so tiny. Good graphical representations are important, especially in problems that are inherently three-dimensional, and we improved the way vectors are represented in three-dimensional drawings. This is particularly important in discussions of magnetism (Chapter 20), the propagation of electromagnetic waves (Chapter 23), and rotational motion (Chapter 9).

A new worked example and a new Concept Check were added to nearly every chapter, giving students additional problem-solving exemplars. All of the end-of-chapter questions and problems were systematically reviewed and edited for clarity, and new reasoning and relationships problems (typically three per chapter) were added.

Chapter	New Worked Examples	New Concept Checks
1	—	1.3 Getting the Units Right
2	2.2 Average and Instantaneous Velocity of a Car	2.5 Finding the Velocity
3	3.2 Car Chase	3.5 Normal Forces
4	4.10 Sliding up a Hill	4.2 Falling Balls
5	5.11 A Double Star System	5.3 Net Force and Circular Motion
6	6.2 Kinetic Energy of an Asteroid	6.5 A 1000-Calorie Hill
7	7.3 Impulse and Playing Golf	7.2 Momentum and Kinetic Energy of Two Particles
		7.9 Explosions and the Center of Mass
8	8.9 Which Stick Falls Faster?	8.5 Carrying Your Share of the Weight
9	9.4 Rolling up a Hill	9.5 Rolling, Rolling, Rolling
10	10.9 How Can a Submarine Float?	10.4 Floating in a Lake
11	11.5 Energy of a Simple Harmonic Oscillator	11.2 Analyzing a Simple Harmonic Oscillator
		11.5 Amplitude and Energy of a Simple Harmonic Oscillator
12	12.7 Bending a Note on a Guitar	12.1 Waves on a String
13	13.5 Two Organ Pipes	13.6 Using Beats to Tune a Guitar
14	14.4 Heating a Bucket of Water	14.4 Heating Your Piano
15	15.5 Physics of a Weather Balloon	15.1 Mass of a Water Molecule
16	16.9 A Magnetic Heat Engine	16.8 Heat Engine or Refrigerator?
17	17.8 Electric Field Between Two Sheets of Charge	17.4 Electric Forces and Neutral Objects
18	18.5 Measuring the Charge on an Electron	18.7 Forces on a Capacitor
19	19.9 An *RC* Circuit in Your Car	19.6 A Resistor Puzzle

Chapter	New Worked Examples	New Concept Checks
20	20.2 Magnetic Force on an Electron 20.5 Magnetic Force on a Wire	20.9 Forces on a Current Loop
21	21.4 Moving Magnets and Induced Currents	21.7 Energy in an *LR* Circuit
22	22.5 Analyzing an *LR* Circuit	22.2 AC Power in the U.S.
23	23.5 Noise on Your Radio	23.7 Polarization of a Wave
24	24.9 Designing the Lenses for a Pair of Glasses	24.2 Using Snell's Law 24.10 Ray Tracing With a Diverging Lens
25	25.7 Rayleigh Criterion for a Marksman	25.4 Color of a *Very* Thin Soap Film
26	26.5 Galileo's Telescope	26.4 Making a Better Microscope
27	27.7 A Remarkable Proton	27.6 The Fastest Proton Ever Known
28	28.4 Photons From Your Cell Phone	28.5 Wavelengths of Electrons and Protons
29	29.6 Filling the 4*f* Subshell	29.4 Electron Configuration of an Ion
30	30.10 Application of a Radioactive Tracer	30.5 Making a New Element
31	31.5 Annihilation of a Positron	31.3 Conserving Lepton Number

Additional Features of this Book

Diagrams with Additional Explanatory Labeling

Every college physics textbook contains line art with labeling. This book adds another layer of labeling that explains the phenomenon being illustrated, much as an instructor would explain a process or relationship in class. This additional labeling is set off in a different style.

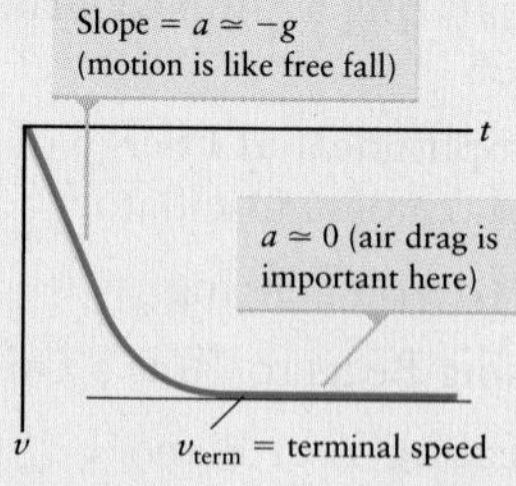

▲ **Figure 3.27** The skydiver's motion is initially like free fall; compare with Figure 3.16A. Eventually, however, air drag becomes as large as the force of gravity and the skydiver reaches her terminal speed v_{term}.

From Chapter 3, page 80

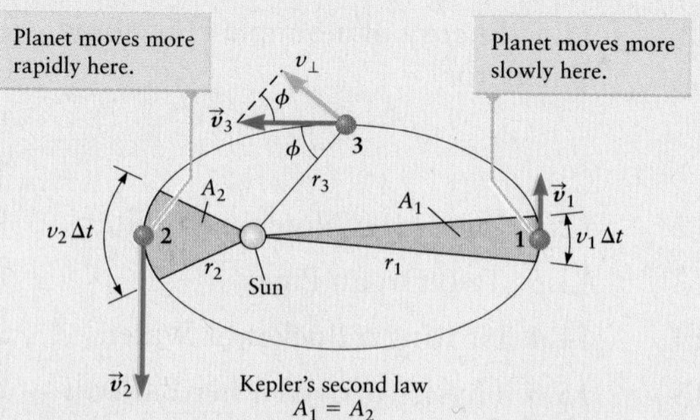

▲ **Figure 9.15** When a planet follows an elliptical orbit, the speed is greatest when the planet is closest to the Sun. According to Kepler's second law, a planet sweeps out equal areas in equal time intervals as it moves about the Sun. Hence, $A_1 = A_2$.

From Chapter 9, page 297

Concept Checks

At various points in the chapter are conceptual questions, called "Concept Checks." These questions are designed to make the student reflect on a fundamental issue. They may involve interpreting the content of a graph or drawing a new graph to predict a relationship between quantities. Many Concept Check questions are in multiple-choice format to facilitate their use in audience response systems. Answers to Concept Checks are given at the end of the book. Full explanations of each answer are given in the *Instructor's Solutions Manual.*

CONCEPT CHECK 6.6 Spring Forces and Newton's Third Law

Figure 6.27 shows two identical springs. In both cases, a person exerts a force of magnitude F on the right end of the spring. The left end of the spring in Figure 6.27A is attached to a wall, while the left end of the spring in Figure 6.27B is held by another person, who exerts a force of magnitude F to the left. Which statement is true?

(1) The spring in Figure 6.27A is stretched half as much as the spring in Figure 6.27B.
(2) The spring in Figure 6.27A is stretched twice as much as the spring in Figure 6.27B.
(3) The two springs are stretched the same amount.

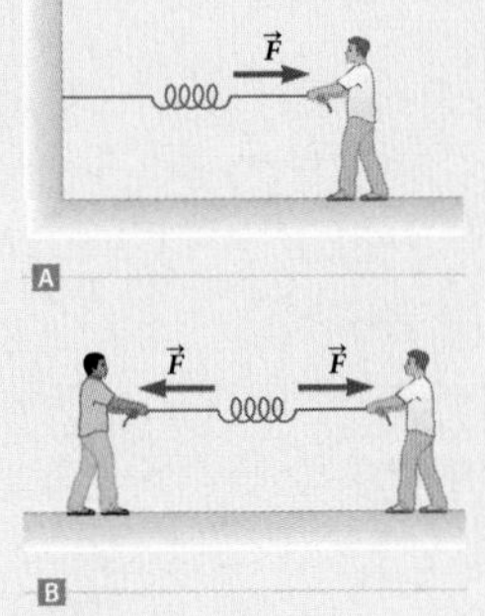

▲ **Figure 6.27** Concept Check 6.6.

From Chapter 6, page 189

Insights

Each chapter contains several special marginal comments called "Insights" that add greater depth to a key idea or reinforce an important message. For instance, Insight 3.3 emphasizes the distinction between weight and mass, and Insight 16.1 explains why diesel engines are inherently more efficient than conventional gasoline internal combustion engines.

Insight 16.1
EFFICIENCY OF A DIESEL ENGINE
A diesel engine is similar to a gasoline internal combustion engine (Example 16.8). One difference is that a gasoline engine ignites the fuel mixture with a spark from a spark plug, whereas a diesel engine ignites the fuel mixture purely "by compression" (without a spark). The compression of the fuel mixture is therefore much greater in a diesel engine, which leads to a higher temperature in the hot reservoir. According to Equation 16.21 and Example 16.8, this higher temperature gives a higher theoretical limit on the efficiency of a diesel engine.

From Chapter 16, page 526

Chapter Summaries

Each concept is described in its own panel, often with an explanatory diagram. This format helps students organize information for review and further study. Concepts are classified into two major groups: (1) *Key Concepts and Principles* and (2) *Applications.*

From Chapter 5, page 158

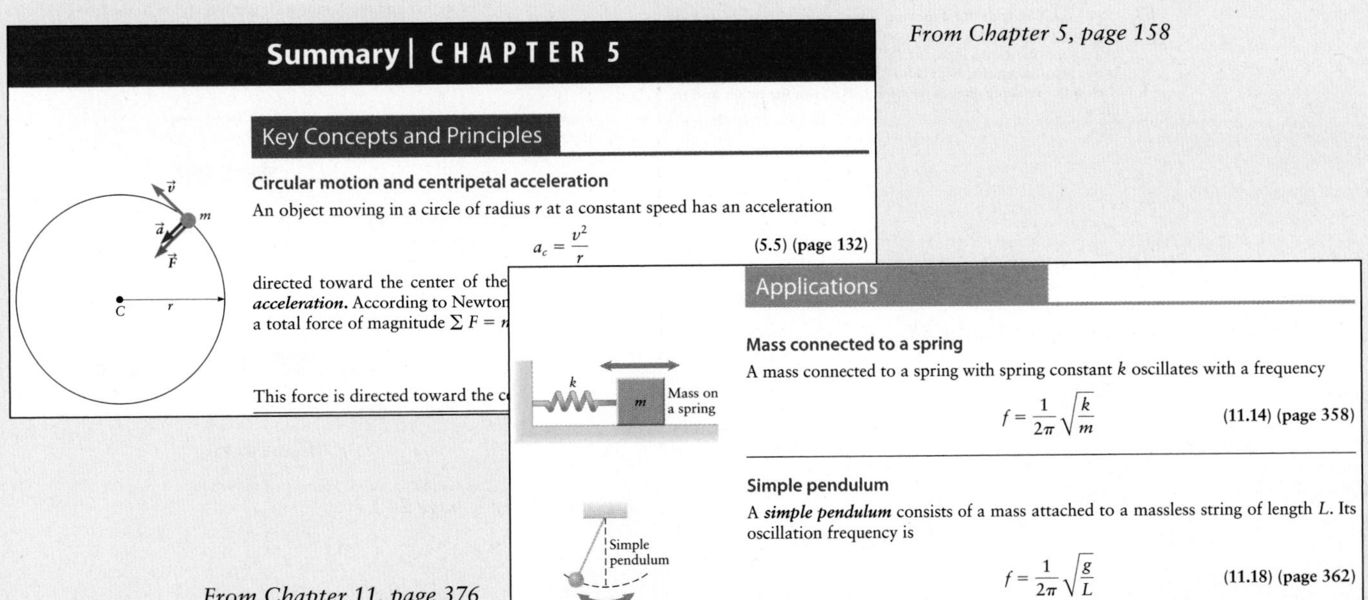

Summary | CHAPTER 5

Key Concepts and Principles

Circular motion and centripetal acceleration
An object moving in a circle of radius r at a constant speed has an acceleration

$$a_c = \frac{v^2}{r} \quad (5.5)\ (\text{page } 132)$$

directed toward the center of the
acceleration. According to Newto
a total force of magnitude $\sum F = n$

This force is directed toward the c

Applications

Mass connected to a spring
A mass connected to a spring with spring constant k oscillates with a frequency

$$f = \frac{1}{2\pi}\sqrt{\frac{k}{m}} \quad (11.14)\ (\text{page } 358)$$

Simple pendulum
A ***simple pendulum*** consists of a mass attached to a massless string of length L. Its oscillation frequency is

$$f = \frac{1}{2\pi}\sqrt{\frac{g}{L}} \quad (11.18)\ (\text{page } 362)$$

From Chapter 11, page 376

End-of-Chapter Questions

Approximately 20 questions at the end of each chapter ask students to reflect on and strengthen their understanding of conceptual issues. These questions are suitable for use in recitation sessions or other group work. Answers to questions designated SSM are provided in the *Student Companion & Problem-Solving Guide*, and all questions are answered in the *Instructor's Solutions Manual.*

22. The author once had a small car with a "hatchback," a rear door that opens upward. One day, he forgot to latch it before driving on the highway, and he found that at high speeds the rear door lifted up as in Figure Q10.22B. Explain why.

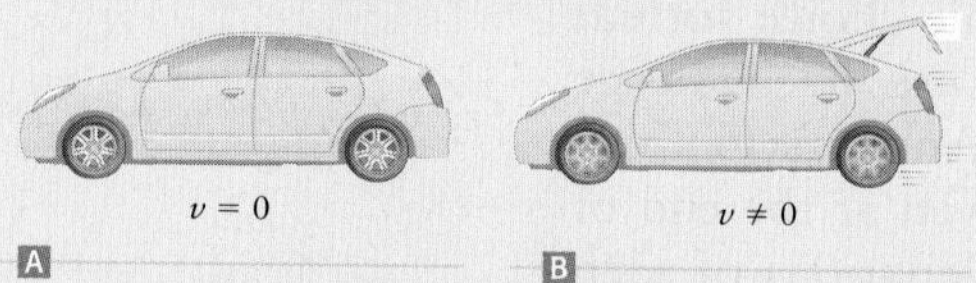

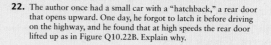

Figure Q10.22

From Chapter 10, page 347

End-of-Chapter Problems

Homework problems are designed to match the examples that are worked throughout each chapter. Most of these problems are grouped according to the matching chapter section. A final list of "Additional Problems" contains problems that bring together ideas from across the chapter or from multiple chapters. Unmarked problems are straightforward, and intermediate and challenging problems are indicated. Problems of special interest to life science students, reasoning and relationship problems R, problems for which a reasoning tutorial exists in Enhanced WebAssign RT, and problems whose solutions appear in the *Student Companion & Problem-Solving Guide* SSM are so indicated. Answers to odd-numbered problems appear at the end of the book.

63. SSM You work for a moving company and are given the job of pulling two large boxes of mass $m_1 = 120$ kg and $m_2 = 290$ kg using ropes as shown in Figure P3.63. You pull very hard, and the boxes are accelerating with $a = 0.22$ m/s². What is the tension in each rope? Assume there is no friction between the boxes and the floor.

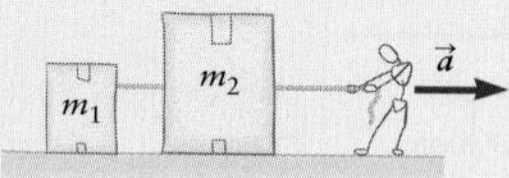

Figure P3.63

From Chapter 3, page 88

74. RT A man ($m_1 = 90$ kg) is standing on a railroad flat car that is 9.0 m long and of unknown mass m_2. The man and the car are initially at rest on a level track (Fig. P7.74A), and the wheels of the car are frictionless. The man then runs to the opposite end of the car and the car moves a distance 1.5 m to the left (Fig. P7.74B). Estimate the mass of the car.

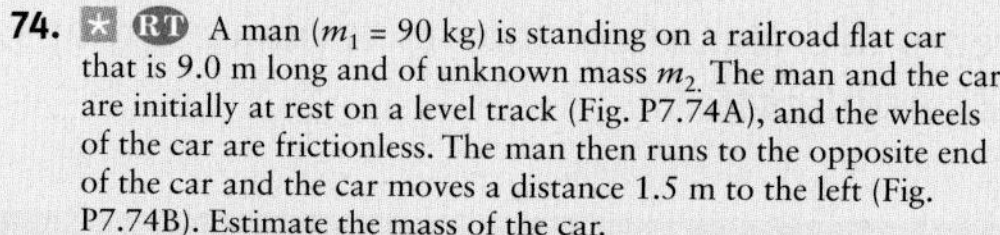

Figure P7.74

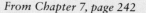

From Chapter 7, page 242

Ancillaries

Using Technology to Enhance Learning

ENHANCED WebAssign Enhanced WebAssign for Giordano's *College Physics: Reasoning and Relationships*, Second Edition

Exclusively from Cengage Learning, Enhanced WebAssign® offers an extensive online program for physics to encourage the practice that's so critical for concept mastery. The meticulously crafted pedagogy and exercises in our proven texts become even more effective in Enhanced WebAssign. Enhanced WebAssign includes the Cengage YouBook, a highly customizable, interactive eBook.

WebAssign includes:

- All of the quantitative end-of-chapter problems
- Reasoning and Relationships tutorials with feedback to address common misconceptions
- Quick Prep: 67 algebra and trigonometry math remediation tutorials
- Concept Checks
- Active Figure animations with assignable questions
- PHET simulations
- The Cengage YouBook

WebAssign has a customizable and interactive eBook, the Cengage YouBook, that lets you tailor the textbook to fit your course and connect with your students. You can remove and rearrange chapters in the table of contents and tailor assigned readings that match your syllabus exactly. Powerful editing tools let you change as much as you'd like—or leave it just like it is. You can highlight key passages or add sticky notes to pages to comment on a concept in the reading, and then share any of these individual notes and highlights with your students, or keep them personal. You can also edit narrative content in the textbook by adding a text box or striking out text. With a handy link tool, you can drop in an icon at any point in the eBook that lets you link to your own lecture notes, audio summaries, video lectures, or other files on a personal website or anywhere on the web. A simple YouTube widget lets you easily find and embed videos from YouTube directly into eBook pages. There is a light discussion board that lets students and instructors find others in their class and start a chat session. The Cengage YouBook helps students go beyond just reading the textbook. Students can also highlight the text, add their own notes, and bookmark the text. Animations play right on the page at the point of learning so that they're not speed bumps to reading but true enhancements.

Please visit **www.webassign.net/brookscole** to view an interactive demonstration of Enhanced WebAssign.

PowerLecture with JoinIn and ExamView for Giordano's *College Physics: Reasoning and Relationships*, Second Edition

PowerLecture™

This DVD provides the instructor with dynamic media tools for teaching. Create, deliver, and customize tests (both print and online) in minutes with ExamView® Computerized Testing. The *Instructor's Solutions Manual*, which includes all solutions to end-of-chapter material, is available as Microsoft® Word documents on the PowerLecture DVD. JoinIn™ "clicker" content helps you turn your lectures into an interactive learning environment that promotes conceptual understanding. Available exclusively for higher education from our partnership with Turning Technologies, JoinIn is the easiest way to turn your lecture into a personal, fully interactive experience for your students! By using JoinIn, you can pose text-specific questions to a large group, gather results, and display students' answers seamlessly. To further facilitate an engaging lecture environment,

Microsoft® PowerPoint® lecture slides and figures from the book are also included on this DVD, as well as animations and movies.

Additional Instructor Resources

The *Instructor's Solutions Manual* was authored by Michael Meyer (Michigan Technological University) and James Heath (Austin Community College). For the second edition, Meyer and Heath have reviewed all of the solutions and provided additional support for many of the solutions.

A *Test Bank* created by Ed Oberhofer (University of North Carolina at Charlotte and Lake-Sumter Community College) is available on the *PowerLecture*™ CDROM as editable electronic files or via the ExamView® test software. The file contains questions in multiple-choice format for all chapters of the text. Instructors may print and duplicate pages for distribution to students.

Student Resources

Student Companion & Problem-Solving Guide by Richard Grant (Roanoke College) will prove to be an essential study resource. For each chapter, it contains a summary of problem-solving techniques (following the text's methodology), a list of frequently-asked questions students often have when attempting homework assignments, selected solutions to end-of-chapter problems, solved Capstone Problems representing typical exam questions, and a set of MCAT review questions with explanations of strategies behind the answers.

Physics Laboratory Manual, Third Edition by David Loyd (Angelo State University) supplements the learning of basic physical principles while introducing laboratory procedures and equipment. Each chapter includes a prelaboratory assignment, objectives, an equipment list, the theory behind the experiment, experimental procedures, graphing exercises, and questions. A laboratory report form is included with each experiment so that the student can record data, calculations, and experimental results. Students are encouraged to apply statistical analysis to their data. A complete *Instructor's Manual* is also available to facilitate use of this lab manual.

Physics Laboratory Experiments, Seventh Edition by Jerry D. Wilson (Lander College) and Cecilia A. Hernández (American River College). This market-leading manual for the first-year physics laboratory course offers a wide range of class-tested experiments designed specifically for use in small to midsize lab programs. A series of integrated experiments emphasizes the use of computerized instrumentation and includes a set of "computer-assisted experiments" to allow students and instructors to gain experience with modern equipment. This option also enables instructors to determine the appropriate balance between traditional and computer-based experiments for their courses. By analyzing data through two different methods, students gain a greater understanding of the concepts behind the experiments. The Seventh Edition is updated with the latest information and techniques involving state-of-the-art equipment and a new Guided Learning feature addresses the growing interest in guided-inquiry pedagogy. Fourteen additional experiments are also available through custom printing.

Acknowledgments

Creating a textbook is an enormous job requiring the assistance of many people. To all these people, I extend my sincere thanks.

Contributors

Raymond Hall of California State University, Fresno, and Richard Grant of Roanoke College contributed many interesting and creative end-of-chapter questions and problems.

Accuracy Reviewers

David Aaron, *South Dakota State University*
David Cole, *Northern Arizona University*
Andrew Cornelius, *University of Nevada—Las Vegas*
Shelly Lesher, *University of Wisconsin—La Crosse*
Mark Lucas, *Ohio University*
Michel Pleimling, *Virginia Tech*
Mahdi Sanati, *Texas Tech University*
Michael Strauss, *University of Oklahoma*
Richard Vallery, *Grand Valley State University*
David Young, *Louisiana State University*

Pre-Revision Reviewers

The following individuals provided helpful comments on the first edition and valuable advice in preparing the revision.

David Aaron, *South Dakota State University*
Gerald Cleaver, *Baylor University*
Christopher Coffin, *Oregon State University*
Alejandro Garcia, *University of Washington*
Eitan Gross, *University of Arkansas*
James Heath, *Austin Community College*
Leo Piilonen, *Virginia Tech*
David Young, *Louisiana State University*

The following individuals provided useful and detailed feedback toward the revision of end-of-chapter questions and problems.

Kevin Haglin, *St. Cloud State University*
Marko Horbatsch, *York University*
Sylvio May, *North Dakota State University*
Michael Meyer, *Michigan Technological University*
David Sokoloff, *University of Oregon*

First Edition Manuscript Reviewers

Jeffrey Adams, *Montana State University*
Anthony Aguirre, *University of California—Santa Cruz*
David Balogh, *Fresno City College*
David Bannon, *Oregon State University*
Phil Baringer, *University of Kansas*
Natalie Batalha, *San Jose State University*
Mark Blachly, *Arsenal Technical High School*
Gary Blanpied, *University of South Carolina*
Ken Bolland, *The Ohio State University*
Scott Bonham, *Western Kentucky University*
Marc Caffee, *Purdue University*
Lee Chow, *University of Central Florida*
Song Chung, *William Patterson University*
Alice Churukian, *Concordia College*
Thomas Colbert, *Augusta State University*
David Cole, *Northern Arizona University*
Sergio Conetti, *University of Virginia*
Gary Copeland, *Old Dominion University*
Doug Copely, *Sacramento City College*
Robert Corey, *South Dakota School of Mines & Technology*
Andrew Cornelius, *University of Nevada—Las Vegas*
Nimbus Couzin, *Indiana University, Southeast*
Carl Covatto, *Arizona State University*
Thomas Cravens, *University of Kansas*
Sridhara Dasu, *University of Wisconsin—Madison*
Timir Datta, *University of South Carolina*
Susan DiFranzo, *Hudson Valley Community College*
David Donnelly, *Texas State University*
Sandra Doty, *The Ohio State University*
Steve Ellis, *University of Kentucky*
Len Finegold, *Drexel University*
Carl Fredrickson, *University of Central Arkansas*
Joe Gallant, *Kent State University, Warren Campus*
Kent Gee, *Brigham Young University*
Bernard Gerstman, *Florida International University*
James Goff, *Pima Community College*
Richard Grant, *Roanoke College*
William Gregg, *Louisiana State University*
James Guinn, *Georgia Perimeter College, Clarkston*
Richard Heinz, *Indiana University—Bloomington*

John Hopkins, *Pennsylvania State University*
Karim Hossain, *Edinboro University of Pennsylvania*
Linda Jones, *College of Charleston*
Alex Kamenev, *University of Minnesota*
Daniel Kennefick, *University of Arkansas*
Aslam Khalil, *Portland State University*
Jeremy King, *Clemson University*
Randy Kobes, *University of Winnipeg*
Raman Kolluri, *Camden County College*
Ilkka Koskelo, *San Francisco State University*
Fred Kuttner, *University of California—Santa Cruz*
Richard Ledet, *University of Louisiana—Lafayette*
Alexander Lisyansky, *Queens College, City University of New York*
Carl Lundstedt, *University of Nebraska—Lincoln*
Donald Luttermoser, *East Tennessee State University*
Steven Matsik, *Georgia State University*
Sylvio May, *North Dakota State University*
Bill Mayes, *University of Houston*
Arthur McGum, *Western Michigan University*
Roger McNeil, *Louisiana State University*
Rahul Mehta, *University of Central Arkansas*
Charles Meitzler, *Sam Houston State University*
Michael Meyer, *Michigan Technological University*
Vesna Milosevic-Zdjelar, *University of Winnipeg*
John Milsom, *University of Arizona*
Wouter Montfrooij, *University of Missouri*
Ted Morishige, *University of Central Oklahoma*
Halina Opyrchal, *New Jersey Institute of Technology*
Michelle Ouellette, *California Polytechnic State University—San Louis Obispo*
Kenneth Park, *Baylor University*
Galen Pickett, *California State University—Long Beach*
Dinko Pocanic, *University of Virginia*
Amy Pope, *Clemson University*
Michael Pravica, *University of Nevada—Las Vegas*
Laura Pyrak-Nolte, *Purdue University*
Mark Riley, *Florida State University*
Mahdi Sanati, *Texas Tech University*
Cheryl Schaefer, *Missouri State University*
Alicia Serfaty de Markus, *Miami Dade College—Kendall Campus*
Marc Sher, *College of William & Mary*
Douglas Sherman, *San Jose State University*

Marllin Simon, *Auburn University*
Chandralekha Singh, *University of Pittsburgh*
David Sokoloff, *University of Oregon*
Noel Stanton, *Kansas State University*
Donna Stokes, *University of Houston*
Carey Stronach, *Virginia State University*
Chun Fu Su, *Mississippi State University*
Daniel Suson, *Texas A&M University—Kingsville*
Doug Tussey, *Pennsylvania State University*
John Allen Underwood, *Austin Community College—Rio Grande Campus*
James Wetzel, *Indiana University-Purdue University—Fort Wayne*
Lisa Will, *Arizona State University*
Gerald T. Woods, *University of South Florida*
Guoliang Yang, *Drexel University*
David Young, *Louisiana State University*
Michael Yurko, *Indiana University-Purdue University—Indianapolis*
Hao Zeng, *State University of New York at Buffalo*
Nouredine Zettili, *Jacksonville State University*

Extra Credit

Special thanks go to Lachina Publishing Services, especially Jennifer Bonnar, Jeanne Lewandowski, and Aaron Kantor, for keeping up with innumerable changes during the production stages of this book. Kathleen M. Lafferty once again edited the copy with diligence. Roy Neuhaus brightened the interior with a new design. Steve McEntee provided valuable updates to the art program. Greg Gambino continued creating the "Stick Dude" art for worked examples and end-of-chapter problems that is so popular with reviewers and students. Dena Digilio Betz once again went the extra mile when needed to find photos from unusual sources. Katie Huha and Martha Hall of PreMediaGlobal assisted in securing permissions.

I would also like to thank all the team at Cengage Learning, including Michelle Julet, Mary Finch, Charles Hartford, Cate Barr, Cathy Brooks, Nicole Mollica, Jack Cooney, Brandi Kirksey, Brendan Killion, Shalice Shah-Caldwell, Diane Gibbons, and Doug Bertke, for their fine work in the development, production, and promotion of this new edition. Special recognition goes to Rebecca Berardy Schwartz for her skillful management of the new reasoning tutorials, and again to Susan Pashos for her unwavering and tireless support. I also thank my wife Pat for her continued patience and encouragement.

No textbook is perfect for every student or for every instructor. It is my hope that both students and instructors will find some useful, stimulating, and even exciting material in this book and that you will all enjoy physics as much as I do.

Nicholas J. Giordano

About the Author

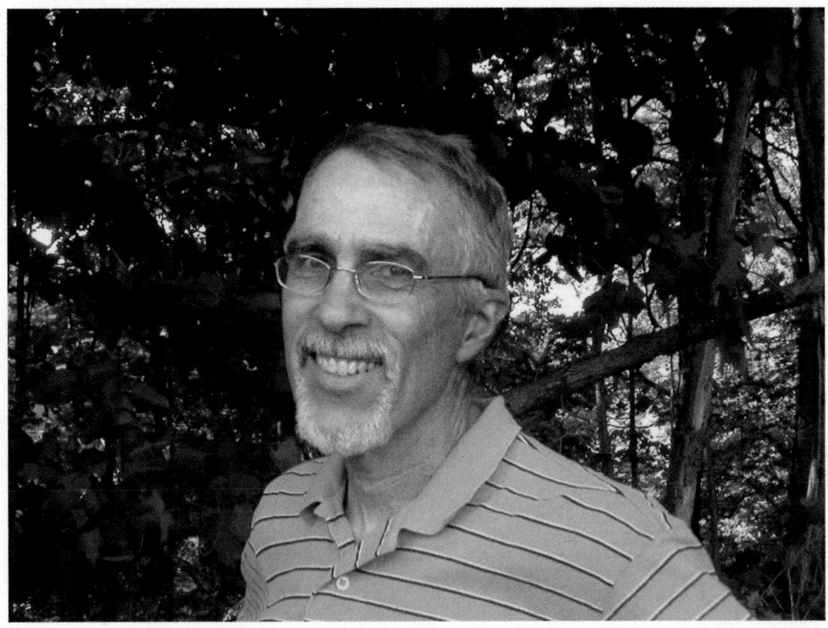

Nicholas J. Giordano obtained his B.S. at Purdue University and his Ph.D. at Yale University. He has been on the faculty at Purdue since 1979, served as an Assistant Dean of Science from 2000 to 2003, and in 2004 was named the Hubert James Distinguished Professor of Physics. His research interests include the properties of nanoscale metal structures, nanofluidics, science education, and biophysics, along with musical acoustics and the physics of the piano. Dr. Giordano earned a Computational Science Education Award from the Department of Energy in 1997, and he was named Indiana Professor of the Year by the Carnegie Foundation for the Advancement of Teaching and the Council for the Advancement and Support of Education in 2004. His hobbies include distance running and restoring antique pianos, and he is an avid baseball fan. He also just finished a book on the physics of the piano.

COLLEGE PHYSICS

Volume 1

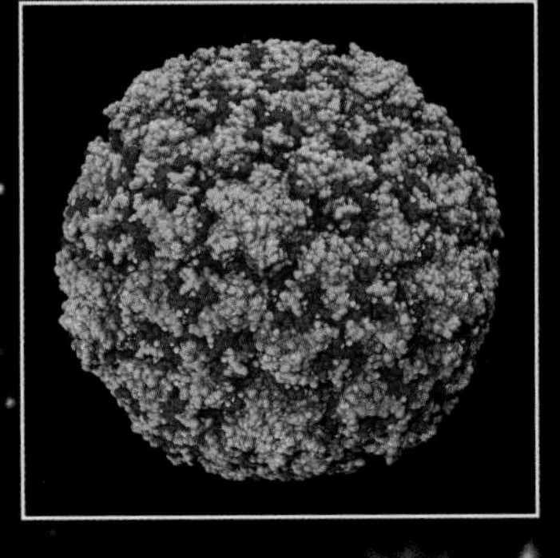

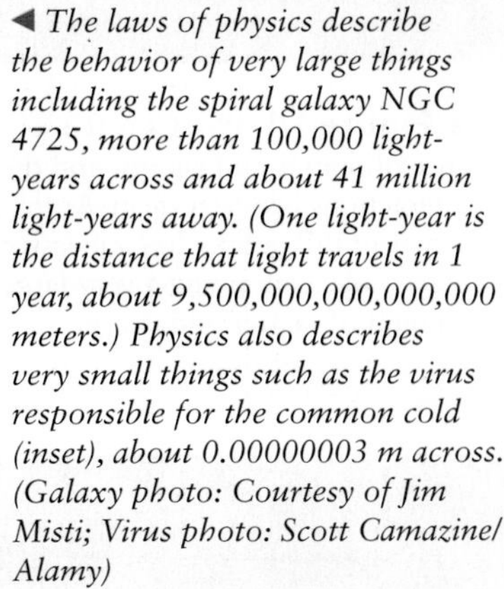

CHAPTER

1

◀ *The laws of physics describe the behavior of very large things including the spiral galaxy NGC 4725, more than 100,000 light-years across and about 41 million light-years away. (One light-year is the distance that light travels in 1 year, about 9,500,000,000,000,000 meters.) Physics also describes very small things such as the virus responsible for the common cold (inset), about 0.00000003 m across. (Galaxy photo: Courtesy of Jim Misti; Virus photo: Scott Camazine/Alamy)*

Introduction

OUTLINE

NASA & The Hubble Heritage Team (STScI/AURA)

▲ **Figure 1.1** Physics is the science of matter and energy, and the interactions between them. At the center of this galaxy lies a massive black hole, containing a very large amount of matter and energy.

© Visual Arts Library (London)/Alamy

▲ **Figure 1.2** Isaac Newton (1642–1727) developed the laws of mechanics we study in the first part of this book. Newton was also a great mathematician and invented much of calculus.

1.1 The Purpose of Physics

This book is about the field of science known as physics. Let's therefore begin by considering what the word *physics* means. One popular definition is

***physics*: the science of matter and energy, and the interactions between them.**

Matter and energy are fundamental to all areas of science; thus, physics is truly a foundational subject. The principles of physics form the basis for understanding chemistry, biology, and essentially all other areas of science. These principles enable us to understand phenomena ranging from the very small (atoms, molecules, and cells) to the very large (planets and galaxies; Fig. 1.1). Indeed, the word *physics* has its origin in the Greek word for "nature." For this reason, an alternative and much broader definition is

***physics*: the study of the natural or material world and phenomena; natural philosophy.**

An important part of this definition is the term *natural philosophy*. In a sense, physics is the oldest of the sciences, and at one time *all* scientists were physicists. Many famous thinkers, including Aristotle, Galileo, and Isaac Newton (Fig. 1.2), were such natural philosophers/physicists.

So, if physics is the science of matter and energy, how does one actually study and learn physics? Our primary objective is to learn how to predict and understand the way matter and energy behave; that is, we want to predict and understand how the universe works. Physics is organized around a collection of ***physical laws***. In the first part of this book, we learn about Newton's laws, which are concerned with the motion of mechanical objects such as cars, baseballs, and planets. Later we'll encounter physical laws associated with a variety of other phenomena including heat, electricity, magnetism, and light. Our job is to learn about these physical laws (sometimes called ***laws of nature***) and how to use them to predict the workings of the universe: how objects move, how electricity flows, how light travels, and more. These physical laws are usually expressed mathematically, so much of our work will involve mathematics. However, good physics is more than just good mathematics; an appreciation of the basic concepts and how they fit together is essential.

In addition to *predicting how* the world works, we would also like to *understand why* it works the way it does. Making predictions requires us to apply the physical laws to a particular situation and work out the associated mathematics to arrive at specific predictions. Understanding the world is more difficult because it involves understanding where the physical laws "come from." This "come from" question is so difficult because physical law comes into being in the following way. Initially, someone formulates a hypothesis that describes all that is known about a particular phenomenon. For example, Newton probably formulated such an initial hypothesis for the laws of motion. One must then show that this hypothesis correctly describes *all* known phenomena. For Newton, it meant that his proposed laws of motion had to account correctly for the motion of apples, rocks, arrows, and all other terrestrial objects. In addition, a successful hypothesis is often able to explain things that were previously not understood. In Newton's case, his hypothesis was able to explain celestial motion (the motion of the planets and the Moon, as sketched in Fig. 1.3), a problem that was unsolved before Newton's time. Only after a hypothesis passes such tests does it qualify as a law or principle of physics.

This process through which a hypothesis becomes a law of physics means that there is no way to prove that such a physical law is correct. It is always possible that a future experiment or observation will reveal a flaw or limitation in a particular "law." This step is part of the scientific process because the discovery of such flaws leads to the discovery of new laws and new insights into nature.

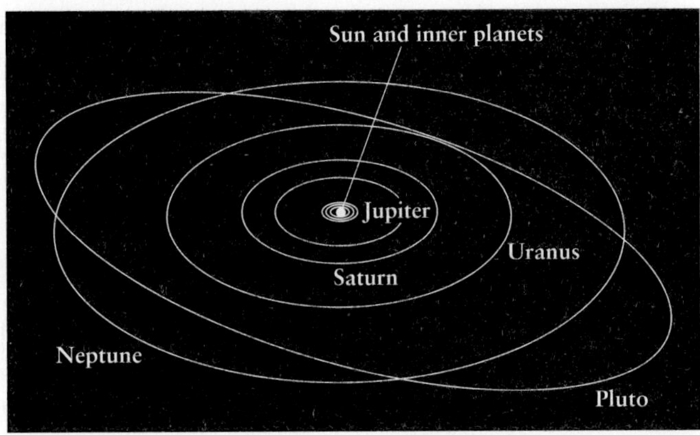

◀ **Figure 1.3** Newton showed that the motion of the planets around our Sun can be explained by the same laws of motion that describe the movement of terrestrial objects, such as baseballs and automobiles. This discovery unified human understanding of the motion of terrestrial and celestial objects.

Although this process of constructing and testing hypotheses can lead us to the laws of physics, it does not necessarily give us an understanding of why these laws are correct. That is, why has nature chosen a particular set of physical laws instead of a different set? Insight into this question can often be gained by examining the form of a physical law and the predictions it leads to. We'll do that as we proceed through this book and in this way get important glimpses into how nature works. The important point is that the problem of *predicting how* the world works is different from *understanding why* it works the way it does. Our goal is to do both.

1.2 Problem Solving in Physics: Reasoning and Relationships

According to our definitions in Section 1.1, physics involves predicting how matter and energy behave in different situations. Making such predictions usually requires that we apply a general physical law to a particular case, a process known as ***problem solving***. Problem solving is an essential part of physics, and we'll spend a great deal of time learning how to do it in many different situations. In some ways, learning the art of problem solving is like learning how to play the piano or hit a golf ball: it takes practice. Just as in playing a musical instrument or learning a sport, certain practices are the keys to good problem solving. Following these practices consistently will help lead to a thorough understanding of the underlying concepts.

Problem solving often involves some mathematical calculations. Many problems we encounter are ***quantitative problems***, which give quantitative information about a situation and require a precise mathematical calculation using that information. An example is to calculate the time it takes an apple to fall from a tree to the ground below, given the initial height of the apple. A problem may also involve the application of a key concept in a nonmathematical way. We use such ***concept checks*** to test your general understanding of a particular physical law or concept and how it is applied. We will also encounter ***reasoning and relationship problems***, which will often require you to identify important information that might be "missing" from an initial description of the problem. For example, you might be asked to calculate the forces acting on two cars when they collide, given only the speed of the cars just before the collision. From an understanding of the relationship between force and motion, you would need to recognize that additional information is required to solve this problem. (In this problem, that additional information is the mass of the cars and the way the car bumpers deform in the collision.) It is your job to use common sense and experience to estimate realistic values of these "missing" quantities for typical cars and then use these values to compute an answer. Because such estimated values vary from case to case (e.g., not all cars have the same mass), an approximate mathematical solution and an approximate numerical answer are usually sufficient for such reasoning and relationship problems.

An ability to deal with these three different types of problems is essential to gaining a thorough understanding of physics and for success in any science- or technology-related field.

PROBLEM SOLVING Problem-Solving Strategies

We'll encounter problems involving different laws of physics, including Newton's laws of mechanics, the laws of electricity and magnetism, and quantum theory. Although these problems involve many different situations, they can all be attacked using the same basic problem-solving strategy.

1. **RECOGNIZE THE** key physics **PRINCIPLES** that are central to the problem. For example, one problem might involve the principle of conservation of energy, whereas another might require Newton's action–reaction principle. The ability to recognize the central principles requires a conceptual understanding of the laws of physics, how they are applied, and how they are interrelated. Such knowledge and skill are obtained from experience, practice, and careful study.

2. **SKETCH THE PROBLEM.** A diagram showing all the given information, the directions of any forces, and so forth is valuable for organizing your thoughts. A good diagram will usually contain a coordinate system to be used in measuring the position of an object and other important quantities.

3. **IDENTIFY THE** important **RELATIONSHIPS** among the known and unknown quantities. For example, Newton's second law gives a relationship between force and motion, and is thus the key to analyzing the motion of an object. This step in the problem-solving process may involve several parts (substeps), depending on the nature of the problem. For example, problems involving collisions may involve steps that aren't necessary for a problem in magnetism. When dealing with a reasoning and relationship problem, one of these substeps may involve identifying the "missing" information or quantities and then estimating their values.

4. **SOLVE** for the unknown quantities using the relationships in step 3.

5. **WHAT DOES IT MEAN?** Does your answer make sense? Take a moment to think about your answer and reflect on the general lessons to be learned from the problem.

1.3 Dealing with Numbers

Scientific Notation

During the course of problem solving, we will often encounter numbers that are very large or very small. ***Scientific notation*** was invented as a convenient way to abbreviate extremely large or extremely small numbers. We can understand how scientific notation works by using it to express some of the numbers found in Table 1.1, which lists some important lengths and distances. As you probably know, lengths and distances can be measured in units of ***meters***. We'll say more about meters and other units of measure in the next section. For now, we can rely on your intuitive notion of length and that 1 meter (abbreviated "m") is approximately the distance from the floor to the doorknob for a typical door. The height of a typical adult male is about 1.8 m, and the diameter of a compact disc (CD) or digital video disc (DVD) is about 0.12 m (Fig. 1.4).

The numerical values in Table 1.1 span a tremendous range. The largest is the distance from Earth to the Sun, which is 150,000,000,000 m. This number is so large and contains so many zeros to the left of the decimal point[1] that in Table 1.1 we have written it in scientific notation. When written in this way, the distance from Earth to the Sun is 1.5×10^{11} m. The exponent has the value 11 because the deci-

[1]In numbers like this one, the decimal point is usually not written explicitly. If we were to include the decimal point, however, this number would be written as 150,000,000,000. m.

TABLE 1.1 Some Common Lengths and Distances

Quantity	Length or Distance in Meters (m)	Quantity	Length or Distance in Meters (m)
Diameter of a proton	1×10^{-15}	Height of the author	1.80
Diameter of a red blood cell*	8×10^{-6}	Height of the Empire State Building	443.2
Diameter of a human hair (removed from the author's head)	5.5×10^{-5}	Distance from New York City to Chicago	1,268,000
Thickness of a sheet of paper	6.4×10^{-5}	Circumference of the Earth	4.00×10^{7}
Diameter of a CD or DVD	0.12	Distance from the Earth to the Sun	1.5×10^{11}

*Red blood cells are not spherical, but are shaped more like flat plates. The value given here is the approximate diameter of the plate.

mal point in the number 150,000,000,000 is 11 places to the right of its location in the number 1.5. Likewise, the distance from New York to Chicago is 1,268,000 m, which is 1.268×10^6 m. Here the decimal point in the number 1,268,000 is six places to the right of its location in the number 1.268. Scientific notation is also a useful way to write very small numbers. For example, the diameter of a hair taken from the author's head is 0.000055 m, which is written as 5.5×10^{-5} m in scientific notation (Table 1.1). Here the exponent has the value -5 because the decimal point in 0.000055 is five places to the left of its location in the number 5.5.

Writing numbers in scientific notation

These rules for expressing a number in scientific notation can be summarized as follows. Move the decimal point in the original number to obtain a new number between 1 and 10. Count the number of places the decimal point has been moved; this number will become the exponent of 10 in scientific notation. If you started with a number greater than 10 (such as 150,000,000,000), the exponent of 10 is positive (1.5×10^{11}). If you started with a number less than 1 (such as 0.000055), the exponent is negative (5.5×10^{-5}).

EXAMPLE 1.1 Population of the Earth

The number of people living on the Earth in 2007 is estimated to have been approximately 6,600,000,000. Express this number in scientific notation.

RECOGNIZE THE PRINCIPLE

To write this number using scientific notation, we must compare the location of the decimal point in the number 6,600,000,000 with its location in the number 6.6.

SKETCH THE PROBLEM

We must move the decimal point nine places to get a number between 1 and 10:

$$\underbrace{6,600,000,000}_{\text{9 places}}$$

IDENTIFY THE RELATIONSHIPS AND SOLVE

We started with a number greater than 10. Using our rules for expressing a number in scientific notion, the population of the Earth in 2007 was

$$6,600,000,000 = \boxed{6.6 \times 10^9}$$

► ***What does it mean?***

The exponent in this answer is 9 because the decimal point in the number 6,600,000,000 is nine places to the right of its location in the number 6.6.

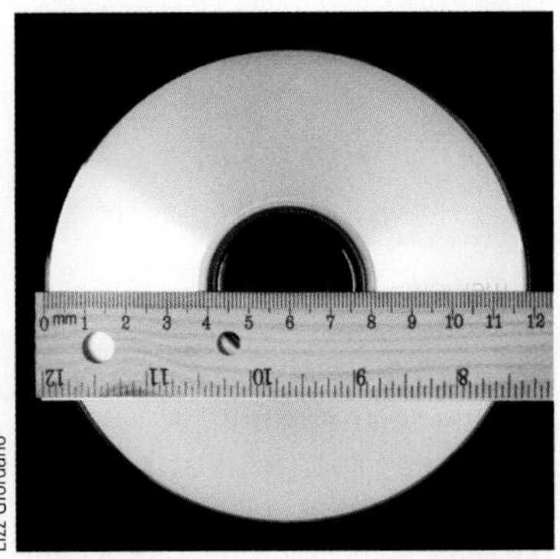

▲ **Figure 1.4** A compact disc has a diameter of approximately 0.12 m.

Significant Figures

Suppose you are asked to measure the width of a standard piece of paper such as a page in this book. The answer is approximately 0.216 meter, as you can confirm for yourself using a ruler. However, you should realize that this value is not exact. Different pieces of paper will be slightly different in size, and your measurement will also have some uncertainty (some experimental error). There is an uncertainty associated with all such measured values, including all the values in Table 1.1, and these uncertainties affect the way we write a numerical value. For our piece of paper, we have written the width using three ***significant figures***. The term *significant* refers to the number of digits that are meaningful with regard to the accuracy of the value. In this particular case, writing the value as 0.216 m implies that the true value lies between 0.210 m and 0.220 m, and that it is likely to be close to 0.216 m. We should not be surprised to find a value of 0.215 m or 0.217 m, though.

When writing a number in scientific notation, the number of digits that are written depends on the number of significant figures. For our piece of paper, we would write 2.16×10^{-1} m and would say that all three digits are significant. There can be some ambiguity when dealing with digits that are zero. For example, the length of a particular race in the sport of track and field is 100 meters (the distance between the starting line and the finish). In terms of significant figures, you might think that this number has only one significant figure, which would imply that the length of the race falls within the rather large range of 0 to 200 m. In this example, however, the distance between the start and finish lines will be measured quite precisely, so we should suspect the value to have at least three significant figures, with the true value lying between 99 and 101 meters. If this accuracy is in fact correct, the value "100 meters" actually contains three significant figures; the zeros here are significant because they are indicative of the accuracy. In such cases, we must determine the number of significant figures from the context of the problem or from other information. This ambiguity does not arise with scientific notation; the length of our race would be written as 1.00×10^{2} m, which indicates explicitly that the zeros are significant.

Significant figures are also important in calculations. As an example, Table 1.1 lists the thickness of a typical sheet of paper as 6.4×10^{-5} m. This value—which has two significant figures—was measured by the author for a piece of paper of the kind used in this book. Suppose a book were to contain 976 such sheets. We could then compute the thickness of the book (without the covers) as

$$\begin{aligned}\text{thickness of book} &= \text{number of sheets of paper} \times \text{thickness of one sheet} \\ &= 976 \times 0.000064 \text{ m} = 0.062464 \text{ m} \qquad (1.1)\end{aligned}$$

Insight 1.1
SIGNIFICANT FIGURES IN NUMBERS LESS THAN ONE
The number of significant figures in the value 0.062464 is *five*. For a number smaller than 1, the zeros immediately to the right of the decimal point are *not* significant figures. Hence, the number 0.000064 has just two significant figures.

The final result in Equation 1.1 is written with five significant figures, as you might read from a calculator used to do this multiplication. Because the number of significant figures indicates the expected accuracy of a value, writing this answer with five significant figures implies that we know the thickness of the book to very high accuracy. However, since the thickness of one page is known with an accuracy of only two significant figures, the final value of this calculation—the thickness of the entire book—will actually have an accuracy of only two significant figures. So, we should round the result in Equation 1.1 to two significant figures:

$$0.062464 \text{ m rounded to two significant figures} = 0.062 \text{ m} \qquad (1.2)$$

Determining the number of significant figures in a calculation involving multiplication or division

Rules for significant figures in calculations involving *multiplication and division.*

1. Use the full accuracy of all known quantities when doing the computation. In Equation 1.1, we used the number of sheets as given to three

significant figures and the thickness per sheet, which is known to two significant figures.

2. At the *end* of the calculation, round the answer to the number of significant figures present in the *least* accurate starting quantity. In Equation 1.1, the least accurate starting value was the thickness of a single sheet, which was known to two significant figures, so we rounded our final answer to two significant figures in Equation 1.2.

Notice that the rounding in Equation 1.2 took place at the *end* of the calculation. Some calculations involve a sequence of several separate computations, and rounding can sometimes cause trouble if it is applied at an intermediate stage of the computation. For example, suppose we want to use the result from Equation 1.2 to find the height of a stack of 12 such books. We could multiply the answer from Equation 1.2 by 12 to get the height of the stack. We thus have

$$\begin{aligned}\text{height of stack} &= 12 \times (0.062 \text{ m}) = 0.744 \text{ m} \\ &= 0.74 \text{ m (rounding to two significant figures)}\end{aligned} \tag{1.3}$$

Here we have rounded our final answer to two significant figures because the starting value—the thickness of one book—is known to two significant figures. On the other hand, if we use the unrounded value of the thickness of one book from Equation 1.1, we get

$$\begin{aligned}\text{height} &= 12 \times (0.062464 \text{ m}) = 0.749568 \text{ m} \\ &= 0.75 \text{ m (rounding to two significant figures)}\end{aligned} \tag{1.4}$$

In the last step in Equation 1.4, we again rounded to two significant figures. Comparing the results in Equations 1.3 and 1.4, we see that the final answers differ by 0.01 m. This difference is due to ***round-off error***, which can happen when we round an answer too soon in the course of a multistep calculation. Such errors can be avoided by carrying along an extra significant figure through intermediate steps in a computation and then performing the final rounding at the very end. In this example, we should keep three significant figures for the answer from Equation 1.1 (carrying an extra digit); using this three-significant-figure value in Equation 1.3 would then have given us the correct answer of 0.75 m.

The procedures we have just described for dealing with significant figures apply to calculations that involve multiplication and division. Computations involving addition or subtraction require a different approach to determine the final accuracy.

Determining the number of significant figures in a calculation involving addition or subtraction

Rule for significant figures in calculations involving *addition or subtraction*.

The location of the least significant digit in the answer is determined by the location of the least significant digit in the starting quantity that is known with the *least accuracy*.

Consider the addition of the numbers 4.52 and 1.2. The least accurate of these numbers is 1.2, so the least significant digit here and in the final answer is one place to the right of the decimal point. Pictorially, we have

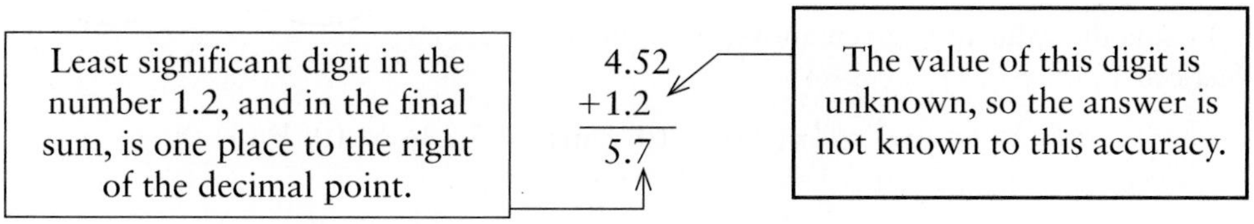

Least significant digit in the number 1.2, and in the final sum, is one place to the right of the decimal point.

$$\begin{array}{r} 4.52 \\ +1.2 \\ \hline 5.7 \end{array}$$

The value of this digit is unknown, so the answer is not known to this accuracy.

Hence, in this example, the final answer has two significant figures.

An example involving subtraction is shown below, where we consider the calculation 4.52 − 1.2. The starting numbers contain three and two significant figures, respectively, and the final answer contains two. Pictorially,

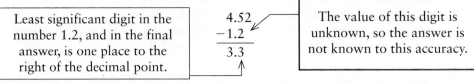

Least significant digit in the number 1.2, and in the final answer, is one place to the right of the decimal point.	4.52 −1.2 3.3	The value of this digit is unknown, so the answer is not known to this accuracy.

As another example, consider the subtraction of two numbers whose difference is very small, such as 4.52 − 4.1. Now we have

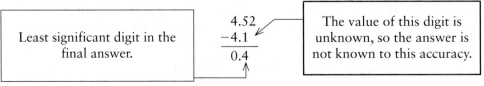

Least significant digit in the final answer.	4.52 −4.1 0.4	The value of this digit is unknown, so the answer is not known to this accuracy.

In this case, the number of significant figures in the final answer is *smaller* than the number of significant figures in either starting number.

Some numbers that appear in a calculation are known exactly. For example, there are 60 seconds in 1 minute. This value is *exact*, so it really should be thought of as 60.000000 . . . in a calculation. The number of significant figures in a calculation with such numbers is then determined by the accuracy (i.e., the number of significant figures) with which other quantities in the calculation are known.

In most of the calculations and problems in this book, we'll round answers to two significant figures.

Insight 1.2

A MISUSE OF SIGNIFICANT FIGURES

It is often quoted that normal human body temperature is 98.6°F. This value is measured with the Fahrenheit temperature scale. This temperature was originally measured in experiments that employed the Celsius scale. When measured in Celsius, it was found that healthy people exhibit temperatures in the range of 36°C to 38°C, with 37°C being typical. So, the value for body temperature on the Celsius scale is 37°C and thus has *two* significant figures. However, when this value was converted to the Fahrenheit scale it was specified as 98.6°F, with *three* significant figures. To be consistent with significant figures (and with the accuracy of the original experimental value and the variation from person to person), this value should actually be rounded to two significant figures, which would give 99°F. Using the value of 98.6°F with the *incorrect* number of significant figures implies that a difference of 0.1°F is significant, but that is not the case at all!

EXAMPLE 1.2 Volume of a Blood Cell

Consider a red blood cell, and for simplicity assume it is spherical with radius $r_1 = 5.0 \times 10^{-6}$ m.

(a) What is the volume of this cell?

(b) Suppose a second red blood cell has a radius $r_2 = 5.1 \times 10^{-6}$ m. What is the *difference* in the volume of the two cells?

RECOGNIZE THE PRINCIPLE

The volume of a sphere of radius r is $V = \frac{4}{3}\pi r^3$. We must evaluate V for both values of the radius, r_1 and r_2. Since r_1 and r_2 are both given with two significant figures, the results for V must also have two significant figures.

SKETCH THE PROBLEM

The radii r_1 and r_2 are almost the same.

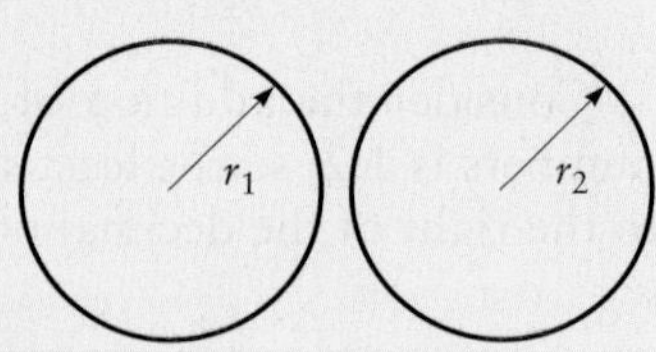

IDENTIFY THE RELATIONSHIPS AND SOLVE

(a) Using the value of r_1 given above, the volume of the first cell V_1 is[2]

$$V_1 = \tfrac{4}{3}\pi r_1^3 = \tfrac{4}{3}\pi(5.0 \times 10^{-6}\ \text{m})^3 = 5.236 \times 10^{-16}\ \text{m}^3$$

[2]Notice that this answer is given in units of m³ = cubic meters. We'll say more about this and other such units below.

The value of r_1 is given to two significant figures, so we must round our final answer to two significant figures:

$$V_1 = \boxed{5.2 \times 10^{-16}\ \text{m}^3}$$

(b) A similar calculation gives the volume of the second cell:

$$V_2 = \tfrac{4}{3}\pi r_2^3 = \tfrac{4}{3}\pi(5.1 \times 10^{-6}\ \text{m})^3 = 5.6 \times 10^{-16}\ \text{m}^3$$

where we have again rounded to two significant figures. The difference in the volumes of the two cells is

$$\Delta V = V_2 - V_1 = (5.6 \times 10^{-16} - 5.2 \times 10^{-16})\ \text{m}^3$$

$$\Delta V = 0.4 \times 10^{-16}\ \text{m}^3 = \boxed{4 \times 10^{-17}\ \text{m}^3}$$

▸ ***What does it mean?***

The given values of r_1 and r_2 contained two significant figures, so the results for V_1 and V_2 also contained two significant figures. However, according to the rule for significant figures in subtraction, the value for ΔV has only one significant figure.

CONCEPT CHECK 1.1 Significant Figures

Give the number of significant figures for all the lengths and distances in Table 1.1.

1.4 Physical Quantities and Units of Measure

To conduct experiments and test physical theories, we must be able to *measure* various physical quantities. Typical quantities of importance in physics include the ***distance*** traveled by a baseball, the ***time*** it takes an apple to fall from a tree, and the ***mass*** of the Earth. These three types of quantities—distance (or equivalently, length), time, and mass—play central roles in physics. In order to measure a particular physical quantity, we must have a ***unit of measure*** for that quantity. Consider, for example, the quantity length. Table 1.1 lists a variety of lengths and distances, with values given in ***meters***, a unit of measure widely used in scientific work. The meter was initially defined in terms of the circumference of the Earth and was subsequently redefined and maintained using a carefully constructed metal bar composed of platinum and iridium (Fig. 1.5). In 1960, the meter was redefined again, this time in terms of the wavelength of light emitted by krypton atoms. Other units, such as inches and feet, can be used to measure length, and scientists involved in metrology (the science of measurement) have determined how different units of length, including meters, inches, and feet, are related.

Courtesy NIST

▲ **Figure 1.5** Prior to 1960, the length of the standard meter was maintained using a platinum–iridium bar like those shown here from the National Institute of Standards and Technology.

It is often necessary to convert the value of a particular length from one unit to another. For example, the width of a page of this book is approximately 0.216 m. We can express this width in inches using the appropriate ***conversion factor***, which relates the two units of interest. For instance, a length of 1 inch (abbreviated "in.") is equal to 2.54×10^{-2} m. The conversion between inches and meters can be expressed in equation form as 1 in. = 2.54×10^{-2} m or in fractional form as

$$\frac{1\ \text{in.}}{2.54 \times 10^{-2}\ \text{m}}$$

If a page is 0.216 m wide, its width in inches is

$$0.216\ \text{m} = 0.216\ \text{m} \times \overbrace{\frac{1\ \text{in.}}{2.54 \times 10^{-2}\ \text{m}}}^{\text{conversion factor}} = 8.50\ \text{in.} \tag{1.5}$$

In terms of only the units, the conversion in Equation 1.5 has the form

$$\text{meters} \times \frac{\text{inches}}{\text{meters}} = \cancel{\text{meters}} \times \frac{\text{inches}}{\cancel{\text{meters}}} = \text{inches} \quad (1.6)$$

Converting from one unit to another

so the "unwanted" units—in this case meters—cancel. The conversion of units always has this form, with the unwanted units canceling to leave the desired unit.

This example of units conversion involves units of length, but the same approach applies to other types of units (e.g., converting from hours to seconds). A set of commonly used conversion factors is given inside the front cover of this book. These conversion factors either are exact numbers or are known with very high precision, so they can generally be treated as exact numbers when determining the number of significant figures in a calculation.

EXAMPLE 1.3 Converting Units

Find the number of inches in 1 mile.

RECOGNIZE THE PRINCIPLE

Inches and miles are two units used to measure length. They are not used often in scientific work, but are often encountered in everyday activities. To convert from miles to inches, we use our knowledge that 1 mile is equal to exactly 5280 feet and that there are exactly 12 inches in 1 foot.

SKETCH THE PROBLEM

No sketch is needed.

IDENTIFY THE RELATIONSHIPS AND SOLVE

The conversion from miles to inches looks like

$$1 \text{ mile} \times \frac{5280 \text{ feet}}{1 \text{ mile}} \times \frac{12 \text{ inches}}{1 \text{ foot}} = \boxed{63{,}360 \text{ inches}} \quad (1)$$

In terms of the units, this conversion has the form

$$1 \text{ mile} = \cancel{\text{mile}} \times \frac{\cancel{\text{feet}}}{\cancel{\text{mile}}} \times \frac{\text{inches}}{\cancel{\text{foot}}} = \text{inches}$$

▶ *What does it mean?*

The answer (Eq. 1) is *exact* because the number of feet in 1 mile and the number of inches in 1 foot are both exact values, determined by the definitions of these units and not by measurement.

▲ **Figure 1.6** Some of the most accurate clocks are based on the properties of cesium atoms and the light they emit. These clocks gain or lose less than 10^{-8} s each day.

Units of Time and Mass

A particularly important physical quantity is ***time***. As you probably know, time is measured in units of seconds, minutes, hours, and so forth. Table 1.2 lists the values of a number of important time intervals. In most scientific work, time is measured in units of seconds (denoted by "s"). The value of the second is based on the frequency of light emitted by cesium atoms. Figure 1.6 shows a cesium atomic clock from the National Institute of Standards and Technology.

A third important physical quantity is ***mass***. Intuitively, mass is related to the amount of material contained in an object; a slightly better definition will be given when we discuss Newton's laws of motion in Chapter 2. Mass can be measured in units called kilograms and in several other units, including grams and slugs. Table 1.3 lists the masses of several common objects in units of kilograms (abbreviated "kg"). By international agreement, the current standard of mass is a piece of a spe-

TABLE 1.2 Some Common Time Intervals

Quantity	Time in Seconds (s)
Time for light to travel 1 meter	3.34×10^{-9}
Time between heartbeats (approximate)	1
Time for light to travel from the Sun to Earth	500
1 day	86,400
1 month (30 days)	2,592,000
Human life span (approximate)	2×10^{9}
Age of the universe	5×10^{17}

TABLE 1.3 Some Common Masses

Object	Mass in Kilograms (kg)
Electron	9.1×10^{-31}
Proton	1.7×10^{-27}
Red blood cell	1×10^{-13}
Mosquito	1×10^{-5}
Typical person	70
Automobile	1200
Earth	6.0×10^{24}
Sun	2.0×10^{30}

cially configured metal alloy composed of platinum and iridium stored in Paris under extremely well controlled conditions (Fig. 1.7).

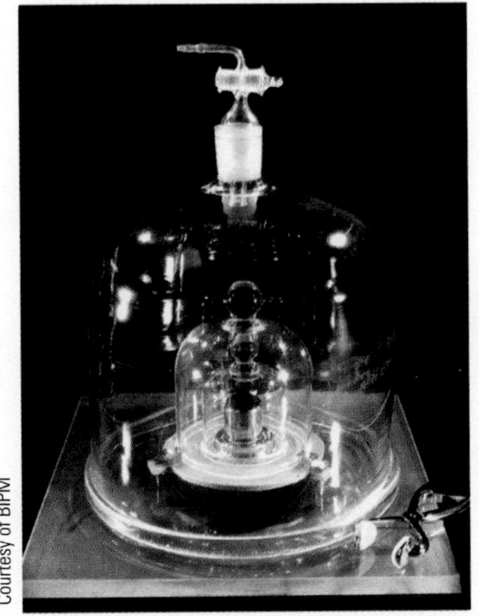
Courtesy of BIPM

▲ **Figure 1.7** The standard kilogram is a cylinder of metal composed of an alloy of platinum and iridium, and is stored in France. This photo shows a copy that is kept in a vacuum inside two separate glass jars. (It is treated very carefully!)

The SI System of Units

To be useful, units of measure must be standardized. Everyone should agree on the length of 1 meter, the length of 1 second, and the amount of mass in 1 kilogram. Without such an agreement, there would be no way to conduct science or commerce since we could not accurately compare quantities measured by different individuals. This standardization has been established by international agreements. One such agreement established the ***Système International d'Unités***, usually referred to as the SI system of units. It employs meters, seconds, and kilograms as the primary units of length, time, and mass. We generally use the SI system of units throughout this book and usually abbreviate meters as m, seconds as s, and kilograms as kg.

Two other systems of units are often encountered. The CGS system uses the units of centimeters (cm) for length, grams (g) for mass, and seconds (s) for time. Another common system, the U.S. customary system, measures length in feet, mass in slugs, and time in seconds. These systems are summarized in Table 1.4.

Powers of 10 and Prefixes

We have already seen how to use scientific notation to work with numbers that are very large or very small. Another way to deal with such numbers is to add a ***prefix*** to the unit of measure. Table 1.5 lists the common prefixes; when attached to a unit, a prefix multiplies the unit by the amount indicated in the table. These prefixes can be applied to any unit of measure. For example, the prefix ***milli*** is equivalent to a factor of 10^{-3}, so 1 millimeter is 10^{-3} m $= 0.001$ m. Another common prefix is ***kilo***, which is equivalent to a factor of $10^3 = 1000$. One kilometer is thus 1000 m, whereas 1 kilosecond is 1000 s. Indeed, you can see that the SI unit of mass contains this prefix since 1 kilogram $=$ 1000 grams.

TABLE 1.4 Units of Length, Mass, and Time

Dimension	SI	CGS	U.S. Customary Units
length	meter (m)	centimeter (cm)	foot (ft)
mass	kilogram (kg)	gram (g)	slug
time	second (s)	second (s)	second (s)

TABLE 1.5 Prefixes That Can Be Attached to Units of Measure

Prefix	Power of 10	Symbol
atto	10^{-18}	a
femto	10^{-15}	f
pico	10^{-12}	p
nano	10^{-9}	n
micro	10^{-6}	μ
milli	10^{-3}	m
centi	10^{-2}	c
deci	10^{-1}	d
deka	10^{1}	da
hecto	10^{2}	h
kilo	10^{3}	k
mega	10^{6}	M
giga	10^{9}	G
tera	10^{12}	T
peta	10^{15}	P
exa	10^{18}	E

EXAMPLE 1.4 Units Conversion and Scientific Notation

How many seconds are in 1 year? Express your answer in scientific notation.

RECOGNIZE THE PRINCIPLE

To compute the number of seconds in 1 year, we use our knowledge that there are 365 days in a year (except for leap years), 24 hours in 1 day, 60 minutes in 1 hour, and 60 seconds in 1 minute.

SKETCH THE PROBLEM

No sketch is needed.

IDENTIFY THE RELATIONSHIPS AND SOLVE

The conversion from years to seconds is

$$1 \text{ year} = 1 \text{ year} \times \frac{365 \text{ days}}{1 \text{ year}} \times \frac{24 \text{ hours}}{1 \text{ day}} \times \frac{60 \text{ minutes}}{1 \text{ hour}} \times \frac{60 \text{ seconds}}{1 \text{ minute}}$$

$$1 \text{ year} = 31{,}536{,}000 \text{ s}$$

Expressed in scientific notation,

$$1 \text{ year} = \boxed{3.1536 \times 10^{7} \text{ s}}$$

► *What does it mean?*

Note that this value is exact (since there are exactly 60 seconds in 1 minute, etc.).

CONCEPT CHECK 1.2 Using Prefixes and Powers of 10

Write the values of all the lengths in Table 1.1 using a prefix from Table 1.5.

1.5 Dimensions and Units

We have discussed the physical quantities length, time, and mass and their units in some detail, but they are only three of the many different types of physical quantities we encounter in physics. In many cases, the units for these other quantities are *derived* from the units of length, time, and mass. In the SI system, *all* the units needed for the study of mechanics can be derived from the ***primary units*** meters, seconds, and kilograms. For example, the volume of an object is measured in units of cubic meters, which is written as m^3. Hence, the unit of volume is a derived unit since it can be expressed in terms of the primary unit (meters). Likewise, the density of an object is equal to its mass divided by its volume, so density is measured in units of

$$\text{density} = \frac{\text{mass}}{\text{volume}} \Rightarrow \frac{\text{kg}}{\text{m}^3} \tag{1.7}$$

The units of density are thus derived from the SI units kilogram and meter.

This distinction between derived units and the primary units of meters, seconds, and kilograms also carries over to our fundamental view of physical quantities. For example, what is length? We all have a clear intuitive notion of length, but it is very difficult to give a formal definition of the term *length* that does not rely on some other physical quantity or idea.[3] Likewise, it is difficult to give a satisfactory formal definition of the quantity we call time. The best we can do is take a set of primary

[3]Defining length is like trying to give a definition of every word in a dictionary. The first definitions have to use words that are not yet defined!

physical quantities as "given" and then use them to build our theories of physics. For our studies of mechanics, we need only three primary physical quantities—length, time, and mass—and we will build all other necessary quantities—such as force, velocity, energy, pressure, and work—from these primary quantities.

In later chapters, we will learn about four more primary units associated with electricity, magnetism, and heat. There are thus seven primary units in the SI system. All other units can be derived from the seven primary units.

Dimensions and Dimensional Analysis

In our calculation of the volume of a blood cell in Example 1.2, we expressed the answer in units of cubic meters. Hence, we calculated both a numerical value and the units of the answer. This pattern is followed for most calculations because most answers are meaningless unless a unit of measure is specified. Calculating and then checking the units of an answer are important parts of the problem-solving process. This checking process, called ***dimensional analysis***, can be carried out in a very general way using the primary physical quantities. For example, the distance from New York City to Chicago has the ***dimensions*** of length and can be measured in units of meters. If we denote these primary dimensional quantities as **L** (length), **T** (time), and **M** (mass), we can express the dimensions of any combination of these quantities in terms of **L**, **T**, and **M**. Suppose the answer for a particular problem can be written as a length divided by a time. The SI units of the answer would then be the units for length (meters) divided by the units for time (seconds), or m/s. We would express the dimensions of this answer as **L/T**. The dimensions of an answer are independent of the particular units used to measure length, time, and so forth.

EXAMPLE 1.5 Using Dimensional Analysis to Check an Answer

Suppose we have performed a calculation of the position of an object (x) as a function of time (t) and arrived at the result

$$x = vt \qquad (1)$$

Use dimensional analysis to find the dimensions of the quantity v.

RECOGNIZE THE PRINCIPLE

The dimensions must be the same on both sides of the equal sign in Equation (1), so the dimensions of x must equal the dimensions of the product vt. If we rearrange the equation to solve for v, we get $v = x/t$. Hence, the dimensions of v must equal the dimensions of x divided by t.

SKETCH THE PROBLEM

No sketch is necessary.

IDENTIFY THE RELATIONSHIPS AND SOLVE

The dimensions of position x (i.e., distance) is **L**, and the dimensions of t is **T**. The dimensions of v must therefore be

$$\text{dimensions of } v = \boxed{\frac{\mathbf{L}}{\mathbf{T}}}$$

► *What does it mean?*

In the SI system, v could be measured in units of meters per second, or m/s. It is always a good idea to check the dimensions of an answer; this check can sometimes reveal errors in a calculation.

CONCEPT CHECK 1.3 Getting the Units Right

The area of a surface can be measured in units of square meters (m^2). Which of the following combinations of units could also be used to measure area? (More than one answer may be correct.)

(a) m · cm (b) ft^3/m^2 (c) $in.^2$ (d) m · ft

1.6 Algebra and Simultaneous Equations

The problem solving we encounter in this book involves three basic types of mathematics: algebra, trigonometry, and vectors. The next few sections are intended as quick refreshers on these topics. A mathematical review is also given in Appendix B.

When solving a problem, we often need to deal with one or two equations containing one or two unknown quantities. Our job is usually to solve for the unknown quantities. Suppose we are given a single equation containing two variables a and T and are asked to find a in terms of T. The equation might be

$$5a = T - 10$$

The goal is to isolate a on one side of the equal sign (usually the left). In this example, we can divide both sides by the constant factor 5 to get the solution

$$a = \frac{T}{5} - 2$$

The numerical value of T might be already known so that the value of a can now be found, or a result expressed in terms of T may be sufficient, depending on the problem.

A slightly more complicated case is with two equations,

$$5a = T - 10 \tag{1.8}$$

$$7a = T + 24 \tag{1.9}$$

A set of equations like this one might arise in a problem in which there are two unknown quantities, a and T. The goal now is to solve for both unknowns. According to the rules of algebra, to solve for two unknown quantities we must have (at least) two equations, so Equations 1.8 and 1.9 should contain all the mathematical information we need to solve for these unknowns. To work out a solution, we can first rearrange Equation 1.8 to obtain a in terms of T:

$$a = \frac{T}{5} - 2 \tag{1.10}$$

Substituting into Equation 1.9 gives

$$7a = T + 24$$

$$7\left(\frac{T}{5} - 2\right) = T + 24$$

$$\frac{7T}{5} - 14 = T + 24 \tag{1.11}$$

We now have a single equation containing only the unknown T. We can find T by moving all the terms involving T to one side to get

$$\frac{7T}{5} - T = 24 + 14$$

$$\frac{7T - 5T}{5} = 38$$

$$\frac{2T}{5} = 38$$

$$T = \frac{5 \times 38}{2} = 95 \qquad (1.12)$$

This value of T can then be inserted back into Equation 1.10 to find the value of a:

$$a = \frac{T}{5} - 2 = \frac{95}{5} - 2 = 17$$

Checking the Units of an Answer

Algebra can be used to solve an equation or a set of equations to find the values of variables such as a and T in the examples above. In physics, these variables represent physical quantities, such as mass, or force, or velocity. Physical quantities possess units, and the algebra we use to compute a and T will also give the value of the associated unit. For example, suppose we have computed the density of an object (represented by the variable ρ) and arrived at the answer

$$\rho = \frac{M}{V} \qquad (1.13)$$

Here, M and V both have units that combine to determine the units of ρ. If we focus only on the units in Equation 1.13, we have the relation

$$\text{units of } \rho = \frac{\text{units of } M}{\text{units of } V}$$

M is the mass of an object so its units are kilograms, and V is a volume with units of cubic meters. The units of ρ are then

$$\text{units of } \rho = \frac{\text{units of } M}{\text{units of } V} = \frac{\text{kg}}{\text{m}^3} \qquad (1.14)$$

Insight 1.3

DIMENSIONS OF DENSITY

In terms of the dimensions, Equation 1.14 reads M/L^3 = mass divided by $(\text{length})^3$.

This answer is indeed the correct unit for density (compare with Equation 1.7).

In Example 1.5, we saw how dimensional analysis can be used to check an answer, and exactly the same approach can be used when calculating and checking units. For example, in our calculation of the density, we might have made an error and obtained the result

$$\rho = \frac{M^2}{V} \quad \textbf{(wrong!!)} \qquad (1.15)$$

The units in this case would work out as

$$\text{units of } \rho = \frac{\text{units of } M^2}{\text{units of } V} = \frac{\text{kg}^2}{\text{m}^3} \quad \textbf{(wrong!!)}$$

Since we know that kg^2/m^3 is not the correct unit for density, this result tells us immediately that there must be an error in the calculation that led to Equation 1.15. It is *always* useful to check the units (and dimensions) of your answers.

1.7 Trigonometry

We often need to deal with angles and the location of an object within a coordinate system. This is most easily done with trigonometry, right triangles, and the

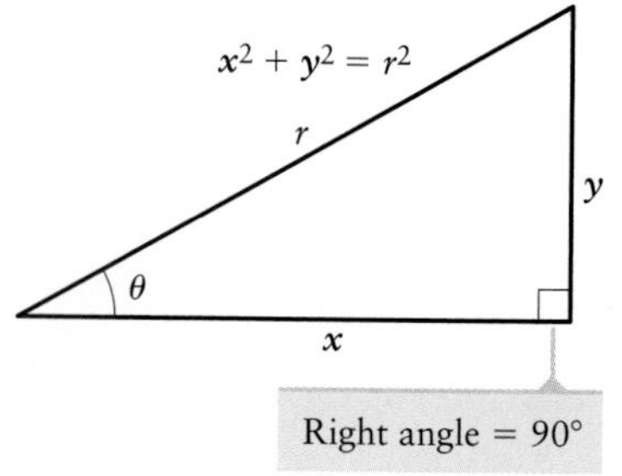

▲ **Figure 1.8** The Pythagorean theorem is a relation between the lengths of the sides of a right triangle.

Pythagorean theorem. Figure 1.8 shows a right triangle with sides x, y, and r, where r is the hypotenuse. According to the Pythagorean theorem,

$$x^2 + y^2 = r^2 \tag{1.16}$$

The trigonometric functions sine, cosine, and tangent are defined as

$$\sin\theta = y/r \tag{1.17}$$

$$\cos\theta = x/r \tag{1.18}$$

$$\tan\theta = y/x \tag{1.19}$$

The Pythagorean theorem (Equation 1.16) implies that

$$\sin^2\theta + \cos^2\theta = 1 \tag{1.20}$$

for any value of the angle θ. Equation 1.20 also leads to a number of other trigonometric relations (called ***identities***), some of which are listed inside the back cover of this book and in Appendix B.

Equations 1.17 through 1.19 tell us how to calculate the sine, cosine, and tangent of an angle from the lengths of the sides of the associated right triangle. We can also use this approach to find the value of the angle. For example, if we know the values of y and r, we can solve for the angle:

$$\sin\theta = y/r$$

$$\theta = \sin^{-1}(y/r) \tag{1.21}$$

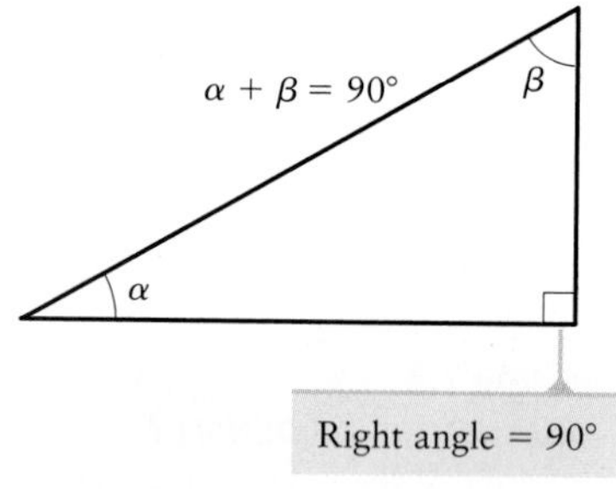

▲ **Figure 1.9** The sum of the three interior angles of a triangle is 180°. Hence, $\alpha + \beta = 90°$.

The function $\sin^{-1}$ is the ***inverse sine*** function, also known as the ***arcsine*** (= arcsin = $\sin^{-1}$). In words, Equation 1.21 says that θ equals the angle whose sine is y/r. Scientific calculators all contain this function (as a single "button"). If the values of y and r are known, you only take the ratio y/r and then compute the arcsine of this value to find θ in Equation 1.21. There are also functions for the inverse cosine ($\cos^{-1}$ = arccos) and inverse tangent ($\tan^{-1}$ = arctan). From Equations 1.18 and 1.19, we have

$$\theta = \cos^{-1}(x/r) \tag{1.22}$$

and

$$\theta = \tan^{-1}(y/x) \tag{1.23}$$

We often need to deal with triangles like the one in Figure 1.9, where we have a right triangle with interior angles α and β. The sum of a triangle's three interior angles is always 180°. For the triangle in Figure 1.9, one of the angles is 90° (because it is a right triangle), so

$$\alpha + \beta = 90° \tag{1.24}$$

Two angles whose sum is 90° are called complementary angles. From the definitions of the sine and cosine functions in Equations 1.17 and 1.18, we have the relations

$$\sin\alpha = \cos\beta$$

$$\cos\alpha = \sin\beta$$

for any pair of complementary angles α and β.

Measuring Angles

The value of an angle is often measured in units of ***degrees***; indeed, we have already used this unit when discussing trigonometry. Another common unit for measuring angles is the ***radian*** (abbreviated "rad"). The conversion between radians and degrees is accomplished just like any other units conversion problem, using the relation

$$360° = 2\pi \text{ rad}$$

or

$$1° = \frac{2\pi}{360} \text{ rad} \tag{1.25}$$

At this point, you might ask why we need two different units—degrees and radians—to measure angles. To explain the mathematical origin of the radian unit, Fig-

An angle of approximately 57° corresponds to 1 radian.

The angle θ here is approximately 1 radian.

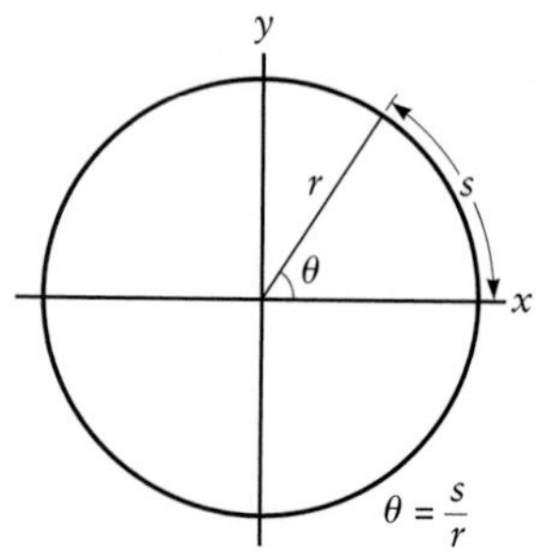

▲ **Figure 1.10** The angle θ equals the ratio of the length s measured along the circular arc, divided by the radius r of the circle. This ratio gives the value of θ in units of radians.

ure 1.10 shows a circle of radius r along with a "radius line" that makes an angle θ with the x axis. The arc length s is measured *along the circle* from the x axis to the point where the radius line meets the circle. The angle θ when measured in radians is equal to the ratio of the arc length to the radius:

$$\theta = \frac{s}{r} \quad (\theta \text{ measured in radians}) \tag{1.26}$$

This connection between the angle and the length measured along a circular arc is very useful in work on circular motion (in which a particle moves along a circular path) and rotational motion (in which an object spins about an axis). The relation between s, r, and θ in Equation 1.26 only holds when s and r are measured in the same units (e.g., both measured in meters) and the angle is measured in radians. Most calculators can be set to work with angles in either degrees or radians, but you should always be clear about which way your calculator is set!

EXAMPLE 1.6 The 3-4-5 Right Triangle and Designing a Roof

The roof of a house has the form of a right triangle with sides of length 3.0 m, 4.0 m, and 5.0 m (Fig. 1.11). Find the interior angles θ and ϕ of the triangle. Express your answers in degrees and radians.

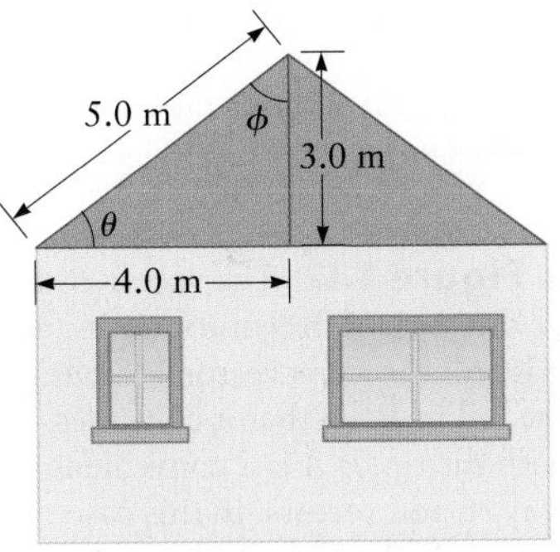

▲ **Figure 1.11** Example 1.6.

RECOGNIZE THE PRINCIPLE

Given the lengths of the sides of a right triangle, we can find the values of the interior angles using the $\sin^{-1}$, $\cos^{-1}$, and $\tan^{-1}$ functions, Equations 1.21 through 1.23. The interior angles θ and ϕ are also complementary (Eq. 1.24).

SKETCH THE PROBLEM

Figure 1.11 shows the problem.

IDENTIFY THE RELATIONSHIPS AND SOLVE

The right triangle in Figure 1.11 has sides $x = 4.0$ m, $y = 3.0$ m, and $r = 5.0$ m. We can compute the value of θ using the inverse sine function ($\sin^{-1}$) in Equation 1.21:

$$\theta = \sin^{-1}(y/r) = \sin^{-1}(3.0/5.0) = \boxed{37°} \tag{1}$$

where we have rounded to two significant figures. To express this value in radians, we convert units:

$$\theta = 37° \times \frac{2\pi \text{ rad}}{360°} = \boxed{0.64 \text{ rad}}$$

The interior angle ϕ is complementary to θ (see Eq. 1.24) and has a value

$$\phi = 90 - \theta = \boxed{53°} = 53° \times \frac{2\pi \text{ rad}}{360°} = \boxed{0.93 \text{ rad}}$$

► ***What does it mean?***

In computing the $\sin^{-1}$ function in Equation (1), you must be sure that you know how your calculator is set and whether it is giving an answer in degrees or radians.

Insight 1.4

INVERSE TRIGONOMETRIC FUNCTIONS

The inverse sine function used in Example 1.6 can be understood as follows. If $\theta = \sin^{-1}(w)$, then w is equal to the sine of the angle θ; that is, $w = \sin\theta$. When computing an inverse trigonometric function, be sure to take note of the units; that is, you should know if your calculator is set for degrees or for radians.

1.8 Vectors

In this book, we deal with two different types of mathematical quantities. One type includes quantities such as time and mass, which are described by a simple number (with a unit, of course) and are called *scalars*. Quantities described by a simple *magnitude*, such as the mass of a car or the number of seconds in 1 year, are scalars. Some physical quantities, however, cannot be described completely in this way. For example, the velocity with which the wind is blowing is a ***vector***. A vector quantity has both a magnitude and a ***direction***. In diagrams and figures, it is often useful to represent a vector quantity by an arrow. The arrow's length indicates the magnitude of the vector (e.g., how fast the wind is blowing), while the arrow's direction indicates the direction of the vector relative to a chosen coordinate system (e.g., the direction in which the wind is blowing). Many quantities in physics are vectors, and we often need to add, subtract, and perform multiplication with vectors.

Vectors have a magnitude and a direction.

These are the same vector $\vec{B}$ even though they are drawn at different locations.

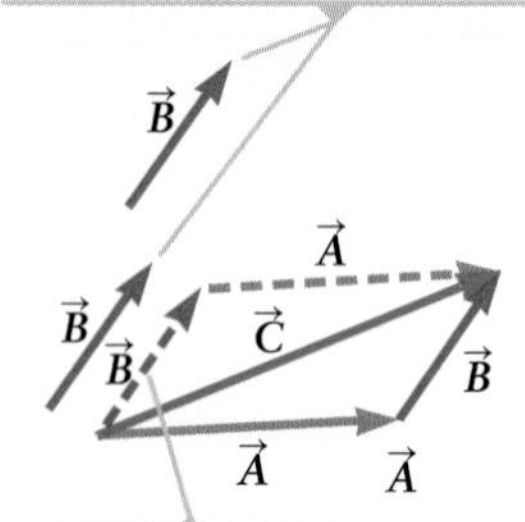

The dashed versions of $\vec{B}$ and $\vec{A}$ show that $\vec{A} + \vec{B} = \vec{B} + \vec{A}$.

For vector addition the order does not matter.

▲ **Figure 1.12** The vector $\vec{B}$ can be drawn in different locations, and it is still the same vector. Drawing the tail of $\vec{B}$ so that it coincides with the tip of $\vec{A}$ is a convenient way to add vectors. In this case, $\vec{C} = \vec{A} + \vec{B}$.

Adding Vectors

Representing vectors as arrows leads to a convenient way to think about the addition of two vectors. Figure 1.12 shows two vectors $\vec{A}$ and $\vec{B}$; here we use a notation in which vector variables are indicated by placing arrows over a boldface symbol. A vector has a magnitude and a direction, so both the lengths and the directions of the arrows in Figure 1.12 are important. However, when we draw a vector in this manner, the location of the arrow is somewhat arbitrary. We can place the arrow in different locations, and as long as it has the same length and the same direction, it is still the same vector. In Figure 1.12, we have drawn $\vec{B}$ several times; in one case, the "tail" of $\vec{B}$ is located at the "tip" of $\vec{A}$. The sum of $\vec{A}$ and $\vec{B}$ is then the vector $\vec{C}$, which connects the tail of $\vec{A}$ with the tip of $\vec{B}$. Mathematically,

$$\vec{C} = \vec{A} + \vec{B} \tag{1.27}$$

A useful way to understand Equation 1.27 is to think of these vectors as displacements or movements. The vector $\vec{A}$ then represents movement from the tail of $\vec{A}$ to the tip of $\vec{A}$, while the vector $\vec{B}$ represents movement from the tail of $\vec{B}$ to the tip of $\vec{B}$. When these two movements are added together, we get the total displacement represented by $\vec{C}$.

Multiplying a Vector by a Scalar and Subtracting Vectors

Before dealing with vector subtraction, it is useful to first consider the multiplication of a vector by a scalar. Scalars are simply numbers like 2 or 5.7. Multiplication of a vector by a scalar changes the vector's length. Figure 1.13 shows a vector $\vec{A}$ multiplied by a scalar K:

$$\vec{B} = K\vec{A} \tag{1.28}$$

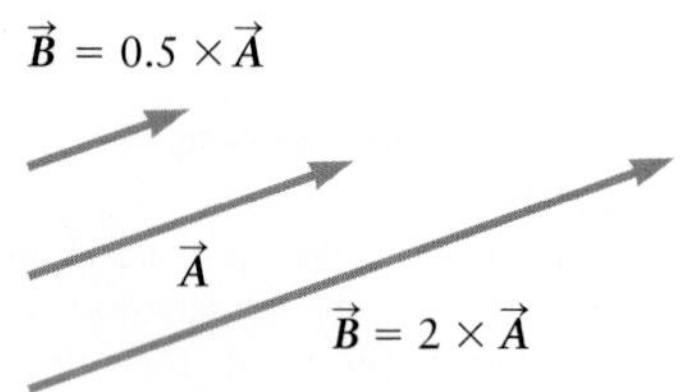

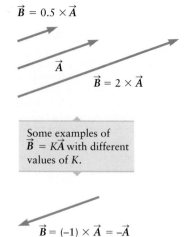

Some examples of $\vec{B} = K\vec{A}$ with different values of K.

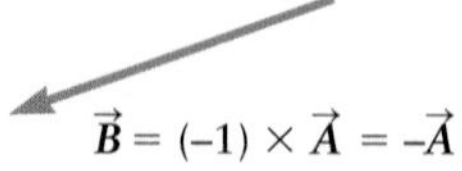

▲ **Figure 1.13** When a vector (such as $\vec{A}$) is multiplied by a scalar, the result is a vector that is parallel to the original vector but with a different length.

If the scalar K is greater than 1 ($K > 1$), the vector $\vec{B}$ is longer than $\vec{A}$, while if K is positive and less than 1, $\vec{B}$ is shorter than $\vec{A}$. If the scalar K is negative, then $\vec{A}$ and $\vec{B}$ are in *opposite directions*.

We can now see how to do vector subtraction. Subtracting the vector $\vec{B}$ from the vector $\vec{A}$ is equivalent to adding the vectors $\vec{A}$ and $-\vec{B}$ (see Fig. 1.14):

$$\vec{A} - \vec{B} = \vec{A} + (-\vec{B}) \tag{1.29}$$

Vectors and Components

When performing calculations involving vector quantities, it is often useful to deal with vectors in their component form. Figure 1.15 shows a vector $\vec{A}$ drawn with its tail at the origin of an $x-y$ coordinate system. This vector has ***components*** along the x and y directions, as shown. The components A_x and A_y are the projections of $\vec{A}$ along x and y.

The vector $\vec{A}$ in Figure 1.15 can be described mathematically in two different ways. The first involves the angle θ this vector makes with the x axis along with the length of $\vec{A}$, which we can denote either by A or by $|\vec{A}|$. The second is to give its components along x and y. According to Figure 1.15, the components of $\vec{A}$ are

$$A_x = A\cos\theta = |\vec{A}|\cos\theta \tag{1.30}$$

$$A_y = A\sin\theta = |\vec{A}|\sin\theta \tag{1.31}$$

With Equations 1.30 and 1.31, we can calculate the components if we know the magnitude (i.e., length) and direction (θ) of a vector. We can also work backward from the components to find the magnitude and direction. Applying some trigonometry to Figure 1.15 gives

$$A = |\vec{A}| = \sqrt{A_x^2 + A_y^2} \tag{1.32}$$

$$\tan\theta = A_y/A_x$$

or

$$\theta = \tan^{-1}(A_y/A_x) \tag{1.33}$$

Hence, if we know the components A_x and A_y, we can find A (the length of the vector) and θ (the direction).

Let's again consider the addition of two vectors, but now work in terms of the components. Figure 1.16 shows two vectors $\vec{A}$ and $\vec{B}$ added graphically, and we have also indicated the values of each vector's components. To add two vectors, we simply add their components. That is, to compute the vector sum $\vec{C} = \vec{A} + \vec{B}$, we add components so that

$$C_x = A_x + B_x \quad \text{and} \quad C_y = A_y + B_y \tag{1.34}$$

Subtraction of two vectors is handled in a similar way. If $\vec{C} = \vec{A} - \vec{B}$, then, in terms of components,

$$C_x = A_x - B_x \quad \text{and} \quad C_y = A_y - B_y \tag{1.35}$$

Multiplication of a vector by a scalar K can also be done using components. If $\vec{B} = K\vec{A}$, we then have (Fig. 1.17)

$$B_x = KA_x \quad \text{and} \quad B_y = KA_y \tag{1.36}$$

In our discussion of vector components, we have so far dealt with only the x and y components. Our results can be generalized to include a third coordinate direction, z, for cases in which we must deal with three-dimensional space. For example, when adding two vectors, Equation 1.34 becomes

$$C_x = A_x + B_x \quad C_y = A_y + B_y \quad C_z = A_z + B_z \tag{1.37}$$

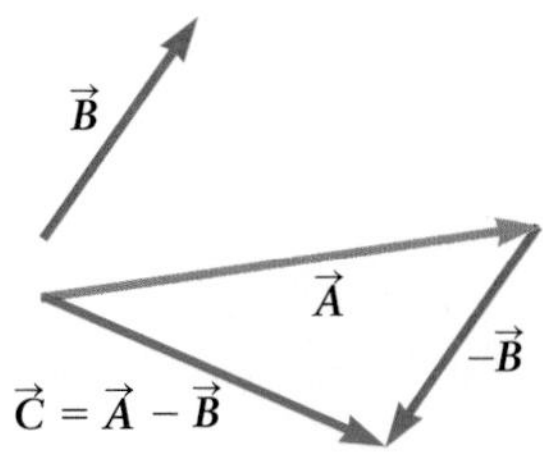

▲ **Figure 1.14** Vector subtraction. Calculating $\vec{A} - \vec{B}$ is equivalent to computing the sum of $\vec{A}$ and $-\vec{B}$.

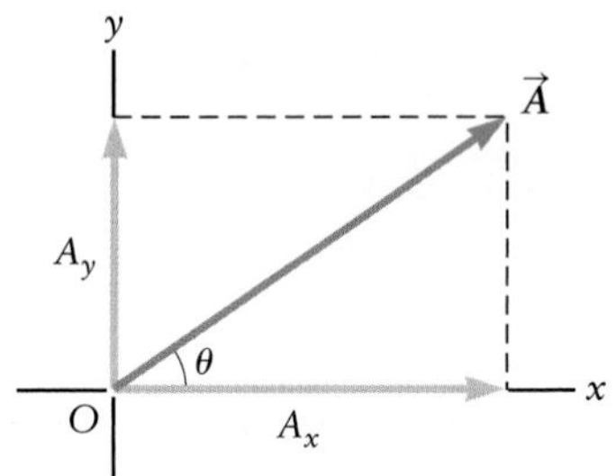

▲ **Figure 1.15** The components of a vector are the projections of the vector onto the coordinate axes. Here we show a vector that lies in the x–y plane, so it has components along the x and y axes. Some vectors have components along the z axis (not shown here).

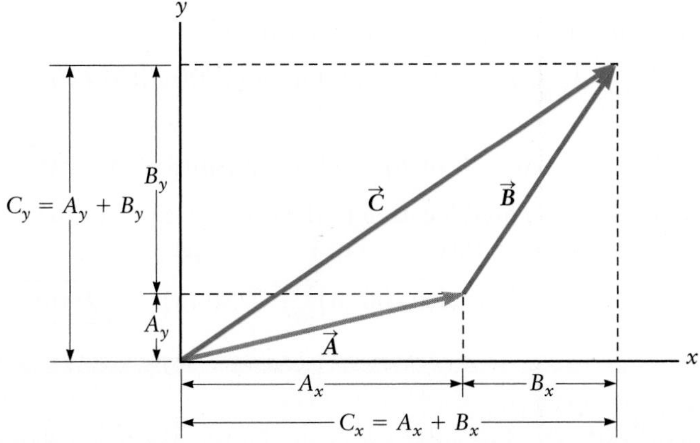

▲ **Figure 1.16** To compute the sum of two vectors, we add their components. Here we add the x components of $\vec{A}$ and $\vec{B}$ to get the x component of $\vec{C}$. We follow the same procedure to find the y component of $\vec{C}$.

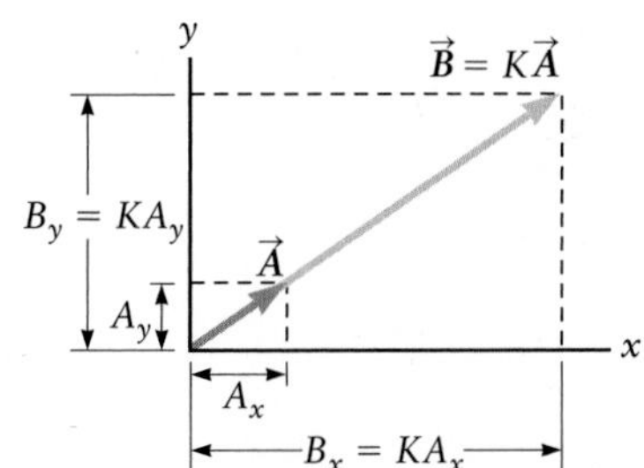

▲ **Figure 1.17** To multiply a vector $\vec{A}$ by a scalar K, we multiply each component of the vector by K.

EXAMPLE 1.7 Hiking and Vectors

A hiker is walking through the woods. He starts from his campsite, which is at the origin in Figure 1.18. He initially travels along the x axis in Figure 1.18A, and his initial displacement of 1200 m takes him to the tip of vector $\vec{A}$; hence, $\vec{A}$ has a length of 1200 m and is parallel to the x direction. He then turns and moves along the path described by vector $\vec{B}$ in Figure 1.18A. This vector has a length of 1500 m and makes an angle $\theta_B = 30°$ with the x axis. When the hiker stops, how far is he from the campsite?

RECOGNIZE THE PRINCIPLE

To find the final location of the hiker, we must add the vectors $\vec{A}$ and $\vec{B}$. We can do so by adding the components of these vectors: $\vec{C} = \vec{A} + \vec{B}$ (Eq. 1.34).

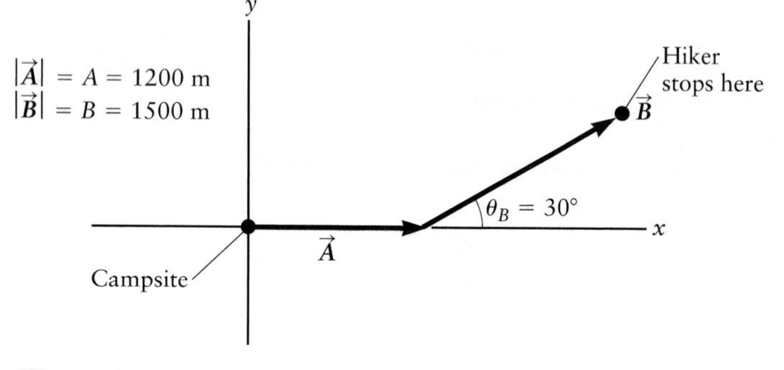

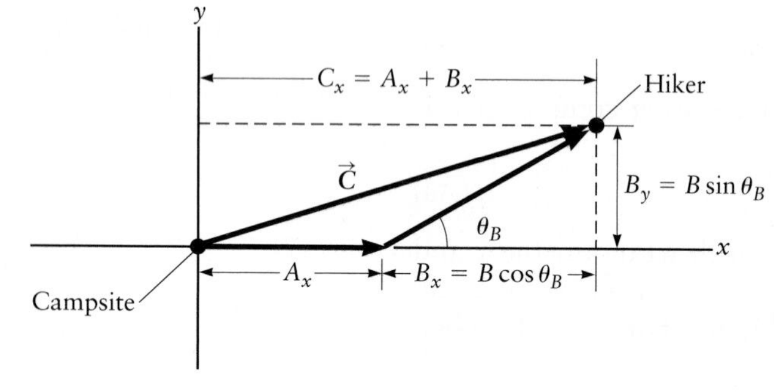

▲ **Figure 1.18** Example 1.7.

SKETCH THE PROBLEM

Figure 1.18 illustrates the problem. Figure 1.18B also shows the components of the various vectors, which will be needed in the solution.

IDENTIFY THE RELATIONSHIPS

According to Figure 1.18B, we can find the distance of the hiker to the campsite (i.e., the origin) if we know the components of the vectors $\vec{A}$ and $\vec{B}$ along the x and y axes. The vector $\vec{A}$ is parallel to x, so

$$A_x = A = 1200 \text{ m} \quad A_y = 0$$

where we again denote the length of the vector $\vec{A}$ by A. For $\vec{B}$, the trigonometry sketched in Figure 1.18B gives

$$B_x = B\cos\theta_B \quad B_y = B\sin\theta_B$$

The hiker's final location is described by the vector $\vec{C}$, which is equal to the sum of the vectors $\vec{A}$ and $\vec{B}$:

$$\vec{C} = \vec{A} + \vec{B}$$

In terms of components,

$$C_x = A_x + B_x \qquad C_y = A_y + B_y$$

SOLVE

Inserting our values for the components of $\vec{A}$ and $\vec{B}$ gives

$$C_x = A_x + B_x = A_x + B\cos\theta_B = 1200 \text{ m} + (1500 \text{ m})\cos(30°) = 2500 \text{ m}$$

and

$$C_y = A_y + B_y = 0 + B\sin\theta_B = (1500 \text{ m})\sin(30°) = 750 \text{ m}$$

The distance from the hiker's final location to the origin is equal to the length of the vector $\vec{C}$. From the Pythagorean theorem (Eq. 1.32), we get

$$C = \sqrt{C_x^2 + C_y^2} = \sqrt{(2500 \text{ m})^2 + (750 \text{ m})^2} = \boxed{2600 \text{ m}}$$

► ***What does it mean?***

Note that we have followed our rules for dealing with significant figures; the lengths of the vectors $\vec{A}$ and $\vec{B}$ and of the angle θ were given to two significant figures, so the answer C was also given with two significant figures. We'll work with vectors and their components in many problems. At this time, it might be valuable to review your trigonometry (see Appendix B).

Summary | CHAPTER 1

What is physics?

Physics is the science of matter and energy, and the interactions between them. The laws of physics enable us to predict and understand the way the universe works.

Problem solving

Problem solving is an essential part of physics. Problems may be quantitative with a precise answer, or they may be conceptual. For some problems, you may need to use common-sense reasoning to estimate the values of important quantities.

Scientific notation and significant figures

Scientific notation is used to express very large and very small numbers. We can also use prefixes to write very large or very small numbers, such as 1 micrometer = 1 μm = 1×10^{-6} meter and 1 megasecond = 1 Ms = 1×10^{6} seconds.

The accuracy of a quantity is reflected in the number of ***significant figures*** used to express its value. The result of a calculation should be expressed with an appropriate number of significant figures.

Units of measure

The ***primary units*** of mechanics involve length, time, and mass. Most scientific work employs the SI system of units, in which length is measured in ***meters***, time is measured in ***seconds***, and mass is measured in ***kilograms***. The values of the standard meter, the standard second, and the standard kilogram are established by international agreement and are a foundation for all scientific work.

Physical quantities and dimensions

We cannot give a definition of the concepts of length, time, and mass, but must instead treat them as "givens." The definitions of all other physical quantities encountered in mechanics can be derived from these three primary quantities. A total of seven primary physical quantities are needed to describe all of physics; the other four are connected with electricity, magnetism, and heat.

The ***dimensions*** of all quantities in mechanics can be expressed in terms of length **L**, mass **M**, and time **T**. Dimensional analysis involves checking that the dimensions of an answer correctly match the dimensions of the quantity being calculated.

The mathematics of physics

Several types of mathematics are used in this book: algebra, trigonometry, and vectors. Algebra is essential for solving systems of equations. When dealing with motion or other problems in physics, we often need to express position or movement in terms of a ***coordinate system*** (usually the x–y–z set of coordinate axes). Trigonometry and vectors are extremely useful in such calculations. A ***vector*** quantity has both a ***magnitude*** and a ***direction***. Vectors can be added graphically or in terms of their ***components***. Appendix B contains a quick review of algebra, trigonometry, and vectors.

Questions

SSM = answer in *Student Companion & Problem-Solving Guide* (X) = life science application

1. Suppose a friend told you that the density of a sphere is $\frac{4}{3}\pi r^4$, where r is the radius of the sphere. Compute the dimensions of this expression and show that it cannot be correct.
2. Discuss the difficulties of giving a definition of the concept of "time" without using any other scientific terms.
3. SSM Which of the following are units of volume?
   cubic meters — acres/m^2
   mm × mi^2 — hours × mm^3/s
   kiloseconds × ft^2 — mm × cm × mm^2/ft
   kg^2 × cm
4. What physical property, process, or situation is described by the following combinations of primary units? (For example, the combination of units kilograms per square meter can be used to measure the mass per unit area, sometimes called the areal density, of a thin sheet of metal.)
   m/s — m/s^2
   m^3/s — m^2/s
   kg/m
5. The standard meter used to be a metal bar composed of platinum and iridium. The current standard is based on the measured speed of light and an optical device called an interferometer. What advantages does the new standard have over the old?
6. SSM Which of the following quantities have the properties of a vector and which have the properties of a scalar?
   mass — density
   velocity — temperature
   displacement (change in position)
7. Give and discuss one advantage and one disadvantage of the SI system compared with the U.S. customary system of units.
8. **Astronomical distances.** Spiral galaxies like our own Milky Way measure approximately 2×10^5 light-years in diameter, where 1 light-year (ly) equals the distance traveled by light in 1 year. Our nearest-neighbor galaxy is the great spiral galaxy Andromeda, which has been determined to be approximately 2.5 million ly away. Construct a scale model to show the sizes of the Milky Way and Andromeda galaxies using two pie tins each 20 cm in diameter. To keep the model to scale, how far apart would these pie tins need to be separated?
9. A football field in the United States is 120 yards long (including the end zones). With how many significant figures do you think this value should be written?

Problems

SSM = solution in *Student Companion & Problem-Solving Guide* (R) = reasoning and relationships problem
★ = intermediate ✪ = challenging (X) = life science application (RT) = reasoning tutorial in Enhanced WebAssign

1.3 DEALING WITH NUMBERS

1. (R) The thickness of a typical piece of paper is 6×10^{-5} m. Suppose a large stack of papers is assembled, reaching to the top floor of the Empire State Building. Approximately how many pieces of paper are in the stack? *Hint:* The Empire State Building has 102 floors. You may want to begin by estimating (based on your experience) the height of a single floor.
2. You take a long walk on a sandy beach. When you return home, you dump the sand out of one of your shoes and find that the pile has a volume of 1.5 cm^3. If the volume of one grain of sand is approximately 0.1 mm^3, estimate the number of grains of sand that were in your shoe.
3. (RT) You decide to walk from Chicago to New York City. Approximately how many steps will you take during this journey?
4. (R) A red blood cell has a radius of approximately 5.0×10^{-6} m. A large number of red blood cells are laid side by side in a line of length 1.0 m. How many cells are in the line?
5. Determine the number of significant figures for all the values listed in Tables 1.2 and 1.3 (see page 11) that are not exact values.
6. A bucket contains several rocks with a combined mass (including the bucket) of 4.55 kg. Another rock of mass 0.224 kg is added to the bucket. What is the total mass of the bucket plus rocks? Be sure to give the correct number of significant figures in your answer.
7. Points *A*, *B*, *C*, and *D* in Figure P1.7 lie along a line. The distance from point *A* to point *B* is 3.45 km, and the distance from *B* to *C* is 5.4 km. What is the distance from *A* to *C*? Be sure to give the correct number of significant figures in your answer.

Figure P1.7 Problems 7 and 8.

8. The distance from point *B* to point *D* in Figure P1.7 is 3.15 km while the distance from *A* to *B* is 3.45 km. What is the distance from *A* to *D*? Be sure to give the correct number of significant figures in your answer.
9. The mass of an object is m = 23 kg, and its volume is V = 0.005 m^3. What is its density, m/V? Be sure to give the correct number of significant figures in your answer.
10. The magnitude of the momentum of an object is the product of its mass m and speed v. If m = 6.5 kg and v = 1.54 m/s, what is the magnitude of the momentum? Be sure to give the correct number of significant figures in your answer.
11. The speed of light is c = 299,792,458 m/s. (a) Write this value in scientific notation. (b) What is the value of c to four significant figures?
12. SSM For each of the following formulas and associated values, find and express the solution with the appropriate number of significant figures and units. (a) The area A of a rectangle of length L = 2.34 m and width w = 1.874 m. (b) The area A of a circle of radius r = 0.0034 m. (c) The volume V of a cylinder with height $h = 1.94 \times 10^{-2}$ m and radius $r = 1.878 \times 10^{-4}$ m. (d) The perimeter P of a rectangle of length L = 207.1 m and width w = 28.07 m.
13. (RT) A hydrogen atom has a diameter of approximately 10^{-10} m. If this atom were expanded to the size of an apple, how big would a real apple be if it were expanded by the same amount? Can you think of an object whose size is comparable to this gigantic apple?
14. (X) (R) (a) Estimate the number of times your heart beats in 1 year. Express your answer in scientific notation. (b) Some people have (jokingly) suggested that each person has only a certain number of heartbeats before he or she dies. The average life expectancy for a person born in the United States in 2004

is 77.9 years. If the average person's heart rate were reduced by 12%, how long would he or she live? Be sure to give the correct number of significant figures in your answer.

15. Write the following values using scientific notation. (a) The radius of the Earth, 6,370,000 m. (b) The distance from the Earth to the Moon, 384,000 km. (c) The mass of the Earth, 5,970,000,000,000,000,000,000,000 kg.

1.4 PHYSICAL QUANTITIES AND UNITS OF MEASURE

1.5 DIMENSIONS AND UNITS

16. Express the distance from Chicago to New York City (Table 1.1, page 5) in (a) miles, (b) inches, (c) kilometers, and (d) micrometers.

17. Ⓡ Calculate the volume of a typical 1000-page textbook. Express your answer in cubic meters (m^3) and in cubic centimeters (cm^3). *Hint*: You will need to use a ruler (or some other means) to find the dimensions of the book.

18. RT What is the volume of a basketball? *Hint*: You will have to estimate, measure, or find a basketball's diameter.

19. SSM A U.S. football field is 120 yards long (including the end zones). How long is the field in (a) meters, (b) millimeters, (c) feet, and (d) inches?

20. The age of the universe is about 5×10^{17} s (Table 1.2, page 11). Express this age in years.

21. Express 5.0 m in units of (a) centimeters, (b) feet, (c) inches, and (d) miles.

22. Express 75 ft in units of (a) meters and (b) millimeters.

23. An object has a mass of 273 g. What is its mass in kilograms?

24. A card table has a top surface area of exactly 1 square meter (1 m^2). What is its area as measured in (a) square centimeters and (b) square millimeters?

25. A bottle has a volume of 1.2 L. (a) What is its volume in cubic centimeters? (b) In cubic meters?

26. Michael Jordan is approximately 6.5 ft tall. What is his height in millimeters?

27. A cardboard box can hold exactly 1 cubic meter of volume (1 m^3). Use the conversion factors 1 m = 100 cm = 1000 mm to find this volume expressed in units of (a) cubic centimeters and (b) cubic millimeters.

28. It takes about 190 s for light to travel from the Sun to the planet Mercury (the planet closest to the Sun) and about 15,000 s for light to travel from the Sun to Neptune (the farthest planet). Suppose it takes light 950 s to travel from the Sun to a particular asteroid. How many minutes does it take?

29. **Sports cars.** A Corvette C4 (circa 1971) came equipped with a huge 454 V8 engine, where the volume of the combustion chambers summed to 454 cubic inches (in.^{3}). In modern engines, this volume is measured in SI units. For example, a 2005 Corvette Z06 touts a 7.0-liter V8 engine. Which engine has the larger combustion volume? Is it surprising that the 2005 Corvette engine is 19% more powerful than the C4's 454 V8?

30. Consider the equation $mgh = \frac{1}{2}mv^2$, where m has units of mass (kilograms), g has units of length/time^2 (m/s^2), h has units of length (meters), and v has units of length/time (m/s). (a) Is this equation dimensionally correct or incorrect? (b) Is the equation $v^2/h = g$ dimensionally correct or incorrect? (c) What are the units of the combination g/v^2?

31. The variable m has dimensions of mass, h and y have dimensions of length, and t has dimensions of time. What are the dimensions of the following quantities? (a) my/t, (b) hy/t^2, (c) $y^3/(ht)$

1.6 ALGEBRA AND SIMULTANEOUS EQUATIONS

32. Consider the equation $5x - 17 = 10x - 27$. Solve for x.

33. The speed v of a bullet satisfies the relation $vt = 95$ m, where $t = 0.25$ s. Find v.

34. SSM ★ Consider the equations

$$5x + 2y = 13 \qquad -3x + 7y = 25$$

(a) Put each equation into the standard format for a straight line ($y = mx + b$) and plot both lines. Use your graph to find the point (x, y) where these lines intersect. (b) Use algebra to find the values of x and y that simultaneously satisfy these two equations. (c) How do your answers to parts (a) and (b) compare?

35. ★ Two well-known organic chemicals are benzene, C_6H_6, and propane, C_3H_8. Benzene has a molecular weight of 78.11 g/mol and propane 44.096 g/mol, where the molecular weight is a useful measure in chemistry and is defined as the mass of 1 mole (Avogadro's number) of molecules or atoms. Determine the molecular weight (in g/mol) of pure hydrogen and pure carbon. *Hint*: The subscripts in the molecular formula are the number of atoms of that type per molecule. Set up two simultaneous equations to solve for the molecular weights.

1.7 TRIGONOMETRY

36. SSM ★ Consider the motion of the hour hand of a clock. What angle does the hour hand make with respect to the vertical 12 o'clock position when it is (a) 3:00, (b) 6:00, (c) 6:30, (d) 9:00, and (e) 11:10? Express each angle in both degrees and radians, and measure angles going clockwise from the vertical axis.

37. A right triangle (Fig. 1.8) has sides of length $x = 4.5$ m and $y = 3.7$ m. (a) Find the value of r (the length of the hypotenuse). (b) Find the values of the interior angles. Express your answers in degrees and radians. (c) Find the sine, cosine, and tangent of the smaller interior angle. (d) Find the sine, cosine, and tangent of the larger interior angle.

38. A right triangle has an interior angle of 25°. What are the values of the other interior angle (the one besides the right angle)?

39. ★ Find values of x, y, and r in Figure 1.8 that will give a right triangle in which one of the triangle's interior angles is 0.70 radian.

40. An angle has a value of 73°. Find its value in radians.

41. An angle has a value of 1.25 radians. Find its value in degrees.

42. A car travels on a circular track (Fig. P1.42) of radius $r = 38$ m. If the car travels a distance of 210 m along the track, through what angle (in rad) does the car move?

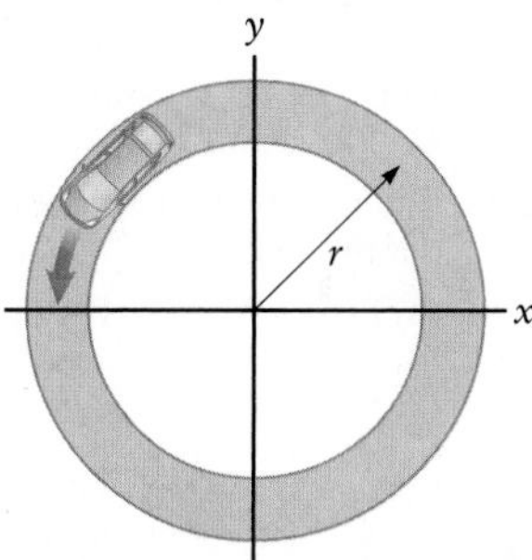

Figure P1.42 Problems 42 and 43.

43. ★ The car in Figure P1.42 travels a distance of 750 m along the track, and the driver finds that she has traveled through an angle of 450°. What is the radius of the track?

44. A ladder of length 2.5 m is leaned against a wall (Fig. P1.44). If the top of the ladder is 1.7 m above the floor, what is the angle the ladder makes with the wall? What angle does it make with the floor?

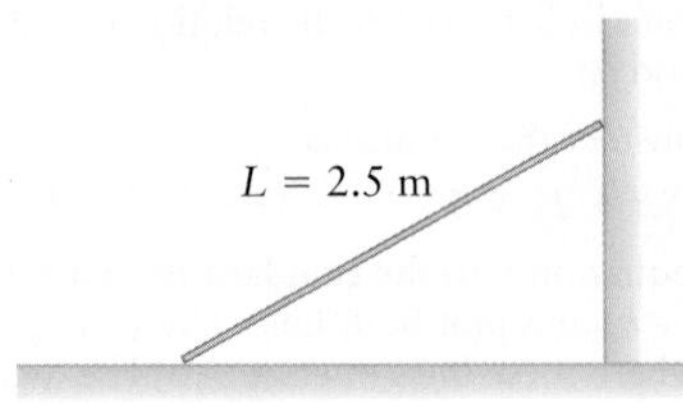

Figure P1.44

45. Steep grade. A mountain road makes an angle $\theta = 8.5°$ with the horizontal direction (Fig. P1.45). If the road has a total length of 3.5 km, how much does it climb? That is, find h.

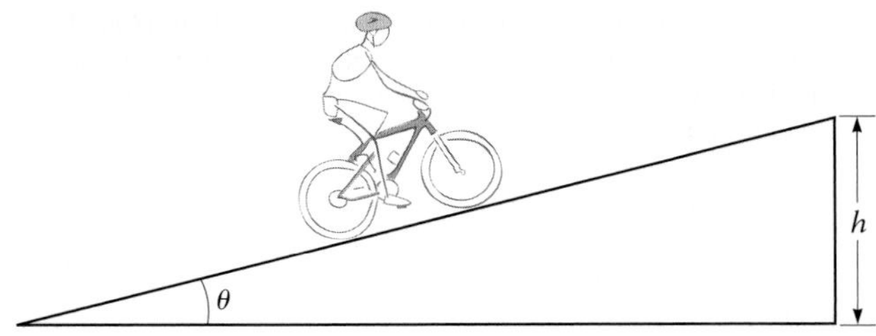

Figure P1.45

1.8 VECTORS

46. Consider the vectors $\vec{A}$ and $\vec{B}$ in Figure P1.46. (a) Draw a picture that graphically constructs the sum of these two vectors $\vec{A} + \vec{B}$ (i.e., similar to what was done in Fig. 1.12). Also construct $\vec{A} - \vec{B}$, $\vec{B} - \vec{A}$, $\vec{A} - 5\vec{B}$, and $-1.5\vec{A}$. (b) Suppose $\vec{A}$ has a length of 2.3 m and $\vec{B}$ has a length of 1.5 m. Use your graph to estimate the lengths of all the vectors in part (a).

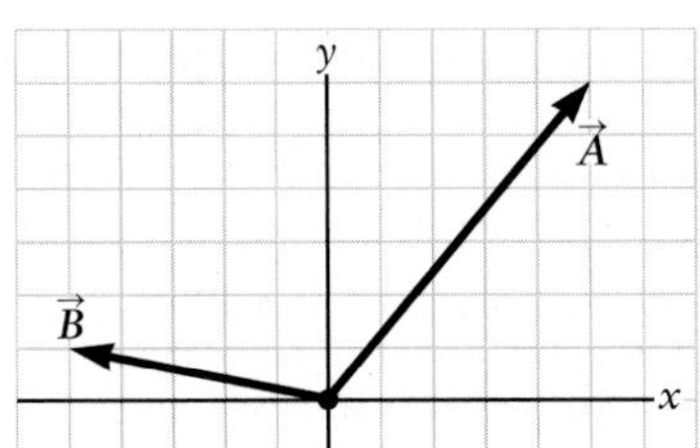

Figure P1.46

47. A vector has components $A_x = 54$ m and $A_y = 23$ m. Find the length of the vector $\vec{A}$ and the angle it makes with the x axis.

48. A vector of length of 2.5 m makes an angle of 38° with the y axis. Find the components of the vector.

49. Which of the vectors in Figure P1.49 have approximately the same magnitude?

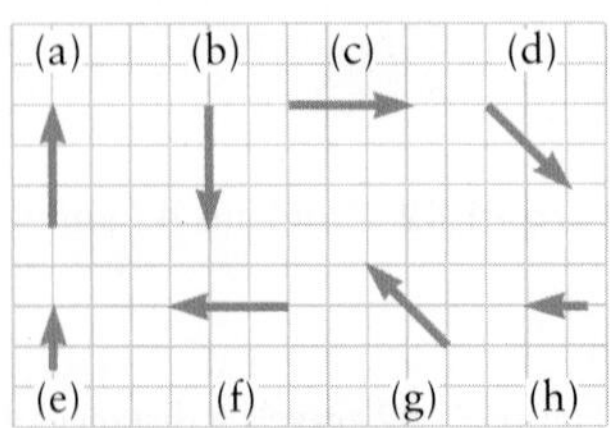

Figure P1.49 Problems 49 and 50.

50. Which of the vectors in Figure P1.49 have approximately the same direction?

51. Which vector in Figure P1.51 might equal $\vec{A} + \vec{B}$?

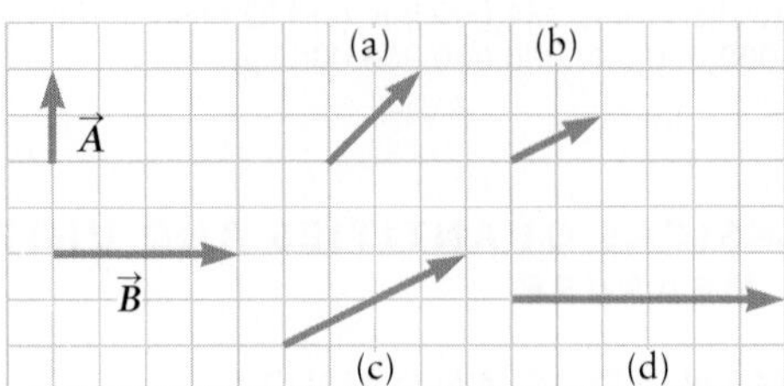

Figure P1.51

52. A vector in the x–y plane has a length of 4.5 m and a y component of –2.7 m. What angle does this vector make with the x axis? Why does this problem have two answers?

53. Consider a map oriented so that the x axis runs east–west (with east being the "positive" direction) and y runs north–south (with north "positive"). A person drives 15 km to the west and then turns and drives 45 km to the south. Find the magnitude and direction of the total movement of the driver. Express the direction as an angle measured counterclockwise from the positive x axis.

54. Consider a map oriented so that the x axis runs east–west (with east being the "positive" direction) and y runs north–south (with north "positive"). A person drives 25 km to the north, turns and drives 75 km to the east, and then turns north and drives for an unknown distance z. If his final position is 95 km from where he started, find z.

55. Consider the vector $\vec{C}$ in Figure P1.55. Construct a drawing that shows the vector $\vec{C}$ along with the vector $4\vec{C}$, $-\vec{C}$, and $-3\vec{C}$.

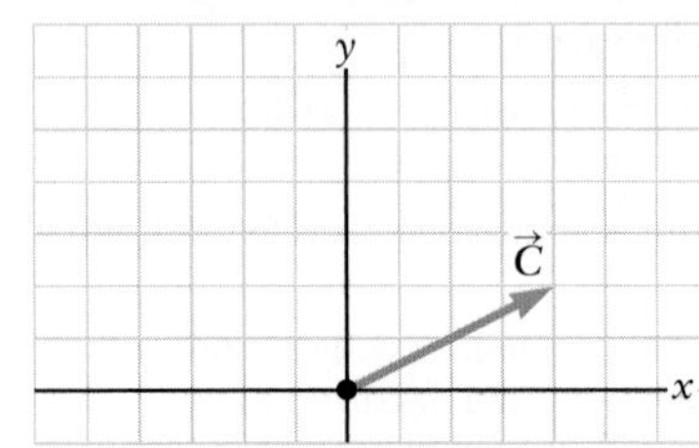

Figure P1.55

56. A vector $\vec{A}$ has a magnitude of 15 (in some unspecified units) and makes an angle of 25° with the x axis, and a vector $\vec{B}$ has a magnitude of 25 and makes an angle of 70° with the x axis. Compute the magnitudes and directions of the vectors:

(a) $\vec{C} = \vec{A} + \vec{B}$ (b) $\vec{C} = \vec{A} - \vec{B}$
(c) $\vec{C} = \vec{A} + 4\vec{B}$ (d) $\vec{C} = -\vec{A} - 7\vec{B}$

57. Find the components of the vector $\vec{C}$ in parts (a) through (d) of Problem 56.

58. A boater travels from some initial point P_1 to a final point P_2 along the dashed line shown in Figure P1.58. Find θ and the total distance traveled.

59. A car starts from the origin and is driven 1.2 km south and then 3.1 km in a direction 53° north of east. Relative to the origin, what is the car's final location? Express your answer in terms of a distance and an angle.

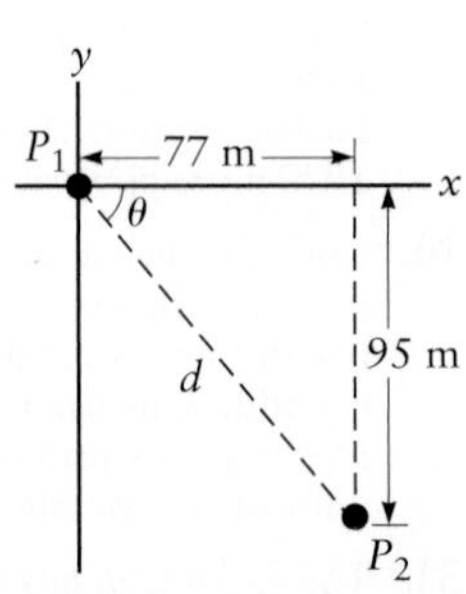

Figure P1.58

60. SSM ★ A mountaineer uses a global positioning system receiver to measure his displacement from base camp to the top of Mount McKinley. The coordinates of base camp are $x = 0$, $y = 0$, and $z = 4300$ m (here z denotes altitude), and those of the top of Mount McKinley are $x = 1600$ m, $y = 4200$ m, and $z = 6200$ m. What is the magnitude of the displacement in going from base camp to the top?

Additional Problems

61. ⓧ RT Estimate the number of times a student's heart will beat during the 4 years she is a college student.

62. SSM ★ Ⓑ **The solar system in perspective.** The diameters of the Earth and the Sun are approximately 1.3×10^7 m and 1.4×10^9 m respectively, and the average Sun–Earth distance is 1.5×10^{11} m. Consider a scale model of the solar system, where the Earth is represented by a peppercorn 3.7 mm in diameter. (a) Calculate the diameter of the Sun in this model. Name an object of this size that could be used in the model. (b) To keep the model in scale, how far would the peppercorn Earth be from the center of the model Sun? (c) Pluto's orbit can take it as far as 5.9×10^{12} m away from the Sun. A grain of sand could represent Pluto in our model. How far away from the model Sun would this grain of sand be?

63. The unit of area called the acre comes from an old system of measurement in which 1 acre is the area of a rectangle whose length is 220 yards and whose width is 22 yards. (a) Determine the number of acres in 1 square mile. (b) How many square meters are in 1 acre? (c) How many square feet are in 1 acre? In 1 square mile?

64. Ⓑ Consider the following three vectors, described here by their lengths and angles with respect to the x axis: $\vec{A}$: 5.0 cm, 30°; $\vec{B}$: 7.5 cm, 0°; and $\vec{C}$: 10.0 cm, 350°. (a) Using the tip-to-tail graphical method for vector addition, show that the order in which these vectors are added does not change the final displacement, $\vec{D}$. That is, show that the following vector equation holds: $\vec{D} = (\vec{A} + \vec{B}) + \vec{C} = \vec{A} + (\vec{B} + \vec{C})$. (b) Use your graph to estimate the length of $\vec{D}$ and the angle it makes with the positive x axis.

65. ★ You wish to replace the flooring in your kitchen with square linoleum tiles that measure 12 inches on a side. If your kitchen has a floor area of 10.7 square yards, how many such tiles will you need to buy?

66. ★ A river flows with a current of 5.0 mi/h directly south. A small boat has a maximum speed in still water (i.e., on a quiet lake) of 10 mi/h. The pilot points the boat due east and sets the engine at full throttle to move the boat at maximum speed. In what direction does the boat actually move? *Hint*: The velocities of the river and the boat in still water can both be described as vectors, each having a magnitude and direction. The total velocity of the boat when traveling in the river is the sum of these two vectors.

67. ★ ⓧ Ⓑ Approximately how many cells are there in the human body? *Hint*: The volume of an average cell is about 1×10^{-16} m^3.

68. ★ ⓧ Ⓑ Approximately how fast does human hair grow? Express your answer in m/s.

69. ★ Ⓑ A typical human baby at birth is about 20 inches as measured from head to toe. An adult is much longer. What is the approximate ratio of the height of an adult female to the length of a baby at birth?

70. SSM ★ ⓧ Ⓑ At room temperature, 1.0 g of water has a volume of 1.0 cm^3. (a) What is the approximate volume of one water molecule? (b) The human body is composed mainly of water. Assuming for simplicity your body is only water (and nothing else), find the approximate number of water molecules in your body.

71. A person stretches a rope of length 55 m between the top of a building to the base of a tree 25 m away (Fig. P1.71). How tall is the building?

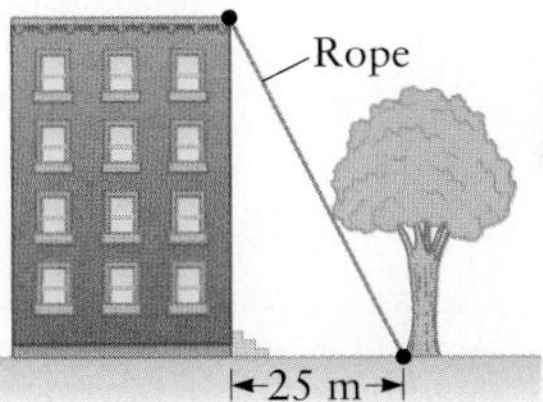

Figure P1.71

72. ★ The radius of Jupiter is 11 times as large as the radius of Earth. What is the ratio of the volume of Jupiter to the volume of Earth?

73. ★ **If Saturn could fit in a bathtub, would it float?** Use the planetary data in Appendix A to calculate the density (mass/volume) of Earth, Mars, Jupiter, and Saturn. What are the differences? Can you explain them? If an object has a density less than 1000 kg/m^3, it will float in water. Would any of these planets float (assuming a big enough container of water!)? (*Note*: Your calculated values of the density should assume spherical planets. The density values in Table A.3 do not make this assumption and are therefore more accurate.)

74. ★ Two vectors have lengths $A = |\vec{A}| = 12$ m and $B = |\vec{B}| = 17$ m, but their directions are unknown. (a) What are the maximum and minimum possible values of $\vec{A} + \vec{B}$? (b) What are the maximum and minimum values of $\vec{A} - \vec{B}$?

75. ★ Ⓑ You are on a road trip, traveling by expressway from one city to another, without stopping and without leaving the expressway. If you obey the speed limit, approximately how far do you travel in 30 minutes? Express your answer in miles and in kilometers.

CHAPTER

2

▶ *The laws of mechanics describe the motion of objects such as this skateboarder. (© Rubberball/Mike Kemp/Getty Images)*

Motion, Forces, and Newton's Laws

OUTLINE

We begin our study of physics with the field known as ***mechanics***. This area of physics is concerned with the motion of objects such as rocks, balloons, cars, water, planets, and skateboarders. To understand mechanics, we must be able to answer two questions. First, what *causes* motion? Second, given a particular situation, *how* will an object move?

The laws of physics that deal with the motion of terrestrial objects were developed over the course of many centuries, culminating with the work of Isaac Newton and the formulation of what are known as Newton's laws of motion. Newton's laws are a cornerstone of physics and are the basis for nearly everything we do in the first part of this book. Even so, it is useful to begin with a brief discussion of some of the ideas that dominated physics before Newton's time. Those ideas originated many centuries ago with the Greek philosopher–scientist Aristotle. Although Aristotle was a leading scientist of his era and was certainly a very deep thinker, we now know that many of his ideas about how objects move were not correct. Why, then, should we now be concerned with such historical (and incorrect) notions about motion? We examine Aristotle's ideas because many students begin this course with some of the *same* incorrect notions. A good way to reach a thorough understanding of the physics of motion is to consider the origins of Aristotle's ideas and identify where they go wrong.

2.1 Aristotle's Mechanics

Aristotle began by identifying two general types of motion: (1) celestial motion, the motion of things like the planets, the Moon, and the stars; and (2) terrestrial motion, the motion of "everyday" objects such as rocks and arrows. We all know that while rocks and other terrestrial objects can move in a variety of ways, they usually come to rest eventually. On the other hand, it appears that celestial objects never come to a stop. According to Aristotle, terrestrial objects move only when something acts on them directly. A rock moves because some other object, such as a person's hand, acts on it to make it move. In contrast, there does not seem (at least to Aristotle) to be anything acting on the Moon to cause its motion. Aristotle thus raised a key point: the motions of celestial and terrestrial objects look very different. A theory of mechanics must explain why.

Aristotle asserted that the "natural" state for a terrestrial object is a state of rest and that this is why terrestrial objects move only when acted on by another object. In modern terminology, Aristotle claimed that an object only moves when acted on by a ***force***. A force is simply a push or a pull, and Aristotle believed that a push or a pull on one object is always produced by a second object. Furthermore, he claimed that a force could only exist if the two objects are in direct contact.

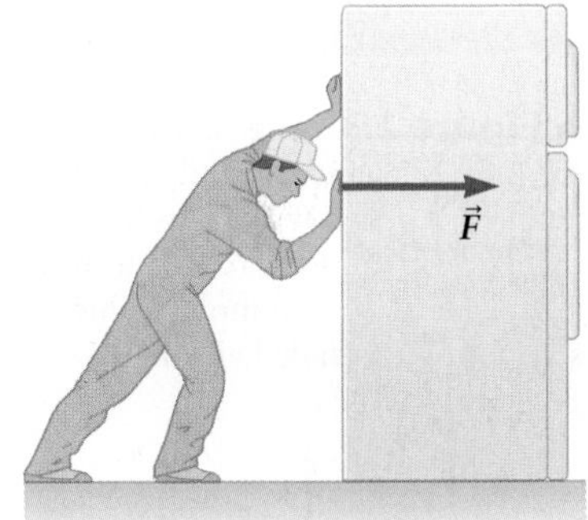

▲ **Figure 2.1** This person is exerting a force, denoted by $\vec{F}$, on the refrigerator.

Hence, Aristotle believed that (1) motion is caused by forces and (2) forces are produced by contact with other objects. These ideas *seem* quite reasonable when we consider something like a refrigerator being pushed along a level floor as in Figure 2.1. Our everyday experience is that the refrigerator only moves while someone is pushing on it. If the person stops pushing, the refrigerator quickly comes to a stop. The person pushing on the refrigerator exerts a force directly on the refrigerator, and the person is in contact with the refrigerator. So far, so good.

A key aspect of mechanics is the notion of force. We have already noted that a force is simply a push or a pull on an object. A force has both a magnitude and a direction, so *force is a vector quantity*, often denoted by $\vec{F}$. The magnitude of a force is the strength of the push or the pull, while the direction of this vector is the direction of the push or the pull.

Force is a *vector* quantity. It has a magnitude and a direction.

We next need to consider how to describe motion. One quantity we normally associate with motion is ***velocity***. Velocity is also a vector quantity and is usually represented by the symbol $\vec{v}$. A careful mathematical definition of velocity is given in the next section. For our discussion of Aristotle's ideas, it is enough to know that the magnitude of $\vec{v}$ is the distance that an object travels (as might be measured in meters) per second. The direction of the vector $\vec{v}$ is the direction of motion (Fig. 2.2).

▲ **Figure 2.2** The velocities of an airplane and a train are vectors, having both magnitude and direction.

Aristotle proposed that force and velocity are directly connected. In the mathematical language of today, he would have written

$$\vec{v} = \frac{\vec{F}}{R} \quad \text{(INCORRECT!)} \tag{2.1}$$

In words, this relationship says that the velocity $\vec{v}$ of an object is proportional to the force $\vec{F}$ that acts on it, with the constant of proportionality being related to a quantity R, which is the resistance to motion. We'll refer to this relationship as "Aristotle's law of motion." We also caution that this relationship is *not* a correct law of physics. It does, however, seem to explain the motion of the refrigerator in Figure 2.1. If no force is exerted on the refrigerator, then $\vec{F} = 0$. According to Equation 2.1, the velocity $\vec{v}$ is then also zero and hence the refrigerator does not move. When a person

© Mike Zarilli/Getty Images

▲ **Figure 2.3** After it leaves the pitcher's hand, the only forces acting on the baseball are a force from gravity and a force from the air through which it moves. This second force is called air drag.

▲ **Figure 2.4** An archer shoots an arrow straight upward. While the arrow is traveling on the upward part of its trajectory, the forces acting on it—from gravity and from air drag—are both directed downward. Hence, these forces are not parallel to the velocity of the arrow while it is traveling upward.

pushes on the refrigerator, $\vec{F}$ is nonzero and Aristotle's theory (Eq. 2.1) predicts that $\vec{v}$ is therefore also nonzero. Since $\vec{v}$ and $\vec{F}$ are both vectors, Equation 2.1 asserts that the velocity and the force are always in the same direction.

The Failures of Aristotle's Ideas about Mechanics

Aristotle's law of motion—Equation 2.1—*seems* to explain the motion of the refrigerator in Figure 2.1, but it does *not* work as well in other situations. Consider the motion of a thrown baseball (Fig. 2.3). After a baseball leaves the thrower's hand, it continues to move until it hits the ground, is struck by a bat, or is caught by someone. This motion is *not* consistent with Equation 2.1. Recall Aristotle's assertion that forces are always caused by direct contact with another object. While the ball is traveling through the air, there does not appear to be another object in contact with it, so according to this reasoning the force on the baseball is zero after it leaves the thrower. If the force is zero, Equation 2.1 says that the ball's velocity should also be zero. However, we know from experience that a thrown baseball continues to move! One way to attempt an escape from this dilemma is to recognize that the ball also experiences a force due to gravity. Unfortunately (at least for Aristotle), the gravitational force does not fit into Aristotle's ideas about forces because there is nothing in "direct" contact with the baseball to produce this force. Even if we overlook this difficulty, we still have to consider that the gravitational force on the ball is directed downward (toward the surface of the Earth) and hence the gravitational force and the velocity are not in the same direction for the ball in Figure 2.3. Figures 2.3 and 2.4 both illustrate situations that contradict Equation 2.1, which predicts that force and velocity are always parallel. It is particularly puzzling (at least for Aristotle) that objects such as baseballs and arrows can be traveling *upward* when the forces acting on them are directed *downward*.

More problems with Aristotle's law of motion can be seen when we consider falling objects. If someone simply drops a baseball, it will fall to the ground. There is nothing in "direct contact" with the ball to cause it to fall. What, then, makes it move from the person's hand? Because the case of falling objects is quite common, Aristotle gave it special attention. He proposed that there is a special force that does not require contact with anything. This force is the ***weight*** of the object, which Aristotle believed was equivalent to what we now call mass.[1] Aristotle also thought that the resistance factor R in Equation 2.1 is a property of the substance through which the object moves. For an object dropped near the Earth's surface, this substance is air. This reasoning leads one to expect that if Equation 2.1 is correct, a heavy object will fall faster than a light one when both are dropped in the same medium. Thanks to experiments performed by Galileo (Fig. 2.5), we know that is *not* the case. Galileo showed that light objects fall at the same rate as heavy objects.

Despite all these difficulties, Aristotle's ideas have a certain appeal. They are based on the notion that an object's velocity is directly proportional to the force acting on the object, which at first glance seems plausible. However, we have seen a number of very simple situations in which this notion fails badly. In the next few sections, we discuss the connection between force and motion in more detail and arrive at the laws of motion discovered by Isaac Newton. We'll then see how Newton's laws overcome the difficulties that frustrated Aristotle.

Although we must reject Aristotle's theory of motion, it is still very instructive to understand where his theory goes wrong and why. Some of the everyday ideas that *you* have brought to this course may be similar to those of Aristotle. Understanding precisely when and why those ideas fail will help you understand and appreciate the correct way to describe motion.

[1]We'll see in the next chapter that mass and weight are *not* the same. Recall that mass is a fundamental physical quantity, as explained in Chapter 1. Weight is defined carefully in Chapter 3.

CONCEPT CHECK 2.1 Force and Motion

Do the following examples of motion appear to be *consistent* or *inconsistent* with Aristotle's law of motion?

(a) A hockey puck sliding along an icy horizontal surface
(b) A car coasting along a level road
(c) A piano that is pushed across a room

2.2 What Is Motion?

It is now time to consider how motion is described and measured in a precise, mathematical sense. The concept of motion is more complicated than you might first guess, and we'll need several quantities—***position***, ***velocity*** (which we have already encountered), and ***acceleration***—to describe it fully.

Let's first consider motion along a straight line, called ***one-dimensional motion***. A hockey puck sliding on a horizontal icy surface is a good example of this type of motion, and the top portion of Figure 2.6A shows what would happen if we took a multiple-exposure photo of such a hockey puck. Figure 2.6A is called a ***motion diagram***, a multiple-exposure photograph or similar sketch that shows the location of an object at regularly spaced instants in time. These exposures are captured at evenly spaced time intervals, and we can use them to construct the graph of the puck's position as a function of time shown in Figure 2.6B. Here we measure position as the distance from the origin on the x axis in Figure 2.6A to the center of the hockey puck. For one-dimensional motion, this distance, which we denote by x, completely specifies the position of the object. Notice in Figure 2.6B that the x axis is now vertical as we plot the position (x) as a function of time (t), which is plotted along the horizontal axis.

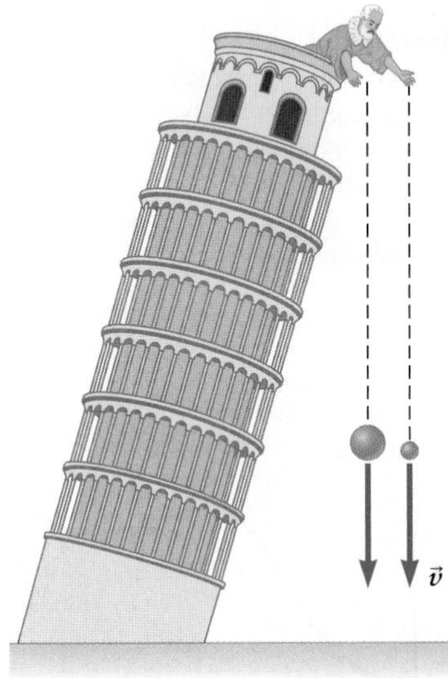

▲ **Figure 2.5** Galileo is reported to have used the Leaning Tower of Pisa in Italy in his studies of falling objects. Although the story may not be strictly accurate, Galileo certainly did conduct experiments showing that light and heavy objects fall at the same rate.

Velocity and Speed

The distance between adjacent dots on the x axis in Figure 2.6A shows how far the puck has moved during each time interval. We have already mentioned that velocity $\vec{v}$ is a vector quantity. The magnitude of $\vec{v}$ is called the ***speed***, the distance traveled per unit of time, while the direction of $\vec{v}$ gives the direction of the motion (in this example, $\vec{v}$ is directed to the right). Speed is a *scalar* quantity; it does not have a direction. In SI units, position is measured in meters and time is measured in seconds, so velocity and speed are both measured in meters per second, or simply m/s.

Speed and velocity are related quantities, but they are *not* the same. Speed tells how fast an object is moving, and it is always a positive quantity (or perhaps zero). The velocity contains this information and in addition tells the *direction* of motion. For the hockey puck in Figure 2.6A, the direction may be positive (motion to the right, toward larger or more positive values of x) or negative (motion to the left, toward smaller or more negative values of x). For one-dimensional motion, the direction of the velocity vector must lie parallel to the x axis. Thus, in cases involving one-dimensional motion, we need only deal with the *component* of the velocity parallel to x. This component can be positive, negative, or zero.

Vector quantities are usually written with arrows overhead, so the velocity vector is generally written as $\vec{v}$. For one-dimensional motion, the velocity vectors have only one component, and we can refer to this component as simply v, without an arrow. The *sign* of v (either positive or negative) then gives the direction of the velocity. Thus, for motion in one dimension,

$$\text{speed} = |v| \quad \text{(one dimension)} \tag{2.2}$$

In words, this expression says that speed is equal to the magnitude of the velocity. For two- or three-dimensional motion,

$$\text{speed} = |\vec{v}| \quad \text{(two or three dimensions)} \tag{2.3}$$

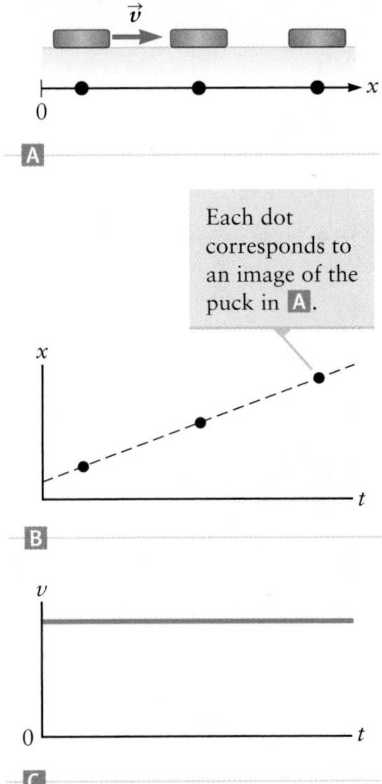

▲ **Figure 2.6** (A) Multiple images of a hockey puck traveling across an icy surface. (B) Plot of the position x of the puck as a function of time. The dots correspond to the images of the puck in part A. (C) Velocity v of the puck as a function of time.

Insight 2.1

SPEED AND VELOCITY

Speed is equal to the *magnitude* of the velocity. For motion in two or three dimensions, velocity is a vector. In these cases, speed is the *magnitude* of the velocity vector $\vec{v}$.

where the vertical bars again indicate that the speed is equal to the magnitude of the velocity vector. In either case (Eq. 2.2 or 2.3), speed is always a positive quantity or zero, but never negative.

The hockey puck in Figure 2.6A is sliding at a constant speed, so its velocity has a constant value as shown in the velocity–time (v–t) graph in Figure 2.6C. In this case, the velocity is positive, which means that the direction of motion is toward increasing values of x (i.e., to the right).

How Is an Object's Velocity Related to Its Position?

Velocity is the change in position per unit time. Since time is measured in seconds, it is natural to think about the position at 1-s time intervals as suggested by the dots in Figure 2.6B. Alternatively, we could consider a particular time interval that begins at time t_1 and ends at time t_2 so that the size of the time interval is $\Delta t = t_2 - t_1$. The change in position during a particular time interval is called the ***displacement.*** For one-dimensional motion, displacement is denoted by $\Delta x = x_2 - x_1$, where x_1 is the ***initial position***, the position at the beginning of the time interval (t_1), and x_2 is the ***final position***, the position at the end of the interval (t_2). The ***average velocity*** during this time interval is

$$v_{\text{ave}} = \frac{x_{\text{final}} - x_{\text{initial}}}{t_{\text{final}} - t_{\text{initial}}} = \frac{x_2 - x_1}{t_2 - t_1}$$

$$v_{\text{ave}} = \frac{\Delta x}{\Delta t} \tag{2.4}$$

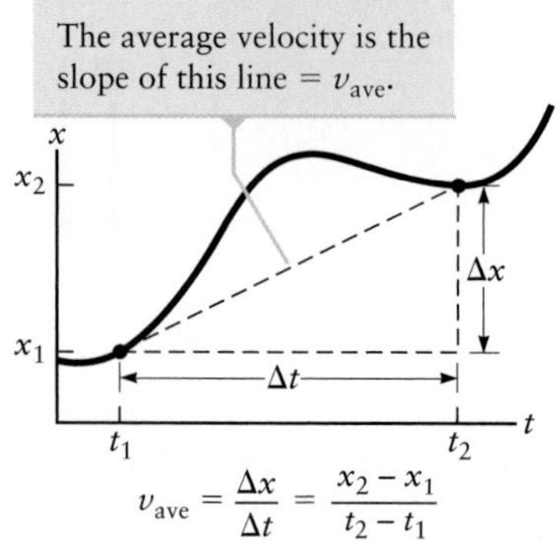

▲ **Figure 2.7** Hypothetical plot of an object's position as a function of time. The average velocity during the time interval from t_1 to t_2 is the slope of the line connecting the two corresponding points on the x–t curve.

According to Equation 2.4, the average velocity is the slope of the line segment that connects the positions at the beginning and end of the time interval. This is illustrated in the hypothetical x–t graph in Figure 2.7.

Another example of motion along a line is the case of a rocket-powered car traveling on a flat road. Let's assume the car is at rest when our clock reads zero. At $t = 0$, the driver turns on the rocket engine and the car begins to move in a straight-line path, along a horizontal axis we denote as x. Figure 2.8A is a motion diagram showing the position of the car at evenly spaced instants in time. The x axis is along the road, and the position of the car at a particular instant is the distance from the origin to the center of the car. The corresponding position–time graph for the car is shown in Figure 2.8B, where the dots mark the car's position at evenly spaced time intervals. In this case, the spacing between dots increases as the car travels. So, the car moves a greater distance during each successive (and equal) time interval and hence the speed of the car increases with time. The car moves toward increasing values of x, so the velocity is again positive and v increases smoothly with time as shown in Figure 2.8C. The precise shape of the v–t curve will depend on the way the engine fires, a problem we will consider in Chapter 3.

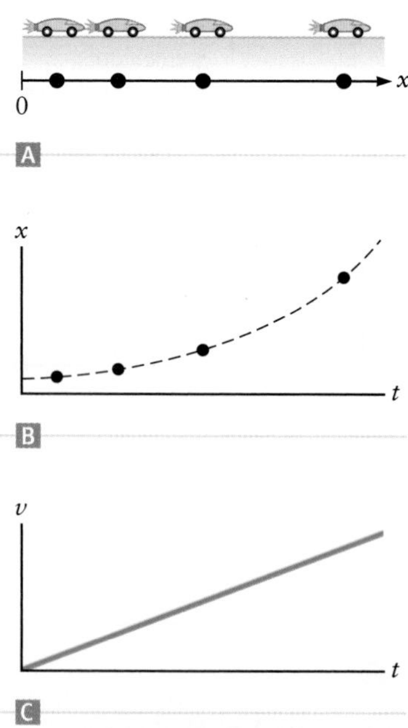

▲ **Figure 2.8** A Motion diagram for a rocket-propelled car traveling along a horizontal road. B Position as a function of time for the rocket-powered car. C Velocity of the car as a function of time.

Average Velocity and Instantaneous Velocity

Figure 2.9A shows the position as a function of time for a hypothetical object moving in a straight line, using dots to mark the position at the beginning and end of a time interval that starts at $t = 1.0$ s and ends at $t = 2.0$ s. According to Equation 2.4, the average velocity during this time interval is just the displacement during the interval divided by the length of the interval. Figure 2.9A shows that this average velocity is the *average slope* of the position–time curve (i.e., the slope of the line connecting the start to the end of the entire interval).

With this approach, though, we lose all details about what happens in the middle of the interval. In Figure 2.9A, the slope of the x–t curve varies considerably as we move through the interval from $t = 1.0$ s to $t = 2.0$ s. If we want to get a more accurate description of the object's motion at a particular instant within this time interval, say at $t = 1.5$ s, it is better to use a smaller interval. How small an interval should we use? Intuitively, we expect that using a smaller interval will give a better measure of

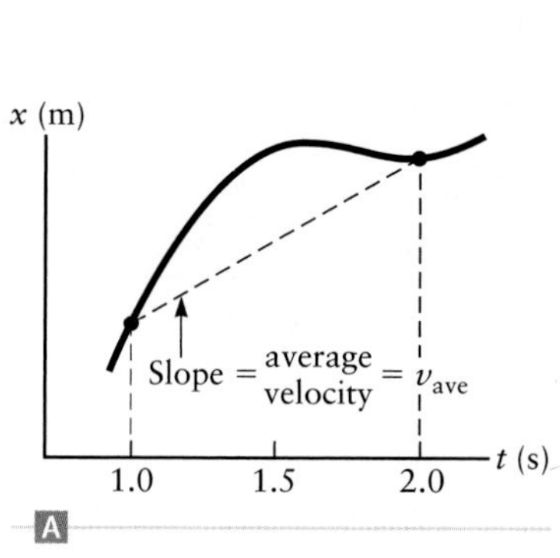

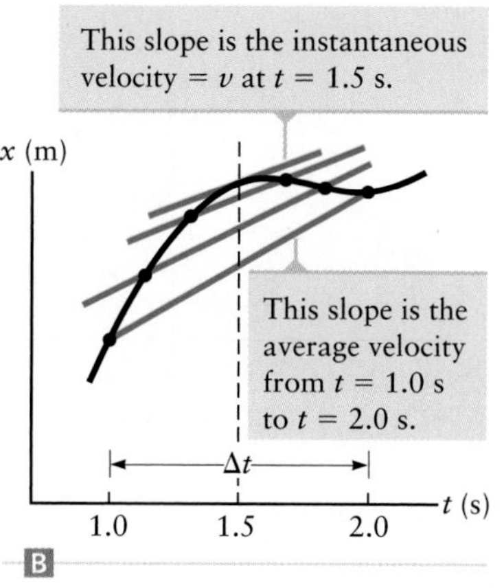

◀ **Figure 2.9** A The average velocity during a particular time interval is the slope of the line connecting the start of the interval to the end of the interval. B The instantaneous velocity at a particular time is the slope of the x–t curve at that time. The instantaneous velocity in the middle of a time interval is not necessarily equal to the average velocity during the interval. The instantaneous velocity is defined as the limit of the slope over a time interval Δt as $\Delta t \to 0$.

the motion at a particular point in time. From Figure 2.9B, we see that as we take ever smaller time intervals we are actually calculating the slope of the position–time curve at the point of interest (here at $t = 1.5$ s). The slope of the position–time curve at the point of interest is called the ***instantaneous velocity***. For one-dimensional motion, the instantaneous velocity v is the *slope of the position–time (x–t) curve* and is given by[2]

$$v = \lim_{\Delta t \to 0} \frac{\Delta x}{\Delta t} \tag{2.5}$$

Definition of instantaneous velocity

For the example shown in Figure 2.9, v is not constant. Rather, it varies with time over the interval from $t = 1.0$ s to $t = 2.0$ s.

The difference between the average and instantaneous values can be understood in analogy with a car's speedometer. The speedometer reading gives your instantaneous speed, the magnitude of your instantaneous velocity at a particular moment in time. If you are taking a long drive, your average speed will generally be different because the average value will include periods at which you are stopped in traffic, passing other cars, and so forth. In many cases, such as in discussions with a police officer, the instantaneous value will be of greatest interest.

The instantaneous velocity gives a mathematically precise measure of how the position is changing at a particular moment, making it much more useful than the average velocity. For this reason, from now on in this book we refer to the instantaneous velocity as simply the "velocity," and we denote it by v as in Equation 2.5.

CONCEPT CHECK 2.2 Estimating the Instantaneous Velocity

Is there an instant in time in Figure 2.9 at which the instantaneous velocity is zero? If so, what is the approximate value of t at which $v = 0$?

EXAMPLE 2.1 Average Velocity of a Bicycle

Consider the multiple images in Figure 2.10, showing a bicyclist moving along a level road. Find the average velocity of the bicyclist during the interval from $t = 2.0$ s to $t = 3.0$ s.

RECOGNIZE THE PRINCIPLE

The average velocity during a particular time interval is the slope of the position–time graph during that interval, $v_{ave} = \Delta x/\Delta t$ (Eq. 2.4). The time interval is given, so we need to deduce the displacement Δx from Figure 2.10.

(continued) ▶

[2]Here the term "lim" (limit) means to take the ratio $\Delta x/\Delta t$ as the quantity Δt approaches zero.

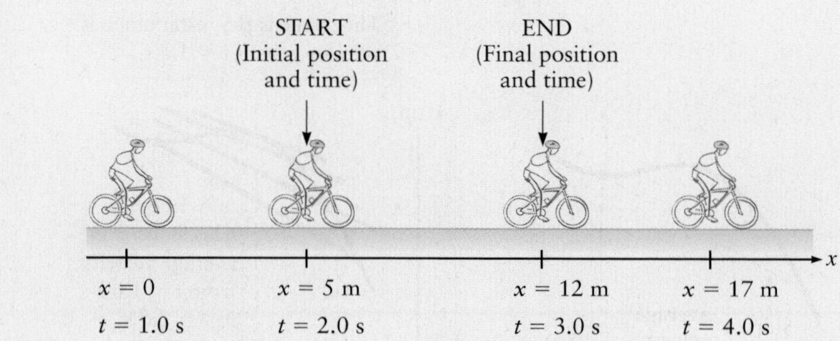

▲ **Figure 2.10** Example 2.1.

Insight 2.2

ALWAYS DRAW A PICTURE

The solution to most problems in physics starts with a picture. Begin by drawing a picture showing all the given information. The picture should show coordinate axes and any other information that seems relevant to the problem.

SKETCH THE PROBLEM

Our first step is to draw a picture that contains all the relevant information; this step is essential for organizing our thoughts and seeing connections. The images in Figure 2.10 form the heart of the picture, but because we want to extract some quantitative information, we have added the x axis, with its origin at the bicyclist's position at $t = 1.0$ s. Using this coordinate axis, we can read off the value of the bicycle's position at the times of interest.

IDENTIFY THE RELATIONSHIPS

The average velocity is the bicyclist's displacement during a particular time interval divided by the length of the interval (Eq. 2.4). We have therefore added arrows to our picture that mark the positions at the start ($t = 2.0$ s) and end ($t = 3.0$ s) of the interval of interest. We have

$$v_{\text{ave}} = \frac{\Delta x}{\Delta t} = \frac{x_{\text{final}} - x_{\text{initial}}}{t_{\text{final}} - t_{\text{initial}}} \tag{1}$$

SOLVE

Inserting values from Figure 2.10 into Equation (1), we get

$$v_{\text{ave}} = \frac{x(t = 3.0\text{ s}) - x(t = 2.0\text{ s})}{t_{\text{final}} - t_{\text{initial}}} = \frac{(12\text{ m} - 5\text{ m})}{(3.0\text{ s} - 2.0\text{ s})} = \boxed{7\text{ m/s}}$$

► ***What does it mean?***

Each bicycle in Figure 2.10 corresponds to a point on the x–t graph, with coordinates given by reading off the values of x and t. Once we had the values of x and t, we then found the average velocity through the relation $v_{\text{ave}} = \Delta x/\Delta t$.

EXAMPLE 2.2 Average and Instantaneous Velocity of a Car

A car is traveling on a long, straight road. It starts at the origin at $t = 0$ and moves with a velocity of $+9.0$ m/s for 30 s. After stopping for 10 s at a crosswalk to allow a pedestrian to cross the road, it then moves at $+6.0$ m/s for another 20 s. (**a**) What is the final position of the car? (**b**) What is the car's average velocity?

RECOGNIZE THE PRINCIPLE

The car's motion can be broken into three separate intervals, with values for the velocity of $+9.0$ m/s, zero, and $+6.0$ m/s, respectively. During each of these intervals the velocity is constant, so the displacement during each interval will be related to the velocity by $\Delta x = v\Delta t$. The final position of the car will be the sum of the displacements during the three intervals, and the average velocity will be the total displacement divided by the total time.

SKETCH THE PROBLEM

Figure 2.11A shows the velocity as a function of time, and Figure 2.11B shows how the displacement varies with time.

IDENTIFY THE RELATIONSHIPS AND SOLVE

(**a**) The displacement during the first interval (between $t = 0$ and 30 s) is $\Delta x_1 = v_1 \Delta t_1$, where $v_1 = +9.0$ m/s and $\Delta t_1 = 30$ s. We thus find

$$\Delta x_1 = v_1 \Delta t_1 = (9.0 \text{ m/s})(30 \text{ s}) = 270 \text{ m}$$

In a similar way, we get $\Delta x_2 = v_2 \Delta t_2 = (0)(10 \text{ s}) = 0$ and $\Delta x_3 = v_3 \Delta t_3 = (6.0 \text{ m/s})(20 \text{ s}) = 120$ m. The total displacement is thus

$$\Delta x_{\text{total}} = \Delta x_1 + \Delta x_2 + \Delta x_3 = (270 + 0 + 120) \text{ m} = \boxed{390 \text{ m}}$$

(**b**) The average velocity during the entire interval is (using Eq. 2.4)

$$v_{\text{ave}} = \frac{\Delta x_{\text{total}}}{\Delta t_{\text{total}}} = \frac{390 \text{ m}}{(30 + 10 + 20) \text{ s}} = \boxed{6.5 \text{ m/s}}$$

► ***What does it mean?***

The average velocity during the entire interval is not equal to the instantaneous velocity during any portion of the motion. The value of the v_{ave} pertains to the entire interval, while the instantaneous velocity describes the motion at a particular instant in time.

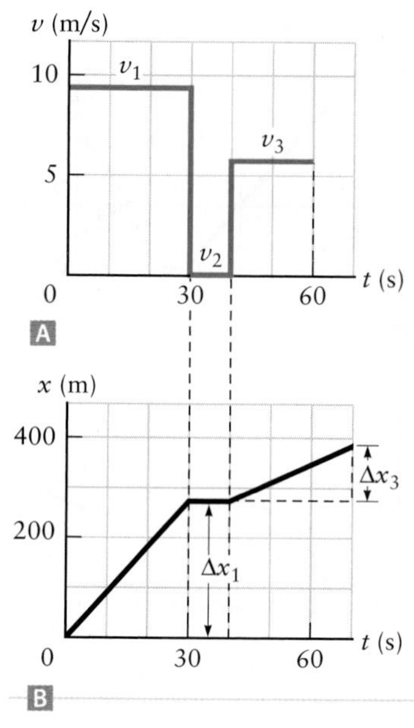

▲ **Figure 2.11** Example 2.3. A Velocity versus time for the car. B Position versus time.

In Example 2.1, we saw how to use observations of the position as a function of time to compute an object's average velocity. It is also very useful to be able to deduce the instantaneous velocity from the position–time behavior. We can do so by estimating the slope of the position–time curve at different values of t and using these estimates to make a qualitative plot of the velocity–time relation. This approach is illustrated in Example 2.3.

EXAMPLE 2.3 Computing Velocity Using a Graphical Method

A hypothetical object moves according to the x–t graph in Figure 2.12A. This object is initially (when t is near t_1) moving to the right, in the "positive" x direction. The object reverses direction near t_2, and it is again moving to the right at the end when t is near t_4. (**a**) Sketch the qualitative behavior of the velocity of the object as a function of time using a graphical approach. (**b**) Estimate the average velocity during the interval between $t_1 = 1.0$ s and $t_2 = 2.5$ s.

RECOGNIZE THE PRINCIPLE

For part (**a**), we want to find the velocity—which means the *instantaneous* velocity—so we need to estimate the slope of the x–t curve as a function of time. For part (**b**), the average velocity over the interval $t = 1.0$ s to $t = 2.5$ s is the slope of the x–t curve during this interval.

SKETCH THE PROBLEM

Figure 2.12B shows the x–t graph with lines drawn tangent to the x–t curve at various instants. The slopes of these tangent lines are the velocities at times $t_1, t_2, \ldots$ in Figure 2.12A.

(continued) ►

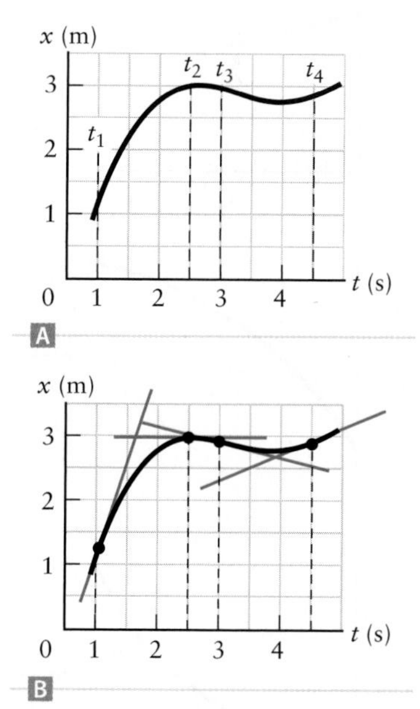

▲ **Figure 2.12** Example 2.3. A Hypothetical position–time graph. The slopes of the tangent lines in B are equal to the velocity at various instants in time.

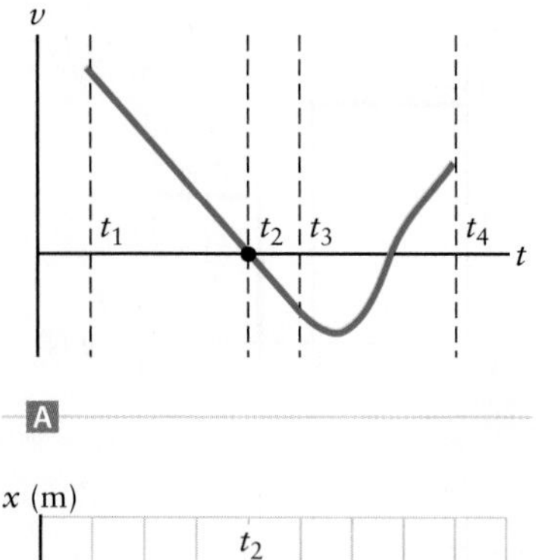

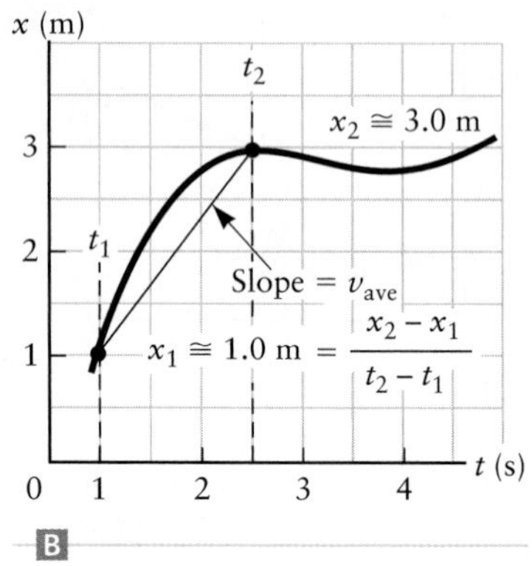

▲ **Figure 2.13** Example 2.3. **A** Qualitative plot of the velocities obtained from the slopes in Figure 2.12B. **B** Calculation of the average velocity during the interval between $t_1 = 1.0$ s and $t_2 = 2.5$ s.

IDENTIFY THE RELATIONSHIPS AND SOLVE

(a) At $t = t_1$, the $x-t$ slope is large and positive, so our result for v in Figure 2.13A is large and positive at $t = t_1$. At $t = t_2$, the $x-t$ slope is approximately zero and hence v is near zero. At $t = t_3$, the object is moving toward smaller values of position x, so the slope of the x–t curve and hence also the velocity are negative. Finally, at $t = t_4$, the object is again moving to the right as x is increasing with time, so v is again positive. After estimating the $x-t$ slope at these places, we can construct the smooth v–t curve shown in Figure 2.13A, which shows the *qualitative* behavior of the velocity as a function of time.

(b) To estimate the average velocity between t_1 and t_2, we refer to Figure 2.13B, which shows a line segment that connects these two points on the position–time graph. The slope of this segment is the average velocity:

$$v_{ave} = \frac{\Delta x}{\Delta t} = \frac{x_2 - x_1}{t_2 - t_1}$$

Reading the values of x_1, x_2, t_1, and t_2 from that graph, we find

$$v_{ave} = \frac{x_2 - x_1}{t_2 - t_1} = \frac{(3.0 \text{ m}) - (1.0 \text{ m})}{(2.5 \text{ s}) - (1.0 \text{ s})} = \boxed{1.3 \text{ m/s}}$$

► ***What does it mean?***

The (instantaneous) velocity is the slope of the position–time graph. To find the qualitative behavior of the velocity, we found approximate values by drawing lines tangent to the x–t curve at several places and estimating their slopes. The value of v at a particular value of t is always equal to the slope of the x–t curve at that time.

CONCEPT CHECK 2.3 The Relation between Velocity and Position

For which of the position–time graphs in Figure 2.14 are the following statements true?

(a) The velocity increases with time.
(b) The velocity decreases with time.
(c) The velocity is constant (does not change with time).

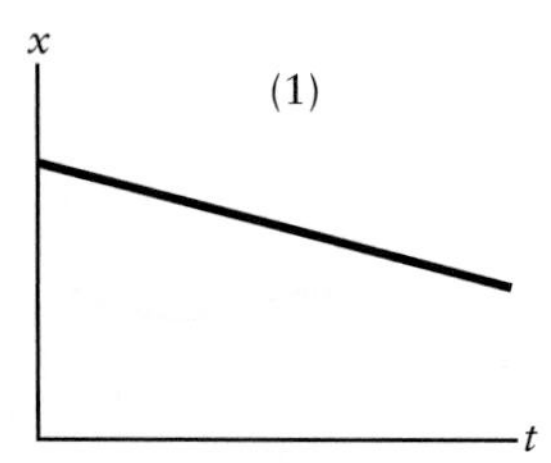

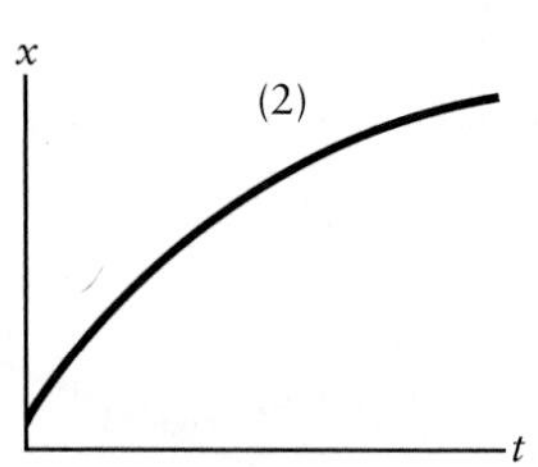

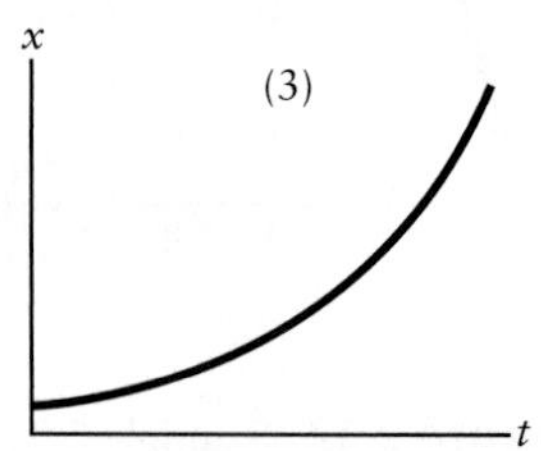

▲ **Figure 2.14** Concept Check 2.3.

Acceleration

We have seen that two quantities—position and velocity—are important for describing the motion of an object. One additional quantity, ***acceleration***, will play a central role in our theory of motion. Acceleration is related to how the velocity changes with time. Consider again our rocket-powered car from Figure 2.8. In Figure 2.15, we resketch the v–t plot that we derived in Figure 2.8C and notice again that the car's velocity increases as time proceeds. Acceleration is defined as the rate at which the velocity is changing. If the velocity changes by an amount Δv over the time interval Δt, the ***average acceleration*** during this interval is

$$a_{ave} = \frac{\Delta v}{\Delta t} \tag{2.6}$$

As with the velocity, we are usually concerned with the acceleration at a particular instant in time, which leads us to consider the acceleration in the limit of very small time intervals. We thus define the ***instantaneous acceleration*** as

$$a = \lim_{\Delta t \to 0} \frac{\Delta v}{\Delta t} \tag{2.7}$$

The instantaneous acceleration a equals the *slope* of the v–t curve at a particular instant in time. The SI unit of acceleration is m/s^2 (meters per second squared).

We have now introduced several quantities associated with motion, including position, displacement, velocity, and acceleration. These quantities are connected in a mathematical sense through Equations 2.5 and 2.7. We also showed important graphical relationships: velocity is the slope of the position–time graph, while acceleration is the slope of the velocity–time graph. You might now be wondering if we'll continue this progression and consider the slope of the acceleration and so forth. The answer is that acceleration is as far as we need to go; x, v, and a are all we need in our formulation of a complete theory of motion. Newton's laws of motion (Section 2.4) will show us why.

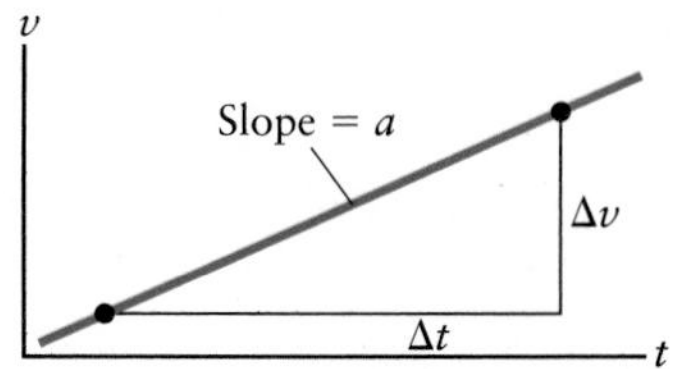

▲ **Figure 2.15** Acceleration is the slope of the velocity–time curve. For the case shown here, in which v varies linearly with time, the average acceleration is equal to the instantaneous acceleration.

EXAMPLE 2.4 Acceleration of a Sprinter

Consider a sprinter (Fig. 2.16A) running a 100-m dash. Figure 2.16B shows the velocity–time graph for the sprinter. Use a graphical approach to calculate the corresponding acceleration–time graph. What is the approximate value of the sprinter's maximum acceleration, and when does it occur?

RECOGNIZE THE PRINCIPLE

Acceleration is the slope of the velocity–time graph, so we must estimate this slope at enough different values of t to be able to make a qualitative plot of the sprinter's acceleration as a function of time.

SKETCH THE PROBLEM

In Figure 2.16B, we have drawn in several lines tangent to the v–t curve at various times. The slopes of these tangent lines give the acceleration.

IDENTIFY THE RELATIONSHIPS AND SOLVE

We have measured approximate values of the slopes of the tangent lines in Figure 2.16B, and the results are plotted in Figure 2.17. This acceleration–time graph is only qualitative (approximate). More accurate results would be possible had we started with a more detailed graph of the velocity.

▶ *What does it mean?*

The shape of the acceleration–time curve in Figure 2.17 is consistent with expectations. The largest value of the acceleration (about 10 m/s^2) occurs at the start of the race when the sprinter is "getting up to speed"; that is where the v–t slope is largest. At the end of the race as the runner crosses the finish line, she slows down and eventually comes to a stop with $v = 0$ at the far right in Figure 2.16B. During that time, her velocity is decreasing with time, so her acceleration is negative.

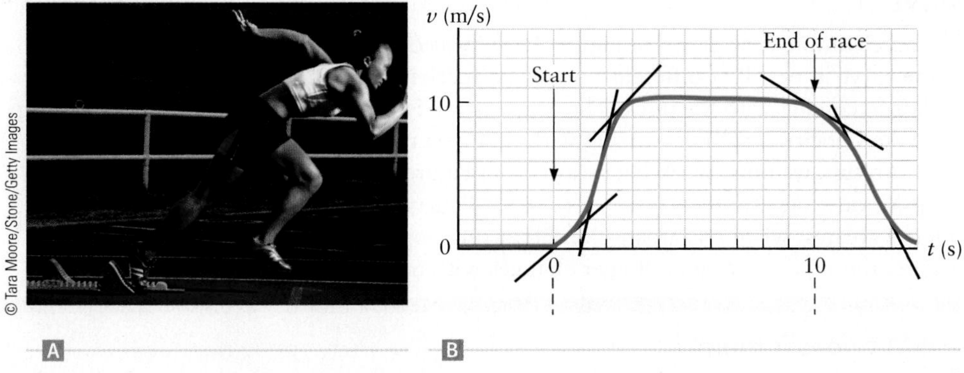

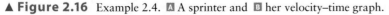

▲ **Figure 2.16** Example 2.4. A A sprinter and B her velocity–time graph.

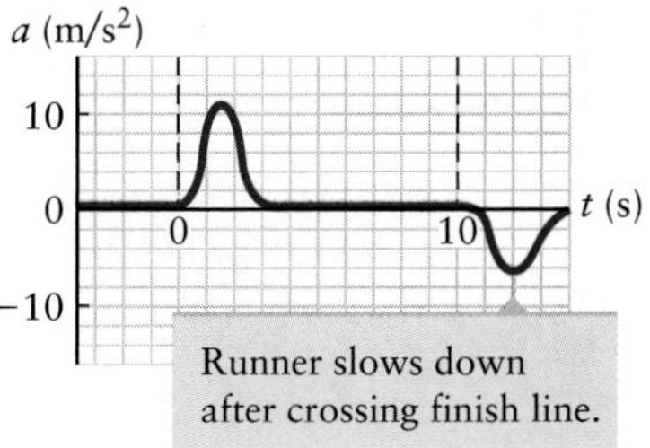

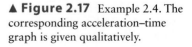

▲ **Figure 2.17** Example 2.4. The corresponding acceleration–time graph is given qualitatively.

The Relation between Velocity and Acceleration

An interesting result in Example 2.4 is that the maximum velocity and the maximum acceleration do *not* occur at the same time. It is tempting to think that if the "motion" is large, both v and a will be large, but this notion is *incorrect*. Acceleration is the slope—the rate of change—of the velocity with respect to time. The time at which the rate of change of velocity is greatest may not be the time at which the velocity itself is greatest.

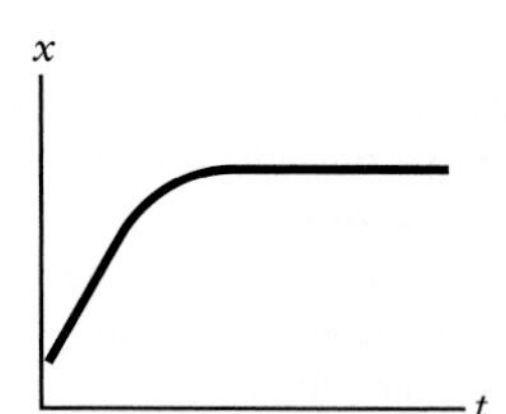

▲ **Figure 2.18** Concept Check 2.4. What type of motion is described by this position–time graph?

CONCEPT CHECK 2.4 Analyzing a Position–Time Graph

The position–time curve of a hypothetical object is shown in Figure 2.18. Which of the following scenarios might be described by this x–t graph?

(a) A car starting from rest when a traffic light turns green.
(b) A car slowing to a stop when a traffic light turns red.
(c) A bowling ball rolling toward the pins.
(d) A runner slowing from top speed to a stop at the end of a race.

EXAMPLE 2.5 Sliding to a Stop

Consider a hockey puck that starts from rest. At $t = 0$, the puck is struck by a hockey stick, which starts the puck into motion with a high velocity. The puck then slides a very long distance before coming to rest. Draw qualitative plots of the puck's position, velocity, and acceleration as functions of time.

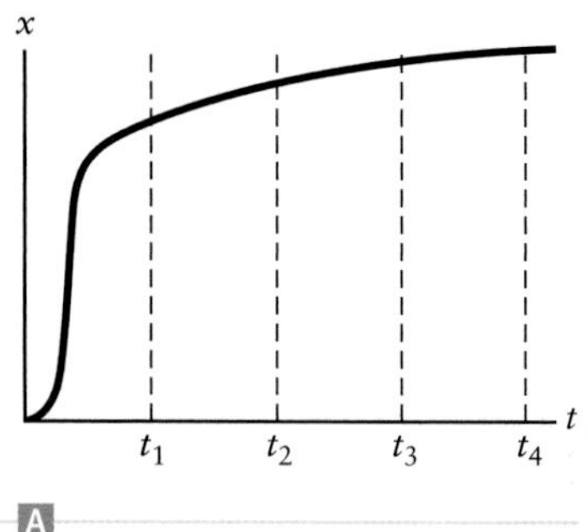

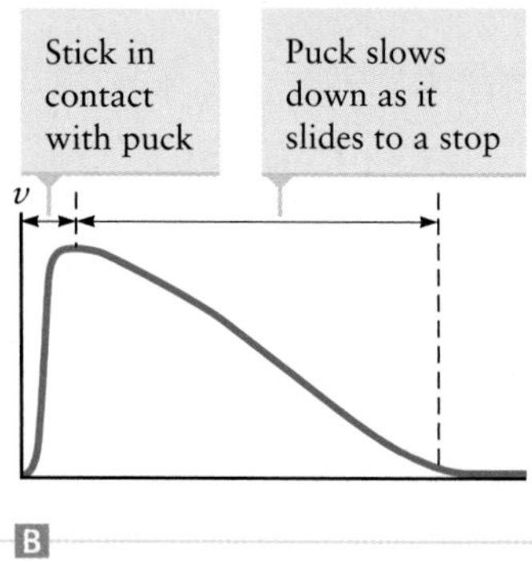

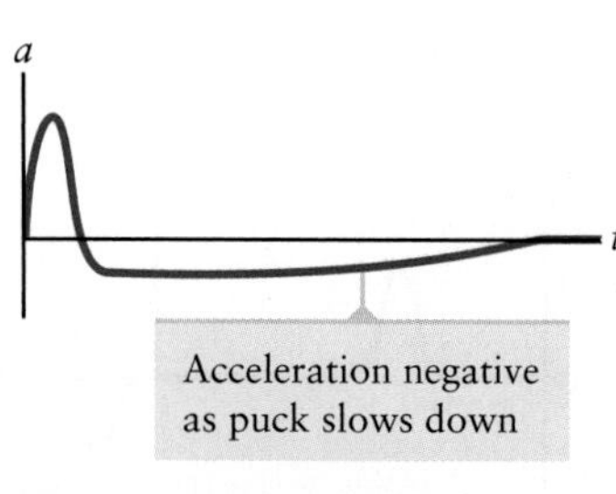

▲ **Figure 2.19** Example 2.5. Motion of a hockey puck as described by graphs of its position, velocity, and acceleration versus time.

RECOGNIZE THE PRINCIPLES

We first use our experience with sliding objects to deduce the qualitative shape of the velocity–time curve. The puck starts from rest, so initially $v = 0$. The velocity then increases quickly to a high value when the stick is in contact with the puck. After the puck leaves the stick, the velocity gradually decreases as the puck slides to a stop. From the v–t curve, we can get the position and acceleration as functions of time.

SKETCH THE PROBLEM

In Figure 2.19, imagine drawing part B first (qualitative behavior of velocity). The puck reaches a high velocity very quickly and then falls to zero at long times.

IDENTIFY THE RELATIONSHIPS

To find the acceleration, we must find the slope of the v–t curve as a function of time. To get the position, we must find an x–t curve whose slope gives this velocity–time graph.

SOLVE

The acceleration–time graph in Figure 2.19C was obtained from estimates of the slope of the curve in part B at different times. The acceleration is *positive* and large when the stick is in contact with the puck. The acceleration is *negative* as the puck slides to a stop since the velocity is then decreasing with time and the v–t slope is negative. To deduce the x–t result, we must work backward from Figure 2.19B so as to make the slope of the position–time graph correspond to the v–t curve. We do so by noting that the x–t slope is greatest when v is greatest and that this slope is small when v is small. Once v reaches zero, the position no longer changes with time.

▸ *What does it mean?*

These graphs of the position and velocity can also be appreciated using the "time-lapsed" sketches in Figure 2.20. These sketches show the location of the puck at

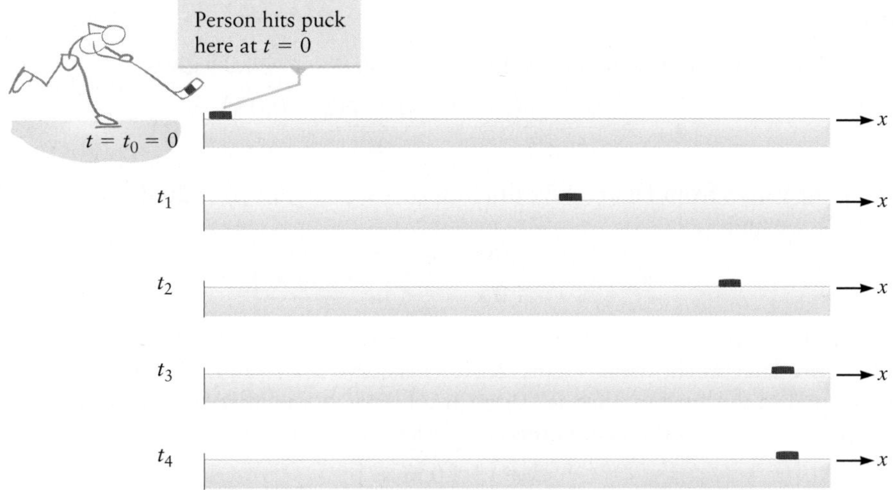

▲ **Figure 2.20** Example 2.5. Time-lapse sketches of the hockey puck's motion.

evenly spaced instants in time, beginning at $t_0 = 0$ when it is struck by the hockey stick and then at $t_1, t_2, \ldots$ until the puck comes to rest at t_4. These times are equally spaced along the time axis (Fig. 2.19A), so the time intervals $\Delta t = t_1 - t_0 = t_2 - t_1, \ldots$ are all equal. The displacements during these intervals are *not* equal. For example, the displacement during the first interval (from t_0 to t_1) is much larger than the displacement during the second interval. In words, the puck moves farther—because it has a higher velocity—during the first interval. Likewise, at the end of the time period, the puck's velocity is small, so the distance traveled between t_3 and t_4 is much smaller than the distance traveled between t_0 and t_1.

EXAMPLE 2.6 Stopping in a Hurry

A driver is in a hurry, and her car is traveling along a straight road with velocity $v_0 = 20$ m/s (about 40 mi/h) when she spots a problem in the road ahead and applies the brakes. The car then slows to a stop according to the velocity–time curve in Figure 2.21. Find the average acceleration during the interval from $t = 0.0$ s to $t = 4.0$ s and the instantaneous acceleration at $t = 2.0$ s.

RECOGNIZE THE PRINCIPLE

The average acceleration is the average slope of the velocity–time curve *during the time interval* of interest. The instantaneous acceleration is the slope of the v–t curve *at the point* of interest.

SKETCH THE PROBLEM

Figure 2.21 describes the problem and contains all the information we need to solve it.

IDENTIFY THE RELATIONSHIPS

From the definition of average acceleration in Equation 2.6, we have

$$a_{ave} = \frac{\Delta v}{\Delta t} \quad (1)$$

We can read the values of Δv and Δt from Figure 2.21.

(continued) ▶

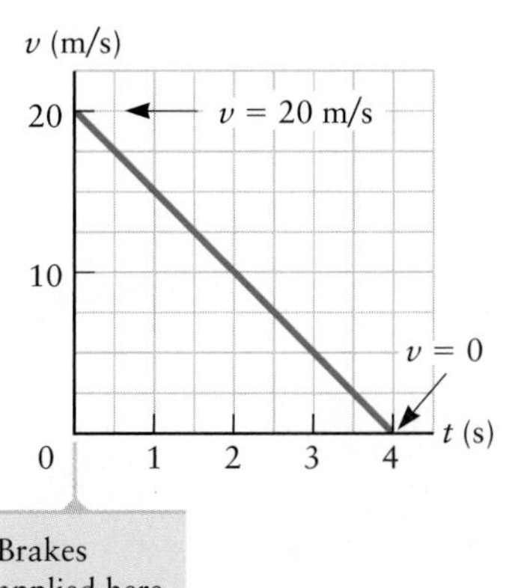

▲ **Figure 2.21** Example 2.6.

SOLVE

Applying Equation (1) over the interval from $t = 0.0$ s to $t = 4.0$ s gives

$$a_{ave} = \frac{\Delta v}{\Delta t} = \frac{v(t = 4.0\text{ s}) - v(t = 0.0\text{ s})}{\Delta t}$$

Inserting the values from Figure 2.21 of $v(4.0\text{ s}) = 0$ and $v(0.0\text{ s}) = 20$ m/s along with $\Delta t = 4.0$ s, we find

$$a_{ave} = \frac{(0 - 20)\text{ m/s}}{4.0\text{ s}} = \boxed{-5.0\text{ m/s}^2}$$

This result is the average acceleration, so it is the average slope of the v–t curve during the entire interval shown in Figure 2.21. This curve is a straight line, so the average slope is equal to the "instantaneous" slope at all times in the interval. The instantaneous slope of the v–t curve is the instantaneous acceleration a; hence

$$a = \boxed{-5.0\text{ m/s}^2}$$

▸ *What does it mean?*

The average and instantaneous acceleration in this example are both negative because the velocity *decreases* during the time interval of interest. The velocity is positive (Fig. 2.21) during the entire interval, so the velocity and acceleration are in opposite directions: the velocity is positive (along the $+x$ direction), while the acceleration is negative (along the $-x$ direction).

CONCEPT CHECK 2.5 Finding the Velocity

Figure 2.22A shows the position–time graph for an object. Which of the curves in Figure 2.22B describes the qualitative behavior of the object's velocity as a function of time?

(a) Curve 1 (b) Curve 2 (c) Curve 3

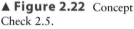

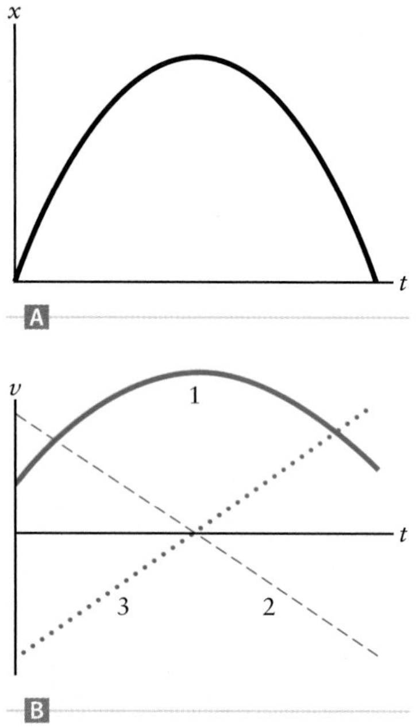

▲ **Figure 2.22** Concept Check 2.5.

2.3 The Principle of Inertia

We spent much of Section 2.1 discussing Aristotle's ideas about motion, and you should now be convinced that his theory of motion has some serious flaws. Nevertheless, it is worthwhile to consider how Aristotle might have explained the motion of the rocket-powered car in Figure 2.8. He probably would have claimed that the car moves by virtue of the force exerted on it by the engine, with a velocity $v = F/R$ as predicted from Equation 2.1. If the rocket engine is then turned off, however, the force would vanish ($F = 0$) and, according to the same argument, the car would stop immediately. Your intuition should tell you that this prediction is not correct; the car would instead continue along, for at least a short period, after the engine is turned off. That is, the car will coast for a while before coming to rest.

The difficulties with Aristotle's ideas about motion all come down to his belief that force and velocity are directly linked. This linkage is expressed in Equation 2.1, which states that if an object has a nonzero velocity, then according to Aristotle there must be a nonzero force acting on the object at that time. However, our example with a rocket-powered car shows that force and velocity are *not* linked in this way because it is possible for the car to have a nonzero velocity along the x direction even when the engine is turned off and the force is zero. The correct connection between force and motion is instead based on a direct linkage between force and *acceleration*. This connection to acceleration is at the heart of Newton's laws of motion.

Before we can really appreciate Newton, however, we must first consider the work of Galileo. Although Galileo did not arrive at the correct laws of motion, his experiments on the motion of terrestrial objects showed that an object can move even if

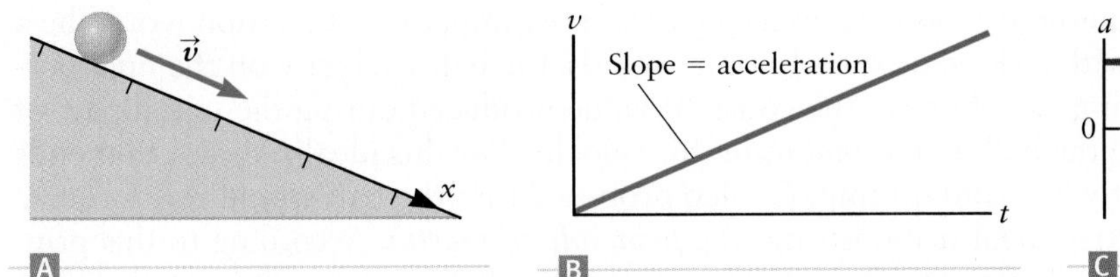

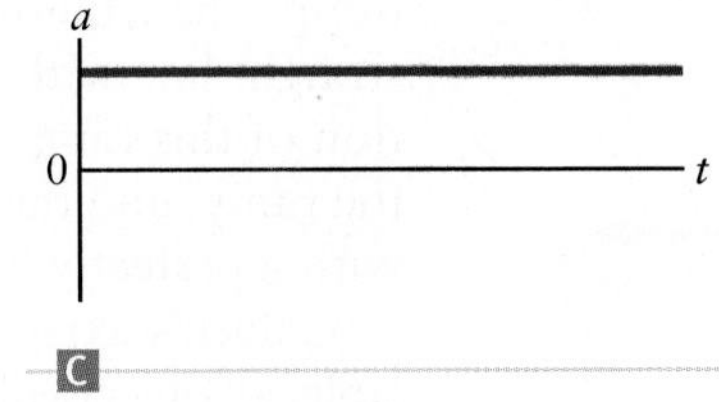

▲ **Figure 2.23** A Ball rolling down a simple incline. B Galileo found that the velocity of the ball increases linearly with time. The slope of this line is the acceleration in C.

there is *zero* total force acting on it. This result led to the discovery of the ***principle of inertia***, which is a cornerstone of Newton's laws.

Galileo's Experiments on Motion

Aristotle was not able to explain how an object could move when there did not appear to be any force acting on it. A good example is the somewhat idealized case of a hockey puck sliding on a very smooth, horizontal, icy surface. Our intuition tells us that the puck will slide a very long way before coming to rest; in fact, if we could somehow make the surface "perfectly" icy so that there is absolutely no friction whatsoever, our intuition suggests that the puck would slide forever.

Galileo did not carry out experiments with a hockey puck on an icy surface. Instead, his experiments involved a ball rolling on an incline. If the ball is very hard, like a billiard ball, and the surface of the incline is also very hard and smooth, the effect of friction on the ball's motion is very small. Hence, this situation is actually quite similar to that of our idealized hockey puck.

Galileo experimented with the motion of a ball on an incline as sketched in Figure 2.23. He found that when a ball is released from rest, its velocity varies with time as shown in Figure 2.23B. This is another example of one-dimensional motion, with the direction of motion denoted by an x axis lying along the incline as shown. The velocity of the ball increases as the ball rolls down the incline, and Galileo found that the magnitude of the velocity increases *linearly* with time. Since acceleration is the slope of the v–t curve, the acceleration is *constant* and *positive*.

Galileo then repeated this experiment, but this time with the ball rolling *up* an incline with the same tilt angle (Fig. 2.24A). When he gave the ball some initial velocity, he found that it rolled up the incline with the v–t curve shown in Figure 2.24B. Again he observed that the velocity varies *linearly* with time, but now the slope is *negative*. In fact, the slope in this case is equal in magnitude to the slope found for the down-tilted incline. Hence, the acceleration was opposite in sign but equal in magnitude to that found when the ball rolled down the incline. Galileo performed this experiment with many inclines and with many different tilt angles, and he always observed that the velocity varied linearly with time, with the slope of the $v-t$ graph being determined by the tilt angle. Moreover, the acceleration (the slope of the v–t relation) when a ball rolled up a particular incline was always equal in magnitude, but opposite in sign, when compared with the acceleration when the ball rolled down the same incline. He then reasoned that if the tilt of the incline were precisely zero (a perfectly level surface), the slope of the v–t line and hence the acceleration would be zero. Galileo therefore asserted that on a level surface the ball would roll with a *constant velocity*.

This result may seem obvious to you, since it means that a ball placed at rest on a horizontal surface will remain motionless, with zero velocity. Such a ball has a constant velocity because $v = 0$ is a constant! It was, however, the genius of Galileo to realize that his experiment implied the result sketched in Figure 2.25. Here, a ball is rolling along a different sort of ramp. The initial portion of the ramp is sloped like the incline in Figure 2.23, but the final portion is perfectly horizontal and extremely long, with only part of it shown here. Galileo realized that for this type of ramp the ball would have a positive velocity when it completes the first portion of the

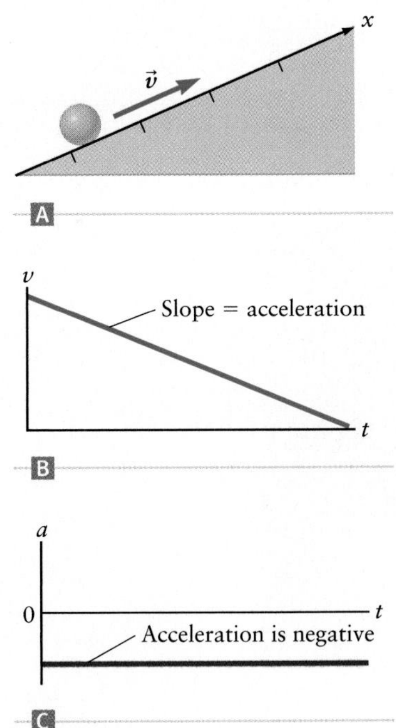

▲ **Figure 2.24** A Ball rolling up an incline with the same angle as in Figure 2.23. B Galileo found that the velocity of the ball decreases linearly with time. C The slope of this line—the acceleration—has the same magnitude, but the opposite sign, as the acceleration when the ball rolls down the incline (Fig. 2.23C).

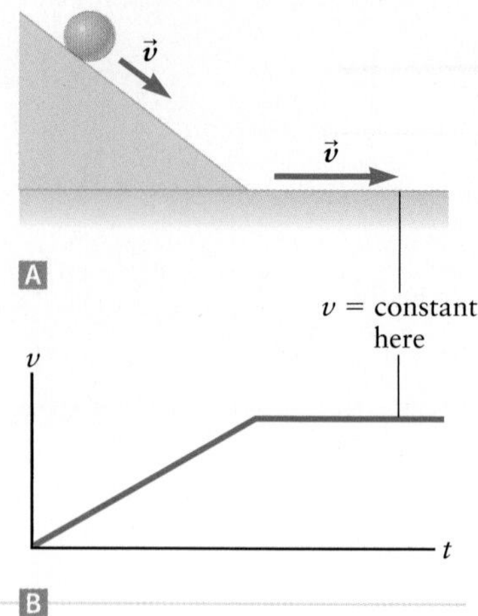

▲ **Figure 2.25** When a ball rolls along this two-part incline, its velocity increases linearly with time while on the initial (sloped) portion, and v is then constant on the final (flat) portion.

ramp. Then, on the second (horizontal) part of the ramp, the v–t relation would be a straight line with a slope of zero. In other words, the ball's velocity on the final portion of this ramp would have the constant value produced during the initial part of the ramp, and the ball would *maintain* this velocity. For this idealized case that ends with a perfectly horizontal ramp, Galileo proposed that the ball would roll *forever*.

Galileo's experiment demonstrates the ***principle of inertia***. According to this principle, an object will *maintain its state of motion*—its velocity—*unless it is acted on by a force*. On a horizontal surface, there is no force in the direction of motion, so the velocity is constant and the ball rolls forever. With his discovery of the principle of inertia, Galileo broke Aristotle's link between velocity and force. Galileo showed that one can have motion (a nonzero velocity) *without* a force. This discovery does not answer the question of exactly how force is linked to motion, however. The answer to that question was provided by Newton. (It is interesting that Newton was born in 1643, the year after Galileo died. In a very direct sense, Newton thus built on the work of Galileo.)

2.4 Newton's Laws of Motion

Newton's laws of motion are three separate statements about how things move.

Newton's First Law

Newton's first law is a careful statement about the principle of inertia that we described in the previous section in connection with Galileo's experiments.

> ***Newton's first law of motion:*** **If the total force acting on an object is zero, the object will maintain its velocity forever.**

If the total force acting on an object is *zero*, the object will move with a constant velocity. In other words, such an object will move with a constant speed along a particular direction and will continue this motion—with the same speed in the same direction—forever, or as long as the total force acting on it is zero.[3] This is another way of stating the principle of inertia, and you can see that it is essentially what Galileo found in his experiments with rolling balls. The total force acting on one of Galileo's balls rolling on a horizontal surface is zero, so in this ideal case the ball will roll forever.

Inertia and Mass

Now that we have introduced the principle of inertia, you should be curious about the term *inertia* and precisely what it means. The inertia of an object is a measure of its *resistance to changes in motion*. This resistance to change depends on the object's ***mass***. It is not possible to give a "first principles" definition of the term *mass*, but your intuitive notion is very useful. The mass of an object is a measure of the amount of matter it contains. Objects that contain a large amount of matter have a larger mass and a greater inertia than objects containing a small amount of matter. The SI unit of mass is the kilogram (kg). Mass is an intrinsic property of an object. For example, the mass m of an object is independent of its location; m is the same on the Earth's surface as on the Moon and in distant space. In addition, an object's mass does not depend on its velocity or acceleration. Although an object's mass does not appear explicitly in Newton's first law, it has an essential role in Newton's second law.

Newton's first law may seem surprising to you. After all, terrestrial objects always come to rest eventually. The difference here, and a key to appreciating Newton's first law, is that an object will move with a constant velocity only in the ideal case that the

[3]This result may be surprising because it may *appear* to contradict your intuition. This result only holds if the total force on an object is *precisely zero*, however, and that can be hard to achieve in practice.

total force is precisely zero. For terrestrial objects, it is very difficult to find such ideal cases because it is difficult to completely eliminate all forces, especially frictional forces. Hence, terrestrial objects come to rest because of the forces that act on them.

Two other hypothetical cases may help make Newton's first law fit with your intuition. Case 1: Imagine a frozen lake with an extremely smooth surface. A hockey puck sliding on such a perfectly flat and icy surface experiences a very tiny frictional force and will therefore slide for a very long distance before coming to rest. If this frictional force could be made to vanish, the puck would slide forever (if the lake were large enough). Of course, actual icy surfaces are not perfect; there will always be a small amount of friction, which would make the puck eventually come to rest. Case 2: Imagine a spaceship that is coasting (i.e., with its engines turned off) someplace in the universe very far from any stars or planets. Such a spaceship would experience only a very small gravitational force (from the nearest stars and planets). If the nearest stars and planets were very far away, this force would be negligible and the spaceship would move with a constant velocity, in accord with Newton's first law.

Newton's Second Law

Newton's second law of motion: **In many situations, several different forces act on an object simultaneously. The total force on the object is the sum of these individual forces, $\vec{F}_{\text{total}} = \sum \vec{F}$. The acceleration of an object with mass m is then given by**

$$\vec{a} = \frac{\sum \vec{F}}{m} \tag{2.8}$$

Newton's second law tells us how an object will move when acted on by a force or by a collection of forces. *This law is our link between force and motion.* The acceleration of an object is directly proportional to the total force that acts on it. Newton's second law, Equation 2.8, is often written in the equivalent form

$$\sum \vec{F} = m\vec{a}$$

Keep in mind that the term $\sum \vec{F}$ in Newton's second law is the *total* force on the object. As you might imagine, in most cases there are several forces acting on an object. So, we have to add them all up, and that resulting vector sum is the force $\sum \vec{F}$ in Newton's second law (see Fig. 2.26). You should also recall that vectors must be added according to the vector arithmetic procedures described in Chapter 1. The direction of the acceleration $\vec{a}$ is then parallel to the direction of the total force $\sum \vec{F}$.

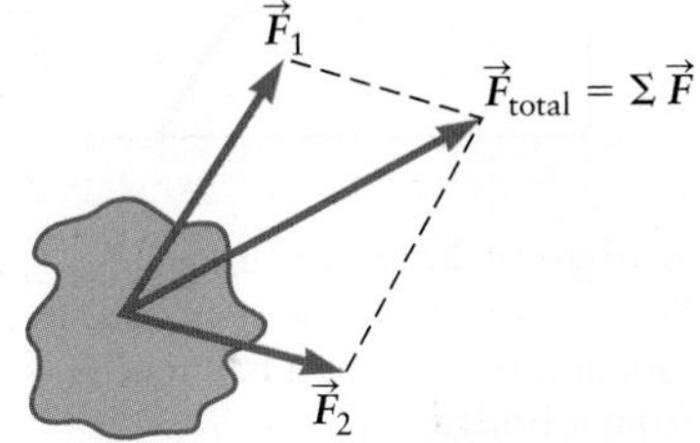

▲ **Figure 2.26** When several forces act on an object, the vector sum of these forces $\sum \vec{F}$ determines the acceleration according to Newton's second law.

In the SI system of units, force is measured in units called newtons (N). We can use Newton's second law to express this unit in terms of the primary SI units. Mass is measured in units of kilograms, while acceleration has the units meters per second squared. In terms of only the units, Equation 2.8 can be rearranged to read

$$\text{force} = \text{mass} \times \text{acceleration}$$

$$\text{newtons} = \text{kg} \times \frac{\text{m}}{\text{s}^2}$$

The value of the newton as a unit of force is therefore

$$1\,\text{N} = 1\,\text{kg} \cdot \text{m/s}^2 \tag{2.9}$$

We'll spend the next dozen chapters or so exploring applications of Newton's second law. Many applications start by determining the total force acting on an object from all sources. The object's acceleration can then be calculated using Equation 2.8. Acceleration is the change in velocity per unit time, so it is possible to use the acceleration to deduce the velocity. In a similar manner, since velocity is the change in position per unit time, one can use the velocity to find the object's position as a function of time. In this way, an object's acceleration, velocity, and position can all be found.

EXAMPLE 2.7 Using Newton's Second Law

A single force of magnitude 6.0 N acts on a stone of mass 1.1 kg. Find the acceleration of the stone.

RECOGNIZE THE PRINCIPLE

The force on the stone and its acceleration are related through Newton's second law,

$$\vec{a} = \frac{\sum \vec{F}}{m} \tag{1}$$

where $\sum \vec{F}$ is the total force on the stone, of magnitude 6.0 N as given.

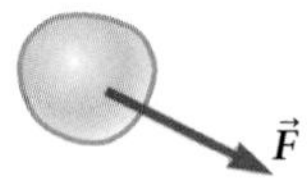

▲ **Figure 2.27** Example 2.7.

SKETCH THE PROBLEM

We begin by drawing a picture, showing all the forces acting on the stone (Fig. 2.27). Since there is only a single force in this example, it is also the total force.

IDENTIFY THE RELATIONSHIPS AND SOLVE

Using Newton's second law, Equation (1), the magnitudes of the acceleration and force are related by

$$a = \frac{\sum F}{m} = \frac{6.0 \text{ N}}{1.1 \text{ kg}} = \frac{6.0 \text{ kg} \cdot \text{m/s}^2}{1.1 \text{ kg}} = \boxed{5.5 \text{ m/s}^2} \tag{2}$$

The direction of the stone's acceleration is parallel to that of the total force, so the direction of $\vec{F}$ in Figure 2.27 gives the direction of $\vec{a}$.

► ***What does it mean?***

Notice how the units in Equation (2) combine, as the factors of kilograms on the top and bottom cancel to give an answer with units of meters per second squared, as expected for acceleration. You should always check the units of an answer in this way. An incorrect result for the units usually indicates an error in the calculation.

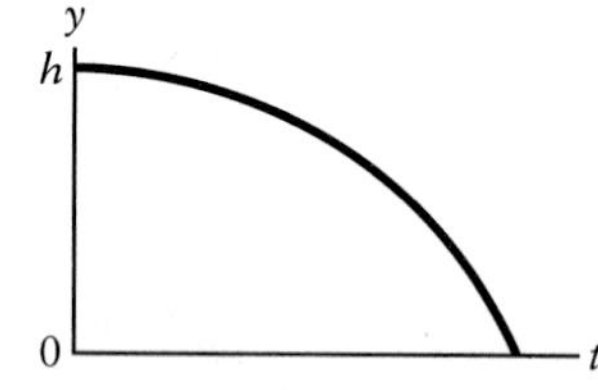

▲ **Figure 2.28** Example 2.8. Position (y) above the ground as a function of time for a ball that falls from a bridge.

EXAMPLE 2.8 Motion of a Falling Object

A ball is dropped from a bridge onto the ground below. The height of the ball above the ground as a function of time is shown in Figure 2.28. Use a graphical approach to find, as functions of time, the qualitative behavior of (**a**) the velocity of the ball, (**b**) the acceleration of the ball, and (**c**) the total force on the ball.

RECOGNIZE THE PRINCIPLE

Velocity is the slope of the position–time curve, so we can find v from the slope of the y-t curve in Figure 2.28. Notice that here we use y to measure the position, taking the place of x in previous examples. Acceleration is the slope of the velocity–time curve, so we can find the behavior of a once we have the behavior of the velocity. We can then find the total force on the ball through Newton's second law, $\sum \vec{F} = m\vec{a}$.

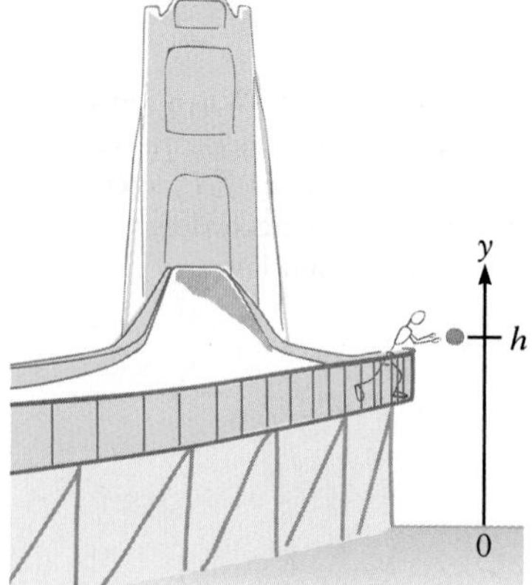

▲ **Figure 2.29** Example 2.8. A ball is dropped from a bridge, starting from rest ($v = 0$) at a height h above the ground. The motion of this ball is described by the y–t graph in Figure 2.28.

SKETCH THE PROBLEM

Figure 2.28 shows how the position of the ball varies with time, and Figure 2.29 shows a picture of the ball as it falls from the bridge. The motion of the ball is one-dimensional, falling directly downward from the bridge to the ground, and its position can be measured by its height y above the ground. Figure 2.29 also shows the y coordinate axis, with its origin at ground level.

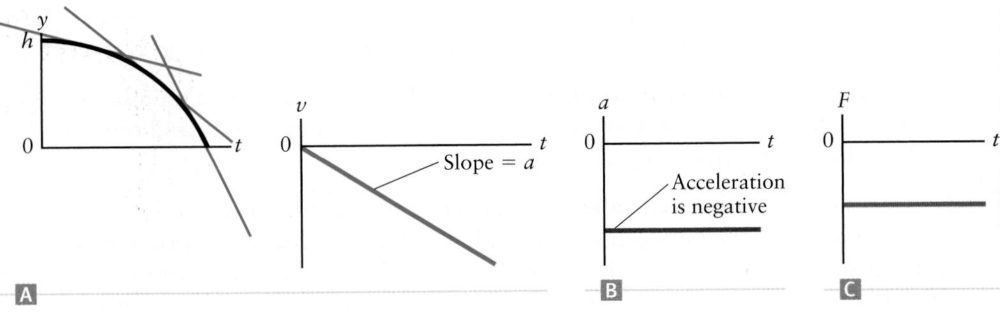

◀ **Figure 2.30** Example 2.8. **A** Position as a function of time for the falling ball from Figure 2.28. The slopes of the tangent lines give the ball's velocity at three instants in time. The velocity of the ball as a function of time is obtained from plotting these slopes. **B** The ball's acceleration is constant. **C** The force on the ball is proportional to the ball's acceleration.

IDENTIFY THE RELATIONSHIPS AND SOLVE

(**a**) Velocity is the change in position per unit time, so the value of v at any particular time is the slope of the y–t graph at that instant. We can obtain this slope graphically by following the approach in Example 2.3 and drawing lines tangent to the y–t curve at various times in Figure 2.30A. Using estimates for the slopes of these lines leads to the qualitative velocity–time graph shown alongside the y–t curve. Note that the velocity is always negative since y decreases monotonically with time. Also, the magnitude of the velocity, which is the speed, becomes larger and larger as the ball falls. (The plot here shows v as a function of time up until just before the ball reaches the ground.)

(**b**) Acceleration is the slope of the v–t curve; that slope is constant in Figure 2.30A, leading to the qualitative acceleration–time graph in Figure 2.30B. The acceleration is negative because the value of v is decreasing (the value of v becomes more negative with time). The magnitude of a is approximately constant during this period.

(**c**) To compute the force on the ball, we use Newton's second law. According to Equation 2.8, the total force is proportional to the acceleration. Even if we do not know the sources of all the forces on the ball, we can still compute the total force from Newton's second law. We can rearrange Newton's second law (Eq. 2.8) as $\sum \vec{F} = m\vec{a}$ and thus arrive at the qualitative force–time graph in Figure 2.30C.

▶ ***What does it mean?***

Given the behavior of the position as a function of time, we can deduce (by estimating slopes) the qualitative behavior of both the velocity and the acceleration as functions of time. The behavior of the force can then be found by using Newton's second law. For this falling ball, the total force is negative (Fig. 2.30C), meaning that the force is directed downward, along the $-y$ direction. This force is just the gravitational force acting on the ball.

Newton's Second Law and the Directions of $\vec{v}$ and $\vec{a}$

Newton's second law can be written as $\sum \vec{F} = m\vec{a}$, so the acceleration of an object $\vec{a}$ is always parallel to the total force $\sum \vec{F}$ acting on the object. However, *the velocity and total force need not be in the same direction.* For example, when an object such as the arrow in Figure 2.31 is fired upward, its velocity is "upward," along the $+y$ direction in the figure. The total force on this object is due almost entirely to gravity (which we'll discuss more in later chapters) and is downward, along $-y$. Hence, in this case, $\vec{v}$ and $\sum \vec{F}$ are in *opposite* directions. While the total force and acceleration are *always* parallel, the direction of the velocity can be different.

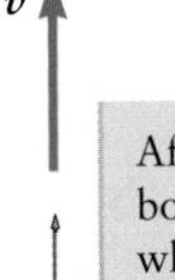

After leaving the bow, $\vec{v}$ is upward while the total force on the arrow and the acceleration of the arrow are both downward.

▲ **Figure 2.31** As the arrow travels upward, its velocity is upward but the total force on the arrow is downward, so $\vec{v}$ and $\sum \vec{F}$ are in opposite directions.

Newton's Third Law

The term $\sum \vec{F}$ appearing in Newton's second law is the total force acting on the object whose motion we are studying or calculating. This force must come from somewhere; in fact, it is *always* produced by other objects. For example, when a

A

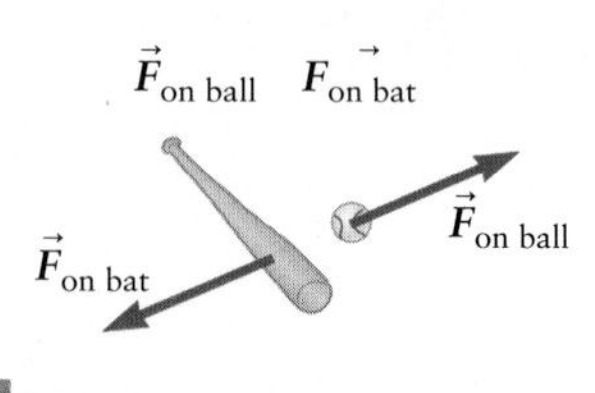

B

▲ **Figure 2.32** A When a bat strikes a baseball, the bat exerts a large force on the ball. B According to Newton's third law, the ball exerts a reaction force—a force of equal magnitude and opposite direction—on the bat.

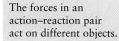

baseball is struck by a bat, the force on the ball is due to the action of the bat. Newton's third law is a statement about what happens to the "other" object, the bat.

> ***Newton's third law of motion:*** **When one object exerts a force on a second object, the second object exerts a force of the same magnitude and opposite direction on the first object.**

Newton's third law is often called the ***action–reaction principle.***

The forces exerted between a baseball bat and a ball (Fig. 2.32) provide an excellent illustration of Newton's third law. When the bat is in contact with the baseball, the ball experiences a force that leads to its acceleration, as can be calculated from Newton's second law. At the same time, the bat experiences a force acting on it that comes from the ball. You may have already encountered this force; it is most noticeable when you hit the ball a little away from the "sweet spot" of the bat. The point of Newton's third law is that forces *always* come in such pairs. The force exerted by the bat on the ball and the force exerted by the ball acting back on the bat are known as an action–reaction pair of forces. According to Newton's third law, these two forces are always *equal in magnitude* and *opposite in direction*, and they must act on *different* objects.

Which Law Do We Use?

Let's now look ahead to how we will actually use Newton's laws to calculate the motion of various objects. Newton's second law tells us how to calculate an object's acceleration. For example, if we want to study the motion of a baseball, we need to calculate the total force on the ball and then insert this $\sum \vec{F}$ into Newton's second law (Eq. 2.8) to find the ball's acceleration. This total force will often have contributions from several sources; for a baseball, there may be a force from the impact with a bat along with other forces. The main points are that the total force $\sum \vec{F}$ in Equation 2.8 is the total force on *just the ball* and that Newton's second law enables us to calculate the acceleration of *just the ball*. On the other hand, Newton's third law tells us that forces always come in pairs. Hence, if we were somehow able to calculate or measure the force exerted by a baseball bat on a ball, Newton's third law tells us that there must be a corresponding reaction force that acts *on the bat*. It is extremely useful to know about this reaction force since it will contribute to the total force *on the bat*.

Newton's three laws of motion are the foundation for nearly everything that we do in the first part of this book. In this section, we have given some background for each of the three laws so that you can appreciate what they tell us about the nature of motion. We'll show in subsequent chapters that Newton's laws contain many other ideas and concepts, such as energy and momentum, that are essential to our thinking about the physical world.

The forces in an action–reaction pair act on different objects.

EXAMPLE 2.9 Action–Reaction

A person pushes a refrigerator across the floor of a room. The person exerts a force $\vec{F}_1$ on the refrigerator. From Newton's third law, we know that $\vec{F}_1$ is part of an action–reaction pair of forces. What is the reaction force to $\vec{F}_1$?

RECOGNIZE THE PRINCIPLE

The force on the refrigerator $\vec{F}_1$ is caused by the action of the person *on the refrigerator*. According to Newton's third law, the reaction force must be equal in magnitude to $\vec{F}_1$ but in the opposite direction. The reaction force must also act on a different object (it cannot act on the refrigerator).

SKETCH THE PROBLEM

Figure 2.33 shows the problem.

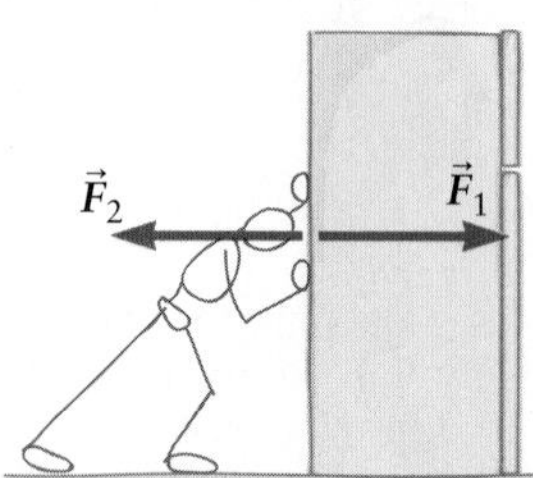

▲ **Figure 2.33** Example 2.9. Newton's third law, the action–reaction principle, applies to people and refrigerators, too. Here the person exerts a force $\vec{F}_1$ on the refrigerator while the refrigerator exerts a force $\vec{F}_2$ on the person.

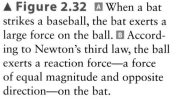

IDENTIFY THE RELATIONSHIPS AND SOLVE

The reaction force $\vec{F}_2$ is the action of the refrigerator *on the person* as shown in Figure 2.33. According to Newton's third law, the forces in an action–reaction pair have equal magnitudes and are in opposite directions, so $\vec{F}_2 = -\vec{F}_1$.

► ***What does it mean?***

Forces *always* come in action–reaction pairs. The two forces in an action–reaction pair always act on *different* objects.

CONCEPT CHECK 2.6 Action–Reaction Force Pairs

Which of the following is *not* an action–reaction pair of forces? (More than one answer may be correct.)

(a) The force exerted by a pitcher on a baseball and the force exerted by the ball when it hits the bat

(b) When you lean against a wall, the force exerted by your hands on the wall and the force exerted by the wall on your hands

(c) In Figure 2.32, the force exerted by the ball on the bat and the force exerted by the bat on the player's hands

2.5 Why Did It Take Newton to Discover Newton's Laws?

Newton's second law tells us that the acceleration of an object is given by $\vec{a} = (\sum \vec{F})/m$, where $\sum \vec{F}$ is the total force acting on the object. In the simplest situations, there may only be one or two forces acting on an object, but in some cases there may be a very large number of forces acting on an object. Multiple forces can make things appear to be very complicated, which is perhaps why the correct laws of motion—Newton's laws—were not discovered sooner.

Forces on a Swimming Bacterium

Figure 2.34 shows a photo of the single-celled bacterium *Escherichia coli*, usually referred to as *E. coli*. An individual *E. coli* propels itself by moving thin strands of protein that extend away from its body (rather like a tail) called flagella. Most *E. coli* possess several flagella as in the photo in Figure 2.34A, but to understand their function, we consider a single flagellum as sketched in Figure 2.34B. A flagellum is fairly rigid, and because it has a spiral shape, one can think of it as a small propeller. An *E. coli* bacterium moves about by rotating this propeller, thereby exerting a force $\vec{F}_w$ on the nearby water. According to Newton's third law, the water exerts a force $\vec{F}_E$ of equal magnitude and opposite direction on the *E. coli* as sketched in Figure 2.34B. One might be tempted to apply Newton's second law with the force $\vec{F}_E$ and conclude that the *E. coli* will move with an acceleration that is proportional to this force. However, this conclusion is incorrect because we have not included the forces from the water on the body of the *E. coli*. These forces are also indicated in Figure 2.34B; to properly describe the total force from the water, we must draw in many force vectors, pushing the *E. coli* in virtually all directions. At the molecular level, we can understand these forces as follows. Water is composed of molecules that are in constant motion and bombard the *E. coli* from all sides. Each time a water molecule collides with the *E. coli*, the molecule exerts a force on the bacterium, much like the collision of the baseball and bat in Figure 2.32. As we saw in that case, the two colliding objects both experience a recoil force, another example of action–reaction forces; hence, the *E. coli* and the water molecule exert forces on each other. An individual *E. coli* is not very large, but a water molecule is much smaller than the bacterium, and the force from one such collision will have only a small effect on the

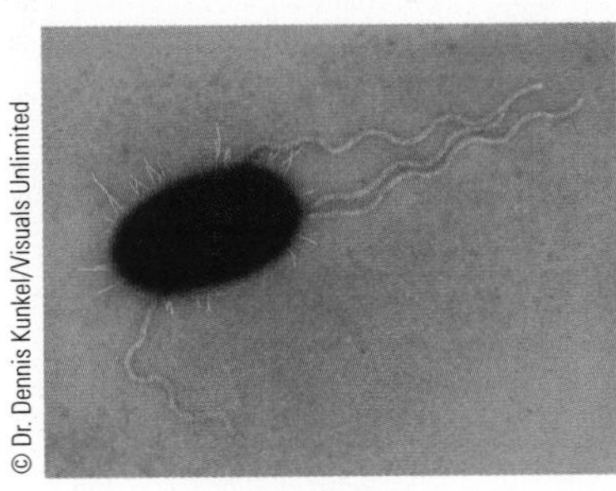

A

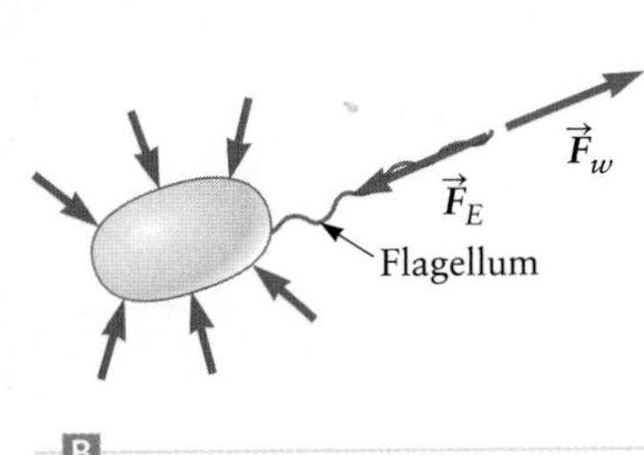

B

▲ **Figure 2.34** A *E. coli* use the action–reaction principle to propel themselves. An individual *E. coli* bacterium is a few micrometers in diameter (1 μm = 0.000001 m). B The flagellum exerts a force $\vec{F}_w$ on the water, and the water exerts a force $\vec{F}_E = -\vec{F}_w$ on the *E. coli*. There are also forces from the water molecules (indicated by the other red arrows) that act on the cell body.

E. coli. However, because there are many water molecules and many such collisions, the sum of the forces from all the water molecule collisions is quite substantial.

Suppose the bacterium is moving toward the left in Figure 2.34B. There will then be more collisions with the water molecules ahead of it (on the left) than with those behind it, resulting in a larger collision force on the bacterium from the left than from the right. This extra force on the front edge of the bacterium is the same type of force you feel when you try to move your body while swimming through water. This resistive force, due to collisions with the water molecules, brings the *E. coli* to a stop when the flagellum stops rotating.

This example with *E. coli* emphasizes again that the force in Newton's second law is the *total* force on the object. It is essential that we account for *all* the forces on an object when applying Newton's second law. Accounting for and calculating all the collision forces on an *E. coli* bacterium is a very complex problem. In fact, we'll learn about other ways to deal with the motion of an *E. coli* and similar problems when we consider the process of *diffusion* in Chapter 15.

2.6 Thinking about the Laws of Nature

In Chapter 1, we described in general terms how ideas and theories can eventually lead to the discovery of a "law" of physics. Let's now discuss this process in a little more detail and consider how Newton's laws came to be.

Discovery of a New Law of Physics

During his lifetime and for many years after, Aristotle's ideas were, in a sense, laws of physics. As we have seen, however, these ideas were not able to explain some important observations, which ultimately led to Galileo's work on the principle of inertia and to Newton. Although we'll never know for sure how Newton arrived at his discoveries, it seems likely that he developed his ideas by first hypothesizing his laws of motion. He then tested his ideas by comparing their predictions with the behavior observed in the experiments of Galileo and others. When Newton found discrepancies, he would rework his theories until eventually they were able to describe correctly the motion of everything that had been studied up to that time. Newton also found that his theories were able to explain the motion of the Moon and the planets. This finding was a tremendous result because prior to Newton there was really no satisfactory theory or explanation of celestial motion. After thus showing that his theory could successfully predict the motion of a wide variety of objects under a wide range of conditions, Newton proposed that they are true "laws" of nature. Other scientists then used Newton's ideas to predict motion in situations that had not yet been studied and tested these predictions with new experiments. When Newton's theories passed these tests, they were accepted as "the" laws of physics, replacing what had come before.

After Newton, What Next?

After Newton's laws were accepted as the "true" laws of motion, what then? Newton's laws are limited to describing how things move. Although this is a broad area of physics, many problems concerning matter and energy are not covered by Newton's laws. Such problems include the behavior of light and electricity, and we require additional theories and laws to describe these phenomena. Also, while we may accept Newton's three laws of motion as the laws of physics that describe mechanics, it is important to continue testing these laws in new situations. In the vast majority of cases, Newton's laws have passed these tests, but around the beginning of the 20th century it was found that Newton's laws do not correctly predict the behavior of electrons and protons, and of atoms and molecules. This monumental discovery eventually led to the development of new laws of physics known as quantum mechanics. What, then, happened to Newton's laws? They were not discarded; instead, physicists realized that Newton's ideas work extremely well for describing motion in what is now known as the ***classical*** regime. This regime includes terrestrial-scale objects, such as rocks and cars, and it even extends to *E. coli* and to planets. Newton's laws break down in the ***quantum*** regime of electrons, protons, and atoms, however, and in that regime a different law of nature—quantum mechanics—must be used.

If Newton's laws break down when applied in certain regimes or to certain types of objects, are we still justified in referring to them as laws of nature? One could take the point of view that a law of nature must always hold true and apply to all situations, with no exceptions. Unfortunately, *no* current law of physics passes this test! All the presently known laws of physics are known to fail or to be inadequate in some regime or another. A more widely accepted viewpoint is that a law of physics must correctly describe all behavior in a particular regime of nature. It is believed that Newton's laws provide an accurate description of all motion in the regime of classical physics, which includes terrestrial objects and extends to things like planets, the Sun, and individual cells.

Although the process of testing and retesting the laws of physics may lead to the discovery of new laws, it is usually found that the law being tested works fine. Such testing, however, can still lead to a better understanding of physics and can reveal important new insights. For example, Newton's three laws do not mention anything about the notion of ***energy***. Nevertheless, we'll see that the concept of energy and many other useful ideas are contained within Newton's laws. Applying Newton's laws to new situations can help us discover such unanticipated results.

Summary | CHAPTER 2

Key Concepts and Principles

Motion

A complete description of an object's motion involves its ***position***, ***velocity***, and ***acceleration***. For motion in one dimension (along a line), position can be specified by a single quantity, x. The ***instantaneous velocity*** is related to changes in position by

$$v = \lim_{\Delta t \to 0} \frac{\Delta x}{\Delta t} \qquad \text{(2.5) (page 31)}$$

The instantaneous velocity is usually referred to as simply the "velocity."

The ***instantaneous acceleration*** is given by

$$a = \lim_{\Delta t \to 0} \frac{\Delta v}{\Delta t} \qquad \text{(2.7) (page 34)}$$

and is usually referred to simply as the "acceleration."

Position, velocity, and acceleration are vector quantities. For motion in two or three dimensions, we must consider how the directions of these quantities relate to our chosen coordinate axes. This topic is explored further in Chapter 4.

Newton's laws of motion

The connection between ***force*** and motion is at the heart of ***mechanics***. A force is a push or a pull. Force is a vector. Given the forces acting on an object, we can calculate how it will move using Newton's three laws of motion.

- ***Newton's first law***: If the total force acting on an object is zero, the object will move with a constant velocity.
- ***Newton's second law***:

$$\vec{a} = \frac{\Sigma \vec{F}}{m} \qquad \text{(2.8) (page 41)}$$

Forces thus cause acceleration.

(Continued)

- ***Newton's third law***: For every action (force), there is a reaction (force) of equal magnitude and opposite direction.

All forces come in ***action–reaction pairs***. The two forces in an action–reaction pair act on different objects.

Applications

The relationships between position, velocity, and acceleration

The ***instantaneous velocity*** at a particular time t is the slope of the position–time curve at that time (Fig. 2.35).

The ***average velocity*** during a particular time interval is equal to the slope of the line connecting the start and end of that interval on the position–time curve (Fig. 2.35).

The ***instantaneous acceleration*** equals the slope of a plot of the instantaneous velocity v as a function of time (Fig. 2.36).

The ***average acceleration*** during a time interval is equal to the slope of the line connecting the start and end of that interval on the velocity–time curve (Fig. 2.36).

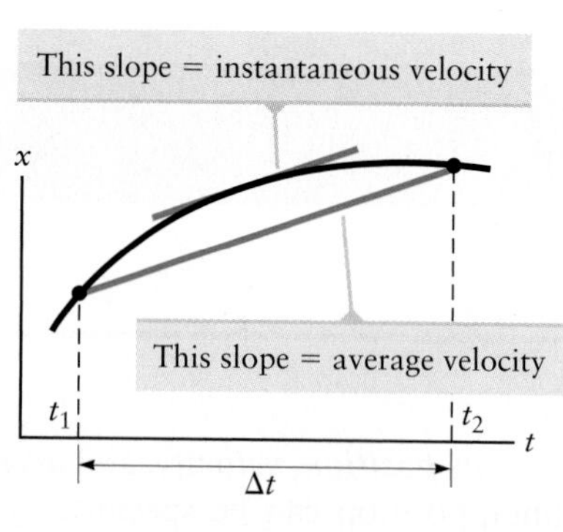

▲ **Figure 2.35**

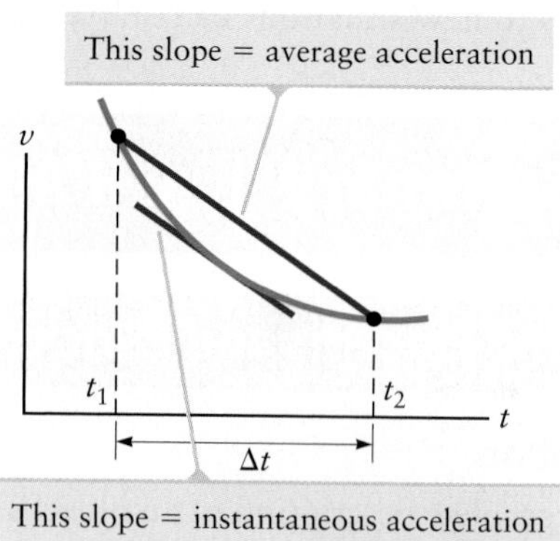

▲ **Figure 2.36**

Questions

SSM = answer in *Student Companion & Problem-Solving Guide* (X) = life science application

1. Make a hypothetical sketch of a velocity–time graph in which the velocity is always positive, but the acceleration is always negative. Give a physical example of such motion.

2. A ball is thrown into the air with an upward velocity of 10 m/s. A short time later, it is caught on its way down, also with a speed of 10 m/s.
(a) Draw the velocity–time graph for this situation.
(b) Draw the acceleration–time graph for this situation.
(c) Many people would describe the ball's motion as "slowing down" until it reaches the top of its flight and then "speeding up" on the downward trip. However, a correct graph in part (b) should show a consistently negative (downward) acceleration. Explain how a negative acceleration can mean both "speeding up" and "slowing down."

3. When a car collides with a wall, a force on the car causes it to stop. Identify an action–reaction pair of forces involving the car.

4. You push on a refrigerator, as in Figure 2.1, but the refrigerator does not move. Hence, even though you are applying a nonzero force, the acceleration is still zero. Explain why this does not contradict Newton's second law.

5. A car is traveling on an icy road that is extremely slippery. The driver finds that she is not able to stop or turn the car. Explain this situation in terms of the principle of inertia (Newton's first law).

6. A car is initially at rest on an icy road that is extremely slippery. The driver finds that he is unable to get the car to drive away because the wheels simply spin when he tries to accelerate. Explain this situation in terms of the principle of inertia (Newton's first law).

7. Give an example of motion for which the average velocity is zero, but the speed is never zero.

8. Give an example of motion for which the average velocity is equal to the instantaneous velocity. *Hint*: Sketch the velocity–time graph.

9. SSM **Abracadabra!** A magician pulls a tablecloth off of a set table with one swift, graceful motion. Amazingly, the fine china, glassware, and silverware are practically undisturbed. Although amazing, this feat is not an illusion. Describe the behavior of the plates and glasses in terms of the principle of inertia (Newton's first law).

10. Give examples of motion matching the following descriptions. (a) The velocity is positive and the acceleration is positive. (b) The velocity is negative and the acceleration is negative. (c) The velocity is positive and the acceleration is negative.

11. A car starts from the origin at $t = 0$. At some later time, is it possible for the car's velocity to be positive but its displacement from the origin to be zero? Explain and give an example.

12. The acceleration of an object that falls freely under the action of gravity near the Earth's surface is negative and constant. (a) Does the object's instantaneous acceleration equal its average acceleration? (b) Draw the corresponding velocity–time graph. (c) Does the instantaneous velocity equal the average velocity? Explain why or why not.

13. Consider again Example 2.8. Draw plots of the position, velocity, and acceleration as functions of time, starting from when the ball is released and ending *after* the ball hits the ground. Indicate the time at which the ball hits the ground.

14. Make a qualitative sketch of the position y as a function of time for the center of a yo-yo (the point at the middle of the axle). Also make sketches of the velocity and acceleration as functions of time. Is the total force on the yo-yo zero or nonzero? Explain how you can tell from your graphs.

15. SSM Figure Q2.15 shows a motion diagram for a rocket-powered car. The photos are taken at 1.0-s intervals. Make qualitative plots of the position, velocity, acceleration, and force on the car as functions of time.

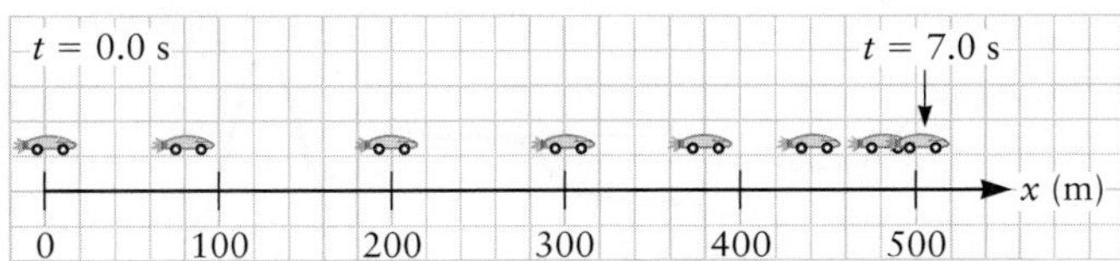

Figure Q2.15

16. A person stands on level ground and throws a baseball straight upward, into the air. (a) Does the person exert a force on the ball while it is in his hand? After it leaves his hand? (b) Does the ball ever exert a force on the person? If so, what is the direction of this force? (c) If your answer to part (b) is yes, why does the person not accelerate?

17. Consider the motion of the Moon as it orbits the Earth. (a) Is the Moon's acceleration zero or nonzero? Explain. (b) If the Moon has a nonzero acceleration, what force is responsible?

18. Consider the motion of a marble as it falls to the bottom of a jar of honey. Experiments show (see also Chapter 3) that the marble moves with a constant velocity. Applying Newton's first law, does that mean that no forces are acting on the marble?

19. According to Newton's first law (the principle of inertia), if there is no force exerted on an object, the object will move with constant velocity. Consider the following examples of motion. Is the velocity of the object constant? What are the forces acting on each object? In cases in which the velocity is constant, explain why Newton's first law applies.
   (a) A hockey puck sliding on an ice-covered surface that is horizontal and frictionless
   (b) A mug of root beer sliding down a bar, whose surface is slippery but not frictionless
   (c) A car skidding on a flat, desert highway

20. Three blocks rest on a table as shown in Figure Q2.20. Identify three action–reaction pairs of forces.

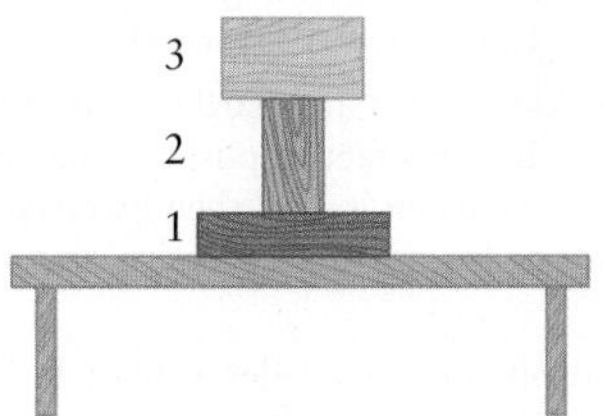

Figure Q2.20

21. Two football players start running at opposite ends of a football field (opposite goal lines), run toward each other, and then collide at the center of the field. They start from rest and are running at top speed when they collide. (a) Draw a graph showing the position as a function of time for both players. (b) Draw a graph showing their velocities as functions of time.

22. A person is riding in a car traveling on a straight level road. What is the direction of the *net* force on the person if the car is (a) speeding up, (b) slowing down, (c) has a constant speed?

Problems

SSM = solution in *Student Companion & Problem-Solving Guide*
★ = intermediate ✪ = challenging ✗ = life science application
Ⓡ = reasoning and relationships problem
RT = reasoning tutorial in Enhanced WebAssign

2.2 WHAT IS MOTION?

1. In SI units, velocity is measured in units of meters per second (m/s). Which of the following combinations of units can also be used to measure velocity?
   (a) cm/s
   (b) cm/s^2
   (c) $m^3/(mm^2 \cdot s^2)$
   (d) km/s
   (e) miles per hour
   (f) km/hour
   (g) miles/cm

2. A typical airplane can fly at a speed of 400 miles per hour. What is its speed in meters per second?

3. In SI units, acceleration is measured in units of meters per second squared (m/s^2). Which of the following combinations of units can also be used to measure acceleration?
   (a) cm/s
   (b) cm/s^2
   (c) $m^3/(mm^2 \cdot s^2)$
   (d) km/s^2
   (e) miles per hour
   (f) miles/(minutes)^2

4. A falling baseball has an acceleration of magnitude 9.8 m/s^2. What is its acceleration in feet per second squared?

5. An elite runner can run 1500 m in 3 minutes and 35 s. What is his average speed? Give your answer to two significant figures.

6. Consider the motion of a sprinter running a 100-m dash. When it is run outdoors, this race is run along a straight-line portion of a track, so it is an example of motion in one dimension. Draw qualitative plots of the position, velocity, and acceleration as functions of time for the sprinter. Start your plots just before the race starts and end them when the runner comes to a complete stop.

7. A jogger maintains a speed of 3.0 m/s for 200 m until he encounters a stoplight, and he abruptly stops and waits 30 s for the light to change. He then resumes his exercise and maintains a speed of 3.5 m/s for the remaining 50 m to his home. (a) What was his average velocity for this entire time interval? (b) What were his maximum and minimum velocities, and how do they compare with this average?

8. A hockey puck that is sliding on an icy surface will eventually come to rest. The (horizontal) force that makes it stop is due to friction between the puck and the ice. Draw qualitative plots of the position, velocity, and acceleration as functions of time for the puck. Pay special attention to the sign of the acceleration.

9. Consider a marble falling through a very thick fluid, such as molasses. Draw qualitative plots of the position, velocity, and acceleration as functions of time for the marble, assuming it is released from rest just above the molasses.

10. A person riding on a skateboard is initially coasting on level ground. He then uses his feet to push on the ground so that he speeds up for a few seconds, and then he coasts again. Draw qualitative plots of his position, velocity, and acceleration as functions of time.

11. **In the drop zone.** Consider a skydiver who jumps from an airplane. Suppose she waits for 1 min before opening her parachute and she lands 4 min after leaving the airplane. Draw qualitative plots of the position, velocity, and acceleration as functions of time for the skydiver, starting from the time she jumps from the plane, assuming (somewhat unrealistically) that she falls straight down. Here, the position y is her vertical height above the ground. Be sure to indicate the time at which she opens the parachute. *Hint*: Consider how the velocity will vary with time after the chute is open for a long time.

12. Consider a skier who coasts up to the top of a hill and then continues down the other side. Draw a qualitative plot of what the skier's speed might look like.

13. SSM Figure P2.13 shows three motion diagrams, where the dots indicate the positions of an object after equal time intervals. Assume left-to-right motion. For each motion diagram, sketch the appropriate position–time, velocity–time, and acceleration–time graphs.

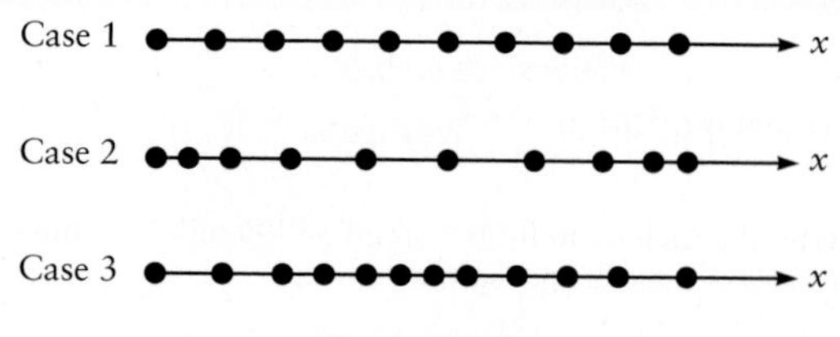

Figure P2.13

14. A bicycle is moving initially with a constant velocity along a level road. The bicyclist then decides to slow down, so she applies her brakes over a period of several seconds. Thereafter, she again travels with a constant velocity. Draw qualitative sketches of her position, velocity, and acceleration as functions of time.

15. Figure P2.15 shows several hypothetical position–time graphs. For each graph, sketch qualitatively the corresponding velocity–time graph.

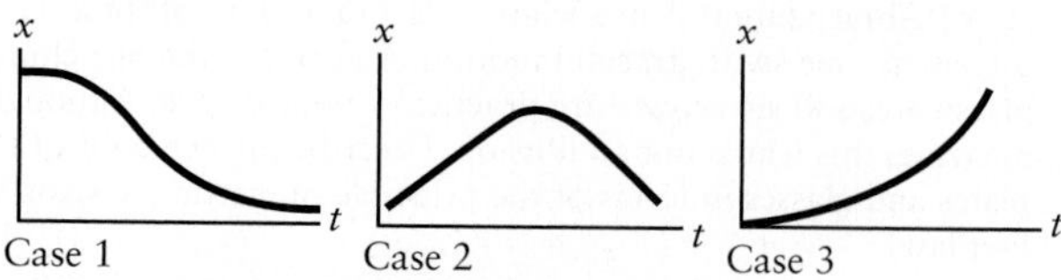

Figure P2.15 Problems 15 and 16.

16. The position–time graphs in Figure P2.15 each describe the possible motion of a particular object. Give at least one example of what the object and motion might be in each case.

17. Figure P2.17 shows several hypothetical velocity–time graphs. For each case, sketch qualitatively the corresponding acceleration–time graph.

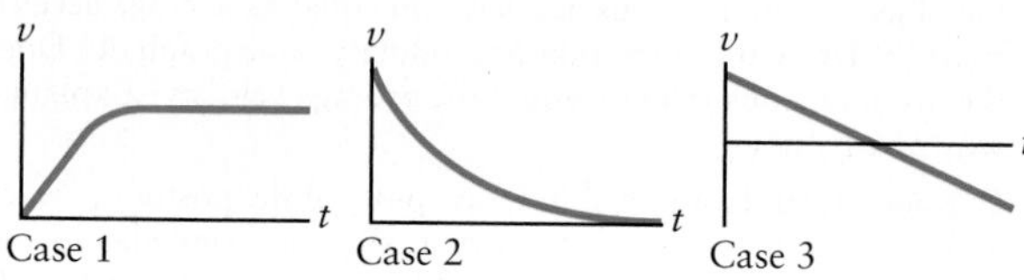

Figure P2.17 Problems 17, 18, and 19.

18. Figure P2.17 shows several hypothetical velocity–time graphs. For each case, sketch qualitatively a possible corresponding position–time graph.

19. Give examples of objects whose motion might be described by the graphs in Figure P2.17.

20. Figure P2.20 shows several hypothetical acceleration–time graphs. For each case, sketch qualitatively a possible corresponding velocity–time graph.

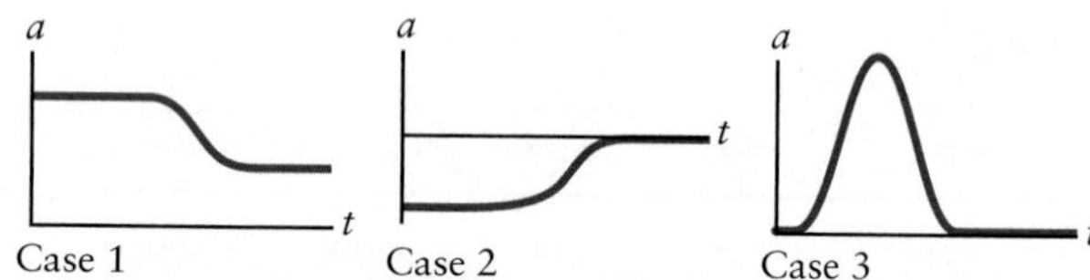

Figure P2.20 Problems 20 and 21.

21. Give examples of objects whose motion is described by the plots in Figure P2.20.

22. Match each of the following examples of motion to one of the position–time graphs in Figure P2.22.
(a) A person at the beginning of a race, starting from rest
(b) A runner near the end of a race, just after crossing the finish line

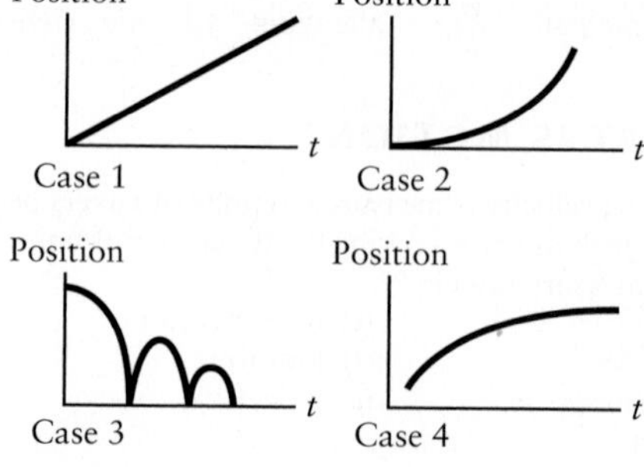

Figure P2.22

(c) A ball dropped from a window, hitting the ground below and then bouncing several times
(d) A bowling ball as it rolls down a lane, just after it leaves the bowler's hand

23. Match each of the examples of motion in Problem 22 to one of the velocity–time graphs shown in Figure P2.23.

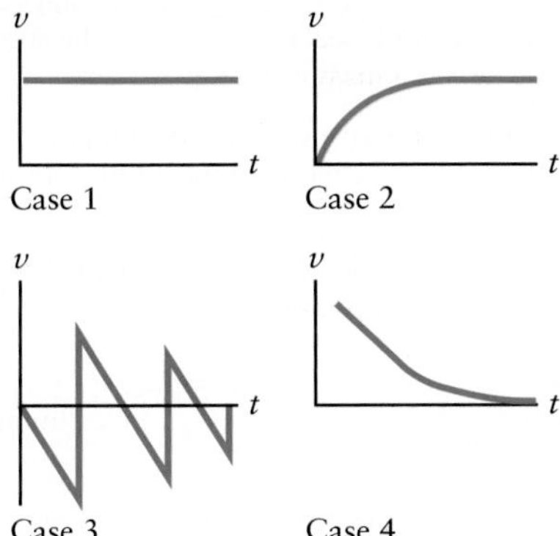

Figure P2.23

24. Consider the position–time graph shown in Figure P2.24. Make a careful graphical estimate of the velocity as a function of time by measuring the slopes of tangent lines. What is an approximate value of the maximum velocity of the object?

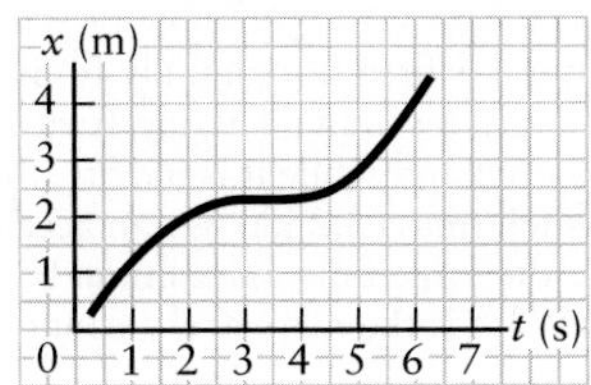

Figure P2.24 Problems 24 and 25.

25. For the object described by Figure P2.24, estimate the average velocity (a) over the interval from $t = 2.0$ s to $t = 4.0$ s and (b) over the interval from $t = 1.0$ s to $t = 5.0$ s.

26. Repeat Problem 24 using the position–time graph in Figure P2.26.

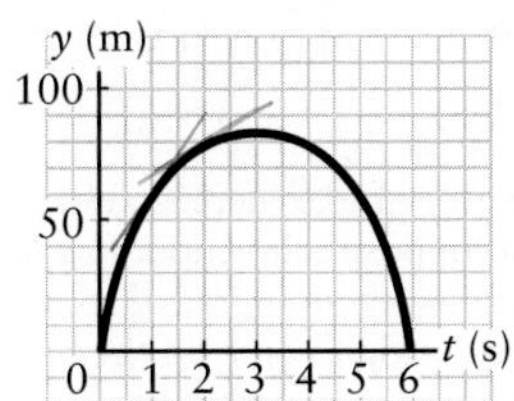

Figure P2.26

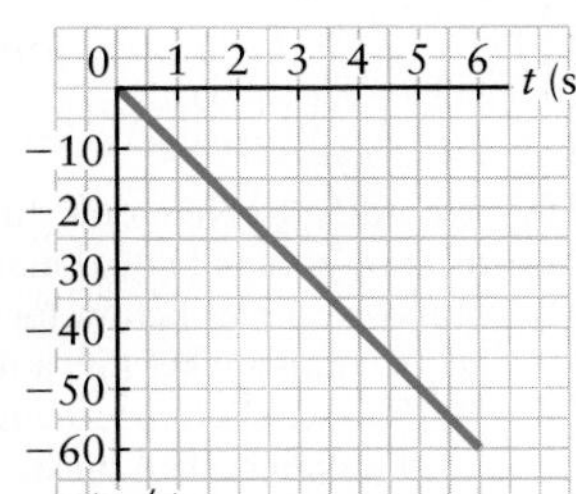

Figure P2.27

27. Figure P2.27 shows the velocity–time curve of a falling brick. Make a careful estimate of the slope to find the acceleration of the brick at $t = 3.0$ s.

28. Using a graphical approach (i.e., by estimating the slope at various points), find the qualitative behavior of the acceleration as a function of time for the object described by the velocity–time graph in Figure P2.28.

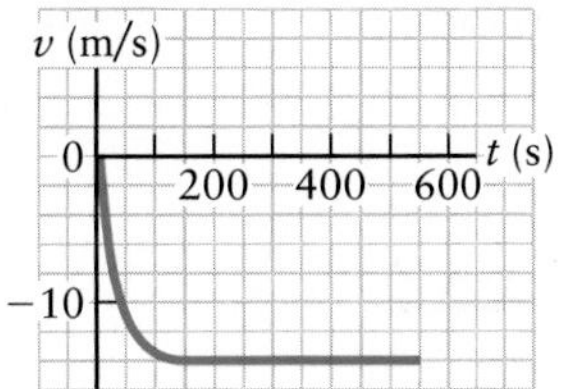

Figure P2.28 Problems 28 and 30.

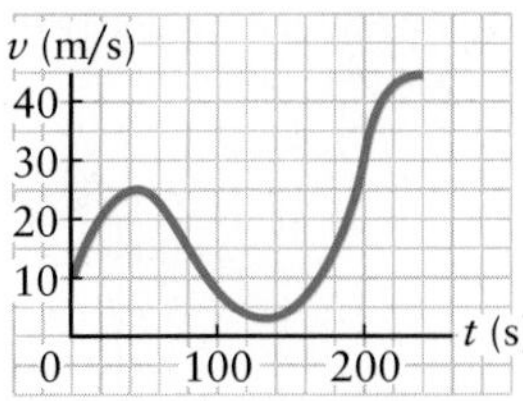

Figure P2.29 Problems 29 and 30.

29. For the object described by the velocity–time graph in Figure P2.29, estimate the average acceleration over the interval from $t = 0$ s to $t = 50$ s and over the interval from $t = 100$ s to $t = 200$ s.

30. Draw a possible position–time graph for an object whose velocity as a function of time is described by (a) Figure P2.28 and (b) Figure P2.29.

31. Draw a graph showing the position x as a function of time for an object whose acceleration is (a) constant and positive, (b) constant and negative, and (c) positive and increasing with time. Assume the object starts at $x = 0$ in parts (a) and (c) and at a positive value of x in part (b).

32. A car travels along a straight, level road. The car begins a distance $x = 25$ m from the origin at $t = 0.0$ s. At $t = 5.0$ s, the car is at $x = 100$ m; at $t = 8.0$ s, it is at $x = 300$ m. Find the average velocity of the car during the interval from $t = 0.0$ s to $t = 5.0$ s and during the interval from $t = 5.0$ s to $t = 8.0$ s.

33. SSM A squirrel falls from a very tall tree. Initially (at $t = 0$), the squirrel is at the top of the tree, a distance $y = 50$ m above the ground. At $t = 1.0$ s, the squirrel is at $y = 45$ m, and at $t = 2.0$ s it is at $y = 30$ m. Estimate the average velocity of the squirrel during the intervals from $t = 0.0$ s to $t = 1.0$ s and from $t = 1.0$ s to $t = 2.0$ s. Use these results to estimate the average acceleration of the squirrel during this time. (No squirrels were harmed during the writing of this problem.)

34. A spacecraft takes off from Cape Canaveral in Florida and orbits Earth 18 times before landing back at Cape Canaveral 24 hours and 15 minutes later. During this time, it moves in a circular orbit with radius 6.7×10^6 m. (a) What is the average speed of the space shuttle during its journey? (b) What is the average velocity?

35. Figure P2.35 shows the velocity as a function of time for an object. (a) What is the average acceleration during the interval from $t = 0$ s to $t = 20$ s? (b) Estimate the instantaneous acceleration at $t = 5$ s, (c) at $t = 10$ s, and (d) at $t = 20$ s.

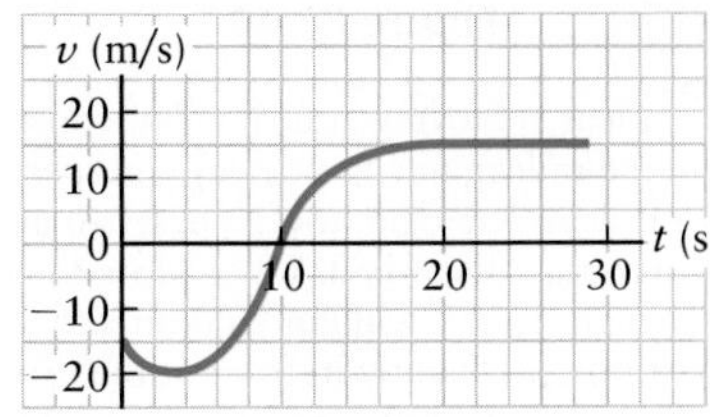

Figure P2.35

36. Figure P2.36 shows the acceleration as a function of time for an object. (a) If the object starts from rest at $t = 0$, what is the velocity of the object as a function of time? (b) If the object instead has a velocity of +40 m/s at $t = 0$, how does your result for part (a) change?

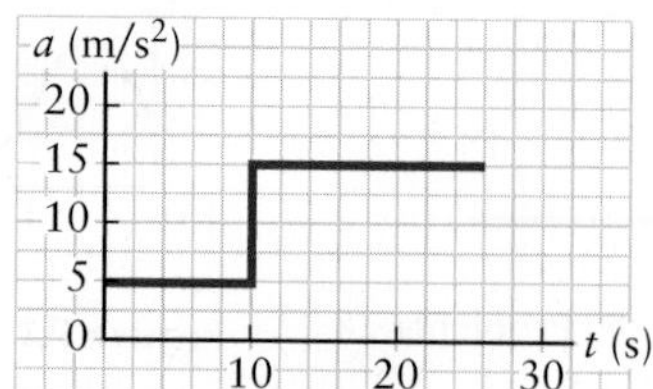

Figure P2.36

37. A rabbit runs in a straight line with a velocity of +1.5 m/s for a period of time, rests for 10 s, and then runs again along the same line at +0.60 m/s for an unknown amount of time. The rabbit travels a total distance of 1200 m, and its average speed is 0.80 m/s. (a) What is the total time the rabbit spends running at 1.5 m/s? (b) How long does it spend running at 0.60 m/s?

2.3 THE PRINCIPLE OF INERTIA

38. SSM Figure 2.24 (page 39) shows one of Galileo's experiments in which a ball rolls up an incline. A ball that is initially rolling up the incline will roll up to some maximum height and then roll back down the incline. Draw qualitative plots of the position and velocity as functions of time for the ball. Take $x = 0$ at the bottom of the ramp.

39. A block is sliding at a particular initial speed v when it encounters the following horizontal surfaces: (a) a grass field, (b) an asphalt street, (c) a hardwood floor, (d) an ice rink (with real, not ideal, ice), and (e) an ideal, frictionless surface. Draw qualitative plots of the velocity as a function of time for each case, with $t = 0$ corresponding to the time at which the block reaches each of these surfaces.

40. **An unsafe way to transport your penguin.** A flatbed truck hauls a block of ice on top of which stands a penguin (see Fig. P2.40). Assume the block of ice is frozen solid to the bed of the truck, but the top surface of the ice is extremely slippery. The truck skids to a stop from an initial velocity of 20 m/s. Describe what happens to the penguin in terms of Newton's first law. How does it differ from what happens to the driver of the truck, who is wearing a seat belt? Draw and compare the velocity–time graphs of the penguin and truck driver.

Figure P2.40

41. A flatbed railcar is moving at a slow but constant velocity. A man stands in the railcar, facing sideways (perpendicular) to the motion of the railcar. The man holds a baseball at arm's length and drops it onto the railcar bed. Where does the principle of inertia predict that the ball will land: (a) directly below the drop point on the railcar, (b) slightly in front of the drop point on the railcar, or (c) slightly behind the drop point on the railcar? Explain.

2.4 NEWTON'S LAWS OF MOTION

42. The person shown in Figure 2.1 (page 27) is pushing on a refrigerator that sits on a level floor. Assume the section of the floor underneath the refrigerator is very slippery so that the only horizontal force on the refrigerator is due to the person. If the force exerted by the person has a magnitude of 120 N and the refrigerator has a mass of 180 kg, what is the magnitude of the refrigerator's acceleration?

43. **Action and reaction.** Consider again the man pushing the refrigerator in Problem 42, but now assume the entire floor, including the portion under the man, is frictionless. Use Newton's third law along with his second law to find the acceleration of the man. Assume he has a mass of 60 kg.

44. An object is found to move with an acceleration of magnitude 12 m/s² when it is subjected to a force of magnitude 200 N. Find the mass of the object.

45. In SI units, force is measured in newtons, with 1 N = 1 kg · m/s². Which of the following combinations of units can also be used to measure force?
(a) g · m/s
(b) g · cm²/s
(c) kg · m⁴/(s² · cm³)
(d) g · cm/s²
(e) kg · miles/(minutes)²

46. In the U.S. customary system of units, mass is measured in units called slugs. Suppose an object has a mass of 15 kg. Use the conversion factors inside the front cover of this book to express the mass of the object in slugs.

47. In the U.S. customary system of units, force is measured in units of pounds (abbreviated lb). Suppose the force on an object is 150 lb. Using the conversion factors inside the front cover of this book, express this force in units of newtons.

48. A force is found to be 240 g · cm/s². Convert this value into units of newtons.

49. SSM A cannon is fired horizontally from a platform (Fig. P2.49). The platform rests on a flat, icy, frictionless surface. Just after the shell is fired and while it is moving through the barrel of the gun, the shell (mass 3.2 kg) has an acceleration of +2500 m/s². At the same time, the cannon has an acceleration of −0.76 m/s². What is the mass of the cannon?

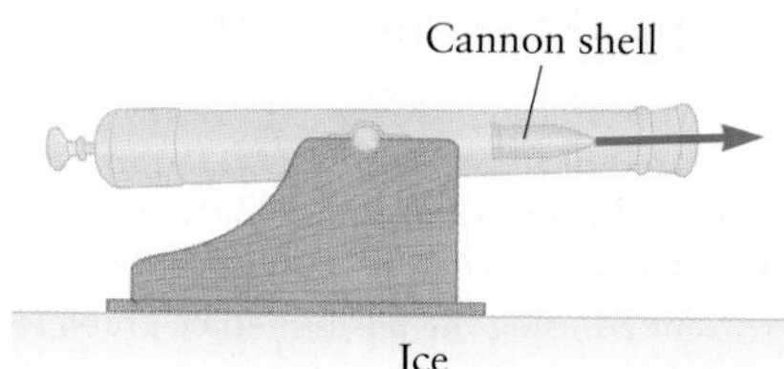

Figure P2.49

50. According to Newton's third law, for every force there is always a reaction force of equal magnitude and opposite direction. In each of the examples below, identify an action–reaction pair of forces.
(a) A tennis racket hits a tennis ball, exerting a force on the ball.
(b) Two ice skaters are initially at rest and in contact. One of the skaters then pushes on the other skater's back, exerting a force on that skater.
(c) A car is moving at high speed and runs into a tree, exerting a force on the tree.
(d) Two cars are moving in opposite directions and collide head-on.
(e) A person leans on a wall, exerting a force on the wall.
(f) A hammer hits a nail, exerting a force on the nail.
(g) A mass hangs by a string tied to a ceiling, with the string exerting a force on the mass.
(h) A bird sits on a telephone pole, exerting a force on the pole.

Additional Problems

51. ✱ Tom has two ways he could drive home from work. He could take Highway 99 for 45 miles with a speed limit of 65 mi/h, or he could take Interstate 5, which would take him a bit out of the way at 57 miles, but with a speed limit of 75 mi/h. (a) Assuming Tom misses the rush hour and drives at an average speed equal to the maximum speed limit, which route gets him home the fastest? How much time does he save? (b) If Tom breaks the law and travels at 75 mi/h on Highway 99, how much time would he save (if he is not stopped to get a speeding ticket) compared with driving at the legal speed limit on Highway 99?

52. SSM ✱ **Throwing heat.** In professional baseball, most pitchers can throw a fastball at a speed of 90 mi/h. For simplicity, assume the pitcher releases the ball just above the center of the pitcher's mound. (a) Given that the regulation distance from the pitcher's mound to home plate is 60.5 ft, how long does it take the ball to reach home plate after the ball leaves the pitcher's hand? (b) It takes (on average) 0.20 s for the batter to get the tip of his bat over home plate. How much time does that give him to react? (c) The average fast pitch in professional women's softball is about 60 mi/h, where the regulation distance from the pitcher's mound to home plate is 40.0 ft. How do the travel time of a pitched ball and reaction time of a batter in softball compare with those in baseball? Keep three significant figures in your calculation.

53. ✱ Figure P2.53 shows the position as a function of time for an object. (a) What is the average velocity during the period from $t = 0.0$ s to $t = 10.0$ s? (b) What is the average velocity between $t = 0.0$ s and $t = 5.0$ s? (c) Between $t = 5.0$ s and $t = 10.0$ s? (d) Explain why and how your answers to parts (a), (b), and (c) are related.

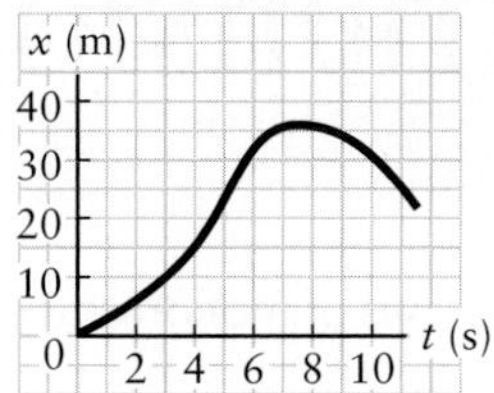

Figure P2.53

54. SSM RT On your vacation, you fly from Atlanta to San Francisco (a total distance of 3400 km) in 4.0 h. (a) Draw a qualitative sketch of how the speed of your airplane varies with time. (b) What is the average speed during your trip? (c) Estimate the top speed during your trip. *Hint*: You reach your top speed about 10 minutes after taking off. (d) What is your average acceleration during the first 10 minutes of your trip? (e) What is the average acceleration during the central hour of your trip?

55. ✪ Figure P2.55 shows the position as a function of time for an apple that falls from a very tall tree. (a) At what time does the apple hit the ground? (b) Use a graphical approach to estimate and plot the acceleration as a function of time. (c) Make a sketch of the velocity of the apple as a function of its height above the ground.

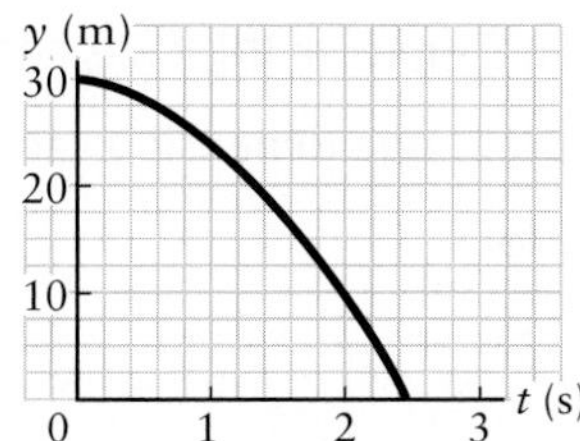

Figure P2.55

56. ✱ ✖ A cheetah runs a distance of 100 m at a speed of 25 m/s and then runs another 100 m in the same direction at a speed of 35 m/s. What is the cheetah's average speed? Explain why the correct answer is *not* just the average of the speeds during the first and last 100 m.

57. **Astronaut frequent-flyer miles.** The average speed of a spacecraft while in orbit is about 8900 m/s. How far does the spacecraft travel during a mission that lasts 7.5 days?

58. ✪ A cat is being chased by a dog. Both are running in a straight line at constant speeds. The cat has a head start of 3.5 m. The dog is running with a speed of 8.5 m/s and catches the cat after 6.5 s. How fast did the cat run?

59. ✪ ✖ **Predator and prey.** The author's cat enjoys chasing chipmunks in the front yard. In this game, the cat sits at one edge of a yard that is $w = 30$ m across (Fig. P2.59), watching as chipmunks move toward the center. The cat can run faster than a chipmunk, and when a chipmunk moves more than a certain distance L from the far end of the yard, the cat knows that it can catch the chipmunk before the chipmunk disappears into the nearby woods. If the cat's top speed is 7.5 m/s and a chipmunk's top speed is 4.5 m/s, find L. Assume the chipmunk and cat move along the same straight-line path.

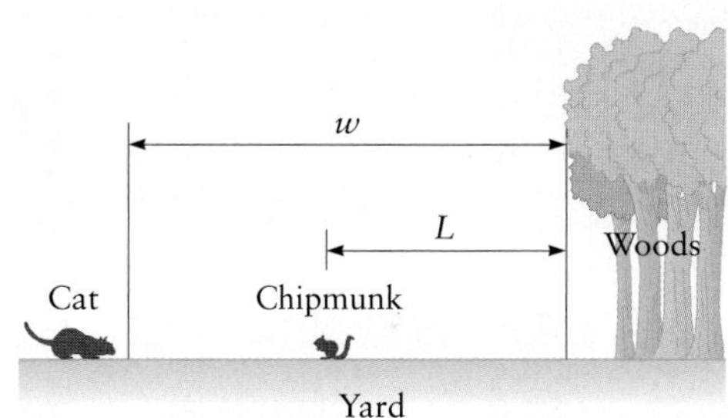

Figure P2.59

60. ✪ A thief is trying to escape from a parking garage after completing a robbery, and the thief's car is speeding ($v = 12$ m/s) toward the door of the parking garage (Fig. P2.60). When the thief is $L = 30$ m from the door, a police officer flips a switch to close the garage door. The door starts at a height of 2.0 m and moves downward at 0.20 m/s. If the thief's car is 1.4 m tall, will the thief escape?

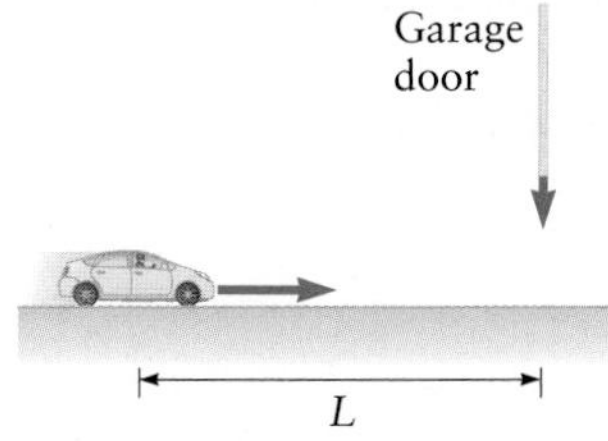

Figure P2.60

CHAPTER

3

▶ *These railroad tracks carry trains across the desert in southern California. They are very straight and very long, so the motion of the trains they carry is indeed one-dimensional. These tracks can be viewed as a coordinate axis complete with "tick marks"! (© Richard T. Norwitz/National Geographic/ Getty Images)*

Forces and Motion in One Dimension

OUTLINE

In Chapter 2, we presented Newton's laws of motion and explained what they mean from a qualitative point of view. We also introduced a number of quantities, including displacement, velocity, and acceleration, that are essential for describing motion. Our next job is to apply Newton's laws to calculate how things move in various situations. In this chapter, we consider motion in one dimension, that is, motion along a straight line. You might wonder why we devote an entire chapter to such an idealized case; after all, the world is not one dimensional, and in most situations, objects move along two- or three-dimensional trajectories. Even so, it is very useful to start with one-dimensional examples. The mathematics is simpler in this case, and the approaches we develop for dealing with motion in one dimension can be applied quite directly to motion in higher dimensions.

3.1 Motion of a Spacecraft in Interstellar Space

Figure 3.1 shows a hypothetical spacecraft traveling somewhere in distant space, traveling along a straight-line path from one galaxy to another. Since the spacecraft is moving along a line, this is an example of one-dimensional motion. To describe this motion, we must first choose a coordinate system with which to measure the displacement, velocity, and acceleration of the spacecraft. The spacecraft's motion is along the line that connects the two galaxies, so we choose that line to be our coordinate axis. We have drawn this axis in Figure 3.1 and labeled it the x axis. We have also placed the origin at the starting galaxy and taken the "positive" direction along the x axis as headed toward the destination.

If our spacecraft is very far from any planets or stars, the force of gravity on it will be very small. To make this example as simple as possible, let's assume this gravitational force is zero. We now want to calculate how the spacecraft will move in two different cases.

In the first case, the engine is turned off so that the spacecraft is "coasting." Since the engine is off and the gravitational force is zero, there are no forces acting on the spacecraft. Newton's first law tells us that when the total force acting on an object is zero, the object has a constant velocity. Hence, our spacecraft moves with a constant velocity; that is, it moves in a straight-line path with a constant speed. Examples of such constant-velocity motion are given in Figures 3.2 and 3.3, which show graphs of how the velocity and position might look as functions of time. We learned in Chapter 2 that displacement and velocity are vectors, having both magnitude and direction. For the cases involving one-dimensional motion that we consider here and throughout this chapter, these vectors all lie along the chosen coordinate axis. In general, we could write these quantities using our usual vector notation with vector arrows, but with one-dimensional motion, we can simplify the notation and drop the vector arrows because we specify direction by the sign of each quantity (+ or −). This notation corresponds to specifying the *components* of the displacement and velocity along the coordinate axis.

Figure 3.2 shows the behavior of the velocity and position of our spacecraft when the velocity is zero, while Figure 3.3 shows the behavior for a constant nonzero velocity. Recall that v is the slope of the position–time curve; in both Figures 3.2 and 3.3, the velocity is constant, so the x–t relations are both linear. Although Newton's first law tells us that the velocity is constant, it does not tell us the value of v. This value is determined by forces applied to the spacecraft at earlier times. In this problem, a force might have been applied to the spacecraft before the engine was turned off, giving a nonzero acceleration prior to $t = 0$ and causing the velocity in Figure 3.3 to be nonzero.

Insight 3.1

ALWAYS START WITH A PICTURE

The first step in attacking any problem involving motion is to draw a picture. This picture should always show the coordinate axes you have chosen to use for the problem.

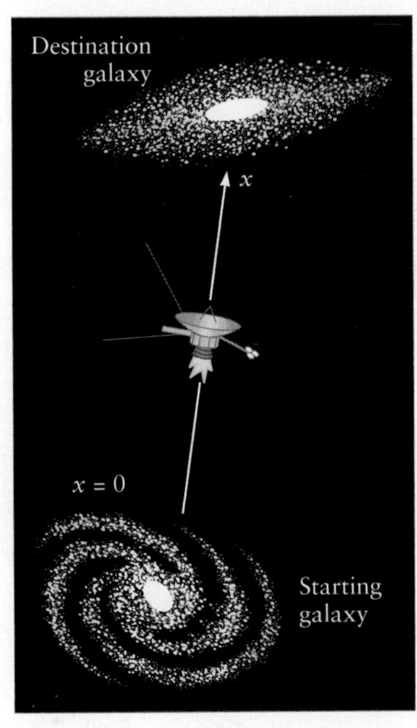

▲ **Figure 3.1** This hypothetical spacecraft is following a one-dimensional trajectory along the x axis.

Motion with a Constant Nonzero Acceleration

Let's next consider what happens when the spacecraft's engine is turned on. There is now a force exerted on the spacecraft, so we need to use Newton's second law to

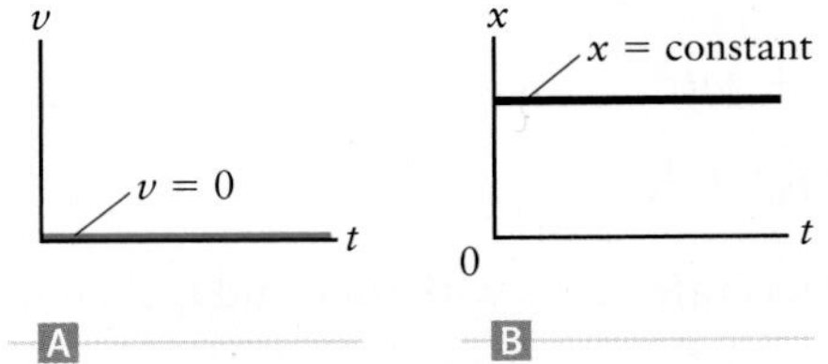

▲ **Figure 3.2** If the spacecraft's velocity is zero, its position x is a constant. This is one example of motion with a constant velocity.

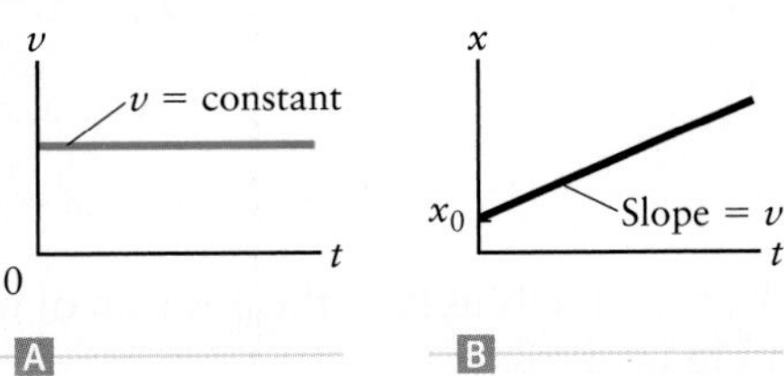

▲ **Figure 3.3** Motion with a constant and nonzero velocity. Since v is equal to the slope of the position–time curve, the graph of x as a function of t is a straight line.

determine the resulting motion. Recall that according to Newton's second law the acceleration is given by $a = \sum F/m$. For our case of one-dimensional motion, both the total force $\sum F$ and the acceleration a are parallel to the x axis in Figure 3.1 (so they are the components of the force and acceleration along x). In this example, there is only a single force (due to the engine); in the simplest case, the engine will be designed to produce a constant force on the spacecraft, so the force term $\sum F$ in Newton's second law is a constant. For simplicity, we also assume the mass m of the spacecraft is constant.[1] We can then use Newton's second law to calculate the acceleration and find that a is also a constant. Since acceleration is the slope of the velocity–time curve, the v–t relation must be a *straight line*. This relationship is shown in Figure 3.4A; we have

Velocity versus time for motion with a constant acceleration

$$v = v_0 + at \tag{3.1}$$

In Equation 3.1, v_0 is the "initial" velocity of the spacecraft, that is, the velocity at $t = 0$; it is useful to think of v_0 as the velocity at the moment the stopwatch used to describe this problem is started. The value of v_0 depends on what happened prior to $t = 0$; for example, one might imagine that the engine was turned on for a while to launch the spacecraft and then turned off before $t = 0$. In most situations (and problems!), we must rely on someone else (such as the person who writes the problem or the pilot of the spacecraft) to provide the value of v_0.

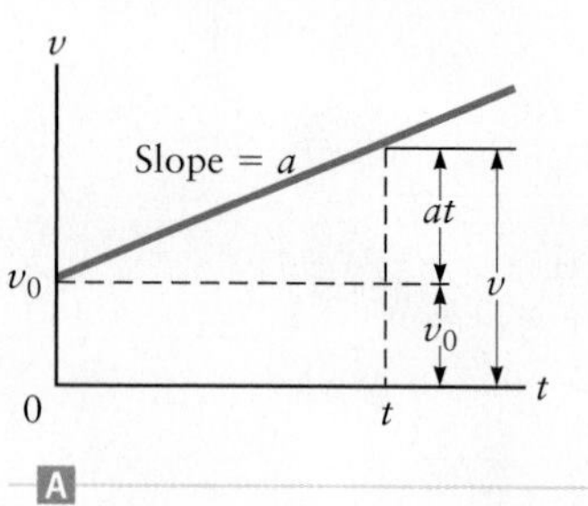

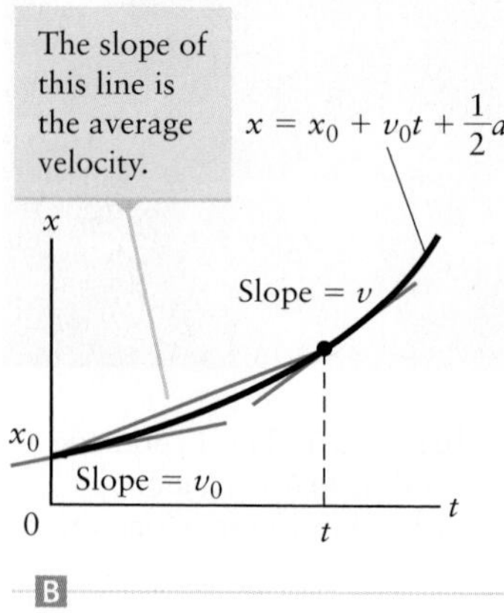

▲ **Figure 3.4** A The slope of the velocity–time graph is equal to the acceleration, so when the acceleration a is constant, the v–t plot is a straight line. B The corresponding graph of position versus time. The slope of the x–t plot at a certain point on the curve is equal to the instantaneous velocity v.

To give a complete description of the spacecraft's motion, we must also derive its position as a function of time. We already have the velocity as a function of time from Equation 3.1, and we know that v is the slope of the x–t curve. To derive the x–t relationship we use the approach shown in Figure 3.4B. In Chapter 2, we learned that the instantaneous velocity v is the slope of the x–t curve at a particular point on the curve, whereas the average velocity over a particular time interval is found by taking the slope of the line that spans the interval. One such time interval is shown in Figure 3.4B. The slope of the line connecting $t = 0$ with some general time t is

$$v_{\text{ave}} = \frac{x - x_0}{t}$$

where x_0 is the value of x at $t = 0$ (the "initial" position). We can also calculate the average velocity from Equation 3.1; since v varies linearly with time, the average velocity during a particular time interval is the average of the instantaneous velocities at the start and end of the interval. Hence,

$$v_{\text{ave}} = \frac{v(t = 0) + v(t)}{2}$$

We can now insert our result for $v(t)$ from Equation 3.1:

$$v_{\text{ave}} = \frac{v(t = 0) + v(t)}{2} = \frac{v_0 + v_0 + at}{2}$$

$$v_{\text{ave}} = v_0 + \tfrac{1}{2}at \tag{3.2}$$

Equating this expression to our previous result $v_{\text{ave}} = (x - x_0)/t$ and doing a little rearranging leads to

$$v_{\text{ave}} = \frac{x - x_0}{t} = v_0 + \tfrac{1}{2}at$$

$$x - x_0 = v_0t + \tfrac{1}{2}at^2$$

Position versus time for motion with a constant acceleration

$$x = x_0 + v_0t + \tfrac{1}{2}at^2 \tag{3.3}$$

This result tells us how the position of our spacecraft varies with time and is sketched in Figure 3.4B.

[1] Most rocket engines work by expelling gas, so the total mass of the spacecraft, including the engine and fuel, will generally decrease with time. Here, however, we ignore that effect.

Relations for Motion with Constant Acceleration

It is useful to recap how we used Newton's second law to analyze the motion of our spacecraft. The first step was to determine the total force on the object. For our spacecraft problem, this force was simply a given; for example, the engine designer might have told us the value. We then used Newton's second law $a = \sum F/m$ to calculate the acceleration a. Because the total force and m were both constants, a is also constant. Acceleration is the slope of the v–t relation; hence, the velocity is a linear function of t. The final step was to deduce the x–t relation; that analysis was sketched in Figure 3.4B and led to Equation 3.3. Keep in mind that these results for the position and velocity of an object with constant acceleration (Eqs. 3.3 and 3.1, respectively) are not "new" laws of physics; both are a direct result of Newton's second law.

It is rare to find situations in real life in which the acceleration is *precisely* constant, so Equations 3.1 and 3.3 will usually not give an exact description of the motion of an object. However, in many interesting cases the acceleration is very nearly constant, and these cases are well described by the constant-acceleration relations in Equations 3.1 and 3.3.

For future reference, Table 3.1 collects these results and also lists one more very useful relationship. Equations 3.1 and 3.3 give the velocity and position as functions of time. We can eliminate t from these two relations to get a single equation that relates x and v without direct reference to t. We first use Equation 3.1 to express t in terms of v and a:

$$v = v_0 + at$$

$$t = \frac{v - v_0}{a}$$

We next insert this result for t into Equation 3.3:

$$x = x_0 + v_0 t + \frac{1}{2}at^2 = x_0 + v_0\left(\frac{v - v_0}{a}\right) + \frac{1}{2}a\left(\frac{v - v_0}{a}\right)^2$$

Collecting and combining terms gives

$$x = x_0 + \frac{v_0 v}{a} - \frac{v_0^2}{a} + \frac{1}{2}\frac{v^2}{a} - \frac{v_0 v}{a} + \frac{1}{2}\frac{v_0^2}{a} = x_0 + \frac{1}{2a}(v^2 - v_0^2)$$

We can now rearrange to reach our desired result:

$$v^2 = v_0^2 + 2a(x - x_0) \tag{3.4}$$

which is also listed in Table 3.1. This relation is interesting because it does not contain time (t); it only involves position, velocity, and acceleration. We'll encounter many cases in which this result is very handy.

TABLE 3.1 Equations for Motion with Constant Acceleration

Equation Number	Mathematical Relation	Variables
3.3	$x = x_0 + v_0 t + \frac{1}{2}at^2$	position, acceleration, and time
3.1	$v = v_0 + at$	velocity, acceleration, and time
3.4	$v^2 = v_0^2 + 2a(x - x_0)$	position, velocity, and acceleration

Note: In cases in which the position variable in a problem is y, the same equations apply with y inserted in place of x.

Relations between x, v, a, and t for motion with a constant acceleration

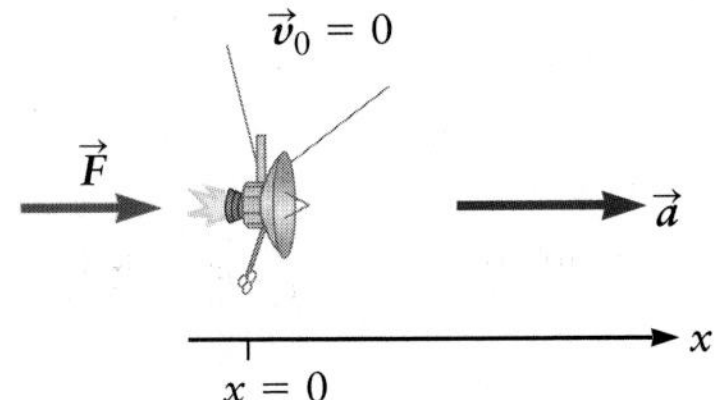

▲ **Figure 3.5** Example 3.1. Sketch of a spacecraft moving in response to a constant force F along the $+x$ direction.

EXAMPLE 3.1 Accelerating into Space

Let's see how the basic relations between x, v, and t in Table 3.1 can be used to calculate the motion of our spacecraft in a particular case. We assume the spacecraft has mass $m = 200$ kg and is initially at rest. The engine is turned on at $t = 0$, and from this time forward there is a constant force $F = 2000$ N on the spacecraft. (**a**) What is the velocity of the spacecraft at $t = 40$ s? (**b**) How far does the spacecraft travel during this time? (**c**) What is the velocity of the spacecraft when it reaches a distance 1000 m from where it started?

RECOGNIZE THE PRINCIPLE

The only force on the spacecraft is the force F from the engine, so this force is equal to $\sum F$ in Newton's second law. The force is constant; hence, the acceleration is also constant and we can find its value using Newton's second law. We can then use our relations for motion with constant acceleration from Table 3.1.

SKETCH THE PROBLEM

Figure 3.5 shows the spacecraft moving along the x direction, with a force F from the engine. The force and therefore also the acceleration are both along the $+x$ direction. We have also noted that the spacecraft is initially at rest (hence $v_0 = 0$) and have chosen the origin for the x axis to be the spacecraft's location when $t = 0$, so $x_0 = 0$.

IDENTIFY THE RELATIONSHIPS

Using Newton's second law along with the given values of the mass and the component of the force along x gives

$$a = \frac{\sum F}{m} = \frac{2000 \text{ N}}{200 \text{ kg}}$$

The unit of force is newtons (N), with $1 \text{ N} = 1 \text{ kg} \cdot \text{m/s}^2$. We thus have

$$a = 10 \, \frac{\text{kg} \cdot \text{m/s}^2}{\text{kg}}$$

Canceling the unit of kilograms that appears in both the numerator and denominator we get

$$a = 10 \text{ m/s}^2$$

We can now use our relations for motion with a constant acceleration from Table 3.1. For part (**a**) of the problem, we are given t and asked to find the velocity. Equation 3.1 involves v, t, and other variables whose values we already know, namely a and v_0, so we can use it to solve for v. Likewise, for parts (**b**) and (**c**) of this problem, we use the relations in Table 3.1 that contain the quantity we want to find, along with the quantities whose values we know.

SOLVE

(**a**) We compute the velocity at $t = 40$ s using Equation 3.1:

$$v = v_0 + at = 0 + (10 \text{ m/s}^2)(40 \text{ s}) = \boxed{400 \text{ m/s}}$$

(**b**) We are given the value of t (= 40 s) and asked to find x. We use Equation 3.3 because it contains x, t, and other quantities we know, namely a and v_0. So,

$$x = x_0 + v_0 t + \tfrac{1}{2}at^2 = 0 + 0 + \tfrac{1}{2}(10 \text{ m/s}^2)(40 \text{ s})^2 = \boxed{8000 \text{ m}}$$

(**c**) To deal with this part of the problem, we note that we are given the value of the position and asked to find the velocity; there is no mention of time. So, we can use Equation 3.4 in Table 3.1 to get

$$v^2 = v_0^2 + 2a(x - x_0) = 0 + 2(10 \text{ m/s}^2)(1000 \text{ m}) = 2.0 \times 10^4 \text{ m}^2/\text{s}^2$$

$$v = \boxed{140 \text{ m/s}}$$

Insight 3.2

WORKING WITH UNITS

In our calculations in Example 3.1, we canceled and combined units as reviewed in Chapter 1. When dealing with the units of force, it is simplest to express the newton in terms of the fundamental units of meters, kilograms, and seconds before canceling and combining units. (Recall that $1 \text{ N} = 1 \text{ kg} \cdot \text{m/s}^2$.) You should always include the units in your calculations and check that they correctly match the quantity being calculated.

▸ ***What does it mean?***
Problem solving with the relations for motion with constant acceleration involves one or more of Equations 3.1, 3.3, and 3.4. Choosing the appropriate relation depends on what quantities are known and what you wish to calculate. It is usually best to choose the equation that contains only one unknown quantity.

EXAMPLE 3.2 Car Chase

A car is traveling at a constant velocity $v = 20$ m/s on a long, straight road. Unfortunately, it is near a school where the speed limit is 10 m/s, and when the car passes a police officer, the officer immediately gives chase. If the police car starts from rest and has a constant acceleration $a_2 = 2.5$ m/s^2, how long does it take the officer to catch the speeder?

RECOGNIZE THE PRINCIPLE

Both cars are moving with a constant acceleration, so we can use the result for the displacement from Table 3.1, $x = x_0 + v_0 t + \frac{1}{2}at^2$, for both. We need to find the value of t at which the displacements of the two cars are equal, indicating that the police car has caught the speeder.

SKETCH THE PROBLEM

Figure 3.6 shows the displacement of both cars, where the speeder is car 1 and the police car is car 2.

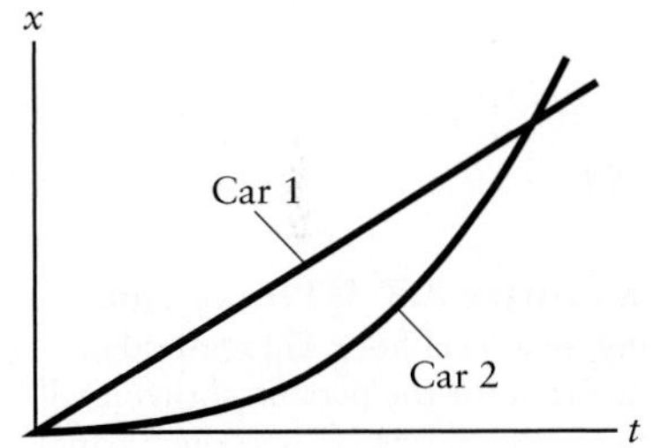

▲ **Figure 3.6** Example 3.2.

IDENTIFY THE RELATIONSHIPS

The acceleration of the speeder is zero ($a_1 = 0$), so it moves with a constant velocity equal to $v_{1,0} = 20$ m/s, while the police car (car 2) has zero initial velocity ($v_{2,0} = 0$) and a nonzero acceleration $a_2 = 2.5$ m/s^2. Both cars start at the origin ($x_{1,0} = x_{2,0} = 0$), so their displacements are

$$x_1 = x_{1,0} + v_{1,0}t + \tfrac{1}{2}a_1t^2 = v_{1,0}t \qquad (1)$$

$$x_2 = x_{2,0} + v_{2,0}t + \tfrac{1}{2}a_2t^2 = \tfrac{1}{2}a_2t^2 \qquad (2)$$

SOLVE

Setting the displacements in Equations (1) and (2) equal ($x_1 = x_2$) and solving for t, we get

$$v_{1,0}t = \tfrac{1}{2}a_2t^2 \qquad (3)$$

$$t = \frac{2v_{1,0}}{a_2} = \frac{2(20 \text{ m/s})}{2.5 \text{ m/s}^2} = \boxed{16 \text{ s}}$$

▸ ***What does it mean?***
Equation (3) has another solution for t, namely $t = 0$, because car 1 initially passes the police car (car 2) at $t = 0$.

3.2 Normal Forces and Weight

We next consider some examples in which the force of gravity plays an important role. Figure 3.7A shows a person standing on the floor of a room. The person is standing still, so her velocity and acceleration are both zero. Since the acceleration is zero and, according to Newton's second law, $\sum \vec{F} = m\vec{a}$, the total force on the person is zero. This statement is true, but it is not the whole story.

In this case there are two forces acting on the person. One is the gravitational force exerted by the Earth; this force is called the ***weight*** of the person and is denoted in Figure 3.7 by the vector $\vec{F}_{grav}$. When an object of mass m is located near the surface of the Earth, the gravitational force on the object is directed downward (toward the Earth) and has a magnitude

Weight is the force of gravity exerted by the Earth on an object.

$$|\vec{F}_{grav}| = mg \tag{3.5}$$

We'll see in Chapter 5 how this result for the force of gravity near the Earth's surface is a consequence of Newton's law of universal gravitation. For now, we note that (1) the value of g is approximately the same for all locations near the surface of the Earth, with $g \approx 9.8\ \text{m/s}^2$; (2) because weight is due to the gravitational attraction of the Earth (or whichever planet or star on which the object is located), the weight of an object will be different if it is taken to another planet or to the Moon; (3) the value of g is independent of the mass of the object, so the weight of an object (Eq. 3.5) is proportional to its mass; and (4) for reasons we'll see shortly, g is commonly referred to as the ***acceleration due to gravity***.

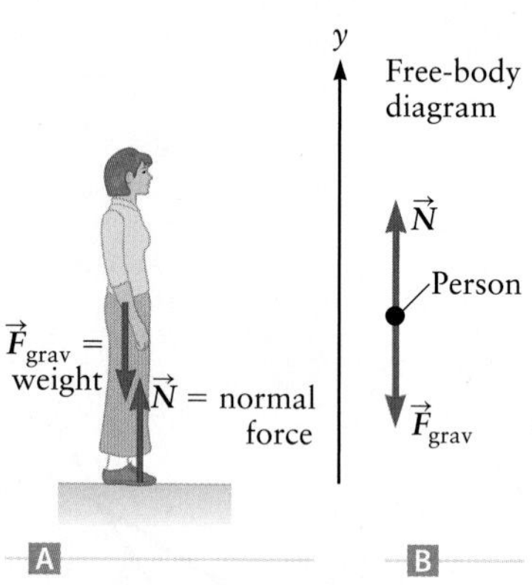

▲ **Figure 3.7** A Person standing on a level floor. B Free-body diagram for the person, showing all the forces acting on her: the normal force $\vec{N}$ and the force of gravity $\vec{F}_{grav}$. In a free-body diagram, the object is often denoted by a "dot."

The weight $\vec{F}_{grav}$ of an object is a force and can thus be measured in units of newtons. It is often convenient to use the coordinate system shown in Figure 3.7, where we follow common convention and label the vertical axis as y. The gravitational force lies along the vertical direction with a component along y given by

$$F_{grav} = -mg \tag{3.6}$$

The negative sign here indicates that the force is in the "negative" y direction, or *down*, toward the center of the Earth.

Besides the gravitational force, the other force acting on the person in Figure 3.7 is exerted by the floor on the bottoms of her feet. This force is called a ***normal force*** because it acts in a direction perpendicular ("normal") to the plane of contact with the floor. Normal forces occur whenever the surfaces of two objects come into contact. Here the normal force is directed upward, as indicated by the upward arrow labeled $\vec{N}$. The person in Figure 3.7 is at rest with an acceleration $a = 0$. Using Newton's second law for components of the force and acceleration along y, we get

$$\sum F = -mg + N = ma = 0$$

Hence, in this case $N = mg$. In this case, the normal force is equal in magnitude and opposite in direction to the person's weight.

CONCEPT CHECK 3.1 Where Is the Reaction?

There are two forces acting on the person in Figure 3.7A: the force of gravity $\vec{F}_{grav}$ and $\vec{N}$, the normal force from the floor. These forces are equal in magnitude and opposite in direction. Are they an action–reaction pair? Explain why or why not. If not, identify the reaction force to $\vec{N}$.

Free-Body Diagrams

Always draw a free-body diagram.

Drawing a picture is an important part of the problem-solving process. This picture should include coordinate axes along with other information such as the values of forces and the initial velocity. In preparation for an analysis using Newton's second law, you should also construct a simplified diagram showing all the forces acting on each object involved in the problem. In Figure 3.7, these forces are the normal force and the force of gravity acting on the person. The diagram that shows them, Figure 3.7B, is called a ***free-body diagram***. The recommended procedures for constructing a free-body diagram are as follows.

PROBLEM SOLVING Constructing Free-Body Diagrams and Applying Newton's Laws

1. **RECOGNIZE** the objects of interest. List all the forces acting on each one.

2. **SKETCH THE PROBLEM.**
   - Start with a drawing that shows all the objects of interest in the problem along with all the forces acting on each one, as in Figure 3.7A.
   - Make a separate sketch showing the forces acting on each object, which is the free-body diagram for that object. For clarity, you can represent each object by a simple dot.
   - Forces in a free-body diagram should be represented by arrows that show the direction of each force. In each free-body diagram, show only the forces acting on that particular object.

3. **IDENTIFY THE RELATIONSHIPS.** A force in a free-body diagram may be known (i.e., its direction and magnitude may be given), or it may be unknown. Represent unknown quantities by a symbol, such as $\vec{N}$ in Figure 3.7. Typically, equations derived from Newton's second law are used to solve for these unknowns.

4. **SOLVE.** The information contained in a free-body diagram can be used in writing Newton's second law for an object. A few algebraic steps will then lead to values for the unknown quantities.

5. Always *consider what your answer means* and check that it makes sense.

Acceleration and Apparent Weight

The normal force acting on the person in Figure 3.7 is equal in magnitude to her weight (mg). This result is often found for objects at rest on a level surface. However, in many situations the normal force on an object is not equal to its weight. Such a situation is sketched in Figure 3.8A, which shows a person standing in a moving elevator. There are again two forces acting on the person (Fig. 3.8B), the force of gravity (the person's weight) and the normal force exerted by the floor of the elevator on the bottoms of his feet. However, this situation is not the same as in our previous example because the elevator in Figure 3.7A has an acceleration $\vec{a}$, which is also the acceleration of the person inside.

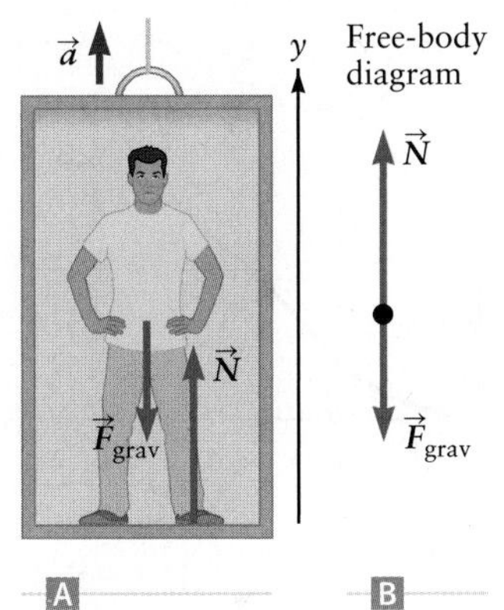

▲ **Figure 3.8** A When a person stands in an elevator, there are only two forces acting on him: $\vec{N}$ and $\vec{F}_{grav}$. B Free-body diagram for the person.

Applying Newton's second law to the motion of the person gives

$$m\vec{a} = \sum \vec{F} = \vec{N} + \vec{F}_{grav}$$

where m is the person's mass. These vectors are along y, so we can write this expression in terms of the y components of the acceleration and force vectors:

$$ma = +N - mg \tag{3.7}$$

The signs here indicate the directions of the normal force and the force of gravity. As usual, these directions are defined by our choice of the coordinate axis (y) in Figure 3.8, which shows that the "positive" direction is upward. This choice for the positive direction also applies to the acceleration. Solving Equation 3.8 for the normal force, we find

$$N = mg + ma$$

Hence, if the acceleration is positive (directed upward in Fig. 3.8), the normal force is greater than the weight mg, while if the acceleration is negative (downward), the normal force is less than mg. In words, when you are in an elevator that is just starting to move to a higher floor, the acceleration is positive (directed upward) and you "feel" heavier. The force exerted by the floor of the elevator on the bottoms of your feet is greater than your true weight mg. Likewise, if you are in an elevator that is just beginning to move to a lower floor, your acceleration is negative (downward), so N is smaller than mg and you feel lighter.

The normal force acting on the bottoms of a person's feet as in Figure 3.8 is called the ***apparent weight.*** Loosely speaking, the only way that you are sensitive to the value of your weight is through the normal force that supports you. Consider (hypothetically) what would happen if the elevator in Figure 3.8 were to malfunction

Insight 3.3
WEIGHT AND MASS
Weight and mass are closely related but they are not the same. For an object on or near the Earth's surface, the weight has a magnitude of $F_{grav} = mg$ (Eq. 3.5), so F_{grav} and m are proportional. However, weight and mass are fundamentally different quantities. The mass of an object is an intrinsic property of the object, whereas the weight of an object is a force whose magnitude varies depending on the location of the object. For example, the magnitude of your weight on Earth is different from its value on the Moon, but your mass is the same wherever you are located.

▲ **Figure 3.9** These young people are experiencing apparent "weightlessness" in a special NASA plane. Notice the padding.

and fall freely down the elevator shaft. This kind of arrangement is similar to the special airplane used by NASA to train astronauts (Fig. 3.9). That airplane flies in a long dive, during which time the plane and the astronauts fall together, so the normal force on the astronauts is zero. Their apparent weight is then zero, and the astronauts feel "weightless." During such a dive, their apparent weight is zero even though their true weight is not zero (and the force of gravity is also *not* zero).

All Forces Come from Interactions

Newton's third law implies that every force on an object arises from an interaction with a second object. This notion that forces come from interactions leads to the concept of an action–reaction pair of forces. Newton also understood that two objects could exert forces on each other even when they are not in direct contact. The gravitational force is a good example. The Earth exerts a gravitational force on objects such as the person in the elevator in Figure 3.8 or the young people in a free-fall dive in Figure 3.9, even though these people are not in contact with the planet. The exertion of forces between objects that are not in direct contact is known as ***action-at-a-distance.*** You probably know that the gravitational force from the Sun is responsible for the Earth's orbital motion. The space between the Sun and the Earth is largely a vacuum, and it is certainly not obvious how this force can make itself felt through such an empty region. Action-at-a-distance is a fundamental aspect of nature that bothered many scientists after it was proposed by Newton, and it took the genius of Einstein to explain how this works for the gravitational force. The notion of action-at-a-distance applies to other forces, including electric and magnetic forces, as we'll discuss later.

Another important idea is contained in the notion of "direct contact" between the floor and the person's feet in Figure 3.7, where it certainly seems like the floor and the person are in contact. But what does this mean at a microscopic scale? We know that the floor and the person's feet are composed of atoms. Can two atoms actually "touch"? Contact forces are a result of electric forces between atoms that are in very close proximity. This atom–atom interaction cannot be described by Newton's laws, but must be treated using quantum mechanics (Chapters 28 and 29).

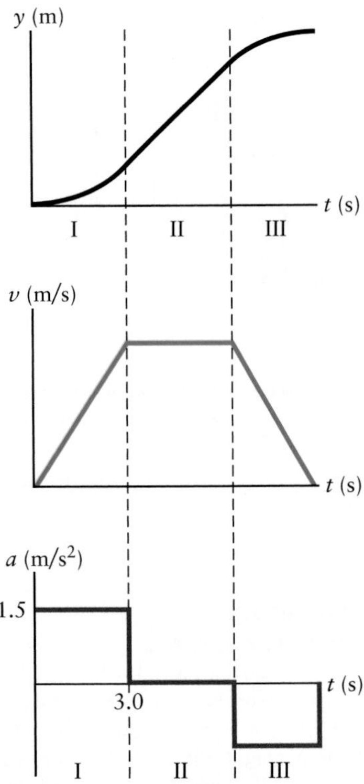

▲ **Figure 3.10** Example 3.3. Position, velocity, and acceleration versus time for an elevator as it travels from the first to the fifth floor of a building. In period I, the elevator is accelerating from rest. In period II, the acceleration is zero and the elevator moves with a constant velocity. During period III, the elevator slows to a stop.

EXAMPLE 3.3 Traveling Up

An elevator begins from rest at the first floor of a building. It starts a journey to the fifth floor by moving with an acceleration of 1.5 m/s^2 for 3.0 s; thereafter, it moves with a constant velocity until it gets near to its stopping point. Find the apparent weight of a person of mass m who is traveling in the elevator at different times during the motion.

RECOGNIZE THE PRINCIPLE

The person's apparent weight will depend on the acceleration. During its journey, the elevator must go through three stages of motion: (I) a period of positive acceleration (with $a = 1.5$ m/s^2), (II) a period of constant velocity (with $a = 0$), and (III) a period of negative acceleration while it slows to a stop.

SKETCH THE PROBLEM

Figure 3.8 shows a person in an elevator and defines our coordinate system; the motion is again along y (the vertical direction). Figure 3.8B is our free-body diagram.

IDENTIFY THE RELATIONSHIPS

Figure 3.10 contains qualitative plots of the position, velocity, and acceleration of the elevator versus time during the three phases of the motion. In each phase, we apply the relations for motion with constant acceleration from Table 3.1.

The apparent weight of a person in the elevator depends on the acceleration, so it is different in periods I, II, and III in Figure 3.10. Using Newton's second law and the components of the forces along y from the free-body diagram in Figure 3.8B gives

$$ma = \sum F = +N + F_{grav} = +N - mg \quad (1)$$

The minus sign in front of the last term indicates (again) that the force of gravity is directed downward.

SOLVE

Rearranging Equation (1) gives the person's apparent weight as $N = ma + mg$. The person's true weight is mg, while his apparent weight during period I is

$$N = m(a + g) = m(1.5\ \text{m/s}^2 + 9.8\ \text{m/s}^2) = m(11.3\ \text{m/s}^2)$$

Comparing N to the true weight, we find

$$\frac{N}{mg} = \frac{m(11.3\ \text{m/s}^2)}{mg} = \frac{11.3\ \text{m/s}^2}{g} = \frac{11.3\ \text{m/s}^2}{9.8\ \text{m/s}^2} \approx 1.2$$

or

$$N = \boxed{1.2(mg)}$$

Hence, during the acceleration period (I) the person's apparent weight N is approximately 20% larger than his true weight.

During period II, the elevator's acceleration is zero, so $N = mg$ and the person's apparent weight is equal to his true weight. During period III, the acceleration is negative. The normal force is again $N = ma + mg$, and since a is now negative, the normal force and thus also the person's apparent weight are smaller than mg. The precise value of N depends on the value of a during period III (Fig. 3.10).

► ***What does it mean?***

The value of the apparent weight depends on an object's acceleration. The apparent weight can be greater than, equal to, or less than the true weight.

What Is "Mass"?

So far, we have only considered the force of gravity on terrestrial objects, and we have seen that this force has a very simple form

$$F_{grav} = -mg \quad (3.8)$$

where (as usual) we take the "positive" direction to be upward. However, this simplicity tends to hide a very mysterious fact. According to Equation 3.8, the force of gravity depends on the mass of the object. We originally encountered mass in connection with the concept of inertia and Newton's laws of motion. According to Newton's second law, the mass of an object determines how it will move in response to forces. Here is the puzzle: if mass is the property of an object that determines how it moves (as calculated using Newton's second law), why does the *same* quantity determine the magnitude of the force of gravity? Even Newton was not able to answer this profound question, which we'll discuss more in Chapters 5 and 27. The quantity called "mass" that enters into Newton's second law is called the ***inertial mass*** of an object, while the "mass" that enters into the gravitational force is referred to as the object's ***gravitational mass***. Physicists believe that the inertial mass is precisely equal to the gravitational mass, which suggests a fundamental connection between gravitation and Newton's laws of motion. We'll explain this connection when we study relativity in Chapter 27.

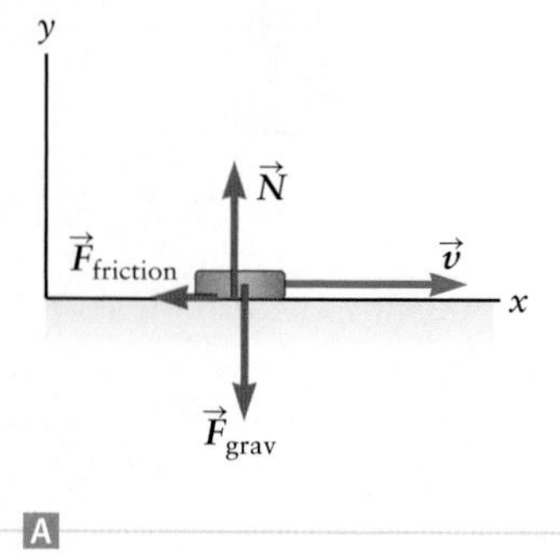

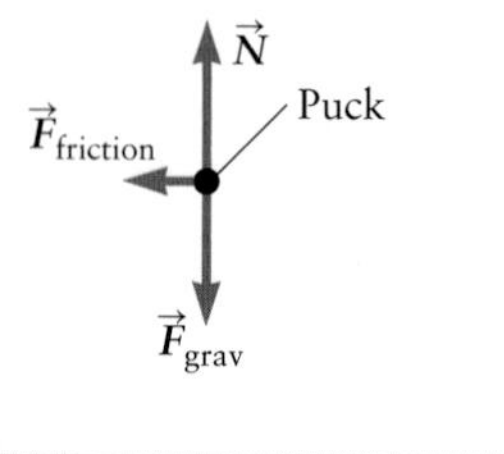

▲ **Figure 3.11** A Three forces act on this hockey puck: the force of gravity (its weight), the normal force from the ice, and the force of friction. B Free-body diagram for the puck.

3.3 Adding Friction to the Mix

One of Aristotle's central beliefs was that terrestrial objects all tend toward a natural state of rest. Indeed, that is what we commonly observe in the world around us because objects generally do spend most of their time at rest relative to us as observers. However, Newton's explanation of this behavior was quite different from the one provided by Aristotle. According to Newton, when an object comes to a state of rest, it does so because a force acts on it, and this force is often due to a phenomenon called friction.

Kinetic Friction

Figure 3.11 shows a hockey puck sliding on an icy surface. While this surface is very slippery, there is a small amount of friction, and we now consider how the force of friction eventually causes the puck to stop.

Figure 3.11B shows a free-body diagram containing all the forces acting on our hockey puck.[2] The puck's motion is along the horizontal direction x, and we assume the puck is sliding toward the right in the $+x$ direction. There is only one force along the horizontal, the force of friction. Because the puck's velocity is along $+x$ (to the right in Fig. 3.11), the force of friction will oppose this motion and hence be directed along $-x$ (to the left). There are also two forces in the vertical direction, y: the weight of the puck (the force of gravity) and the normal force exerted by the icy surface acting on the puck. This example is our first one involving forces acting in more than one direction, and we have not yet discussed how this affects the use of Newton's laws. For now, we appeal to your intuition and argue that the two forces directed along y, the normal force and the weight of the puck, will cancel each other. This is very similar to the cancellation we found in connection with Figure 3.7. We thus have $N - mg = 0$, or $N = mg$ for the hockey puck.

If the motion of our puck is along x, why did we need to mention forces along y? The reason is that the normal force is intimately connected with the force of friction. Experiments show that for a sliding object, the magnitudes of these two forces are related by

Force of kinetic friction

$$F_{\text{friction}} = \mu_K N \tag{3.9}$$

The quantity μ_K is called the ***coefficient of kinetic friction***. The frictional force in Equation 3.9 describes cases in which two surfaces are in motion (slipping) with respect to each other. As a "coefficient," μ_K is a pure number without dimensions or units, and its value depends on the surface properties. For a hockey puck on an icy surface, μ_K is relatively small and the frictional force is very small; the value of μ_K might typically be $\mu_K = 0.05$. The coefficient of kinetic friction for two rough surfaces would be much larger, with $\mu_K = 0.3$ being a common value. Table 3.2 lists coefficients of kinetic and static friction (discussed below) for some common materials. These measured values of μ_K show that the frictional force depends on the nature of the surfaces that are in contact.

Note that the magnitude of the frictional force in Equation 3.9 depends on the normal force. If the normal force is increased—say, by piling a large additional mass on top of our hockey puck—the frictional force is larger than if the extra mass is not present. We must emphasize that Equation 3.9 is not a "law" of nature; it is merely an approximate relation that has been found to work well in a wide variety of cases. To derive a fundamental law or theory of friction would require us to consider in detail the atomic interactions that occur when two surfaces are in contact. This problem is very complicated and is currently a topic of much research.

[2]Actually, we have omitted the force from air drag (sometimes called air resistance), but that will usually be very small for a hockey puck.

TABLE 3.2 Typical Values for the Coefficients of Kinetic Friction and Static Friction for Some Common Materials

Surfaces	μ_K	μ_S	Surfaces	μ_K	μ_S
rubber on dry concrete	0.8	0.9	ski on snow	0.05	0.10
rubber on wet concrete	0.5	0.6	glass on glass	0.4	0.9
rubber on glass	0.7	0.9	Teflon on Teflon	0.04	0.04
hockey puck on ice	0.05	0.10	metal on metal	0.5	0.6
wood on wood	0.3	0.5	bone on bone (with joint fluid)	0.015	0.02

Notes: The values for metal on metal are only approximate. The coefficients in all cases depend on the smoothness of the surfaces.

Analyzing Motion in the Presence of Friction

Suppose we know the value of μ_K and that the hockey puck in Figure 3.11 is given some initial velocity v_0. How long will the puck slide before coming to rest? We can apply Newton's second law along with the free-body diagram in Figure 3.11B to calculate the puck's acceleration. In this problem, the motion in which we are interested is along x, so we need to focus on the forces along the x direction. Writing Newton's second law for motion along x, we get

$$ma = \sum F_x = F_{\text{friction}} = -\mu_K N$$

All these terms involve components of acceleration and force along the horizontal (x) direction. The negative sign in the last term indicates that the frictional force is directed along $-x$, because friction acts to oppose the motion of the puck. In words, the velocity in Figure 3.11 is to the right, so the frictional force acts to the left. The normal force in this case is equal to mg, and we have

$$ma = -\mu_K N = -\mu_K mg$$
$$a = -\mu_K g$$

Since μ_K and g are both constants, the puck's acceleration is constant, so we can apply our general relations for motion with constant acceleration in Table 3.1. However, we must decide which one of these relations to use. In this problem, we are asked to find when the puck stops, and we'll know that the puck has come to a stop when $v = 0$. We therefore need a relation that involves v and t. Examining Table 3.1, we see that Equation 3.1 involves these quantities. Using this relation, we have

$$v = v_0 + at = v_0 - \mu_K gt = 0$$

and solving for t gives

$$t = \frac{v_0}{\mu_K g} \tag{3.10}$$

It is always a good idea to work out the units of an answer, even if no numbers are given. The coefficient of friction is dimensionless, so the units of the term on the right in Equation 3.10 are $(\text{m/s})/(\text{m/s}^2) = \text{s}$, as expected because the result is a time. Notice also that we did not need to know the mass of the puck; the value of m canceled when calculating the acceleration.

Another question that we might have asked is, "How far will the puck slide before coming to a stop?" We could find the answer in two different ways: (1) by using the value of t found in Equation 3.10 and inserting it into Equation 3.3 in Table 3.1 or (2) by using Equation 3.4. Let's use the second approach. When the puck stops, its velocity is $v = 0$, and inserting this into Equation 3.4 gives

$$v^2 = v_0^2 + 2a(x - x_0) = 0$$

We can then solve for the distance traveled:

$$2a(x - x_0) = -v_0^2$$

$$(x - x_0) = -\frac{v_0^2}{2a} = \frac{v_0^2}{2\mu_K g}$$

Static Friction

The case of the sliding hockey puck in Figure 3.11 involves surfaces that are in motion (slipping) relative to each other, so the frictional force in such cases is referred to as kinetic friction. Situations in which the relevant surfaces are not slipping involve what is known as ***static friction***. An example involving static friction is given in Figure 3.12, which shows a refrigerator at rest on a level floor. This situation is similar to the hockey puck in Figure 3.11, but with an additional force exerted by a person pushing on the refrigerator. We know intuitively that when the force exerted by the person is small, the refrigerator will not move. Since the acceleration is zero, applying Newton's second law along the horizontal direction leads to

$$ma = \sum F_x = F_{\text{push}} + F_{\text{friction}} = 0$$

In this case, the force exerted by the person and the force of friction cancel. The force of static friction is sufficiently strong that no relative motion (no slipping) occurs. In terms of magnitudes, we have

$$|F_{\text{push}}| = |F_{\text{friction}}|$$

However, the value of F_{push} can vary and the refrigerator will still remain at rest. That is, we can push a lot or a little or not at all, without the refrigerator starting into motion. In all these cases, the frictional force exactly cancels F_{push}. The only way this can happen is if the magnitude of the frictional force varies depending on the value of F_{push}, as described by

Force of static friction

$$|F_{\text{friction}}| \leq \mu_S N \qquad (3.11)$$

where μ_S is the ***coefficient of static friction.***

The term *static* again means that the two surfaces (the floor and the bottom of the refrigerator) are not moving relative to each other. According to Equation 3.11, the magnitude of the force of static friction can take any value up to a maximum of $\mu_S N$. If F_{push} in Figure 3.12 is small, the force of static friction is small and will precisely cancel F_{push} so that the total horizontal force is zero. If F_{push} is increased, the force of static friction increases and again precisely cancels F_{push}. However, the magnitude of the static friction force has an upper limit of $\mu_S N$.

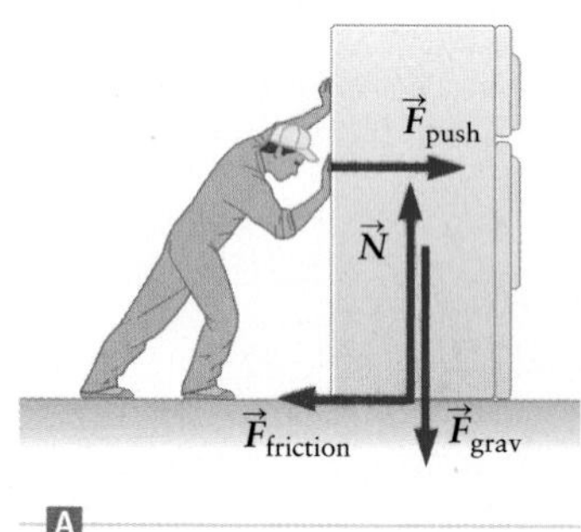

Free-body diagram

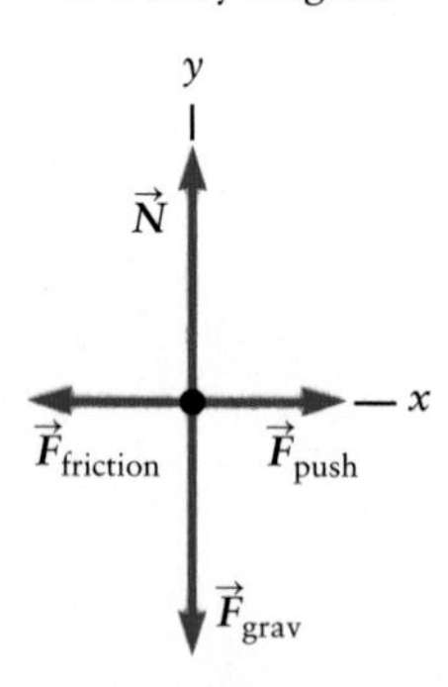

▲ **Figure 3.12** A Four forces act on this refrigerator. The force of static friction acts along the horizontal and opposes the movement of the refrigerator relative to the floor. B Free-body diagram for the refrigerator. Here we assume the force F_{push} is small enough that the refrigerator does not tip.

Comparing Kinetic Friction and Static Friction

To describe friction fully, we must consider both the kinetic and static cases, with two different coefficients of friction. In the kinetic case, the two surfaces are slipping relative to each other and the frictional force has the simple value $F_{\text{friction}} = \mu_K N$ (Eq. 3.9). The static case is a bit more complicated; here, the frictional force varies so as to cancel the other force(s) on the object, up to a maximum value of $\mu_S N$ (Eq. 3.11). For a given combination of surfaces, it usually happens that $\mu_S > \mu_K$ (see Table 3.2), so the maximum force of static friction is greater than the force of kinetic friction. This confirms an experiment you have probably done yourself. The force required to start something into motion—that is, the minimum value of F_{push} required to just barely move the refrigerator in Figure 3.12—is larger than the force required to keep it moving subsequently at a constant velocity.

EXAMPLE 3.4 Sliding Away

A wooden crate sits on the floor of a flatbed truck that is initially at rest, and the coefficient of static friction between the crate and the floor of the truck is $\mu_S = 0.25$. The driver of the truck wants to accelerate to a velocity of 20 m/s within 4.0 s. If she does so, will the crate slide off the back of the truck? Assume the truck moves with a constant acceleration.

RECOGNIZE THE PRINCIPLE

Figure 3.13A shows that the only horizontal force on the crate is the force of static friction. The static frictional force is not opposing any other force. Instead, static friction is opposing the tendency of the crate to slide along the floor of the truck. That is, static friction opposes any *relative motion* of the two surfaces. If there is no relative motion and the crate does not slip, it will be moving along with the truck. Hence, in this example the force of friction is what causes the crate to accelerate! Since there is an upper limit to the magnitude of the static frictional force (Eq. 3.11), there is an upper limit to this horizontal force on the crate, giving an upper limit to the acceleration that static friction can provide to the crate. If this acceleration limit is smaller than the truck's actual acceleration, the crate will not be able to keep up with the truck and the crate will slide off the back.

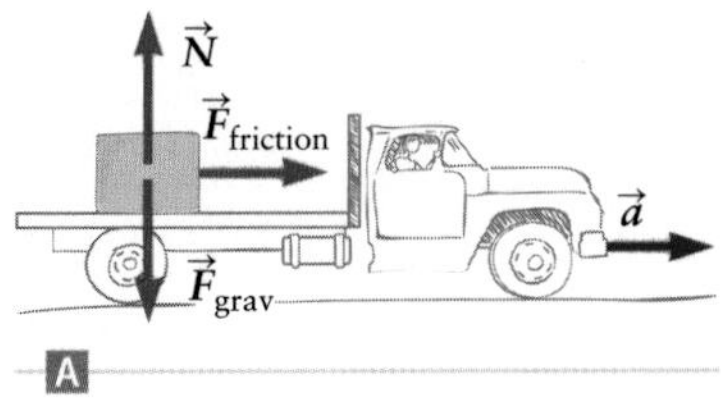

Free-body diagram

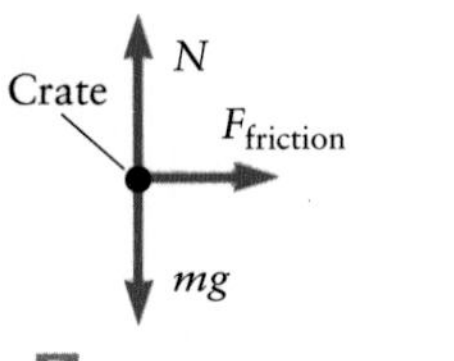

▲ **Figure 3.13** Example 3.4. A The force of friction on the crate is directed to the right, in the direction of motion of the truck and the crate. The frictional force acts to prevent the crate from slipping relative to the truck. B Free-body diagram for the crate.

SKETCH THE PROBLEM

Figure 3.13B shows a free-body diagram for the crate. The vertical forces on the crate are its weight (directed downward) and the normal force from the floor of the truck (directed upward). As the truck accelerates, there is also a horizontal force on the crate caused by static friction from the floor of the truck.

IDENTIFY THE RELATIONSHIPS

Using Newton's second law, the force of static friction leads to a maximum acceleration of the crate in the x direction of

$$a_{\text{crate, max}} = \frac{F_{\text{friction, max}}}{m} = \frac{\mu_S N}{m} \tag{1}$$

Calculating the normal force between the crate and floor of the truck is done in the same way as with the hockey puck in Figure 3.11. Hence, the magnitude of the normal force acting on the crate is $N = mg$, where m is the mass of the crate. Substituting into Equation (1) leads to

$$a_{\text{crate, max}} = \frac{\mu_S N}{m} = \frac{\mu_S mg}{m} = \mu_S g \tag{2}$$

To calculate the acceleration of the truck as determined by the driver, we note that the truck is undergoing motion with constant acceleration and choose Equation 3.1 from Table 3.1 because this relation involves velocity, acceleration, and time. The truck starts from rest, so $v_0 = 0$. Inserting this into Equation 3.1 gives

$$v = v_0 + a_{\text{truck}}t = a_{\text{truck}}t$$

$$a_{\text{truck}} = \frac{v}{t} \tag{3}$$

SOLVE

We now need to calculate the value of the truck's acceleration in Equation (3) and compare it with the maximum possible acceleration of the crate in Equation (2). Solving for a_{truck}, we find

$$a_{\text{truck}} = \frac{v}{t} = \frac{20 \text{ m/s}}{4.0 \text{ s}} = 5.0 \text{ m/s}^2$$

Evaluating Equation (2) for the maximum possible acceleration of the crate gives

$$a_{\text{crate, max}} = \mu_S g = (0.25)(9.8 \text{ m/s}^2) = 2.5 \text{ m/s}^2$$

The acceleration of the truck is thus greater than the maximum possible frictional acceleration of the crate, so the crate will *not* be able to keep up with the truck and it will slide off the back.

(continued) ▶

► ***What does it mean?***

Normally, friction either prevents an object from moving or brings it to a stop. However, in this case friction is responsible for the motion of the crate because friction always opposes the relative motion of the surfaces in contact.

CONCEPT CHECK 3.2 Trucks and Crates

The driver of the truck in Example 3.4 is in a hurry and wants to drive at the highest speed possible to her destination without losing the crate. Is there a maximum speed at which she can drive without having the crate slide off the truck?

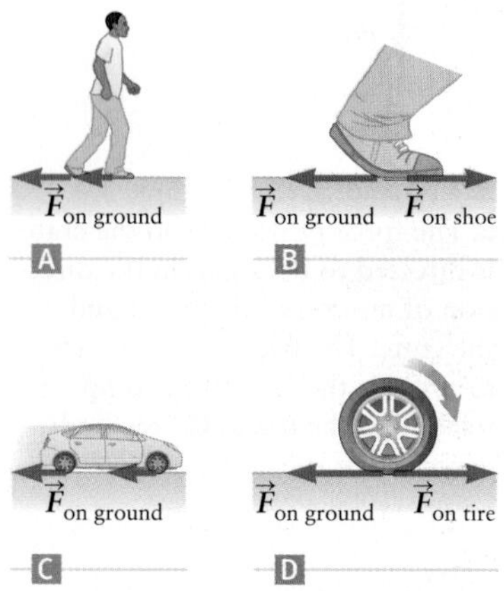

▲ **Figure 3.14** Frictional forces are essential for walking and rolling.

The Role of Friction in Walking and Rolling

It is interesting that the force of static friction on the crate in Figure 3.13 actually *causes* the crate's motion. This happens in other cases too. Figure 3.14A shows a person walking on level ground. As the person "pushes" off during each step, the bottoms of his shoes exert a force $\vec{F}_{\text{on ground}}$ on the ground. If the shoes do not slip, this force is due to static friction since the shoes do not move relative to the ground. According to Newton's third law, there is a reaction force $\vec{F}_{\text{on shoe}}$ on the person's shoes (Fig. 3.14B), and this force "propels" a person as he walks. If the surface were extremely slippery and there were no frictional force, then $\vec{F}_{\text{on ground}}$ and $\vec{F}_{\text{on shoe}}$ would both be zero and the person would slip when he attempted to move. Hence, the force of friction makes walking and running possible.

A similar set of forces is found when a wheel rolls along level ground. If a car's tire does not slip, friction between the tire and the ground leads to a force $\vec{F}_{\text{on ground}}$ on the ground (Fig. 3.14C). There is a reaction force on the tire $\vec{F}_{\text{on tire}}$ (Fig. 3.14D), and this force propels the car forward. Friction thus also plays a key role in rolling motion.

3.4 Free Fall

In Chapter 2, we discussed how several quantities, including velocity and acceleration, are needed when describing the motion of an object. Indeed, the relationship and differences between velocity and acceleration are absolutely crucial for understanding motion, as we now illustrate for a type of motion known as ***free fall***. An example of free fall is the motion of a ball that is thrown upward, reaches some maximum height, and then falls back down to Earth. After the ball leaves the thrower's hand, it follows the familiar trajectory sketched in Figure 3.15. Let's analyze this trajectory and consider how the acceleration, velocity, and position of the ball all vary with time.

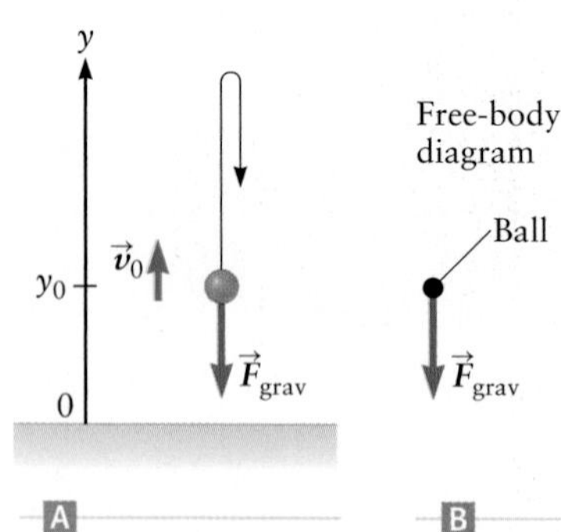

▲ **Figure 3.15** A ball is thrown upward beginning at an initial height y_0 with initial velocity $\vec{v}_0$. After leaving the thrower's hand, the ball is in free fall and the only force acting on the ball is the force of gravity. Even though this force is directed downward, the ball still moves up initially due to the initial velocity imparted by the thrower.

We begin with the acceleration. Before the ball leaves the thrower's hand, there are forces from gravity and from her hand acting on the ball, and these forces lead to an upward acceleration as she gives the ball an upward velocity. In Figure 3.15, we take $t = 0$ to be the instant just *after* the ball leaves her hand. From that instant until the ball hits the ground, the only forces acting on the ball are the force of gravity and a force due to air drag. In many cases, the force from air drag is very small, so we'll ignore it here.

We choose a coordinate system that measures the position as the height y of the ball above the ground, with the origin chosen to be at ground level. Using Newton's second law, we have

$$a = \frac{F_{\text{grav}}}{m} = \frac{-mg}{m} = -g \tag{3.12}$$

In words, the negative sign here means that the force of gravity is directed downward, in the $-y$ direction. We can again use our relations for motion with constant acceleration (Eqs. 3.1 and 3.3):

$$v = v_0 + at = v_0 - gt \tag{3.13}$$

$$y = y_0 + v_0 t + \tfrac{1}{2}at^2 = y_0 + v_0 t - \tfrac{1}{2}gt^2 \tag{3.14}$$

Sketches of how v and y vary with time are shown in Figure 3.16. The velocity varies linearly with time with a slope of $-g$ and an intercept equal to v_0, the velocity of the ball at $t = 0$ (just after it leaves the thrower's hand). The position varies quadratically with t, with a positive intercept y_0 that is the height of the ball when it leaves the thrower's hand.

The sketches in Figure 3.16 reveal some very important features of freefall. Initially, the ball's velocity is *positive*, whereas its acceleration is *negative*. That is, the velocity vector is directed *upward*, and the acceleration vector is directed *downward*. In fact, the ball's acceleration is downward during the *entire* time (Fig. 3.16C and Eq. 3.12). The force on the ball and therefore its acceleration are both negative; they are both directed along $-y$ throughout the motion. Since the acceleration is the slope of the v–t curve, the v–t relation must have a negative and constant slope throughout as shown in Figure 3.16A. Although this slope is always negative, the velocity itself can have a value that is positive, negative, or zero. The velocity is positive while the ball is on the upward (initial) part of its trajectory and negative while it is on the downward (final) part. The velocity is zero, shown by the v–t curve crossing through zero in Figure 3.16A, at the instant the ball reaches its highest point along y (the highest point on its trajectory).

To understand the behavior of the displacement, recall again that v is the slope of the y–t curve, so while the ball is on the way up and v is positive, the y–t curve has a positive slope. The ball reaches its highest point when the velocity is zero, and the y–t slope at that instant is zero. While the ball is on the way down, the velocity is negative, so the slope of the y–t curve is negative.

An *extremely* important lesson from this problem is that the acceleration and velocity can be in *different* directions; that is, they can have different signs. Depending on the situation, they can have either the same signs (both positive or both negative) or one can be positive while the other is negative.

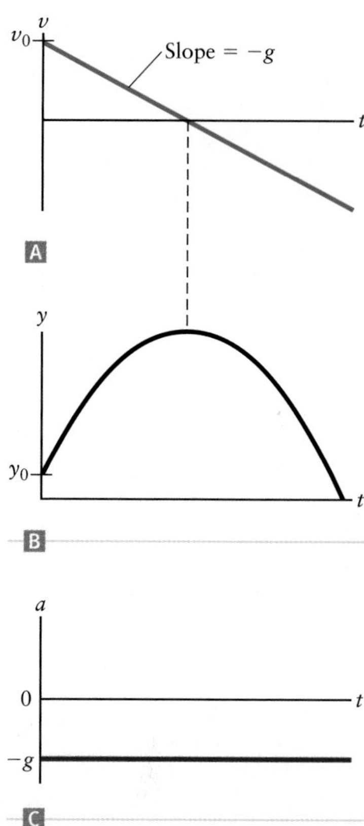

▲ **Figure 3.16** A For a ball in free fall, the acceleration is constant, so the velocity varies linearly with time. The slope of the v–t line is $-g$. B Height of the ball y versus time. The ball reaches its maximum height when $v = 0$. C The acceleration of the ball is constant and nonzero ($a = -g$) during the entire time, even at the highest point on the trajectory.

CONCEPT CHECK 3.3 The Direction of Velocity and Acceleration

For which of the following situations is the velocity parallel to the acceleration, and for which are $\vec{v}$ and $\vec{a}$ in opposite directions?

(a) The hockey puck sliding to a stop in Figure 3.11.
(b) The crate on the flatbed truck in Figure 3.13.
(c) An apple falling from a tree.

EXAMPLE 3.5 Maximum Height of a Projectile

Consider the motion of the ball in Figure 3.15 and assume the ball has an initial speed of 12 m/s (about 25 mi/h). Calculate (**a**) the time it takes for such a ball to reach its maximum height, (**b**) the maximum height reached by the ball, (**c**) the time the ball spends in the air, and (**d**) the ball's velocity just prior to hitting the ground. For simplicity, assume the initial position of the ball is at ground level.

RECOGNIZE THE PRINCIPLE

Since the acceleration is constant for free fall, we can use the relations from Table 3.1 or, equivalently, Equations 3.13 and 3.14. For each part of the problem, we select the

(continued) ▶

relation that contains the quantity we want to find along with quantities that we know. For part (a), we know that the velocity is zero when the ball reaches the maximum height, so we select Equation 3.13. We attack the other parts of the problem in a similar way.

SKETCH THE PROBLEM

Figure 3.15 contains all the information we need, including a free-body diagram. This free-body diagram applies at *all* points on the trajectory. While the ball is traveling through the air, the only force exerted on it is the force of gravity. (We again assume the force exerted by air drag is very small and can be neglected.)

IDENTIFY THE RELATIONSHIPS

The ball starts at ground level, so $y_0 = 0$. The initial velocity of the ball is $v_0 = 12$ m/s.

(a) At the top of the trajectory, the velocity is $v = 0$. From Equation 3.13, we get

$$v_{top} = v_0 - gt_{top} = 0$$

$$t_{top} = \frac{v_0}{g}$$

(b) Once we know the time needed to get to the top of the trajectory, we can find the ball's position at the top of its path using Equation 3.14:

$$y_{max} = y_0 + v_0 t_{top} - \tfrac{1}{2}gt_{top}^2$$

SOLVE

(a) Inserting the given value of v_0 into our result for t_{top} yields

$$t_{top} = \frac{12\text{ m/s}}{9.8\text{ m/s}^2} = \boxed{1.2\text{ s}}$$

(b) Substituting the value we just found for t_{top} into the equation for y_{max} gives

$$y_{max} = 0 + (12\text{ m/s})(1.2\text{ s}) - \tfrac{1}{2}(9.8\text{ m/s}^2)(1.2\text{ s})^2 = \boxed{7.3\text{ m}}$$

IDENTIFY THE RELATIONSHIPS

(c) When the ball reaches the ground, its position will be $y = 0$. Using Equation 3.14, we have

$$y = y_0 + v_0 t - \tfrac{1}{2}gt^2 = 0 \qquad (1)$$

We set $t = t_{ground}$ and take $y_0 = 0$:

$$y = 0 + v_0 t_{ground} - \tfrac{1}{2}gt_{ground}^2 = 0$$

$$t_{ground} = \frac{2v_0}{g}$$

(d) To find the velocity of the ball just before it lands, we use Equation 3.13 because it contains v and t. We want the velocity at time $t = t_{ground}$.

SOLVE

(c) Inserting the given value for v_0 into our result for t_{ground} yields

$$t_{ground} = \frac{2(12\text{ m/s})}{9.8\text{ m/s}^2} = \boxed{2.4\text{ s}}$$

(d) Substituting the value we just found for t_{ground} into Equation 3.13 for the velocity gives

$$v = v_0 - gt_{ground} = 12\text{ m/s} - (9.8\text{ m/s}^2)(2.4\text{ s}) = \boxed{-12\text{ m/s}}$$

► ***What does it mean?***
In this example, we assumed the initial position of the ball was $y_0 = 0$. A nonzero value would be more realistic, but would complicate the mathematics in part (c) since solving Equation (1) with a nonzero value of y_0 would require us to solve a quadratic equation. (See Problem 53 at the end of this chapter.)

Motion of a Dropped Ball

Example 3.5 gave us an interesting result. The ball's speed just before it hit the ground is equal to the ball's initial speed (although the ball's *velocity* is positive at the start and negative at the end). This result is no accident; it is true because the position y is a symmetric function with respect to the time at which the ball reaches its highest point (see Fig. 3.16B). The time spent on the way up is equal to the time spent falling back to the ground. We can see why by comparing the results for t_{top} and t_{ground} in Example 3.5, where we found

$$t_{top} = \frac{v_0}{g} \quad \text{and} \quad t_{ground} = \frac{2v_0}{g}$$

Hence, t_{ground} is precisely twice t_{top}. This symmetry is an important property of motion with constant acceleration, although it is not found if the acceleration varies with time. For example, when air drag is important, the acceleration is not constant and the position–time relation is no longer symmetric with respect to t_{top}.

EXAMPLE 3.6 Symmetry of Free Fall

Consider an apple that drops from a tree, and assume the apple falls from a height that is the same as the maximum height reached in the previous example. Calculate (**a**) how long it takes the apple to reach the ground and (**b**) its velocity the instant before it hits the ground.

RECOGNIZE THE PRINCIPLE

Since the apple is in free fall, the only force on it is due to gravity and the acceleration is constant. We can thus (yet again) use the expressions for motion with a constant acceleration in Table 3.1.

SKETCH THE PROBLEM

Figure 3.17 shows the problem. The apple starts at $y_0 = 7.3$ m (the maximum height reached by the ball in Example 3.5), with an initial velocity of zero. It undergoes free fall until it reaches the ground.

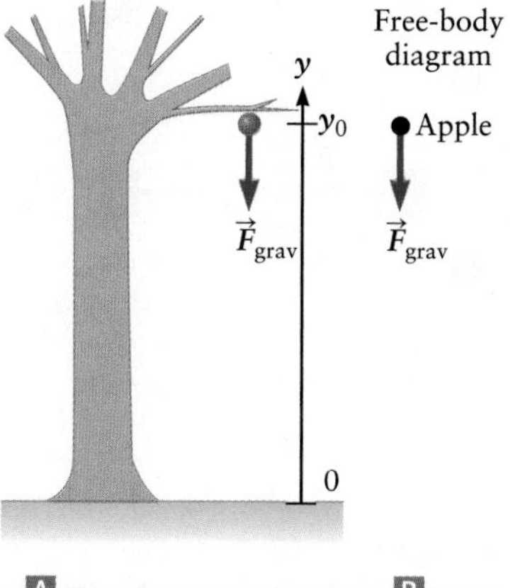

▲ **Figure 3.17** Example 3.6. The acceleration of an apple during free fall is the same as for the thrown ball in Figure 3.15, but the initial velocity ($v = 0$) is different.

IDENTIFY THE RELATIONSHIPS

(**a**) The apple reaches the ground when $y = 0$, so to calculate when the apple reaches this point we choose the relation that contains both position and time (Eq. 3.14 or its equivalent, Eq. 3.3). Applying Equation 3.14, we have

$$y = y_0 + v_0 t - \tfrac{1}{2}gt^2 \qquad (1)$$

The initial velocity is $v_0 = 0$ because the apple starts from rest.

SOLVE

(**a**) Substituting $v_0 = 0$ and $y = 0$ into Equation (1) and solving for t gives

$$y = y_0 - \tfrac{1}{2}gt_{ground}^2 = 0$$

$$t_{ground} = \sqrt{\frac{2y_0}{g}} = \sqrt{\frac{2(7.3\text{ m})}{9.8\text{ m/s}^2}} = \boxed{1.2\text{ s}}$$

(continued) ►

IDENTIFY THE RELATIONSHIPS

(**b**) We now need to evaluate the velocity at the value of t found in part (a). For the apple's velocity the instant before it hits the ground, we have (Eq. 3.13)

$$v = v_0 - gt_{\text{ground}} = -gt_{\text{ground}}$$

where again $v_0 = 0$.

SOLVE

(**b**) Inserting the value of t from part (a) gives

$$v = -(9.8 \text{ m/s}^2)(1.2 \text{ s}) = \boxed{-12 \text{ m/s}}$$

▸ ***What does it mean?***

This result for the velocity is the same as we found for the thrown ball in Example 3.5, part (d). That is no accident: the motion *on the way down* is precisely the same in the two cases.

CONCEPT CHECK 3.4 Free Fall

Consider an acorn as it falls from the top of a tall tree. Take the y axis to be in the vertical direction, with "up" being the $+y$ direction. Which of the following statements is true?

(a) The velocity is negative, the acceleration is negative, and the speed is increasing.

(b) The velocity is negative, the acceleration is positive, and the speed is decreasing.

(c) The velocity is negative and the speed is negative.

3.5 Cables, Strings, and Pulleys: Transmitting Forces from Here to There

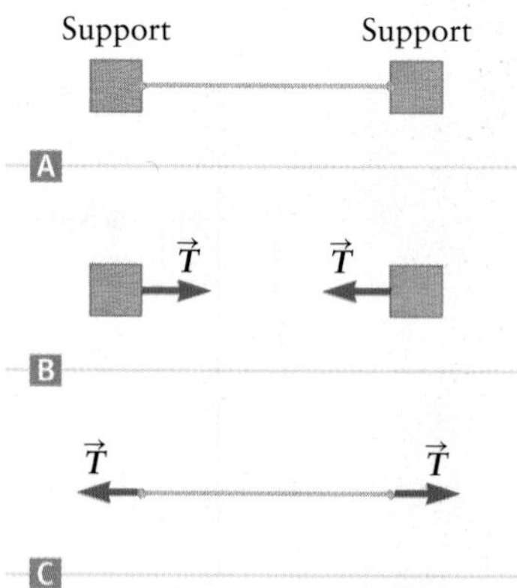

▲ **Figure 3.18** A A rope stretched between two supports. B The rope exerts a force of magnitude T on both supports, the one on the left and the one on the right. C According to Newton's third law, each support exerts a force of magnitude T on the rope.

Strings, cables, and ropes exert a force on the objects they are connected to due to their ***tension***. To understand these tension forces, we begin by considering an elevator supported by a cable. To choose a cable that is strong enough to support the elevator, you need to understand the forces acting on the cable. When a cable (or rope or string, etc.) is pulled tight, it exerts a force on whatever it is connected to. This force is illustrated in Figure 3.18; the ends of the rope both exert a force of magnitude T on the supports to which they are connected (Fig. 3.18B). At the same time, Newton's third law (the action–reaction principle) tells us that the supports both exert a reaction force of magnitude T on the ends of the rope (Fig. 3.18C). In words, T is equal to the tension in the rope.

The reasoning in Figure 3.18 also applies to the elevator in Figure 3.19; the cable exerts an upward force T on the elevator compartment and a separate (downward) force of magnitude T on whatever it is connected to at the top of the elevator shaft.

Tension Forces

To deal with the motion of the elevator compartment in Figure 3.19, we start with the free-body diagram in Figure 3.19B. There are two forces acting on the compartment, the force of gravity and a force from the cable that has a magnitude T (recall that T is the tension in the cable). If the elevator has an acceleration a along the vertical direction and m is the mass of the compartment plus passengers, we can apply Newton's second law for motion along the vertical direction to write

$$ma = \sum F = F_{\text{grav}} + T = -mg + T$$

The tension in the cable is thus

$$T = mg + ma$$

This answer is quite reasonable; if the elevator is at rest and the acceleration is zero, the force exerted by the cable must only support the weight of the elevator compartment and $T = mg$. If a is positive, a larger tension is required.

We next consider the forces acting on the cable. We have just seen that the cable exerts an upward force T on the elevator. According to Newton's third law—the action–reaction principle—the elevator must exert a force of equal magnitude and opposite direction on the cable. This is the origin of the downward force T acting on the cable in the free-body diagram in Figure 3.19C. There is a second force T_C acting on the cable, which comes from its connection to the top of the elevator shaft (to a motor, etc., located there). Usually the mass of a cable is very small when compared with the mass of other parts of the system, so for simplicity we will assume the cable is massless. If the mass of the cable is zero, the force of gravity on the cable will also be zero. Applying Newton's second law to our elevator cable in Figure 3.19C and assuming its mass is zero, we find

$$m_{\text{cable}}a = (0)a = +T_C - T = 0$$

$$T_C = T$$

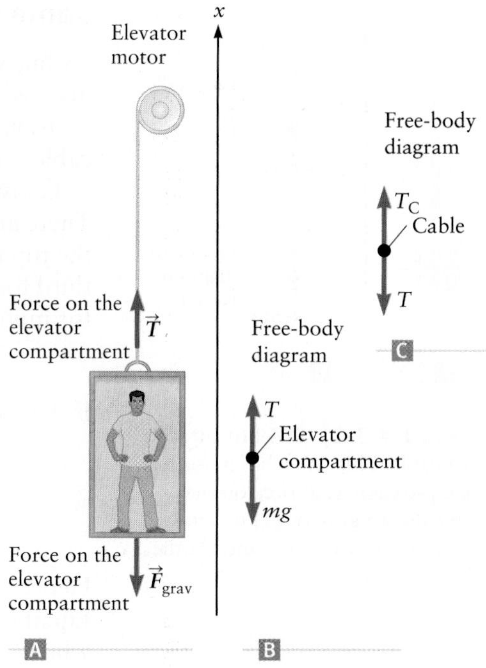

▲ **Figure 3.19** A When an elevator is lifted by a cable, there are two forces on the elevator compartment, the gravitational force and the tension from the cable. This tension force is part of an action–reaction pair; the other force in the pair acts downward on the cable. B Free-body diagram for the elevator compartment. C Free-body diagram for the cable.

Hence, for this cable, and for all such massless cables, the forces acting on the two ends are equal and both equal T, the tension in the cable. Put another way, for a massless cable the tension is the same at all points along the cable.

Let's now take on the more complicated situation in Figure 3.20, which shows a hanging spider holding onto its dinner (a bug). Here we have two "cables," both composed of spider silk, and their tensions (denoted T_1 and T_2) will be different. Figure 3.20 also shows free-body diagrams for the spider, the bug, and the lower piece of silk. The tension in the lower piece of silk is T_2. This silk is attached to the spider, so there is a force of magnitude T_2 directed downward on the spider, and there is also a corresponding reaction force T_2 directed upward on the silk. The silk exerts a separate force of magnitude T_2 directed upward on the bug, and there is a reaction force T_2 acting downward on the silk. There are similar forces due to the tension T_1 in the upper piece of silk.

The spider and bug have masses m_{spider} and m_{bug} and a common acceleration a. Let's now find the tension in each piece of silk, assuming the silk is massless. We begin by applying Newton's second law to the bug in Figure 3.20. The free-body diagram for the bug shows that the forces in this case are just the force of gravity and the tension T_2, so from Newton's second law for motion along y we have

$$m_{\text{bug}}a = \sum F = F_{\text{grav, bug}} + T_2 = -m_{\text{bug}}g + T_2$$

$$T_2 = m_{\text{bug}}(g + a)$$

We next write Newton's second law for the spider:

$$m_{\text{spider}}a = \sum F = F_{\text{grav, spider}} + T_1 - T_2 = -m_{\text{spider}}g + T_1 - T_2$$

$$T_1 = m_{\text{spider}}(g + a) + T_2$$

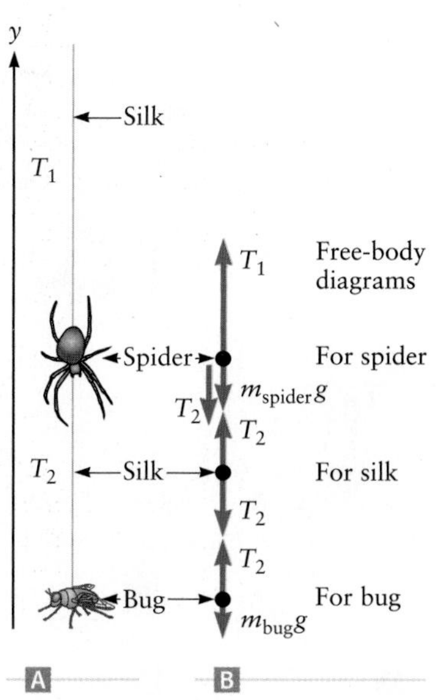

▲ **Figure 3.20** A A spider hanging and holding onto a bug using a strand of silk. B Free-body diagrams showing the forces acting on the spider, the bug, and the lower pieces of silk. There are several action–reaction force pairs in this problem.

Inserting our result for T_2 leads to

$$T_1 = m_{\text{spider}}(g + a) + m_{\text{bug}}(g + a) = (m_{\text{spider}} + m_{\text{bug}})(g + a)$$

This result for T_1 could also have been obtained by realizing that because the spider and the bug are connected and move together, we can think them as just a single object of mass $(m_{\text{spider}} + m_{\text{bug}})$. Applying Newton's second law to this composite object leads to the same result.

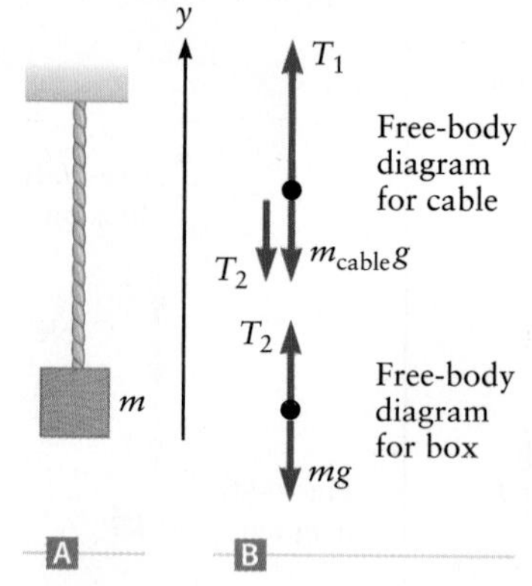

▲ **Figure 3.21** A A box attached to a cable. This cable has a nonzero mass, so there is a force on the cable due to gravity. B Free-body diagrams for the cable and for the hanging box.

Some Cables Are Not Massless

So far, we have assumed cables and ropes are massless. This assumption is often justified since the mass of a realistic cable or rope is usually very small compared with that of other parts of the system. Let's now see how to deal with cases when a cable's mass is not zero.

Consider the cable of mass m_{cable} in Figure 3.21, holding up a box of mass m. There are now three forces acting on the cable: one tension force from the support at the top (T_1), a second tension force from the box tied to the bottom end (T_2), and a third force due to the weight of the cable. Writing Newton's second law for the cable for motion along the y direction gives

$$m_{\text{cable}}a = \sum F = +T_1 - T_2 - m_{\text{cable}}g$$

If the cable is not accelerating, then $a = 0$ and

$$T_1 = T_2 + m_{\text{cable}}g \tag{3.15}$$

so

$$T_1 > T_2$$

Hence, we now have two different values for the tension in the cable. According to Equation 3.15, the tension T_1 at the top is larger than the tension at the bottom, which makes sense because the tension at the top must support the weight of the cable. We can also see what is required for a cable to be approximated as "massless." From Equation 3.15, the difference in the tensions at the top and bottom is equal to the weight of the cable. A cable can thus be approximated as massless if its weight is small compared to either tension.

Using Pulleys to Redirect a Force

Cables and ropes are an efficient way to transmit force from one place to another and are often used together with a device called a pulley. A simple pulley is shown in Figure 3.22A; it is just a wheel free to spin on an axle through its center, and it is arranged so that a rope or cable runs along its edge without slipping. For simplicity, we assume both the rope and the pulley are massless. Typically, a person pulls on one end of the rope so as to lift an object connected to the other end. Here the pulley changes the direction of the force associated with the tension in the rope as illustrated in Figure 3.22B, which shows the rope "straightened out" (i.e., with the pulley removed). In either case—with or without the pulley in place—the person exerts a force F on one end of the rope, and this force is equal to the tension. The tension is the same everywhere along this massless rope, so the other end of the rope exerts a force of magnitude T on the object. A comparison of the two arrangements in Figure 3.22—one with the pulley and one without—suggests the tension in the rope is the same in the two cases, and in both cases the rope transmits a force of magnitude $T = F$ from the person to the object.

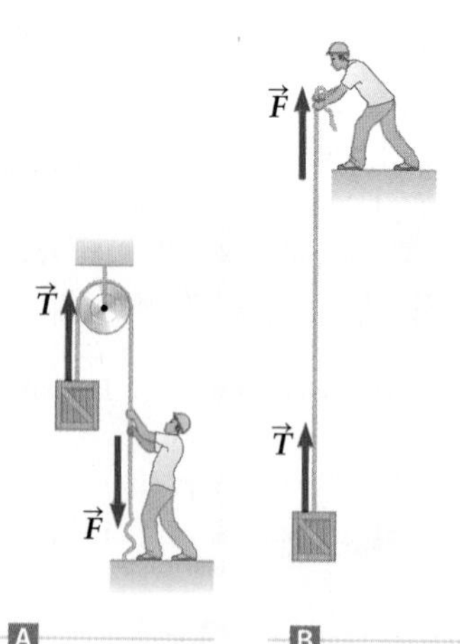

▲ **Figure 3.22** A A simple pulley. If the person exerts a force F on a massless string, there is a tension T in the string, and this tension force can be used to lift an object. B The string "straightened out." The pulley simply redirects the force.

CONCEPT CHECK 3.5 Normal Forces

Consider the man pulling downward on the rope in Figure 3.22A. Is the normal force of the floor on the man (a) greater than, (b) less than, or (c) equal to the man's weight?

Amplifying Forces

A pulley can do much more than simply redirect a force. Figure 3.23 shows a pulley configured as a block and tackle, a device used to lift heavy objects. To analyze this case, we have to think carefully about the force between the string and the surface of the pulley. We have already mentioned that the rope does not slip along this surface. There is a frictional force between the rope and the surface of the pulley wherever the two are in contact, which is all along the bottom half of the pulley in Figure 3.23. Since the rope does not slip relative to the pulley, the rope exerts a force on the pulley as if the rope and pulley were actually attached to each other. First consider the part of the rope held by the person: it exerts an upward force on the pulley, equal in magnitude to the tension T (Fig. 3.23B). The same thing happens with the section of the rope on the left that is tied to the ceiling; this part of the rope also exerts a force on the pulley, and this force is again equal to the tension. Hence, there is a force $2T$ acting upward on the pulley. The interesting point is that the person exerts a force of only T on the rope, but the block and tackle converts it into a force of magnitude $2T$ *on the pulley*. This force on the pulley can then be used to lift an object attached to the pulley, such as the wooden crate that hangs from the pulley's axle in Figure 3.23A. A block and tackle can thus *amplify forces*. This particular arrangement enables the person to lift twice as much weight as would be possible if he used the simpler pulley arrangement in Figure 3.22.

The block and tackle in Figure 3.23 amplifies forces by a factor of two, and more complex block-and-tackle devices can amplify an applied force by even greater factors. This result might disturb you a little because it appears to violate an intuitive notion that "you can't get something for nothing." Although this statement is not a formal law of physics, you might still wonder what Newton would have thought. A key feature of the block and tackle in Figure 3.23 is that while it amplifies the force by a factor of two, the displacement of the object is *reduced* by a factor of two. If the person pulls the end of the rope upward through a distance L in Figure 3.23A, the body of the pulley and hence also the mass attached at the bottom move a distance of only $L/2$. In fact, this reduction of the displacement is a feature of all block and tackles. While it is possible to amplify a force by a factor that can be much larger than unity, the corresponding displacement is always reduced by the same factor. You do *not* get something for nothing.

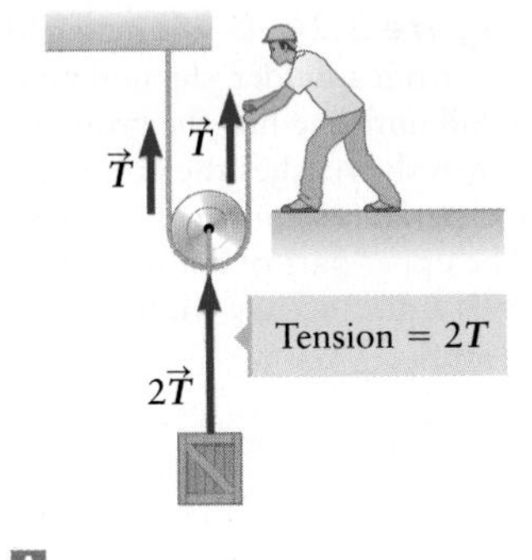

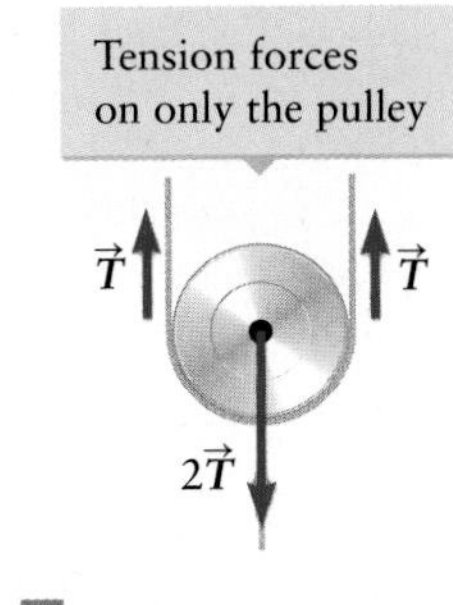

▲ **Figure 3.23** A A block-and-tackle arrangement. The upward force on the pulley, from both sides of the rope, is $2T$. This force can be used to lift an object attached to the pulley. B Close-up of rope in contact with the pulley.

3.6 Reasoning and Relationships: Finding the Missing Piece

You may have the impression that physics always involves a lot of precise mathematics. However, it is sometimes very useful to solve a problem approximately, perhaps because some important quantities (such as the initial velocity of an object) are not known precisely or because an approximate answer is sufficient. The reasoning involved in an approximate analysis can often give clear insight into a problem, without the "distraction" of a lot of mathematics. In this section, we practice such ***reasoning and relationships problems***.

Dealing with these problems involves two challenges that have not been present in the problems we have faced so far. First, we may need to identify important information that is "missing" from an initial description of the problem. For example, you might be asked to calculate the forces acting on two cars when they collide, given only the speed of the cars just before the collision. From an understanding of the relationship between force and motion, you would have to *recognize that additional information is needed* to solve this problem. In this problem, that information is the mass of the cars and the way the car bumpers deform in the collision. (We'll actually consider this problem later in this section.) It would then be your job to use common sense to *make reasonable estimates* of these "missing" quantities for typical cars and use your values to compute an answer. Because such estimated values vary from case to case (not all cars have the same mass or the same type of bumper), an approximate mathematical solution and an approximate numerical answer is usually sufficient. Let's now consider a few problems of this kind.

Jumping off a Ladder

Consider the case of a child who jumps off a ladder or playground structure for fun. We want to find the approximate force between the child and the ground when she

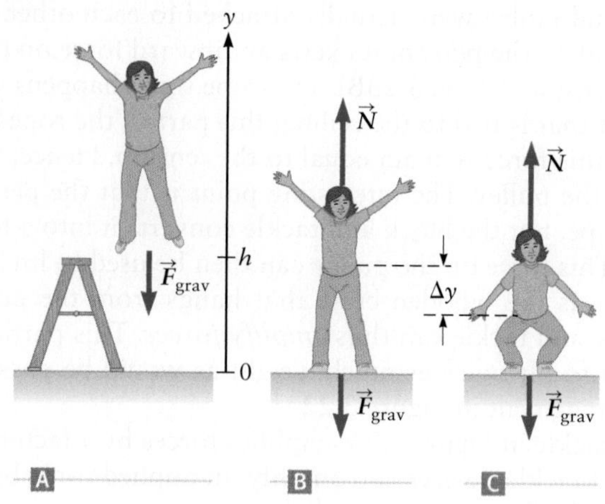

► **Figure 3.24** A When a child jumps from a ladder, she undergoes free fall until she hits the ground. B When she reaches the ground, she experiences a normal force N. C The upper part of her body then travels a distance Δy while coming to a stop.

hits the ground and what strategy she should follow to minimize this force so as to make the landing as "soft" as possible. This type of problem would be of interest to someone who designs playgrounds or to a doctor who deals with children's injuries. The same sort of problem arises in various sports and activities, such as basketball and skydiving. For example, a skydiver might want to know how much force her legs will have to withstand on landing.

As usual, we begin by making a drawing for the problem, Figure 3.24. The initial part of the motion is just free fall as the child falls from the ladder to the ground. When she strikes the ground, the ground exerts a force on the child, which causes her to stop. Let's first deal with the initial motion, the free-fall phase. For simplicity, we assume the child steps off the ladder with an initial speed of zero ($v_0 = 0$). While she is falling, her motion is described by our expressions for constant acceleration in Table 3.1, and we use Equation 3.4 with an acceleration $a = -g$ to calculate her velocity just before she hits the ground. We choose the origin ($y = 0$) to be at ground level as in Figure 3.24A and the child's initial position to be $y_0 = h$, where h is the height of the ladder. We can thus calculate v_{land}, the speed of the child just before she hits the ground at $y = 0$:

$$v_{land}^2 = v_0^2 + 2a(y - y_0) = 0 + 2(-g)(0 - h) = 2gh$$

$$v_{land} = \sqrt{2gh} \tag{3.16}$$

This result is her *speed* at the end of the free-fall period; her velocity will, of course, be directed downward.

When the child reaches the ground, she begins the second phase of her motion. There are now two forces acting on her (Fig. 3.24B): gravity and the normal force from the ground N, a quantity that we want to calculate. The precise value of N and how it varies with time depends on how the child lands. To get an approximate value, we *assume* N is constant during this period. This is only an approximation to the true behavior, but it should lead to a good approximate value for the average normal force. Our goal is obtain a value for N that is accurate to within an order of magnitude (i.e., to within a factor of 10).

Identifying the "Missing" Information and Making an Estimate. We now need to consider in a little more detail the child's motion just after hitting the ground. During this time, her legs will flex at the knees, and the upper portion of her body travels a distance Δy while coming to a stop. We know intuitively that if she lands with stiff knees (very little bending), the normal force will be much higher and will act over a shorter time than if her knees flex a lot. The maximum flex possible is roughly the distance from her feet to her knees, which is about $\Delta y \approx 0.3$ m for an average child. We can use Equation 3.4 again to describe her motion while she comes to a stop.

Since the distance traveled during the deceleration phase is $y - y_0 = -\Delta y$ (see Figs. 3.24B and C), we have

$$v^2 = v_0^2 + 2a(y - y_0) = v_0^2 + 2a(-\Delta y) \tag{3.17}$$

Here v is the speed at the end of the deceleration phase, which is zero because the child comes to a stop. In Equation 3.17, v_0 is the speed at the *start* of the deceleration phase and is thus equal to her speed at the *end* of the free-fall phase, v_{land} from Equation 3.16. With these values, we can use Equation 3.17 to calculate the acceleration during this part of the motion. Once we have the acceleration, we can use Newton's second law to find the normal force N exerted by the ground.

Rearranging Equation 3.17 to solve for the acceleration, we find

$$v^2 = 0 = v_0^2 + 2a(y - y_0) = v_{\text{land}}^2 + 2a(-\Delta y)$$

$$a = \frac{v_{\text{land}}^2}{2\,\Delta y} \tag{3.18}$$

This is the acceleration of the child just after she hits the ground and before she comes to a complete stop. Using Newton's second law during this period gives

$$ma = \sum F = F_{\text{grav}} + N = -mg + N$$

$$N = mg + ma$$

Although it seems natural to take m to be the mass of the child, it may actually be more accurate to let m be the mass of only her upper body because that is what is decelerating. However, because we are aiming for an approximate answer we'll let m be her total mass. Inserting the result for the acceleration from Equation 3.18, we can solve for the normal force acting on the child:

$$N = mg + ma = mg + \frac{mv_{\text{land}}^2}{2\,\Delta y}$$

According to Equation 3.16, $v_{\text{land}}^2 = 2gh$. Therefore,

$$N = mg + \frac{mgh}{\Delta y} = mg\left(1 + \frac{h}{\Delta y}\right) \tag{3.19}$$

For a typical "experiment" using a ladder of height $h = 2$ m and a child with $\Delta y = 0.3$ m, Equation 3.19 predicts

$$N = mg\left(1 + \frac{h}{\Delta y}\right) = mg\left(1 + \frac{2}{0.3}\right) \approx mg \times 8$$

The normal force is thus about eight times the weight (mg) of the child.

We emphasize that this is only an approximate solution. In our calculation, we assumed the stopping acceleration is constant, and in practice that may not be exactly the case. Another key to the analysis was our estimate of the stopping distance Δy. Our estimate was $\Delta y = 0.3$ m; a value of 0.03 m (about 1 inch) seems much too small, whereas a value of 3 m is unrealistically large. Our value of Δy should be acccurate to better than a factor of 10, which should lead to similar accuracy in the final answer. Here and in most reasoning and relationships problems, our goal is to get the correct order of magnitude for the answer.

Besides giving a numerical estimate of the magnitude of the impact force, Equation 3.19 also provides insight on how to minimize that force. We see that a larger stopping distance Δy leads to a smaller value of N. This is why playgrounds and playground equipment are made of soft materials and why the steering column of a car is designed to collapse on impact.

The example in Figure 3.24 is intended to give some insight into how to deal with reasoning and relationships problems. A general strategy is presented on page 78.

PROBLEM SOLVING Dealing with Reasoning and Relationships Problems

1. **RECOGNIZE THE PRINCIPLE.** Determine the key physics ideas that are central to the problem and that connect the quantity you want to calculate with the quantities you know. In the examples found in this section, this physics involves motion with constant acceleration.

2. **SKETCH THE PROBLEM.** Make a drawing that shows all the given information and everything else you know about the problem. For problems in mechanics, your drawing should include all the forces, velocities, and so forth.

3. **IDENTIFY THE RELATIONSHIPS.** Identify the important physics relationships; for problems concerning motion with constant acceleration, they are the relationships between position, velocity, and acceleration in Table 3.1. For many reasoning and relationships problems, values for some of the essential unknown quantities may not be given. You must then use common sense to make reasonable estimates for these quantities. Don't worry or spend time trying to obtain precise values of every quantity (such as the amount that the child's knees flex in Fig. 3.24). Accuracy to within a factor of 3 or even 10 is usually fine because the goal is to calculate the quantity of interest to within an order of magnitude (a factor of 10). Use the Internet, the library, and (especially) your own intuition and experiences.

4. **SOLVE.** Since an exact mathematical solution is not required, cast the problem into one that is easy to solve mathematically. In the examples in this section, we were able to use the results for motion with constant acceleration.

5. Always *consider what your answer means* and check that it makes sense.

As is often the case, practice is a very useful teacher.

EXAMPLE 3.7 Ⓡ Cars and Bumpers and Walls

Consider a car of mass 1000 kg colliding with a rigid concrete wall at a speed of 2.5 m/s (about 5 mi/h). This impact is a fairly low-speed collision, and the bumpers on a modern car should be able to handle it without significant damage to the car. Estimate the force exerted by the wall on the car's bumper.

RECOGNIZE THE PRINCIPLE

The motion we want to analyze starts when the car's bumper first touches the wall and ends when the car is stopped. To treat the problem approximately, we *assume* the force on the bumper is constant during the collision period, so the acceleration is also constant. We can then use our expressions from Table 3.1 to analyze the motion. Our strategy is to first find the car's acceleration and then use it to calculate the associated force exerted by the wall on the car from Newton's second law.

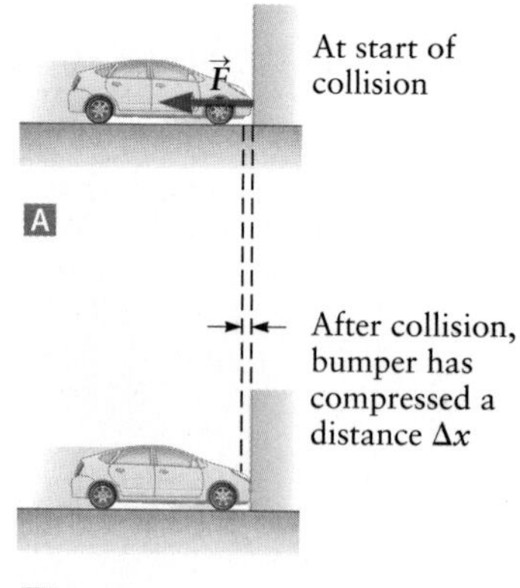
At start of collision
$\vec{F}$
A
After collision, bumper has compressed a distance Δx
B

▲ **Figure 3.25** Example 3.7. A When a car collides with a wall, the wall exerts a force F on the bumper. This force provides the acceleration that stops the car. B During the collision and before the car comes to rest, the bumper deforms by an amount Δx. The car travels this distance while it comes to a complete stop.

SKETCH THE PROBLEM

Figure 3.25 shows a sketch of the car along with the force exerted by the wall on the car. There are also two vertical forces—the force of gravity on the car and the normal force exerted by the road on the car—but we have not shown them because we are concerned here with the car's horizontal (x) motion, which we can treat using $\sum F = ma$ for the components of force and acceleration along x.

IDENTIFY THE RELATIONSHIPS

To find the car's acceleration, we need to estimate either the stopping time or the distance Δx traveled during this time. Let's take the latter approach. We are given the initial velocity ($v_0 = 2.5$ m/s) and the final velocity ($v = 0$). Both of these quantities are in Equation 3.4:

$$v^2 = v_0^2 + 2a(x - x_0) \tag{1}$$

We can use this to find the acceleration, provided we can make an estimate of $\Delta x = x - x_0$. Intuition tells us that for this low-speed collision, the bumper will deform only a little. We'll guess that a deformation of $\Delta x \approx 0.1$ m is a reasonable estimate and expect it to be correct to within an order of magnitude. We do not suggest that you try this experiment at home, but you have likely witnessed it! A value $\Delta x = 0.01$ m $= 1$ cm seems too small, whereas $\Delta x = 1$ m is much too large; our estimate of 0.1 m is somewhere in the middle.

SOLVE

The final velocity is $v = 0$ because the car comes to rest at the end of the collision. Inserting this into Equation (1) leads to

$$v^2 = v_0^2 + 2a(\Delta x) = 0$$

$$a = -\frac{v_0^2}{2\,\Delta x}$$

The acceleration is negative because it is directed opposite to the initial velocity. Using $v_0 = 2.5$ m/s and our estimate $\Delta x = 0.1$ m, the magnitude of the acceleration is

$$|a| = \frac{v_0^2}{2\,\Delta x} = \frac{(2.5\text{ m/s})^2}{2(0.1\text{ m})} = 30\text{ m/s}^2$$

Applying Newton's second law for the mass $m = 1000$ kg, this acceleration requires a force of magnitude

$$|F| = m|a| = \boxed{3 \times 10^4\text{ N}}$$

▸ *What does it mean?*

A key to this problem was our estimate of the stopping distance Δx. Using some intuition and common sense, we estimated Δx with an accuracy of about an order of magnitude; this level of accuracy is usually sufficient for a reasoning and relationships problem.

CONCEPT CHECK 3.6 Force on a Car Bumper

How large is the force exerted by the wall on the car bumper in Example 3.7 compared with the weight of the car?

3.7 Parachutes, Air Drag, and Terminal Speed

We have spent considerable time discussing problems in which the acceleration is constant. Such cases are very useful because the mathematics of how the position and velocity vary with time is fairly simple and also because these cases are a good approximation to many real-life examples. In this section, we consider some interesting and important cases in which the acceleration is *not* constant.

In our discussions of the motion of a ball through the air, we claimed that the force of air drag can often be neglected, with the excuse that it is generally small compared with the force of gravity for objects moving slowly through our atmosphere. However, as the speed of an object increases, the magnitude of the air drag force increases rapidly. You have probably felt the force of air drag by holding your hand outside the window of a moving car. This force is barely noticeable at low speeds, but can be large at highway speeds. The drag force also depends on how you hold your hand; that is, the force depends on the exposed surface area of the moving

Insight 3.4

DRAG FORCES DEPEND ON THE SHAPE OF THE OBJECT

According to Equation 3.20, the force due to air drag is given by $F_{\text{drag}} = \frac{1}{2}\rho A v^2$. The drag force thus depends on the size of the object through its frontal area A. This expression for the drag force is only approximate. In a more accurate treatment, the drag force also depends on the aerodynamics of the object. For example, a smoothly shaped object like a race car will have a smaller drag force than a truck with a boxy shape having the same area and speed. This difference can be accounted for through the improved relation for F_{drag}:

$$F_{\text{drag}} = \tfrac{1}{2}C_D\rho A v^2$$

where C_D is called the "drag coefficient." The value of C_D depends on the aerodynamic shape. For a "normal" shape, such as a boxy truck, $C_D \approx 1$, but for more streamlined shapes, the drag coefficient can be much smaller than 1.

object. Hence, the drag force is generally most significant for objects moving at high speeds, or with large surface areas perpendicular to the direction of motion, or both.

To get a rough idea of the importance of air drag, let's consider again the motion of a ball. To be specific, consider a baseball hit directly upward with an initial speed of 45 m/s, which is typical of what a professional player can do (about 100 mi/h). If we ignore air drag, we can use results from Section 3.4 for the motion of a thrown ball (Fig. 3.15 and Eq. 3.14) and predict that the ball will spend approximately 9.2 s in the air. For a real ball, the correct answer is about half as long. (We encourage you to observe this situation for yourself at a future baseball game.) The error is due to neglect of the force from air drag.

A high-accuracy calculation of the force due to air drag is extremely complicated, and there is no simple general "law" or formula that can be used. For common cases such as a baseball moving with speed v through the air, a good estimate of the magnitude of the drag force is

Force due to air drag

$$F_{drag} = \tfrac{1}{2}\rho A v^2 \tag{3.20}$$

where ρ is the density of the air and A is the exposed area of the moving object; that is, A is the cross-sectional area perpendicular to the direction of motion. The direction of the drag force is always opposite to the direction of the velocity. Equation 3.20 is only approximate, but it is generally useful for speeds between about 1 m/s and 100 m/s.

The form of Equation 3.20 makes good intuitive sense. The drag force is proportional to the density of the fluid through which the object is moving. Here this fluid is air, but if we were considering a fluid such as water, the density and hence the drag force would increase. The drag force also increases as the exposed cross-sectional area A of the object is made larger (larger parachutes are better than small ones) and increases rapidly with the speed of the object.

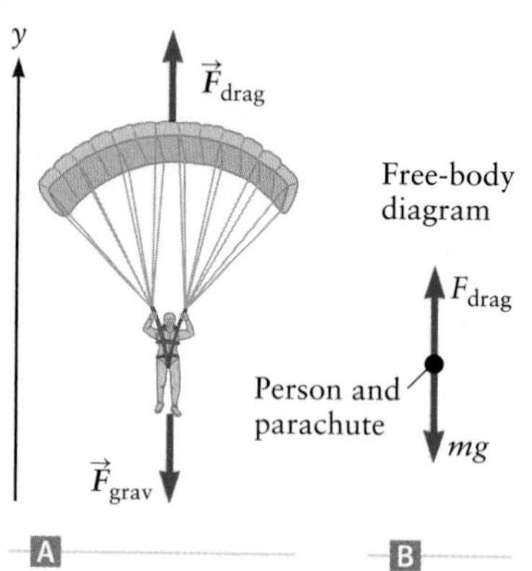

▲ **Figure 3.26** A There are two forces acting on this skydiver: the force of gravity and an upward force due to air drag. B Free-body diagram for the skydiver.

Skydiving and Air Drag

Consider a skydiver who uses a parachute (Fig. 3.26). The free-body diagram shows two forces acting on the skydiver, the force of gravity (her weight) and the force of air drag. Using Newton's second law for motion along y, we find

$$ma = \sum F = F_{grav} + F_{drag}$$

Inserting the expression from Equation 3.20 for F_{drag} gives

$$ma = -mg + \tfrac{1}{2}\rho A v^2 \tag{3.21}$$

There is no simple way to solve Equation 3.21 to get the velocity as a function of time, but we can understand the general behavior as sketched in Figure 3.27. We assume the skydiver opens her parachute immediately after exiting the plane at $t = 0$, so the drag force is acting on her throughout the jump. When she leaves the plane, her velocity in the vertical direction is initially very small, so we have $v \approx 0$ in Equation 3.21, which leads to

$$ma = -mg + \tfrac{1}{2}\rho A v^2 \approx -mg$$
$$a \approx -g$$

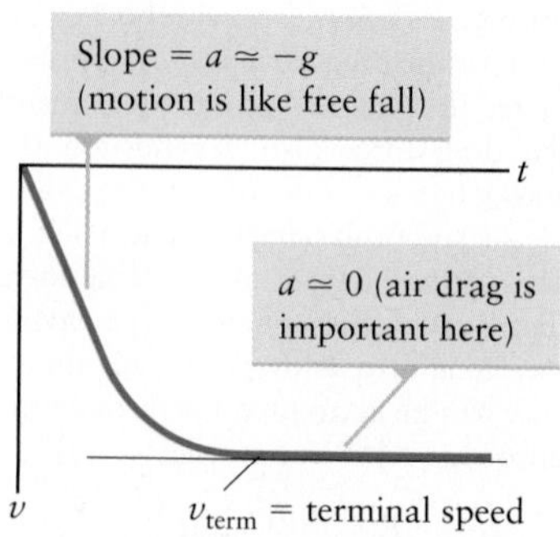

▲ **Figure 3.27** The skydiver's motion is initially like free fall; compare with Figure 3.16A. Eventually, however, air drag becomes as large as the force of gravity and the skydiver reaches her terminal speed v_{term}.

Hence, the initial motion is approximately the same as if no air drag were present. We can then use our standard result for motion with a constant acceleration (Eq. 3.1), with $v_0 = 0$ and $a = -g$, to get $v \approx -gt$. The skydiver's speed thus increases with time. However, as this speed increases she will eventually reach a time when the drag force is not small, and we must then include the effect of air drag. In fact, the speed will eventually become large enough that F_{drag} is nearly equal in magnitude, and opposite in direction, to the force of gravity so that the total force on the skydiver is very close to zero. From Newton's second law, we know that her acceleration is then also approximately zero and her speed is constant as sketched in Figure 3.27. This constant speed is known as the ***terminal speed***, and the skydiver maintains this speed for the remainder of her jump. (The word *terminal* refers to the end of the skydiver's jump and not to the end of something else.) When the skydiver reaches her terminal speed, the total force is zero and she is moving at a constant speed equal to v_{term}, and

$$\sum F = 0 = F_{grav} + F_{drag} = -mg + \tfrac{1}{2}\rho A v_{term}^2$$

Solving for the magnitude of v_{term}, we find

$$v_{term} = \sqrt{\frac{2mg}{\rho A}} \tag{3.22}$$

For a skydiver of mass 60 kg with a parachute of area 100 m^2, and using the density of air ($\rho = 1.3$ kg/m^3), we find a terminal speed $v_{term} = 3.0$ m/s. This is approximately 7 mi/h, which should be an acceptably gentle landing.

The message here is that air drag *cannot* always be neglected. The drag force increases rapidly with velocity. Air drag is generally small for slowly moving objects, but it is very important in some cases.

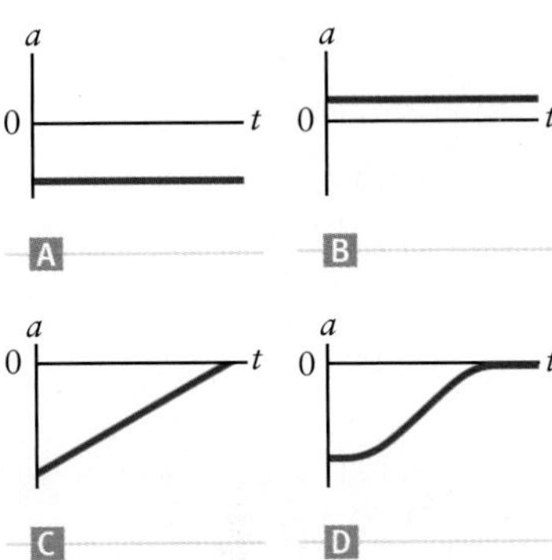

▲ **Figure 3.28** Concept Check 3.7.

CONCEPT CHECK 3.7 Acceleration of a Skydiver

Figure 3.27 shows a skydiver's velocity as a function of time. Which of the plots in Figure 3.28 shows the qualitative behavior of the skydiver's acceleration as a function of time?

(a) Plot (A) (b) Plot (B) (c) Plot (C) (d) Plot (D)

EXAMPLE 3.8 Ⓡ Air Drag on a Car

Estimate the drag force on a car moving at 50 mi/h and at 150 mi/h and compare it to the weight of a typical car with $m = 1000$ kg. The density of air is $\rho = 1.3$ kg/m^3. *Hint*: Treat this as a reasoning and relationships problem and consider what essential information is missing from the description of the problem.

RECOGNIZE THE PRINCIPLE

The drag force depends on the car's speed v, the density of the air ρ, and the frontal area A (the cross-sectional area perpendicular to the direction of motion, Fig. 3.29). We are given v and ρ, but not A. Hence, A is the "missing" information, and we must estimate a typical car's frontal area.

SKETCH THE PROBLEM

The problem is shown in Figure 3.29.

IDENTIFY THE RELATIONSHIPS

The frontal area A depends on the car's size and shape. When viewed from the front, a typical car is about 2 m wide and 1.5 m tall, so we estimate A to be approximately 3 m^2.

SOLVE

The magnitude of the drag force is given by Equation 3.20,

$$F_{drag} = \tfrac{1}{2}\rho A v^2$$

To evaluate F_{drag}, it is convenient to express the speed in units of meters per second. For 50 mi/h, we find

$$50\ \frac{\text{mi}}{\text{h}} = 50\ \frac{\text{mi}}{\text{h}} \times \frac{1600\ \text{m}}{1\ \text{mi}} \times \frac{1\ \text{h}}{3600\ \text{s}} = 22\ \text{m/s}$$

(continued) ▶

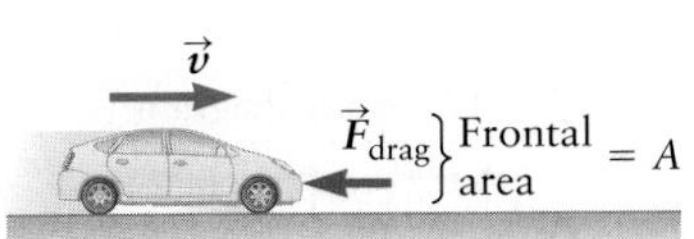

▲ **Figure 3.29** Example 3.8. The drag force on a car depends on the frontal area A of a cross section perpendicular to the car's motion. The drag force always opposes the velocity.

Inserting this value into our expression for F_{drag} along with our values of ρ and A leads to

$$F_{drag} = \tfrac{1}{2}\rho A v^2 = \tfrac{1}{2}(1.3 \text{ kg/m}^3)(3 \text{ m}^2)(22 \text{ m/s})^2 = \boxed{1000 \text{ N}}$$

The mass of the car is 1000 kg, so its weight is $mg = 9800$ N, which is about 10 times larger than this drag force. Repeating the calculation for a speed of 150 mi/h (= 67 m/s), we find $F_{drag} = \boxed{10{,}000 \text{ N}}$. This is about a factor of 10 larger than the result at 50 mi/h; the difference can be traced to the strong dependence of the drag force on the speed. At 150 mi/h, the air drag force on a car is approximately equal to the car's weight.

▸ ***What does it mean?***

These estimates for the drag force ignore the aerodynamics of the car because our simple expression for the drag force (Eq. 3.20) does not account for the effect of an object's aerodynamic shape. (See Insight 3.4.) For this reason, Equation 3.20 provides only an approximate value of F_{drag}. Even so, our results show that the air drag force on a car is quite substantial at high speeds. That is why a car's fuel efficiency decreases significantly at speeds greater than about 50 mi/h.

3.8 Life as a Bacterium

In this section, we consider the motion of a small object in a liquid, such as the *E. coli* bacterium discussed in Chapter 2. Bacteria do not move as fast as baseballs or skydivers, and the expression for the drag force in Equation 3.20 is not very accurate for a swimming *E. coli*. For this case—and for other situations involving small things moving slowly (less than about 1 m/s or so)—the drag force is described much more accurately by an expression worked out in the 19th century by George Stokes. He applied Newton's laws to this problem and showed that the drag force on a spherical object moving slowly through a fluid is given by

Drag force in a liquid

$$\vec{F}_{drag} = -Cr\vec{v} \tag{3.23}$$

where C is a constant that depends on the properties of the fluid and r is the radius of the object. For water, $C \approx 0.02 \text{ N}\cdot\text{s/m}^2$. The negative sign in Equation 3.23 indicates that the drag force is always directed opposite to the velocity. Strictly speaking, Equation 3.23 applies only to spherical objects. Of course, most bacteria are not spherical (!), but we can usually approximate their shape at least roughly in this way.

Let's now consider how *E. coli* move about in a fluid. Recall from Section 2.5 that an *E. coli* possesses one or more flagella that it rotates to propel itself (Fig. 3.30). Writing Newton's second law for motion along the horizontal direction in Figure 3.30, we have

$$ma = \sum F = F_{flagellum} + F_{drag}$$
$$ma = F_{flagellum} - Crv \tag{3.24}$$

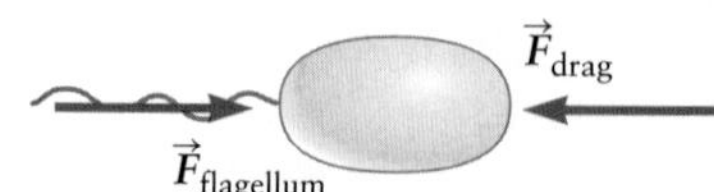

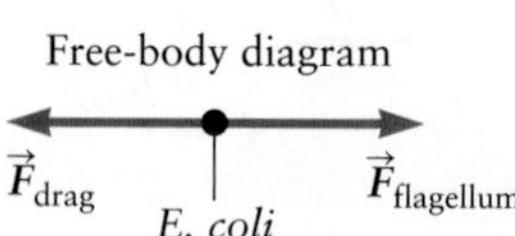

▲ **Figure 3.30** The two dominant forces on an *E. coli* are from its flagellum (propeller) and from the drag force. These forces are in opposite directions.

The drag force on a bacterium is very large compared to its weight, so these microbes cannot move very rapidly. In fact, their acceleration is always so small that the two terms on the right-hand side of Equation 3.24 are both much larger in magnitude than ma in any realistic situation. In other words, the forces on the bacterium are both large, but they are in opposite directions and they almost exactly cancel. To a very good approximation, we can therefore simply set the right-hand side equal to zero and solve for the velocity, with the result

$$v = \frac{F_{flagellum}}{Cr} \tag{3.25}$$

Hence, in this case when the flagellum stops turning and the force stops, the bacterium stops moving immediately; it does not coast!

It is common for thrown balls and sliding hockey pucks to coast since the frictional and drag forces they encounter are usually small. However, for a bacterium in water the drag force is substantial and leads to very different behavior. In either case, though, the object's motion is described correctly and accurately by Newton's laws.

Summary | CHAPTER 3

Key Concepts and Principles

Forces and motion

A force is a push or a pull exerted by one object on another. Force is a ***vector*** quantity.

All calculations of motion start with Newton's second law. For one-dimensional motion, the forces and acceleration are directed along a line, and Newton's second law reads

$$\sum F = ma$$

A ***free-body diagram*** is a very useful first step in the application of Newton's second law.

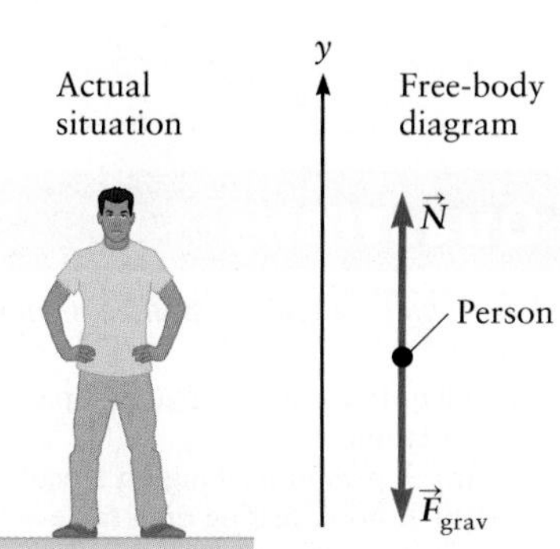

Applications

Motion with constant acceleration

A common and extremely important case is motion with constant acceleration. When a is constant, the position, velocity, and acceleration are related by

$$x = x_0 + v_0 t + \tfrac{1}{2}at^2 \qquad \text{(3.3) (page 56)}$$

$$v = v_0 + at \qquad \text{(3.1) (page 56)}$$

$$v^2 = v_0^2 + 2a(x - x_0) \qquad \text{(3.4) (page 57)}$$

Common types of forces

Gravity
Near the surface of the Earth, the gravitational force on an object of mass m is directed along the vertical direction and is given by

$$F_{\text{grav}} = -mg \qquad \text{(3.8) (page 63)}$$

The negative sign here indicates that F_{grav} is downward. This force is also called the ***weight*** of the object. When an object is moving freely under the action of gravity, the motion is called ***free fall*** and the acceleration is $a = -g$.

Normal force
When two surfaces are in contact, there is a normal force between them. This force is perpendicular to the contact surface.

Friction
When two surfaces are in contact, a frictional force opposes motion of the surfaces relative to each other. If the surfaces are slipping relative to each other, the opposing force is ***kinetic friction*** given by

$$F_{\text{friction}} = \mu_K N \qquad \text{(3.9) (page 64)}$$

where N is the normal force. If the surfaces are not slipping, the resistive force is ***static friction*** given by

$$|F_{\text{friction}}| \le \mu_S N \qquad \text{(3.11) (page 66)}$$

(Continued)

Drag forces

When an object moves through a fluid such as air or water, there is a drag force arising from contact between the fluid molecules and the moving object. We consider drag forces in two regimes:

- For motion through the air, the magnitude of the drag force is described approximately by

$$F_{\text{drag}} = \tfrac{1}{2}\rho A v^2 \qquad \text{(3.20) (page 80)}$$

  This regime is appropriate for skydivers, cars, and baseballs.
- For motion of cells and bacteria through water, the drag force is given by

$$\vec{F}_{\text{drag}} = -Cr\vec{v} \qquad \text{(3.23) (page 82)}$$

Questions

SSM = answer in *Student Companion & Problem-Solving Guide*

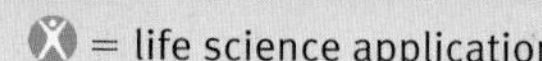
= life science application

1. Identify an action–reaction pair of forces in each of the following situations.
(a) A person pushing on a wall
(b) A book resting on a table
(c) A hockey puck sliding across an icy surface
(d) A car accelerating from rest (*Hint*: Consider the tires.)
(e) An object undergoing free fall in a vacuum
(f) A basketball player jumping to dunk a basketball
(g) A person throwing a baseball

2. Give an example of motion in which the acceleration and the velocity are in opposite directions.

3. Give an example of motion in one dimension for an object that starts at the origin at $t = 0$. At some time later, the displacement from the origin is zero but the velocity is not zero. Draw the corresponding position–time and velocity–time graphs.

4. Two objects are released simultaneously from the same height. The objects are both spherical with the same radius, but their masses differ by a factor of two. (a) Ignoring air drag, which object strikes the ground first? Or do they strike at the same time? Explain. (b) Including air drag, which object strikes the ground first? Or do they strike at the same time? Explain. (c) Find two household objects that might be used to demonstrate part (b).

5. Two mountain climbers are suspended by a rope as shown in Figure Q3.5. Which rope is most in danger of breaking? That is, which rope has the greatest tension?

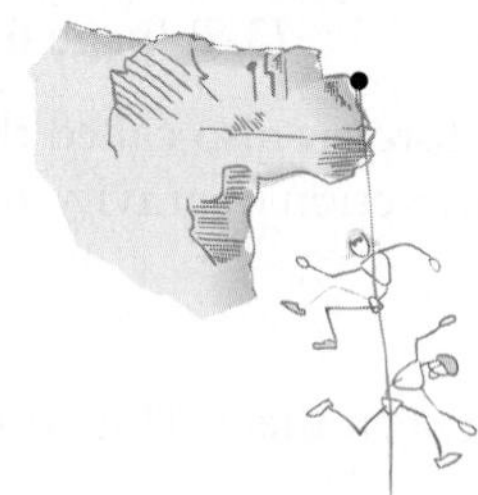

Figure Q3.5

6. You are riding in a car that starts from rest, accelerates for a short distance, and then moves with constant velocity. Explain why you feel a force from the back of your seat only while the car is accelerating and not while you are moving with a constant velocity.

7. Two balls are thrown from a tall bridge. One is thrown upward with an initial velocity $+v_0$, while the other is thrown downward with an initial velocity $-v_0$. Which one has the greater speed just before it hits the ground?

8. SSM (a) Suppose a tire rolls without slipping on a horizontal road. Explain the role friction plays in this motion. What two surfaces are involved in this frictional force? Is it static friction or kinetic friction?

(b) Suppose a racecar driver wants to get his car started very quickly and "burns rubber" as he leaves the starting line. Is he exploiting static friction or kinetic friction?

(c) Normally, the coefficient of kinetic friction between two surfaces is smaller than the coefficient of static friction for the same two surfaces. Assuming that is true for the case of friction between a car tire and a road, explain why the car will accelerate faster if the driver does not "burn rubber."

9. The lower piece of silk in Figure 3.20 is acted on by two forces, $+T_2$ at the upper end and $-T_2$ at the lower end. These two forces are equal in magnitude and opposite in direction. Are they an action–reaction force pair? Explain why or why not.

10. Devise a block-and-tackle arrangement that amplifies the applied force by a factor of three.

11. Imagine that you are a passenger on the International Space Station. While on the station, you are "weightless" and so is everything else, including wrenches and furniture. Even though everything appears weightless, explain why mass still matters. Why does it still take longer for a given force to move a massive object a certain distance than to move an object with a very small mass the same distance?

12. Imagine a skydiver who waits a long time before opening her parachute. For simplicity, assume she moves along a straight line. (a) In what direction is the skydiver moving (what is the direction of her velocity) just before she opens her parachute? (b) Suppose that when she finally opens her parachute, she is traveling faster than her terminal speed with the parachute open. What is the direction of the skydiver's acceleration just after she opens her parachute?

13. Explain why the air bags in a car reduce the forces on a passenger in the event of an accident.

14. For the books in Figure Q3.14, there is a force F from the table supporting the book on the bottom. Identify two action–reaction force pairs for each of the following objects: (a) the table,

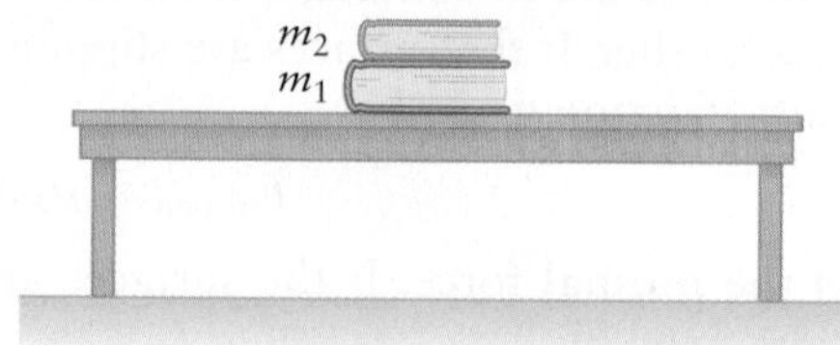

Figure Q3.14 Question 14 and Problems 30 and 31.

(b) the book on top, (c) the book on the bottom, and (d) the Earth.

15. The bacterium in Figure 3.30 experiences a force due to drag from the surrounding fluid. Does the reaction force to F_{drag} act on (a) the flagellum, (b) the body of the bacterium, or (c) the fluid?

16. SSM Two balls of the same diameter are dropped simultaneously from a very tall bridge. One ball is solid lead, and the other is hollow plastic and has a much smaller mass than the solid lead ball. Use a free-body diagram to explain why the solid lead ball reaches the ground first. *Hint*: Include the air drag force in your analysis.

17. Consider a string with one end tied to a tall ceiling and the other end hanging freely. Explain why the tension at the bottom of the string is smaller than the tension at the top.

18. In our discussion of the block and tackle in Figure 3.23, we claimed that when the right end of the string is lifted through a distance L, the body of the pulley is lifted a distance $L/2$. Give a geometric argument to prove this claim.

19. An astronaut measures her mass and her weight on Earth and again when she reaches the Moon. Which one of these quantities changes as a result of this trip and which one does not? Explain.

20. One end of a string is tied to the ceiling of an elevator, and the other end is tied to a rock (Fig. Q3.20). The elevator is moving in such a way that the tension in the string is zero. (a) What is the acceleration of the rock? (b) Explain why you cannot find the velocity of the rock, even though you can find its acceleration.

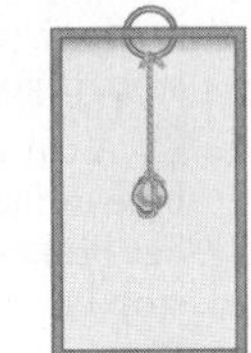

Figure Q3.20

Problems

SSM = solution in *Student Companion & Problem-Solving Guide*

★ = intermediate ✪ = challenging ⊗ = life science application

Ⓡ = reasoning and relationships problem

RT = reasoning tutorial in 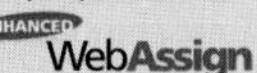

3.1 MOTION OF A SPACECRAFT IN INTERSTELLAR SPACE

1. A spacecraft that is initially at rest turns on its engine at $t = 0$. If its mass is $m = 3000$ kg and the force from the engine is 45 N, what is the acceleration of the spacecraft?

2. A hockey puck moves on an icy surface that is frictionless, with a constant speed of 30 m/s. How long does it take the puck to travel the length of the hockey rink (60 m)?

3. An ice skater moves without friction on a frozen pond. While traveling at 8.0 m/s, she finds that it takes 17 s to travel the length of the pond. How long is the pond?

4. An object moves with a constant acceleration of 4.0 m/s². If it starts with an initial speed of 30 m/s, how long does it take to reach a velocity of 250 m/s?

5. A car has a velocity of 10 m/s at $t = 7.0$ s. It then accelerates uniformly and reaches a velocity of 42 m/s at $t = 12.0$ s. What is its acceleration during this period?

6. ★ A constant force of 400 N acts on a spacecraft of mass 8000 kg that has an initial velocity of 30 m/s. How far has the spacecraft traveled when it reaches a velocity of 5000 m/s?

7. ★ A rocket-powered sled of mass 3500 kg travels on a level snow-covered surface with an acceleration of +3.5 m/s² (Fig. P3.7). What are the magnitude and direction of the force on the sled?

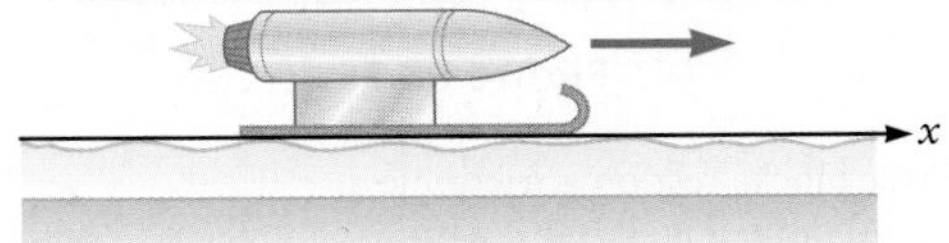

Figure P3.7

8. ★ Your car has a dead battery. It is initially at rest, and you push it along a level road with a force of 120 N, finding that it reaches a velocity of 2.0 m/s in 50 s. What is the mass of the car? Ignore friction.

9. ★ You are a newly graduated astronaut preparing for your first trip into space. Plans call for your spacecraft to reach a velocity of 500 m/s after 2.4 min. If your mass is 75 kg, what force will be exerted on your body? Assume the acceleration is constant.

10. ★ ⊗ **Pulling *g*'s.** Suppose again you are the astronaut in Problem 9. When most people are subjected to an acceleration greater than about $5 \times g$, they will usually become unconscious ("black out"). Will you be in danger of blacking out?

11. SSM ★ An object with an initial velocity of 12 m/s accelerates uniformly for 25 s. (a) If the final velocity is 45 m/s, what is the acceleration? (b) How far does the object travel during this time?

12. ★ A hockey puck has an initial velocity of 50 m/s and a final velocity of 35 m/s. (a) If it travels 35 m during this time, what is the acceleration? (b) If the mass of the puck is 0.11 kg, what is the horizontal force on the puck?

13. ✪ An elevator is moving at 1.2 m/s as it approaches its destination floor from below. When the elevator is a distance h from its destination, it accelerates with $a = -0.50$ m/s², where the negative sign indicates a downward vertical direction. ("Up" is positive.) Find h.

14. ★ An airplane must reach a speed of 200 mi/h to take off. If the runway is 500 m long, what is the minimum value of the acceleration that will allow the airplane to take off successfully?

15. ✪ A drag racer is able to complete the 0.40-km course (0.25 mi) in 6.1 s. (a) If her acceleration is constant, what is a? (b) What is her speed when she is halfway to the finish line?

16. ⊗ Consider a sprinter who starts at rest, accelerates to a maximum speed v_{max}, and then slows to a stop after crossing the finish line. Draw qualitative plots of the acceleration, velocity, and position as functions of time. Indicate the location of the finish line on the x–t plot.

17. ★ Draw a qualitative plot of the total force acting on the ball in Figure 3.15 (page 68) as a function of time. Begin your plot while the ball is still in the thrower's hand and end it after the ball comes to rest back on the ground.

18. ✪ A barge on a still lake is moving toward a bridge at 10.0 m/s. When the bridge is 40.0 m away, the pilot slows the boat with a constant acceleration of −0.73 m/s². (a) Use Equation 3.3 to find the time it takes the barge to reach the bridge. Note that you will obtain two answers! Do both calculated times correspond to possible physical situations? (b) Find the final velocity for each time using Equation 3.1. (c) For each set of answers for t and v, sketch plots of velocity versus time and position versus time, and use them to describe the two situations.

19. ★ To measure the height of a tree, you throw a rock directly upward, with a speed just fast enough that the rock brushes against the uppermost leaves and then falls back to the ground. If the rock is in the air for 3.6 s, how tall is the tree?

20. Your car is initially traveling at a speed of 25 m/s. As you approach an intersection, you spot a dog in the road 30 m ahead and immediately apply your brakes. (a) If you stop the instant before you reach the dog, what was the magnitude of your acceleration? (b) If your velocity is positive while you are slowing to a stop, is your acceleration positive or negative?

21. A car is traveling at 25 m/s when the driver spots a large pothole in the road a distance 30 m ahead. She immediately applies her brakes. If her acceleration is $-27\ \text{m/s}^2$, does she manage to stop before reaching the pothole?

22. A bullet is fired upward with a speed v_0 from the edge of a cliff of height h (Fig. P3.22). (a) What is the speed of the bullet when it passes by the cliff on its way down? (b) What is the speed of the bullet just before it strikes the ground? (c) If the bullet is instead fired downward with the same initial speed v_0, what is its speed just before it strikes the ground? Express your answers in terms of v_0, h, and g. Ignore air drag. Assume the bullet is fired straight up in (a) and (b) and straight down in (c).

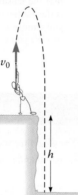

Figure P3.22 Problems 22 and 23.

23. SSM A ball is thrown upward with a speed of 35 m/s from the edge of a cliff of height $h = 15$ m (similar to Fig. P3.22). (a) What is the speed of the ball when it passes by the cliff on its way down to the ground? (b) What is the speed of the ball when it hits the ground? Ignore air drag. Assume the ball is thrown straight up.

24. A rock is dropped from a tree of height 25 m into a lake (depth 5.0 m) below. After entering the water, the rock floats gently down through the water at a constant speed of 1.5 m/s to the bottom of the lake. What is the total elapsed time?

25. You are a passenger ($m = 110$ kg) in an airplane that is accelerating on the runway, beginning to take off. The force between your back and your seat is 400 N. Starting from rest, how far does the plane travel as it accelerates to a takeoff speed of 130 m/s?

26. A motivated mule can accelerate an empty cart of mass $m = 180$ kg from rest to 5.0 m/s in 10 s. If the cart is loaded with 540 kg of wood, how long will it take the mule to get the cart to 5.0 m/s? Assume a constant acceleration. Also assume the mule exerts the same force as when the cart is empty.

3.2 NORMAL FORCES AND WEIGHT

27. A person has a weight of 500 N. What is her mass?

28. A book of mass 3.0 kg sits at rest on a horizontal table. What is the magnitude of the normal force exerted by the table on the book?

29. **Under siege.** Twenty soldiers hold a long beam of heavy wood against a fortified castle door and attempt to push it open (Fig. P3.29). If each soldier is able to supply a forward force on the beam of 80 N, what normal force does the castle door exert on the beam if the door does not open?

Figure P3.29

30. Two books of mass $m_1 = 8.0$ kg and $m_2 = 5.5$ kg are stacked on a table as shown in Figure Q3.14. Find the normal force acting between the table and the bottom book.

31. The table in Figure Q3.14 is now sitting in an elevator, with $m_1 = 9.5$ kg and $m_2 = 2.5$ kg. The normal force between the table and the bottom book is 70 N. Find the magnitude and direction of the elevator's acceleration.

32. A grand piano with three legs has a mass of 350 kg and is at rest on a level floor. (a) Draw a free-body diagram for the piano. Show the force of the floor on each leg as a separate force in your diagram. Assume the weight of the piano is distributed equally among the three legs. (b) What is the total force of the piano on the floor?

33. Your friend's car has broken down, and you volunteer to push it to the nearest repair shop, which is 2.0 km away. You carefully move your car so that the bumpers of the two cars are in contact and then slowly accelerate to a speed of 2.5 m/s over the course of 1 min. (a) If the mass of your friend's car is 1200 kg, what is the normal force between the two bumpers? (b) If you then maintain a speed of 2.5 m/s, what is the total time (including the time the cars are accelerating) it takes you to reach the repair shop?

34. SSM A tall strongman of mass $m = 95$ kg stands upon a scale while at the same time pushing on the ceiling in a small room. Draw a free-body diagram of the strongman (Fig. P3.34) and indicate all normal forces acting on him. If the scale reads 1100 N (about 240 lb), what is the magnitude of the normal force that the ceiling exerts on the strongman?

Figure P3.34

35. A bodybuilder configures a leg-press apparatus (Fig. P3.35) to a resistance of 1000 N (about 230 lb). She pushes the weight to her full extension and comes to rest. (a) What is the normal force on each of her feet? (b) Assume she is able to move the push-plate of the leg-press machine at an acceleration of $0.50\ \text{m/s}^2$ for the first half of the displacement. What is the normal force on each foot during this period of acceleration? Assume the resistive force is due to a mass with weight of 1000 N hanging on the cable of the leg-press apparatus.

© CORBIS RF/Alamy

Figure P3.35

36. Three blocks rest on a frictionless, horizontal table (Fig. P3.36), with $m_1 = 10$ kg and $m_3 = 15$ kg. A horizontal force $F = 110$ N is applied to block 1, and the acceleration of all three blocks is found to be $3.3\ \text{m/s}^2$. (a) Find m_2. (b) What is the magnitude of the normal force between blocks 2 and 3?

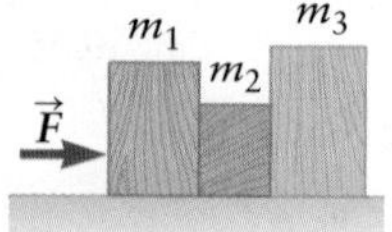

Figure P3.36

3.3 ADDING FRICTION TO THE MIX

37. A car is moving with a velocity of 20 m/s when the brakes are applied and the wheels lock (stop spinning). The car then slides to a stop in 40 m. Find the coefficient of kinetic friction between the tires and the road.

38. A hockey puck slides with an initial speed of 50 m/s on a large frozen lake. If the coefficient of kinetic friction between the puck and the ice is 0.030, what is the speed of the puck after 10 s?

39. Your moving company runs out of rope and hand trucks, so you are forced to push two crates along the floor as shown in Figure P3.39. The crates are moving at constant velocity, their masses are m_1 = 45 kg and m_2 = 22 kg, and the coefficients of kinetic friction between both crates and the floor are 0.35. Find the normal force between the two crates.

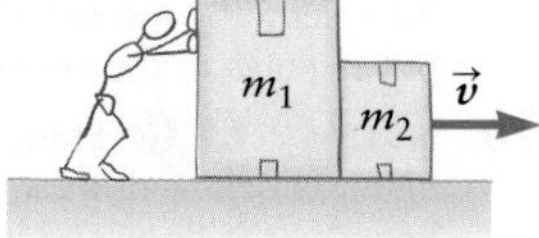

Figure P3.39

40. You are given the job of moving a refrigerator of mass 100 kg across a horizontal floor. The coefficient of static friction between the refrigerator and the floor is 0.45. What is the minimum force that is required to just set the refrigerator into motion?

41. SSM The coefficient of kinetic friction between a refrigerator (mass 100 kg) and the floor is 0.20, and the coefficient of static friction is 0.25. If you apply the minimum force needed to get the refrigerator to move, what will the acceleration then be?

42. After struggling to move the refrigerator in Problem 41, you finally apply enough force to get it moving. What is the minimum force required to keep it moving with a constant velocity? Assume $\mu_K = 0.20$.

43. You are trying to slide a refrigerator across a horizontal floor. The mass of the refrigerator is 200 kg, and you need to exert a force of 350 N to make it just begin to move. (a) What is the coefficient of static friction between the floor and the refrigerator? (b) After it starts moving, the refrigerator reaches a speed of 2.0 m/s after 5.0 s. What is the coefficient of kinetic friction between the refrigerator and the floor?

44. A driver makes an emergency stop and inadvertently locks up the brakes of the car, which skids to a stop on dry concrete. Consider the effect of rain on this scenario. If the coefficients of kinetic friction for rubber on dry and wet concrete are $\mu_K(\text{dry}) = 0.80$ and $\mu_K(\text{wet}) = 0.50$, how much farther would the car skid (expressed in percentage of the dry-weather skid) if the concrete were instead wet?

45. Jeff Gordon (a racecar driver) discovers that he can accelerate at 4.0 m/s^2 without spinning his tires, but if he tries to accelerate more rapidly, he always "burns rubber." (a) Find the coefficient of friction between his tires and the road. Assume the force from the engine is applied to only the two rear tires. (b) Have you calculated the coefficient of static friction or kinetic friction?

46. **Antilock brakes.** A car travels at 65 mi/h when the brakes are suddenly applied. Consider how the tires of a moving car come in contact with the road. When the car goes into a skid (with wheels locked up), the rubber of the tire is moving with respect to the road; otherwise, when the tires roll, normally the point where the tire contacts the road is stationary. Compare the distance required to bring the car to a full stop when (a) the car is skidding and (b) when the wheels are not locked up. Assume the coefficients of friction between the tires and the road are $\mu_K = 0.80$ and $\mu_S = 0.90$. (c) How much farther does the car go if the wheels lock into a skidding stop? Give your answer as a distance in meters and as a percent of the nonskid stopping distance. (d) Can antilock brakes make a big difference in emergency stops? Explain.

47. A hockey puck slides along a rough, icy surface. It has an initial velocity of 35 m/s and slides to a stop after traveling a distance of 95 m. Find the coefficient of kinetic friction between the puck and the ice.

3.4 FREE FALL

48. A rock is dropped from a very tall tower. If it takes 4.5 s for the rock to reach the ground, what is the height of the tower?

49. A baseball is hit directly upward with an initial speed of 45 m/s. Find the velocity of the ball when it is at a height of 40 m. Is there one correct answer for v or two? Explain why.

50. A squirrel is resting in a tall tree when it slips from a branch that is 50 m above the ground. It is a very agile squirrel and manages to land safely on another branch after only 0.50 s. What is the height of the branch it lands on?

51. **Basketball on the Moon.** If LeBron James can jump 1.5 m high on Earth, how high could he jump on the Moon (assume an indoor court), where g = 1.6 m/s^2?

52. An apple falls from a branch near the top of a tall tree. If the branch is 12 m above the ground, what is the apple's speed just before it hits the ground?

53. SSM A ball is thrown directly upward with an initial velocity of 15 m/s. If the ball starts at an initial height of 3.5 m, how long is the ball in the air? Ignore air drag.

54. Two children are playing on a 150-m-tall bridge. One child drops a rock (initial velocity zero) at t = 0. The other waits 1.0 s and then throws a rock downward with an initial speed v_0. If the two rocks hit the ground at the same time, what is v_0?

55. A rock is dropped from a tall bridge into the water below. If the rock begins with a speed of zero and has a speed of 12 m/s just before it hits the water, what is the height of the bridge as measured from the surface of the water?

56. A roofing tile falls from rest off the roof of a building. An observer from across the street notices that it takes 0.43 s for the tile to pass between two windowsills that are 2.5 m apart. How far is the sill of the upper window from the roof of the building?

57. You are standing at the top of a deep, vertical cave and want to determine the depth of the cave. Unfortunately, all you have is a rock and a stopwatch. You drop the rock into the cave and measure the time that passes until you hear the rock hitting the floor of the cave far below. If the elapsed time is 8.0 s, how deep is the cave? *Hints*: (1) Sound travels at a constant speed of 340 m/s. (2) Consider both the time the rock undergoes free fall and the time it takes the sound to travel back up the cave. During the free-fall phase, the rock starts from rest, moves with a constant acceleration, and lands at the bottom of the cave. During the second period, sound travels at a constant velocity back up the cave.

58. Your friend is an environmentalist who is living in a tree for the summer. You are helping provide her with food, and you do so by throwing small packages up to her tree house. If her tree house is 30 m above the ground, what is the minimum (initial) speed you must use when throwing packages up to her?

59. You are standing across the street from a tall building when the top of the building (h = 80 m) is hit by lightning and a brick is knocked loose. You see the lightning strike and immediately see that the brick will fall to hit a person standing at the base of the building. You then run toward the person and push him out of the path of the brick, the instant before the brick reaches him. If you are initially 25 m from the building, how fast do you have to run?

3.5 CABLES, STRINGS, AND PULLEYS: TRANSMITTING FORCES FROM HERE TO THERE

60. A cable attached to a block of mass 12 kg pulls the block along a horizontal floor at a constant velocity. If the tension in the cable is 5.0 N, what is the coefficient of kinetic friction between the block and the floor?

61. A crate of mass 55 kg is attached to one end of a string, and the other end of the string runs over a pulley and is held by a person as in Figure 3.22A. If the person pulls with a force of 85 N, what is the crate's acceleration?

62. A car of mass 1200 kg is lowered onto a junk pile at the end of the cable of a large crane. If the car is accelerating downward at 0.20 m/s^2, what is the tension in the cable?

63. SSM You work for a moving company and are given the job of pulling two large boxes of mass m_1 = 120 kg and m_2 = 290 kg using ropes as shown in Figure P3.63. You pull very hard, and the boxes are accelerating with a = 0.22 m/s^2. What is the tension in each rope? Assume there is no friction between the boxes and the floor.

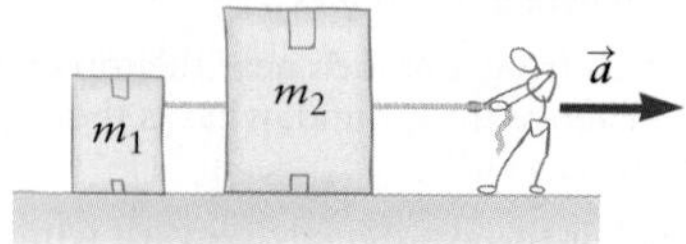

Figure P3.63

64. **In traction.** When a large bone such as the femur is broken, the two pieces are often pulled out of alignment by the complicated combination of tension and compression forces that arise from the muscles and tendons in the leg (see the X-ray image in Fig. P3.64A). To realign the bones and allow proper healing, these forces must be compensated for. A method called traction is often employed. If a total tension force of 400 N is applied to the leg as depicted in Figure P3.64B to realign the parts of the femur, how much mass m must be attached to the bottom pulley?

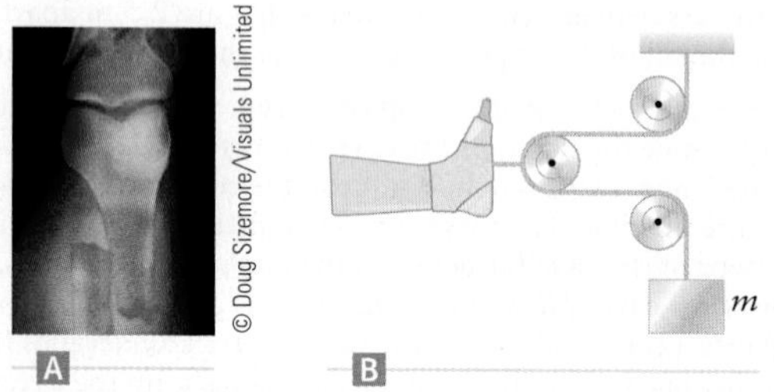

Figure P3.64

65. A mass with M = 102 kg is attached to the bottom of a block-and-tackle pulley system as depicted in Figure P3.65. How much tension force is needed to keep the mass at its current position?

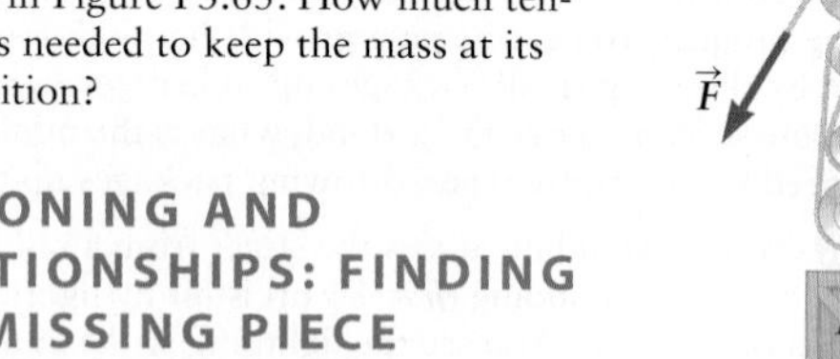

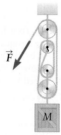

Figure P3.65

3.6 REASONING AND RELATIONSHIPS: FINDING THE MISSING PIECE

66. What is the approximate speed at which the force of air drag on a car is equal to the weight of the car? *Hint*: Start by estimating the mass and the frontal area of the car.

67. A bullet of mass 10 g leaves the barrel of a rifle at 300 m/s. Assuming the force on the bullet is constant while it is in the barrel, find the magnitude of this force. *Hint*: Start by estimating the length of the barrel.

68. When an airplane takes off, it accelerates on the runway starting from rest. What is the magnitude of this runway acceleration for a passenger jet? *Hint*: You will need to find (or estimate) one or more of the following quantities: the takeoff velocity, the distance traveled on the runway, and the time spent on the runway.

69. Calculate the terminal speed for a pollen grain falling through the air using the drag force Equation 3.20. Assume the pollen grain has a diameter of 1 μm and a density of 0.2 g/cm^3. If this grain is released from the top of a tree (height 10 m), estimate the time it will take to fall to the ground. *Hint*: The pollen grain will reach its terminal speed very quickly and will have this speed for essentially the entire motion. Your answer will explain why pollen stays in the air for a very long time.

70. SSM **Hang time.** LeBron James decides to jump high enough to dunk a basketball. What is the approximate force between the floor and his feet while he is jumping from the floor? *Hint*: Start by estimating the distance he moves while in contact with the floor and the height he must jump for his hands to reach just above the rim.

3.7 PARACHUTES, AIR DRAG, AND TERMINAL SPEED

71. Consider a skydiver who lands on the ground with a speed of 3 m/s. What is the approximate force on the skydiver's legs when she lands?

72. Calculate the terminal speed for a baseball. A baseball's diameter is approximately d = 0.070 m, and its mass is m = 0.14 kg. Express your answer in meters per second and miles per hour.

73. Hail forms high in the atmosphere and can be accelerated to a high speed before it reaches the ground. Estimate the terminal speed of a spherical hailstone that has a diameter of 2.0 cm. *Hint*: The mass of a piece of hail that has a volume of 1 cm^3 is about 1 g.

74. Calculate the force of air drag on a hockey puck moving at 30 m/s. A hockey puck is approximately 3.0 cm tall and 8.0 cm in diameter.

75. The following items are dropped from an airplane. Rank them in order from lowest terminal speed to highest and justify your ranking.

(a) Bowling ball
(b) Beach ball
(c) Spear or javelin (pointing downward)
(d) Watermelon
(e) Cantaloupe
(f) Apple

76. A Styrofoam ball of radius 28 cm falls with a terminal speed of 5.0 m/s. What is the mass of the ball?

77. SSM You are a secret agent and find that you have been pushed out of an airplane without a parachute. Fortunately, you are wearing a large overcoat (as secret agents often do). Thinking quickly, you are able to spread out and hold the overcoat so that you increase your overall area by a factor of four. If your terminal speed would be 43 m/s without the coat, what is your new terminal speed with the coat? Could you survive impact with the ground? How about over a lake?

78. Calculate the drag force on a bullet that is 5.0 mm in diameter and moving at a speed of 600 m/s. If the mass of the bullet is 10 g, compare this force to the weight of the bullet and to your own weight. Assume the drag force on the bullet is given by Equation 3.20.

3.8 LIFE AS A BACTERIUM

79. SSM The force exerted on a bacterium by its flagellum is 4×10^{-13} N. Find the velocity of the bacterium in water. Assume a size r = 1 μm.

80. Calculate the terminal speed for a bacterium in water. Use the expression for the drag force in Equation 3.23. Assume the bacterium has a mass $m = 4 \times 10^{-15}$ kg, a drag coefficient $C = 0.02$ N · s/m^2 and can be approximated as a sphere of radius r = 1 μm. How long does it take a bacterium to fall from the top of a lake of depth 5.0 m to the bottom?

Additional Problems

81. RT A textbook rests at the edge of a desk of average height. (a) If the book gently tips off the desk, what is its approximate speed when it hits the floor? (b) If the height of the desk were doubled, by what factor would the book's speed at impact with the floor increase?

82. The kick experienced when firing a rifle can be explained by Newton's third law. A .22-caliber rifle has a mass $M = 5.2$ kg, and a bullet with a mass $m = 3.0$ g leaves the barrel of the gun at a velocity of 320 m/s. (a) If the bullet starts from rest and leaves the gun barrel after $t = 0.010$ s, what was the acceleration of the bullet? Assume the bullet's acceleration is constant while it travels along the barrel. (b) What was the force on the bullet? (c) What was the magnitude of the force exerted on the gun? (d) What acceleration did the gun experience? (e) Compare the ratio of M and m to the ratio of the accelerations of each object.

83. **What is your reaction time?** The following simple method can be employed to determine reaction time. A partner holds a meterstick by pinching it at the top and letting it hang vertically. To measure your reaction time, place your thumb and forefinger just below the base of the meterstick, ready to pinch it when it falls. Without signaling, your partner releases the meterstick; it accelerates due to gravity at a rate of 9.8 m/s^2, and you grab it as fast as possible. (a) If your thumb pinches the meterstick at the 45-cm mark, what was your reaction time? Using a similar calculation, one can calibrate a "Grab-it Gauge" such as that shown in Figure P3.83. (b) Calculate the distance to draw each line from the bottom starting point for reaction times of 0.14, 0.16, 0.18, 0.20, and 0.22 s.

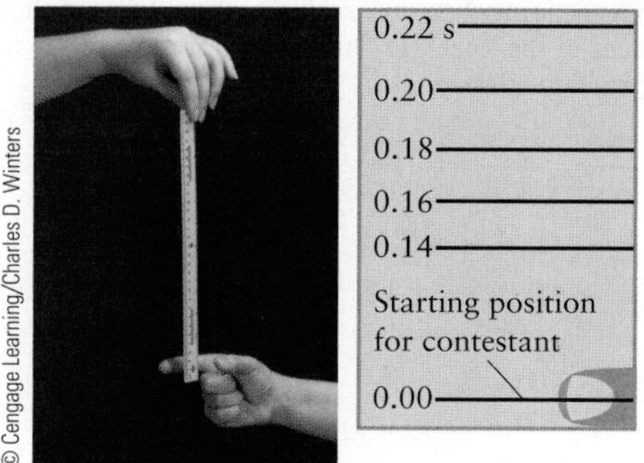

Figure P3.83

84. **Deer in the headlights.** There are two important time intervals to consider when coming to an emergency stop while driving. The first is the driver's reaction time to get a foot on the brake pedal, and the second is the time it takes to slow the car to a stop. Consider a car moving at 30 m/s (about 65 mi/h) when the driver sees a deer in the road ahead and applies the brakes. (a) If the driver's reaction time is 1.1 s, how far does the car travel before the brakes are applied? (b) If the deer is 100 m away when the driver sees it, what acceleration is needed to stop the car without hitting the deer? (c) If the concrete streets are wet, will the car be able to stop without hitting the deer? (d) If the car cannot stop in time, how fast will it be going when it strikes the deer? If the car can stop in time, how far away from the deer will it come to rest?

85. A boy pushes a 3.1-kg book against a vertical wall with a horizontal force of 40 N. What is the minimum coefficient of friction that will keep the book in place without sliding?

86. An impish young lad stands on a bridge 10 m above a lake and drops a water balloon on a boat of unsuspecting tourists. Although the boat is traveling at a speed of 7.5 m/s, the boy manages to land the balloon right on the deck of the boat. How far away from the base of the bridge was the boat when the boy released the balloon? Assume he just lets the balloon go without throwing it (i.e., he simply drops it).

87. RT Two mischievous children drop water balloons from a bridge as depicted in Figure P3.87. Each water balloon is approximately 30 cm in diameter, and in this figure the red balloon is about 1.8 m below the railing of the bridge. What is the time interval between when the first balloon was let go and the second balloon was dropped? Assume both balloons were let go just above the bridge railing. Take measurements directly from the figure and scale appropriately.

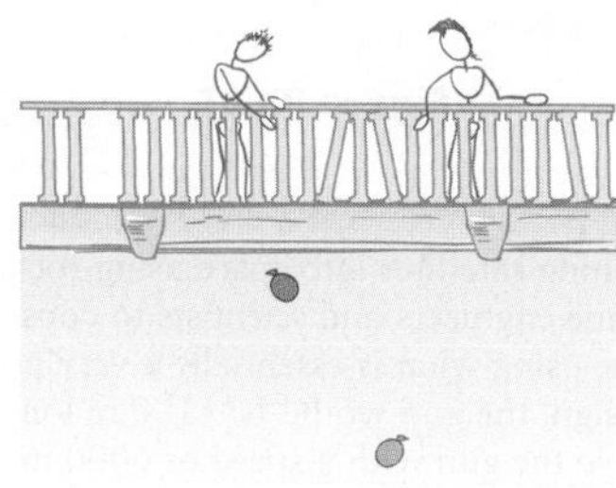

Figure P3.87

88. A subway train is designed with a maximum acceleration of +0.20 m/s^2, which allows for both passenger safety and comfort. (a) If subway stations are 1.2 km apart, what is the maximum velocity that can be obtained between stations? (b) How long does it take to travel between two stations? (Do not include time spent stopped at a station.) (c) The train stops for a total of 45 s at each station. What is the overall average velocity of the train from station to station?

89. A spring scale indicates that a helium balloon tied to it produces a tension of 0.20 N in the string. The string is then cut, and the balloon rises until it comes to rest on the ceiling. (a) Draw a free-body diagram of the balloon on the ceiling. (b) What is the normal force exerted by the ceiling on the balloon?

90. R **Slapshot.** In professional hockey, a slapshot is a type of shot in which a player literally "slaps" the puck with his stick. The puck may start with a very low speed and can leave the stick with a speed as high as 45 m/s (100 mi/h). Assuming the puck starts from rest, what is the average force exerted by the hockey stick on the puck? A hockey puck has a mass of about 0.17 kg. *Hint*: For approximately what distance is the stick in contact with the puck?

91. A car is outfitted with a flat piece of plywood mounted vertically on its front bumper. As seen in Figure P3.91, a block of wood is simply placed in front of the car just as the car begins to accelerate. (a) If the coefficient of static friction between the block and the plywood is $\mu_S = 0.90$, what acceleration is needed to keep the block from falling? (b) Safety concerns limit the maximum speed of the car to 50 m/s (about 110 mi/h). How long can the car keep the block from falling this way? (c) If the mass of the wooden block is doubled, how does the answer to part (a) change?

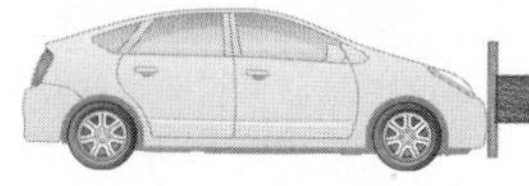

Figure P3.91

92. Two spheres, one of balsa wood (diameter 40 cm) and one of steel (diameter 10 cm), have the same mass. (a) How many times greater is the terminal speed of the steel sphere than that of the

wooden sphere? (b) The balsa wood sphere is now trimmed down so that it has the same diameter as the steel sphere. Find the ratio of the terminal speed of the steel sphere to that of the wooden sphere of mass 10 kg.

93. A block of mass $M_1 = 3.0$ kg rests on top of a second block of mass $M_2 = 5.0$ kg, and the second block sits on a surface that is so slippery that the friction can be assumed to be zero (see Fig. P3.93). (a) If the coefficient of static friction between the blocks is $\mu_S = 0.21$, how much force can be applied to the top block without the blocks slipping apart? (b) How much force can be applied to the bottom block for the same result?

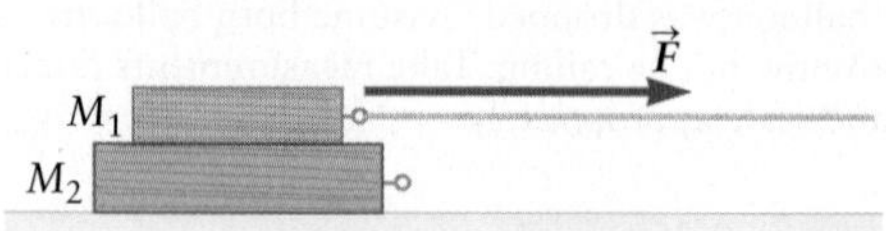

Figure P3.93

94. Ⓑ Launching satellites into space using rockets is expensive, prompting some engineers and scientists to consider firing satellites into space using what is essentially a very powerful gun. In one recent design, the gun would be 1100 m long and the satellite would leave the gun with a speed of 6000 m/s. (a) What is the acceleration of the satellite while it moves though the gun (assume a is constant)? (b) Do you think that a human astronaut could survive such a launch?

95. A bullet of mass $m = 10$ g and velocity 300 m/s is shot into a block of wood that is firmly attached to the ground. The bullet comes to rest in the wood at a depth of 8.1 cm. (a) What was the acceleration of the bullet? (b) How long did it take the bullet to come to rest once it entered the wood? (c) What was the force exerted by the wood on the bullet?

96. Ⓡ Consider a small sailboat with a triangular sail of height 10 m and width at the base of 5.0 m. (a) Assuming a wind speed of 15 mi/h relative to the boat and a wind direction perpendicular to the sail, estimate the force exerted by the wind on the sail. (b) If the sailboat is moving with a constant speed, what is the drag force due to the water? (c) Suppose the speed of the wind relative to the boat is doubled to 30 mi/h. By what factor does the speed of the boat *relative to the water* increase? *Hint*: Assume as in Stokes's expression (Eq. 3.23) that the drag force due to the water is proportional to the speed of the boat relative to the water.

97. SSM **High dive.** The cliff-divers of Acapulco are famous for diving from steep cliffs that overlook the ocean into places where the water is very shallow. (a) Suppose a cliff-diver jumps from a cliff that is 25 m above the water. What is the speed of the diver just before he enters the water? Ignore air drag. (b) If the water is 4.0 m deep, what is the acceleration of the diver after he enters the water? Assume this acceleration is constant and it begins at the moment his hands enter the water.

98. Ⓡ The pedestrian walkway on the Golden Gate Bridge is about 75 m above the water below. This bridge is (unfortunately) a popular spot for some unhappy people, who attempt to jump off. Ignore air drag and calculate (a) the time it takes an object to fall from the bridge to the water and (b) the speed of the object just before it hits the water. Express your answer to part (b) in meters per second and miles per hour. (c) Use Equation 3.22 to calculate the terminal speed of a person. Will air drag make it significantly more likely that the person will survive the impact with the water?

99. Ⓧ Ⓡ The surfaces where bones meet are lubricated by a fluidlike substance, which makes the coefficient of friction between bones very small (see Table 3.2). What is the approximate lateral (i.e., horizontal) force required to make the bones in a typical knee joint slide across each other? For simplicity, assume the surfaces in the knee are flat and horizontal, and consider an adult of average mass.

CHAPTER

4

◀ *These Adelie penguins diving from an iceberg demonstrate motion in two dimensions. In this chapter, we learn how to describe and predict their motion. (© Joseph Van Os, Getty Images)*

Forces and Motion in Two and Three Dimensions

In Chapter 3, we considered how to apply Newton's laws to deal with motion in one dimension (along a line). We now build on our understanding of motion in one dimension and deal with forces and motion in two and three dimensions. Newton's laws are again our foundation, and we continue to describe motion in terms of displacement, velocity, and acceleration. Now, though, we must allow for the vector nature of force, displacement, velocity, and acceleration. Recall that vector quantities have both a magnitude and a direction. For one-dimensional motion, we chose a positive direction at the start of a problem and used an algebraic sign (+ or −) to denote the directions of the displacement, velocity, and acceleration. For motion in two or three dimensions, we must express the directions of these vectors according to our chosen coordinate axes. Once that is done, we can use the same principles and problem-solving techniques we developed in Chapter 3.

OUTLINE

4.1 Statics

We begin with Newton's second law in vector form,

Newton's second law

$$\sum \vec{F} = m\vec{a} \tag{4.1}$$

where $\sum \vec{F}$ is the total force acting on the object. When applying Newton's second law, we generally start by determining all the individual forces acting on the object and construct a free-body diagram, showing these forces in graphical form. These individual forces must be added *as vectors* to determine the total force $\sum \vec{F}$ acting on the object. Once the total force is known, Equation 4.1 can be used to compute the object's acceleration, which then leads to its velocity and displacement.

Conditions for Translational Equilibrium

Let's first consider problems in which the velocity and acceleration are both zero. This area of mechanics is known as ***statics***, and in such situations we say that an object is in ***translational equilibrium***. The problem of when an object is or is not in translational equilibrium is very important for many engineering applications. (Imagine being on a bridge or in a building that is not in translational equilibrium, but is accelerating instead!) To simplify our terminology, we often drop the word *translational* and refer to such objects as being in "static equilibrium" or just "equilibrium."

From Equation 4.1, we see that if the acceleration is zero, the total force on the object $\sum \vec{F}$ must also be zero. Since there may be several separate forces $\vec{F}_1, \vec{F}_2, \vec{F}_3, \ldots$ acting on an object, the sum of all these forces must obey

$$\sum \vec{F} = \vec{F}_1 + \vec{F}_2 + \vec{F}_3 + \cdots = 0 \tag{4.2}$$

which is known as the ***condition for translational equilibrium***.[1]

Let's apply the condition for translational equilibrium to a situation we encountered in Chapter 3. Figure 4.1 shows a person attempting to push a refrigerator across a level floor.[2] The person is applying only a small force, so the refrigerator is at rest with zero acceleration. There are four forces acting on the refrigerator. Two of these forces are in the vertical direction (y): the force of gravity and the normal force acting between the floor and the refrigerator. Two other forces are acting in the horizontal direction (x): the force exerted by the person and the force of static friction.

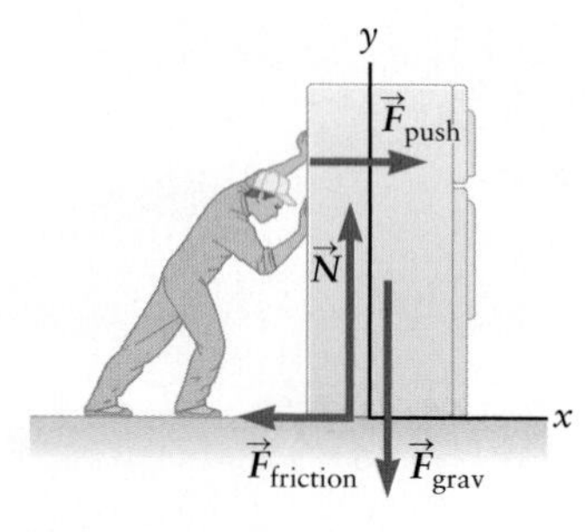

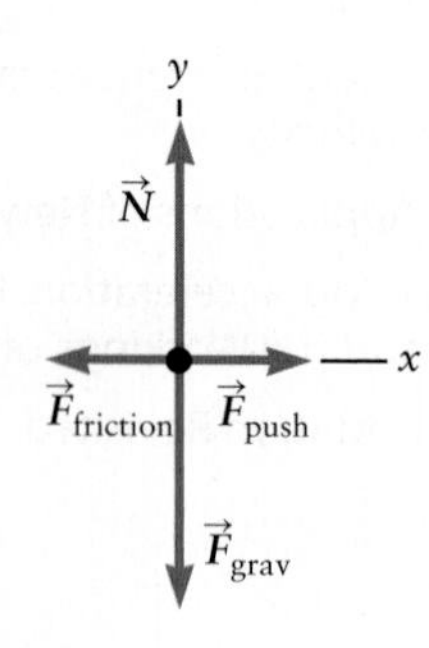

▲ **Figure 4.1** A Four forces are acting on this refrigerator. Two of them, the normal force $\vec{N}$ exerted by the floor and the force of gravity $\vec{F}_{grav}$, lie along the vertical direction (y). The other two forces, $\vec{F}_{push}$ exerted by the person and the force of friction $\vec{F}_{friction}$, lie along the horizontal direction (x). B Free-body diagram for the refrigerator.

To apply Equation 4.2 to this case, we first draw a free-body diagram (Fig. 4.1B). This diagram also shows our coordinate system, with the usual horizontal and vertical axes denoted x and y. Our next step is to express the forces in terms of their components along x and y. In this example, that work is already done for us since each of the four forces is directed along either x or y. We can then proceed to apply Equation 4.2. In this case, Equation 4.2 actually contains two relations, one for the x components of the forces and another for the y components:

$$\sum F_x = 0 \quad \text{and} \quad \sum F_y = 0 \tag{4.3}$$

Hence, for the refrigerator to be in equilibrium, the sum of all the forces along x must be zero and the sum of all forces along y must be zero. From Figure 4.1, we have $N - mg = 0$ for the y components and $F_{push} - F_{friction} = 0$ for the x components. These results are identical to what we derived in Section 3.3. This analysis could then be the first step in finding the minimum value of F_{push} needed to just break the equilibrium condition and accelerate the refrigerator.

Figure 4.2 shows a case in which the forces do not all align with the x or y directions. Here a sled is stuck in the snow, and a child is attempting to extract it by pull-

[1]When we discuss objects in translational equilibrium, we will also assume their velocity is zero.

[2]For simplicity, we assume the refrigerator does not tip.

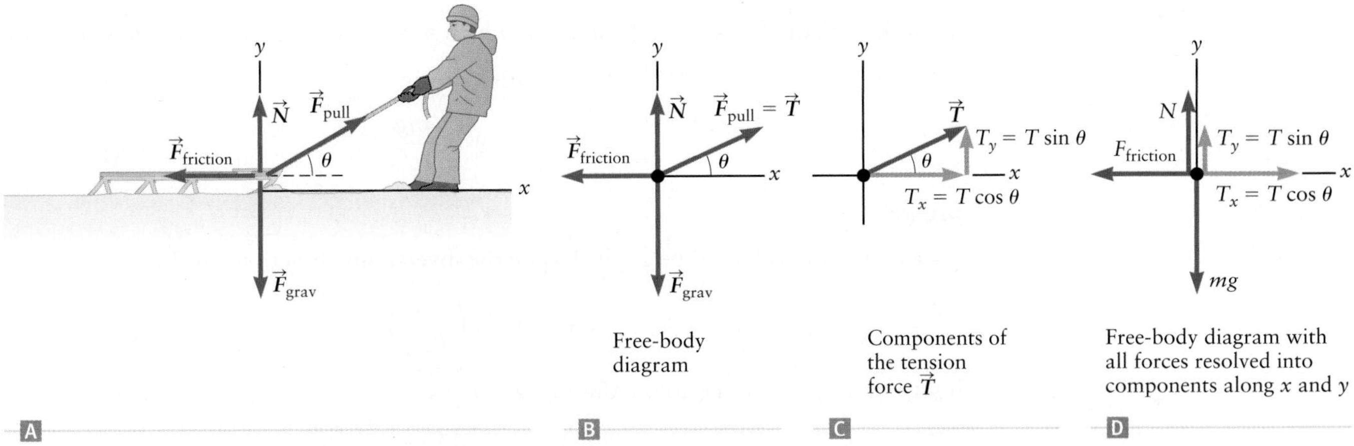

Free-body diagram

Components of the tension force $\vec{T}$

Free-body diagram with all forces resolved into components along x and y

▲ **Figure 4.2** A The force exerted by the rope on the sled has components along the horizontal and the vertical directions. B Free-body diagram for the sled. C Resolving the tension force into x and y components. D Modified free-body diagram that can be used to write equations for the force components in both the x and y directions.

ing with a rope that makes an angle θ with respect to the x axis. If the child pulls with a force $\vec{F}_{pull}$, the force on the sled due to the tension in the rope is $\vec{T} = \vec{F}_{pull}$. The free-body diagram in Figure 4.2B shows that several of the forces are parallel to our chosen coordinate axes. The normal force is parallel to y, the force of gravity is along $-y$, and the force of friction is parallel to the x axis. However, the force exerted by the rope on the sled has components along both x and y. These components are shown in Figure 4.2C, where we also show how the vector $\vec{T}$ and its components along x and y form a right triangle with components $T_x = T\cos\theta$ along x and $T_y = T\sin\theta$ along y.

In Figure 4.2D, we have redrawn the free-body diagram with $\vec{T}$ replaced by its components, so all the forces in the free-body diagram are now parallel to the coordinate axes. We next apply the conditions for equilibrium in two dimensions (Eq. 4.3). For the forces along x,

$$\sum F_x = T_x - F_{friction} = T\cos\theta - F_{friction} = 0$$

whereas for the forces along y,

$$\sum F_y = N - mg + T_y = N - mg + T\sin\theta = 0 \tag{4.4}$$

In Example 4.1, we show how this line of analysis can be used.

EXAMPLE 4.1 Trying to Move a Sled

Suppose the sled in Figure 4.2 has a mass of 12 kg and the child is exerting a force of magnitude $F_{pull} = T = 200$ N. Find the angle θ at which the normal force N becomes zero. What is happening physically when $N = 0$?

RECOGNIZE THE PRINCIPLE

We need to apply the conditions for static equilibrium to the sled. Given the mass of the sled and the magnitude of F_{pull}, these conditions cannot be satisfied if the angle θ is too large.

SKETCH THE PROBLEM

Figure 4.2 shows all the forces on the sled and their components along x and y.

IDENTIFY THE RELATIONSHIPS

From Equation 4.4, for the forces along the vertical (y),

$$\sum F_y = +N - mg + T\sin\theta = N - mg + F_{pull}\sin\theta = 0$$

(continued) ▶

We want to find the value of the angle at which $N = 0$. Inserting this condition leads to

$$F_{\text{pull}} \sin\theta = mg$$

$$\sin\theta = \frac{mg}{F_{\text{pull}}}$$

SOLVE

We can find the value of the angle θ using the inverse sine function ($\sin^{-1}$):

$$\theta = \sin^{-1}\left(\frac{mg}{F_{\text{pull}}}\right)$$

Inserting the given values of m and F_{pull} gives us

$$\theta = \sin^{-1}\left(\frac{mg}{F_{\text{pull}}}\right) = \sin^{-1}\left[\frac{(12\text{ kg})(9.8\text{ m/s}^2)}{200\text{ N}}\right]$$

$$\theta = \sin^{-1}(0.59) = \boxed{36°}$$

▸ *What does it mean?*

The normal force is due to contact between the sled and the ground. When the child pulls at an angle of 36°, the normal force goes to zero, indicating that contact between the sled and the ground is being lost.

CONCEPT CHECK 4.1 Pulling on the Sled

What will happen in Example 4.1 if the angle θ is made even larger than 36°?

(a) The sled will not move.
(b) The sled will be lifted off the ground.
(c) The rope will break.

A

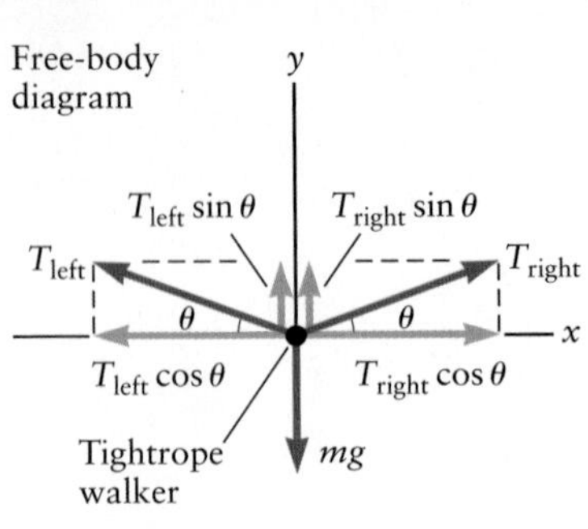

B

▲ **Figure 4.3** A Both sections of the rope exert a tension force on the portion at the center where the tightrope walker is standing. B Free-body diagram for a tightrope walker located at the rope's center point.

A Tightrope Walker in Equilibrium

Figure 4.3 shows an interesting equilibrium situation in which several forces are involved. The photo in part A shows a person standing at the middle of a tightrope. Suppose the walker has a mass $m = 60$ kg and the tension in the rope is $T = 800$ N. What angle θ does the rope make with the x axis?

We construct, as usual, a free-body diagram (Fig. 4.3B). Since the tightrope walker and the rope are assumed to be at rest, we can apply our conditions for translational equilibrium to the small piece of rope on which the tightrope walker stands. Imagine the rope is composed of two separate pieces, left and right, tied together with a small knot, with the tightrope walker standing on the knot. These two pieces of rope have tensions T_{right} and T_{left}. The walker is located at the midpoint of the rope, so you should suspect that these tensions are equal. (In the end, we'll set both of them equal to T.) However, when analyzing the equilibrium conditions, it is useful to distinguish between the tensions on both sides of the knot.

The free-body diagram in Figure 4.3B shows the forces acting on the knot. There are tension forces from the left and right sections of the rope as well as a downward force $\vec{F}_{\text{grav}}$ whose magnitude is equal to the weight of the tightrope walker (mg). Our next step is to pick a coordinate system; we choose the usual x–y coordinate axes along the horizontal and vertical directions, respectively. We then express all the forces in terms of their components along x and y as shown in Figure 4.3B. With these components, we can apply the conditions for translational equilibrium in Equation 4.3 to get

$$\sum F_x = +T_{\text{right}} \cos\theta - T_{\text{left}} \cos\theta = 0 \qquad (4.5)$$

$$\sum F_y = +T_{\text{right}} \sin\theta + T_{\text{left}} \sin\theta - mg = 0 \qquad (4.6)$$

From Equation 4.5,

$$T_{\text{right}} \cos\theta = T_{\text{left}} \cos\theta$$

and hence

$$T_{\text{right}} = T_{\text{left}}$$

The tensions in the two sides of the rope are thus equal (as we had already expected), and we can write $T_{\text{right}} = T_{\text{left}} = T$, where T is "the" tension in the rope. From Equation 4.6, we see that there are two vertical components of the tension forces, one from the left side of the rope and one from the right. These forces act to support the tightrope walker. Inserting $T_{\text{right}} = T_{\text{left}} = T$ into Equation 4.6 gives

$$+T_{\text{right}} \sin\theta + T_{\text{left}} \sin\theta = 2T\sin\theta = mg$$

We can now solve for θ:

$$\sin\theta = \frac{mg}{2T}$$

$$\theta = \sin^{-1}\left(\frac{mg}{2T}\right)$$

Inserting the given values of m and T leads to

$$\theta = \sin^{-1}\left(\frac{mg}{2T}\right) = \sin^{-1}\left[\frac{(60\text{ kg})(9.8\text{ m/s}^2)}{2(800\text{ N})}\right] = 22°$$

PROBLEM SOLVING Plan of Attack for Problems in Statics

1. **RECOGNIZE THE PRINCIPLE.** For an object to be in static equilibrium, the sum of all the forces on the object must be zero. This principle leads to Equation 4.2, which can be applied to calculate any unknown forces in the problem.

2. **SKETCH THE PROBLEM.** It is usually a good idea to show the given information in a picture, which should include a coordinate system. Figures 4.1 through 4.3 and the following examples provide guidance and advice on choosing coordinate axes.

3. **IDENTIFY THE RELATIONSHIPS.**
   - Find all the forces acting on the object that is (or should be) in equilibrium and construct a free-body diagram showing all the forces on the object.
   - Express all the forces on the object in terms of their components along x and y.
   - Apply the conditions $\sum F_x$ and $\sum F_y = 0$.
   - Forces along the z direction, if any, should also be included in your free-body diagram, and the additional condition $\sum F_z = 0$ must be applied.

4. **SOLVE.** Solve the equations resulting from step 3 for the unknown quantities. The number of equations must equal the number of unknown quantities.

5. Always *consider what your answer means* and check that it makes sense.

EXAMPLE 4.2 Stuck in the Mud

Your car is stuck in the mud, and you ask a friend to help you pull it free using a cable. You tie one end of the cable to your car and then pull on the other end with a force of 1000 N. Unfortunately, the car does not move. Your friend then suggests you tie the other end of the cable to a tree as shown from above in Figure 4.4. Although you are

(continued) ▶

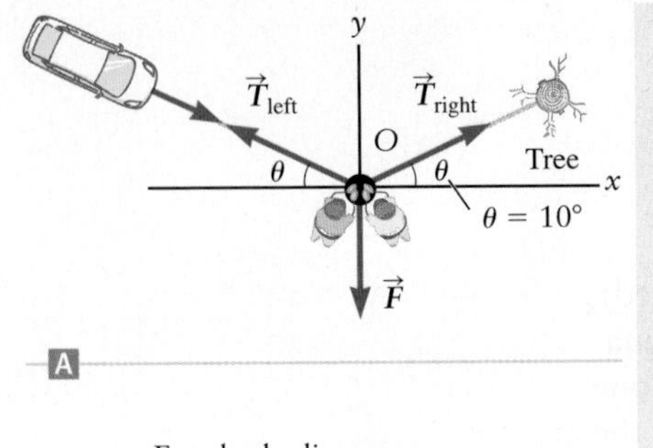

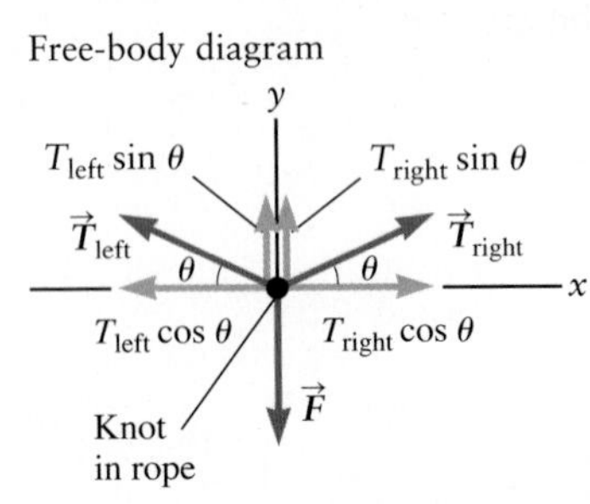

▲ **Figure 4.4** Example 4.2. A Pulling on a car. B Free-body diagram for the "imaginary knot" at the center of the cable. These drawings show top views of the predicament. (*Note*: The angle θ is not drawn to scale.)

skeptical that your friend's idea will help, you try it anyway and find that when a force $\vec{F}$ with the same magnitude (1000 N) is applied to the middle of the cable in the direction shown in Figure 4.4, you are able to pull the car free. Why does this work? Assume an angle $\theta = 10°$ as indicated in Figure 4.4.

RECOGNIZE THE PRINCIPLE

Our objective is to calculate the tension in the cable given the force $\vec{F}$ exerted by you and your friend as shown in Figure 4.4. The value we find for the tension T will *not* necessarily be the same as the magnitude of $\vec{F}$. We imagine the car is just on the verge of moving, so we can apply our conditions for equilibrium to this situation. We need to think carefully in selecting the object to which we should apply these conditions. We could apply them to the car, but that would not help calculate T. However, if we consider the forces acting on the cable at point O (the point at which you and your friend exert your force), we can then find the tension in terms of the applied force $\vec{F}$. This approach is similar to the problem of the tightrope walker in Figure 4.3, where we considered the forces applied to an imaginary knot in the rope.

SKETCH THE PROBLEM

We follow the steps listed in the Plan of Attack for Problems in Statics and apply the conditions for static equilibrium to the cable at point O in Figure 4.4A. Part B of the figure shows a free-body diagram with the forces acting on the knot at point O.

IDENTIFY THE RELATIONSHIPS

Three forces act on the cable at point O: the force $\vec{F}$ and the tension forces from each side of the cable. Choosing the x–y coordinate system shown in the figure, we calculate the components of these three forces along x and y, and then apply Equation 4.3. Along y,

$$\sum F_y = +T_{right} \sin\theta + T_{left} \sin\theta - F = 0 \qquad (1)$$

Since we have a single continuous cable and the angles on the two sides are equal, the tensions in the left and right portions of the cable are the same, $T_{right} = T_{left} = T$. Inserting this result into Equation (1) leads to

$$2T \sin\theta - F = 0 \qquad (2)$$

SOLVE

Rearranging Equation (2) to solve for T, we find

$$T = \frac{F}{2\sin\theta}$$

Inserting the value of θ (Fig. 4.4) gives

$$T = \frac{F}{2\sin\theta} = \frac{F}{2\sin(10°)} \approx \boxed{2.9 \times F}$$

▶ *What does it mean?*

The tension in the cable is thus *larger* than F! By tying the cable to a tree, T is larger than it would have been had you simply pulled on one end of the cable with the same force F. This arrangement essentially *amplifies* the force applied to the car. Recall that we encountered another example of amplifying forces in Chapter 3 when we discussed the block-and-tackle device. We'll say more about amplifying forces when we discuss work and energy in Chapter 6.

Static Equilibrium and Frictional Forces

Consider the situation in Figure 4.5 in which a car is parked on a steep hill. Imagine it is winter, when the surface of the hill might be slippery, and we want to determine if we can park the car safely or if it will instead slide down the hill. That is, we want to know if the car will be in equilibrium. Following our general strategy in the Plan of Attack for Problems in Statics, Figure 4.5B shows the free-body diagram for the car. This diagram contains three forces: the force from gravity $\vec{F}_{\text{grav}}$, the normal force $\vec{N}$ exerted by the surface, and the force of friction $\vec{F}_{\text{friction}}$, which is directed up the hill. Our next step is to choose a coordinate system. It might seem natural to choose coordinate axes that are along the horizontal and the vertical. Although this choice is a reasonable one, let's instead choose axes that are parallel and perpendicular to the plane defined by the hill as shown in Figure 4.5. With this choice of coordinate axes, the normal force is purely along y, whereas the frictional force is along x; we only have to worry about finding the components of the gravitational force.[3] Another reason for choosing x to be parallel to the hill is that were the car to move (as might happen in future problems), its velocity and acceleration would be purely along x. The trigonometry needed to find the components of the gravitational force $\vec{F}_{\text{grav}}$ along x and y is shown in Figure 4.5B.

Using the forces and components from Figure 4.5B, we apply the conditions for translational equilibrium. Along x,

$$\sum F_x = F_{\text{grav}} \sin\theta - F_{\text{friction}} = 0$$

where we have taken the $+x$ direction to be downward along the hill. The magnitude of the force of gravity is $F_{\text{grav}} = mg$; inserting this we get

$$mg \sin\theta - F_{\text{friction}} = 0 \tag{4.7}$$

Along y,

$$\sum F_y = N - F_{\text{grav}} \cos\theta = N - mg\cos\theta = 0$$

which leads to

$$N = mg\cos\theta \tag{4.8}$$

Unlike many of our previous problems, the normal force here is *not* equal to mg. The value of N depends on the angle of the hill and will be much smaller than the weight if θ is large.

If we know the coefficient of friction for the car's tires with the hill and also the angle of the hill, we can determine if the car will slip. From Equation 4.7, we find that the minimum frictional force required to keep the car from slipping is

$$F_{\text{friction}}(\text{required}) = mg\sin\theta \tag{4.9}$$

This result is static friction, and we know from Chapter 3 that the force of static friction can be as large as $\mu_S N$ but no larger. If the required force of friction in Equation 4.9 is greater than $\mu_S N$, the car will slip. On the other hand, if the required frictional force in Equation 4.9 is equal to $\mu_S N$, the car will just barely be in equilibrium. In this case,

$$\mu_S N = \mu_S mg\cos\theta = F_{\text{friction}}(\text{required}) = mg\sin\theta$$

where we have used the result for N from Equation 4.8. Solving for θ gives

$$\mu_S mg\cos\theta = mg\sin\theta$$

$$\frac{\sin\theta}{\cos\theta} = \frac{\mu_S mg}{mg}$$

$$\tan\theta = \mu_S \tag{4.10}$$

[3]Using coordinate axes oriented along the vertical and horizontal directions would lead to precisely the *same* answer as found with the axes chosen in Figure 4.5, although the algebra corresponding to Equations 4.7 and 4.8 would be a little different.

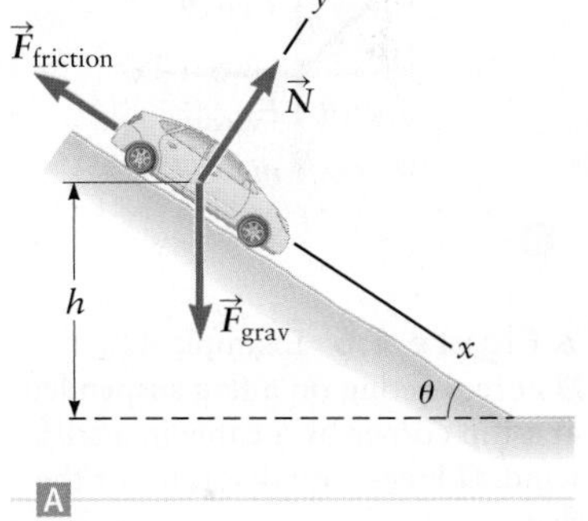

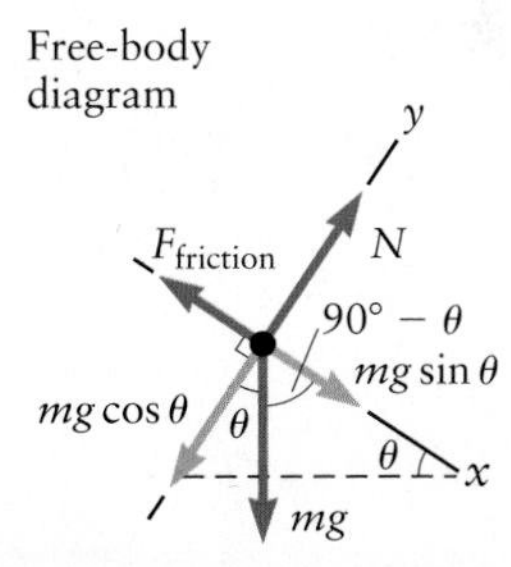

▲ **Figure 4.5** A Three forces are acting on the car. For the car to be in translational equilibrium, the three forces must add to zero. We choose the x and y axes to be parallel and perpendicular (respectively) to the plane defined by the hill. B We then express all the forces in terms of their components along these two coordinate directions. Notice that the magnitude of the force of gravity is mg.

Let's estimate the value of θ for a car just barely in equilibrium on a snowy street. The coefficient of static friction for rubber tires with snow varies with the way the snow is compacted, but is typically in the neighborhood of $\mu_S \approx 0.2$. Inserting this value into Equation 4.10 leads to $\theta \approx 10°$. This angle corresponds to a steep but realistic street, so our result suggests that a car parked on such a street might not always be in equilibrium under snowy conditions.

EXAMPLE 4.3 Supporting a Flag

A flag of mass $m = 3.5$ kg is suspended by a single cable attached at the top corner as shown in Figure 4.6A. Suppose a stiff breeze exerts a horizontal force of 75 N on the flag. Find the angle θ at which the cable hangs from the flagpole. For simplicity, assume the flag stays flat as sketched in Figure 4.6A.

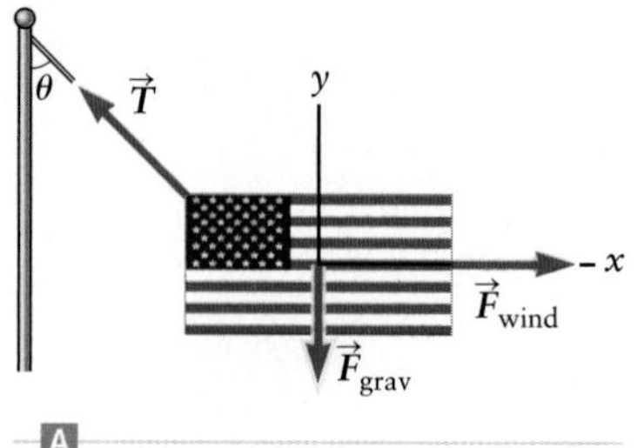

Free-body diagram

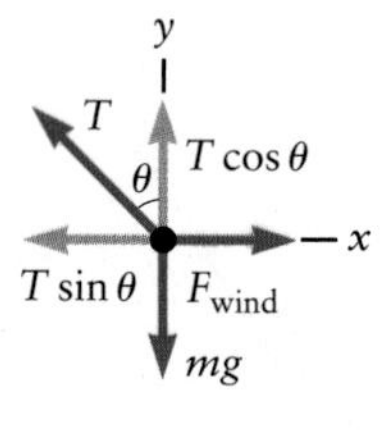

▲ **Figure 4.6** Example 4.3. A Forces acting on a flag suspended at a top corner by a cable in a stiff wind. B Free-body diagram for the flag.

RECOGNIZE THE PRINCIPLE

We want the flag to be in static equilibrium. The unknowns in this problem are the magnitude of the tension T and the angle θ. Since we have two unknowns, we must use the conditions for equilibrium in two dimensions (Eq. 4.3) to obtain two equations to solve the problem.

SKETCH THE PROBLEM

Figure 4.6 shows the forces acting on the flag: the tension force $\vec{T}$ from the cable, the force $\vec{F}_{wind}$ from the wind, and the force of gravity $\vec{F}_{grav}$. The picture shows our choice of coordinate axes with x and y along the horizontal and vertical, respectively. The free-body diagram shows the components of all the forces along x and y.

IDENTIFY THE RELATIONSHIPS

Writing the condition for equilibrium for the force components along x gives us

$$\sum F_x = F_{wind} - T\sin\theta = 0$$

$$T\sin\theta = F_{wind} \tag{1}$$

Along y,

$$\sum F_y = T\cos\theta - F_{grav} = 0$$

$$T\cos\theta = F_{grav} = mg \tag{2}$$

SOLVE

We can solve for θ by taking the ratio of Equations (1) and (2):

$$\frac{T\sin\theta}{T\cos\theta} = \frac{F_{wind}}{mg}$$

$$\tan\theta = \frac{F_{wind}}{mg}$$

Inserting the given values of the force exerted by the wind and of m leads to

$$\theta = \tan^{-1}\left(\frac{F_{wind}}{mg}\right) = \tan^{-1}\left[\frac{75\text{ N}}{(3.5\text{ kg})(9.8\text{ m/s}^2)}\right] = \boxed{65°}$$

► ***What does it mean?***

Our work on this problem began with a general picture together with a free-body diagram. The same approach will be useful when we apply Newton's second law to calculate how objects move, beginning with projectiles in Section 4.2.

Insight 4.1

STATICS IN THREE DIMENSIONS

All the examples in this section have been two dimensional, with vector forces in an x–y plane. Many situations can be treated in this way by choosing the x–y plane to match the geometry of the problem. Problems in which the force vectors lie in three-dimensional space can be treated using the same basic approach. We use the two conditions for equilibrium in Equation 4.3 and add a corresponding relation for the forces along the z direction, $\sum F_z = 0$.

4.2 Projectile Motion

We are now ready to consider objects that are in motion and learn how to calculate their displacement, velocity, and acceleration. In this section, we use ***projectile motion*** to illustrate the basic ideas. Projectile motion is the motion of a baseball, arrow, rock, or similar object that is thrown (i.e., *projected*), typically through the air. For the simplest type of projectile motion, the only force acting on the projectile is the force of gravity, and we ignore (at least for now) the force from air drag.

It is convenient to use the coordinate system sketched in Figure 4.7, with x along the horizontal and y along the vertical direction. The force of gravity has components

$$F_{\text{grav},\,x} = 0 \qquad F_{\text{grav},\,y} = -mg \tag{4.11}$$

Writing Newton's second law (Eq. 4.1) in component form, we have

$$\sum F_x = ma_x \qquad \sum F_y = ma_y \tag{4.12}$$

For the case of simple projectile motion, the only force is due to gravity, so Equation 4.11 gives the total force. Inserting this into Newton's second law (Eq. 4.12) gives

$$\sum F_x = 0 = ma_x$$
$$a_x = 0 \tag{4.13}$$

and

$$\sum F_y = -mg = ma_y$$
$$a_y = -g \tag{4.14}$$

Hence, the acceleration is constant along both x and y. In fact, the component of the acceleration along x is zero.

Now comes a crucial point: the motions along x and y are *independent* of each other. According to Equation 4.12, the acceleration along x is determined by the force along x, while the acceleration along y is determined by the force along y. For the gravitational force, the components of the force along x and y are independent of each other (Eq. 4.11), so the accelerations along x and y are also independent. What's more, according to Equation 4.13 the motion of a projectile along x is completely equivalent to one-dimensional motion along the x axis with an acceleration $a_x = 0$. Hence, the results found in Chapter 3 for motion in one dimension can be applied to give the x component of the position and the x component of the velocity for a projectile. Likewise, according to Equation 4.14, the motion of our projectile along the y direction is equivalent to one-dimensional motion along y with an acceleration $a_y = -g$. Hence, we can again use our results for one-dimensional motion to calculate the y component of the position and the y component of the velocity of a projectile. In fact, the motion along y is the same as simple free fall, which we studied in Chapter 3. As a result, *projectile motion is simply two cases of motion with constant acceleration*: one along x and one along y. The relationships among displacement, velocity, acceleration, and time given in Table 3.1 for constant acceleration thus apply directly to projectile motion.

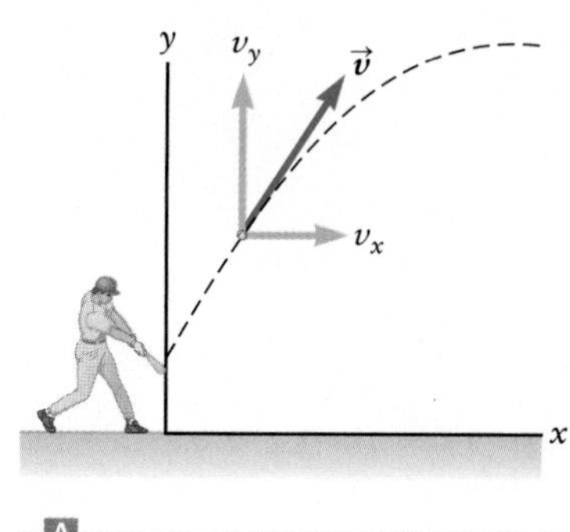

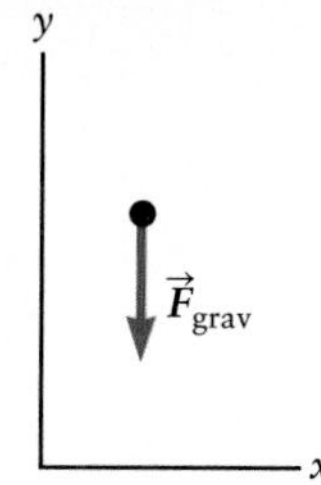

▲ **Figure 4.7** A The dashed line shows the trajectory of a thrown or batted baseball. B Free-body diagram for the ball. Throughout the entire course of the motion, the force on the ball $\vec{F}_{\text{grav}}$ is directed downward with a magnitude equal to mg.

Rolling Off a Cliff

An example of projectile motion is shown in Figure 4.8, which shows a car rolling off a cliff. Suppose the car leaves the cliff with a velocity of 10 m/s directed along the horizontal. If the cliff has a height $h = 20$ m, where and when will the car land? To attack this problem, we begin, as usual, with a picture that establishes our coordinate system and contains the starting information. In Figure 4.8, we choose x

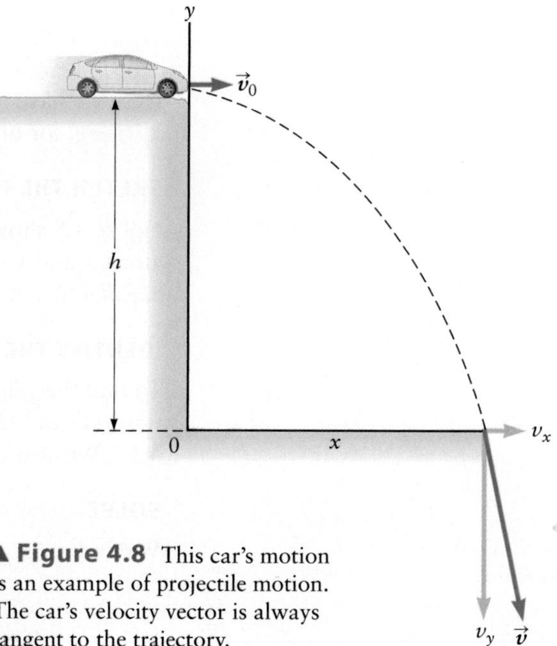

▲ **Figure 4.8** This car's motion is an example of projectile motion. The car's velocity vector is always tangent to the trajectory.

and y to be horizontal and vertical, with the origin at the bottom of the cliff. We then apply our relations for motion with constant acceleration from Table 3.1. Here we are given information on where the car starts and are asked to find where and when it lands, so we choose the relation that involves position and time. Writing this equation twice, once for the horizontal motion and once for the vertical motion, we have

$$x = x_0 + v_{0x}t + \tfrac{1}{2}a_x t^2 = 0 + v_{0x}t + 0 = v_{0x}t \tag{4.15}$$

$$y = y_0 + v_{0y}t + \tfrac{1}{2}a_y t^2 = h + 0 - \tfrac{1}{2}gt^2 = h - \tfrac{1}{2}gt^2 \tag{4.16}$$

where the initial position of the car is $x_0 = 0$ and $y_0 = h$ (the height of the cliff). The initial velocity components are $v_{0x} = 10$ m/s and $v_{0y} = 0$ since the car is moving horizontally at the moment it leaves the cliff. Notice again that the acceleration along x is zero (so $a_x = 0$ and v_x is constant) and that $a_y = -g$. Equations 4.15 and 4.16 give the position of the car as a function of time. This position is a vector quantity and has components x and y that vary with time. We can calculate when the car lands by computing the value of t when the car reaches $y = 0$ because this point is ground level (the bottom of the cliff). From Equation 4.16, we have

$$y = h - \tfrac{1}{2}gt^2 = 0$$

$$t = \sqrt{\frac{2h}{g}} = \sqrt{\frac{2(20\text{ m})}{9.8\text{ m/s}^2}} = 2.0\text{ s} \tag{4.17}$$

To find where the car lands, we use this value of t in Equation 4.15:

$$x = v_{0x}t = (10\text{ m/s})(2.0\text{ s}) = 20\text{ m}$$

EXAMPLE 4.4 Driving Off a Cliff

For the car in Figure 4.8, find the x and y components of the velocity and also the speed of the car just before it hits the ground.

RECOGNIZE THE PRINCIPLE

We use the relations for motion with constant acceleration from Table 3.1. We are interested in the velocity and have to deal with two equations, one for the horizontal component of the velocity v_x and another for the vertical component v_y. So,

$$v_x = v_{0x} + a_x t = v_{0x} \tag{1}$$

$$v_y = v_{0y} + a_y t = v_{0y} - gt \tag{2}$$

Here we have also inserted values for the components of the acceleration, $a_x = 0$ and $a_y = -g$, for our projectile.

SKETCH THE PROBLEM

Figure 4.8 shows the trajectory of the car and also the velocity and its components along x and y just before the car hits the ground. Notice that the direction of $\vec{v}$ is *parallel* (i.e., tangent) to the trajectory curve.

IDENTIFY THE RELATIONSHIPS

To find the velocity when the car reaches the ground, we evaluate v_x and v_y in Equations (1) and (2) at the value of t at which the car hits the ground as found in Equation 4.17. We also use the given values of the initial velocity, $v_{0x} = 10$ m/s and $v_{0y} = 0$.

SOLVE

We find

$$v_x = v_{0x} = \boxed{10\text{ m/s}}$$

and

$$v_y = v_{0y} - gt = 0 - (9.8 \text{ m/s}^2)(2.0 \text{ s}) = \boxed{-20 \text{ m/s}}$$

The speed v of the car is the magnitude of the velocity vector:

$$v = \sqrt{v_x^2 + v_y^2} = \sqrt{(10 \text{ m/s})^2 + (-20 \text{ m/s})^2} = \boxed{22 \text{ m/s}}$$

▶ What does it mean?

Notice that we have calculated the velocity and speed the instant *before* the car strikes the ground. When it contacts the ground, there will be a normal force exerted by the ground and also a frictional force, and the car's behavior will no longer be described by simple projectile motion.

Independence of the Vertical and Horizontal Motion of Projectiles

The independence of a projectile's vertical and horizontal motions leads to some interesting results. In the example with the car in Figure 4.8, this independence means that the time it takes the car to reach the ground (Eq. 4.17) is independent of the motion along x. This time is the same if the car is initially traveling very fast (large v_{0x}) or very slow, or even if the car is just nudged from the cliff ($v_{0x} \approx 0$).

This result is demonstrated in the time-lapse photos shown in Figure 4.9 showing the motion of two falling balls that are released from the same height at the same time. The ball on the left is simply dropped, while the ball on the right is given an initial velocity along the horizontal direction x. The ball on the left falls directly downward, whereas the ball on the right follows a parabolic path characteristic of projectile motion. Although the balls have very different velocities along the horizontal direction, at each "flashpoint" in the photo the two balls are always at the same height, which shows that their displacements and velocities along y are the same. This result confirms that the motion along y does *not* depend on the velocities along the horizontal direction, in accordance with Equations 4.13 through 4.16.

CONCEPT CHECK 4.2 Falling Balls

When the balls in Figure 4.9 reach the ground, which one has the larger speed?

(a) The red ball does.
(b) The yellow ball does.
(c) They have the same speed when they reach the ground.

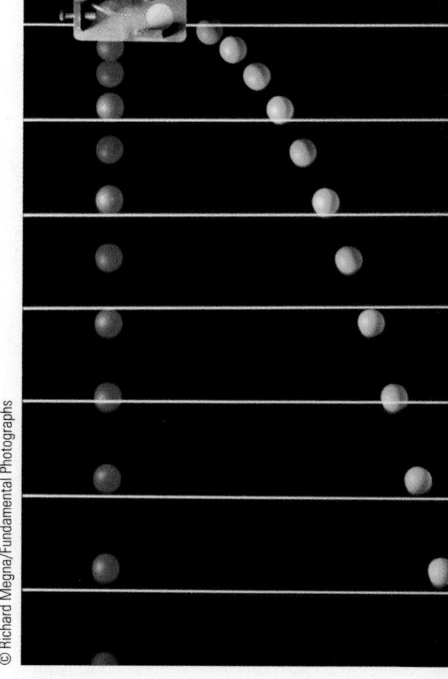

▲ **Figure 4.9** Time-lapse photos of two falling balls. The ball on the left falls straight downward, whereas the ball on the right has a nonzero component of velocity along the horizontal direction. The two balls strike the ground at the same time.

The independence of the motion along x and y for a projectile is central to the following hypothetical situation. A monkey has escaped from a zoo and is hiding in a tree. The zookeeper wants to capture the monkey without hurting him, so the zookeeper plans to shoot a tranquilizer dart at the monkey (Fig. 4.10). This very clever monkey is watching closely as the rifle is aimed at him. When the monkey observes a flash of light from the rifle (indicating that it has been fired), the monkey releases his grip on the tree and begins to fall. The monkey does this so that he will fall some distance vertically while the dart is traveling on its way to him. The monkey believes his falling will cause the dart to pass over his head and thereby miss him, but will the dart really miss the monkey?

Figure 4.10 shows the monkey and his predicament. Part A of the figure shows what would happen if we could (hypothetically) turn off gravity for the duration of the problem. That is, we assume $g = 0$. With this assumption, the acceleration of both the dart and the monkey are zero along both x and y, and both the dart and the monkey move with constant velocities. The dart thus moves in a straight line that takes it from the rifle to the monkey (as sketched in Fig. 4.10A), while the monkey

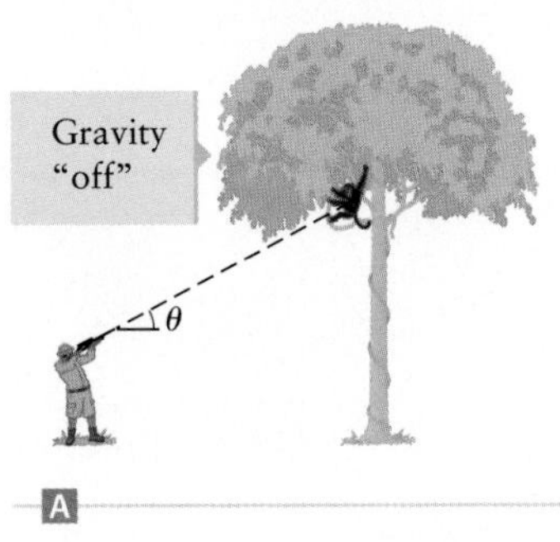

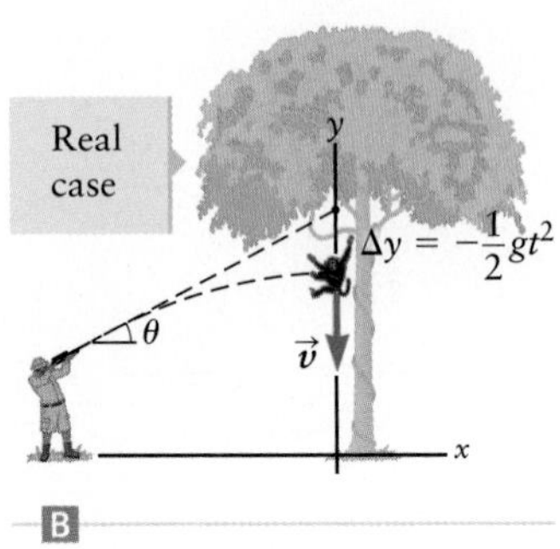

▲ **Figure 4.10** A If gravity were "turned off," this dart would follow a straight-line path to the monkey. B When gravity is turned on (which is usually the case!), the dart and the monkey are both accelerated downward and they fall the same distance Δy. The dart thus hits the monkey, but at a location below the monkey's initial position.

does not move at all since his initial velocity is zero. Hence, with gravity turned off, the dart would hit the monkey, as you should have expected.

Now consider the case with gravity turned on. The trajectories of the dart and the monkey in this case are shown in Figure 4.10B. They have the same acceleration: $a_x = 0$, $a_y = -g$, so the vertical motions of the dart and the monkey are both described by

$$y = y_0 + v_{0y}t - \tfrac{1}{2}gt^2$$

In words, we can write this result for y as

$$y = \overbrace{y_0 + v_{0y}t}^{\substack{\text{displacement} \\ \text{without gravity}}} \overbrace{-\tfrac{1}{2}gt^2}^{\substack{\text{displacement} \\ \text{due to gravity}}} \tag{4.18}$$

Equation 4.18 emphasizes that the effect of gravity on y is the *same* for the dart and the monkey. The trajectories of both fall a distance $\Delta y = -\frac{1}{2}gt^2$ below the trajectories in the absence of gravity in Figure 4.10A. Hence, gravity causes the dart and the monkey to fall the same distance, so their trajectories still intersect. The dart *will* hit the monkey. The point of this story is that in projectile motion, the motion along the vertical (y) is *independent* of the horizontal motion. The dart and the monkey fall the same amount due to the acceleration caused by gravity, even though their horizontal velocities are very different.

Projectile Motion and Target Practice

Figure 4.11 shows a sharpshooter firing a bullet horizontally at a target a distance L away. From our previous arguments, we know that because of the acceleration caused by the gravitational force, the bullet falls a distance $\Delta y = -\frac{1}{2}gt^2$ below the straight-line trajectory it would have followed were gravity (hypothetically) turned off (Eq. 4.18).

It is interesting to work out the value of Δy for a typical case. On a shooting range, the distance to the target might be 100 m or more, so let's take $L = 100$ m. The speed of the bullet when it leaves the rifle is typically in the range of 400 m/s to 800 m/s; we'll assume a value in the middle of this range and take $v_{0x} = 600$ m/s (about 1200 mi/h) for our calculation. To compute Δy, we must first calculate the time it takes the bullet to reach the target. Assuming simple projectile motion (so that there is no force due to air drag), we can use Equation 4.15 to find

$$x = v_{0x}t = L$$

which leads to

$$t = \frac{L}{v_{0x}} = \frac{100 \text{ m}}{600 \text{ m/s}} = 0.17 \text{ s}$$

For Δy, we then get

$$\Delta y = -\tfrac{1}{2}gt^2 = -\tfrac{1}{2}(9.8 \text{ m/s}^2)(0.17 \text{ s})^2 = -0.14 \text{ m}$$

You may now wonder how a sharpshooter can ever hit the bull's-eye. If he aims directly at the center of the target, the bullet in this example will pass 0.14 m = 14 cm below the bull's-eye. The resolution of this "paradox" is that a sharpshooter does not actually aim directly at the target. The sighting telescope on a rifle is calibrated so that when the rifle is "sighted" on the bull's-eye, the barrel is in reality

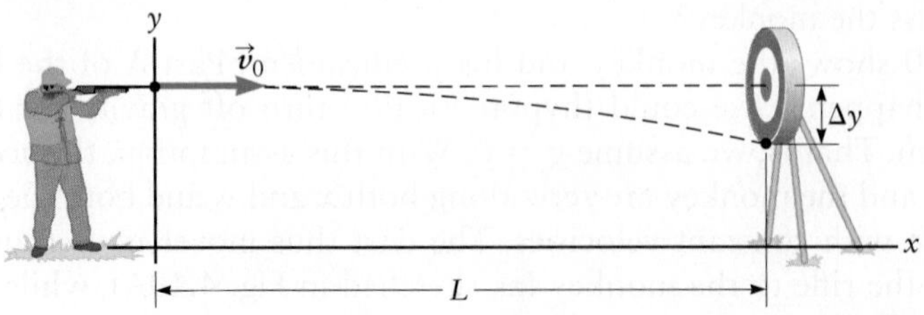

► **Figure 4.11** This bullet's initial velocity is horizontal. Gravity, however, causes the bullet to "fall" a distance Δy while it travels to the target.

aimed a distance Δy *above* the target. Since the value of Δy depends on the distance to the target and the speed of the bullet, the sighting telescope must be adjusted for the particular target distance and bullet in use. Hence, the rifle "knows" physics (or, rather, the rifle maker and sharpshooter do).

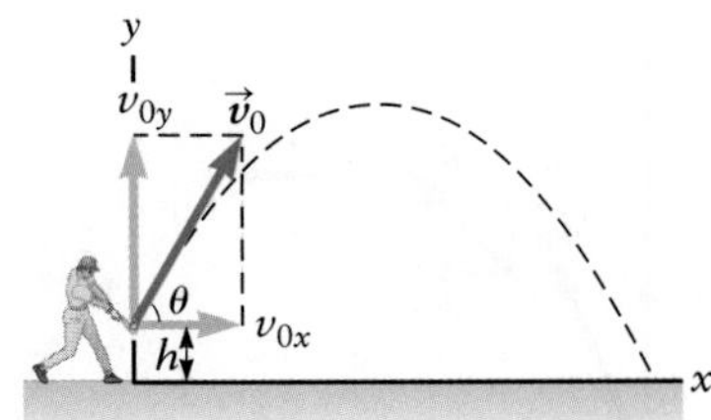

▲ **Figure 4.12** Trajectory of a batted baseball.

Motion of a Baseball: Calculating the Trajectory and the Velocity

Let's now consider the motion of a batted baseball. Our projectile—the baseball—starts from some initial position with a specified initial velocity, and we want to calculate its trajectory. The problem is sketched in Figure 4.12. The ball has an initial position with $x_0 = 0$ and $y_0 = h$, where h is the height of the ball when it leaves the bat. The ball is hit with an initial speed v_0 at an angle θ with respect to the horizontal, so the initial components of its velocity are $v_{0x} = v_0 \cos\theta$ and $v_{0y} = v_0 \sin\theta$. Our relations for motion with constant acceleration then have the form

$$x = x_0 + v_{0x}t + \tfrac{1}{2}a_x t^2 = 0 + v_{0x}t + 0 = v_0(\cos\theta)t \tag{4.19}$$

and

$$y = y_0 + v_{0y}t + \tfrac{1}{2}a_y t^2 = h + v_{0y}t - \tfrac{1}{2}gt^2 = h + v_0(\sin\theta)t - \tfrac{1}{2}gt^2 \tag{4.20}$$

To completely describe the motion, we also need the velocity:

$$v_x = v_{0x} + a_x t = v_0 \cos\theta \tag{4.21}$$

$$v_y = v_{0y} + a_y t = v_0 \sin\theta - gt \tag{4.22}$$

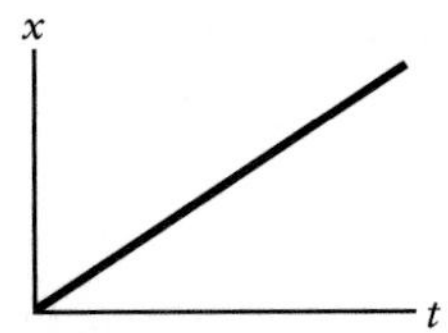

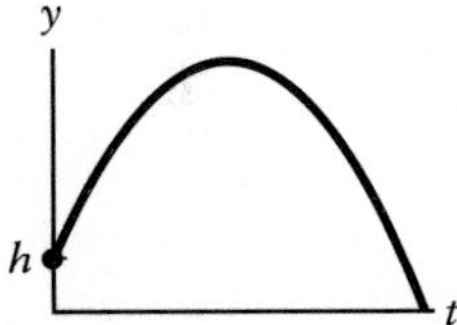

▲ **Figure 4.13** Plots of the x and y coordinates of a baseball as it travels along its trajectory.

Figures 4.13 and 4.14 show the components of the ball's position and velocity as functions of time. The results for the x direction are similar to what we would find for the dart and bullet in Figures 4.10 and 4.11. Since there is no acceleration along the horizontal direction, v_x is constant and x varies linearly with t. For the motion along y, we have an acceleration of $-g$; hence, y varies quadratically with t, and the path followed by the baseball is a parabola in the x–y plane as sketched in Figure 4.12. This path is symmetric in the following sense: the trajectory followed on the way toward the highest point is a mirror image of the trajectory on the way down. Moreover, the time the ball spends traveling from some initial height (such as h) to the top is the same as the time it spends traveling back down to that same height.

It is instructive to show the velocity components at different points along the trajectory as plotted in Figure 4.15. This is just another way of displaying the results contained in Equations 4.21 and 4.22, and in Figure 4.14. The velocity also displays a symmetry like that found for the trajectory. For example, the value of v_y at some point on the way up (i.e., at some particular height) is equal in magnitude but opposite in sign compared with v_y when the ball is at the same height on the way down. In other words, if a baseball begins its motion with speed v_0 at an initial height $y = h$, it will have the same speed when it is at this height on the way back down.

CONCEPT CHECK 4.3 Trajectory of a Baseball

For the baseball in Figure 4.12, at what point(s) on its trajectory is

(a) the acceleration zero?
(b) the vertical component of the velocity zero?
(c) the force on the ball zero?
(d) the speed zero?

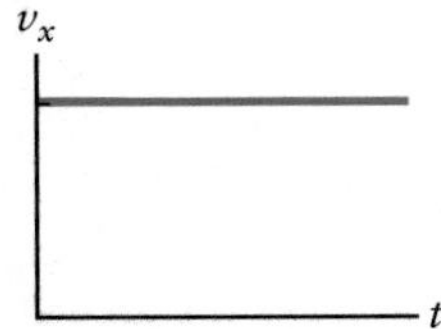

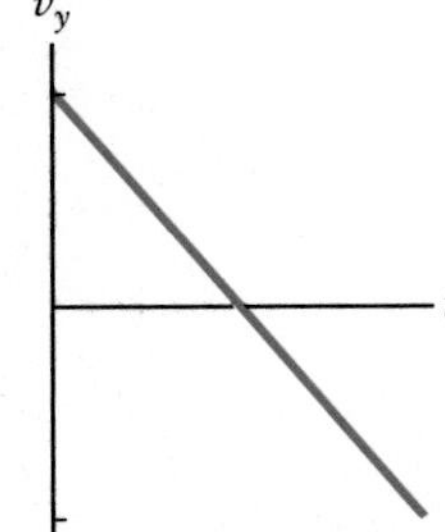

▲ **Figure 4.14** Plots of the velocity components v_x and v_y as a baseball travels along its trajectory.

Motion of a Baseball: Analyzing the Results

Let's now consider a few quantitative examples involving the baseball's trajectory in Figures 4.12 through 4.15. To be somewhat realistic, we assume the ball has an initial speed of 45 m/s (about 100 mi/h) with $\theta = 30°$ and it starts at an initial height of $h = 1.0$ m.

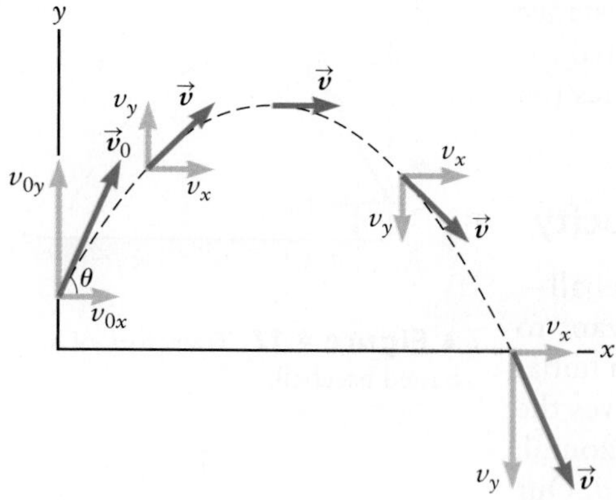

▲ **Figure 4.15** Components of the velocity at various points along the trajectory of a baseball. Compare with m Figure 4.14.

Top of the Trajectory (Maximum Height). Suppose we want to find both the height and the speed of the ball when it reaches the uppermost point on its trajectory. We can find the time at which it reaches this point by knowing that $v_y = 0$ at that instant. Inserting this into Equation 4.22 gives

$$v_y = 0 = v_0 \sin\theta - gt_{\text{top}}$$

$$t_{\text{top}} = \frac{v_0 \sin\theta}{g} \tag{4.23}$$

Using Equation 4.20, the ball's height at this moment is

$$y_{\text{top}} = h + v_0(\sin\theta)t_{\text{top}} - \tfrac{1}{2}gt_{\text{top}}^2 \tag{4.24}$$

Inserting our initial conditions into these results, we get

$$t_{\text{top}} = \frac{v_0 \sin\theta}{g} = \frac{(45\ \text{m/s})\sin(30°)}{9.8\ \text{m/s}^2} = 2.3\ \text{s}$$

and

$$y_{\text{top}} = h + v_0(\sin\theta)t_{\text{top}} - \tfrac{1}{2}gt_{\text{top}}^2$$

$$y_{\text{top}} = (1.0\ \text{m}) + (45\ \text{m/s})\sin(30°)(2.3\ \text{s}) - \tfrac{1}{2}(9.8\ \text{m/s}^2)(2.3\ \text{s})^2 = 27\ \text{m}$$

The ball's speed at this moment will be

$$v_{\text{top}} = \sqrt{v_x^2 + v_y^2} = \sqrt{v_x^2 + 0} = v_x = v_0 \cos\theta \tag{4.25}$$

since $v_y = 0$ at the top. Indeed, we could have obtained this result for the speed directly from Figure 4.14. Evaluating Equation 4.25, we find

$$v_{\text{top}} = v_0 \cos\theta = (45\ \text{m/s})\cos(30°) = 39\ \text{m/s}$$

Landing (End of Trajectory). To find the velocity of the baseball just before it hits the ground, we can use the fact that ground level is at $y = 0$. Using Equation 4.20, we get

$$y_{\text{lands}} = 0 = h + (v_0 \sin\theta)t_{\text{lands}} - \tfrac{1}{2}gt_{\text{lands}}^2 \tag{4.26}$$

which gives us a quadratic equation to solve for t_{lands}. The solution[4] is $t_{\text{lands}} = 4.64$ s. (Here we are keeping three significant figures to minimize rounding errors in the next calculation.) To find the speed at this moment, we first need to find the component of the velocity along y, which we can obtain from Equation 4.22:

$$v_{\text{lands}, y} = v_0 \sin\theta - gt_{\text{lands}}$$

$$v_{\text{lands}, y} = (45\ \text{m/s})\sin(30°) - (9.8\ \text{m/s}^2)(4.64\ \text{s}) = -23\ \text{m/s}$$

We can now compute the speed using $v_{\text{lands}} = \sqrt{v_{\text{lands}, x}^2 + v_{\text{lands}, y}^2}$. The component v_x is constant (see Eq. 4.21 and Fig. 4.14), so $v_{\text{lands}, x} = v_x = v_0 \cos\theta = 39$ m/s as calculated above. Inserting these results we find

$$v_{\text{lands}} = \sqrt{v_{\text{lands}, x}^2 + v_{\text{lands}, y}^2} = \sqrt{(39\ \text{m/s})^2 + (-23\ \text{m/s})^2} = 45\ \text{m/s}$$

We can use these results to check for the symmetry of the trajectory mentioned earlier. For a symmetric trajectory, t_{lands} should be twice the time it takes to reach the top of the trajectory found from Equation 4.23, and the speed just before the ball reaches the ground should be equal to the initial speed. To be strictly correct, we should notice that in this example the trajectory of the ball does not end at its initial

[4]Question 9 at the end of this chapter asks you to use the quadratic formula to solve for t_{lands}.

height because it starts a small distance above the ground at $y_0 = h = 1.0$ m but lands at $y = 0$. Here, however, h is sufficiently small that to within the two significant figures used in most of our numerical evaluations, t_{lands} is equal to twice the time it takes to reach the peak of the trajectory (t_{top}) and the speed at landing is equal to the initial speed. In Example 4.5, we show that the time of flight is exactly symmetric for objects projected from ground level.

Range of a Projectile. In our example with the baseball, we assumed the ball was hit with a particular value of the initial angle θ. It is interesting to consider how the trajectory varies as a function of θ. If you were a baseball coach, you might want to instruct your players on the value of θ they should use to get the ball to travel as far as possible. (This would be applied physics at its best.) This horizontal distance traveled by the ball is called the ***range***. To solve this problem, we must calculate the distance traveled by a projectile as a function of angle θ and then find the value of θ that maximizes this distance. For simplicity, let's assume the baseball starts from ground level so that $y_0 = h = 0$. The time the baseball spends in the air can be found from Equation 4.26:

$$y_{\text{lands}} = 0 = h + (v_0 \sin\theta)t_{\text{lands}} - \tfrac{1}{2}gt_{\text{lands}}^2 = (v_0 \sin\theta)t_{\text{lands}} - \tfrac{1}{2}gt_{\text{lands}}^2$$

$$t_{\text{lands}} = \frac{2v_0 \sin\theta}{g} \tag{4.27}$$

The ball then lands at the location

$$x_{\text{lands}} = v_{0x}t_{\text{lands}} = v_0(\cos\theta)\frac{2v_0 \sin\theta}{g} = \frac{2v_0^2 \sin\theta \cos\theta}{g}$$

and x_{lands} is also equal to the range. Using the trigonometric identity $\sin\theta\cos\theta = \frac{1}{2}\sin(2\theta)$ from Appendix B, we find

$$x_{\text{lands}} = \text{range} = \frac{v_0^2 \sin(2\theta)}{g} \tag{4.28}$$

For a given value of the initial speed v_0, the largest value of the range will be found when $\sin(2\theta) = 1$ since that is the largest possible value of the sine function. So, $2\theta = 90°$, or $\theta = 45°$ gives the maximum range. This result applies to all projectiles that start and end at the same height, provided that the only force acting on the projectile is the force of gravity. When the force from air drag is significant, the range is no longer given by Equation 4.28.

EXAMPLE 4.5 Symmetry of Projectile Motion

Consider a projectile that starts and ends at ground level. Show that the time spent traveling to the point of maximum height y_{top} is equal to the time spent moving from y_{top} back to the ground.

RECOGNIZE THE PRINCIPLE

We want to prove this result in the general case, so we return to Equations 4.23 and 4.27. We need to compare t_{top} and t_{lands} and show that t_{lands} is exactly twice t_{top}.

SKETCH THE PROBLEM

Figures 4.13, 4.14, and 4.16 describe the problem. For this example, the initial height is $y_0 = 0$ because the projectile starts at ground level. We have already mentioned that the trajectory of a projectile is a symmetric function of time, which is why the y–t graph in Figure 4.16 is drawn as a symmetric parabola.

(continued) ▶

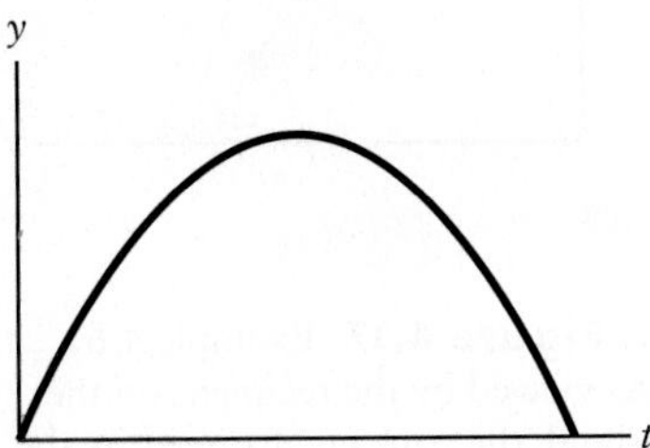

▲ **Figure 4.16** Example 4.5. Vertical position y versus t for a projectile that starts at ground level.

IDENTIFY THE RELATIONSHIPS

From Equation 4.23, we have

$$t_{top} = \frac{v_0 \sin\theta}{g} \tag{1}$$

Applying Equation 4.27 to find when the projectile lands gives

$$t_{lands} = \frac{2v_0 \sin\theta}{g} \tag{2}$$

SOLVE

By comparing Equations (1) and (2), we see that $t_{lands} = 2t_{top}$. Therefore, the time spent traveling down is *precisely* equal to the time spent traveling to the top.

► ***What does it mean?***

This result confirms our earlier claim that the trajectory for a projectile that begins and ends at the same height is symmetric.

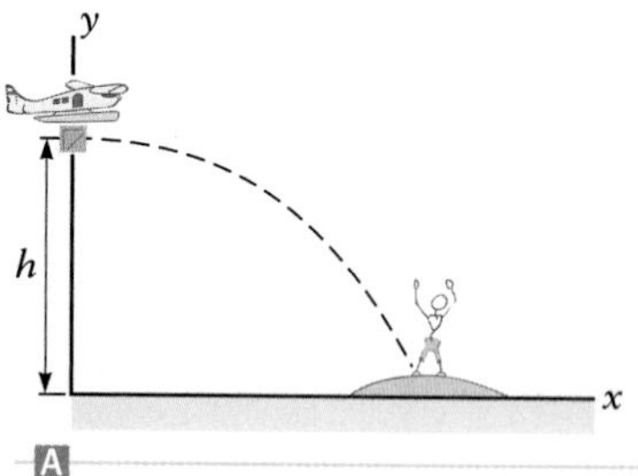

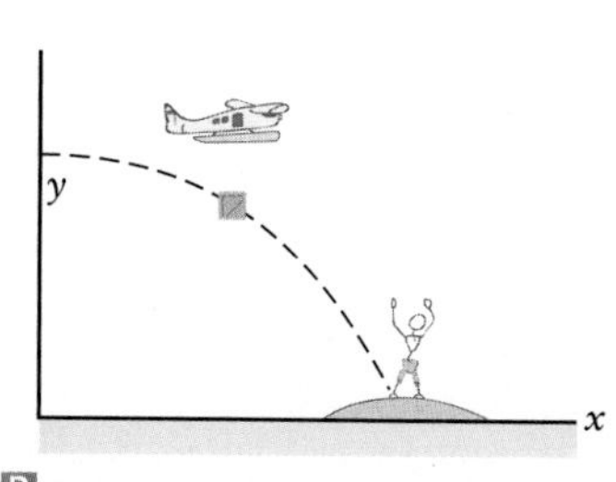

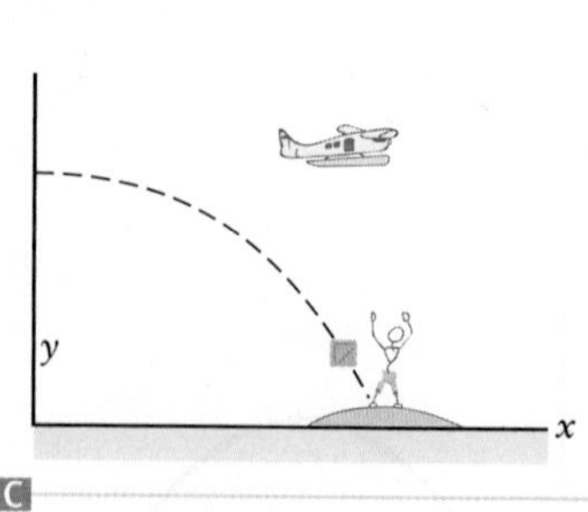

▲ **Figure 4.17** Example 4.6. As viewed by the recipient on the island, the package dropped by the airplane is a projectile that has an initial velocity equal to the velocity of the airplane. The package then follows the parabolic trajectory as sketched.

EXAMPLE 4.6 Dropping a Payload

An airplane is carrying relief supplies to a person stranded on an island. The island is too small to land on, so the pilot decides to drop the package of supplies as she flies horizontally over the island. **(a)** Find the time the package spends in the air. **(b)** If the airplane is flying horizontally at an altitude h at speed v_{plane}, where should the pilot release the package?

RECOGNIZE THE PRINCIPLE

For this projectile to hit the target, the pilot must release the package *before* she is directly over the island. We first compute how long it takes the package to fall a distance h and land on the island. This answer will give us the time the package spends in the air. We can then calculate how far along x the package travels during this time, and that will tell us how far in advance of the island the pilot should release the package.

SKETCH THE PROBLEM

The trajectory of the package is sketched in Figure 4.17. Just after its release (Fig. 4.17A), the package has a velocity along the horizontal equal to the speed of the airplane. Since the package is a projectile, the horizontal (x) component of its velocity is constant while the package falls, so the velocity of the package along the x direction is always the same as the airplane's velocity. The package is therefore always directly under the airplane as it falls to the island as shown in Figure 4.17. This example again shows the independence of the horizontal and vertical motions for a projectile.

IDENTIFY THE RELATIONSHIPS AND SOLVE

(a) The time it takes the package to reach the ground can be calculated from the displacement along y:

$$y = y_0 + v_{0y}t + \tfrac{1}{2}a_y t^2$$

The initial height of the package is $y_0 = h$, and at the instant the package is released, $v_{0y} = 0$. The package's acceleration is $a_y = -g$. We want to find when the package reaches $y = 0$, so we have

$$y = h - \tfrac{1}{2}gt_{lands}^2 = 0$$

Solving for the time gives

$$\boxed{t_{\text{lands}} = \sqrt{\frac{2h}{g}}}$$

(b) During the time the package is in the air, the airplane and the package both travel a horizontal distance

$$x = x_0 + v_{0x}t = v_{\text{plane}}t_{\text{lands}} = \boxed{v_{\text{plane}}\sqrt{\frac{2h}{g}}}$$

▸ ***What does it mean?***
The pilot should release the package when she is this distance (horizontally) from the island.

4.3 A First Look at Reference Frames and Relative Velocity

In Example 4.6, we considered the motion of the airplane and the falling package from the point of view of an observer on the ground. According to such an observer, the airplane is moving along the $+x$ direction with a velocity $\vec{v}_{\text{plane}}$ and the package moves along the parabolic path shown in Figure 4.17. Another way to view the problem is from the point of view of the pilot. According to her, the package simply falls straight down to Earth below (Fig. 4.18), while the island is moving along $-x$; that is, the island moves (!) relative to the airplane. Although the trajectories viewed by these two observers are different, the time the package takes to fall is the *same* from either point of view.

This is an example of using two different reference frames. A ***reference frame*** is an observer's choice of coordinate system for making measurements. In this example, one reference frame is at rest relative to the ground and the other reference frame (containing the airplane and its pilot) is moving with a constant velocity relative to the ground. Newton's laws give a correct description of the motion in any reference frame that moves with a constant velocity. We discuss what happens in other types of reference frames in Section 4.5.

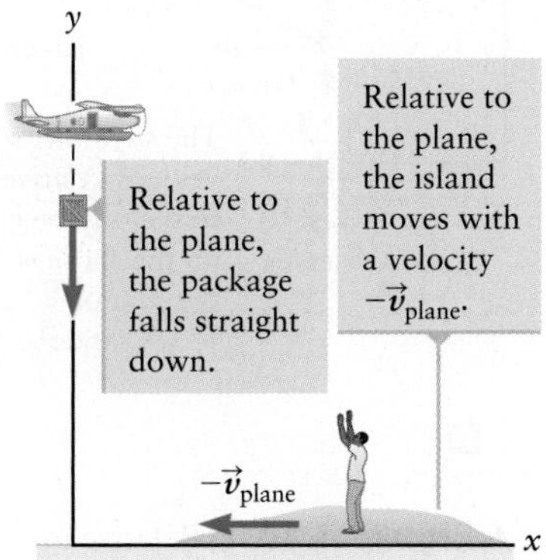

▲ **Figure 4.18** As viewed by the pilot of the airplane, the package falls directly downward because the package is always directly beneath the airplane. When viewed in the pilot's reference frame, the island is moving with a velocity $-\vec{v}_0 = -\vec{v}_{\text{plane}}$.

Relative Velocity

When we discuss or calculate the velocity of an object such as the package in Figures 4.17 and 4.18, we are always considering the velocity *relative* to a particular coordinate system or observer; that is, we consider the velocity in a particular reference frame. So, given $\vec{v}$ in one reference frame, how do we find the velocity in a different reference frame? Let's first consider this problem for motion in one dimension.

Figure 4.19A shows two cars traveling at constant velocities along the x direction. According to an observer (and reference frame) at rest on the sidewalk, the cars have velocities $\vec{v}_1$ and $\vec{v}_2$. Now consider a reference frame at rest with respect to car 1; this reference frame is defined by the coordinate axes x' and y' in Figure 4.19B that "travel along" with car 1. What is the velocity of car 2 in this new reference frame with axes x' and y'? The velocity of car 2 *relative to car 1* is just $\vec{v}_2' = \vec{v}_2 - \vec{v}_1$. Furthermore,

$$\text{velocity of car 2 relative to observer} = \text{velocity of car 2 relative to car 1} + \text{velocity of car 1 relative to observer}$$

This is the general way velocities in different reference frames—that is, the velocities seen by different observers—are related.

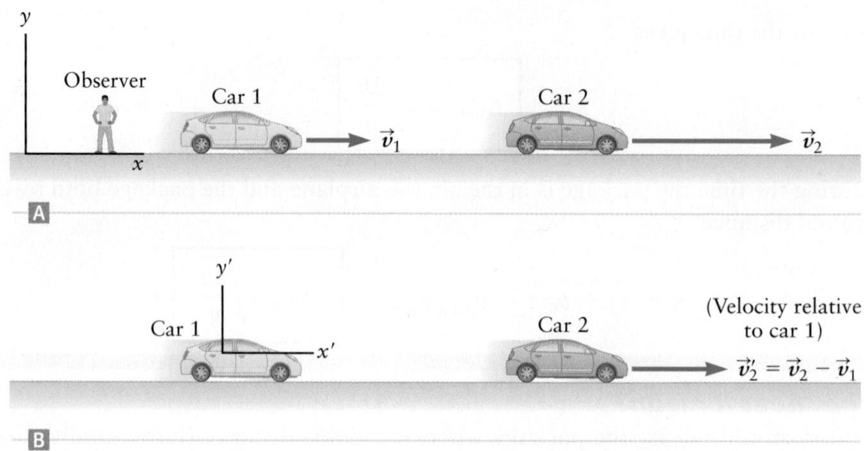

▶ **Figure 4.19** A Two cars moving along a straight road (i.e., in one dimension). B The velocity of car 2 relative to car 1 is $\vec{v}_2 - \vec{v}_1$.

The same ideas concerning relative velocities apply for motion in two dimensions. Figure 4.20 shows two people playing in a river: one is floating along with the current, while the other is swimming from one riverbank to the other. Relative to an observer on the shore, the water in this river moves at a velocity $\vec{v}_{\text{current}}$ along the x direction, and this current affects the motion of the two people in the river. The person floating along with the current (Fig. 4.20A) moves with a velocity equal to the velocity of the water. In terms of the velocity components in Figure 4.20A,

$$v_{\text{floater}, x} = v_{\text{current}}$$
$$v_{\text{floater}, y} = 0$$

The person in Figure 4.20B is swimming so as to travel across the river. Let's assume this swimmer is moving relative to the water with a velocity of magnitude v_0 in the y direction. Since the swimmer is carried along with the current, she will also have a nonzero velocity along x, and this velocity component will be equal to v_{current}. In words,

$$\begin{pmatrix}\text{velocity of swimmer relative} \\ \text{to observer on shore}\end{pmatrix} = \begin{matrix}\text{velocity of swimmer relative to water} \\ + \text{ velocity of water relative to observer}\end{matrix}$$

In terms of components, we have

$$v_{\text{swimmer}, x} = v_{\text{current}} \tag{4.29}$$
$$v_{\text{swimmer}, y} = v_0$$

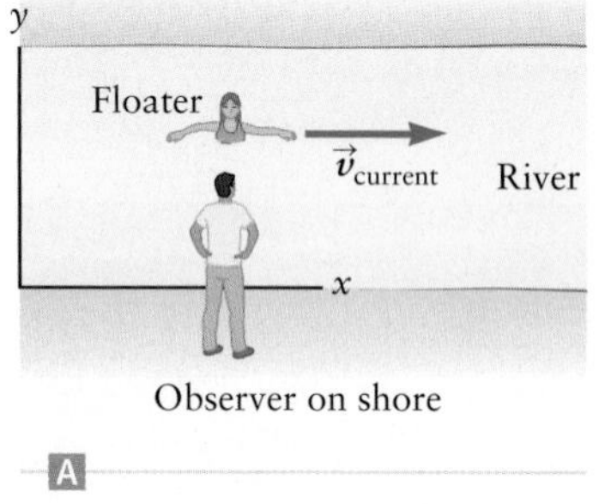

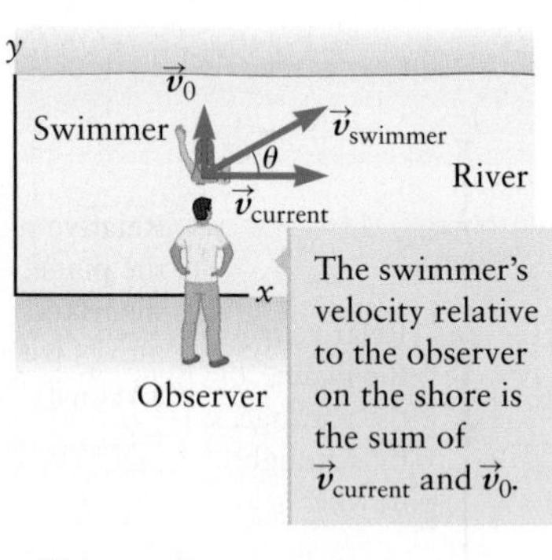

▲ **Figure 4.20** A Relative to an observer on the shore, a person floating along in a river has a velocity equal to the velocity of the river's current, $\vec{v}_{\text{current}}$. B As viewed by the observer on the shore, this swimmer's velocity is equal to the sum of the swimmer's velocity relative to the water plus the velocity of the river's current.

According to an observer on the shore in Figure 4.20B, the swimmer's velocity is the sum of the two components in Equation 4.29, so the speed of the swimmer is

$$v_{\text{swimmer}} = \sqrt{v_{\text{swimmer}, x}^2 + v_{\text{swimmer}, y}^2} = \sqrt{v_{\text{current}}^2 + v_0^2} \tag{4.30}$$

Moreover, *relative to this observer*, the swimmer does not travel directly across the river, but instead moves at an angle θ. From Figure 4.20B, this angle is given through the relation

$$\tan\theta = \frac{v_{\text{swimmer}, y}}{v_{\text{swimmer}, x}} \tag{4.31}$$

The swimmer's velocity is thus different for an observer on the shore (Eqs. 4.30 and 4.31) than for the observer who floats along with the water current in Figure 4.20A.

EXAMPLE 4.7 Swimming Across a River

Suppose the river's current in Figure 4.21 has a speed $v_{current} = 2.0$ m/s. To travel directly across the river *as viewed by an observer on the shore*, at what velocity (along the direction of the river) should the person swim?

RECOGNIZE THE PRINCIPLE

Let the components of the swimmer's velocity *relative to the water* be v_x and v_y (see Fig. 4.21). The velocity of the swimmer relative to an observer on the shore is equal to the velocity of the swimmer relative to the water *plus* the velocity of the water relative to the observer. These velocity vectors are shown in Figure 4.21, which also shows their vector sum. Relative to the observer on the shore, we thus have

$$v_{swimmer,\,x} = v_{current} + v_x \tag{1}$$

$$v_{swimmer,\,y} = v_y$$

To swim directly across the river, we must have $v_{swimmer,x} = 0$.

SKETCH THE PROBLEM

Figure 4.21 shows the problem, including the components of the swimmer's velocity.

IDENTIFY THE RELATIONSHIPS AND SOLVE

Inserting $v_{swimmer,x} = 0$ into Equation (1), we find

$$v_{current} + v_x = 0$$

$$v_x = -v_{current}$$

Using the given value of $v_{current}$, we get

$$v_x = \boxed{-2.0 \text{ m/s}} \tag{2}$$

► ***What does it mean?***

The negative sign in Equation (2) means that the swimmer must swim against the current. Notice that the value of v_y is not important; as long as it is not zero, the swimmer will be able to swim across the river.

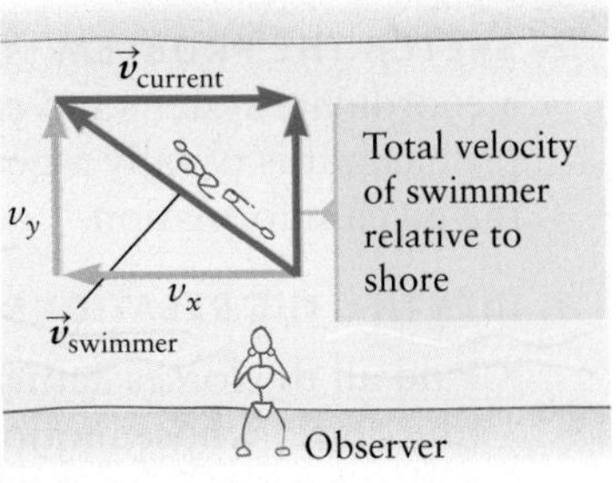

▲ **Figure 4.21** Example 4.7.

4.4 Further Applications of Newton's Laws

We have now worked through a number of applications of Newton's second law, both in one dimension (in Chapter 3) and in our work on projectile motion in Section 4.2. With that experience, we can formulate a general strategy for attacking such problems. We will apply this problem-solving strategy as we explore a number of additional applications of Newton's laws in the rest of this chapter.

PROBLEM SOLVING with Newton's Second Law

1. **RECOGNIZE THE PRINCIPLE.** To use Newton's second law to find the acceleration, velocity, and position of an object, we need to consider all the forces acting on an object and compute the total force. For an object in static equilibrium, the total force must be zero; now, for an object that is not in equilibrium, the total force is equal to $m\vec{a}$.

(continued) ►

2 **SKETCH THE PROBLEM.** Your picture should define a coordinate system and contain all the forces in the problem. It is usually a good idea to also show all the given information.

3. **IDENTIFY THE RELATIONSHIPS.**

- Find all the forces acting on the object of interest (the object whose motion you wish to describe) and construct a free-body diagram.
- Express all the forces in terms of their components along x and y.
- Apply Newton's second law (Eq. 4.1) in component form:

$$\sum F_x = ma_x \quad \text{and} \quad \sum F_y = ma_y$$

- If the acceleration is constant along x or y (or both), you can apply the kinematic equations from Table 3.1.

4. **SOLVE.** Solve the equations resulting from step 3 for the unknown quantities in terms of the known quantities. The number of equations must equal the number of unknown quantities.

5. Always *consider what your answer means* and check that it makes sense.

Traveling Down a Hill

A typical problem involving Newton's second law is sketched in Figure 4.22, which shows a sled traveling down a snowy hillside. If the sled starts from rest at the top of the hill, how long does it take to reach the bottom? For simplicity, we ignore friction between the sled and the snow and assume the hill makes a constant angle θ with the horizontal. There are then only two forces acting on the sled, the force of gravity and the normal force exerted by the hill's surface.

Following our problem-solving strategy, we next add a coordinate system to our diagram. We know that the sled's motion is along the hill, so we choose x to be directed parallel to the hill, with x increasing downward. The y axis is then perpendicular to the hill's surface. We next draw a free-body diagram (Fig. 4.22B), which shows the components of the forces along x and y. The normal force is already along y, but the force of gravity on the sled has components along both x and y. Using some trigonometry, we can get those components (Fig. 4.22B). We now apply Newton's second law for motion along x:

$$\sum F_x = mg \sin \theta = ma_x \tag{4.32}$$

Applying Newton's second law along y, we have

$$\sum F_y = N - mg \cos \theta = ma_y \tag{4.33}$$

Since the sled is moving purely along the x direction, the acceleration along y must be zero. Hence, $a_y = 0$, and we could use Equation 4.33 to calculate the magnitude of the normal force, N. However, in this particular problem we are interested in the motion along the hillside (along x), so all we need is the acceleration a_x. We can get this acceleration by solving Equation 4.32:

$$a_x = g \sin \theta \tag{4.34}$$

This acceleration is constant, and we can use our relation for motion with constant acceleration:

$$x = x_0 + v_{0x}t + \tfrac{1}{2}a_x t^2$$

The sled starts at $x_0 = 0$ and begins from rest, so $v_{0x} = 0$. Inserting the acceleration from Equation 4.34 gives

$$x = \tfrac{1}{2}a_x t^2 = \tfrac{1}{2}g(\sin \theta)t^2 \tag{4.35}$$

If the hill has a vertical height h, the distance to the bottom of the hill as measured along the slope is $h/\sin \theta$ (see Fig. 4.22A). Inserting this into Equation 4.35, we find

$$x = \tfrac{1}{2}g(\sin \theta)t^2 = \frac{h}{\sin \theta}$$

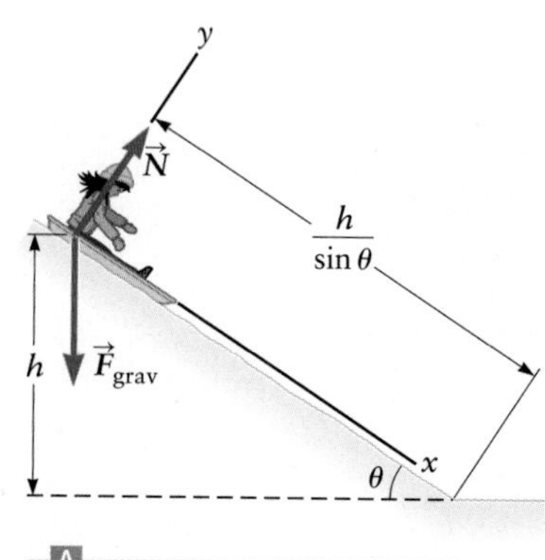

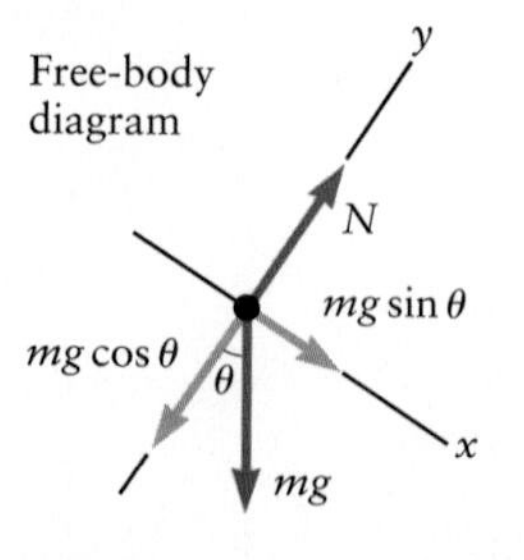

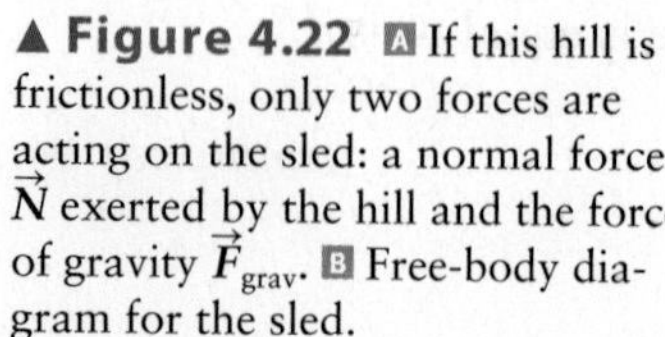
▲ **Figure 4.22** A If this hill is frictionless, only two forces are acting on the sled: a normal force $\vec{N}$ exerted by the hill and the force of gravity $\vec{F}_{grav}$. B Free-body diagram for the sled.

when the sled is at the bottom of the hill. The time it takes to reach the bottom is then

$$t_{\text{bottom}} = \frac{\sqrt{2h/g}}{\sin\theta} \tag{4.36}$$

EXAMPLE 4.8 Sledding Down the Hill

For the sled in Figure 4.22, find the velocity when it reaches the bottom of the hill.

RECOGNIZE THE PRINCIPLE

We wish to find the components of the sled's velocity along the x and y directions as defined by the coordinate axes in Figure 4.22. Here, the x direction is parallel to the hill and y is perpendicular. From Newton's second law, we can find the acceleration along x (Eq. 4.34) and then apply the relations for motion with a constant acceleration (Table 3.1) to get the velocity.

SKETCH THE PROBLEM

Figure 4.22 describes the problem.

IDENTIFY THE RELATIONSHIPS

The sled is moving along the hill (along x), so the component of the velocity along y (i.e., perpendicular to the hill) is zero; that is,

$$\boxed{v_y = 0}$$

We already calculated the acceleration along x in Equation 4.34, where we found $a_x = g\sin\theta$. The acceleration is constant, so the velocity along x is again given by one of our relations for motion with constant acceleration,

$$v_x = v_{0x} + a_x t$$

Inserting the result for a_x from Equation 4.34 gives

$$v_x = v_{0x} + a_x t = g(\sin\theta)t \tag{1}$$

SOLVE

The sled reaches the bottom at t_{bottom}, which we found in Equation 4.36. Inserting t_{bottom} into Equation (1) leads to

$$v_{x,\,\text{bottom}} = g(\sin\theta)t_{\text{bottom}} = g\sin\theta\,\frac{\sqrt{2h/g}}{\sin\theta} = \boxed{\sqrt{2gh}}$$

► ***What does it mean?***

While the speed at the bottom depends on the height h of the hill, it is *independent* of the angle θ of the hill. We show in Chapter 6 that this is a general property of the gravitational force and is connected with the notion of potential energy.

EXAMPLE 4.9 Towing a Car

Your car breaks down, and you ask a friend to help you move it to the nearest repair shop. You need to move your car up a hill, so your friend is going to use his car to pull your vehicle using a rope as shown in Figure 4.23A. If the angle of the hill is $\theta = 5.0°$ and you pull your car ($m = 1200$ kg) with a constant velocity, what is the tension in the rope? Assume all friction can be neglected.

(continued) ►

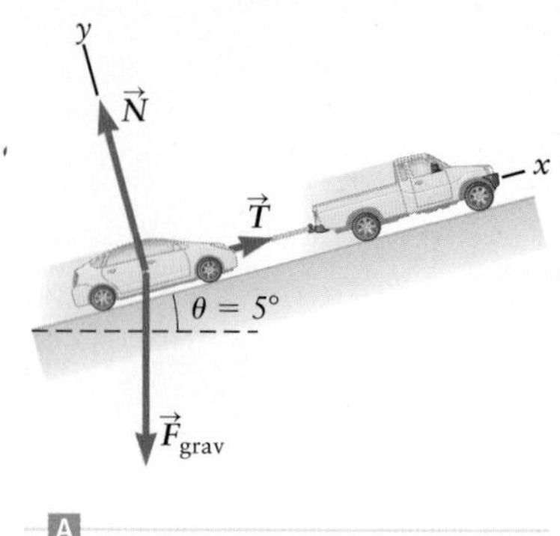

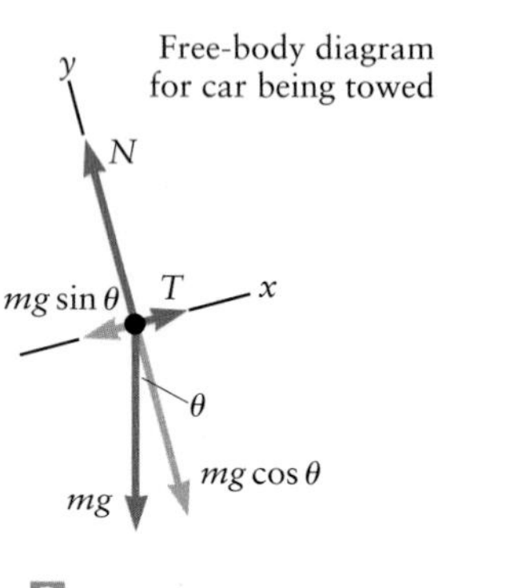

▲ **Figure 4.23** Example 4.9. A Three forces are acting on this car: a normal force $\vec{N}$ exerted by the road, the force of gravity $\vec{F}_{grav}$, and a force due $\vec{T}$ to the cable. B Free-body diagram for the towed car.

RECOGNIZE THE PRINCIPLE

We can use Newton's second law to calculate the motion of the cars (Fig. 4.23A). Since the cars are moving with constant velocity, the acceleration along the hill (the x direction) is $a_x = 0$; hence, the sum of all the forces along the hill must be zero. From this we can solve for the tension T in the rope.

SKETCH THE PROBLEM

Figure 4.23B shows a free-body diagram for the car being towed. The forces acting on it are the tension from the rope, the force of gravity, and the normal force exerted by the hill's surface. The figure also shows the components of these forces along the hill and perpendicular to the hill. The $+x$ direction is along the hill's surface and up the incline, while y is perpendicular to x.

IDENTIFY THE RELATIONSHIPS

Writing Newton's second law for motion along the x direction (along the hill), we have

$$\sum F_x = -mg \sin\theta + T = ma_x$$

Since the velocity is constant, a_x is zero, which leads to

$$-mg \sin\theta + T = 0$$

$$T = mg \sin\theta$$

SOLVE

Inserting the given values of m and θ gives

$$T = (1200 \text{ kg})(9.8 \text{ m/s}^2)\sin(5.0°) = \boxed{1000 \text{ N}}$$

▶ ***What does it mean?***

This tow rope only needs to withstand 1000 N (about 250 lb) to pull the car at a 5.0° angle without breaking, which is much less than the weight of the car.

EXAMPLE 4.10 Sliding Up a Hill

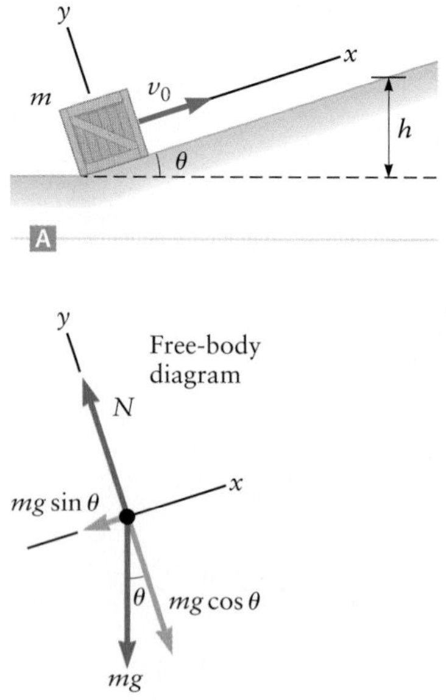

▲ **Figure 4.24** Example 4.10.

Consider a box of mass m that slides *up* a hill as sketched in Figure 4.24A. The hill is frictionless and makes an angle θ with the horizontal, and the box is given an initial speed $v_0 = 5.0$ m/s. What is the maximum height reached by the box?

RECOGNIZE THE PRINCIPLE

Following our problem-solving strategy for Newton's second law, we begin by noting that since the hill is frictionless, there are only two forces on the box, the force of gravity (magnitude mg) and the normal force (N) from the hill. These are the forces we will use in Newton's second law to find the acceleration, velocity, and position of the box. When the box reaches its maximum height, its velocity will be zero.

SKETCH THE PROBLEM

Figure 4.24 shows the problem, including a free-body diagram. We choose a coordinate system with y perpendicular to the hill and x along it, with the $+x$ direction being uphill.

IDENTIFY THE RELATIONSHIPS

We use the free-body diagram and the components of the forces along x and y in Figure 4.24B to write Newton's second law (Eq. 4.1) for both the x and y direction:

$$\sum F_x = -mg \sin\theta = ma_x \qquad (1)$$

$$\sum F_y = N - mg \cos\theta = ma_y$$

We are only concerned with motion along the x direction and can get the acceleration along x from Equation (1):

$$a_x = -g \sin \theta \tag{2}$$

This acceleration is constant, so we can apply our usual relations from Table 3.1. We want to find the displacement when the velocity v_x of the box is zero, and the relation containing both of these quantities is

$$v_x^2 = v_0^2 + 2a_x(x - x_0) \tag{3}$$

with a_x given in Equation (2).

SOLVE

The block starts at the origin, so $x_0 = 0$. The velocity of the box will be zero when it reaches its highest point; setting $v_x = 0$ in Equation (3) and solving for the displacement, we find

$$0 = v_0^2 + 2a_x(x - 0)$$

$$x = -\frac{v_0^2}{2a_x} = \frac{v_0^2}{2g \sin \theta}$$

If the maximum height of the block is h (Fig. 4.24A), we can use some trigonometry to write

$$\frac{h}{x} = \sin \theta$$

and solving for h gives

$$h = x \sin \theta = \frac{v_0^2}{2g \sin \theta} \sin \theta = \frac{v_0^2}{2g}$$

$$h = \frac{v_0^2}{2g} = \frac{(5.0 \text{ m/s})^2}{2(9.8 \text{ m/s}^2)} = \boxed{1.3 \text{ m}}$$

▶ ***What does it mean?***

The maximum height of the block depends only on its initial speed and is independent of the angle θ. In fact, the maximum height here is the same as for a projectile thrown straight up! We will explain why that happens when we consider mechanical energy in Chapter 6.

Adding the Frictional Force

In the previous examples involving motion along a slope, we have assumed friction is negligible. Now let's see how to include it. Returning to our problem of a sled sliding down a snowy hill (Fig. 4.22), we add friction between the sled and the hill. The revised free-body diagram is shown in Figure 4.25, which contains a frictional force of magnitude $F_{\text{friction}} = \mu_K N$ directed up the incline (along $-x$). This force is kinetic friction because the sled is slipping relative to the hill. To calculate the frictional force, we must first find the normal force N, which we can get from Equation 4.33:

$$\sum F_y = N - mg \cos \theta = ma_y$$

Since the acceleration along y (perpendicular to the hill) is zero,

$$\sum F_y = N - mg \cos \theta = 0$$

$$N = mg \cos \theta \tag{4.37}$$

We next apply Newton's second law for motion along the hill:

$$\sum F_x = mg \sin \theta - F_{\text{friction}} = mg \sin \theta - \mu_K N = ma_x$$

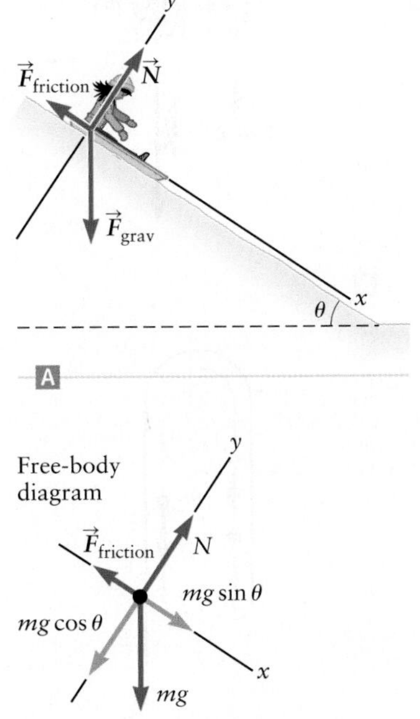

▲ **Figure 4.25** For this sled, there is a frictional force $\vec{F}_{\text{friction}}$ from the sled slipping on the snow. Compare with Figure 4.22.

The acceleration down the incline thus depends on the normal force. Using the result for the normal force from Equation 4.37 leads to

$$a_x = g\sin\theta - \frac{\mu_K N}{m} = g\sin\theta - \frac{\mu_K mg\cos\theta}{m} = g(\sin\theta - \mu_K\cos\theta) \quad (4.38)$$

Let's now estimate quantitatively the effect of friction. Suppose we have a hill with $\theta = 10°$, which is a typical value for a reasonably steep sledding hill. Without friction, $\mu_K = 0$ in Equation 4.38 and we get $a_x = g\sin\theta = 1.7\ \text{m/s}^2$. A typical value for the coefficient of kinetic friction between a wooden ski and snow is $\mu_K \approx 0.05$, and using this value in Equation 4.38 we find $a_x \approx 1.2\ \text{m/s}^2$. Hence, friction reduces the acceleration (and therefore also the speed at any moment) by about 30%.

Pulleys and Cables

In Chapter 3, we found that pulleys could be used to change the direction of a force and also to amplify forces. Let's now consider the associated motion in more detail. Figure 4.26 shows a common situation of two crates connected by a rope that passes over a pulley. For simplicity, we assume the rope is very light and the pulley is frictionless and massless. We want to find how the crates move; that is, we want to calculate their acceleration.

As usual, we need to apply Newton's second law. The forces on each crate are shown in Figure 4.26. Recall that for an ideal (massless) string, the tension is the same throughout the string, so the magnitudes of the tension forces acting on the two crates are the same. We next need to establish a coordinate system, and for this problem the choice requires some thought. Because the two crates are connected by the string, they must move together; otherwise, the string would break. Therefore, the magnitudes of their displacements, velocities, and accelerations must be the same. With that in mind, we choose the x coordinate direction to be *parallel to the string* as sketched in Figure 4.26B. This coordinate axis *follows the string* as it loops around the pulley. It may be useful to think of this coordinate "axis" as a sort of "railroad track" along which the two crates move. If we take the "positive" direction for this track to be upward for m_1, it will then be downward for m_2 as shown in the figure. Although the "positive" directions are thus opposite for the two crates, the motion is still essentially one dimensional, so we can use our familiar results for problems involving one-dimensional motion. This choice of a coordinate axis whose direction is different for the two crates will make our mathematical solution of Newton's second law a bit simpler because now the acceleration of the two crates will be the same from the start.

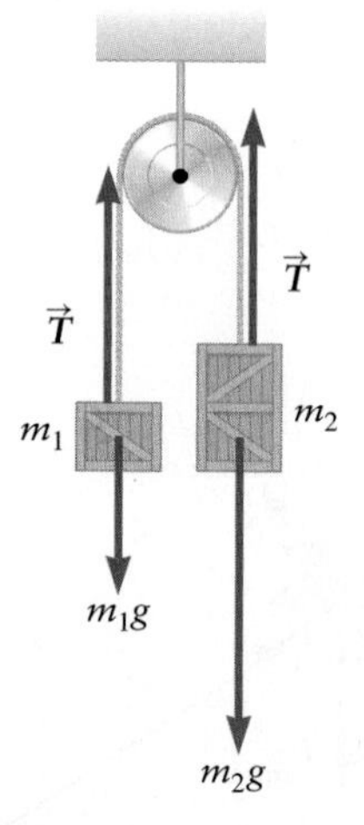

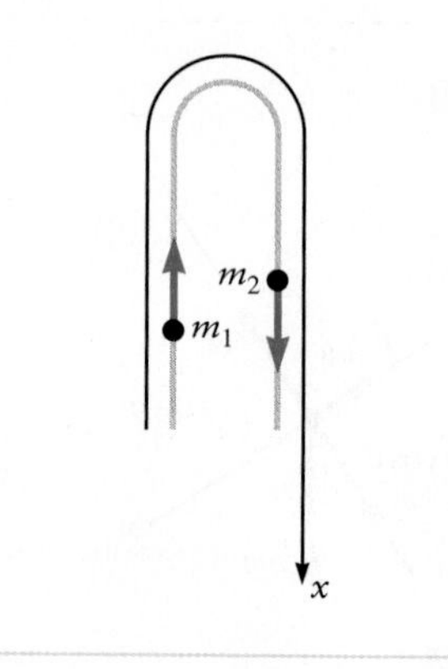

▲ **Figure 4.26** A Forces on two crates connected by a rope-and-pulley system. B The coordinate axis (the x axis) follows the string as it bends around the pulley.

To find the acceleration, we first write Newton's second law for m_1. Since the forces are all along the string direction x, the acceleration will also be along this direction. We denote the total force on m_1 as $\sum F_1$, so Newton's second law for this crate reads

$$\sum F_1 = m_1 a = +T - m_1 g \quad (4.39)$$

where we have written the acceleration as simply a (because the acceleration is the same for m_1 and m_2). The forces on the right side of Equation 4.41 are $+T$ (since the tension force is directed along the $+x$ direction for m_1) and $-m_1 g$ (since the force of gravity is along $-x$). For m_2,

$$\sum F_2 = m_2 a = -T + m_2 g \quad (4.40)$$

Here the tension term is $-T$ because the tension force on m_2 is directed along $-x$ for this crate. The sign is reversed with respect to Equation 4.39 because the x direction is reversed for m_2. We now have two equations and two unknowns—the acceleration and the tension—so we can solve for both a and T. To solve for the acceleration, we add Equations 4.39 and 4.40, which eliminates T and gives

$$m_1 a + m_2 a = +T - m_1 g - T + m_2 g$$
$$(m_1 + m_2)a = -m_1 g + m_2 g$$
$$a = \frac{(m_2 - m_1)g}{m_1 + m_2} \quad (4.41)$$

To solve for the tension, we insert this result for a back into Equation 4.39 to find

$$T = m_1 a + m_1 g = m_1(a + g)$$

$$T = m_1\left[\frac{(m_2 - m_1)g}{m_1 + m_2} + g\right] = m_1 g\left[\frac{m_2 - m_1 + m_2 + m_1}{m_1 + m_2}\right]$$

$$T = \frac{2m_1 m_2 g}{m_1 + m_2}$$

If you are uneasy about our choice of the x direction in this problem, look again at Equation 4.41. There we see that if $m_2 > m_1$, the acceleration is positive and the block with the larger mass (m_2) moves downward as expected. On the other hand, if $m_2 < m_1$, Equation 4.41 tells us that the acceleration is negative, which simply means that m_2 now accelerates upward, again as we should expect. The *physical answer* does not depend on how we choose the "positive" direction.

EXAMPLE 4.11 More Crates and a Pulley

A crate sits on a frictionless table and is connected to a second crate by a string that passes over a pulley as shown in Figure 4.27. If the pulley is frictionless and massless and the string is also massless, find the acceleration of the crates and the tension in the string.

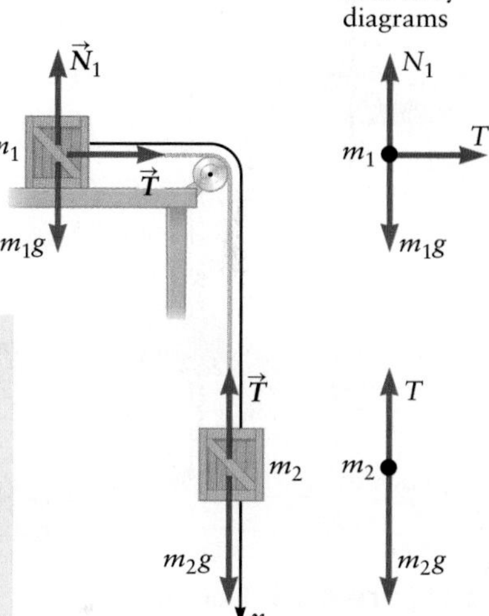

▲ **Figure 4.27** Example 4.11. Another arrangement of crates and a pulley. We assume there is no frictional force between the table and crate 1. Notice that we have chosen the x direction to follow the string.

RECOGNIZE THE PRINCIPLE

We follow our problem-solving strategy for applying Newton's laws. To find the acceleration of the crates, we must consider all the forces acting on them. Our sketch (Fig. 4.27) shows all these forces. We write Newton's second law for each crate and solve for the unknown quantities.

SKETCH THE PROBLEM

Figure 4.27 shows the forces acting on both crates and indicates our choice of coordinate system. As in Figure 4.26, we take the x direction to follow the string around the pulley. That is, we choose the $+x$ direction to be horizontal and to the right for the crate on the table, and down along the string for the crate that is suspended in midair.

IDENTIFY THE RELATIONSHIPS

Using the forces shown in Figure 4.27, Newton's second law for the crate on the table reads

$$\sum F_x = m_1 a = +T \tag{1}$$

For the crate that hangs from the string, Newton's second law gives us

$$\sum F_x = m_2 a = m_2 g - T \tag{2}$$

SOLVE

To find the acceleration, we eliminate T by adding Equations (1) and (2):

$$m_1 a + m_2 a = +T + m_2 g - T = m_2 g$$

$$a = \boxed{\frac{m_2 g}{m_1 + m_2}}$$

Solving for T then gives, from Equation (1),

$$m_1 a = +T$$

$$T = m_1 a = \boxed{\frac{m_1 m_2 g}{m_1 + m_2}}$$

(continued) ▶

▶ What does it mean?

Our approach to this problem follows our usual pattern. We start with a picture and construct free-body diagrams for the objects of interest (the two crates). We then write Newton's second law for each crate and solve for the unknowns (a and T).

EXAMPLE 4.12 Crates with Friction

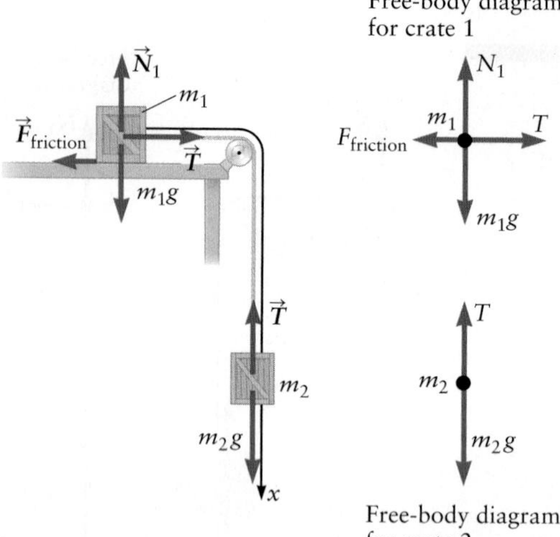

▲ **Figure 4.28** Example 4.12. We add friction to the system in m Figure 4.27, so there is now a frictional force between crate 1 and the table.

Suppose there is a frictional force between the crate m_1 and the table in Example 4.11 (Fig. 4.27). Write the new Newton's second law equation for this crate's acceleration along the horizontal direction.

RECOGNIZE THE PRINCIPLE

We follow the same approach as taken in Example 4.11. Now our sketch (Fig. 4.28) contains an additional force due to friction on crate 1. This frictional force has a magnitude of $\mu_K N_1$, where N_1 is the normal force between crate 1 and the table.

SKETCH THE PROBLEM

Figure 4.28 shows the new situation along with free-body diagrams for both crates. This frictional force acts in the $-x$ direction since it opposes the motion of crate 1.

IDENTIFY THE RELATIONSHIPS AND SOLVE

When we add the frictional force, Newton's second law for the crate on the table becomes

$$\boxed{\sum F_x = m_1 a = +T - \mu_K N_1}$$

▶ What does it mean?

In this problem we have an additional unknown, the normal force N_1. This normal force can be found by considering the forces on m_1 along the y direction, leading to $N_1 = m_1 g$.

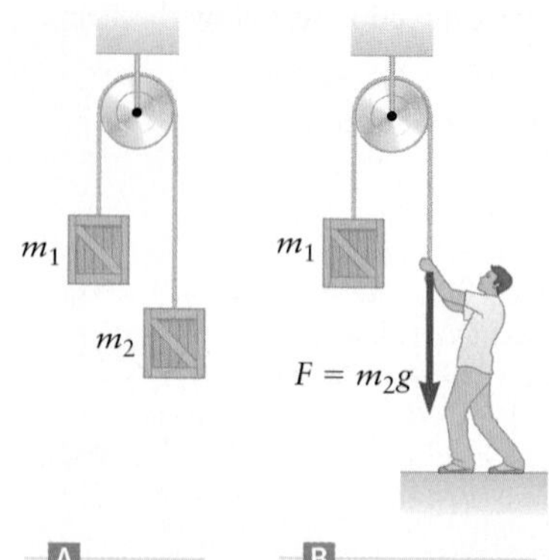

▲ **Figure 4.29** Concept Check 4.4. Do these two systems have the same acceleration?

CONCEPT CHECK 4.4 Tension in a String

Consider the two pulley arrangements in Figure 4.29. In both cases, a crate of mass m_1 is attached to one end of a string. In Figure 4.29A, the other end of the string is tied to a crate of mass m_2, so there is a force from gravity of magnitude $F_{2,\,grav} = m_2 g$ acting on this crate. In Figure 4.29B, crate m_2 is replaced by a person who pulls on the string, and we suppose he pulls with a force *equal* to $F_{2,\,grav} = m_2 g$. Will the acceleration of m_1 be the same in these two cases?

4.5 Detecting Acceleration: Reference Frames and the Workings of the Ear

Suppose you are traveling in an airplane initially flying at a constant horizontal velocity and the pilot decides to accelerate slightly. You immediately sense this acceleration. How did you do it? The human ear contains structures that act as sensitive acceleration sensors, and we can understand how they work using Newton's second law.

Before we consider the workings of an actual ear, let's first consider a simple mechanical device you could build to detect and measure an airplane's acceleration. Our device, shown in Figure 4.30, consists of a rock tied to one end of a string, with the other end of the string fastened to the airplane's ceiling. When the acceleration is zero, the string hangs vertically, which is consistent with a Newton's law analysis. Two forces act on the rock—the force of gravity and the tension force from the string—and in Figure 4.30A these forces are both along the vertical (y). The acceleration along y is zero, so these forces must add to zero with the result $T = mg$. There are no forces along the horizontal; the acceleration along x is zero, and the rock moves with a constant velocity.

When the airplane is accelerating along the horizontal (x), the rock will have the same acceleration as the airplane and the string hangs at an angle θ as sketched in Figure 4.30B. The value of θ depends on the airplane's acceleration. Writing Newton's laws for motion along x and y leads to (Fig. 4.30C)

$$\sum F_x = ma_x = T\sin\theta$$

$$\sum F_y = ma_y = T\cos\theta - mg = 0 \quad (4.42)$$

since the acceleration along y is again zero. We can use this relation for a_y to find the tension in the string,

$$T\cos\theta = mg$$

$$T = \frac{mg}{\cos\theta}$$

and then get the acceleration along x:

$$a_x = \frac{T\sin\theta}{m} = \frac{\left(\dfrac{mg}{\cos\theta}\right)\sin\theta}{m} = g\tan\theta \quad (4.43)$$

Whenever the airplane has a nonzero acceleration, the angle θ is nonzero and the string hangs at an angle from the vertical. By measuring this angle, we can use Equation 4.43 to find the value of the acceleration a_x.

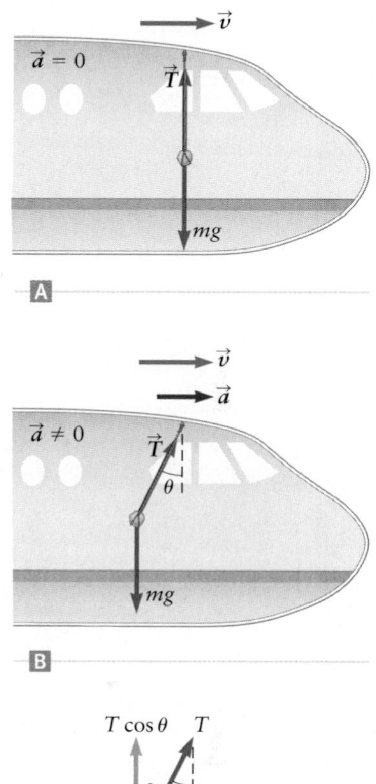

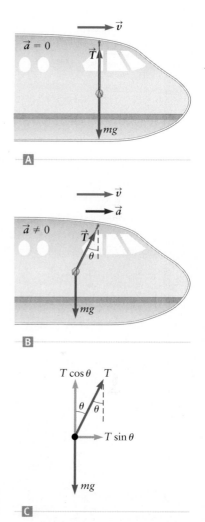

▲ **Figure 4.30** A rock hanging by a string acts as an accelerometer, a device that measures acceleration. A When the system (here an airplane cabin) moves with a constant velocity, the string hangs vertically. B When the system is accelerating, the string hangs at a nonzero angle θ, and the value of θ depends on the acceleration. C Vertical and horizontal components of the forces on the rock.

CONCEPT CHECK 4.5 Effect of Mass on an Accelerometer

How will the angle of the accelerometer string in Figure 4.30 depend on the *mass* of the rock?

(a) The angle is larger for a rock with a larger mass.
(b) The angle is independent of the mass of the rock.
(c) The angle is smaller for a rock with a larger mass.

The Accelerometer in Your Ear

Your ears contain a structure that acts very much like the simple accelerometer in Figure 4.30. A region within the ear called the utricle contains hair cells that project up into a gelatinous layer filled with a very thick fluid (Fig. 4.31). This gelatinous layer also contains some small masses called otoliths (also known as "ear stones"). When your head—that is, the utricle—accelerates along the horizontal, the gelatinous layer lags behind a small amount in the same way the string in Figure 4.30B hangs behind when the airplane accelerates. This displacement of the gelatinous layer causes the hair cells to deflect (just like the string in Fig. 4.30B), and the hair cells send signals to the brain when they are bent in this way. Hence, the brain is "told" that the ear is being accelerated. Although the ear is certainly more complex than a rock on a string, the accelerometers in Figures 4.30 and 4.31 are remarkably similar.

Inertial Reference Frames

Let's analyze the accelerometer in Figure 4.30 once more, this time from a slightly different point of view. In Equation 4.42, we applied Newton's second law from the point of view of an observer outside the airplane, a person who might have viewed the rock and string through a window as the airplane flew past. Now imagine how things look to an observer sitting inside the airplane. When the acceleration is zero and the string hangs vertically as in Figure 4.30A, an observer inside the airplane would see two forces acting vertically on the rock: gravity and the tension in the string. This observer would say that the acceleration along y is zero and hence that $T = mg$. The observer inside the airplane would also say that there is no force along x and there is also no acceleration along x, so Newton's second law works just fine. Hence, when the airplane moves with a constant velocity, an observer inside the airplane would be in complete agreement with the observer outside the airplane. Both would find that Newton's second law describes things perfectly.

The situation is different when the airplane has a nonzero acceleration relative to an observer outside the airplane. Free-body diagrams for this case are shown in Figure 4.32. Our two observers—one outside the airplane and one inside—would draw the same free-body diagrams. However, they would disagree when they use these free-body diagrams to apply Newton's second law:

$$\sum F_x = T \sin \theta = ma_x \tag{4.44}$$

The outside observer would say that the acceleration a_x is nonzero and that Newton's second law works fine since θ is nonzero (Fig. 4.32A). The observer inside the airplane would also see that the string hangs at an angle and hence would conclude that there is a force $T \sin \theta$ along the x direction. According to this observer, though, there is *no acceleration along* x. That is, *as viewed by an observer riding inside the airplane*, the rock is simply at rest; there is no velocity or acceleration along the x direction *relative to this observer*. The observer inside the airplane would therefore say that there is a force along x but no acceleration along x, so the left-hand side of Equation 4.44 is not zero, whereas the right-hand side is zero. Our inside observer would thus be tempted to conclude that Newton's second law does not work inside airplanes!

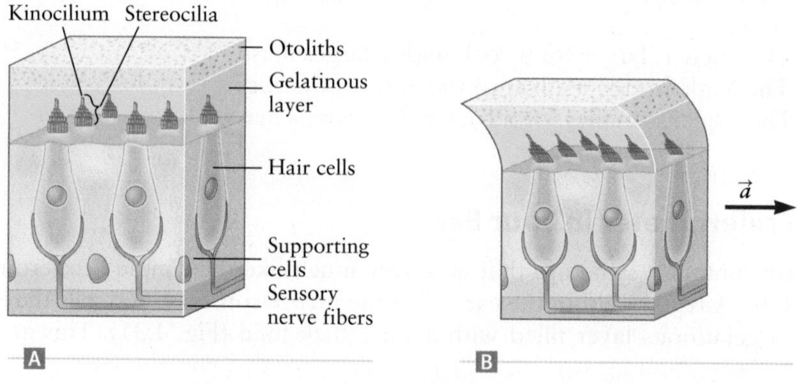

▶ **Figure 4.31** A Hair cells within the utricle in a human ear. These cells act like the accelerometer in Figure 4.30. B When your head is accelerated, the hair cells are deflected from the vertical, producing a signal that is sent to your brain.

The correct interpretation of this situation is that the fault lies not with Newton, but with the observer inside the airplane who is viewing things in an *accelerating reference frame*. Newton's second law can only be applied in nonaccelerated reference frames, also called ***inertial reference frames***. An inertial reference frame can be moving, but it must be moving with a constant velocity relative to other inertial frames. Suppose an observer in one inertial reference frame (frame 1) conducts a test of Newton's second law by measuring the forces on an object and the acceleration of the object. He would then (in agreement with Newton's second law) find that

$$\vec{F}_{\text{measured}} = m\vec{a}_{\text{observed}} \tag{4.45}$$

Now consider a second observer who is moving at a constant velocity relative to frame 1, so his reference frame (frame 2) is an inertial frame. Observer 2 would measure the same horizontal force as observer 1 because he will observe the tension in the string in Figure 4.32 to be the same. Observer 2 will also measure the *same* acceleration as observer 1; since acceleration is the rate of change of velocity, adding the constant velocity of frame 2 does not change the acceleration. Hence, Equation 4.45 would still be satisfied. However, if an observer in a *noninertial* reference frame (frame 3) were to conduct the same test, he would measure the same force but a *different* acceleration. As in the case of the airplane in Figure 4.32, the acceleration in frame 3 would differ from that found in frame 1 by an amount equal to the acceleration of the noninertial frame. Hence, to the observer in frame 3, it would appear that Newton's laws fail. Again, Newton's laws can only be used in inertial reference frames.

The problem of inertial and noninertial reference frames has a deep connection to Newton's laws and to other theories of motion. We'll explore this connection further when we discuss gravitation and relativity in Chapter 27.

Observer outside airplane: inertial frame

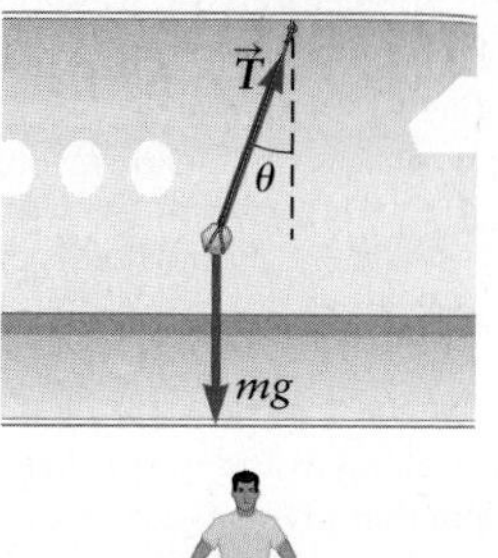

A

Observer inside airplane: noninertial frame

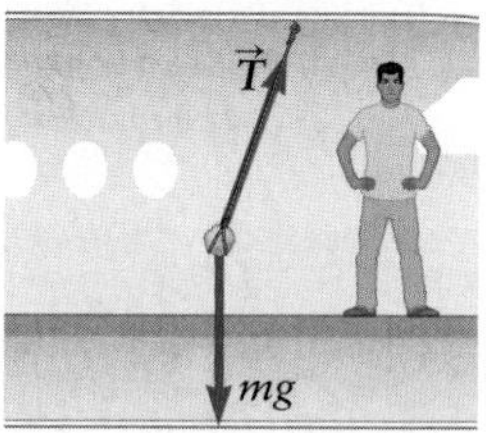

B

▲ **Figure 4.32** A An observer in an inertial reference frame (i.e., an observer at rest, outside the airplane) finds that there is a horizontal component to the total force (because the string hangs at an angle) and also measures that the rock is accelerating, in agreement with Newton's second law. B An observer inside the airplane accelerates along with the rock. According to this noninertial observer, the rock's acceleration is zero since the rock is at rest relative to this observer. However, this observer still finds that there is a horizontal component to the total force on the rock. According to this observer, there is a horizontal force on the rock but no horizontal acceleration, so this observer claims that Newton's second law does not work!

CONCEPT CHECK 4.6 **Newton's Laws and Reference Frames**

Give an example showing that Newton's first law—the principle of inertia—does not hold in a noninertial reference frame.

4.6 Projectile Motion Revisited: The Effect of Air Drag

In Section 4.2, we discussed "ideal" projectile motion in which the only force acting on a projectile is the force of gravity. Real projectiles generally also experience a force due to air drag, and in some cases this drag force can have a very significant effect. Let's consider the drag force in a few cases and estimate its effects.

We begin with the motion of a batted baseball and calculate how far it would travel in the absence of air drag. We worked through this problem in Section 4.2, where we saw that the range of the ball is (Eq. 4.28)

$$x_{\text{range}} = \text{range} = \frac{v_0^2 \sin(2\theta)}{g}$$

We also showed that for a given value of the initial speed v_0, the maximum range occurs when $\theta = 45°$. A skilled baseball player can set a ball into motion with an initial speed of $v_0 = 45$ m/s (about 100 mi/h). Inserting this value into our relation for x_{range} gives a range of 210 m, or about 680 ft. This distance is far longer than a batted baseball has ever traveled. (For a typical baseball field, the distance from home plate to the centerfield fence is about 120 m, and only the longest home runs travel this far.) Our calculation must have omitted something important: the force of air drag.

Insight 4.2

APPLYING NEWTON'S SECOND LAW IN THREE DIMENSIONS

In our applications of Newton's second law, we have dealt with many examples, such as projectiles, that move in two dimensions. For such cases, Newton's second law can be written as

$$\sum F_x = ma_x \quad \text{and} \quad \sum F_y = ma_y$$

and we can deal with forces and acceleration along x and y. To deal with a problem that involves forces or motion in three dimensions, we must also use

$$\sum F_z = ma_z$$

The basic approach to such problems is the same as in two dimensions.

In Chapter 3, we saw that for an object like a baseball, a bullet, a cannon shell, or an airplane, the magnitude of the air drag force is given approximately by

$$F_{\text{drag}} = \tfrac{1}{2}\rho A v^2 \tag{4.46}$$

where ρ is the density of air (1.3 kg/m^3), A is the cross-sectional area of the object, and v is its speed. (See Insight 3.4 on page 79 for more on the drag force.) For a baseball[5] moving at 45 m/s, Equation 4.46 gives $F_{\text{drag}} = 5.4$ N. The baseball's weight is approximately $mg = 1.4$ N, so the drag force here is much *greater* than the force of gravity on the ball. We certainly cannot ignore the drag force in this case.

Although we have a mathematical expression for the drag force in Equation 4.46, it is difficult to work out a simple expression for the trajectory of a projectile (such as a baseball) when this force is included. However, such trajectories can be readily calculated with computer simulation methods, so we can describe how drag affects the shape of the trajectory and the range. We have already seen that for an ideal projectile (i.e., one without air drag), the maximum range occurs when the initial velocity makes an angle of 45° with the horizontal. Drag can substantially alter this result, and in either direction. For a baseball, the maximum range is found with a much smaller angle. The "best" angle depends on the initial speed of the ball and is typically around 35°. On the other hand, for a powerful artillery gun, the artillery shell can travel very high into the atmosphere where the density of the air is much less than at ground level, which greatly reduces the drag force on the shell. To achieve maximum range, these guns are usually aimed well above 45°, and the artillery shell spends a large amount of time at high altitudes where the air density is lower than it is near sea level.

Effect of Air Drag on a Bicycle

Air drag is very important for vehicles such as cars and bicycles. Indeed, most of us have felt the force from air drag when riding a bicycle even at rather low speeds. Consider the bicyclist in Figure 4.33 riding downhill on a frictionless bicycle so that the only forces along the incline are due to gravity and air drag. If the slope of the hill is $\theta = 5.0°$, how fast will the bicycle coast?

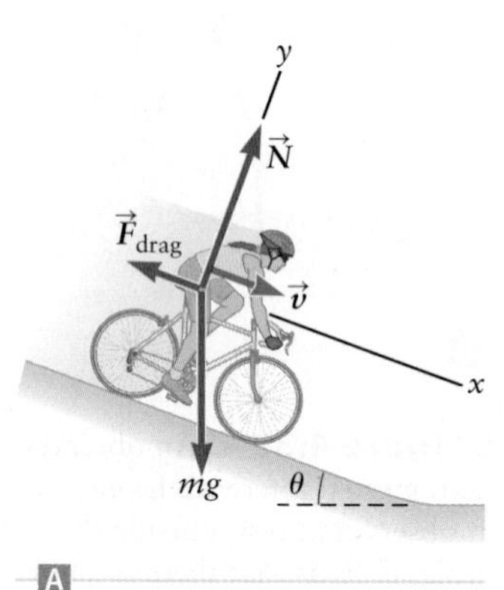

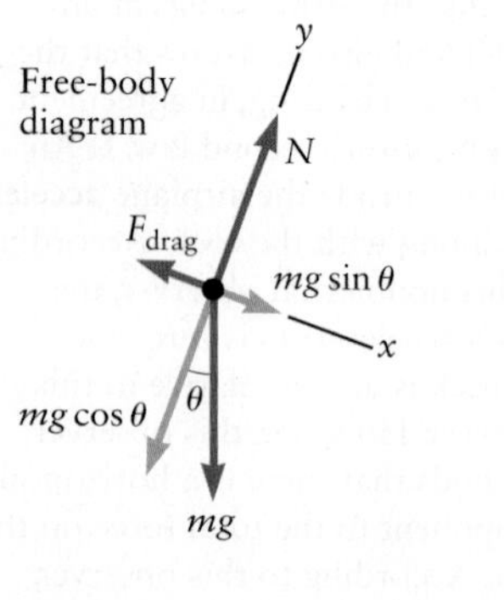

▲ **Figure 4.33** A When a bicycle coasts down a long hill, the bicycle eventually reaches its terminal velocity. The forces acting on the bicycle are the normal force $\vec{N}$ exerted by the hill's surface, the force of gravity $\vec{F}_{\text{grav}}$, and the drag force $\vec{F}_{\text{drag}}$. B Free-body diagram with forces resolved along x and y.

Figure 4.33 shows the forces acting on the bicycle. We take the x direction to be along the incline. Writing Newton's second law for motion along x, we get

$$\sum F_x = F_{\text{grav},x} + F_{\text{drag}}$$

The component of the gravitational force along x is

$$F_{\text{grav},x} = mg \sin\theta \tag{4.47}$$

Using Equation 4.46, the drag force is

$$F_{\text{drag}} = -\tfrac{1}{2}\rho A v_x^2 \tag{4.48}$$

where the negative sign indicates that F_{drag} is directed up the incline (along $-x$). When the bicyclist is coasting at her terminal velocity, v is constant, so the total force along x must be zero. Using Equations 4.47 and 4.48, we thus find

$$\sum F_x = mg\sin\theta - \tfrac{1}{2}\rho A v_x^2 = 0$$

$$v_x = \sqrt{\frac{2mg\sin\theta}{\rho A}}$$

To get a feeling for the value of this terminal (coasting) velocity, we consider a bicycle plus rider of mass 100 kg, with an approximate frontal area of 1 m^2. For a hill with slope angle $\theta = 5.0°$, a moderately steep hill, we have a speed

$$v_x = \sqrt{\frac{2mg\sin\theta}{\rho A}} = \sqrt{\frac{2(100\text{ kg})(9.8\text{ m/s}^2)\sin(5.0°)}{(1.3\text{ kg/m}^3)(1\text{ m}^2)}} = 10\text{ m/s}$$

[5]The frontal area of the baseball is $A = \pi r^2$, where the radius of a baseball is $r = 3.6$ cm.

(Here we keep just one significant figure, since the area is only estimated with this accuracy.) This is about 25 mi/h. How does this answer compare with your experience?

Summary | CHAPTER 4

Key Concepts and Principles

Translational equilibrium

When the total force on an object is zero, Newton's second law tells us that the acceleration is also zero. In other words, all the vector components of $\vec{F}_{\text{total}} = \sum \vec{F}$ and $\vec{a}$ are zero. Such an object is in ***translational equilibrium***.

Analyzing motion in two and three dimensions

We use Newton's second law to calculate the acceleration of an object:

$$\sum \vec{F} = m\vec{a} \qquad \text{(4.1) (page 92)}$$

The acceleration can then be used to find the velocity and displacement.

Inertial and noninertial reference frames

A ***reference frame*** is an observer's choice of coordinate system, including an origin, for making measurements. An ***inertial reference frame*** is one that moves with a constant velocity, whereas a ***noninertial*** reference frame is one that is accelerating. Newton's laws are only obeyed in inertial reference frames. They are not obeyed in noninertial reference frames.

Applications

Projectile motion

In many cases, the force of air drag is small. The force on a projectile near the Earth's surface, such as a baseball, is then just the force of gravity, and the acceleration has a constant magnitude g and is directed downward. The displacement and velocity for such a projectile are described by the relations for motion with constant acceleration.

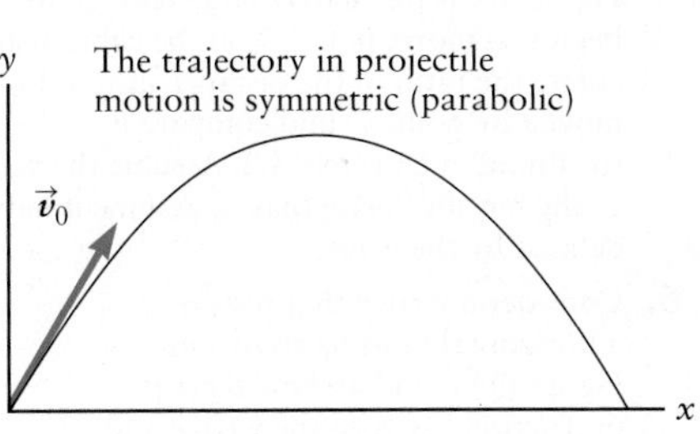

The trajectory in projectile motion is symmetric (parabolic)

For projectile motion, the vertical and horizontal motions are independent. Two objects can have very different horizontal velocities, but they still fall at the same rate.

The trajectory for simple projectile motion starting from the origin is symmetric in two ways:

- It is symmetric in time. The time spent traveling to the point of maximum height is equal to the time spent falling back down to the initial height.
- It is symmetric in space. The trajectory has a parabolic shape.

How the body detects acceleration

The ear contains a sensitive acceleration detector that enables the brain to know when the head is accelerating.

Questions

SSM = answer in *Student Companion & Problem-Solving Guide* = life science application

1. At what angle should a ball be thrown so that it has the maximum range? Ignore air drag.

2. Two children are having a snowball fight, and one is trying to land a snowball on a friend who is hiding behind a wall (Fig. Q4.2). The child on the left can only throw snowballs with one value of the initial speed and finds that for one throwing angle α that is less than 45° he is able to have his snowball land on his friend. Show that there is a corresponding throwing angle β greater than 45° that yields the same value of the range and find a general relation between α and β. *Hint:* Consider the range of a projectile as calculated in Equation 4.28 (i.e., ignoring the effect of air drag), and assume that the snowballs in Figure Q4.2 start and land at the same height above the ground.

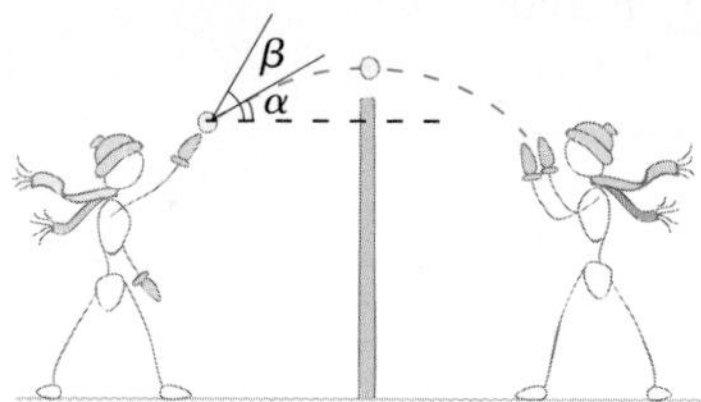

Figure Q4.2

3. An object has an initial velocity $\vec{v}_i$ and a final velocity $\vec{v}_f$ as sketched in Figure Q4.3. Sketch the approximate direction of the acceleration $\vec{a}$ of the object during this time interval.

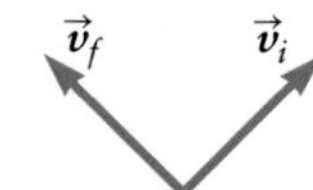

Figure Q4.3

4. During a particular time interval, an object has an acceleration $\vec{a}$ in the direction shown in Figure Q4.4. If $\vec{v}_i$ is the initial velocity of the object, draw a vector $\vec{v}_f$ that might be the velocity of the object at the end of the interval.

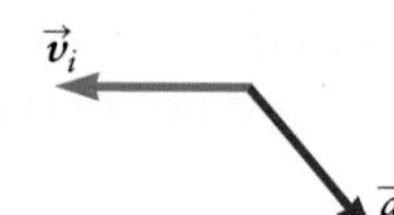

Figure Q4.4

5. The arrangement in Example 4.2 (p. 95) makes it possible to amplify a force. (Compare this example to the block and tackle in Fig. 3.23.) Amplifying forces in this way comes at a price. To appreciate this price, imagine that the force on the car in Figure 4.4 is just barely large enough to move the car. If the car begins to move, point O on the cable also begins to move. Calculate the ratio of the distance moved by the car to the distance moved by point O and compare it to the force amplification factor found in Example 4.2. Assume the car moves in the direction of the tension force; that is, assume it moves along the direction defined by the cable.

6. Consider a wedge that rests on a horizontal table as shown in Figure Q4.6 and assume there is no friction between the wedge and the table. A force is now applied vertically as shown in the figure. You might expect this force will cause the wedge to move to the left. Explain why the wedge does not move, even though there is no friction between it and the table.

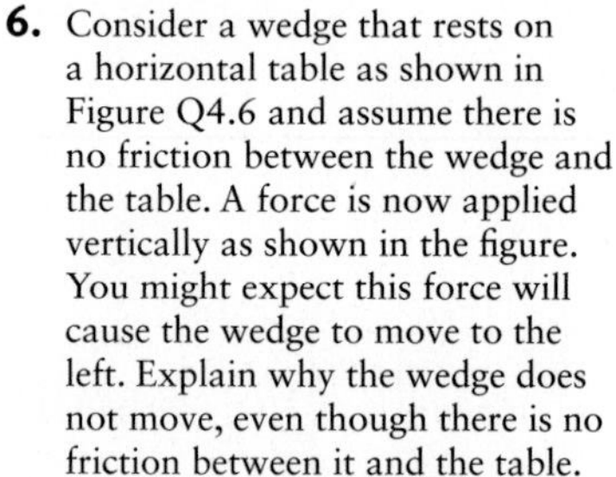

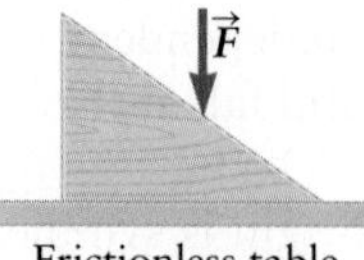

Frictionless table

Figure Q4.6

7. SSM Consider again the wedge in Question 6, but now assume a block is placed onto it as shown in Figure Q4.7. There is again no friction between the wedge and the table, and there is also no friction between the block and the wedge. Explain why in this case the wedge *will* be accelerated to the left. How does this situation differ from that in Question 6? *Hint*: Compare the free-body diagrams for the wedge in the two cases.

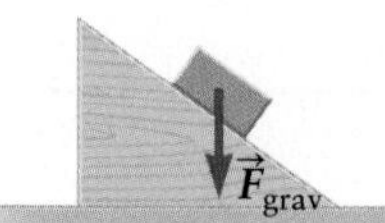

Frictionless table

Figure Q4.7

8. The rock-on-a-string accelerometer in Figure 4.30 can be used to measure acceleration along a horizontal direction. Discuss how you could use it to also measure the vertical component of the acceleration.

9. **Time of flight.** Equation 4.26 gives a relation for the time at which a batted baseball lands. Use the quadratic formula to find the solutions for t_{lands}. (See Appendix B if you don't remember the quadratic formula.) Since you are solving a quadratic equation, there are *two* solutions for t_{lands}. Find numerical values for both solutions and give a physical interpretation of the results. *Hint:* Consider how the problem would be changed if the ball was initially directed "backward" along the trajectory in Figure 4.12. Assume $h = 1.0$ m, $v_0 = 45$ m/s, and $\theta = 30°$.

10. Explain why the string in Figure 4.30 hangs vertically if the velocity is constant. Consider both an airplane moving horizontally and one moving with a nonzero velocity component along y.

11. Explain why you cannot run unless there is a frictional force between your feet and the ground.

12. **Peel out!** The author has a small pickup truck. He finds that it is much easier to "burn rubber" (i.e., spin the back wheels so that they slip relative to the road surface) when the truck is empty than when it is carrying a heavy load. Explain why.

13. Give an example in which the magnitude of the instantaneous velocity is always larger than the average velocity.

14. Give an example in which the acceleration is perpendicular to the velocity.

15. Give an example of motion in which the instantaneous velocity is zero but the acceleration is not zero.

16. A rock is thrown up in the air in such a way that its speed is zero at the top of its trajectory. Where does the rock land?

17. Give three examples in which the force of friction on an object has the same direction as the velocity of the object.

18. Ball 1 is thrown with an initial speed v_0 at an angle θ relative to the horizontal (x) axis. Ball 2 is thrown at the same angle, but with an initial speed of $2v_0$. (a) If ball 1 is in the air for a time t_1, how long is ball 2 in the air? (b) If ball 1 reaches a maximum height h_1, what is the maximum height reached by ball 2 in terms of h_1?

19. Ball 1 rolls off the edge of a table and falls to the floor below, while ball 2 is dropped from the same height (Fig. Q4.19). Which ball reaches the ground first? Explain why your answer does, or does not, depend on the velocity of ball 1 just before it leaves the table.

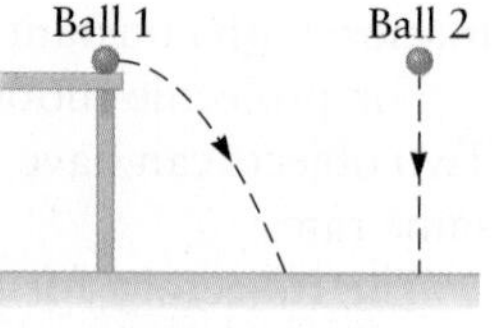

Figure Q4.19

20. SSM Two balls are thrown into the air with the same initial speed, directed at the same initial angle with respect to the horizontal. Ball 1 has a mass five times the mass of ball 2, and the force of air drag is negligible.
 (a) Which ball has the larger acceleration as it moves through the air?
 (b) Which ball lands first?
 (c) Which ball reaches the greatest height?
 (d) For which ball is the force of gravity larger at the top of the trajectory?

Problems

SSM = solution in *Student Companion & Problem-Solving Guide*
★ = intermediate ✪ = challenging ⓧ = life science application
Ⓡ = reasoning and relationships problem
RT = reasoning tutorial in Enhanced WebAssign

4.1 STATICS

1. Two ropes are attached to a skater as sketched in Figure P4.1 and exert forces on her as shown. Find the magnitude and direction of the total force exerted by the ropes on the skater.

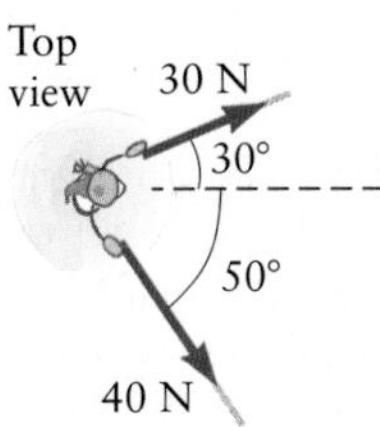

Figure P4.1

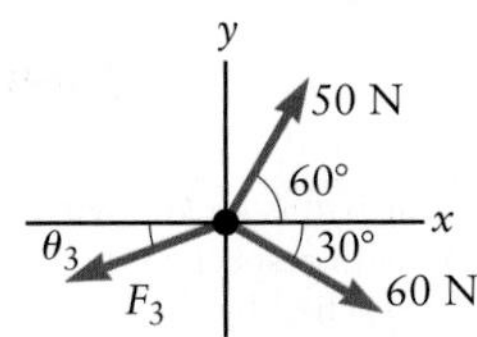

Figure P4.2

2. The three forces shown in Figure P4.2 act on a particle. If the particle is in translational equilibrium, find F_3 (the magnitude of force 3) and the angle θ_3.

3. SSM ★ Several forces act on a particle as shown in Figure P4.3. If the particle is in translational equilibrium, what are the values of F_3 (the magnitude of force 3) and θ_3 (the angle that force 3 makes with the x axis)?

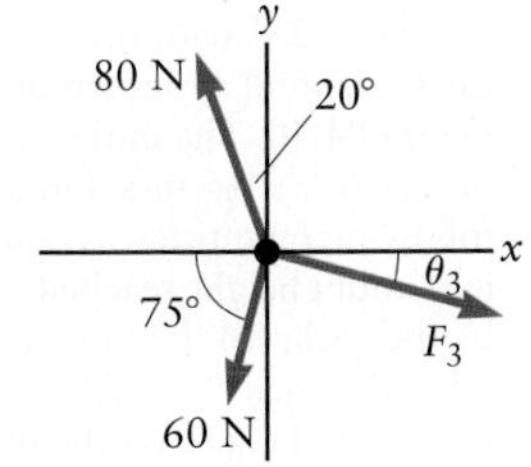

Figure P4.3

4. A man is lazily floating on an air mattress in a swimming pool (Fig. P4.4). (a) Draw a free-body diagram for the man and for the mattress. (b) Identify the reaction forces for all the forces in your free-body diagrams in part (a). (c) If the mass of the man is 110 kg and the mass of the mattress is 2.5 kg, what is the upward force of the water on the mattress? What is the force that the man exerts on the mattress? *Note*: Keep three significant figures in your answer.

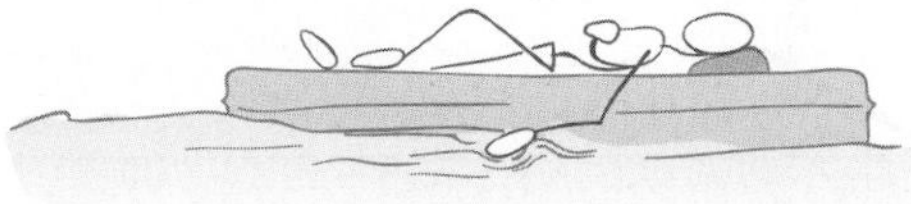

Figure P4.4

5. A person leans against a wall (Fig. P4.5). Draw a free-body diagram for the person. Assume there is no frictional force between the wall and the person.

Figure P4.5

6. ★ The sled in Figure 4.2 is stuck in the snow. A child pulls on the rope and finds that the sled just barely begins to move when he pulls with a force of 25 N, with the rope at an angle of 30° with respect to the horizontal. (a) Sketch the sled and all the forces acting on it. Also choose a coordinate system. (b) Determine the components of all the forces on the sled along the coordinate axes. (c) Write the conditions for static equilibrium along your two coordinate directions. (d) If the sled has a mass of 12 kg, what is the coefficient of friction between the sled and the snow? (e) Is this the coefficient of static friction or the coefficient of kinetic friction? (f) If the child continues to pull on the sled and it has an acceleration of 0.30 m/s^2, find the coefficient of kinetic friction between the sled and the snow.

7. ★ A system of cables is used to support a crate of mass m = 45 kg as shown in Figure P4.7. Find the tensions in all three cables.

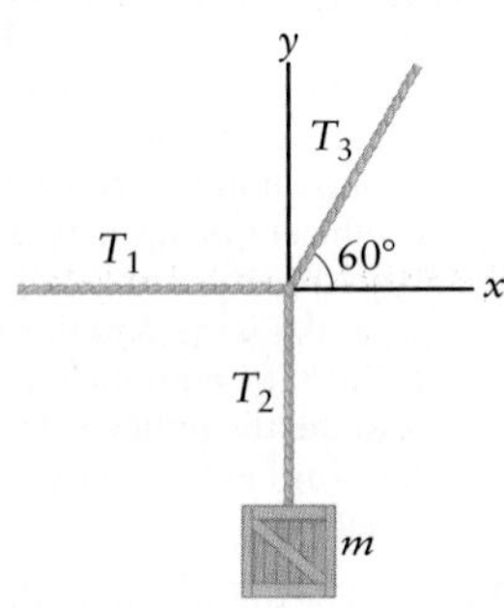

Figure P4.7

8. **Balancing act.** The tightrope walker in Figure P4.8 gets tired and decides to stop for a rest. During this rest period, she is in translational equilibrium. She stops at middle of the rope and finds that both sides of the rope make an angle θ = 15° with the horizontal. (a) Sketch the walker and the rope. Show all the forces acting on the point of the rope on which the walker stands (call this point P). Also include a coordinate system. (b) Determine the components of all the forces on point P along the coordinate axes. (c) Write the conditions for static equilibrium along your two coordinate directions. (d) If the mass of the tightrope walker is 60 kg, what is the tension in the rope?

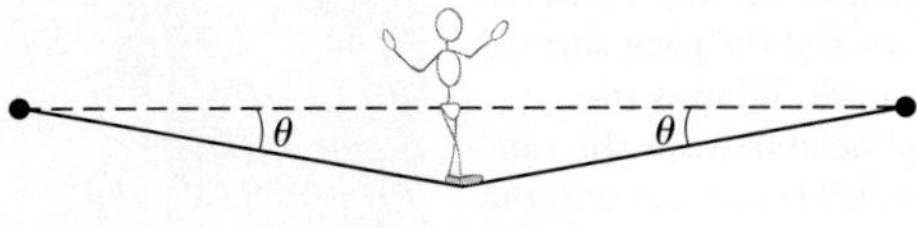

Figure P4.8

9. A flag of mass 2.5 kg is supported by a single rope as shown in Figure P4.9. A strong horizontal wind exerts a force of 12 N on the flag. Find the tension in the rope and the angle θ the rope makes with the horizontal.

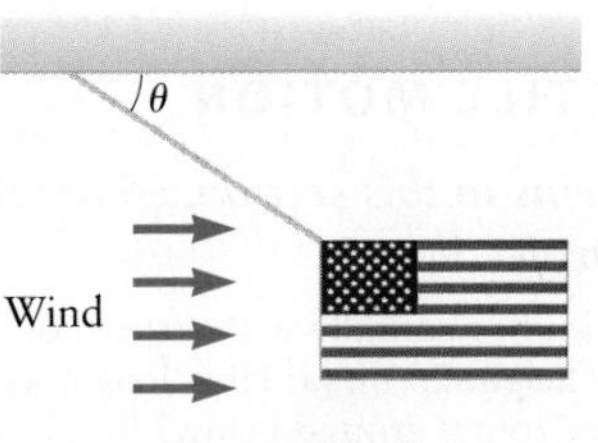

Figure P4.9

10. ★ A car of mass 1400 kg is parked on a very slippery hillside (Fig. P4.10). To keep it from sliding down the hill, the owner attaches a cable. (a) Sketch all the forces on the car. Include coordinate axes in your sketch. (b) Determine the components of all the forces on the car along the coordinate axes. (c) Write the conditions for static equilibrium along your two coordinate directions. (d) If there is no frictional force between the road and the tires, what is the tension in the cable?

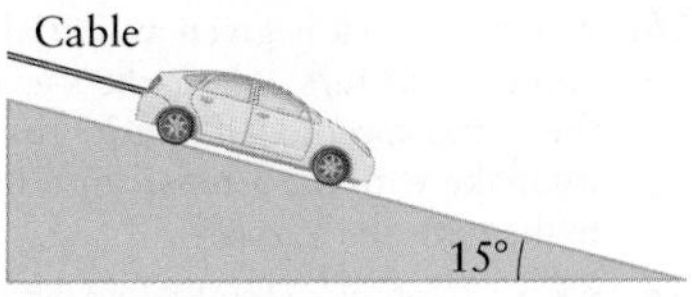

Figure P4.10

11. A crate is placed on an inclined board as shown in Figure P4.11. One end of the board is hinged so that the angle θ is adjustable. If the coefficient of static friction between the crate and the board is $\mu_S = 0.30$, what is the value of θ at which the crate just begins to slip?

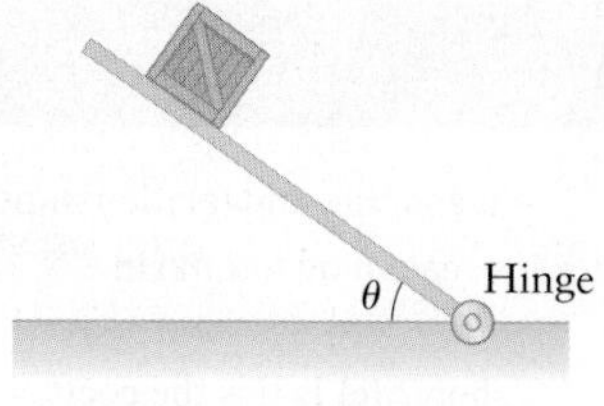

Figure P4.11

12. Two blocks of mass $m_1 = 45$ kg and $m_2 = 12$ kg are connected by a massless string that passes over a pulley as shown in Figure P4.12. The coefficient of static friction between m_1 and the table is $\mu_S = 0.45$. (a) Will this system be in static equilibrium? Assume the pulley is frictionless. (b) Find the tension in the string.

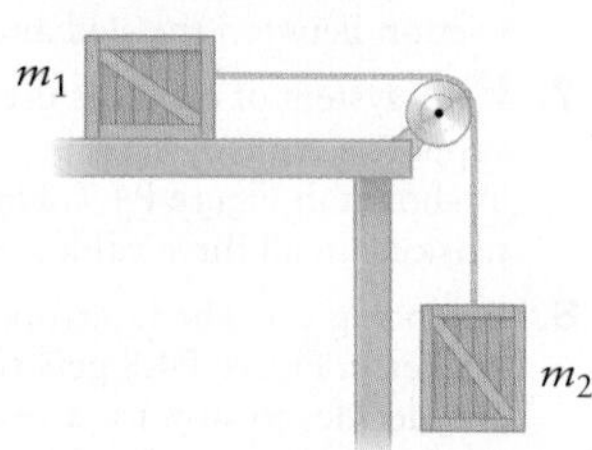

Figure P4.12 Problems 12, 13, and 44.

13. For the system in Problem 12 and Figure P4.12, how large can m_2 be made without the system starting into motion?

14. **Stemming a chimney.** A rock climber of mass 60 kg wants to make her way up the crack between two rocks as shown in Figure P4.14. The coefficient of friction between her shoes and the rock surface is $\mu_S = 0.90$. What is the smallest normal force she can apply to both surfaces without slipping? Assume the rock walls are vertical. *Hint*: Why is the normal force between the climber and the rock on the left equal to the normal force between her and the rock on the right?

Figure P4.14

4.2 PROJECTILE MOTION

For all the problems in this section, ignore the force due to air drag on the projectile.

15. A rock is thrown horizontally with a speed of 20 m/s from a vertical cliff of height 25 m. (a) How long does it take the rock to reach the horizontal ground below? (b) How far will it land from the base of the cliff? (c) What is the velocity (magnitude and direction) of the rock just before it hits the ground?

16. A hockey puck is given an initial velocity such that $v_x = 12$ m/s and $v_y = 18$ m/s, where the x–y plane is horizontal. (a) What is the initial speed of the puck? (b) What angle does the initial velocity make with the x axis? (c) What angle does the initial velocity make with the y axis?

17. A quarterback is asked to throw a football to a receiver who is 35 m away. What is the minimum speed the football must have when it leaves the quarterback's hand? Assume the ball is caught at the same height as it is thrown.

18. Which of the graphs in Figure P4.18 might be a plot of the vertical component of the velocity of a projectile that is thrown from the top of a building?

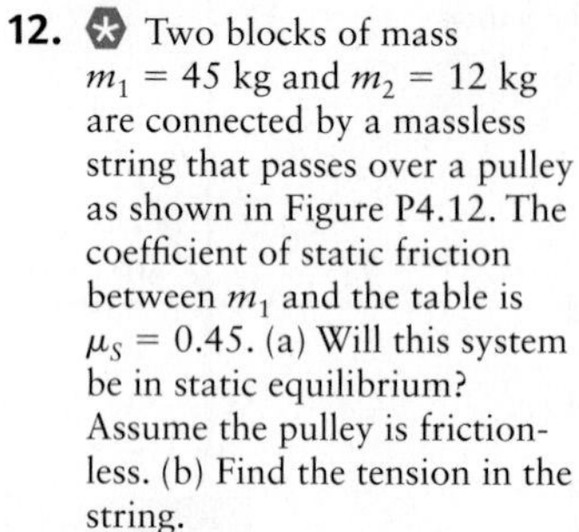
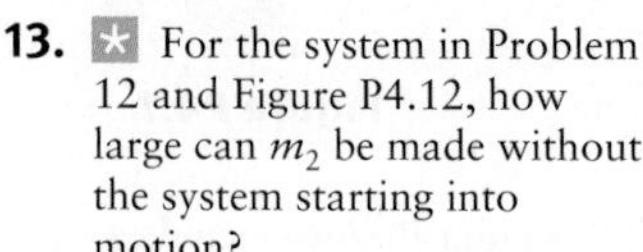

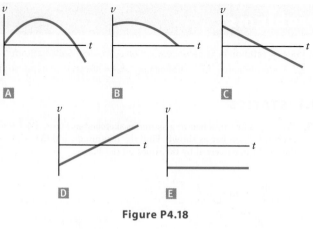

Figure P4.18

19. A soccer ball is kicked with an initial speed of 30 m/s at an angle of 25° with respect to the horizontal. Find (a) the maximum height reached by the ball and (b) the speed of the ball when it is at the highest point on its trajectory. (c) Where does the ball land? That is, what is the range of the ball? Assume level ground.

20. Consider a rock thrown off a bridge of height 75 m at an angle $\theta = 25°$ with respect to the horizontal as shown in Figure P4.20. The initial speed of the rock is 15 m/s. Find the following quantities: (a) the maximum height reached by the rock, (b) the time it takes the rock to reach its maximum height, (c) the place where the rock lands, (d) the time at which the rock lands, and (e) the velocity of the rock (magnitude and direction) just before it lands.

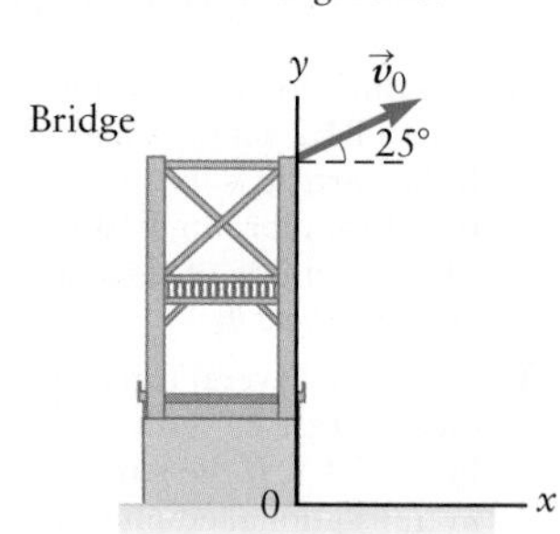

Figure P4.20

21. A football player wants to kick a ball through the uprights as shown in Figure P4.21. The ball is kicked from a distance of 30 m (he's playing metric football) with a velocity of magnitude 25 m/s at an angle of 30°. Will the ball make it over the crossbar (height 3.1 m)?

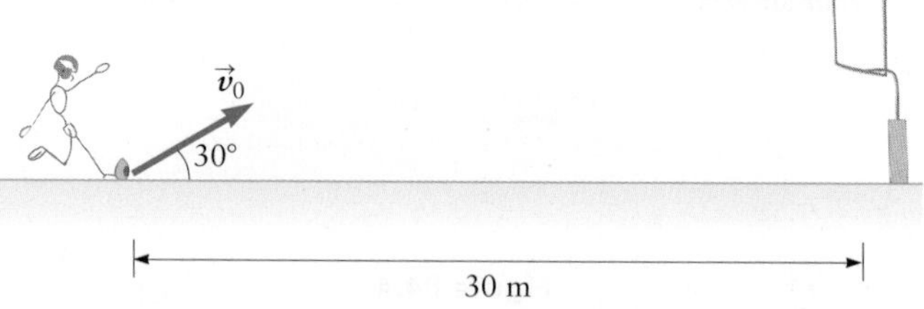

Figure P4.21

22. SSM An airplane is flying horizontally with a constant velocity of 200 m/s at an altitude of 5000 m when it releases a package. (a) How long does it take the package to reach the ground? (b) What is the distance between the airplane and the package when the package hits the ground? (c) What is the distance between the airplane and the package's landing spot, as measured along the x direction, when the package is released? Ignore air drag.

23. A horizontal rifle is fired at a target 50 m away. The bullet falls vertically a distance of 12 cm on the way to the target. (a) What is the speed of the bullet just after it leaves the rifle? (b) What is the speed of the bullet just before it hits the target?

24. A bullet of mass 0.024 kg is fired horizontally with a speed of 75 m/s from a tall bridge. If the bullet is in the air for 2.3 s, how far from the base of the bridge does it land?

25. Consider the game of baseball. A pitcher throws a ball to the catcher at a speed of 100 mi/h (45 m/s). If the velocity of the ball is horizontal when it leaves the pitcher's hand, how far (vertically) will it fall on the way to the catcher? The distance from the pitcher to the catcher is 60.5 ft. Express your answer in meters and in feet.

26. A batted baseball is hit with a speed of 45 m/s starting from an initial height of 1 m. Find how high the ball travels in two cases: (a) a ball hit directly upward and (b) a ball hit at an angle of 70° with respect to the horizontal. Also find how long the ball stays in the air in each case.

27. A baseball is hit with an initial speed of 45 m/s at an angle of 35° relative to the x (horizontal) axis. (a) Find the speed of the ball at $t = 3.2$ s. (b) What angle does $\vec{v}$ make with the x axis at this moment? (c) While the ball is in the air, at what value of t is the speed smallest? Ignore air drag.

28. An archer lines up an arrow on the horizontal exactly dead center on a target as drawn in Figure P4.28. She releases the arrow, and it strikes the target a distance $h = 7.6$ cm below the center. If the target is 10 m from the archer, (a) how long was the arrow in the air before striking the target? (b) What was the initial velocity of the arrow as it left the bow? Neglect air drag.

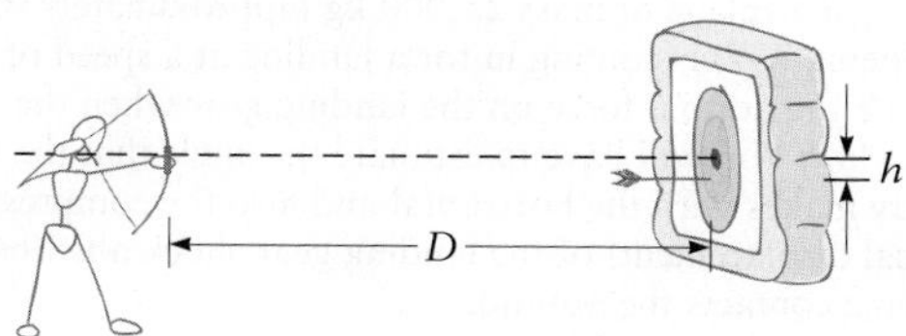

Figure P4.28

29. Ⓡ Consider the problem of kicking a soccer ball past a goalkeeper into the goal (Fig. P4.29). You are 25 m away from the goal and kick the ball at an angle of 30° with respect to the horizontal, and the ball just passes over the goalkeeper's hands. Find the initial speed of the ball. Also, calculate how long it takes the ball to get from you to the goal. *Hint*: You will have to estimate the height of the goalkeeper.

Figure P4.29

30. A ball is thrown straight up and rises to a maximum height of 24 m. At what height is the speed of the ball equal to half its initial value? Assume the ball starts at a height of 2.0 m above the ground.

31. Two rocks are thrown off a cliff. One rock (1) is thrown horizontally with a speed of 20 m/s. The other rock (2) is thrown at an angle θ relative to the horizontal with a speed of 30 m/s. While the two rocks are in the air, rock 2 is always directly above 1. Find θ.

32. A bullet is fired from a rifle with speed v_0 at an angle θ with respect to the horizontal axis (Fig. P4.32) from a cliff that is a height h above the ground below. (a) Calculate the speed of the bullet when it strikes the ground. Express your answer in terms of v_0, h, g, and θ. (b) Explain why your result is independent of θ.

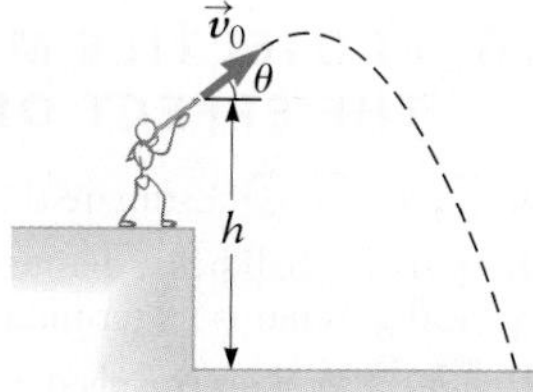

Figure P4.32

33. A high jumper can run horizontally with a top speed of 10.0 m/s. (a) If he can "convert" this velocity to the vertical direction when he leaves the ground, what is the theoretical limit on the height of his jump? (b) How does your result to part (a) compare with the current high-jump record of approximately 2.5 m? Can you explain the difference?

34. A baseball is thrown with an initial velocity of magnitude v_0 at an angle of 60° with respect to the horizontal (x) direction. At the same time, a second ball is thrown with the same initial speed at an angle θ with respect to x. If the two balls land at the same spot, what is θ?

4.3 A FIRST LOOK AT REFERENCE FRAMES AND RELATIVE VELOCITY

35. A swimmer is able to swim at a speed of 0.90 m/s in still water. (a) How long does it take the swimmer to go a distance of 1500 m in still water? Call this time t_1. (b) The swimmer then decides to swim the same distance in a river having a current with a speed of 0.50 m/s. How long does this take if she swims with the current? Call this time t_2. (c) How long does it take her to swim the same distance against the current? Call this time t_3. (d) Explain why $t_1 - t_2$ is not equal to $t_3 - t_1$.

36. Consider a swimmer who wants to swim directly across a river as in Example 4.7. If the speed of the current is 0.30 m/s and the swimmer's speed relative to the water is 0.60 m/s, how long will it take her to cross a river that is 15 m wide?

37. Suppose the swimmer in Figure 4.21 has a swimming speed relative to the water of 0.45 m/s and the current's speed is 2.5 m/s. If it takes the swimmer 200 s to cross the river, how wide is the river?

38. An airplane has a velocity relative to the air of 200 m/s in a westerly direction. If the wind has a speed relative to the ground of 60 m/s directed to the north, what is the speed of the airplane relative to the ground?

39. Consider again the airplane in Problem 38, but now suppose the wind is directed in an easterly direction. How long does it take the airplane to travel a distance between two cities that are 300 km apart?

40. SSM An airplane flies from Boston to San Francisco (a distance of 5000 km) in the morning and then immediately returns to Boston. The airplane's speed relative to the air is 250 m/s. The wind is blowing at 50 m/s from west to east, so it is "in the face" of the airplane on the way to San Francisco and it is a tailwind on the way back. (a) What is the average speed of the airplane relative to the ground on the way to San Francisco? (b) What is the average speed relative to the ground on the way back to Boston? (c) What is the average speed for the entire trip? (d) Why is the average of the average speeds for the two legs of the trip not equal to the average speed for the entire trip?

41. **Round trip.** An airplane flies from Chicago to New Orleans (a distance of 1500 km) in the morning and then immediately returns to Chicago. The airplane's speed relative to the air is 250 m/s. The wind is blowing at 50 m/s from west to east, so it is perpendicular to the line that connects the two cities. (a) What is the average speed of the airplane relative to the ground on the way to New Orleans? (b) What is the average speed relative to the ground on the way back to Chicago? (c) What is the average speed for the entire trip?

42. A 35-m-wide river flows in a straight line (the x direction) with a speed of 0.25 m/s. A boat is rowed such that it travels directly across the river (along y). If the boat takes exactly 4 minutes to cross the river, what is the speed of the boat relative to the water?

4.4 FURTHER APPLICATIONS OF NEWTON'S LAWS

43. Two children pull a sled of mass 15 kg along a frictionless surface as shown in Figure P4.43. (a) Find the magnitude and direction of the sled's acceleration. (b) How long does it take the sled to reach a speed of 10 m/s?

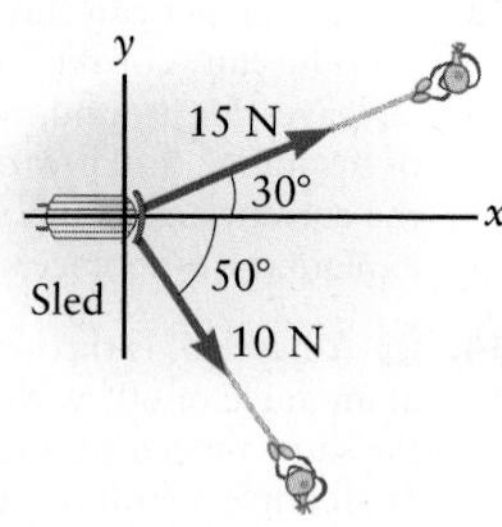

Figure P4.43

44. Consider the crates in Problem 12 (Fig. P4.12). Assume there is no friction between m_1 and the table. (a) Sketch all the forces on the crates. Include a coordinate system. (b) Determine the components of the forces along the coordinate axes. (c) Write Newton's second law for motion along both coordinate directions. (d) Solve the equations from part (c) to find the acceleration of the crates and the tension in the string.

45. ✪ When a car is traveling at 25 m/s on level ground, the stopping distance is found to be 22 m. This distance is measured by pushing hard on the brakes so that the wheels skid on the pavement. The same car is driving at the same speed down an incline that makes an angle of 8.0° with the horizontal. What is the stopping distance now, as measured along the incline?

46. A block slides down an inclined plane that makes an angle of 40° with the horizontal. There is friction between the block and the plane. (a) Sketch all the forces on the block. Include a coordinate system. (b) Determine the components of the forces along the coordinate axes. (c) Write Newton's second law for motion along both coordinate directions. (d) If the acceleration of the block is 2.4 m/s², what is the coefficient of kinetic friction between the block and the plane?

47. ✱ A sled is pulled with a horizontal force of 20 N along a level trail, and the acceleration is found to be 0.40 m/s². An extra mass $m = 4.5$ kg is placed on the sled. If the same force is just barely able to keep the sled moving, what is the coefficient of kinetic friction between the sled and the trail?

48. Two crates are connected by a massless rope that passes over a pulley as shown in Figure 4.26. If the crates have mass 35 kg and 85 kg, what is their acceleration? If the system begins at rest, with the more massive crate a distance 12 m above the floor, how long does it take the more massive crate to reach the floor? Assume the pulley is massless and frictionless.

49. ✱ A hockey puck is sliding down an inclined plane with angle $\theta = 15°$ as shown in Figure P4.49. If the puck is moving with a constant speed, what is the coefficient of kinetic friction between the puck and the plane?

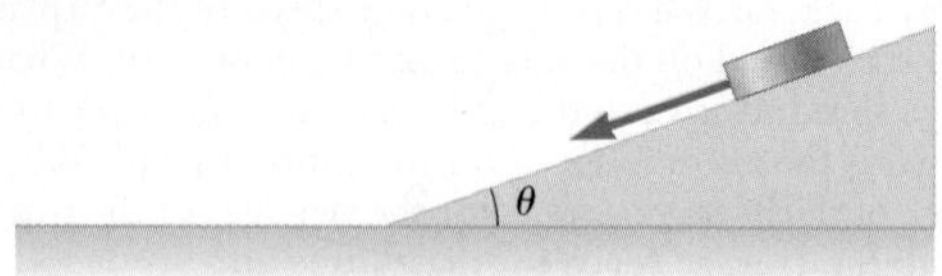

Figure P4.49

50. ✪ RT Consider a baseball player who wants to catch the fly ball in Problem 26(b). Estimate the force exerted by the ball on the player's glove as the ball is coming to rest. *Hint*: You will have to estimate the acceleration of the ball, and to do so it is useful to estimate the distance the ball travels while coming to rest in the glove. The mass of a baseball is 0.14 kg.

51. Two forces are acting on an object. One force has a magnitude of 45 N and is directed along the $+x$ axis. The other force has a magnitude of 30 N and is directed along the $+y$ axis. What is the direction of the object's acceleration? Express your answer by giving the angle between $\vec{a}$ and the x axis.

52. ✱ A car is traveling down a hill that makes an angle of 20° with the horizontal. The driver applies her brakes, and the wheels lock so that the car begins to skid. The coefficient of kinetic friction between the tires and the road is $\mu_K = 0.45$. (a) Find the acceleration of the car. (b) How long does the car take to skid to a stop if its initial speed is 30 mi/h (14 m/s)?

53. SSM ✱ A skier is traveling at a speed of 40 m/s when she reaches the base of a frictionless ski hill. This hill makes an angle of 10° with the horizontal. She then coasts up the hill as far as possible. What height (measured vertically above the base of the hill) does she reach?

54. Two crates of mass $m_1 = 35$ kg and $m_2 = 15$ kg are stacked on the back of a truck (Fig. P4.54). The frictional forces are strong enough that the crates do not slide off the truck. Assume the truck is accelerating with $a = 1.7$ m/s². Draw free-body diagrams for both crates and find the values of all forces in your diagrams.

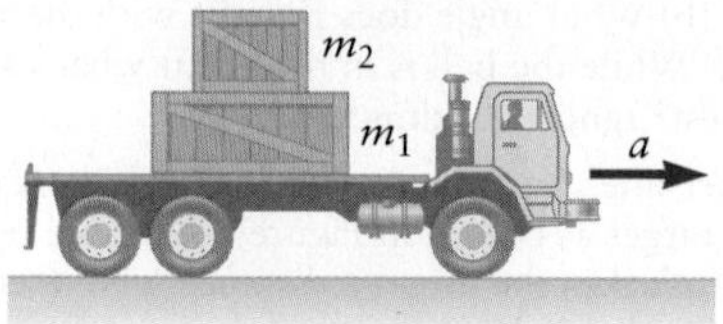

Figure P4.54

55. ✪ B An airplane of mass 25,000 kg (approximately the size of a Boeing 737) is coming in for a landing at a speed of 70 m/s. Estimate the normal force on the landing gear when the airplane lands. *Hint*: You will have to estimate the angle that the landing velocity makes with the horizontal and also the compression (vertical displacement) of the landing gear shock absorbers when the plane contacts the ground.

4.5 DETECTING ACCELERATION: REFERENCE FRAMES AND THE WORKINGS OF THE EAR

56. Consider the rock-on-a-string accelerometer in Figure 4.30. What is the accelerometer angle if the airplane has a horizontal acceleration of 1.5 m/s²?

57. ✪ Consider the accelerometer design shown in Figure P4.57, in which a small sphere sits in a hemispherical dish. Suppose the dish has an acceleration of a_x along the horizontal and a small amount of friction causes the sphere to reach a stable position relative to the dish as shown in the figure. Find the position of the sphere (i.e., find the angle θ) as a function of a_x.

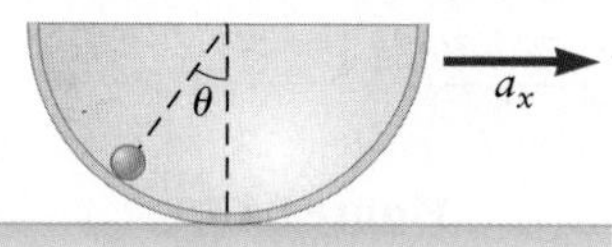

Figure P4.57

58. SSM ✪ RT Consider a commercial passenger jet as it accelerates on the runway during takeoff. This jet has a rock-on-a-string accelerometer installed (as in Fig. 4.30). Estimate the angle of the string during takeoff.

4.6 PROJECTILE MOTION REVISITED: THE EFFECT OF AIR DRAG

59. SSM ✱ B Estimate the terminal velocity for a golf ball.

60. A soccer ball has a diameter of about 22 cm and a mass of about 430 g. What is its terminal velocity?

61. ✱ B A skier travels down a steep, frictionless slope of angle 20° with the horizontal. Assuming she has reached her terminal velocity, estimate her speed.

62. Consider the motion of a bicycle with air drag included. We saw how to deal with the motion on a hill in connection with Figure 4.33. Now assume the bicycle is coasting on level ground and is being pushed along by a tailwind of 4.5 m/s (about 10 mi/h). If the bicycle starts from rest at $t = 0$, what is the acceleration of the bicycle at that moment? Assume there are no frictional forces (e.g., from the bicycle's tires or bearings, or any other source) opposing the motion of the bicycle.

63. What is the approximate magnitude of the drag force on an airplane (such as a commercial jet) traveling at 250 m/s? Consider only the drag force on the fuselage of the airplane (and not on the wings).

Additional Problems

64. SSM Consider the system of blocks in Figure P4.64, with $m_2 = 5.0$ kg and $\theta = 35°$. If the coefficient of static friction between block 1 and the inclined plane is $\mu_S = 0.25$, what is the largest mass m_1 for which the blocks will remain at rest?

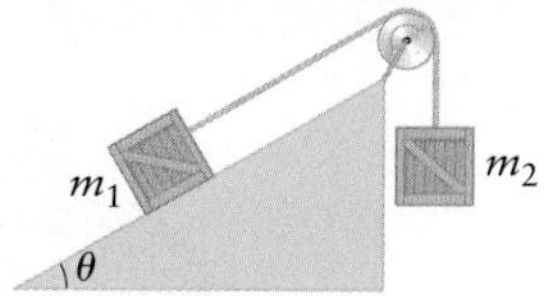

Figure P4.64

65. A water skier (Fig. P4.65) of mass $m = 65$ kg is pulled behind a boat at a constant speed of 25 mi/h. If the tension in the horizontal rope is 1000 N (approximately 225 lb), what are the magnitude and direction of the force the water exerts on the skier's ski?

© SuperStock/Alamy

Figure P4.65

66. An abstract sculpture is constructed by suspending a sphere of polished rock along the vertical with four cables connected to a square wooden frame 1.2 m on a side as shown in Figure P4.66. If the stone has mass $m = 8.0$ kg and the bottom two cables are under a tension of 25 N each, what is the tension in each top cable?

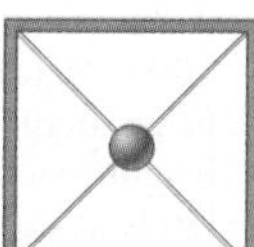
Figure P4.66

67. An airplane pulls a 25-kg banner on a cable as depicted in Figure P4.67. When the airplane has a constant cruising velocity, the cable makes an angle $\theta = 20°$ with the horizontal. (a) What is the drag force exerted on the banner by the air? (b) The weather changes abruptly and the pilot finds herself in a 20-minute rainstorm. After the storm, she finds that the cable now makes an angle $\theta_{water} = 30°$. Assuming the velocity and drag force are the same after the storm, find the mass of the water absorbed by the banner's fabric during the storm.

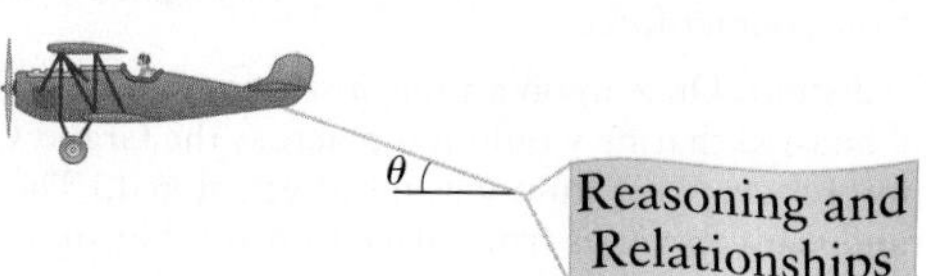

Figure P4.67

68. A piece of wood with mass $m = 2.4$ kg is held in a vise sandwiched between two wooden jaws as shown in Figure P4.68. A blow from a hammer drives a nail that exerts a force of 450 N on the wood. If the coefficient of static friction between the wood surfaces is 0.67, what minimum normal force must each jaw of the vise exert on the wood block to hold the block in place?

Figure P4.68

69. **Stomp rocket.** A "stomp rocket" is a toy projectile launcher illustrated in Figure P4.69 (left), which uses a blast of air to propel a dart made of plastic. The air blast is produced by jumping on a plastic bladder as shown the figure on the right. It is found that when a 190-lb father jumps on the bladder, the rocket dart will fly on average a horizontal distance of 500 ft when launched at an angle of 45°. (a) If the rocket dart is instead pointed straight up and the same father jumps on the launcher, what would be the maximum height obtained? (b) What is the initial velocity of the rocket dart as it leaves the launch tube? Would you recommend the use of safety goggles? Ignore air resistance.

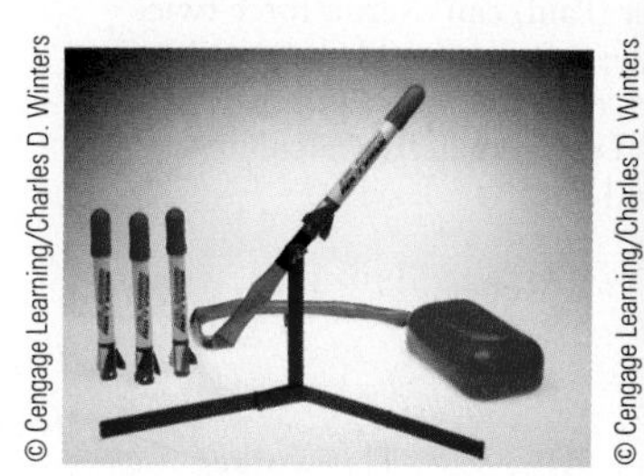
© Cengage Learning/Charles D. Winters

© Cengage Learning/Charles D. Winters

Figure P4.69

70. You are a serious basketball player and want to use physics to improve your free-throw shooting. Do an approximate calculation of the minimum speed the ball must have to travel from your hand to the basket in a successful free throw. You will have to estimate or find several of the quantities indicated in Figure P4.70, including the distance x from your hand to the basket, the height h of the ball when it leaves your hand, and the height y of the basket. Graph how the initial velocity changes as a function of launch angle to find the minimum initial speed of the ball and the particular launch angle.

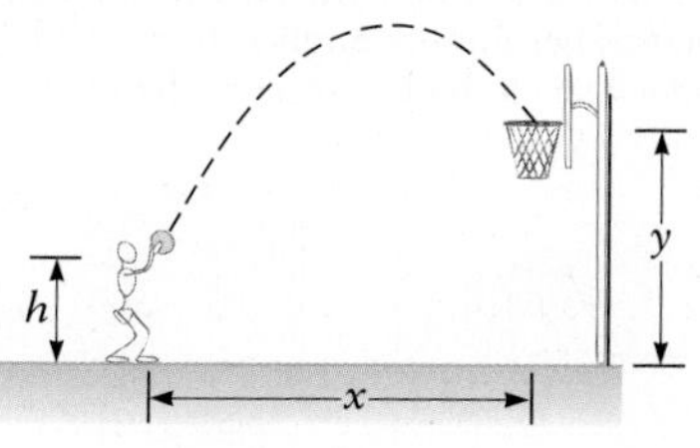

Figure P4.70

71. Consider once again the swimmer in Example 4.7. Assume she can swim at a velocity of 0.30 m/s and the river is 15 m wide. She needs to get across the river as quickly as possible. (a) What direction should her velocity vector make relative to the water to get her across in the shortest time? (b) What is her overall velocity as measured from the shore? (c) What is the direction of her total velocity as seen from the shore? (d) How long does it take her to cross the river? (e) How far downstream does she end up?

72. A bartender slides a mug of root beer with mass $m = 2.6$ kg down a bar top of length $L = 2.0$ m to an inattentive patron who lets the mug fall a height $h = 1.1$ m to the floor. The bar top (Fig. P4.72) is smooth, but it still has a coefficient of kinetic friction $\mu_K = 0.080$. (a) If the bartender gave the mug an initial velocity of 2.5 m/s, at what distance D from the bottom of the bar will the mug hit the floor? (b) What is the mug's velocity (magnitude and direction) as it impacts the floor? (c) Draw velocity–time graphs for both the x and y directions for the mug, from the time the bartender lets go of the mug to the time it hits the floor. Ignore air drag.

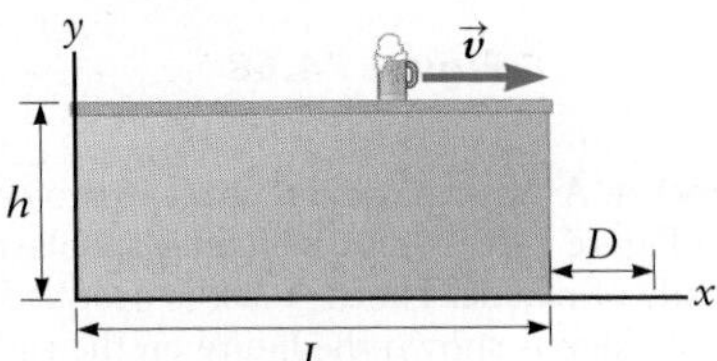

Figure P4.72

73. Two workers must slide a crate designed to be pushed with a rigid rod (on the left in Fig. P4.73) and at the same time pulled with a rope (on the right). The crate has mass $m = 45$ kg, the angle $\theta = 25°$, and the coefficient of kinetic friction between the floor and crate is 0.56. One of the workers (named Joe) can exert a force of 160 N, while the other (Paul) can exert a force twice that large. If we want to move the crate as quickly as possible, is it better to have the stronger worker (Paul) push and the weaker worker (Joe) pull, or vice versa? Does it matter? Calculate the acceleration for both cases to find out.

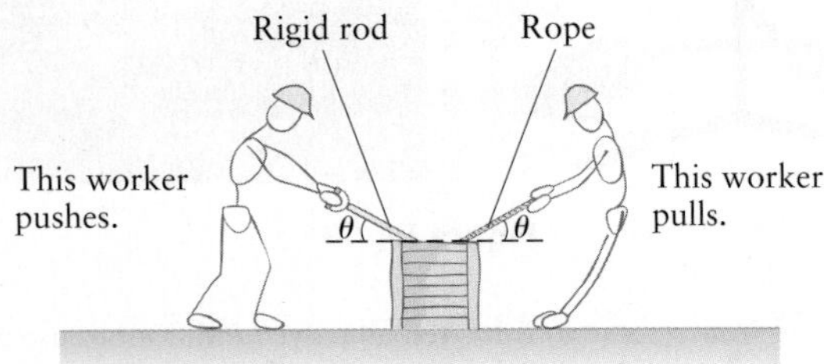

Figure P4.73

74. **Inner ear.** A student constructs a model of the utricle of the ear by attaching wooden balls ($m = 0.080$ kg) by strings to the bottom of a fish tank and then submerging them in water so that they float with the strings in a vertical direction. She finds that the wooden balls float with a buoyant force that exerts a tension of 0.20 N on each string as shown in Figure P4.74 (left). She then has her mother take her (and the model) on a drive. She notes that as her mother applies the gas pedal, the strings make an angle $\theta = 16°$ with respect to the vertical as shown in Figure P4.74 (right). (a) What was the acceleration of the car? (b) If the car started from rest, how long would it take to get to freeway speed of 30 m/s (about 65 mi/h)? During the acceleration, the water also makes an angle with the horizontal. Are the wooden balls even needed? Explain. Ignore any forces due to the pressure in the water.

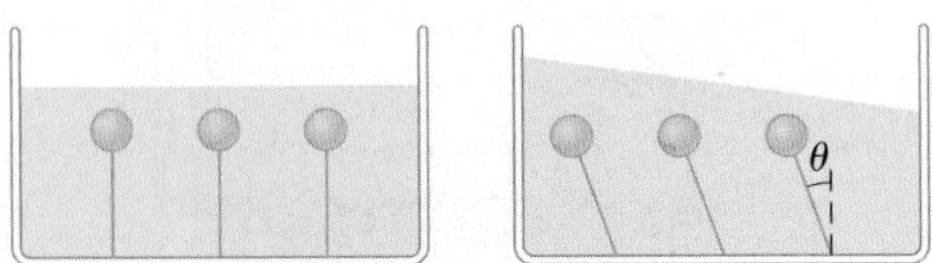

Figure P4.74

75. A vintage sports car accelerates down a slope of $\theta = 17°$ as depicted in Figure P4.75. The driver notices that the strings of the fuzzy dice hanging from his rearview mirror are perpendicular to the roof of the car. What is the car's acceleration?

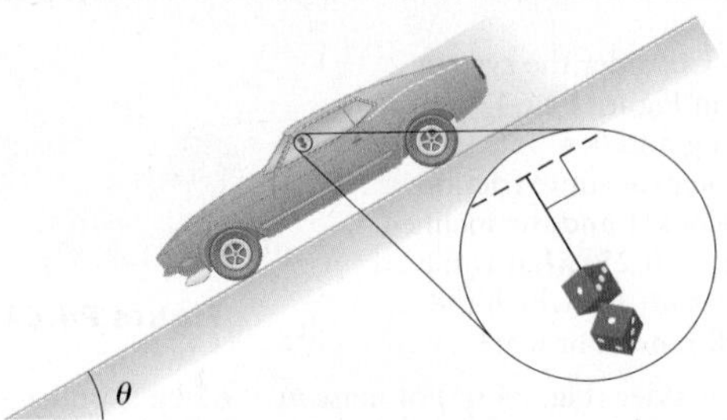

Figure P4.75

76. A spacecraft is traveling through space parallel to the x direction with an initial velocity of magnitude $v_0 = 3500$ m/s. At $t = 0$, the pilot turns on the engines, producing an acceleration of magnitude $a = 150$ m/s^2 along the y direction. Find the speed and the angle the velocity makes with the x direction at (a) $t = 15$ s and (b) 45 s.

77. A juvenile delinquent wants to break a window near the top of a tall building using a rock. The window is 30 m above ground level, and the base of the building is 20 m from his hiding spot behind a bush. Find the minimum speed the rock must have when it leaves his hand. Also find the corresponding launch angle of the rock. *Hint:* Graph how the initial velocity changes as a function of launch angle to find the minimum value needed and the corresponding launch angle. Ignore air drag.

78. A cannon shell is fired from a battleship with an initial speed of 700 m/s. (a) What is the maximum range of the shell? (b) For the firing angle that gives the maximum range, what is the maximum height reached by the shell? Ignore air drag and assume the shell lands at the same height as it is launched.

79. SSM Two blocks of mass m_1 and m_2 are sliding down an inclined plane (Fig. P4.79). (a) If the plane is frictionless, what is the magnitude of the contact force between the two masses? (b) If the coefficient of kinetic friction between m_1 and the plane is $\mu_1 = 0.25$ and between m_2 and the plane is $\mu_2 = 0.25$, what is the contact force between the two masses? (c) If $\mu_1 = 0.15$ and $\mu_2 = 0.25$, what is the contact force?

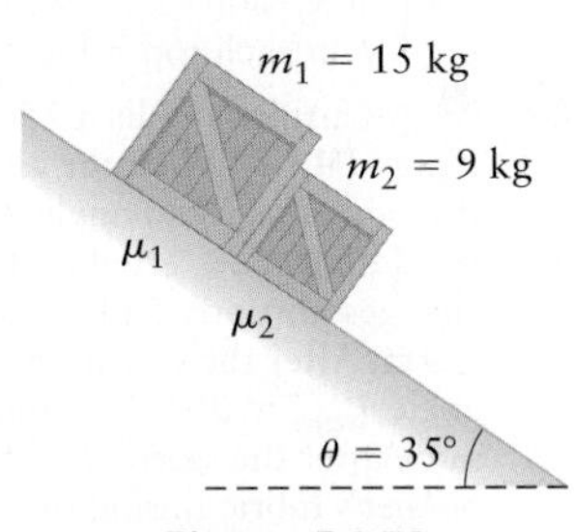

Figure P4.79

80. **Daredevil stunt.** Once upon a time, a stuntman named Evel Knievel boasted that he would jump across the Grand Canyon on a motorcycle. (This stunt was never attempted.) The width of the Grand Canyon varies from place to place, but suppose Mr. Knievel attempted to jump across at a point where the width is about 5 mi and the two rims are at the same altitude. What minimum initial speed would be required to make it safely across the canyon? Express your answer in meters per second and miles per hour. Ignore air drag.

81. RT Although Evel Knievel never succeeded in jumping over the Grand Canyon (see Problem 80), he was famous for jumping (with the help of his motorcycle) over, among other things, 14 large trucks. This jump covered an approximate distance of

135 ft. What was Mr. Knievel's minimum initial velocity for this jump? Ignore air drag.

82. A person travels on a ski lift (Fig. P4.82). (a) If the support strut on the ski lift makes an angle $\theta = 15°$, what is the horizontal acceleration of the person? (b) If the person plus the lift chair have a combined mass $m = 120$ kg, what is the tension force along the support strut? Assume the ski lift is moving horizontally. For simplicity, when considering forces involving the support strut, you can treat the strut in the same way as you would treat a massless cable.

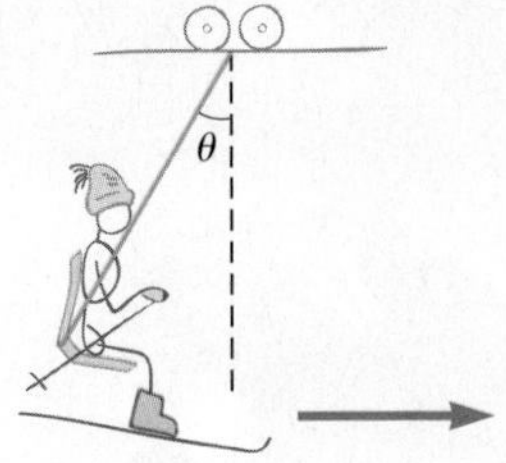

Figure P4.82

83. A block slides up a frictionless, inclined plane with $\theta = 25°$ (Fig. P4.83). (a) If the block reaches a maximum height $h = 4.5$ m, what was the initial speed of the block? (b) Now consider a second inclined plane with the same tilt angle and for which the coefficient of kinetic friction between the block and the plane is $\mu_K = 0.15$. What initial speed is now required for the block to reach the same maximum height?

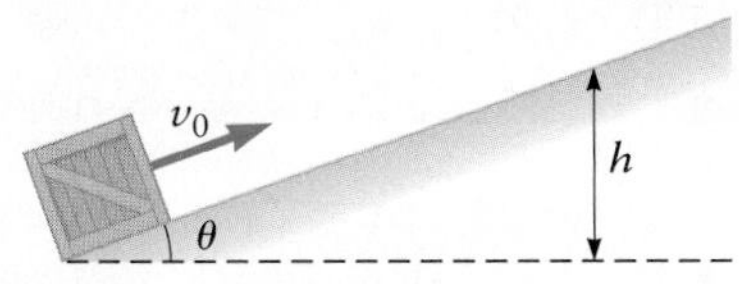

Figure P4.83

84. SSM A golf ball is hit with an initial velocity of magnitude 60 m/s at an angle of 65° with respect to the horizontal (x) direction. At the same time, a second golf ball is hit with an initial speed v_0 at an angle 35° with respect to x. If the two balls land at the same time, what is v_0?

85. A car travels along a level road with speed $v = 25$ m/s (about 50 mi/h). The coefficient of kinetic friction between the tires and the pavement is $\mu_K = 0.55$. (a) If the driver applies the brakes and the tires "lock up" so that they skid along the road, how far does the car travel before it comes to a stop? (b) If the same car is traveling downhill along a road that makes an angle $\theta = 12°$, how much does the car's stopping distance increase? (c) What is the stopping distance on an uphill road with $\theta = 12°$?

86. A person riding a bicycle on level ground at a speed of 10 m/s throws a baseball forward at a speed of 15 m/s relative to the bicycle at an angle of 35° relative to the horizontal (x) direction. (a) If the ball is released from a height of 1.5 m, how far does the ball travel horizontally, as measured from the spot where it is released? (b) How far apart are the bicycle and the ball when the ball lands? Ignore air drag.

87. R Two blocks of mass $m_1 = 2.5$ kg and $m_2 = 3.5$ kg rest on a double inclined plane with equal angles (Fig. P4.87). The blocks are connected by a string that passes over a pulley, and the blocks are in motion. Consider two different scenarios: (1) there is a coefficient of friction μ_K between m_1 and the plane, and there is no friction between m_2 and the plane; and (2) there is a coefficient of friction μ_K between m_2 and the plane, and there is no friction between m_1 and the plane. In which scenario is the acceleration of the blocks larger?

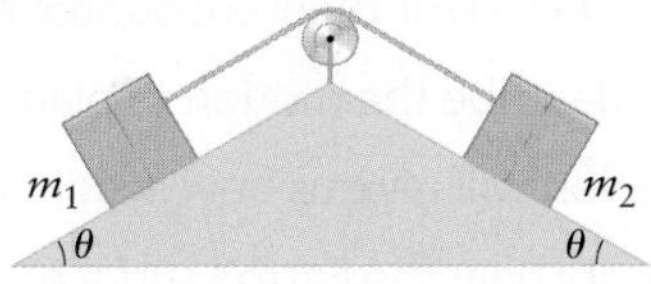

Figure P4.87

88. A canoe travels across a 40-m-wide river that has a current of speed 5.0 m/s. The canoe has a speed of 0.80 m/s relative to the water (Fig. P4.88). The person paddles the canoe so as to cross the river as fast as possible. (a) What is the canoe's velocity relative to an observer on shore? Give the magnitude and direction of the velocity vector. (b) How far downstream does the canoe travel as it crosses from one side of the river to the other?

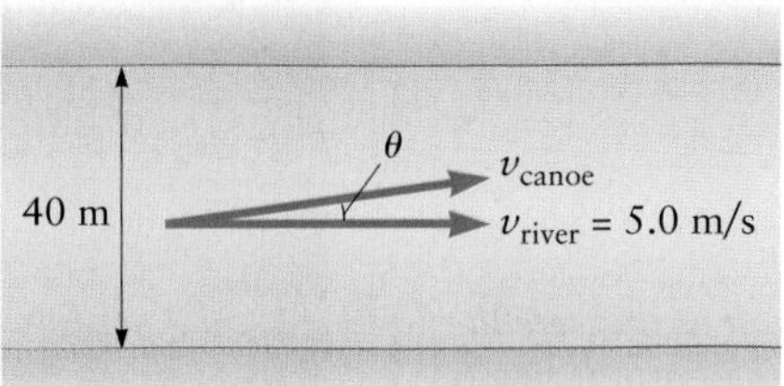

Figure P4.88

89. RT **Hanging out.** (a) A block of mass $m = 20$ kg hangs from two cables attached to a ceiling as shown in Figure P4.89A. What is the approximate tension in each cable? *Hint:* The two cables make the same angle with the vertical direction, but you will have to estimate this angle from the figure. (b) The two cables are now rearranged so that they make different angles with the vertical direction as shown in Figure P4.89B. Estimate the tensions in the two cables.

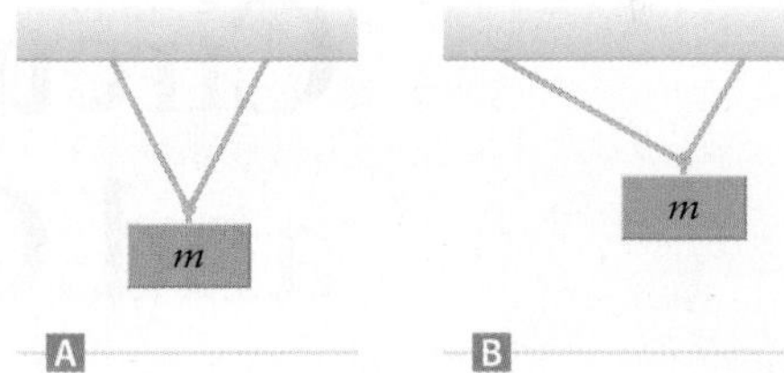

Figure P4.89

90. A projectile is fired uphill as sketched in Figure P4.90. If $v_0 = 150$ m/s and $\theta = 30°$, what is L?

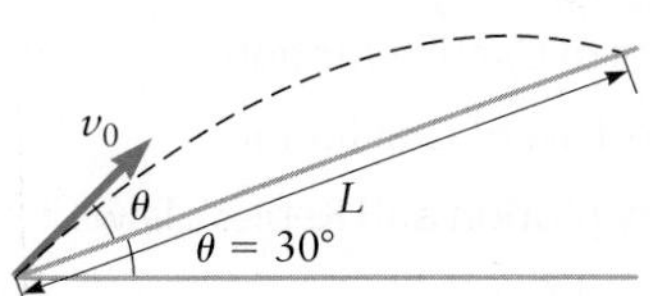

Figure P4.90

91. R When airplanes land or take off, they always travel along a runway in the direction that is "into" the wind because the "lift force" on an airplane wing depends on the speed of the airplane relative to the air (v_{rel}). We'll see in Chapter 10 that the lift force is proportional to v_{rel}^2. Consider a case in which a runway is parallel to the x axis and the wind velocity is $v_{wind} = +20$ mi/h (directed along $+x$). (a) In what direction should the airplane travel to take off (along $+x$ or $-x$)? (b) Estimate the speed of a typical airplane (e.g., a Boeing 737) relative to the runway as it takes off. (c) Find the ratio of the lift force for an airplane that travels into the wind to the lift force when it travels in the opposite direction. (d) Is this factor large enough to make a significant difference?

92. R **Soccer on a windy day.** A professional soccer player can kick a ball with a speed of about 30 m/s (about 75 mi/h). Does air drag play a significant role in the trajectory of the ball? Give a reason for your answer. *Hint*: Compare the force of air drag to the force of gravity on the ball.

CHAPTER

5

▶ *The motion of this satellite as it orbits the Earth is nearly circular and also involves the gravitational force. We study both of these topics in this chapter. (Erik Simonsen/ Photographer's Choice/Getty Images)*

Circular Motion and Gravitation

OUTLINE

In the past few chapters, we studied the connection between forces and motion, and we learned how to apply Newton's laws in a variety of situations. In many cases, those situations involved motion in one dimension. While we have also studied two- and three-dimensional motion, most of those cases could be treated as a combination of one-dimensional problems. For example, in our work on projectile motion, we were able to treat the horizontal and vertical motions as essentially separate one-dimensional problems. In this chapter, we consider a different type of motion, called **circular motion**, in which the acceleration is far from constant and that *cannot* be reduced to a one-dimensional problem. Circular motion is found in many situations, such as a car traveling around a turn, a roller coaster traveling near the top or bottom of its track, a centrifuge, and the Earth orbiting the Sun. Our studies of orbital motion and the forces that make it possible will also lead us to explore the gravitational force in more detail. We'll then encounter Kepler's laws of orbital motion and see how they describe the motion of planets, moons, and satellites. These studies of gravitation will give us new insights into the quantity g, the acceleration due to gravity near the Earth's surface.

5.1 Uniform Circular Motion

Figure 5.1 shows a top-down view of a person running around a circular track. For simplicity, let's assume our runner is moving with a constant speed so that the magnitude of her velocity is constant. The direction of her velocity $\vec{v}$, however, changes as she moves around the track since the velocity vector is always directed tangent to the circle. Her velocity is thus not constant, even though her speed is constant. Circular motion at constant speed is known as ***uniform circular motion.***

The runner in Figure 5.1 is moving at a constant speed v, so the time it takes her to travel once around the track (to run one complete lap) is equal to the circumference of the track divided by v. If the track has a radius r, the circumference is $2\pi r$ and the time to complete one lap is

$$T = \frac{2\pi r}{v} \tag{5.1}$$

This time T is called the ***period*** of the motion.

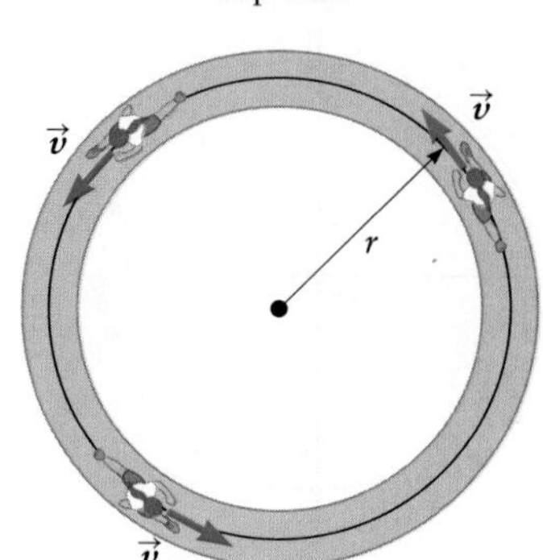

▲ **Figure 5.1** This runner is moving with a constant speed around a circular track. While her speed is constant, her velocity is not constant. The direction of $\vec{v}$ changes as she moves around the track.

EXAMPLE 5.1 A Bug on a Compact Disc

A bug is sitting on the edge of a compact disc of radius $r = 6.0$ cm (Fig. 5.2). The bug undergoes uniform circular motion as the CD spins. **(a)** If the bug traverses this circle exactly six times in precisely 1 s, what is the period of the motion? **(b)** What is the bug's speed?

RECOGNIZE THE PRINCIPLE

The period T is the time it takes to travel once around the circle, so we can find T from the given information. We can then use the relation between period and v in Equation 5.1 to find the bug's speed.

SKETCH THE PROBLEM

Figure 5.2 shows the problem. The bug moves in a circle of radius r.

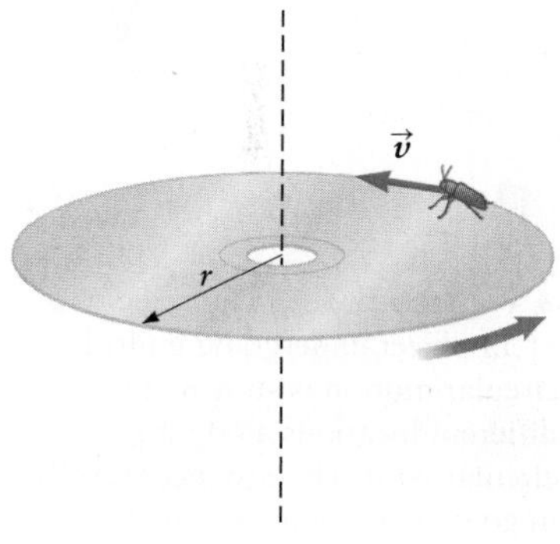

▲ **Figure 5.2** Example 5.1. The bug sitting on this CD is undergoing uniform circular motion.

IDENTIFY THE RELATIONSHIPS AND SOLVE

(a) Since the bug completes six trips each 1 s, we have

$$T = \frac{1.0 \text{ s}}{6 \text{ trips around circle}} = \frac{0.17 \text{ s}}{1 \text{ trip}}$$

The period is the time to complete one trip around the circle, so the bug's period is

$$T = \boxed{0.17 \text{ s}}$$

(b) The bug is undergoing uniform circular motion, so its speed v is constant and equal to the total distance traveled divided by the time. A distance d equal to six times the circumference is traveled in 1 s, so $d = 6 \times (2\pi r)$ while $t = 1.0$ s. Hence,

$$v = \frac{d}{t} = \frac{6(2\pi)(0.060 \text{ m})}{1.0 \text{ s}} = \boxed{2.3 \text{ m/s}}$$

► ***What does it mean?***

The period of circular motion is equal to the time it takes to move around the circle once. Hence, T does not depend on the bug's location on the CD; the period is the same for a bug at the edge of the CD or near the center. The speed of the bug, however, does depend on its location (the value of r). The speed is largest at the edge (a large value of r) and decreases as the bug moves toward the center.

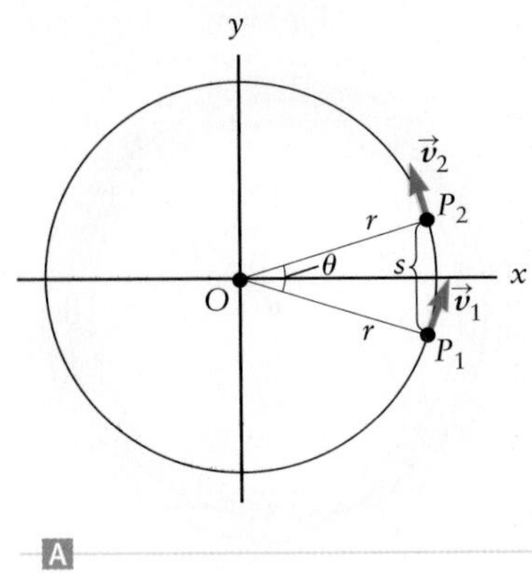

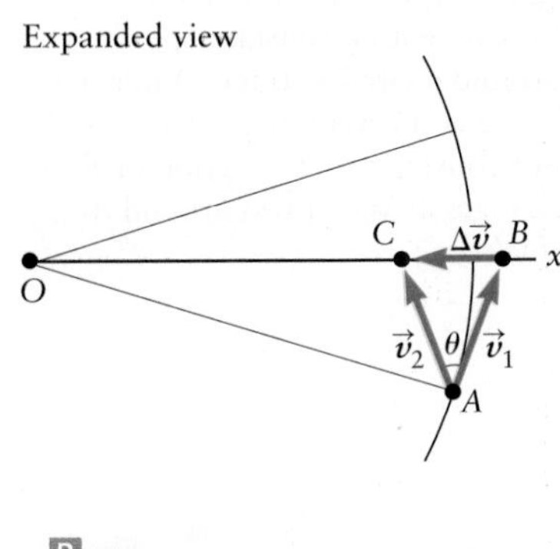

▲ **Figure 5.3** A The velocity $\vec{v}$ of an object undergoing uniform circular motion is shown at two different locations along the circular path. The distance traveled in going from point P_1 (time t_1) to point P_2 (time t_2) is s. B Expanded view of the velocity vectors at P_1 and P_2. The difference $\Delta\vec{v}$ is directed toward the center of the circle.

Centripetal Acceleration

Recall from Chapter 2 that the acceleration $\vec{a}$ of an object is proportional to the change in $\vec{v}$ over the course of a time interval Δt:

$$\vec{a} = \lim_{\Delta t \to 0} \frac{\Delta\vec{v}}{\Delta t} \tag{5.2}$$

For an object undergoing uniform circular motion, the velocity is changing with time, so $\Delta\vec{v}$ is definitely not zero. Uniform circular motion therefore involves a *nonzero acceleration.*

To calculate this acceleration, let's consider how the velocity of the runner in Figure 5.1 changes as she moves around the track. Figure 5.3A shows her velocity vectors at two nearby locations P_1 and P_2, and Figure 5.3B shows the corresponding change in velocity $\Delta\vec{v}$ over a short time interval Δt. We can use the geometry in Figure 5.3B to estimate the average acceleration $\vec{a}_{\text{ave}} = \Delta\vec{v}/\Delta t$ during this interval. If Δt is small, the instantaneous acceleration (Eq. 5.2) is approximately equal to $\vec{a}_{\text{ave}}$. We can therefore write

$$\vec{a} \approx \vec{a}_{\text{ave}} = \frac{\Delta\vec{v}}{\Delta t} \tag{5.3}$$

Let's first consider the direction of $\vec{a}$. During the time interval $\Delta t = t_2 - t_1$, our runner travels a distance $s = v\,\Delta t$ along the circumference of the circle. (Recall that the speed v is the magnitude of both $\vec{v}_1$ and $\vec{v}_2$.) The velocity vectors $\vec{v}_1$ and $\vec{v}_2$ are both tangent to the circle and hence are perpendicular to the radius lines drawn at P_1 and P_2 in Figure 5.3A. In Figure 5.3B, we have redrawn the velocity vectors so that they have a common "tail point." When drawn in this way, the velocity vectors $\vec{v}_1$ and $\vec{v}_2$ form two sides of the triangle labeled ABC. The third side of this triangle is $\Delta\vec{v} = \vec{v}_2 - \vec{v}_1$ and is directed toward the center of the circle. This result is true at all points along our runner's circular path. Since the acceleration vector is parallel to $\Delta\vec{v}$ (see Eq. 5.3), the acceleration of an object undergoing uniform circular motion is always directed *toward the center of the circle.*

To compute the magnitude of this acceleration, we again make use of triangle ABC in Figure 5.3B. Two sides of this triangle are formed by $\vec{v}_1$ and $\vec{v}_2$, so these sides are both of length v, while the other side of this triangle has a length Δv. Triangle ABC has the same interior angles as the triangle defined by O, P_1, and P_2 in Figure 5.3A, so these triangles are similar. Two sides of triangle OP_1P_2 are along the radius of the circle and are thus of length r, while the other side (between points P_1 and P_2) has a length $s = v\Delta t$, and when Δt is small, this arc approaches a straight line. Triangles ABC and OP_1P_2 are similar, so the ratios of their corresponding sides are equal; hence,

$$\frac{\Delta v}{v} = \frac{s}{r} = \frac{v(\Delta t)}{r}$$

$$\Delta v = \frac{v^2(\Delta t)}{r} \tag{5.4}$$

According to Equation 5.3, the magnitude of the acceleration is equal to the ratio $\Delta v/\Delta t$, so we can rearrange the result in Equation 5.4 to get

$$a_c = \frac{\Delta v}{\Delta t} = \frac{v^2(\Delta t)/r}{\Delta t}$$

Magnitude of the centripetal acceleration

$$a_c = \frac{v^2}{r} \tag{5.5}$$

This acceleration, called the ***centripetal acceleration***,[1] is usually denoted by the symbol a_c. Although this derivation started with the average acceleration in Equation 5.3, the result becomes exact for a very small time interval $\Delta t \to 0$; hence, Equation 5.5 gives the instantaneous centripetal acceleration.

[1]The adjective *centripetal* means "center seeking."

Circular Motion and Forces

The result for the acceleration in Equation 5.5 applies to any object undergoing circular motion. Such an object has an acceleration of magnitude $a_c = v^2/r$, and this acceleration is directed toward the center of the circle. From Newton's second law, we know that accelerations are caused by forces, so for an object undergoing uniform circular motion, we have

$$\sum \vec{F} = m\vec{a}$$

In terms of magnitudes,

$$\sum F = ma_c = \frac{mv^2}{r} \tag{5.6}$$

Force required for uniform circular motion

which is the force required to make an object of mass m travel with speed v in a circle of radius r. This force must be directed toward the center of the circle.

As an example, consider a rock tied to a string and suppose a person is twirling the string so that the rock moves in a circle. For simplicity, let's assume this experiment is being done by an astronaut in distant space (Fig. 5.4), so the only force on the rock comes from the string (i.e., all gravitational forces are negligible). This rock is undergoing circular motion; hence, its motion is described by Equation 5.6. The only force comes from the tension T in the string, so

$$\sum F = T = \frac{mv^2}{r}$$

For the rock to travel in a circle, the tension must have this value. (Note that T here is the tension in the string, not the period of the motion.)

When an object undergoes circular motion, there must be a force of magnitude mv^2/r acting on it. For the rock in Figure 5.4A, this force is the tension in the string, but in other situations, the force might be due to gravity, friction, or some other source. Without such a force, the object *cannot* undergo circular motion.

What happens if the string in Figure 5.4 suddenly breaks? The force on the rock would then be zero, and according to Newton's first law, the rock would then move away in a straight-line path—that is, with a constant velocity (Fig. 5.4B). The rock does not move radially outward, nor does it "remember" its circular trajectory. The only way the rock can move in a circle is if there is a force that makes it do so, and that force is provided by the string.

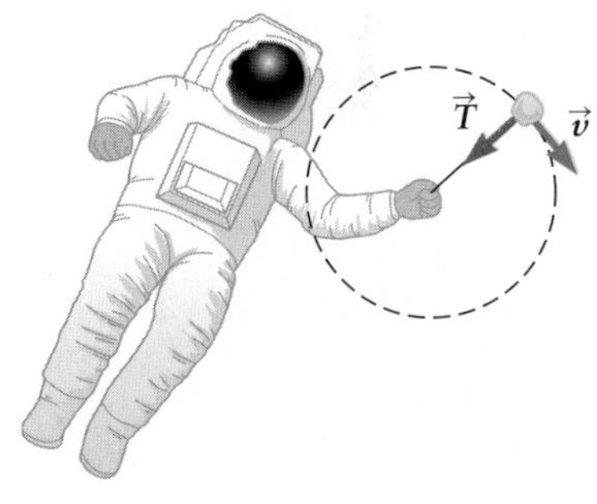

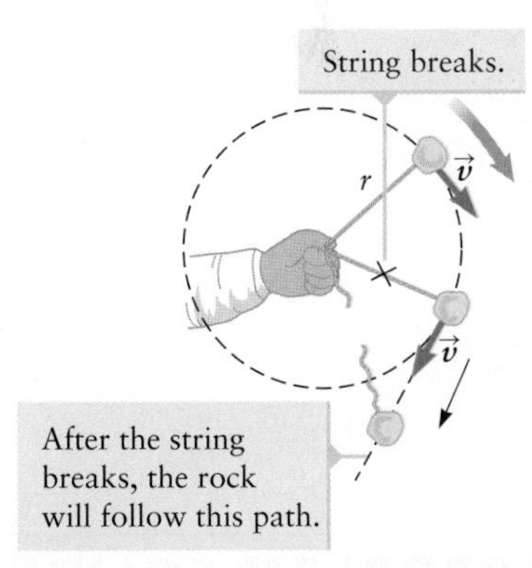

▲ Figure 5.4 A An astronaut in distant space twirls a rock on a string. The only force on the rock is due to the tension in the string (all gravitational forces are negligible). B If $T = mv^2/r$, the rock will undergo uniform circular motion. If the string breaks, the rock will move along a straight line (obeying Newton's first law).

CONCEPT CHECK 5.1 Velocity and Acceleration in Circular Motion

Consider the rock in Figure 5.4A as it undergoes uniform circular motion (before the string breaks). Which of the following statements is correct? (More than one statement may be correct.)

(a) The direction of $\vec{a}$ changes as the rock moves around the circle.
(b) The direction of $\vec{v}$ changes as the rock moves around the circle.
(c) $\vec{a}$ and $\vec{v}$ are always perpendicular.
(d) $\vec{a}$ and $\vec{v}$ are always parallel.

EXAMPLE 5.2 Centripetal Acceleration of a Compact Disc

Suppose the bug in Example 5.1 has a mass $m = 5.0$ g and sits on the edge of a compact disc of radius 6.0 cm. The CD is spinning such that the bug travels around its circular path three times per second. Find **(a)** the centripetal acceleration of the bug and **(b)** the total force on the bug. Also identify the *origin* of the force that enables the bug to move in a circle.

(continued) ▶

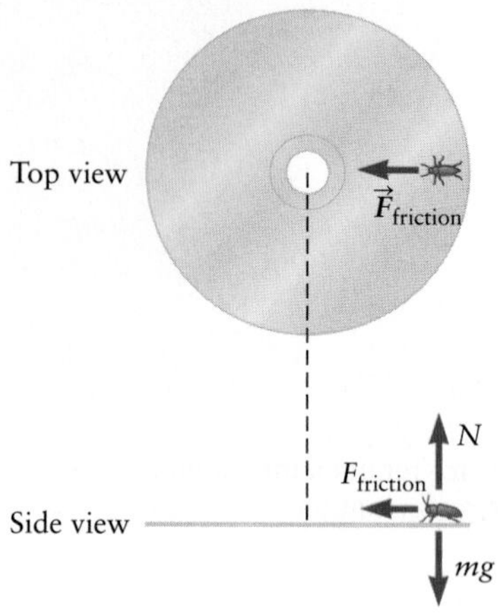

▲ **Figure 5.5** Example 5.2. There are three forces acting on the bug on this CD. The force responsible for the centripetal acceleration is provided by the force of static friction and is directed toward the center of the circle.

RECOGNIZE THE PRINCIPLE

Because the bug's acceleration along the vertical direction is zero, the normal force N and the force of gravity mg must cancel; hence, $N = mg$. The bug is undergoing circular motion, so we know a third force must provide the force required to produce the centripetal acceleration a_c. This third force keeps the bug from slipping relative to the CD and is just the frictional force. To make the bug undergo uniform circular motion, this force must be directed toward the center of the circle.

SKETCH THE PROBLEM

The forces acting on our bug are shown at the bottom of Figure 5.5.

IDENTIFY THE RELATIONSHIPS

To find the centripetal acceleration, we need to determine the speed of the bug, which we can find from its period and the radius of its circular path. We can then compute the magnitude of a_c and the force needed to produce it.

SOLVE

(a) The bug traverses the circle three times in 1 s, so it travels a distance equal to three times the circumference each second. Its speed is thus

$$v = \frac{3 \times (2\pi r)}{t} = \frac{3(2\pi)(6.0 \text{ cm})}{1.0 \text{ s}} = 110 \text{ cm/s} = 1.1 \text{ m/s}$$

From Equation 5.5, the centripetal acceleration is

$$a_c = \frac{v^2}{r} = \frac{(1.1 \text{ m/s})^2}{0.060 \text{ m}} = \boxed{20 \text{ m/s}^2}$$

(b) The required force is (Eq. 5.6)

$$F = ma_c = (5.0 \text{ g})(20 \text{ m/s}^2) = (5.0 \times 10^{-3} \text{ kg})(20 \text{ m/s}^2)$$

$$F = 0.10 \text{ kg} \cdot \text{m/s}^2 = \boxed{0.10 \text{ N}}$$

This force is produced by friction between the bug's feet and the surface of the CD.

► ***What does it mean?***

Even though the bug is moving, the force in part (**b**) is static friction because the bug is not slipping relative to the CD's surface.

Now that we have analyzed some examples of circular motion, it is time to outline a general way to approach such problems.

PROBLEM SOLVING Analyzing Circular Motion

1. **RECOGNIZE THE PRINCIPLE.** An object can move in a circle only if there is a force $F = mv^2/r$ directed toward the center of the circle.

2. **SKETCH THE PROBLEM.** Make a drawing that shows the path followed by the object of interest. This drawing should identify the circular part of the path, the radius of this circle, and the center of the circle.

3. **IDENTIFY THE RELATIONSHIPS.**
   - Find all the forces acting on the object; as in our applications of Newton's laws in Chapters 3 and 4, a free-body diagram is often very useful.
   - Using your drawing and free-body diagram, find the components of the forces that are directed *toward the center* of the circle and find the components perpendicular to this direction.

- Apply Newton's second law $\sum F = ma$ for motion toward the center of the circle and (if necessary) in the perpendicular direction. The total force directed toward the center of the circle is (Eq. 5.6)

$$\sum F_{\text{center}} = ma_c = \frac{mv^2}{r}$$

4. **SOLVE FOR THE QUANTITIES OF INTEREST.** For example, the centripetal acceleration is (Eq. 5.5)

$$a_c = \frac{v^2}{r}$$

5. Always *consider what your answer means* and check that it makes sense.

When discussing the centripetal acceleration and the associated forces, it is often convenient to use a coordinate system in which the positive direction is toward the center of the circle. The "positive" direction changes as the object moves around the circle (as for the bug in Fig. 5.5). An advantage of this approach is that the centripetal acceleration is always positive, which can eliminate some extra minus signs in your equations.

Centripetal Acceleration of a Turning Car: What Are the Forces?

The difficulties faced by the bug in Example 5.2 are similar to those encountered by the driver of a car. While cars do not usually travel in perfect circles, a car making a turn or rounding a bend in a road is traveling in an approximate circle. Over a short distance, any curved path can be approximated by a circle, and the radius of this circle is called the ***radius of curvature***. Let's analyze the motion of a car that travels along a short section of such a circle as sketched in Figure 5.6. As it travels around a bend in this road, the car travels in a circular path, so there must be a force on the car directed toward the center of the circle (point C in Figure 5.6A). Continuing with our problem-solving strategy for circular motion (step 2), the sketch in Figure 5.6A shows the circular path followed by the car as the dashed curve. The center of this circle is at point C, and its radius is r. For step 3, the sketch in Figure 5.6B shows all the forces on the car along with a free-body diagram. There are two forces along the vertical direction, the force of gravity and the normal force exerted by the road on the car, while the only horizontal force is the force of friction F_{friction} from the road on the car. This force is static friction because we assume the tires are not slipping relative to the road. Applying Newton's second law along y, we get

$$\sum F_y = +N - mg = ma_y$$

Since the acceleration along y is zero, this result leads to $N = mg$. The only force directed toward the center of the circle is F_{friction}, and inserting it into Equation 5.6 gives

$$F_{\text{friction}} = \frac{mv^2}{r}$$

Given the values of m, v, and r, we can now calculate the force required to make the turn successfully, without leaving the road.

Notice that the required force increases as the car's speed increases. There is an upper limit on the force of static friction; hence, there is a maximum speed at which the car can make the turn without slipping. The maximum frictional force is $F_{\text{friction,max}} = \mu_S N = \mu_S mg$, so the condition for this maximum speed is

$$\frac{mv^2}{r} = F_{\text{friction, max}} = \mu_S mg$$

Solving for the maximum speed gives

$$v = \sqrt{\mu_S g r} \tag{5.7}$$

Let's put in some realistic numbers to see how the result in Equation 5.7 compares with our everyday experience. For a typical intersection in a residential

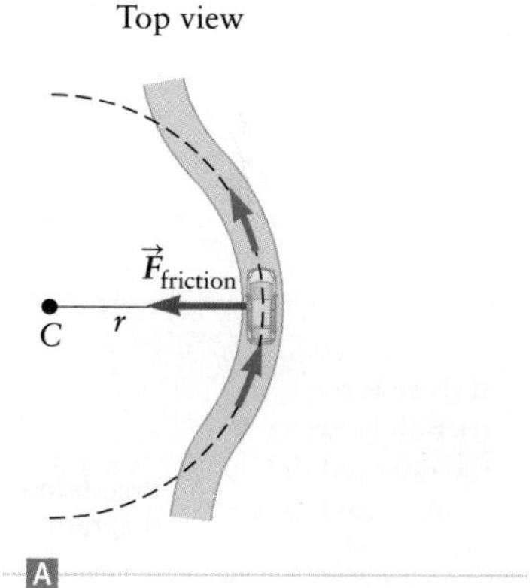

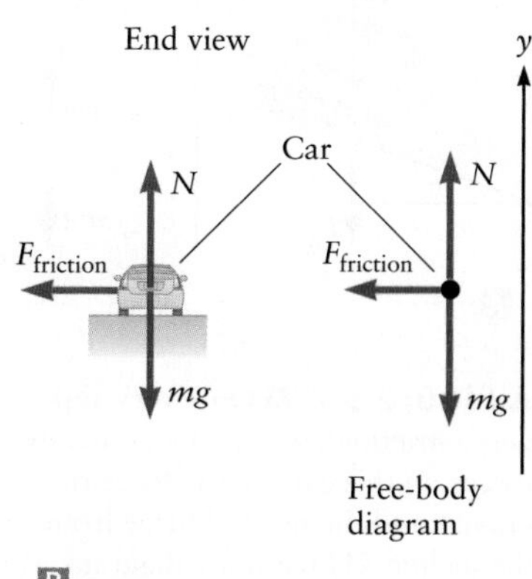

▲ **Figure 5.6** This car moves in an approximately circular path as it travels along a curve in the road. Friction between the car's tires and the road provides the force that causes the centripetal acceleration. **A** Top view. **B** End view showing forces on the car and a free-body diagram.

neighborhood, the effective radius of a turn might be $r = 10$ m, and the coefficient of friction between rubber and a road surface is typically $\mu_S = 0.6$. Inserting these values into Equation 5.7 gives

$$v = \sqrt{\mu_S g r} = \sqrt{0.6(9.8 \text{ m/s}^2)(10 \text{ m})} = 8 \text{ m/s}$$

which is approximately 15 mi/h. Hence, our physics does indeed give a good description of "real-life" driving.

CONCEPT CHECK 5.2 High-Speed Driving

Consider two vehicles, a compact car and a large truck, that are both traveling around the turn in the road in Figure 5.6 and assume the coefficient of friction between the tires and the road is the same for both. Which vehicle can drive through the turn at the largest speed without slipping?

(a) The car
(b) The truck
(c) The nonslipping speed is the same.

A Car on a Banked Turn: Analyzing the Forces

The maximum speed for making a turn successfully can be increased by banking the turn as shown in Figure 5.7. Let's analyze this situation, but for simplicity, we assume there is no friction between the tires and the road. (We'll add friction to this problem in Example 5.3.) We want to calculate the speed at which a car can successfully travel through such a turn without slipping off the road.

We again follow our problem-solving strategy for circular motion. The car travels in a circle, so there must be a force $F = ma_c$ directed toward the center of its circular path. For step 2, the car is shown in Figure 5.7A. The overall circular path is the same as for the car in Figure 5.6, and we again take r to be the radius of curvature. For step 3, the forces on the car are shown in Figure 5.7, which also gives a free-body diagram. We have assumed there is no friction, so the only forces on the car are due to gravity and the normal force exerted by the road on the car. We determine the components of the force along the vertical and along the direction toward the center of the circle (horizontal). Due to the banking, there is now a component of the normal force directed toward the center of the circle. From Figure 5.7B, the component of the force directed toward the center of the circle is $N \sin\theta$, and this must be equal to the mass of the car times the centripetal acceleration. For a car of mass m traveling at a speed v, we have

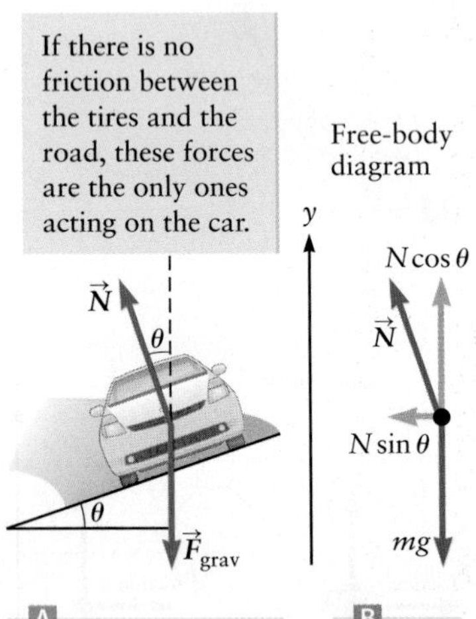

▲ **Figure 5.7** A On a very slippery—frictionless—incline, the only forces on this car are the force of gravity and the normal force from the incline. B Free-body diagram with forces resolved into horizontal and vertical components. The horizontal component of the normal force is $N\sin\theta$. This is the only horizontal force on the car, and it provides the centripetal acceleration that enables the car to move in a circular path as it travels around a banked curve.

$$F = ma_c = \frac{mv^2}{r} = N\sin\theta$$

which can be solved for the speed of the car:

$$v = \sqrt{\frac{Nr\sin\theta}{m}} \tag{5.8}$$

At this speed, the car will just be able to negotiate the turn without sliding up or down the banked road. To complete our calculation of the speed v, we must know the value of the normal force, which we can determine by considering the forces along the vertical direction in Figure 5.7B. Since the acceleration along the vertical direction is zero, the total force along y must be zero:

$$\sum F_y = N\cos\theta - mg = 0$$

Hence,

$$N = \frac{mg}{\cos\theta}$$

Inserting this result for N into Equation 5.8, we find

$$v = \sqrt{\frac{Nr \sin\theta}{m}} = \sqrt{\frac{mg}{\cos\theta}\frac{r\sin\theta}{m}} = \sqrt{gr\tan\theta}$$

When $\theta = 0$, this road is flat, and the result for the maximum speed is then zero because $\tan(0) = 0$. Therefore, you cannot turn on a very icy (frictionless) unbanked road without slipping, as you probably already knew.

EXAMPLE 5.3 Traveling through a Banked Turn with Friction

Consider again the problem of a car making a turn on a banked road, but now let's add friction to the situation. If the coefficient of static friction between the tires and the road is μ_S, what is the maximum speed at which the car can safely negotiate a turn of radius r with a banking angle θ?

RECOGNIZE THE PRINCIPLE

Because the car is traveling in a circle, the total force on the car must be $\sum F = mv^2/r$ directed toward the center of the circle. We must now include the force of friction when computing $\sum F$.

SKETCH THE PROBLEM

Figure 5.8A shows our car along with all the forces—the force of gravity, the normal force from the road, and the force of static friction—acting on it. Since we want to calculate the maximum safe speed, the force of friction will be directed *down* the incline so as to oppose any slipping of the car that would otherwise be directed up the plane.

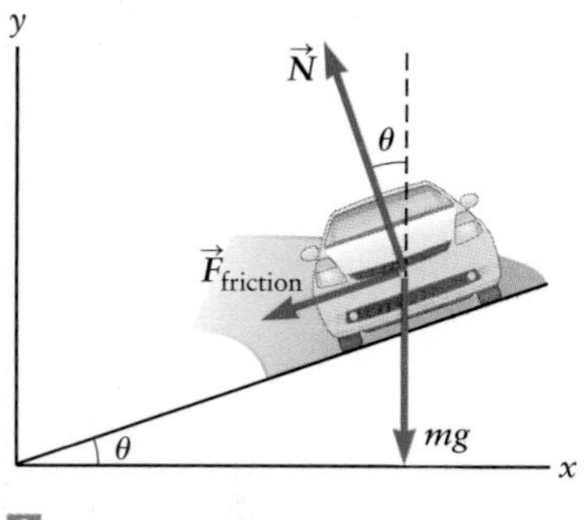

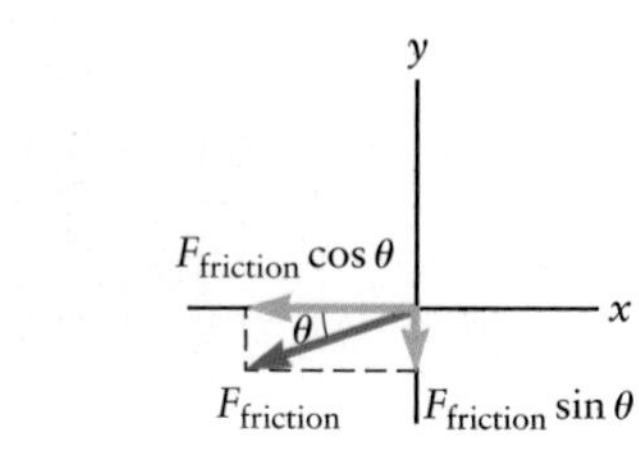

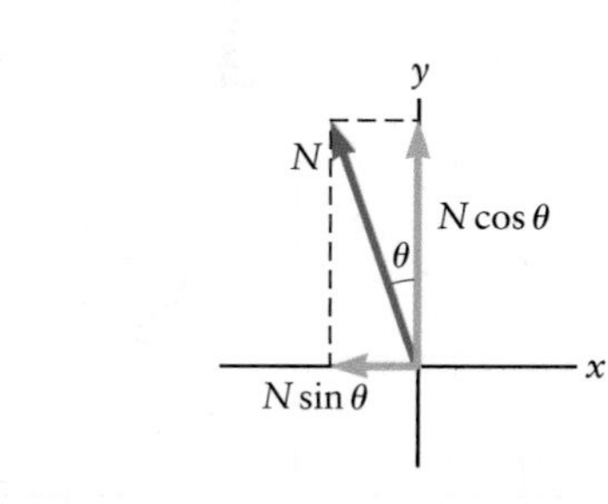

▲ **Figure 5.8** Example 5.3. A When friction is added to the road surface in Figure 5.7, there is an additional force (static friction) parallel to the incline. The horizontal and vertical components of B the frictional force and C the normal force are shown. The total horizontal force on the car is the sum of the horizontal components of the frictional force and the normal force.

IDENTIFY THE RELATIONSHIPS

The components of the forces are shown in parts B and C of Figure 5.8. We have two unknown quantities, the speed of the car and the normal force. We can get two relations by applying Newton's second law along the vertical and horizontal directions, just as we did for the case without friction. We can then solve for N and v.

SOLVE

The car is not slipping up or down the incline, so the acceleration along y is zero. Hence, the total force along y must be zero, and from Figure 5.8, we have

$$\sum F_y = 0 = +N\cos\theta - F_{friction}\sin\theta - mg$$

If the car is on the verge of slipping, the frictional force will have a magnitude of $F_{friction} = \mu_S N$. Hence,

$$N\cos\theta - \mu_S N\sin\theta - mg = 0$$

and we can rearrange to find the normal force

$$N = \frac{mg}{\cos\theta - \mu_S\sin\theta}$$

The total force along the horizontal direction provides the centripetal acceleration; in the coordinate system in Figure 5.8, we find

$$\sum F_x = -N\sin\theta - F_{friction}\cos\theta = -N\sin\theta - \mu_S N\cos\theta$$

(continued) ▶

The minus signs indicate that these forces are along the $-x$ direction. However, the center of the circle is also in this direction, so we have

$$N \sin\theta + \mu_S N \cos\theta = ma_c = \frac{mv^2}{r}$$

Solving for v then gives

$$v = \sqrt{\frac{Nr(\sin\theta + \mu_S \cos\theta)}{m}}$$

Inserting our result for the normal force leads to

$$v = \sqrt{\frac{mg}{\cos\theta - \mu_S \sin\theta}\,\frac{r(\sin\theta + \mu_S\cos\theta)}{m}} = \boxed{\sqrt{gr\frac{\mu_S\cos\theta + \sin\theta}{\cos\theta - \mu_S\sin\theta}}}$$

► ***What does it mean?***

For an automobile racetrack, a banking of 20° is common and the tracks are quite large; the radius of the turn could be 200 m. Inserting these values along with a coefficient of friction of $\mu_S = 0.6$, we find $v = 50$ m/s, which is approximately 110 mi/h; this is a typical speed in auto racing, so our model seems reasonable. Obtaining an even higher speed for this value of the track radius would require a larger coefficient of friction, a larger banking angle, or both.

5.2 Examples of Circular Motion

In all our examples so far, we have assumed the speed of the object undergoing circular motion is constant. In this case, the centripetal acceleration is the total acceleration. In many situations, however, the speed is not constant, resulting in *nonuniform* circular motion. In such cases, the total acceleration has two components as indicated in Figure 5.9. One component is directed along the circumference, that is, tangent to the circle. The magnitude of the tangential acceleration is nonzero only when v is changing with time. The other component of the acceleration is directed toward the center of the circle and is just the centripetal acceleration we derived in Equation 5.5. Even when v is not constant, our results from Section 5.1 for the centripetal acceleration and the associated force are still valid. Whenever an object is moving in a circle, there must be a component of the acceleration equal to a_c directed toward the center of the circle, so there must also be an associated force equal to mv^2/r. Hence, we can use the results from Section 5.1 to analyze a great many situations involving circular motion; we are *not* limited to situations in which the speed is constant.

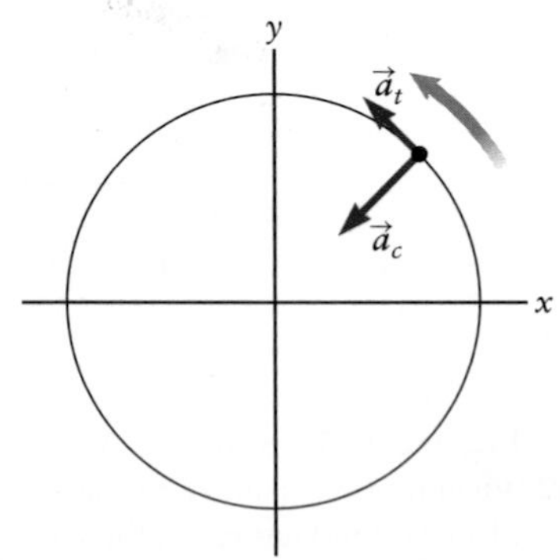

▲ **Figure 5.9** In general, an object traveling in a circle might not move with a constant speed. If the speed is not constant, there is a nonzero tangential acceleration $\vec{a}_t$ directed tangent to the circle. There is still a centripetal acceleration $\vec{a}_c$ directed toward the center of the circle, with magnitude $a_c = mv^2/r$.

Twirling a Rock on a String: What Is the Tension in the String?

Figure 5.10A shows a simple experiment in circular motion you have probably performed yourself: a rock tied to string is twirled in a vertical circle. The force of gravity acting on the rock makes it very difficult to keep the speed constant. Intuitively, we know that the speed will tend to be highest when the rock is at the bottom of the circle and lowest when the rock is at the top. Because the speed varies with time, the tangential acceleration is nonzero. Even so, the centripetal acceleration, the component of the acceleration directed toward the center of the circle, is still equal to $a_c = v^2/r$, where v is the speed at the point of interest.

Let's assume we know the speed v of the rock and consider how to calculate the tension T in the string. The answer will depend on the rock's location, so let's first deal with the case when the rock is at the bottom of the circle (Fig. 5.10B). Here the two forces on the rock, gravity and tension, are in opposite directions. These two forces together must add up to produce the centripetal acceleration

$$\sum F_{\text{center}} = +T_{\text{bottom}} - mg = \frac{mv^2}{r}$$

Solving for the tension gives

$$T_{\text{bottom}} = m\left(\frac{v^2}{r} + g\right) \tag{5.9}$$

Hence, the tension when the rock is at the bottom of the circle is large enough to overcome the gravitational force (mg) and also provide the required centripetal acceleration (mv^2/r).

When the rock is at the top of the circle as shown in Figure 5.10C, the gravitational force is parallel to the tension. The total force directed toward the center of the circle is now (recall that we take the positive direction to be toward the center of the circle)

$$\sum F_{\text{center}} = T_{\text{top}} + mg$$

The centripetal acceleration is also directed toward the center, so

$$\sum F_{\text{center}} = T_{\text{top}} + mg = \frac{mv^2}{r}$$

which leads to

$$T_{\text{top}} = m\left(\frac{v^2}{r} - g\right) \tag{5.10}$$

Comparing the results for T_{bottom} and T_{top} we see that if the speed v is the same at the top and bottom, the tension is smaller at the top. From Equation 5.10, we can also see that there is a minimum value of the speed that will keep the string taut. The tension in a string cannot be negative because a string cannot "push" on an object. As v is made smaller and smaller, however, the tension T_{top} in Equation 5.10 will eventually become zero. That will happen when

$$T_{\text{top}} = 0 = m\left(\frac{v^2}{r} - g\right)$$

$$\frac{v^2}{r} = g$$

$$v = \sqrt{gr} \tag{5.11}$$

If the speed is made smaller than this value, the string will become slack and circular motion is no longer possible.

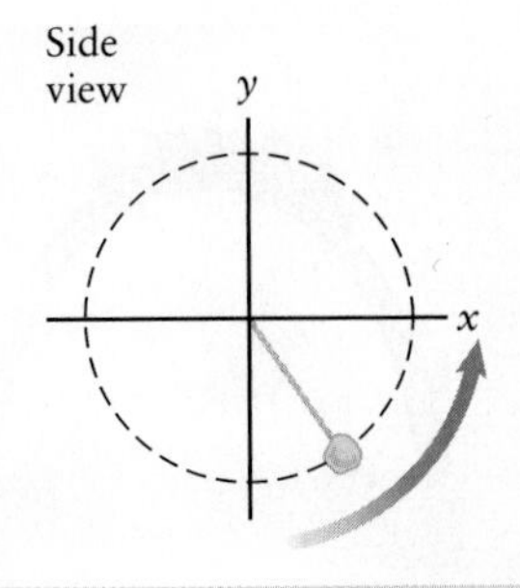

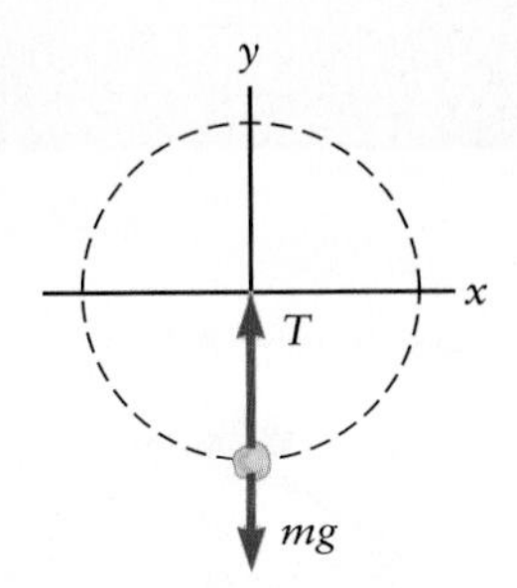

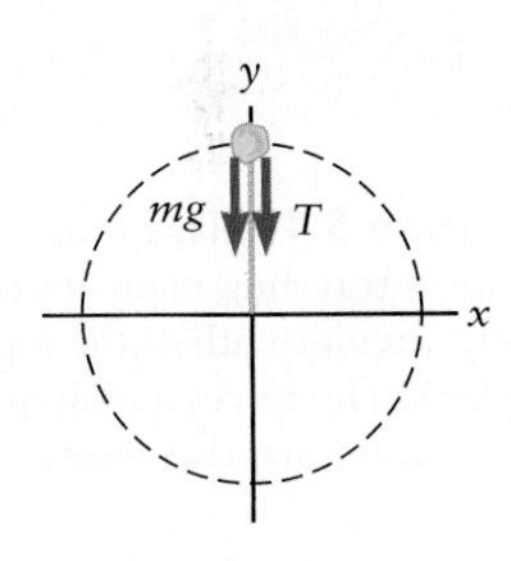

▲ **Figure 5.10** A This rock is moving in a vertical circle. There are two forces acting on it: the force of gravity and the tension from the string. B Forces on the rock when it is at the bottom of the circle. C Forces on the rock when it is at the top of its circular path.

Circular Motion and Amusement Park Activities: Maximum Speed of a Roller Coaster

The analysis of the rock's motion in Figure 5.10 can be applied to various amusement park activities. For example, when a roller coaster is near the minimum or maximum points on its track, its path is approximately circular. At these points, there must be an acceleration of magnitude v^2/r directed toward the center of the circle, where r is the track's radius of curvature. The roller coaster in Figure 5.11 is at the highest point on the track; let's calculate the maximum speed it can have without leaving the track. (This calculation would presumably be of interest to the riders.) Figure 5.11 shows the circular path followed by the roller coaster as well as the forces in the problem, gravity and the normal force exerted by the track on the roller coaster. The gravitational force is toward the center of the circle while the normal force is directed upward and is thus away from the center of the circle. The total force directed toward the center of the circle is

$$\sum F_{\text{center}} = mg - N$$

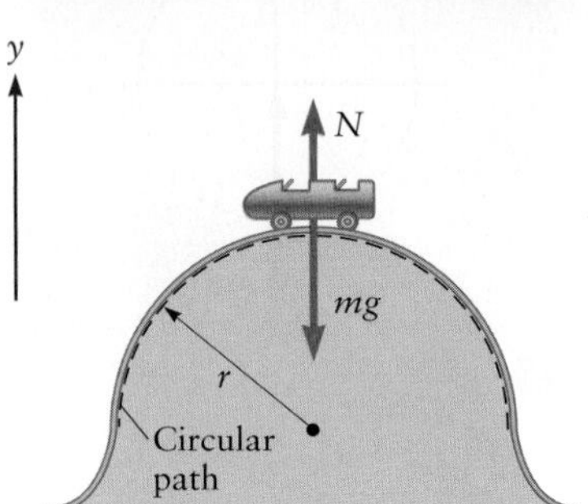

▲ **Figure 5.11** This roller coaster is traveling in an approximately circular path at the top of the track. The forces acting on the roller coaster are also shown.

which is related to the centripetal acceleration by (Eq. 5.6)

$$\sum F_{\text{center}} = mg - N = ma_c = \frac{mv^2}{r} \tag{5.12}$$

To find the maximum safe speed, we first rearrange Equation 5.12 to get the normal force:

$$N = m\left(g - \frac{v^2}{r}\right)$$

From this expression, we can see that as the speed v increases, the normal force N decreases. Eventually, we will reach a value of v at which N is zero:

$$N = m\left(g - \frac{v^2}{r}\right) = 0$$

which occurs when the speed satisfies

$$g - \frac{v^2}{r} = 0$$

$$v = \sqrt{gr} \tag{5.13}$$

If the speed is increased beyond this value, the value of N at the top of the track would have to be *negative*. Since a normal force can only push (but not pull), this scenario is impossible. Such a negative solution for N tells us that the roller coaster will leave the track if its speed exceeds $\sqrt{gr}$. Notice also that the normal force is the apparent weight of the roller coaster, and it will be your apparent weight if you are a passenger. Hence, when the speed is $v = \sqrt{gr}$, you will feel "weightless."

You might notice that our result (Eq. 5.13) for the speed at which the roller coaster leaves the track is the same as that found for the rock-on-a-string example in Figure 5.10C. There we saw (Eq. 5.11) that the string will go slack when the speed at the top is less than $v = \sqrt{gr}$. It is no accident that these two results are the same; in both cases, we are dealing with circular motion (so the acceleration is v^2/r for both), and in both cases, we have special situations in which the only force is gravity (because the tension in the string is zero and the normal force from the track is zero). In that sense, these problems are the "same."

EXAMPLE 5.4 Apparent Weight on a Roller Coaster

Consider the roller coaster in Figure 5.12 at the lowest point on its track, where the radius of curvature is $r = 20$ m. If the apparent weight of a passenger on the roller coaster is 3.0 times her true weight, what is the speed of the roller coaster?

RECOGNIZE THE PRINCIPLE

The apparent weight is the normal force on the object (or person) undergoing circular motion. So, we need to find the value of v at which $N = 3.0 \times mg$.

SKETCH THE PROBLEM

Figure 5.12 shows the circle followed by the roller coaster; also shown are the forces on the roller coaster, gravity and the normal force. Since the roller coaster is at the low point of the track, the normal force is directed toward the center of the circular arc defined by the track, and gravity is downward, away from the center.

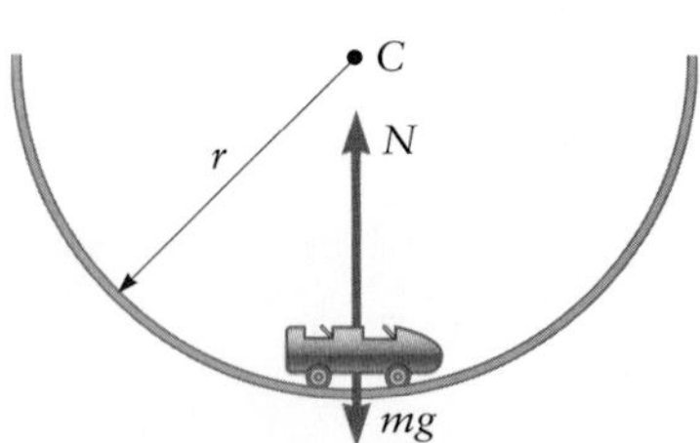

▲ **Figure 5.12** Example 5.4. Forces on a roller coaster at the bottom of its track.

IDENTIFY THE RELATIONSHIPS

We apply Newton's second law to relate the normal force to the speed of the roller coaster. Applying our relation for circular motion (Eq. 5.6), we have

$$\sum F_{\text{center}} = +N - mg = ma_c = \frac{mv^2}{r}$$

The apparent weight is three times the true weight, so $N = 3.0 \times mg$ and

$$N - mg = 3.0(mg) - mg = 2.0(mg) = \frac{mv^2}{r}$$

SOLVE

Solving for v, we find

$$v = \sqrt{2.0(gr)}$$

For a track of radius of $r = 20$ m, this result gives

$$v = \sqrt{2.0(gr)} = \sqrt{2.0(9.8\ \text{m/s}^2)(20\ \text{m})} = \boxed{20\ \text{m/s}}$$

▶ ***What does it mean?***

Our result for v is about 45 mi/h, which is a typical roller-coaster speed. Notice the similarity of this problem to the rock-on-a-string example in Figure 5.10B. The normal force on the roller coaster plays the role of tension in the string.

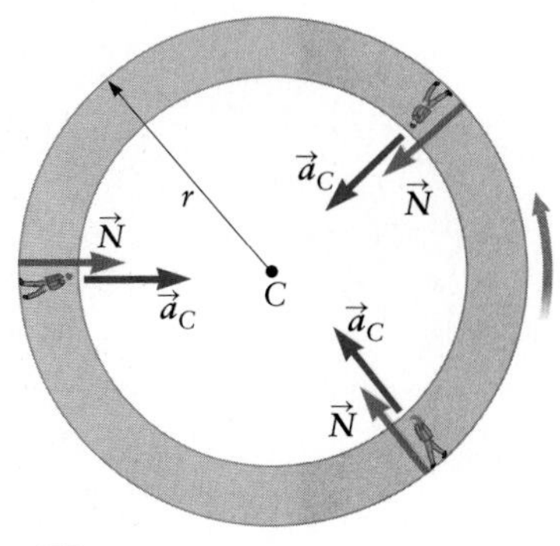

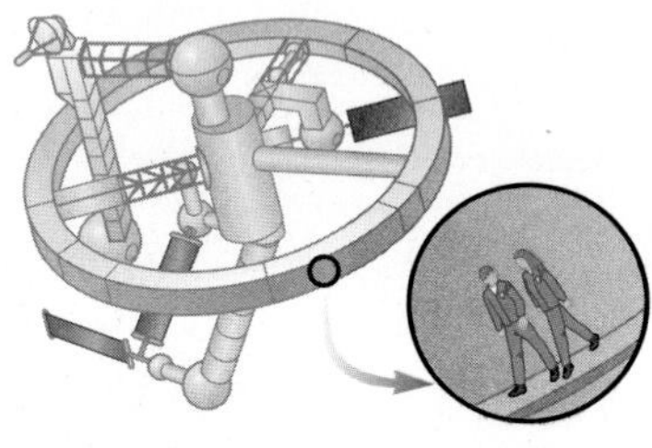

▲ **Figure 5.13** A circular space station that rotates about its central axis (C) generates "artificial gravity" for its passengers.

"Artificial Gravity" and a Rotating Space Station

A practical use of circular motion is sketched in Figure 5.13, which shows a hypothetical space station that uses circular motion to produce "artificial gravity" for its inhabitants. The station spins about a central axis so that the edge of the station, where the passengers spend most of their time, undergoes circular motion. The passengers in Figure 5.13 all experience a centripetal acceleration directed toward the center of the station (i.e., toward the center of the circle). This acceleration is produced by the normal force from the station floor on each passenger. For each passenger, $N = mv^2/r$, where v is the speed of the station edge and r is the radius of the station.

The passengers would have a slightly different interpretation of the situation. They would still be aware of the normal force acting on the bottoms of their shoes, but to them this force would be like the gravitational force they feel when on the surface of the Earth. In fact, if $N = mv^2/r = mg$, this force of "artificial gravity" would be the same as the real gravitational force on the Earth and the passengers would feel right at "home"!

EXAMPLE 5.5 Designing a Space Station

Consider a rotating space station similar to the one in Figure 5.13. If the radius of the station is $r = 40$ m, how many times per minute must the station rotate to produce a force due to "artificial gravity" equal to 30% of that found on the Earth?

RECOGNIZE THE PRINCIPLE

Figure 5.13A shows the circular path that the passengers follow. The only force is the normal force N, directed toward the center of the circle. We can therefore apply Equation 5.6 to get

$$\sum F = N = \frac{mv^2}{r} \tag{1}$$

We want the normal force to be 30% of the gravitational force on the Earth, which means $N = 0.30 \times mg$. We can use this expression together with Equation (1) to find the required speed v.

SKETCH THE PROBLEM

Figure 5.13 describes the problem and shows the forces on a passenger.

(continued) ▶

IDENTIFY THE RELATIONSHIPS AND SOLVE

Inserting the desired value of N into Equation (1) gives

$$N = 0.30(mg) = \frac{mv^2}{r}$$

and solving for v we find

$$v = \sqrt{0.30(gr)} \tag{2}$$

The time T for one revolution is equal to the distance traveled divided by the speed. The distance traveled is just the circumference of the station, so

$$T = \frac{2\pi r}{v} = \frac{2\pi r}{\sqrt{0.30(gr)}} = 2\pi\sqrt{\frac{r}{0.30g}} = 2\pi\sqrt{\frac{40 \text{ m}}{0.30(9.8 \text{ m/s}^2)}} = 23 \text{ s}$$

The number of revolutions per minute (rpm) is then

$$\frac{1 \text{ revolution}}{23 \text{ s}} \times \frac{60 \text{ s}}{\text{min}} = \boxed{2.6 \text{ rpm}}$$

▸ ***What does it mean?***

The speed in Equation (2) is independent of the person's mass. Hence, all passengers (regardless of their mass) experience an "artificial gravity" that is 30% of their weight on the Earth.

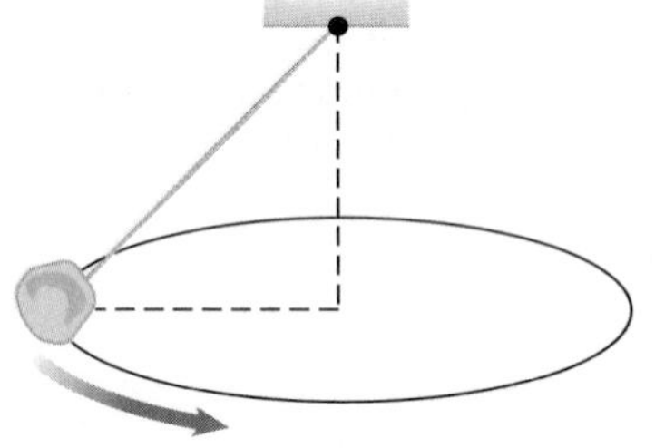

▲ **Figure 5.14** Concept Check 5.3.

CONCEPT CHECK 5.3 Net Force and Circular Motion

The rock in Figure 5.14 is suspended by a string tied to the ceiling and is undergoing uniform circular motion in a horizontal plane. What is the direction of the *net force* on the rock?

(a) It is along the string.
(b) It is downward (from gravity).
(c) It is toward the center of the circle.
(d) The net force is zero.

Physics of a Centrifuge

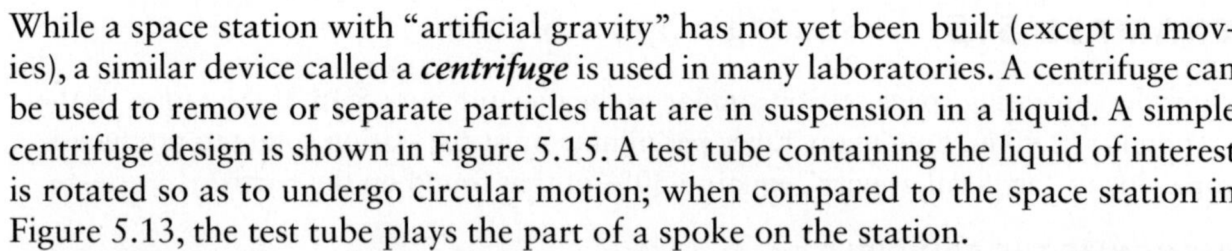

While a space station with "artificial gravity" has not yet been built (except in movies), a similar device called a ***centrifuge*** is used in many laboratories. A centrifuge can be used to remove or separate particles that are in suspension in a liquid. A simple centrifuge design is shown in Figure 5.15. A test tube containing the liquid of interest is rotated so as to undergo circular motion; when compared to the space station in Figure 5.13, the test tube plays the part of a spoke on the station.

The liquid in Figure 5.15 might be blood containing a particular type of cell. How long will it take to separate the cells from the blood? That is, how long will it take a cell to move from an initial location near the center of the centrifuge tube to the outer end of the tube?

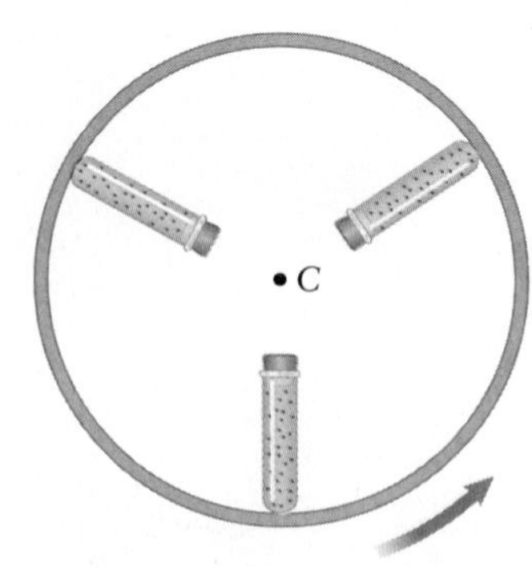

▲ **Figure 5.15** A centrifuge rotates at an extremely high rate about its axis at C, producing a large centripetal acceleration for the contents of the centrifuge. Here, the contents are a liquid that might contain cells in suspension. The result is that the cells move to the outer end of the tube.

Consider a cell that is already located at the end of the test tube so that it is in contact with the bottom of the tube. The circular motion of this cell is just like the motion of a passenger in the space station in Figure 5.13; hence, there is a normal force $N = mv^2/r$ exerted by the end of the tube on the cell. The cell thus experiences a force of "artificial gravity" $F_{AG} = mv^2/r$ much like the artificial gravity felt by a passenger on the space station. What's more, this "effective force" due to artificial gravity acts on the cell even when it is not at the end of the test tube, and this force will cause a cell to move toward the bottom (the outer end). When viewed by an observer moving along with the test tube, it seems as if the cell is simply falling "vertically" to the bottom of the tube in response to the force of artificial gravity (Fig. 5.16A).

We have called F_{AG} an "effective force" because things appear different when viewed by an observer who is not rotating with the centrifuge. From the point of view of a stationary observer, the motion of the cell appears as shown in Figure 5.16B. A cell that is initially located away from the end of the test tube moves in an expanding arclike trajectory since there is nothing to provide the force required to make it move in a circle (until it reaches the end of the tube). The cell would move in a straight line (compare to the rock in Fig. 5.4) if not for the walls of the test tube and the drag force from the liquid.

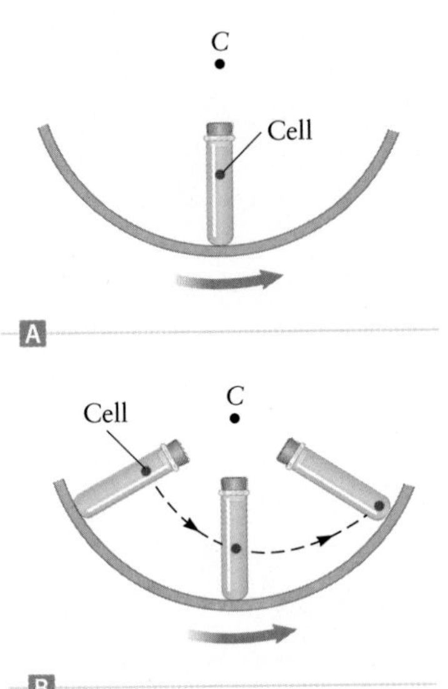

▲ **Figure 5.16** A A cell in the centrifuge moves as if there is a force due to "artificial gravity" acting on the cell. B When viewed by an observer who is not spinning with the centrifuge, the cell moves in a path that spirals outward as indicated by the dashed line.

Inertial and Noninertial Reference Frames Applied to a Centrifuge

While the motion of the cell in a centrifuge may be interpreted differently by different observers (as in Fig. 5.16), all observers will agree that the cell moves outward along the test tube, eventually coming to rest at the end of the tube. We have seen, however, that one observer (the cell) will deduce the presence of a force F_{AG}, whereas the other (stationary) observer will not. Which interpretation is correct? Is there actually a force F_{AG} or not? We mentioned in Chapter 4 that Newton's laws of motion should be used only by observers in *inertial reference frames*. You will recall that an inertial reference frame is one that moves with a constant velocity. In this example, the stationary observer is in an inertial reference frame, while Figure 5.16A corresponds to a noninertial reference frame. Only the stationary observer can use Newton's second law, and that observer's conclusion is correct: there is *no force* pushing the cell outward along the test tube. For this reason, F_{AG} as observed by the cell—that is, by the noninertial observer—is called a ***fictitious force***. Observers in rotating, and hence noninertial, reference frames often describe motion in terms of such fictitious forces.

Why should we ever want to use a noninertial reference frame? Why not stick to inertial frames, where we know that Newton's laws can be used? The answer is that sometimes motion appears simpler when viewed in a noninertial frame. For example, with our cell in a centrifuge, the path seen by a noninertial observer (Fig. 5.16A) is very simple. The cell just undergoes free fall to the bottom of the test tube, and it moves *as if* there were a force F_{AG} directed along the "vertical" (i.e., along the tube). This motion is certainly simpler than the curved path seen by an inertial observer as sketched in Figure 5.16B. It is thus sometimes useful to analyze motion from a noninertial reference frame. However, in such an analysis one must always account for the fictitious forces, such as F_{AG}, because these forces are necessary to make the inertial and noninertial results agree.

To calculate how long it will take for the cell in Figure 5.16 to be removed from the solution, let's take the point of view of the cell as it falls to the bottom of the test tube due to the force of artificial gravity, $F_{AG} = mv^2/r$. We must recognize that another force on the cell opposes its fall: the drag force we encountered in Chapter 3. There we saw that for a spherical object moving slowly through a liquid, there is a drag force described by Stokes's law (Eq. 3.23),

$$\vec{F}_{\text{drag}} = -Cr_{\text{cell}}\vec{v}_{\text{cell}} \tag{5.14}$$

where r_{cell} is the cell's radius and v_{cell} is the cell's speed along the test tube. Notice that v_{cell} is *not* the same as the speed v of the end of the test tube (v is the speed of the outer rim of the spinning centrifuge). Here, C is a constant that depends on the viscosity (the resistance to flow) of the fluid. For blood, C has a value of approximately 0.075 kg/(m·s). The drag force becomes larger as the speed v_{cell} of the cell increases, and the situation is very similar to the problem of a falling skydiver we encountered in Chapter 3. There we found that the skydiver reaches a terminal speed at which the drag force balances the force of gravity. For the cell in a centrifuge, the drag force in Equation 5.14 will balance F_{AG} and the cell will move at a constant speed along the test tube. This balance condition is

$$F_{AG} + F_{\text{drag}} = F_{AG} - Cr_{\text{cell}}v_{\text{cell}} = 0$$

Insight 5.1

USING DIFFERENT REFERENCE FRAMES

It is often useful to analyze a situation from different viewpoints. In Chapter 4, we took this approach when we considered the behavior of the mass-on-a-string accelerometer in a moving airplane. These different viewpoints are called ***reference frames***. A reference frame is simply a coordinate system, but it is necessary to specify the motion of the coordinate axes. For the cell in a centrifuge (Fig. 5.15), one reference frame moves with the cell as it undergoes circular motion (Fig. 5.16A). Hence, these coordinate axes rotate along with the arms of the centrifuge. This problem can also be analyzed from the point of view of an observer who is watching the centrifuge from the "outside" (Fig. 5.16B).

Since $F_{AG} = mv^2/r$, we have

$$\frac{mv^2}{r} = Cr_{\text{cell}}v_{\text{cell}}$$

Solving for the speed along the test tube gives

$$v_{\text{cell}} = \frac{mv^2}{Cr_{\text{cell}}r} \tag{5.15}$$

It is important to distinguish the tangential speed of the centrifuge (v) from the settling speed of the cell (v_{cell}), and the radius of the cell (r_{cell}) from the radius of the centrifuge (r).

CONCEPT CHECK 5.4 Inertial and Noninertial Reference Frames

We have seen that an observer in an inertial reference frame can use Newton's laws, while observers in a noninertial frame must include fictitious forces. Identify the following reference frames as either inertial or noninertial and explain your answers.

(a) A car traveling on a level road at a constant velocity
(b) A car traveling up a steep mountain road at a constant velocity
(c) A child sitting on a rotating merry-go-round
(d) An apple falling from a tree

EXAMPLE 5.6 Operation of a Centrifuge

Calculate the settling speed v_{cell} of a cell of radius 10 μm and mass 10^{-5} mg ($= 10^{-11}$ kg) in blood using Equation 5.15. Assume the centrifuge has a radius of 10 cm and a rotation rate of 3000 revolutions per minute. How long must the centrifuge run to make the cell settle out? Assume $C = 0.075$ kg/(m · s) for blood.

RECOGNIZE THE PRINCIPLE

The radius and mass of a cell are never known precisely and vary from cell to cell. Moreover, Stokes's relation for the drag force, Equation 5.14, applies exactly only for a spherical object, and cells are usually not precisely spherical! Even so, we can use Equation 5.15 to get an approximate value for the speed at which a cell moves. The only unknown quantity in Equation 5.15 is v, the speed of the centrifuge. We can find v from the period of the centrifuge, along with the relation between period and v in Equation 5.1.

SKETCH THE PROBLEM

Figure 5.17 shows the problem from the point of view of the cell. According to this noninertial observer, the force of artificial gravity ($\vec{F}_{AG}$) causes an acceleration a_c "downward" along the centrifuge tube.

IDENTIFY THE RELATIONSHIPS AND SOLVE

Our centrifuge makes 3000 revolutions in 1 min, so the time to make 1 revolution is $T = 1$ min/3000 $= 0.020$ s, and the circumferential speed is

$$v = \frac{2\pi r}{T} = \frac{2\pi(0.10 \text{ m})}{0.020 \text{ s}} = 31 \text{ m/s}$$

Inserting this in Equation 5.15 gives

$$v_{\text{cell}} = \frac{mv_c^2}{Cr_{\text{cell}}r} = \frac{(1 \times 10^{-11} \text{ kg})(31 \text{ m/s})^2}{[0.075 \text{ kg/(m} \cdot \text{s)}](1.0 \times 10^{-5} \text{ m})(0.10 \text{ m})} = \boxed{0.1 \text{ m/s}}$$

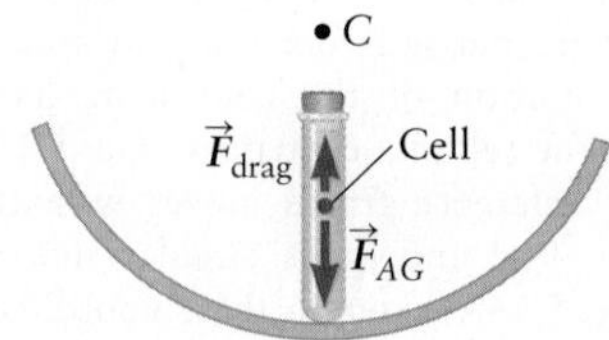

▲ **Figure 5.17** Example 5.6. A cell in a centrifuge moves as if there is a force of artificial gravity acting "downward" on the cell.

The length of the centrifuge tube will be limited by the radius of the centrifuge (Fig. 5.15). As a rough estimate, we can assume the tube length L is about half the radius of the centrifuge, so the cell in this case must travel a distance of at most $L = 5$ cm $= 0.05$ m. Using the value found for v_{cell}, the time required for the cell to move this distance is

$$t = \frac{L}{v_{cell}} = \frac{0.05 \text{ m}}{0.1 \text{ m/s}} = \boxed{0.5 \text{ s}}$$

▸ ***What does it mean?***

A centrifuge with a similar speed is used to separate blood cells from plasma. Note that our analysis used the value of v as measured at the end of the centrifuge tube, while most cells will begin their motion closer to the center of the centrifuge. Our calculation will thus overestimate F_{AG} and v_{cell}. Even allowing for this, and for variations in cell size, our analysis shows that it only takes a few seconds of spinning to completely separate the cells from the plasma.

5.3 Newton's Law of Gravitation

The orbital motions of planets and moons are circular (to a good approximation) in many cases, and the force responsible for this circular motion is gravity. We have already encountered the force of gravity as it acts on objects near the Earth's surface. Now we need to consider the gravitational force in more general situations.

Newton's three laws of motion are the foundation of our study of mechanics. However, Newton discovered another law of nature, the law of gravitation, that is perhaps just as important as his laws of motion. Newton's law of gravitation plays a key role in physics for two reasons. First, it allows us to calculate and understand the motions of a wide variety of objects, including baseballs, apples, and planets. Second, Newton's application of his law of gravitation to the motion of planets and moons was the first time that physics was applied (successfully) to describe the motion of the solar system. His work showed for the first time that the laws of physics apply to *all* objects and had a profound effect on how people viewed the universe.

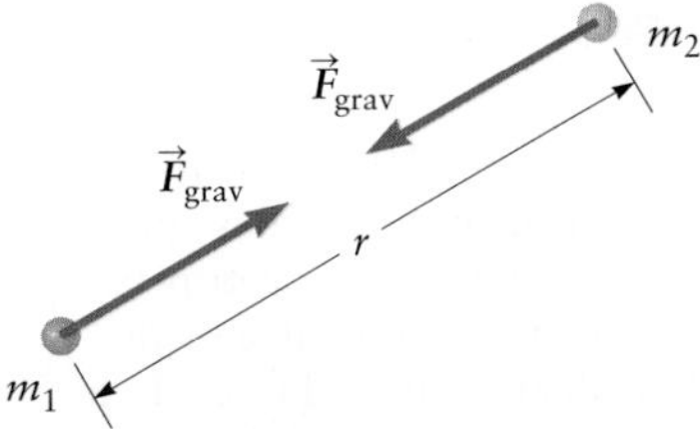

▲ **Figure 5.18** The gravitational force between two point masses m_1 and m_2 that are separated by a distance r is given by Equation 5.16. Notice that here the distance r is not the radius of a circle.

Newton's law of gravitation: **There is a gravitational attraction between *any* two objects. If the objects are point masses m_1 and m_2, separated by a distance r (Fig. 5.18), the magnitude of the gravitational force is**

$$F_{grav} = \frac{Gm_1m_2}{r^2} \tag{5.16}$$

Newton's law of gravitation

where $G = 6.67 \times 10^{-11}\ \text{N} \cdot \text{m}^2/\text{kg}^2$ is a constant of nature known as the universal gravitational constant. The gravitational force is *always attractive.* Every mass attracts every other mass.

The gravitational force law, Equation 5.16, is "symmetric." That is, the magnitude of the gravitational force exerted by mass 1 on mass 2 is equal to the magnitude of the gravitational force exerted by mass 2 on mass 1. Since the forces are both attractive, this result is precisely what we would expect from Newton's third law; the two gravitational forces are an *action–reaction pair* because they are equal in magnitude and opposite in direction and they act on different members of the pair of objects.

Gravitation and the Orbital Motion of the Moon

The gravitational force is responsible for the motion of planets, moons, asteroids, and comets. For example, the Moon follows an approximately circular path as it orbits the Earth (Fig. 5.19). This circular motion requires a force, which is provided by gravity. As a check, we can use Equation 5.6 to calculate the force required to make the Moon move in its circular orbit and then compare it with the gravitational

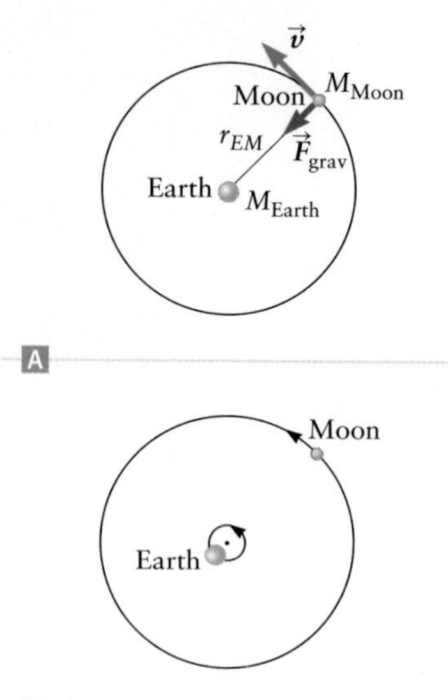

▲ **Figure 5.19** A The Moon's orbit around the Earth is approximately circular. The centripetal acceleration is provided by the Earth's gravitational force. To a good approximation, the Earth in this picture can be assumed to be fixed in space. B In a more accurate description, the Earth also moves in a circular "orbit" due to the gravitational force of the Moon on the Earth. This sketch is not to scale; the center of this orbit is inside the Earth.

force calculated from Newton's law of gravitation, Equation 5.16. The force required to make the Moon move in a circle is (from Eq. 5.6)

$$F = \frac{M_{\text{Moon}} v^2}{r_{EM}}$$

where M_{Moon} is the mass of the Moon and r_{EM} is the distance from the Earth to the Moon (the radius of the Moon's orbit). To find the speed v of the Moon, we notice that it completes one orbit and travels a distance of $2\pi r_{EM}$ in approximately $T =$ 27.3 days;[2] hence,

$$v = \frac{2\pi r_{EM}}{27.3 \text{ days}} = \frac{2\pi r_{EM}}{T}$$

The required force is thus

$$F = \frac{M_{\text{Moon}} v^2}{r_{EM}} = \frac{M_{\text{Moon}}(2\pi r_{EM}/T)^2}{r_{EM}} = \frac{4\pi^2 M_{\text{Moon}} r_{EM}}{T^2} \tag{5.17}$$

Inserting the known values (see Table 5.1) $M_{\text{Moon}} = 7.35 \times 10^{22}$ kg and $r_{EM} =$ 3.85×10^8 m along with $T = 27.3$ days $= 2.36 \times 10^6$ s, we find

$$F = 2.0 \times 10^{20} \text{ N} \tag{5.18}$$

This is the magnitude of the force needed to make the Moon follow its observed circular orbit. To confirm that this force is actually provided by gravity, let's calculate the gravitational force exerted by the Earth on the Moon from Equation 5.16 and show that it is indeed equal to the result for F in Equation 5.18. The mass of the Earth is (from Table 5.1) $M_{\text{Earth}} = 5.98 \times 10^{24}$ kg, and inserting the other values given above into Equation 5.16 gives

$$F_{\text{grav}} = \frac{G M_{\text{Earth}} M_{\text{Moon}}}{r_{EM}^2} = 2.0 \times 10^{20} \text{ N} \tag{5.19}$$

which does agree with the expected value of F in Equation 5.18.

When he was developing his theories of motion and gravitation, Newton almost certainly carried out these same calculations and obtained the results in Equations 5.18 and 5.19. That the force of gravity on the Moon is precisely equal to the force required to make the Moon move in its circular orbit must have been very strong evidence for Newton that his theories were indeed correct.

Insight 5.2

ORBITAL MOTION OF THE MOON AND THE EARTH

In our analysis of the Moon's orbital motion, we have assumed the Moon orbits around a "fixed" Earth. That is, we have assumed the Earth does not move at all as the Moon moves in a circle of radius r_{EM}. Although that is a good approximation, it is only an approximation because it ignores two aspects of the Earth's motion. (1) The Earth orbits the Sun. That orbital motion is much slower than the Moon's orbital speed, so it still makes sense (it is mathematically extremely accurate) in Equation 5.17 to treat the Earth as fixed. (2) The Earth also "orbits" the Moon. That is, as the Moon moves in its circular orbit, the Earth moves in a corresponding circular orbit as shown in Figure 5.19B. The radius of the Earth's circular orbit is much smaller than that of the Moon. In fact, the center of the Earth's circle is inside the Earth(!) and is located at a spot called the ***center of mass*** that we'll discuss in Chapter 7.

Applying Newton's Law of Gravitation: Calculating the Value of g

In Chapter 3, we introduced the "constant" g and learned that the gravitational force on an object near the Earth's surface has the magnitude $F_{\text{grav}} = mg$. Let's now see how this result is contained in the universal law of gravitation. Strictly speaking, Equation 5.16 applies only to the case of two "point" masses, two objects that are very small compared to the distance between them. This assumption works for our calculation of the gravitational force between the Earth and the Moon, but we certainly cannot consider the Earth to be a point mass in relation to an object on its surface. To appreciate the problem, consider the force exerted by the Earth's gravity on a person standing on the surface as sketched in Figure 5.20A. We would like to use Equation 5.16 to calculate this force, but what value should we use for the distance r? Some parts of the Earth are very close to the person (just beneath his feet), whereas other parts are quite far away (as much as twice the radius of the Earth). The answer is that when dealing with a spherical object, one can usually calculate the gravitational force it exerts on another object *as if* all the sphere's mass were located at its center. (See Insight 5.3.)

[2]In our examples involving planetary and satellite motion, we will carry three significant figures through the calculation (one more than usual in this book) to avoid problems due to rounding errors.

TABLE 5.1 Solar System Data: Properties of Several Objects in the Solar System, Including the Planets and Two of the Largest Dwarf Planets (Pluto and Eris)

Name	Mean Orbital Radius ($\times 10^{11}$ m)	Orbital Period (years)	Radius of Object ($\times 10^{6}$ m)	Mass ($\times 10^{24}$ kg)	Orbital Eccentricity
Mercury	0.579	0.240	2.44	0.330	0.21
Venus	1.08	0.615	6.05	4.87	0.007
Earth	1.50	1.00	6.37	5.98	0.017
Mars	2.28	1.88	3.39	0.644	0.093
Jupiter	7.78	11.9	71.5	1900	0.048
Saturn	14.3	29.4	60.3	568	0.054
Uranus	28.7	83.8	25.6	86.6	0.047
Neptune	45.0	164	24.8	102	0.009
Pluto	59.1	248	1.14	0.0131	0.25
Eris	100	560	2.4	—	0.44
Earth's Moon	3.85×10^{8} m	27.3 days	1.74	0.0735	0.055
Sun			6.96×10^{8} m	1.99×10^{30} kg	

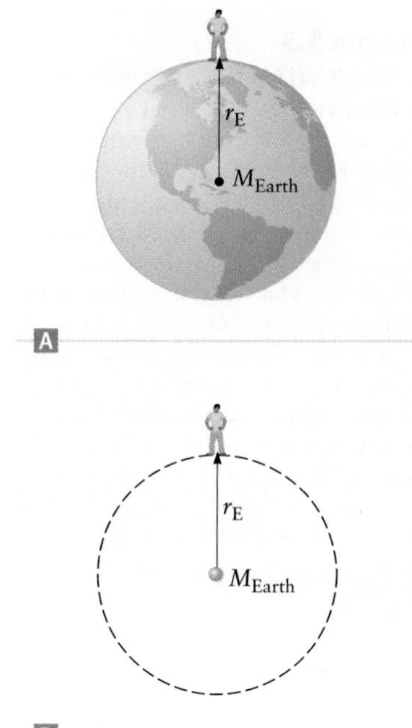

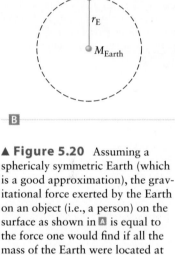

▲ **Figure 5.20** Assuming a sphericaly symmetric Earth (which is a good approximation), the gravitational force exerted by the Earth on an object (i.e., a person) on the surface as shown in A is equal to the force one would find if all the mass of the Earth were located at its center as shown in B.

The Earth's shape is very close to spherical, and the gravitational force exerted by the Earth on the person in Figure 5.20 can be calculated as if all the Earth's mass is at its center as sketched in Figure 5.20B. We can therefore use Equation 5.16 to calculate this force, with the separation r equal to the distance from the person to the center of the Earth; this distance is just the Earth's radius, r_E. We thus have

$$F_{\text{grav}} = \frac{GM_{\text{Earth}}M_{\text{person}}}{r_E^2} \tag{5.20}$$

This force is the weight of the person, which we have denoted by $M_{\text{person}}g$, where g is the "acceleration due to the Earth's gravity." Hence, we can use Equation 5.20 to calculate the value of g. Examining Equation 5.20, we see that it has the form $M_{\text{person}}g$ provided that g is given by

$$g = \frac{GM_{\text{Earth}}}{r_E^2} \tag{5.21}$$

The value of g is a function of only the Earth's mass and radius, and the universal gravitational constant G. The value of g is thus the same for all terrestrial objects near the Earth's surface. Inserting the known values of M_{Earth} and r_E (Table 5.1) along with G, we find

g and Newton's law of gravitation

$$g = \frac{GM_{\text{Earth}}}{r_E^2} = \frac{(6.67 \times 10^{-11}\ \text{N}\cdot\text{m}^2/\text{kg}^2)(5.98 \times 10^{24}\ \text{kg})}{(6.37 \times 10^{6}\ \text{m})^2}$$

$$g = 9.8\ \text{m/s}^2$$

which is the value we have already been using for g.

This calculation shows us where the value of g "comes from," and it also shows that g is not really a constant. The Earth happens to be approximately spherical, so all objects on the surface of the planet (i.e., all terrestrial objects) are approximately the same distance from the center and hence all have approximately the same value of g. However, we see from Equation 5.21 that g—and hence the weight of an object—will change if the object is moved farther from the Earth (e.g., by climbing a mountain). It also shows that the weight of an object will be different on another planet or on the Moon than it is on the Earth.

Insight 5.3
THE GRAVITATIONAL FORCE FROM THE EARTH

When dealing with a spherically shaped object, the gravitational force on a mass located outside the object can usually be calculated by assuming *all* the object's mass is located at its center. That is true as long as the density of the object is spherically symmetric. Hence, the result applies if the object has a uniform (constant) density. It also applies when the density varies with depth, as long as the density depends only on distance from the center. The result thus applies to a more realistic model of the Earth, consisting of a core with a different density than the crust.

EXAMPLE 5.7 What Would You Weigh on the Moon?

What is your weight on the Moon? Compare it with your weight on the Earth.

RECOGNIZE THE PRINCIPLE

Your weight on the Moon is just the Moon's gravitational force when you are located on its surface. Hence, we need to evaluate Equation 5.20 using the mass of the Moon and the radius of the Moon in place of those quantities for the Earth. (These data are listed in Table 5.1.)

SKETCH THE PROBLEM

This problem is described by Figure 5.20, but with the Earth replaced by the Moon.

IDENTIFY THE RELATIONSHIPS AND SOLVE

The author has a mass of approximately $M_{\text{author}} = 70$ kg; inserting this value into Equation 5.20 gives

$$F_{\text{grav}} = \frac{GM_{\text{moon}}M_{\text{author}}}{r_M^2} = \frac{(6.67 \times 10^{-11}\ \text{N} \cdot \text{m}^2/\text{kg}^2)(7.35 \times 10^{22}\ \text{kg})(70\ \text{kg})}{(1.74 \times 10^6\ \text{m})^2}$$

$$F_{\text{grav}} = 110\ \text{N}$$

The author's weight on the Earth is $M_{\text{author}}g = (70\ \text{kg})g = 690$ N, so his weight on the Moon is approximately one-sixth of his weight on the Earth.

► ***What does it mean?***

The *ratio* of the weight on the Moon to the weight on the Earth is *independent* of the mass of the object because both are proportional to the mass of the object. All objects weigh less on the Moon by a factor of approximately one-sixth.

EXAMPLE 5.8 The Force of Gravity in a Very Tall Building

When this book was written, the tallest building in the world was the Burj Khalifa building in Dubai (Fig. 5.21A), with the top being a distance $h = 830$ m above the bottom. Find the ratio of your weight at the top of the building to your weight at ground level.

RECOGNIZE THE PRINCIPLE

At the top of the Burj Khalifa building, your distance from the center of the Earth is $r_E + h$, where r_E is the Earth's radius (Fig. 5.21B). Hence, the force of gravity (your weight) is slightly smaller than when you are on the ground. Note that because h is much smaller than r_E, we must carry extra significant figures in this calculation.

SKETCH THE PROBLEM

We want to compare the force of gravity at two locations as sketched in Figure 5.21B. The force of gravity depends on the distance from the center of the Earth.

IDENTIFY THE RELATIONSHIPS

At ground level, you are a distance r_E from the center of the Earth, so your weight is

$$F_{\text{grav}}\ (\text{ground level}) = \frac{GM_{\text{Earth}}M_{\text{person}}}{r_E^2}$$

At the top of the building, you are a distance $r_E + h$ from the center of the Earth, so

$$F_{\text{grav}}\ (\text{top}) = \frac{GM_{\text{Earth}}M_{\text{person}}}{(r_E + h)^2}$$

The ratio of these two forces is

$$\frac{F_{\text{grav}}\text{ (top)}}{F_{\text{grav}}\text{ (ground level)}} = \frac{GM_{\text{Earth}}M_{\text{person}}/(r_E + h)^2}{GM_{\text{Earth}}M_{\text{person}}/r_E^2} = \left(\frac{r_E}{r_E + h}\right)^2$$

SOLVE

Inserting values for the Earth's radius and the height of the Burj Khalifa building, we find

$$\frac{F_{\text{grav}}(\text{top})}{F_{\text{grav}}(\text{ground level})} = \left(\frac{r_E}{r_E + h}\right)^2 = \left[\frac{6.37 \times 10^6 \text{ m}}{(6.37 \times 10^6 + 830)\text{ m}}\right]^2 = \boxed{0.99974}$$

► *What does it mean?*

This result is independent of the mass of the person; all objects weigh less at the top of the Burj Khalifa building by approximately 0.026%. Sensitive instruments called gravimeters can measure changes as small as 1 part in 10^9 in the gravitational force, so this difference can be easily measured.

CONCEPT CHECK 5.5 Gravity on Another Planet

You travel to another planet in our solar system and find that your weight is twice as large as it is on Earth. If the radius of this planet is twice the Earth's radius, is the mass of the planet (a) two times the Earth's mass, (b) four times the Earth's mass, (c) half of the Earth's mass, or (d) eight times the Earth's mass?

Measuring *G*: The Cavendish Experiment

The first precision measurement of the force of gravity between two terrestrial objects was carried out in a famous experiment by Henry Cavendish, who measured the gravitational force between two lead spheres. This experiment is very difficult for the following reason. The Cavendish spheres each had a diameter of 1 m and a mass of approximately 5900 kg. We have already seen that the gravitational force exerted by a spherical mass can be calculated as if all the mass is at the center. With that in mind, let's calculate the gravitational force between two point particles of mass 5900 kg, separated by a distance $r = 1.0$ m. Using Equation 5.16 with $m_1 = m_2 = 5900$ kg gives

$$F_{\text{grav}} = \frac{Gm_1m_2}{r^2} = \frac{(6.67 \times 10^{-11}\text{ N}\cdot\text{m}^2/\text{kg}^2)(5900\text{ kg})(5900\text{ kg})}{(1.0\text{ m})^2}$$

$$F_{\text{grav}} = 2.3 \times 10^{-3}\text{ N} \quad (5.22)$$

This force is a quite small, only a little larger than the weight of a mosquito! That's why a high-precision measurement of the gravitational force between two terrestrial objects is a very challenging experiment.

Cavendish did his work more than 200 years ago, but the same basic experimental design is still used today to study gravitational forces. The Cavendish apparatus (Fig. 5.22) uses a "dumbbell" with two spheres each of mass m_1 at the ends, suspended at the middle by a very thin wire. Another pair of spheres, each of mass m_2, is then brought very close to the dumbbell masses, and the force of gravity causes the dumbbell rod to rotate. The rotation angle θ in Figure 5.22 depends on the force, so by measuring θ, Cavendish was able to determine the gravitational force. Since the values of m_1 and m_2 and their separation can also be measured, Cavendish was able to deduce the value of G from Equation 5.16.

The quantity G is our first encounter with a *constant of nature*. Laws of physics, such as Newton's law of gravitation, often contain such constants, and the only way to know their values is through experimentation. It is interesting that Cavendish's method for measuring G is still the basis for most experimental studies of gravitation.

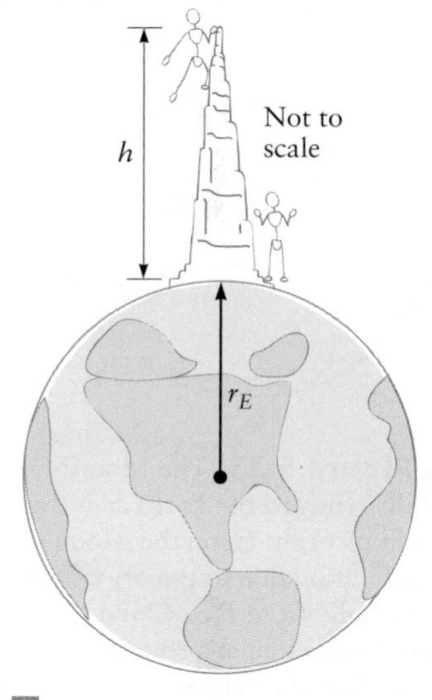

▲ **Figure 5.21** Example 5.8. **A** The Burj Khalifa building in Dubai is the tallest building in the world. **B** A person at the top of the Burj Khalifa Building is a distance $r_E + h$ from the center of the Earth, where h is the height of the building.

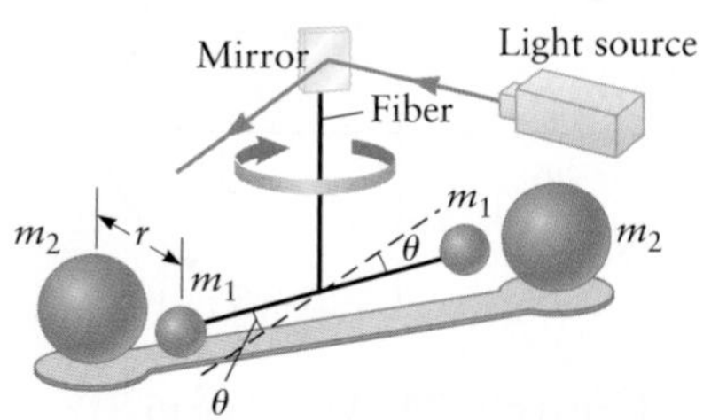

▲ **Figure 5.22** Cavendish used an apparatus like this one to measure the force of gravity between terrestrial objects.

Insight 5.4

GRAVITY IS A WEAK FORCE

In our calculations of the gravitational force between two terrestrial-scale objects, such as the two spheres in the Cavendish experiment, we found that the force of gravity is much weaker than typical normal and tension forces. In later chapters, we'll see that in most cases gravity is also much weaker than electric, magnetic, and nuclear forces. For this reason, gravitation is considered the weakest of the fundamental forces in nature.

Newton's Apple

Newton's discovery of the law of gravitation had a profound impact on our view of the universe. To appreciate this impact, it is useful to recall the allegory of Newton's apple. The story has several versions, but the general idea is that Newton was sitting under an apple tree, thinking about the motion of the Moon, when an apple fell from the tree and struck him on the head. According to the story, this jolt led to the "discovery" of gravity. The point of this story is not simply that falling apples undergo the motion we have called "free fall." Rather, it is that the acceleration of an apple falling from a tree and the acceleration of the Moon as it moves in a circular orbit around the Earth are *basically the same* (Fig. 5.23). They are caused by the same force (gravity), which has the same direction (toward the center of the Earth), and if the Moon and the apple were at the same distance from the Earth, their accelerations would be equal. This connection is not obvious because for circular motion the acceleration is perpendicular to the velocity, whereas the acceleration of the apple is parallel to $\vec{v}$. It took the genius of Newton to make this connection.

Prior to Newton, it was widely believed that the motion of celestial objects, such as the Moon, is fundamentally different from the motion of terrestrial objects, such as apples. Newton showed that the motion of falling apples and the motion of the Moon are caused by the same force and governed by the same laws of motion. So, one can use Newton's laws to understand the motion of the solar system and beyond, a regime that had previously been "off-limits" to such scientific study.

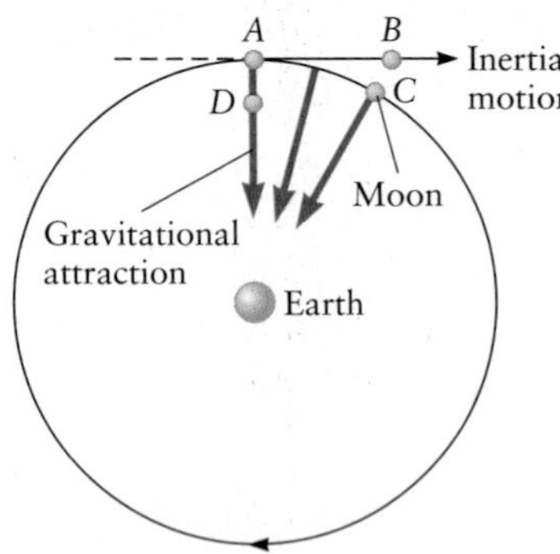

▲ **Figure 5.23** The Moon "falls" toward the Earth as it travels in its orbit. Here, the Moon's acceleration due to gravity causes it to "fall" from *B* to *C* and thereby travel in a circular orbit. If the Moon's initial velocity were zero, it would simply fall from *A* to *D*, in the same way that Newton's apple fell from its tree.

5.4 Planetary Motion and Kepler's Laws

Of the many important astronomers prior to the time of Newton, one of the most famous is Johannes Kepler (1571–1630). During and prior to his lifetime, a number of astronomers recorded the positions of the Moon, planets, comets, and so forth as functions of time and showed that bodies in the solar system move in an extremely precise fashion. Kepler studied these results very carefully and found that the motion of the Moon and planets could be described by what are now known as ***Kepler's laws*** of planetary motion. Kepler's laws are not "laws of nature" in the sense of Newton's laws of motion or Newton's law of gravitation. Rather, Kepler's laws are mathematical rules that describe motion in the solar system. These rules were inferred by Kepler from the available astronomical observations, but he was not able to give a scientific explanation or derivation. It remained for Newton to show how his three fundamental laws of motion together with his law of gravitation explain Kepler's laws.

Kepler's First Law

Kepler's first law is a statement about the shapes of orbits. Aristotle and many other early scientists believed that orbits are circular. The main reason for this belief was that circles are the most "perfect" shape for such a curve, and early thinkers believed nature must be perfect. By Kepler's time, however, it was well established[3] that while planetary orbits are approximately circular, they are definitely not perfect circles. Kepler showed that all the planetary orbits are *elliptical*. Furthermore, for the motion of planets around the Sun, the Sun is not at the center of the ellipse; instead, it is at one of the foci of the ellipse as illustrated in Figure 5.24. This discovery must have been quite a shock because the off-center placement of the Sun would seem to violate nature's tendency for symmetry.

For the planets known to Kepler, as well for the Moon, the orbits deviate only a small amount from perfect circles (by much less than the orbit in Fig. 5.24). For example, the distance between the Earth and the Sun varies by only ±3% during the

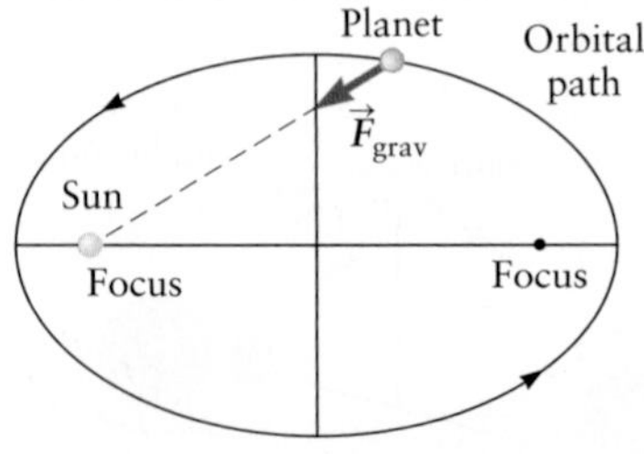

▲ **Figure 5.24** According to Kepler's first law, the planetary orbits are elliptical. An ellipse has two foci, which are offset from the center of the ellipse. The Sun is located at one of the foci.

[3]At that time, only the six innermost planets in our solar system had been discovered: Mercury, Venus, Earth, Mars, Jupiter, and Saturn.

course of an orbit. Nevertheless, the deviation from a perfect circle is certainly real, and it had been accurately measured by Kepler's time.

Kepler's first law is the statement that planetary orbits are elliptical. This statement also applies to the Moon because the Moon is really just a "planet" belonging to the Earth, and Kepler's first law also applies to artificial satellites that have been launched into orbit around the Earth. Because a circle is a special case of an ellipse, Kepler's first law also allows circles, but real orbits generally deviate at least a small amount from a perfect circular shape. Soon after he developed his law of gravitation, Newton was able to use it to derive Kepler's first law and thus explain the shapes of planetary orbits.[4]

> ***Kepler's first law of planetary motion:*** **Planets move in elliptical orbits.**

Most comets have highly elliptical orbits. Perhaps the best known case is Halley's comet. A scale drawing of the orbit of Halley's comet is shown in Figure 5.25. This figure includes the orbit of the dwarf planet Pluto, which is also noticeably elliptical. In fact, Pluto's orbit is such that it actually spends a substantial amount of time inside Neptune's orbit.

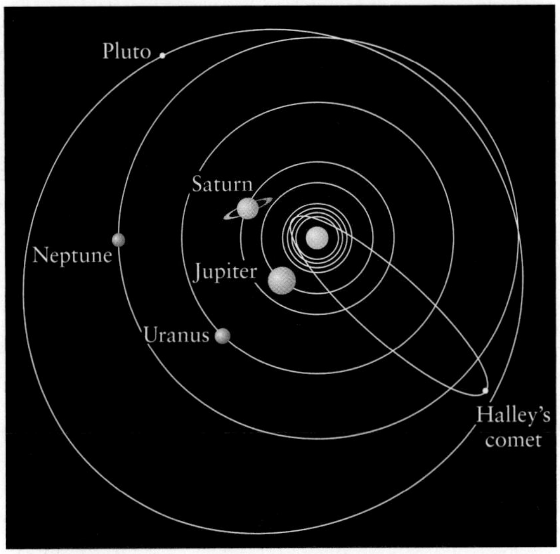

Pluto
Saturn
Neptune
Jupiter
Uranus
Halley's comet

▲ **Figure 5.25** Scale drawing of the planetary orbits in our solar system. The orbits of most planets are nearly circular. The orbit of the dwarf planet Pluto deviates substantially from a circle. The orbit of Halley's comet, which is highly elliptical, is also shown.

CONCEPT CHECK 5.6 Acceleration of a Comet

Sketch the direction of the acceleration vector for Halley's comet at various points in its orbit in Figure 5.25.

Kepler's Second Law

Kepler's second law concerns the speed of a planet as it moves around its orbit. This law applies to a planet in our solar system moving about our Sun as well as to a planet in another solar system moving about its sun. For a perfectly circular orbit, this speed is constant. However, for an elliptical orbit, the speed is smallest when the planet is farthest from its sun, whereas v is largest when the planet is nearest its sun (Fig. 5.26). To understand Kepler's second law, it is useful to consider a line drawn from the planet to its sun. This line moves along with the planet and sweeps out area as the planet moves. According to Kepler's second law, this line sweeps out area at a constant rate. Let A_1 be the area swept out in time Δt when the planet is near its sun and A_2 be the amount of area swept out in the same amount of time Δt when the planet is somewhere else in its orbit. Kepler's second law then states that $A_1 = A_2$. For this statement to be true, the planet must (as we have already noted) speed up when it is nearest its sun. We'll discuss Kepler's second law again in Chapter 9, where we'll see that it is closely connected with the ***angular momentum*** of the planet.

> ***Kepler's second law of planetary motion:*** **A line connecting a planet to its sun sweeps out equal areas in equal times as the planet moves around its orbit.**

v is smallest here.
sun
A_1
A_2
v is largest here.

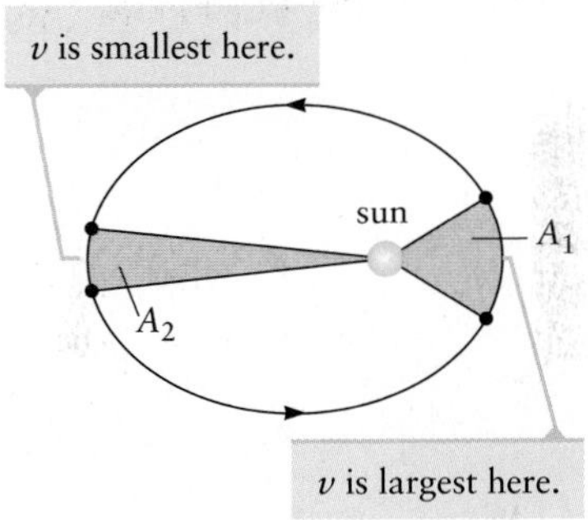

▲ **Figure 5.26** Kepler's second law is called the "equal areas" law. A line that extends from the sun to a planet sweeps out equal areas during equal time intervals as a planet moves around its orbit. If the time required for the planet to sweep out area A_1 is equal to the time associated with area A_2, the areas will be equal.

Kepler's Third Law

Kepler's third law relates the timing of an orbit to the size of the orbit. It is simplest to derive Kepler's third law for the special case of a perfectly circular orbit, although it also applies to elliptical orbits. For a circular orbit, the speed of the planet is

[4]Nonclosed trajectories are also possible. For example, objects that pass near the Sun once and never return move along either a parabolic or a hyperbolic path. These trajectories, including the elliptical orbits of the planets, are ***conic sections.*** Kepler's first law is sometimes worded to reference these nonclosed trajectories. The only closed trajectories—that is, the only true orbits—are elliptical, however.

constant. If the orbit has a radius r, the time it takes to complete one orbit equals the distance traveled (the circumference) divided by the speed:

$$T = \frac{2\pi r}{v} \tag{5.23}$$

Since the gravitational force of the sun on the planet is equal to ma_c, where a_c is the centripetal acceleration, we get (compare with Eqs. 5.16 and 5.17)

$$F_{\text{grav}} = \frac{GM_{\text{sun}}M_{\text{planet}}}{r^2} = ma_c = \frac{M_{\text{planet}}v^2}{r} \tag{5.24}$$

From Equation 5.23, we have $v = 2\pi r/T$, and substituting for v gives

$$\frac{M_{\text{planet}}v^2}{r} = \frac{M_{\text{planet}}(2\pi r/T)^2}{r} = \frac{4\pi^2 M_{\text{planet}}r}{T^2} \tag{5.25}$$

We can now combine Equations 5.24 and 5.25 to get a relation between the orbital time (the period T) and the orbital radius r. We find

$$\frac{GM_{\text{sun}}M_{\text{planet}}}{r^2} = \frac{4\pi^2 M_{\text{planet}}r}{T^2}$$

and solving for T gives

$$T^2 = \left(\frac{4\pi^2}{GM_{\text{sun}}}\right)r^3 \tag{5.26}$$

which is Kepler's third law. In words, Equation 5.26 states that the square of the orbital period is proportional to the cube of the orbital radius. This result does not depend on the mass of the planet. It applies to all orbiting bodies, including planets, comets, and spacecraft. Jupiter and Saturn, for example, possess many moons, and the orbital periods and radii of these moons must obey Equation 5.26, with M_{sun} replaced by the mass of the appropriate planet.

Kepler's third law: **The square of the period of an orbit is proportional to the cube of the orbital radius.**

EXAMPLE 5.9 Neptune's Orbit

The Earth completes one orbit about the Sun in 1 year and has an orbital radius of 1.50×10^{11} m (see Table 5.1). If the orbital radius of Neptune is 4.50×10^{12} m, what is the period of Neptune's orbit?

RECOGNIZE THE PRINCIPLE

We could solve this problem by simply evaluating Equation 5.26 using the given radius of Neptune's orbit. Here, though, we take a different approach by applying Equation 5.26 first to Neptune and next to the Earth, and then computing the ratio of their orbital periods. Notice that the orbits of the Earth and Neptune are both very close to circular.

SKETCH THE PROBLEM

No sketch is needed.

IDENTIFY THE RELATIONSHIPS

Applying Equation 5.26 to Neptune, we have

$$T^2_{\text{Neptune}} = \left(\frac{4\pi^2}{GM_{\text{Sun}}}\right)r^3_{\text{Neptune}}$$

with a similar result for the Earth. Taking the ratio gives

$$\frac{T^2_{\text{Neptune}}}{T^2_{\text{Earth}}} = \frac{\left(\dfrac{4\pi^2}{GM_{\text{Sun}}}\right)r^3_{\text{Neptune}}}{\left(\dfrac{4\pi^2}{GM_{\text{Sun}}}\right)r^3_{\text{Earth}}} = \frac{r^3_{\text{Neptune}}}{r^3_{\text{Earth}}} \quad (1)$$

SOLVE

Inserting the orbital radius of Neptune and of the Earth from Table 5.1 gives

$$\frac{T^2_{\text{Neptune}}}{T^2_{\text{Earth}}} = \left(\frac{4.50 \times 10^{12}\text{ m}}{1.50 \times 10^{11}\text{ m}}\right)^3 = 2.70 \times 10^4$$

$$T^2_{\text{Neptune}} = 2.70 \times 10^4 \times T^2_{\text{Earth}}$$

$$T_{\text{Neptune}} = \boxed{160 \times T_{\text{Earth}}}$$

Since $T_{\text{Earth}} = 1$ year, it takes Neptune approximately $\boxed{160 \text{ years}}$ to complete one orbit.

▸ ***What does it mean?***

Neptune thus takes more than a century to complete one orbit! When doing a calculation, it is sometimes mathematically simpler to use the ratio of two similar quantities. In this example, the factors of G and M_{Sun} canceled when the ratio was taken in Equation (1), simplifying the calculation.

Satellite Orbits around the Earth

Many satellites, including most that carry astronauts, are in what is often called a "low Earth" orbit. You may wonder why any orbit would be called "low," but the reason for this expression becomes clear when we calculate the radius of a typical orbit. We can do so using Kepler's third law (Eq. 5.26) if we know the period of the orbit. The period for a satellite in low Earth orbit, such as the International Space Station, is approximately 90 min, so $T = (90\text{ min})(60\text{ s/min}) = 5400\text{ s}$. Kepler's third law, with the mass of the Earth as that of the central body, gives

$$T^2 = \left(\frac{4\pi^2}{GM_{\text{Earth}}}\right)r^3$$

Solving for the radius of the orbit, we find

$$r = \left(\frac{GM_{\text{Earth}}T^2}{4\pi^2}\right)^{1/3} \quad (5.27)$$

Inserting values for the mass of the Earth and T, we get

$$r = \left(\frac{GM_{\text{Earth}}T^2}{4\pi^2}\right)^{1/3} = \left[\frac{(6.67 \times 10^{-11}\text{ N}\cdot\text{m}^2/\text{kg}^2)(5.98 \times 10^{24}\text{ kg})(5400\text{ s})^2}{4\pi^2}\right]^{1/3}$$

$$r = 6.66 \times 10^6\text{ m} \quad (5.28)$$

The radius of the Earth is $r_E = 6.37 \times 10^6$ m, so these satellites orbit at a height above the Earth's surface that is only 5% of the Earth's radius. This orbit is illustrated in the scale drawing in Figure 5.27. Even though it may seem low when compared to the Earth's radius, the tallest mountain on the Earth (Mount Everest) is only about 0.1% of r_E above sea level, so the satellites are in no danger of colliding with any mountains!

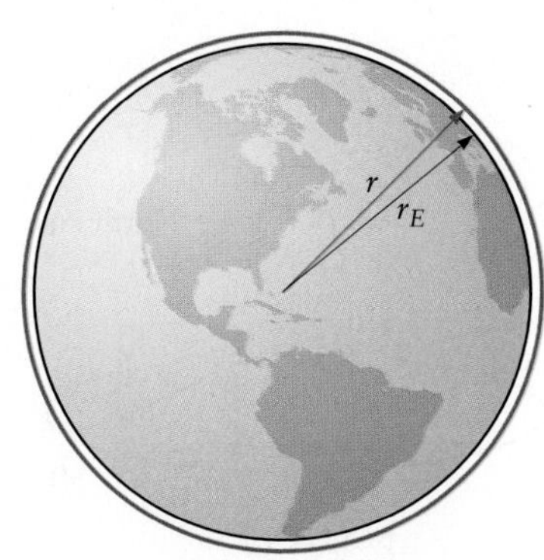

▲ **Figure 5.27** Scale drawing of a "low Earth" orbit. The orbit is shown in red. On this scale, it is barely distinguishable from the surface of the Earth.

EXAMPLE 5.10 Geosynchronous Orbits

Earth-orbiting satellites used for transmitting telephone and television signals travel in ***geosynchronous*** orbits. These orbits have a period of 1 day, so these satellites move in synchrony with the Earth's rotation and are thus always at the same position in the sky relative to a person on the ground. Sending signals to and from these satellites is thus greatly simplified because an antenna such as the author's satellite TV dish can be aligned only once and then needs no further adjustment. Calculate the orbital radius of a satellite in a geosynchronous orbit.

RECOGNIZE THE PRINCIPLE

We have already done a very similar problem in our discussion of a satellite in low Earth orbit. We can thus apply Equation 5.27 to the geosynchronous case using a period $T = 1$ day.

SKETCH THE PROBLEM

No sketch is needed.

IDENTIFY THE RELATIONSHIPS AND SOLVE

The period of a geosynchronous orbit is 1 day. Hence, the period $T = 1$ day = (1 day)(24 h/day)(3600 s/h) = 86,000 s. Inserting this result into Equation 5.27 gives

$$r = \left(\frac{GM_{\text{Earth}}T^2}{4\pi^2}\right)^{1/3}$$

$$= \left[\frac{(6.67 \times 10^{-11}\ \text{N} \cdot \text{m}^2/\text{kg}^2)(5.98 \times 10^{24}\ \text{kg})(8.6 \times 10^4\ \text{s})^2}{4\pi^2}\right]^{1/3}$$

$$r = \boxed{4.2 \times 10^7\ \text{m}}$$

► ***What does it mean?***

The radius of the Earth is 6.37×10^6 m, so the radius of a geosynchronous orbit is about seven times larger than the radius of the Earth. It is thus much larger than the low Earth orbit in Figure 5.27, which is why much more fuel is required to launch a satellite into geosynchronous orbit than into low Earth orbit.

Kepler's Laws, Putting a Satellite into Orbit, and the Origin of the Solar System

Kepler's three laws of planetary motion apply to all types of gravitationally produced orbital motion, including the motion of planets and comets about the Sun and the motion of moons and satellites about a planet. Indeed, one can also think of a freely falling apple as an example of gravitationally produced motion. It is fascinating that such a variety of motion can be produced by a single force. These different types of motion result from the different ways in which these objects are initially set into motion. The dropped apple is released from rest, so its initial velocity is zero. For the Moon to follow a nearly circular orbit, it must have been given the proper "initial" velocity at some point in the distant past. For planets such as the Earth and Jupiter to be moving in orbits about the Sun that are now approximately circular, it was necessary that the planets be set into motion with the proper velocity. This problem brings us to some very interesting questions concerning the origin and evolution of the solar system. It is now believed that the solar system was originally a rotating mass of gas and that this gas gradually condensed to form the planets. The rotational motion of the original gas cloud then led to the approximately circular orbits we now observe.

The problem of setting up an orbit of the desired shape is also encountered when a satellite is launched from the Earth. NASA usually launches satellites into equatorial orbits that are approximately parallel to the Earth's equator as sketched in Figure 5.28. We have always considered or computed orbital speeds as measured with respect to a

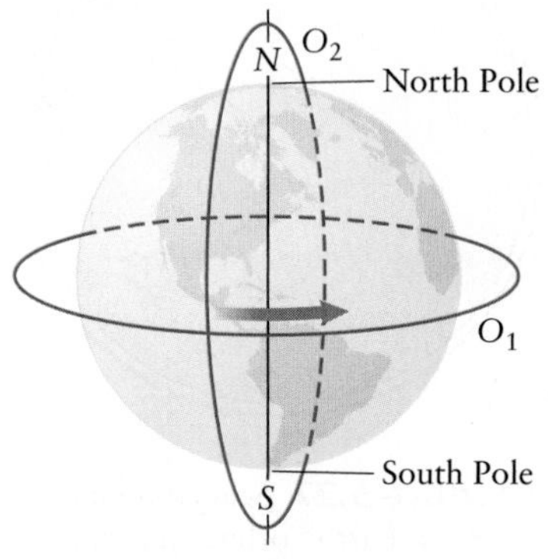

▲ **Figure 5.28** It is possible to place a satellite into an equatorial orbit (O_1) or a polar orbit (O_2). A polar orbit takes a satellite over the Earth's North and South Poles.

stationary observer. The Earth's surface is not stationary, however; rather, it moves with a substantial speed v_R due to the Earth's rotation. When a satellite is launched toward the east in an equatorial orbit, it starts with a speed v_R and the rocket engines then add to this speed as the satellite is put into orbit. If the satellite were launched into a different orbit—say, into an equatorial orbit to the west—it would not be able to take advantage of v_R and the launch would require more fuel (and therefore more money).

CONCEPT CHECK 5.7 Apparent Weight and Earth's Rotational Motion

A person at the equator moves in a circle due to the Earth's rotational motion. How does this circular motion affect the person's apparent weight?

(a) The circular motion causes the apparent weight to be *larger* than it would be if the Earth were not rotating.
(b) The circular motion causes the apparent weight to be *smaller* than it would be if the Earth were not rotating.
(c) There is no effect on the apparent weight.

EXAMPLE 5.11 A Double Star System

A double star system is one in which two stars orbit around each other due to their mutual gravitational attraction. In the simplest case, the stars have the same mass m and move in circular orbits with the same radius r and speed v (Fig. 5.29), with the center of each orbit at the point midway between the two stars. Find the orbital period T of such a double star system. Express your answer in terms of r, m, and G.

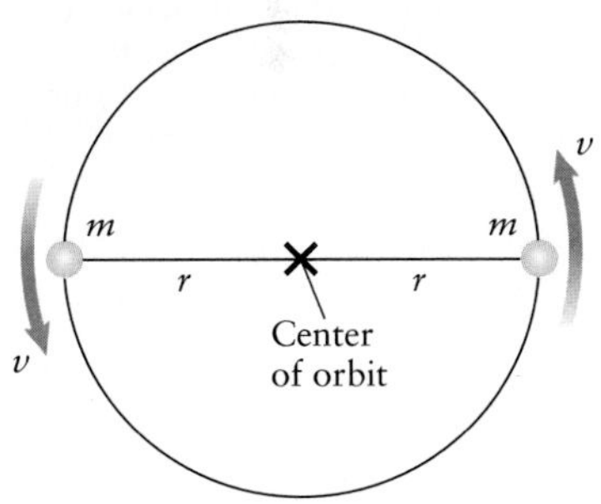

▲ **Figure 5.29** Example 5.11. A double star system.

RECOGNIZE THE PRINCIPLE

The gravitational force on one of the stars can be calculated from Newton's law of universal gravitation. This force provides the centripetal acceleration of that star. We can use Newton's second law (Eq. 5.6) to find the orbital period, following the approach we used in deriving Kepler's third law (Eq. 5.26).

SKETCH THE PROBLEM

Figure 5.29 shows the problem.

IDENTIFY THE RELATIONSHIPS

Both stars have mass m and are separated by a distance $2r$, so the gravitational force on one of the stars is

$$F_{\text{grav}} = \frac{Gmm}{(2r)^2} = \frac{Gm^2}{4r^2} \tag{1}$$

The force required to make a star move in a circular orbit is mv^2/r, where the speed is equal to the circumference of an orbit divided by the period, $v = 2\pi r/T$. Setting this force equal to the gravitational force in Equation (1) leads to

$$\frac{Gm^2}{4r^2} = \frac{mv^2}{r} = \frac{m(2\pi r/T)^2}{r} = \frac{4\pi^2 mr}{T^2} \tag{2}$$

SOLVE

Solving Equation (2) for the period, we get

$$\frac{Gm^2}{4r^2} = \frac{4\pi^2 mr}{T^2}$$

$$T^2 = \frac{16\pi^2 r^3}{Gm}$$

$$T = \boxed{4\pi\sqrt{\frac{r^3}{Gm}}}$$

(continued) ▶

▶ ***What does it mean?***
The period thus depends on both the mass of the stars and the orbital radius. In their studies of double star systems, astronomers use measurements of this period to learn about both of these quantities.

Courtesy of Nova Scotia Economic & Rural Development & Tourism (both)

▶ **Figure 5.30** An example of tides. The ocean level can vary substantially between periods of low and high tide. There are generally two high tides every 24 h, although in some regions factors such as the shape of the ocean basin lead to only one high tide and one low tide each day.

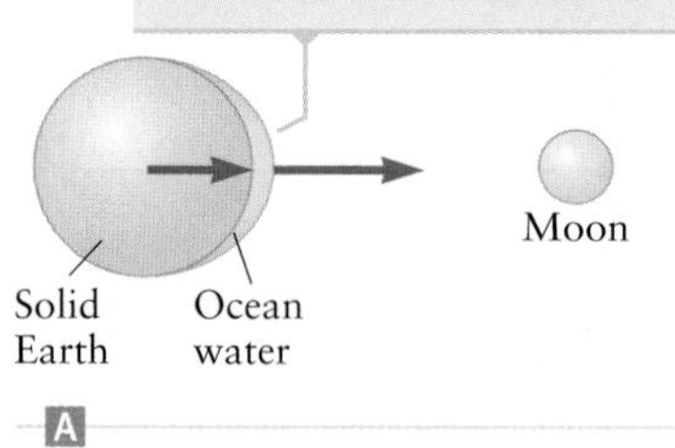

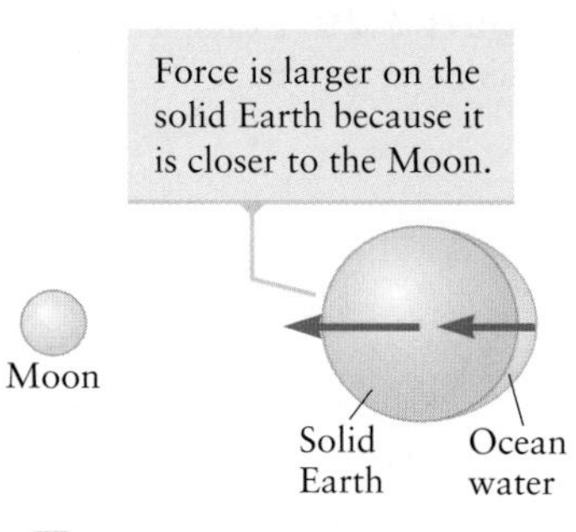

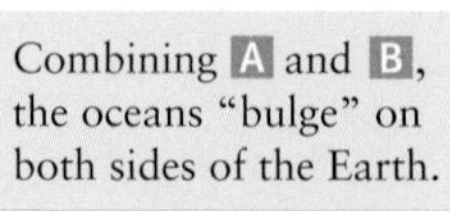

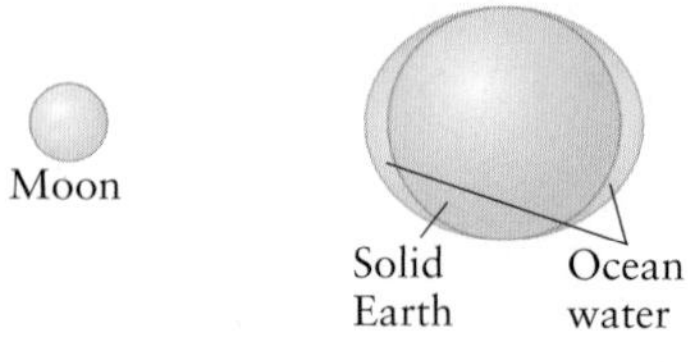

▲ **Figure 5.31** The Moon's gravitational force on the oceans is responsible for the tides. (Not to scale.)

5.5 Moons and Tides

The Origin of Tides

You have probably noticed that the level of the Earth's oceans fluctuates up and down every day; these fluctuations are called tides (Fig. 5.30). Tides are due to the Moon's gravitational force on the oceans. Although the Moon exerts a gravitational force on both the ocean water and the solid Earth, the gravitational force decreases with distance, so the resulting acceleration is slightly higher for the water nearest the Moon (Fig. 5.31A). Thus, the water nearest the Moon "falls toward" the Moon slightly faster than does the solid Earth. The result is a "bulge" in the ocean, which is just a high tide.

Figure 5.31 explains why there is a high tide when the Moon is overhead. However, there are actually *two* high tides every day in most parts of the world, one when the Moon is overhead and another 12 h later when it is on the opposite side of the Earth. The origin of this second high tide is explained in Figure 5.31B. Because the gravitational force of the Moon decreases with increasing distance, the acceleration of the solid Earth is larger than the acceleration of the ocean on the far side of the Earth. As a result, the solid Earth "falls toward" the Moon faster than the ocean water on the far side and the ocean now bulges away from the Moon, producing another high tide.

The Sun also affects the tides, although its effect is smaller than the Moon's. When the Sun and the Moon are aligned and are on the same side of the Earth, their gravitational forces add, giving a higher high tide than produced by the Moon alone.

5.6 Deep Notions Contained in Newton's Law of Gravitation

The Inverse Square Law

A key feature of the gravitational force is the manner in which it varies with distance. According to Equation 5.16, the gravitational force between two objects falls off as the square of the distance between them:

$$F_{\text{grav}} = \frac{Gm_1m_2}{r^2} \propto \frac{1}{r^2} \tag{5.29}$$

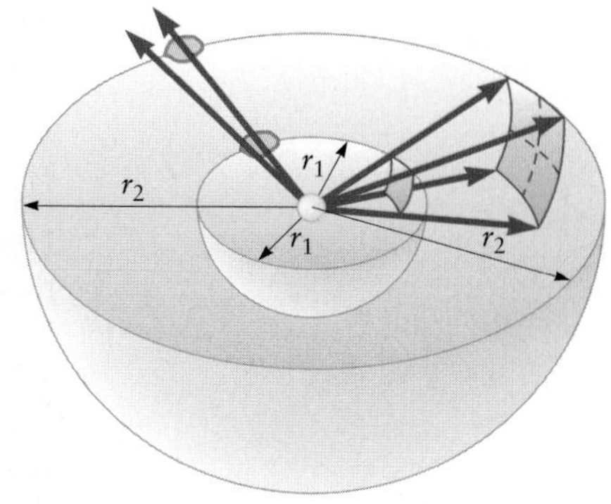

▲ **Figure 5.32** Lines of gravitational force. According to the field line picture, gravitational force acts along lines that emanate from all objects. The gravitational force exerted by one object on another is proportional to the number of lines intersecting the second object. This number falls as $1/r^2$ and thus accounts for the inverse square force law in Equation 5.29. Here, the object shown in blue (which might be the Earth) is located at two different distances from the central object in yellow (which might be the Sun). The number of lines intercepted by the blue mass is smaller when the separation increases because the field lines emanate radially and expand into a larger area (tan squares).

Mathematically, Equation 5.29 is called an "inverse square" law. A number of other forces in nature, such as the force between two electric charges, fall off as the square of the distance and are thus also described by inverse square laws. Why do many natural forces follow this pattern? Imagine that an object—for example, the Sun—possesses gravitational "field lines" that emanate radially outward from it as sketched in Figure 5.32. We also imagine that the number of these lines is proportional to the mass of the object. When these lines intersect another object, such as the Earth, there is a force on that object directed parallel to the lines and hence toward the original object; for example, there is a force on the Earth directed toward the Sun.

Because the force lines emanate in three-dimensional space, the number of lines over a given area—that is, the number of lines that intercept the Earth in Figure 5.32—falls off with distance as $1/r^2$. Hence, this picture explains the inverse square dependence in Equation 5.29. This result implies that gravity follows an inverse square law because we live in a three-dimensional space. It also means that we should expect other forces described by a field line picture to have the same inverse square dependence, which does seem to be the way nature works.

Do such field lines really exist? So far, no one has devised a way to "see" the lines; the best we can do is to observe the resulting force that the lines are presumed to cause. The field line picture explains the general form of Newton's law of gravitation, so there is strong evidence in favor of this model. However, while the field line picture is very attractive, it leads to other questions. The field lines that emanate from the Sun in Figure 5.32 and eventually reach the Earth must travel through the nearly perfect vacuum of space. Other observations indicate that the gravitational force is felt even when a perfect vacuum separates two objects. How can "something," like a gravitational field line, exist in a vacuum? Does a field line have a mass? If the Sun were to suddenly move, how fast would the corresponding change in its gravitational field be felt on the Earth?

The gravitational force is an example of what is called ***action at a distance***. Newton's theory of gravitation tells us that action at a distance does indeed occur, but it does not tell us *how* it occurs. The field line picture was invented to answer this "how" question, but it does not provide a complete answer. For some forces, such as the electric force, there is a theoretical explanation of the nature of the associated field lines. Einstein's theories of relativity answer many of these questions in the case of the gravitational field, as we'll describe in Chapter 27.

Gravitation and Mass

A very important feature of the gravitational force law is that F_{grav} is linearly proportional to the mass m of each object. This quantity m is sometimes referred to as the ***gravitational mass*** of an object. We first encountered the concept of mass in Newton's second law of motion ($\sum \vec{F} = m\vec{a}$), and the m in Newton's second law is often called the ***inertial mass*** of the object. As far as physicists can tell, the gravitational mass of an object is *precisely equal* to the inertial mass. That $m_{\text{grav}} = m_{\text{inertial}}$ suggests a deep connection between gravitation and inertia (and hence motion). Newton did not understand why there should be a connection; it was not explained until the work of Einstein, whose theories showed why the inertial and gravitational masses are the same.

Einstein also considered the effect of a gravitational field on the motion of light. Consider how we would apply Newton's second law to the motion of a "particle" of mass m_p that experiences a gravitational force caused by another object, such as the Sun. We would write Newton's second law as

$$m_p a_p = \frac{GM_{\text{Sun}} m_p}{r^2}$$

where a_p is the acceleration of the particle and r is the distance from it to the Sun. Strictly speaking, the mass m_p on the left side is the inertial mass of the particle, whereas the mass m_p on the right side is the gravitational mass. Because they are equal, however, we can cancel these factors to get

$$a_p = \frac{GM_{\text{Sun}}}{r^2} \tag{5.30}$$

Hence, the acceleration is independent of the mass of the particle. We have seen this result many times before (e.g., in free fall), but there is a new point to make. Given that the acceleration is independent of the mass, does this result also apply to a "particle" that has no mass? As we'll see in Chapter 28, light can be described in terms of particles called ***photons*** that have no mass. Even though a photon's mass is zero, Equation 5.30 suggests that it is still accelerated by gravity. That is indeed the case, although a correct theoretical description of this acceleration requires Einstein's general theory of relativity. We'll say more about this acceleration and about the equivalence of gravitational and inertial mass in Chapter 27.

Summary | CHAPTER 5

Key Concepts and Principles

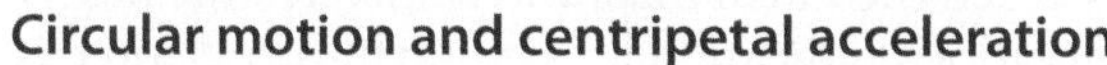

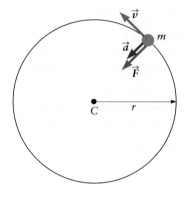

Circular motion and centripetal acceleration

An object moving in a circle of radius r at a constant speed has an acceleration

$$a_c = \frac{v^2}{r} \quad \text{(5.5) (page 132)}$$

directed toward the center of the circle. The quantity a_c is called the ***centripetal acceleration.*** According to Newton's second law, this acceleration must be caused by a total force of magnitude $\sum F = ma_c$, so

$$\sum F = \frac{mv^2}{r} \quad \text{(5.6) (page 133)}$$

This force is directed toward the center of the circle.

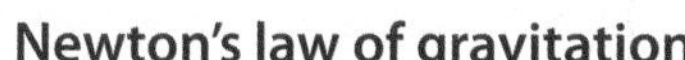

Newton's law of gravitation

There is a gravitational force

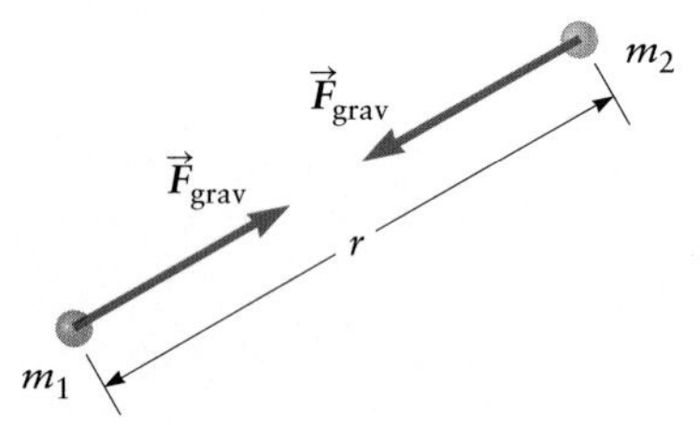

$$F_{\text{grav}} = \frac{Gm_1m_2}{r^2} \quad \text{(5.16) (page 145)}$$

between any two objects. This force is always attractive.

Newton's law of gravitation is an example of an ***inverse square law.*** This $1/r^2$ dependence suggests a field line model of gravity and tells us something about the geometry of the universe. Gravitation is also an example of "action at a distance." Other forces, including electric forces, exhibit this property.

Applications

There are many examples of circular motion, including roller coasters, cars traveling on a curved road, and centrifuges. In all cases, the total force on an object undergoing uniform circular motion must equal $F = mv^2/r$ and be directed toward the center of the circle.

The motion of the Moon as it orbits the Earth and of the planets as they orbit the Sun are examples of (nearly) circular motion. The force in these cases is due to gravitation. Near the Earth's surface, the magnitude of the gravitational force on an object of mass m is $F_{\text{grav}} = mg$ and is directed toward the center of the Earth.

Kepler's laws

Kepler deduced three laws of planetary motion: (1) planetary orbits are elliptical, (2) a planetary orbit sweeps out equal areas in equal times, and (3) the square of the orbital period is proportional to the cube of the average orbital radius. These laws apply to planets orbiting a sun and also to satellites and moons orbiting a planet.

Kepler's laws, and hence the motions of planets, moons, comets, and other objects in the solar system, are all explained by Newton's law of gravitation, together with Newton's laws of motion. The circular motion of the Moon and the free fall of an apple look different, but they are due to the same force.

Questions

SSM = answer in *Student Companion & Problem-Solving Guide* ⊗ = life science application

1. Give an example of motion in which (a) the magnitude of the instantaneous velocity is always larger than the average velocity and (b) the instantaneous velocity is never parallel to the instantaneous acceleration.
2. SSM In Example 5.3, we considered a car traveling on a banked turn with friction. Draw free-body diagrams for the car when the speed is low and when the speed is high, and explain why they are different. *Hint*: Consider the direction of the frictional force in the two cases.
3. In a reference listing found on the Internet, it is stated that $g = 9.80665$ m/s^2. Discuss why it is not correct to think that the "exact" value of g can be given with this accuracy. Indeed, is there an "exact" value of g?
4. Consider the Cavendish experiment in Figure 5.22. When he designed this experiment, Cavendish had to decide how large to make the spheres. If they are made larger, they will have a larger mass, which, according to Equation 5.16, will make the gravitational attraction larger and the force therefore easier to measure. If the spheres are made larger, however, the distance between their centers will necessarily increase, which makes the force smaller. Assuming the spheres in Figure 5.22 all have the same radii, suppose the value of r is increased by a factor of two. Will that change increase or decrease the force, and by what factor? Assume that the spheres all have the same constant density.
5. Explain why a geosynchronous satellite cannot remain directly overhead for an observer in Boston. *Hint*: Consider first an observer at the North Pole.
6. Kepler's second law, the statement that a planet sweeps out equal areas in equal times, can be derived by a geometrical argument. To see how one might construct such a geometrical proof, consider the simpler case of a planet moving with constant velocity as shown in Figure Q5.6. The points A, B, C, . . . are spaced at equal time intervals. Show that this planet obeys Kepler's second law; that is, show that it sweeps out equal areas in equal times. *Hint*: Calculate the areas of triangles OAB, OBC, and so on.

Figure Q5.6

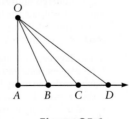

7. It is sometimes claimed by astrologers (but not by astronomers!) that because the Moon dramatically affects the seas of the world, as evidenced by the tides, the Moon must also affect individual people because more than 60% of an average adult's mass is water. Does that claim make sense? Use Newton's law of universal gravitation to make an estimate of the forces involved to justify your answer.
8. What force makes it possible for a car to move along a curved road? A straight, flat road?
9. Astronomers have discovered that some stars orbit around regions of space that contain large amounts of cosmic dust. This dust provides the gravitational force that causes the stars to move in a circular orbit. Suppose the density of dust is constant in a certain region and several stars orbit through this dust. Find a relation between the orbital period T and orbital radius R for these stars and show that in this case T is independent of R. Compare your result to Kepler's third law. Studies of these stars give astronomers a way to determine the distribution of cosmic dust in the universe.
10. Reconnaissance satellites (often called spy satellites) travel in very low orbits so that their cameras can take photos of objects on the Earth's surface with the highest possible resolution. Explain why such satellites are often launched into polar orbits; when in these orbits, satellites travel over both the north and south poles of the Earth.
11. It is believed that much of the mass in the universe is carried by what is called dark matter, matter that does not emit enough visible light or other radiation to be detected by conventional telescopes. However, dark matter does exert a gravitational force on other objects in the universe. Explain how dark matter might be detected and studied through the observation of "normal" matter (such as conventional stars).
12. When a planet orbits around a star, the star also moves in an "orbit." Since it is much more massive than the planet, the star's orbital radius r_{star} is much smaller than that of the planet r_{planet}. Work out how the ratio r_{star}/r_{planet} depends on the ratio of the masses. Discuss how this effect could be used to detect the presence of planets in distant solar systems. For simplicity, assume circular orbits.
13. NASA uses a specially equipped airplane (called the "Vomit Comet") to provide a simulated zero-gravity environment for training and experiments. This airplane flies in a long, parabolic path. Explain how a passenger can feel "weightless" near the top of the parabola.
14. You are a prospector looking for gold by taking high-precision measurements of the acceleration due to gravity, g, at different points on the Earth's surface. In one region, you find that g is slightly higher than its average value. Are you standing over what might be a deposit of gold or over an underground lake? Explain.
15. How does your weight on a ship in the middle of the ocean compare with your weight when you are standing on solid ground? Explain why they are not the same.

16. The Sun exerts an overall force on the Earth many times greater than that of the Moon. How can it then be that the ocean tides are primarily due to the influence of the Moon and, to a much lesser extent, the Sun?

17. SSM The difference in the gravitational force is only about 10% less on an object that is in a low Earth orbit than it is for the same object on the ground. Why is it that an astronaut in orbit experiences weightlessness?

18. An astronaut on the peak of a mountain on the Moon fires a rifle along the horizontal direction. Is it possible, given a sufficient initial speed for the bullet, that the bullet might hit her in the back? Explain how it could happen.

19. **Pluto's mass.** In 1978, it was discovered that Pluto had a moon of its own. The moon was given the name Charon (now known to be one of three; see Fig. Q5.19). After the discovery of this moon, the hitherto unknown mass of Pluto was calculated to a precision of less than 1%. How did the discovery of Charon allow the mass to be determined?

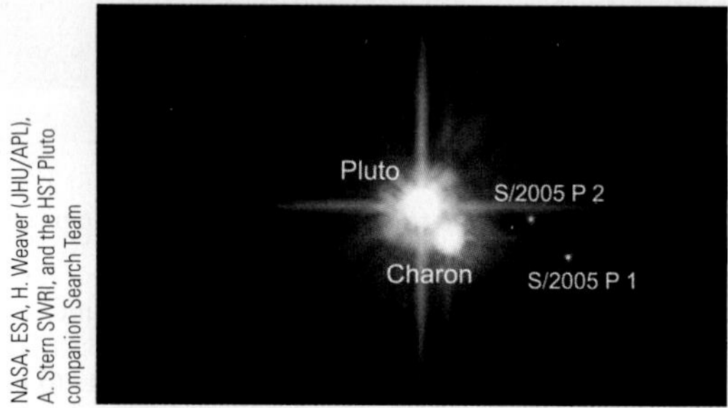

Figure Q5.19 Pluto and its three moons imaged by the Hubble Space Telescope.

20. **A coffee centrifuge?** In one popular demonstration, a full cup of hot coffee is placed on a platform suspended by strings to the lecturer's hand as seen in Figure Q5.20. With some practice, the platform, coffee and all, can be made to rotate in a vertical circle. How does the coffee stay in the cup? If the rotation of the coffee and cup is sustained for some time, what would happen to any grinds that happen to be mixed in the coffee?

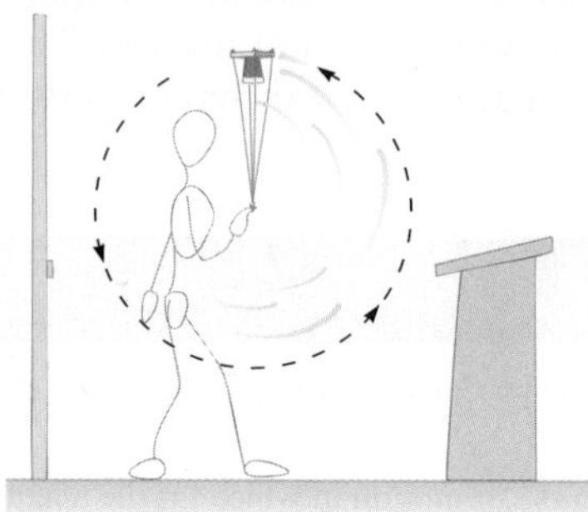

Figure Q5.20 Question 20 and Problem 78.

Problems

SSM = solution in *Student Companion & Problem-Solving Guide*
★ = intermediate ✪ = challenging ⓧ = life science application
Ⓡ = reasoning and relationships problem
RT = reasoning tutorial in Enhanced WebAssign

5.1 UNIFORM CIRCULAR MOTION

1. A bicycle wheel of radius 0.30 m is spinning at a rate of 60 revolutions per minute. (a) What is the centripetal acceleration of a point on the edge of the wheel? (b) What is the period of the wheel's motion?

2. For the bicycle wheel in Problem 1, what is the centripetal acceleration of a point that is 0.10 m from the outer edge of the wheel? Explain why this value is different from the answer to part (a) of Problem 1.

3. ★ RT The Earth has a radius of 6.4×10^6 m and completes one revolution about its axis in 24 h. (a) Find the speed of a point at the equator. (b) Find the speed of New York City.

4. ★ Ⓡ In the game of baseball, a pitcher throws a curve ball with as much spin as possible. This spin makes the ball "curve" on its way to the batter. In a typical case, the ball spins at about 30 revolutions per second. What is the maximum centripetal acceleration of a point on the edge of the baseball?

5. A jogger is running around a circular track of circumference 400 m. If the jogger has a speed of 12 km/h, what is the magnitude of the centripetal acceleration of the jogger?

6. The hard disk in a laptop computer contains a small disk that rotates at a rate of 5000 rpm. If this disk has a radius of 2.0 cm, what is the centripetal acceleration of a point at the edge of the disk?

7. Suppose the circular track in Figure 5.1 has a radius of 100 m and the runner has a speed of 5.0 m/s. (a) What is the period of the motion? (b) If the radius of the track were reduced to 50 m and the runner maintained this speed, by what factor would the runner's centripetal acceleration change?

8. What is the acceleration of the Moon as it moves in its circular orbit around the Earth? *Hint*: You will find some useful data in Table 5.1.

9. ★ Consider points on the Earth's surface as sketched in Figure P5.9. Because of the Earth's rotation, these points undergo uniform circular motion. Compute the centripetal acceleration of (a) a point at the equator, and (b) at a latitude of 30°.

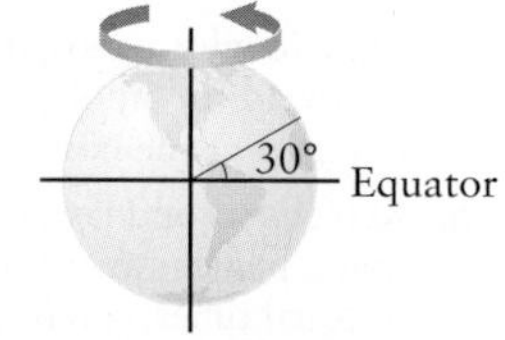

Figure P5.9

10. In the days before compact discs and MP3 players (ancient history!), music was recorded in scratches in the surface of vinyl-coated disks called records. In a typical record player, the record rotated with a period of 1.8 s. Find the centripetal acceleration of a point on the edge of the record. Assume a radius of 15 cm.

11. SSM ★ A compact disc spins at 2.5 revolutions per second. An ant is walking on the CD and finds that it just begins to slide off the CD when it reaches a point 3.0 cm from the CD's center. (a) What is the coefficient of friction between the ant and the CD? (b) Is this the coefficient of static friction or kinetic friction?

12. ⓧ When a fighter pilot makes a very quick turn, he experiences a centripetal acceleration. When this acceleration is greater than about $8 \times g$, the pilot will usually lose consciousness ("black out"). Consider a pilot flying at a speed of 900 m/s (about

2000 mi/h) who wants to make a very sharp turn. What is the minimum radius of curvature he can take without blacking out?

13. At a practice for a recent automobile race, officials found that the drivers were nearly "blacking out," which led to cancellation of the race. The cars were traveling at about 240 mi/h, and the track was approximately 1.5 mi long. Find the centripetal acceleration during a turn and compare it with the physiological limit of approximately $8 \times g$ discussed in Problem 12. Assume the track was circular.

14. The Daytona 500 stock car race is held on a track that is approximately 2.5 mi long, and the turns are banked at an angle of 31°. It is currently possible for cars to travel through the turns at a speed of about 180 mi/h. Assuming these cars are on the verge of slipping into the outer wall of the racetrack, find the coefficient of static friction between the tires and the track.

15. Consider again the problem of a car traveling along a banked turn. Sometimes roads have a "reversed" banking angle. That is, the road is tilted "away" from the center of curvature of the road. If the coefficient of static friction between the tires and the road is $\mu_S = 0.50$, the radius of curvature is 15 m, and the banking angle is 10°, what is the maximum speed at which a car can safely navigate such a turn?

5.2 EXAMPLES OF CIRCULAR MOTION

16. Consider the motion of a rock tied to a string of length 0.50 m. The string is spun so that the rock travels in a vertical circle as shown in Figure P5.16. The mass of the rock is 1.5 kg, and it is twirling at constant speed with a period of 0.33 s.

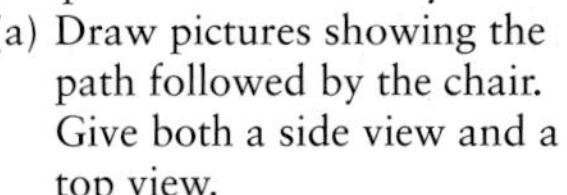

Figure P5.16
Problems 16 and 17.

(a) Draw free-body diagrams for the rock when it is at the top and when it is at the bottom of the circle. Your diagrams should include the tension in the string, but the value of T is not yet known.

(b) What is the total force on the rock directed toward the center of the circle?

(c) Find the tension in the string when the rock is at the top and when it is at the bottom of the circle.

17. Consider the motion of the rock in Figure P5.16. What is the minimum speed the rock can have without the string becoming "slack"?

18. A stone of mass 0.30 kg is tied to a string of length 0.75 m and is swung in a horizontal circle with speed v. The string has a breaking-point force of 50 N. What is the largest value v can have without the string breaking? Ignore any effects due to gravity.

19. The track near the top of your favorite roller coaster has a circular shape with a diameter of 20 m. When you are at the top, as in Figure 5.11, you feel as if your weight is only one-third your true weight. What is the speed of the roller coaster?

20. A roller-coaster track is designed so that the car travels upside down on a certain portion of the track as shown in Figure P5.20. What is the minimum speed the roller coaster can have without falling from the track? Assume the track has a radius of curvature of 30 m.

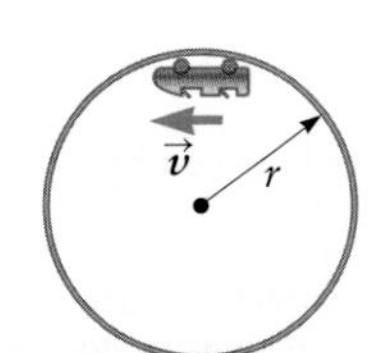

Figure P5.20

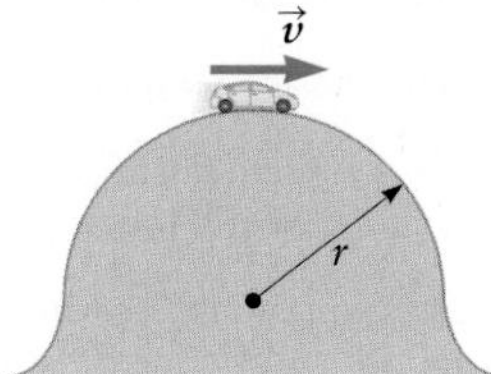

Figure P5.21

21. SSM A car of mass 1000 kg is traveling over the top of a hill as shown in Figure P5.21. (a) If the hill has a radius of curvature of 40 m and the car is traveling at 15 m/s, what is the normal force between the hill and the car at the top of the hill? (b) If the driver increases her speed sufficiently, the car will leave the ground at the top of the hill. What is the speed required to make that happen?

22. On a popular amusement park ride, the rider sits in a chair suspended by a cable as shown in Figure P5.22. The top end of the cable is tied to a rotating frame that spins, hence moving the chair in a horizontal circle with $r = 10$ m. The ride makes one complete revolution every 10 s.

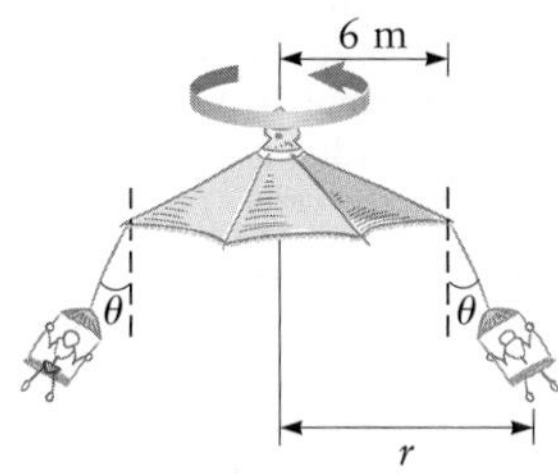

Figure P5.22

(a) Draw pictures showing the path followed by the chair. Give both a side view and a top view.

(b) In your diagrams in part (a), indicate all the forces on the chair. Also draw a free-body diagram for the chair.

(c) Find the components of the forces in the vertical direction and in the direction toward the center of the chair's circular path. Express your answers in terms of the tension in the cable, the angle θ, and the mass m of the chair.

(d) Apply Newton's second law in both the vertical and radial directions. What is the acceleration of the chair along y?

(e) Find the angle θ the cable makes with the vertical.

23. Consider a roller coaster as it travels near the bottom of its track as sketched in Figure P5.23. At this point, the normal force on the roller coaster is three times its weight. If the speed of the roller coaster is 20 m/s, what is the radius of curvature of the track?

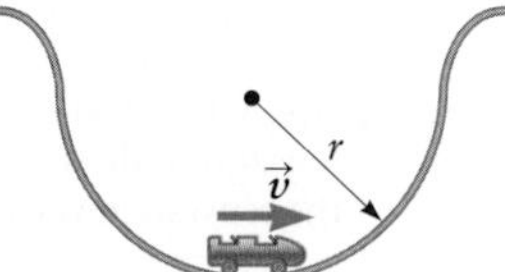

Figure P5.23

24. A rock is tied to a string and spun in a circle of radius 1.5 m as shown in Figure P5.24. The speed of the rock is 10 m/s.

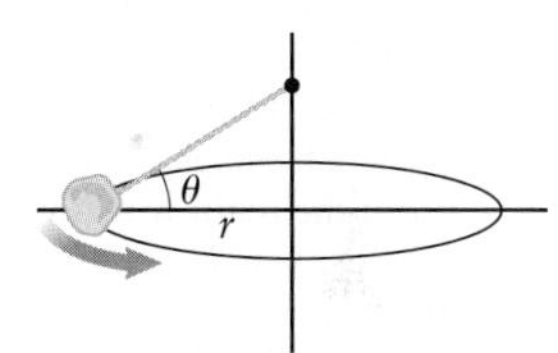

Figure P5.24

(a) Draw a picture giving both a top view and a side view of the motion of the rock.

(b) What are all the forces acting on the rock? Add them to your pictures in part (a). Then draw a free-body diagram for the rock.

(c) What is the total force on the rock directed toward the center of its circular path? Express your answer in terms of the (unknown) tension T in the string.

(d) Apply Newton's second law along both the vertical and the horizontal direction and find the angle θ the string makes with the horizontal.

25. A coin is sitting on a record as sketched in Figure P5.25. It is found that the coin slips off the record when the rotation rate is 0.30 rev/s. What is the coefficient of static friction between the coin and the record?

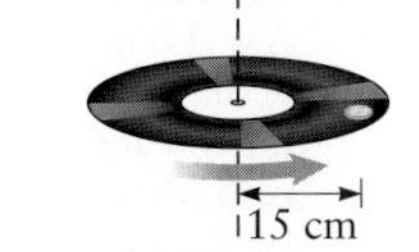

Figure P5.25

26. **Spin out!** An interesting amusement park activity involves a cylindrical room that spins about a vertical axis (Fig. P5.26). Participants in the "ride" are in contact with the wall of the room, and the circular motion of the room results in a normal force from the wall on the riders. When the room spins sufficiently fast, the floor is retracted and the frictional force from the wall keeps the people "stuck" to the wall. Assume the room

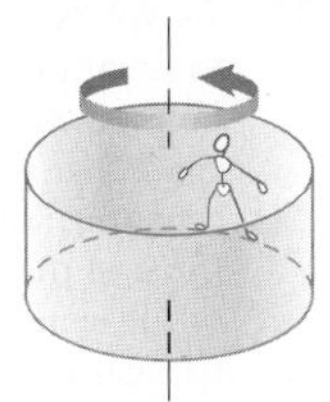

Figure P5.26

has a radius of 2.0 m and the coefficient of static friction between the people and the wall is $\mu_S = 0.50$.
(a) Draw pictures showing the motion of a "rider." Give both a side view and a top view.
(b) What are all the forces acting on a rider? Add them to your pictures in part (a). Then draw a free-body diagram for a rider.
(c) What are the components of all the forces directed toward the center of the circle (the radial direction)?
(d) Apply Newton's second law along both the vertical and the radial directions. Find the minimum rotation rate for which the riders do not slip down the wall.

27. Consider the circular space station in Figure 5.13. Suppose the station has a radius of 15 m and is designed to have an acceleration due to "artificial gravity" of $g/2$. Find the speed of the rim of the space station.

28. A child of mass $m = 50$ kg sits at the end of a rope of length $L = 3.2$ m. The other end of the rope is fastened to a ceiling in a gymnasium, and the child travels so that he moves in a horizontal circle of radius $r = 2.50$ m as sketched in Figure P5.28.

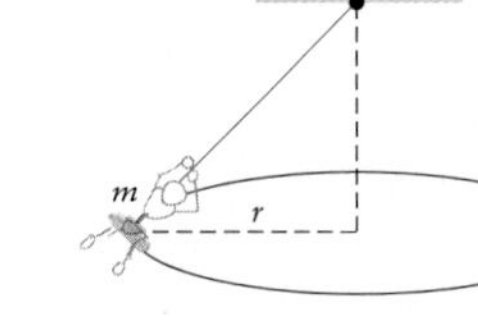

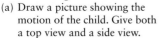

Figure P5.28

(a) Draw a picture showing the motion of the child. Give both a top view and a side view.
(b) What are all the forces on the child? Add them to your pictures in part (a). Then draw a free-body diagram for the child.
(c) Find the vertical and horizontal components of all the forces on the child.
(d) Apply Newton's second law in both the vertical (y) and the horizontal directions. What is the acceleration along y? Find the tension in the rope.

29. A car of mass 1700 kg is traveling without slipping on a flat, curved road with a radius of curvature of 35 m. If the car's speed is 12 m/s, what is the frictional force between the road and the tires?

30. RT Consider a Ferris wheel in which the chairs hang down from the main wheel via a cable. The cable is $L = 2.0$ m long, and the radius of the wheel is 12 m (Fig. P5.30). When a chair is in the orientation shown in the figure (the "3 o'clock" position), the cable attached to the chair makes an angle $\theta = 20°$ with the vertical. Find the speed of the chair. For simplicity, assume that at this moment the chair is moving in a circle of radius equal to the radius of the wheel.

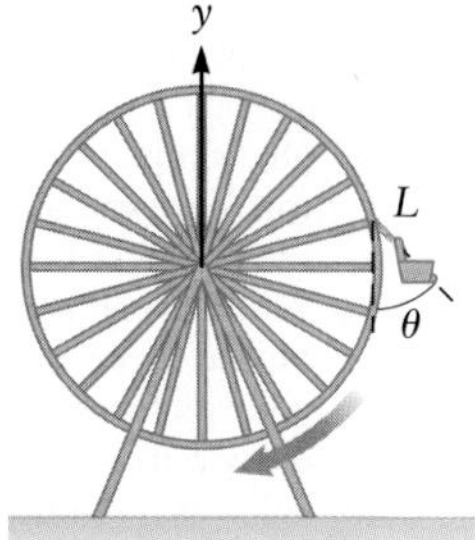

Figure P5.30

31. A rock of mass $m = 1.5$ kg is tied to a string of length $L = 2.0$ m and is twirled in a vertical circle as shown in Figure 5.10. The speed v of the rock is constant; that is, it is the same at the top and the bottom of the circle. If the tension in the string is zero when the rock is at its highest point (so that the string just barely goes slack), what is the tension when the rock is at the bottom?

32. B We saw in Example 5.6 how a centrifuge can be used to separate cells from a liquid. To increase the rate at which objects can be separated from solution, it is useful to make the centrifuge's speed as large as possible. If you want to design a centrifuge of diameter 1.5 m to have a force of 100,000 times the force of the Earth's gravity, what is the speed of the outer edge of the centrifuge? Such a device is called an *ultra*centrifuge. Centrifuges of this type are currently being used by certain countries (including Iran) to produce what is called enriched uranium for use in nuclear power plants or weapons. Naturally occurring uranium contains two isotopes (which have slightly different masses), only one of which is useful in these applications (see Chapter 30). Ultracentrifuges like the one considered in this problem are used to process the gas UF_6 and thereby separate the two uranium isotopes.

33. B A centrifuge can be used to separate DNA molecules from solution. Estimate how long it will take the centrifuge in Example 5.6 to separate a DNA molecule from water. Assume the centrifuge tube is 2.0 cm long. For this case, the drag coefficient in Stokes's law (Eq. 5.14) is $C = 0.020$ N · s/m^2.

34. SSM NASA has built centrifuges to enable astronauts to train in conditions in which the acceleration is very large. The device in Figure P5.34 shows one of these "human centrifuges." If the device has a radius of 8.0 m and attains accelerations as large as $5.0 \times g$, what is the rotation rate?

NASA

Figure P5.34

5.3 NEWTON'S LAW OF GRAVITATION

35. Two small objects of mass 20 kg and 30 kg are a distance 1.5 m apart. What is the gravitational force of one of these objects on the other?

36. If the masses of the objects in Problem 35 are both increased by a factor of $\sqrt{5}$, by what factor does the gravitational force change? Do not use a calculator to solve this problem!

37. Three lead balls of mass $m_1 = 15$ kg, $m_2 = 25$ kg, and $m_3 = 9.0$ kg are arranged as shown in Figure P5.37. Find the total gravitational force exerted by balls 1 and 2 on ball 3. Be sure to give the magnitude and the direction of this force.

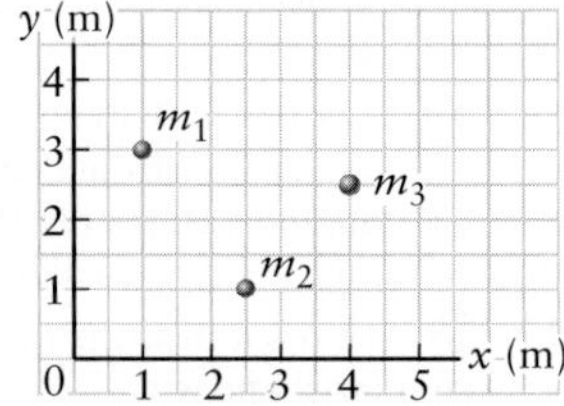

Figure P5.37

38. **Travel and lose pounds!** Your apparent weight is the force you feel on the bottoms of your feet when you are standing. Due to the Earth's rotation, your apparent weight is slightly more when you are at the South Pole than when you are at the equator. What is the ratio of your apparent weights at these two locations? Carry three significant figures in your calculation.

39. Find the gravitational force of the Sun on the Earth.

40. Calculate the acceleration due to gravity on the surface of Mars.

41. RT Estimate the gravitational force between two bowling balls that are nearly touching.

42. RT Suppose the bowling balls in Problem 41 are increased in size (radius) by a factor of two, but their density does not change. By what factor does the gravitational force change? *Hint*: When the radius is changed, both the mass and the separation will change.

43. SSM ★ When a spacecraft travels from the Earth to the Moon, both the Earth and the Moon exert a gravitational force on the spacecraft. Eventually, the spacecraft reaches a point where the Moon's gravitational attraction overcomes the Earth's gravity. How far from the Earth must the spacecraft be for the gravitational forces from the Moon and the Earth to just cancel?

44. Find the ratio of your weight on the Earth to your weight on the surface of the Sun.

45. ★ Some communications and television towers are much taller than any buildings. These towers have been used to study how the Earth's gravitational force varies with distance from the center of the Earth. Calculate the ratio of the acceleration due to the Earth's gravity at the top of a tower that is 600 m tall to the value of g at the Earth's surface. *Hint*: Keep four significant figures in your calculation.

46. Suppose the density of the Earth was somehow reduced from its actual value to 1000 kg/m^3 (the density of water). Find the value of g, the acceleration due to gravity, on this new planet. Assume the radius does not change.

47. You are an astronaut (m = 95 kg) and travel to a planet that is the same radius and mass as the Earth, but it has a rotational period of only 2 h. What is your apparent weight at the equator of this planet?

48. ★ In Section 5.4, we showed that the radius of a geosynchronous orbit about the Earth is 4.2×10^7 m, compared with the radius of the Earth, which is 6.4×10^6 m. By what factor is the force of gravity smaller when you are in geosynchronous orbit than when you are on the Earth's surface?

5.4 PLANETARY MOTION AND KEPLER'S LAWS

49. ★ Saturn makes one complete orbit of the Sun every 29.4 years. Calculate the radius of the orbit of Saturn. *Hint*: It is a very good approximation to assume this orbit is circular.

50. ★ The region of the solar system between Mars and Jupiter contains many asteroids that orbit the Sun. Consider an asteroid in a circular orbit of radius 5.0×10^{11} m. Find the period of the orbit.

51. SSM ★ A newly discovered asteroid is found to have a circular orbit, with a radius equal to 27 times the radius of the Earth's orbit. How long does this asteroid take to complete one orbit around the Sun?

52. ✪ In recent years, a number of nearby stars have been found to possess planets. Suppose the orbital radius of such a planet is found to be 4.0×10^{11} m, with a period of 1100 days. Find the mass of the star.

53. ✪ Mars has two moons, Phobos and Deimos. It is known that the larger moon, Phobos, has an orbital radius of 9.4×10^6 m and a mass of 1.1×10^{16} kg. Find its orbital period.

54. ★ In our derivation of Kepler's laws, we assumed the only force on a planet is due to the Sun. In a real solar system, however, the gravitational forces from the other planets can sometimes be important. Calculate the gravitational force of Jupiter on the Earth and compare it to the magnitude of the force from the Sun. Do the calculations for the cases when Jupiter is both closest to and farthest from the Earth (Fig. P5.54).

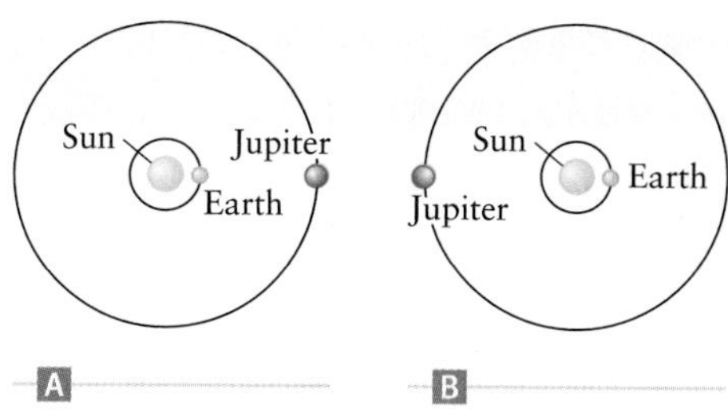

Figure P5.54

55. ★ What is the speed of a satellite in a geosynchronous orbit about the Earth? Compare it with the speed of the Earth as it orbits the Sun.

5.5 MOONS AND TIDES

56. ★ Ⓡ **Syzygy.** We have seen that normal tides are due to the gravitational force exerted by the Moon on the Earth's oceans. When the Moon, Sun, and Earth are aligned as shown in Figure P5.56, the magnitude of the tide increases due to the gravitational force exerted by the Sun on the oceans (at the times of the "new" Moon and the "full" Moon during the course of a month). Calculate the ratio of the gravitational force of the Sun to that of the Moon on the oceans. This "extra" force from the Sun does make a difference!

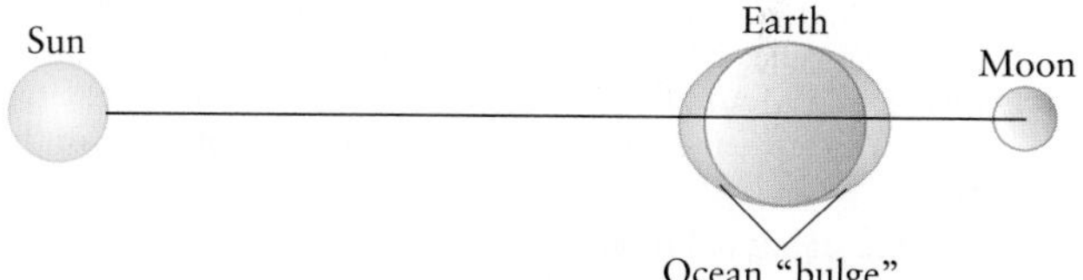

Figure P5.56

57. ★ Ⓡ In Figure 5.31, we saw that tides on the Earth are due to the variation of the Moon's gravitational force with distance. Find the approximate ratio of the Moon's gravitational force on two portions of the ocean, one nearest the Moon and one on the opposite side of the Earth.

58. ★ Your weight is due to the gravitational attraction of the Earth. The Moon, though, also exerts a gravitational force on you, and when it is overhead, your weight decreases by a small amount. Calculate the effect of the Moon on your weight. Express your result as a percentage change for the cases of the Moon overhead and the Moon on the opposite side of the Earth.

59. SSM ★ During an eclipse, the Sun, Earth, and Moon are arranged in a line as shown in Figure P5.59. There are two types of eclipse: (a) a lunar eclipse, when the Earth is between the Sun and the Moon, and (b) a solar eclipse, when the Moon is between the Sun and the Earth. Calculate the percentage change in your weight when going from one type of eclipse to the other.

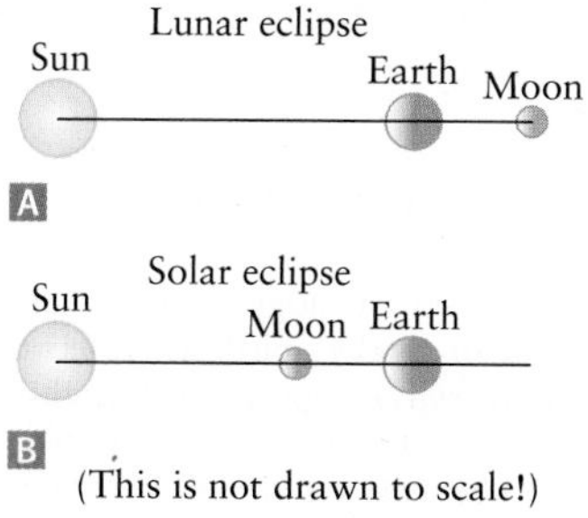

Figure P5.59

Additional Problems

60. In the film *Mission to Mars* (released in 2000), the spacecraft (Fig. P5.60) features a rotating section to provide artificial gravity for the long voyage. A physicist viewing a scene from the interior of the spacecraft notices that the diameter of the rotating portion of the ship is about five times the height of an astronaut walking in that section (or about 10 m). Later, in a scene showing the spacecraft from the exterior, she notices that the living quarters of the ship rotate with a period of about 30 s. Did the movie get the physics right? Compare the centripetal acceleration of a 1.7-m-tall astronaut at his feet to that at his head. Compare these accelerations to g.

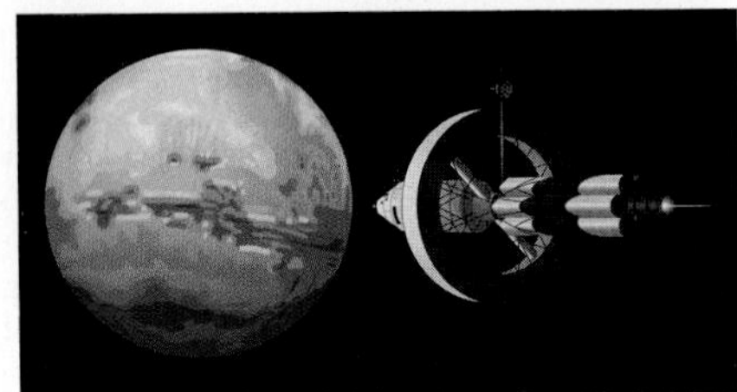

Figure P5.60 Not to scale.

61. Will your apparent weight at the top of Mount Everest (altitude 8850 m = 29,035 ft) be more or less than at sea level at the same latitude (27.98° N)? What is the ratio of your apparent weight on Mount Everest to that at sea level? For simplicity, consider only the effect of altitude and ignore the spinning of the Earth.

62. A man stands 6.0 ft tall at sea level on the North Pole as shown in Figure P5.62. (a) What is the difference in the value of g (the gravitational acceleration) between his head and his feet? (b) The man is now put in a space suit and transported to a location one Earth radius away from a black hole of mass equal to 20 times the mass of our Sun. If the man's feet are pointing in the direction of the black hole, what is the difference in the gravitational acceleration between his head and his feet? Would this difference in acceleration be harmful, just noticeable, or somewhere in between?

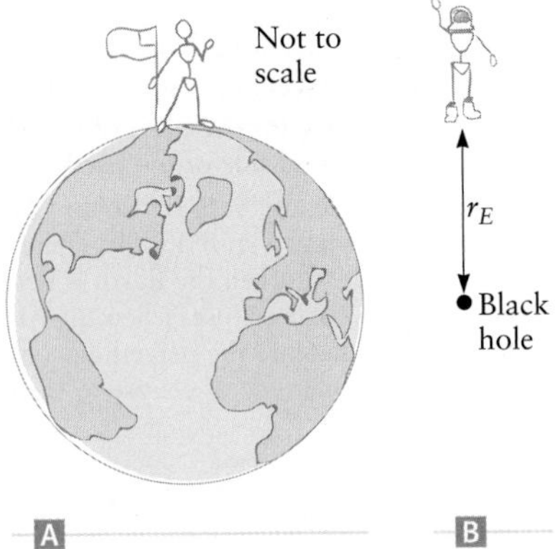

Figure P5.62

63. RT Proponents of astrology claim that the positions of the planets at the time of a baby's birth will affect the life of that person in important ways. Some assert that this effect is due to gravity. Use Newton's law of gravity to examine this claim. Calculate the ratio of the maximum force exerted on the baby by the planet Mars to the force of gravity exerted by the mass of the doctor's head as she delivers the baby. Assume Mars is at its closest approach to the Earth at the time. You will need to estimate the masses of and distance between the doctor's head and the baby.

64. SSM R An ancient and deadly weapon, a ***sling*** consists of two braided cords, each about half an arm's length long, attached to a leather pocket. The pocket is loaded with a projectile made of lead, carved rock, or clay and made to swing in a vertical circle as shown in Figure P5.64. The projectile is released by letting go of one end of the cord. (a) If a Roman soldier can swing the sling at a rate of 7.5 rotations per second, what is the maximum range of his 100-g projectile? (Ignore air drag.) (b) What is the maximum tension in each cord during the rotation?

Figure P5.64 The proper use of a sling.

65. A popular circus act features daredevil motorcycle riders encased in the "Globe of Death" (Fig. P5.65), a spherical metal cage of diameter 16 ft. (a) A rider of mass 65 kg on a 125-cc (95-kg) motorcycle keeps his bike horizontal as he rides around the "equator" of the globe. What coefficient of friction is needed between his tire and the cage to keep him in place? (b) How many loops will the rider make per second? (c) The same rider performs vertical loops in the globe. What force does the cage need to withstand at the top and the bottom of the rider's loop? Assume a speed of 20 mi/h for both tricks.

Figure P5.65 The "Globe of Death."

66. RT **Asteroid satellite.** While on its way to Jupiter in 1993, the *Galileo* spacecraft made a flyby of asteroid Ida. Images captured of Ida (Fig. P5.66) showed that the asteroid has a tiny moon of its own, since given the name Dactyl. Measurements found Ida to be about 56 × 24 × 21 km (35 × 15 × 13 mi) in size, and

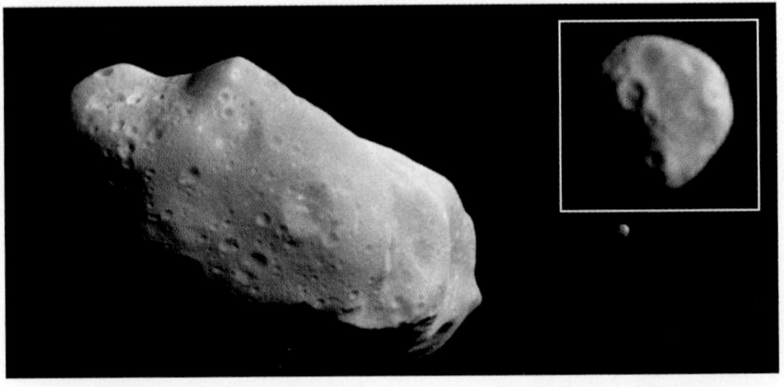

Figure P5.66 Ida and Dactyl from the *Galileo* spacecraft.

Dactyl's orbital period and radius are approximately 27 h and 95 km, respectively. From these data, determine Ida's approximate mass and density. *Hint*: When calculating Ida's centripetal acceleration and the force of gravity on Ida, it is safe to assume both Ida and Dactyl are spherical.

67. **The death spiral.** An Olympic pair figure-skating routine features an element called the death spiral shown in Figure P5.67. In this routine, the male skater swings his female partner in a circle. If the rotation rate is three-fourths a rotation per second, estimate the tension in the arms that the skaters' grip must withstand when performing this element. *Hint*: Approximate the female skater as a point mass located at her waist. It may be useful to take measurements directly from Figure P5.67 and scale appropriately.

Figure P5.67

68. Consider a hypothetical extrasolar world, planet Tungsten, that has twice the radius of the Earth and twice its density. (a) What is the acceleration due to gravity on the surface of planet Tungsten? (b) An interstellar astronaut lands on the equator of this planet and finds that his apparent weight matches his weight on the Earth. What is the period of rotation of planet Tungsten?

69. The movie *2001: A Space Odyssey* (released in 1968) features a massive rotating space station of radius 100 m, similar to the one in Figure 5.13B. (a) What period of rotation is needed to provide an artificial gravity of g at the rim? (b) At what speed is the rim moving? (c) What is your apparent weight if you run along the rim at 4.2 m/s opposite the rotation direction? (d) What is your apparent weight if you instead run in the direction of rotation? (e) In which direction would you run (either with or against the rotation) to get the best workout? Would it matter?

70. An astronaut stands on the surface of Vesta, which, with an average radius of 270 m, makes it the third largest object in the asteroid belt. The astronaut picks up a rock and drops it from a height of 1.5 m. He times the fall and finds that the rock strikes the ground after 3.2 s. (a) Determine the acceleration due to gravity at the surface of Vesta. (b) Find the mass of Vesta. (c) If the astronaut can jump to a height of 82 cm while wearing his space suit on the Earth, how high could he jump on Vesta? Assume any rotation of Vesta can be ignored.

71. The International Space Station orbits at an average height of 350 km above sea level. (a) Determine the acceleration due to gravity at that height and find the orbital velocity and the period of the space station. (b) The Hubble Space Telescope orbits at 600 km. What is the telescope's orbital velocity and period?

72. **Oil exploration.** When searching for gold, measurements of g can be used to find regions within the Earth where the density is larger than that of normal soil. Such measurements can also be used to find regions in which the density of the Earth is smaller than normal soil; these regions might contain a valuable fluid (oil). Consider a deposit of oil that is 300 m in diameter and just below the surface of the Earth. For simplicity, assume the deposit is spherical. Estimate the change in the acceleration due to gravity on the surface above this deposit. Assume the density of the oil is 800 kg/m^3 and the density of normal soil and rock is 2000 kg/m^3. *Note*: Companies that search for valuable minerals actually use this method.

73. SSM A rock of mass m is tied to a string of length L and swung in a horizontal circle of radius r. The string can withstand a maximum tension T_{max} before it breaks. (a) What is the maximum speed v_{max} the rock can have without the string breaking? (b) The speed of the rock is now increased to $3v_{max}$. The original single string is then replaced by N pieces that are all identical to the original string. What is the minimum value of N required so that the strings do not break? Ignore the force of gravity on the rock.

74. Lake Baikal in Siberia is the deepest lake on the Earth, with a maximum depth of 1600 m. Consider the weight of twin brothers with the same mass. One twin is in a rowboat in the middle of Lake Baikal, while his brother is standing on solid ground a few kilometers from the lake. What is the ratio of the weight of the twin in the boat to the weight of his brother on solid ground?

75. Experiments have shown that riders in a car begin to feel uncomfortable while traveling around a turn if their acceleration is greater than about $0.40 \times g$. Use this fact to calculate the minimum radius of curvature for turns at (a) 10 m/s (appropriate for driving in town) and (b) 30 m/s (highway driving).

76. In Insight 5.2, we discussed how, because of the force of gravity from the Moon, the Earth moves in an orbit around a point that lies between it and the Moon. (a) Find the radius of the Earth's orbit. (b) Does the center of the Earth's orbit lie inside or outside the Earth itself?

77. **Gravitational tractor.** In some science fiction stories, a "tractor beam" is used to pull an object from one point in space to another. That may not be just science fiction. It has been proposed that a "gravitational tractor" could be used to "tow" an asteroid. In theory, this tractor could be used to deflect asteroids that would otherwise collide with the Earth. A typical asteroid that could significantly damage life on the Earth might have a radius of 100 m and density 2000 kg/m^3. The "tractor" would be just a very massive spacecraft; in current designs, it would have a mass of about 2×10^4 kg. (a) What is the maximum gravitational force the tractor could exert on the asteroid? (b) What would be the acceleration of the asteroid? (c) If the tractor stayed near the asteroid for 1 year, what would be the deflection of the asteroid? (d) Consider an asteroid initially headed straight for the Earth with a speed of 3×10^4 m/s. If the tractor first comes close to the asteroid when it is 6×10^{11} m from the Earth (near the orbit of Jupiter), will the asteroid hit or miss the Earth? For simplicity, ignore the gravitational force of the Earth on the asteroid. *Note*: NASA is seriously considering the development of such a gravitational tractor.

78. Consider the coffee centrifuge in Question 20 (Fig. Q5.20). If this cup is moving in a circle with constant speed, what is the approximate minimum speed that will keep the coffee in the cup?

CHAPTER

6

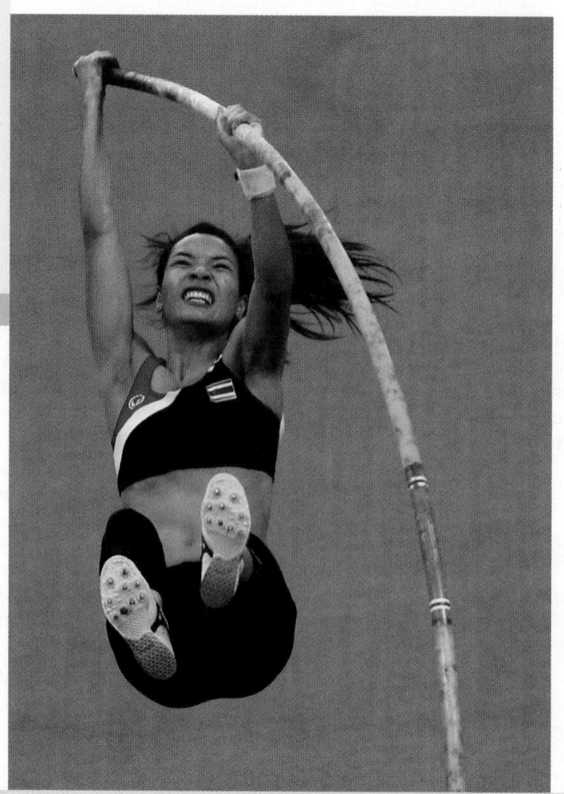

This pole vaulter uses her kinetic and potential energies in making her vault. In this chapter, we explore the connection between mechanical energy and motion. (© Mark Dadswell/Getty Images)

Work and Energy

OUTLINE

So far in this book, our discussions of motion have been based on very direct applications of Newton's laws. For example, to predict the motion of an object, we used Newton's second law ($\sum \vec{F} = m\vec{a}$) to calculate the acceleration; from there, we worked out the velocity and position and how they vary with time. Such direct applications of Newton's laws can take us a long way, but there is a lot more to mechanics than simply forces and acceleration. In this and the next few chapters, we will explore some very important concepts and principles that are, in a sense, "hidden" just beneath the surface of Newton's second law. This chapter is based on the concepts of ***work*** and ***energy***, and we'll see how Newton's second law leads to definitions of these quantities. We will also consider the concept of energy as applied to an individual particle and to systems of particles or objects. We'll find that the total energy of an isolated system is constant with time, which will lead us to the general principle of ***conservation of energy***. This principle has an important role in many fields, including engineering, physiology, and physics.

6.1 Force, Displacement, and Work

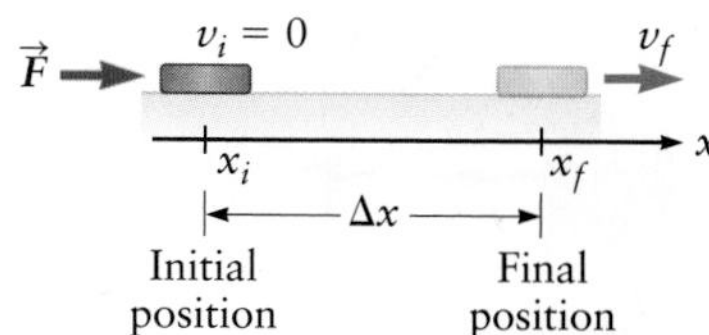

▲ **Figure 6.1** When a force $\vec{F}$ acts on this hockey puck, the puck accelerates. We also say that this force does work on the puck.

According to Newton's second law ($\sum \vec{F} = m\vec{a}$), the acceleration of a particle is proportional to the total applied force. A large force thus gives a large acceleration, while a small force gives a small acceleration. Given enough time, though, a small acceleration can still produce a large velocity and large displacement, so there is a sort of trade-off between force and time. Our intuition tells us that a small force acting for a sufficiently long time can have the same effect as a large force acting over a shorter time, but how long a time is required? The notion of "work" gives a way to answer this and other similar questions.

Consider a hockey puck of mass m placed on a frictionless, horizontal surface as sketched in Figure 6.1. The puck is initially at rest, and we want to consider how a constant force of magnitude F applied horizontally will set the puck into motion. Since the force is constant, we can use our results for motion with constant acceleration from Chapter 3. Applying Equation 3.4 and denoting the initial and final velocities by v_i and v_f, respectively, we can write

$$v_f^2 = v_i^2 + 2a\Delta x \tag{6.1}$$

where Δx is the displacement. The puck starts from rest, so $v_i = 0$, giving $v_f^2 = 2a\Delta x$. Using Newton's second law, we also have $F = ma$. Putting it all together leads to

$$v_f^2 = 2a\Delta x = \frac{2}{m}(ma)\Delta x = \frac{2}{m}(F\Delta x)$$

which we can rearrange to read

$$F\Delta x = \tfrac{1}{2}mv_f^2 \tag{6.2}$$

In words, this result means that if we want to accelerate an object that starts from rest to a particular velocity, we can exert a large force over a short distance or a small force over a large distance. As long as the product of force and displacement is the same, the object will reach the same final velocity. The product $F\,\Delta x$ in Equation 6.2 is called ***work***, and the work done by the applied force F on the hockey puck in Figure 6.1 is

$$W = F\,\Delta x \tag{6.3}$$

In SI units, we have

$$\text{work} = \text{force} \times \text{displacement} = \text{newtons} \cdot \text{meters} = \text{N} \cdot \text{m}$$

The unit N · m is called a joule (abbreviated "J").

The definition of work given in Equation 6.3 only applies for one-dimensional motion in which a constant force F is applied along the direction of motion. In two or three dimensions, we must consider that force $\vec{F}$ and displacement $\Delta\vec{r}$ are both vectors. In this case (Fig. 6.2), the work done on the particle is

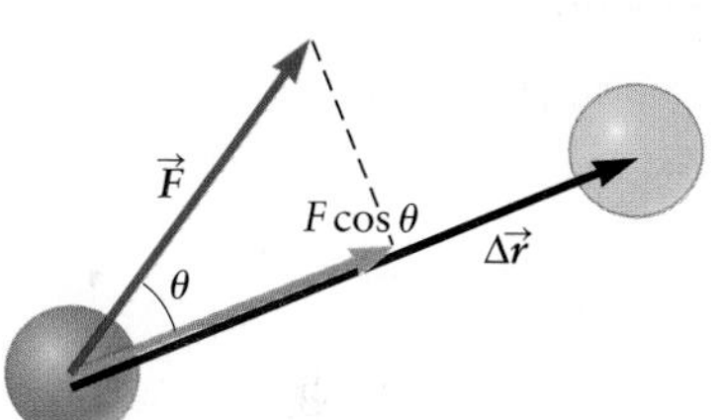

▲ **Figure 6.2** The force $\vec{F}$ acts on this particle while the particle moves through a displacement $\Delta\vec{r} = \vec{r}_f - \vec{r}_i$. The work done by the force is given by $W = F(\Delta r)\cos\theta$.

Definition of work

$$W = F(\Delta r)\cos\,\theta \tag{6.4}$$

where θ is the angle between the force and the particle's displacement. The factor F in Equation 6.4 denotes the magnitude of the force, while Δr is the magnitude of the displacement. Although force and displacement are both vectors, work is a scalar. Also, W can be positive, negative, or zero and does not have a direction.

W Depends on the Direction of the Force Relative to the Displacement

Figure 6.2 shows that the factor $F\cos\theta$ in Equation 6.4 is equal to the component of the force along the direction of the displacement. Hence, the work W equals the component of the force *along* the displacement multiplied by the magnitude of the displacement. In Figure 6.2, this component of the force is parallel to $\Delta\vec{r}$, and W is positive. A similar case in one dimension is shown in Figure 6.3A, in which the angle

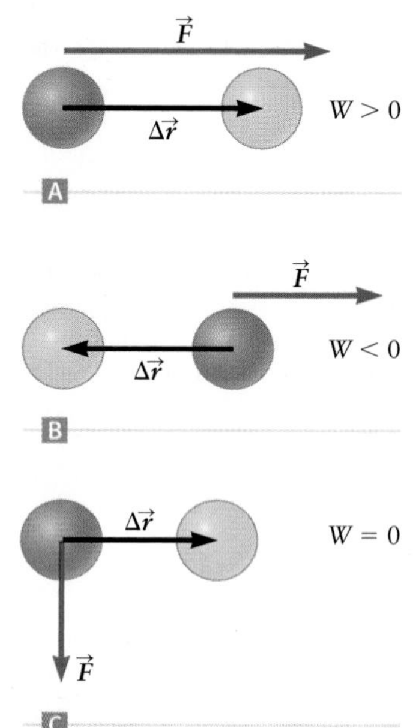

▲ **Figure 6.3** A The work done on an object is positive when the applied force and the object's displacement are parallel. B W is negative if the force and displacement are antiparallel. C When the force is perpendicular to the displacement, as in uniform circular motion, W is zero.

between the applied force and the displacement is zero (which was also the case for the hockey puck in Fig. 6.1). The cos θ factor in Equation 6.4 is then equal to unity, because $\cos(0) = 1$, and this expression is equivalent to the relation for work in one dimension (Eq. 6.3).

Figure 6.3B shows a situation in which the force on an object and the object's displacement are in opposite directions; that is, $\vec{F}$ and $\Delta\vec{r}$ are *anti*parallel. This object might be a hockey puck sliding across a surface with friction, since the force due to friction is directed opposite to the displacement. We then have $\theta = 180°$, giving $\cos\theta = -1$ in Equation 6.4, and the work done by friction on the puck is *negative*. Another important case is shown in Figure 6.3C, in which the force is perpendicular to the displacement. This situation often arises because, as we learned in Chapter 5, the force responsible for uniform circular motion is perpendicular to the direction of motion. When $\vec{F}$ and $\Delta\vec{r}$ are perpendicular, $\theta = 90°$ and $\cos\theta = 0$ in Equation 6.4; in this case, the work done on the object is $W = 0$.

The relationship between force, displacement, and work are central to this chapter and can be summarized as follows.

Key concepts: The relationship between force, displacement, and work

- Work is done *by* a force $\vec{F}$ acting *on* an object.
- The work W depends on the force acting on the object and on the object's displacement, according to Equation 6.4.
- The value of W depends on the direction of $\vec{F}$ relative to the object's displacement.
- W may be positive, negative, or zero, depending on the angle θ between the force and the displacement. W is a scalar; it is *not* a vector.
- If the displacement is zero (the object does not move), then $W = 0$, even though the force may be very large.

How Physics Uses the Term *Work*

The term *work* is used in everyday conversation. It is crucial to see how the everyday definition of this term differs from the "physics definition" in Equations 6.3 and 6.4. One important difference is that W (the "physics" work) can be negative. An example with $W < 0$ is sketched in Figure 6.3B, where the value of W is negative because the force and displacement are in opposite directions. As a result, the final speed of the object is less than its initial speed. In general, we can say that if $W > 0$, an object will "speed up"; if $W < 0$, it will "slow down."

Although we have not yet defined the concept of energy, we'll soon see how the motion of an object is related to its energy. Situations in which the force *increases* the energy of the object correspond to a positive value of W. Conversely, it is possible for a force to *reduce* the energy of an object (as with friction and the hockey puck in Fig. 6.3B); in such cases, the work done on the object is negative.

CONCEPT CHECK 6.1 Force, Displacement, and Work

Figure 6.3 shows hypothetical cases in which the force is (a) parallel, (b) antiparallel, and (c) perpendicular to the displacement. Identify which case applies to the following situations.

(1) An apple falling from a tree
(2) A satellite in geosynchronous orbit around the Earth
(3) A car skidding to a stop on a horizontal road

EXAMPLE 6.1 Towing a Car

A tow truck pulls a car of mass 1200 kg to a repair shop several miles away. As part of this journey, the tow truck and car must travel on a highway, and when entering the

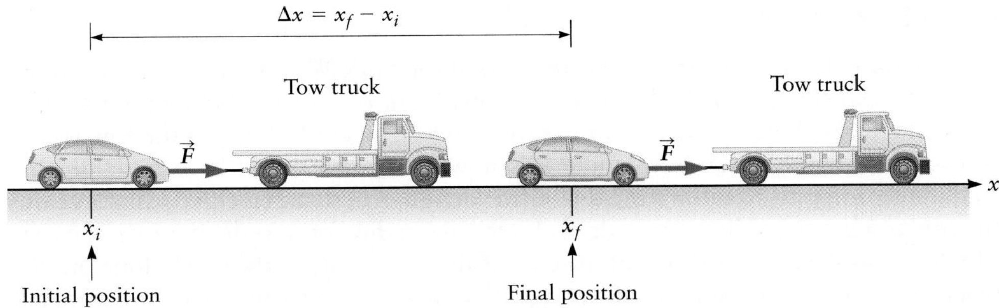

◀ **Figure 6.4** Example 6.1. The work done in towing this car along a straight road is equal to $F\,\Delta x$.

highway they accelerate uniformly as they merge into traffic. They have an initial speed of 10 m/s when they enter the merge ramp and a final speed of 30 m/s when they leave the merge lane and enter the highway. The length of the merge lane is 20 m. Find the work done on the car.

RECOGNIZE THE PRINCIPLE

To calculate W, we need to know the displacement of the car, the force on the car, and the angle between the force and the displacement. The displacement is equal to the length of the merge lane, whereas the force is directed along the lane; hence, $\theta = 0$ in Equation 6.4. We can obtain the force on the car from the acceleration together with Newton's second law.

SKETCH THE PROBLEM

Figure 6.4 shows the problem. The car and tow truck have a constant acceleration along the direction of the road. The displacement and the force are both directed along the merge lane, which we can take as the x direction.

IDENTIFY THE RELATIONSHIPS

To calculate the acceleration, we use the relations for motion with constant acceleration from Chapter 3. We denote the initial velocity as v_i and the initial position as x_i. From Equation 3.4, we have

$$v_f^2 = v_i^2 + 2a(x_f - x_i)$$

The initial speed is given as $v_i = 10$ m/s and the final speed as $v_f = 30$ m/s, and the displacement during this time is $(x_f - x_i) = 20$ m. Solving for the acceleration, we get

$$a = \frac{v_f^2 - v_i^2}{2(x_f - x_i)} = \frac{(30\text{ m/s})^2 - (10\text{ m/s})^2}{2(20\text{ m})} = 20\text{ m/s}^2$$

We now use Newton's second law to find the force needed to produce this acceleration:

$$\sum F = ma = (1200\text{ kg})(20\text{ m/s}^2) = 2.4 \times 10^4\text{ N}$$

Here the total force is equal to the total horizontal force, which we denote by F.

SOLVE

The work done on the car is equal to F times the displacement:

$$W = F\,\Delta x = F(x_f - x_i) = (2.4 \times 10^4\text{ N})(20\text{ m})$$

$$W = 4.8 \times 10^5\text{ N}\cdot\text{m} = \boxed{4.8 \times 10^5\text{ J}}$$

Notice that the units of work are joules (J).

► ***What does it mean?***

This result is the work done by the tow truck on the car. Notice that W is positive and that the car's final speed is greater than its initial speed.

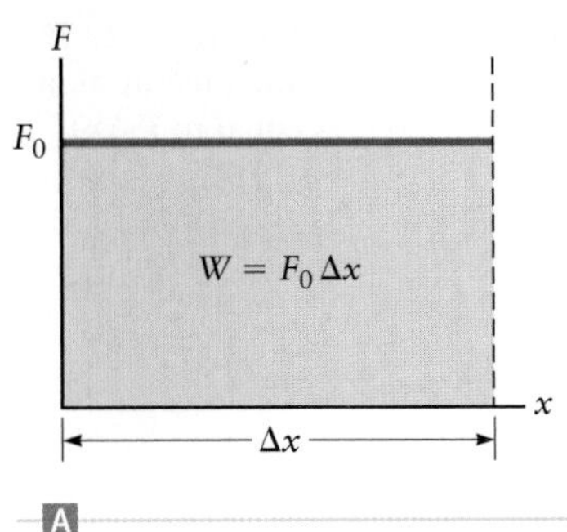

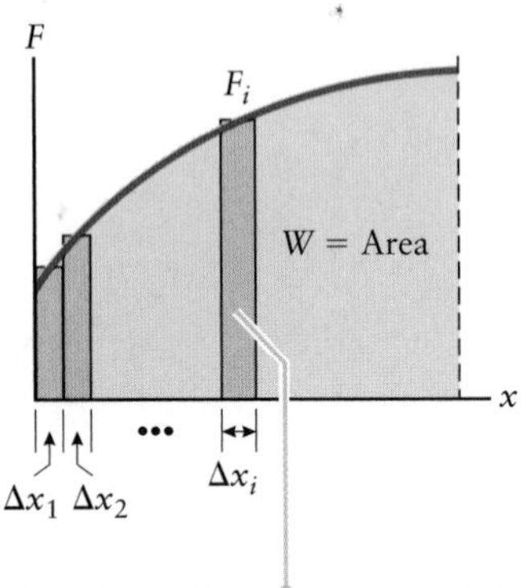

The work done during each small step Δx_i is equal to $F\Delta x_i$, which is just the area of a shaded box.

The total work done is the total area under the curve.

▲ **Figure 6.5** Work is equal to the area under a graph of force versus displacement. A The force is constant and equals the area of the shaded rectangle. B The force is not constant, but the work done by this force is still equal to the area under the curve of F versus x.

What Does the Work?

In Example 6.1, we found that a certain amount of work $W = 4.8 \times 10^5$ J was done, and we could say that the tow truck does this work *on the car*. We might also ask about the work done *by the car on the tow truck*. The work done on the tow truck is also given by Equation 6.4, where F is now the force exerted by the car on the truck. According to Newton's third law (the action–reaction principle), the force on the tow truck is equal in magnitude and opposite in direction to the force on the car. The force on the truck is thus opposite to its displacement, so the work done on the truck is *negative*. In general, when an agent applies a force to an object and does an amount of work W *on that object*, the object will do an amount of work equal to $-W$ "back" *on the agent*.

We can use the same ideas to discuss the work done in cases in which several forces, from different agents, act on an object. It is often useful to talk about the work done by each separate agent, that is, the work done separately by each force, which can be calculated using Equation 6.4, with F being the force exerted by the particular agent of interest.

Graphical Analysis and Work Done by a Variable Force

In our discussions of work, we have so far assumed the force is constant. To see how to deal with situations in which the force is not constant, it is useful to plot the force as a function of the displacement. When F is constant (Fig. 6.5A), this graph is simply a horizontal line. For simplicity, let's assume we have a case of one-dimensional motion, so the force and displacement are both along the x axis. The work done is then equal to the force times the displacement, and W is equal to the area under this force–displacement plot. Now consider a case in which the force is not constant as sketched in Figure 6.5B. We can (mathematically) think of the motion as a sequence of displacements Δx_1, Δx_2, Δx_3, and so forth, with the force approximately constant during each of these displacements. For each displacement, we can calculate the work using Equation 6.3 and then sum the results to get the total work done. We can see from Figure 6.5B that this process is equivalent to calculating the area under the entire force–displacement curve. Hence, when dealing with a nonconstant force, the work W is equal to the area under the force–displacement graph. In practice (see Problems 13 and 14 at the end of this chapter), this area can be estimated by dividing the area under the force–displacement graph into a series of rectangles as in Figure 6.5B and then computing the areas of these rectangles.

6.2 Kinetic Energy and the Work–Energy Theorem

In our derivation of Equation 6.2, we assumed the object started from rest so that its initial velocity was $v_i = 0$. Let's consider that problem again, but without making this assumption. Returning to Equation 6.1, we have

$$v_f^2 = v_i^2 + 2a\Delta x = v_i^2 + 2a(x_f - x_i) \tag{6.5}$$

where the initial velocity v_i is not zero and the displacement is $\Delta x = x_f - x_i$, with x_f being the final position. Notice again that we are using the subscript i to denote the initial position and initial velocity and subscript f to denote the final values. Rearranging Equation 6.5 gives

$$a(x_f - x_i) = \frac{v_f^2 - v_i^2}{2} \tag{6.6}$$

We now want to calculate the work done on the object as it moves from the initial position x_i to the final position x_f. Multiplying both sides of Equation 6.6 by m and using the definition of work $W = F\Delta x$ (Equation 6.3) leads to

$$W = ma(x_f - x_i) = m\left(\frac{v_f^2 - v_i^2}{2}\right)$$

$$W = \tfrac{1}{2}mv_f^2 - \tfrac{1}{2}mv_i^2 \qquad (6.7)$$

The expression $\frac{1}{2}mv^2$ that appears on the right-hand side of Equation 6.7 is an *extremely* important quantity. When an object of mass m has speed v, it has energy $\frac{1}{2}mv^2$ due to its motion. We call this ***kinetic energy*** (***KE***):

$$KE = \tfrac{1}{2}mv^2 \qquad (6.8)$$

Kinetic energy

"Energy" is a new concept for us in mechanics, and we can use Equation 6.7 to gain some understanding of it. Equation 6.7 tells us that the kinetic energy of an object can be changed by doing work on the object. If we write the change in kinetic energy as ΔKE, Equation 6.7 can be written

$$W = \tfrac{1}{2}mv_f^2 - \tfrac{1}{2}mv_i^2 = KE_f - KE_i$$

$$W = \Delta KE \qquad (6.9)$$

Work–energy theorem

The work done on an object is thus equal to the *change* in its kinetic energy. This relation tells us how work, and hence also force and displacement, are connected to the kinetic energy of an object. Equation 6.9 is called the ***work–energy theorem***. According to this result, the units of work and energy are the same, so energy is measured in joules. Another commonly used unit of energy is the calorie, with 1 cal equal to 4.186 J.

EXAMPLE 6.2 Kinetic Energy of an Asteroid

Asteroids are constantly bombarding the Earth and often explode as they heat up while passing through the atmosphere. Consider an asteroid with radius $r = 5$ m and a mass of 5×10^6 kg. Such asteroids have a speed of about $v = 10^4$ m/s when they enter the Earth's atmosphere. What is the kinetic energy of this asteroid? Is that a lot of energy?

RECOGNIZE THE PRINCIPLE

The mass and speed of the asteroid are given, so we simply need to evaluate its kinetic energy using Equation 6.8.

SKETCH THE PROBLEM

No figure is necessary.

IDENTIFY THE RELATIONSHIPS AND SOLVE

The kinetic energy is given by

$$KE = \tfrac{1}{2}mv^2$$

Inserting the given values of m and v, we find

$$KE = \tfrac{1}{2}mv^2 = \tfrac{1}{2}(5 \times 10^6)(10^4 \text{ m/s})^2 = \boxed{3 \times 10^{14} \text{ J}}$$

Here we have kept only one significant figure in our answer because the initial values were only given to that accuracy. This is a lot of energy; it is about five times the energy released by the atomic bomb that was dropped on Hiroshima, Japan, in 1945.

(continued) ▶

▶ ***What does it mean?***

The hypothetical asteroid we have considered here is not very large—about the size of a small bus—but it carries a *lot* of kinetic energy and can thus do a lot of damage. Data obtained from the U.S. Air Force Defense Support Program suggest that an asteroid explosion with this much energy occurs more than once a year! Fortunately, such explosions usually take place in the upper atmosphere and do little damage on the Earth.

EXAMPLE 6.3 Kinetic Energy of a Falling Object

A rock of mass m is dropped from the top of a tall building of height h. Find **(a)** the kinetic energy and **(b)** the speed of the rock just before it reaches the ground.

RECOGNIZE THE PRINCIPLE

The only force on the rock is the gravitational force, and as shown in the coordinate system of Figure 6.6, this force and the rock's displacement are both along the $-y$ direction. We have $F_{grav} = -mg$ where the negative sign indicates that the direction of this force is downward. This is a case of one-dimensional motion with the force parallel to the displacement, so we can apply Equation 6.3 to find the work done on the rock. We can then use the work–energy theorem to find the final kinetic energy and speed of the rock.

SKETCH THE PROBLEM

Figure 6.6 shows the problem. We have also drawn in a coordinate system along with a free-body diagram for the rock.

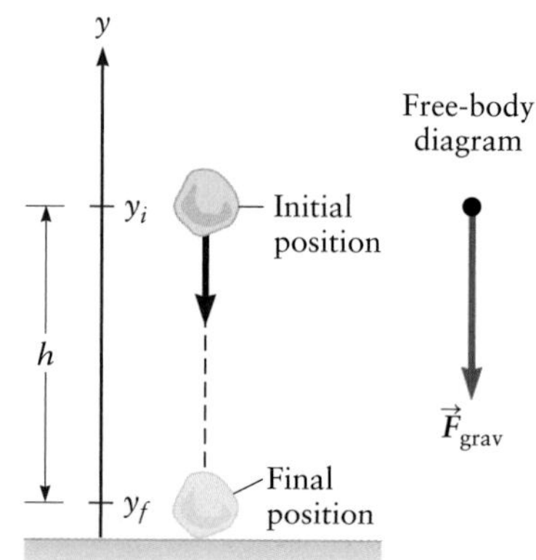

▲ **Figure 6.6** Example 6.3. When a rock falls vertically, the gravitational force is parallel to the rock's displacement. The work done by gravity on the rock is thus positive, with $W = mgh$.

IDENTIFY THE RELATIONSHIPS

The initial position of the rock is at $y_i = h$ and the final location is at ground level, $y_f = 0$, so the displacement of the rock is

$$\Delta y = y_f - y_i = 0 - h = -h$$

The work done on the rock is equal to the force times the displacement (Eq. 6.3):

$$W = F_{grav}\,\Delta y = (-mg)(-h) = mgh$$

Using the work–energy theorem (Eq. 6.9), we can write

$$W = \Delta KE = KE_f - KE_i$$

SOLVE

(a) The rock is initially at rest ($v_i = 0$), so its initial kinetic energy is $KE_i = \frac{1}{2}mv_i^2 = 0$, which leads to

$$W = mgh = \Delta KE = KE_f - KE_i = KE_f - 0$$

$$KE_f = \boxed{mgh} \quad (1)$$

Since the force and displacement are parallel (both downward), W is positive.

(b) The final speed v_f of the rock is related to its final kinetic energy by (Eq. 6.8):

$$KE_f = \tfrac{1}{2}mv_f^2$$

Inserting the result for KE_f from Equation (1) and solving for v_f gives

$$mgh = \tfrac{1}{2}mv_f^2$$

$$v_f^2 = 2gh$$

$$v_f = \boxed{\sqrt{2gh}} \quad (2)$$

▶ ***What does it mean?***
These results could also have been obtained using the methods for dealing with forces and acceleration that we developed in Chapter 3. Because the acceleration is constant, the displacement and velocity are related by (Eq. 3.4)

$$v_f^2 = v_i^2 + 2a(y_f - y_i) \quad (3)$$

The acceleration of the rock is $a = -g$, and (as we have already noted) $v_i = 0$, $y_i = h$, and $y_f = 0$. Inserting into Equation (3) gives

$$v_f = \sqrt{2gh}$$

This result is precisely the same as we found using the work–energy approach (Eq. 2). For this particular problem, both approaches are straightforward to carry out, but we'll soon encounter cases that are much easier to solve with the work–energy method.

CONCEPT CHECK 6.2 Dependence of the Kinetic Energy on Mass and Speed

Consider a rock of mass m traveling at speed v. Suppose the mass is doubled and the speed is cut in half. Which of the following statements is true?
(a) The kinetic energy is the same.
(b) The kinetic energy increases.
(c) The kinetic energy decreases.

EXAMPLE 6.4 Speed of a Skier

Consider a person of mass m skiing down an extremely icy, frictionless ski slope. This specially constructed ski slope has the shape of an inclined plane of angle α as shown in Figure 6.7. If the skier starts from an initial height h, what is the work done by gravity on the skier while she skis to the bottom of the slope? Use this result to find the speed of the skier when she reaches the bottom, assuming she starts from rest.

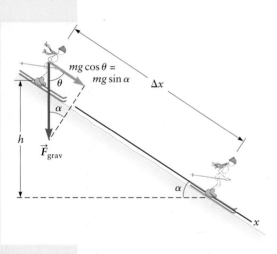

▲ **Figure 6.7** Example 6.4. The gravitational force is not parallel to this skier's displacement. The work done by gravity on the skier is equal to the component of the gravitational force along the slope multiplied by the skier's displacement.

RECOGNIZE THE PRINCIPLE

From Equation 6.4 and Figure 6.2, the work done by gravity on the skier equals the component of the gravitational force along the slope (parallel to the displacement) multiplied by the displacement. The force makes an angle θ with the displacement (see Fig. 6.7). This angle is *not* the same as the angle of the incline α.

SKETCH THE PROBLEM

Figure 6.7 shows the problem. Because the motion of the skier is along the incline, we take the coordinate axis to be parallel to the slope, with the positive x direction downward, parallel to the displacement of the skier.

IDENTIFY THE RELATIONSHIPS

The work done by the gravitational force is (Eq. 6.4)

$$W = F_{grav}\,\Delta x \cos\theta \quad (1)$$

The angle θ the force makes with the direction of the skier's displacement is related to the angle of the slope by $\theta = 90° - \alpha$ because they are two of the interior angles of a

(continued) ▶

right triangle. The displacement of the skier Δx is the distance traveled along the slope. From the trigonometry in Figure 6.7, we have

$$\frac{h}{\Delta x} = \sin \alpha$$

$$\Delta x = \frac{h}{\sin \alpha}$$

SOLVE

Inserting this result for Δx into Equation (1) and using $\theta = 90° - \alpha$, we find

$$W = F_{\text{grav}} \, \Delta x \cos \theta = mg\left(\frac{h}{\sin \alpha}\right)\cos(90° - \alpha)$$

This result can be simplified if we use the trigonometric identity $\cos(90° - \alpha) = \sin \alpha$:

$$W = mg \frac{h}{\sin \alpha} \cos(90° - \alpha) = mg \frac{h}{\sin \alpha} \sin \alpha = \boxed{mgh} \quad (2)$$

To find the final speed of the skier, we use the work–energy theorem and that her initial speed and hence also her initial kinetic energy are zero:

$$W = KE_f - KE_i = \tfrac{1}{2}mv_f^2$$

Using our result for W (= mgh) and solving for v_f, we find

$$W = mgh = \tfrac{1}{2}mv_f^2$$

$$v_f = \boxed{\sqrt{2gh}}$$

► ***What does it mean?***

Our result for the work W done by gravity in Equation (2) has *precisely* the same functional form as the work done on the rock in Example 6.3. The work done by the gravitational force on a rock that falls vertically through a height h (Example 6.3) is precisely equal to the work done by gravity on a skier who moves along a slope through the same vertical height h. In the next section, we'll see that this result is no accident and that it has some profound consequences.

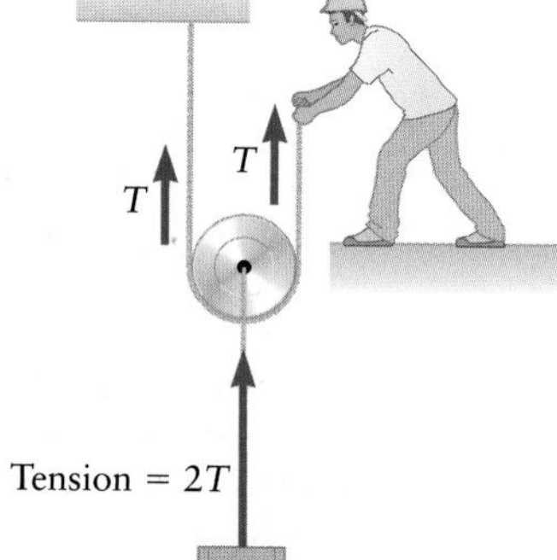

▲ **Figure 6.8** The person applies a force T to the rope. The block and tackle amplifies this force, and the total force applied to the crate by the rope is $2T$. However, when he moves the end of the rope a distance L, the crate moves a distance of only $L/2$.

Work, Energy, and Amplifying Forces

In Chapter 3, we discussed the block-and-tackle device in Figure 6.8, and showed that it amplifies forces by a factor of two. If a person applies a force F to the rope, the tension in the rope is $T = F$. Because the pulley is suspended by two portions of the rope, the upward force on the pulley is $2T$ and the pulley exerts a total force $2T$ on the object to which it is connected, that is, the crate in Figure 6.8. Let's now consider how this process of force amplification affects the work done by the person. If the person lifts his end of the rope through a distance L, the pulley will be raised by half that amount, that is, a distance of $L/2$. You can see why by noticing that when the person pulls his end of the rope up a distance L, the rope on the left side of the pulley will be shortened by an amount $L/2$, and this is also the distance that the pulley moves upward.

When the pulley moves upward by a distance $L/2$, the crate is displaced by the same amount. The work done *by the pulley on the crate* is equal to the total force of the pulley on the crate ($2T$) multiplied by the displacement of the crate, which is $L/2$:

$$W_{\text{on crate}} = 2T(L/2) = TL$$

At the same time, the work done *by the person on the rope* is equal to the force that he exerts on the rope (T) multiplied by the displacement of the rope, which is L, so

$$W_{\text{on rope}} = FL = TL$$

Thus, the work done on the rope is precisely equal to the work done on the crate.

There are several important messages from this example.

1. The work done by the person is effectively "transferred" to the crate.
2. Forces can be amplified, but work cannot be increased in this way. The force exerted by the person is increased by the block and tackle such that the force on the crate is amplified by a factor of two. However, the work is *not* amplified.
3. We'll see that this result—that work cannot be amplified—is a consequence of the principle of conservation of energy.

These points are all in accord with our observation in Chapter 3 that you "cannot get something for nothing." There, we also observed that pulleys can amplify a force, but they always do so by decreasing the associated displacement. Now we can go a bit further. According to the work–energy theorem, $W = \Delta KE$, which suggests that we can "convert" or "trade" work to kinetic energy and vice versa. The fact that work cannot be amplified suggests that this trade does not increase the amount of energy that is available. These qualitative ideas are leading us to a very important fundamental principle of physics, the principle of conservation of energy.

6.3 Potential Energy and Conservation of Energy

In Examples 6.3 and 6.4, we calculated the work done by the gravitational force in two different situations. In one case (Example 6.3), described by the path in Figure 6.9A, a rock of mass m fell vertically from a height h to ground level ($y = 0$), and we found that the work done by gravity on the rock is $W = mgh$. In the second case (Example 6.4), a skier of mass m skied down a slope, starting from an initial height h above the bottom of the slope, and we found that the work done by gravity is given by $W = mgh$. So, the work done by gravity was precisely the same, even though the rock and the skier followed very different paths. In fact, when an object of mass m follows *any* path that moves through a vertical distance h, the work done by the gravitational force is *always* equal to $W = mgh$. This amazing fact is connected with the notion of ***potential energy***.

Figure 6.9 shows several different paths that might be followed by an object as it moves between the same initial and final positions. The work done by the gravitational force as an object moves from a particular initial location (i) to a particular final location (f) is completely *independent of the path* the object takes to move from i to f. This statement is true for all conceivable paths and can be understood as follows. An object near the Earth's surface has a potential energy (PE) that depends only on the object's height h. The PE depends only on where the object is located and does not depend on how the object got there. This potential energy is a result of the gravitational force between the Earth and the object. Strictly speaking, both the force and the potential energy are properties of the "system" that is composed of the Earth and the object. Although we will often speak of the gravitational potential energy of an object, it should be understood that this PE is a property of the object and Earth together. A general feature of potential energy is that it is always a property of the *system* of particles or objects that interact through the underlying force.

The potential energy associated with a particular force is related to the work done by that force on an object as the object moves from one position to another. If this work is W, the change in the potential energy ΔPE is defined as

$$\Delta PE = PE_f - PE_i = -W \tag{6.10}$$

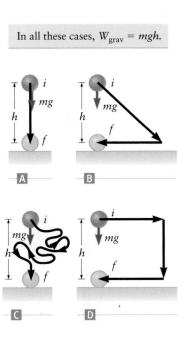

In all these cases, $W_{\text{grav}} = mgh$.

▲ **Figure 6.9** The work done by gravity is independent of the path taken. The initial and final points are the same in parts A, B, C, and D, so the work done is the same in all four cases. In addition, the work done by gravity on objects near the Earth's surface, W_{grav}, depends only on the change in height h.

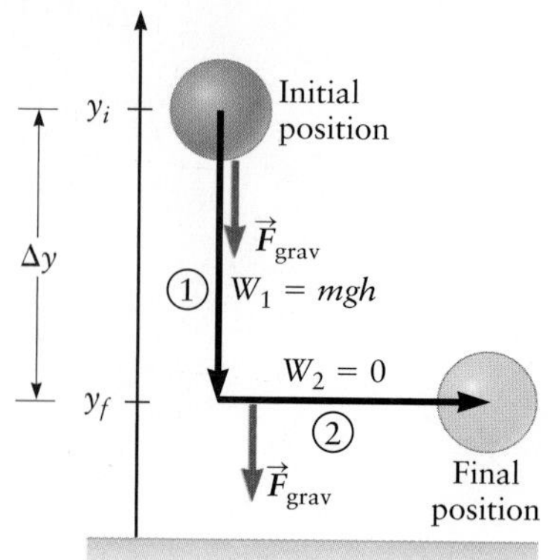

▲ **Figure 6.10** The work done by gravity along part 1 of this path is $W_1 = mgh$, whereas the work along part 2 is $W_2 = 0$ because the gravitational force and the object's displacement are perpendicular along this part of the path.

Since W is a scalar, potential energy is also a scalar. For the case of the gravitational force, we can thus find the change in the potential energy by calculating the work done in moving from an initial height y_i to a final height y_f. The work done is independent of the path, so we might as well use the simple path sketched in Figure 6.10 for the calculation of W. This path moves from our initial point to the final point in two steps. In the first step, we start at y_i and move downward (vertically) to the height y_f, while the second step moves horizontally to the final point. The work done by gravity during the second step is zero because the gravitational force and the object's displacement are perpendicular. We are thus left only with the work done during the first step, which is

$$W = F_{grav}\,\Delta y$$

The force is $F_{grav} = -mg$ (since the gravitational force is downward, along the $-y$ direction), and the displacement is $\Delta y = y_f - y_i$, so

$$W = F_{grav}\,\Delta y = (-mg)(y_f - y_i) \tag{6.11}$$

Using our relation for potential energy in Equation 6.10, the change in the potential energy is

$$\Delta PE = PE_f - PE_i = -W$$
$$\Delta PE = mg(y_f - y_i) \tag{6.12}$$

Another way to write this result is to say that the initial potential energy is $PE_i = mgy_i$ and the final potential energy is $PE_f = mgy_f$. Or, we could simply say that the gravitational potential energy of the object when it is at a height y is

Gravitational potential energy for an object near the Earth's surface

$$PE_{grav} = mgy \tag{6.13}$$

Our derivation of Equation 6.13 used the fact that $F_{grav} = -mg$, so this result for the gravitational potential energy applies only for objects near the surface of the Earth. In Section 6.4, we'll consider the gravitational potential energy of planets, moons, and other objects in the solar system.

Potential Energy Is Stored Energy

The result for the gravitational potential energy in Equation 6.13 tells us that the potential energy increases as the object is moved higher and decreases when the object is moved lower, which agrees with our intuition. If an object is moved to a greater height, it is capable of producing a greater kinetic energy if it were to fall back to a lower height, where its potential energy is smaller. This example also shows that potential energy is *stored energy*. By "allowing" an object to fall, the energy stored as gravitational potential energy can be converted to kinetic energy. We can increase the gravitational PE of an object by increasing its height as sketched in Figure 6.11. As the object moves higher, the work done *by gravity* is negative because the gravitational force is directed downward while the displacement of the object is upward. Now, the object moves higher because there is another force acting on it. We call this an "external" force F_{ext}, and it might be produced by your hand as you lift the object. This external force is directed upward, so the work done by F_{ext} on the object is positive. In fact, the work done by this external force, W_{ext}, is equal in magnitude but opposite in sign compared to the work done by gravity. Hence, we have stored an amount of energy W_{ext} as gravitational potential energy. We can recover this energy by letting the object fall back down to its initial height, thus gaining kinetic energy.

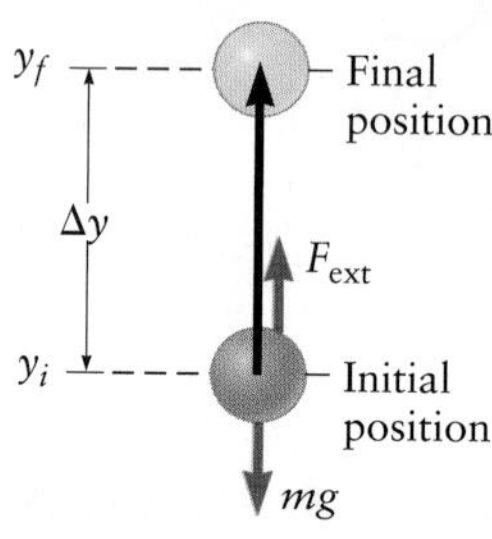

▲ **Figure 6.11** An external agent (perhaps a person's hand) lifts an object by exerting a force F_{ext} on the object. This agent does an amount of work $W_{ext} = F_{ext}\,\Delta y$ on the object. This energy is stored as potential energy of the object.

Potential Energy and Conservative Forces

We have seen that the work done by the gravitational force depends only on the initial and final locations and not on the path taken to get from one point to the other. We say that the work done by gravity is *independent of the path* taken. This property allowed us to define a potential energy function that depends only on the height of the object being studied. Many other forces besides gravity have the property that the work done is independent of the path, and they all have a corresponding potential energy function. Such forces are called ***conservative forces.*** All conservative forces can be used to store energy as potential energy. Other types of forces do not have potential energy functions and are called ***nonconservative forces.*** We'll spend the rest of this section learning how to apply and understand the gravitational potential energy in Equation 6.13. We then explore how these ideas apply to other types of conservative and nonconservative forces in subsequent sections.

We have introduced several new and important concepts connected with potential energy, and it is worth summarizing them now.

Key Facts about Potential Energy and Conservative Forces

- Potential energy is a result of the force(s) that act on an object. Since forces always come from the interaction of two objects, *PE* is a property of the objects (the "system") involved in the force.
- Potential energy is energy that an object or system has by virtue of its *position.*
- Potential energy is *stored energy*; it can be converted to kinetic energy.
- Potential energy is a scalar; its value can be positive, negative, or zero, but it does not have a direction.
- For forces that are associated with a potential energy (and hence have a potential energy function), the work done is independent of the path taken. Such forces are called *conservative forces.*
- Some forces, such as friction and air drag, are *nonconservative forces.* These forces do not have potential energy functions and cannot be used to store energy. The work done by a nonconservative force depends on the path taken by the object of interest.

Potential Energy, the Work–Energy Theorem, and Conservation of Energy

Let's now reexamine the work–energy theorem and consider how potential energy fits into our ideas about work and energy. We first wrote the work–energy theorem as (compare with Eq. 6.9)

$$W = \Delta KE = KE_f - KE_i \tag{6.14}$$

Here, W is the work done by *all* the forces acting on the object of interest. In many cases, some of or all these forces are associated with a potential energy. If so, the potential energy associated with a particular force is related to work done by that force, with $\Delta PE = -W$ (Eq. 6.10). If *all* the work is done by conservative forces such as gravity, then in Equation 6.14 we can replace the work done by that force with the negative of the change in the potential energy, and we have

$$W = -\Delta PE = -(PE_f - PE_i) = KE_f - KE_i$$

which can be rearranged to read

$$KE_i + PE_i = KE_f + PE_f \tag{6.15}$$

Conservation of mechanical energy

This version of the work–energy theorem applies to situations in which all the forces are conservative forces, and it contains potential energy and kinetic energy together. In words, Equation 6.15 says that the sum of the potential and kinetic energies of an object is *conserved.* That is, the sum of the potential and kinetic energies when the object is at its initial location (i) is equal to the sum of the potential and kinetic energies at the final location (f). This sum of the potential and kinetic energies is called the ***total mechanical energy.***

Equation 6.15 is thus a statement of the ***conservation of mechanical energy.*** In our derivation, we assumed only conservative forces are acting on the object of interest, so that all the forces involved possess a potential energy function. Frictional forces do not satisfy this requirement, and we'll learn how to deal with them in Section 6.5. Nevertheless, in many situations all the forces in a problem are connected with potential energy. In these cases, the principle of conservation of mechanical energy in Equation 6.15 is a very powerful tool for understanding, analyzing, and predicting motion.

Conservation of Mechanical Energy and the Speed of a Snowboarder

Let's apply the conservation of energy relation in Equation 6.15 to analyze the motion of a snowboarder who is sliding down an extremely icy (frictionless) hill (Fig. 6.12). If she starts from a height $h = 20$ m above the bottom of the hill with initial speed $v_i = 15$ m/s, how fast will she be traveling at the bottom of the hill?

The only forces acting on the snowboarder are the force of gravity and the normal force from the hill. (Recall that the hill is very slippery and the frictional force is zero.) The normal force is perpendicular to the displacement of the snowboarder, so it does no work on her. The only remaining force is gravity, which is a conservative force, so the mechanical energy of the snowboarder is conserved. As the snowboarder moves down the hill, her potential energy is converted to kinetic energy, resulting in a larger velocity at the bottom of the hill. Applying Equation 6.15, we have

$$KE_i + PE_i = KE_f + PE_f$$

$$\tfrac{1}{2}mv_i^2 + mgy_i = \tfrac{1}{2}mv_f^2 + mgy_f \tag{6.16}$$

where v_f is her velocity at the bottom of the hill and y_f is her height at the bottom. If we place the origin of the y axis at the bottom of the hill, then $y_i = h$ and $y_f = 0$. Inserting these values into Equation 6.16 and solving for v_f, we find

$$\tfrac{1}{2}mv_i^2 + mgh = \tfrac{1}{2}mv_f^2 + mg(0)$$

$$\tfrac{1}{2}v_i^2 + gh = \tfrac{1}{2}v_f^2$$

$$v_f = \sqrt{v_i^2 + 2gh} = \sqrt{(15 \text{ m/s})^2 + 2(9.8 \text{ m/s}^2)(20 \text{ m})} = 25 \text{ m/s}$$

The final speed v_f does *not* depend on the angle of the hill; it only depends on the height of the hill. That is because the gravitational potential energy depends only on the snowboarder's height, and changes in this potential energy are independent of the path she follows. So, the final speed will be the same for the two different hills in parts A and B of Figure 6.12, provided they have the same height.

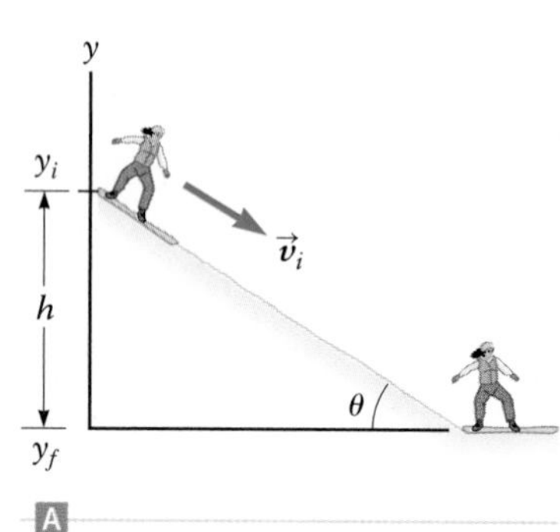

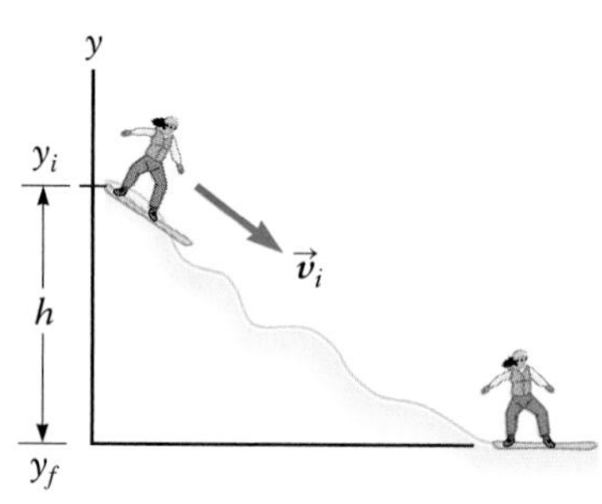

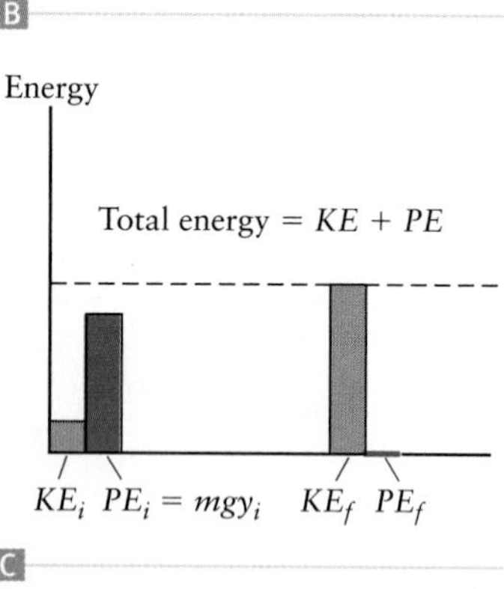

▲ **Figure 6.12** A and B These two hills have the same height h, so even though they have very different shapes, the work done by gravity in the two cases is the same: W[hill A] = W[hill B]. The change in the potential energy is therefore the same in the two cases ($\Delta PE_A = \Delta PE_B$). C These bar charts show the initial and final kinetic and potential energies of the snowboarder.

Charting the Energy

A useful way to illustrate the conservation of energy is with bar charts. The bar charts in Figure 6.12C show the kinetic and potential energies of the snowboarder at the start and the end of the slope. These charts show how the initial potential energy of the snowboarder is converted to kinetic energy. Notice also how the sum of the kinetic and potential energies is the same at the start and end, which is just another way to represent the conservation of energy.

Why Is the Principle of Conservation of Energy Useful?

Using conservation of energy principles made the calculation with the snowboarder in Figure 6.12 very straightforward. For the hill with the simple shape in Figure 6.12A, we could have done this problem in Chapter 4 by applying Newton's second law ($\sum \vec{F} = m\vec{a}$). Because this hill is just a ramp, the component of the gravitational force along the hill is constant; the acceleration is therefore also constant, so we could have used our relations for motion with constant acceleration to find the final velocity. We did not really need to use conservation of energy principles as expressed by Equation 6.15 to solve this problem. However, most snowboard hills do not have such a simple shape. The hill in Figure 6.12B has a much more interesting and realistic shape, as it undulates several times as the snowboarder travels to the bottom. Because the hill's angle is not constant, the acceleration is definitely not constant and the methods we employed in Chapter 4 will not work here. However, this complicated hill can easily be handled with the conservation of energy approach. To find the final velocity of the snowboarder, we only need to know the height of the hill; the shape of the hill does not have any effect on the final velocity, and our calculation thus applies to any hill, regardless of its shape. This example illustrates the power of the conservation of energy approach; this approach will allow us to tackle many problems that cannot be handled using the methods of Chapter 4.

PROBLEM SOLVING Applying the Principle of Conservation of Mechanical Energy

Most applications of conservation of energy ideas can be attacked using the following steps.

1. **RECOGNIZE THE PRINCIPLE.** Start by finding the object (or system of objects) whose mechanical energy is conserved.
2. **SKETCH THE PROBLEM,** showing the initial and final states of the object. This sketch should also contain a coordinate system, including an origin, with which to measure the potential energy.
3. **IDENTIFY THE RELATIONSHIPS.** Find expressions for the initial and final kinetic and potential energies. One or more of these energies may involve unknown quantities.
4. **SOLVE.** Equate the initial and final mechanical energies and solve for the unknown quantities.
5. Always *consider what your answer means* and check that it makes sense.

This approach can only be used when mechanical energy is conserved, that is, when all the forces that do work on an object are conservative forces. We'll consider how to extend this approach to deal with nonconservative forces such as friction in Section 6.5.

EXAMPLE 6.5 Roller-Coaster Physics

Consider the roller coaster in Figure 6.13. The car starts at position A with an initial height $h_A = 20$ m and speed $v_A = 15$ m/s. Find (**a**) the speed of the roller coaster when it reaches the top of the track at position B, where $h_B = 25$ m, and (**b**) the speed of the roller coaster when it reaches the bottom of the track, where $h_C = 0$. (**c**) Make a qualitative sketch of the roller-coaster car's kinetic energy and potential energy as a function of position along the track. Also make a corresponding energy bar chart for the system at points A, B, and C. Ignore the rotational motion of the wheels and assume all frictional drag on the roller coaster is negligible.

RECOGNIZE THE PRINCIPLE

We follow the steps outlined in the problem-solving strategy on conservation of energy. Step 1: The object of interest is the roller-coaster car; its mechanical energy is conserved

(continued) ▶

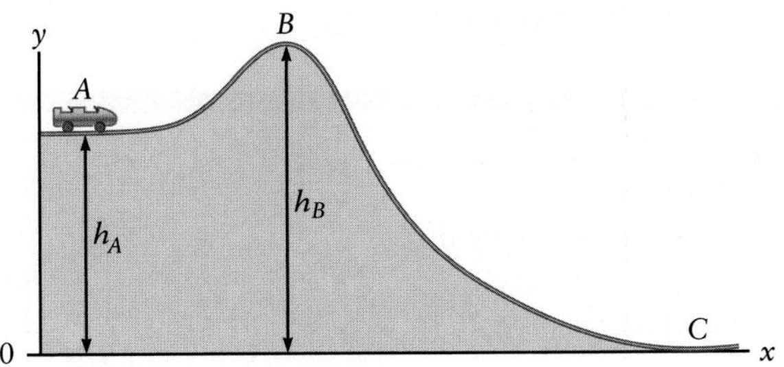

◀ **Figure 6.13** Example 6.5.

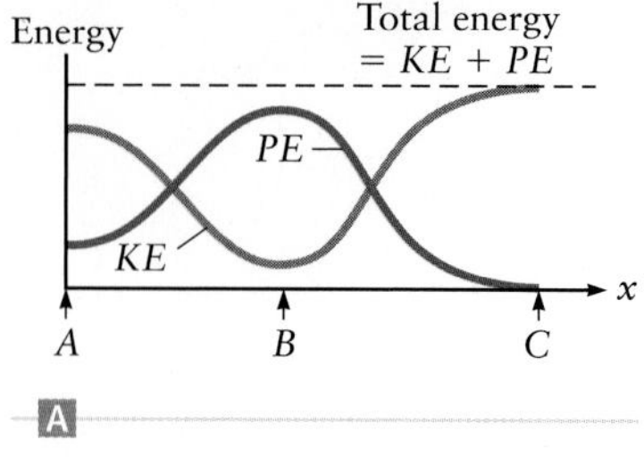

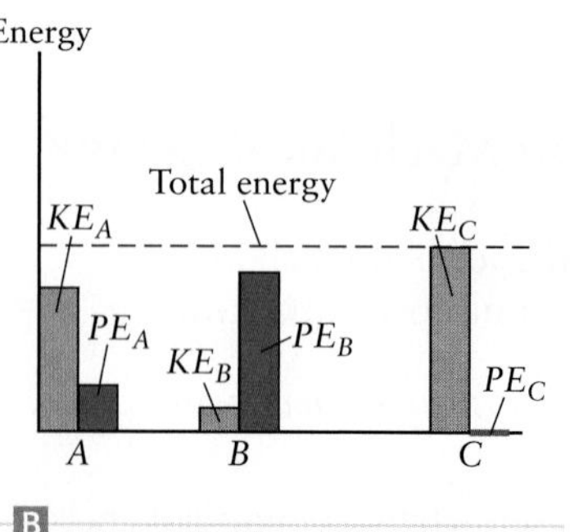

▲ **Figure 6.14** Behavior of the kinetic and potential energies of the roller-coaster car in Example 6.5.

because the only force that does work on the car is gravity, which is a conservative force.

SKETCH THE PROBLEM

Step 2: Figure 6.13 shows the initial position of the car along with the other positions of interest (points B and C). We take the origin of the y axis to be at the bottom of the track.

IDENTIFY THE RELATIONSHIPS

Step 3: We next apply the conservation of energy relation (Eqs. 6.15 and 6.16) and then (step 4) solve for the speed at B and C.

SOLVE

(**a**) Applying the conservation of energy relation (Eq. 6.15) at locations A and B gives

$$KE_A + PE_A = KE_B + PE_B$$

$$\tfrac{1}{2}mv_A^2 + mgh_A = \tfrac{1}{2}mv_B^2 + mgh_B$$

We now see that the mass of the roller coaster cancels. Solving for v_B, we find

$$\tfrac{1}{2}v_A^2 + gh_A = \tfrac{1}{2}v_B^2 + gh_B$$

$$v_B^2 = v_A^2 + 2gh_A - 2gh_B = v_A^2 + 2g(h_A - h_B)$$

$$v_B = \sqrt{v_A^2 + 2g(h_A - h_B)}$$

Inserting the given values of v_A, h_A, and h_B leads to

$$v_B = \sqrt{(15 \text{ m/s})^2 + 2(9.8 \text{ m/s}^2)(20 \text{ m} - 25 \text{ m})} = \boxed{11 \text{ m/s}}$$

(**b**) We again use Equation 6.15, but now we apply it at locations A and C, which leads to

$$v_C = \sqrt{v_A^2 + 2g(h_A - h_C)} = \sqrt{(15 \text{ m/s})^2 + 2(9.8 \text{ m/s}^2)(20 \text{ m} - 0)} = \boxed{25 \text{ m/s}}$$

(**c**) Figure 6.14A shows the qualitative behavior of the kinetic and potential energies as functions of x, the position along the track, while bar charts showing the kinetic and potential energies at points A, B, and C are given in Figure 6.14B. The kinetic energy is largest at the lowest point on the track (C), while the potential energy is largest when the roller-coaster car is at the highest point (B). The sum $KE + PE$ is the total mechanical energy of the car and is constant throughout.

► ***What does it mean?***

The value of the final velocity v_C is independent of the shape of the roller-coaster track and depends only on the initial and final heights. However, this result assumes the roller coaster will actually make it past the top of the track. If the height at location B were made much larger, the roller coaster would not make it to the top of the track and hence would not be able to travel to location C.

CONCEPT CHECK 6.3 Roller Coasters and Conservation of Energy

Consider the roller coaster in Example 6.5. How would you use conservation of energy ideas to calculate the value of h_B for which the roller coaster would barely make it over the highest point of the track?

EXAMPLE 6.6 Driving through a Loop-the-Loop Track

The author's daughter likes to relax by driving fast. One particularly relaxing case involves loop-the-loop tracks like the one shown in Figure 6.15. Although few roads for automobiles are constructed with this shape, it is a common design for roller coasters.

The middle of the track has a circular shape with radius $r = 15$ m, and the car enters and leaves along the horizontal portions at ground level. If a driver wants to coast all the way through without losing contact with the track when she reaches the top, what is the smallest velocity she can have when she enters the track at point A? As was the case with the roller coaster in Figure 6.13, ignore the rotational motion of the car's wheels and assume all friction is negligible.

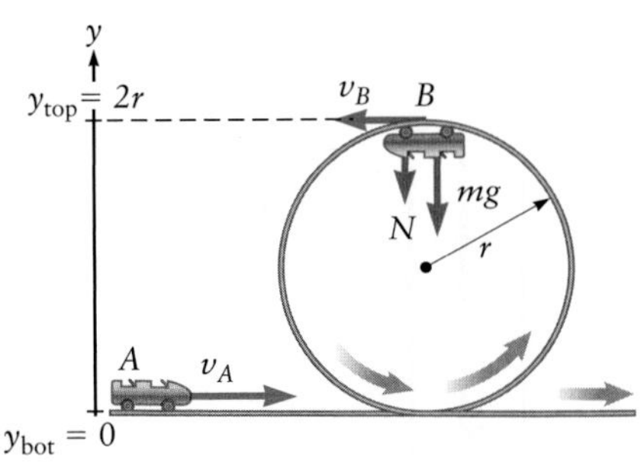

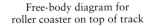

Free-body diagram for roller coaster on top of track

▲ **Figure 6.15** Example 6.6. A car enters this loop-the-loop track from the left at point A and coasts all the way through. The middle portion of the track has a circular shape with radius r.

RECOGNIZE THE PRINCIPLE

The only force that does work on the car is gravity (because the normal force exerted by the track is always perpendicular to the car's path), so the car's mechanical energy is conserved. We therefore follow the problem-solving strategy for problems in which energy is conserved as applied (step 1) to the car. Step 2: Figure 6.15 shows the car at the initial point of its trip (point A). You should suspect that the most dangerous part of the trip will be when the car is upside down at the top of the track (point B), so we take this point as the final position. We can then follow steps 3 and 4 of the problem-solving strategy (see the solution below) to find the speed v_B of the car at the top of the track. However, just knowing v_B will not tell us if the car will stay on the track. For the car to stay on the track at point B, the normal force N exerted by the track on the car must always be greater than zero. If the car has a very high speed at the top of the track, then N will be large. As this speed is reduced, N decreases. Eventually, if v_B is too small, the normal force is zero, signaling that the car has just lost contact. We can use the condition that $N = 0$ to find the minimum velocity at the top and then use conservation of energy principles to find the corresponding velocity at the bottom of the track at point A.

SKETCH THE PROBLEM

Figure 6.15 shows the car at its initial and final positions, along with a free-body diagram corresponding to the final position at B.

IDENTIFY THE RELATIONSHIPS

There are two forces acting on the car at point B—the normal force exerted by the track and the gravitational force—and both are directed downward. For an object undergoing uniform circular motion, the total force must be directed toward the center of the circle and have a magnitude mv^2/r. We thus have

$$\frac{mv_B^2}{r} = N + mg$$

Since the car is just barely staying on the track, the normal force will be very small, so we can take $N = 0$. Solving for v_B then gives

$$v_B = \sqrt{gr}$$

This is the speed at the top that will just keep the car on the track. We now use the conservation of energy condition to work back and find the corresponding speed at the bottom of the track. Using conservation of mechanical energy, we have

$$PE_A + KE_A = PE_B + KE_B$$

$$mgy_A + \tfrac{1}{2}mv_A^2 = mgy_B + \tfrac{1}{2}mv_B^2 \quad (1)$$

If we take $y = 0$ at the bottom of the track, then $y_A = 0$ and $y_B = 2r$ (see Fig. 6.15).

SOLVE

Inserting these values of y_A and y_B and our result for v_B into Equation (1) gives

$$\tfrac{1}{2}mv_A^2 = mgy_B + \tfrac{1}{2}mv_B^2 = mg(2r) + \tfrac{1}{2}m(gr) = \tfrac{5}{2}mgr$$

(continued) ▶

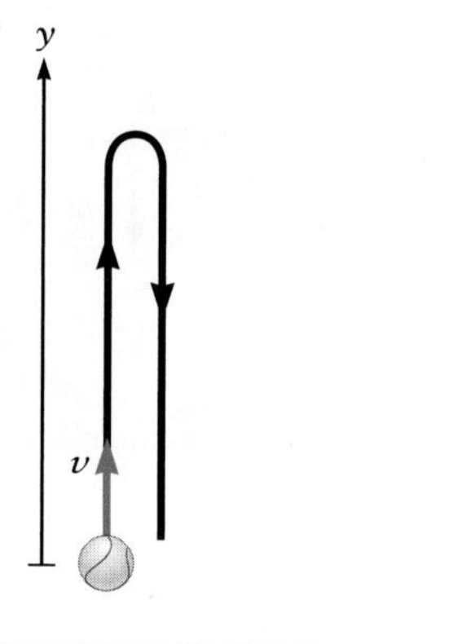

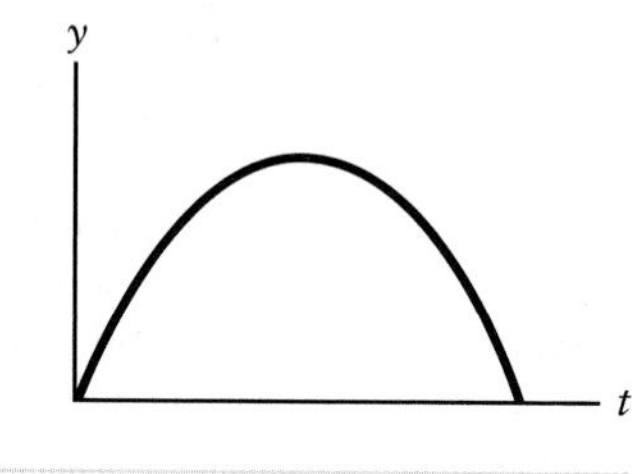

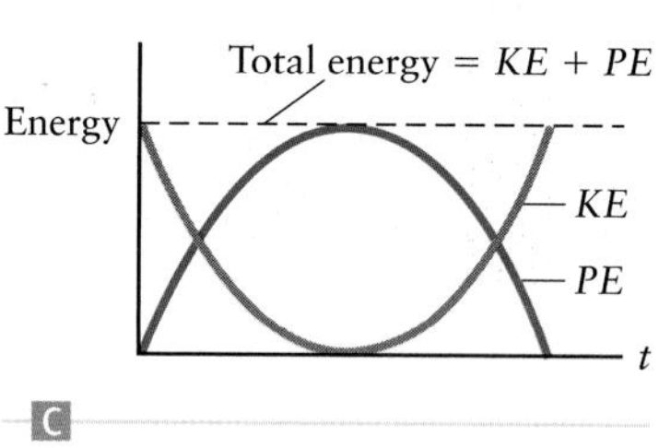

▲ **Figure 6.16** A When a ball is thrown upward (as a simple projectile), the height of the ball varies parabolically with time as sketched in B, the position versus time graph. C The *KE*, *PE*, and total mechanical energy vary with time as shown. Since the total mechanical energy is conserved, the sum $KE + PE$ is constant.

and solving for v_A gives

$$v_A^2 = 5gr$$

$$v_A = \sqrt{5gr}$$

We then insert the value of the radius of the track to get

$$v_A = \sqrt{5gr} = \sqrt{5(9.8 \text{ m/s}^2)(15 \text{ m})} = \boxed{27 \text{ m/s}}$$

► ***What does it mean?***

This speed is approximately 60 mi/h, which is a typical speed for a roller coaster. This also suggests that this stunt should be possible for an ordinary automobile, but the author's daughter has (supposedly) not yet attempted it.

Projectile Motion and Conservation of Energy

It is instructive to apply conservation of energy principles to projectile motion and consider how the kinetic and potential energies vary as an object moves along its trajectory. Figure 6.16A shows an example in which an object, such as a baseball, is thrown upward and then returns to the Earth. When the position y of the ball is plotted as a function of time t, we find the familiar parabolic relation between y and t sketched in Figure 6.16B. The behavior of the ball's energy as a function of time is shown in Figure 6.16C. Since the gravitational potential energy is $PE = mgy$, the potential energy also varies parabolically with time. Figure 6.16C shows the kinetic energy and the total mechanical energy, which is just the sum $PE + KE$. The total mechanical energy is *conserved*, so the sum $KE + PE$ is constant and does not change during the course of the motion. The kinetic energy thus varies as shown, in an "inverted" parabolic manner.

Only Changes in Potential Energy Matter

In our discussion of the gravitational potential energy, we found that $PE_{\text{grav}} = mgh$, where h is the height of the object (Eq. 6.13). However, in our calculations involving potential energy, we have always ended up computing changes in the potential energy as an object moves from an initial position to a final position. Recall the conservation of energy condition (Eq. 6.15):

$$KE_i + PE_i = KE_f + PE_f$$

When this relation is used to compute either the final kinetic energy or the change in the kinetic energy, the answer depends only on the potential energy difference $\Delta PE = PE_f - PE_i$. Indeed, our initial definition of potential energy in Equation 6.10 was in terms of the work done in moving an object from one place to another, and this original definition only referred to *changes* in potential energy.

The idea that only changes in potential energy are physically relevant can be appreciated if we consider the variation of the potential energy as a ball is released from a height h above the ground and falls to ground level (Fig. 6.17). If we use a

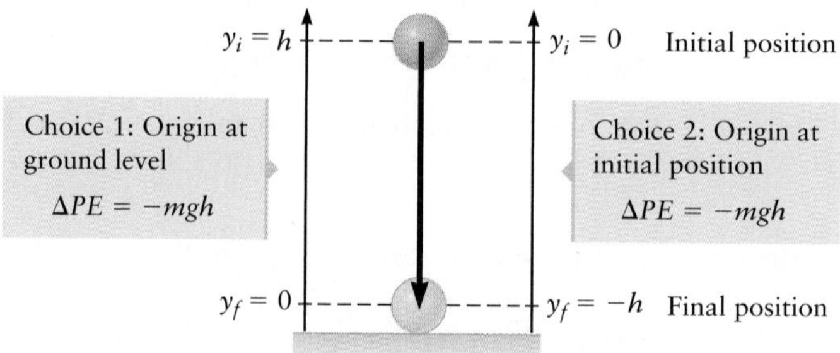

► **Figure 6.17** Only changes in the potential energy matter. Here, we consider the change in the gravitational potential energy using two different choices for the origin on the vertical axis. On the left, we take the origin at ground level, while on the right, the origin is at the initial position of the object. The change in the potential energy is the same; it does not depend on the choice of the origin.

coordinate axis with the origin at ground level (the coordinate axis on the left in Fig. 6.17), we have $y_i = h$ and $y_f = 0$, so the change in potential energy is

$$\Delta PE = PE_f - PE_i = mgy_f - mgy_i = mg(0 - h) = -mgh \quad (6.17)$$

On the other hand, we could just as well have chosen the origin of our coordinate system to be at the initial position of the ball (the axis shown on the right in Fig. 6.17). In this case, we would have $y_i = 0$ and $y_f = -h$, and the change in potential energy would be

$$\Delta PE = PE_f - PE_i = mgy_f - mgy_i = mg(-h - 0) = -mgh \quad (6.18)$$

The change in the potential energy in Equations 6.17 and 6.18 is *precisely the same*, and we can use either coordinate system to calculate the speed of the ball just before it reaches the ground or in any other calculation. Choosing a different origin for our coordinate system is very similar to adding or subtracting a constant from the potential energy at all locations. The results obtained from the application of the conservation of energy condition, Equation 6.15, depend only on changes in the potential energy, never on your choice of the "zero level" of the potential energy.

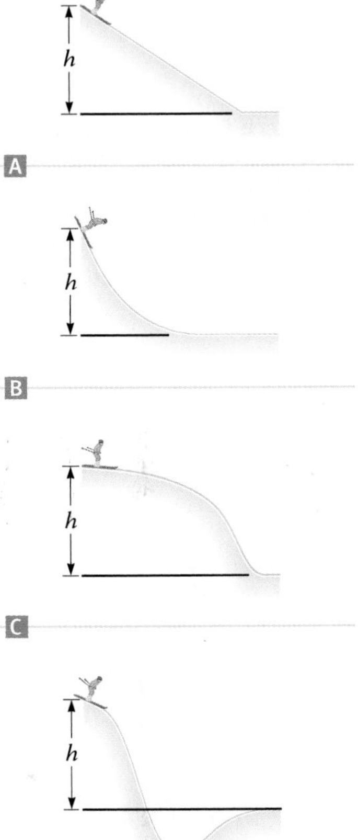

▲ **Figure 6.18** Concept Check 6.4.

CONCEPT CHECK 6.4 Skiing Down a Hill

Four expert skiers travel down the ski trails shown in Figure 6.18. All have the same initial height and the same final height at the bottom, on the far right in each sketch, and there is no friction on any of the trails. Which of the following statements are true? (More than one answer may be correct.) Assume air drag can be neglected.

(1) The skiers all arrive at the bottom at the same time.
(2) The skiers all arrive at the bottom with the same speed.
(3) The skier in Figure 6.18C probably arrives at the far right last.

CONCEPT CHECK 6.5 A 1000-Calorie Hill

The author's local bicycling club often rides along a particularly long and steep stretch of road the cyclists have named "Thousand Calorie Hill." This name suggests that the change in potential energy, and hence the amount of energy expended, in riding up this hill is 1000 Calories. The Calorie is a unit used to measure the energy in food, with 1 Calorie = 1000 calories (lowercase "c") = 4186 J. Do you think this name could be a reasonable one for any biking hill? *Hint*: Estimate the change in gravitational potential energy of a person who climbs the steepest hill in your neighborhood.

6.4 More Potential Energy Functions

In the last section, we used the potential energy associated with the gravitational force near the Earth's surface to illustrate the key ideas connected with potential energy. Let's now consider the potential energy functions associated with some other forces.

Gravitational Potential Energy in the Solar System

For objects near the surface of the Earth, the force of gravity exerted by the Earth is $F_{grav} = -mg$, and in the last section we showed that the potential energy associated with this force is $PE_{grav} = mgy$ (Eq. 6.13). In Chapter 5, we introduced Newton's law of gravitation, which describes the more general case of the gravitational force that exists between two objects separated by *any* distance. According to Newton's law of

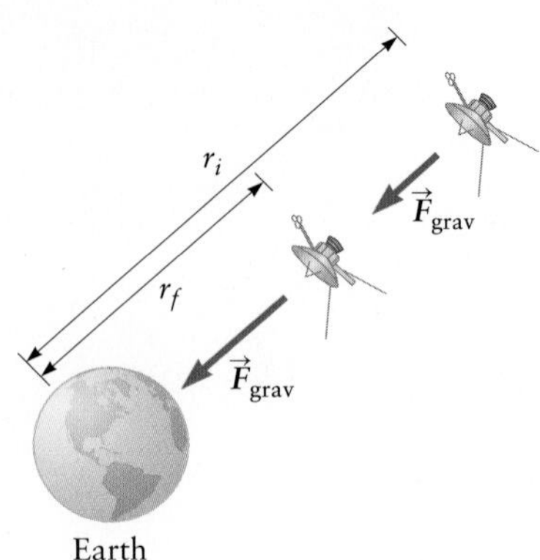

▲ **Figure 6.19** The potential energy associated with the gravitational force exerted by the Earth on this spacecraft changes as the separation r changes. This potential energy is given by Equation 6.20.

gravitation, the gravitational force between two objects of mass m_1 and m_2 separated by a distance r is (Eq. 5.16)

$$F_{grav} = -\frac{Gm_1m_2}{r^2} \tag{6.19}$$

The negative sign here indicates that the force is attractive. In a typical example, the mass m_1 might be the Earth, and m_2 could be a spacecraft that is either leaving the Earth or approaching the Earth as sketched in Figure 6.19. If this spacecraft is initially a distance r_i from the Earth and moves closer to a final distance r_f, the gravitational force does work on the spacecraft. According to the connection between work and potential energy (Eq. 6.10), there is a change in the gravitational potential energy of the spacecraft. While we will not give a detailed derivation here, the gravitational potential energy of two objects separated by a distance r is

$$PE_{grav} = -\frac{Gm_1m_2}{r} \tag{6.20}$$

The negative sign here means that the potential energy is lowered as the objects are brought closer together. The change in potential energy is then

$$\Delta PE_{grav} = -\frac{Gm_1m_2}{r_f} - \left(-\frac{Gm_1m_2}{r_i}\right) = -\frac{Gm_1m_2}{r_f} + \frac{Gm_1m_2}{r_i}$$

Gravitational Potential Energy: Launching a Satellite into Space

The result for the gravitational potential energy in Equation 6.20 can be used to analyze the motion of planets, moons, and satellites, including a calculation of the ***escape speed*** of a satellite. Suppose you want to launch a projectile of mass m_p into distant space. NASA currently does so by attaching a rocket engine to the projectile. However, several scientists have considered the possibility of simply "firing" the projectile from a very large cannon. (This possibility was described in a story by Jules Verne in 1865.) What is the minimum initial velocity required so that a projectile will completely escape from the Earth? To calculate this velocity, we ignore the drag force due to the atmosphere, so the only force acting on the projectile after it leaves the cannon is the gravitational force. This force is conservative, so we can apply the principle of conservation of energy to the projectile's motion. The instant after it leaves the cannon, the projectile is just above the Earth's surface with speed v_i at distance r_i from the Earth's center (Fig. 6.20A), and v_i is the unknown quantity we wish to calculate. The projectile eventually travels very far from the Earth; if it just barely escapes, the final speed is $v_f \approx 0$ and the final distance is very large, so $r_f \approx \infty$ (Fig. 6.20B).

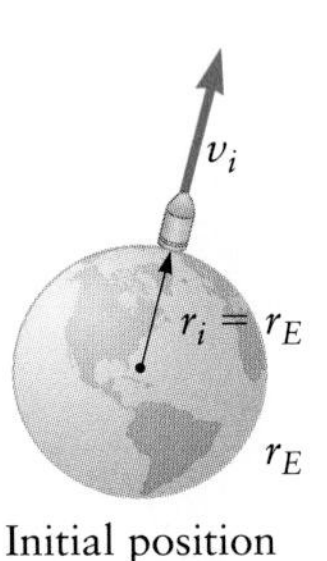

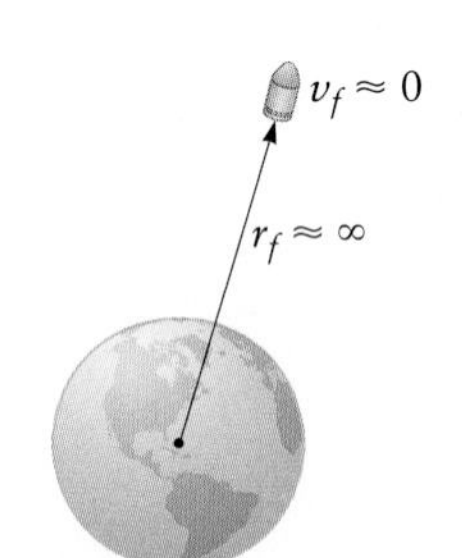

▲ **Figure 6.20** Calculating the escape velocity. A Initial position and speed of the projectile. B Final position and speed.

We now apply the conservation of energy condition with the potential energy given by Equation 6.20. Let M_{Earth} be the mass of the Earth and m_p the mass of the projectile. We then have

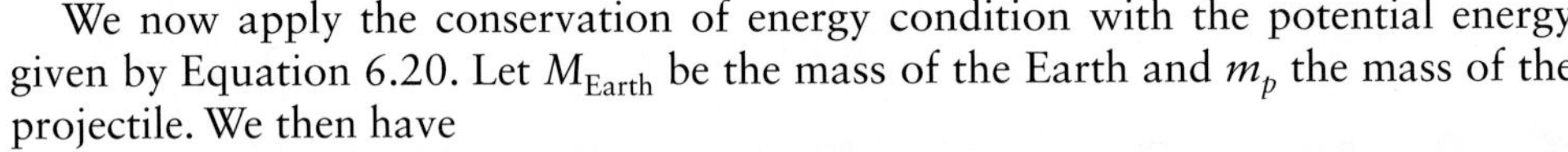

$$KE_i + PE_i = KE_f + PE_f$$

$$\frac{1}{2}m_pv_i^2 - \frac{GM_{Earth}m_p}{r_i} = \frac{1}{2}m_pv_f^2 - \frac{GM_{Earth}m_p}{r_f} \tag{6.21}$$

The projectile begins on the surface of the Earth, so it is initially a distance $r_i = r_E$ from the center of the Earth, where r_E is the radius of the Earth. Since $r_f \approx \infty$, the final potential energy is

$$PE_f = -\frac{GM_{Earth}m_p}{r_f} \approx 0$$

Inserting these results along with the final speed $v_f \approx 0$ into the conservation of energy relation Equation 6.21, we find

$$\frac{1}{2}m_p v_i^2 - \frac{GM_{\text{Earth}}m_p}{r_E} = \frac{1}{2}m_p v_f^2 - \frac{GM_{\text{Earth}}m_p}{r_f} = 0$$

$$\frac{1}{2}m_p v_i^2 = \frac{GM_{\text{Earth}}m_p}{r_E}$$

$$v_i = \sqrt{\frac{2GM_{\text{Earth}}}{r_E}} \tag{6.22}$$

which is called the escape speed. This is the minimum speed an object at the Earth's surface must have (ignoring air drag) to completely escape from the gravitational attraction of the Earth. Inserting values for the mass and radius of the Earth gives

$$v_i = \sqrt{\frac{2GM_{\text{Earth}}}{r_E}} = \sqrt{\frac{2(6.67 \times 10^{-11}\ \text{N} \cdot \text{m}^2/\text{kg}^2)(6.0 \times 10^{24}\ \text{kg})}{6.4 \times 10^6\ \text{m}}} = 1.1 \times 10^4\ \text{m/s}$$

which is approximately 24,000 mi/h.

Escape velocity

The escape speed is different for objects that are fired from different planets or from the Moon because that speed depends on the mass and radius of the planet or the Moon. Of course, to launch a real satellite, NASA should not rely on our simple calculation because the effect of air drag on such a projectile as it moves through the atmosphere cannot be ignored.

Insight 6.1

GRAVITATIONAL POTENTIAL ENERGY

We have now derived two different expressions for the gravitational potential energy. The result $PE_{\text{grav}} = mgy$ (Eq. 6.13) is the potential energy associated with the Earth's gravitational force for objects near its surface. The result $PE_{\text{grav}} = -Gm_1m_2/r$ (Eq. 6.20) is the potential energy associated with the gravitational force between any two objects separated by a distance r. Equation 6.20 is required for problems involving motion in the solar system.

Elastic Forces and Potential Energy: Springs

Hooke's law

One important type of potential energy is associated with springs and other elastic objects. A simple spring is a tight coil of wire as sketched in Figure 6.21. In its "relaxed" state with no force applied to its end, the spring rests as shown in Figure 6.21A. We might now pull on the spring with a force F_{pull}, causing the spring to stretch as shown in Figure 6.21B. When it is stretched, the spring itself exerts a force F_{spring} that opposes the stretching. Likewise, if you were to push on the spring with a force F_{push} so as to compress it (Fig. 6.21C), the spring exerts a force back on you in opposition. The force exerted by the spring when it is either stretched or compressed is

$$F_{\text{spring}} = -kx \tag{6.23}$$

where x is the amount the end of the spring is displaced from its equilibrium position, which is assumed to be at $x = 0$. The quantity k is called the ***spring constant***, and it has units of newtons per meter. The value of k depends on the properties of the spring (the number of coils, the stiffness of the wire, etc.) and is thus different for different springs. The negative sign in Equation 6.23 indicates that the force exerted by the spring always *opposes* the displacement of the end of the spring. If the end of the spring is pulled toward $+x$, the force exerted by the spring is directed along $-x$, whereas if the spring is pushed toward $-x$, the spring exerts a force along $+x$.

Equation 6.23, called ***Hooke's law***, provides an accurate description of spring forces in a wide variety of situations. Hooke's law is not a "law" of physics in the sense of Newton's laws. Instead, Hooke's law is an empirical relation that is accurate for most springs. Springs like the simple coils of wire sketched in Figure 6.21 are rather idealized; real springs can be far more complex. Hooke's law, however, applies to a variety of elastic objects. For example, a diving board is a type of spring: it exerts a force on a diver given by Equation 6.23, with x equal to the distance the diving board is deflected (Fig. 6.22A). A tennis racket also acts like a spring: the force exerted by the racket on a tennis ball is again given by an expression like Equation 6.23, with x equal to the deflection of the strings (Fig. 6.22B).

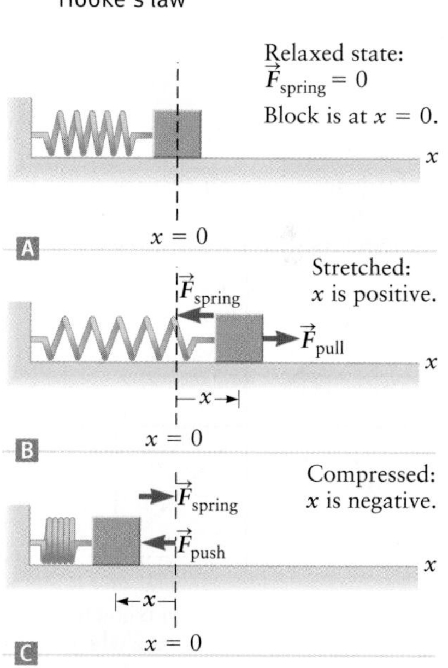

▲ **Figure 6.21** The force exerted by a spring is always opposite to the displacement of the end of the spring. A When a spring is unstretched and uncompressed, the force exerted by the spring is zero. B This spring is stretched by pulling it to the right, so the force exerted by the spring is to the left. C The spring is compressed, and the force exerted by the spring is to the right.

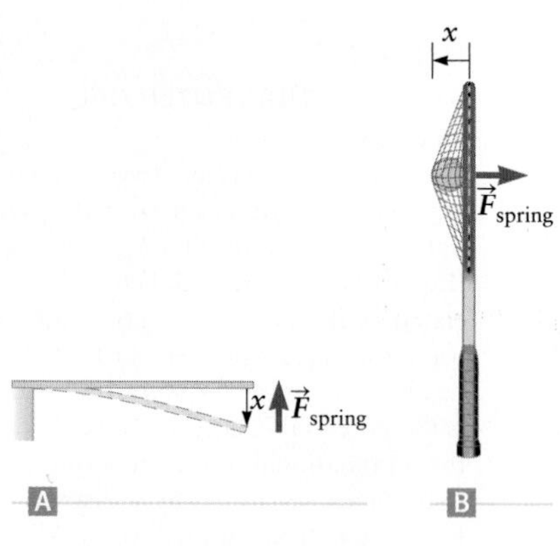

▲ **Figure 6.22** Objects such as A a diving board and B a tennis racket obey Hooke's law (Eq. 6.23), with the force proportional to the displacement x.

Potential Energy Stored in a Spring

According to Hooke's law, the force exerted by a spring is proportional to x, the amount the spring is stretched or compressed. To calculate the potential energy associated with this force, we return to the definition of potential energy (Eq. 6.10),

$$\Delta PE = -W$$

with W now the work done by the Hooke's law force (Eq. 6.23). When you pull or push on a spring to stretch it or compress it as in part B or C of Figure 6.21, W is the work done *by the spring* on your hand.

The force F_{spring} is not a constant, but varies as the displacement changes. In Figure 6.5, we showed that the work done by a variable force is equal to the area under the force–displacement curve. This curve for a spring is shown in Figure 6.23. We assume the spring is initially in its relaxed state (not stretched or compressed), so $x_i = 0$. If the spring is then stretched an amount $x = x_s$, the work done by the spring is the area under the curve $F_{\text{spring}} = -kx$. This area is just a triangle of height $F_s = kx_s$, and the area of this triangle is $\frac{1}{2}F_s x_s$ (see Fig. 6.23). Inserting the expression for F_{spring} from Hooke's law, the area of the triangle is $\frac{1}{2}F_s x_s = \frac{1}{2}kx_s^2$. We must now be careful to get the correct sign for the work done by the spring. In this example, the spring is being stretched, so according to Figure 6.21B the force exerted by the spring is directed to the left, whereas the displacement of the end of the spring is to the right. Since the force and the displacement are in opposite directions, the work done by the spring is negative, and

$$W = -\tfrac{1}{2}kx_s^2$$

Here, W is the work done by the spring as it is *stretched* an amount x_s starting from its relaxed state. From Figure 6.23, we see that the work done in *compressing* the spring is equal to the area under a similar triangular region, the triangle in the upper left in Figure 6.23. The work done by the spring when it is compressed is again negative because the force exerted by the spring is directed opposite to the displacement of the end of the spring. (See Fig. 6.21C.) Hence, if a spring is stretched or compressed an amount x, the work done by the spring is

$$W_{\text{spring}} = -\tfrac{1}{2}kx^2$$

Using the definition of potential energy, we then get

$$PE_{\text{spring}} = \tfrac{1}{2}kx^2 \tag{6.24}$$

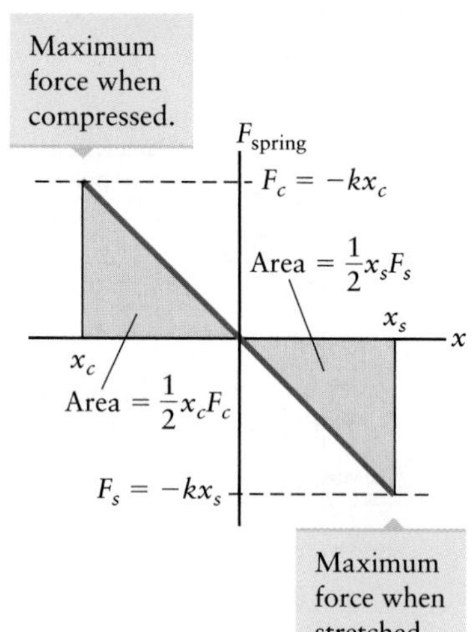

▲ **Figure 6.23** The work done by a variable force is equal to the area under the force–displacement graph. For a spring, this area is a triangle. When the spring is stretched, $x = x_s$ and W is the area of the triangle at the lower right. When the spring is compressed, $x = x_c$ and W is the area of the triangle at the upper left.

Spring Forces and Potential Energy: A Recap

Our results for the elastic (spring) force in Equation 6.23 and potential energy in Equation 6.24 are collected in Figure 6.24, which shows how F_{spring} and PE_{spring} vary as a spring is stretched and compressed. The force exerted by the spring always opposes the displacement, so F_{spring} can be either positive or negative, depending on whether the spring is stretched or compressed. The elastic potential energy is smallest ($PE_{\text{spring}} = 0$) when a spring is in its relaxed state ($x = 0$), and PE_{spring} always increases as a spring is either stretched or compressed.

EXAMPLE 6.7 Force Exerted by a Bungee Cord

A bungee cord is a kind of spring, and the force exerted by the cord on a person attached to it is given by $F_{\text{spring}} = -kh$, where h is the amount the cord is stretched (Hooke's law, Eq. 6.23). Consider a person of mass $m = 75$ kg attached to a bungee cord with spring constant $k = 50$ N/m. When the jump is completed, the person is at rest, suspended by the cord (Fig. 6.25B). How much is the cord stretched at that time?

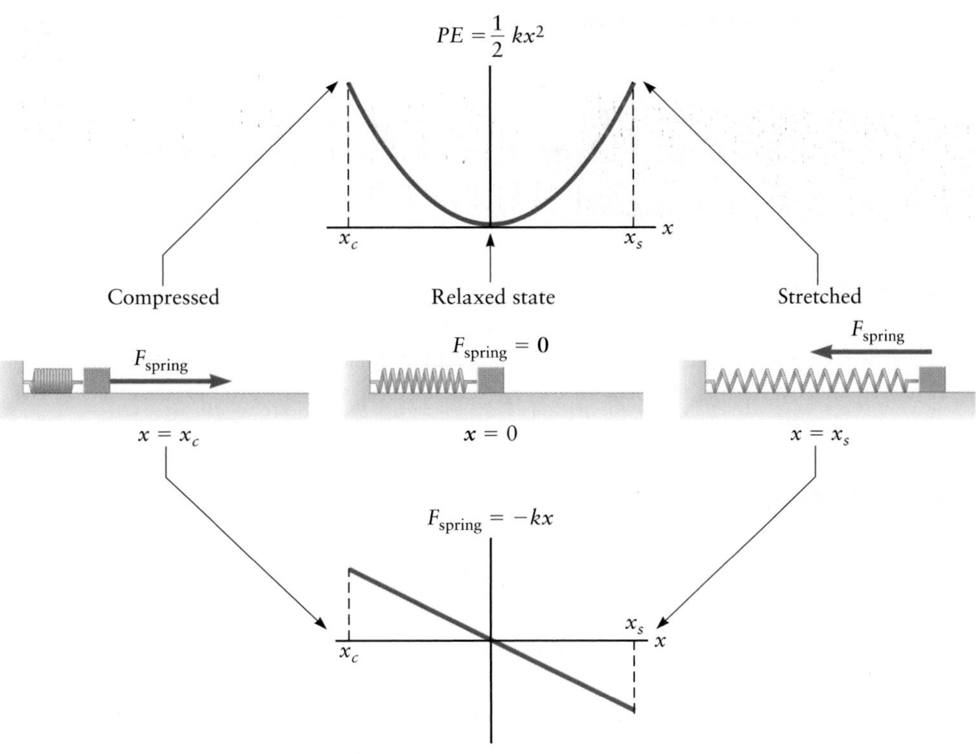

◄ **Figure 6.24** Variation of the potential energy and force as a spring is compressed and stretched.

RECOGNIZE THE PRINCIPLE

At the end of the jump and after the person has stopped oscillating up and down, the person is at rest, so his acceleration is zero. Hence, according to Newton's second law, the total force must be zero.

SKETCH THE PROBLEM

Figure 6.25B shows the forces acting on the person when the cord is stretched: the force of gravity and the spring force from the bungee cord.

IDENTIFY THE RELATIONSHIPS

Applying Newton's second law for motion in the vertical direction for the person in Figure 6.25B, we have

$$\sum F = kh + F_{grav} = kh - mg = ma = 0 \quad (1)$$

The spring force is $+kh$, where h is the amount the bungee cord is stretched; the positive sign is here because this force is directed upward.

SOLVE

Solving Equation (1) for h, we get

$$kh = mg$$

$$h = \frac{mg}{k} = \frac{(75\text{ kg})(9.8\text{ m/s}^2)}{50\text{ N/m}} = \boxed{15\text{ m}}$$

► ***What does it mean?***

This result should be of interest to the bungee cord company. For safety reasons, it would want to be sure that the customer is suspended well above the ground.

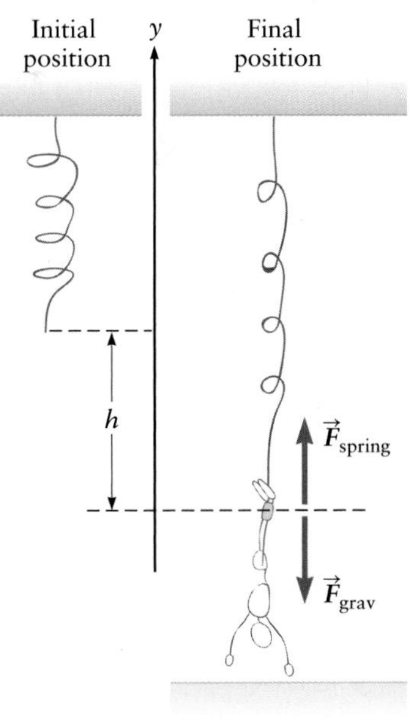

▲ **Figure 6.25** Example 6.7. A The bungee cord in its relaxed state (neither stretched or compressed). B The bungee jumper is hanging at rest, suspended by the cord.

EXAMPLE 6.8 Ⓡ Spring Constant of an Archer's Bow

An archer's bow behaves like a spring and is described by Hooke's law. The force exerted by the bow on the arrow is proportional to the displacement of the bowstring. Suppose the archer is able to shoot an arrow of mass $m = 0.050$ kg with a velocity of 100 m/s just as it leaves the bow. What is the approximate spring constant of her bow?

RECOGNIZE THE PRINCIPLE

We want to apply the conservation of energy condition to the combination of the bow plus arrow. The initial state has the bowstring pulled back a distance x_i and the arrow at rest. The final state has the bow in its relaxed state ($x_f = 0$) and the arrow leaving the bow with velocity $v_f = 100$ m/s.

SKETCH THE PROBLEM

The problem is shown in Figure 6.26A.

IDENTIFY THE RELATIONSHIPS

Writing the conservation of energy condition with the elastic potential energy of the bow, we have

$$KE_i + PE_i = KE_f + PE_f$$

$$\tfrac{1}{2}mv_i^2 + \tfrac{1}{2}kx_i^2 = \tfrac{1}{2}mv_f^2 + \tfrac{1}{2}kx_f^2 \quad (1)$$

SOLVE

Inserting our values for the initial arrow velocity ($v_i = 0$) and final bow stretch ($x_f = 0$) into Equation (1) leaves us with an equation involving k and the initial bow stretch x_i. Solving for the spring constant k, we find

$$0 + \tfrac{1}{2}kx_i^2 = \tfrac{1}{2}mv_f^2 + 0$$

$$k = \frac{mv_f^2}{x_i^2} \quad (2)$$

We are given the mass of the arrow and its final speed v_f. To find the spring constant k, we also need to know x_i, the amount the bow stretches, but this information was not given. Relying on intuition and experience, we estimate $x_i \approx 0.1$ m. A value of

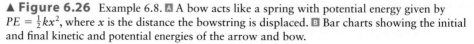

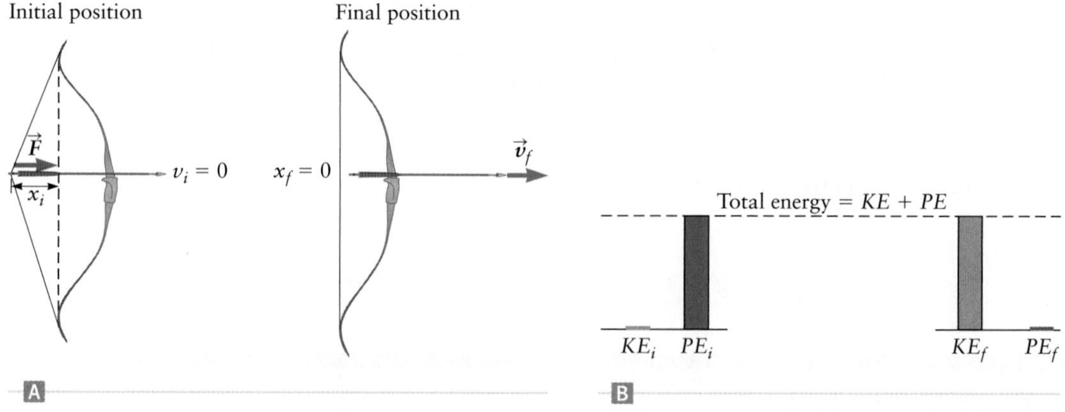

▲ **Figure 6.26** Example 6.8. A A bow acts like a spring with potential energy given by $PE = \tfrac{1}{2}kx^2$, where x is the distance the bowstring is displaced. B Bar charts showing the initial and final kinetic and potential energies of the arrow and bow.

$x = 0.01$ m = 1 cm seems much too small, whereas $x = 1$ m is too large. Our estimate is in the middle. Evaluating Equation (2) for k then gives

$$k = \frac{mv_f^2}{x_i^2} = \frac{(0.050 \text{ kg})(100 \text{ m/s})^2}{(0.1 \text{ m})^2} = \boxed{5 \times 10^4 \text{ N/m}}$$

▶ *What does it mean?*

Notice how we followed the general steps in our strategy for dealing with conservation of energy problems. We began with a picture and identified the initial and final states of the bow plus arrow. We then wrote the conservation of energy relation, Equation (1), and solved for the quantity of interest (k).

Figure 6.26B shows bar charts for the initial and final kinetic and potential energies. The initial potential energy is converted into the final kinetic energy, and the total energy $KE + PE$ is conserved.

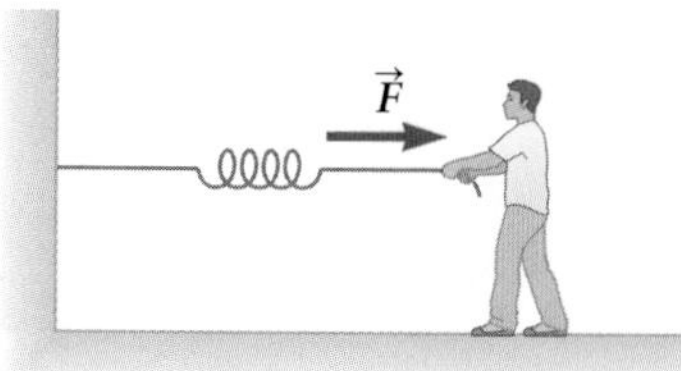

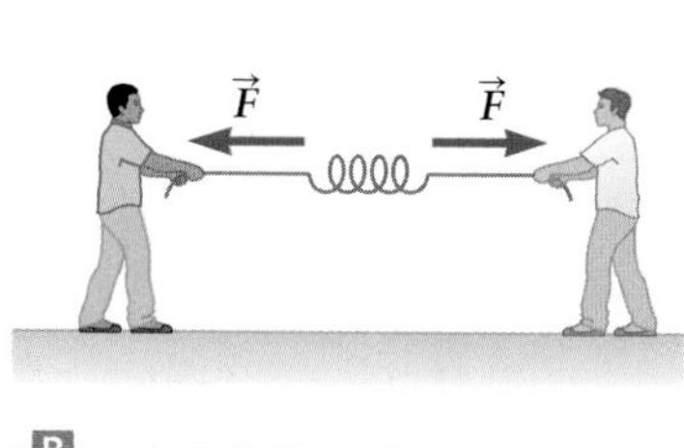

▲ **Figure 6.27** Concept Check 6.6.

CONCEPT CHECK 6.6 Spring Forces and Newton's Third Law

Figure 6.27 shows two identical springs. In both cases, a person exerts a force of magnitude F on the right end of the spring. The left end of the spring in Figure 6.27A is attached to a wall, while the left end of the spring in Figure 6.27B is held by another person, who exerts a force of magnitude F to the left. Which statement is true?

(1) The spring in Figure 6.27A is stretched half as much as the spring in Figure 6.27B.
(2) The spring in Figure 6.27A is stretched twice as much as the spring in Figure 6.27B.
(3) The two springs are stretched the same amount.

Total Potential Energy with Multiple Forces

In all our problems so far, there has only been one type of force and hence only one type of potential energy function to deal with. Often, however, several different forces—such as a gravitational force and a spring force—may be acting together on an object. In these cases, we must be sure to account for the total potential energy from all forces when applying the principle of conservation of energy. For example, if both gravity near the Earth's surface and a spring force are acting on an object, the total potential energy is

$$PE_{\text{total}} = PE_{\text{grav}} + PE_{\text{spring}} = mgh + \tfrac{1}{2}kx^2$$

This type of situation is illustrated in the next example.

EXAMPLE 6.9 Potential Energy from a Spring Plus Gravity

Consider again the bungee jumper from Example 6.7. He is initially standing on the bungee platform with the cord unstretched (Fig. 6.28A). He then jumps from the platform and travels downward (Fig. 6.28B). How much is the cord stretched when the jumper reaches his lowest point? The mass of the jumper is $m = 75$ kg, and the spring constant of the bungee cord is $k = 50$ N/m.

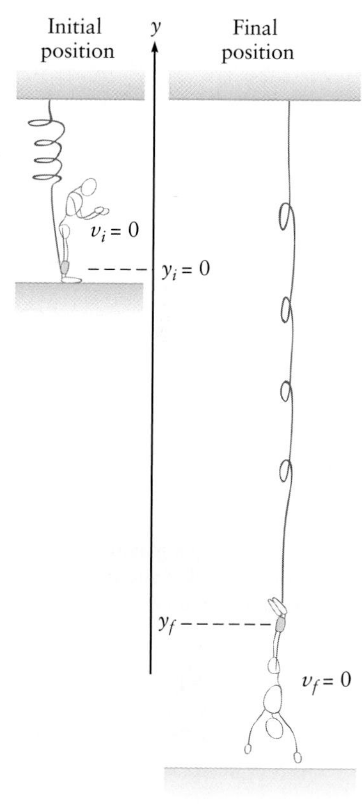

▲ **Figure 6.28** Example 6.9. **A** This bungee jumper is initially at rest with the cord unstretched. **B** At his lowest point, the bungee jumper's position is a distance y_f below the starting point.

RECOGNIZE THE PRINCIPLE

We again apply the principle of conservation of energy, but now we have to deal with the potential energy associated with both gravity and the bungee cord. Initially, the cord is in its relaxed state, so all the potential energy is from gravity. In the final

(continued) ▶

configuration, the cord is stretched, so we have to include the contributions to the potential energy from both this "spring" and from gravity.

SKETCH THE PROBLEM

Figure 6.28 describes the problem. This sketch shows both the initial and final states of the system.

IDENTIFY THE RELATIONSHIPS

According to the conservation of energy condition,

$$KE_i + PE_{\text{spring}, i} + PE_{\text{grav}, i} = KE_f + PE_{\text{spring}, f} + PE_{\text{grav}, f}$$

where we have written the gravitational and cord (i.e., spring) potential energies as separate terms. The jumper begins at rest ($v_i = 0$) and also ends at rest ($v_f = 0$), so the initial and final kinetic energies are both zero. The cord is initially relaxed, so the initial potential energy of the cord is zero. We take the origin ($y = 0$) at the initial position of the jumper, so the initial gravitational potential energy is also zero. Using the coordinate system in Figure 6.28, the final position of the jumper is at $y = y_f$. Inserting these values into the conservation of energy relation leads to

$$PE_{\text{spring}, i} + PE_{\text{grav}, i} = PE_{\text{spring}, f} + PE_{\text{grav}, f}$$

$$0 + 0 = \tfrac{1}{2}ky_f^2 + mgy_f$$

SOLVE

Solving for y_f, we get

$$\tfrac{1}{2}ky_f^2 = -mgy_f$$

$$y_f = -\frac{2mg}{k} = -\frac{2(75\ \text{kg})(9.8\ \text{m/s}^2)}{50\ \text{N/m}} = \boxed{-29\ \text{m}}$$

► ***What does it mean?***

The value of y_f is negative because the final position of the jumper is below the starting point at $y = 0$. Comparing this result with Example 6.7, we see that the jumper in Figure 6.28B is at a lower point than the person in Figure 6.25B. What are the velocity and acceleration of the jumper when he reaches the lowest point in Figure 6.28B? Is this lowest point a point of static equilibrium?

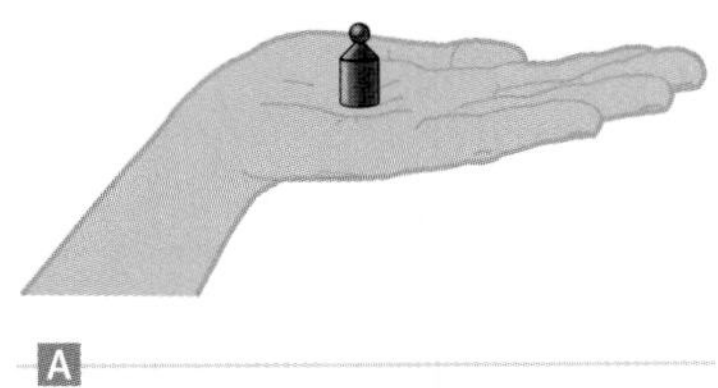

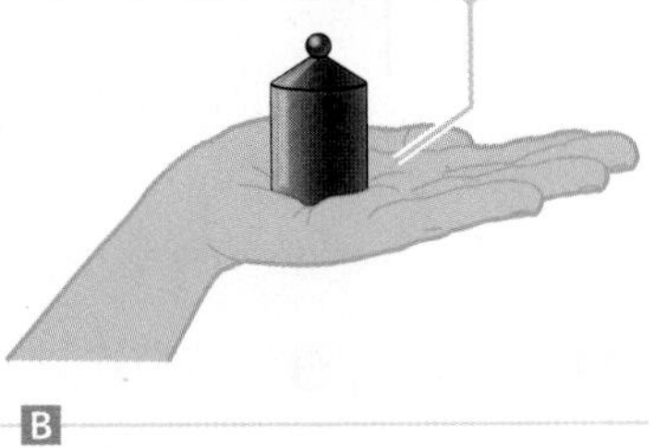

When holding a massive object, your skin and muscles are compressed like a spring.
B

▲ **Figure 6.29** When you support an object with your hand, your muscles deform much like a spring, producing an elastic force on the object.

Elastic Forces and the "Feeling" of Holding a Heavy Object

According to the definition of mechanical work (Eq. 6.3), W is zero if the displacement is zero, even though the force on the object might be very large. This definition is quite different from our "everyday" concept of work. Consider what happens when you hold an object in your hand (Fig. 6.29A). If the object is at rest, the displacement is zero, and the work done by your hand on the object is zero. Hence, $W = 0$ if the object has a very small mass or even a very large mass. However, you also know intuitively that supporting a very massive object will eventually make your hand "tired," and in everyday language we would say that holding something requires an expenditure of "energy" by your hand.

If the "physics" work done on the object is zero, however, why do your muscles get tired? One reason is that your muscles consist of many parallel fibers, and when they apply a force, these fibers are constantly slipping relative to one another. Muscles also contain small motorlike molecules that pull and move each fiber relative to its neighbors to counteract this slipping. (We say more about these motors in Section 6.8.) Because the molecular motors exert a force on the muscle fibers as they move along a fiber, the motors do a nonzero amount of "physics" work, and the associated energy comes from chemical reactions in the muscle. You sense this expenditure of chemical energy in your muscles, and that is why you must exert "energy" to support an object.

6.5 Conservative versus Nonconservative Forces and Conservation of Energy

In our derivation of the potential energy associated with gravity near the surface of the Earth (Eq. 6.13), we mentioned that W is independent of the path taken by the object. This path independence is an essential characteristic of the work done by a conservative force. The corresponding change in the potential energy depends only on the object's initial and final positions. Changes in the potential energy, and hence in the value of W, are independent of the path taken. Many forces satisfy this requirement of path independence, including gravitation and spring (Hooke's law) forces. As already mentioned, such forces are called conservative forces.

The work done by a conservative force is independent of the path.

Although potential energy is an extremely useful concept, some forces—called ***nonconservative forces***—cannot be associated with a potential energy. Unlike conservative forces, the work done by a nonconservative force *does* depend on the path taken. In this section, we'll deal with nonconservative forces.

The work done by a nonconservative force does depend on the path taken.

The Work Done by Friction Depends on the Path

Friction is a good example of a nonconservative force. Consider the work done by friction as an object moves along the path sketched in Figure 6.30, where a crate is pushed across a table from an initial location x_i to a final location x_f. This can be done by pushing the crate directly from x_i to x_f (path A in Fig. 6.30) or by taking an indirect path (path B). Since we are dealing with kinetic friction, $F_{\text{friction}} = -\mu_K N$, where N is the normal force between the table and the crate. The work done by friction along path A is

$$W_A = F_{\text{friction}}\,\Delta x = -\mu_K N(x_f - x_i)$$

Here, W_A is negative because friction opposes the motion. We can calculate the work done in moving along path B in the same way. The work done is

$$W_B = -\mu_K N\ (\text{length of path } B)$$

The length of path B is much longer than that of path A (see Fig. 6.30), so W_A and W_B are certainly not equal. Hence, the work done by friction depends on the path taken by the object. From Equation 6.10, it is therefore impossible to have a single value for the change in a corresponding potential energy, so we conclude that such a potential energy cannot exist. It is thus impossible to have a potential energy associated with the frictional force. Other nonconservative forces have the same property: the work done by a nonconservative force depends on the path taken by the object, so it is not possible to have a corresponding potential energy function.

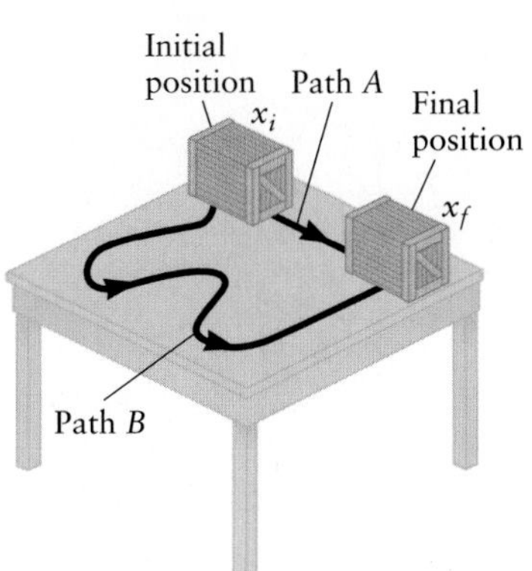

▲ **Figure 6.30** The work done by friction is larger for path B than it is for path A because path B is longer. Since the work done by friction depends on the path taken, friction is a nonconservative force.

The Work–Energy Theorem Revisited: Including Nonconservative Forces

If we cannot define potential energy for a nonconservative force, how do we think about conservation of energy in the presence of nonconservative forces? To answer this question, it is simplest to return to our original version of the work–energy theorem (Eq. 6.9),

$$W_{\text{total}} = \Delta KE = KE_f - KE_i \tag{6.25}$$

which tells us that the change in kinetic energy is equal to the *total* work done on the object. This total work can be due to several different forcess. Let's split it into two terms, one for the work done by all the conservative forces and another for the work done by the nonconservative forces:

$$W_{\text{total}} = W_{\text{con}} + W_{\text{noncon}}$$

According to the definition of potential energy (Eq. 6.10), the work done by the conservative forces is related to the change in the potential energy:

$$-W_{\text{con}} = \Delta PE = PE_f - PE_i$$

Collecting these results and inserting them into Equation 6.25 leads to

Work–energy theorem including nonconservative forces

$$KE_i + PE_i + W_{\text{noncon}} = KE_f + PE_f \tag{6.26}$$

which looks just like our previous conservation of energy result (Eq. 6.15) but with an additional term on the left for the work done by nonconservative forces. Equation 6.26 is our general conservation of energy/work–energy theorem with nonconservative forces included. The sum of the kinetic and potential energies is the total ***mechanical energy*** of an object. According to Equation 6.26, the final mechanical energy ($KE_f + PE_f$) is equal to the initial mechanical energy ($KE_i + PE_i$) plus the work done by any nonconservative forces that act on the object.

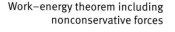

EXAMPLE 6.10 Work Done by Friction on a Skier

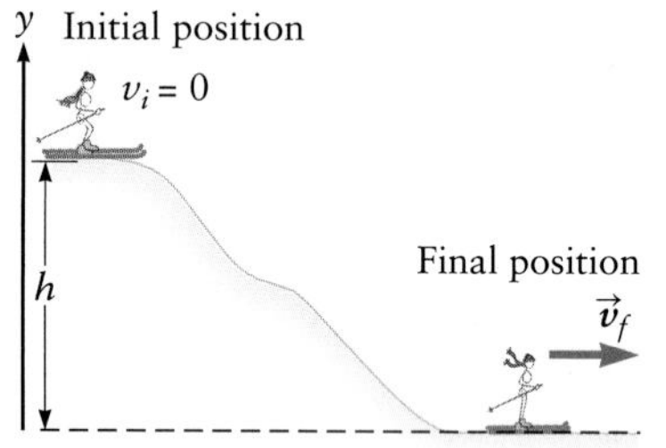

▲ **Figure 6.31** Example 6.10.

A skier of mass $m = 60$ kg starts from rest at the top of a ski hill of height $h = 50$ m (Fig. 6.31). When she reaches the bottom, she has speed $v_f = 25$ m/s. How much work is done by friction on the skier as she travels down the hill?

RECOGNIZE THE PRINCIPLE

Since the ski hill does not have a simple shape (it is not a simple inclined plane), we cannot calculate the work done by friction using the definition of work in Equation 6.3 ($W_{\text{friction}} = F_{\text{friction}}\,\Delta x$). Friction is the only nonconservative force in the problem, however, so $W_{\text{friction}} = W_{\text{noncon}}$, and we can find the work done by friction using the conservation of energy relation

$$KE_i + PE_i + W_{\text{friction}} = KE_f + PE_f \tag{1}$$

SKETCH THE PROBLEM

Figure 6.31 shows the initial and final states of the skier.

IDENTIFY THE RELATIONSHIPS

We first rearrange Equation (1):

$$W_{\text{friction}} = KE_f + PE_f - (KE_i + PE_i) \tag{2}$$

In Equation (2), the potential energy is due to gravity; choosing the origin of the vertical axis at the bottom of the hill, we get $PE_i = mgh$ and $PE_f = 0$. The skier starts from rest, so the initial kinetic energy is zero and the final kinetic energy is $\frac{1}{2}mv_f^2$.

SOLVE

Inserting all this information into Equation (2) leads to

$$W_{\text{friction}} = \tfrac{1}{2}mv_f^2 - mgh = \tfrac{1}{2}(60\text{ kg})(25\text{ m/s})^2 - (60\text{ kg})(9.8\text{ m/s}^2)(50\text{ m})$$

$$W_{\text{friction}} = \boxed{-1.1 \times 10^4\text{ J}}$$

► ***What does it mean?***

The work done by friction is negative because the frictional force opposes the skier's motion along the hill. Friction thus causes the final speed of the skier to be less than it would have been without friction.

Conservation of Energy of a System

We have mentioned the concept of "conservation of energy" several times, in several different contexts. We now consider the principle of conservation of energy from a very general point of view. Suppose a system of particles or objects exert forces on one another as they move about. These forces may be conservative (perhaps gravity and elastic forces) or nonconservative (friction or air drag). While there are many forces involving particles within this system, let's assume there are no forces on any of these particles from outside the system so that the total energy of the system will be conserved. As the particles move around, we would expect some potential energy to be converted to kinetic energy or vice versa. It is also possible for mechanical energy (i.e., elastic potential energy or kinetic energy) to be converted to another form of energy such as heat energy, electrical energy, or chemical energy that we describe later in this book. Nevertheless, the total energy of the entire system will be conserved.

Of course, some agent from outside this system may exert a force on one of the particles within the system. If this outside agent does an amount of work W_{ext} on the particle, the total energy of the system will change by an amount W_{ext}. This change in the system's energy might be positive or negative, depending on the sign of W_{ext}. In this way, energy can be exchanged between an external agent and the system. If a certain amount of energy is added to the system, the same amount of energy must be removed from the agent, so we again arrive at the general principle of ***conservation of total energy***.

6.6 The Nature of Nonconservative Forces: What Is Friction Anyway?

When all the forces acting on an object are conservative forces, the total mechanical energy is conserved:

$$KE_i + PE_i = KE_f + PE_f$$

Intuitively, the idea that energy is conserved should not be surprising. If energy were not conserved, it would be possible to do some very incredible things—such as generate limitless amounts of energy and construct mechanical perpetual motion machines—and physicists have found no evidence that such things are possible.

We have also seen that when a nonconservative force such as friction is present, the principle of conservation of energy can be expressed as (Eq. 6.26)

$$KE_i + PE_i + W_{noncon} = KE_f + PE_f$$

or

$$\underbrace{(KE_f + PE_f)}_{\text{final mechanical energy}} - \underbrace{(KE_i + PE_i)}_{\text{initial mechanical energy}} = W_{noncon} \quad (6.27)$$

In most situations, the work done by friction is negative because the frictional force is usually directed opposite to the displacement. So, Equation 6.27 tells us that in such cases the final mechanical energy is less than the initial mechanical energy. A simple illustration of this situation is given in Figure 6.32, which shows a block sliding along a rough surface. The block has some initial kinetic energy, so the initial mechanical energy is not zero. Friction causes the block to come to a stop, so the final kinetic energy and the final mechanical energy are both zero. Hence, the mechanical energy of the block is not conserved; by the end of this process, the mechanical energy of the block has changed by an amount W_{noncon}. Let's now consider precisely where the energy associated with W_{noncon} goes.

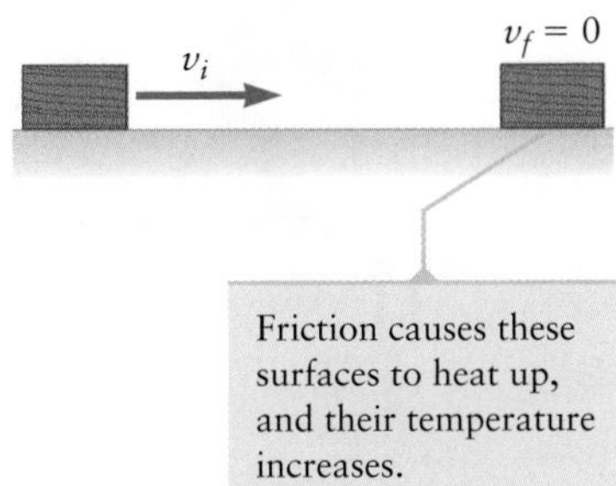
v_i
$v_f = 0$
Friction causes these surfaces to heat up, and their temperature increases.

▲ **Figure 6.32** When friction does work on a sliding block, the block's mechanical energy is converted into heat energy, resulting in a temperature increase at the surfaces of the block and the floor.

For the sliding block in Figure 6.32, the energy associated with W_{noncon} goes into heating up the surface of the block and the surface of the floor. This increase in the temperature of the two surfaces can be understood in terms of the motion of the atoms at the surfaces. All substances are composed of atoms in constant motion. The chemical bonds that hold the atoms together are analogous to springs, which are stretched and compressed as the atoms move and vibrate. When a substance is heated and its temperature is increased, these vibrations become larger in size and the atoms move faster. Since they move faster, the atoms have an increased kinetic energy, and there is also an increase in the potential energy of the chemical bonds/springs. Hence, the initial mechanical energy of the block in Figure 6.32 does not disappear, but shows up as an increase in the energy of the atoms at the surfaces. Even in the presence of friction, the *total* energy is still conserved.

This conclusion leads to another, and deeper, question. If the energy W_{noncon} has simply gone into the mechanical energy (kinetic and potential) of the atoms, is it possible to somehow return all this energy back to its original form? That is, can we convert this added mechanical energy of the heated atoms back into the mechanical energy of the block? We'll see in Chapter 16 that it is possible to transfer *some* of the energy from the heated atoms into kinetic energy of the block. However, it is *not* possible to transfer *all* this energy back to the block.

This energy that flows in connection with a temperature difference is called ***heat***. Heat plays a central role in the branch of physics called ***thermodynamics***, which we will explore in Chapters 14 through 16. Although mechanical energy is easily converted completely into heat as with the sliding block in Figure 6.32, it is not nearly as simple to convert heat energy back into mechanical energy. After all, if the block in Figure 6.32 were initially at rest and we then heated it, we would not expect the block to begin sliding along the floor!

These observations also suggest why friction is called a "nonconservative" force. Conservative forces are associated with potential energy, and we have seen that potential energy is stored energy that can be "retrieved" and then converted freely into other forms of energy. The heat energy produced by friction cannot be retrieved in this manner, and it cannot all be transferred back to other forms of energy.

6.7 Power

In Section 6.2, we found that a block and tackle is able to amplify the force that can be applied to an object. However, this amplification comes at the cost of a reduction in the object's displacement. For the particular block and tackle in Figure 6.8, the force amplification factor is two and the displacement is reduced by the same factor. The work done on the object is therefore the *same* as the work done by the person as he pulls on the rope. We can view the person and the crate in Figure 6.8 as two "systems"; the block and tackle "transfers" energy from the person to the crate, and according to the principle of conservation of energy, the energy expended by the person—that is, the work done by the person—must equal the energy gained by the crate.

More complicated block-and-tackle arrangements can give much greater force amplification factors, so in principle a person could use such a device to lift an extremely massive object. While the force amplification factor can be very large, the displacement is always "deamplified" by the same factor. Hence, it may take a very long time to lift a massive object through an appreciable distance. Time enters into work–energy ideas through the concept of ***power***. Consider the person in Figure 6.33 who is using a rope to lift a crate very slowly. Work is done on the crate as it moves from the floor to a height h, which changes the potential energy of the crate by an amount mgh. This energy comes from the person as he pulls on the other end of the rope and does an amount of work $W = mgh$ on the rope. If this work is expended during a time t, the average power exerted by the person is defined as

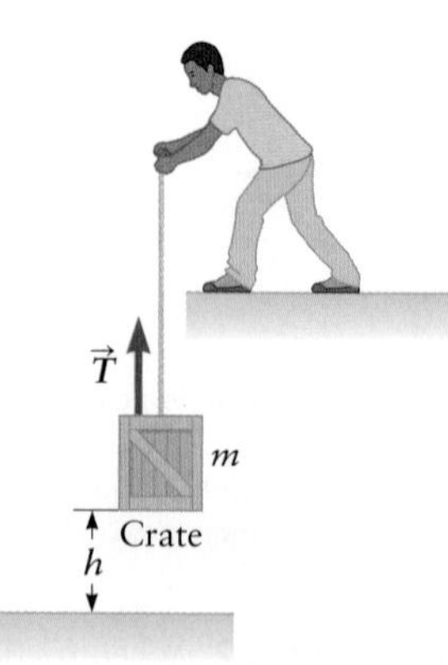

▲ **Figure 6.33** This person is expending a certain amount of energy per unit time as he lifts the crate.

$$P_{ave} = \frac{W}{t} \tag{6.28}$$

TABLE 6.1 Typical Values for the Power Output and Consumption of Some Common Appliances and Devices

Device	Power Output or Consumption
Lightbulb	20 W
Video game console	200 W
Automobile engine (small car)	100 hp = 7.5×10^4 W
Automobile engine (race car)	700 hp = 5.2×10^5 W
Smart phone (when talking)	1 W
Elite bicycle racer	400 W

The units of power are joules per second; this unit has been given the name watt (W) after James Watt (1736–1819), a developer of the steam engine. So,

$$1 \text{ W} = 1 \text{ J/s}$$

According to Equation 6.28, power is the *rate* at which work is done. Since work is a way to transfer energy from one system to another, power is also equal to the energy output per unit time of a device. For example, the specifications for a motor or engine often include the maximum power the engine is able to produce. For historical reasons, the power of an automobile is usually given in terms of an odd unit called "horsepower." Presumably, 1 horsepower (hp) is equal to the power that a typical ("standard"?) horse is able to expend. You can convert between watts and horsepower using the relation

$$1 \text{ hp} = 745.7 \text{ W} = 745.7 \text{ J/s}$$

In addition to mechanical devices, the concept of power also applies to chemical and electrical processes and devices. Indeed, there is a power output or input associated with many of the electrical devices in your house. For example, a typical lightbulb is rated at 20 W = 20 J/s. This rating means that the lightbulb consumes 20 J of electrical energy for every second it is turned on. Likewise, devices such as computers and DVD players also have a "power rating." Table 6.1 lists the power output and consumption of a number of common devices. The distinction between power "consumption" and power "output" is important. For example, a lightbulb consumes a certain amount of electrical energy (for which you pay the utility company), and it outputs a certain amount of energy in the form of visible light along with a certain amount of energy as heat.

EXAMPLE 6.11 Comparing Mechanical and Electrical Energy

Consider a lightbulb rated for a power $P_{\text{bulb}} = 20$ W. This lightbulb consumes a certain amount of electrical energy in 1 second of use. Suppose we want to generate the same amount of energy by dropping the lightbulb from height h to ground level. If the mass of the lightbulb is $m = 0.050$ kg, what is h?

RECOGNIZE THE PRINCIPLE

We apply conservation of energy principles to the lightbulb. The change in potential energy of the lightbulb is $\Delta PE = -mgh$. This energy must equal (in magnitude) the energy used by the lightbulb in 1 s.

(continued) ▶

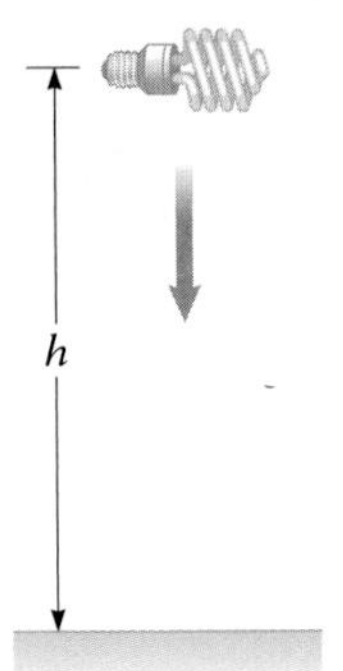

▲ **Figure 6.34** Example 6.11. A falling lightbulb.

SKETCH THE PROBLEM

Figure 6.34 describes the problem.

IDENTIFY THE RELATIONSHIPS

Using the definition of power (Eq. 6.28), the energy consumed by the lightbulb in time t is

$$W = P_{\text{bulb}}t$$

Setting this equal to the change in potential energy when the lightbulb falls gives

$$mgh = P_{\text{bulb}}t$$

SOLVE

Solving for h, we find

$$h = \frac{P_{\text{bulb}}t}{mg} = \frac{(20\ \text{W})(1.0\ \text{s})}{(0.050\ \text{kg})(9.8\ \text{m/s}^2)} = \boxed{41\ \text{m}}$$

► ***What does it mean?***

This result is a surprisingly large height: what we normally consider to be a rather small electrical power rating (of the lightbulb) corresponds to a substantial amount of mechanical energy. Imagine what h would be if the lightbulb were turned on for 1 day!

Power and Velocity

Consider a "device" such as a motor or a person that is moving an object from one place to another. This device is expending an amount of power given by Equation 6.28. For one-dimensional motion with the force parallel to the displacement Δx, we can write this relation as

$$P_{\text{ave}} = \frac{W}{t} = \frac{F\,\Delta x}{t}$$

Since the average velocity is equal to $v_{\text{ave}} = \Delta x/t$, we have

$$P_{\text{ave}} = Fv_{\text{ave}}$$

This relationship between power and velocity also holds for the instantaneous velocity v and the instantaneous power P:

$$P = Fv \tag{6.29}$$

If the device, such as a motor, is capable of expending a certain amount of power, Equation 6.29 shows that there is a trade-off between force and velocity. That is, with a fixed value of P, a motor can exert a large force (as might be required to move a heavy object) while moving an object at a small velocity, or the motor can exert a small force (as needed to move a light object) while moving the object at a large velocity. Such trade-offs involving force and velocity are quite common.

Power, Force, and Efficiency

The trade-off between force and velocity in Equation 6.29 limits the *maximum* possible force consistent with Newton's laws and conservation of mechanical energy. *Real* motors and other devices usually do not attain this theoretical limit, however. For example, an automobile engine might stop running ("stall") at a very low velocity, or there might be some frictional forces in the problem. To account for this behavior, we define the efficiency ε by

$$F = \varepsilon\frac{P}{v} \tag{6.30}$$

The efficiency cannot be larger than unity. We'll see an example involving efficiency in the next section, when we discuss the operation of molecular motors.

6.8 Work, Energy, and Molecular Motors

In Section 6.7, we discussed how our ideas about work, energy, and power can be used to understand the behavior of motors and similar devices. The same ideas apply to the molecular motors that transport materials within and between cells. Several different types of molecular motors have been discovered, one example of which is sketched in Figure 6.35. This motor is based on a molecule called myosin that moves along long filaments composed of actin molecules.

Myosin is believed to move by "walking" along a filament in steps, much like a person walks on two legs. This process takes place in muscles, where the filaments form bundles. A myosin molecule is also attached to its own muscle filament, so as it walks, the myosin drags one muscle filament relative to another. The operation of your muscles is produced by these molecular motors.

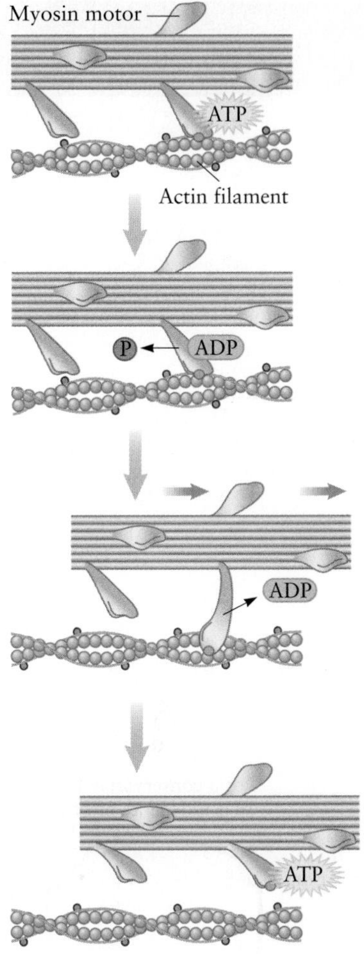

▲ **Figure 6.35** Some molecular motors move by "walking" along long strands of a protein called actin. These motors are the subject of much current research. We can use work–energy principles to understand their behavior.

Calculating the Force Exerted by a Molecular Motor

The precise biochemical reactions involved in the myosin walking motion are not completely understood. However, we do know that each step has a length of approximately 5 nm (5×10^{-9} m) and that the energy for this motor comes from a chemical reaction in which the molecule ATP reacts to form ADP. (ATP is shorthand for adenosine triphosphate, and ADP is adenosine diphosphate.) The energy released when a single ATP molecule breaks down is $E_{ATP} \approx 5 \times 10^{-20}$ J. With this information, we can use work–energy ideas to calculate the maximum force this motor can achieve.

The energy E_{ATP} is stored as chemical potential energy. One ATP molecule is consumed during each step, so the molecular motor uses an amount of potential energy E_{ATP} during each step. During the course of a single step, the potential energy of the motor decreases by E_{ATP} while it does an amount of work W_{motor}. Energy is conserved, so

$$W_{motor} = E_{ATP}$$

The work done by the motor is also equal to the motor force times the length of a single step Δx:

$$W_{motor} = F_{motor}\,\Delta x = E_{ATP} \tag{6.31}$$

$$F_{motor} = \frac{E_{ATP}}{\Delta x}$$

Inserting the values of E_{ATP} and Δx given above, we find

$$F_{motor} = \frac{E_{ATP}}{\Delta x} = \frac{5 \times 10^{-20}\ \text{J}}{5 \times 10^{-9}\ \text{m}} = 1 \times 10^{-11}\ \text{N}$$

for the force generated by the motor. This force is 10 piconewtons, which may seem like a very small force. However, it is actually very substantial compared with other "molecular-scale" forces such as the weight of a typical molecule. Indeed, the weight of a single ATP molecule is about 1×10^{-23} N, which is *many* orders of magnitude smaller than the force exerted by this molecular motor!

Our calculation of F_{motor} assumes all the ATP potential energy is converted to work by the motor, that is, that the efficiency of the motor is $\varepsilon = 1$ (a "perfect" motor). Real motors will have a smaller efficiency as some of the chemical potential energy will go into vibrations, other molecular motions of the actin filament, and so forth. For this reason, our value for F_{motor} is an upper limit on the force that can be produced by a myosin motor. Evidence suggests that real molecular motors operate with a typical efficiency $\varepsilon > 0.5$, so our result for F_{motor} provides a fairly good description of real myosin motors.

Summary | CHAPTER 6

Key Concepts and Principles

F

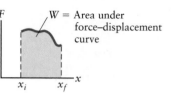

W = Area under force–displacement curve

x_i x_f x

Work

When a force acts on an object, the force does ***work*** on the object:

$$W = F(\Delta r)\cos\theta \qquad \textbf{(6.4) (page 167)}$$

The work done by an applied force can be interpreted in a graphical manner as the area under the force–displacement curve.

Kinetic energy

Kinetic energy is energy of motion and is given by

$$KE = \tfrac{1}{2}mv^2 \qquad \textbf{(6.8) (page 171)}$$

Work and kinetic energy are related through the ***work–energy theorem***

$$W = \Delta KE \qquad \textbf{(6.9) (page 171)}$$

where W is the work done on the object.

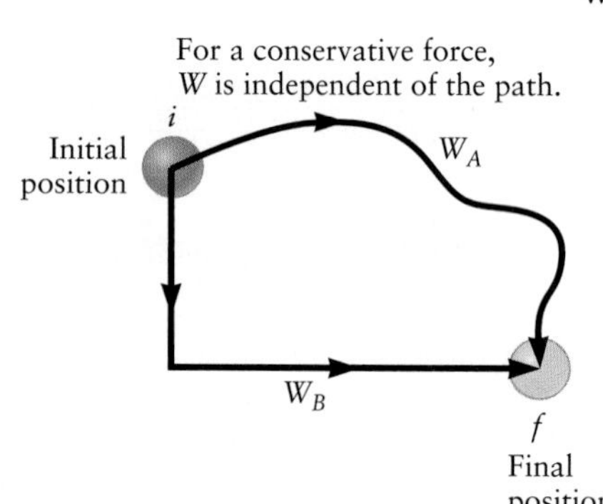

Potential energy

Potential energy is stored energy associated with a particular force or forces. The potential energy associated with a force is related to the work done by this force:

$$\Delta PE = PE_f - PE_i = -W \qquad \textbf{(6.10) (page 175)}$$

The work done by a ***conservative force*** is independent of the path.

Conservation of energy

The principle of conservation of energy can be written as

$$KE_i + PE_i + W_{\text{noncon}} = KE_f + PE_f \qquad \textbf{(6.26) (page 192)}$$

where W_{noncon} is the work done on the system by nonconservative forces.

Applications

Hooke's law

The force exerted by a spring or similar elastic object is

$$F = -kx \qquad \textbf{(6.23) (page 185)}$$

where k is the ***spring constant*** and x is the amount that the spring is stretched or compressed.

Potential energy functions

Gravitational potential energy for an object near the Earth's surface:

$$PE_{\text{grav}} = mgy \qquad \textbf{(6.13) (page 176)}$$

Gravitational potential energy of two masses (general case):

$$PE_{grav} = -\frac{Gm_1m_2}{r} \qquad (6.20) \text{ (page 184)}$$

Elastic potential energy:

$$PE_{spring} = \tfrac{1}{2}kx^2 \qquad (6.24) \text{ (page 186)}$$

Questions

SSM = answer in *Student Companion & Problem-Solving Guide* ⊗ = life science application

1. Consider an object undergoing uniform circular motion. We know from Chapter 5 that circular motion is caused by a force that is directed toward the center of the circle. What is the work done by this force during circular motion?
2. Explain why kinetic energy can never be negative, but the potential energy can be positive, negative, or zero.
3. A heavy crate is pushed across a level floor at a constant velocity. What is the total work done on the crate by all forces? How does your answer change if the crate is being pushed up a hill?
4. Consider a satellite in an elliptical orbit around the Earth. What is the sign (positive or negative) of the power associated with the gravitational force of the Earth on the satellite? *Hint*: The answer depends on the position of the satellite along its orbit. Where in the orbit is the power zero?
5. Consider an apple that falls from a branch to the ground below. At what moment is the kinetic energy of the apple largest? At what moment is the gravitational potential energy largest?
6. SSM For the bungee jumper in Example 6.9 (page 189), plot the following quantities as a function of time: (a) the gravitational potential energy of the jumper, (b) the kinetic energy of the jumper, (c) the potential energy associated with the bungee cord, and (d) the total mechanical energy.
7. The work done by friction is usually negative, as in the case of an object that slides to a stop. The work done by friction does not have to be negative, however. Give an example in which the work done by friction is positive.
8. Motor 1 does 20 J of work in 10 s, while motor 2 does 5 J of work in 1 s. Which motor produces the greatest power?
9. According to Equation 6.29, power is related to force and velocity by $P = Fv$. Is it possible for P to be negative? Give an example and explain what it means for P to be negative.
10. Give two examples in which the work done on an object by friction is negative.
11. Give two examples in which a nonzero force acts on an object, but the work done by that force is zero.
12. Explain how a mosquito can have a greater kinetic energy than a baseball.
13. A hockey puck slides along a level surface, eventually coming to rest. Is the energy of the hockey puck conserved? Is the total energy of the universe conserved? Discuss where the initial kinetic energy of the puck "goes."
14. A spring is initially compressed. It is then released, sending an object flying off the spring. Discuss this process in terms of conservation of energy. What and "where" is the initial energy, and where is the energy found at the end?
15. Tarzan starts at rest near the top of a tall tree. He then swings on a vine and reaches the base of the tree, where he jumps off the vine and comes to rest on the ground. Discuss and compare Tarzan's total energy at (a) the top of the tree, (b) just before he jumps off the vine at the bottom, and (c) when he is standing on the ground. Is Tarzan's total energy the same in all three places? Explain how total energy is conserved in going from (a) to (b) to (c).
16. During the course of each day, you move from place to place and do work on various objects. Where does the energy for these everyday activities come from? Explain.
17. SSM When a rubber ball is dropped onto a concrete floor, it bounces to a height that is slightly lower than its initial height. Compare the ball's (a) initial mechanical energy to (b) its energy just before hitting the ground, to (c) the total energy just after bouncing off the ground, and to (d) the total energy when it reaches it final height. Is the ball's mechanical energy conserved? Explain and discuss using the principle of conservation of total energy.
18. Construct bar charts showing the kinetic energy and potential energy of the projectile in Figure 6.16. Show these energies when the projectile has just left the ground, when it is at the highest point of its trajectory, and just before it returns to the ground. Explain how these charts illustrate the principle of conservation of mechanical energy.
19. Construct bar charts showing the kinetic energy and potential energy for the bungee jumper in Example 6.9 at various points during his jump. Explain how these charts illustrate the principle of conservation of mechanical energy.

Problems

SSM = solution in *Student Companion & Problem-Solving Guide* Ⓡ = reasoning and relationships problem
★ = intermediate ✪ = challenging ⊗ = life science application RT = reasoning tutorial in Enhanced WebAssign

6.1 FORCE, DISPLACEMENT, AND WORK

1. A sphere of mass 35 kg is attached to one end of a rope as shown in Figure P6.1. It is found that the rope does an amount of work $W = 550$ J in pulling the sphere upward through a distance 1.5 m. Find the tension in the rope.

Figure P6.1 Problems 1 and 2.

2. ★ For the sphere in Problem 1, find the work done by the force of gravity on the sphere.
3. A crate of mass 50 kg is pushed across a level floor by a person. If the person exerts a force of 25 N in the horizontal direction and moves the crate a distance of 10 m, what is the work done *by the person*?
4. SSM ★ A hockey puck of mass 0.25 kg is sliding along a slippery frozen lake, with an initial speed of 60 m/s. The coefficient

of friction between the ice and the puck is $\mu_K = 0.030$. Friction eventually causes the puck to slide to a stop. Find the work done by friction.

5. A crate of mass 24 kg is pushed up a frictionless ramp by a person as shown in Figure P6.5. Calculate the work done by the person in pushing the crate a distance of 20 m as measured along the ramp. Assume the crate moves at constant velocity.

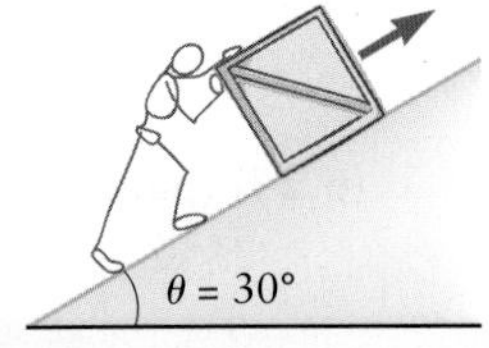

Figure P6.5
Problems 5, 6, and 7.

6. ✱ Consider again the crate in Problem 5, but now include friction. Assume the coefficient of kinetic friction between the crate and the ramp is $\mu_K = 0.25$. What is the work done by the person as he pushes the crate up the ramp?

7. ✱ Find the work done by gravity on the crate in Problem 6.

8. ✱ A block of mass $m = 5.0$ kg is pulled along a rough horizontal surface by a rope as sketched in Figure P6.8. The tension in the rope is 40 N, and the coefficient of kinetic friction between the block and the surface is $\mu_K = 0.25$. (a) If the block travels a distance of 4.5 m along the surface, what is the work done by the rope? (b) Find the work done by friction on the block.

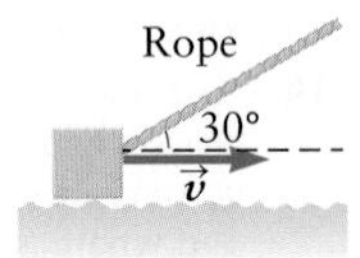

Figure P6.8

9. A car ($m = 1200$ kg) is traveling at an initial speed of 15 m/s. It then slows to a stop over a distance of 80 m due to a force from the brakes. How much work is done on the car? Assume the acceleration is constant.

10. A snowboarder of mass 80 kg slides down the trail shown in Figure P6.10. The first part of the trail is a ramp that makes an angle of 20° with the horizontal, and the final portion of the trail is flat. Find the work done by gravity on the snowboarder as she travels from the beginning to the end of this trail.

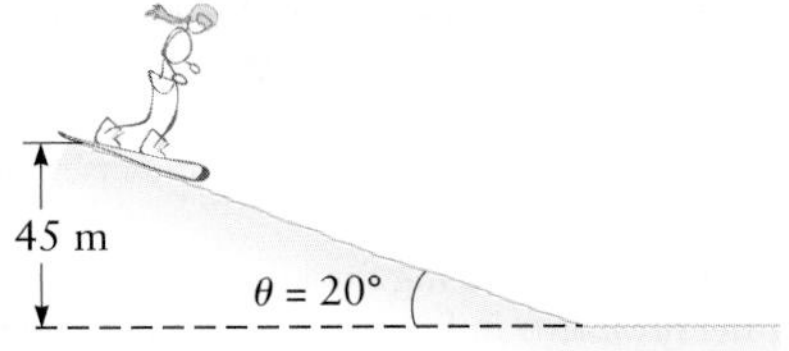

Figure P6.10

11. Two railroad cars, each of mass 2.0×10^4 kg, are connected by a cable to each other, and the car in front is connected by a cable to the engine as shown in Figure P6.11. The cars start from rest and accelerate to a speed of 1.5 m/s after 1 min. (a) Find the work done by cable 1 on the car in the back. (b) Find the work done by cable 1 on the car in front. (c) Find the work done by cable 2 on the car in front.

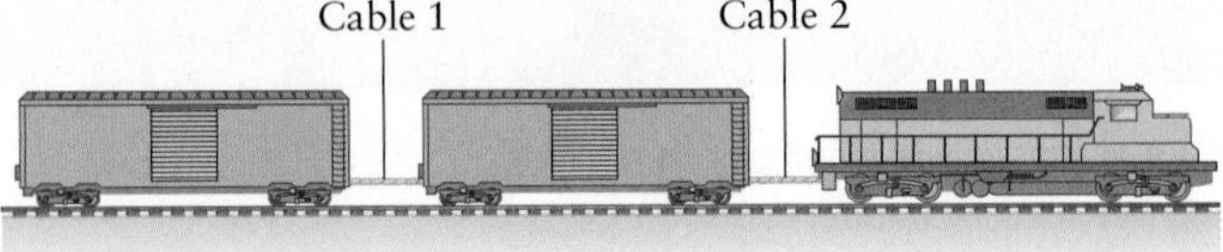

Figure P6.11

12. A person pushes a broom at an angle of 60° with respect to the floor (Fig. P6.12). If the person exerts a force of 30 N directed along the broom handle, what is the work done by the person on the broom as he pushes it a distance of 5.0 m?

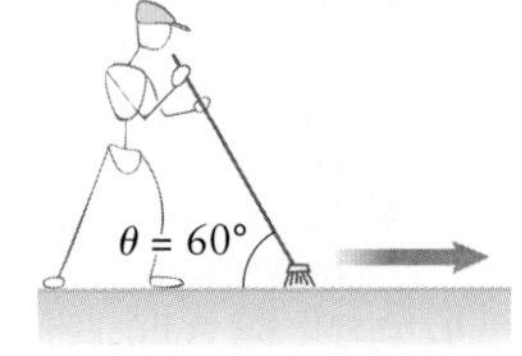

Figure P6.12

13. A car is pushed along a long road that is straight, horizontal, and parallel to the x direction. (a) The horizontal force on the car varies with x as shown in Figure P6.13. What is the work done on the car by this force? (b) What is the work done by gravity on the car?

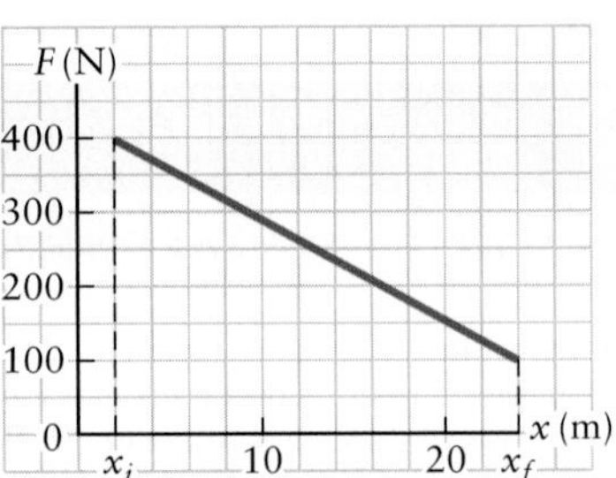

Figure P6.13

14. The force on an object varies with position as shown in Figure P6.14. Estimate the work done on the object as it moves from the origin to $x = 5.0$ m. Assume the motion is one dimensional and the force is along the x direction.

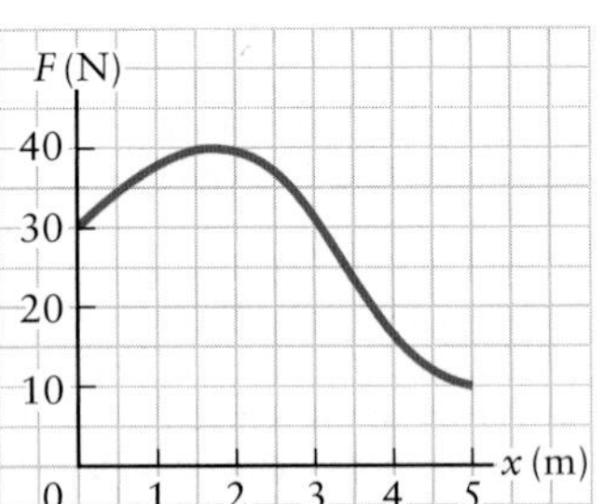

Figure P6.14

6.2 KINETIC ENERGY AND THE WORK–ENERGY THEOREM

15. A tomato of mass 0.22 kg is dropped from a tall bridge. If the tomato has a speed of 12 m/s just before it hits the ground, what is the kinetic energy of the tomato?

16. Two objects have the same kinetic energy. One has a speed that is 2.5 times the speed of the other. What is the ratio of their masses?

17. A car of mass 1500 kg is initially traveling at a speed of 10 m/s. The driver then accelerates to a speed of 25 m/s over a distance of 200 m. Calculate the change in the kinetic energy of the car.

18. ✱ A rock of mass 0.050 kg is thrown upward with an initial speed of 25 m/s. Find the work done by gravity on the rock from the time it leaves the thrower's hand until it reaches the highest point on its trajectory. *Hint*: You do not have to calculate the rock's maximum height.

19. A horizontal force of 15 N pulls a block of mass 3.9 kg across a level floor. The coefficient of kinetic friction between the block and the floor is $\mu_K = 0.25$. If the block begins with a speed of 8.0 m/s and is pulled for a distance of 12 m, what is the final speed of the block?

20. You are pushing a refrigerator across the floor of your kitchen. You exert a horizontal force of 300 N for 7.0 s, during which time the refrigerator moves a distance of 3.0 m at constant velocity. (a) What is the total work (by all forces) done on the refrigerator? (b) What is the work done by friction?

21. An archer is able to fire an arrow (mass 0.020 kg) at a speed of 250 m/s. If a baseball (mass 0.14 kg) is given the same kinetic energy, what is its speed?

22. B A baseball pitcher can throw a baseball at a speed of 50 m/s. If the mass of the ball is 0.14 kg and the pitcher has it in his hand over a distance of 2.0 m, what is the average force exerted by the pitcher on the ball?

23. RT A softball pitcher can exert a force of 100 N on a softball. If the mass of the ball is 0.19 kg and the pitcher has it in her hand over a distance of 1.5 m, what is the speed of the ball when it leaves her hand?

24. ★ Consider a skydiver of mass 70 kg who jumps from an airplane flying at an altitude of 1500 m. With her parachute open, her terminal velocity is 8.0 m/s. (a) What is the work done by gravity and (b) what is the average force of air drag during the course of her jump?

25. Consider a small car of mass 1200 kg and a large sport utility vehicle (SUV) of mass 4000 kg. The SUV is traveling at the speed limit ($v = 35$ m/s). The driver of the small car travels so as to have the same kinetic energy as the SUV. Find the speed of the small car.

26. B A truck of mass 9000 kg is traveling along a level road at an initial speed of 30 m/s. The driver then shifts into neutral so that the truck is coasting, and the truck gradually slows to a final speed of 12 m/s. (a) Find the total work done on the truck. (b) What force(s) do this work on the truck? (There may be more than one.) (c) The final kinetic energy of the truck is less than its initial kinetic energy. Where does this energy go?

27. ★ You use an elevator to travel from the first floor ($h = 0$) to the fourth floor of a building ($h = 12$ m). If you start from rest and you end up at rest on the fourth floor, what is the work done by gravity on you? Assume you have a mass of 70 kg.

28. SSM ★ A car of mass $m = 1500$ kg is pushed off a cliff of height $h = 24$ m (Fig. P6.28). If the car lands a distance of 10 m from the base of the cliff, what was the kinetic energy of the car the instant after it left the cliff?

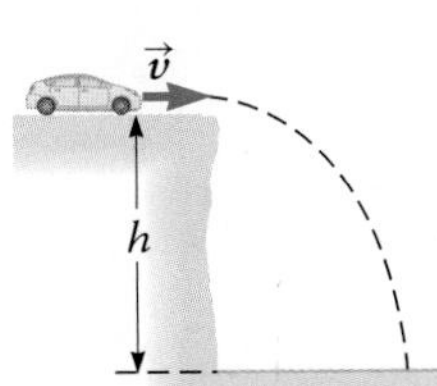

Figure P6.28

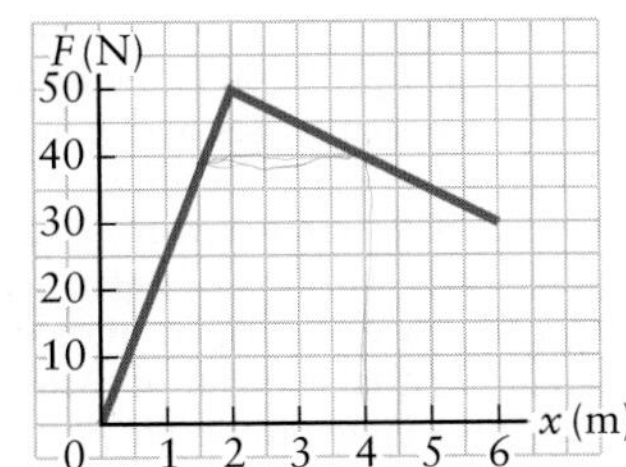

Figure P6.29 Problems 29 and 30.

29. ★ The force on an object varies with position as shown in Figure P6.29. The object begins at rest from the origin and has a speed of 5.0 m/s at $x = 4.0$ m. What is the mass of the object? Assume one-dimensional motion with the force along the direction of motion.

30. Suppose the object in Figure P6.29 has a mass of 2.5 kg and a speed of 9.0 m/s at $x = 0$. What is its speed at $x = 5.0$ m?

6.3 POTENTIAL ENERGY AND CONSERVATION OF ENERGY

31. A skier of mass 110 kg travels down a frictionless ski trail. (a) If the top of the trail is a height 200 m above the bottom, what is the work done by gravity on the skier? (b) Find the velocity of the skier when he reaches the bottom of the ski trail. Assume he starts from rest.

32. Consider a roller coaster that moves along the track shown in Figure P6.32. Assume all friction is negligible and ignore the kinetic energy of the roller coaster's wheels. (a) Is the mechanical energy of the roller coaster conserved? (b) Add a coordinate system to the sketch in Figure P6.32. (c) If the roller coaster starts from rest at location A, what is its total mechanical energy at point A? (d) If v_B is the speed at point B, what is the total mechanical energy at point B? (e) Find the speed of the roller coaster when it reaches locations B and C.

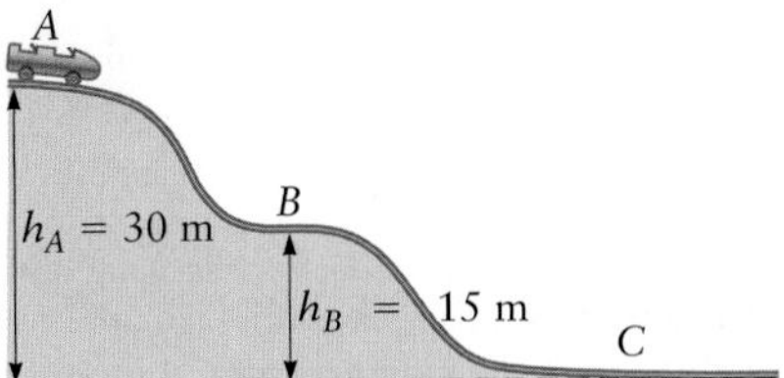

Figure P6.32 Problems 32 and 34.

33. Suppose the ski trail in Problem 31 is *not* frictionless. (a) Find the work done by gravity on the skier in this case. (b) If the skier has a speed of 30 m/s at the bottom of the hill, what is the work done by friction on the skier?

34. Consider again the roller coaster in Figure P6.32, but now assume the roller coaster starts with a speed of 12 m/s at point A. Find the speed of the roller coaster when it reaches locations B and C. Again assume friction is negligible and ignore the kinetic energy of the wheels.

35. ★ Mount McKinley (also called Denali) is the tallest mountain in North America, with a height of 6200 m above sea level. If a person of mass 120 kg walks from sea level to the top of Mount McKinley, how much work is done by gravity on the person?

36. ✪ The roller coaster in Figure P6.36 starts with a velocity of 15 m/s. One of the riders is a small girl of mass 30 kg. Find her apparent weight when the roller coaster is at locations B and C. At these two locations, the track is circular, with the radii of curvature given in the figure. The heights at points A, B, and C are $h_A = 25$ m, $h_B = 35$ m, and $h_C = 0$. Assume friction is negligible and ignore the kinetic energy of the wheels.

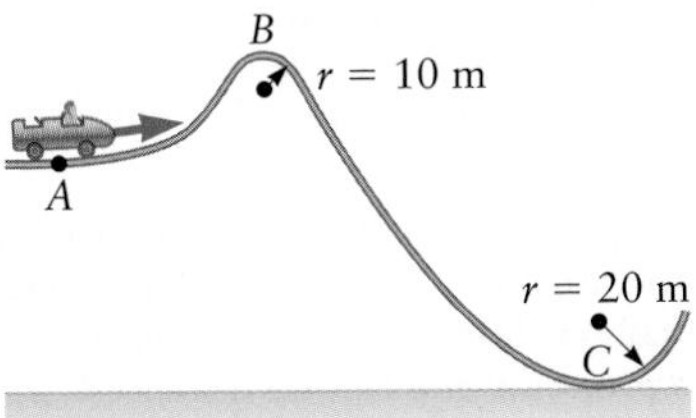

Figure P6.36

37. SSM ★ A roller coaster (Fig. P6.37) starts at the top of its track (point A) with a speed of 12 m/s. If it reaches point B traveling at 16 m/s, what is the vertical distance (h) between A and B? Assume friction is negligible and ignore the kinetic energy of the wheels.

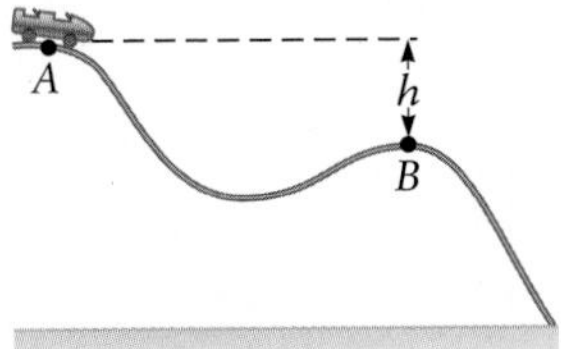

Figure P6.37

38. A skateboarder starts at point A on the track (Fig. P6.38) with a speed of 15 m/s. Will he reach point B? Assume friction is negligible and ignore the kinetic energy of the skateboard's wheels.

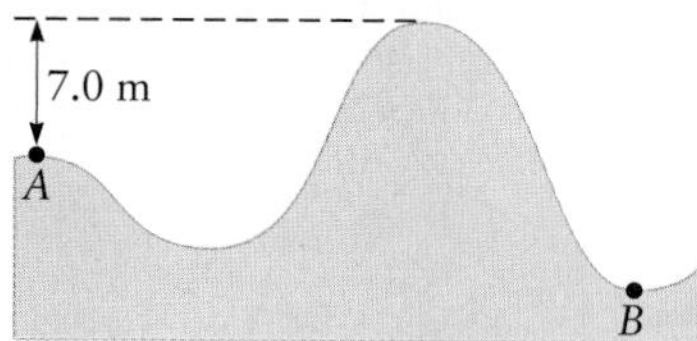

Figure P6.38

39. **Frontside air.** A skateboarder is practicing on the "half-pipe" shown in Figure P6.39, using a special frictionless skateboard. (You can also ignore the kinetic energy of the skateboard's wheels.) (a) If she starts from rest at the top of the half-pipe, what is her speed at the bottom? (b) If the skateboarder has mass $m = 55$ kg, what is her apparent weight at the bottom of the half-pipe? (c) What speed does she then have when she reaches the top edge on the other side of the half-pipe? (d) Now suppose she has a speed of 11 m/s at the bottom of the half-pipe. What is the highest point she can reach? *Hint*: This point may be above the edge of the half-pipe.

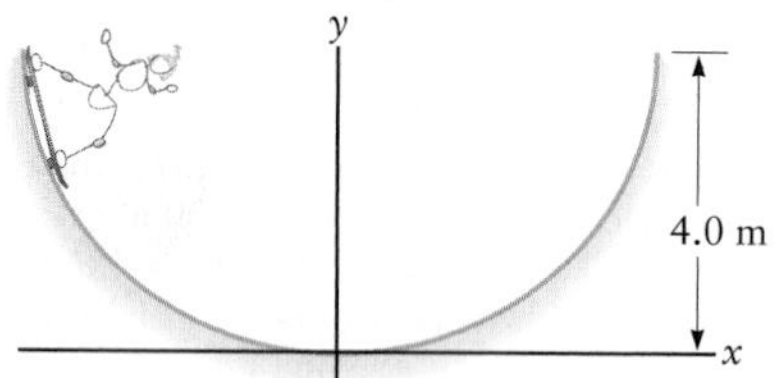

Figure P6.39 Problems 39 and 40.

40. Consider again the skateboarder in Figure P6.39. (a) Assuming she starts at the top of the half-pipe, make qualitative sketches of her kinetic energy and potential energy as a function of x in the half-pipe. Start your plots when she is at the far left in Figure P6.39 and end them when she reaches the far right. Also assume friction is negligible and ignore the kinetic energy of the wheels. (b) Repeat part (a), but now assume friction is not negligible.

41. A hockey puck of mass 0.25 kg starts from rest and slides down the frictionless ramp shown in Figure P6.41. The bottom end of the ramp is horizontal. After it leaves the ramp, the puck travels as a projectile and lands a distance L from the ramp. (a) Is the puck's mechanical energy at the top of the ramp equal to its mechanical energy when it leaves the ramp and is in the air? Explain why or why not. (b) Add a coordinate system to the sketch in Figure P6.41. Where is a convenient place to put the origin of the vertical (y) axis? Do you have to put it there? (c) What is the initial mechanical energy of the puck? (d) What is the mechanical energy of the puck the instant it leaves the ramp? (e) What is the speed of the puck the instant it leaves the ramp? (f) How far from the base of the ramp does the puck land?

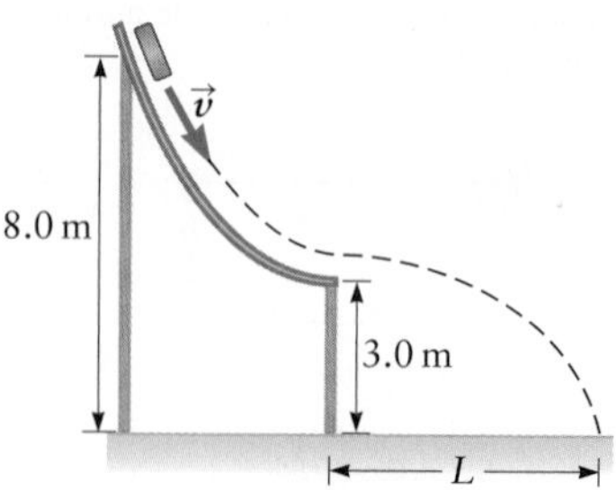

Figure P6.41

42. SSM A rock of mass 12 kg is tied to a string of length 2.4 m, with the other end of the string fastened to the ceiling of a tall room (Fig. P6.42). While hanging vertically, the rock is given an initial horizontal velocity of 2.5 m/s. (a) Add a coordinate system to the sketch in Figure P6.42. Where is a convenient place to choose the origin of the vertical (y) axis? (b) What are the initial kinetic and potential energies of the rock? (c) If the rock swings to a height h above its initial point, what is its potential energy? What is its total mechanical energy at that point? (d) How high will the rock swing? Express your answer in terms of the angle θ that the string makes with the vertical when the rock is at its highest point. (e) Make qualitative sketches of the kinetic energy and potential energies as functions of height h.

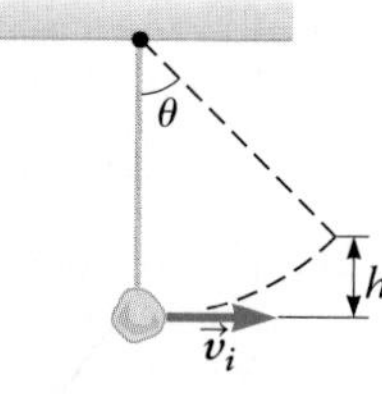

Figure P6.42

43. Consider the rock and string in Figure P6.43. The string is fastened to a hinge that allows it to swing completely around in a vertical circle. If the rock starts at the lowest point on this circle and is given an initial speed v_i, what is the smallest value of v_i that will allow the rock to travel completely around the circle without the string becoming slack at the top? Assume the length of the string is 1.5 m.

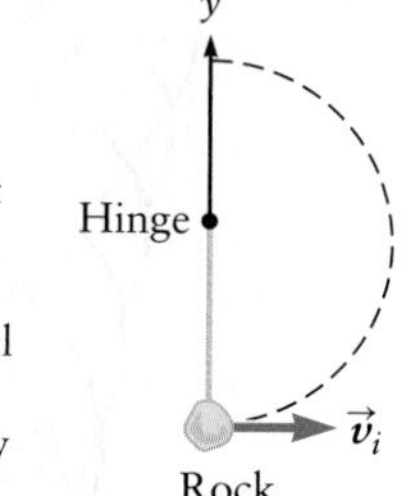

Figure P6.43

44. RT A rock climber ($m = 90$ kg with gear included) starts at the base of a cliff and climbs to the top ($h = 35$ m). He then walks along the plateau at the top for a distance $L = 40$ m. Find the work done by gravity and the change in gravitational potential energy of the rock climber.

45. Two crates of mass $m_1 = 40$ kg and $m_2 = 15$ kg are connected by a massless rope that passes over a massless, frictionless pulley as shown in Figure P6.45. The crates start from rest. (a) Add a coordinate system to Figure P6.45. (b) What are the initial kinetic and potential energies of each crate? (c) If the crate on the left (m_1) moves downward through a distance h, what is the change in the potential energy of m_1 and m_2? Express your answers in terms of h. (d) What is the speed of the crates after they have moved a distance $h = 2.5$ m?

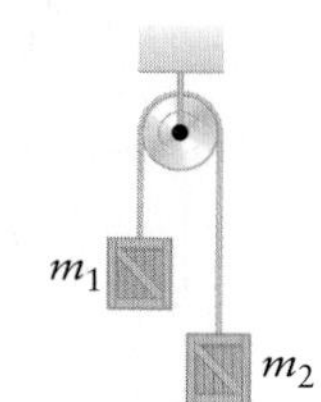

Figure P6.45 Problems 45 and 46.

46. Consider again the crates in Problem 45, but now assume the rope has a mass of 2 kg. If the tops of the crates start at the same initial height and the rope has a length of 10 m, what is the final speed after the crates have moved a distance of 2.5 m?

47. Sketch how the potential energy, kinetic energy, and total mechanical energy vary with time for an apple that drops from a tree.

48. (a) Sketch how the potential energy, kinetic energy, and total mechanical energy vary with time for a cannon shell that is fired from a large cannon. Ignore air drag and assume the cannon is aimed straight upward. (b) Repeat part (a), but now with the cannon aimed at 45° above the horizontal direction.

49. A rock of mass 3.3 kg is tied to a string of length 1.2 m. The rock is held at rest as

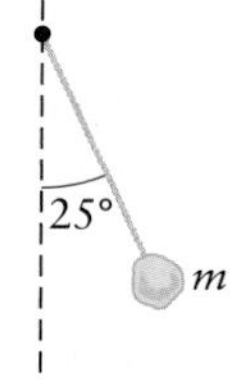

Figure P6.49

shown in Figure P6.49 so that the string is initially tight, and then it is released. (a) Find the speed of the rock when it reaches the lowest point of its trajectory. (b) What is the maximum tension in the string?

50. ★ A ball of mass 1.5 kg is tied to a string of length 6.0 m as shown in Figure P6.50. The ball is initially hanging vertically and is given an initial velocity of 5.0 m/s in the horizontal direction. The ball then follows a circular arc as determined by the string. What is the speed of the ball when the string makes an angle of 30° with the vertical?

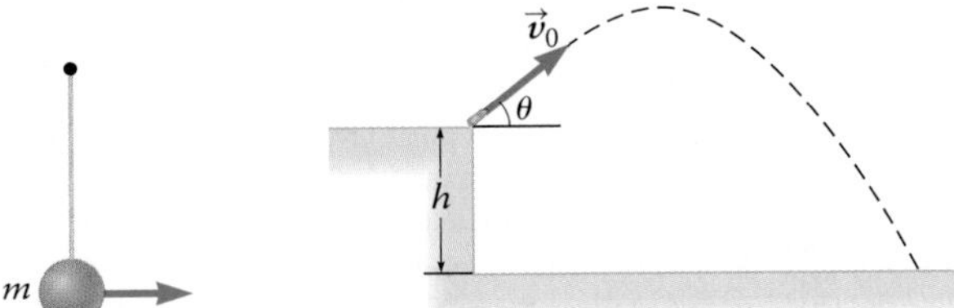

Figure P6.50 Figure P6.51

51. A bullet is fired from a rifle with speed v_0 at angle θ with respect to the horizontal axis (Fig. P6.51) from a cliff at height h above the ground below. (a) Use conservation of energy principles to calculate the speed of the bullet when it strikes the ground. Express your answer in terms of v_0, h, g, and θ, and ignore air drag. (b) Explain why your result is independent of the angle θ.

6.4 MORE POTENTIAL ENERGY FUNCTIONS

52. Calculate the escape speed for an object on the Moon.

53. An engineer at NASA decides to save money on fuel by launching satellites from a very tall mountain. (a) Calculate the escape speed for a satellite launched from an altitude of 8000 m above sea level (a mountain comparable to Mount Everest). (b) Do you think this method would be a good way for NASA to save fuel?

54. SSM ★ Calculate the velocity needed for an object starting at the Earth's surface to just barely reach a satellite in a geosynchronous orbit. Ignore air drag and assume the object has a speed of zero when it reaches the satellite.

55. Sketch how the potential energy, kinetic energy, and total mechanical energy vary with position for a projectile that is fired from the Earth's surface. Assume the projectile's initial speed is equal to the escape speed and ignore air drag.

56. ★ Consider a spacecraft that is to be launched from the Earth to the Moon. Calculate the minimum velocity needed for the spacecraft to just make it to the Moon's surface. Ignore air drag from the Earth's atmosphere. *Hint*: The spacecraft will not have zero velocity when it reaches the Moon.

57. ★ A space station is orbiting the Earth. It moves in a circular orbit with a radius equal to twice the Earth's radius. A supply satellite is designed to travel to the station and dock smoothly when it arrives. If the supply satellite is fired as a simple projectile from the Earth's surface, what is the minimum initial speed required for it to reach the space station and have a speed equal to the station's speed when it arrives? Ignore both air drag and the rotational motion of the Earth.

58. A mass and spring are arranged on a horizontal, frictionless table as shown in Figure P6.58. The spring constant is $k = 500$ N/m, and the mass is 4.5 kg. The block is pushed against the spring so that the spring is compressed an amount 0.35 m, and then it is released. Find the velocity of the mass when it leaves the spring.

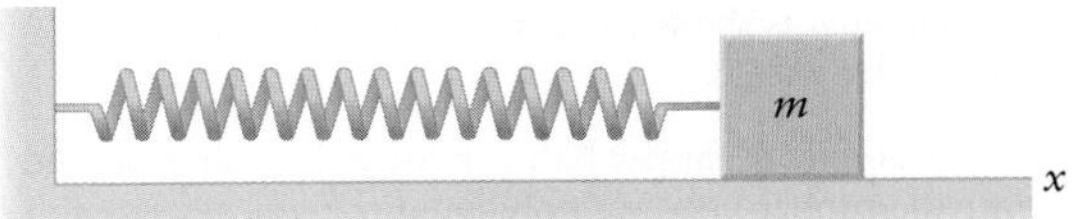

Figure P6.58

59. A block is dropped onto a spring with $k = 30$ N/m. The block has a speed of 3.3 m/s just before it strikes the spring. If the spring compresses an amount 0.12 m before bringing the block to rest, what is the mass of the block?

60. A block of mass 35 kg is sitting on a platform as shown in Figure P6.60. The platform sits on a spring with $k = 2000$ N/m. The mass is initially at rest. (a) Add a coordinate system to this sketch. Where is a convenient place to choose the origin of the vertical (y) axis? (b) If the platform is depressed a distance 0.50 m and held there, what is the mechanical energy of the system? (c) The platform is then released, and the mass moves upward and eventually flies off the platform. Assume the spring returns to its relaxed state (unstretched and uncompressed) after the mass leaves the platform. What are the kinetic energy, gravitational potential energy, and elastic potential energy of the system when the mass reaches its highest point? (d) How high will the block go? Measure this distance from the height of the uncompressed platform.

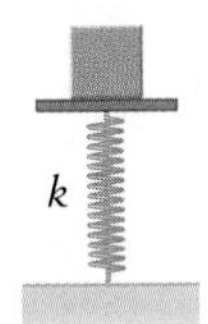

Figure P6.60

61. An archer's bow can be treated as a spring with $k = 3000$ N/m. If the bow is pulled back a distance 0.12 m before releasing the arrow, what is the kinetic energy of the arrow when it leaves the bow?

62. For the arrow in Problem 61, what is the work done on the bow as the bowstring is pulled back into position?

63. ★ RT The struts on a car are fancy springs. Because one is attached to each wheel, your car is supported by four of these springs. Estimate the spring constant for one of these springs.

64. ★ R A tennis racket can be treated as a spring. The displacement of the spring (what we commonly call x) is the displacement of the strings at the center of the racket. Estimate the spring constant of this spring. Assume a typical tennis ball has a mass of about 57 g and a speed of 50 m/s when it leaves the racket.

65. ★ RT A diving board acts like a spring and obeys Hooke's law (Fig. 6.22A). Estimate the spring constant of a diving board.

66. ★ RT A tennis ball is a flexible, elastic object. If a person of average size stands on a tennis ball, the ball will compress to about half its original (noncompressed) diameter. What is the approximate spring constant of a tennis ball?

6.5 CONSERVATIVE VERSUS NONCONSERVATIVE FORCES AND CONSERVATION OF ENERGY

67. SSM ★ A tennis ball ($m = 57$ g) is projected vertically with an initial speed of 8.8 m/s. (a) If the ball rises to a maximum height of 3.7 m, how much kinetic energy was dissipated by the drag force of air resistance? (b) How much higher would the ball have gone in a vacuum?

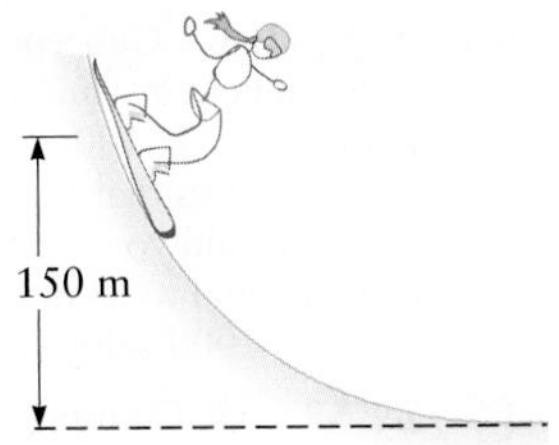

68. ★ A snowboarder of mass 80 kg travels down the slope of height 150 m shown in Figure P6.68. If she starts from rest at the top and has a velocity of 12 m/s when she reaches the

150 m

Figure P6.68

bottom, what is the work done on her by friction? *Note*: Figure not drawn to scale.

69. **Big bounce.** A Super Ball is a toy ball made from the synthetic rubber polymer polybutadiene vulcanized with sulfur. Manufactured by Wham-O since 1965, these toys have a "super" elastic property such that they bounce to 90% of the height from which they are dropped. A Super Ball of mass 100 g is dropped from a height of 2.0 m. How much kinetic energy is dissipated by the collision with the ground? A lump of clay of equal mass is dropped from a similar height and sticks to the ground without a bounce. Where did the kinetic energy go in the case of the clay? What about in the case of the Super Ball?

70. Consider the skateboarder in Figure P6.70. If she has a mass of 55 kg, an initial velocity of 20 m/s, and a velocity of 12 m/s at the top of the ramp, what is the work done by friction on the skateboarder? Ignore the kinetic energy of the skateboard's wheels.

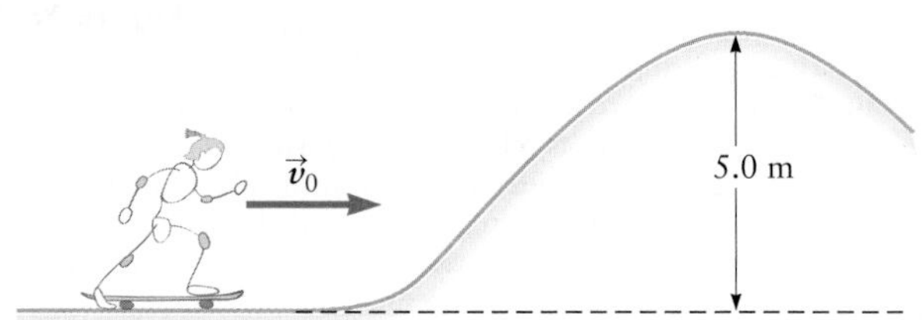

Figure P6.70

71. A skier of mass 100 kg starts from rest at the top of the ski slope of height 100 m shown in Figure P6.71. If the total work done by friction is -3.0×10^4 J, what is the skier's speed when he reaches the bottom of the slope?

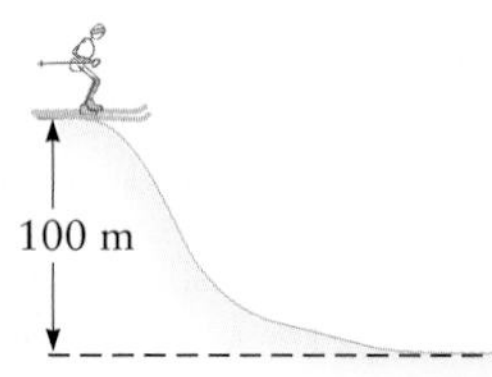

Figure P6.71

6.7 POWER

72. Every year, there is a foot race to the top of the Empire State Building. The vertical distance traveled (using the stairs!) is about 440 m. The current record for this race is about 9 min 30 s. If the record holder has a mass of 60 kg, what was his average power output during the race?

73. An electric motor is rated to have a maximum power output of 0.75 hp. If this motor is being used to lift a crate of mass 200 kg, how fast (i.e., at what speed) can it lift the crate? *Hint*: Assume the only other force on the crate is the force of gravity.

74. In Example 6.11, we compared the gravitational potential energy of a 20-W lightbulb to the electrical energy consumed when it is turned on. Repeat that calculation, but now assume the lightbulb is turned on for one day. Find the height h through which the lightbulb would have to fall for the change in gravitational potential energy to equal the electrical energy consumed by the lightbulb. Compare this answer to the height of Mount Everest. Assume you can use the relation $PE_{grav} = mgh$.

75. The main frictional force on a good bicycle is the force of air drag (Eq. 3.20). Consider a bicyclist coasting at a constant velocity on a road that is sloped downward at 5.0° with respect to the horizontal. (a) Find her approximate terminal speed. (b) Approximately how much work is done by air drag as she travels a distance of 5.0 km? *Hint*: She moves at a constant speed for essentially the entire trip. (c) What is the power "output" associated with the air drag force?

76. The rate of energy use in a typical house is about 2.0 kW. If the kinetic energy of a car (mass 1400 kg) is equal to the total energy used by the house in 1 min, what is the car's speed?

77. RT A typical car has a power rating of 150 hp. Estimate the car's maximum speed on level ground. Assume air drag is much larger than any frictional forces.

6.8 WORK, ENERGY, AND MOLECULAR MOTORS

78. Consider a molecular motor that has an efficiency of 50%. Find the maximum force the motor can produce. Assume a step size of 4 nm. Also assume the energy consumed for each step is 5×10^{-20} J (as given for an ATP-powered motor in Section 6.8).

79. Suppose a molecular motor consumes 3 ATP molecules in every step. If this motor has a step size of 8.0 nm, what is the maximum force the motor could exert?

80. Molecular motors are often studied by attaching them to small plastic spheres. These spheres enable the motor's motion to be observed with a microscope and are much larger than the motor. Consider a molecular motor to which a plastic sphere of radius 20 nm is attached. As the motor moves through water, a drag force given by Equation 3.23 acts on the sphere. If the molecular motor is able to produce a power output of 1×10^{-17} W, what is the speed of the motor as it pulls the sphere through water?

81. Consider a molecular motor that consumes the energy from 100 ATP molecules per second. What is the power output of this motor?

82. If the motor in Problem 81 takes a step of length 6.0 nm for each ATP molecule it consumes, what are the speed of the motor and the force it is able to produce?

83. SSM In Section 6.8, we calculated the force produced by a myosin molecular motor and found that it is approximately 10 piconewtons. Compare this force to the weight of a typical amino acid.

Additional Problems

84. SSM **Conceptualizing units.** A small burrito, like those bought at your local fast-food restaurant, has a mass of approximately 100 g and contains about 300 C. (The capital "C" indicates that this unit is a "food Calorie.") From what height would you need to drop the burrito to give it a kinetic energy equal to its dietary "energy"? Note that 1 food Calorie is equal to 1000 calories = 4186 J. Ignore air drag.

85. RT **Counting calories.** The chemical potential energy in foods is measured in units called Calories (notice the capital C). This "food Calorie" is equal to 1000 calories. For example, a typical apple contains approximately 75 Calories, which is actually 75 *kilo*calories. If all the energy in an apple were converted to gravitational potential energy, how high would you be lifted?

86. Repeat Problem 85, but replace the apple with a typical fast-food hamburger (approximately 400 C).

87. SSM A doughnut contains roughly 350 C (1.5×10^6 J) of potential energy locked in chemical bonds. Find the ratio of the potential energy in the doughnut to that in an equivalent volume of TNT (trinitrotoluene), which has a density of 1.65 g/cm³ and releases 4.7×10^6 J per kg of explosive. Aren't you glad the doughnut does not release all its potential energy at once?

88. One gallon of gasoline contains 3.1×10^7 calories of potential energy that are released during combustion. If 1 gal of gasoline can provide the force that moves a car through a displacement of 25 mi, what is the average force produced by the gasoline?

89. Tarzan (m = 74 kg) commutes to work swinging from vine to vine. He leaves the platform of his tree house and swings on the end of a vine of length L = 8.0 m. (a) If the platform is 1.9 m above the lowest point in the swing, what is the tension in the vine at the lowest point in the swing? (b) Tarzan again takes to his morning commute, but this time a monkey of mass m_M = 23 kg hitches a ride by jumping onto Tarzan's back. If a vine can withstand a maximum tension of 1200 N, will it snap under the tension of the added passenger? If so, at what angle with respect to the vertical does the vine break?

90. A toy gun shoots spherical plastic projectiles by means of a spring. A typical projectile has mass m = 25 g. The spring used has spring constant k = 15 N/m, and when put under load, it is displaced an amount Δx = 6.0 cm as shown in Figure P6.90. The barrel of the gun exerts a slight frictional force of magnitude $F_{friction}$ = 0.074 N on the pellet as it moves down the barrel from its starting point a total length L = 15 cm. If the toy gun is fired in a horizontal position, (a) at what position measured from the starting point does the pellet reach maximum velocity? *Hint*: Consider an equilibrium condition between the spring force and the friction force on the pellet before it leaves the spring. (b) What is the maximum speed achieved by the pellet as it is fired from the gun? (c) At what speed does the pellet leave the barrel?

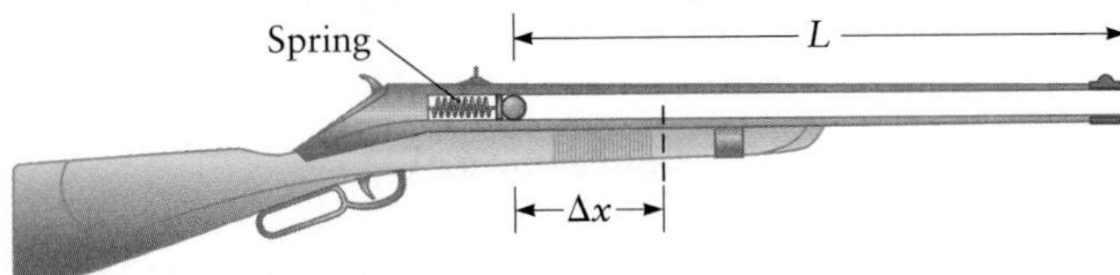

Figure P6.90

91. A pole vaulter of mass 70 kg can run horizontally with a top speed of 10.0 m/s. The current record height for the pole vault is about 6.2 m. (a) For the vaulter to clear this height, approximately how much energy must be stored in the pole just before he leaves the ground? Assume the vaulter's speed is zero just prior to taking off and that you can treat him as a point particle. (b) The energy in part (a) is the maximum energy stored in the pole, so we denote it by PE_{max}. How does PE_{max} compare to the maximum kinetic energy KE_{run} of the pole vaulter before he takes off? (c) You should find that PE_{max} is greater than KE_{run}, which suggests the vaulter is able to go higher than expected based on our simple analysis. Why does our simple analysis underestimate the maximum height the vaulter can clear? Did we overlook some source of kinetic or potential energy? Is the point particle assumption valid?

92. A spring is mounted at angle θ = 35° on a frictionless incline as illustrated in Figure P6.92. The spring is compressed by 15 cm where it is allowed to propel a mass of 5.5 kg up the incline. (a) If the spring constant is 550 N/m, how fast is the mass moving when it leaves the spring? (b) To what maximum distance from the starting point will the mass rise up the incline?

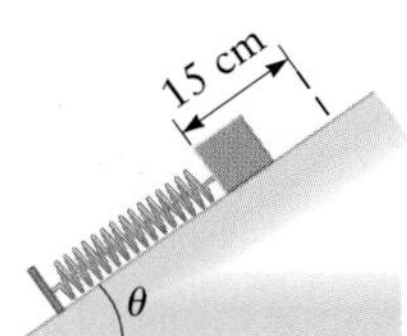

Figure P6.92 Problems 92 and 93.

93. Consider the system in Problem 92 and assume the coefficient of kinetic friction between the mass and the incline is 0.17. (a) At what position, measured from the starting point, does the maximum velocity occur? (b) What is the maximum velocity attained by the mass at this point? (c) What is now the maximum distance from the starting point that the mass will rise up the incline?

94. **Spring shoes.** Tae-Hyuk Yoon of South Korea is the inventor of a novel method of human locomotion: the Poweriser spring boot shown in Figure P6.94. Promotional materials claim that an 80-kg man can jump up to 2 m high while wearing these devices. Estimate the spring constant of the leaf springs on the Poweriser. The maximum compression can be estimated from the figure.

Figure P6.94 Poweriser spring boots in use.

95. A chairlift rises in elevation 600 m (about 2000 ft) up a ski slope and has a total of 200 chairs evenly spaced every 12 m (about 40 ft) apart. On a busy weekend, the ski lift is at full capacity (two people per seat), and skiers come off the lift every 10 s. Assuming a skier has a mass of 80 kg (including clothes and equipment) on average, what is the power output required for the ski lift to operate?

96. A 70-kg skydiver reaches a terminal velocity of 50 m/s. At what rate is the drag force dissipating his kinetic energy? How does this rate compare with his rate of change in potential energy?

97. The world record for the highest-altitude skydive was made by Joseph Kittinger in 1960. Kittinger jumped from a high-altitude balloon (Fig. P6.97) at a height of 102,800 ft (31,330 m) above sea level and, according to reports, attained speeds up to 624 mi/h (279 m/s) allowed by the thin air of the stratosphere. Calculate his maximum speed (i.e., neglect air drag) for his 14,500 ft (4420 m) of free fall. How does this speed compare with the reported value?

Figure P6.97

98. The NEAR spacecraft flew by the asteroid Eros in 1997 (Fig. P6.98). The flyby allowed the craft to measure the asteroid's density at $2.7\ g/cm^3$. Assume most asteroids have similar composition and you personally can jump 1.0 m high on the Earth. (a) What is the radius of the largest spherical asteroid you could literally jump off (in other words, that you could jump from with an initial speed equal to the escape speed of the asteroid)? (b) Eros has a mass of 6.7×10^{15} kg. Could you jump off Eros? If not, how far from the surface would you go before descending back to the asteroid? For simplicity, assume Eros is spherical.

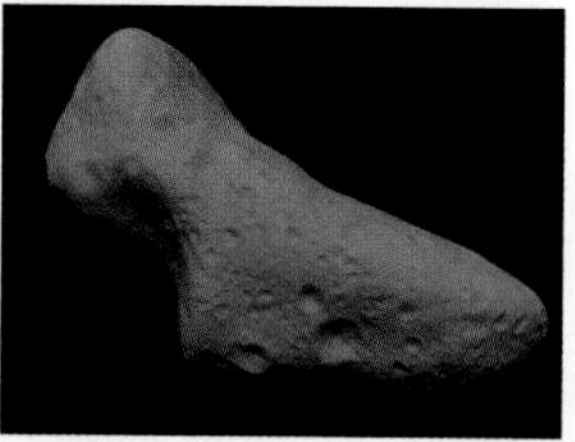
NASA/JHUAPL/NLR

Figure P6.98 Image of Eros asteroid from NEAR encounter.

99. **Aerobic workout.** After a 5-min workout on a Climb Max stair machine, the readout panel (Fig. P6.99) reports that the 75-kg user burned 19.7 Calories (see Problem 85) and climbed a total of 180 steps. (a) If each step is 15 cm in height, how much would the potential energy of the user have changed if she were actually climbing a stairway? (b) What was her rate of change in potential energy? (c) At what rate was her metabolism consuming energy? (d) What is her efficiency in converting the energy in the chemical bonds in her food into potential energy? (e) She really didn't change her vertical displacement on this machine, so where did that energy go?

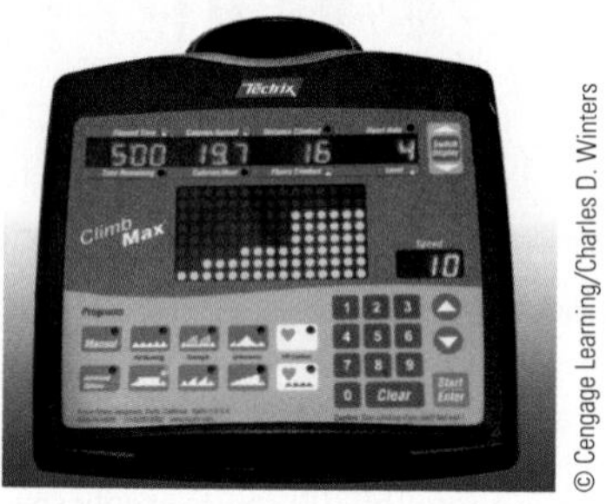

© Cengage Learning/Charles D. Winters

Figure P6.99 Results from a workout.

100. The expression for the force due to air drag described in Chapter 3 (Eq. 3.20) ignores the aerodynamic shape of an object. Some objects, such as a race car, are shaped so as to minimize the drag force. This can be accounted for by adding a factor called the drag coefficient C_D. The drag force then has a magnitude

$$F_{\text{drag}} = \tfrac{1}{2} C_D \rho A v^2$$

A boxy car might have a drag coefficient $C_D \approx 1.0$, whereas a race car might have $C_D \approx 0.25$. If all else is the same (the power produced by the engine, etc.) and there is no other source of friction or drag, what is the ratio of the race car's top speed to the boxy car's top speed?

101. SSM For a car moving with speed v, the force of air drag is proportional to v^2. If the power output of the car's engine is doubled, by what factor does the speed of the car increase?

102. **The Roche limit.** Some "bodies" in the solar system are a collection of rocks held together solely or mainly by their gravitational attraction. When such an object comes close to a planet, so-called tidal forces will break the object apart. Consider an asteroid composed of two pieces, each of mass $m = 1 \times 10^{16}$ kg, and each spherical with diameter $d = 2 \times 10^4$ m. Suppose this object approaches the Earth as shown in Figure P6.102. The two pieces of the asteroid are at slightly different distances from the Earth, so the Earth's gravitational force on the two pieces will be slightly different. At what distance R from the Earth will the Earth's gravitational force pull this asteroid apart? This distance is called the ***Roche limit,*** and it determines when asteroids or moons near planets such as Saturn break apart to form rings around the planet.

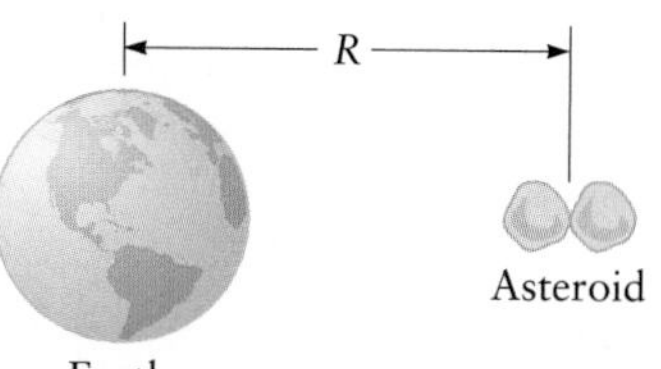

Figure P6.102

103. Collisions of commercial aircraft with birds are a serious safety hazard, especially when a plane is taking off or landing. Consider a typical goose that collides with an airplane taking off at 70 m/s (about 150 mi/h) and compare the impact with that of a 500-kg mass (about half the size of a small car) dropped from a height of 3 m. Which impact—the goose hitting the plane or the mass hitting the ground—will likely do more damage? Or will it be about the same? *Hint*: Compare the kinetic energy of the mass just before it hits the ground with the kinetic energy of a goose moving at 70 m/s.

CHAPTER 7

◀ *Collisions are important in many everyday situations. In this chapter, we will learn how to analyze different types of collisions. (TRL Ltd./Photo Researchers, Inc.)*

Momentum, Impulse, and Collisions

OUTLINE

Until now, we have treated all objects as "point particles," objects whose mass is located at a single point in space. The point particle assumption is very useful and is the correct way to deal with many situations. However, some important kinds of motion cannot be dealt with in this way. For example, objects such as baseballs and boomerangs can spin or rotate as they move along their trajectory. To deal with such motion, we must consider the ball or boomerang as an extended object, taking size and shape into account. In this and the next two chapters, we show how Newton's laws can be used to treat the motion of such extended objects. Conceptually, we will imagine that our baseball is composed of many very small "pieces." If the pieces are small enough, we can treat each as a point particle using the methods developed in previous chapters. The different pieces of an object exert forces on one another—when you push on one side of a baseball, the rest of the ball moves, too—so we must also deal with the forces that inevitably exist in such a *system* of interacting particles. We first encountered the notion of a system in Chapter 6, where we found that the total energy of a closed system (i.e., a system that is not acted on by any outside forces) is conserved. In this chapter, we introduce another property called ***momentum*** and discover that the total momentum of a closed system of particles is also conserved. This principle of ***conservation of momentum*** provides a very powerful tool in the analysis of motion in a wide variety of situations. It will lead us to the concept of ***center of mass*** and provide a general way to deal with collisions between particles.

7.1 Momentum

The variables velocity and acceleration are needed to describe the motion of a single particle. When dealing with a collection of particles, it is useful to define an additional quantity, ***momentum***. The momentum $\vec{p}$ of a single particle of mass m moving with a velocity $\vec{v}$ is defined as

The momentum of a particle depends on its mass and velocity.

$$\vec{p} = m\vec{v} \tag{7.1}$$

The momentum of a particle thus is along the same direction as the velocity. Notice also that $\vec{p}$ is proportional to the mass of the particle. We'll soon see how this relation for the momentum arises from Newton's second law.

EXAMPLE 7.1 The Momentum of a Bullet

Consider a bullet that has a mass of 10 g and a speed of 600 m/s. (**a**) Find the momentum of the bullet. (**b**) Find the speed of a baseball (m = 0.14 kg) that has the same momentum as the bullet.

RECOGNIZE THE PRINCIPLE

The magnitude of the momentum can be found directly from Equation 7.1, given the mass and speed of the objects of interest. The goal of this example is to get a sense of the momentum of two "everyday" objects.

SKETCH THE PROBLEM

No figure is necessary.

IDENTIFY THE RELATIONSHIPS

The speed and mass of the bullet are both given, so we can apply Equation 7.1,

$$\vec{p} = m\vec{v}$$

SOLVE

(**a**) In terms of magnitudes, the momentum of the bullet is

$$p = (0.010 \text{ kg})(600 \text{ m/s}) = \boxed{6.0 \text{ kg} \cdot \text{m/s}}$$

The direction of $\vec{p}$ is parallel to the velocity of the bullet.

(**b**) If a baseball has the same momentum as this bullet, we have

$$\vec{p}_{\text{ball}} = \vec{p}_{\text{bullet}}$$

In terms of the magnitudes, we then get

$$m_{\text{ball}}v_{\text{ball}} = m_{\text{bullet}}v_{\text{bullet}}$$

Hence, the speed of the ball is

$$v_{\text{ball}} = \frac{m_{\text{bullet}}v_{\text{bullet}}}{m_{\text{ball}}}$$

Inserting the numerical values given above, we find

$$v_{\text{ball}} = \frac{m_{\text{bullet}} v_{\text{bullet}}}{m_{\text{ball}}} = \frac{(0.010 \text{ kg})(600 \text{ m/s})}{0.14 \text{ kg}} = \boxed{43 \text{ m/s}}$$

▶ ***What does it mean?***

This speed is approximately 100 mi/h and is about the speed that can be reached by a professional baseball pitcher. The momentum of a particle is the *product* of its mass and velocity, so a particular value of $\vec{p}$ can be obtained with a small mass moving at a high velocity or with a large mass moving with a smaller velocity. The momentum carried by a baseball thrown by a major-league pitcher is about the same as the momentum of a bullet!

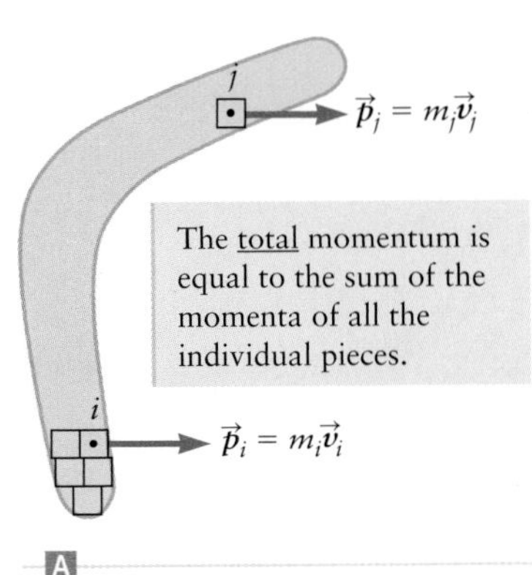

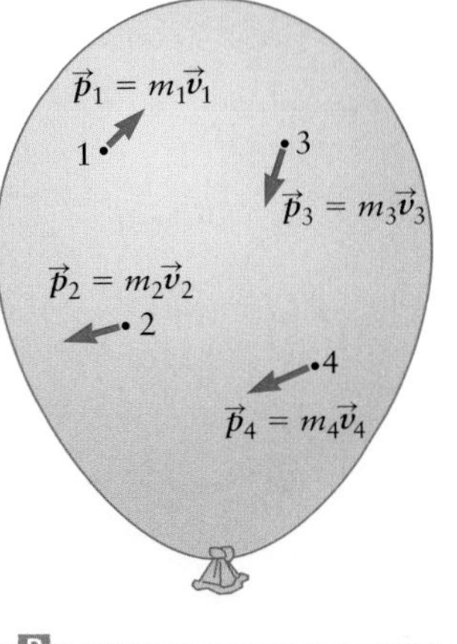

▲ **Figure 7.1** **A** A solid object such as a boomerang can be thought of as a system composed of very small pieces, that is, a collection of point particles. **B** The gas in a balloon is a system of molecules. The total momentum is the sum of the individual momenta of all the molecules.

Momentum of a System of Particles

To find the total momentum of a system of particles, we take the sum of the momenta of all the individual pieces of the system. For example, if we want to find the total momentum of a boomerang, we can imagine it as a collection of many small pieces as illustrated in Figure 7.1A. Each piece can be treated as a point particle, and the momentum of each piece is given by Equation 7.1. The total momentum of a collection of such pieces is then

$$\vec{p}_{\text{total}} = \Sigma_i\, \vec{p}_i = \Sigma_i\, m_i \vec{v}_i \tag{7.2}$$

where the sums are over all the individual particles (pieces) in the system. Such a system of particles may be the pieces of a solid object, such as a boomerang, or pieces of a nonsolid object, such as the molecules in a gas (Fig. 7.1B). In either case, the total momentum of the system is given by Equation 7.2.

EXAMPLE 7.2 Momentum of Two Cars

Two cars are on course for a collision in one dimension (Fig. 7.2). The small car has a mass of 1200 kg and a speed of 20 m/s, and the large car has a mass of 2000 kg and a speed of 15 m/s in the opposite direction. What is the total momentum of the two cars?

RECOGNIZE THE PRINCIPLE

We can find the momentum of each car using Equation 7.1 and then add the results to get the total momentum. We must include the sign of the velocity (+ or −) of each car when calculating their momenta.

SKETCH THE PROBLEM

Figure 7.2 describes the problem. The cars are both moving along the x axis.

IDENTIFY THE RELATIONSHIPS

To calculate the total momentum, we can treat each car as a point particle. We then have (Eq. 7.2)

$$\vec{p}_{\text{total}} = \Sigma_i\, \vec{p}_i = \Sigma_i\, m_i\vec{v}_i = m_1\vec{v}_1 + m_2\vec{v}_2$$

where m_1 and m_2 are the masses of the cars and $\vec{v}_1$ and $\vec{v}_2$ are the cars' velocities. In this example, the motion is one dimensional, so we can work in terms of the x components of $\vec{p}$ and $\vec{v}$. From Figure 7.2, we have $v_1 = +20$ m/s for the smaller car and $v_2 = -15$ m/s for the larger car. Here, the signs *do* matter.

(continued) ▶

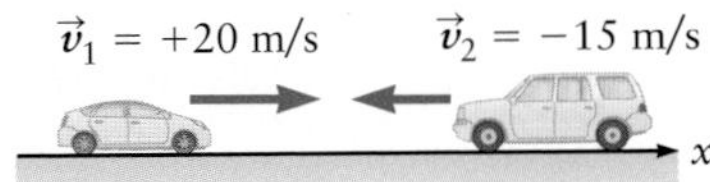

▲ **Figure 7.2** Example 7.2.

SOLVE

The total momentum of our system of two cars is thus

$$p_{\text{total}} = m_1 v_1 + m_2 v_2$$

$$p_{\text{total}} = (1200 \text{ kg})(20 \text{ m/s}) + (2000 \text{ kg})(-15 \text{ m/s}) = \boxed{-6000 \text{ kg} \cdot \text{m/s}}$$

► *What does it mean?*

Even though the smaller car has a higher speed, the magnitude of the momentum of the larger car is greater because of its greater mass. For this reason, the total momentum is in the $-x$ direction, that is, toward the left in Figure 7.2.

CONCEPT CHECK 7.1 Momentum and Kinetic Energy

Consider a rock of mass m moving with a speed v. Suppose the mass of the rock is doubled, while the speed is cut in half. Which of the following statements is true?

(a) The momentum of the rock is the same, but its kinetic energy is smaller.
(b) The momentum and kinetic energy of the rock are both unchanged.
(c) The momentum of the rock is greater, and its kinetic energy is smaller.

CONCEPT CHECK 7.2 Momentum and Kinetic Energy of Two Particles

Two particles of different mass have the same kinetic energy. Which one has the larger momentum?

(a) The particle with the smaller mass
(b) The particle with the larger mass
(c) They have the same momentum.

7.2 Force and Impulse

Suppose a force acts on a particle of mass m. Let's consider how this force affects the momentum of the particle. If $\vec{F}$ is the total force acting on the particle, from Newton's second law we then have $\vec{F} = m\vec{a}$. For the moment, we assume the force and hence also the acceleration are constant. During the course of a very short time interval Δt, the acceleration is related to changes in the velocity according to $\vec{a} = \Delta\vec{v}/\Delta t$. Combining these two facts gives

$$\vec{F} = m\vec{a} = m\frac{\Delta\vec{v}}{\Delta t} = m\frac{\vec{v}_f - \vec{v}_i}{\Delta t} \tag{7.3}$$

Here, we imagine that the particle has a velocity $\vec{v}_i$ just before the force is applied and that after a time Δt the velocity is $\vec{v}_f$. From the definition of momentum in Equation 7.1, the initial momentum of the particle is $\vec{p}_i = m\vec{v}_i$ and the final momentum is $\vec{p}_f = m\vec{v}_f$. Inserting this all into Equation 7.3 and rearranging gives

$$\vec{F}\,\Delta t = m(\vec{v}_f - \vec{v}_i) = \vec{p}_f - \vec{p}_i = \Delta\vec{p} \tag{7.4}$$

where $\Delta\vec{p}$ is the change in the momentum. The product $\vec{F}\,\Delta t$ is called the ***impulse***. According to Equation 7.4, when a force acts on a particle, the change in momentum of the particle is equal to the impulse:

Impulse theorem

$$\text{impulse} = \vec{F}\,\Delta t = \Delta\vec{p} \tag{7.5}$$

Equation 7.5 is called the ***impulse theorem***, and it follows directly from Newton's laws. Impulse is a vector, so its direction is parallel to the total force on the particle. Because impulse is the product of force and time, a particular value of the impulse can be obtained with a small force acting for a long time or with a large force acting for a short time.

Impulse Associated with a Variable Force

In our derivation of the relation between impulse and momentum in Equation 7.5, we assumed the force acts for a very short time Δt and the force is constant during this time interval as in Figure 7.3A. In many cases involving impulse and momentum, however, the force is not constant. Such a case is illustrated in Figure 7.4, which shows a tennis ball being struck by a racket. The force exerted by the racket on the ball has the form sketched in Figure 7.3B: the magnitude of the force grows from zero to some maximum value and then decreases to zero as the ball leaves the racket. For the simpler case of a constant force, the magnitude of the impulse is simply $F\,\Delta t$ (Eq. 7.5), which is the area under the corresponding force–time curve in Figure 7.3A. When the force is not constant, we can calculate the impulse by dividing the entire time period into many small intervals and using Equation 7.5 for each interval as illustrated in Figure 7.3B. The total impulse is then the sum of the impulses during each interval and is equal to the area under the complete force–time curve:

$$\text{impulse} = \text{area under the force–time curve} = \Delta\vec{p} \tag{7.6}$$

In both cases—with a constant force as in Figure 7.3A or a force that varies with time as in Figure 7.3B—the impulse is equal to the change in the momentum of the object. Hence, the impulse theorem we initially derived in Equation 7.5 for the case of a constant force also applies for the impulse produced by a force that varies with time.

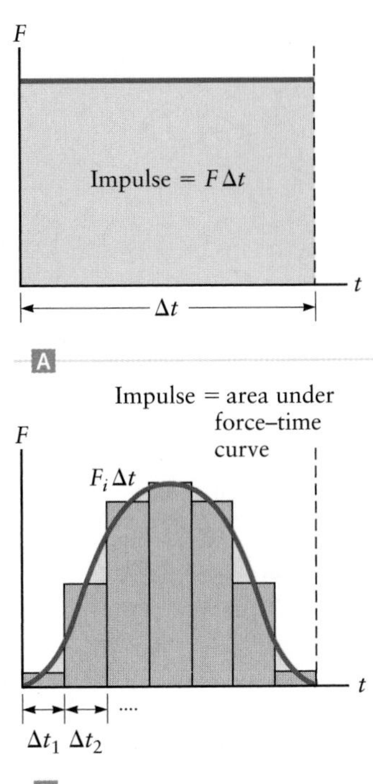

▲ **Figure 7.3** Impulse is equal to the area under the force−time graph. A Impulse for a constant force. B When the force varies with time, the impulse during a time interval Δt_i is approximately equal to the area of a rectangular bar of height F_i.

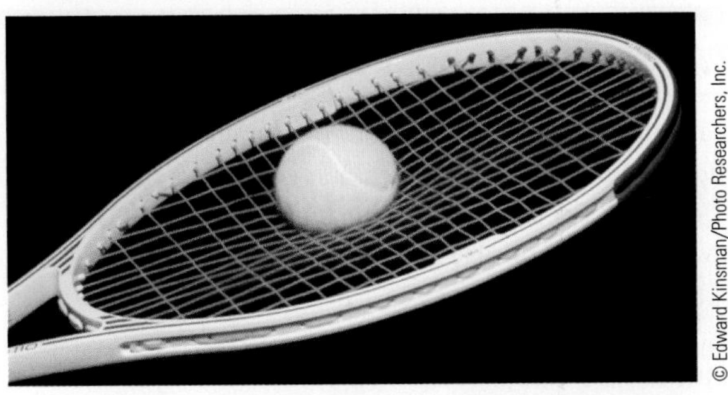

▲ **Figure 7.4** The force exerted by a tennis racket on a ball is usually not constant. In such cases, the impulse is still equal to the area under the force−time graph as illustrated in Figure 7.3B.

EXAMPLE 7.3 Impulse and Playing Golf

A golf ball ($m = 0.046$ kg) is hit from a tee (Fig. 7.5). If the ball has speed $v_f = 40$ m/s just after it is hit, find (a) the magnitude of the impulse imparted by the golf club on the ball and (b) the work done by the club on the ball.

RECOGNIZE THE PRINCIPLE

From the impulse theorem, the impulse imparted to the ball is equal to the change in momentum. We can find the change in momentum since the initial and final velocities are known. The work done on the ball is equal to the change in its kinetic energy.

SKETCH THE PROBLEM

Figure 7.5 shows the ball just after it leaves the tee, when its speed is v_f.

(continued) ▶

▲ **Figure 7.5** Example 7.3.

IDENTIFY THE RELATIONSHIPS AND SOLVE

(**a**) Using the impulse theorem (Eq. 7.5), we have

$$\text{impulse} = \Delta\vec{p} = \vec{p}_f - \vec{p}_i = m\vec{v}_f - m\vec{v}_i$$

The ball is initially at rest, so $\vec{v}_i = 0$. The magnitude of the impulse is thus

$$\text{impulse} = mv_f$$

Inserting the given mass and final speed, we get

$$\text{impulse} = mv_f = (0.046 \text{ kg})(40 \text{ m/s}) = \boxed{1.8 \text{ kg} \cdot \text{m/s}}$$

(**b**) According to the work–energy theorem, the work done on the ball equals the change in kinetic energy. The initial kinetic energy is zero, so

$$W = \Delta KE = \tfrac{1}{2}mv_f^2 - \tfrac{1}{2}mv_i^2 = \tfrac{1}{2}mv_f^2$$

Inserting the given values of m and v_f gives

$$W = \tfrac{1}{2}mv_f^2 = \tfrac{1}{2}(0.046 \text{ kg})(40 \text{ m/s})^2 = \boxed{37 \text{ J}}$$

► ***What does it mean?***

The initial momentum is zero, so the direction of the impulse is parallel to the final velocity $\vec{v}_f$ in Figure 7.5.

EXAMPLE 7.4 Impulse and a Home Run

Consider a baseball player who hits a baseball as sketched in Figure 7.6. The ball travels from the pitcher toward the batter with a speed of 45 m/s (about 100 mi/h). The batter then hits it directly back at the pitcher with a speed of 50 m/s. If the ball has a mass of 0.14 kg, what is the impulse imparted by the bat to the baseball?

RECOGNIZE THE PRINCIPLE

We are given the initial and final velocities of the baseball, so we can find the impulse using Equation 7.5. We must be careful to use the correct signs (+ and −) for the velocities.

SKETCH THE PROBLEM

Figure 7.6 shows the problem. The initial velocity is along $-x$ because the ball is moving to the left, and the final velocity is positive (to the right along $+x$).

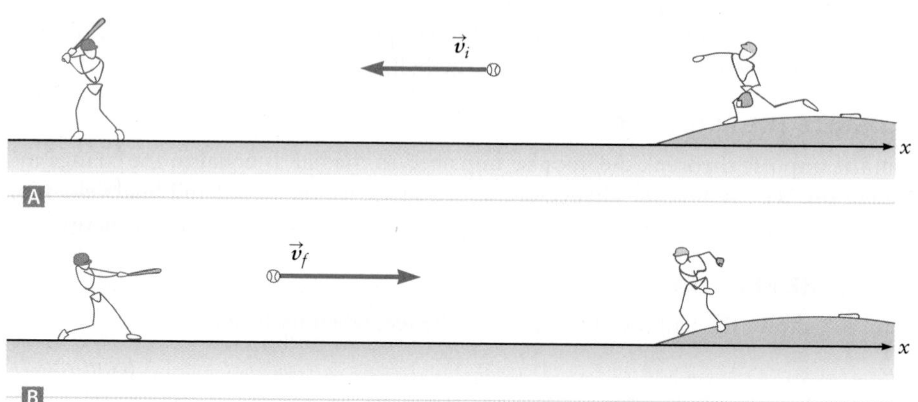

► **Figure 7.6** Example 7.4. **A** The ball has some initial momentum as it travels to the batter and **B** some final value of momentum after being hit by the bat. The impulse imparted to the baseball equals the change in its momentum. Here, $\Delta\vec{p}$ is equal to the final momentum minus the initial momentum.

IDENTIFY THE RELATIONSHIPS

We begin with the impulse theorem:

$$\text{impulse} = \Delta\vec{p} = m\vec{v}_f - m\vec{v}_i \quad (1)$$

This problem is one dimensional since the initial and final velocities are both along the line that extends from the batter to the pitcher, the x direction in Figure 7.6. The force exerted by the bat on the ball is also parallel to this direction. We can thus work simply in terms of the components of the velocity and momentum vectors along x. Using the coordinate system in Figure 7.6, we have $v_i = -45$ m/s and $v_f = +50$ m/s.

SOLVE

Inserting our values of m, v_i, and v_f into Equation (1), the impulse is

$$\Delta p = mv_f - mv_i = (0.14 \text{ kg})(50 \text{ m/s}) - (0.14 \text{ kg})(-45 \text{ m/s}) = \boxed{13 \text{ kg} \cdot \text{m/s}}$$

▶ ***What does it mean?***

The initial velocity and the initial momentum are both in the $-x$ direction. The bat then imparts a large positive impulse to the ball, so the final momentum of the ball is positive, directed toward the right in Figure 7.6B.

Impulse and the Average Force

For cases such as the force exerted by a tennis racket on a ball in Figure 7.4, it is usually very difficult to calculate the precise form of the force–time curve, so it is also difficult to apply Equation 7.6 to find the impulse. However, analyzing the impulse in a graphical manner is very useful in a different way. Often, the interaction time—the length of time during which the force is nonzero—is very short. For example, when a baseball is hit by a bat, high-speed photography (Fig. 7.7) shows that the bat and ball are in contact for approximately $\Delta t = 0.001$ s. We can approximate the force–time curve by assuming F is constant during this short interaction time and write

$$\text{impulse} = \vec{F}_{\text{ave}}\,\Delta t = \Delta\vec{p} \quad (7.7)$$

where $\vec{F}_{\text{ave}}$ is the *average* interaction force as indicated graphically in Figure 7.8. The value of F_{ave} is chosen so that the area under the rectangular region bounded by F_{ave} in Figure 7.8 is equal to the area under the true force–time curve. In many cases, this average force is all we are interested in. For example, if we assume the baseball in Example 7.4 is in contact with the bat for a time $\Delta t = 0.001$ s, the average force is related to the change in momentum by

$$\vec{F}_{\text{ave}}\,\Delta t = \Delta\vec{p}$$

We can then use the result for the impulse found in Example 7.4 to calculate the size of this force. In terms of the magnitudes, we get

$$F_{\text{ave}} = \frac{\Delta p}{\Delta t} = \frac{13 \text{ kg} \cdot \text{m/s}}{0.001 \text{ s}} \approx 1 \times 10^4 \text{ N} \quad (7.8)$$

The force on the baseball is thus *extremely* large. The weight of the ball is $mg = (0.14 \text{ kg})g = 1.4$ N, so the force exerted by the bat in Equation 7.8 is about 10,000 times the ball's weight! Such collision forces are often very large because the corresponding interaction times are usually very short.

▲ **Figure 7.7** A baseball is in contact with a bat for approximately 1 ms (= 0.001 s). During this short time, the force exerted by the bat on the ball is very large.

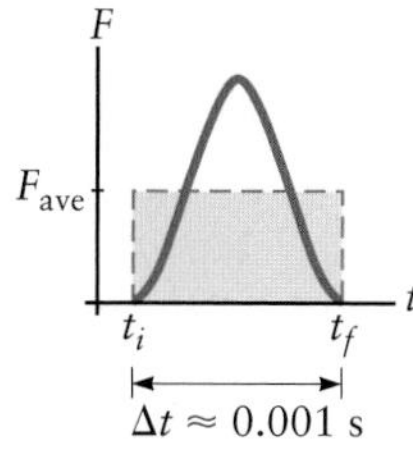

▲ **Figure 7.8** We can find the average force associated with an impulse by approximating the force–time relation by a constant (F_{ave}) during the time interval over which the force acts. The dashed horizontal line shows the (constant) average force over this time interval.

Minimizing Collision Forces

In Section 3.6, we considered a child jumping from a tall ladder to the ground below and analyzed the forces on the child's legs during the landing. We can also view that example as an impulse problem. The child has a certain initial momentum just before she hits the ground, and her final momentum is zero because she comes to rest. The

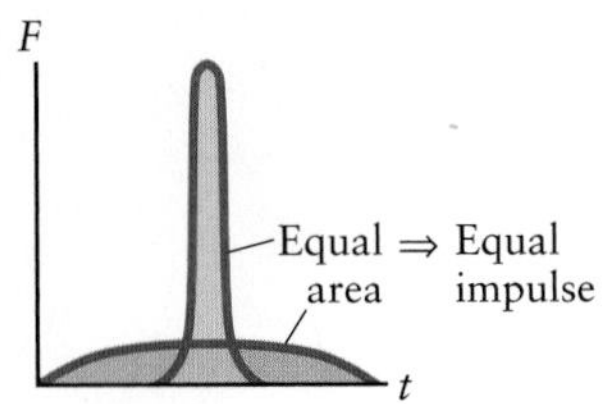

▲ **Figure 7.9** Impulse is equal to the area under the force–time curve. A particular value of the impulse can be produced by a small force acting for a long time or by a large force acting for a short time.

impulse exerted by the ground on the child can be calculated in the same way we found the impulse on the batted baseball in Example 7.4. In Chapter 3, we discussed strategies for minimizing the force on the child, and we found that it is safest for her to allow her legs to bend and flex as much as possible. By doing so, she is making the interaction time Δt as long as possible. Our understanding of impulse and force (Eq. 7.7) also explains why increasing Δt will reduce the average force F_{ave}. The effect of Δt on the force is illustrated graphically in Figure 7.9. A particular value of the impulse (the area under the force–time curve) can be obtained with a large force acting for a short time or with a small force acting for a long time. There are many other examples of this relationship between force and collision time. For example, the air bags in your car are designed to take advantage of this trade-off. When your car stops suddenly, as in an accident, the air bags inflate and fill the region between you (the driver) and the steering wheel and front windshield. Your collision with an air bag then involves a much longer interaction time than if you were to collide with the steering column or windshield. This longer collision time leads to a smaller force, and a safer accident.

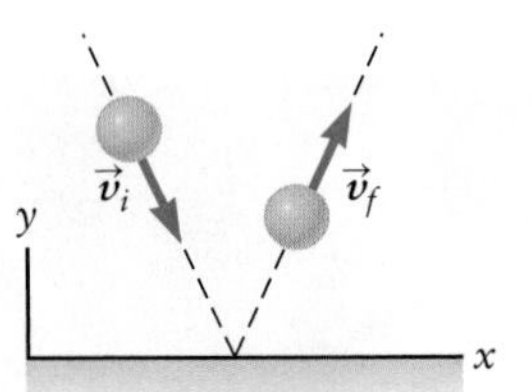

▲ **Figure 7.10** Concept Check 7.3.

CONCEPT CHECK 7.3 Impulse on a Bouncing Ball

Consider the collision of a bouncing ball with a floor as sketched in Figure 7.10. What is the direction of the impulse imparted by the floor to the ball?

(a) Horizontal and to the right in Figure 7.10, $\Delta\vec{p} = \rightarrow$

(b) Vertical and upward, $\Delta\vec{p} = \uparrow$

(c) At 45° from the vertical direction, $\Delta\vec{p} = \nearrow$

7.3 Conservation of Momentum

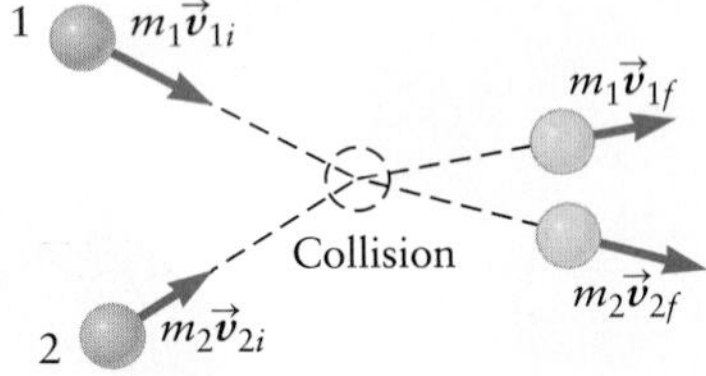

▲ **Figure 7.11** When two objects collide, the total momentum just after the collision is equal to the total momentum just before the collision. Momentum is conserved.

It is useful to apply our ideas about momentum and the impulse theorem to the motion of a system of particles. Consider the system comprised of just two colliding point particles sketched in Figure 7.11. For the moment, we ignore any rotational or spinning motion. With this assumption, let's use the notion of impulse to analyze how a collision affects the subsequent motion of the two particles.

When two objects collide, there is an associated collision force. Let $\vec{F}_1$ be the force exerted by object 2 on object 1 and $\vec{F}_2$ be the force exerted by object 1 on object 2. These forces are an action–reaction pair, so according to Newton's third law, they must be equal in magnitude and opposite in direction. Hence,

$$\vec{F}_1 = -\vec{F}_2 \tag{7.9}$$

If this collision takes place over a time interval Δt, the impulse imparted to object 1 is $\vec{F}_1\,\Delta t$. According to the impulse theorem, this impulse equals the change in momentum of object 1, so we have

$$\Delta\vec{p}_1 = \vec{p}_{1f} - \vec{p}_{1i} = \vec{F}_1\,\Delta t \tag{7.10}$$

Here, $\vec{p}_{1i}$ is the (initial) momentum of object 1 just before the collision and $\vec{p}_{1f}$ is the (final) momentum just after the collision. Likewise, the impulse imparted to object 2 is $\vec{F}_2\,\Delta t$, which equals the change in the momentum of object 2:

$$\Delta\vec{p}_2 = \vec{p}_{2f} - \vec{p}_{2i} = \vec{F}_2\,\Delta t \tag{7.11}$$

Since $\vec{F}_1 = -\vec{F}_2$ and the interaction times Δt are the same in Equations 7.10 and 7.11, the impulse imparted to object 1 is equal in magnitude but opposite in sign to the impulse imparted to object 2. We thus have

$$\Delta\vec{p}_1 = \vec{F}_1\,\Delta t = -\vec{F}_2\,\Delta t = -\Delta\vec{p}_2$$

$$\Delta\vec{p}_1 = -\Delta\vec{p}_2 \tag{7.12}$$

We can also write Equation 7.12 as

Conservation of momentum in a collision

$$\Delta\vec{p}_1 + \Delta\vec{p}_2 = \Delta\vec{p}_{total} = 0 \tag{7.13}$$

Hence, the collision does *not* change the *total* momentum of the two particles. Whatever momentum is lost or gained by one particle in the collision is gained or lost by the other. The *total momentum of the system is conserved.*

Conservation of Momentum for a System of Many Particles

Our derivation of the conservation of momentum in Equation 7.13 was based on a simple system of just two particles undergoing a single collision. We can apply the same argument to a system containing a very large number of particles, in which the particles undergo many collisions with one another. Further, let's assume this system is *closed*, meaning that there are no external forces exerted on any of the particles; that is, the only forces on a particle are exerted by other particles within the system. These internal forces are due to collisions between particles in the system; each collision can be viewed as in Figure 7.11, and the sum of the momenta of the two particles is the same before and after the collision (Eq. 7.13). The same argument applies to all the collisions between particles in the system, so we conclude that the total momentum of the entire system is conserved.

This result is very similar to what we found when we considered the total energy of a closed system in Chapter 6. Recall that the particles in a closed system can exchange energy among one another in various ways, but the particles do not gain energy from or lose energy to objects outside the system. The total energy of the system is thus conserved. The same notion applies to the total momentum. The particles in a closed system do not experience any forces from objects external to the system. The momentum gained or lost by one particle in a collision is lost or gained by other particles in the system, and the total momentum of the system is conserved.

This analysis has so far assumed the system is composed of a collection of particles that are traveling more or less independently, except for occasional collisions. It also applies to a solid object such as a boomerang, a football, or a planet. We can think of a solid object as being composed of many point particles; these individual particles are subject to strong forces from other particles in the system that hold the object together. These forces will always come in action–reaction pairs, so for any two interacting particles, we always have $\vec{F}_1 = -\vec{F}_2$, just as in Equation 7.9. The same reasoning then tells us that any momentum lost by one particle will be gained by the other. Hence, we again conclude that the total momentum of a system, in this case a solid object, is conserved, provided there are no external forces acting on the object.

Momentum Conservation and External Forces

We have just shown that for a closed system of particles, the interaction forces between particles do not change the total momentum of the system. These interactions cause some particles to lose momentum while others gain momentum, but the total momentum of the system stays the same. However, these arguments apply only to the internal forces, that is, the forces that act between particles within the system. There may also be forces acting from outside the system, called external forces. As an example, consider again the two-particle system in Figure 7.11 and assume these two particles are asteroids in outer space. If these two asteroids are near a star, the gravitational force from the star is an external force on this two-asteroid system. This external force causes the asteroids to accelerate toward the star and hence changes the total momentum of the two asteroids.

In practice, it is difficult to find situations in which the external forces on a system are exactly zero. For example, when dealing with any system of terrestrial objects, the Earth's gravitational force is always present as an external force. However, our analysis of the collision between a baseball and a bat (Eq. 7.8) showed that collision forces are often much larger than this gravitational force. For the baseball–bat collision, the collision force was about 10,000 times larger than the weight of the ball. Thus, while the total momentum of the bat–baseball system is not conserved precisely during this collision, it will change by only about 0.01% (1 part in 10,000).

For this reason, the notion of momentum conservation is still a very useful way to analyze most collisions.

7.4 Collisions

In this section, we apply the principle of momentum conservation to analyze several different types of collisions. In general, a collision changes the particle velocities; the velocities just after the collision are different from the velocities just before the collision. The kinetic energy of a particle is proportional to the square of its speed, so the kinetic energy of any particular particle is also changed by the collision. Collisions fall into two general types, depending on what happens to the total kinetic energy of the two particles involved. In an ***elastic collision***, the system's kinetic energy is conserved: the total kinetic energy of the two particles after the collision is equal to their total kinetic energy before the collision. In contrast, ***inelastic collisions*** are collisions in which some kinetic energy is lost. In both types of collisions, *momentum is conserved.*

The term *elastic* can help us understand how and why a collision affects the kinetic energy. For example, an extremely elastic ball (such as a Super Ball) is compressed in a collision, and this compression stores energy in the ball just as energy is stored in a compressed spring. In an ideal Super Ball, all this potential energy is turned back into kinetic energy when the ball uncompresses ("springs back") at the end of the collision. Collisions involving such objects will therefore tend to be elastic. On the other hand, consider a collision involving a ball composed of soft putty or clay. Such a ball does not "spring back" at the end of the collision, causing the kinetic energy after the collision to be smaller than the initial kinetic energy. The collision is thus inelastic.

We next analyze a few examples of these different types of collisions and show how conservation principles regarding momentum (for all collisions) and kinetic energy (for elastic collisions) can be used to predict the motions of two colliding particles. Our analyses will be guided by the following general approach.

PROBLEM SOLVING Analyzing a Collision

1. **RECOGNIZE THE PRINCIPLE.** The momentum of a system is conserved only when the external forces are zero. For a collision between two particles, the system is just the two particles. When external forces are zero or negligible, one can use the principle of conservation of momentum to analyze a collision.
2. **SKETCH THE PROBLEM.** Make a sketch of the system, showing the coordinate axes and (where possible) the initial and final velocities of the particles in the system.
3. **IDENTIFY THE RELATIONSHIPS.**
   - Express the conservation of momentum condition (Eq. 7.13) for the system.
   - Is kinetic energy conserved?

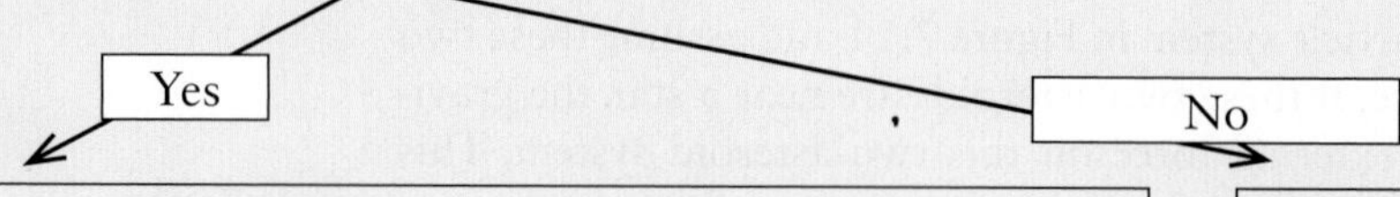

Elastic collision

4. **SOLVE.**
   - Express the conservation of kinetic energy for both particles.
   - Use the equations describing conservation of momentum and kinetic energy to solve for unknown quantities, such as the final velocities.

Inelastic collision

4. **SOLVE** for unknown quantities using the equations derived from the conservation of momentum.

5. Always *consider what your answer means* and check that it makes sense.

Elastic Collisions in One Dimension

Let's apply the above strategy for dealing with collisions to an elastic collision in one dimension involving particles with masses m_1 and m_2 that are subject to no external forces. Step 1: Because the external forces are zero, the momentum of the system (the two colliding particles) will be conserved. Step 2 is the sketch in Figure 7.12, which shows the particles and their velocities before and after the collision. The particles are moving along the x axis before the collision with velocities v_{1i} and v_{2i}; since this problem is one dimensional, we only need to be concerned with the x components of the velocities. After the collision, the velocities are v_{1f} and v_{2f}. The total momentum of the two particles is conserved, so we have

$$\overbrace{m_1v_{1i} + m_2v_{2i}}^{\text{initial momentum}} = \overbrace{m_1v_{1f} + m_2v_{2f}}^{\text{final momentum}} \tag{7.14}$$

This is an elastic collision, so the total kinetic energy of the system is also conserved, which gives

$$\overbrace{\tfrac{1}{2}m_1v_{1i}^2 + \tfrac{1}{2}m_2v_{2i}^2}^{\text{initial kinetic energy}} = \overbrace{\tfrac{1}{2}m_1v_{1f}^2 + \tfrac{1}{2}m_2v_{2f}^2}^{\text{final kinetic energy}} \tag{7.15}$$

In a typical collision, the initial velocities are known, and we have two equations—Equations 7.14 and 7.15—to solve for the two unknowns, the final velocities v_{1f} and v_{2f}.

Insight 7.1

ALWAYS BEGIN BY IDENTIFYING THE SYSTEM

When applying the principle of conservation of momentum, it is important to first identify the system. It is usually best to choose the system so that all the important forces act between different parts of the system and the external forces are zero or at least small. If the external forces are zero, the total momentum of the system will be conserved exactly.

Elastic collision in one dimension

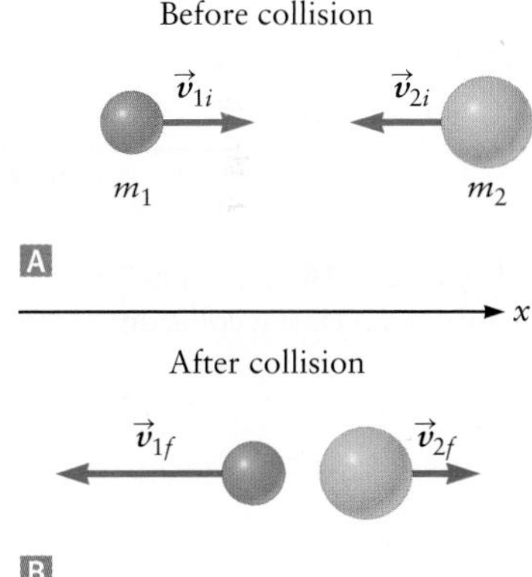

▲ **Figure 7.12** In an elastic collision, both momentum and kinetic energy are conserved. A Velocities just before a one-dimensional collision. B Velocities just after the collision.

A Collision between Two Billiard Balls

Let's now use Equations 7.14 and 7.15 to analyze a one-dimensional elastic collision between two billiard balls, assuming one of the balls is initially at rest and the other has an initial speed v_0 as sketched in Figure 7.13A. Because our problem is one dimensional, the collision is "head-on" and the initial and final velocities are all along the same direction. For simplicity we'll ignore all effects of spin in this problem. In the game of billiards, all balls have the same mass m. Inserting $v_{1i} = v_0$ and $v_{2i} = 0$ into our relations for the momentum (Eq. 7.14) and the kinetic energy (Eq. 7.15), we find

$$mv_0 = mv_{1f} + mv_{2f} \tag{7.16}$$

$$\tfrac{1}{2}mv_0^2 = \tfrac{1}{2}mv_{1f}^2 + \tfrac{1}{2}mv_{2f}^2 \tag{7.17}$$

Solving for v_{2f} in Equation 7.16 gives

$$v_{2f} = v_0 - v_{1f} \tag{7.18}$$

and inserting this into Equation 7.17, we get

$$\tfrac{1}{2}mv_0^2 = \tfrac{1}{2}mv_{1f}^2 + \tfrac{1}{2}m(v_0 - v_{1f})^2$$

Canceling the factors of $\frac{1}{2}m$ and solving for v_{1f} leads to

$$v_0^2 = v_{1f}^2 + (v_0 - v_{1f})^2 = v_{1f}^2 + v_0^2 - 2v_0v_{1f} + v_{1f}^2$$

Subtracting v_0^2 from each side and collecting terms gives

$$0 = v_{1f}^2 - 2v_0v_{1f} + v_{1f}^2 = 2v_{1f}^2 - 2v_0v_{1f}$$

$$2v_{1f}^2 = 2v_0v_{1f} \tag{7.19}$$

This equation has two solutions. One solution is $v_{1f} = 0$, and using Equation 7.18 then gives $v_{2f} = v_0$. Physically, ball 1 comes to a complete stop after the collision and ball 2 moves away with a velocity of v_0 afterward (Fig. 7.13B). This solution is familiar to many pool players.

The second solution of Equation 7.19 is $v_{1f} = v_0$, and using Equation 7.18 leads to $v_{2f} = 0$. Hence, with this solution, ball 2 stays at rest and ball 1 simply passes through it. You can think of this as a collision between two transparent balls. While you may not have expected this solution for a collision problem, it certainly does

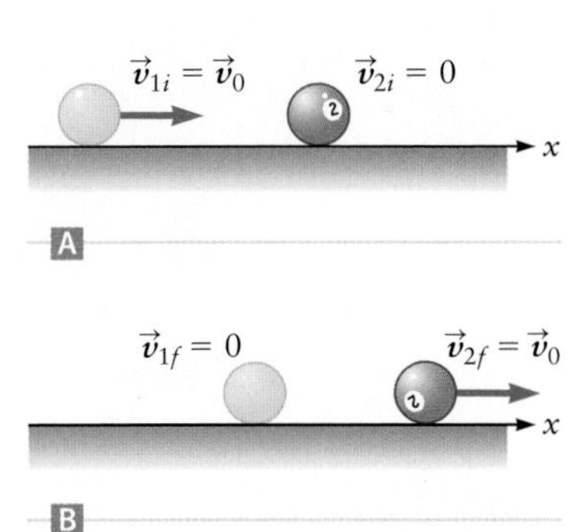

▲ **Figure 7.13** Elastic collision in one dimension between two billiard balls. A One of the balls is initially at rest. B After the collision, the other ball is at rest.

conserve both kinetic energy and momentum. This solution would only apply if the collision force between the two balls is zero, and that does not occur in any real collision between two objects.

EXAMPLE 7.5 Elastic Collision between Balls of Different Mass

Consider again an elastic collision between two balls with one initially at rest (Fig. 7.13A), but this time assume the balls have different masses. Repeat the analysis of this collision and find the final velocities of both balls.

RECOGNIZE THE PRINCIPLE

We can again apply conservation of momentum (Eq. 7.14) and conservation of kinetic energy (Eq. 7.15). Our goal is to understand how the relative masses of the particles affect the final velocities. We follow the steps outlined in the problem-solving box on analyzing a collision. Step 1: The external forces are zero, so the momentum of the colliding particles will be conserved.

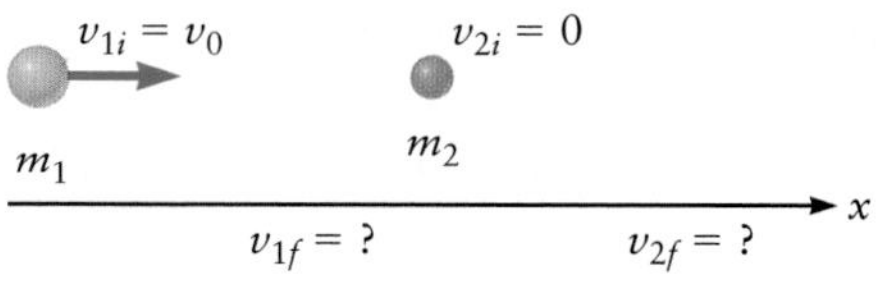

SKETCH THE PROBLEM

Step 2: Figure 7.14 shows the problem. We denote the masses as m_1 and m_2. One of them is initially at rest, so the velocities before the collision are $v_{1i} = v_0$ and $v_{2i} = 0$. The final velocities (denoted v_{1f} and v_{2f}) are unknowns that we wish to find. This information is all collected in Figure 7.14.

▲ **Figure 7.14** Example 7.5. Analyzing an elastic collision.

IDENTIFY THE RELATIONSHIPS

We continue with our problem-solving strategy. Step 3: The conservation of momentum condition gives

$$\overbrace{m_1v_{1i} + m_2v_{2i}}^{\text{initial momentum}} = m_1v_0 = \overbrace{m_1v_{1f} + m_2v_{2f}}^{\text{final momentum}} \tag{1}$$

We are given that kinetic energy is conserved, so we have an elastic collision.

SOLVE

Step 4: Setting the initial and final kinetic energies equal, we find

$$\overbrace{\tfrac{1}{2}m_1v_{1i}^2 + \tfrac{1}{2}m_2v_{2i}^2}^{\text{initial } KE} = \tfrac{1}{2}m_1v_0^2 = \overbrace{\tfrac{1}{2}m_1v_{1f}^2 + \tfrac{1}{2}m_2v_{2f}^2}^{\text{final } KE} \tag{2}$$

We next rearrange the momentum conservation relation (Eq. 1) to solve for the final velocity of mass 2:

$$v_{2f} = \frac{m_1(v_0 - v_{1f})}{m_2} \tag{3}$$

Substituting this result into the conservation of kinetic energy expression, Equation (2):

$$\frac{1}{2}m_1v_0^2 = \frac{1}{2}m_1v_{1f}^2 + \frac{1}{2}m_2\left[\frac{m_1(v_0 - v_{1f})}{m_2}\right]^2$$

We can now solve for the final velocity of mass 1 and then use Equation (3) to find the final velocity of mass 2. After some algebra, we find two solutions. Solution 1 is

$$v_{1f} = v_0 \quad \text{and} \quad v_{2f} = 0$$

In this solution, there is no collision at all because mass 1 simply passes through mass 2; that is, the final velocity of mass 1 is equal to its initial velocity, while mass 2 is at rest before and after the collision. This particular solution is only possible if the collision force between the two balls is zero, which does not occur in a real collision. There is a second solution for the final velocities that does describe a real collision. It is

$$v_{1f} = \left(\frac{m_1 - m_2}{m_1 + m_2}\right)v_0 \quad \text{and} \quad v_{2f} = \left(\frac{2m_1}{m_1 + m_2}\right)v_0$$

► ***What does it mean?***

These results for the final velocities of the two balls correspond to an actual collision, and we explore some of their consequences in the end-of-chapter problems. One important case is when m_2 is very small compared with m_1, which leads to

$$v_{1f} = \left(\frac{m_1 - m_2}{m_1 + m_2}\right)v_0 \approx \left(\frac{m_1 - 0}{m_1 + 0}\right)v_0 = v_0$$

Hence, the final velocity of a very massive particle (m_1) is approximately unchanged by the collision with a very light particle (m_2). When a car collides with a mosquito, the car's velocity barely changes.

The Power of Conservation Principles

In dealing with the elastic collisions in Figures 7.12 through 7.14, we started with the conditions for conservation of momentum and conservation of kinetic energy. For a one-dimensional collision, this process leads to two equations (Eqs. 7.14 and 7.15). In a typical problem, we might know the initial velocities of the two objects and be asked to find their final velocities. Hence, the two equations are all we need to solve for these two unknowns. This result is remarkable: it means that we can work out the outcome of the collision without knowing *anything* about the force that acts between the two objects during the collision. For an elastic collision in one dimension, the conditions for conservation of kinetic energy and momentum *completely* determine the results. This example illustrates that conservation laws are an extremely powerful way to get from general principles to specific results.

Inelastic Collisions in One Dimension

We have just analyzed some collisions in which kinetic energy was conserved. However, in many collisions the total kinetic energy after a collision is smaller than the kinetic energy just prior to the collision. These collisions are called *inelastic*. Although some portion of the kinetic energy of the two colliding objects is "lost" in such a collision, the *total* energy of the universe is still conserved. This "lost" kinetic energy goes into other forms of energy, such as potential energy and heat energy. Although kinetic energy is not conserved in an inelastic collision, the total momentum is still conserved. Hence, for a one-dimensional inelastic collision, we can still use the conservation of momentum relation

$$m_1v_{1i} + m_2v_{2i} = m_1v_{1f} + m_2v_{2f} \tag{7.20}$$

However, this equation does not give enough information to allow us to solve for the final velocities of both objects. We have only one relation, but two unknown quantities, v_{1f} and v_{2f}. So, to analyze an inelastic collision fully, we need additional information.

Completely Inelastic Collisions

The simplest type of inelastic collision is one in which the two objects stick together after the collision so that the objects have the same final velocity. This is called a ***completely inelastic collision***. A good example of a completely inelastic collision is one in which two cars lock bumpers. Figure 7.15A shows two cars of mass m_1 and m_2 coasting on a straight, one-dimensional road with velocities v_{1i} and v_{2i}. The cars collide and lock bumpers, and they have the same velocity v_f after the collision (Fig. 7.15B). The cars are moving in a horizontal direction (x), and if they are coasting, there are no external forces on the cars in this direction. Hence, the total momentum along x is conserved and we can apply our problem-solving strategy for dealing with collisions. Writing the condition for conservation of momentum along x gives

$$m_1v_{1i} + m_2v_{2i} = m_1v_f + m_2v_f \tag{7.21}$$

In this case, we have only one unknown, the final velocity v_f, so the conservation of momentum relation in Equation 7.21 is all we need. Solving for v_f, we have

$$m_1v_{1i} + m_2v_{2i} = (m_1 + m_2)v_f$$

Completely inelastic collision in one dimension

$$v_f = \frac{m_1v_{1i} + m_2v_{2i}}{m_1 + m_2} \tag{7.22}$$

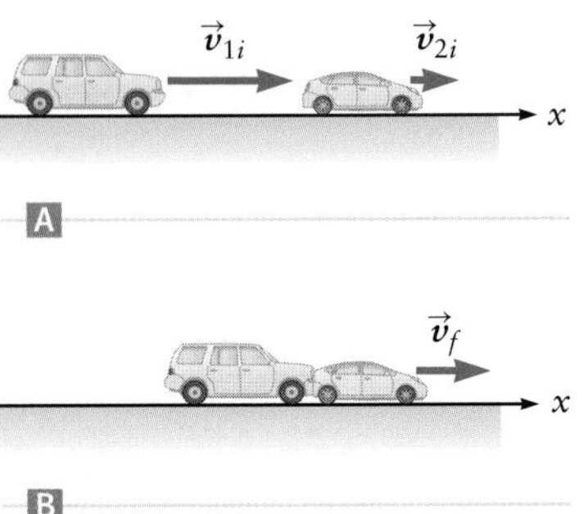

▲ **Figure 7.15** A Two cars on a collision course in one dimension. B In a completely inelastic collision, the two objects stick together after the collision.

EXAMPLE 7.6 A Completely Inelastic Collision between Two Cars

Two cars, an SUV of mass 2500 kg and a compact model of mass 1200 kg, are coasting along a straight road. Their initial velocities along this coordinate direction are 40 m/s and 10 m/s, respectively, so the SUV is overtaking the compact car. When they collide, the cars lock bumpers. Find the cars' velocity just after the collision.

RECOGNIZE THE PRINCIPLE

The cars "stick together" after the collision, so this collision is completely inelastic, and we can use our problem-solving strategy for collisions. Step 1: We first note that the two cars are the system. Since they are both coasting, there are no external forces along the horizontal direction (x), and the total momentum along x is conserved.

SKETCH THE PROBLEM

Step 2: The collision is described in Figure 7.15. It is a one-dimensional collision, so the given initial velocities are the velocity components along our coordinate direction (x), which we choose as parallel to the road.

IDENTIFY THE RELATIONSHIPS

Step 3: The cars have masses $m_1 = 2500$ kg (the SUV) and $m_2 = 1200$ kg (the compact car), with initial velocities $v_{1i} = 40$ m/s and $v_{2i} = 10$ m/s (Fig. 7.15A). The final velocity of the two cars together is unknown, and we denote it by v_f. Continuing with our problem-solving strategy, conservation of momentum leads to the results in Equations 7.21 and 7.22. This collision is inelastic, so the kinetic energy of the two cars is not conserved.

SOLVE

Inserting the given values of the masses and initial velocities into Equation 7.22 leads to

$$v_f = \frac{m_1v_{1i} + m_2v_{2i}}{m_1 + m_2} = \frac{(2500\text{ kg})(40\text{ m/s}) + (1200\text{ kg})(10\text{ m/s})}{2500\text{ kg} + 1200\text{ kg}} = \boxed{30\text{ m/s}}$$

▶ *What does it mean?*

The final velocity lies between the two initial velocities, but v_f is closer to the initial velocity of the larger car because it had a larger fraction of the initial momentum.

CONCEPT CHECK 7.4 Hitting a Parked Car

A careless driver is coasting at 20 m/s (about 40 mi/h) along a straight, level road in a car of mass 1400 kg when she collides with a parked car. The parked car is in neutral, and its brakes are off. The two cars lock bumpers and move off together. If the final velocity is 10 m/s, what is the mass of the parked car?

(a) 700 kg (b) 1400 kg (c) 2800 kg

Inelastic Collisions: What Happens to the Kinetic Energy?

We have already mentioned that kinetic energy is not conserved in an inelastic collision. Let's compare the initial and final kinetic energies for the specific case in Example 7.6. The kinetic energy before the collision is

$$KE_i = \tfrac{1}{2}m_1v_{1i}^2 + \tfrac{1}{2}m_2v_{2i}^2$$

Inserting the values of the masses and velocities from Example 7.6 leads to

$$KE_i = \tfrac{1}{2}m_1v_{1i}^2 + \tfrac{1}{2}m_2v_{2i}^2 = \tfrac{1}{2}(2500 \text{ kg})(40 \text{ m/s})^2 + \tfrac{1}{2}(1200 \text{ kg})(10 \text{ m/s})^2$$
$$KE_i = 2.1 \times 10^6 \text{ J}$$

The cars stick together after the collision, so they have the same final velocity v_f. The final kinetic energy is therefore

$$KE_f = \tfrac{1}{2}m_1v_{1f}^2 + \tfrac{1}{2}m_2v_{2f}^2 = \tfrac{1}{2}(m_1 + m_2)v_f^2$$

Inserting the value of the final velocity from Example 7.6 gives

$$KE_f = \tfrac{1}{2}(m_1 + m_2)v_f^2 = \tfrac{1}{2}(2500 \text{ kg} + 1200 \text{ kg})(30 \text{ m/s})^2 = 1.7 \times 10^6 \text{ J}$$

The final kinetic energy is thus about 20% smaller than the initial kinetic energy. This "missing" energy would be converted to several other forms, including heat and sound.

Now consider a completely inelastic "head-on" collision involving two objects. For example, suppose two cars of equal mass are moving with equal speeds, but in opposite directions. In this case, $v_{1i} = -v_{2i}$ and $m_1 = m_2 = m$. Inserting these values into the result for v_f for a completely inelastic collision (Eq. 7.22) gives

$$v_f = \frac{m_1v_{1i} + m_2v_{2i}}{m_1 + m_2} = \frac{m(v_{1i} - v_{1i})}{2m} = 0$$

Hence, the two objects come to a complete stop after the collision, and the final kinetic energy is zero. *All* the initial kinetic energy is converted into other forms of energy.

Collisions in Two Dimensions

Dealing with collisions in two dimensions involves the same basic ideas as in one dimension, except that the final velocity of each object involves two unknowns: the two components of the velocity vector. Hence, for two colliding objects, there are usually four unknowns (two velocity components for each object). The momentum conservation relation then becomes

$$\vec{p}_{1i} + \vec{p}_{2i} = \vec{p}_{1f} + \vec{p}_{2f}$$

where $\vec{p}_{1i}$ is the initial momentum of object 1, $\vec{p}_{1f}$ is its final momentum, and so forth. It is useful to write this relation in terms of components:

$$p_{1ix} + p_{2ix} = p_{1fx} + p_{2fx} \quad \text{and} \quad p_{1iy} + p_{2iy} = p_{1fy} + p_{2fy} \tag{7.23}$$

To analyze a two-dimensional collision, we can again follow the basic approach outlined in our problem-solving strategy for analyzing a collision, using the momentum conservation relations in Equation 7.23. We must then include whatever additional information is available. This additional information might be that the

▶ **Figure 7.16** This crater (known as Meteor Crater) was produced when an asteroid struck the Earth in eastern Arizona about 49,000 years ago. This crater is about 1.2 km across.

collision is elastic and hence that total kinetic energy is conserved, or we might have some information regarding the final velocity of one of the objects.

A Collision in Two Dimensions: A Rocket, an Asteroid, and Saving the Earth

Our planet is constantly being bombarded with asteroids. These asteroids are usually very small and disintegrate completely as they enter the Earth's atmosphere. On occasion, however, a very large asteroid reaches the Earth's surface and causes a great deal of damage. It is believed that such a collision formed Meteor Crater, an impact crater 1.2 km in diameter located in eastern Arizona (Fig. 7.16). It has been proposed that a similar collision involving a much larger asteroid led to the extinction of the dinosaurs, and NASA is concerned about what might happen to life on the Earth when another large asteroid strikes the Earth at some future time. Scientists have therefore considered how to use rockets to deflect incoming asteroids before they reach the Earth. The collision between such a rocket and an asteroid is sketched in Figure 7.17. For simplicity, we assume the rocket hits the asteroid and then becomes embedded in it, so this collision is completely inelastic. Given the masses of the rocket and asteroid and their initial velocities, we want to calculate the common final velocity of the two. From that result, we can determine if the asteroid will subsequently strike the Earth.

We can analyze this collision using our problem-solving strategy for dealing with collisions. Step 1: The system consists of the two colliding "particles," the rocket and the asteroid. The external forces on this system are due to gravity from the Sun and the Earth, which will be much smaller than the collision force when the rocket hits the asteroid. Hence, to very good accuracy, the momentum of this system will be conserved. Step 2: We choose the coordinate system in Figure 7.17 with the initial velocity of the asteroid in the $+y$ direction, aimed directly toward the Earth. The asteroid's initial velocity is then $v_{ax} = 0$, $v_{ay} = v_{a0}$. The rocket approaches the asteroid with an initial velocity in the $+x$ direction, so the rocket's initial velocity is $v_{rx} = v_{r0}$, $v_{ry} = 0$. The rocket "sticks to" the asteroid after the collision, so the rocket and asteroid have the same final velocity, and we denote the components of this final velocity by v_{fx} and v_{fy}. Step 3: We can now write the momentum conservation relations for our system of the rocket plus the asteroid. For the momentum along the x direction, we have

$$\overbrace{m_a v_{ax} + m_r v_{rx}}^{\text{initial momentum along } x} = m_r v_{r0} = \overbrace{(m_a + m_r) v_{fx}}^{\text{final momentum along } x}$$

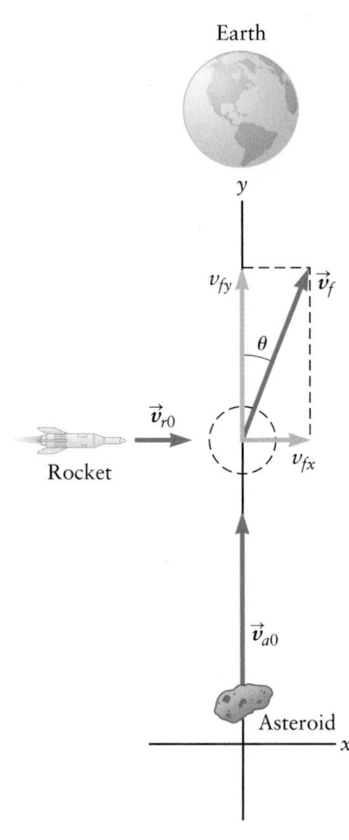

▲ **Figure 7.17** The collision between a rocket and an asteroid could be used to deflect an asteroid that is heading toward the Earth. If the rocket embeds (sticks) in the asteroid, the collision would be completely inelastic.

The corresponding result for the momentum along y is

$$\overbrace{m_a v_{ay} + m_r v_{ry}}^{\text{initial momentum along } y} = m_a v_{a0} = \overbrace{(m_a + m_r) v_{fy}}^{\text{final momentum along } y}$$

Step 4: Solving for the components of the final velocity, we find

$$v_{fx} = \frac{m_r v_{r0}}{m_a + m_r} \tag{7.24}$$

$$v_{fy} = \frac{m_a v_{a0}}{m_a + m_r} \tag{7.25}$$

We can now calculate the final speed using $v_f = \sqrt{v_{fx}^2 + v_{fy}^2}$ and express the final direction in terms of the angle $\theta = \tan^{-1}(v_{fx}/v_{fy})$ shown in Figure 7.17.

Now let's put in some realistic values to see if this asteroid deflection plan can actually work. A rocket of reasonable size has an approximate mass $m_r = 2 \times 10^6$ kg, and a large asteroid has mass $m_a = 1 \times 10^{18}$ kg. If we assume the approach velocity of the rocket is similar to the velocity attained when NASA sent the Apollo spacecrafts to the Moon, then $v_{r0} = 1000$ m/s. We also assume the asteroid's initial velocity is equal to the Earth's orbital velocity relative to the Sun, which is approximately 3×10^4 m/s. Inserting all these values into Equations 7.24 and 7.25 leads to

$$v_{fx} = \frac{m_r v_{r0}}{m_a + m_r} = \frac{(2 \times 10^6 \text{ kg})(1000 \text{ m/s})}{(1 \times 10^{18} + 2 \times 10^6) \text{ kg}} = 2 \times 10^{-9} \text{ m/s} \tag{7.26}$$

$$v_{fy} = \frac{m_a v_{a0}}{m_a + m_r} = \frac{(1 \times 10^{18} \text{ kg})(3 \times 10^4 \text{ m/s})}{(1 \times 10^{18} + 2 \times 10^6) \text{ kg}} = 3 \times 10^4 \text{ m/s}$$

These values for the final velocity of the asteroid plus rocket show that asteroid deflection is a very tough problem. The value of v_{fx} is *extremely* small because the mass of the asteroid is much, much larger than the mass of the rocket. In addition, v_{fy} is essentially equal to the initial velocity of the asteroid, again because the rocket is much smaller than the asteroid. The real test of success (or failure) can be determined by calculating the angle through which the asteroid is deflected. That angle is

$$\theta = \tan^{-1}\left(\frac{v_{fx}}{v_{fy}}\right) = \tan^{-1}\left(\frac{2 \times 10^{-9}}{3 \times 10^4}\right)$$

$$\theta = 7 \times 10^{-14} \text{ rad} = 4 \times 10^{-12} \text{ degrees} \tag{7.27}$$

This angle is very small. It is not easy to deflect an asteroid.

CONCEPT CHECK 7.5 Saving the Earth

The deflection angle in the rocket–asteroid collision in Figure 7.17 (and Eq. 7.27) is very small. Which of the following changes would increase the deflection angle? More than one choice may apply.

(a) Increase the mass of the rocket.
(b) Decrease the mass of the rocket.
(c) Increase the speed of the rocket.
(d) Decrease the speed of the rocket.
(e) Pick a slower asteroid.
(f) Options (a), (c), and (e) would all work.
(g) Options (b) and (d) would both work.

EXAMPLE 7.7 Ⓡ Forces in an Automobile Collision

Two cars approach an intersection as shown schematically in Figure 7.18. One of the drivers ignores a stop sign, resulting in a right-angle collision with the cars locked together after the collision. What is the approximate average force between the two cars during the collision? Assume they each have mass $m = 1000$ kg and they both are traveling with an initial speed of 20 m/s (about 45 mi/h).

RECOGNIZE THE PRINCIPLE

This problem requires us to combine what we know about force and impulse together with the principle of conservation of momentum. The geometry is similar to the rocket–asteroid problem in Figure 7.17, and we can take the same approach to calculating the final velocity $\vec{v}_f$ of the two cars. From $\vec{v}_f$, we can use the impulse theorem to get the impulse exerted by one car on the other. If we then estimate the collision time, we can get the average force exerted by one car on the other during the collision.

(continued) ▶

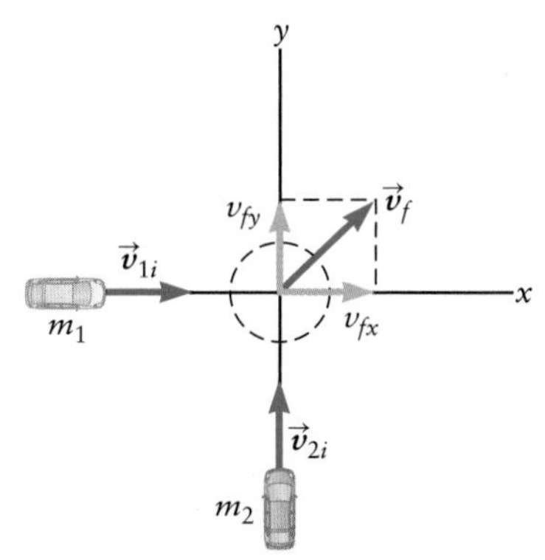

▲ **Figure 7.18** Example 7.7.

SKETCH THE PROBLEM

Figure 7.18 shows the problem. The initial velocities are along both x (car 1) and y (car 2), so we will have to account for momentum conservation in two dimensions.

IDENTIFY THE RELATIONSHIPS

Noticing the directions of $\vec{v}_{1i}$ and $\vec{v}_{2i}$ in Figure 7.18, we can write the components of the initial velocities of the two cars as $v_{1x} = 20$ m/s, $v_{1y} = 0$, and $v_{2x} = 0$, $v_{2y} = 20$ m/s. The collision is a completely inelastic right-angle collision, just as with the rocket and asteroid in Figure 7.17, so we can modify Equations 7.24 and 7.25 to get

$$v_{fx} = \frac{mv_{1x}}{m+m} = \frac{v_{1x}}{2}$$

$$v_{fy} = \frac{mv_{2y}}{m+m} = \frac{v_{2y}}{2}$$

Inserting the given values for the initial velocities, we find

$$v_{fx} = v_{fy} = 10 \text{ m/s} \tag{1}$$

We can now use the impulse theorem to relate the collision force exerted on one of the cars to the change in the momentum of that car.

SOLVE

Writing the impulse theorem (Eq. 7.5) for the x component of the momentum of car 2, we have

$$\text{impulse on car 2 along } x = \Delta p_{2x}$$

Expressing the impulse along x in terms of the average force along x (Eq. 7.7) gives

$$F_{\text{ave},\,x}\,\Delta t = \Delta p_{2x}$$

The initial momentum of car 2 along x is zero, which leads to

$$F_{\text{ave},\,x}\,\Delta t = \Delta p_{2x} = mv_{fx} - mv_{ix} = mv_{fx}$$

$$F_{\text{ave},\,x} = \frac{mv_{fx}}{\Delta t}$$

For our final step, we need to know the collision time Δt. We estimate a value of approximately $\Delta t = 0.1$ s. As usual, we aim here for accuracy to within a factor of 10. A value of $\Delta t = 1$ s seems too long, whereas 0.01 s seems too short. We found v_{fx} in Equation (1). Using that result, we get

$$F_{\text{ave},\,x} = \frac{mv_{fx}}{\Delta t} = \frac{(1000 \text{ kg})(10 \text{ m/s})}{(0.1 \text{ s})} = \boxed{1 \times 10^5 \text{ N}}$$

► ***What does it mean?***

This force is quite large, as we should expect; cars are massive objects. Notice also that this is just the x component of the collision force. From the symmetry of this problem in terms of x and y, however, you should suspect that the y component of the collision force equals $F_{\text{ave},\,x}$.

CONCEPT CHECK 7.6 How Big Is the Collision Force?

In Example 7.7, we estimated the collision force in an automobile accident. Which of the following statements is correct?

(a) The magnitude of the collision force is much larger than a car's weight.
(b) The magnitude of the collision force is much smaller than a car's weight.
(c) The magnitude of the collision force is approximately equal to a car's weight.

7.5 Using Momentum Conservation to Analyze Inelastic Events

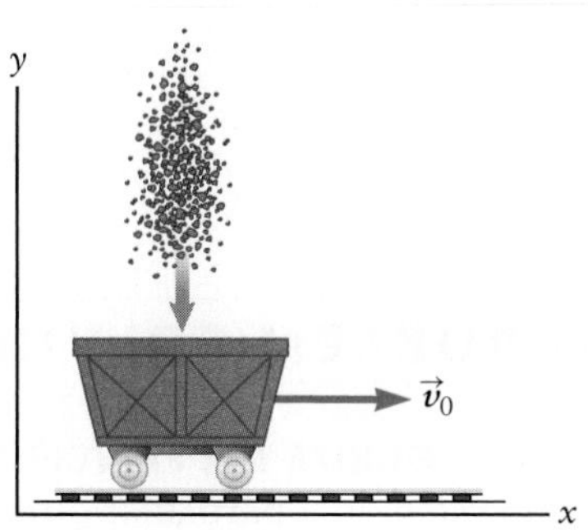

▲ **Figure 7.19** This system consists of the railroad car and the load of gravel dropped into it. The momentum of this system along the x direction is conserved.

Momentum is the product of mass and velocity. In all our examples so far, the mass has been constant (that is, unchanging). When the momentum of an object or system with constant mass is conserved, the velocity is constant. In some situations, though, the mass can change with time, and it is interesting to consider how the principle of conservation of momentum applies in such cases.

Figure 7.19 shows an empty railroad car coasting at speed v_0 along a level track. The track is straight and lies along the x direction. A load of gravel is then dropped into the car as it moves along. Given the masses of the empty car m_c and the load of gravel m_g, what is the speed of the railroad car after the gravel is dropped into it?

We can treat the problem as a "collision" involving the gravel and the railroad car and apply the principle of momentum conservation. The railroad car moves on a straight track, so its motion is one dimensional, along the x direction in Figure 7.19. If we assume there is no frictional force on the car (from bearings in the wheels, etc.), there are no external forces along x, and the component of the momentum along this direction is conserved. The initial momentum of the car along x is $m_c v_0$. Prior to landing in the car, the gravel has no momentum along x, so the total initial momentum of the car plus the gravel along x is also $m_c v_0$. After the gravel is added to the car, the car plus gravel has a velocity v_f, so the momentum conservation relation has the form

$$\underbrace{m_c v_0}_{\substack{\text{initial momentum} \\ \text{along } x}} = \underbrace{(m_c + m_g)v_f}_{\substack{\text{final momentum} \\ \text{along } x}} \tag{7.28}$$

Solving for the final velocity gives

$$v_f = \frac{m_c v_0}{m_c + m_g}$$

The value of v_f is always smaller than v_0, so the railroad car will always slow down after the gravel is dropped into it.

Our analysis so far has only considered momentum along x; we should also consider what happens to the momentum along y during this collision. Your first expectation may be that the momentum along y should also be conserved. Before the collision, the gravel has momentum along the $-y$ direction, whereas the car has no momentum along y. After the collision, the motion is solely along x, so the final momentum along y is zero. Hence, the momentum of the gravel plus the car along y is *not* conserved in this example. Why not? When the gravel hits the car, a vertical force is exerted by the car on the gravel, bringing the gravel to a stop along the y direction in the car. A corresponding normal force exerted by the railroad track on the car prevents the car from falling through the track and moving along y. The force exerted by the track on our system (the car plus gravel) is an external force; it imparts an impulse to the system and hence produces a change in the system's momentum along y. For this reason, the momentum of this system along y is not conserved. The momentum of a system is conserved *only* if the forces from outside the system are negligible. In this example, there is an external force along the y direction, so the momentum of the system along this direction is not conserved. However, the momentum of the system along the x direction is still conserved since there are no external forces along x.

Applying the Principle of Conservation of Momentum to Inelastic Events

The principle of momentum conservation can be applied to inelastic events, such as the railroad car in Figure 7.19, using a problem-solving strategy similar to the one we have used for dealing with collisions. That strategy was introduced with collisions in mind, so it is useful to state a new version written specifically for inelastic events. The basic ideas are the same: whenever the external forces on a system are zero or negligible, the total momentum of the system is conserved.

PROBLEM SOLVING Analyzing an Inelastic Event

1. **RECOGNIZE THE PRINCIPLE.** The momentum of a system in a particular direction is conserved only when the net external force in that direction is zero or negligible.
2. **SKETCH THE PROBLEM.** Make a sketch of the system, including a coordinate system, and use the given information to determine (where possible) the initial and final velocities of the system's components.
3. **IDENTIFY THE RELATIONSHIPS.**
   - Express the conservation of momentum condition for the direction(s) identified in step 1.
   - Use any given information to determine the increase or decrease of the kinetic energy.
4. **SOLVE** for the unknown quantities, such as the final velocities or the force exerted on one of the particles in the system.
5. Always *consider what your answer means* and check that it makes sense.

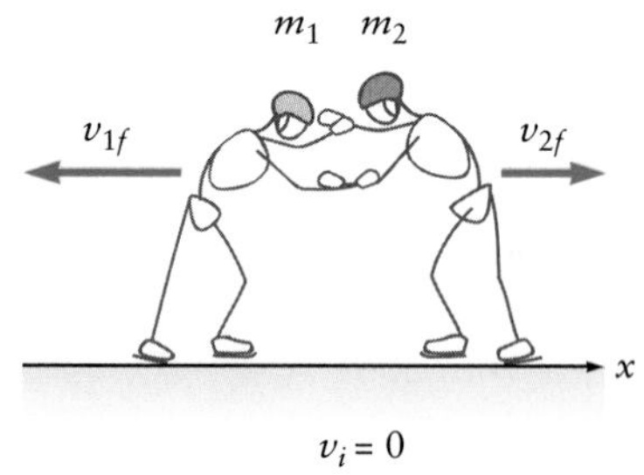

▲ **Figure 7.20** Example 7.8.

EXAMPLE 7.8 Momentum on Ice

Two angry children are fighting while standing on a very slippery icy surface (Fig. 7.20). Their masses are $m_1 = 25$ kg and $m_2 = 40$ kg, and they are initially at rest. The larger child pushes the smaller child, who then slides away to the left with a speed of 1.5 m/s. What is the final velocity of the larger child?

RECOGNIZE THE PRINCIPLE

We follow the strategy outlined in the problem-solving box on inelastic events. Step 1: The two children are the "system." The ice is very slippery, so the external forces in the horizontal (x) direction are negligible and the total momentum along x is conserved.

SKETCH THE PROBLEM

Continuing with our problem-solving strategy (step 2), the problem is sketched in Figure 7.20, which shows the initial and final velocities. Initially, the children are at rest, so $v_{1i} = v_{2i} = 0$, and we denote their final velocities v_{1f} and v_{2f}.

IDENTIFY THE RELATIONSHIPS

Step 3: The total initial momentum of this system along x is zero, and the final momentum along x is $m_1v_{1f} + m_2v_{2f}$. The final momentum must equal the initial momentum:

$$\overbrace{0}^{\text{initial momentum}} = \overbrace{m_1v_{1f} + m_2v_{2f}}^{\text{final momentum}}$$

SOLVE

Solving for the final velocity of child 2 gives

$$v_{2f} = -\frac{m_1v_{1f}}{m_2} = -\frac{(25\text{ kg})(-1.5\text{ m/s})}{40\text{ kg}} = \boxed{0.94\text{ m/s}}$$

► ***What does it mean?***

The final velocity of child 1 is to the left, so his velocity is negative, which leads to a positive velocity (to the right) for child 2.

Inelastic Processes Are Similar to Collisions

Most inelastic processes are very similar to collisions. Consider again the two children in Example 7.8. In Figure 7.21A, the children are initially at rest, whereas after pushing on each other they are traveling in opposite directions (Fig. 7.21B). A closely related situation is sketched in parts C and D of Figure 7.21. Here, the children are initially moving toward each other; then they collide and stick together, which is just a completely inelastic collision. Comparing these two scenarios, we see that the inelastic event in which the children push each other apart is just like a completely inelastic collision in "reverse." In both scenarios, the total momentum is conserved.

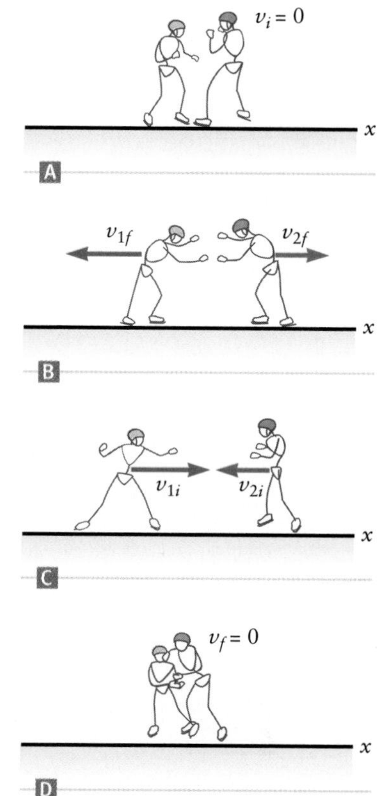

▲ **Figure 7.21** A and B: An inelastic event in which two children on ice push on each other and move apart. C and D: An inelastic collision in which the two children come together and "collide."

CONCEPT CHECK 7.7 Children and Forces in an Inelastic Event

Consider again the two children in Example 7.8, who push on each other as they fight. If the push lasts for 0.50 s, what is the magnitude of the average force exerted by the larger child on the smaller one?

(a) 25 N (b) 38 N (c) 75 N (d) Zero

Splitting Asteroids

Let's take another look at the daunting task of deflecting an asteroid that threatens the Earth. In our previous analysis, we tried to deflect the asteroid by arranging for an inelastic collision with a rocket (Fig. 7.17) and found that it is difficult to deflect an asteroid very much with a rocket of reasonable size. We will therefore try a different approach in which a large bomb splits the asteroid into two pieces as illustrated in Figure 7.22. This is another example of an inelastic event, and we can attack it using our problem-solving strategy for inelastic events. The two pieces of the asteroid are the system, and because the external forces are very small (we assume we are far from the Sun, so its gravitational force is small), the momentum of this system is conserved. For simplicity, we assume the two pieces are of equal mass m and choose the coordinate axis x to pass through the center of the asteroid (Fig. 7.22A). Before the explosion, the x component of the velocity is zero, so the momentum along x is also zero. Since there are no external forces on the system (the asteroid plus the bomb), the final momentum along x is also zero. Hence, the two pieces will separate after the explosion with velocities that are equal in magnitude and opposite in direction. We therefore take $\pm v_f$ to be the x components of the velocity of the two pieces of the asteroid after the explosion (Fig. 7.22B).

To calculate v_f, we need to know something about the final kinetic energy. Suppose an energy E_{exp} is released in the explosion. If we are optimistic, we might hope all this energy goes into the final kinetic energy.

$$\tfrac{1}{2}mv_f^2 + \tfrac{1}{2}mv_f^2 = mv_f^2 = E_{\text{exp}}$$

Here, there are two terms equal to $\tfrac{1}{2}mv_f^2$ because the asteroid is split into two pieces, each of mass m. We can now solve for the final velocity and find

$$v_f = \sqrt{\frac{E_{\text{exp}}}{m}} \tag{7.29}$$

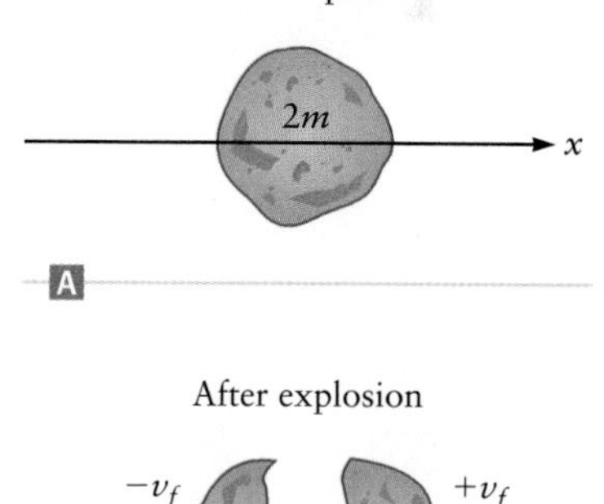

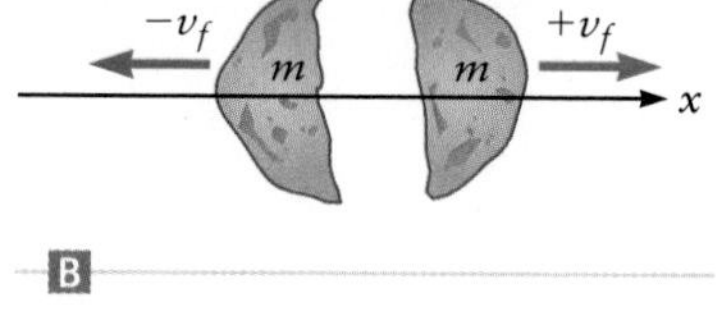

▲ **Figure 7.22** When this asteroid is split into two pieces by an explosion, the explosion releases energy, so the kinetic energy of the system—the pieces of the asteroid—is larger after the explosion. However, the forces from outside the system are zero, so the total momentum is still conserved.

To decide if this approach will succeed, we need to estimate the value of v_f with some realistic numbers. The mass of a typical large asteroid is 1×10^{18} kg, so if the asteroid is split into two equal-sized pieces, we would have $m = 5 \times 10^{17}$ kg. We must next consider the energy we can obtain from the bomb because it will be providing the final kinetic energy of the two asteroid pieces. The energy released by one of the atomic bombs that was dropped in World War II was approximately 1×10^{17} J. Let's assume we can increase that value by a factor of 10 with a contemporary thermonuclear explosive, so $E_{exp} = 1 \times 10^{18}$ J. Inserting these values into Equation 7.29 gives

$$v_f = \sqrt{\frac{E_{exp}}{m}} = \sqrt{\frac{1 \times 10^{18}\text{ J}}{5 \times 10^{17}\text{ kg}}} \approx 1\text{ m/s}$$

which is *much* larger than the lateral velocity we found for a collision with a rocket (Eq. 7.26). Hence, splitting the asteroid with a bomb is the better approach for achieving a significant deflection. Even in this case, though, v_f is not very large, so the task is still not easy. Notice that the momentum of our asteroid system is conserved since the *external* forces on it are zero. Although the bomb's explosion certainly does exert forces on the asteroid, the bomb is part of the system, so these are *internal* forces. The kinetic energy is *not* conserved because the energy of the bomb goes into the final kinetic energy of the system, the two pieces of the asteroid.

7.6 Center of Mass

In the previous sections, we distinguished between *internal* and *external* forces, that is, between forces that act between the particles in aa system and forces that come from outside the system. This distinction is important for understanding the motion of a composite object, that is, an object composed of many point particles such as the boomerang in Figure 7.1A. Before we attempt to deal with an object as complicated as a boomerang, let's consider the simpler case of two point masses connected by a light rod (Fig. 7.23A). We can write Newton's second law for the two combined masses as

$$\sum \vec{F} = \vec{F}_1 + \vec{F}_2 = m_1\vec{a}_1 + m_2\vec{a}_2 \tag{7.30}$$

where $\vec{F}_1$ is the total force acting on m_1 and $\vec{F}_2$ is the total force acting on m_2. A portion of the force acting on m_1 will come from its interaction with m_2 and is an internal force. The rest of the force on m_1 will come from outside the system and is thus an external force. The same argument applies to the force on m_2. Let's therefore write the total force on the left-hand side of Equation 7.30 as a sum of the external and internal forces:

$$\overbrace{\sum \vec{F}}^{\text{total force}} = \overbrace{\sum \vec{F}_{ext}}^{\text{external forces}} + \overbrace{\sum \vec{F}_{int}}^{\text{internal forces}}$$

The internal forces always come in action–reaction pairs; for the example in Figure 7.23A, the force exerted by m_1 on m_2 is equal in magnitude and opposite in sign to the force exerted by m_2 on m_1. Hence, when we consider the entire system, the internal forces will cancel so that

$$\sum \vec{F}_{int} = 0 \tag{7.31}$$

While we have derived this result for a system of just two particles, Equation 7.31 is true for a system with any number of particles. The internal forces always come in action–reaction pairs, so the sum of all the internal forces is always zero. Thus, Equation 7.30 becomes

$$\sum \vec{F}_{ext} = m_1\vec{a}_1 + m_2\vec{a}_2$$

If we denote the total mass of the system as $m_1 + m_2 = M_{tot}$, we have

$$\sum \vec{F}_{ext} = m_1\vec{a}_1 + m_2\vec{a}_2 = M_{tot}\vec{a}_{CM} \tag{7.32}$$

where $\vec{a}_{CM}$ is the acceleration of what is called the ***center of mass*** of the system. Equation 7.32 has precisely the same form as Newton's second law for a point particle. In words, it says that the total external force ($\sum \vec{F}_{ext}$) on a system of particles is equal to the mass of the system (M_{tot}) multiplied by a new quantity called the acceleration of the center of mass.

What Is the Center of Mass and How Is It Useful?

So what is the "center of mass"? According to Equation 7.32, the acceleration of the center of mass is related to the accelerations of all the pieces of the system. We can rearrange Equation 7.32 to find $\vec{a}_{CM}$ for a system of just two particles moving in one dimension. Solving for $\vec{a}_{CM}$ gives

$$\vec{a}_{CM} = \frac{m_1\vec{a}_1 + m_2\vec{a}_2}{M_{tot}}$$

This relation suggests that we can express the position of the center of mass along the x coordinate direction in terms of the positions of all the pieces of the system:

$$x_{CM} = \frac{m_1x_1 + m_2x_2}{M_{tot}} \tag{7.33}$$

Here, x_1 and x_2 are the positions of the two particles measured along the x axis in Figure 7.23A, and x_{CM} is the position of the center of mass along x. If the two particles have equal masses m, according to Equation 7.33 the center of mass is at

$$x_{CM} = \frac{m_1x_1 + m_2x_2}{M_{tot}} = \frac{mx_1 + mx_2}{2m} = \frac{x_1 + x_2}{2}$$

The center of mass of this system is thus at the point midway between the two particles. On the other hand, if one particle is more massive than the other, the center of mass is closer to the one with greater mass. Intuitively, you can think of the center of mass as the "balance point"; that is, the center of mass is the point on the connecting rod in Figure 7.23B where the system would be "in balance."

This result for the center of mass for two particles can be generalized to a system of many particles in two or three dimensions. The result is

$$x_{CM} = \frac{\sum_i m_ix_i}{\sum_i m_i} = \frac{\sum_i m_ix_i}{M_{tot}}$$
$$y_{CM} = \frac{\sum_i m_iy_i}{\sum_i m_i} = \frac{\sum_i m_iy_i}{M_{tot}} \tag{7.34}$$

Center of mass

with a similar expression for z_{CM}. Here, the sums are over all the particles (i) in the system. The total mass of the system is just

$$M_{tot} = \sum_i m_i$$

When using Equation 7.34 to find the location of the center of mass, you must first choose a coordinate system (with an origin). The values of x_{CM} and y_{CM} then refer to the same coordinate system.

For a collection of several masses, computing the center of mass requires a sum involving the positions and masses of the objects. For example, if there are three particles of equal mass m arranged in plane as shown in Figure 7.24, the coordinates of the center of mass are given by

$$x_{CM} = \frac{\sum_i m_ix_i}{\sum_i m_i} = \frac{m_1x_1 + m_2x_2 + m_3x_3}{m_1 + m_2 + m_3} = \frac{m(0) + m(0) + m(L)}{3m} = \frac{L}{3}$$

and

$$y_{CM} = \frac{\sum_i m_iy_i}{\sum_i m_i} = \frac{m_1y_1 + m_2y_2 + m_3y_3}{m_1 + m_2 + m_3} = \frac{m(L) + m(0) + m(0)}{3m} = \frac{L}{3}$$

The calculation of the center of mass of a solid object, such as a baseball bat or a boomerang, requires a sum over all the pieces of the object; that is, a sum over all the individual "point particles" that make up the object. Such a calculation can be

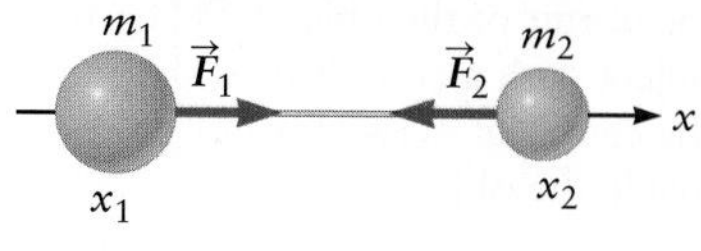

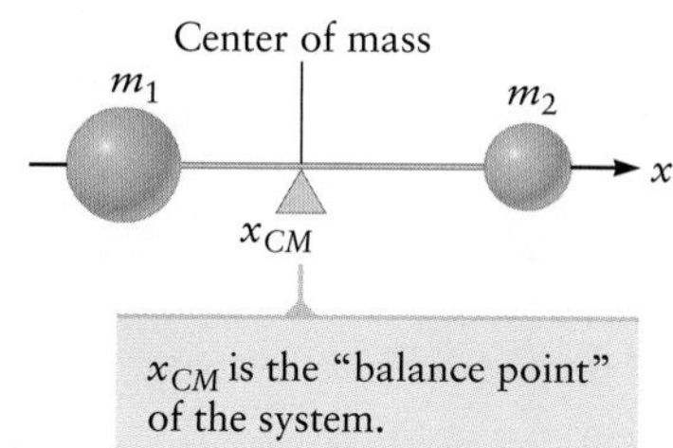

▲ **Figure 7.23** A A system consisting of two particles connected by a massless rod. The forces exerted by each particle on the other (through the rod) are internal forces. B The center of mass of the two point particles is the "balance" point along a line that connects them.

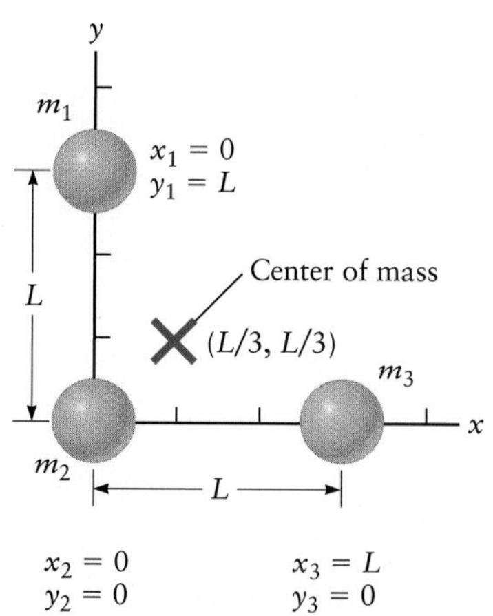

▲ **Figure 7.24** The center of mass of a collection of particles is given by Equation 7.34.

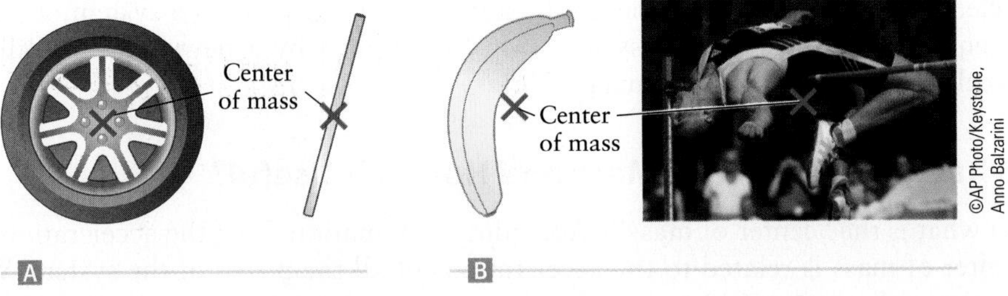

▶ **Figure 7.25** Approximate location of the center of mass of various objects. **A** For a symmetric object, the center of mass is located at the center of symmetry (i.e., the center of the object). **B** For an object with a complicated shape, the center of mass may not lie inside the object.

complicated, especially if the mass is not distributed uniformly or if the shape is complex. Even so, the intuitive notion of "balance point" can be very helpful in locating the center of mass. Thus, for a symmetric object such as a simple rod or wheel, the center of mass is at the center of symmetry of the object. Notice also that the center of mass need not be located within an object. The center of mass of a banana may not be inside the banana (Fig. 7.25B).

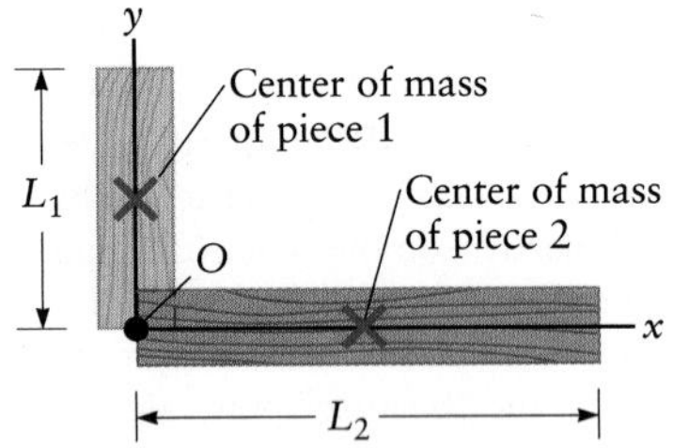

▲ **Figure 7.26** Example 7.9. This bracket is composed of two separate pieces. Each piece is a simple board, one lies along x and one along y, and they overlap near the origin.

EXAMPLE 7.9 Center of Mass of a Bracket

You work for a carpenter and are building a bracket composed of two straight sections of wood as sketched in Figure 7.26. You want to show your boss that physics can be useful, so you decide to calculate the position of the center of mass of the bracket. If the two sections of wood have lengths $L_1 = 1.2$ m and $L_2 = 2.0$ m and masses $m_1 = 0.50$ kg and $m_2 = 1.0$ kg, what is the location of the center of mass of the entire bracket?

RECOGNIZE THE PRINCIPLE

We could find the center of mass of this system of particles by applying the definition in Equation 7.34 and summing over all the pieces that make up the bracket. However, a simpler approach uses the notion of the center of mass as a balance point. With that in mind, we first find the center of mass of each straight section of the bracket. We can treat these two pieces as particles of mass m_1 and m_2 located at their respective centers of mass and then compute the location of the overall center of mass of this two-particle system.

SKETCH THE PROBLEM

Figure 7.26 shows the problem.

IDENTIFY THE RELATIONSHIPS

For each piece of wood, we use the balance-point notion of center of mass to tell us that the center of mass of each piece will be at each respective center as indicated in Figure 7.26. Thus, the center of mass of piece 1 is at

$$x_{1,\,CM} = 0 \quad \text{and} \quad y_{1,\,CM} = \frac{L_1}{2} = 0.60 \text{ m}$$

and for piece 2, we have

$$x_{2,\,CM} = \frac{L_2}{2} = 1.0 \text{ m} \quad \text{and} \quad y_{2,\,CM} = 0$$

SOLVE

We can now treat this bracket as a system of two particles of mass m_1 and m_2 at these two locations and use the relations for the center of mass coordinates in Equation 7.34 to find the center of mass of the entire bracket:

$$x_{CM} = \frac{m_1 x_{1,\,CM} + m_2 x_{2,\,CM}}{m_1 + m_2} = \frac{(0.50\text{ kg})(0) + (1.0\text{ kg})(1.0\text{ m})}{(0.50\text{ kg}) + (1.0\text{ kg})} = \boxed{0.67\text{ m}}$$

$$y_{CM} = \frac{m_1 y_{1,\,CM} + m_2 y_{2,\,CM}}{m_1 + m_2} = \frac{(0.50\text{ kg})(0.60\text{ m}) + (1.0\text{ kg})(0)}{(0.50\text{ kg}) + (1.0\text{ kg})} = \boxed{0.20\text{ m}}$$

► ***What does it mean?***

Notice that the center of mass of the bracket is outside the bracket. When an object has a simple shape (e.g., a wheel, a ball, or a straight piece of wood), the center of mass is at the center of symmetry of the object. For more complicated shapes, Equation 7.34 must be used to find the center of mass.

CONCEPT CHECK 7.8 Finding the Center of Mass

A sledgehammer has a massive metal head attached to a much lighter wooden handle (Fig. 7.27). Which location is most likely to be the sledgehammer's center of mass?

(a) As shown in the left drawing, midway along the handle
(b) As shown in the center drawing, on the metal head
(c) As shown in the right drawing, slightly below the head

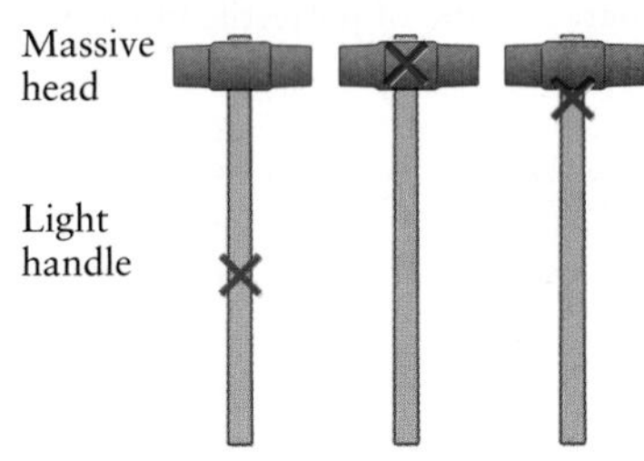

▲ **Figure 7.27** Concept Check 7.8.

Motion of the Center of Mass

Now that we have seen how to calculate the location of the center of mass, let's consider how it *moves*. Figure 7.28 shows two ice skaters; they are initially very close together, so they form a system with the center of mass initially located as shown in part A. We also assume the skaters begin at rest, so their center of mass is initially at rest. If the skaters then push very forcefully on each other, they will move apart as sketched in Figure 7.28B.

If the ice is very slippery (i.e., frictionless), the skaters will move with constant velocities after they separate since there are no forces on either skater along the horizontal (x) direction. Moreover, the *external* force on this system along x is zero the entire time, so the x component of the total momentum of the two skaters is conserved. The initial velocities of the two skaters are zero, so the initial momentum of the system (before the push) is zero and the final momentum must also be zero. The final momentum of the system can be written in terms of the final velocities of the two skaters (see Fig. 7.28), which gives

$$p_{\text{final}} = m_1 v_{1f} + m_2 v_{2f} = 0 \tag{7.35}$$

Since the skaters move with constant velocities and they both start at the origin (being initially very close together), their positions after the push are given by $x_1 = v_{1f}t$ and $x_2 = v_{2f}t$. We have

$$m_1 v_{1f} + m_2 v_{2f} = m_1 \frac{x_1}{t} + m_2 \frac{x_2}{t}$$

Combining this with Equation 7.35, we find

$$m_1 \frac{x_1}{t} + m_2 \frac{x_2}{t} = 0$$

If we now cancel the factor of t, we obtain

$$m_1 x_1 + m_2 x_2 = 0 \tag{7.36}$$

According to Equation 7.34, the center of mass of the two skaters is given by

$$x_{CM} = \frac{m_1 x_1 + m_2 x_2}{m_1 + m_2}$$

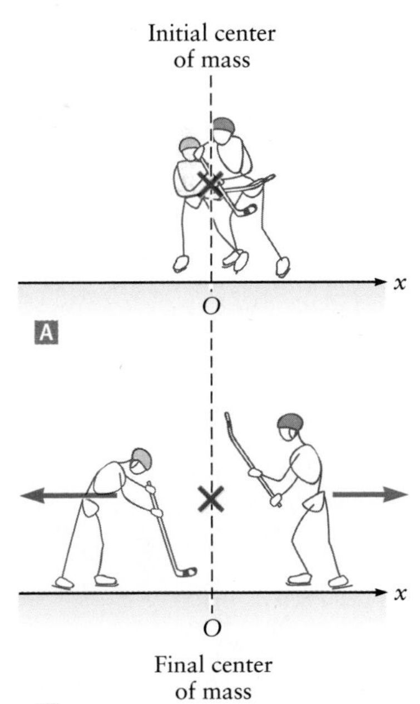

▲ **Figure 7.28** A These two skaters are initially at rest and there are no external forces along the horizontal direction (x). B After they push on each other, the skaters move apart but the center of mass does not move.

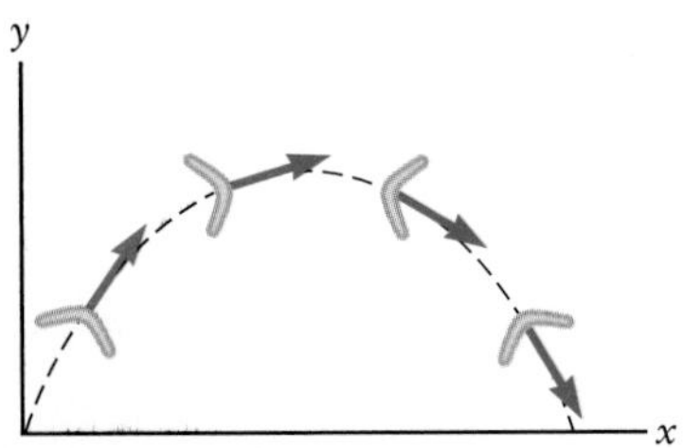

▲ **Figure 7.29** An object such as a boomerang may rotate or tumble when it is thrown, but the center of mass follows a parabolic trajectory characteristic of projectile motion.

Using Equation 7.36, the numerator in this expression for x_{CM} is equal to zero. So, the center of mass of the two skaters is at $x_{CM} = 0$. Moreover, this result applies at all values of t. Hence, even though the skaters move after the push, their center of mass does not move. It stays in the same location, before and after the push.

We could have anticipated this result from our original derivation of the center of mass with Equation 7.32, where we found

$$\sum \vec{F}_{\text{ext}} = M_{\text{tot}}\vec{a}_{CM} \tag{7.37}$$

which is simply Newton's second law written for the center of mass. In words, Equation 7.37 says that a system of particles moves as if all the mass (M_{tot}) is located at the center of mass and that the center of mass motion is caused *only by the external forces* acting on the system.

Translational Motion of a System

Consider the motion of a complicated object such as a boomerang (Fig. 7.29). A boomerang has an asymmetrical shape, which causes it to spin in a complicated manner when thrown. This complicated motion can be viewed as a combination of ***translational motion*** of the center of mass together with ***rotational motion*** as the boomerang spins about an axis that extends through its center of mass. Translational motion is often referred to as ***linear motion***. Here the term *linear* does not mean motion along a straight line; rather, it means motion in which the center of mass moves along a path such as the parabolic trajectory we found in projectile motion. Translational motion is different from rotational motion, in which an object spins or rotates about a particular axis. According to Equation 7.37, the translational motion of any system of particles is described by Newton's second law as applied to an equivalent particle of mass M_{tot}. This equivalent particle is located at the center of mass, and for the purposes of the translational motion, we can treat the motion as if all the mass were located at the center of mass. We can then use everything we know about the motion of a point particle to analyze and predict the motion of more complicated objects such as boomerangs. For example, if the force due to air drag is small, the center of mass of the boomerang will move in a simple parabolic trajectory as calculated in Chapter 4. Even though the boomerang may appear to spin in a very complicated manner as it moves through the air, the center of mass motion will be *precisely the same* as that of a simple point particle.

The concept of center of mass allows us to deal with the translational motion of any complicated object. According to Equation 7.37, the center of mass always moves as if all the mass were located at the center of mass of the object. The rotational motion of the object can then be treated separately as we'll do in Chapters 8 and 9.

CONCEPT CHECK 7.9 Explosions and the Center of Mass

A bomb is initially located at the origin ($x = 0$). The bomb then explodes, breaking into two pieces with masses m and $3m$. The pieces fly away in opposite directions along the x axis. Some time later, the smaller piece is at $x = -6.0$ m. If all friction is negligible, where is the larger piece at that moment?

(a) $x = 2.0$ m (b) $x = 3.0$ m (c) $x = 12$ m (d) $x = 18$ m

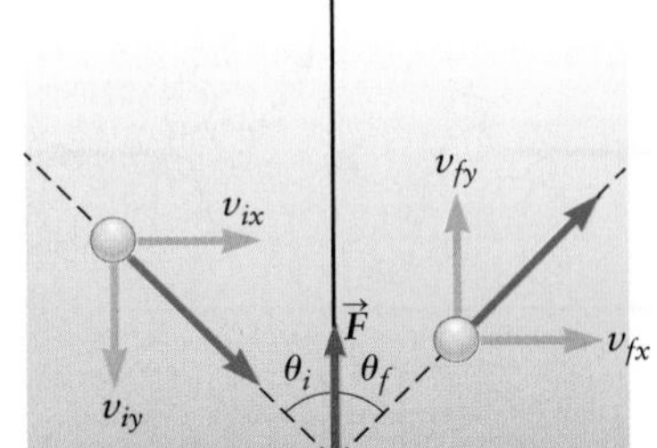

Top view of a cue ball bouncing from the edge of a pool table.

▲ **Figure 7.30** When this cue ball bounces from a wall, the force $\vec{F}$ of the wall on the ball is along y, so the momentum of the ball along x is conserved. Here, $\vec{F}$ is a normal force directed perpendicular to the wall.

7.7 A Bouncing Ball and Momentum Conservation

Consider a pool player who wants to bounce the cue ball off a "rail" at the edge of the table as sketched in Figure 7.30. A good player must be able to predict how a ball will bounce off the rail. The ball approaches the rail at a certain angle θ_i (Fig. 7.30), and we want to calculate the outgoing angle θ_f. We can attack this problem using the principle of momentum conservation.

When the ball collides with the rail, the force exerted by the rail changes the momentum of the ball. For a typical ball and table, the cue ball has nearly the same speed after the collision as it had before hitting the rail, so for simplicity let's assume the initial and final speeds are equal. The final kinetic energy of the ball will therefore equal the initial kinetic energy, so we could also think of this collision as an elastic process. The force of the rail on the cue ball is just a normal force and is directed perpendicular to the rail.[1] Using the coordinate system in Figure 7.30, this normal force is along the y direction and will therefore impart an impulse to the ball in the y direction. There is no force on the ball along the x direction, so the momentum of the ball along x is conserved, and we can write

$$mv_{ix} = mv_{fx}$$

Hence, the initial and final components of the velocity along x are equal:

$$v_{ix} = v_{fx} \tag{7.38}$$

Since the initial and final kinetic energies of the ball are equal, we also have

$$\tfrac{1}{2}mv_i^2 = \tfrac{1}{2}mv_f^2$$

$$\tfrac{1}{2}m\sqrt{v_{ix}^2 + v_{iy}^2} = \tfrac{1}{2}m\sqrt{v_{fx}^2 + v_{fy}^2} \tag{7.39}$$

which leads to

$$\sqrt{v_{ix}^2 + v_{iy}^2} = \sqrt{v_{fx}^2 + v_{fy}^2}$$

$$v_{ix}^2 + v_{iy}^2 = v_{fx}^2 + v_{fy}^2 \tag{7.40}$$

From Equation 7.38, we have $v_{ix} = v_{fx}$, so the terms involving these quantities in Equation 7.40 cancel, leaving

$$v_{iy}^2 = v_{fy}^2$$

which gives

$$v_{fy} = \pm v_{iy}$$

We thus have two solutions for the final y component of the velocity, and we have to decide which one describes the reflection in Figure 7.30. The cue ball bounces from the rail,[2] so the y component of the final velocity is directed opposite to that component of the initial velocity. The solution that describes the reflection is therefore $v_{fy} = -v_{iy}$. The direction of the cue ball's final velocity in Figure 7.30 is then

$$\theta_f = \tan^{-1}\left|\frac{v_{fx}}{v_{fy}}\right| = \tan^{-1}\left|\frac{v_{ix}}{v_{iy}}\right| = \theta_i$$

The outgoing angle is thus equal to the incoming angle. This result, called a ***specular reflection***, is no surprise to an experienced pool player. This kind of reflection will come up again when we study the reflection of light from a mirror in Chapter 24.

7.8 The Importance of Conservation Principles in Physics

Conservation of momentum is the second example of a ***conservation principle*** we have encountered in this book; conservation of energy was a key part of Chapter 6. Conservation principles are important for several reasons. First, they allow us to analyze problems in a very general and powerful way. For example, we saw that for a one-dimensional elastic collision, the combination of conservation of momentum and conservation of kinetic energy completely determines the outcome of the

[1]Things can be more complicated if the ball has sidespin, which can lead to a force on the ball along x.

[2]The other solution corresponds to a ball that passes through the rail, much like some of the solutions we saw in Section 7.4 in which two particles "pass through" each other instead of colliding.

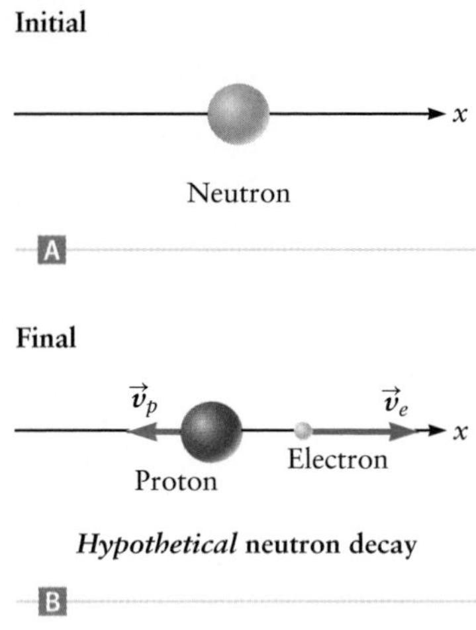
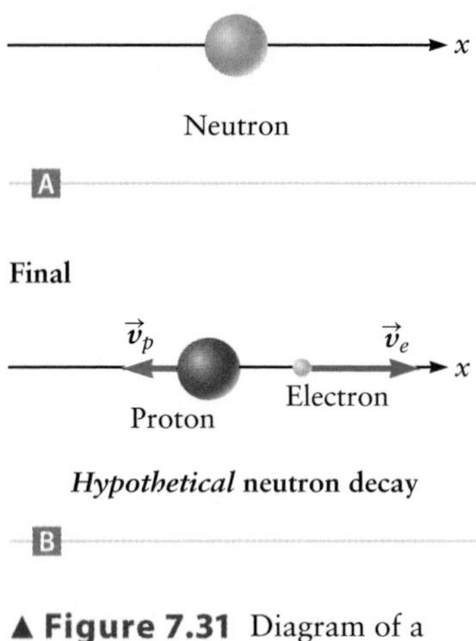

▲ **Figure 7.31** Diagram of a hypothetical decay of a neutron into a proton and an electron. A The neutron is initially at rest and then B decays into a proton and an electron. If there are only two outgoing particles, they must be emitted in opposite directions; otherwise, momentum cannot be conserved. This leads to a single value for the speed of the outgoing electron. Neutrons do not decay in this way.

collision. Analyzing the collision in this way shows that the outcome is a result of these general conservation principles rather than the details of the interaction forces.

Second, conservation principles are extremely general statements about the physical world. While we derived conservation of momentum starting from Newton's laws, this principle is much more general than that. In later chapters, we will study situations in which Newton's laws fail, such as the quantum regime of atoms and subatomic particles, and the astrophysical regime of massive stars and black holes. Although Newton's laws fail in those situations, momentum is still conserved. Indeed, the principle of conservation of momentum is believed to be an exact law of physics, valid in all regimes.

Third, careful tests of conservation principles can sometimes lead to new discoveries. The discovery of an elementary particle called the ***neutrino*** was based quite closely on ideas from this chapter. A subatomic particle called the ***neutron*** was discovered in the early 1930s and was soon recognized as an important component of the atomic nucleus. It was also found that an isolated neutron—that is, a neutron removed from a nucleus—is an unstable particle. When the decay of an isolated neutron was first studied, the details of the decay process were very puzzling. It was initially believed that a neutron decays to leave behind just a proton and an electron:

$$\text{neutron} \rightarrow \text{proton} + \text{electron} \qquad \text{(incomplete!)} \tag{7.41}$$

We have labeled this reaction as "incomplete" for the following reason. This decay process, in which one particle (the neutron) decays to form two other particles (a proton and an electron), is very similar to the exploding asteroid in Figure 7.22. The decaying neutron releases a certain amount of energy E_{decay} (analogous to the bomb that splits the asteroid), which goes into the kinetic energy of the reaction products. There are no external forces acting on the system, so we also know that momentum must be conserved. With conservation of momentum together with knowledge of the reaction energy, we can calculate the final velocities of the proton and electron, just as we calculated the final speeds of the asteroid's pieces in Figure 7.22B.

Let's now analyze the neutron decay reaction in a little more detail.[3] To conserve momentum, the proton and electron in Figure 7.31 must travel away in opposite directions, along a common axis we call x. The problem is thus one dimensional, and we only need to find the components of the proton and electron velocities along x. For simplicity, we assume the neutron was at rest prior to its decay, so the total momentum of the proton plus electron after the collision must be zero. If the electron's velocity along x is v_e and the proton's velocity along x is v_p, conservation of momentum along x gives

$$m_e v_e + m_p v_p = 0 \tag{7.42}$$

The conservation of energy condition can be written as

$$\tfrac{1}{2} m_e v_e^2 + \tfrac{1}{2} m_p v_p^2 = E_{\text{decay}} \tag{7.43}$$

We thus have two conservation conditions (two equations) and two unknowns, the final velocities of the electron and the proton. Hence, these final velocities are fully determined: their precise values are just the solution of these two equations. If Equation 7.41 is really the correct decay reaction, all the electrons must have the same outgoing speed.

When the neutron decay reaction was studied, a very different result was observed. The speed of the outgoing electron was *not* always a single, precise value as expected from the analysis in Equations 7.42 and 7.43. Instead, a range of speeds was observed. This result was very puzzling, and some physicists suggested that either conservation of momentum or conservation of energy might not apply to subatomic particles.

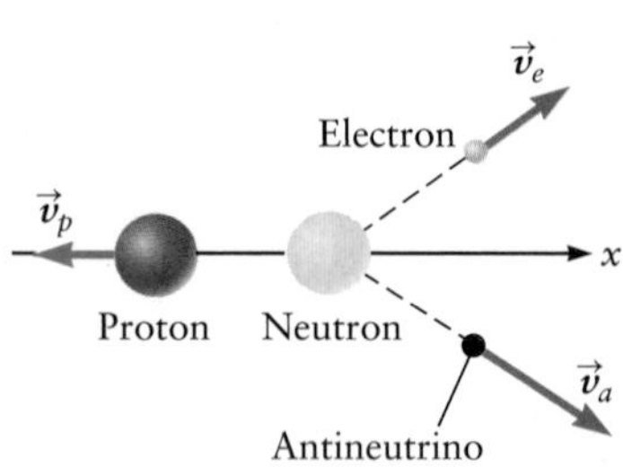

▲ **Figure 7.32** Diagram of the actual decay of a neutron. There are three outgoing particles: a proton, an electron, and an antineutrino. The speed of the outgoing electron is not the same for every decay.

[3]We will learn in Chapter 27 that our Newton's law expressions for momentum and kinetic energy begin to break down when applied to the motion of neutrons, protons, electrons, and neutrinos moving at high speeds. However, these expressions are still a useful approximation and will take us to the correct conclusions regarding the decay of the neutron.

The correct explanation was not this drastic, however. It was eventually found that a neutron actually decays into *three* particles (Fig. 7.32):

neutron → proton + electron + antineutrino (correct!)

The third particle was an entirely new type of particle called an ***antineutrino***. Adding the contributions of the antineutrino to the final momentum and kinetic energy in Equations 7.42 and 7.43 leads to a range of possible values for the final velocity of the electron, as found in the experiments. In fact, this experiment, together with its analysis in terms of conservation laws, marked the discovery of the antineutrino and led to the observation of a closely related particle called the neutrino. The point of this story is that a very careful test of conservation principles led to an entirely unexpected discovery.

Summary | CHAPTER 7

Key Concepts and Principles

Momentum

The ***momentum*** of a particle is given by

$$\vec{p} = m\vec{v} \quad (7.1)$$ **(page 208)**

The total momentum of a system of particles is equal to the sum of the momenta of the individual particles.

Impulse theorem

When a constant force acts on an object during a time interval Δt, it imparts an ***impulse*** to the object equal to $\vec{F}\,\Delta t$. The impulse is equal to the area under the force–time curve. The impulse imparted to an object is equal to the change in the momentum of the object:

$$\text{impulse} = \vec{F}\,\Delta t = \Delta\vec{p} \quad (7.5)$$ **(page 210)**

Conservation of momentum

If there is no external force on a system of particles, the total momentum of the system is constant. In such a case, momentum is ***conserved***.

Applications

Collisions

The two objects involved in a collision can be thought of as a "system." During a collision, forces external to the system are usually very small and the total momentum of the system after the collision is equal to the total momentum of the system before the collision. The momentum is conserved.

- In an ***elastic collision***, both momentum and kinetic energy are conserved.
- In an ***inelastic collision***, momentum is conserved but kinetic energy is not conserved.
- In a ***completely inelastic collision***, the two objects stick together after the collision. Momentum is still conserved.

(Continued)

Center of mass

Most real objects have a size and shape and can be thought of as being composed of many separate pieces. Such extended objects possess a ***center of mass***, and the motion of this point is very simple. For any object, no matter how complicated, the center of mass moves in response to the total external force acting on the object. This motion follows Newton's laws as if all the mass were located at the center of mass. For a system of two particles, the x coordinate of the center of mass is given by

$$x_{CM} = \frac{m_1 x_1 + m_2 x_2}{m_1 + m_2} \qquad (7.33)\ (\text{page } 229)$$

with a similar expression for y_{CM} and z_{CM}. For a system of many particles, the result is

$$x_{CM} = \frac{\sum_i m_i x_i}{\sum_i m_i} = \frac{\sum_i m_i x_i}{M_{\text{tot}}} \qquad (7.34)\ (\text{page } 229)$$

where M_{tot} is the total mass of the system.

Questions

SSM = answer in *Student Companion & Problem-Solving Guide* ⓧ = life science application

1. SSM A bomb that is initially at rest breaks into several pieces of approximately equal mass, two of which are shown with their velocity vectors in Figure Q7.1. Use conservation of momentum to determine if there might be other pieces of the bomb not shown in the figure. Assume there is only one missing piece and estimate the direction it is traveling after the explosion.

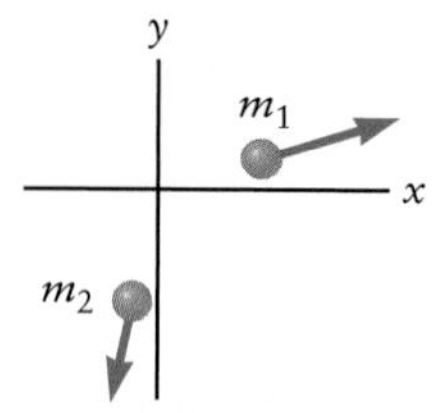

Figure Q7.1

2. Ⓡ The boxes in Figure Q7.2 all have the same mass and size. What is the approximate location of the center of mass of the two boxes in case 1, the three boxes in case 2, and the four boxes in case 3?

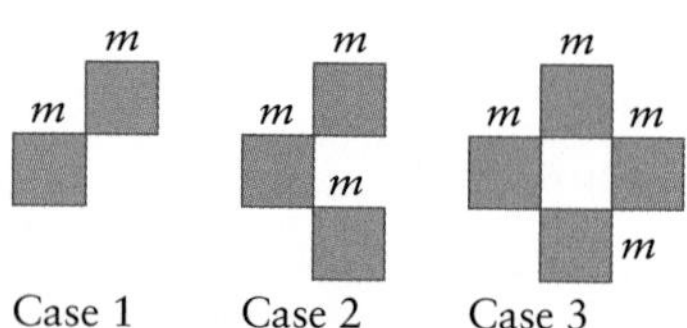

Figure Q7.2

3. Ⓡ What is the approximate location of the center of mass of the boomerang in Figure Q7.3?

4. Explain why the center of mass of an object (such as the boomerang in Fig. Q7.3) may not be "inside" the object. We discussed how the center of mass can be thought of as the "balance point" of an object. How might you use the idea of a balance point to locate the center of mass in cases in which the center of mass is outside the object?

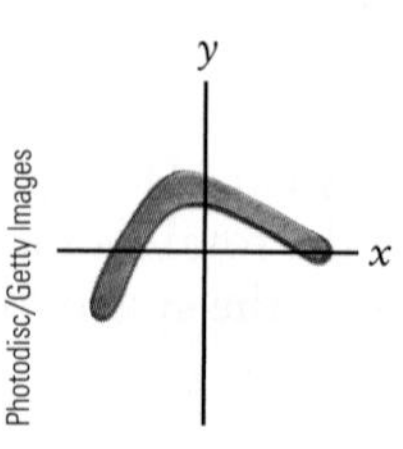

Figure Q7.3 Questions 3 and 4.

5. If the magnitude of the momentum of an object is increased by a factor of three, by what factor is the kinetic energy changed? Assume the mass of the object is unchanged.

6. Two objects of different mass have the same kinetic energy. Which one has the larger momentum?

7. A baseball has a certain momentum p_{ball} and a certain kinetic energy KE_{ball}. One golf ball (ball 1) is given a momentum equal to p_{ball}, and a second golf ball (ball 2) is given a kinetic energy equal to KE_{ball}. Which golf ball has the higher speed?

8. Consider the momentum of an object that is undergoing projectile motion. Is its momentum conserved? If the answer is no, is any component of the momentum conserved? Explain.

9. In which case is the momentum change of a baseball largest, (a) a pitch caught by a catcher, (b) a baseball thrown by a pitcher, or (c) a baseball hit by a batter? Explain your answer.

10. **Deep impact?** Consider the collision between an asteroid and a rocket as discussed in connection with Figure 7.17. We have already calculated the deflection angle that results from the collision with the rocket (Eq. 7.27), but we need to also determine if this deflection will actually save the Earth. If we assume the collision takes place near the Moon, will the asteroid miss the Earth? Ignore the effect of the Earth's gravitational force on the asteroid.

11. Finish the analysis of the force in the automobile collision in Example 7.7. Find the average collision force along y and the magnitude of the total collision force.

12. A cue ball bounces off the rail of a pool table as shown in Figure Q7.12. The momentum of the ball changes. (a) What is the source of the impulse? (b) That is, what is the source of the force? (c) Explain why the answers to (a) and (b) are closely(!) related.

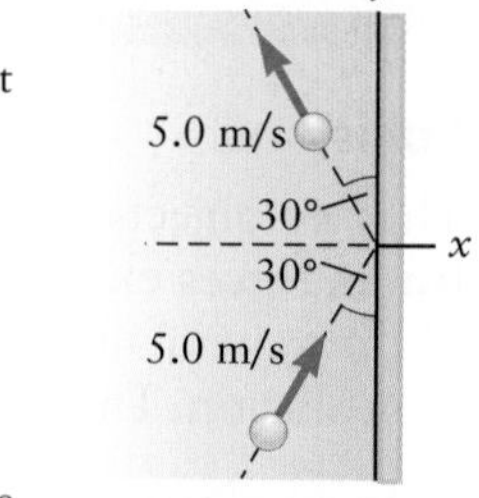

Figure Q7.12 Question 12 and Problems 8 and 9.

13. Two crates are on an icy, frictionless, horizontal surface. One of the crates is then given a push and collides with the other. If the crates stick together after the collision, which of the following quantities is conserved in the collision, (a) horizontal momentum, (b) kinetic energy, (c) both, or (d) neither?

14. SSM A tennis ball and a ball of soft clay are dropped from the same height onto the floor below. The force from the floor

produces an impulse on each ball. If the balls have the same mass, which impulse is larger? Explain.

15. When you fire a gun, you experience a recoil force due to the impulse produced by the bullet on the barrel of the gun. Is it possible to design a gun in which the recoil force is zero? If so, explain how you could do it. Why do you think guns are not made in this way?

16. Two masses m_1 and m_2 have the same momentum. If the masses are not equal, which statement(s) are correct?
(a) The velocities are the same.
(b) The velocities are in the same direction.
(c) The kinetic energies are the same.

17. Ⓧ In the event of a two-car, head-on collision, would you rather be riding in a heavy car (large mass) or a small car? Justify your answer in terms of the impulse and force on a car during a collision.

18. Ⓧ Explain why a padded dashboard makes a car safer in the event of an accident.

19. An astronaut is working in distant space (outside his spaceship and far from any planets or stars) and floating freely when he accidently throws a wrench. (a) Is the astronaut's momentum conserved (i.e., is his momentum the same before and after he throws the wrench)? (b) Is the wrench's momentum conserved? (c) Is there a system in this situation whose momentum is conserved?

20. As an apple falls from a tree, its speed increases, so the magnitude of its momentum also increases. Explain why this is consistent with the principle of conservation of momentum.

Problems

SSM = solution in *Student Companion & Problem-Solving Guide*
★ = intermediate ✪ = challenging Ⓧ = life science application
Ⓡ = reasoning and relationships problem
RT = reasoning tutorial in ENHANCED WebAssign

7.1 MOMENTUM

1. What is the magnitude of the momentum of a baseball (m = 0.14 kg) traveling at a speed of 100 mi/h? Express your answer in SI units.

2. Consider two cars, one a compact model (m = 800 kg) and the other an SUV (m = 2500 kg). The SUV is traveling at 10 m/s, while the compact car is traveling at an unknown speed v. If the two cars have the same momentum, what is v?

3. Consider two cars that are on course for a head-on collision. If they have masses m_1 = 1200 kg and m_2 = 1800 kg and are both traveling at 30 m/s, what is the magnitude of the total momentum?

4. SSM Two particles of mass m_1 = 1.2 kg and m_2 = 2.9 kg are traveling as shown in Figure P7.4. What is the total momentum of this system? Be sure to give the magnitude and direction of the momentum.

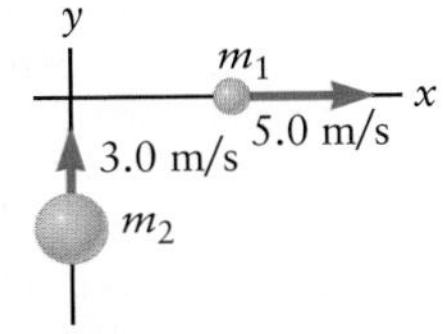

Figure P7.4

7.2 FORCE AND IMPULSE

5. ★ **Slap shot!** A hockey player strikes a puck that is initially at rest. The force exerted by the stick on the puck is 1000 N, and the stick is in contact with the puck for 5.0 ms (0.0050 s). (a) Find the impulse imparted by the stick to the puck. (b) What is the speed of the puck (m = 0.12 kg) just after it leaves the hockey stick?

6. A constant force of magnitude 25 N acts on an object for 3.0 s. What is the magnitude of the impulse?

7. A rubber ball (mass 0.25 kg) is dropped from a height of 1.5 m onto the floor. Just after bouncing from the floor, the ball has a speed of 4.0 m/s. (a) What are the magnitude and direction of the impulse imparted by the floor to the ball? (b) If the average force of the floor on the ball is 18 N, how long is the ball in contact with the floor? Ignore the force of gravity during the collision of the ball with the floor.

8. ★ Ⓡ For the cue ball in Figure Q7.12, make a qualitative sketch of how the force exerted by the rail on the ball varies with time. Be sure to plot both the x and y components of the force and indicate when the ball is in contact with the rail.

9. ★ Ⓡ Consider again the cue ball in Question 12 and Figure Q7.12. (a) What is the impulse imparted to the ball? Be sure to give the magnitude and direction. (b) What is the change in momentum of the ball? Be sure to give the magnitude and direction.

10. ★ A rubber ball is dropped and bounces vertically from a horizontal concrete floor. If the ball has a speed of 3.0 m/s just before striking the floor and a speed of 2.5 m/s just after bouncing, what is the average force of the floor on the ball? Assume the ball is in contact with the floor for 0.12 s and the mass of the ball is 0.15 kg.

11. SSM ★ A baseball player hits a baseball (m = 0.14 kg) as shown in Figure P7.11. The ball is initially traveling horizontally with a speed of 40 m/s. The batter hits a fly ball as shown, with speed v_f = 55 m/s. (a) What are the magnitude and direction of the impulse imparted to the ball? (b) If the ball and bat are in contact for a time of 8.0 ms, what is the magnitude of the average force of the bat on the ball? Compare this answer to the weight of the ball. (c) What is the impulse imparted to the bat?

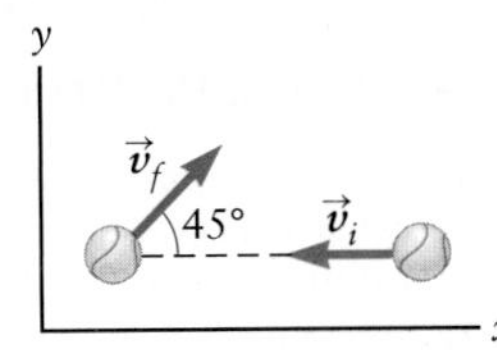

Figure P7.11 Problems 11 and 12.

12. The baseball in Problem 11 is replaced by a very flexible rubber ball of equal mass. (a) Does the contact time increase or decrease? (b) If the contact time changes by a factor of 50 but the initial and final velocities are the same, by what factor does the force of the bat on the ball change?

13. ★ An egg and a tomato of equal mass are dropped from the roof of a three-story building onto the sidewalk below. The sidewalk imparts an impulse to both as they splatter. (a) Is the impulse imparted to the egg greater than, less than, or equal to the impulse imparted to the tomato? (b) Is the average force exerted on the egg greater than, less than, or equal to the force on the tomato?

For Problems 14 through 18, you will need to estimate the collision time, that is, the time two objects are in contact during a collision. For comparison, the collision time for a baseball–bat collision is approximately 0.001 s, while the time for a tennis racket–ball collision is about 0.01 s. In general, hard things produce short contact times and soft things lead to long contact times.

14. ★ RT Consider an archer who fires an arrow. The arrow has a mass of 0.030 kg and leaves the bow with a horizontal velocity of 80 m/s. (a) What is the impulse imparted to the arrow? Give both the magnitude and direction of the impulse. (b) What is the approximate average force of the bow string on the arrow?

15. ★ RT Consider a tennis player who hits a serve at 140 mi/h (63 m/s). Estimate the average force exerted by the racket on the ball ($m = 57$ g).

16. ★ RT Consider the problem of returning a serve in tennis. A serve is hit at a speed of 63 m/s and is hit back to the server with a speed of 40 m/s. (a) What is the impulse imparted to the ball? (b) What is the approximate average force on the ball ($m = 57$ g)? For simplicity, assume one-dimensional motion; that is, assume the serve and the return travel along the same line.

17. SSM ★ R A golf ball is hit from the tee and travels a distance of 300 yards. Estimate the magnitude of the impulse imparted to the golf ball. Ignore air drag in your analysis.

18. ★ ✖ R **Uppercut!** A boxer hits an opponent on the chin and imparts an impulse of 500 N · s. Estimate the magnitude of the average force.

19. RT Suppose the force on an object varies with time as shown in Figure P7.19. Estimate the impulse imparted to the object. Assume the motion is one dimensional.

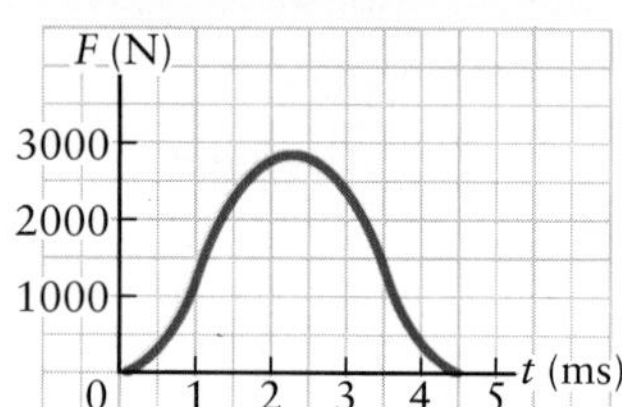

Figure P7.19

7.3 CONSERVATION OF MOMENTUM

7.4 COLLISIONS

20. ★ Two objects of mass m and $3m$ undergo a completely inelastic collision in one dimension. If the two objects are at rest after the collision, what was the ratio of their speeds before the collision?

21. Two particles of mass $m_1 = 1.5$ kg and $m_2 = 3.5$ kg undergo a one-dimensional head-on collision as shown in Figure P7.21. Their initial velocities along x are $v_{1i} = 12$ m/s and $v_{2i} = -7.5$ m/s. The two particles stick together after the collision (a completely inelastic collision). (a) Find the velocity after the collision. (b) How much kinetic energy is lost in the collision?

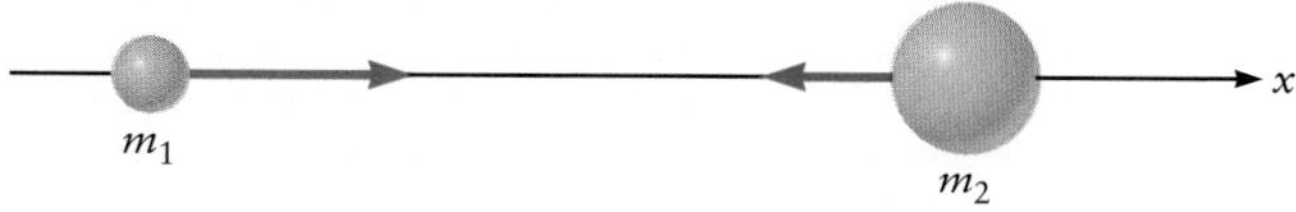

Figure P7.21 Problems 21 and 23.

22. Two skaters are studying collisions on an ice-covered (frictionless) lake. Skater 1 ($m_1 = 85$ kg) is initially traveling with a speed of 5.0 m/s, and skater 2 ($m_2 = 120$ kg) is initially at rest. Skater 1 then "collides" with skater 2, and they lock arms and travel away together. (a) Identify a "system" whose momentum is conserved. (b) Draw a sketch of your system in part (a) before the collision. Include a coordinate system and identify the initial velocities of the different parts of the system. (c) Express the conservation condition for your system. (d) Does your system undergo an elastic collision or an inelastic collision? How do you know? (e) Solve for the final velocity of the two skaters.

23. The particles in Figure P7.21 ($m_1 = 1.5$ kg and $m_2 = 3.5$ kg) undergo an elastic collision in one dimension. Their velocities before the collision are $v_{1i} = 12$ m/s and $v_{2i} = -7.5$ m/s. Find the velocities of the two particles after the collision.

24. Two hockey pucks approach each other as shown in Figure P7.24. Puck 1 has an initial speed of 20 m/s, and puck 2 has an initial speed of 15 m/s. They collide, and some glue on one of the pucks causes them to stick together. (a) If the two pucks form a "system," is the momentum of this system along x or y conserved? (b) Find the components along x and y of the initial velocities of both pucks. (c) Express the conservation of momentum condition for motion along x and y. (d) Is this collision elastic or inelastic? (e) Find the final velocity of the two pucks after the collision. (f) What fraction of the initial kinetic energy is lost in the collision?

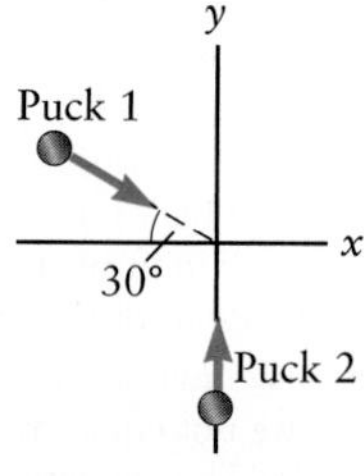

Figure P7.24 Problems 24 and 25.

25. SSM ★ Consider again the collision between two hockey pucks in Figure P7.24, but now they do not stick together. Their speeds before the collision are $v_{1i} = 20$ m/s and $v_{2i} = 15$ m/s. It is found that after the collision one of the pucks is moving along x with a speed of 10 m/s. What is the final velocity of the other puck?

26. ✚ In Example 7.5, we considered an elastic collision in one dimension in which one of the objects is initially at rest. Notice that mass 2 is the one that is initially at rest. Use the results for v_{1f} and v_{2f} to answer the following problems. (a) Assume $m_1 = 10m_2$ so that the incoming object is much more massive than the one at rest. Find v_{1f}/v_{1i} and v_{2f}/v_{1i}. Explain why the final velocity of the smaller object is much larger (in magnitude) than the final velocity of the bigger object. (b) Assume $m_1 = m_2/10$. Find v_{1f}/v_{1i} and v_{2f}/v_{1i}. Explain why the final velocity of the smaller object is opposite its initial direction.

27. ★ Two billiard balls undergo an elastic collision as shown in Figure P7.27. Ball 1 is initially traveling along x with a speed of 10 m/s, and ball 2 is at rest. After the collision, ball 1 moves away with a speed of 4.7 m/s at an angle $\theta = 60°$. (a) Find the speed of ball 2 after the collision. (b) What angle does the final velocity of ball 2 make with the x axis?

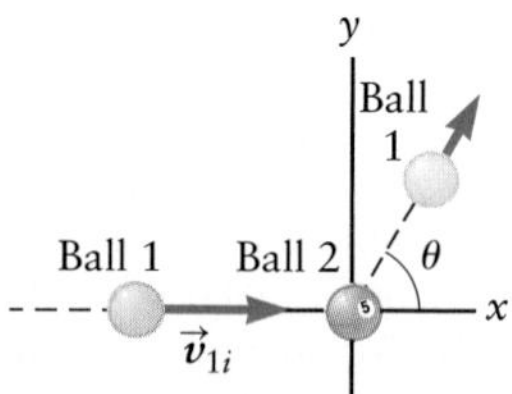

Figure P7.27

28. Consider an elastic collision in one dimension that involves objects of mass 2.5 kg and 4.5 kg. The larger mass is initially at rest, and the smaller one has an initial velocity of 12 m/s. Find the velocities of the two objects after the collision.

29. ★ Two hockey players are traveling at velocities of $v_1 = 12$ m/s and $v_2 = -18$ m/s when they undergo a head-on collision. After the collision, they grab each other and slide away together with a velocity of -4.0 m/s. Hockey player 1 has a mass of 120 kg. What is the mass of the other player?

30. SSM ★ Two cars of equal mass are traveling as shown in Figure P7.30 just before undergoing a collision. Before the collision, one of the cars has a speed of 18 m/s along $+x$ and the other has a speed of 25 m/s along $+y$. The cars lock bumpers and then slide away together after the collision. What are the magnitude and direction of their final velocity?

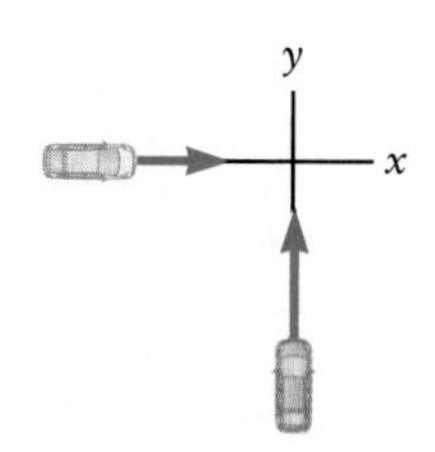

Figure P7.30

31. Two cars collide head-on and lock bumpers on an icy (frictionless) road. One car has a mass of 800 kg and an initial speed of 12 m/s and is moving toward the north. The other car has a mass of 1200 kg. If the speed of the cars after the collision is 4.5 m/s and they are moving north, what was the initial velocity of the large car?

7.5 USING MOMENTUM CONSERVATION TO ANALYZE INELASTIC EVENTS

32. A railroad car ($m = 3000$ kg) is coasting along a level track with an initial speed of 25 m/s. A load of coal is then dropped into the car as sketched in Figure 7.19. (a) Treat the railroad car plus coal

as a system. Is the momentum of this system along the horizontal (x) or vertical (y) conserved? (b) What are the velocities of the car and the coal before the collision along the direction found in part (a)? (c) Express the conservation of momentum condition for the system. (d) If the final speed of the car plus coal is 20 m/s, what is the mass of the coal?

33. A baseball pitcher (m = 80 kg) is initially standing at rest on an extremely slippery, icy surface. He then throws a baseball (m = 0.14 kg) with a horizontal velocity of 50 m/s. What is the recoil velocity of the pitcher?

34. A military gun is mounted on a railroad car (m = 1500 kg) as shown in Figure P7.34. There is no frictional force on the car, the track is horizontal, and the car is initially at rest. The gun then fires a shell of mass 30 kg with a velocity of 300 m/s at an angle of 40° with respect to the horizontal. Find the recoil velocity of the car.

Figure P7.34

35. An open railroad car of mass 2500 kg is coasting with an initial speed of 14 m/s on a frictionless, horizontal track. It is raining, and water begins to accumulate in the car. After some time, it is found that the speed of the car is only 11 m/s. How much water (in kilograms) has accumulated in the car?

36. The M79 grenade launcher was first used by the U.S. Army in 1961. It fires a grenade with an approximate mass $m \approx 7.0$ kg at a speed of 75 m/s. (a) If a soldier fires the M79 horizontally while standing on a very slippery surface, what is the recoil speed of the soldier? Take the mass of the soldier plus M79 to be 100 kg. (b) The barrel of the M79 is 36 cm long. Estimate the average recoil force on the soldier.

37. A skateboarder (m = 85 kg) takes a running jump onto a friend's skateboard that is initially at rest as sketched in Figure P7.37. The friend is standing on the skateboard (mass of friend plus skateboard = 110 kg). After landing on the skateboard, the velocity of the board plus the two skateboarders is 3.0 m/s. What was the horizontal component of the velocity of the jumping skateboarder just before he landed on the skateboard?

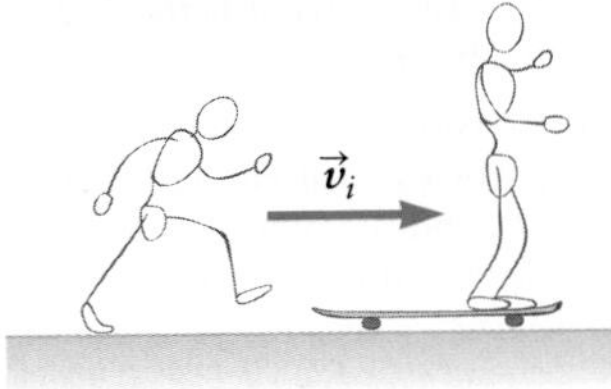

Figure P7.37

38. A bullet of mass 15 g is fired with an initial speed of 300 m/s into a wooden block that is initially at rest. The bullet becomes lodged in the block, and the bullet and block then slide together on the floor for a distance of 1.5 m before coming to rest (Fig. P7.38). If the coefficient of friction between the block and the floor is 0.40, what is the mass of the block?

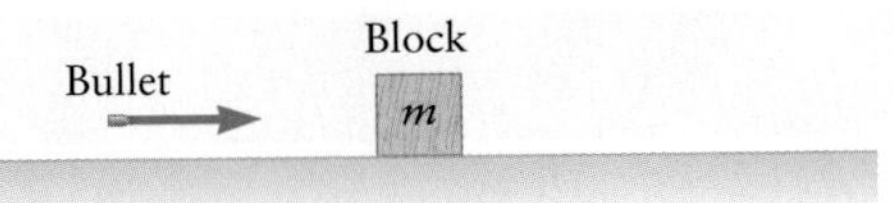

Figure P7.38

39. An airplane is sent on a rescue mission to help a stranded explorer near the South Pole. The plane cannot land, but instead drops a supply package of mass m = 60 kg while it is flying horizontally with speed v_p = 120 m/s. The supply package drops onto a waiting sled that is initially at rest (Fig. P7.39). After the package lands in the sled, the speed of the sled plus package is found to be 30 m/s. What is the mass of the sled? Ignore air drag.

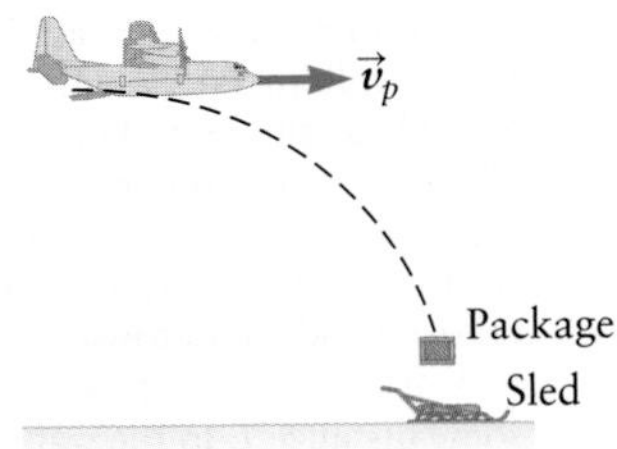

Figure P7.39

40. SSM A railroad car containing explosive material is initially traveling south at a speed of 5.0 m/s on level ground. The total mass of the car plus explosives is 3.0×10^4 kg. An accidental spark ignites the explosive, and the car breaks into two pieces, which then roll away along the same track. If one piece has a mass of 2.0×10^4 kg and a final speed of 2.5 m/s toward the south, what is the final speed of the other piece?

7.6 CENTER OF MASS

41. Find the center of mass of the two particles in Figure P7.41.

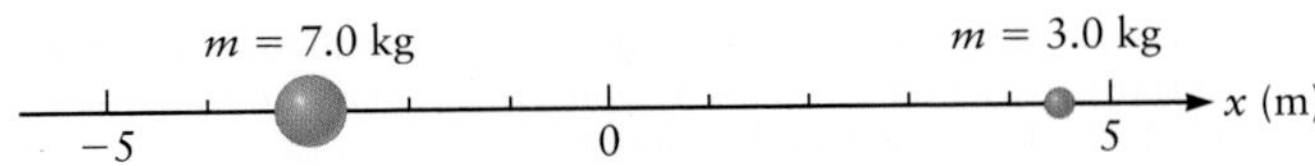

Figure P7.41

42. Find the center of mass of the four particles in Figure P7.42. Their masses are m_1 = 12 kg, m_2 = 18 kg, m_3 = 7.9 kg, and m_4 = 23 kg.

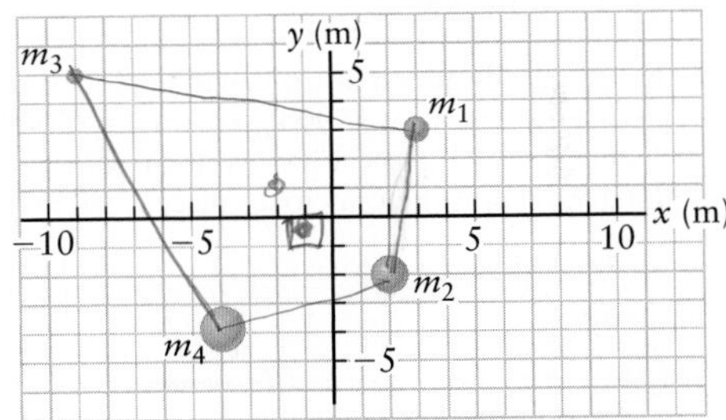

Figure P7.42

43. Two objects are moving along the x axis with velocities of 35 m/s (object 1) and −25 m/s (object 2). (a) If the center of mass has a velocity of +10 m/s, which object has the greater mass? (b) What is the ratio of their masses?

44. SSM Consider the motion of the two ice skaters in Figure 7.28 and assume they have masses of 60 kg and 100 kg. If the larger skater has moved a distance of 12 m from his initial position, where is the smaller skater?

45. RT Figure P7.45 shows several arrangements of blocks. In cases 1 through 3, the blocks all have the same mass, but in case 4, one of the blocks has a mass twice that of the other two blocks.

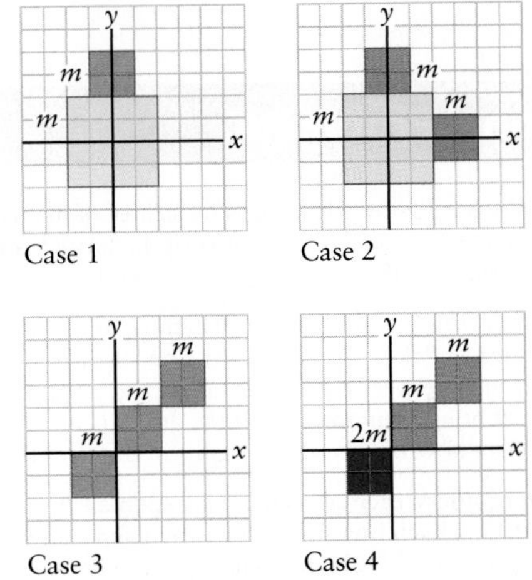

Figure P7.45

Find the approximate center of mass for each arrangement. Assume the grid lines are spaced 1 m apart in both the x and y directions.

46. Estimate the location of the center of mass of the thin metal plate in Figure P7.46. Assume the grid lines are spaced 1 cm apart in both the x and y directions.

Figure P7.46

47. Consider the Sun and the Earth as a "system" of two "particles." The center of mass of this system lies a distance L from the center of the Sun. Find L and compare it to the radius of the Sun.

48. A person of mass 90 kg standing on a frictionless, horizontal surface fires a bullet of mass 40 g in a horizontal direction. After the bullet has traveled a distance of 50 m, what distance has the person moved?

49. Consider the motion of an apple and the Earth as the apple falls to the ground. If the apple begins at the top of a tree of height 15 m, how much does the Earth move?

7.7 A BOUNCING BALL AND MOMENTUM CONSERVATION

50. A bomb that is initially at rest at the origin explodes and breaks into two pieces of mass $m_1 = 1500$ kg and $m_2 = 2500$ kg as shown in Figure P7.50. If piece 1 has a velocity of +35 m/s directed along x, what is the velocity of piece 2 after the explosion?

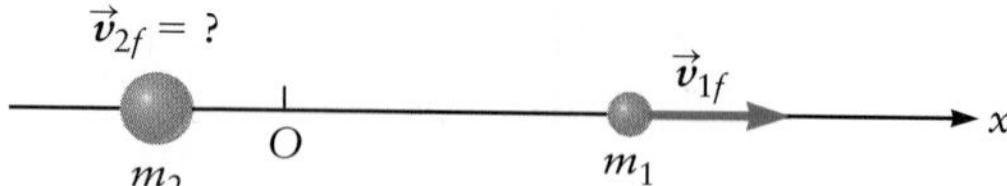

Figure P7.50

51. SSM A hand grenade is thrown with speed $v_0 = 30$ m/s, as sketched in Figure P7.51, just prior to exploding. It breaks into two pieces of equal mass after the explosion. One piece has a final velocity of 40 m/s along x. Find the velocity of the other piece after the explosion.

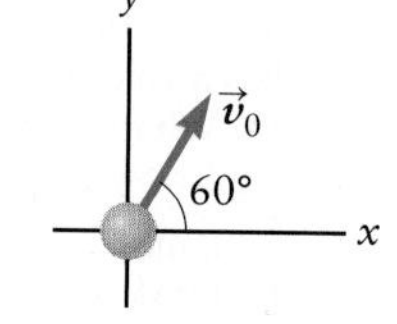

Figure P7.51

52. **The ballistic pendulum.** A bullet of mass $m = 30$ g is fired into a wooden block of mass $M = 5.0$ kg as shown in Figure P7.52. The block is attached to a string of length 1.5 m. The bullet is embedded in the block, causing the block to then swing as shown in the figure. If the block reaches a maximum height $h = 0.30$ m, what was the initial speed of the bullet?

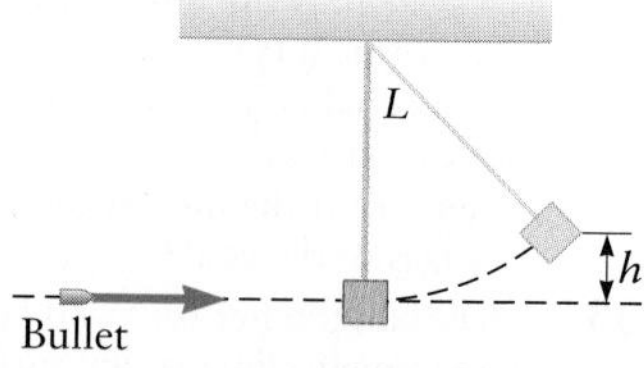

Figure P7.52

53. A rock of mass 30 g strikes a car windshield while the car is traveling on a highway. Estimate the force of the rock on the windshield. *Hint*: You will need to estimate the collision time (i.e., the time that the rock is in contact with the windshield). For comparison, the collision time for a baseball–bat collision is approximately 0.001 s, while the time for a tennis racket–ball collision is about 0.01 s.

54. Two particles of mass 2.3 kg and mass 4.3 kg that are free to move on a horizontal track are initially held at rest so that they compress a spring as shown in Figure P7.54. The spring has a spring constant $k = 400$ N/m and is compressed 0.12 m. Find the final velocity of each particle.

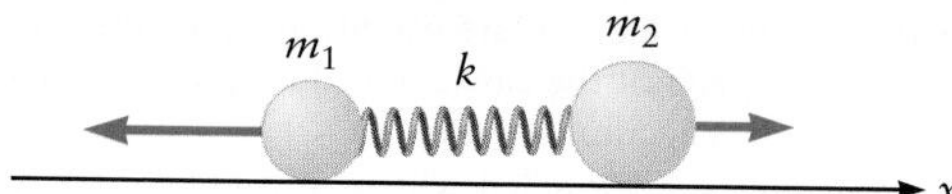

Figure P7.54

55. A tennis ball of mass 0.12 kg is dropped vertically onto a hard concrete floor from a height of 1.5 m. The ball then bounces up to some maximum height h. If the ball bounces up to a height $h = 1.0$ m, what is the approximate average force of the floor on the ball?

56. Consider a reaction that involves the creation of two elementary particles as in Figure P7.56. One particle is a proton of mass m_p, and the other has mass $m_2 = m_p/1800$. If the proton leaves the reaction with speed v_p, what is the final speed of the other particle? Express your answer in terms of v_p.

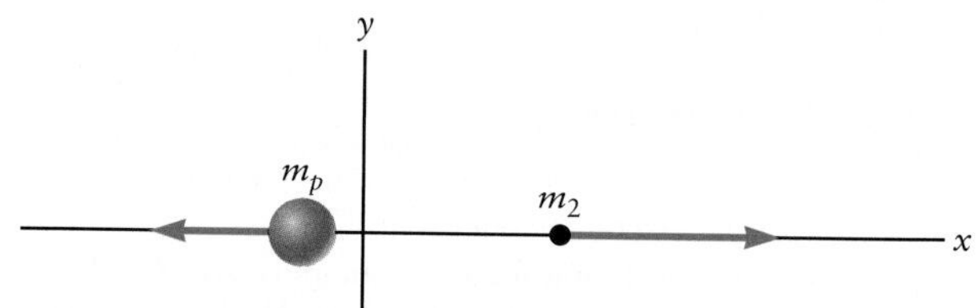

Figure P7.56

Additional Problems

57. **Finding the barycenter.** How far is the center of mass of the Earth–Moon system from the center of the Earth? Is it below the surface of the Earth?

58. A father ($m_F = 75$ kg) and his daughter ($m_D = 35$ kg) stand on a flat, frozen lake of negligible friction. They hold a 10-m-long rope stretched between them. The father and daughter then pull the rope to bring them together. If the father is initially standing at the origin, how far from the origin will they meet?

59. Twin brothers, each of mass 50 kg, sit in a symmetric canoe of mass 40 kg. Each boy sits 1.5 m away from the center of the canoe as they toss a basketball (mass = 0.60 kg) back and forth. (a) How much does the center of the canoe move as the ball moves from one brother to the other? (b) How much would the canoe shift on each throw if instead they used a medicine ball of mass 11 kg?

60. A block of mass 3.5 kg is initially at rest on a wedge of mass 20 kg, height 0.30 m, and incline angle $\theta = 35°$ as shown in Figure P7.60. There is no friction between the wedge and the floor. Starting at the top of the incline, the block is released and slides toward the bottom of the wedge. At the same time, the wedge "recoils" and slides some distance L to the right. Find L when the block has reached the bottom of the wedge. *Hint*: The center of mass of a right triangle of height h is a distance $h/3$ above the base of the triangle.

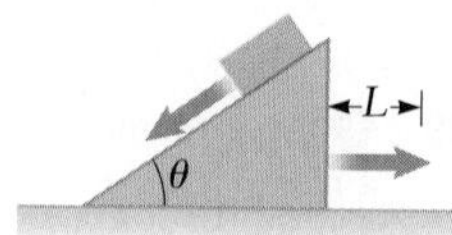

Figure P7.60

61. Consider three scenarios for a one-dimensional elastic collision in which a "bullet" is fired at a stationary target. For each scenario, determine the final velocities of both the bullet, v_{1f}, and the target, v_{2f}, in terms of the bullet's initial velocity, v_0. Assume both the bullet and target are on a horizontal surface for which friction is negligible. Case (a): Both the bullet and target are bowling balls; $m_1 = m_2 = 7.25$ kg. Case (b): The target is one of the bowling balls, but the bullet is a Styrofoam ball of the same diameter; $m_1 = 140$ g and $m_2 = 7.25$ kg. Case (c): Now make the Styrofoam ball the target and the bowling ball the bullet; $m_2 = 140$ g and $m_1 = 7.25$ kg. Compare parts (b) and (c) to the following extremes: (d) when the target is the Earth and the bullet is a tennis ball and (e) when the target is a frozen pea and the bullet is a truck's windshield.

62. **Fun with a Super Ball.** A Super Ball has a coefficient of restitution $\alpha = 0.90$. This means that the y components of the ball's velocity just before and after hitting the floor are related by $v_{fy} = -\alpha v_{if}$. It is dropped from a height of 1.8 m such that it goes under a table of height 70 cm. The ball bounces back and forth between the underside of the table and the ground (Fig. P7.62). How many total times will the ball bounce off the bottom of the table?

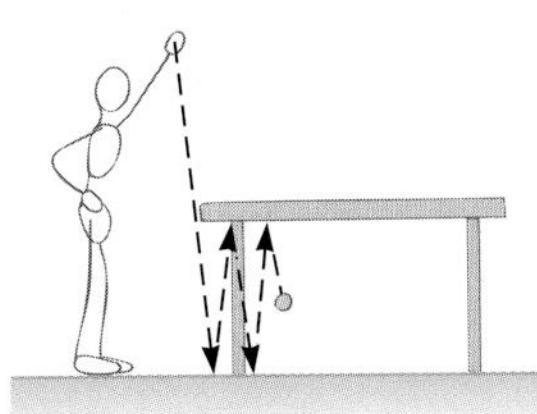

Figure P7.62

63. RT Consider again the discussion in Section 7.2 concerning air bags and impulse for a car colliding with a tree. The car's velocity of 40 mi/h is reduced to zero very quickly, but the head of a passenger is still moving at 40 mi/h (about 18 m/s). Calculate the average force exerted on the passenger's head coming to rest for the case (a) when it is stopped by the dashboard in 5.0 ms and (b) when it is stopped by an air bag that compresses in 45 ms. (c) Air bags must deploy quickly. If the distance between the passenger's head and the dashboard is 60 cm and it takes the air bag 45 ms to safely bring the passenger to rest, how much time is there for the air bag to inflate?

64. **Crumple zones.** A life-saving development in automobile manufacture is the invention of crumple zones, areas of the body and frame of a car deliberately made to collapse in a collision (Fig. P7.64) such that the passenger compartment will not. Introduced in 1955 on the Heckflosse made by Mercedes, crumple zones became mandatory on all cars sold in the United States in 1967. For a head-on collision with a tree at 40 mi/h (about 18 m/s), find the ratio of the average force on a car with a stiff frame (stop time of 0.010 s) to that on a car with a crumple zone (stop time of 0.25 s).

© Volvo Car Corporation

Figure P7.64 Front crumple zone on a Volvo.

65. A projectile of mass 10 kg is launched at an angle of 55° and an initial speed of 87 m/s. Just as the object reaches the maximum height in its trajectory, a small explosion along the horizontal blows off the projectile's back portion equal to one-fourth its original mass. The smaller piece lands exactly at the launch point as shown in Figure P7.65. How far from the launch point does the larger portion land?

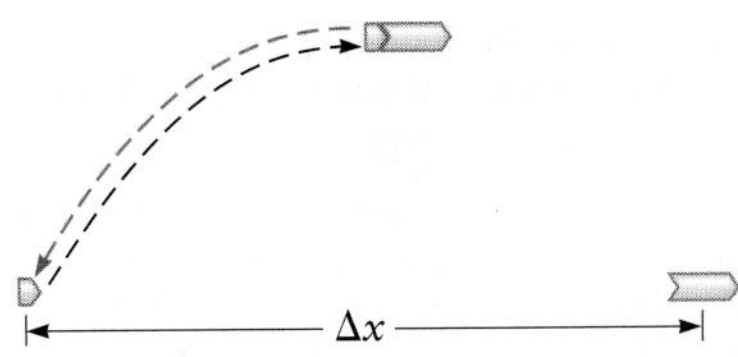

Figure P7.65

66. A cart of mass $m_1 = 10$ kg slides down a frictionless ramp and is made to collide with a second cart of mass $m_2 = 20$ kg, which then heads into a vertical loop of radius 0.25 m as shown in Figure P7.66. (a) Determine the height h from which cart 1 would need to start to make sure that cart 2 completes the loop without leaving the track. Assume an elastic collision. (b) Find the height needed if instead the more massive cart is allowed to slide down the ramp into the smaller cart.

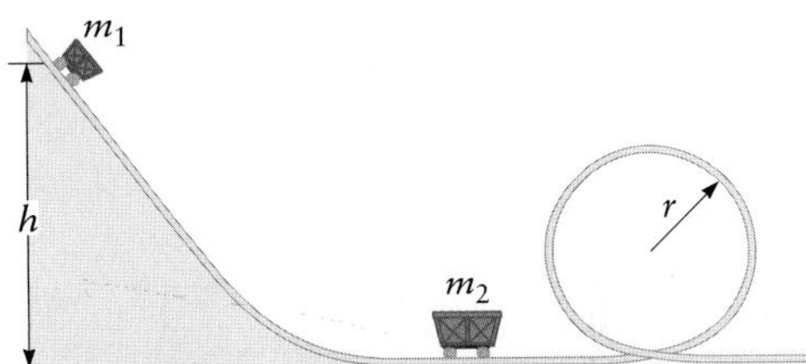

Figure P7.66 Problems 66 and 67.

67. Again consider the track and carts in Figure P7.66 with $m_1 = 10$ kg and $m_2 = 20$ kg, but this time the carts stick together after the collision. (a) Find the height h from which cart 1 would need to start to make sure that both carts complete the loop without leaving the track. (b) Find the height needed if instead the more massive cart is made to collide with the smaller cart.

68. A block of wood ($M_{block} = 5.2$ kg) rests on a horizontal surface for which the coefficient of kinetic friction is 0.28. A bullet ($m_{bullet} = 32$ g) is shot along the horizontal direction and quickly becomes stuck in the block. The block plus bullet then slide 87 cm before they come to rest. What was the initial speed of the bullet before impact?

69. A firefighter directs a horizontal stream of water toward a fire at an angle of 40° measured from the horizontal. The fire hydrant and hose deliver 95 gallons per minute (1 gal/min = 6.31×10^{-4} m^3/s), and the nozzle projects the water with a speed of 7.5 m/s. How much horizontal force must the firefighter exert on the hose to keep it stationary?

70. R A machine gun directs a flow of bullets at a steel target mounted on a cart with negligible friction and of mass 600 kg. In this case, the gun is an Uzi, which uses 9-mm (8.0-g) bullets and can fire at a rate of 10 rounds/s, and the bullets have an average muzzle velocity of 400 m/s. (a) Assuming each bullet's ricochet off the target can be approximated by an elastic collision, with what final velocity would the cart be moving if all 50 rounds in the Uzi's magazine were fired at the maximum rate at the target? (b) How fast would the cart be moving if the target were made of soft wood such that all 50 bullets became embedded? (c) The Uzi is now firmly attached to the cart while at rest. The trigger is then pulled, and all 50 rounds are discharged horizontally into the air. What is the final speed of the cart?

71. SSM Rocket engines work by expelling gas at a high speed as illustrated in Figure P7.71. We wish to design an engine that expels an amount of gas $m_g = 50$ kg each

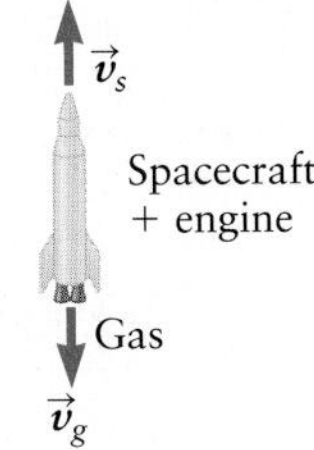

Figure P7.71

second, and we want the engine to exert a force $F = 20{,}000$ N on a rocket. At what speed v_g should the gas be expelled? Assume the mass of the rocket is very large.

72. Comet Shoemaker–Levy struck Jupiter in 1994. Just before the collision, the comet broke into several pieces. The total mass of all the pieces was estimated to be 3.8×10^8 kg, and they all had a velocity relative to Jupiter of 60 km/s. How much energy was deposited on Jupiter?

73. ★ R Figure P7.73 shows two stars with masses m_1 and m_2 moving in circular orbits (the dashed circles) due to their mutual gravitational attraction. What is the approximate value of the mass ratio m_1/m_2?

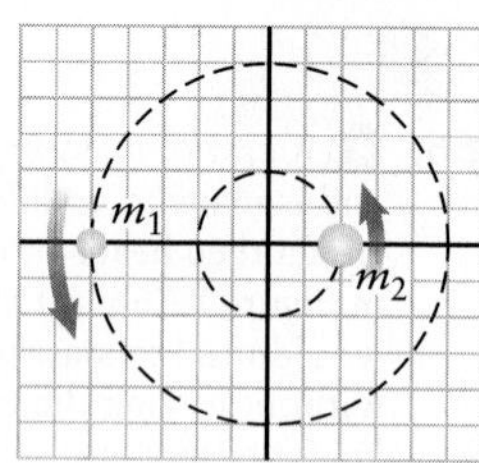

Figure P7.73

74. ★ RT A man ($m_1 = 90$ kg) is standing on a railroad flat car that is 9.0 m long and of unknown mass m_2. The man and the car are initially at rest on a level track (Fig. P7.74A), and the wheels of the car are frictionless. The man then runs to the opposite end of the car and the car moves a distance 1.5 m to the left (Fig. P7.74B). Estimate the mass of the car.

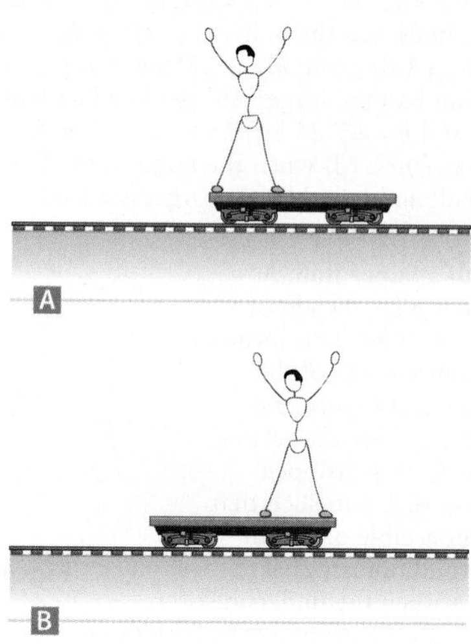

Figure P7.74

CHAPTER

8

◀ *Rotating objects have played a major role in science and technology for many years. Each of these windmills can provide the electrical power for about 500 homes. (Jon Boyes/ Getty Images)*

Rotational Motion

In previous chapters, we learned how to apply Newton's laws to deal with translational motion in many different situations, and in this work, we generally assumed our objects were "point particles." A point particle is one in which all the mass is located at an infinitesimal point in space. Real objects like soccer balls, cars, and windmills are definitely not point particles. For these objects, mass is distributed over a certain region of space; that is, they have a certain size and shape. Although the point particle treatment is the correct way to describe the *translational* motion of all objects, this approach does not allow us to deal with the ***rotational motion*** of an object. Our job in this chapter and Chapter 9 is to learn how to use Newton's laws to describe and analyze rotational motion. To do so, we must treat objects more realistically by accounting for their size and shape.

OUTLINE

8.1 Describing Rotational Motion

Top view of a rod rotating in the x–y plane. The rotation axis z is perpendicular to the plane of this figure.

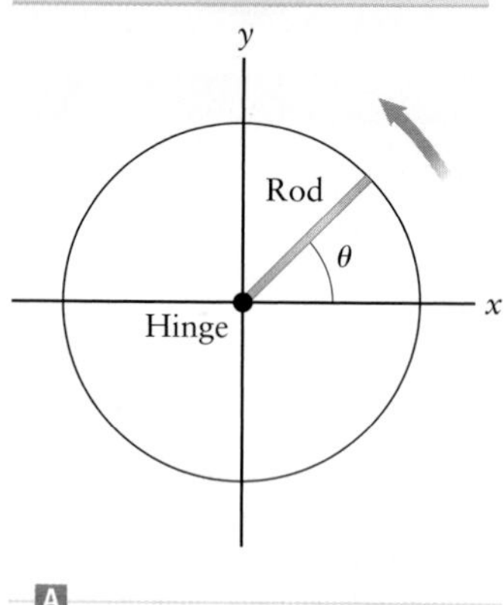

Perspective view of a CD in the x–y plane. The rotation axis is along z.

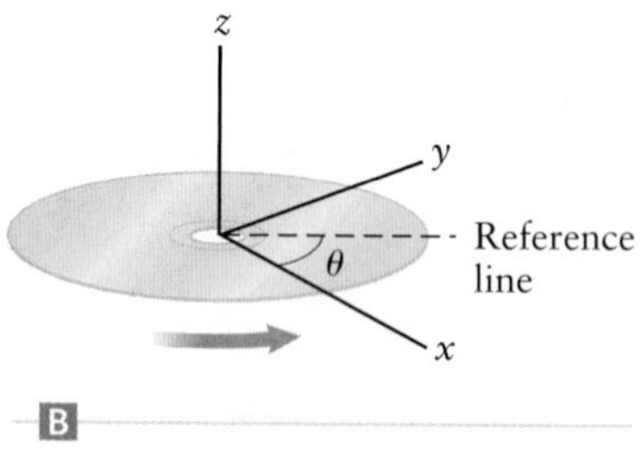

▲ **Figure 8.1** For both A a rotating rod and B a spinning compact disc, the angular position is described by the angle θ. In both of these examples, the rotation axis is the z coordinate axis.

In our studies of translational motion in Chapters 2 and 3, we introduced the concepts of position, velocity, and acceleration. Because translational motion usually involves motion along a trajectory or line in space, it is commonly referred to as ***linear motion***. Hence, the translational velocity and acceleration are often referred to as the linear velocity and linear acceleration. To deal with rotational motion, we now need to define the analogous rotational quantities: ***angular position***, ***angular velocity***, and ***angular acceleration***. We encountered angular position in Chapter 5 when we studied circular motion. Circular motion is related to rotational motion: in both cases, an object or a point on an object moves in a circular path. A simple particle undergoing circular motion as in Chapter 5 thus closely resembles a point on a rotating wheel or cylinder.

For problems involving rotational motion, the first step is to identify the ***rotation axis***. Figure 8.1 shows two different objects undergoing rotational motion. The rod in Figure 8.1A is hinged at one end so that it can rotate about the z axis. The angle θ gives the angular position of the rod, that is, the angle the rod makes with the x axis. Figure 8.1B shows a compact disc spinning about an axis passing through its center, and we again take this rotation axis to be along the z direction. The angular position of the CD can be specified by the angle θ that a reference line on the CD makes with the x axis. For both the hinged rod and the CD, the angle θ tells us all we need to know about the object's angular position.

Figure 8.2 shows the motion of our hinged rod in a little more detail. Suppose the moving end of the rod travels a distance s along its circular path. At the same time, the rod sweeps out an angle θ as sketched in Figure 8.2. If the rod has length r, the distance s and the angle θ are related by

$$\theta = \frac{s}{r} \tag{8.1}$$

where the angle θ is measured in ***radians*** (rad). Of course, angles can also be measured in degrees, and we use both units throughout this book. An angle of 360° is equivalent to 2π radians (one complete circle). (The measurement of angles using units of degrees and radians was discussed in Chapter 1. You might want to review that material now.)

CONCEPT CHECK 8.1 Measuring Angles in Degrees and Radians

What is the approximate value of the angle θ between the rod and the x axis in Figure 8.2?

(a) 90°	(b) 60°	(c) 0°
(d) π	(e) $\pi/2$	(f) $\pi/3$
(g) both values (a) and (e)		(h) both values (b) and (f)

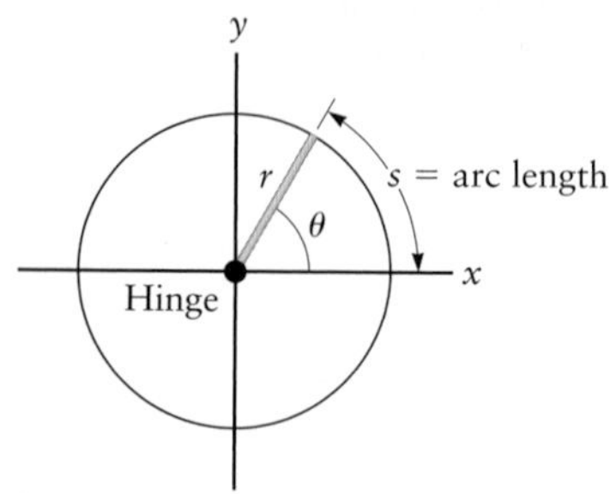

▲ **Figure 8.2** The arc length s measured along the circle is proportional to the angle $\theta = s/r$, measured in radians.

Angular Velocity and Acceleration

To describe the rotational motion of a hinged rod or a CD, we need a variable that describes how the angular position θ is changing with time. That variable is the ***angular velocity***, denoted by ω (lowercase Greek omega). The angular velocity ω is a measure of how rapidly the angle θ is changing with time. If we are interested in the rotational motion over a time interval Δt, we can define the average angular velocity as

$$\omega_{\text{ave}} = \frac{\Delta\theta}{\Delta t} \tag{8.2}$$

where $\Delta\theta$ is the change in the angle θ during this time interval. We can also define an instantaneous angular velocity given by

$$\omega = \lim_{\Delta t \to 0} \frac{\Delta\theta}{\Delta t} \tag{8.3}$$

In words, ω is the slope of the $\theta-t$ curve, which can be calculated by considering the change $\Delta\theta$ during a time interval Δt as this interval is allowed to shrink to zero ($\Delta t \to 0$). As with linear motion, the instantaneous angular velocity equals the average angular velocity when ω is constant. Angular velocity ω is usually measured in units of radians per second (rad/s), but can also be expressed in terms of revolutions per minute (rpm), where one revolution about the rotation axis corresponds to an angular change of 360°.

Because ω is called the angular *velocity*, you should expect it to be a vector quantity, just as the linear velocity $\vec{v}$ is a vector. All vectors have both a magnitude and a direction. The magnitude of the angular velocity is given by Equation 8.3; how do we determine its direction? The direction of ω is determined by the rotation axis. Figure 8.3 shows a rotating circular disc lying in the x–y plane with a rotation axis along $+z$. If the angle θ increases with time, the angular velocity is positive according to the definition in Equation 8.3. Hence, a counterclockwise rotation corresponds to a positive value of ω. A clockwise rotation would then give a negative value for ω.

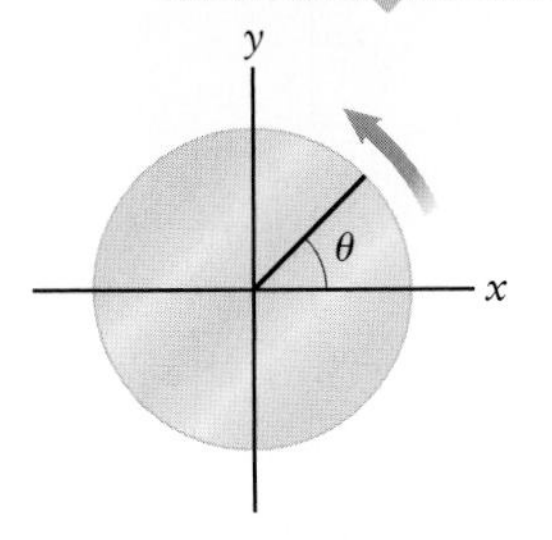

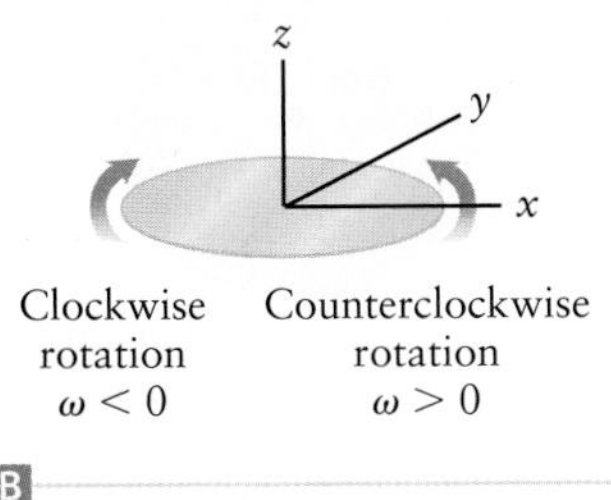

▲ **Figure 8.3** An increasing value of θ corresponds to counterclockwise motion. Hence, the angular velocity is positive if the object is rotating counterclockwise.

Acceleration is a very important quantity when dealing with translational motion, and the same is true with rotational motion. We now define the ***angular acceleration***, denoted by α (lowercase Greek alpha). The average angular acceleration during a time interval Δt is related to changes in ω by

$$\alpha_{\text{ave}} = \frac{\Delta\omega}{\Delta t} \tag{8.4}$$

while the instantaneous angular acceleration is defined by

Definition of angular acceleration

$$\alpha = \lim_{\Delta t \to 0} \frac{\Delta\omega}{\Delta t} \tag{8.5}$$

Thus, α is the slope of the $\omega-t$ curve and has units of rad/s^2. As with linear motion, the instantaneous angular acceleration equals the average angular acceleration when α is constant.

A good example of rotational motion is a ceiling fan (Fig. 8.4). Suppose the fan is spinning (i.e., rotating) at a constant angular velocity of 85 rpm. Let's calculate the angular velocity in rad/s and also find the total rotation angle after exactly 5 minutes.

To convert from units of rpm to rad/s, we know that one complete revolution corresponds to 2π radians, so calculating ω gives (to two significant figures)

$$\omega = 85\,\frac{\text{rev}}{\text{min}} \times \frac{2\pi\text{ rad}}{1\text{ rev}} \times \frac{1\text{ min}}{60\text{ s}} = 8.9\text{ rad/s}$$

In this example, ω is constant, so the instantaneous angular velocity equals the average angular velocity. We can therefore rearrange Equation 8.3 to get the angular displacement $\Delta\theta$:

$$\Delta\theta = \omega\,\Delta t$$

Inserting our value for ω and using $\Delta t = 5\text{ min} = 300\text{ s}$, we have

$$\Delta\theta = \omega\,\Delta t = \left(\frac{8.9\text{ rad}}{\text{s}}\right)(300\text{ s}) = 2700\text{ rad}$$

Note that $\Delta\theta = 2700\text{ rad}$ corresponds to $(2700\text{ rad}) \times (1\text{ rev}/2\pi\text{ rad}) = 430$ revolutions.

▲ **Figure 8.4** When this ceiling fan is turned on, the blades undergo rotational motion.

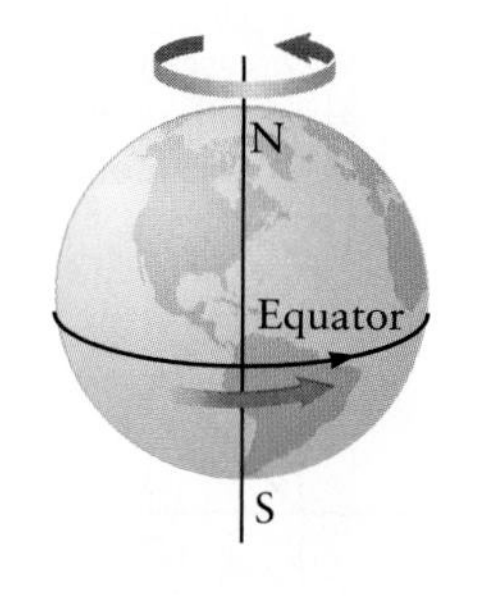

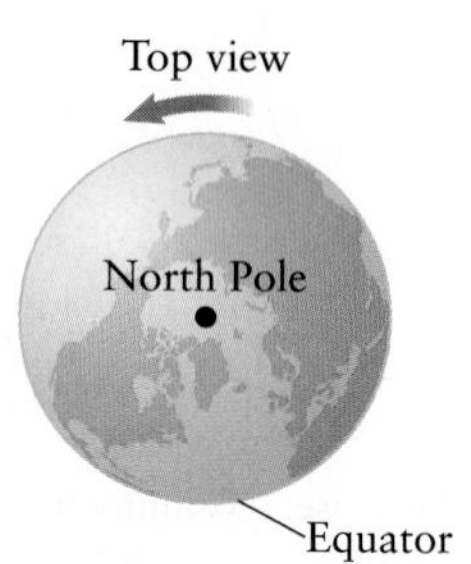

▲ **Figure 8.5** Example 8.1. The Earth undergoes rotational motion as it spins about its axis.

EXAMPLE 8.1 Angular Velocity of the Earth

The Earth rotates about the axis that connects the North and South Poles as sketched in Figure 8.5A. Find the Earth's angular velocity in rad/s.

RECOGNIZE THE PRINCIPLE

The angular velocity of the Earth is constant, so $\omega = \omega_{ave}$, which can be calculated from the angle $\Delta\theta$ through which the Earth rotates in a given amount of time Δt using Equation 8.2 or 8.3.

SKETCH THE PROBLEM

Figure 8.5B shows a top view of how the Earth's equator rotates (compare with Figs. 8.1 through 8.3).

IDENTIFY THE RELATIONSHIPS

The Earth completes one rotation ($\Delta\theta = 1$ rev $= 2\pi$ rad) every day, so its angular velocity is simply 1.0 rev/day.

SOLVE

Expressing 1.0 rev/day in rad/s, we find

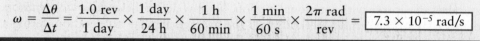

$$\omega = \frac{\Delta\theta}{\Delta t} = \frac{1.0 \text{ rev}}{1 \text{ day}} \times \frac{1 \text{ day}}{24 \text{ h}} \times \frac{1 \text{ h}}{60 \text{ min}} \times \frac{1 \text{ min}}{60 \text{ s}} \times \frac{2\pi \text{ rad}}{\text{rev}} = \boxed{7.3 \times 10^{-5} \text{ rad/s}}$$

▶ ***What does it mean?***

The Earth's angular velocity is constant, so its average angular velocity equals the instantaneous angular velocity. Also, a point on the Earth's equator moves in a circle with constant speed and thus undergoes uniform circular motion as discussed in Chapter 5.

▲ **Figure 8.6** Example 8.2. The information on a DVD is stored in a very long spiral "track."

EXAMPLE 8.2 Angular Acceleration of a DVD

A DVD used to view a movie does not spin with a constant angular velocity. Instead, it spins most rapidly when images are being read from regions nearest the rotation axis (the inner "tracks" near the center of the disc) and slowest when the images are played from regions near the edge of the disc (the outer "tracks"). If the angular velocity decreases uniformly from $\omega = 1500$ rpm to 630 rpm as the DVD is scanned from the innermost to the outermost track, what is the angular acceleration? Assume the DVD contains a movie that lasts 110 min.

RECOGNIZE THE PRINCIPLE

We are given that the angular velocity decreases uniformly, which means that the angular acceleration is constant; hence, $\alpha = \alpha_{ave}$. We can compute α from the change in ω during the 110 min it takes to play the DVD.

SKETCH THE PROBLEM

Figure 8.6 shows the tracks on a DVD. These tracks form one very long "spiral" that begins near the center of the disc and winds out to the edge.

IDENTIFY THE RELATIONSHIPS

The angular velocity ω decreases from 1500 rpm to 630 rpm as the DVD player moves from the inside to the outside tracks. Because α is constant, we can use Equation 8.4

with a time interval that is not infinitesimally small. The angular acceleration is then given by

$$\alpha = \alpha_{\text{ave}} = \frac{\Delta\omega}{\Delta t} = \frac{\omega_f - \omega_i}{\Delta t}$$

SOLVE

The initial angular velocity is $\omega_i = 1500$ rpm and the final angular velocity is $\omega_f = 630$ rpm, which leads to

$$\alpha = \frac{\omega_f - \omega_i}{\Delta t} = \frac{(630 - 1500)\text{ rpm}}{110\text{ min}} = -7.9\text{ rev/min}^2$$

We now need to convert the units and express the answer in rad/s²:

$$\alpha = \frac{-7.9\text{ rev}}{\text{min}^2} \times \frac{2\pi\text{ rad}}{\text{rev}} \times \left(\frac{1\text{ min}}{60\text{ s}}\right)^2 = \boxed{-1.4 \times 10^{-2}\text{ rad/s}^2}$$

▸ *What does it mean?*

The information on a DVD is stored in a very "tight" spiral (Fig. 8.6), so at any particular point on the disc, the path containing the video information is extremely close to being a perfect circle. During one revolution, the length of each of these approximately circular paths is just the circle's circumference, a distance smaller near the center of the DVD than at regions near the edge. The amount of information stored in a circular track is proportional to the length of a track, so a circle near the center contains less information than a circle near the edge. To compensate for this difference, the DVD player spins faster when reading images and sound from near the center since the information must be processed and displayed at a constant rate.

Angular and Centripetal Acceleration Are Different

The angular acceleration α defined in Equation 8.5 is *different* from the centripetal acceleration associated with circular motion discussed in Chapter 5. This difference can be appreciated from Figure 8.2. The free end of the rod moves like a particle moving around the circle with a constant linear speed v. This particle is undergoing uniform circular motion, and its angular position θ increases at a constant rate. Hence, the particle's angular velocity is constant, and a constant value for ω means the angular acceleration $\alpha = 0$ (according to Eq. 8.5). At the same time, the particle is undergoing uniform circular motion, so from Chapter 5 we know that it also has centripetal acceleration $a_c = v^2/r$ (Eq. 5.5). This centripetal acceleration refers to the linear motion (and the linear acceleration) of the particle, whereas the angular acceleration α is concerned with the associated angular motion. Hence, the centripetal acceleration a_c is *not* zero, even though the angular acceleration α does equal zero.

The Period of Rotational Motion

When an object such as a compact disc, a DVD, or the Earth is rotating with a particular value of ω, the angular velocity is the *same* for all points on the object. This is one property that makes ω such a useful quantity for describing rotational motion. While the angular velocity is the same for all points, the linear velocity can be different for different points on the rotating object. To illustrate, we consider again the motion of a compact disc as it rotates about its axis in Figure 8.7. If we focus on the motion of a general point P on the CD, we can relate the magnitude of the linear velocity v of this point to the angular velocity ω. If the angular velocity is constant, the CD moves through ω radians every second. Since one revolution corresponds to 2π radians, the CD moves through $\omega/(2\pi)$ complete revolutions every second. Equivalently, we can

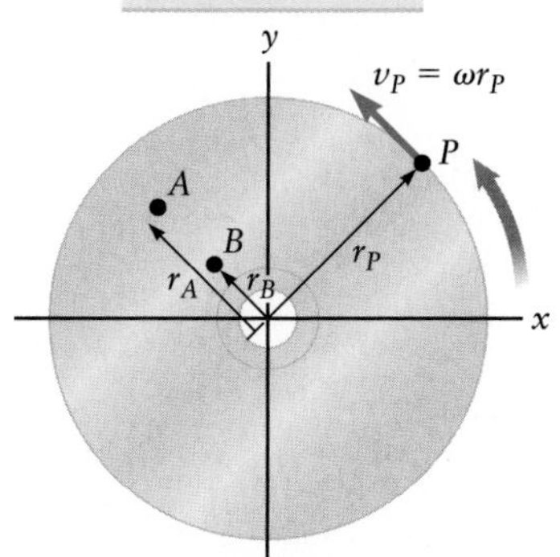

▲ Figure 8.7 When an object such as this compact disc undergoes rotational motion, the linear velocity of a point on the object depends on its distance from the rotation axis, but ω is the same for all points on the object.

say that each complete revolution takes $2\pi/\omega$ seconds. The time required to complete one revolution is called the ***period*** of the motion, denoted by T:

Period and angular velocity

$$\text{period} = T = \frac{2\pi}{\omega} \tag{8.6}$$

We encountered the notion of period in Chapter 5 in connection with circular motion. The relation between period and angular velocity in Equation 8.6 assumes ω is constant over the time it takes to complete one revolution (one period).

The Connection between Linear and Rotational Motion

Point P on our CD in Figure 8.7 lies on a circle of radius r_P. As the CD rotates, this point completes one revolution of the circle during each period of the motion, so it travels a distance $2\pi r_P$ every T seconds. Its velocity thus has a magnitude of $2\pi r_P/T$. Using Equation 8.6, we have

$$v_P = \frac{2\pi r_P}{T} = \frac{2\pi r_P}{2\pi/\omega} = \omega r_P$$

This result applies just as well to any point on the CD if r_P is replaced by the distance r between that point and the rotation axis. The *linear* velocity of a point a distance r from the rotation axis is related to the *angular* velocity by

Relation between linear speed and angular speed

$$v = \omega r \tag{8.7}$$

Equation 8.7 gives a very general and useful relation between the linear velocity and the angular velocity. To be precise, this relation really only involves the linear and angular *speeds* because here we have not accounted for the directions of these two types of velocities.

Let's compare the linear speeds of points A and B on the CD sketched in Figure 8.7. These points are located distances r_A and r_B, respectively, from the axis of the CD. Using Equation 8.7, we find $v_A = \omega r_A$ and $v_B = \omega r_B$. Point A is farther from the axis (so $r_A > r_B$), so the linear speed of point A will be *greater* than that of point B, even though they are both attached to the same CD. Physically, the reason for this difference is the different values of the radius. Point A is farther from the rotation axis, so it travels on a larger circle than point B. The motions of the two points have the same period, however, so point A must travel faster to complete its circle in the same amount of time (T). While the linear speeds at points A and B are thus *not* equal, the angular velocities at these two points are the *same*.

The relationship between the angular and translational speeds in Equation 8.7 can be extended to the accelerations. For linear motion, we showed in Chapter 2 that when the acceleration is constant, the instantaneous acceleration is related to changes in velocity by (Eqs. 2.6 and 2.7):

$$a = \frac{\Delta v}{\Delta t}$$

If a is constant, the angular acceleration is also constant, so we can write (from Eq. 8.5)

$$\alpha = \frac{\Delta \omega}{\Delta t}$$

The ratio a/α is therefore

$$\frac{a}{\alpha} = \frac{\Delta v/\Delta t}{\Delta \omega/\Delta t} = \frac{\Delta v}{\Delta \omega}$$

From Equation 8.7, we can write $\omega = v/r$, which leads to $\Delta\omega = \Delta v/r$. We thus get

$$\frac{a}{\alpha} = \frac{\Delta v}{\Delta \omega} = \frac{\Delta v}{\Delta v/r} = r$$

The *angular* acceleration α and the *linear* acceleration a of a point a distance r from the rotation axis are therefore related by

$$a = \alpha r \tag{8.8}$$

Relation between linear acceleration and angular acceleration

CONCEPT CHECK 8.2 Angular Velocity and Period

Consider two merry-go-rounds (Fig. 8.8) whose radii differ by a factor of two. If they have the same angular velocities, how do their periods compare?

(a) The larger merry-go-round has a shorter period.
(b) The larger merry-go-round has a longer period.
(c) The periods are the same.

▲ **Figure 8.8** Concept Check 8.2.

EXAMPLE 8.3 Ⓡ Speed of a Merry-Go-Round

Estimate the angular velocity of a typical merry-go-round and find the corresponding linear speed of a point on its outer edge.

RECOGNIZE THE PRINCIPLE

The angular velocity is related to the period of the motion (Eq. 8.6), and it is also related to the linear speed v of a point on a rotating object (the merry-go-round) through Equation 8.7.

SKETCH THE PROBLEM

Figure 8.9 shows a top view of a rotating merry-go-round. A point on the edge moves in a circle with r equal to the radius of the merry-go-round.

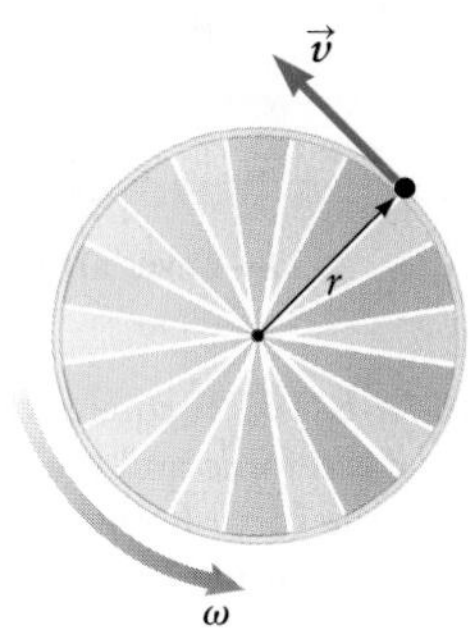

▲ **Figure 8.9** Example 8.3.

IDENTIFY THE RELATIONSHIPS

According to Equation 8.6, the angular velocity and period are related by

$$\omega = \frac{2\pi}{T}$$

To find ω, we must therefore estimate the value of T for a typical merry-go-round. The linear speed is related to ω by

$$v = \omega r$$

Hence, to find v, we must know the radius r of the merry-go-round. The value of r is not given, so we must also estimate r for a typical case.

SOLVE

A typical merry-go-round (Fig. 8.8) might have a radius of 3 m and move at a rate such that it completes one revolution in about 5 s. Hence, the period of the motion is $T = 5$ s. Working to one significant figure (to match our estimates of r and T), these values lead to

$$\omega = \frac{2\pi}{T} = \frac{2\pi}{5 \text{ s}} = \boxed{1 \text{ rad/s}}$$

The linear speed of a point on the edge of the merry-go-round is then

$$v = \omega r = (1 \text{ rad/s})(3 \text{ m}) = \boxed{3 \text{ m/s}}$$

(continued) ▶

▶ What does it mean?

The angular velocity is the same for all points on the merry-go-round, but the linear speed depends on r, the distance from the rotation axis. Notice also how the units for v here work out to be m/s. As we discussed in Chapter 1 (see also Appendix B), radians are a dimensionless unit, so they cancel automatically.

8.2 Torque and Newton's Laws for Rotational Motion

Our next job is to explore the connection between force and rotational motion. In particular, we want to know how forces give rise to angular acceleration; this approach is very similar to the one we took in Chapter 2 when we dealt with the linear (i.e., translational) motion of a particle.

Torque and Lever Arm

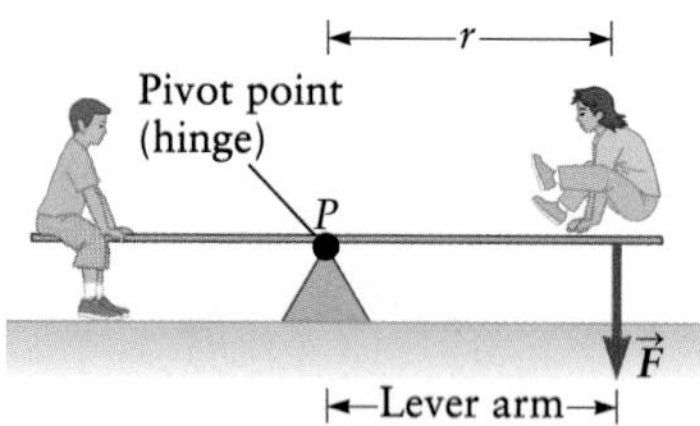

▲ Figure 8.10 Torque is the product of the force and the lever arm. Here the force $\vec{F}$ produces a torque of magnitude Fr.

Figure 8.10 shows a child sitting on one end of a seesaw while his friend prepares to jump onto the opposite side. The seesaw is supported near the center by a hinge that allows it to rotate about its point of support P. Hence, both ends of the seesaw, and indeed all points along it, move on circular arcs as the seesaw rotates. When the child jumps on board, her weight gives a force on the seesaw, and when the seesaw is horizontal (as drawn here), this force is directed perpendicular to the seesaw. This force makes the seesaw rotate, causing an angular acceleration of the seesaw. To predict the angular acceleration we must consider both the magnitude of this force and *where* it is applied to the seesaw.

From our playground experiences, we know that applying a force far from the support point P is more effective at producing rotation than applying the same force close to P. We therefore define a quantity called ***torque***, which equals the product of the applied ***force*** and the ***distance*** it is applied from the support point. Since the seesaw is able to rotate about this point, point P is often called the ***pivot point***. In the simplest cases, such as this seesaw, the force is directed *perpendicular* to a line connecting its point of application to the pivot point. We then define the ***lever arm*** as the distance between the pivot point and the place where the force acts. Later, we'll describe how the lever arm is defined in more general situations. The torque in this simple case is given by

$$\tau = Fr \tag{8.9}$$

where r is the length of the lever arm. Torque is the fundamental "force-type" quantity with regard to rotational motion; it is analogous to force in translational motion. The units of τ are $N \cdot m$ (force times distance).

We can use the seesaw in Figure 8.10 to understand why torque depends on both the force and the lever arm. For a force of a given magnitude, it is easier to rotate the seesaw if r is large than if r is small. That is, a given force produces a larger angular acceleration when it is applied farther from the pivot point.

Relating Torque and Angular Acceleration

Our next job is to derive the relation between torque and the angular motion it produces. Consider Figure 8.11, which shows a hinged rod to which a ball of mass m is attached at one end. The other end of the rod at point P is held in place, but the hinge allows the rod to rotate about P. If a force $\vec{F}_{\text{applied}}$ is applied to the ball, the rod and ball will rotate. Let's now apply Newton's second law to analyze this motion.

For simplicity, we assume the rod is extremely light (massless), so all the rotating mass is concentrated in the ball. Applying Newton's second law to the ball, we have

$$\sum \vec{F} = m\vec{a}$$

where $\sum \vec{F}$ is the total force. There are two forces acting on the ball: our applied force $\vec{F}_{\text{applied}}$ and another force $\vec{F}_{\text{rod}}$ from the rod. We thus have

$$\sum \vec{F} = \vec{F}_{\text{applied}} + \vec{F}_{\text{rod}} = m\vec{a} \tag{8.10}$$

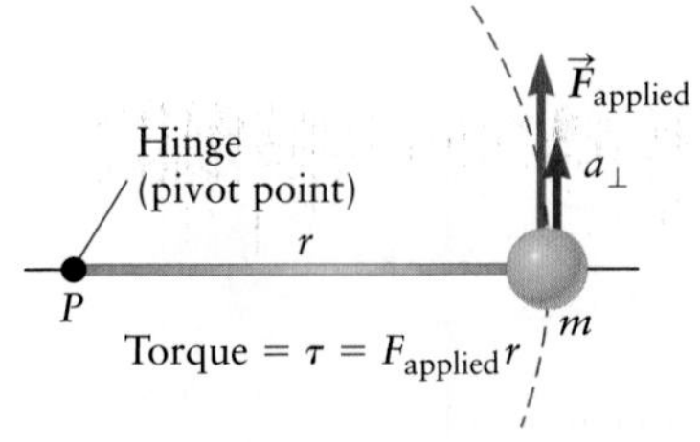

▲ **Figure 8.11** Only a perpendicular force can make an object rotate. The force $\vec{F}_{\text{applied}}$ accelerates this ball in the perpendicular direction.

Let's begin with the case in which $\vec{F}_{\text{applied}}$ is directed perpendicular to the rod as shown in Figure 8.11. The rod is rigid, so the ball will move in a circular arc along the direction of $\vec{F}_{\text{applied}}$. The force $\vec{F}_{\text{rod}}$ is along the direction of the rod, perpendicular to the ball's circular path. As a result, $\vec{F}_{\text{rod}}$ does not contribute to the ball's angular acceleration. *Only forces with a component perpendicular to the rod can contribute to the angular acceleration.* With that in mind, consider only the perpendicular components in Equation 8.10 of the force on the ball and its acceleration along this circular path. We call this acceleration $a_\perp$ because it denotes the acceleration perpendicular to the rod as in Figure 8.11. Likewise, we denote the perpendicular component of the applied force as $F_\perp$. The components of Equation 8.10 along this perpendicular direction then read

$$F_\perp = ma_\perp$$

The component $a_\perp$ is a linear (translational) acceleration, and we saw in Section 8.1 that this linear acceleration is proportional to the angular acceleration of the ball as it moves along its circular path. Using that result (Eq. 8.8) here leads to

$$F_\perp = ma_\perp = m\alpha r$$

where r is the distance from the hinge to the ball. Multiplying both sides by r and doing a little rearranging gives

$$F_\perp r = (m\alpha r)r = (mr^2)\alpha \tag{8.11}$$

Comparing with Equation 8.9, you should recognize that the left-hand side of Equation 8.11 is the torque on the ball; it is the product of the force and the lever arm r. The right-hand side contains mass and (angular) acceleration, so Equation 8.11 looks a lot like Newton's second law. Indeed, it *is* Newton's second law for rotational motion, and it is usually written as

Newton's second law for rotational motion

$$\sum \tau = I\alpha \tag{8.12}$$

where $\sum \tau$ is the total torque, that is, the sum of all the different torques acting on the object. Here we have introduced a new quantity I called the ***moment of inertia***. In rotational motion, I plays the role that mass does in translational motion. For the system considered here, consisting of a ball attached to a massless hinged rod, we can compare Equations 8.11 and 8.12 to find that the moment of inertia is

$$I = mr^2 \quad \text{(massless hinged rod/massive ball)} \tag{8.13}$$

Newton's Second Law for Rotational Motion and the Analogy with Translational Motion

Newton's second law for rotational motion, $\sum \tau = I\alpha$, and the associated definitions of torque and moment of inertia have many applications. Before exploring them, let's review and recap three major points.

First, *torque* plays the role of force in rotational motion. Torque depends on the magnitude of the force and on *where* the force is applied relative to the pivot point. There may be multiple forces acting on an object; hence, there may be multiple lever arms and multiple torques, all involving a single pivot point.

Second, the *moment of inertia* enters into rotational motion in the same way that mass enters into translational motion. We found (Eq. 8.13) that the moment of inertia of a ball at the end of a massless rod of length r is $I = mr^2$. For an object

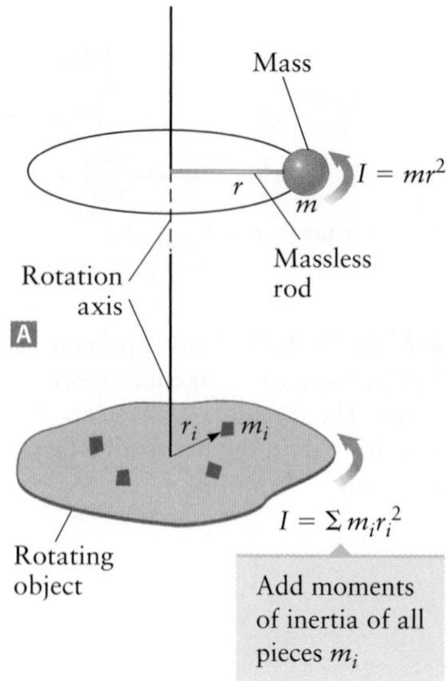

▲ **Figure 8.12** A Moment of inertia of a single point mass rotating about an axis: $I = mr^2$. B The moment of inertia of an extended object is the sum of the moments of inertia of all the individual pieces of the object. The moment of inertia of each piece can be calculated as in part A.

TABLE 8.1 Quantities Used to Describe Linear Motion and Rotational Motion

Linear Motion	Rotational Motion
Linear position $= x$	Angular position $= \theta$
Linear velocity $= v$	Angular velocity $= \omega$
Linear acceleration $= a$	Angular acceleration $= \alpha$
Force $= F$	Torque $= \tau$
Mass $= m$	Moment of inertia $= I$

composed of many pieces of mass m_i located at distances r_i from the pivot point, the total moment of inertia *of the object* is just the sum of the moments of inertia of the individual pieces (Fig. 8.12); hence,

$$I = \sum_i m_i r_i^2 \tag{8.14}$$

The moment of inertia of an object thus depends on the object's mass and on how that mass is distributed relative to the rotation axis. We'll apply Equation 8.14 and consider the moment of inertia for various types of objects in Section 8.4.

Third, Newton's second law for translational motion leads to *Newton's second law for rotational motion*:

$$\sum \tau = I\alpha$$

We'll spend the rest of this chapter exploring how to apply this relation.

The similarities between translational and rotational motion are reinforced in Table 8.1, where we list some of the quantities used to describe each.

Torques and Lever Arms Revisited: A More General Definition

We have defined torque for cases in which the applied force is purely in the perpendicular direction (Figs. 8.10 and 8.11), but, in general, the force may be applied in a nonperpendicular direction. One such example is shown in Figure 8.13, again for our hinged rod with a ball attached at the end. Here the applied force $\vec{F}_{\text{applied}}$ acts at an angle ϕ with respect to the rod. Since the ball is attached to the rod, the ball is again restricted to move along the circular arc sketched in Figure 8.13. Only forces directed along this arc (perpendicular to the rod) contribute to the torque. For this reason, we can ignore the force of the rod on the ball when calculating the torque; this force does not contribute to the angular acceleration. The perpendicular force is $F_\perp = F_{\text{applied}} \sin \phi$, and the torque in this case is

Definition of torque

$$\tau = F_\perp r = (F_{\text{applied}} \sin \phi) r$$

$$\tau = F_{\text{applied}} r \sin \phi \tag{8.15}$$

Here $F_\perp$ is the component of the force that is perpendicular to the rod and r is the distance between the pivot point and the point where the force acts. Equation 8.15 is our general definition of torque, and it applies even when the force is not directed perpendicular to the lever arm. When the force is directed perpendicular to the lever

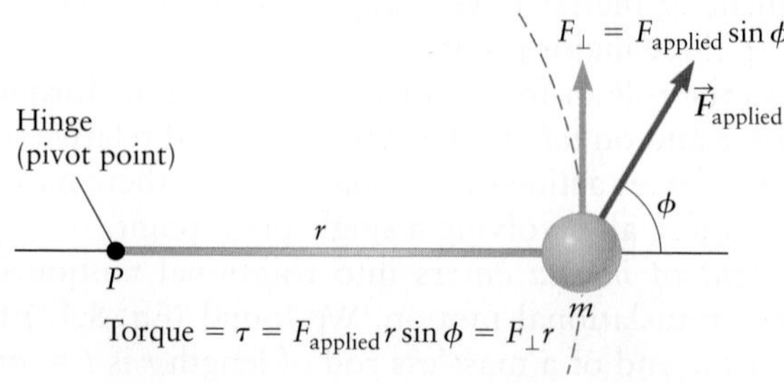

► **Figure 8.13** Only the perpendicular component of $\vec{F}_{\text{applied}}$ contributes to the torque. Any force components parallel to the rod do not contribute to the torque.

arm, $\phi = 90°$; hence, $\sin\phi = 1$ and $F_\perp = F_{applied}$. When the force is parallel to the lever arm, $\phi = 0$ and the torque is zero. This conclusion makes sense because a force directed parallel to the lever arm ($\phi = 0$) cannot cause an object to rotate.

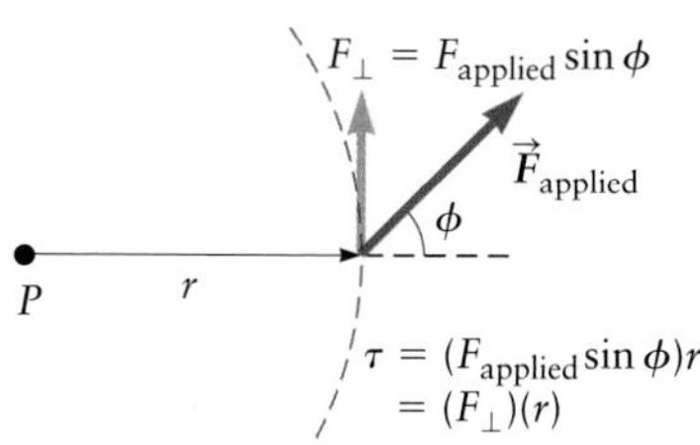

Two Ways to Think about Torque

The general relation for the torque in Equation 8.15 can be illustrated graphically in two different ways. The perpendicular component of the applied force is $F_\perp = F_{applied} \sin\phi$, so we can group the terms in Equation 8.15 as

$$\tau = \overbrace{(F_{applied} \sin\phi)}^{F_\perp} r = F_\perp r \tag{8.16}$$

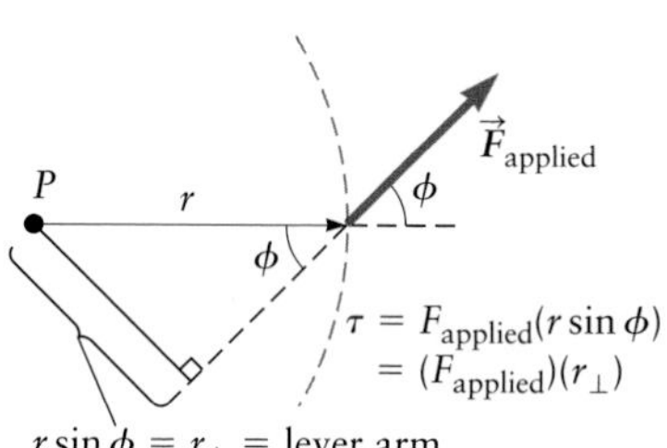

which is illustrated in Figure 8.14A; in words, this says that torque equals the product of the perpendicular component of the force ($F_\perp$) and the distance from the pivot point to the point where the force acts on the object (r). This product of perpendicular force and distance is equivalent to our definition of torque in the simpler cases in Figures 8.10 and 8.11. The dashed line in Figure 8.14A also shows that you can find the angle ϕ by extending the "radius line" beyond the point where the force acts; ϕ is the angle between this extension and the force vector.

On the other hand, we can group the terms in Equation 8.15 as

$$\tau = F_{applied} \overbrace{(r \sin\phi)}^{r_\perp} = F_{applied} r_\perp \tag{8.17}$$

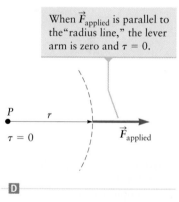

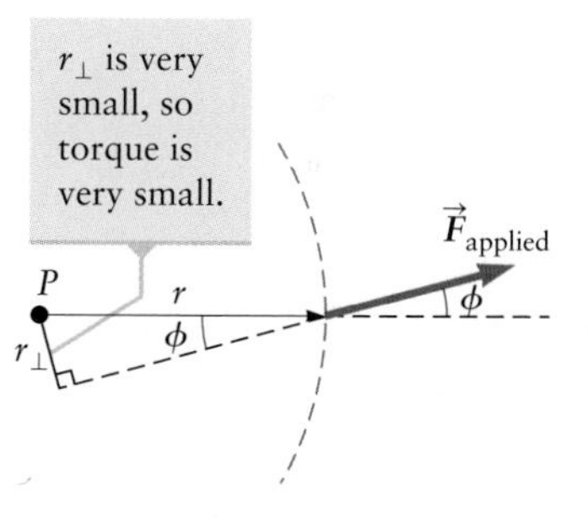

which is illustrated in parts B and C of Figure 8.14; in words, this equation says that torque equals the product of the applied force ($F_{applied}$) and the perpendicular distance $r_\perp$. This perpendicular distance is the distance from the line defined by the force vector (parts B and C of Fig. 8.14) to the rotation axis as measured on a line that is perpendicular to both. The perpendicular distance $r_\perp = r \sin\phi$ is the lever arm in the general case of a force applied at an angle ϕ to the rotation axis. These two ways of writing the torque (Eqs. 8.16 and 8.17) are completely equivalent. The general expression for the lever arm is

$$r_\perp = r \sin\phi \tag{8.18}$$

When the force is purely perpendicular and $\phi = 90°$, this expression reduces to just r as in the cases we considered in Figures 8.10 and 8.11. On the other hand, if $\phi = 0$, the lever arm is zero, corresponding to a force applied parallel to the "radius line" (Fig. 8.14D). Such a parallel force cannot cause an object to rotate and thus cannot produce a torque.

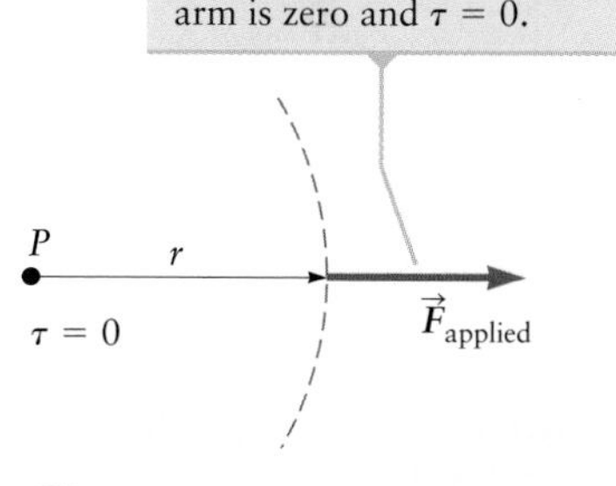

▲ **Figure 8.14** Torque is given by $\tau = F_{applied}(r \sin\phi) = F_\perp r = F_{applied} r_\perp$.

CONCEPT CHECK 8.3 Forces on a Torque Wrench

A torque wrench is a special type of wrench that limits the magnitude of the torque it is applying to a bolt, helping the user avoid breaking a bolt or stripping a thread. The torque wrench in Figure 8.15 has a long handle, and we suppose one user applies a force $F_1 = 50$ N at a distance of 10 cm from the head of the wrench. If a second user applies a force F_2 at a point 25 cm from the head, what value of F_2 will produce the same torque? Assume forces F_1 and F_2 are both applied perpendicular to the wrench handle.

(a) 10 N (b) 20 N (c) 50 N (d) 100 N (e) 125 N

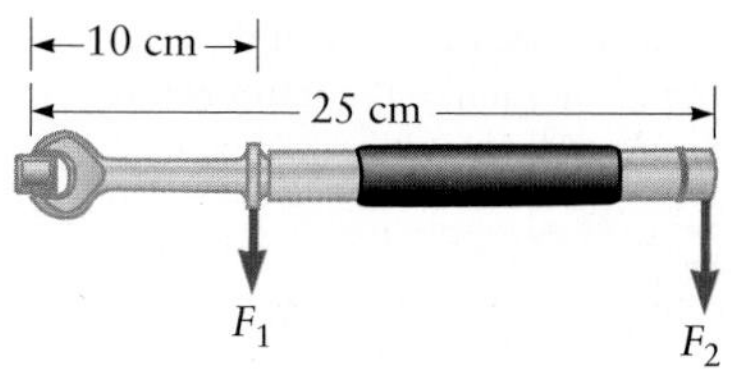

▲ **Figure 8.15** Concept Check 8.3. Forces on a torque wrench.

Center of Gravity, Center of Mass, and the Direction of Torque

Consider the motion of the hands of a large outdoor clock such as one might find near the top of a clock tower as represented very schematically in Figure 8.16. Let's calculate the torque on one clock hand due to the force of gravity. For simplicity, we first assume (unrealistically) that all the mass m is located at the end of the hand, which makes the clock hand just like the ball attached to a massless rod in Figures 8.11 and 8.13. Using Equation 8.15, the magnitude of the torque is

$$|\tau| = F_{\text{grav}} r \sin \phi$$

Figure 8.16 shows the clock hand at two different positions. When the hand is at "3 o'clock" (i.e., horizontal), the gravitational force is perpendicular to the hand and $\phi = 90°$. If L is the length of the clock hand, we have

$$|\tau| = F_{\text{grav}} r \sin \phi = mgL \sin(90°)$$

$$|\tau| = mgL \tag{8.19}$$

which is the *magnitude* of the torque. We must also pay attention to its *direction*. When we have a single rotation axis, as is the case here, the direction of the torque is specified by its sign. Recall our convention for the positive and negative directions for rotations in Figure 8.3. Those conventions also give the positive and negative directions for τ. A *positive* torque is one that would produce a counterclockwise rotation, whereas a *negative* torque would produce a clockwise rotation. In this example, with the clock hand at 3 o'clock, the torque from gravity is negative since this torque would, if it acted alone, cause the clock hand to move downward and hence rotate clockwise. If the clock hand were instead at the 9 o'clock position (Fig. 8.16B), the torque would have the same magnitude as found in Equation 8.19 (because the angle ϕ is again a right angle), but τ would now be positive because it would cause a counterclockwise rotation of the hand.

Having all the mass located at the end of the clock hand is rather unrealistic, even for a physics problem, so let's now make the more reasonable assumption that the mass is distributed uniformly along the clock hand. This situation is shown in Figure 8.17, with the clock hand located at 3 o'clock. To calculate the torque, we want to apply our relation for τ in Equation 8.15, but we immediately run into a problem. Torque involves the product of force and lever arm, and in this case, with the force in a purely perpendicular direction, the lever arm lies along the clock hand. The length of the lever arm depends on *where* the force acts. The gravitational force acts wherever there is mass, so it acts at all points along the hand. We could imagine the hand to be broken up into many infinitesimally small pieces and then add up the torques on each piece to get the total torque, but a much more convenient approach is to use the concept of ***center of gravity***.

When calculating the torque due to the gravitational force, one can assume all the force acts at a single location called the object's center of gravity. For our clock hand and for all other examples we encounter in this book, the center of gravity of an object is located at its center of mass,[1] defined in Chapter 7. For our uniform

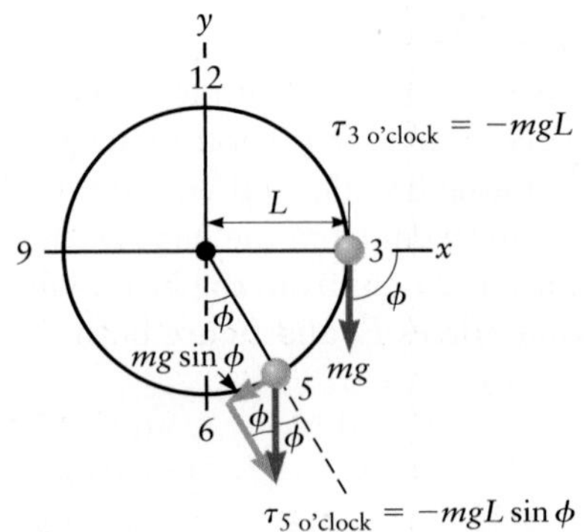

Clockwise direction = negative torque

A

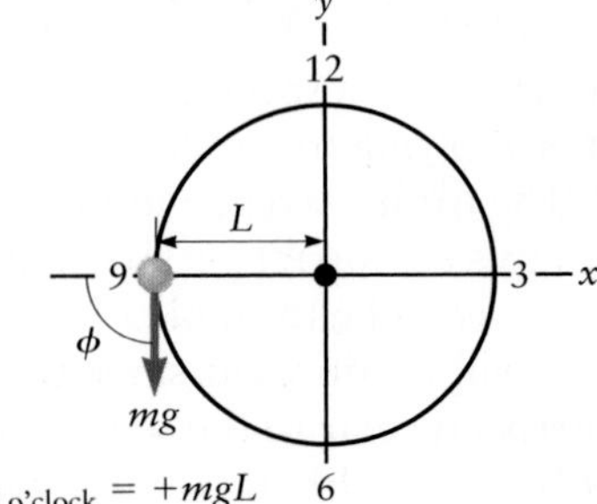

Counterclockwise direction = positive torque

B

▶ Figure 8.16 A When all mass of this clock hand is at the end, the torque due to gravity is $\tau = -mgL \sin \phi$. When the clock hand is at 3 o'clock, $\phi = 90°$, whereas when the hand is at 5 o'clock, this angle is smaller, so the magnitude of the torque is smaller. B When the clock hand is at 9 o'clock, the torque is positive because it would (if acting alone) make the hand rotate counterclockwise.

[1]The center of gravity and the center of mass of an object are usually at the same point, but that is not the case when the force of gravity varies with position, as in planetary motion. We do not treat such cases in this text.

clock hand, the center of mass is at the center of the hand. The length of the lever arm is thus $L/2$ as indicated in Figure 8.17, and the torque with the clock hand at 3 o'clock is

$$\tau = -mgr \sin\phi = -mg\left(\frac{L}{2}\right)\sin(90°) = -\frac{mgL}{2}$$

because again the angle $\phi = 90°$. The negative signs here account for gravity's tendency to make the clock hand rotate clockwise, which, according to our convention, is the negative direction.

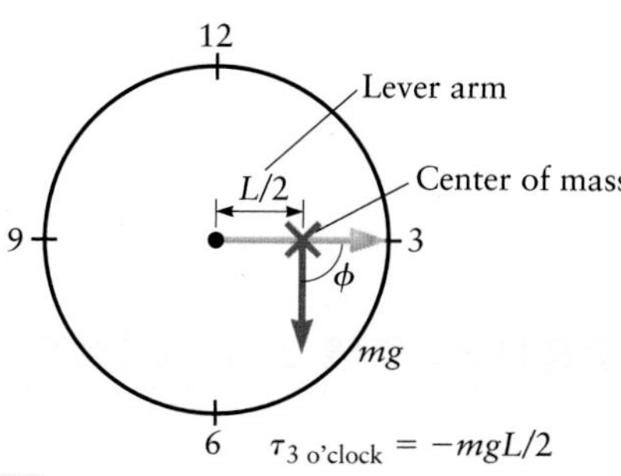

▲ **Figure 8.17** When the mass of the clock hand is distributed uniformly, the torque is calculated by letting the force of gravity act at the center of mass.

CONCEPT CHECK 8.4 Determining the Sign of the Torque

Figure 8.18 shows several rotating objects and the forces applied to them. According to our convention for finding the sign of τ, which of these torques are positive and which are negative?

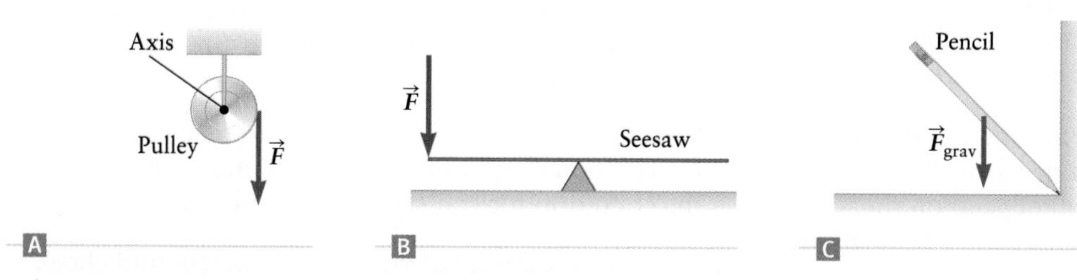

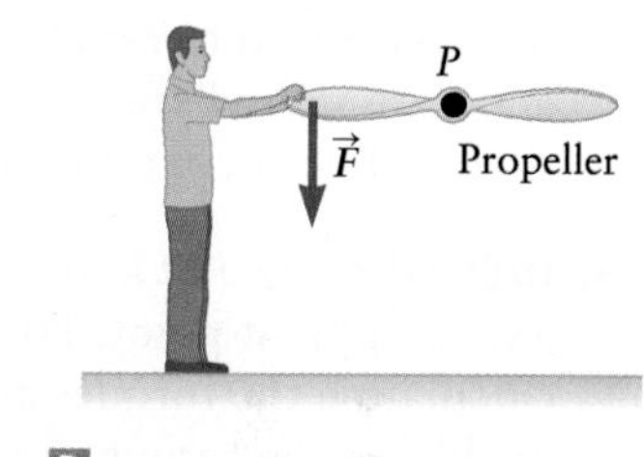

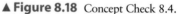

▲ **Figure 8.18** Concept Check 8.4.

8.3 Rotational Equilibrium

We showed in Chapter 4 that for an object to be in equilibrium with respect to its translational motion, the linear acceleration $\vec{a}$ must be zero. According to Newton's second law, $\sum \vec{F} = m\vec{a}$, so if an object is in translational equilibrium, the total force must be zero, $\sum \vec{F} = 0$. Also, since the acceleration is zero, an object in translational equilibrium will have a constant velocity $\vec{v}$. If that velocity is zero, we say that the object is in static equilibrium with respect to its translational motion.

We now want to extend these ideas to the notion of ***rotational equilibrium***. An object that is in equilibrium with regard to both its translational and its rotational motion must have a linear acceleration of zero *and* an angular acceleration that is zero. The condition that the total force is zero, $\sum \vec{F} = 0$, is *not* sufficient to guarantee that both these requirements are satisfied. See, for instance, Figure 8.19, which shows two forces applied to an airplane propeller. The forces $\vec{F}_1$ and $\vec{F}_2$ are applied horizontally, and we suppose they are equal in magnitude but opposite in direction, so their vector sum is zero. Although the condition for translational equilibrium along x is thus satisfied, this propeller will not be in rotational equilibrium. Even though the sum of these two forces is zero, they still produce a net torque on the propeller and lead to an angular acceleration in the clockwise direction.

For an object to be in complete equilibrium, the angular acceleration must be zero. From Newton's second law for rotational motion (Eq. 8.12), the total torque must be zero:

$$\sum \tau = \tau_1 + \tau_2 + \tau_3 + \cdots = 0 \tag{8.20}$$

For the problems we consider in this book, all the torques in this equation refer to a particular rotation axis. The same ideas can be applied to the rotation about two or three different axes simultaneously, but we leave that for a more advanced course.

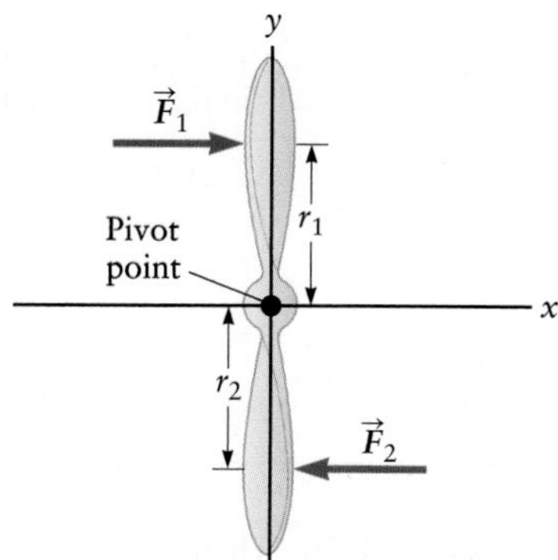

▲ **Figure 8.19** Although the horizontal forces on this propeller may add up to zero, they will still produce a nonzero torque. Hence, the propeller will not be in rotational equilibrium.

To apply the condition for rotational equilibrium in Equation 8.20, we must calculate the total torque on an object from all the applied forces. The following procedure describes how to calculate the torque and apply the conditions for equilibrium in any situation.

PROBLEM SOLVING Applying the Conditions for Translational and Rotational Equilibrium

1. **RECOGNIZE THE PRINCIPLE.** An object is in translational static equilibrium if its linear acceleration and linear velocity are both zero. An object is in rotational static equilibrium if its angular acceleration and angular velocity are both zero.

2. **SKETCH THE PROBLEM.** Make a drawing showing the object of interest along with all the forces that act on it and where those forces are applied. This sketch should contain a set of coordinate axes.

3. **IDENTIFY THE RELATIONSHIPS.** Find the rotation axis and pivot point. They are necessary for calculating the torques and will depend on the problem. The torque calculations can be broken into several steps:
   - For each force, first determine the lever arm, the perpendicular distance from the pivot point to the point at which the force acts.
   - Calculate the magnitude of the torque using $\tau = Fr \sin \phi$ (Eq. 8.15). Here, F is the magnitude of the force.
   - Determine the sign of τ. If this force *acting alone* would produce a *counterclockwise* rotation, the torque is *positive.* If the force would cause a *clockwise* rotation, the torque is *negative.*
   - Add the torques from each force to get the total torque. Be sure to include the proper sign for each individual torque.

4. **SOLVE** for the unknowns by applying the condition for rotational equilibrium $\sum \tau = 0$ and, if necessary, the condition for translational equilibrium $\sum \vec{F} = 0$.

5. Always *consider what your answer means* and check that it makes sense.

When we applied Newton's laws to the translational motion of a point particle (in Chapters 3 and 4), we began with a free-body diagram showing all the forces acting on the particle. When dealing with rotational motion, we also start with a force diagram, but we cannot simply show the object as a point. We must consider all the forces on the object and the *places on the object where these forces act.* This information is needed to determine the torque associated with each force.

Choosing a pivot point is also an essential part of dealing with torques and rotational motion, especially in equilibrium problems. In many cases, such as the seesaw in Figure 8.10, there is a "natural" choice for the pivot point, but sometimes there may be more than one plausible choice. Fortunately, for an object in rotational equilibrium, you may choose *any* spot to be the pivot point without affecting the final answer. However, some choices will lead to a simpler torque equation. Since torque is equal to the product of force and lever arm, forces whose lever arms are zero will not contribute to the torque. These forces will then not appear in the torque equation. This is illustrated in Example 8.4.

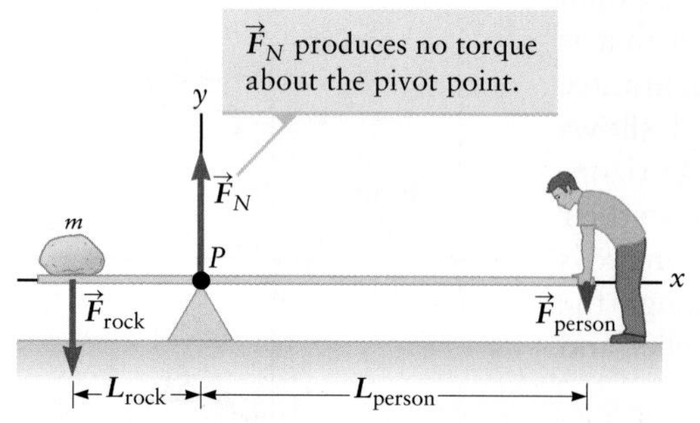

▲ **Figure 8.20** There are three forces acting on the lever (assuming it is very light so that its weight can be ignored). The normal force F_N exerted by the support acts at the pivot point P, so this force does not produce a torque about P.

Rotational Equilibrium of a Lever: Amplifying Forces

Figure 8.20 shows a lever being used to lift a very heavy rock of mass m. Let's calculate the minimum force needed to just begin lifting the rock and then find how it compares to the rock's weight.

Since the rock is just barely being lifted from the ground, both its acceleration and its angular acceleration are essentially zero and we can apply our procedure for dealing with objects in rotational equilibrium to the lever. Figure 8.20 serves as our drawing (step 2 in our problem-solving procedure). It shows all the forces on the lever and the places at which they act. For simplicity, we assume the lever is much lighter than the rock, so we can ignore the force of gravity on the lever. That still leaves three forces acting on the lever: the force $\vec{F}_{\text{rock}}$ from the rock on the lever (the weight of the rock), the force $\vec{F}_{\text{person}}$ applied by the person on the other end of the lever, and the force $\vec{F}_N$ of the support on the lever (at point P). We next (step 3) identify the pivot point: we choose the pivot point to be the spot at which the lever is supported (point P in Fig. 8.20), with the rotation axis perpendicular to the plane of the picture. The lever arm associated with $\vec{F}_{\text{rock}}$ is L_{rock}, while the lever arm for the torque associated with $\vec{F}_{\text{person}}$ is L_{person}. The lever arm for the force from the support is zero because this force acts at the pivot point.

Using the results from Figure 8.20 for the lever arms and forces, the condition for rotational equilibrium is

$$\sum \tau = \tau_{\text{rock}} + \tau_{\text{person}} + \tau_N = 0$$

$$+|F_{\text{rock}}|L_{\text{rock}} - |F_{\text{person}}|L_{\text{person}} + 0 = 0 \tag{8.21}$$

The signs in front of these terms come from our convention for positive and negative torques. The force from the rock has a magnitude $|F_{\text{rock}}|$ and is directed downward, so it gives a counterclockwise (positive) torque (hence the positive sign in front of this term in Eq. 8.21). To balance this torque, the force exerted by the person on the lever's other end must be directed downward, with a magnitude $|F_{\text{person}}|$. This force produces a clockwise (negative) torque (hence the negative sign in front of that term in Eq. 8.21). Since F_{rock} is due to the weight of the rock, $|F_{\text{rock}}| = mg$. Inserting into Equation 8.21 leads to

$$mgL_{\text{rock}} - |F_{\text{person}}|L_{\text{person}} = 0 \tag{8.22}$$

We want to find the force F_{person} that is just sufficient to balance the rock on the lever, and we can find its value by solving Equation 8.22 (step 4):

$$|F_{\text{person}}| = mg\frac{L_{\text{rock}}}{L_{\text{person}}}$$

This expression gives the magnitude of F_{person}, but we also need to know its sign (i.e., direction). From Figure 8.20, we can see that the person exerts a downward force on the lever (as we also assumed in our calculation of the torque), so F_{person} must be negative. We get

$$F_{\text{person}} = -mg\frac{L_{\text{rock}}}{L_{\text{person}}} \tag{8.23}$$

What does this result mean? Equation 8.23 shows that the magnitude of the force exerted by the person (F_{person}) can be smaller than the weight of the rock. If the lever arm extending to the person is longer than the lever arm to the rock (i.e., if $L_{\text{person}} > L_{\text{rock}}$), this simple lever will *amplify* the force exerted by the person. Of course, this result is probably what you expected if you have ever used a screwdriver to pry something open.

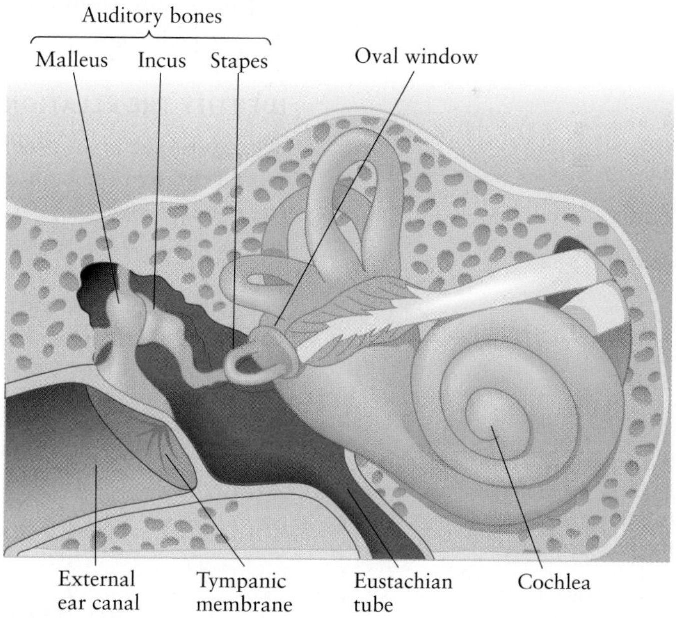

Auditory bones
Malleus Incus Stapes
Oval window
External ear canal
Tympanic membrane
Eustachian tube
Cochlea

▲ **Figure 8.21** The incus bone in the ear acts as a lever as it transmits forces to the inner ear. The pivot point for the incus lever is near where the incus meets the malleus.

Amplification of Forces in the Ear

Many mechanical devices, including the human ear, use a lever to amplify forces. As you may know, three tiny bones help carry sound through the middle ear as shown in Figure 8.21. One of them, the incus, is supported by a sort of hinge so that it is able to transmit forces from one end to the other, just like the lever in Figure 8.20. The analysis given above applies also to the forces at the ends of the incus. The lever arm from this hinge to the end in contact with the stapes bone is about a factor of three shorter than the end connected to the malleus bone. By Equation 8.23, the lever associated with the incus bone amplifies the forces associated with a sound vibration by the same factor of three.

EXAMPLE 8.4 Painters and Physics

A housepainter is using a board of length $L = 3.0$ m that sits on two supports as a scaffold (Fig. 8.22). The painter wants to reach very far to one side of the board so that he can paint as much area as possible without moving the scaffolding. (**a**) How far can the painter in Figure 8.22 walk to the right-hand side of the board before it tips? The mass of the painter is 80 kg, and the mass of the board is 30 kg. (**b**) What is the force of the right-hand support on the board when the board just begins to tip? Assume the painter is always standing straight up, so his center of mass is always above his feet.

RECOGNIZE THE PRINCIPLE

The board does not quite tip, so it is always in rotational equilibrium. The total torque must therefore equal zero. The board is also in translational equilibrium, so the total force on the board must also equal zero.

SKETCH THE PROBLEM

Figure 8.22 shows all the forces on the board: the forces exerted by both supports F_R and F_L, the force of gravity on the board F_B, and the force of gravity on the painter F_P. The figure also shows where each force acts, information that is needed to calculate the associated torques.

IDENTIFY THE RELATIONSHIPS

We choose the pivot point to be at P because that is the point around which the board would rotate if the painter moves too far to the right. When the painter is standing as far to the right as possible at a distance x from the pivot point, the board will just barely begin to tip, which means that F_L, the force exerted by the left-hand support, is *zero*.

We next compute the torques using the information in Figure 8.22 and apply the condition for rotational equilibrium $\sum \tau = 0$. Using the lever arms from Figure 8.22

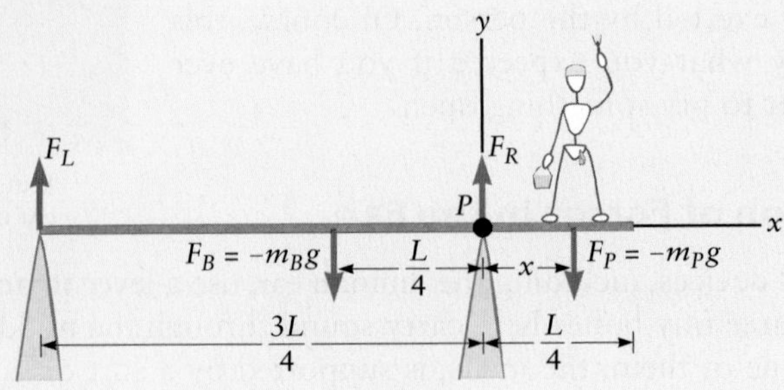

▶ **Figure 8.22** Example 8.4. How far to the right from point P can this painter go without tipping the board?

and $F_B = -m_B g$ (the weight of the board) and $F_P = -m_P g$ (the weight of the painter), we have

$$\Sigma\tau = 0 = -F_L\left(\frac{3}{4}L\right) + F_R(0) + m_B g\left(\frac{L}{4}\right) - m_P g(x) \tag{1}$$

The signs of these torque terms are determined using the convention that a counterclockwise rotation is in the positive direction. For example, the last term on the right in Equation (1) is the torque from the weight of the painter; this torque would produce a clockwise rotation and thus is negative.

The next step is to write the condition for translational equilibrium $\Sigma \vec{F} = 0$. The forces on the board are all along the y direction. The force condition is then

$$\Sigma F = F_L + F_R + F_B + F_P = 0 \tag{2}$$

Notice that some of these terms, such as $F_B = -m_B g$ and $F_P = -m_P g$, are themselves negative. Inserting $F_L = 0$ and our values of F_B and F_P into Equation (2) gives

$$F_R - m_B g - m_P g = 0 \tag{3}$$

SOLVE

(**a**) We can now solve Equation (1) to find x, the location of the painter. Inserting $F_L = 0$ because the board is on the verge of tipping, we find

$$m_B g\left(\frac{L}{4}\right) - m_P g(x) = 0$$

$$x = \frac{m_B g(L/4)}{m_P g} = \frac{m_B L}{4m_P}$$

Inserting the given values of the masses of the board and painter along with L gives

$$x = \frac{m_B L}{4m_P} = \frac{(30\text{ kg})(3.0\text{ m})}{4(80\text{ kg})} = \boxed{0.28\text{ m}}$$

The painter should thus remain a distance less than 0.28 m from the pivot point.

(**b**) To find the force from the right-hand support, we rearrange Equation (3) and get

$$F_R = m_B g + m_P g = (m_B + m_P)g = (80\text{ kg} + 30\text{ kg})(9.8\text{ m/s}^2)$$

$$F_R = \boxed{1100\text{ N}}$$

▶ ***What does it mean?***

Since F_L is zero, the right-hand support must support the entire weight of the system. Hence, the force from the right-hand support must equal the weight of the board plus the weight of the painter, as found in the answer to part (**b**).

CONCEPT CHECK 8.5 Carrying Your Share of the Weight

The two workers in Figure 8.23 are carrying a long, heavy steel beam. Which one is exerting a larger force on the object? How can you tell?

(a) The worker on the left (b) The worker on the right

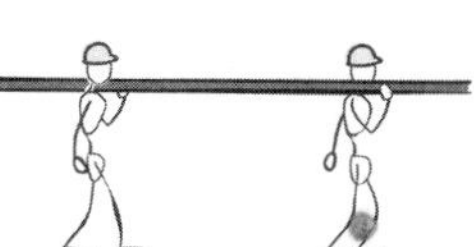

▲ **Figure 8.23** Concept Check 8.5.

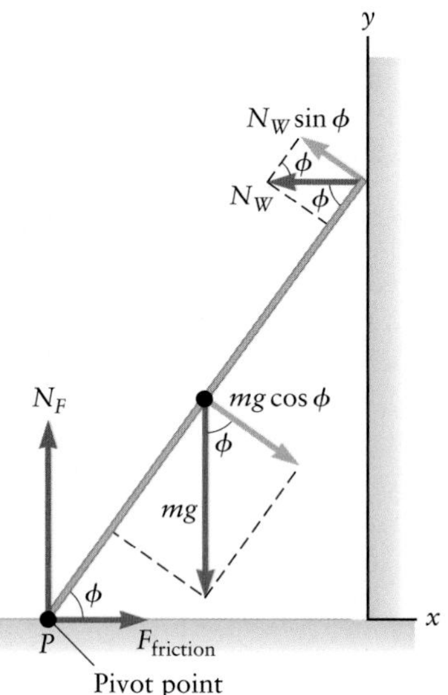

▲ **Figure 8.24** Example 8.5. Forces acting on a leaning ladder. We assume there is no friction between the wall and the top end of the ladder.

EXAMPLE 8.5 Leaning on a Wall

Consider the stability of a ladder of mass m and length L leaning against a wall as shown in Figure 8.24. A painter wants to be sure the ladder will not slip across the floor. He knows that the frictional force exerted by the floor on the bottom end of the ladder is needed to keep the ladder in equilibrium, and he also knows that the coefficient of static friction between the ladder and the floor is $\mu_S = 0.50$. What is the angle ϕ at which the ladder will just become unstable and slide across the floor? For simplicity, assume there is no frictional force between the ladder and the wall.

RECOGNIZE THE PRINCIPLE

The ladder is in equilibrium, so we can apply the general conditions for static equilibrium $\sum \tau = 0$ and $\sum \vec{F} = 0$. The angle ϕ plays a key role: if we take the bottom of the ladder (point P) as our pivot point, the torque on the ladder due to gravity increases as ϕ becomes smaller because the lever arm increases. At some critical value of the angle, this torque will be large enough that the ladder will start to move.

SKETCH THE PROBLEM

Figure 8.24 shows all the forces acting on the ladder along with their components along x and y. There are normal forces exerted by the floor (N_F) and the wall (N_W), the force of friction exerted by the floor ($F_{friction}$), and the ladder's weight (mg). The force from the floor acts on the end of the ladder at point P, while the force of gravity acts at the ladder's center of mass.

IDENTIFY THE RELATIONSHIPS

Let's choose the base of the ladder at P as the pivot point. By doing so, we make the torques due to the forces from the floor (N_F and $F_{friction}$) zero because these forces act at P and hence their lever arms are zero. Using the lever arms and angles from Figure 8.24, the torque condition is

$$\sum \tau = 0 = +N_W L \sin\phi - mg\left(\frac{L}{2}\right)\sin(90^\circ - \phi)$$

$$0 = +N_W L \sin\phi - mg\left(\frac{L}{2}\right)\cos(\phi) \tag{1}$$

where we have used the general definition of torque (Eq. 8.15) along with the trigonometric identity $\sin(90^\circ - \phi) = \cos(\phi)$. The appearance of the cosine term in Equation (1) can be understood from the expression for torque in Equation 8.16. The torque due to the ladder's weight is the product of $L/2$ (the distance from the center of mass to the pivot point) and the component of the force of gravity *perpendicular* to the ladder. Figure 8.24 shows that this component involves $\cos\phi$.

Since the ladder is in static equilibrium, the sum of all forces along both x and y must be zero, which leads to

$$\sum F_x = 0 = -N_W + F_{friction} \tag{2}$$

$$\sum F_y = 0 = N_F - mg = 0 \tag{3}$$

SOLVE

From Equation (3), we can solve for the normal force from the floor and get $N_F = mg$. The maximum frictional force is $F_{friction} = \mu_S N_F$, so we have

$$F_{friction} = \mu_S N_F = \mu_S mg$$

and, according to Equation (2), this result is also equal to N_W. Inserting into Equation (1) leads to

$$+N_W L \sin\phi - mg\left(\frac{L}{2}\right)\cos\phi = 0$$

$$\mu_S mgL \sin\phi = mg\left(\frac{L}{2}\right)\cos\phi$$

$$\tan\phi = \frac{1}{2\mu_S}$$

Using the given value $\mu_S = 0.50$, we find

$$\tan\phi = \frac{1}{2\mu_S} = \frac{1}{2(0.50)} = 1.0 \quad (4)$$

$$\phi = \boxed{45°}$$

▸ *What does it mean?*

For a smaller angle, the frictional force would not be able to keep the ladder in equilibrium and the ladder would slip. A sensible painter would choose a steeper angle than 45°.

CONCEPT CHECK 8.6 Adding Friction between the Ladder and the Wall

Suppose we add friction between the wall and the upper end of the ladder in Example 8.5. How would that change the result for ϕ?

(a) The minimum safe value of ϕ would increase.
(b) The minimum safe value of ϕ would decrease.
(c) The minimum safe value of ϕ would stay the same.

Pushing on a Crate: When Will It Tip?

Another example of an object in static equilibrium is shown in Figure 8.25A. A cubical crate of length L sits at rest on a level floor as a person pushes on one side of the crate. If the person applies a very large force of magnitude F_{person} horizontally at the top edge and the frictional force $F_{friction}$ at the bottom edge is large, the crate will tip as sketched in Figure 8.25B. Let's consider how to calculate the value of F_{person} at which the crate just begins to tip.

When the crate is on the verge of tipping, it is still in static equilibrium, so the total torque and the total force are both zero. There are four forces acting on the crate: (1) the force of gravity, which acts at the center of mass; (2) the force of static friction, which acts along the bottom surface of the crate; (3) the normal force exerted by the floor, which acts at the surface in contact with the floor; and (4) the force applied by the person, F_{person}. Any tipping will involve rotation about the point P located at the corner of the crate, so let's choose that as our pivot point. The corresponding lever arms are indicated in Figure 8.25A; the lever arm associated with the frictional force is zero, so it will not contribute to the torque. If the crate is on the verge of rotating about P, the normal force exerted by the floor will then be acting along a line passing through P and its lever arm is also zero. The total torque on the crate is then

$$\sum \tau = 0 = \underbrace{mg(L/2)}_{\text{gravity}} - \underbrace{F_{person}(L)}_{\text{person}} + \underbrace{N(0)}_{\text{normal force}} + \underbrace{F_{friction}(0)}_{\text{friction}} = 0 \quad (8.24)$$

where the corresponding lever arm factors are all given in parentheses. The lever arm for the gravitational force is $L/2$ because this force acts at the center of mass, which is in the middle of the crate, a distance $L/2$ from the crate's side. The lever

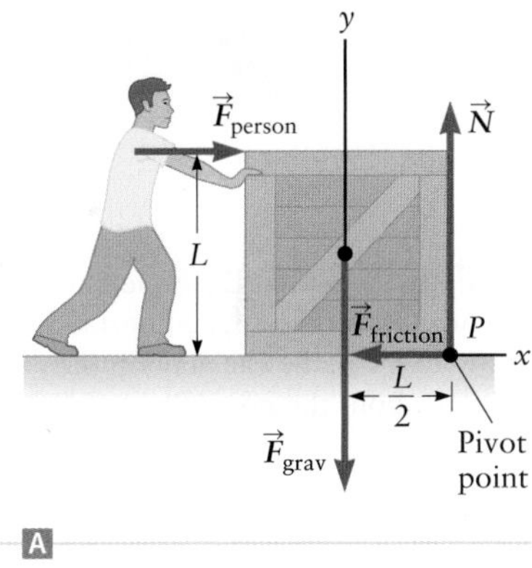

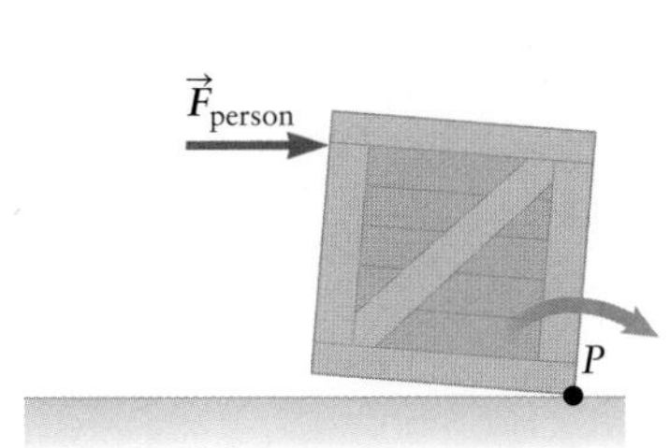

▲ Figure 8.25 A Forces on a crate. As the person pushes harder and harder, the crate will eventually tip and B rotate about a pivot point P at the corner of the crate.

arm for the torque from F_{person} is L because the person is pushing at the top edge of the crate. The signs in Equation 8.24 can again be understood using our convention for positive and negative torque. For example, a positive value of the applied force (F_{person} directed to the right, along $+x$) gives a clockwise rotation and hence a *negative* torque. We can solve Equation 8.24 for F_{person} and find

$$mg(L/2) = F_{person}(L)$$

$$F_{person} = \frac{mg}{2}$$

which is half the weight of the crate. If F_{person} exceeds this value, the crate will tip.

This problem of a tipping crate is very similar to the problem of a car rolling over in a traffic accident. This subject is of great interest to automobile designers and is explored in the next example.

EXAMPLE 8.6 Ⓡ Cars and Static Equilibrium

Figure 8.26 shows the forces acting on a car in a collision. The force $\vec{F}_{collision}$ comes from another car, and for simplicity we assume this force is directed horizontally and acts at the center of the body of the car. Estimate the magnitude of $\vec{F}_{collision}$ for which the car in Figure 8.26 will just roll over. Express the answer as a ratio $|\vec{F}_{collision}|/mg$, where m is the mass of the car, and use the dimensions of a typical car in your calculation.

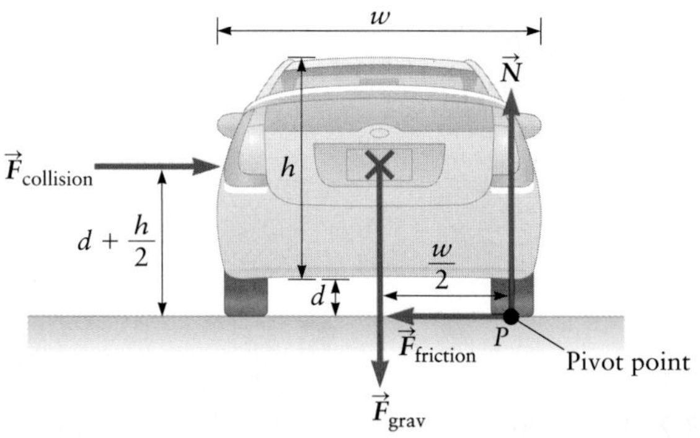

▲ **Figure 8.26** Example 8.6. If a force $\vec{F}_{collision}$ is applied horizontally to a car, the car may tip (i.e., rotate about the point P).

RECOGNIZE THE PRINCIPLE

We wish to calculate the minimum force $\vec{F}_{collision}$ that will just barely cause the car to tip. This problem is very similar to the tipping crate in Figure 8.25. When the car tips, it will rotate about point P in Figure 8.26, so we take that as the pivot point.

SKETCH THE PROBLEM

Figure 8.26 shows all the forces on the car, including the force of friction between the tires and road ($F_{friction}$) and the force of gravity ($F_{grav} = mg$). It also shows the points at which these forces act.

IDENTIFY THE RELATIONSHIPS

Although this problem is similar to the tipping crate in Figure 8.25, there are some important differences. Because the car sits on four wheels, we must account for the height of the wheels when finding the car's center of mass. Indeed, we are not given any information about the dimensions of the car's body (w and h in Fig. 8.26) or the wheels (d), so we'll have to estimate their values. First, though, let's express the conditions for equilibrium in terms of w, h, and d.

We treat the car as a simple "box" of width w and height h that is a distance d off the ground and ignore the mass in the wheels. There are four forces on the car as indicated in Figure 8.26. The condition for rotational equilibrium is

$$\sum \tau = 0 = \overbrace{-F_{collision}\left(d + \frac{h}{2}\right)}^{\text{collision force}} + \overbrace{mg\left(\frac{w}{2}\right)}^{\text{gravity}} + \overbrace{N(0)}^{\text{normal force}} + \overbrace{F_{friction}(0)}^{\text{friction}} \qquad (1)$$

where the lever arm factors are given in parentheses. The term involving $F_{collision}$ is the torque produced by the collision force. The lever arm in this case is the distance from the center of the car to the ground ($d + h/2$) since this force acts at the middle of the car, at its center of mass. This force is directed to the right (along $+x$) and would, if it acted alone, produce a clockwise rotation; hence, this torque is *negative*.

SOLVE

We can rearrange Equation (1) to get

$$F_{\text{collision}}\left(d + \frac{h}{2}\right) = mg\left(\frac{w}{2}\right)$$

Dividing $F_{\text{collision}}$ by the weight of the car leads to

$$\boxed{\frac{F_{\text{collision}}}{mg} = \frac{w}{2d + h}} \qquad (1)$$

► *What does it mean?*

If $F_{\text{collision}}/mg$ is small, a relatively small force is enough to cause a rollover accident, whereas a large ratio means that a larger impact force is needed to cause a rollover. Equation (1) shows that a large value of d—that is, a large ground clearance—will make a rollover more likely. For a typical compact car that sits relatively low to the ground, we estimate[2] $w = 170$ cm, $d = 20$ cm, and $h = 120$ cm, which gives the ratio

$$\frac{F_{\text{collision}}}{mg} = \frac{170 \text{ cm}}{2(20 \text{ cm}) + 120 \text{ cm}} = 1.1 \quad \text{(compact car)}$$

For an SUV, we estimate $w = 170$ cm, $d = 40$ cm, and $h = 140$ cm, which leads to

$$\frac{F_{\text{collision}}}{mg} = \frac{170 \text{ cm}}{2(40 \text{ cm}) + 140 \text{ cm}} = 0.77 \quad \text{(SUV)}$$

Hence, according to this measure, the SUV is more likely to roll over. Of course, real accidents are more complicated than this simple model, but current government rollover ratings are based in part on a similar calculation.

8.4 Moment of Inertia

In Section 8.2, we introduced the rotational form of Newton's second law:

$$\sum \tau = I\alpha \qquad (8.25)$$

In all our work on rotational equilibrium, $\alpha = 0$, so the right-hand side of Equation 8.25 was zero and we did not have to worry about the moment of inertia I. To tackle cases in which the angular acceleration is not zero, we must consider the moment of inertia in more detail.

Equation 8.14 defines the moment of inertia for a general object composed of many pieces of mass m_i as

$$I = \Sigma_i\, m_i r_i^2 \qquad (8.26)$$

where each piece i is a distance r_i from the axis of rotation. The moment of inertia of an object depends on its mass *and* on how this mass is distributed with respect to the rotation axis. The value of I for a particular object thus depends on the choice of rotation axis, as illustrated using the simple "object" in Figure 8.27. Two point particles, both with mass m, are connected by a very light (massless) rod of length L. Figure 8.27A shows a rotation about the center of the rod. Applying Equation 8.26, the moment of inertia in this case is

$$I_a = \Sigma_i\, m_i r_i^2 = m_1 r_1^2 + m_2 r_2^2 \qquad (8.27)$$

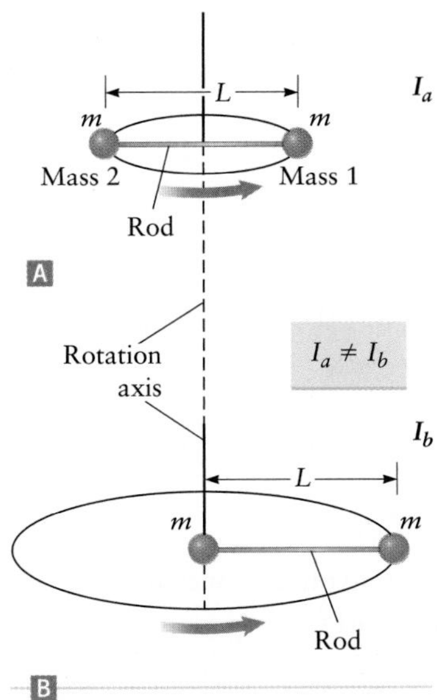

▲ **Figure 8.27** The moment of inertia depends on the choice of rotation axis.

[2]These estimates of w, d, and h were obtained from the specifications of actual vehicles.

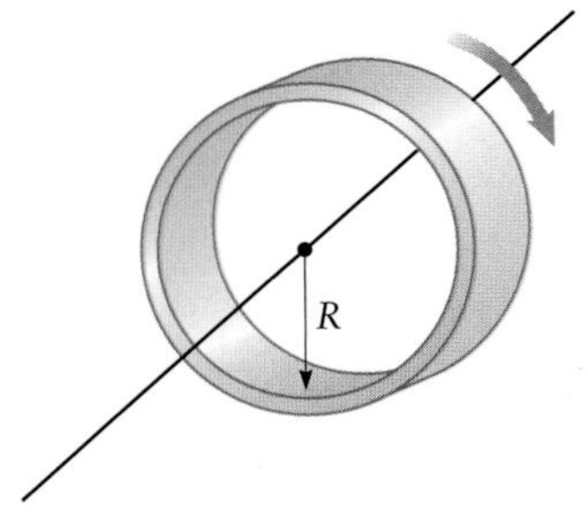

▲ **Figure 8.28** For a simple hoop rotating about its geometric center, all the mass is a distance R from the rotation axis.

Both particles have mass m; hence, $m_1 = m_2 = m$, and they are both a distance $L/2$ from the rotation axis, so $r_1 = r_2 = L/2$. Inserting this information into Equation 8.27 leads to

$$I_a = m(L/2)^2 + m(L/2)^2 = mL^2/2 \tag{8.28}$$

This object can also rotate about an axis that passes through one end of the rod (Fig. 8.27B). The moment of inertia in this case is again given by Equation 8.27, but now $r_2 = 0$ (because the axis runs through mass 2) while $r_1 = L$. The moment of inertia is thus

$$I_b = m_1 r_1^2 + m_2 r_2^2 = m(L)^2 + m(0)^2 = mL^2 \tag{8.29}$$

The moments of inertia I_a and I_b (Eqs. 8.28 and 8.29) are different, showing that the moment of inertia of an object does indeed depend on the choice of axis.

Using Equation 8.26 to find the moment of inertia of a complicated object can be difficult. Fortunately, though, the results for a variety of common shapes are known, and some are listed in Table 8.2. For example, the moment of inertia of a solid sphere of mass m and radius R is $I_{\text{sphere}} = 2mR^2/5$ for an axis that passes through the center of the sphere. This result applies to *all* solid spheres. Likewise, the moment of inertia of a disc or solid cylinder rotating about its central axis is $I_{\text{disc}} = mR^2/2$, where m is the total mass and R is the radius. The result is different from the moment of inertia of a hoop, which is $I_{\text{hoop}} = mR^2$ for an axis perpendicular to the plane of the hoop and passing through its center as indicated in Table 8.2. This case with a hoop is particularly simple because all the mass is a distance R from the rotation axis (Fig. 8.28). In this case, the sum in Equation 8.26 can be evaluated as

$$I_{\text{hoop}} = \sum_i m_i r_i^2 = \left(\sum_i m_i\right) R^2 \tag{8.30}$$

TABLE 8.2 Moment of Inertia for Some Common Objects

Object	Shape	Object	Shape
Hoop $I = mR^2$	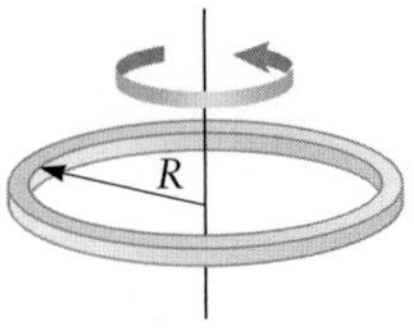	Rod pivoted at center $I = \frac{1}{12}mL^2$	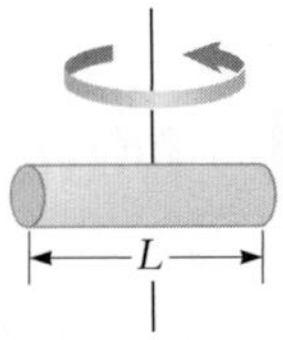
Solid sphere $I = \frac{2}{5}mR^2$	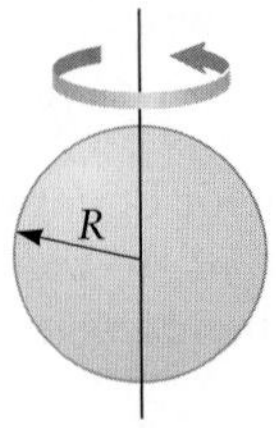	Pulley/cylinder/disc $I = \frac{1}{2}mR^2$	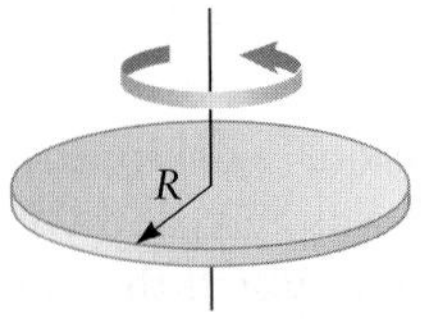
Spherical shell $I = \frac{2}{3}mR^2$	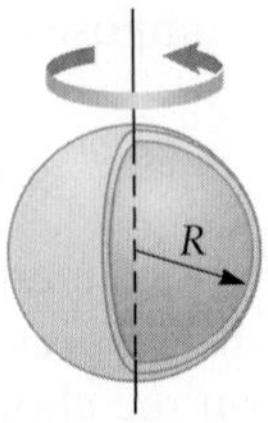	Wheel or hollow cylinder $I = \frac{1}{2}m(R_{\text{max}}^2 + R_{\text{min}}^2)$	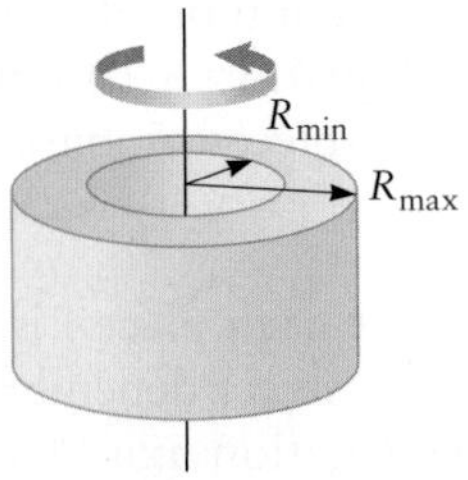
Rod pivoted at one end $I = \frac{1}{3}mL^2$	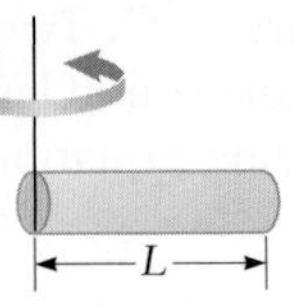	Solid square plate with axis perpendicular to plate $I = \frac{1}{6}mL^2$	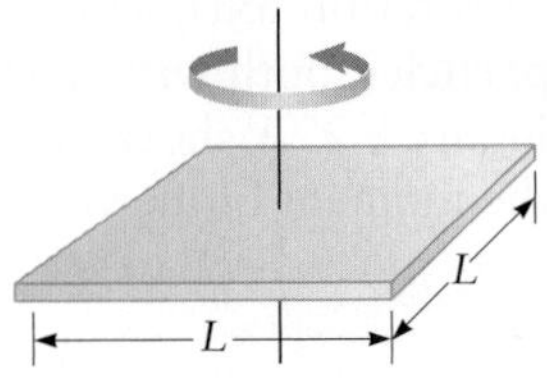

Note: In each case, m is the total mass of the object.

All the mass of the hoop is at the same distance R from the rotation axis at the center, so we can pull this factor out of the summation. The sum $\Sigma_i m_i$ is the total mass m of the hoop, so Equation 8.30 becomes $I_{\text{hoop}} = mR^2$.

Table 8.2 also lists the moment of inertia for a uniform rod of length L and shows how the moment of inertia depends on the axis of rotation, just as we found for the simple object in Figure 8.27. If a rod rotates about a perpendicular axis that runs through the center of the rod, then $I_{\text{rod center}} = mL^2/12$, whereas if the rotation axis runs through one end of the rod, the moment of inertia is $I_{\text{rod end}} = mL^2/3$.

As an example, let's calculate the moment of inertia of a compact disc. A typical CD has a mass of about 9.0 g and a diameter of 12 cm. If we ignore the relatively small area of the hole in its center, a CD is just a disc of mass 0.0090 kg and radius 0.060 m. Inserting these values into the expression for a disc in Table 8.2 gives

$$I_{\text{disc}} = \frac{mR^2}{2} = \frac{(0.0090\text{ kg})(0.060\text{ m})^2}{2} \tag{8.31}$$

$$I_{\text{disc}} = 1.6 \times 10^{-5}\text{ kg}\cdot\text{m}^2 \tag{8.32}$$

This is the moment of inertia for rotations about an axis that runs through the center and is perpendicular to the plane of the disc. This is the axis appropriate for a CD player.

EXAMPLE 8.7 Moment of Inertia of a Dirty CD

A rather large bug of mass 5.0 g manages to land on the edge of your favorite video-game CD (mass 9.0 g and radius 6.0 cm). How does that affect the CD's moment of inertia? (Ignore the hole in the center of the CD.)

RECOGNIZE THE PRINCIPLE

To calculate the total moment of inertia of the bug plus the CD, we start with the general definition of the moment of inertia in Equation 8.26:

$$I_{\text{total}} = \Sigma_i m_i r_i^2$$

We now split this expression into two terms, one for the clean CD and one for the contribution of the bug:

$$I_{\text{total}} = \Sigma_i m_i r_i^2 = (\Sigma_i m_i r_i^2)_{\text{CD}} + (\Sigma_i m_i r_i^2)_{\text{bug}}$$

We calculated the first term (for the clean CD) in Equation 8.32. The bug (the second term) acts just as a point mass, as treated in connection with the object in Figure 8.11.

SKETCH THE PROBLEM

The CD, bug included, is shown in Figure 8.29.

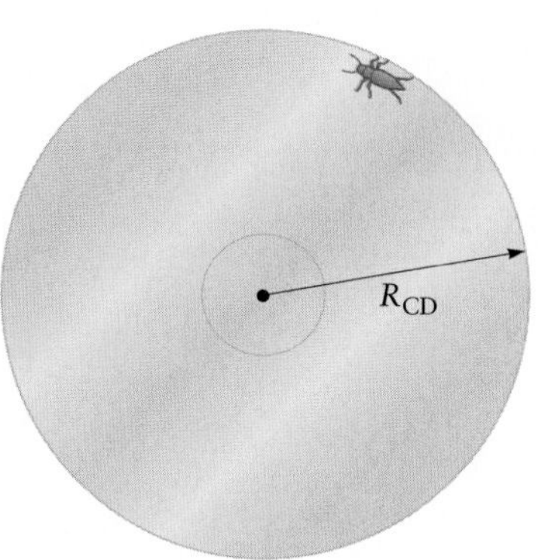

▲ **Figure 8.29** Example 8.7. The total moment of inertia of this "dirty" compact disc is the sum of the moment of inertia of the clean CD plus the moment of inertia of the bug.

IDENTIFY THE RELATIONSHIPS

A clean CD is simply a disc; its moment of inertia is $I_{\text{CD}} = m_{\text{CD}}R_{\text{CD}}^2/2$ and was calculated in Equations 8.31 and 8.32. For the contribution of the bug, we notice that all the bug's mass is located at the edge of the CD, so according to Equation 8.26 $I_{\text{bug}} = m_{\text{bug}}r_{\text{bug}}^2 = m_{\text{bug}}R_{\text{CD}}^2$. Combining these results gives

$$I_{\text{total}} = \frac{m_{\text{CD}}R_{\text{CD}}^2}{2} + m_{\text{bug}}R_{\text{CD}}^2$$

SOLVE

Inserting all the given values, we find

$$I_{\text{total}} = \frac{(0.0090\text{ kg})(0.060\text{ m})^2}{2} + (0.0050\text{ kg})(0.060\text{ m})^2 = \boxed{3.4 \times 10^{-5}\text{ kg}\cdot\text{m}^2}$$

(continued) ▶

▸ ***What does it mean?***
Comparing this result with the moment of inertia of a clean CD (Eq. 8.32), we see that the bug increases the moment of inertia by about a factor of two. This increase could well affect the way your videogame player is able to spin a CD. It pays to keep your CDs clean.

8.5 Rotational Dynamics

We are now ready to tackle some problems involving rotational dynamics. The basis for all our calculations will be

$$\sum \tau = I\alpha \tag{8.33}$$

We can predict the rotational motion of an object by calculating both the total torque and the moment of inertia and using Equation 8.33 to find the angular acceleration α. We can then find the angular velocity and angular displacement of the object. Often, the angular acceleration is constant, as in cases in which the applied torque is constant. The results for ω and θ then have particularly simple forms. In fact, these mathematical forms are already quite familiar to us. You should recall that α is the slope of the ω–t curve (see Eq. 8.5). If the angular acceleration has the constant value α, then ω must vary linearly with time:

$$\omega = \omega_0 + \alpha t \tag{8.34}$$

Equation 8.34 should remind you of the behavior of the linear velocity v as a function of time in situations in which the linear acceleration a is constant, as we studied in Chapter 3. Likewise, for this case of constant angular acceleration, the angular displacement has the familiar form

Rotational motion with constant angular acceleration

$$\theta = \theta_0 + \omega_0 t + \tfrac{1}{2}\alpha t^2 \tag{8.35}$$

We can also combine Equations 8.34 and 8.35 to get a relation that does not involve time; the result is

$$\omega^2 = \omega_0^2 + 2\alpha(\theta - \theta_0) \tag{8.36}$$

These relations, Equations 8.34 through 8.36, are very similar to the results for linear velocity and displacement we encountered for translational motion when the linear acceleration is constant. They are extremely useful, so we have collected them together in Table 8.3.

Angular Motion of a Compact Disc

Let's now use the relations in Table 8.3 to tackle a few problems involving rotational motion. Consider yet again the motion of a compact disc. Suppose a CD is inserted into the optical drive of a computer at $t = 0$ and accelerates uniformly to reach its operational angular velocity of 500 rpm in 5.0 s. Let's calculate (a) the angular acceleration of the CD, (b) the total angular displacement of the CD during this period, and (c) the torque the optical drive must exert on the CD.

TABLE 8.3 Kinematic Relations for Constant Angular Acceleration α and Corresponding Relations for Linear Motion with Constant Acceleration a

Rotational Motion	Equation Number	Translational Motion
$\theta = \theta_0 + \omega_0 t + \frac{1}{2}\alpha t^2$	8.35	$x = x_0 + v_0 t + \frac{1}{2}at^2$
$\omega = \omega_0 + \alpha t$	8.34	$v = v_0 + at$
$\omega^2 = \omega_0^2 + 2\alpha(\theta - \theta_0)$	8.36	$v^2 = v_0^2 + 2a(x - x_0)$

To find the angular acceleration, we note that the initial angular velocity is zero ($\omega_0 = 0$) and that the final angular velocity is

$$\omega = 500 \text{ rpm} = 500 \frac{\text{rev}}{\text{min}} \times \frac{2\pi \text{ rad}}{1 \text{ rev}} \times \frac{1 \text{ min}}{60 \text{ s}} = 52 \text{ rad/s}$$

Inserting this result into Equation 8.34 and solving for α gives

$$\alpha = \frac{\omega - \omega_0}{t} = \frac{(52 \text{ rad/s}) - 0}{5.0 \text{ s}} = 10 \text{ rad/s}^2 \tag{8.37}$$

The angular displacement during this interval can be obtained from Equation 8.35,

$$\theta = \theta_0 + \omega_0 t + \tfrac{1}{2}\alpha t^2 = 0 + 0 + \tfrac{1}{2}(10 \text{ rad/s}^2)(5.0 \text{ s})^2 = 130 \text{ rad}$$

The torque that must be supplied by the CD player can be found from Equation 8.33,

$$\sum \tau = I\alpha$$

We have already calculated the moment of inertia of a CD (Eq. 8.32). Using that result together with the angular acceleration from Equation 8.37, we find

$$\sum \tau = I\alpha = (1.6 \times 10^{-5} \text{ kg} \cdot \text{m}^2)(10 \text{ rad/s}^2) = 1.6 \times 10^{-4} \text{ N} \cdot \text{m}$$

EXAMPLE 8.8 Ⓡ Motion of a Frisbee

You and a friend are playing catch with a Frisbee (Fig. 8.30). Estimate the total angular displacement of the Frisbee as it travels from you to your friend. Assume the Frisbee's angular velocity is constant.

RECOGNIZE THE PRINCIPLE

We are given that the Frisbee moves with a constant angular velocity as it travels through the air; this makes sense since we know from experience that when a Frisbee reaches its destination, it is usually still spinning quite rapidly. If ω is constant, the angular acceleration $\alpha = 0$. We can then use the relations in Table 8.3 to find the total angular displacement.

SKETCH THE PROBLEM

The problem is shown in Figure 8.30. The top of the figure shows the rotation axis of the Frisbee and the angular displacement θ.

IDENTIFY THE RELATIONSHIPS

We can find the total angular displacement from Equation 8.35 provided we can estimate the initial angular velocity ω_0 and the time the Frisbee spends in the air. (We already know that $\alpha = 0$.) Based on experience, we estimate that the Frisbee spends about 5 s in the air. It is a bit more difficult to estimate the Frisbee's angular velocity, but a reasonable guess is 10 rev/s. This value should be correct to within an order of magnitude; a value of 1 rev/s seems much too slow, and 100 rev/s would be quite fast. (The maximum rotation rate for most car engines is about 100 rev/s, and we don't expect a Frisbee to rotate that rapidly.)

SOLVE

Our estimate for the angular velocity is thus

$$\omega_0 \approx 10 \text{ rev/s} = 10 \frac{\text{rev}}{\text{s}} \times \frac{2\pi \text{ rad}}{\text{rev}} = 60 \text{ rad/s}$$

(continued) ▶

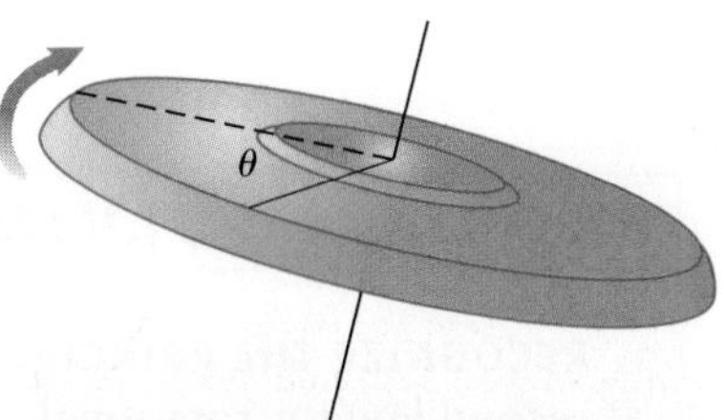

▲ **Figure 8.30** Example 8.8. A Frisbee spins about its axis with an angular velocity that is approximately constant during the course of a "throw." This rotational motion is independent of the Frisbee's translational motion.

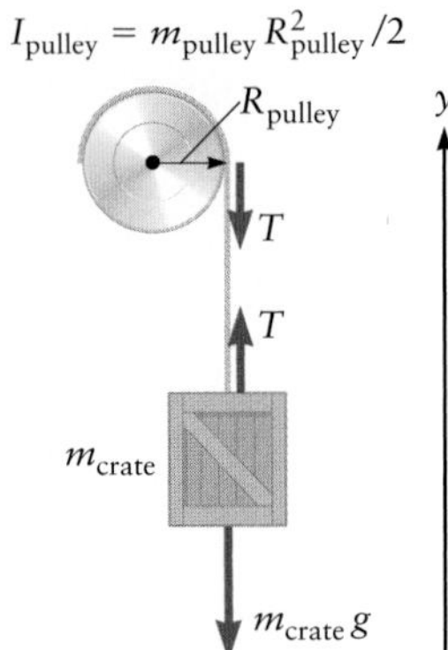

▲ **Figure 8.31** The torque on this pulley is due to the tension in the rope. The rotation axis is the axle of the pulley, so the torque equals the tension multiplied by the radius of the pulley.

(Notice that our value for ω_0 contains only one significant figure because it is only an estimate.) Using Equation 8.35, in a time $t = 5$ s the Frisbee will undergo an angular displacement of

$$\Delta\theta = \theta - \theta_0 = \omega_0 t = (60 \text{ rad/s})(5 \text{ s}) = \boxed{300 \text{ rad}}$$

► ***What does it mean?***

One revolution corresponds to $2\pi \approx 6.3$ rad, so 300 rad is about 50 revolutions.

Pulling on a Pulley: Real Pulleys with Mass

In previous chapters, we discussed pulleys and explained how they can be used to amplify forces. However, our previous treatments have assumed extremely light (massless) pulleys. Now we are ready to deal with the more realistic case of a pulley with mass.

Figure 8.31 shows a crate suspended by a very light (massless) rope. The rope is wound around a pulley, so as the crate is accelerated by the force of gravity, the pulley rotates. We now want to calculate the motion of the crate and pulley.

To calculate the linear acceleration of the crate and the angular acceleration of the pulley, we apply Newton's second law for linear motion ($\sum \vec{F} = m\vec{a}$) and Newton's second law for rotational motion ($\sum \tau = I\alpha$, Eq. 8.33). We take an approach similar to the one we used to deal with problems involving rotational equilibrium.

PROBLEM SOLVING Applying Newton's Second Law for Rotational Motion

1. **RECOGNIZE THE PRINCIPLE.** According to Newton's second law for rotational motion ($\sum \tau = I\alpha$), torque causes an angular acceleration. To find α, we must therefore find the total torque ($\sum \tau$) and the moment of inertia I.
2. **SKETCH THE PROBLEM.** Make a drawing showing all the objects of interest along with the forces that act on them and where those forces are applied. This sketch should contain a set of coordinate axes (x and y) for translational motion.
3. **IDENTIFY THE RELATIONSHIPS.** Determine the rotation axis and pivot point for calculating the torques for objects that may rotate. Then
   - Find the total torque on the objects that are undergoing rotational motion. Calculate the torque using the method employed in our work on rotational statics. This torque will then be used in Newton's second law for rotational motion, $\sum \tau = I\alpha$.
   - Calculate the sum of the forces on the objects that are undergoing linear motion for use in Newton's second law, $\sum \vec{F} = m\vec{a}$.
   - Check for a relation between linear acceleration and angular acceleration. There will usually be a connection between a and α through a relation such as Equation 8.8.
4. **SOLVE** for the quantities of interest (which might be a, α, or both) using Newton's second law for linear motion ($\sum \vec{F} = m\vec{a}$) and rotational motion ($\sum \tau = I\alpha$).
5. Always *consider what your answer means* and check that it makes sense.

Motion of a Pulley and Crate: Example of Combined Translational and Rotational Motion

We demonstrate the above problem-solving approach by applying it to calculate the crate's translational motion and the pulley's rotational motion in Figure 8.31. This figure serves as our sketch; it shows all the forces on the crate (gravity and the force from tension in the rope). Since the same rope is attached to the pulley, the torque on the pulley is also due to this tension. (We omit the force of gravity on the pulley because it is balanced by the force from the pulley's support, and neither of these forces produces a torque on the pulley.) The pulley rotates about its axle, so we pick that as the rotation axis.

To find the torque on the pulley, we first note that the tension in the rope will make the pulley rotate clockwise. According to our convention, the torque is negative. The force acts at the edge of the pulley, so the lever arm equals R_{pulley}. We thus have

$$\sum \tau = -TR_{\text{pulley}} = I_{\text{pulley}}\alpha \tag{8.38}$$

The pulley is a disc and from Table 8.2 its moment of inertia is $I_{\text{pulley}} = m_{\text{pulley}}R^2_{\text{pulley}}/2$.

We next write Newton's second law for the translational motion of the crate. Taking the $+y$ direction as the "positive" direction for forces, we have

$$\sum F_y = +T - m_{\text{crate}}g = m_{\text{crate}}a \tag{8.39}$$

Here, a is the acceleration of the crate along the y direction.

There are three unknowns—the tension T in the rope, the acceleration a of the crate, and the angular acceleration α of the pulley—but only two equations (Eqs. 8.38 and 8.39). We therefore need one more relation among these unknowns. We now recognize that the linear acceleration of the crate equals the linear acceleration of a point on the edge of the pulley. We have already discussed this situation in connection with Eq. 8.8 and found that a and α are related by

$$a = \alpha R_{\text{pulley}} \tag{8.40}$$

Equation 8.40 is actually a relation between the magnitudes of the linear and angular accelerations. In this problem, we must also be careful about the signs of a and α. Equation 8.40 treats these signs correctly for the crate and the pulley in Figure 8.31: a negative acceleration of the crate (i.e., a negative value of a) corresponds to a clockwise angular acceleration of the pulley and hence a negative value of α.

We can now solve these three relations (Eqs. 8.38 through 8.40) to find the accelerations of the crate and pulley. We first use Equation 8.40 to eliminate a from Equation 8.39:

$$T - m_{\text{crate}}g = m_{\text{crate}}a = m_{\text{crate}}\alpha R_{\text{pulley}} \tag{8.41}$$

Equation 8.38 can be simplified by inserting the expression for I_{pulley}

$$-TR_{\text{pulley}} = I_{\text{pulley}}\alpha = \tfrac{1}{2}m_{\text{pulley}}R^2_{\text{pulley}}\alpha$$

and canceling one factor of R_{pulley}:

$$T = -\tfrac{1}{2}m_{\text{pulley}}R_{\text{pulley}}\alpha$$

Using this result, we can eliminate T from Equation 8.41:

$$\overbrace{-\tfrac{1}{2}m_{\text{pulley}}R_{\text{pulley}}\alpha}^{T} - m_{\text{crate}}g = m_{\text{crate}}\alpha R_{\text{pulley}}$$

Now we can solve for α:

$$\alpha(m_{\text{crate}}R_{\text{pulley}} + \tfrac{1}{2}m_{\text{pulley}}R_{\text{pulley}}) = -m_{\text{crate}}g$$

$$\alpha = \frac{-m_{\text{crate}}g}{m_{\text{crate}}R_{\text{pulley}} + \tfrac{1}{2}m_{\text{pulley}}R_{\text{pulley}}} = \frac{-g/R_{\text{pulley}}}{1 + [m_{\text{pulley}}/(2m_{\text{crate}})]}$$

This result for the angular acceleration of the pulley can be used in Equations 8.40 and 8.39 to find the linear acceleration of the crate and the tension. Since α is constant, the angular velocity and displacement of the pulley are given by the relations for motion with a constant angular acceleration in Table 8.3.

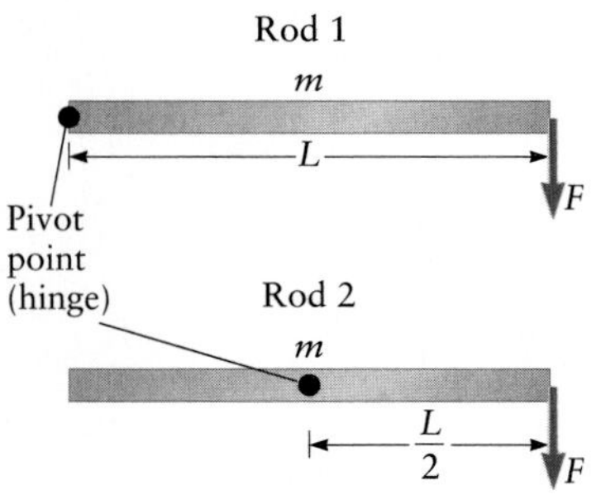

▲ **Figure 8.32** Concept Check 8.7.

CONCEPT CHECK 8.7 Angular Acceleration of a Rotating Rod

A rod of mass m and length L is hinged at one end, and a force F is applied at one end as shown in Figure 8.32 for rod 1. An identical rod is mounted on an axle so that it rotates about its center, and a force F is applied to its end as shown for rod 2 in Figure 8.32. If the forces are the same, which rod has the greater angular acceleration? What is the ratio of the angular accelerations?

(a) The angular accelerations are the same, $\alpha_1/\alpha_2 = 1$.
(b) Rod 1 has the larger angular acceleration, with $\alpha_1/\alpha_2 = 2$.
(c) Rod 1 has the larger angular acceleration, with $\alpha_1/\alpha_2 = 4$.
(d) Rod 2 has the larger angular acceleration, with $\alpha_2/\alpha_1 = 2$.
(e) Rod 2 has the larger angular acceleration, with $\alpha_2/\alpha_1 = 4$.

EXAMPLE 8.9 Which Stick Falls Faster?

Two uniform flat sticks are held at the same angle and with one end of each stick resting in the corner of a wall as shown in Figure 8.33A. The sticks have the same mass m, but lengths L and $2L$. If the sticks are released simultaneously, which one hits the floor first?

RECOGNIZE THE PRINCIPLE

We can use Newton's law for rotational motion to calculate the angular acceleration of each stick. The stick for which the magnitude of α is larger will have the larger angular speed at all times and thus will reach the floor first.

SKETCH THE PROBLEM

Figure 8.33B shows the gravitational force acting on the shorter stick and has the information we need to calculate the associated torque.

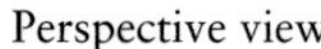

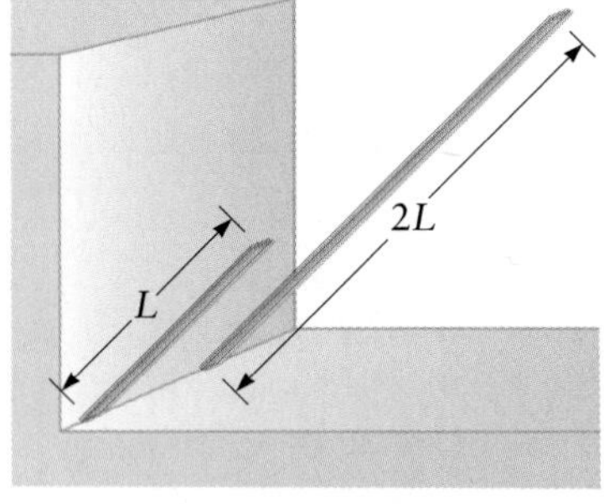

Enlarged side view of shorter rod

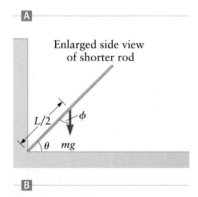

L/2 φ θ mg

B

▲ **Figure 8.33** Example 8.9.

IDENTIFY THE RELATIONSHIPS

Let's calculate α for the stick of length L. The only torque τ is from gravity. Using Newton's second law for rotational motion, we have $\tau = I\alpha$, where the moment of inertia of the stick is $I = \frac{1}{3}mL^2$ (from Table 8.2, modeling the stick as a rod). The stick will rotate clockwise in Figure 8.33B, so both τ and α will be negative. The torque can be calculated by assuming the force of gravity acts at the center of mass of the stick, a distance $L/2$ from the end. The torque is thus (Fig. 8.33B and Eq. 8.15)

$$\tau = Fr\sin\phi = -mg\left(\frac{L}{2}\right)\cos\theta$$

Inserting this result into Newton's second law for rotational motion gives

$$\tau = -mg\left(\frac{L}{2}\right)\cos\theta = I\alpha = \frac{1}{3}mL^2\alpha$$

and we can solve for the angular acceleration:

$$\alpha = -\frac{3g\cos\theta}{2L} \quad (1)$$

SOLVE

The angular acceleration for the longer stick can be found by inserting $2L$ in place of L in Equation (1). The magnitude of the angular acceleration will thus be *larger* for the *shorter* stick, and **the shorter stick will reach the floor first**.

> **▸ What does it mean?**
> You may recall from Chapter 3 that Galileo showed that balls of different mass fall at the same rate, so you might have guessed that the two sticks in Figure 8.33A would reach the floor at the same time. For rotational motion, however, the free end of the longer stick has farther to travel, so it reaches the floor last.

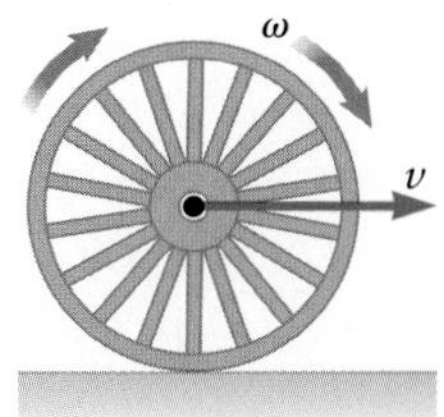

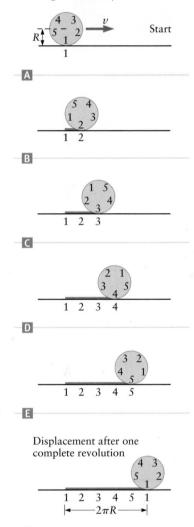

▲ **Figure 8.34** A rolling object undergoes translational motion with a center of mass velocity v and simultaneous rotational motion with angular velocity ω.

8.6 Combined Rotational and Translational Motion

The combination of rotational and translational motion arises in many common situations, including a rolling wheel and the motion of a baseball bat.

Rolling Motion

An extremely common situation that combines translation and rotation is ***rolling motion.*** Figure 8.34 shows a simple wheel as it rolls along the ground. Here the center of a wheel is moving horizontally with a constant velocity of magnitude v, but a point on the edge of the wheel is certainly not moving with a constant velocity. In fact, if the wheel is rolling without slipping, the point of the wheel in contact with the ground is *at rest* during the instant it is in contact.

As a wheel rolls, it rotates about an axis that runs along its axle, and this rotational motion is described by an angular velocity ω. To understand how ω is related to the linear velocity v of the center of the wheel, consider the series of "snapshots" of the wheel in Figure 8.35. Points on the wheel's rim are labeled, and corresponding labels where these points touch the ground as the wheel rolls are also shown. The wheel begins (Fig. 8.35A) with point 1 in contact with the ground, and when it completes one full revolution, this point is again touching the ground (Fig. 8.35F). Since the wheel is always in contact with the ground and does not slip, the distance traveled *as measured along the ground* must equal the *circumference* of the wheel, $2\pi R$. This is also the distance traveled by the center of the wheel, so if the rotation period is T, the speed of the center is

$$v = \frac{2\pi R}{T} \tag{8.42}$$

At the same time, the wheel has completed one full revolution (2π rad), so its angular velocity is

$$\omega = \frac{2\pi}{T}$$

Combing this with Equation 8.42 gives

$$v = \omega R \quad \text{(rolling motion)} \tag{8.43}$$

This relation connects the linear velocity of the center of the wheel with its angular velocity and is true for any object that rolls without slipping. It is interesting that this relation (Eq. 8.43) has the same form as Equation 8.7, but even though these equations are related, the ideas are slightly different. The result in Equation 8.7 involves the angular velocity of a rotating object and the linear speed of a *point on the edge* of the object. The result for rolling motion in Equation 8.43 connects the angular velocity with the velocity of the *center* (i.e., the axis) of a rolling object. The arguments contained in Figure 8.35 can also be used to relate the acceleration of the center of a rolling object to its angular acceleration:

$$a = \alpha R \quad \text{(rolling motion)} \tag{8.44}$$

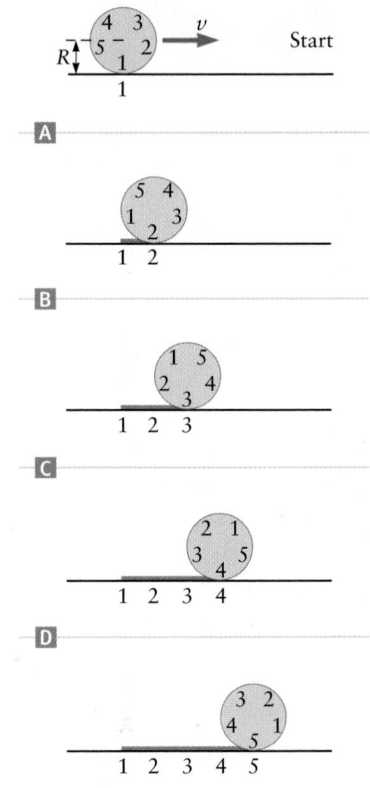

▲ **Figure 8.35** Successive "snapshots" as a wheel rolls to the right. A wheel that rolls without slipping travels a linear distance of $2\pi R$ during each complete revolution of the wheel. Each complete revolution occurs in a time equal to the period, T. This leads to the relation $v = \omega R$ between the translational velocity of the center of the wheel and the wheel's angular velocity.

It is interesting that friction plays an essential role in rolling motion. Indeed, an object cannot roll without it! It may seem a bit strange to claim that an object cannot move *without* friction, but consider a car that is initially at rest on an icy, frictionless surface. If you try to drive away, the wheels will spin freely because there is no frictional force to prevent them from slipping relative to the ice. On normal pavement (*with* friction), the force of friction between the road and the surface of the wheels enables the car to accelerate when the wheels turn. Usually, the wheels do not slip relative to the pavement, so this is static friction; if the driver tries to accelerate very quickly, however, the wheels may slip relative to the pavement bringing kinetic friction into play. Hence, for rolling motion to occur, friction is *required*.

▲ **Figure 8.36** Concept Check 8.8. On some cars, the front and rear wheels have different radii. How are the angular velocities of the two different-sized wheels related?

CONCEPT CHECK 8.8 Angular Velocity of Rolling Wheels

Consider a car whose rear wheels are much larger than the front wheels as shown in Figure 8.36. If the car is traveling up a hill, which wheels will have a larger angular velocity?

(a) The front wheels will.
(b) The rear wheels will.
(c) The front and the rear wheels will have the same angular velocity.

EXAMPLE 8.10 Ⓡ Angular Velocity of an Automobile Wheel

Consider a car traveling on a highway at a speed of 50 mi/h (about 23 m/s). Estimate the angular velocity of the car's wheels.

RECOGNIZE THE PRINCIPLE

A car's wheels are rolling without slipping (we hope!), so we can apply Equation 8.43 to relate the linear velocity v and the angular velocity ω.

SKETCH THE PROBLEM

This problem is illustrated in Figures 8.34 and 8.35.

IDENTIFY THE RELATIONSHIPS

The velocity of the center of a wheel equals the overall speed of the car, $v = 23$ m/s. The angular velocity of a wheel can then be obtained using the relation $v = \omega R$ (Eq. 8.43). This relation involves the radius R of the wheel, so we need to estimate the value of R. For a typical car, we estimate $R \approx 40$ cm $= 0.40$ m.

SOLVE

Using these values gives

$$\omega = \frac{v}{R} = \frac{23\text{ m/s}}{0.4\text{ m}} = \boxed{60\text{ rad/s}}$$

▶ ***What does it mean?***

This angular velocity corresponds to

$$\omega = (60\text{ rad/s}) \times (1\text{ rev}/2\pi\text{ rad}) = 10\text{ rev/s}$$

so the wheels undergo approximately 10 revolutions each second.

Sweet Spot of a Baseball Bat

The general motion of any object can be described as a combination of linear and rotational motion. The center of mass plays an important role in problems involving such combined motions. The linear motion of an object can be calculated from Newton's second law,

$$\sum \vec{F} = m\vec{a}$$

with $\vec{a}$ being the acceleration of the center of mass of the object. As far as the linear (i.e., translational) motion is concerned, the object moves as if it were a point particle with all its mass located at the center of mass. The rotational motion then follows from

$$\sum \tau = I\alpha$$

An interesting example of this principle involves a baseball bat and the problem of how best to hit a baseball or softball. You may know from experience that if you hit a ball off the very end of a bat or very near your hands, you feel a significant and often uncomfortable vibration at your hands. There is, however, a particular spot on the bat called the "sweet spot," and if the ball strikes the bat at that location, your hands feel essentially no vibration. A ball hit from the sweet spot seems to rebound effortlessly from the bat and, all else being equal, will travel farther than a ball that hits the bat away from this location. Suppose a baseball-playing friend wants to use physics to calculate the location of the sweet spot on his favorite bat. How would you help him?

Figure 8.37 shows the problem. As the player swings, the bat undergoes rotational motion about the batter's hands at point P. In practice, it is not quite "pure" rotational motion because a batter will often move the bat handle forward a small amount while he swings. Since this point P usually moves very little compared with the other end of the bat, it is a good approximation to assume P is fixed while the bat simply rotates about P. When the ball strikes the bat, the ball exerts a force F_b on the bat due to the ball–bat collision, producing a torque on the bat. In Figure 8.37A, this force F_b is shown acting at the sweet spot, a distance d from the center of mass of the bat.

The special property of the sweet spot is that the force F_b does not cause the handle of the bat to move. If the batter holds the bat *very lightly*, the force exerted by his hands on the bat is approximately zero and the only significant horizontal force on the bat is the force F_b exerted by the ball. This force will cause the bat's center of mass to move to the left in Figure 8.37B and at the same time will make the bat rotate. If the bat's translational and rotational motions are related in just the right way (by Eqs. 8.43 and 8.44), the bat will undergo a sort of rolling motion with point P just like the point of a wheel in contact with the ground. This behavior is found only if the force is applied at the sweet spot. In this case, no force is needed by the player's hands to keep the end of the bat at P from moving.

To find the location of the sweet spot (the value of d in Fig. 8.37A), we need to calculate the linear acceleration of the center of mass a_{CM} and the angular acceleration α of the bat. Newton's second law for the bat's translational motion reads

$$\sum F = F_b = ma_{CM} \tag{8.45}$$

We consider only motion along the direction of F_b, so Equation 8.45 involves just the components of the force and acceleration along this direction. If we assume a bat with a uniform cross section (a simple rod, another physics simplification), the center of mass is at the center of the bat. The force exerted by the ball on the bat also produces a torque on the bat. To analyze the rotational motion, let's *not* use point P in Figure 8.37 as the pivot point. Instead, let's take the pivot point to be at the bat's center of mass so that the lever arm is d in Figure 8.37A. Newton's second law for the rotational motion is then

$$\sum \tau = F_b d = I\alpha \tag{8.46}$$

For point P on the bat handle in Figure 8.37 to stay at rest, the bat must rotate in the same way as the rolling wheel in Figure 8.35, which happens only if the linear speed of the center of mass "matches" the angular velocity according to $\omega = v_{CM}/R$.

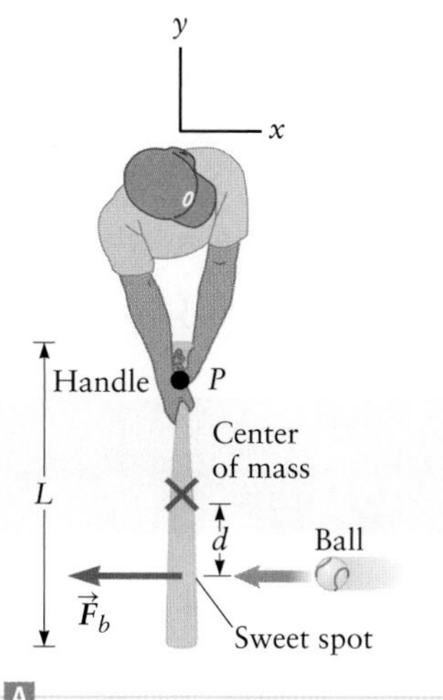

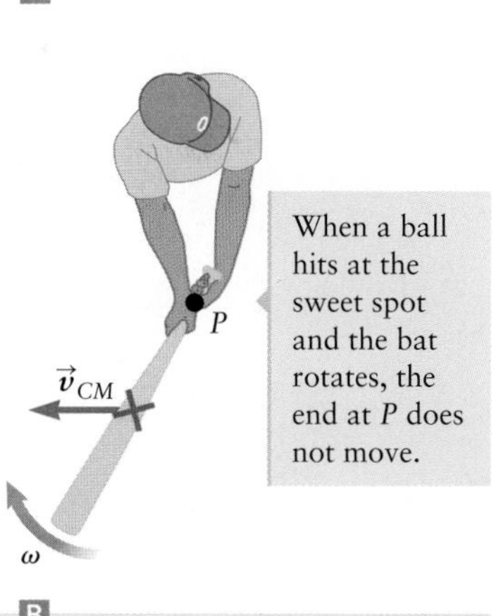

▲ **Figure 8.37** A When a baseball hits at the sweet spot of the bat, the velocity at the handle (point P) is zero. B The handle's velocity will be zero if the velocity v_{CM} of the center of mass and the angular velocity ω are related by $v_{CM} = \omega R = \omega(L/2)$, which will only be true if the linear acceleration and the angular acceleration are related by $a_{CM} = \alpha(L/2)$.

Here, R is the distance from the center of mass to point P, so $R = L/2$, where L is the length of the bat, and

$$\omega = \frac{v_{CM}}{R} = \frac{v_{CM}}{L/2}$$

If the velocities are related in this way, the angular and linear accelerations will also be related in the same way (see Eq. 8.44):

$$\alpha = \frac{a_{CM}}{L/2} \tag{8.47}$$

If we model the bat as a rod pivoted about its center, its moment of inertia from Table 8.2 is $I = mL^2/12$. Inserting our results for I and α into Equation 8.46 leads to

$$\sum \tau = F_b d = I\alpha = \frac{mL^2}{12}\frac{a_{CM}}{(L/2)} = \frac{mLa_{CM}}{6}$$

Combining this result with Equation 8.45 gives

$$F_b d = (ma_{CM})d = \frac{mLa_{CM}}{6}$$

which we can solve to find the location of the sweet spot:

$$d = \frac{L}{6}$$

Hence, the sweet spot for a uniform bat is located a distance $L/6$ from the center of the bat. A real bat is not uniform, but its sweet spot can be found in a similar way.

Summary | CHAPTER 8

Key Concepts and Principles

Angular displacement, velocity, and acceleration

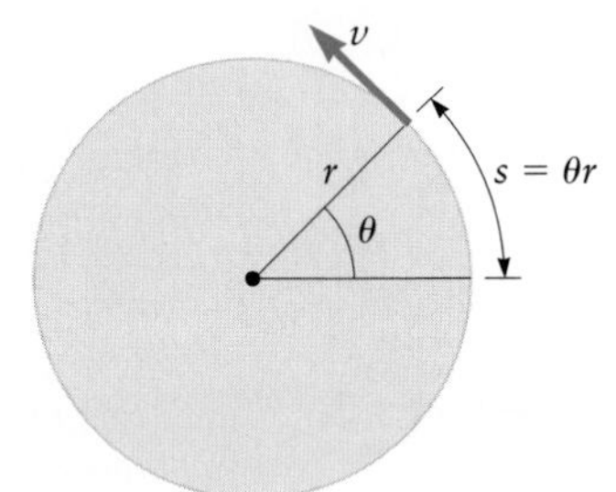

Rotational motion is described by an ***angular displacement*** θ (measured in radians), an ***angular velocity*** ω (in rad/s), and an ***angular acceleration*** α (in rad/s²). They are related to the linear motion of a point on the object by

$$\theta = \frac{s}{r}$$ **(8.1) (page 244)**

$$v = \omega r$$ **(8.7) (page 248)**

$$a = \alpha r$$ **(8.8) (page 249)**

Torque

Torque plays the role of force in cases of rotational motion. The torque produced by an applied force is

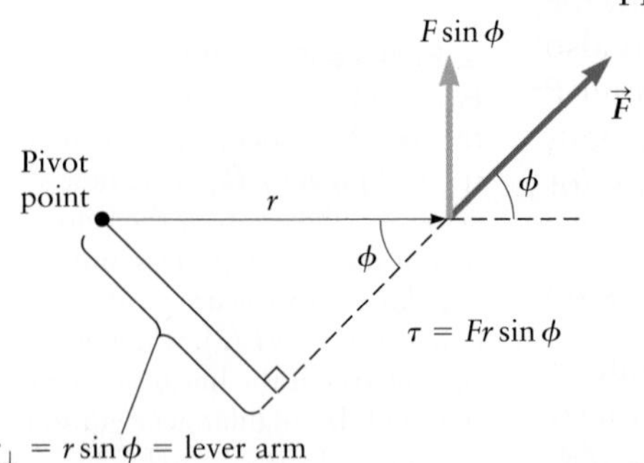

$$\tau = Fr\sin\phi$$ **(8.15) (page 252)**

This relation for the torque can be written in two equivalent ways. One way is

$$\tau = F_\perp r$$ **(8.16) (page 253)**

where $F_\perp$ is the component of the force perpendicular to the line defined by r, the distance from the rotation axis to the point on the object where the force acts. The torque can also be written as

$$\tau = Fr_\perp$$ **(8.17) (page 253)**

where F is the applied force and

$$r_{\perp} = r \sin \phi \qquad (8.18)\ (\text{page } 253)$$

The distance $r_{\perp}$ is called the ***lever arm***; it is the distance from the rotation axis to the line of action of the force, measured along a line that is perpendicular to both.

Newton's second law for rotational motion

$$\sum \tau = I\alpha \qquad (8.12)\ (\text{page } 251)$$

The ***moment of inertia*** I of an object depends on its mass and shape. For a collection of particles of mass m_i, the moment of inertia is

$$I = \sum_i m_i r_i^2 \qquad (8.14)\ (\text{page } 252)$$

where r_i is the distance from the rotation axis to m_i. The moment of inertia is proportional to the total mass and depends on its distribution relative to the rotation axis. Values of I for some common shapes are listed in Table 8.2.

Applications

Rotational equilibrium

An object is in translational equilibrium and rotational equilibrium if its linear acceleration and angular acceleration are both zero. For that to be true, both the total force and the total torque on the object must be zero.

Rotational motion

Calculations of the rotational dynamics start with Newton's second law for rotational motion,

$$\sum \tau = I\alpha \qquad (8.12)\ (\text{page } 251)$$

When the angular acceleration is constant, the solutions for the angular displacement and velocity as functions of time are (Table 8.3)

$$\theta = \theta_0 + \omega_0 t + \tfrac{1}{2}\alpha t^2 \qquad (8.35)\ (\text{page } 266)$$

$$\omega = \omega_0 + \alpha t \qquad (8.34)\ (\text{page } 266)$$

$$\omega^2 = \omega_0^2 + 2\alpha(\theta - \theta_0) \qquad (8.36)\ (\text{page } 266)$$

These relations are very similar to the results we found for linear motion with constant linear acceleration in Chapter 3.

Rolling motion

Rolling motion is an example of combined translational and rotational motion. If an object rolls without slipping, the angular velocity and angular acceleration are related to the linear velocity and acceleration of the center of mass by

$$v = \omega R \qquad (8.43)\ (\text{page } 271)$$

$$a = \alpha R \qquad (8.44)\ (\text{page } 271)$$

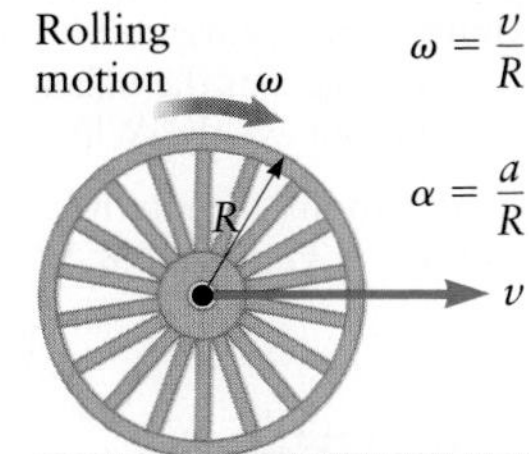

Questions

SSM = answer in *Student Companion & Problem-Solving Guide* ⓧ = life science application

1. ⓧ Explain why placing your hands behind your head rather than placing them on your stomach makes sit-ups more difficult.

2. Two disks, each of mass m, have radii R and $2R$ and are mounted on axles that pass through their centers. (a) Which disk has the larger moment of inertia? (b) If a force of magnitude F is applied at the edge of each disk as shown in Figure Q8.2, which disk will experience the larger torque? (c) Which disk will have the larger angular acceleration?

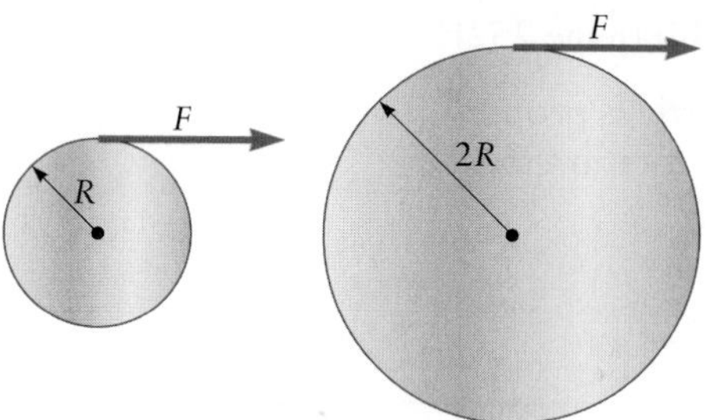

Figure Q8.2

3. A hoop and a disk have the same radius and the same moment of inertia. What is the ratio of their masses?

4. According to Table 8.2, the moment of inertia for a rod that rotates about an axis perpendicular to the rod and passing through one end is $I = ML^2/3$; if the axis passes through the center of the rod, then $I = ML^2/12$. Give a physical explanation for this difference in terms of the way the mass of the rod is distributed with respect to the axis in the two cases.

5. ⓧ The tree in Figure Q8.5 experiences a large torque due to the force of gravity, yet it is in static equilibrium. What additional torque balances the torque due to gravity?

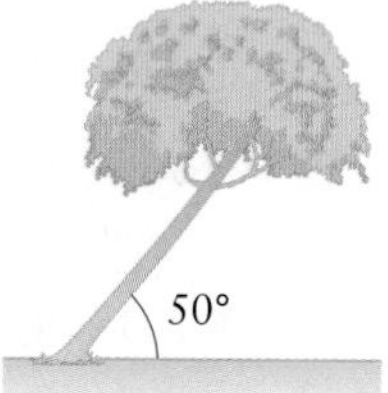

Figure Q8.5 Question 5 and Problem 23.

6. In our analysis of a rollover car accident in Example 8.6, we treated the car in a very simplified way. Develop a more realistic model of a car with special attention to the height of its center of mass and use it to compute the force required to tip over the car. (See Fig. 8.26.)

7. SSM Two golfers team up to win a golf tournament and are awarded a solid gold golf club. To divide their winnings, they balance the club on a finger and then cut the club into two pieces at the balance point with a hacksaw. Was the winning gold split evenly? Why or why not?

8. A mechanic's toolbox contains the two wrenches and two screwdrivers shown in Figure Q8.8. While working on her vintage car, the mechanic comes across a stubborn and rusted bolt. Which wrench would be the most useful in breaking the bolt free? Why is one better than the other? She also encounters a screw that is similarly stuck tight. Which screwdriver should she choose? Is one better than the other? Why?

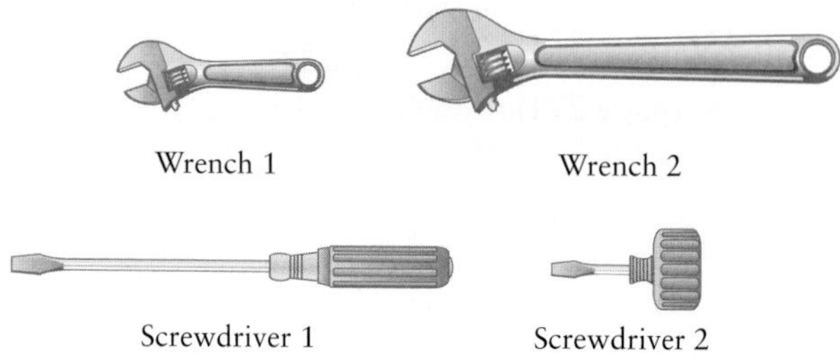

Figure Q8.8

9. A snap of the wrist will usually detach a paper towel from a roll. Why does that almost always work for a full roll, yet an almost empty roll often gives much more than a single sheet?

10. Two metersticks, one with a weight attached to its end, are held with one of their ends in the corner formed by the floor and a wall as depicted in Figure Q8.10. If the ends of the metersticks are let go at the same time and start with the same angle with respect to the floor, which one will hit the floor first? Why?

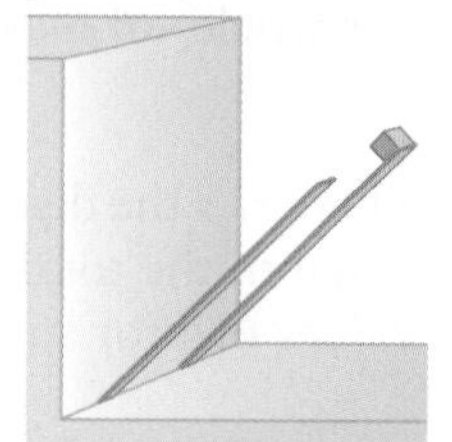

Figure Q8.10

11. **Three types of levers.** Levers can be classified as first, second, or third class, depending on how the load force ($\vec{F}_L$) and effort force ($\vec{F}_E$) are configured with respect to the pivot point or fulcrum as shown in Figure Q8.11. Determine the lever classification of the following items: (a) a crowbar, (b) a pair of pliers, (c) a clawed hammer pulling a nail, (d) a wheelbarrow (Fig. P8.37), (e) a seesaw (Fig. 8.10), (f) a fishing pole, (g) a light switch, (h) a gate with closing spring attached to hinge, and (i) biceps raising the forearm (Fig. P8.79).

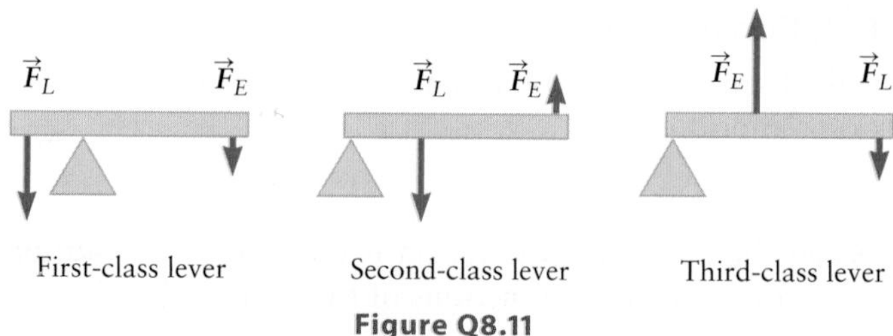

Figure Q8.11

12. In about 200 BCE, Archimedes made the statement, "Give me a lever long enough and a place to stand, and I will move the Earth." Assume he can exert a force equal to his weight on the Earth (about 900 N). Design a lever that would enable him to give the Earth an acceleration of 1 m/s^2.

13. A tennis racket possesses a sweet spot in close analogy to that of a baseball bat. One might expect that rackets are designed so that the sweet spot is in the center of the strings. Somewhat surprisingly, that is not the case. The approximate location of a racket's sweet spot can be found from the simple experiment sketched in Figure Q8.13. The racket is suspended by a string tied to the handle and is then struck (gently) by a ball or a light blow. When this force is exerted at the sweet spot, the handle will not rebound. Perform this experiment yourself and show that the sweet spot is rather far from the center of the strings. Explain how you could add or subtract mass from different spots on the racket to move the sweet spot to this desired location.

Figure Q8.13

14. ⓧ A tree has a large branch that grows horizontally out from the trunk. If the branch doubles in length and doubles in diameter, by what factor does the torque on the branch due to gravity increase?

15. SSM A tree has two large branches that grow out horizontally from the trunk in opposite directions. One branch has length L and diameter d, whereas the other has length $L/2$. If the magnitude of the torque due to gravity is the same on the two branches, what is the diameter of the shorter branch?

16. A force of magnitude F is applied to a rod at various spots and at various angles (Fig. Q8.16). In which case, (1), (2), or (3), is the magnitude of the torque largest? Smallest?

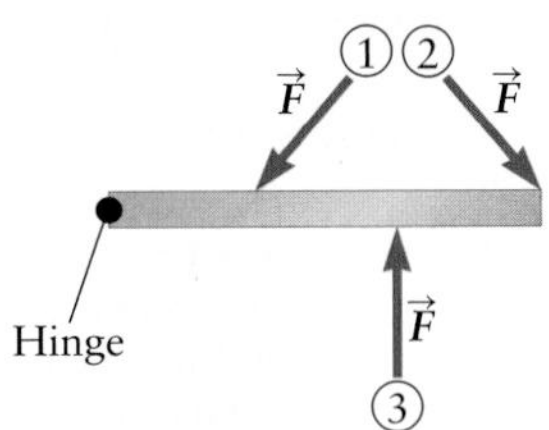

Figure Q8.16

17. Bodybuilders do an exercise called "curls" in which they lift a weight by bending at the elbow. Explain why the torque on the elbow is largest when the arm is horizontal.

18. Explain why it is easier (it requires a smaller force) to open a large jar than a small one.

19. Consider a wheel that rolls without slipping such as the wheel in Figure 8.34. We have seen that slipping is prevented by the force of static friction between the wheel and the pavement. Can this force do work on the wheel?

20. In Section 8.6, we found that for a uniform baseball bat, the sweet spot is a distance $L/6$ from the center of the bat. A real bat is not uniform; it is much thinner in the handle than in the barrel (the end of the bat farthest from point P in Fig. 8.37). Estimate how that affects the location of the sweet spot.

Problems

SSM = solution in *Student Companion & Problem-Solving Guide*
★ = intermediate ✪ = challenging ✕ = life science application
R = reasoning and relationships problem
RT = reasoning tutorial in ENHANCED WebAssign

8.1 DESCRIBING ROTATIONAL MOTION

1. An angle has a value of 85°. What is the angle in radians?

2. The shaft on an engine turns through an angle of 45 rad. (a) How many revolutions does the shaft make? (b) What is this angle in degrees?

3. An old-fashioned (vinyl) record rotates at a constant rate of 33 rpm. (a) What is its angular speed in rad/s? (b) If the record has a radius of 18 cm, what is the linear speed of a point on its edge?

4. The angular speed of the shaft of a car's engine is 360 rad/s. How many revolutions does it complete in 1 hour?

5. Consider again the record in Problem 3. If it plays music for 20 min, how far does a point on the edge travel in meters?

6. The tachometer in the author's car has a maximum reading of 8000 rpm. (A tachometer is a gauge that shows the angular speed of a car's engine.) Express this angular speed in rad/s.

7. A car engine is initially rotating at 200 rad/s. It is then turned off and takes 3.0 s to come to a complete stop. What is the angular acceleration of the engine? Assume α is constant.

8. ★ What is the angular speed of the Earth's center of mass as it orbits the Sun?

9. ★ What is the angular speed with respect to the Earth's rotation axis of a person standing (a) at the equator? (b) At a latitude of 45°?

10. ★ Mercury spins about its axis with a period of approximately 58 days. What is the rotational angular speed of Mercury?

11. ★ Mercury orbits the Sun with a period of approximately 88 days. What is the angular speed of Mercury's orbital motion?

12. ★ An adult is riding on a carousel and finds that he is feeling a bit nauseated. He then decides to move from the outer edge of the carousel to a point a distance r_2 from the center so that his linear speed is reduced by a factor of 2.5. If the carousel has a constant angular speed of 1.0 rad/s and a radius of 3.3 m, what is r_2?

13. SSM ★ What are the angular velocity and the period of the second hand of a clock? If the linear speed of the end of the hand is 5.0 mm/s, what is the clock's radius?

14. Construct a graph of the angular displacement of the minute hand of a clock as a function of time. The axes of your graph should include quantitative scales (i.e., numbers and units). What is the slope of the relation between θ and t?

15. Consider a race car that moves with a speed of 200 mi/h. If it travels on a circular race track of circumference 2.5 mi, what is the car's angular speed?

8.2 TORQUE AND NEWTON'S LAWS FOR ROTATIONAL MOTION

16. A diver stands at rest at the end of a massless diving board as shown in Figure P8.16. (a) If the mass of the diver is 120 kg and the board is 4.0 m long, what is the torque due to gravity on the diving board with a pivot point at the fixed end of the board? (b) According to our convention for positive and negative torques, is the torque on the diving board positive or negative?

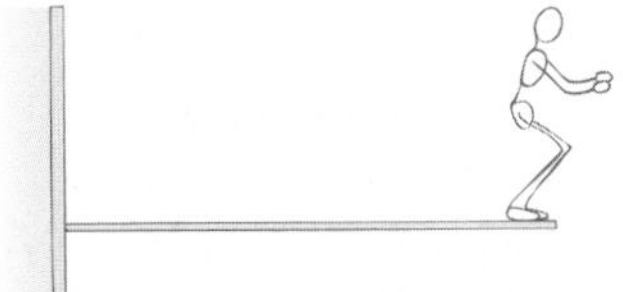

Figure P8.16 Problems 16 and 17.

17. ★ Consider again the diver in Figure P8.16. Assume the diving board now has a mass of 30 kg. Find the total torque due to gravity on the diving board. Assume the mass of the board is uniformly distributed.

18. SSM ★ Consider the clock in Figure 8.17. Calculate the magnitude and sign of the torque due to gravity on the hour hand of the clock at 4 o'clock. Assume the hand has a mass of 15 kg, a length of 1.5 m, and the mass is uniformly distributed.

19. ★ A rod of length 3.8 m is hinged at one end, and a force of magnitude $F = 10$ N is applied at the other (Fig. P8.19). (a) If the magnitude of

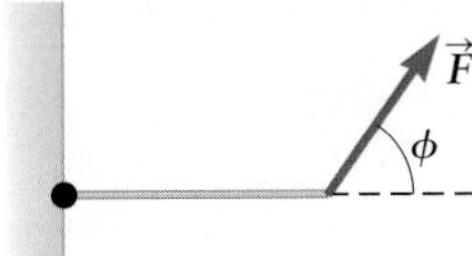

Figure P8.19

the torque associated with this force is 18 N·m, what is the angle ϕ? (b) What is the sign of the torque?

20. A person carries a long pole of mass 12 kg and length 4.5 m as shown in Figure P8.20. Find the magnitude of the torque on the pole due to gravity.

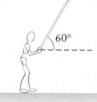

21. If the length of the pole in Problem 20 is increased by a factor of three and its mass is increased by a factor of two, by what factor does the torque change?

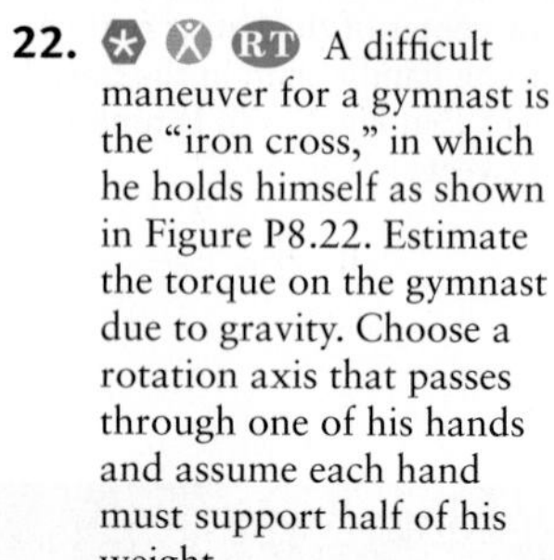

Figure P8.20
Problems 20 and 21.

22. RT A difficult maneuver for a gymnast is the "iron cross," in which he holds himself as shown in Figure P8.22. Estimate the torque on the gymnast due to gravity. Choose a rotation axis that passes through one of his hands and assume each hand must support half of his weight.

Figure P8.22

23. SSM R A tree grows at an angle of 50° to the ground as shown in Figure Q8.5. If the tree is 25 m from its base to its top and has a mass of 500 kg, what is the approximate magnitude of the torque on the tree due to the force of gravity? Take the base of the tree as the pivot point. (The answer reveals one reason trees need roots.)

24. RT A baseball player holds a bat at the end of the handle. If the bat is held horizontally, what is the approximate torque due to the force of gravity on the bat, with a pivot point at the batter's hands?

8.3 ROTATIONAL EQUILIBRIUM

25. RT A uniform pole of length 2.0 m and mass 4.5 kg hangs horizontally from two cables as shown in Figure P8.25. What are the approximate tensions in the two cables?

Figure P8.25

26. A uniform wooden plank of mass 12 kg rests on two supports as shown in Figure P8.26. The plank is at rest, so it is in translational equilibrium and rotational equilibrium. (a) Make a sketch showing all the forces on the plank. Be sure to show where the forces act. (b) Choose the left-hand support for the pivot point and calculate the torque from each of the forces in part (a). (c) Express the conditions for translational and rotational equilibrium. (d) Solve to find the forces exerted by the two supports on the plank.

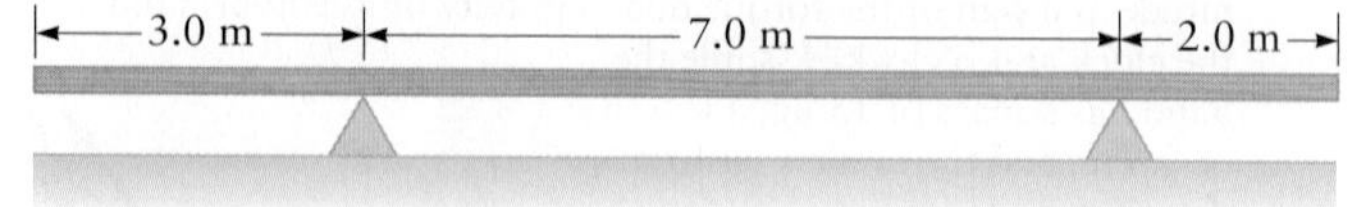

Figure P8.26

27. Suppose a crate of mass 7.5 kg is placed on the plank in Figure P8.27 at a distance 3.9 m from the left end. If the plank has a mass of 12 kg, find the forces exerted by the two supports on the plank.

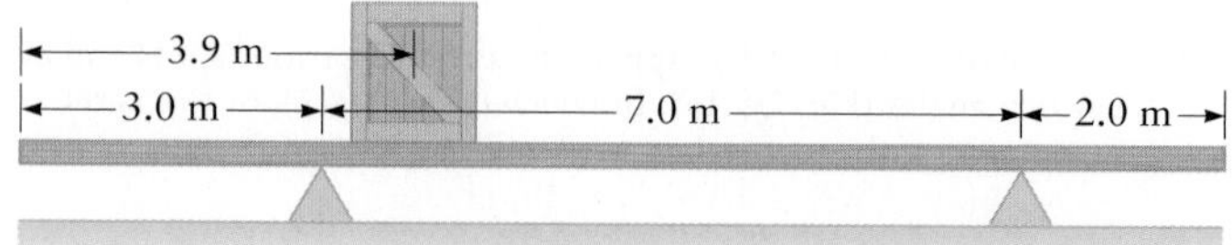

Figure P8.27

28. The seesaw in Figure P8.28 is 4.5 m long. Its mass of 20 kg is uniformly distributed. The child on the left end has a mass of 14 kg and is a distance of 1.4 m from the pivot point while a second child of mass 30 kg stands a distance L from the pivot point, keeping the seesaw at rest. (a) Is the seesaw in rotational equilibrium? (b) Make a sketch showing all the forces on the seesaw. Be sure to show where each force acts. (c) Express the conditions for translational and rotational equilibrium for the seesaw. (d) Solve to find L, the location of the child on the right.

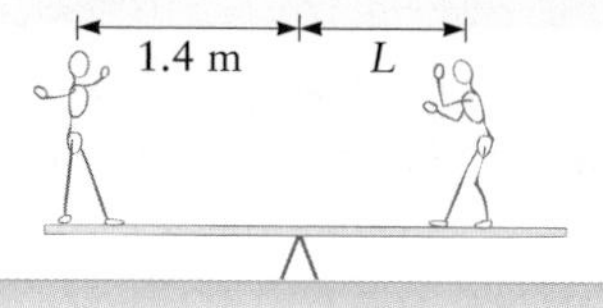

Figure P8.28

29. Consider the flagpole in Figure P8.29. If the flagpole has a mass of 20 kg and length 10 m and the angle the cable makes with the pole is $\phi = 25°$, what are the magnitude and direction of the force exerted by the hinge (at point P) on the flagpole? Assume the mass of the pole is distributed uniformly.

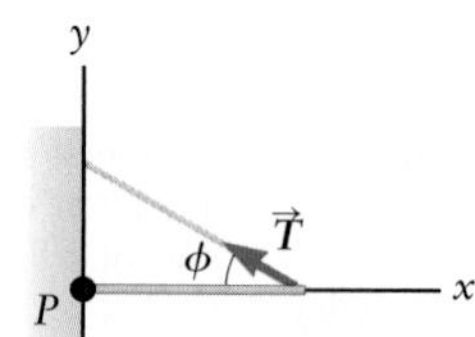

Figure P8.29

30. A painter is standing on the ladder (mass 40 kg and length 2.5 m) in Figure P8.30. There is friction between the bottom of the ladder and the floor with $\mu_S = 0.30$, but there is no friction between the ladder and the wall. The painter has mass 70 kg and is standing a distance of 0.60 m from the top. (a) Make a sketch showing all the forces on the ladder. Be sure to show where each force acts. (b) Express the conditions for translational and rotational equilibrium for the ladder. (c) If $\theta = 60°$, use the results from part (b) and determine whether or not this ladder will slip.

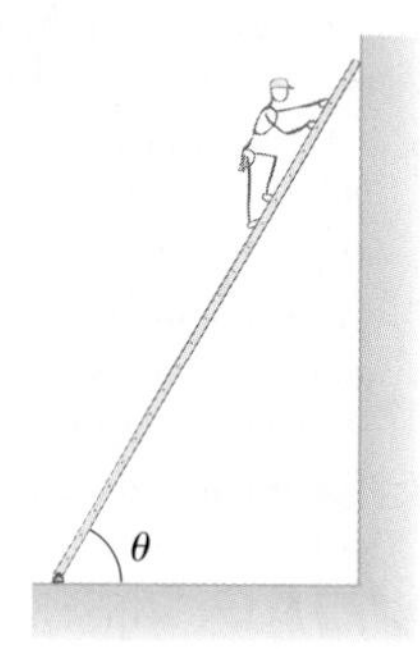

Figure P8.30
Problems 30, 31, and 32.

31. Consider again the ladder in Problem 30. What is the sign of the torque on the ladder due to the force from the wall?

32. R Consider again the ladder in Figure P8.30 and imagine two scenarios. In both scenarios the ladder is in rotational equilibrium. (1) There is no friction between the ladder and the floor, and there is a frictional force F_W between the wall and the ladder. (2) There is no friction between the ladder and the wall, and there is a frictional force F_F between the floor and the ladder. Which force is larger, F_W or F_F? *Hint*: Consider the value of the angle θ in Figure P8.30.

33. Consider the ladder in Example 8.5 and assume there is friction between the vertical wall and the ladder, with $\mu_S = 0.30$. Find the angle ϕ at which the ladder just begins to slip.

34. A solid cube of mass 40 kg and edge length 0.30 m rests on a horizontal floor as shown in Figure P8.34. A person then pushes on the upper edge of the cube with a horizontal force of magnitude F. At what value of F will the cube start to tip? Assume the frictional force from the floor is large enough to prevent the cube from sliding.

Figure P8.34 Problems 34, 35, and 36.

35. Consider again the cube in Problem 34 (Fig. P8.34), but now assume the force is applied along a horizontal line that is 0.20 m above the floor. At what value of F will the crate begin to tip?

36. Repeat Problem 34, but assume the force F is applied at the corner of the cube and at an angle of 30° above the horizontal direction. At what value of F will the crate begin to tip?

37. RT Wheelbarrows are designed so that a person can move a massive object more easily than if he simply picked it up. If the person using the wheelbarrow in Figure P8.37 is able to apply a total vertical (upward) force of 200 N on the handle, what is the approximate mass of the heaviest object he could move with the wheelbarrow?

© Burke/Triolo/Brand X Pictures/Jupiterimages

Figure P8.37 Problem 37 and Question 11.

38. R One way to put a ladder in place on a wall is to "walk" the ladder to the wall. One end of the ladder is held fixed (perhaps at the base of the wall), and a person slides his hands along the ladder as he walks toward the fixed end as sketched in Figure P8.38. Anyone who has raised a ladder in this way knows that the amount of force the person must apply grows larger initially as the person moves toward the end that is fixed. Assuming the force from the person is perpendicular to the ladder, find the force the person must apply to hold the ladder at (a) an angle of 10° and (b) an angle of 20°.

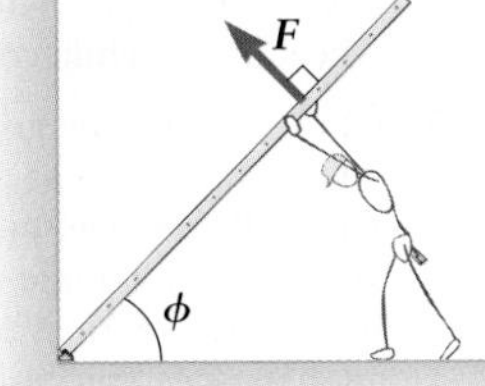

Figure P8.38 Problems 38 and 39.

39. R Consider again the ladder in Figure P8.38. We saw in Problem 8.38 that the force supplied by the person increases with increasing angle ϕ as the ladder is initially lifted off the ground. When the ladder is sufficiently high (the angle ϕ is sufficiently large), though, it becomes easier to raise the ladder higher (i.e., the necessary applied force becomes smaller). Find the angle ϕ at which that happens.

40. SSM A flagpole of length 12 m and mass 30 kg is hinged at one end, where it is connected to a wall as sketched in Figure P8.40. The pole is held up by a cable attached to the other end. Find the tension in the cable.

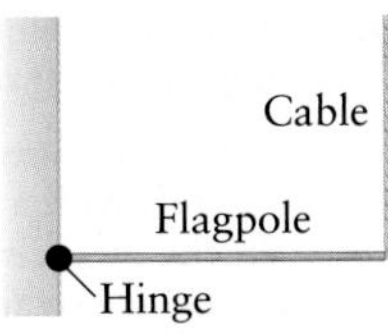

Figure P8.40

8.4 MOMENT OF INERTIA

41. Two particles of mass $m_1 = 15$ kg and $m_2 = 25$ kg are connected by a massless rod of length 1.2 m. Find the moment of inertia of this system for rotations about the following pivot points: (a) the center of the rod, (b) the end at m_1, (c) the end at m_2, and (d) the center of mass. Assume the rotation axis is perpendicular to the rod.

42. SSM Four particles with masses $m_1 = 15$ kg, $m_2 = 25$ kg, $m_3 = 10$ kg, and $m_4 = 20$ kg sit on a very light (massless) metal sheet and are arranged as shown in Figure P8.42. Find the moment of inertia of this system with the pivot point (a) at the origin and (b) at point P. Assume the rotation axis is parallel to the z direction, perpendicular to the plane of the drawing.

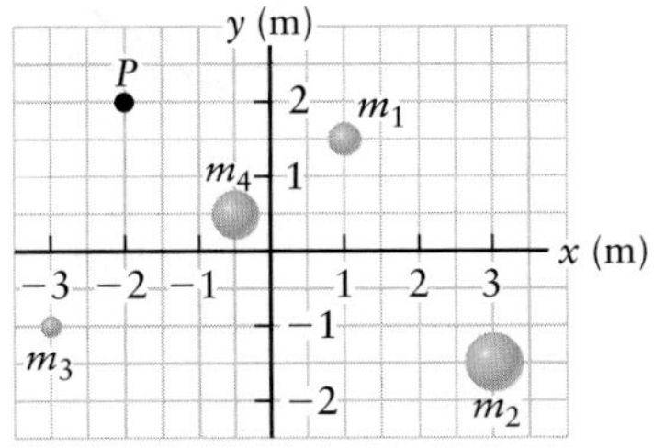

Figure P8.42

43. Consider a merry-go-round that has the form of a disc with radius 5.0 m and mass 100 kg. If three children, each of mass 20 kg, sit on the outer edge of the merry-go-round, what is the total moment of inertia?

44. Which has a larger moment of inertia, a disc of mass 20 kg or a hoop of mass 15 kg with the same radius?

45. RT Estimate the moment of inertia of a bicycle wheel.

46. If the mass of a wheel is increased by a factor of 2 and the radius is increased by a factor of 1.5, by what factor is the moment of inertia increased? Model the wheel as a solid disc.

47. Consider two cylinders with the same density and the same length. If the ratio of their diameters is 1.5/1, what is the ratio of their moments of inertia?

48. R In our examples with CDs (such as Example 8.7), we usually ignored the hole in the center. Calculate the moment of inertia of a CD, including the effect of the hole. For a CD of radius 6.0 cm, estimate the percentage change in the moment of inertia due to a hole of radius 7 mm.

49. The moment of inertia for a square plate of mass M and length L that rotates about an axis perpendicular to the plane of the plate and passing through its center is $ML^2/6$ (Fig. P8.49A). What is the moment of inertia of the same plate when it is rotated about an axis that lies along one edge of the plate (Fig. P8.49B)?

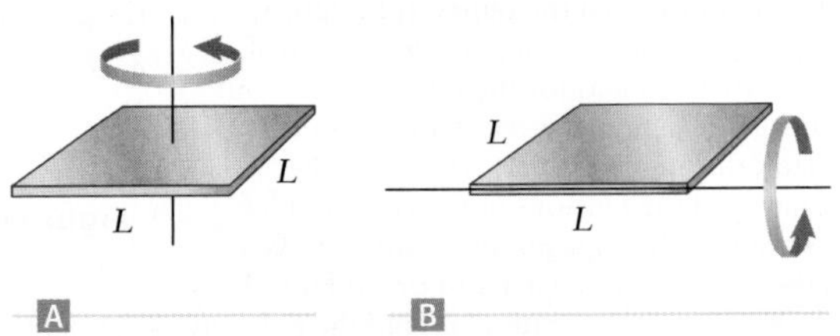

Figure P8.49

50. Estimate the moment of inertia of a baseball bat for the following cases.

(a) Assume the bat is a wooden rod of uniform diameter and take the pivot point to be (i) in the middle of the bat, (ii) at the end of the bat, and (iii) 5 cm from the end of the handle, where a batter who "chokes up" on the bat would place his hands. *Hint*: Consider the bat as two rods held together at the pivot point.

(b) Assume a realistic bat, which is narrower at the handle than at the opposite end, and consider the three locations from part (a).

8.5 ROTATIONAL DYNAMICS

51. Figure P8.51 shows the angular displacement of an object as a function of time. What is the approximate angular velocity of the object?

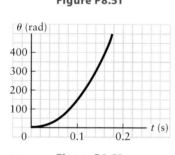

Figure P8.51

52. SSM Figure P8.52 shows the angular displacement of an object as a function of time. (a) What is the approximate angular velocity of the object at $t = 0$? (b) What is the approximate angular velocity at $t = 0.10$ s? (c) Estimate the angular acceleration at $t = 0.050$ s.

Figure P8.52

53. A ceiling fan of radius 45 cm runs at 90 rev/min. How far does the tip of the fan blade travel in 1 hour?

54. Construct a plot of the angular displacement of a CD as a function of time. Use information from this chapter to attach quantitative labels (numbers and units) to your graph. Take $t = 0$ to be the time when the CD player is first turned on. End your graph when the CD player is turned off and the CD stops spinning.

55. A potter's wheel of radius 0.20 m is turned on at $t = 0$ and accelerates uniformly. After 60 s, the wheel has an angular velocity of 2.0 rad/s. Find the angular acceleration and the total angular displacement of the wheel during this time.

56. Construct a graph of the angular velocity of a car wheel as a function of time. (a) Assume the wheel starts from rest and moves with a constant (center of mass) velocity. (b) Assume the car starts from rest and accelerates uniformly. (c) Assume the car comes to a stop uniformly at a traffic light.

57. Two crates of mass 5.0 kg and 9.0 kg are connected by a rope that runs over a pulley of mass 4.0 kg as shown in Figure P8.57. (a) Make a sketch showing all the forces on both crates and the pulley. (b) Express Newton's second law for the crates (translational motion) and for the pulley (rotational motion). The linear acceleration a of the crates, the angular acceleration α of the pulley, and the tensions in the right and left portions of the rope are unknowns. (c) What is the relation between a and α? (d) Find the acceleration of the crates. (e) Find the tensions in the right and left portions of the rope.

Figure P8.57

58. SSM Two crates of mass $m_1 = 15$ kg and $m_2 = 25$ kg are connected by a cable that is strung over a pulley of mass $m_{\text{pulley}} = 20$ kg as shown in Figure P8.58. There is no friction between crate 1 and the table. (a) Make a sketch showing all the forces on both crates and the pulley. (b) Express Newton's second law for the crates (translational motion) and for the pulley (rotational motion). The linear acceleration a of the crates, the angular acceleration α of the pulley, and the tensions in the right and left portions of the rope are unknowns. (c) What is the relation between a and α? (d) Find the acceleration of the crates. (e) Find the tensions in the horizontal and vertical portions of the rope.

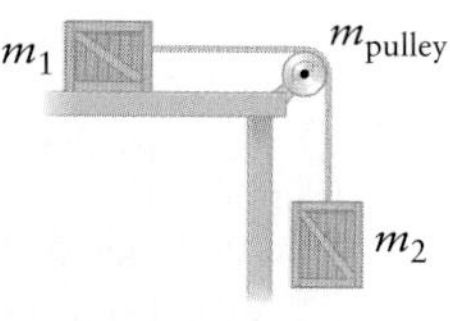

Figure P8.58
Problems 58 and 59.

59. For the system in Figure P8.58, if the coefficient of friction between crate 1 and the table is $\mu_K = 0.15$, what is the acceleration?

60. RT Consider the DVD in Example 8.2. This DVD player is turned on at $t = 0$ and very quickly starts to play a video so that the DVD has an angular velocity of 1500 rpm just a few seconds after $t = 0$. The DVD then plays for 2 hours, during which time it has constant angular acceleration and a final angular velocity of 630 rpm. Estimate the total distance traveled by a point on the edge of the DVD. Assume the DVD has a radius of 6.0 cm.

61. Consider again the DVD in Problem 60. As was explained in Example 8.2, the data on the DVD are encoded in a long spiral "track," where the spacing between each turn of the track is 0.74 μm and the inner and outer radii of the program area are about 25 mm and 58 mm, respectively. Estimate the length of this spiral.

62. You like to swim at a nearby lake. On one side of the lake is a cliff, and the top of the cliff is 6.5 m above the surface of the lake. You dive off the cliff doing somersaults with an angular speed of 2.2 rev/s. How many revolutions do you make before you hit the water? Assume your initial center of mass velocity is horizontal and you begin from a standing position. Do you enter the water head first or feet first?

63. A Ferris wheel is moving at an initial angular velocity of 1.0 rev/30 s. If the operator then brings it to a stop in 3.0 min, what is the angular acceleration of the Ferris wheel? Express your answer in rad/s^2. Through how many revolutions will the Ferris wheel move while coming to a stop?

64. A board of length 1.5 m is attached to a floor with a hinge at one end as shown in Figure P8.64. The board is initially at rest and makes an angle of 40° with the floor. If the board is then released, what is its angular acceleration?

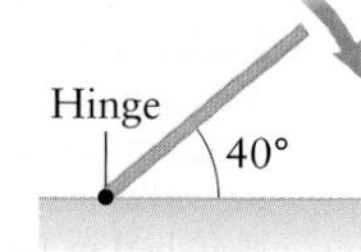

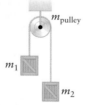

Figure P8.64

8.6 COMBINED ROTATIONAL AND TRANSLATIONAL MOTION

65. A car starts from rest and then accelerates uniformly to a linear speed of 15 m/s in 40 s. If the tires have a radius of 25 cm, what are the magnitudes of (a) the average linear acceleration of a tire, (b) the angular acceleration of a tire, (c) the angular displacement of a tire from $t = 0$ to 40 s, and (d) the total distance traveled by a point on the edge of a tire?

66. A car is traveling with a speed of 20 m/s. If the tires have an angular speed of 62 rad/s, what is the radius of the tires?

67. SSM Consider a tennis ball that is hit by a player at the baseline with a horizontal velocity of 45 m/s (about 100 mi/h). The ball travels as a projectile to the player's opponent on the opposite baseline, 24 m away, and makes 25 complete revolutions during this time. What is the ball's angular speed?

68. A cylinder rolls without slipping down an incline that makes a 30° angle with the horizontal. What is the acceleration of the cylinder?

69. A string is rolled around a cylinder ($m = 4.0$ kg) as shown in Figure P8.69. A person pulls on the string, causing the cylinder to roll without slipping along the table. If there is a tension of 30 N in the string, what is the acceleration of the cylinder?

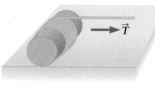

Figure P8.69

70. A truck accelerates from rest to a speed of 12 m/s in 5.0 s. If the tires have a radius of 40 cm, how many revolutions do the tires make during this time?

71. Ⓡ The wheel of an adult's bicycle rolls a distance of 3.0 km. Approximately how many revolutions does the wheel make during this journey?

72. A car has a top speed of 70 m/s (about 150 mi/h) and has wheels of radius 30 cm. If the car starts from rest and reaches its operating speed after 9.0 s, what is the angular acceleration of the wheels? Assume the car accelerates uniformly.

73. The wheels on a special bicycle have different radii R_f and R_b (Fig. P8.73). The bicycle starts from rest at $t = 0$ and accelerates uniformly, reaching a linear speed of 10 m/s at $t = 9.0$ s. (a) What is the bicycle's linear acceleration? (b) At $t = 4.5$ s, the angular speed of the back wheel is 11 rad/s, whereas the angular speed of the front wheel is greater by a factor of three. Find R_b and R_f.

Figure P8.73

Additional Problems

74. Two bricks of uniform density are stacked as shown in Figure P8.74. What is the maximum overhang distance d that can be achieved? *Hint*: Consider how the torque on each brick's center of mass must be in equilibrium about the balance point (the edge of the ledge).

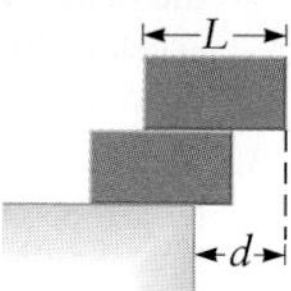

Figure P8.74 Problems 74 and 75.

75. Consider again the problem of overhanging bricks in Figure P8.74. Find the maximum overhang distance d for systems of (a) three bricks and (b) four bricks. (c) How many bricks are needed for the top brick to have an overhang of $d = 2L$?

76. SSM Three gears of a mechanism are meshed as shown in Figure P8.76 and are rotating with constant angular velocities. The ratio of the diameters of gears 1 through 3 is 3.5/1.0/2.0. A torque of 20 N · m is applied to gear 3. If gear 2 has a radius of 10 cm, what is the torque on gears 1 and 2?

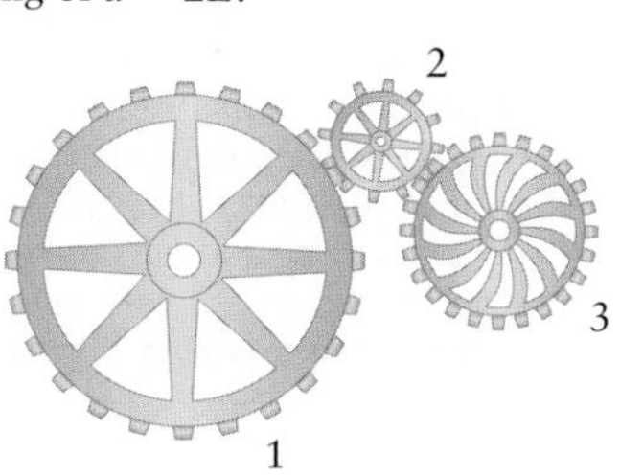

Figure P8.76

77. A chain drive (Fig. P8.77) transfers the rotational motion of a tractor transmission to a grain elevator. The drive shaft places a counterclockwise torque on the small sprocket of radius 2.2 cm so that it can overcome the resistance in the mechanism of the auger, which puts a clockwise torque of 200 N · m on the larger sprocket of radius 17 cm. If the gears turn at a constant rate, (a) what is the tension in the chain and (b) what is the torque applied to the small sprocket by the drive shaft?

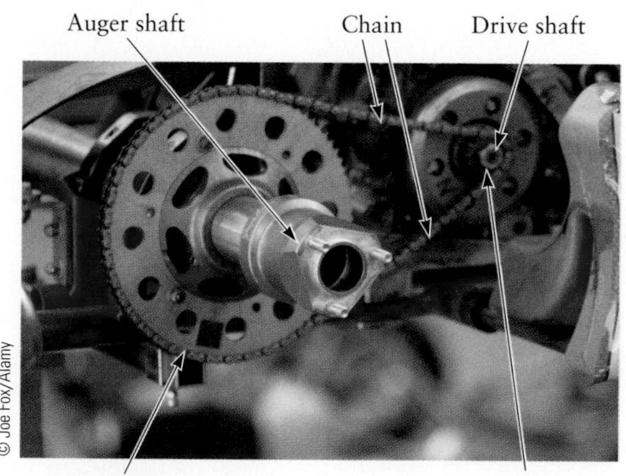

Figure P8.77

78. Bicycles like the one shown in the photo in Figure P8.78 typically have 24 possible gear ratios (three gears on the front and eight on the back). Suppose the diameters of the largest and smallest sprockets are 17 cm and 9.0 cm on the front cluster and 11 cm and 5.0 cm on the rear cluster. The inset drawing in Figure P8.78 gives an expanded view showing just one sprocket on the back and one on the front. The crank (Fig. P8.78, inset) measures 18 cm from the center of the pedal shaft axis to the center of the front sprocket, and the back wheel (including the tire) has a diameter of 68 cm. Assume the cyclist is traveling at constant speed. If a downward force of 320 N is applied to the pedal when the crank is in the horizontal position, what is the horizontal force the rear wheel exerts on the road for (a) the lowest and (b) the highest gear combinations? Also find the torque on the rear axle in both cases. Assuming rolling without slipping, how far does the bike move forward for each full pedal stroke (each full 360° rotation) for (c) the lowest and (d) the highest gear ratios?

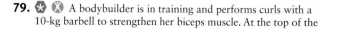

Figure P8.78

79. A bodybuilder is in training and performs curls with a 10-kg barbell to strengthen her biceps muscle. At the top of the stroke, her arm is configured as in Figure P8.79, with her forearm at 35° with respect to the horizontal. Approximate her forearm and hand as a uniform rod of mass 1.4 kg and length 32 cm (from pivot point at the elbow to center of barbell). The biceps muscle is attached 3.1 cm from the pivot point at the elbow joint and produces a force that is directed 85° away from the forearm. What force must the biceps muscle exert to keep the weight at its current position?

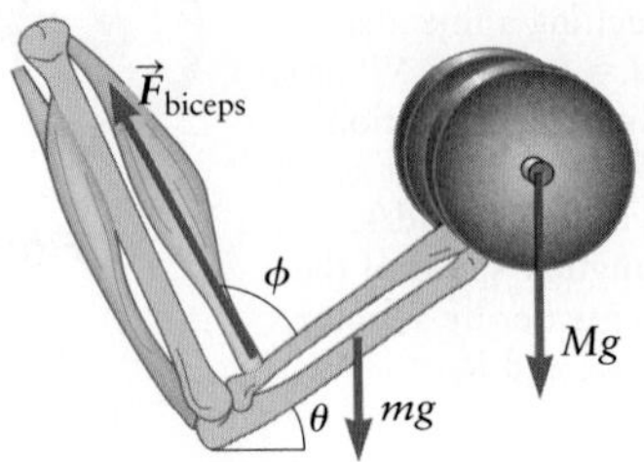

Figure P8.79 Problem 79 and Question 11.

80. Runner's gait. The amount of torque required by an external applied force (other than gravity) to move a leg forward is less than that required to move it backward if the leg is bent as shown in Figure P8.80. The lengths of the vectors $\vec{r}_1$, $\vec{r}_2$, and $\vec{r}_3$ indicate the position of the center of mass of the upper leg, lower leg, and foot from the axis of rotation. The ratio of the mass of the upper leg/lower leg/foot is approximately 5/3/1. Find the ratio of the moment of inertia for the leg moving forward to that of the leg moving backward. Approximate the three parts of the leg as point masses.

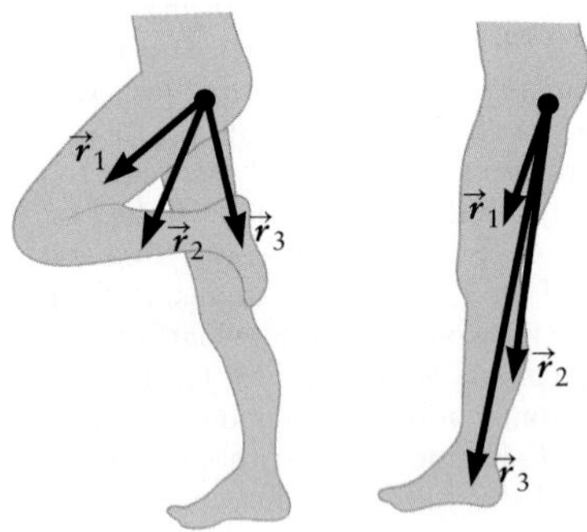

Figure P8.80

81. A wheel of mass 11 kg is pulled up a step by a horizontal rope as depicted in Figure P8.81. If the height of the step is equal to one-third the radius of the wheel, $h = \frac{1}{3}R$, what minimum tension is needed on the rope to start the wheel moving up the step? *Hint*: The wheel will start to move just as the normal force on the very bottom of the wheel becomes zero.

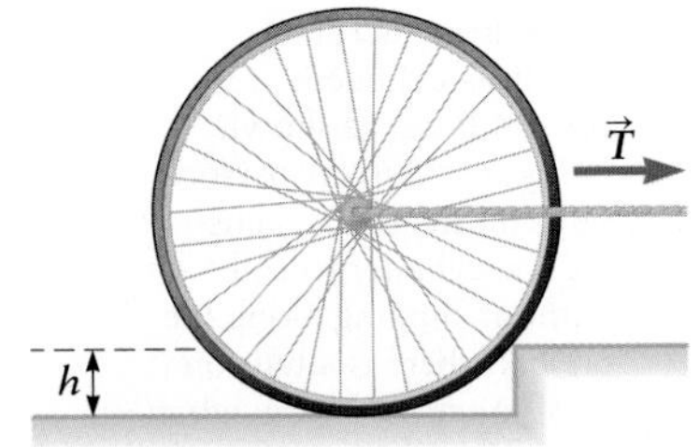

Figure P8.81

82. The end of a pencil of mass m and length L rests in a corner as the pencil makes an angle θ with the horizontal (x) direction (Fig. P8.82). If the pencil is released, it will rotate about point O with an acceleration α_1. Suppose the length of the pencil is increased by a factor of two but its mass is kept the same; the pencil would then have an angular acceleration α_2 when released. Find the ratio α_1/α_2.

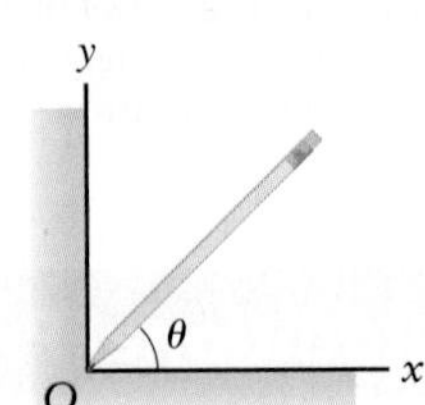

Figure P8.82

83. A long, *nonuniform* board of length 8.0 m and mass m = 12 kg is suspended by two ropes as shown in Figure P8.83. If the tensions in the ropes are $mg/3$ (on the left) and $2mg/3$ (on the right), what is the location of the board's center of mass?

Figure P8.83

84. Consider again the problem of a tipping car in Example 8.6. This time, instead of applying a force F to the car, assume the car is traveling around a curve on a level road. Let the radius of curvature of the turn be r and assume the car's speed v is just barely fast enough to make the car's inside wheels lift off the ground. (a) Find v. Express your answer in terms of r and the variables w, d, and h from Figure 8.26. (b) Use realistic values to get a numerical estimate for v for r = 80 m. *Hint*: Use the car's center of mass as your pivot point.

CHAPTER

9

Energy and Momentum of Rotational Motion

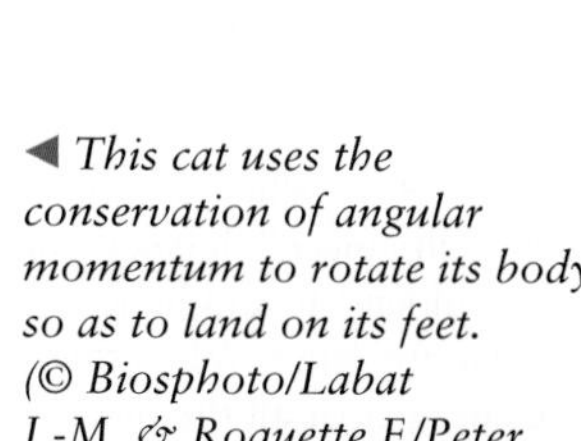

◀ *This cat uses the conservation of angular momentum to rotate its body so as to land on its feet. (© Biosphoto/Labat J.-M. & Roquette F./Peter Arnold, Inc./Photolibrary)*

In Chapter 8, we introduced the notion of torque and saw that torques give rise to rotational motion in much the same way that forces give rise to linear motion. In this chapter, we consider the kinetic energy associated with rotation and derive a work–energy theorem that relates torque and rotational kinetic energy. We then explore situations involving the conservation of mechanical energy, this time with the rotational kinetic energy included. We also introduce the idea of angular momentum, the rotational analog of linear momentum. In Chapter 7, we saw that the general principle of momentum conservation (for linear momentum) has a fundamental place in physics. Angular momentum has a similar role: the conservation of angular momentum is important in many situations, ranging from planetary orbits to the structure of atoms (as we'll see in Chapter 29) and the periodic table, and it also explains the mystery of how cats are almost always able to land on their feet.

OUTLINE

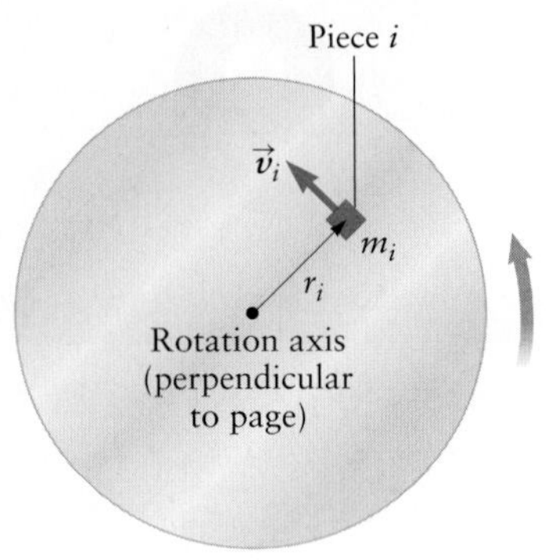

▲ **Figure 9.1** When an object rotates, each "piece" (labeled here as i) has velocity $\vec{v}_i$ and hence kinetic energy $\frac{1}{2}m_i v_i^2$. When the kinetic energies of all the pieces are added up, the total equals $\frac{1}{2}I\omega^2$.

9.1 Kinetic Energy of Rotation

In Chapter 6, we learned that the kinetic energy of a point particle of mass m moving with a linear speed v is

$$KE = \tfrac{1}{2}mv^2 \tag{9.1}$$

We now wish to generalize Equation 9.1 to include the contribution of rotation to the kinetic energy of an extended object. We can view such objects as being composed of many small pieces, and each piece can be thought of as a point particle with a kinetic energy given by Equation 9.1. The total kinetic energy can be found by adding up the kinetic energies of all the pieces. If each piece has mass m_i and moves with linear speed v_i, the total kinetic energy is

$$KE = \textstyle\sum_i \tfrac{1}{2} m_i v_i^2 \tag{9.2}$$

This is the total kinetic energy of any moving object, including objects that are undergoing rotational motion. This result can be simplified if we assume the pieces in Equation 9.2 are part of a rigid object undergoing simple rotational motion. For an object rotating about a fixed axis with an angular velocity ω, we have, from Chapter 8 (Eq. 8.7), $v_i = \omega r_i$ where r_i is the distance of the ith piece from the rotation axis (Fig. 9.1). Because this object is rigid, all pieces have the same angular velocity, and we can rearrange Equation 9.2 to get

$$KE_{\text{rot}} = \textstyle\sum_i \tfrac{1}{2} m_i v_i^2 = \sum_i \tfrac{1}{2} m_i(\omega r_i)^2 = \sum_i \tfrac{1}{2} m_i r_i^2 \omega^2$$

$$KE_{\text{rot}} = \tfrac{1}{2}\left(\textstyle\sum_i m_i r_i^2\right)\omega^2 \tag{9.3}$$

In Equation 9.3, we pulled the factor of ω^2 out of the summation since ω is the same for all pieces of the object. The term in parentheses in Equation 9.3 is the moment of inertia I (recall that $I = \sum_i m_i r_i^2$, Eq. 8.14). We thus have

$$KE_{\text{rot}} = \tfrac{1}{2}I\omega^2 \tag{9.4}$$

This kinetic energy is due to the rotational motion of an extended object. You should see, however, that KE_{rot} has as its origin the translational kinetic energies of all the individual pieces of the object, as given by Equation 9.1.

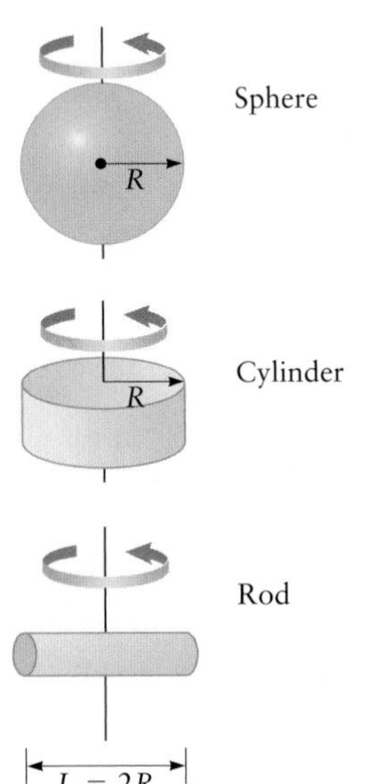

▲ **Figure 9.2** Concept Check 9.1.

CONCEPT CHECK 9.1 Rotational Kinetic Energy

A sphere of radius R, a cylinder of radius R, and a rod of length $L = 2R$ all have the same mass, and all three are rotating about an axis through their centers (Fig. 9.2). If the objects all have the same angular velocity, which one has the greatest rotational kinetic energy, (a) the sphere, (b) the cylinder, or (c) the rod?

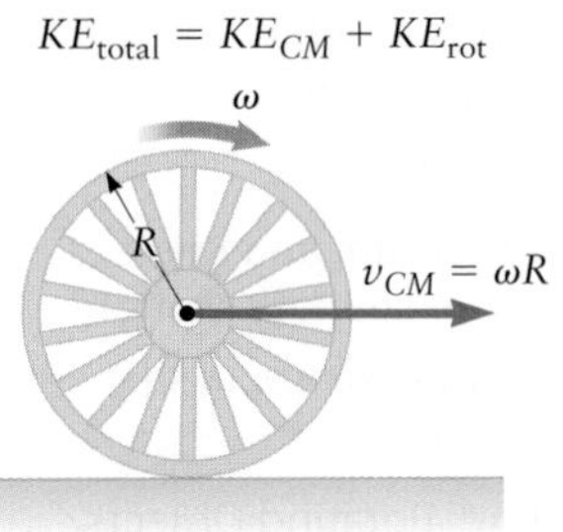

▲ **Figure 9.3** For an object such as a rolling wheel, the total KE is the sum of the translational KE associated with the center of mass velocity ($\frac{1}{2}mv_{\text{CM}}^2$) and the KE of rotation about the center of mass ($\frac{1}{2}I\omega^2$). For a wheel that rolls without slipping, $\omega = v_{\text{CM}}/R$.

The Total Kinetic Energy of an Object Is the Sum of the Rotational and Translational Kinetic Energies

The rotational kinetic energy of an object undergoing "pure" rotation about a fixed axis is $KE_{\text{rot}} = \frac{1}{2}I\omega^2$ (Eq. 9.4). Rolling objects such as the wheel in Figure 9.3 undergo both translational and rotational motions. In this case, the *total* kinetic energy has contributions from both the rotational motion and the overall translational motion. In practice, it is usually desirable to use an axis that passes through the center of mass. For an object of mass m whose center of mass moves with a linear speed v_{CM}, the total kinetic energy is then given by

$$KE_{\text{total}} = \overbrace{\tfrac{1}{2}mv_{\text{CM}}^2}^{\text{translational } KE} + \overbrace{\tfrac{1}{2}I\omega^2}^{\text{rotational } KE} \tag{9.5}$$

which is the total kinetic energy of *any* moving object, including rolling wheels and spinning Frisbees. The first term on the right of the equal sign in Equation 9.5 is the kinetic energy of translation, whereas the second term is the rotational kinetic energy. In Equation 9.5, ω is the angular velocity about a rotation axis that passes through the center of mass.

EXAMPLE 9.1 Energy of a Wind Farm

A promising and environmentally friendly method of generating electricity is a wind turbine (Fig. 9.4A). Wind turns the propellers, which in turn drive an electrical generator that produces electricity. A typical propeller unit has three blades (usually composed of a lightweight carbon fiber composite material) that are about 30 m long and rotate at 30 rpm. Calculate the rotational kinetic energy of this propeller unit. Approximate each blade as a uniform rod of mass 300 kg.

RECOGNIZE THE PRINCIPLE

The total kinetic energy of the propeller unit equals the sum of the rotational kinetic energies of the three blades. The kinetic energy of each blade can be found from Equation 9.4, using the moment of inertia for a uniform rod rotating about an axis that passes through one end, $I_{rod} = \frac{1}{3}mL^2$, where m is the mass of a blade and L is its length (Table 8.2).

SKETCH THE PROBLEM

The problem is sketched in Figure 9.4B. The three blades have the same angular velocity.

IDENTIFY THE RELATIONSHIPS

The total rotational kinetic energy of the propeller unit is

$$KE_{rot} = \tfrac{1}{2}I_{tot}\omega^2 \qquad (1)$$

where I_{tot} is the total moment of inertia of all three blades. We treat each blade as a uniform rod that rotates about an axis through one end. The total moment of inertia of three propeller blades is thus

$$I_{tot} = 3(mL^2/3) = mL^2 = (300\text{ kg})(30\text{ m})^2 = 2.7 \times 10^5\text{ kg}\cdot\text{m}^2$$

where we have inserted the given values for m and L. The angular velocity is

$$\omega = 30\,\frac{\text{rev}}{\text{min}} \times \frac{2\pi\text{ rad}}{\text{rev}} \times \frac{1\text{ min}}{60\text{ s}} = 3.1\text{ rad/s}$$

SOLVE

Inserting these values into our expression for the rotational kinetic energy (Eq. 9.4) gives

$$KE_{rot} = \tfrac{1}{2}I_{tot}\omega^2 = \tfrac{1}{2}(2.7 \times 10^5\text{ kg}\cdot\text{m}^2)(3.1\text{ rad/s})^2 = \boxed{1.3 \times 10^6\text{ J}}$$

▶ ***What does it mean?***

If this much energy can be extracted each second, the power generated by the wind turbine will be

$$P = \frac{KE_{rot}}{\Delta t} = \frac{1.3 \times 10^6\text{ J}}{1\text{ s}} = 1.3 \times 10^6\text{ W} = 1.3\text{ MW}$$

which is enough energy to power about 400 homes.

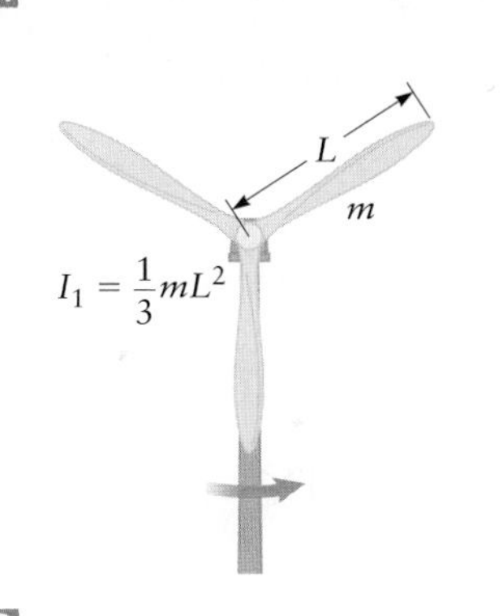

▲ **Figure 9.4** Example 9.1.

Rolling Motion and the Distribution of Kinetic Energy

Rolling objects have both rotational and translational kinetic energy. Let's consider how that energy is distributed. For example, if you are designing the wheels for a skateboard and want the skateboard to go as fast as possible, you will want to minimize the fraction of kinetic energy that goes into the rotational motion of each wheel, since this will leave more energy for the translational motion of the skateboard.

A rolling wheel has translational kinetic energy $\frac{1}{2}mv_{CM}^2$ and rotational kinetic energy $\frac{1}{2}I\omega^2$. If the wheel has the shape of a disk, the moment of inertia is $I = \frac{1}{2}mR^2$, where m is the wheel's mass and R is its radius (Table 8.2). Combining all these results, we find

$$KE_{\text{tot}} = \frac{1}{2}mv_{CM}^2 + \frac{1}{2}I\omega^2 = \frac{1}{2}mv_{CM}^2 + \frac{1}{2}\left(\frac{mR^2}{2}\right)\omega^2 = \frac{1}{2}mv_{CM}^2 + \frac{1}{4}mR^2\omega^2$$

In Chapter 8, we found that a wheel's angular velocity and center of mass velocity are related by $\omega = v_{CM}/R$ (Eq. 8.43). Inserting this information leads to

$$KE_{\text{tot}} = \frac{1}{2}mv_{CM}^2 + \frac{1}{4}mR^2\omega^2 = \frac{1}{2}mv_{CM}^2 + \frac{1}{4}mR^2\left(\frac{v_{CM}}{R}\right)^2$$

$$KE_{\text{tot}} = \overbrace{\tfrac{1}{2}mv_{CM}^2}^{\text{translational } KE} + \overbrace{\tfrac{1}{4}mv_{CM}^2}^{\text{rotational } KE} = \overbrace{\tfrac{3}{4}mv_{CM}^2}^{\text{total}} \tag{9.6}$$

The wheel's rotational kinetic energy is thus $\frac{1}{4}mv_{CM}^2$; this is half as large as the translational kinetic energy and one third of the total kinetic energy. These fractions are *independent* of the wheel's radius since all factors of R have canceled in deriving Equation 9.6. The only assumption we made is that the moment of inertia is given by $I = mR^2/2$, so our results are true for any object that has the shape of a wheel or disk. One might have expected that for a smaller wheel (a smaller radius R), less of the total kinetic energy would go into rotation, but that is not the case. The *fraction* of the total kinetic energy associated with rotation is *independent* of R. The only way to reduce this fraction is to change the moment of inertia, that is, to change the wheel's shape so as to reduce its moment of inertia.

Notice that the distribution of the kinetic energy in Equation 9.6 applies only to the wheels. The rest of a skateboard does not rotate, and the kinetic energy of that part of the skateboard comes just from translational motion ($KE = \frac{1}{2}mv_{CM}^2$).

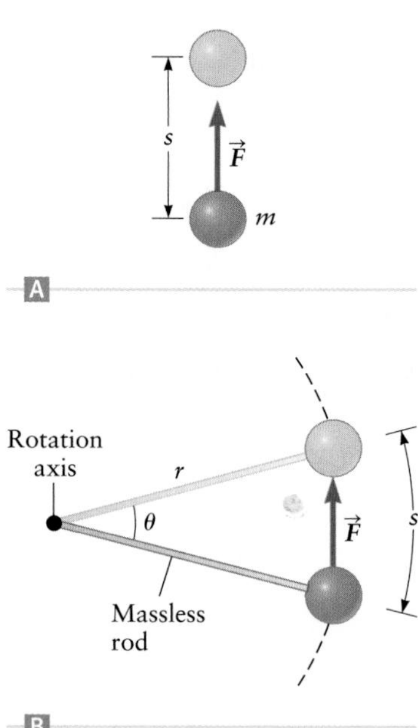

▲ **Figure 9.5** A When this mass moves a distance s while acted on by the force F, the work done is $W = Fs$. B If the mass is attached to a light (massless) rod so that the mass and the rod now rotate, the work done is also equal to $\tau\theta$.

Torque and Rotational Kinetic Energy: Rotational Version of the Work–Energy Theorem

In Chapter 6, we introduced the notion of ***work*** and showed that the work done on an object equals the change in the object's translational kinetic energy. Let's now derive a similar result involving torque and rotational kinetic energy.

Consider a one-dimensional situation in which a force of magnitude F pushes an object of mass m through a distance s as sketched in Figure 9.5A. Since the force here is parallel to the displacement, the work equals the product of the force and the displacement; hence, $W = Fs$. Applying the work–energy theorem (Eq. 6.9), W is equal to the change in kinetic energy of the object:

$$W = Fs = \Delta KE = \tfrac{1}{2}mv_f^2 - \tfrac{1}{2}mv_i^2 \tag{9.7}$$

where v_i and v_f are the object's initial and final speeds. Now suppose this object is held by a very light (massless) rod so that the object rotates as sketched in Figure 9.5B, with a rotational radius of r. If the angle θ is very small, the force F produces a torque $\tau = Fr$ about this axis. If the object moves a linear distance s, there is a corresponding angular displacement $\theta = s/r$ according to the definitions of angles and radians (Eq. 1.26). Inserting a factor of r in the denominator and numerator of Equation 9.7, we get

$$W = Fs = \overbrace{(Fr)}^{\tau}\overbrace{\left(\frac{s}{r}\right)}^{\theta}$$

$$W = \Delta KE = \tau\theta \tag{9.8}$$

Because it is moving along a circle, the object is undergoing rotational motion with an angular velocity related to its linear velocity by $v_i = \omega_i r$ and $v_f = \omega_f r$ for the initial and final velocities, respectively. Inserting into Equation 9.7 gives

$$\Delta KE = \tfrac{1}{2}mv_f^2 - \tfrac{1}{2}mv_i^2 = \tfrac{1}{2}m(\omega_f r)^2 - \tfrac{1}{2}m(\omega_i r)^2$$

$$\Delta KE = \tfrac{1}{2}(mr^2)\omega_f^2 - \tfrac{1}{2}(mr^2)\omega_i^2$$

The term mr^2 is just the moment of inertia of the object (from Eq. 8.13), so the change in kinetic energy is

$$\Delta KE = \tfrac{1}{2}I\omega_f^2 - \tfrac{1}{2}I\omega_i^2$$

Combining this expression with Equation 9.8 gives

$$W = \tau\theta = \Delta KE = \tfrac{1}{2}I\omega_f^2 - \tfrac{1}{2}I\omega_i^2 \quad (9.9)$$

Work–energy theorem for rotational motion

This result is the ***work–energy theorem*** for rotational motion. It says that the product of the torque τ and the angular displacement θ equals the change in the rotational kinetic energy.

EXAMPLE 9.2 Physics on the Playground

A child is given the job of getting a merry-go-round up to an enjoyable speed by pushing on its edge as she runs along the outer circumference as sketched in Figure 9.6. If she pushes with a force of 30 N and the merry-go-round starts from rest, how far will she have to run to get the merry-go-round to rotate at a rate of 0.20 rev/s (which corresponds to an angular velocity of 1 rev every 5.0 s = ω_f = 1.3 rad/s)? Assume the merry-go-round is a disk of mass m = 100 kg and radius R = 2.0 m. Express your answer in terms of the child's angular displacement during her run.

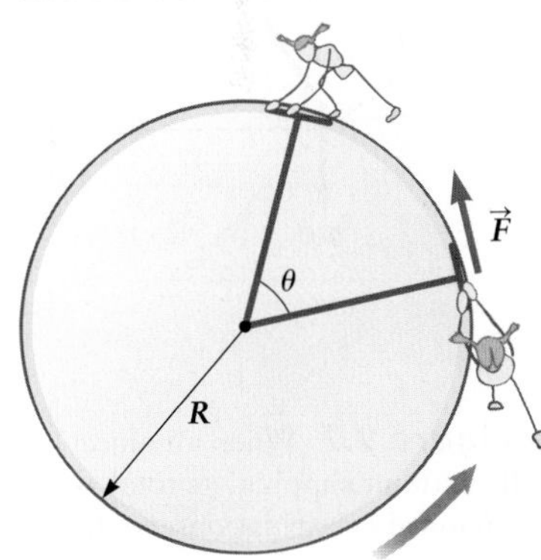

▲ **Figure 9.6** Example 9.2. This child pushes on the merry-go-round with a force F while it moves through an angular displacement θ.

RECOGNIZE THE PRINCIPLE

Since the force and radius of the merry-go-round are given, we can find the applied torque. The initial and final angular velocities are also given, so we can calculate the change in the rotational kinetic energy. The rotational work–energy theorem (Eq. 9.9) then relates the change in kinetic energy to the angular displacement, which is what we want to find.

SKETCH THE PROBLEM

Figure 9.6 shows the problem.

IDENTIFY THE RELATIONSHIPS

The torque exerted by the child is $\tau = FR$ since the associated lever arm is the radius R of the merry-go-round. (See Fig. 9.6.) If the child travels an angular displacement θ as she pushes the merry-go-round, the work done is $W = \tau\theta$. The initial kinetic energy is zero (because the merry-go-round starts from rest), whereas the final kinetic energy is $\tfrac{1}{2}I\omega_f^2$. Inserting all these results into the work–energy theorem (Eq. 9.9) leads to

$$W = \tau\theta = FR\theta = \Delta KE = \tfrac{1}{2}I\omega_f^2$$

$$\theta = \frac{I\omega_f^2}{2FR}$$

SOLVE

The merry-go-round is a disk, so its moment of inertia is $I = mR^2/2$. (See Table 8.2.) Inserting this equation and the given values of F, R, and ω_f gives

$$\theta = \frac{I\omega_f^2}{2FR} = \frac{(mR^2/2)\omega_f^2}{2FR} = \frac{mR\omega_f^2}{4F} = \frac{(100\text{ kg})(2.0\text{ m})(1.3\text{ rad/s})^2}{4(30\text{ N})} = \boxed{2.8\text{ rad}}$$

(continued) ▶

▶ What does it mean?
This answer is less than half of a revolution (since 1 rev corresponds to $2\pi \approx$ 6.2 rad), so the child will not have to run very far.

9.2 Conservation of Energy and Rotational Motion

In Chapter 6, we showed that if all the forces that do work on an object are conservative forces, the total mechanical energy of the object is conserved. In such cases

Conservation of total mechanical energy

$$\underbrace{KE_i + PE_i}_{\text{initial mechanical energy}} = \underbrace{KE_f + PE_f}_{\text{final mechanical energy}} \tag{9.10}$$

Here KE denotes the *total* kinetic energy (including translation and rotation) given by Equation 9.5 and PE is the object's potential energy.

Consider a solid ball (a sphere) that starts from rest and rolls down a hill as sketched in Figure 9.7. Let's apply the principle of conservation of energy to calculate its speed when it reaches the bottom. In this example, there are three forces acting on the ball: the normal force exerted by the hill, the force of friction between the ball and the hill, and the force of gravity. The normal force does not do any work on the ball because this force is always directed perpendicular to the ball's displacement. The force of friction is necessary for the ball to roll without slipping. However, friction does not do any work on the ball because the contact point where the ball meets the surface does not slip. Hence, friction does not play a role in the conservation of energy condition (Eq. 9.10) when an object rolls without slipping (see also Insight 9.1). The only force that does work on the ball in Figure 9.7 is the force of gravity.

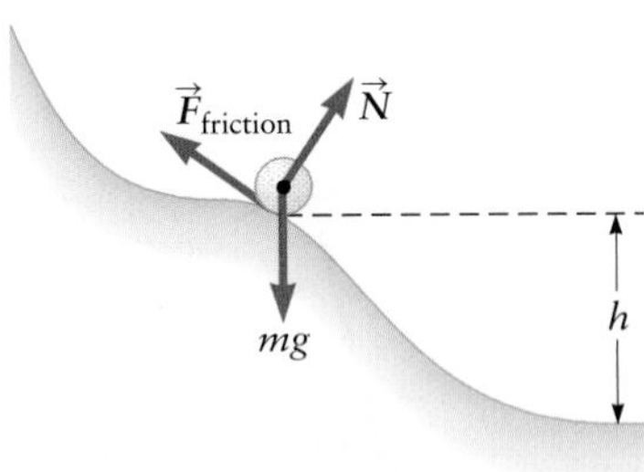

▲ **Figure 9.7** When an object rolls without slipping, its total mechanical energy is conserved.

We saw in Chapter 6 that the work done by gravity equals the change in gravitational potential energy $PE_{\text{grav}} = mgh$. Hence, the conservation of mechanical energy condition for the ball becomes

$$KE_i + PE_i = KE_f + PE_f$$
$$\tfrac{1}{2}mv_i^2 + \tfrac{1}{2}I\omega_i^2 + mgh_i = \tfrac{1}{2}mv_f^2 + \tfrac{1}{2}I\omega_f^2 + mgh_f \tag{9.11}$$

where the subscript i denotes the ball's initial state (at the top of the hill) and the subscript f denotes the ball's final state (at the bottom). The speeds v_i and v_f in Equation 9.11 are the center of mass speeds of the ball, which (until this point) we have denoted by v_{CM}. (To avoid the excessive use of subscripts, in the rest of this chapter we often drop the subscript "CM" in places where it should be obvious.)

The ball started from rest, so $v_i = \omega_i = 0$. The ball is also rolling without slipping, so $\omega_f = v_f/R$, where R is the ball's radius. Inserting this all into Equation 9.11 and taking $h_f = 0$ and $h_i = h$ (the height of the hill), we find

$$\overbrace{\tfrac{1}{2}mv_i^2}^{0} + \overbrace{\tfrac{1}{2}I\omega_i^2}^{0} + \overbrace{mgh_i}^{mgh} = \tfrac{1}{2}mv_f^2 + \tfrac{1}{2}I\omega_f^2 + \overbrace{mgh_f}^{0}$$
$$mgh = \tfrac{1}{2}mv_f^2 + \tfrac{1}{2}I\omega_f^2 = \tfrac{1}{2}mv_f^2 + \tfrac{1}{2}I\left(\frac{v_f}{R}\right)^2 \tag{9.12}$$

Insight 9.1
FRICTION DOES NO WORK ON A ROLLING OBJECT
Consider an object that rolls without slipping, such as the ball in Figure 9.7. Slipping is prevented by the frictional force between the surface of the hill and the ball. This frictional force acts at the point of contact, and because the ball is not slipping, this point is always at rest. Work is the product of force and displacement. Since the contact point is at rest, its displacement is zero, and the work done by the frictional force is zero in such cases.

Inserting the moment of inertia of a solid sphere ($I = 2mR^2/5$) from Table 8.2 leads to

$$mgh = \frac{1}{2}mv_f^2 + \frac{1}{2}I\left(\frac{v_f}{R}\right)^2 = \frac{1}{2}mv_f^2 + \frac{1}{2}\frac{2mR^2}{5}\left(\frac{v_f}{R}\right)^2$$
$$mgh = \frac{1}{2}mv_f^2 + \frac{1}{5}mv_f^2 = \frac{7}{10}mv_f^2$$

Solving for the speed v_f at the bottom, we have

$$v_f = \sqrt{\frac{10gh}{7}} \tag{9.13}$$

In getting this result, we have relied only on the fact that the mechanical energy of the ball is conserved. We did not need to know anything about the hill other than its height h.

Our application of the principle of conservation of energy to the rolling ball in Figure 9.7 is similar to the way we used conservation of energy principles in Chapter 6. Our general strategy for dealing with these problems is as follows.

PROBLEM SOLVING Applying the Principle of Conservation of Energy to Problems of Rotational Motion

1. **RECOGNIZE THE PRINCIPLE.** The mechanical energy of an object is conserved only if all the forces that do work on the object are conservative forces. Only then can we apply the conservation of mechanical energy condition in Equation 9.10.
2. **SKETCH THE PROBLEM.** Always use a sketch to collect your information concerning the initial and final states of the system.
3. **IDENTIFY THE RELATIONSHIPS.** Find the initial and final kinetic and potential energies of the object. They will usually involve the initial and final linear and angular velocities, and the initial and final heights, as for the rolling ball in Figure 9.7. You will usually also need to find the moment of inertia.
4. **SOLVE** for the unknown quantities using the conservation of energy condition,
$$KE_i + PE_i = KE_f + PE_f$$
The kinetic energy terms must account for both the translational and the rotational kinetic energies (Eq. 9.5):
$$KE_{\text{total}} = \underbrace{\tfrac{1}{2}mv_{CM}^2}_{\text{translational } KE} + \underbrace{\tfrac{1}{2}I\omega^2}_{\text{rotational } KE}$$
5. Always *consider what your answer means* and check that it makes sense.

CONCEPT CHECK 9.2 Rolling versus Sliding

Consider a ball that rolls down a hill as compared to an object (such as a skier) that slides down a frictionless hill of the same height. Which of the following statements is correct regarding their speeds at the bottom of the hill?

(a) The ball has a higher linear speed because it does not slip.
(b) The sliding object has a higher speed because the hill is frictionless.
(c) The ball and skier have the same speed at the bottom of the hill.
(d) The ball has a smaller linear speed because some of its kinetic energy goes into rotational motion.

EXAMPLE 9.3 Motion of a Yo-Yo

Consider the yo-yo in Figure 9.8. The yo-yo starts from rest and then moves downward in the manner you have probably practiced at least a few times. Find the speed of the yo-yo after it has moved 50 cm below its starting location. Assume the yo-yo is a solid disk with the string wound around the outer edge.

RECOGNIZE THE PRINCIPLE

There are two forces acting on the yo-yo: the force exerted by the string and the force of gravity. The force exerted by the string plays a role very similar to that of friction for the rolling ball in Figure 9.7, and the string does no work on the yo-yo. That is, the

(continued) ▶

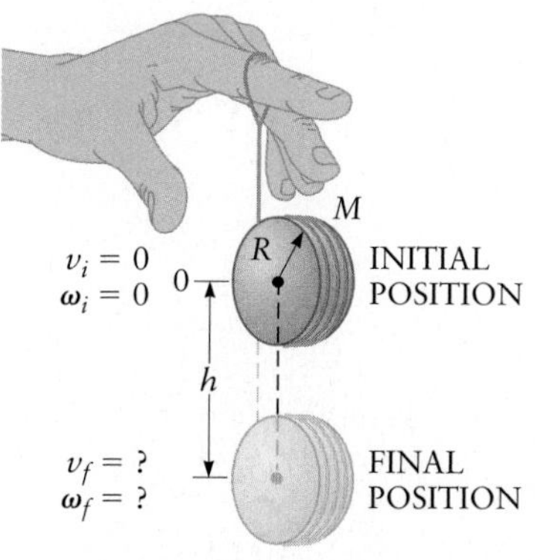

▲ **Figure 9.8** Example 9.3. A yo-yo traveling down a string.

yo-yo "rolls down" the string in the same way that the ball rolls without slipping down the hill. The only force that does work on the yo-yo is the force of gravity. Since gravity is a conservative force, we can apply the principle of conservation of energy.

SKETCH THE PROBLEM

Figure 9.8 shows the initial and final positions of the yo-yo.

IDENTIFY THE RELATIONSHIPS

The conservation of energy relation for the yo-yo is

$$KE_i + PE_i = KE_f + PE_f$$

The kinetic energy has contributions from both translational and rotational motion:

$$\tfrac{1}{2}mv_i^2 + \tfrac{1}{2}I\omega_i^2 + PE_i = \tfrac{1}{2}mv_f^2 + \tfrac{1}{2}I\omega_f^2 + PE_f$$

The potential energy is due to gravity; taking the starting height to be $h_i = 0$, the final height is $h_f = -h$ ($= -0.50$ m as given above). This yo-yo is a disk of mass m and radius R, so its moment of inertia is $I = mR^2/2$ (from Table 8.2). Inserting all this into the conservation of energy relation leads to

$$\frac{1}{2}mv_i^2 + \frac{1}{2}\left(\frac{mR^2}{2}\right)\omega_i^2 = \frac{1}{2}mv_f^2 + \frac{1}{2}\left(\frac{mR^2}{2}\right)\omega_f^2 + mg(-h) \tag{1}$$

The yo-yo starts from rest, so $v_i = \omega_i = 0$. The yo-yo is also (as already noted) "rolling down" the string, so the final linear and angular velocities are related by $\omega_f = v_f/R$.

SOLVE

Putting all this information into Equation (1), we find

$$0 = \frac{1}{2}mv_f^2 + \frac{1}{2}\left(\frac{mR^2}{2}\right)\left(\frac{v_f}{R}\right)^2 - mgh = \frac{1}{2}mv_f^2 + \frac{1}{4}mv_f^2 - mgh$$

$$0 = \frac{3}{4}mv_f^2 - mgh$$

$$\frac{3}{4}mv_f^2 = mgh$$

$$v_f = \sqrt{\frac{4gh}{3}}$$

Inserting the given value of h, we find

$$v_f = \sqrt{\frac{4gh}{3}} = \sqrt{\frac{4(9.8 \text{ m/s}^2)(0.50 \text{ m})}{3}} = \boxed{2.6 \text{ m/s}}$$

► ***What does it mean?***

Notice that v_f does not depend on either the mass or the radius of the yo-yo. A large yo-yo and a small yo-yo will move at the same speed; that is, they will "fall" at the same rate. Hence, Galileo could have used yo-yos in his experiments at the Leaning Tower of Pisa (Chapter 2).

CONCEPT CHECK 9.3 Designing a Faster Yo-Yo

Suppose you want to design and market a faster yo-yo. That is, you want to make a yo-yo that will have a higher speed than that found in Example 9.3. How could you do it?

(a) Make the yo-yo radius larger.
(b) Make the yo-yo radius smaller.
(c) Add mass near the axis so that the yo-yo is shaped more like a sphere.
(d) Hollow out the center so that the yo-yo is shaped like a cylindrical shell.

EXAMPLE 9.4 Rolling up a Hill

A bowling ball rolls without slipping up an incline as shown in Figure 9.9. If the ball has speed $v_i = 12$ m/s at the base of the incline, what height h_f on the incline will the bottom of the ball reach?

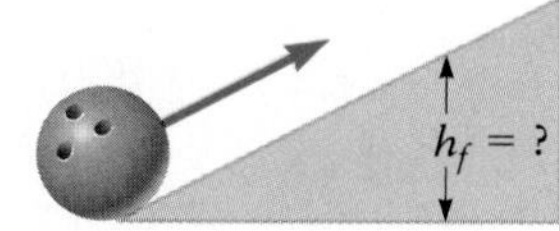

▲ **Figure 9.9** Example 9.4.

RECOGNIZE THE PRINCIPLE

The only force that does work on the ball is gravity, which is a conservative force. The mechanical energy of the ball is therefore conserved, and the initial energy is equal to the final energy.

SKETCH THE PROBLEM

Figure 9.9 shows the problem.

IDENTIFY THE RELATIONSHIPS

We can write

$$KE_i + PE_i = KE_f + PE_f \qquad (1)$$

where the subscripts i and f denote the initial state (with the ball at the bottom of the incline) and the final state (with the ball at its highest point), respectively. If we measure height from the bottom of the ramp, $h_i = 0$, and h_f is the quantity we wish to find. The initial kinetic energy is

$$KE_i = \frac{1}{2}mv_i^2 + \frac{1}{2}I\omega_i^2$$

where m is the mass of the ball and its moment of inertia is $I = \frac{2}{5}mR^2$. The radius of the ball is R, and its initial angular velocity is $\omega_i = v_i/R$. When the ball reaches its highest point, its speed is zero, so the final kinetic energy is zero. Inserting all this into Equation (1) leads to

$$KE_i + PE_i = \frac{1}{2}mv_i^2 + \frac{1}{2}\left(\frac{2}{5}mR^2\right)\left(\frac{v_i}{R}\right)^2 + mg(0) = KE_f + PE_f = 0 + mgh_f \qquad (2)$$

SOLVE

Solving Equation (2) for the final height gives

$$gh_f = \frac{1}{2}v_i^2 + \frac{1}{5}v_i^2 = \frac{7}{10}v_i^2$$

$$h_f = \frac{7v_i^2}{10g} = \frac{7(12 \text{ m/s})^2}{10(9.8 \text{ m/s}^2)} = 10 \text{ m}$$

► ***What does it mean?***

The maximum height of a projectile that does not have any rotational kinetic energy would be $h_f = v_i^2/(2g)$, so a rolling ball will reach higher than a projectile with the same initial speed because the ball has some initial rotational kinetic energy that is converted to potential energy at the highest point.

9.3 Angular Momentum

In Chapter 7, we defined the *linear momentum* for a point particle of mass m moving with velocity $\vec{v}$ as

$$\vec{p} = m\vec{v} \qquad (9.14)$$

We showed that the momentum of a system of particles is conserved provided there is no net external force on the system. It should not surprise you to find that a rotating object has a property called angular momentum. In nearly all the rotational motion examples we have discussed so far, the angular velocity has been defined with respect to a single rotation axis that does not change direction during the motion. In such cases, the ***angular momentum*** is given by

Angular momentum for rotation about a fixed axis

$$L = I\omega \tag{9.15}$$

This equation looks very much like Equation 9.14: the moment of inertia I is analogous to mass, and angular velocity is analogous to linear velocity. One difference is that the linear momentum $\vec{p}$ is a vector, whereas L in Equation 9.15 is not. This difference is due to our simplified treatment of rotations. In Section 9.4, we'll generalize Equation 9.15 to a vector form with a true angular velocity vector when we discuss gyroscopes. In this section, we assume the rotation axis keeps a fixed direction so that we can use the scalar (nonvector) form in Equation 9.15.

So far, we have always associated an angular velocity with objects that rotate about a fixed axis, but the notion of angular momentum also applies to more general cases. Figure 9.10 shows a particle moving freely through space. Even though this particle is not constrained to move in a rotational fashion, it can still have angular momentum. For example, if we choose a particular pivot point P, the particle will at any given instant be moving tangent to the circular arc of radius r sketched in Figure 9.10. At this instant, the motion of the particle is the same as that of an object that rotates along this circular arc. We can therefore define an angular velocity $\omega = v_\perp/r$, where $v_\perp$ is the component of the particle's velocity perpendicular to the radius of the arc and r is the distance from the particle to the point P. The moment of inertia is given by $I = mr^2$ (Eq. 8.13), so using the definition of angular momentum (Eq. 9.15), we have

$$L = I\omega = mr^2\left(\frac{v_\perp}{r}\right)$$

$$L = mrv_\perp \tag{9.16}$$

This result for L applies to any moving particle. We can always choose a hypothetical pivot point and rotation axis, even if the particle is not moving in a complete closed path about the axis. The angular momentum with respect to that axis is then proportional to $v_\perp$, the component of the velocity perpendicular to the line that connects the particle to the pivot point. The result for angular momentum in Equation 9.16 will be important when we wish to calculate the total angular momentum of a system of particles.

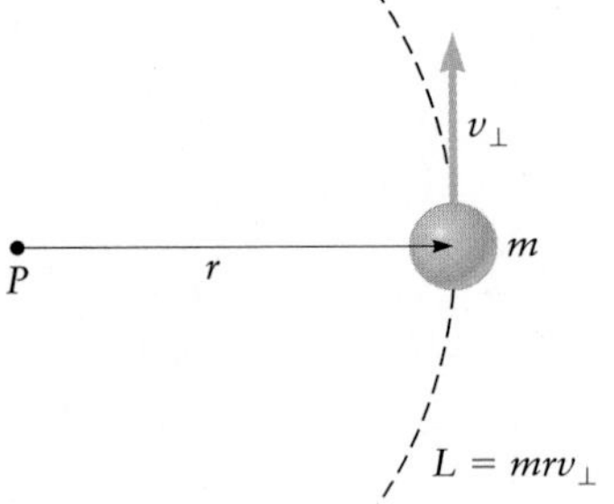

▲ **Figure 9.10** The angular momentum of a moving particle about a rotation axis that passes through point P is $L = mrv_\perp$, where $v_\perp$ is the component of the velocity perpendicular to the line that connects the particle to the pivot point P. Here the rotation axis is perpendicular to the plane of the drawing.

Conservation of Angular Momentum and a Spinning Skater

A rotating object with an angular momentum $L = I\omega$ will maintain its angular momentum provided no external torques act on the object. We say that such an object's angular momentum is *conserved*. This notion of angular momentum conservation also applies to a system of objects. If there are no *external* torques on a system, the total angular momentum of the system will be conserved.

A familiar example in which angular momentum conservation plays a key role is the motion of the figure skater in Figure 9.11. The skater is initially spinning about a vertical axis with an angular velocity ω_i (Fig. 9.11A). If she then pulls her arms and legs in very close to the midline of her body, we know that her spin rate will increase so that her final angular velocity ω_f is much larger than ω_i. This behavior can be understood in terms of the skater's angular momentum.

During the time she is spinning, no torques are acting on the skater. The only forces acting on her are gravity and the normal force from the ice, and they do not exert a torque on the skater. (We assume the ice is very slippery, so the frictional force is negligible.) The skater's angular momentum is therefore conserved:

$$L_i = L_f$$

Her initial angular momentum is $L_i = I_i\omega_i$, where the subscript i indicates the skater's initial state. When she pulls her arms and legs close to her midline, she is redistributing her mass so that it is closer to the axis of rotation, changing her moment of inertia to a new value I_f. Recall that the moment of inertia is equal to a sum over all the individual pieces of an object:

$$I = \sum mr^2$$

If an object changes shape so that some of the individual pieces are moved closer to the axis of rotation, the values of r for these pieces are reduced and the overall moment of inertia decreases. That is precisely what happens with the skater in Figure 9.11B. When she pulls her arms and legs in close to her rotation axis, these "pieces" have smaller values of r and her moment of inertia becomes smaller. Her total angular momentum is conserved; thus,

$$L_i = I_i\omega_i = L_f = I_f\omega_f \tag{9.17}$$

Pulling her arms in close to her body reduces her moment of inertia, so $I_f < I_i$ and hence $\omega_f > \omega_i$. She spins *faster*. This increase in her angular velocity depends on how much she can reduce her moment of inertia; that is, it depends on how closely she can pull her arms and legs in line with the rest of her body.

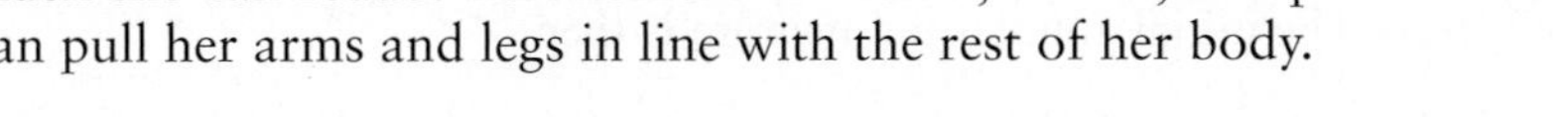

▲ **Figure 9.11** There are no external torques acting on the skater, so her angular momentum is conserved. Her angular velocity increases when she pulls her arms and legs close to her body, but her angular momentum stays the same.

Problem Solving with Angular Momentum

We saw in Chapter 7 that linear momentum and momentum conservation are extremely useful when analyzing collisions and other situations that involve a system of particles. The same is true of angular momentum. Typically, a system such as the skater in Figure 9.11 is moving in some manner with an initial angular momentum. There is then some sort of "change" in the system (such as the skater pulling in her arms and legs) that leaves the system in a final state of motion. If the external torque on the system is zero, the total angular momentum must be conserved. Hence, $L(\text{final}) = L(\text{initial})$, which we can use to find the final angular velocity or some other quantity. The general approach to such problems is as follows.

PROBLEM SOLVING Applying the Principle of Conservation of Angular Momentum

1. **RECOGNIZE THE PRINCIPLE.** If the external torque on a system is zero, the angular momentum is conserved.
2. **SKETCH THE PROBLEM.** Always use a sketch to collect your information concerning the initial and final states of the system.
3. **IDENTIFY THE RELATIONSHIPS.**
   - The system of interest and its initial and final states depend on the problem. The system might be a single object (such as the skater in Figure 9.11) or a collection of objects.
   - Express the initial and final angular velocities and moments of inertia.
   - Apply any information concerning the initial and final mechanical energies. (Such information will only be available in some cases.)
4. **SOLVE** for the quantities of interest using the principle of conservation of angular momentum,

$$L_i = I_i\omega_i = L_f = I_f\omega_f$$

5. Always *consider what your answer means* and check that it makes sense.

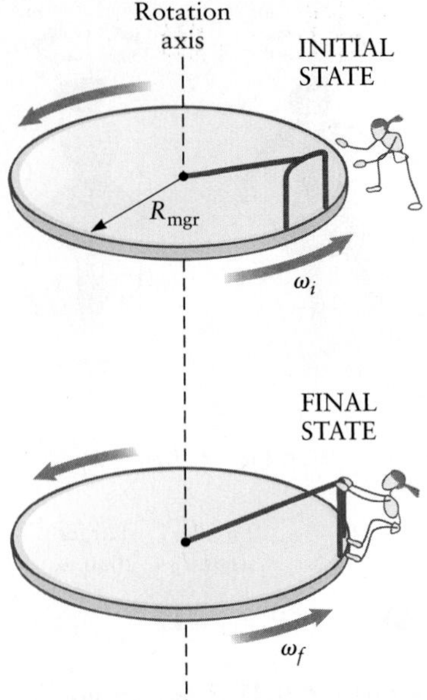

▲ **Figure 9.12** Example 9.5. A child steps onto a rotating merry-go-round.

EXAMPLE 9.5 Angular Momentum on the Playground

A merry-go-round of mass m_{mgr} and radius R_{mgr} is rotating freely as sketched in Figure 9.12. A child then steps onto the edge of the merry-go-round and notices that the angular velocity of the merry-go-round has decreased slightly. If the child's mass is m_c, what is the ratio of the final angular velocity to the initial angular velocity?

RECOGNIZE THE PRINCIPLE

The merry-go-round and the child are a "system," and although there are external forces acting on them (such as gravity), for a well-oiled merry-go-round the torque on this system will be negligible and the angular momentum will be conserved.

SKETCH THE PROBLEM

Figure 9.12 shows the system just before the child steps onto the merry-go-round (the initial state) and just after (the final state).

IDENTIFY THE RELATIONSHIPS

Let I_{mgr} be the moment of inertia of the merry-go-round and ω_i be its angular velocity before the child gets on. Also let I_c be the moment of inertia of the child, and her initial angular velocity is zero. The initial angular momentum is then

$$L_i = I_{mgr}\omega_i + I_c(0) = I_{mgr}\omega_i$$

We can approximate the merry-go-round as a large disk, with moment of inertia $I = \frac{1}{2}mR^2$, so we have $I_{mgr} = \frac{1}{2}m_{mgr}R_{mgr}^2$. The child has a mass m_c and is located a distance R_{mgr} from the axis of rotation, so her moment of inertia is $I_c = \sum mr^2 = m_c R_{mgr}^2$. After the child steps onto the merry-go-round, the angular momentum of the system is

$$L_f = I_{mgr}\omega_f + I_c\omega_f$$

where ω_f is the (common) final angular velocity. Both the child and the merry-go-round have the same final angular velocity.

SOLVE

Angular momentum is conserved, so $L_i = L_f$, which gives

$$L_i = I_{mgr}\omega_i = L_f = I_{mgr}\omega_f + I_c\omega_f = (I_{mgr} + I_c)\omega_f$$

Solving for the final angular velocity, we find

$$\omega_f = \frac{I_{mgr}\omega_i}{I_{mgr} + I_c}$$

The ratio of the initial and final angular velocities is then

$$\frac{\omega_f}{\omega_i} = \frac{I_{mgr}}{I_{mgr} + I_c} \tag{1}$$

Inserting the results for the moments of inertia into Equation (1) gives

$$\frac{\omega_f}{\omega_i} = \frac{I_{mgr}}{I_{mgr} + I_c} = \frac{m_{mgr}R_{mgr}^2/2}{(m_{mgr}R_{mgr}^2/2) + m_c R_{mgr}^2} = \boxed{\frac{m_{mgr}}{m_{mgr} + 2m_c}}$$

▶ ***What does it mean?***
This example is similar to an inelastic collision between two objects as we encountered in Chapter 7. The two "objects"—the merry-go-round and the child—stick together and move with the same angular final velocity, just as the two objects in a completely inelastic collision stick together after colliding.

EXAMPLE 9.6 Jumping off a Merry-Go-Round

Suppose the child in Example 9.5 is initially standing on the edge of the merry-go-round and both are initially at rest. The child then jumps off the merry-go-round, which begins to rotate. If she jumps in a direction tangent to the edge of the merry-go-round as in Figure 9.13A and has a speed of 3.0 m/s while she is in the air, what is the final angular velocity of the merry-go-round? Assume the mass of the child is $m_c = 30$ kg, the merry-go-round has mass $m_{\text{mgr}} = 100$ kg, and the merry-go-round's radius is $R_{\text{mgr}} = 2.0$ m. Also assume the merry-go-round is frictionless.

RECOGNIZE THE PRINCIPLE

Just as in Example 9.5, the external torques on the merry-go-round and child in Figure 9.13 are negligible. Hence, the angular momentum of this system is conserved.

SKETCH THE PROBLEM

Figure 9.13 shows the initial and final states of the system. In Figure 9.13B, the child has just jumped off the merry-go-round and is still moving through the air.

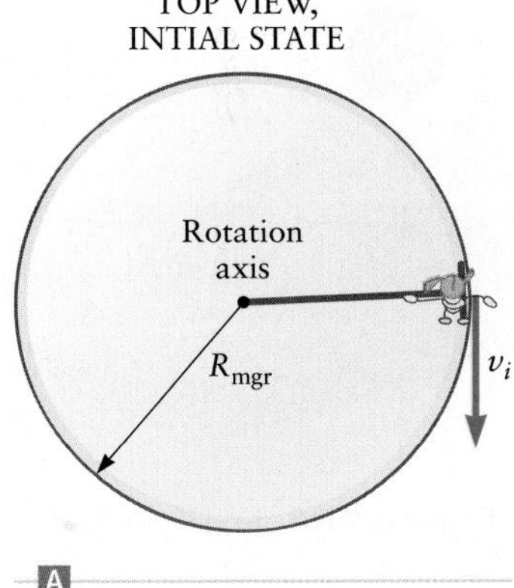

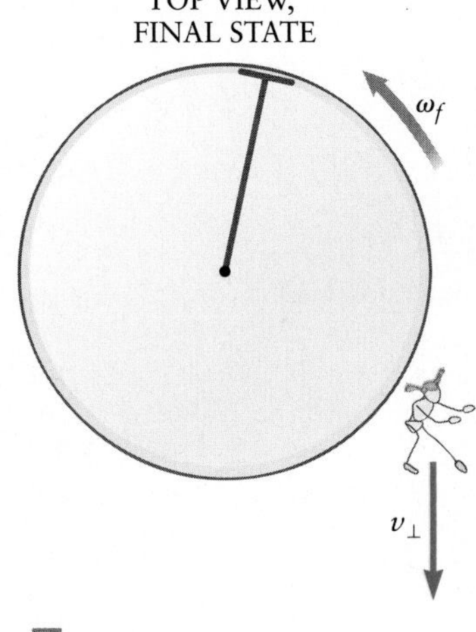

▲ **Figure 9.13** Example 9.6. A child jumps off a stationary merry-go-round. The final angular momentum of the child is $m_c R_{\text{mgr}} v_\perp$.

IDENTIFY THE RELATIONSHIPS

The initial state of the system is just prior to when the child jumps; everything is at rest at that moment, so the initial angular momentum is zero. The final state is the instant just after the child has jumped and before she reaches the ground. *When the child is moving through the air as a projectile, she still has angular momentum* as described in Figure 9.10 and Equation 9.16, which we must account for when we compute the final angular momentum of the system.

If the final angular velocity of the merry-go-round is ω_{mgr}, its final angular momentum is $I_{\text{mgr}}\omega_{\text{mgr}}$, where the moment of inertia of the merry-go-round is as described in Example 9.5. The final angular momentum of the child can be found from Equation 9.16. The distance from the child to the rotation axis is R_{mgr}, and her velocity perpendicular to the radius of the merry-go-round is $v_\perp$ (= 3.0 m/s as given in the problem statement). The child's final angular momentum is thus

$$L_c = m_c R_{\text{mgr}} v_\perp$$

Combining all these results, the final angular momentum of the system is the sum of the child's angular momentum plus that of the merry-go-round:

$$L_f = L_{\text{mgr}} + L_c = I_{\text{mgr}}\omega_{\text{mgr}} + m_c R_{\text{mgr}} v_\perp$$

Inserting the moment of inertia of the merry-go-round ($I_{\text{mgr}} = \frac{1}{2}m_{\text{mgr}}R_{\text{mgr}}^2$) leads to

$$L_f = \tfrac{1}{2}m_{\text{mgr}}R_{\text{mgr}}^2\omega_{\text{mgr}} + m_c R_{\text{mgr}} v_\perp$$

SOLVE

The final angular momentum equals the initial angular momentum, which is zero, so

$$\tfrac{1}{2}m_{\text{mgr}}R_{\text{mgr}}^2\omega_{\text{mgr}} + m_c R_{\text{mgr}} v_\perp = 0$$

Solving for the final angular velocity of the merry-go-round leads to

$$\omega_{\text{mgr}} = -\frac{m_c R_{\text{mgr}} v_\perp}{m_{\text{mgr}}R_{\text{mgr}}^2/2} = -\frac{2m_c v_\perp}{m_{\text{mgr}}R_{\text{mgr}}}$$

(continued) ▶

The negative sign here indicates that the final angular velocities of the child and merry-go-round are in opposite directions. Inserting the given values of the various quantities, we find

$$\omega_{mgr} = -\frac{2m_c v_\perp}{m_{mgr}R_{mgr}} = -\frac{2(30\text{ kg})(3.0\text{ m/s})}{(100\text{ kg})(2.0\text{ m})} = \boxed{-0.90\text{ rad/s}}$$

▸ *What does it mean?*

An important point of this example is that an object moving in a straight line (here, the child after she jumps from the merry-go-round) can still have angular momentum with respect to a particular rotation axis.

CONCEPT CHECK 9.4 Where Does the Torque Come From?

In Example 9.6, we treated the merry-go-round plus child as a system (Fig. 9.13) whose angular momentum is conserved. The merry-go-round is initially at rest, so its initial angular momentum is zero, but at the end it is rotating, so its final angular momentum is nonzero. The angular momentum of the merry-go-round alone thus changes, which can only happen if there is a torque on the merry-go-round. What force provides this torque?

(a) The force of gravity on the merry-go-round
(b) The force of air drag on the merry-go-round
(c) The force produced by the child when she jumps off
(d) The force of friction at the axle of the merry-go-round

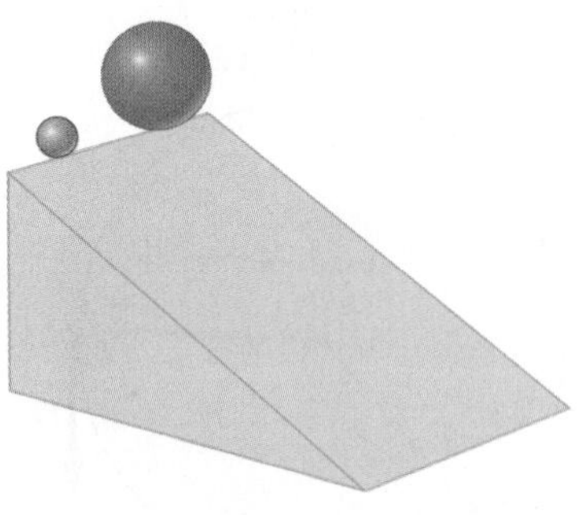

▲ **Figure 9.14** Concept Check 9.5.

CONCEPT CHECK 9.5 Rolling, Rolling, Rolling

Two balls with the same mass but different radii roll down a ramp (Fig. 9.14). If the balls both start from rest, which one has the larger angular momentum when they reach the bottom of the ramp?

Angular Momentum and Kinetic Energy

Let's now return to the problem of the spinning ice skater in Figure 9.11 and consider how her rotational kinetic energy varies during the spin. Our skater began with her arms and legs extended away from her body's rotation axis and then pulled them in close to her body, increasing her angular velocity. In terms of her angular momentum, we wrote (Eq. 9.17)

$$L_i = I_i\omega_i = L_f = I_f\omega_f$$

which can be rearranged to give

$$\omega_f = \frac{I_i}{I_f}\omega_i \tag{9.18}$$

Since her final moment of inertia (with her arms and legs in) is smaller than her initial moment of inertia (arms and legs out), $\omega_f > \omega_i$ and she spins faster at the end.

Rotational kinetic energy is $KE = \frac{1}{2}I\omega^2$, so the skater's final rotational kinetic energy is $KE_f = \frac{1}{2}I_f\omega_f^2$. Inserting ω_f from Equation 9.18 and using the skater's initial kinetic energy of $KE_i = \frac{1}{2}I_i\omega_i^2$ leads to

$$KE_f = \frac{1}{2}I_f\omega_f^2 = \frac{1}{2}I_f\left(\frac{I_i}{I_f}\omega_i\right)^2 = \frac{I_i}{I_f}\left(\frac{1}{2}I_i\omega_i^2\right)$$

$$KE_f = \frac{I_i}{I_f}KE_i \tag{9.19}$$

Because the skater's initial moment of inertia is larger than her final moment of inertia ($I_i > I_f$), Equation 9.19 shows that her final kinetic energy is *greater* than her initial kinetic energy. Even though her angular momentum is conserved, her kinetic energy is *not conserved*.

Where does the additional kinetic energy of the skater come from? That is, what force does work on the skater? This force comes from the skater herself as she pulls on her arms. Her arms undergo a displacement as they are pulled inward, and the work done (W) equals the product of this displacement and the component of the force parallel to the displacement. From the work–energy theorem, W equals the increase in the skater's kinetic energy.

9.4 Angular Momentum and Kepler's Second Law of Planetary Motion

In Chapter 5, we learned about Kepler's three laws of planetary motion. According to Kepler's first law, planets follow elliptical orbits about the Sun, and his second law states that a planet moving about its orbit sweeps out equal areas in equal times. If we observe the motion of the planet during two separate time intervals of equal duration Δt, we can consider the areas A_1 and A_2 swept out during these intervals as sketched in Figure 9.15. During an interval Δt, the planet travels a distance $v\,\Delta t$. When the planet is far from the Sun (at point 1 in Fig. 9.15), its speed is less than when the planet is close to the Sun (at point 2). However, Kepler's second law states that the areas that are swept out are the same; the areas A_1 and A_2 in Figure 9.15 are thus equal. This equality holds for all points along the orbit; the area swept out by the planet in a time Δt is the same for all positions along the orbit. We'll now show that Kepler's second law is intimately connected with the angular momentum of the planet.

Angular Momentum of an Orbiting Planet

The planet orbiting the Sun in Figure 9.15 has angular momentum, just like the moving particle in Figure 9.10. Applying Equation 9.16, the planet's angular momentum is given by

$$L_{\text{planet}} = m_{\text{planet}} r v_{\perp} \tag{9.20}$$

where r is the distance from the planet to the Sun and $v_{\perp}$ is the component of the velocity perpendicular to the line connecting the planet and the Sun. As the planet moves around its orbit, both r and $v_{\perp}$ change but the planet's angular momentum stays constant for the reason shown in Figure 9.16. The planet is now our system, and the only force on this system is the force of gravity from the Sun. The gravitational

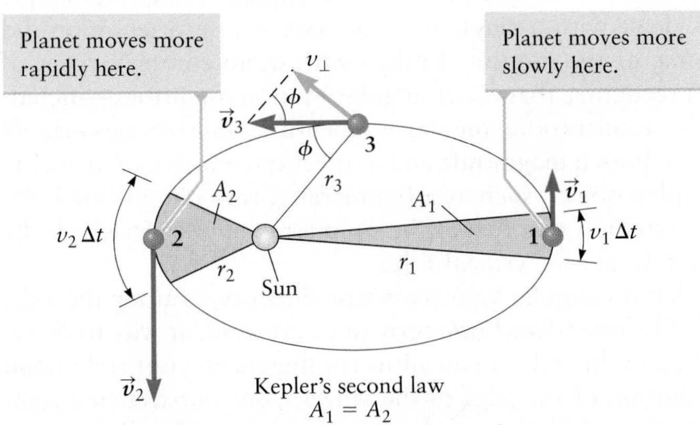

▲ **Figure 9.15** When a planet follows an elliptical orbit, the speed is greatest when the planet is closest to the Sun. According to Kepler's second law, a planet sweeps out equal areas in equal time intervals as it moves about the Sun. Hence, $A_1 = A_2$.

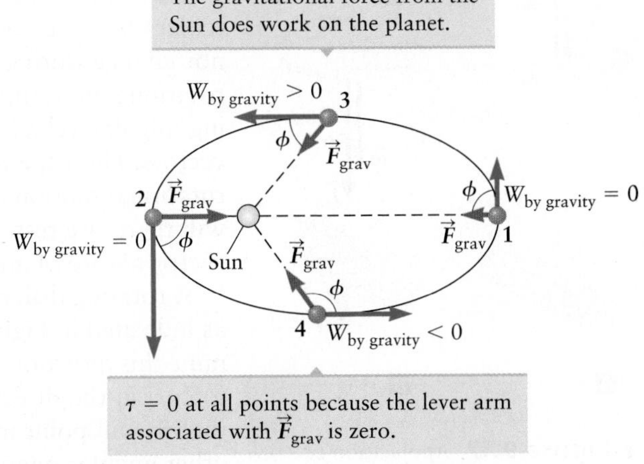

▲ **Figure 9.16** The force of gravity on the planet is directed along the orbital radius. This force cannot exert a torque on the planet, but it can do work on the planet.

force is always directed along the radial line that runs from the planet to the Sun. The axis for the planet's rotational motion passes through the Sun, so the torque due to the gravitational force is zero. The angular momentum of the planet is conserved.

To see how conservation of angular momentum of the planet leads to Kepler's second law, consider the locations along the orbit where the planet is farthest and closest to the Sun, at points 1 and 2 in Figure 9.15. At these positions, the planet's velocity is perpendicular to the line between the Sun and the planet, so $v_\perp = v_1$ at point 1 and $v_\perp = v_2$ at point 2. Using Equation 9.20, we get

$$L_1 = m_{\text{planet}} r_1 v_1 = m_{\text{planet}} r_2 v_2 = L_2 \tag{9.21}$$

$$r_1 v_1 = r_2 v_2 \tag{9.22}$$

The regions A_1 and A_2 swept out by the planet in Figure 9.15 are approximately triangular,[1] so the area A_1 is given by

$$A_1 = \tfrac{1}{2}(\text{base})(\text{height}) = \tfrac{1}{2}\,(v_1 \Delta t)\,(r_1)$$

Using Equation 9.22, we now have

$$A_1 = \tfrac{1}{2}(\Delta t) r_1 v_1 = \tfrac{1}{2}(\Delta t) r_2 v_2 = A_2$$

The areas A_1 and A_2 are thus equal; this relation is Kepler's second law.

We have already noted that because the gravitational force from the Sun acts along the planet's orbital radius, this force cannot exert a torque on the planet. This result is true for all points along the orbit. Gravity, however, *can* do work on the planet. The work done during a time interval Δt is equal to

$$W_{\text{by gravity}} = Fd \cos \phi$$

where d is the displacement during this interval and the angle ϕ is indicated in Figure 9.16. At positions 1 and 2, this angle is 90°, so $\cos \phi = 0$ and the work done is zero, but at other places along the orbit, such as locations 3 and 4, the work done is not zero. The work done by gravity causes the planet to speed up, thus increasing its kinetic energy as it approaches the Sun. As the planet moves farther from the Sun during the course of its orbit, the work done by gravity is negative (at point 4 in Fig. 9.16), causing the planet to slow down.

9.5 The Vector Nature of Rotational Motion: Gyroscopes

To this point, we have treated rotational quantities such as angular velocity as simple scalars because we always dealt with a single rotation axis whose orientation did not change during the course of the motion. To deal with more complex types of rotational motion, we must recognize that many angular motion quantities—including angular velocity, angular acceleration, angular momentum, and torque—are all vectors. These quantities have both a magnitude and a direction, which can make the rotational motion of a complex object, such as a boomerang, very complicated. We will therefore restrict our discussion to a few of the simplest examples in which the vector nature of angular motion plays a crucial role.

A rotating disk or wheel has an angular velocity whose direction is along the axle, as indicated in Figure 9.17. The ***right-hand rule*** provides a convenient way to determine this direction. According to this rule, if you allow the fingers on your right hand to curl in the direction of motion of the edge of the wheel, your outstretched right thumb will point in the direction of $\vec{\omega}$, the angular velocity vector. The direction of other angular quantities such as the angular momentum $\vec{L}$ is also given by the right-

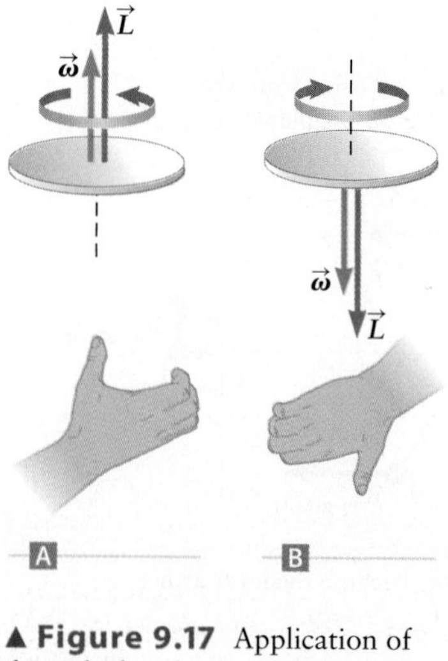

▲ **Figure 9.17** Application of the right-hand rule to determine the direction of the angular velocity $\vec{\omega}$ and angular momentum $\vec{L}$ of a rotating disk or wheel.

[1]This approximation for the area becomes more and more accurate as the time interval Δt is made smaller and smaller.

hand rule. For the cases we consider in this book, the angular momentum is parallel to the angular velocity, with

$$\vec{L} = I\vec{\omega} \tag{9.23}$$

When we say that angular momentum is conserved, we mean that both the *magnitude* and *direction* of $\vec{L}$ are conserved. That is the principle behind the operation of a ***gyroscope.*** There are many ways to build a gyroscope; one design (Fig. 9.18) employs a spinning wheel mounted on a special frame that allows the axle to rotate freely. Because of the way the wheel is mounted in this frame, the torque on the wheel is zero, even when the frame is moved or rotated. Hence, even if the gyroscope frame is moved or rotated, the wheel's orientation does not change because the angular momentum vector is conserved: $\vec{L}$ of the wheel maintains this constant direction. The orientation of a gyroscope thus provides a "direction finder," a tool extremely useful to pilots and navigators. All airplanes have gyroscopes so that pilots can be certain of their orientation and also the direction in which their plane is moving. Of course, modern aeronautical gyroscopes are more complicated than the one in Figure 9.18, but the principle is the same.

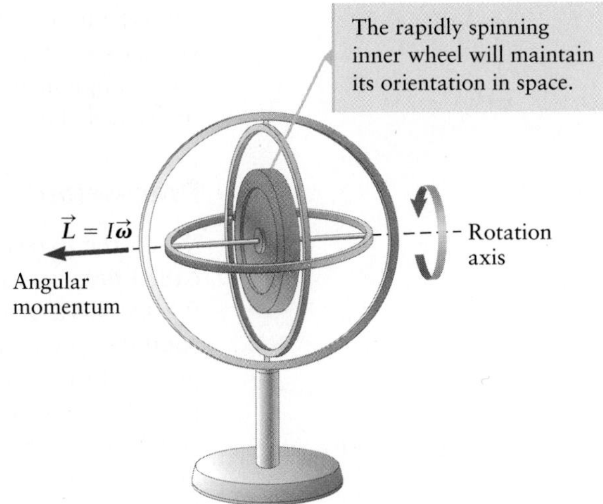

▲ **Figure 9.18** The rotating wheel in this gyroscope is mounted on a system of axles that are free to rotate independently, even when the base of the gyroscope is moved or rotated. Conservation of angular momentum keeps the direction of the inner wheel's axis fixed in space.

The Earth as a Gyroscope

The Earth itself acts as a gyroscope, rotating once each day about an axis that runs between the geographic North and South Poles. This rotational motion is often referred to as "spin," and it produces the "spin angular momentum" of the Earth, which should be distinguished from the orbital angular momentum we discussed in connection with Kepler's second law (Fig. 9.15). The Earth moves in a nearly circular orbit around the Sun, and the Earth's rotational axis is not perpendicular to the orbital plane. Instead, the Earth's spin rotation axis is tilted approximately 23.5° away from the normal to this plane as shown in Figure 9.19. Applying the right-hand rule to the Earth's rotational motion in Figure 9.19 shows that the Earth's spin angular momentum is directed along its rotation axis and points in the direction of the North Pole. There are no external torques on the Earth; its spin angular momentum is therefore conserved, and the rotation axis remains tilted at a fixed angle with respect to the orbital plane. So, for part of each year, the northern half of the Earth tilts toward the Sun, and during the other part of the year, the southern half tilts toward the Sun. The part of the Earth that tilts toward the Sun receives slightly more energy from sunlight, leading to the seasons we observe.

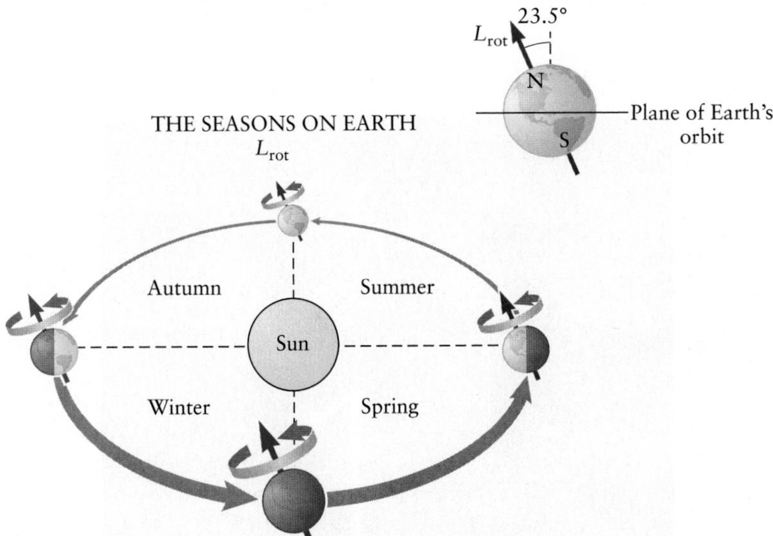

▲ **Figure 9.19** The Earth's spin angular momentum $\vec{L}_{rot}$ lies along a direction parallel to its north–south axis. The vector $\vec{L}_{rot}$ is tilted at an angle of approximately 23.5° away from the normal to the Earth's orbital plane. The seasons indicated in the figure are those in the northern hemisphere.

Angular Momentum and the Stability of a Spinning Wheel

If you have a bicycle wheel that is not mounted on a bicycle frame, you can perform an experiment that demonstrates gyroscopic motion. A nonrotating wheel is very hard to "balance" on its edge and will quickly fall over if left alone. A rolling wheel, however, is much more stable and can travel for quite a long distance before it eventually tips over and falls to one side. The extra stability of a rolling wheel is caused by its angular momentum. According to the right-hand rule, the angular momentum $\vec{L}$ of the wheel is directed along the rotation axis (i.e., it is horizontal). If the external torque on the rolling wheel were exactly zero, then $\vec{L}$ would remain in this direction forever and the wheel would never fall over. In reality, however, there is always some small external torque, mainly from friction, causing the wheel to fall eventually. Even so, its angular momentum makes a rolling wheel much more stable than a nonrotating wheel. This experiment is one you *should* try at home!

Precession

When the external torque on a rotating object is nonzero, it can lead to an effect called ***precession***. A popular device that demonstrates this effect is shown in Figure 9.20A; it is a bicycle wheel mounted on an axle, with one end of the axle resting on a rotatable pivot attached to a vertical pole. As you might expect, this "system" (composed of the wheel plus axle) can be extremely unstable; if the wheel is not rotating, the wheel and its axle will immediately fall over. On the other hand, if the wheel is spinning about the axle, its stability is greatly enhanced. The angular momentum of the spinning wheel prevents it from falling over, even though the force of gravity exerts a large external torque on the system.

We can calculate the torque on the system of the wheel plus axle by assuming the entire gravitational force acts at the center of mass (Chapter 8). Most of the mass of the system in Figure 9.20A is in the wheel, so the system's center of mass is close to the center of the wheel as shown in the figure. To compute the direction of the gravitational torque $\vec{\tau}$ we use another version of the right-hand rule. The procedure is sketched in Figure 9.20B. (1) Start from the pivot point, which in this case is point P on the rotatable pivot at the base of the axle. With P as an origin, place the fingers of your right hand along the rotation axis of the wheel, that is, along the vector $\vec{r}$ in Figure 9.20B. Point your fingers toward the center of the wheel, which is the place where the force $\vec{F}_{grav}$ acts. (2) Curl your fingers in the direction of the force $\vec{F}_{grav}$. Your outstretched right thumb then points in the direction of $\vec{\tau}$ as indicated in Figure 9.20B.

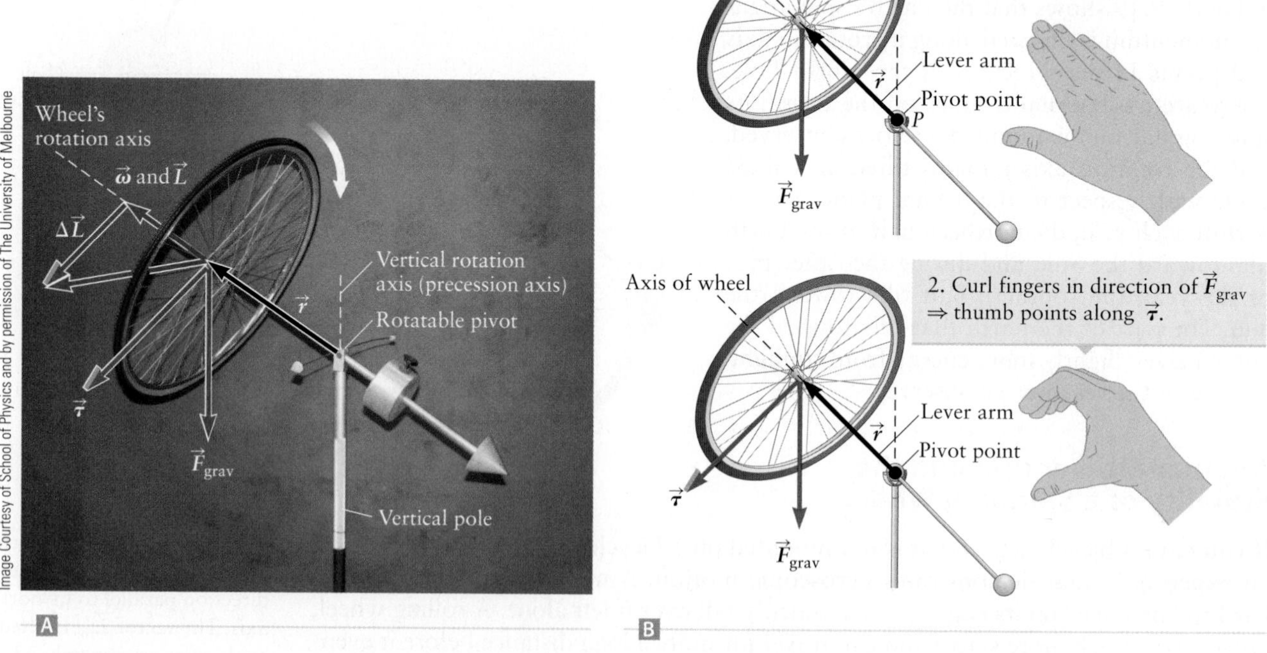

▲ **Figure 9.20** A There is a torque on this bicycle wheel due to the force of gravity. The direction of the torque is given by applying the right-hand rule. This torque makes the axis of the wheel precess. Note that the change in angular momentum $\Delta\vec{L}$ is parallel to $\vec{\tau}$. B Application of the right-hand rule to find the direction of the torque due to gravity.

We must next consider how this torque affects the angular momentum of this system (wheel plus axle). In Chapter 7, we showed that an applied force $\vec{F}$ changes the linear momentum $\vec{p}$ of an object. This leads to the *impulse theorem*, involving force and impulse (Eq. 7.5):

$$\vec{F}\,\Delta t = \Delta\vec{p}$$

An analogous result is true for rotational motion: an applied torque produces a change in the angular momentum. We have

$$\vec{\tau}\,\Delta t = \Delta\vec{L} \tag{9.24}$$

so if a constant torque $\vec{\tau}$ acts for a time Δt, there is a change $\Delta\vec{L}$ in the angular momentum. Applying this result to the wheel plus axle in Figure 9.20, we find that the torque due to the gravitational force is perpendicular to the rotation axis of the wheel, which is also the direction of the angular momentum vector $\vec{L}$. This torque causes the wheel's angular momentum to turn in the direction of $\vec{\tau}$ as indicated in Figure 9.20A. However, as the rotation axis of the wheel turns, the direction of the torque also changes. In fact, $\vec{\tau}$ and hence also $\Delta\vec{L}$ from Equation 9.24 are *always perpendicular* to $\vec{L}$. As a result, the wheel plus axle rotate continuously about a vertical rotation axis (the pole in Fig. 9.20). This movement is called *precession* and is an example of the type of motion that can occur when the angular momentum and the applied torque are not parallel.

9.6 Cats and Other Rotating Objects

In this section, we discuss two interesting examples of rotational motion based on the principle of conservation of angular momentum.

Rotating Cats

A widely known piece of folklore is the claim that a falling cat always lands on its feet. Although we certainly do not want to encourage the unethical treatment of animals, you have probably observed that a falling cat is usually able to rotate its body so as to land feet first, even if it begins its fall with its feet pointing upward. This example of rotational motion is shown in Figure 9.21 and in the opening photo for this chapter. The cat in Figure 9.21A is not rotating when it begins its fall, so its initial angular momentum is zero. Since there are no external torques on the cat while it is falling, its angular momentum must stay constant and hence is always zero during the course of its fall. How can the cat manage to rotate even though it has no angular momentum?

Cats accomplish this feat by changing their shape while they are falling. To understand this process, consider separately the motion of the cat's front section (near its head) and its hindquarters (near its tail). In Figure 9.21B, the cat has pulled its front legs close to its body; just as for the figure skater in Figure 9.11, this reduces the moment of inertia of the cat's front section (I_{front} is small). At the same time, the cat's tail and rear legs are extended away from its body, so the moment of inertia of the rear part of its body is large (I_{rear} is large). The cat then "swivels" at its middle as shown in Figure 9.21C, with the front half of the body rotating in one direction and the hindquarters in the *opposite* direction. The total angular momentum is still zero, but because the front and rear moments of inertia are different, the angular velocities of the front and rear parts of the cat's body are different. Since $I_{\text{rear}} > I_{\text{front}}$, we know from the example with the skater that $\omega_{\text{rear}} < \omega_{\text{front}}$. As a result, the rotation angle for the front half is much larger than for the hindquarters. Figure 9.21D shows the cat's upper body rotating by approximately 180°, whereas the lower body has rotated much less.

The cat then repeats the same process in parts D and E of Figure 9.21, pulling its hind legs and tail close to its body while extending its front legs. Then $I_{\text{rear}} < I_{\text{front}}$, and the cat can now swivel at the middle with $\omega_{\text{rear}} > \omega_{\text{front}}$. This swivel allows the

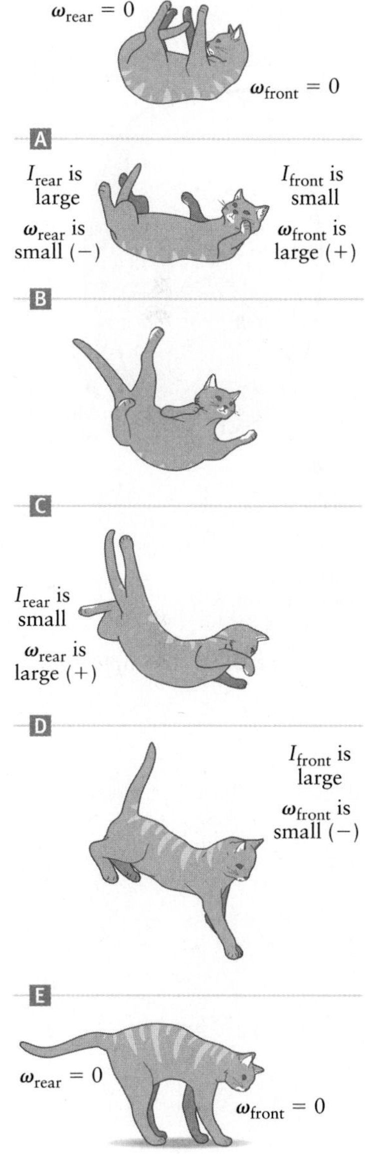

▲ **Figure 9.21** The angular momentum of this cat is conserved as it falls to the ground. The cat is still able to rotate its body so as to land safely, however.

cat to rotate its hindquarters through approximately 180° as in Figure 9.21D while rotating the front part of its body very little. The cat is then able to extend its legs and land safely in Figure 9.21F.

The point of this example is not that cats have a keen understanding of angular momentum, but rather that a system whose total angular momentum is zero can still rotate in some interesting ways. All we needed to explain this cat's motion was the concept that angular momentum is the product of the moment of inertia and the angular velocity.

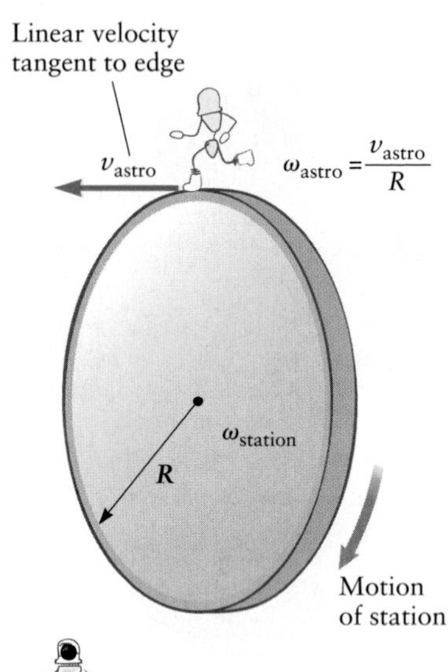

▲ **Figure 9.22** Example 9.7.

EXAMPLE 9.7 An Astronaut's Treadmill

A circular space station (Figure 9.22) is initially at rest (i.e., not rotating), when an astronaut begins his daily exercise in which he runs laps around the station with linear speed $v_{astro} = 3.0$ m/s. This is the speed that would be measured by a stationary observer outside the station and is directed *tangentially*, along the edge of the station. Find the resulting angular velocity of the station. The mass of the astronaut is 80 kg, the mass of the station is 1.0×10^5 kg, and (for simplicity) assume the station has the shape of a simple disk with radius $R = 30$ m.

RECOGNIZE THE PRINCIPLE

The astronaut and the space station form a system. There are no external torques, so its angular momentum must be conserved. If part of the system starts to rotate in one direction (i.e., the astronaut runs in one direction), another part of the system must rotate in the opposite direction. We can use the principle of conservation of angular momentum to find how the angular velocities of the astronaut and the station are related.

SKETCH THE PROBLEM

Figure 9.22 shows the problem. The astronaut and station rotate in opposite directions.

IDENTIFY THE RELATIONSHIPS

The station and astronaut are initially at rest relative to the observer in Figure 9.22, so the system's total angular momentum is zero throughout the astronaut's workout. Let $\omega_{station}$ be the angular velocity of the station when the astronaut is running. As he runs, the astronaut's motion is similar to that of the child on (or near) the edge of the merry-go-round in Figure 9.13 (compare also with Fig. 9.10). We can therefore find the astronaut's angular momentum L_{astro} from Equation 9.16, which gives

$$L_{astro} = m_{astro} R v_{astro}$$

The total angular momentum of the astronaut plus the station is then

$$L_{total} = L_{astro} + L_{station} = m_{astro} R v_{astro} + I_{station}\omega_{station} = 0 \quad (1)$$

We approximate the station as a disk, so its moment of inertia (from Table 8.2) is

$$I_{station} = I_{disk} = \tfrac{1}{2} m_{station} R^2$$

SOLVE

Using this result in Equation (1) leads to

$$0 = m_{astro} R v_{astro} + \tfrac{1}{2} m_{station} R^2 \omega_{station}$$

$$\omega_{station} = -\frac{m_{astro} R v_{astro}}{m_{station} R^2/2} = -\frac{2 m_{astro} v_{astro}}{m_{station} R}$$

Inserting the values of the various quantities gives

$$\omega_{station} = -\frac{2(80\text{ kg})(3.0\text{ m/s})}{(1.0 \times 10^5\text{ kg})(30\text{ m})} = \boxed{-1.6 \times 10^{-4}\text{ rad/s}}$$

▶ ***What does it mean?***
The angular velocity of the astronaut is

$$\omega_{\text{astro}} = \frac{v_{\text{astro}}}{R} = \frac{3.0 \text{ m/s}}{30 \text{ m}} = +0.10 \text{ rad/s}$$

The angular velocity of the station is much smaller because the station is much more massive than the astronaut and has a larger moment of inertia. Notice that ω_{station} is negative because the station rotates in the direction opposite that of the astronaut.

CONCEPT CHECK 9.6 Rotational Motion When the Total Angular Momentum Is Zero

Explain what aspects of the motion in Example 9.7 are similar to the rotating cat in Figure 9.21.

Angular Momentum and Motorcycles

As a final example of angular momentum conservation, we consider the motion of a motorcycle stunt rider who jumps through the air as shown in Figure 9.23. Let's examine the combined translational and rotational motions of the system consisting of the motorcycle plus the rider. In both jumps shown in Figure 9.23, the system's center of mass moves as a projectile, following a parabolic trajectory as we established in Chapter 4. The rotational motions are different in the two cases, however. In Figure 9.23A, the motorcycle frame maintains a fixed angular orientation throughout the jump, and the motorcycle lands on its rear wheel. Throughout the jump, the angular velocity of the system (the motorcycle and rider) is zero, so the angular displacement of the motorcycle frame is also zero. This example shows conservation of angular momentum: the angular momentum of the entire system, and also of the motorcycle frame, is constant and zero throughout the jump.

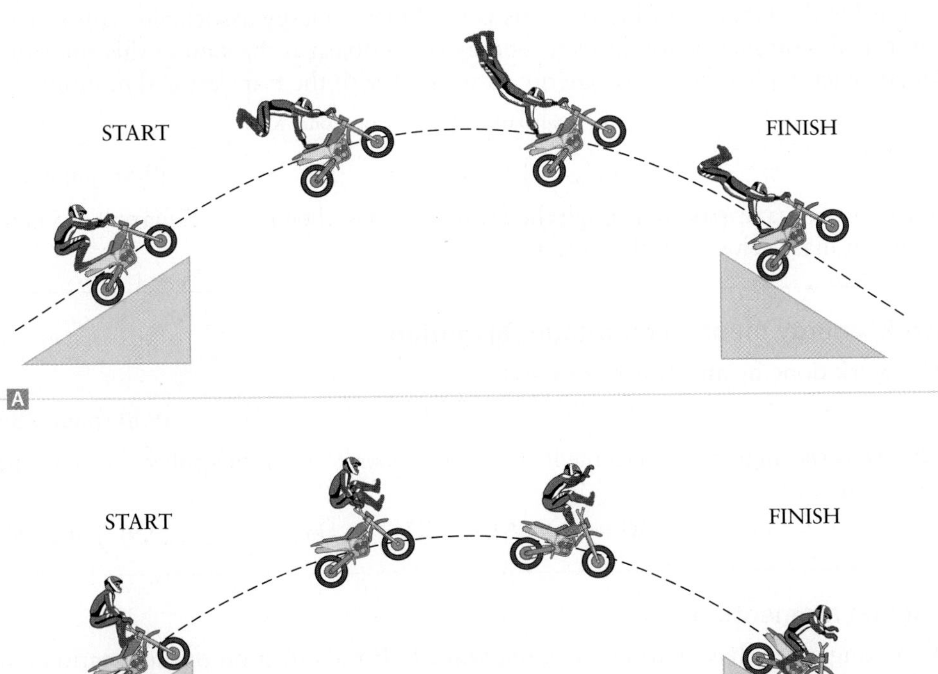

◀ **Figure 9.23** A In this jump, the orientation of the motorcycle frame changes very little during the course of the jump. B Here the rider reduced the angular velocity of the rear wheel after leaving the ground (by adjusting the throttle). To conserve angular momentum about the axis that runs along the rear axle, the frame then began to rotate counterclockwise.

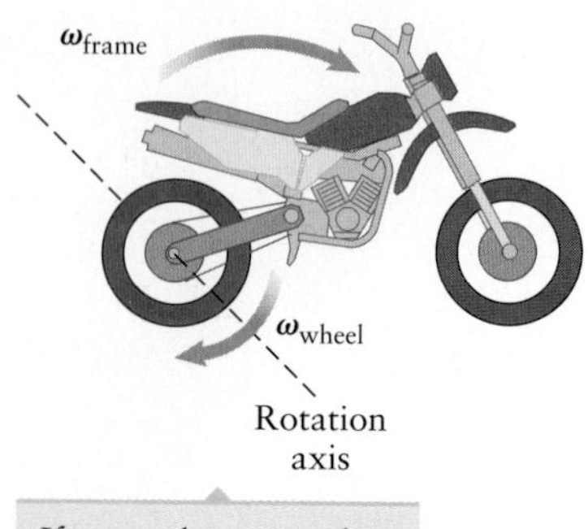

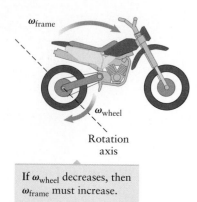

▲ **Figure 9.24** The total angular momentum of a motorcycle is conserved during a jump. The total angular momentum has contributions from the wheels and the frame.

In the jump shown in Figure 9.23B, the motorcycle again begins with zero angular velocity, but then *during the jump* the motorcycle frame rotates. This rotation allows the motorcycle to land simultaneously on both wheels, with the motorcycle frame "sloping down," leading to a gentler (and presumably safer) landing since the landing is usually onto a downward-sloping ramp. How did the rider cause the angular velocity of the motorcycle frame to change from its value of zero at the start of the jump to a nonzero value during the jump?

The answer to this question is well known to stunt riders: during the jump, they adjust the throttle so as to change the angular velocity of the rear wheel as it rotates about its axle, thus changing the wheel's angular momentum. There are no external torques on the system, so the total angular momentum of the system must remain constant. The total angular momentum is equal to the sum of the angular momentum of the wheels plus the angular momentum of the frame plus the angular momentum of the rider. In this case, we can take the rotation axis to be the rear axle, which is perpendicular to the plane of Figure 9.24. If the angular momentum of the rear wheel is decreased (by adjusting the throttle), the remaining angular momentum must increase. This causes the frame to rotate clockwise about the rear axle as in Figures 9.23B and 9.24, and is another nice application of physics!

Summary | CHAPTER 9

Key Concepts and Principles

Rotational kinetic energy

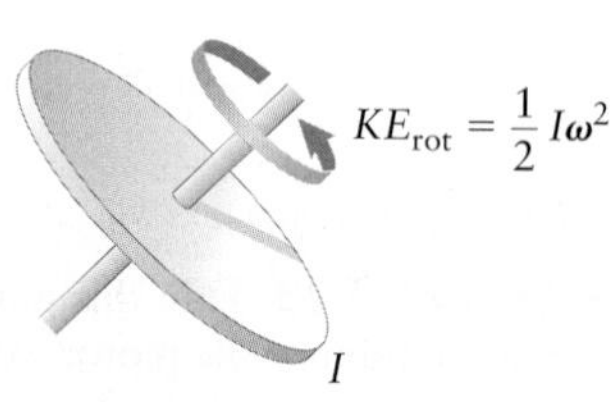

The ***rotational kinetic energy*** of an object is

$$KE_{rot} = \tfrac{1}{2}I\omega^2 \qquad (9.4)\ (\text{page } 284)$$

where I is the moment of inertia. This is the kinetic energy associated with just the rotational motion. The total kinetic energy of an object is the sum of this rotational kinetic energy plus the kinetic energy associated with the translational motion:

$$KE_{total} = \underbrace{\tfrac{1}{2}mv_{CM}^2}_{\text{translational } KE} + \underbrace{\tfrac{1}{2}I\omega^2}_{\text{rotational } KE} \qquad (9.5)\ (\text{page } 284)$$

If the rotation axis passes through the center of mass, then v_{CM} in Equation 9.5 is the center of mass velocity of the object.

Work–energy theorem for rotational motion

The work done by an applied torque is

$$W = \tau\theta \qquad (9.8)\ (\text{page } 286)$$

where θ is the angular displacement. The work–energy theorem applied to rotational motion is

$$W = \tau\theta = \Delta KE = \tfrac{1}{2}I\omega_f^2 - \tfrac{1}{2}I\omega_i^2 \qquad (9.9)\ (\text{page } 287)$$

Angular momentum

A rotating object has an ***angular momentum*** L. If the direction of the rotation axis is fixed, then L is given by

$$L = I\omega \qquad (9.15)\ (\text{page } 292)$$

If the external torque on a system is zero, the total angular momentum of the system is conserved.

Vector nature of angular momentum

The direction of the angular velocity $\vec{\omega}$ is given by the ***right-hand rule***, and the angular momentum vector $\vec{L}$ is

$$\vec{L} = I\vec{\omega} \qquad (9.23)\ (\text{page } 299)$$

When the fingers of your right hand curl in the direction of motion, your outstretched right thumb points along $\vec{\omega}$ and $\vec{L}$.

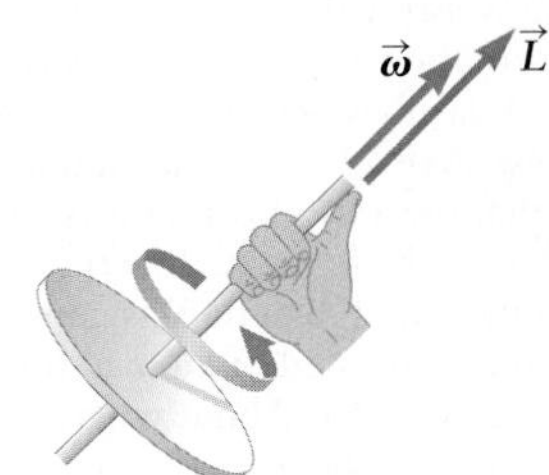

Applications

Conservation of energy

Conservation of energy principles can be applied to rotational motion. When all the forces that do work on an object are conservative forces, the total mechanical energy is conserved. When the potential energy is due to gravity, as for an object that rolls up or down an incline, conservation of mechanical energy leads to

$$KE_i + PE_i = KE_f + PE_f$$

$$\tfrac{1}{2}mv_i^2 + \tfrac{1}{2}I\omega_i^2 + mgh_i = \tfrac{1}{2}mv_f^2 + \tfrac{1}{2}I\omega_f^2 + mgh_f \qquad (9.11)\ (\text{page } 288)$$

The kinetic energy given here is the total kinetic energy, including the rotational contribution.

Conservation of angular momentum

Angular momentum and its conservation play important roles in the motion of many objects, including figure skaters, cats, motorcycles, and gyroscopes.

Questions

SSM = answer in *Student Companion & Problem-Solving Guide* ⊗ = life science application

1. A solid wood ball is rotating about an axis that passes through its center. If its angular speed is doubled, (a) by what factor does its rotational kinetic energy change? (b) By what factor does its angular momentum change?
2. A pitcher throws a baseball (mass $m = 0.14$ kg, $r = 7.4$ cm) with linear speed v and angular velocity ω. Which of the following statements is correct? Explain your answer. (a) The total kinetic energy of the baseball can be calculated knowing only m and v. (b) The total kinetic energy of the baseball can be calculated knowing only m, r, and ω. (c) To calculate the total kinetic energy of the baseball, we need to know only v and ω.
3. Two identical spheres start from rest at the top of a long ramp. One ball rolls down the ramp without slipping, while the other slides down without rolling. Which one reaches the bottom first? Explain your answer.
4. Two objects having different shapes start from rest from the top of a long ramp. Both then roll without slipping down the ramp, and one reaches the bottom before the other. Which object has the smaller moment of inertia? Explain your answer.
5. SSM Consider the translational and rotational kinetic energies of a disc that rolls without slipping. Show that the ratio of these energies is independent of the size (the radius) of the wheel.
6. You are given two objects and a ramp. Explain how you could measure the ratio of the moments of inertia of the two objects. Assume the two objects are both round (so that both roll), have the same mass and radius, and are solid (no holes in the center, etc.). Consider only objects that meet these criteria and with the shapes in Table 8.2. *Hint*: Both objects can roll without slipping down the ramp, and you have the equipment needed to measure the speeds of the objects as they roll.
7. Tightrope walkers often carry a long pole (Fig. Q9.7). Explain why.
8. Is it possible for the total force on a system to be nonzero but the torque to be zero? Give an example.
9. You are given the job of designing a new high-tech skateboard. For it to travel as fast as possible, you want to minimize the kinetic energy in its wheels. All else being equal (i.e., the mass and radius of the wheels), is it better to make the wheels in the shape of a disk, a sphere, or a conventional wheel with spokes?

Figure Q9.7

10. Explain why the rim brakes on a bicycle wheel are located at the outer edge of the wheel instead of near the hub. *Hint*: Consider the work–energy theorem (Eq. 9.9) and how the magnitude of the

torque produced by a given force varies when the force is applied at different distances from the axis of rotation.

11. The flywheel within a car engine is a form of gyroscope, and when rotating at high rpm, it carries significant angular momentum. (A car flywheel is just a disk attached to the engine's main rotating shaft.) In most cases, the engine of a race car is mounted such that the axis of rotation points in the forward direction and $\vec{L}$ is in the forward direction. Because angular momentum (magnitude and direction) is conserved, what happens to the car as it rounds the end of an oval track in a counterclockwise direction (turning left)? About what axis will the car tend to twist? If instead the engine were mounted such that the axis of rotation was parallel to the axles of the car, along which axis would the car then tend to twist? *Hint*: Consider the gyroscope and the rotation of precession. In auto racing, the cars typically go around the oval track in a counterclockwise direction as viewed from above. Would race cars be engineered differently if races were conducted in a clockwise direction?

12. Discuss how and if the moment of inertia of the Earth associated with its orbital motion around the Sun changes as the Earth moves through its orbit. How are these changes connected to changes in the Earth's speed?

13. Many kilograms of asteroids land on the Earth each year. Discuss how they affect the length of a day.

14. Consider a child who is jumping on a trampoline as shown in Figure Q9.14. Explain how the child can have zero angular momentum, but still rotate to either the right or the left while she is in the air.

15. The cat in Figure 9.21 undergoes free fall due to the force of gravity. Explain why we did not consider gravity when we discussed how the cat in Figure 9.21 managed to land on its feet.

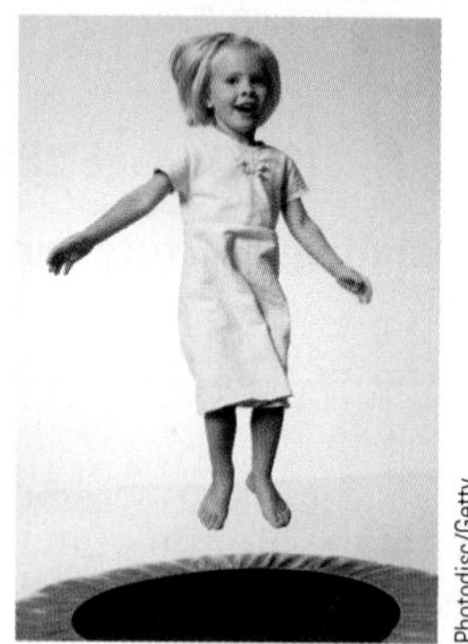
Photodisc/Getty

Figure Q9.14

16. The barrel of a rifle contains spiral grooves that impart some spin to a bullet as it leaves the barrel. Explain why this spin improves the rifle's accuracy.

17. SSM (a) Why do most small helicopters have a rotor on their tail (Fig. Q9.17)? (b) Many large helicopters have two large rotors. Why do these helicopters not need a tail rotor? Do you think the two large propellers rotate in the same direction? Explain.

U.S. Navy/Mass Communication Specialist Seaman Jon Dasbach

Figure Q9.17

18. (life science) Why is it possible for a gymnast to perform more rotations during an airborne somersault when she is in a tuck position than when her body is straight (called the layout position)?

19. Some early spacecrafts used tape recorders to store data. These recorders have spools of magnetic tape that are rotated in spindles to move the tape past "heads" that record or read data. Explain why a tape recorder can affect a spacecraft's motion.

20. When a spinning figure skater pulls her arms close to her body (Fig. 9.11), she spins faster. If she then pushes her arms away from her body, her angular velocity decreases. Her rotational kinetic energy also decreases. Explain why her kinetic energy decreases. What force or torque does the corresponding work?

Problems

SSM = solution in *Student Companion & Problem-Solving Guide*
★ = intermediate ✪ = challenging Ⓧ = life science application
Ⓡ = reasoning and relationships problem
RT = reasoning tutorial in Enhanced WebAssign

9.1 KINETIC ENERGY OF ROTATION

1. A baseball (mass 0.14 kg, radius 3.7 cm) is spinning with an angular velocity of 60 rad/s. What is its rotational kinetic energy?

2. If the baseball in Problem 1 has a linear speed of 45 m/s, what is the ratio of its rotational and translational kinetic energies?

3. Consider a quarter rolling down an incline. What fraction of its kinetic energy is associated with its rotational motion? Assume it rolls without slipping. *Hint*? You do not need to know the mass or radius of a quarter to solve this problem.

4. ★ Ⓡ Estimate the maximum rotational kinetic energy of a yo-yo. Use a rotation axis that passes through the yo-yo's center.

5. ★ A wheel of mass 0.50 kg and radius 45 cm is spinning with an angular velocity of 20 rad/s. You then push your hand against the edge of the wheel, exerting a force F on the wheel as shown in Figure P9.5. If the wheel comes to a stop after traveling 1/4 of a turn, what is F?

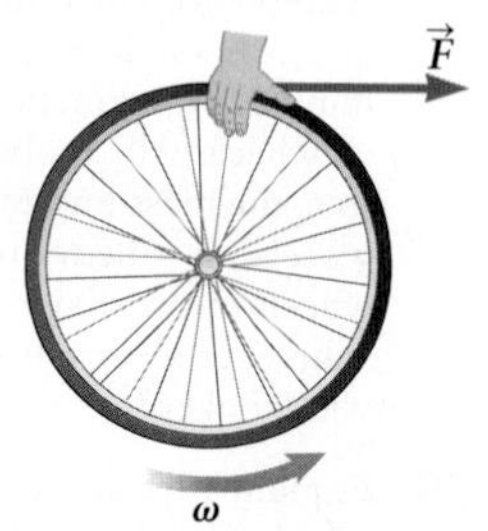

Figure P9.5

6. ★ RT It has been proposed that large flywheels could be used to store energy. Consider a flywheel made of concrete (density 2300 kg/m^3) in the shape of a solid disk, with a radius of 10 m and a thickness of 2.0 m. If its rotational kinetic energy is 100 MJ (1.0×10^8 J), what is the angular velocity of the flywheel?

7. ★ RT Consider a wheel on a racing bicycle for an adult. If the wheel has a mass of 0.40 kg and an angular speed of 15 rad/s, what is the rotational kinetic energy of the wheel?

8. ★ Ⓡ An automobile wheel has a mass of 18 kg and a diameter of 0.40 m. What is the total kinetic energy of one wheel when the car is traveling at 20 m/s?

9. ★ Ⓡ Estimate the rotational kinetic energy of an airplane propeller.

10. ★ Two balls (solid spheres) have the same radius and the same rotational kinetic energy. If the ratio of their masses is 3:1, what is the ratio of their angular velocities?

11. ★ RT Estimate the fraction of a bicycle's total kinetic energy associated with the rotational motion of the wheels. Consider the kinetic energy of the bicycle and not that of the rider.

12. SSM What is the rotational kinetic energy of the Earth as it spins about its axis?

13. What is the ratio of the rotational kinetic energy of the Earth to the rotational kinetic energy of the Moon as they spin about their axes?

9.2 CONSERVATION OF ENERGY AND ROTATIONAL MOTION

14. Two crates of mass $m_1 = 15$ kg and $m_2 = 9.0$ kg are connected by a rope that passes over a frictionless pulley of mass $m_p = 8.0$ kg and radius 0.20 m as shown in Figure P9.14. (a) What force(s) can do work on the masses and the pulley? (b) Identify a system whose mechanical energy will be conserved. (c) The crates are released from rest. Crate 1 falls a distance of 2.0 m, at which time its speed is v_f. Make a sketch showing the initial and final states of the system. (d) What are the total initial and final kinetic energies of the system? Express your answers in terms of v_f. (e) What are the initial and final potential energies of the system? (f) What is the final speed v_f of the crates? (g) What is the final angular velocity of the pulley?

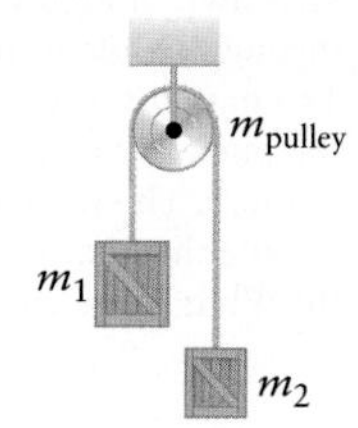

Figure P9.14 Problems 14, 16, and 17.

15. A marble (radius 1.0 cm and mass 8.0 g) rolls without slipping down a ramp of vertical height 20 cm. What is the speed of the marble when it reaches the bottom of the ramp?

16. For the system of two crates and a pulley in Figure P9.14, what fraction of the total kinetic energy resides in the pulley?

17. Consider again the system in Figure P9.14. Use the result from Problem 9.14 to find the acceleration of one of the crates. *Hint*: Is the acceleration constant?

18. Consider a hoop of mass 3.0 kg and radius 0.50 m that rolls without slipping down an incline. The hoop starts at rest from a height $h = 2.5$ m above the bottom of the incline and then rolls to the bottom. (a) What forces act on the hoop? Which of these forces do work on the hoop? (b) Is the mechanical energy of the hoop conserved? (c) Make a sketch showing the hoop at the top and the bottom of the incline. (d) What is the total initial mechanical energy of the hoop? Take the zero of potential energy to be at the bottom of the incline. (e) If the hoop's speed at the bottom of the incline is v_f, what is its mechanical energy? Express your answer in terms of v_f. (f) Solve for v_f.

19. A marble rolls on the track shown in Figure P9.19, with $h_B = 25$ cm and $h_C = 15$ cm. If the marble has a speed of 2.0 m/s at point A, what is its speed at points B and C?

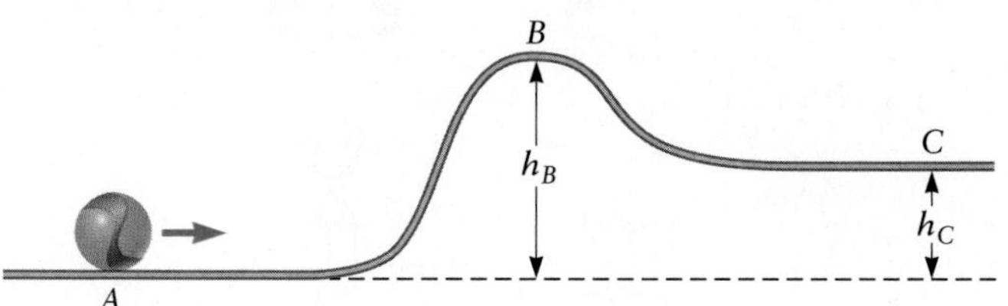

Figure P9.19

20. A yo-yo falls through a distance of 0.50 m as its string unwinds. If it starts from rest, what is its speed?

21. A sphere rolls down the loop-the-loop track shown in Figure P9.21, starting from rest at a height h above the bottom. The ball travels around the inside of the circular portion of the track (radius $r = 5.0$ m). It is found that the ball is just barely able to travel around the inside of the track without losing contact at the top. Find h.

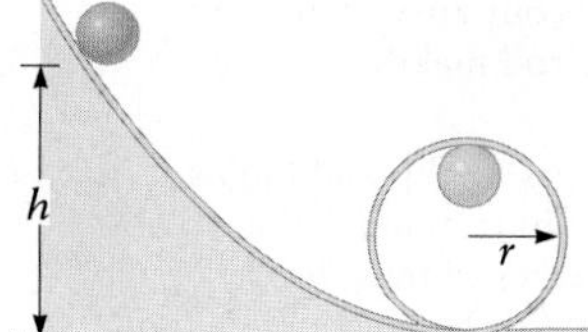

Figure P9.21

22. A metal rod of length $L = 3.0$ m is free to rotate about a frictionless hinge at one end (Fig. P9.22). The rod is initially at rest and oriented horizontally and then released so that it swings to the bottom. (a) What are all the forces acting on the rod? Which force or forces do work on the rod as it swings? (b) Is the mechanical energy of the rod conserved? (c) Make a sketch showing the rod in its initial orientation and when it is at the bottom of its swing. (d) What is the initial mechanical energy of the rod? Choose the zero of potential energy so that the final *PE* is zero. (e) If the speed of the end of the rod when it is at its lowest point is v_f, what is the final mechanical energy of the rod? Express your answer in terms of v_f. (f) Find v_f.

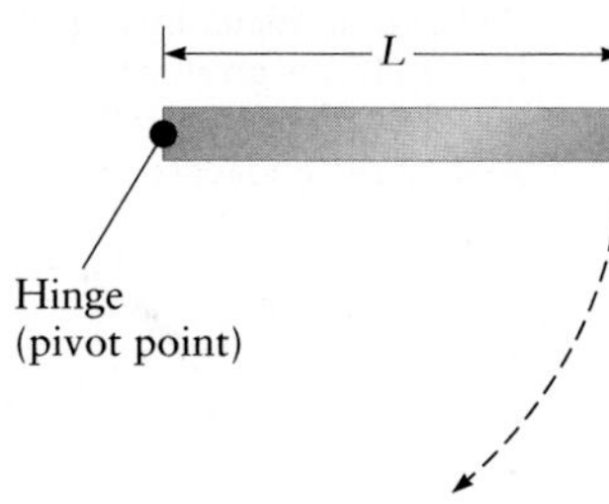

Figure P9.22 Problems 22, 23, and 25.

23. Consider the rod in Figure P9.22, but now assume it starts from rest in the "up" position (i.e., pointing vertically upward from the hinge). If the rod is then given a very small push, what is the linear speed of the end of the rod when it reaches the bottom of its circular path?

24. A ball rolls without slipping down the track shown in Figure P9.24, starting from rest at a height $h_1 = 20$ m as shown. The ball is traveling horizontally when it leaves the bottom of the track, which has a height $h_2 = 9.0$ m. Find where the ball hits the ground; that is, find L.

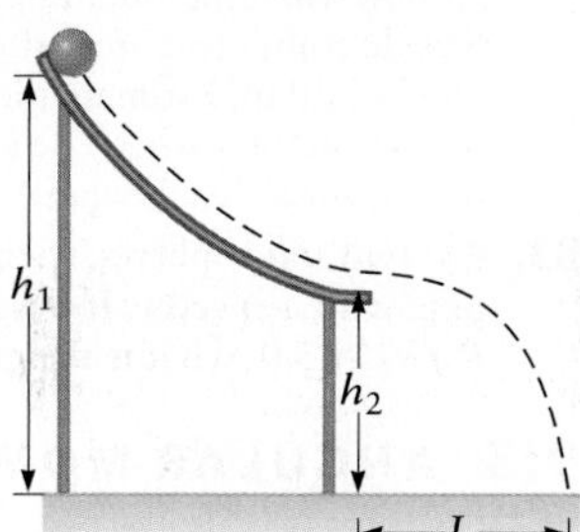

Figure P9.24

25. Repeat Problem 9.22(f) (i.e., find v_f), but now assume the rod is initially oriented an angle of 30° above the horizontal.

26. A pencil (mass 10 g and length 15 cm) is initially balanced so that it is sitting vertically on a flat table as shown in Figure P9.26. If the pencil then falls, what is its angular velocity just before it strikes the tabletop? Assume the mass of the pencil is distributed uniformly and the end of the pencil in contact with the table does not slip.

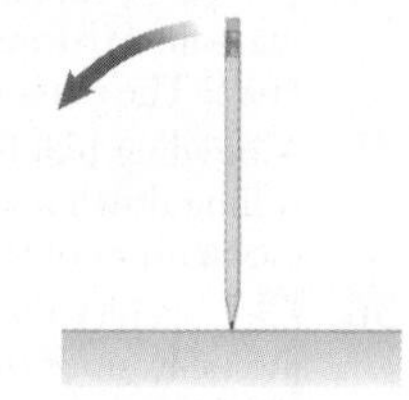

Figure P9.26 Problems 26, 27, and 36.

27. Suppose the table in Figure P9.26 is extremely slippery so that there is no friction between the table and the end of the pencil. As a result, the pencil slips as it tips over. Let $\omega_{\text{no friction}}$ denote the angular velocity of the pencil just before it hits the table, and ω_{friction} denote the angular velocity found in Problem 26. What is the ratio $\omega_{\text{no friction}}/\omega_{\text{friction}}$?

28. A rod of mass 4.0 kg and length 1.5 m hangs from a hinge as shown in Figure P9.28. The end of the rod is then given a "kick" so that it is moving at a speed of 5 m/s. How high will the rod

swing? Express your answer in terms of the angle the rod makes with the vertical.

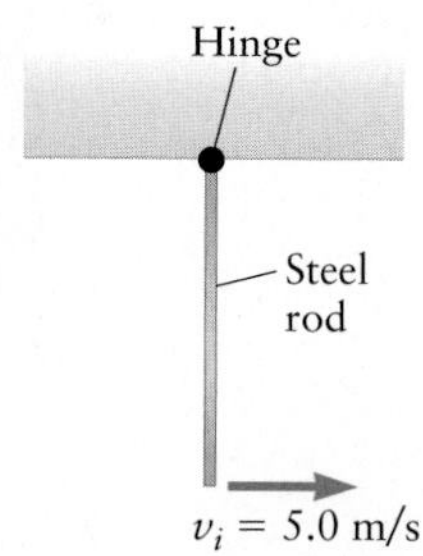

Figure P9.28

29. A cylinder of mass 7.0 kg and radius 0.25 m rolls without slipping along a level floor. Its center of mass has a speed of 1.5 m/s. Find (a) the kinetic energy of translation and (b) the kinetic energy of rotation.

30. Ⓡ An empty bookcase (total mass 10 kg) is accidentally tipped over (Fig. P9.30). If it is given only a very gentle initial push, what is the speed of the top edge of the bookcase just before it strikes the floor?

Figure P9.30

31. SSM ✶ A bucket filled with dirt of mass 20 kg is suspended by a rope that hangs over a pulley of mass 30 kg and radius 0.25 m (Fig. P9.31). Everything is initially at rest, but someone is careless and lets go of the pulley and the bucket then begins to move downward. What is the speed of the bucket when it has fallen a distance of 2.5 m?

Figure P9.31

32. ✪ Ⓡ A bicycle is moving at a speed of 15 m/s when the rider applies her brakes. The bicycle comes to a stop after moving forward another 20 m. Estimate the force of one of the bicycle brake pads on the rim of a wheel. Be sure to include that there are four brake pads (two for each wheel).

33. ✶ Two solid spheres, each of mass m, are rolling without slipping with a speed v. If the spheres have radii R_1 and R_2 with $R_1/R_2 = 3.0$, what is the ratio of their total kinetic energies?

9.3 ANGULAR MOMENTUM

Note that problems 34, 35, 36, 37, 41, and 45 are concerned with the magnitude *of the angular momentum, not its sign.*

34. ✶ RT Estimate the angular momentum of a DVD that is playing your favorite video. Assume an angular speed of 1000 rpm. *Hint*? The mass of a DVD is about 15 g.

35. A bowling ball has a mass of 6.0 kg and a radius of 22 cm. If it is rolling down a lane with a speed of 9.0 m/s, what is the angular momentum of the ball?

36. ✶ Consider the pencil in Problem 26. What is the angular momentum of the pencil just before it strikes the table?

37. ✶ Ⓡ Estimate the angular momentum of a figure skater of mass 50 kg who is spinning at 300 rpm (Fig. 9.11). Assume her arms and legs are pulled in very close to her rotation axis.

38. A child ($m = 40$ kg) is playing on a merry-go-round ($m = 200$ kg, $R = 2.0$ m) that is initially at rest. The child then jumps off in a direction tangent to the edge of the merry-go-round as shown in Figure P9.38. The child has a speed of 5.0 m/s just before she lands on the ground. (a) Identify a system for which the total external torque is zero. (b) What is the initial angular momentum of your system? (c) Make a sketch showing your system just before and just after the child jumps off the merry-go-round. (d) If the final angular velocity of the merry-go-round is ω_f, what is the final angular momentum of the merry-go-round and of the child? Express your answers in terms of ω_f. (e) What is the final angular velocity of the merry-go-round?

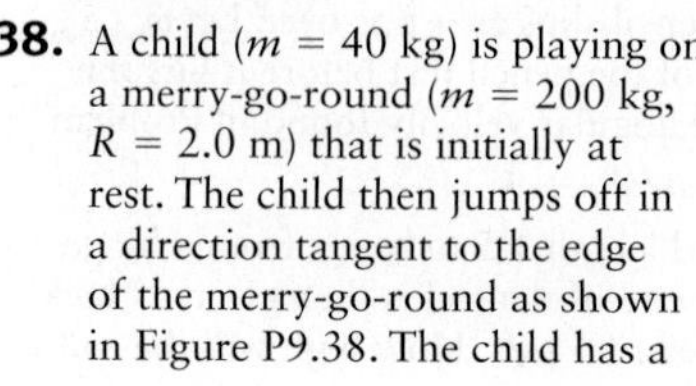
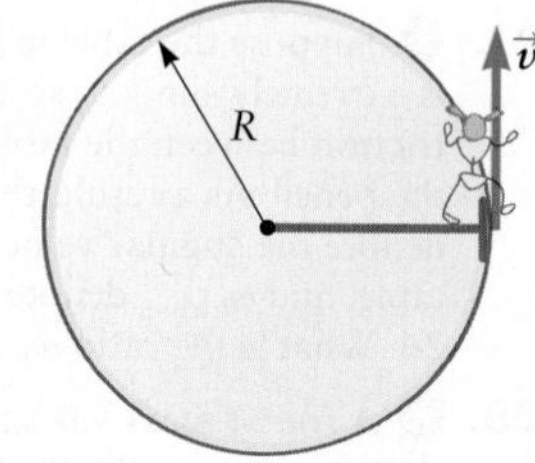

Figure P9.38

39. ✪ A puck (mass $m_1 = 0.50$ kg) slides on a frictionless table as shown in Figure P9.39. The puck is tied to a string that runs through a hole in the table and is attached to a mass $m_2 = 1.5$ kg. The mass m_2 is initially at height $h = 1.5$ m above the floor with the puck traveling in a circle of radius $r = 0.40$ m with a speed of 1.5 m/s. The force of gravity then causes mass m_2 to move downward a distance 0.15 m. (a) What is the new speed of the puck? (b) What is the change in the kinetic energy of the puck?

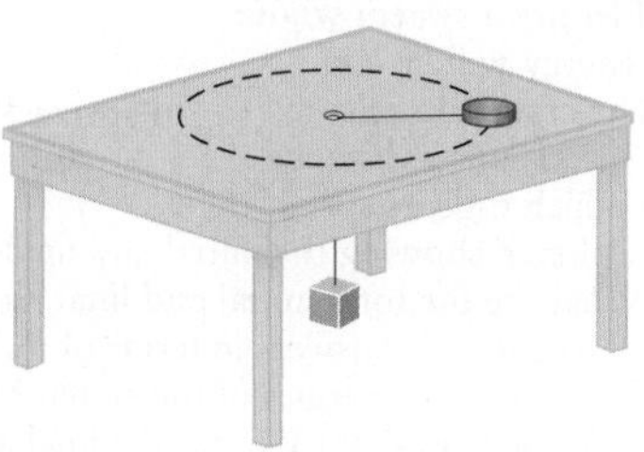
Figure P9.39

40. ✶ Ⓡ A figure skater begins a spin at an angular velocity of 200 rpm with her arms and legs out away from her body (Fig. 9.11A). She then pulls her arms and legs in close to her body, and her angular velocity increases. Estimate her final angular velocity.

41. A bug of mass 3.0 g is sitting at the edge of a CD of radius 8.0 cm. If the CD is spinning at 300 rpm, what is the angular momentum of the bug?

42. A child of mass 50 kg jumps onto the edge of a merry-go-round of mass 150 kg and radius 2.0 m that is initially at rest as sketched in Figure 9.12. While in the air (during her jump), the child's linear velocity in the direction tangent to the edge of the merry-go-round is 10 m/s. What is the angular velocity of the merry-go-round plus child after the child jumps onto the merry-go-round?

43. SSM ✶ Ⓡ Consider a person who is sitting on a frictionless rotating stool as in Figure P9.43. The person initially has his arms outstretched and is rotating with an angular speed of 5.0 rad/s. He then pulls his arms close to his body. (a) Estimate his final angular speed. (b) Estimate the kinetic energy before and after the person pulls his arms into his body.

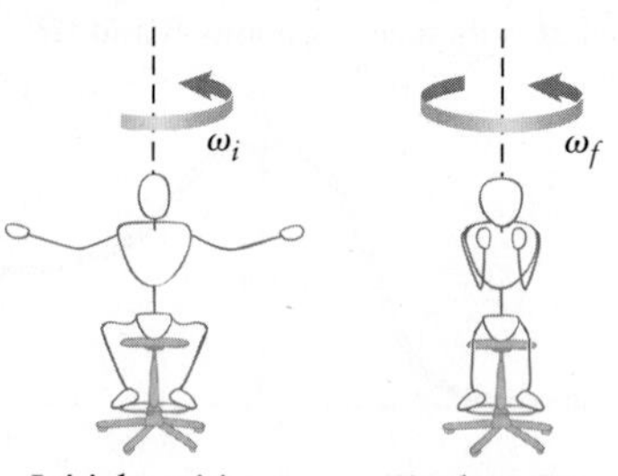

Figure P9.43 Problems 43 and 44.

44. ✶ Ⓡ Repeat Problem 43, but now assume the person is holding a 5.0-kg mass in each hand.

45. ✶ RT Estimate the angular momentum of an airplane propeller that is spinning at 2000 rpm.

46. Two disks are located on an axle as shown in Figure P9.46. The lower disk is initially spinning at 50 rad/s, and the upper one is not spinning. The upper disk then falls onto the lower disk, and they stick together. (a) Is there a torque on the lower disk? On the upper disk? (b) Identify a system for which the external torque is zero so that the angular momentum is conserved. (c) What is the initial angular momentum of your system? (d) What is the final angular momentum of your system? Express your answer in terms of ω_f, the final angular velocity of the two disks. (e) Find ω_f.

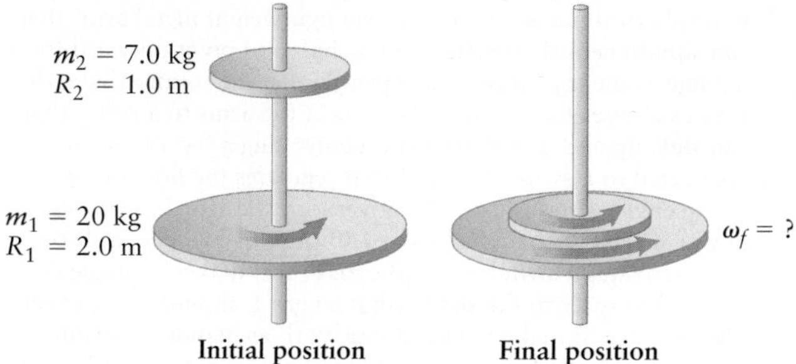

Figure P9.46

47. A particle of mass 3.0 kg moves with a horizontal velocity of 20 m/s as shown in Figure P9.47. (a) What is the angular momentum of the particle about an axis that runs through point P and is directed out of the page? (b) What is L about an axis that runs through Q? (c) Explain why the answers to parts (a) and (b) are the same (or different).

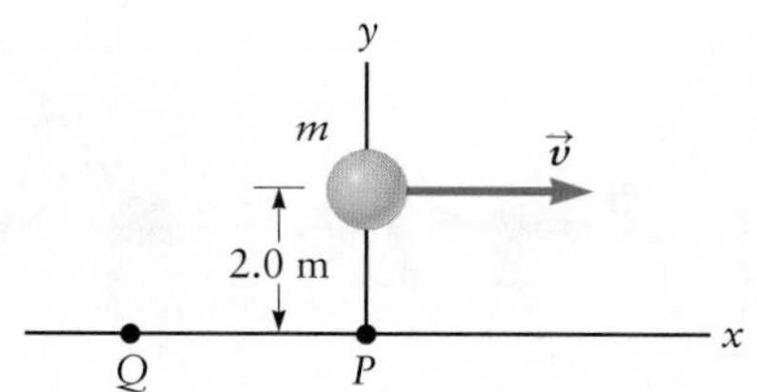

Figure P9.47 Problems 47 and 53.

48. SSM ★ A meteor with a volume of 1.0 km^3 strikes the Earth at the equator as shown in Figure P9.48, and all the fragments stick to the surface. (a) What is the magnitude of the change in the angular momentum of the Earth? (b) What is the change in the length of the day? *Hint*: Assume the average density of the meteor is 1.0×10^4 kg/m^3.

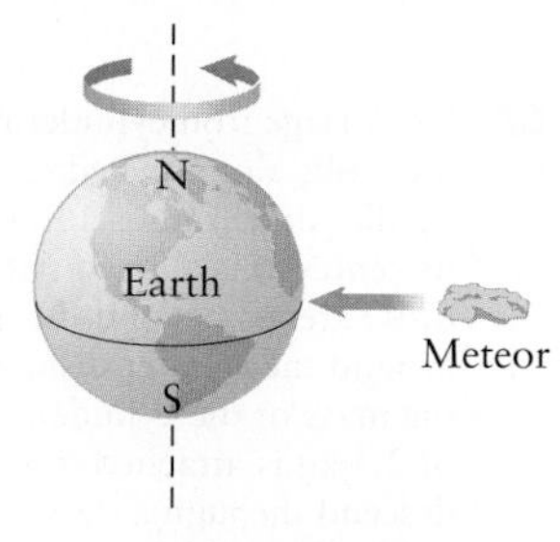

Figure P9.48

49. A DVD (radius 6.0 cm) is spinning freely with an angular velocity of 1200 rpm when a bug drops onto and sticks to the DVD a distance 4.5 cm from the center. If the DVD slows to 800 rpm, what is the ratio of the bug's mass to the DVD's mass? (Ignore the effect of the hole in the center of the DVD.)

9.4 ANGULAR MOMENTUM AND KEPLER'S SECOND LAW OF PLANETARY MOTION

50. SSM Halley's comet moves about the Sun in a highly elliptical orbit. At its closest approach, it is a distance of 8.9×10^{10} m from the Sun and has a speed of 54 km/s. When it is farthest from the Sun, the two are separated by 5.3×10^{12} m. Find the comet's speed at that point in its orbit.

51. RT Estimate the angular momentum associated with the Moon's orbital motion around the Earth.

52. What is the ratio of the angular momentum of the Moon due to its orbital motion around the Earth to the Earth's orbital angular momentum around the Sun?

9.5 THE VECTOR NATURE OF ROTATIONAL MOTION: GYROSCOPES

53. For the particle in Problem 47, what is the direction of the angular velocity of the particle?

54. Use the right-hand rule to find the direction of the orbital angular momentum of the Earth as it orbits the Sun as sketched in Figure 9.19.

55. Consider the rotational motion of the Earth as it spins about the axis that runs from the North Pole to the South Pole. Find the magnitude and direction of the angular momentum $\vec{L}$ associated with the spinning Earth. Assume the Earth is a sphere of uniform density.

56. SSM ★ A ball of mass 3.0 kg is tied to a string of length 2.0 m. The other end of the string is fastened to a ceiling, and the ball is set into circular motion as shown in Figure P9.56. If $\theta = 30°$, what is the magnitude of the angular momentum of the ball with respect to the vertical axis that passes through the center of the circle (shown dashed in the figure)?

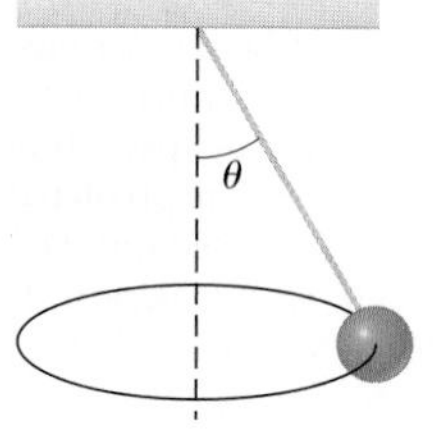

Figure P9.56

57. Consider the rotational motion of the wheels of a car when the car is in motion. Use a drawing to show the direction of the angular momentum of each wheel.

58. Consider the yo-yo in Figure 9.8. What is the direction of the angular momentum vector of the yo-yo?

9.6 CATS AND OTHER ROTATING OBJECTS

59. SSM ✪ A child of mass 35 kg stands at the edge of a merry-go-round of mass 140 kg and radius 2.5 m, and both are initially at rest. The child then walks along the edge of the merry-go-round until she reaches a point opposite her starting point as measured on the ground (point B in Fig. P9.59). How far does the child walk as measured relative to the merry-go-round?

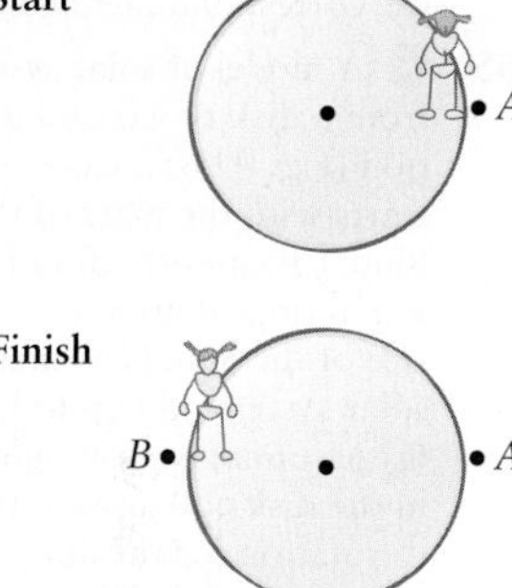

Figure P9.59

60. ★ A boy of mass m_{boy} stands at the edge of a merry-go-round of radius $R = 3.5$ m and mass m_{mgr}, and both are initially at rest. The boy then walks along the edge of the merry-go-round. After walking a distance of 21 m relative to the merry-go-round, the boy finds that the merry-go-round has rotated through an angle of 20°. Find the ratio of m_{boy} to m_{mgr}.

61. ★ ✖ RT Consider the rotating diver shown in Figure P9.61. Suppose she has an angular speed of 60 rpm = 6.3 rad/s when she has her arms and legs pulled close to her body as shown (called the tuck position). Estimate her angular speed just before she enters the water, when her arms and legs are fully extended.

Figure P9.61

Additional Problems

62. The flywheel of an automobile engine is a solid disk with a mass of 8.7 kg and a radius of 22 cm and is rotating at 4000 rpm. If the entire rotational kinetic energy of the flywheel were converted to gravitational potential energy, to what height would the flywheel be lifted? Assume the flywheel is a uniform disk.

63. **Good approximation!** Often when we apply equations to calculate moment of inertia, we are making a useful approximation, such as when we approximate a wagon wheel as a disk without a hole in the center. A manufactured item exists that is extraordinarily spherical, where the approximation of a uniform sphere is astonishingly close. Gravity Probe B is a satellite-borne experiment that uses ultraprecise gyroscopes to detect the warping of space–time due to the rotation of the Earth's gravitational field as predicted by Einstein. Each gyroscope uses a sphere made of fused quartz (density 2.2 g/cm^3), 38 mm in diameter, which is so spherical that were it scaled to the size of the Earth, the difference between hills and valleys would be less than 4 m! The sphere of a gyroscope spins at 10,000 rpm about an axis through its center. (a) What is the kinetic energy of the sphere? (b) What is the angular momentum of the sphere?

64. SSM **Better approximation!** There exists an item in nature that is extraordinarily spherical, where the approximation of a uniform sphere is astonishingly close. A pulsar is a rotating neutron star, the remnant of a supernova of a star between six and eight solar masses. A typical pulsar rotates at a rate of 600 rpm, has an average mass of twice that of the Sun, and is only 20 km in diameter, or about as big across as Washington, D.C. (a) What is the rotational kinetic energy of an average pulsar? (b) What is the corresponding angular momentum?

65. A model of solar system formation is that the Sun condensed from a disk of gas and dust through mutual gravitational attraction (Fig. P9.65). Consider a uniform disk of gas and dust that starts with the mass of the Sun and has a radius equal to that of Pluto's mean orbital radius. The theory is that this disk had an initial tiny rotation ($T = 250{,}000$ years) and collapsed to the size of the Sun. (The mass of all planets and other objects in the solar system add up to less than 0.2% of the mass of the Sun.) (a) Assuming we can ignore the mass of the planets (all the mass in the disk collapses to the solid spherical Sun), what period of rotation for the Sun is predicted by conservation of angular momentum? (b) This model turns out to be faulty; the Sun rotates about once in every 27 days. Repeat your conservation of angular momentum calculation, this time including the angular momentum of Jupiter and Saturn at their appropriate orbital radii. What is the new predicted period of rotation for the Sun? (c) Explain why planets with such (relatively) small masses can have angular momenta that are significant fractions of the angular momentum of the Sun.

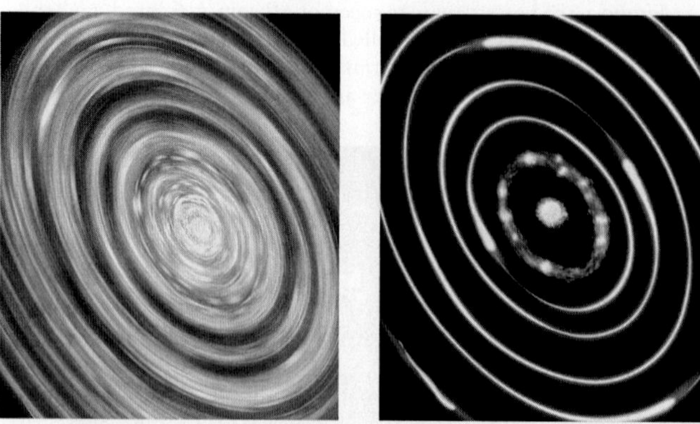

Figure P9.65

66. The mechanical governor shown in Figure P9.66 was invented by James Watt (Chapter 6) to regulate and limit, through feedback, a steam engine's maximum speed. The device consists of two spherical masses connected via lightweight metal arms that can simultaneously rotate about a shaft and pivot outward from a hinge at the top, allowing, depending on the rotation rate, the masses to rise. Two more rods connect the arms to a collar that can slide up and down the shaft, controlling a lever that can be connected to a switch or valve that regulates the flow of fuel to the engine. In this case, the governor is designed to cut off power when the shaft rotates at 150 rpm (or higher). (a) If the lever activates a switch when the mass arm makes an angle $\phi = 75°$ with respect to the shaft, what length L should the connecting rod have to make the governor work as planned? Assume the connecting rods are made of very lightweight material. (b) If each mass is 250 g, approximately how much rotational kinetic energy does the governor have at the time of cutoff? (c) What is the angular momentum of the governor at this time? How would using spheres of larger mass affect the governor's response time to changes in rotational speed?

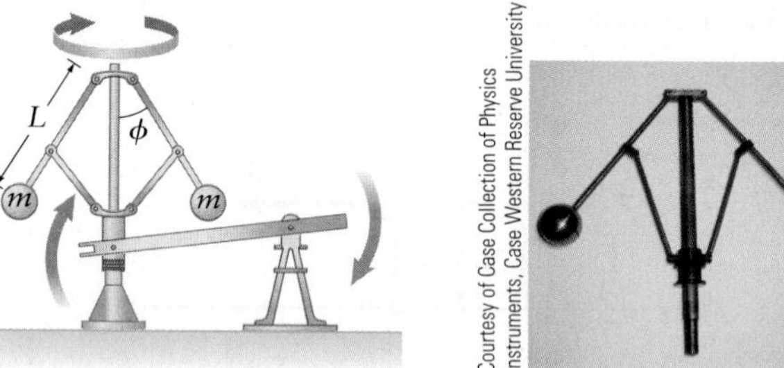

Figure P9.66

67. A large iron cylinder of mass 270 kg rests so that its circular face rolls along the edge of a table as shown in Figure P9.67. A smaller disk (radius 11 cm) is fastened to the circular face with its center axis aligned with the large cylinder such that the small disk extends beyond the edge of the table. A cable is wrapped around the smaller disk, which has half the radius and one-fourth the mass of the cylinder. With the entire assembly at rest, a mass of 2.3 kg is attached to the free end of the cable and allowed to descend through a distance $h = 35$ cm. (a) Calculate the moment of inertia and the total mass of the entire assembly. (b) What

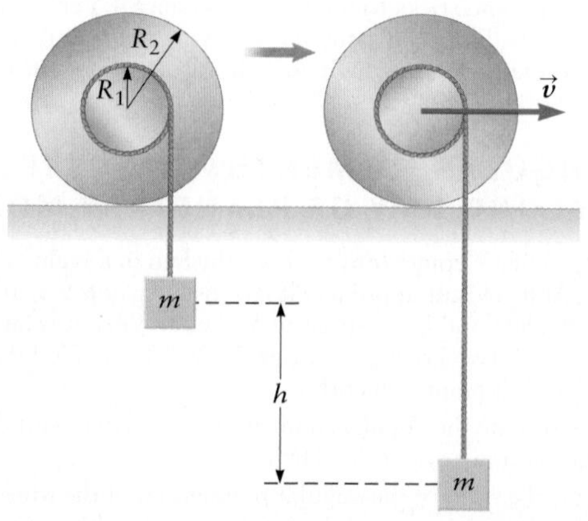

Figure P9.67

is the total kinetic energy of the system after the attached mass descends through a height h? (c) If the entire assembly rolls without slipping, what is its velocity at this time? (d) How far has the system moved along the horizontal?

68. Skee Ball amusement. The popular arcade sport of Skee Ball involves bowling a wooden ball 3 in. (7.6 cm) in diameter down an alley 9 ft long, where the ball is then launched off a short ramp as shown in Figure P9.68. Highest points are awarded to those who can launch the ball such that it lands in the topmost cylindrical hole; landing in lower and larger holes is worth fewer points. The launch ramp is 8.5 cm high and slopes at an angle $\theta = 40°$. From the top of the launch ramp, the location of the score cups for the 30-point hole and 50-point hole are $(x_{30}, y_{30}) = (25\text{ cm}, 6.0\text{ cm})$ and $(x_{50}, y_{50}) = (42\text{ cm}, 12\text{ cm})$, respectively, as shown in Figure P9.68. (a) If the ball rolls without slipping, what initial horizontal velocity must the ball have to make it into the 30-point hole? (b) An expert makes a 50-point score with one ball. What were the initial linear and angular velocities of the ball? (c) Is it likely that the ball actually does not slip while it rolls?

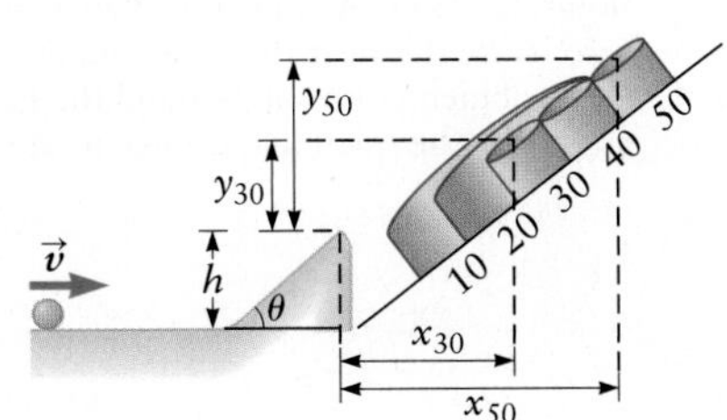

Figure P9.68

69. Extinction. Sixty-seven million years ago, a large asteroid impacted the Earth. It is widely accepted in the field of paleontology that this impact played a large role in a mass extinction that included all dinosaurs and 60% of the species living at the time. Such collisions could also alter fundamental aspects of the Earth's orbit and rotation. Consider an impact of a similar-sized asteroid of mass 8.9×10^{15} kg and a velocity with respect to the Earth's surface of 90,000 km/h. If the asteroid were to strike the Earth at the equator as shown in Figure P9.69, (a) what is the change in the angular velocity of the Earth? (b) What is the maximum possible change in the length of a day?

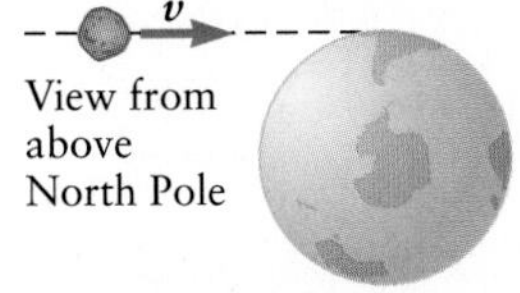

Figure P9.69

70. Some passenger jets have one engine attached to each wing. The core of a jet engine is a rotating turbine that spins at a rate of 9,000 rpm. (a) Estimate the angular momentum of the airplane due to the rotation of the turbines in both engines. These turbines are cylindrical, are usually made of titanium alloys, and can be modeled as a cylinder 1.0 m in diameter and 4.0 m long. Assume the components of each engine are rotating in a counterclockwise direction as seen from the front of the aircraft so that a significant amount of angular momentum is pointing along the forward direction (in the positive direction along the roll axis). (b) If the pilot executes a left turn, about which axis (***roll***, ***pitch***, or ***yaw*** as seen in Fig. P9.70) would the angular momentum of the engines cause the plane to rotate? (c) About which axis would a rotation occur if the plane flew upward in a vertical circle?

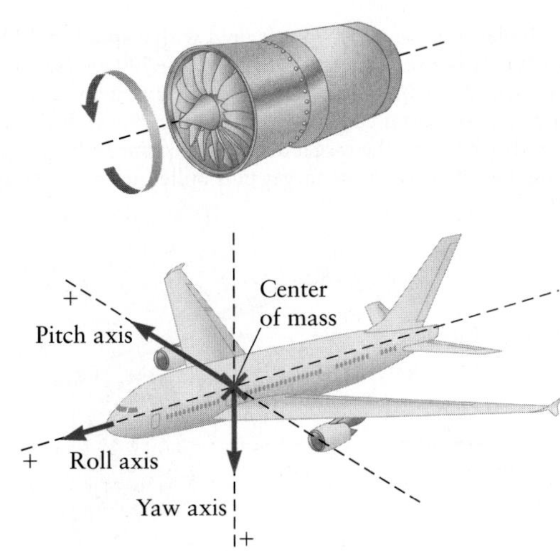

Figure P9.70

71. The hard drive in a laptop computer or some MP3 players contains one or more platters (spinning disks), each having an angular speed of typically 10,000 rpm, a radius of 4.0 cm, and a mass of about 50 g. (a) What is the kinetic energy of one platter? (b) What is the angular momentum of one platter? (c) Can you notice this angular momentum when you rotate your computer or MP3 player? Explain why or why not.

72. The spinning platter of a computer hard disk can act as a sort of gyroscope. Some early portable MP3 players contained a single hard disk, and the angular momentum of the disk could be "felt" when the MP3 player was turned or rotated. It is theoretically possible to make an MP3 player using two identical hard disks mounted in such a way that the angular momentum of the disks is not felt when the player is rotated. In which arrangement in Figure P9.72 are the spinning disks mounted in this way?

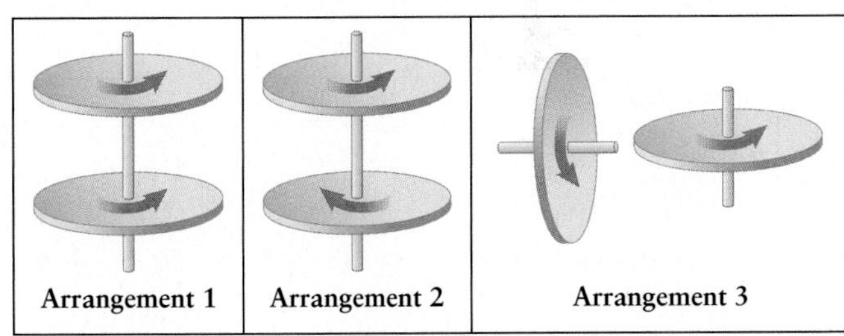

Figure P9.72

73. Estimate the moment of inertia of a basketball. *Hint?* Table 8.2 (page 264) may be helpful.

74. A mass m_1 is connected to a pulley of radius R and mass m_2 as shown in Figure P9.74. The mass m_1 starts from rest and then falls through a distance h. (a) Use conservation of energy to calculate the linear speed of m_1. (b) What is the final angular velocity of the pulley? (c) What is the final angular momentum of the pulley? (d) What was the source of the torque that produced this angular momentum?

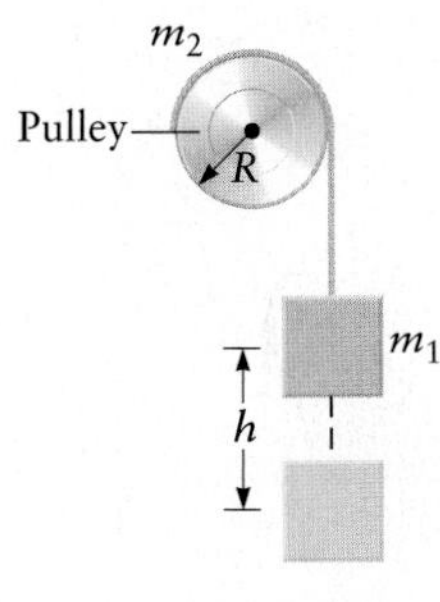

Figure P9.74

75. A bullet of mass $m_b = 25$ g is fired with a speed of 250 m/s at a target that is a sheet of metal (mass $m_t = 500$ g). The target is square with sides of length $L = 20$ cm. The target is hinged along its top edge so that it can swing freely (Fig. P9.75). If the bullet hits in the center of the target and then becomes stuck, what is the angular velocity of the target plus bullet just after the bullet hits?

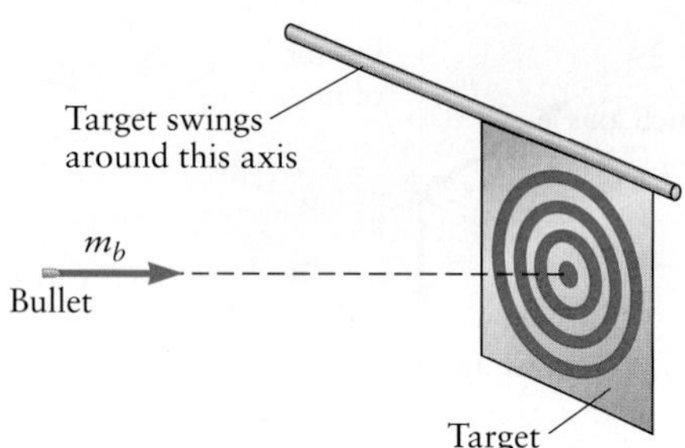

Figure P9.75

76. Ⓡ Estimate the kinetic energy of the spinning top in Figure P9.76. Assume it is made of wood and is rotating with $\omega = 30$ rad/s. *Hint*: You will need to estimate the radius of the top from the figure.

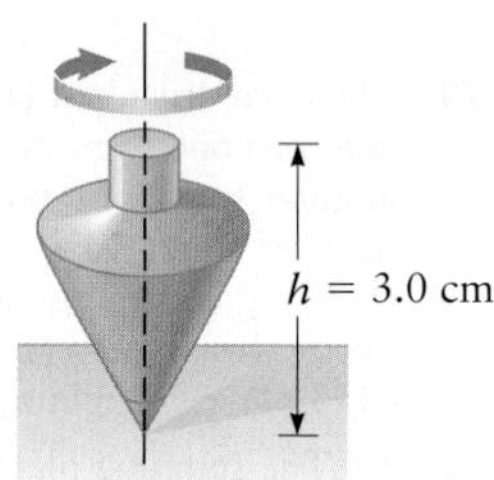

Figure P9.76

77. A figure skater (Fig. 9.11) has an initial angular velocity of 3.8 rad/s. She then pulls her arms in, and her angular velocity increases to 9.5 rad/s. What is the ratio of her final kinetic energy to her initial kinetic energy?

78. An object is rolling without slipping, with a translational kinetic energy of KE_{trans} and a rotational kinetic energy of KE_{rot}. If $KE_{trans}/KE_{rot} = 3/2$, what might this object be, (a) a hoop, (b) a spherical shell, (c) a solid cylinder, or (d) a solid sphere?

79. A solid sphere and a solid cylinder, both of mass m and radius R, are rolling without slipping with speed v. Find (a) the ratio of the angular momentum of the sphere to that of the cylinder and (b) the ratio of the total kinetic energy of the sphere to that of the cylinder.

80. **Electrical power produced by a wind turbine.** The force of the wind on the blades of a wind turbine (Fig. 9.4A) is proportional to v^2, where v is the speed of the wind (Chapter 3). The angular velocity of the turbine is also proportional to the wind speed. (a) Show that the mechanical power exerted by the wind on a turbine varies as $P \propto v^\alpha$ and find the value of α. (b) At some promising sites for wind energy, the wind speed averages about 20 mi/h (about 9 m/s). If a wind turbine at this site produces 1.0 MW with a wind speed of 20 mi/h, what power would it produce when the wind speed is 25 mi/h? Your answer should show why it is important to place wind turbines at places where the wind speed is fastest.

81. RT The ice cap at the North Pole has an area of about 6×10^6 km^2 and an average thickness of about 25 m. If all this ice melts, what will be the approximate effect on the Earth's rotation rate? *Hint*? Assume the angular momentum of the ice cap is zero (since it is near the pole) and the water from the melted ice is distributed uniformly around the Earth's surface. Express your answer as the fractional change in the rotation rate.

CHAPTER 10

◀ *In this chapter, we use Newton's laws to describe and predict the motion of fluids, such as the water in Victoria Falls or the air above it. (The Africa Image Library/Alamy)*

Fluids

In previous chapters, we learned how to apply Newton's laws to the motion of solid objects, first dealing with point particles and then with extended objects that may rotate. In this chapter, we turn our attention to the behavior of ***fluids***. A fluid may be either a liquid or a gas. We say that a fluid *flows* from one place to another, and the shape of a fluid varies according to its container. To describe the mechanics of fluids, we need several new quantities, but our analysis is still based on Newton's laws of motion. Understanding the mechanics of fluids is certainly an important topic, since most of the Earth's surface is covered with a liquid (water), and the atmosphere is a gas (air). Moreover, the motion of fluids, including the flow of blood in the body and the movement of water within plants, is essential to most living things.

OUTLINE

10.1 Pressure and Density

We want to apply Newton's second law to analyze and predict the behavior of a fluid. Recall that Newton's second law is

$$\sum \vec{F} = m\vec{a}$$

When we apply this relation to describe the motion of a particle, m is the mass of the particle, $\vec{a}$ is its acceleration, and $\sum \vec{F}$ is the total force acting on the particle. To treat the motion of a fluid, what should we use for the mass m? If we want to describe the water in a river, should m be the mass of the entire river? The idea of using a single acceleration variable $\vec{a}$ to describe the entire river does not make sense either since different parts of the river might be moving in different directions.

We must therefore rewrite Newton's second law in a form designed with fluids in mind. We need to define some new quantities to use in place of our familiar variables force and mass. The first of these new quantities is ***pressure***. The pressure in a fluid is connected with the force that the fluid exerts on a particular surface. This concept is illustrated in Figure 10.1A, which shows a submarine with a window. Seawater exerts a force on the window. This force is related to the pressure P in the seawater by

Definition of pressure

$$P = \frac{F}{A} \tag{10.1}$$

where F is the magnitude of the force exerted by the seawater on the window and A is the area of the window. This force is *perpendicular* to the window's surface. The units of pressure correspond to force divided by area, which in SI units is newtons/square meter (N/m^2). This unit is called the ***pascal*** (abbreviated Pa). In the U.S. customary system of units, pressure is measured in pounds per square inch (lb/in.^{2}). You may have encountered that unit when you last checked the pressure in an automobile or bicycle tire.

Although force is a vector quantity, pressure is not. When an object is submerged in a static fluid, the fluid exerts a force on the object that is always directed perpendicular to the object's surface. As a result, there is never any doubt about the direction of the force in Equation 10.1. This point is illustrated in Figure 10.1B, which shows a small, sealed box filled with a liquid at pressure P. The fluid within the box produces an outward force on *all* the sides of the box. If the box is small, the pressure will be approximately constant throughout the box.[1] For any particular wall of the box (area A), the total force on the wall is equal to $F = PA$.

The pressure in the Earth's atmosphere near sea level has the approximate value

Value of atmospheric pressure

$$P_{\text{atm}} = 1.01 \times 10^5 \text{ Pa} \tag{10.2}$$

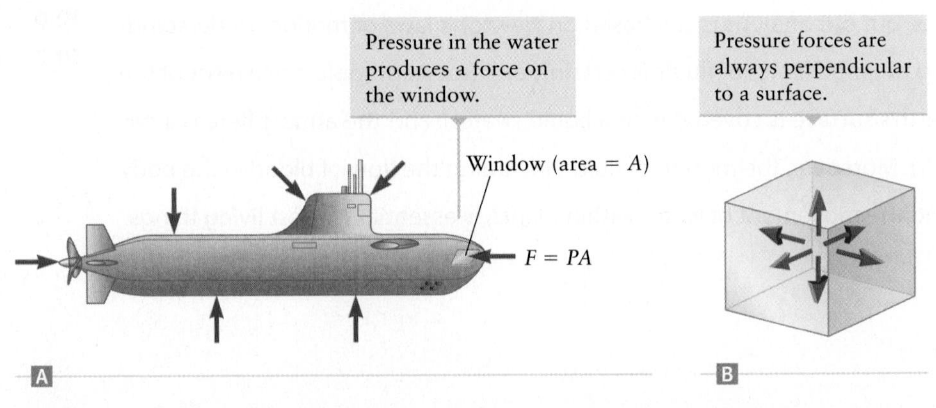

Pressure in the water produces a force on the window.
Window (area = A)
F = PA
Pressure forces are always perpendicular to a surface.
A
B

► **Figure 10.1** A The pressure P in the surrounding water produces a force on a submarine's shell. This force is everywhere directed perpendicular to the surface of the submarine. The force on a submarine window is $F = PA$, where A is the window's area. B When a container is filled with a fluid at a pressure P, the fluid produces an outward force that is perpendicular to the walls of the container.

[1]The pressure in a fluid actually varies from place to place within the container according to depth as we'll see in Section 10.2. For a small box, however, this variation will be small and P will be nearly constant.

This value of P_{atm} is approximate because the pressure in the atmosphere varies in response to changes in the weather and other factors (such as altitude).

Atmospheric Pressure

Atmospheric pressure (Eq. 10.2) corresponds to surprisingly large forces for areas you encounter every day. For example, a sheet of paper (like a page of this book) has an area of about 500 cm^2 = 0.05 m^2. Rearranging Equation 10.1, the force on one side of this sheet of paper is

$$F = PA = P_{atm}A = (1.01 \times 10^5 \text{ Pa})(0.05 \text{ m}^2)$$

$$F = 5 \times 10^3\left(\frac{\text{N}}{\text{m}^2}\right)\text{m}^2 = 5 \times 10^3 \text{ N}$$

which is a very substantial force; it is equal to the combined weight of several large people! You might at first be skeptical of this result and ask why a sheet of paper does not fold or bend under such a large force. The answer is that the air exerts a force on *both sides* of the sheet of paper as sketched in Figure 10.2. The pressure in the air on both sides is equal to P_{atm}. These forces are in opposite directions, so the total force on the paper is zero. This cancellation occurs in many situations, including the force due to air pressure on your body.

$F = P_{atm}A$

$F = P_{atm}A$

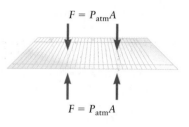

▲ **Figure 10.2** The atmosphere produces a downward force of magnitude $P_{atm}A$ on this sheet of paper, where A is the area of the sheet. The air underneath the sheet produces a force of the same magnitude directed upward.

Vacuum and the Magdeburg Experiment

The balance of forces associated with air pressure can be altered by removing air from inside an enclosed volume with a device called a vacuum pump. This was first demonstrated 400 years ago by Otto von Guericke. He constructed two metal hemispheres as shown in Figure 10.3A; when fitted together, they form a Magdeburg sphere, so named in honor of the town where von Guericke lived. One hemisphere contained a small tube connected to a valve through which air was removed from inside the two hemispheres by a pump that von Guericke invented. The region inside where the pressure was zero, or very small, is known as a ***vacuum***.

To understand von Guericke's experiment, let's analyze the somewhat simpler case of two Magdeburg plates sketched in Figure 10.3B. The air outside produces an inward force equal to $P_{atm}A$, where A is the plate's area. Because there is no air on the inside, $P_{inside} \approx 0$ and there is no outward force from air pressure. There will be a similar inward force $P_{atm}A$ on the other plate, and the net result is a substantial force holding the plates together. The force required to pull the plates apart is calculated in Example 10.1.

A

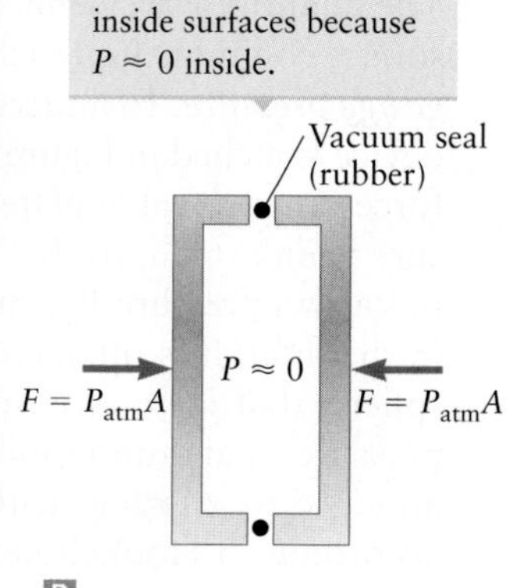

There is no force on the inside surfaces because $P \approx 0$ inside.
Vacuum seal (rubber)
$P \approx 0$

B

◀ **Figure 10.3** A Otto von Guericke demonstrated the force associated with atmospheric pressure using this apparatus. He made two tight-fitting hemispheres and then removed the air from the inside using a vacuum pump, which he invented. Von Guericke employed hemispheres about 1 m in diameter and used horses to (attempt to) pull them apart! B A similar von Guericke apparatus, using two flat plates in place of the hemispheres.

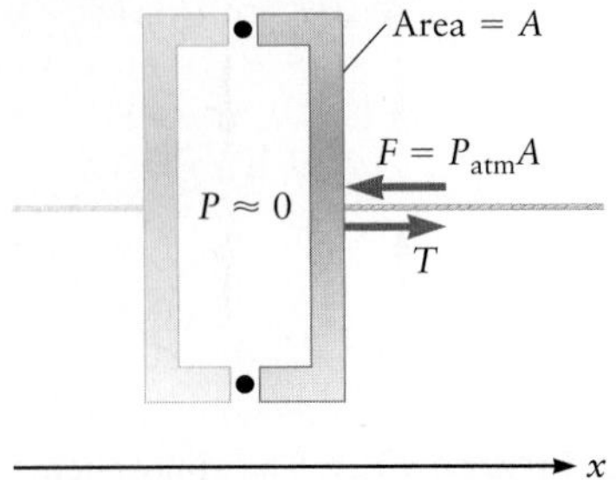

▲ **Figure 10.4** Example 10.1.

EXAMPLE 10.1 Force in the Magdeburg Demonstration

Suppose the Magdeburg plates in Figure 10.3B have an area of $A = 1.0 \text{ m}^2$. What force is required to pull the plates apart?

RECOGNIZE THE PRINCIPLE

This problem involves an application of the relation between pressure, area, and force. The force due to air pressure on one plate is $F = P_{atm}A$ (Eq. 10.1).

SKETCH THE PROBLEM

Figure 10.4 shows the horizontal forces acting on one of the plates.

IDENTIFY THE RELATIONSHIPS

A cable with tension T is attached to the Magdeburg plate on the right in Figure 10.4. There are two forces: a force $P_{atm}A$ to the left due to atmospheric pressure and a force T to the right. (We can ignore the weight of the plate because it will be much smaller than the other two forces.) If the plates are just barely being pulled apart, we can take the acceleration of the plate on the right to be zero.

SOLVE

Applying Newton's second law gives

$$\sum F = +T - P_{atm}A = ma = 0$$

where these forces are along the horizontal. This leads to

$$T = P_{atm}A = (1.01 \times 10^5 \text{ Pa})(1.0 \text{ m}^2) = \boxed{1.0 \times 10^5 \text{ N}}$$

▶ ***What does it mean?***

This force is the combined weight of about 10 cars! It's no wonder the painting in Figure 10.3A shows a sizable team of horses pulling on the Magdeburg plates.

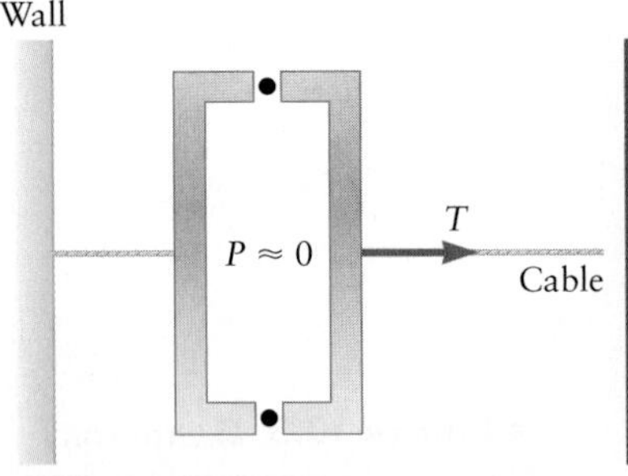

▲ **Figure 10.5** Concept Check 10.1.

CONCEPT CHECK 10.1 Magdeburg Plates and Newton's Third Law

When we calculated the force required to pull apart the Magdeburg plates in Example 10.1, we actually calculated the force exerted by the atmosphere on *one* of the plates and found that $F = 1.0 \times 10^5$ N. Suppose the other plate is attached to a rigid wall as shown in Figure 10.5. Now what tension T is required in the cable on the right to pull apart the plates?

(a) 1.0×10^5 N (b) 2.0×10^5 N (c) 0

Gauge Pressure versus Absolute Pressure

The quantity P we defined in Equation 10.1 is often referred to as the ***absolute*** pressure. A common approach to measuring pressure leads to a related quantity called the ***gauge pressure***. To understand the difference, consider the simple pressure-measuring device sketched in Figure 10.6. This device determines the pressure by measuring the force on a movable plate. One side of this plate is exposed to the fluid whose pressure is unknown, so the force on this side of the plate is equal to the product of the unknown pressure P_{abs} and the plate's area. Notice that P_{abs} is the *absolute* pressure in the fluid. It is often convenient to expose the other side of the plate to the atmosphere, that is, to atmospheric pressure. For example, a gauge used to measure the pressure in an automobile tire is constructed in this way. The movable plate is also attached to a spring (with spring constant k) exerting a force $-kx_{spring}$ on the plate, according to Hooke's law (Chapter 6). Here x_{spring} is the displacement of the plate from a reference position where the spring is unstretched and uncompressed. The

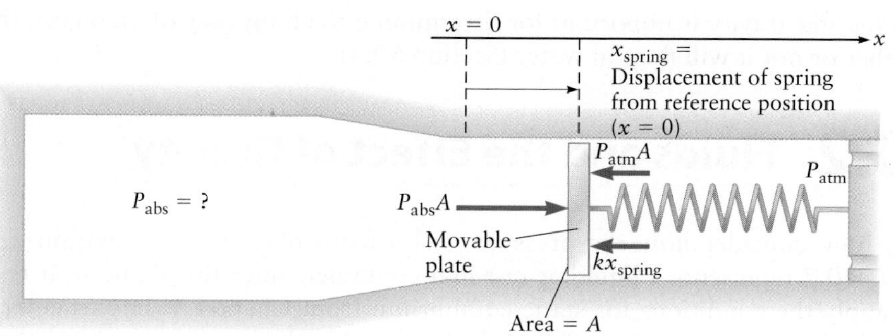

Figure 10.6 Simple design for a pressure gauge. The unknown pressure P_{abs} on the left leads to a force $F = P_{abs}A$ on a movable plate inside the gauge. The other side of the plate is attached to a spring and is exposed to the atmosphere. The deflection of the spring (x) is proportional to the difference in the pressures $P_{abs} - P_{atm}$. This difference is the gauge pressure.

total force on the plate inside the gauge is due to the pressures on both sides plus the force exerted by the spring. If we take the $+x$ direction to be to the right in the figure, we have

$$\sum F = +P_{abs}A - P_{atm}A - kx_{spring}$$

If the gauge plate is in equilibrium, the total force must be zero and we get

$$P_{abs}A - P_{atm}A - kx_{spring} = 0$$

$$x_{spring} = (P_{abs} - P_{atm})\left(\frac{A}{k}\right)$$

The deflection of the gauge x_{spring} is thus proportional to the *difference* between the pressure one wants to measure (P_{abs}) and atmospheric pressure. This difference is called the gauge pressure, $P_{gauge} = P_{abs} - P_{atm}$. Hence, the absolute pressure is related to the gauge pressure by

Gauge pressure

$$P_{abs} = P_{gauge} + P_{atm}$$

Many common pressure gauges work this way and thus measure the gauge pressure. For the remainder of our work in this chapter, the term *pressure* means the absolute pressure unless we specifically note otherwise.

CONCEPT CHECK 10.2 Pressure in an Automobile Tire

The pressure in an automobile tire is usually quoted (in the United States) in units of pounds per square inch (lb/in.^{2}), and in a typical tire the gauge pressure is $P_{gauge} = 25$ lb/in.^{2}. What is the value of the gauge pressure in pascals?

(a) 2.7×10^3 Pa (b) 1.0×10^5 Pa (c) 1.7×10^5 Pa (d) 110 Pa

Density

Another quantity we need to describe fluids is the ***density***, denoted by ρ (lowercase Greek rho). The density of a substance is equal to its mass per unit volume,

$$\rho = \frac{M}{V} \tag{10.3}$$

where M is the mass of a sample of the substance and V is the sample volume. This definition of ρ assumes the density is constant throughout the volume V.

The density of a substance can vary according to temperature and may also depend on pressure. However, for most liquids—including water and oil—density is approximately independent of pressure. Such fluids are called ***incompressible***. The densities of some common substances are listed in Table 10.1.

Although most liquids are incompressible, that is not the case for gases. The density of a gas depends strongly on pressure; that is, gases are ***compressible***, and we'll show in Section 10.3 how that affects their behavior. The densities of some common gases at atmospheric pressure are listed in Table 10.1.

The SI unit of density is the kilogram per cubic meter, or kg/m^3 (= mass/volume). Another way to express the density of a substance is in terms of its ***specific gravity***, which is the ratio of the density of the substance to the density of water. We'll see

TABLE 10.1 Densities of Some Common Solids, Liquids, and Gases[a]

Substance	Density (kg/m^3)
SOLIDS	
Ice (at 0°C)	917
Aluminum	2,700
Lead	11,300
Platinum	21,500
Gold	19,300
Tungsten	19,300
Steel	7,800
Concrete	2,000
Bone	1,500–2,000
Polystyrene	100
Glass	2,500
Wood (spruce)	400
Wood (balsa wood)	120
Wood (oak)	750
LIQUIDS	
Water (0°C)	999.8
Water (4°C)	1,000.0
Water (20°C)	999.9
Seawater	1,025
Ethanol	790
Mercury	13,600
Oil	700–900
GASES	
Air (0°C)	1.29
Air (20°C)	1.20
Air (100°C)	0.84
Helium (0°C)	0.18
Hydrogen (0°C)	0.090
Carbon dioxide (25°C)	1.80

[a]Unless otherwise noted, all values are for room temperature and pressure.

that specific gravity is important for determining the buoyancy of an object, that is, whether or not it will float in water (Section 10.4).

10.2 Fluids and the Effect of Gravity

Let's now consider how the pressure varies from place to place within a fluid. Figure 10.7 represents a liquid at rest in a container. Since this liquid is at rest, we can apply the conditions for static equilibrium from Chapter 4. In particular, consider the forces acting on a region of the fluid within the imaginary cubical "box" in Figure 10.7A. The system is at rest, so we can treat the fluid cube as a "simple" mass. For this cubical box, each wall (or face) has edge length h and area $A = h^2$.

We now apply our condition for translational equilibrium to the mass of fluid within the box. Because this mass is at rest, the total force on it must be zero. There are contributions to the total force from the pressure outside the box. These forces are all directed inward on the walls of the box as indicated in Figure 10.7. The forces along the horizontal directions are equal in magnitude and opposite in direction, so they cancel. However, the forces on the top and bottom of the box do not cancel. Although these forces are in opposite directions, they cannot be equal in magnitude because the pressure force at the bottom must support both the pressure force at the top and the weight of the fluid in the box. Hence, the pressure at the bottom must be greater than the pressure at the top. In general, *pressure always increases as one goes deeper into a fluid.*

To calculate how pressure depends on depth, note that the force on the bottom of the box is upward with $F_{bot} = P_{bot}A$, where P_{bot} is the pressure at the bottom face of the box. The force on the top of the box is $F_{top} = -P_{top}A$, where the negative sign indicates that this force is downward. There is also a gravitational force on the fluid mass due to its weight. This force is along the vertical with $F_{grav} = -mg$, where m is the fluid mass inside the box. The total force along y is then

$$\sum F_y = P_{bot}A - P_{top}A - mg = 0 \tag{10.4}$$

This total force is zero because the system is in translational equilibrium. From the definition of density (Eq. 10.3), the mass of the fluid in the box is $m = \rho V$, and the volume of the box is equal to the area A times the box height h, so $V = Ah = h^3$ and

$$m = \rho V = \rho A h = \rho h^3$$

Inserting this into Equation 10.4 leads to

$$P_{bot}A - P_{top}A - mg = P_{bot}h^2 - P_{top}h^2 - \rho h^3 g = 0$$

$$P_{bot} - P_{top} = \rho g h$$

We can also write this result as

$$P = P_0 + \rho g h \tag{10.5}$$

Here P_0 is the pressure at a "reference point" in the fluid and P is the pressure at a depth h relative to the reference point as sketched in Figure 10.8. We have derived Equation 10.5 with the use of a cubical box, but this result is not limited to fluid "boxes." In words, pressure increases as one goes to greater depths because a particular section of fluid must support the weight of the fluid above it.

When a fluid has a surface open to the atmosphere, it is usually convenient to take the reference point at this surface. In this case, $P_0 = P_{atm}$ as shown in Figure 10.8B. No matter where the reference point is chosen (Fig. 10.8A or B), a positive value of h in Equation 10.5 corresponds to a location that is *deeper* than the reference location.

Equation 10.5 is a general relation that tells how the pressure varies with depth in an *incompressible* fluid since our derivation assumed the density of the fluid is constant; that is, ρ does not vary with depth in the fluid. Equation 10.5 can therefore be applied only to liquids, not to gases. We'll see how to deal with compressible fluids (gases) after we consider a few examples involving liquids.

◀ **Figure 10.7** A An imaginary cubical box enclosing a region of fluid. Pressure from the fluid pushes on each face of the box. The pressure forces on the sides of the box (magnitude P_sA) cancel. B Free-body diagram for the box.

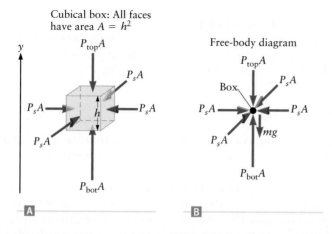

EXAMPLE 10.2 Pressure in the Marianas Trench

The Marianas Trench, a region at the bottom of the Pacific Ocean, is believed to be the deepest spot on the ocean floor. In 1960, a U.S. Navy submersible vehicle went to a spot in this trench that is 10,700 m below sea level. Suppose this submarine had a tiny window of area 1.0 cm^2 so that sailors could enjoy the view. Find the force on this window due to the water pressure.

RECOGNIZE THE PRINCIPLE

The force on the window depends on the pressure through $F = PA$. We thus need to find the pressure in the Marianas Trench, which we can do using the dependence of pressure on depth (Eq. 10.5).

SKETCH THE PROBLEM

Figure 10.8B describes the problem, with the submarine at point B.

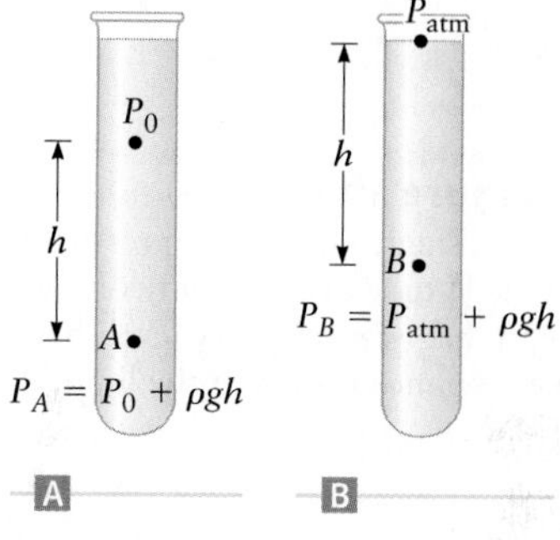

▲ **Figure 10.8** Illustration of the relation $P = P_0 + \rho gh$. In A, the reference point is at a particular point inside the liquid, whereas in B, the reference point is at the surface in contact with the atmosphere, so $P_0 = P_{atm}$. In both cases, h is the depth relative to the reference point.

IDENTIFY THE RELATIONSHIPS

We take the ocean surface as the reference point for calculating the pressure using Equation 10.5. Therefore, $P_0 = P_{atm}$, the depth is $h = +10{,}700$ m, and the density of seawater is given in Table 10.1.

SOLVE

From Equation 10.5 and using $P_0 = P_{atm}$, we find the pressure to be

$$P = P_{atm} + \rho gh = (1.01 \times 10^5 \text{ Pa}) + (1025 \text{ kg/m}^3)(9.8 \text{ m/s}^2)(10{,}700 \text{ m})$$
$$P = 1.1 \times 10^8 \text{ Pa}$$

which is about 1000 times atmospheric pressure. The force on the submarine's window is

$$F = PA = (1.1 \times 10^8 \text{ Pa})(1.0 \text{ cm}^2) = (1.1 \times 10^8 \text{ N/m}^2)(1.0 \times 10^{-4} \text{ m}^2)$$
$$F = \boxed{1.1 \times 10^4 \text{ N}}$$

▶ ***What does it mean?***

This force is approximately the weight of a small automobile. The result explains why submarines do not have large windows.

CONCEPT CHECK 10.3 Pressure at the Bottom of a Swimming Pool

The pressure increases as one goes deeper within a liquid, and in Example 10.2 we found that the pressure increase can be quite large in deep locations in the oceans. This pressure change can also be quite noticeable in a swimming pool.

A and C are at same depth so $P_A = P_C$.

B and D are at same depth so $P_B = P_D$.

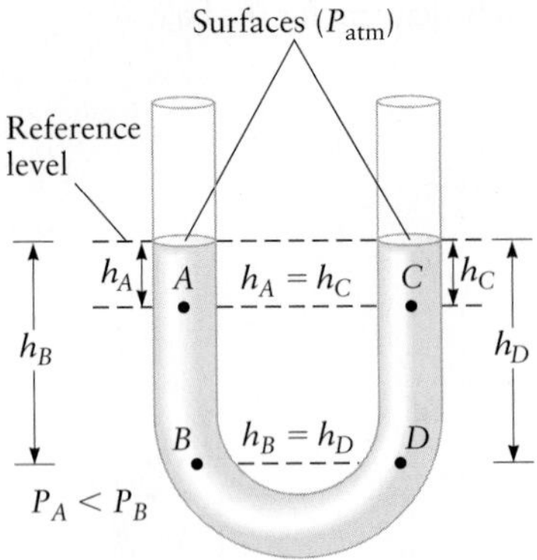

▲ **Figure 10.9** This U-tube contains a single liquid. The pressure in the U-tube depends only on the depth below a reference level; here the reference level is at the liquid's surface.

When you dive to the bottom of a pool, you can sense the pressure increase with your ears. For a pool with a depth of 3.0 m, what is the difference between the pressure at the bottom of the pool and atmospheric pressure?
(a) 2.9×10^4 Pa (b) 3.0×10^5 Pa (c) 3.0×10^3 Pa

Pressure in a U-Tube

Figure 10.9 shows a liquid contained in a glass tube that has a "U" shape; this container is called, not surprisingly, a U-tube, and is very handy for a variety of fluid experiments.

In a U-tube, different regions of the fluid can be widely separated in space. However, this separation does not affect our general expression for how the pressure varies with depth, Equation 10.5. According to that result, the pressure in the liquid in Figure 10.9 depends only on the depth h below a reference level. The pressures are the *same* at two spots on different sides of the U-tube that are at the *same* depth below the surface; hence, $P_A = P_C$ and $P_B = P_D$ in Figure 10.9. This result is true as long as the U-tube is filled with only one type of liquid. Example 10.3 shows how to analyze a U-tube containing two different liquids.

EXAMPLE 10.3 Two Liquids in a U-Tube

The U-tube in Figure 10.10 has both ends open to the atmosphere. The tube is filled with two liquids of different densities. These liquids are immiscible; that is, they do not mix. If liquid 1 has a higher density than liquid 2, the level of liquid 2 is higher than the level of liquid 1 by some distance d. Suppose liquid 1 is water ($\rho_1 = 1000$ kg/m^3); liquid 2 is a type of oil with $\rho_2 = 700$ kg/m^3 and is filled to a height $y_R = 0.10$ m above the oil-water boundary (Fig. 10.10). Find d.

RECOGNIZE THE PRINCIPLE

For the U-tube containing a single type of fluid in Figure 10.9, pressure depends only on depth, so at any particular depth below the surface, the pressure on the two sides of the tube is the same. This result is true only when the U-tube is filled with a single liquid so that the density is constant throughout, a key assumption in the derivation of Equation 10.5. In Figure 10.10, the density changes as we pass from one liquid to the other, so we *cannot* use Equation 10.5 to compare the pressures in the different liquids. We can, however, use Equation 10.5 to compare the pressures at different points within the same connected piece of liquid. Consider the pressure at point B in Figure 10.10. This point is inside liquid 1 and is just below the boundary that separates the two liquids. Because we consider point B to be at the same depth as point A (also located in liquid 1, but on the other side of the U-tube), the pressures at points A and B are the same.

We next consider point C; this point is in liquid 2, just above the boundary between the liquids. Because points B and C are extremely close together, the pressures at these two locations must be the same. You can see how that is true by imagining a thin (massless) plastic sheet stretched along the boundary between liquids 1 and 2. Because the system is at rest, this sheet is in translational equilibrium; hence, the total force on the sheet must be zero. That can only be true if the pressure is the same on the top and bottom surfaces of the sheet. We have thus concluded that the pressure is the same at points A, B, and C; that is, $P_A = P_B = P_C$. This condition will allow us to find the value of d.

SKETCH THE PROBLEM

Figure 10.10 describes the problem.

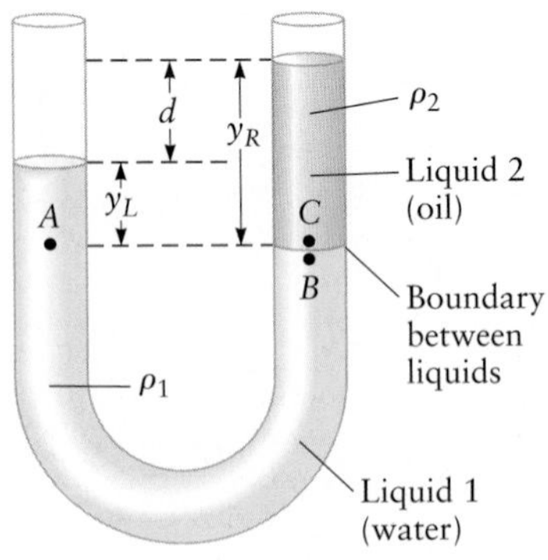

▲ **Figure 10.10** Example 10.3. When a U-tube contains more than one type of liquid, one must apply Equation 10.5 separately within each liquid region. Points B and C are on opposite sides of the boundary between the fluids. These points are extremely close together and are at the same height as point A.

IDENTIFY THE RELATIONSHIPS

In liquid 1, the pressure at the surface is atmospheric pressure since this surface is in contact with the atmosphere; let's take this surface as our reference point in applying Equation 10.5. If y_L is the distance from the point A to the surface, we get

$$P_A = P_{atm} + \rho_1 g y_L$$

Likewise, in liquid 2,

$$P_C = P_{atm} + \rho_2 g y_R$$

where y_R is the distance from point C to the surface of liquid 2. Combining these results and using $P_A = P_C$ gives

$$P_A = P_{atm} + \rho_1 g y_L = P_C = P_{atm} + \rho_2 g y_R$$

$$\rho_1 g y_L = \rho_2 g y_R$$

$$y_L = y_R \frac{\rho_2}{\rho_1}$$

The difference in heights of the two surfaces is then

$$d = y_R - y_L = y_R\left(1 - \frac{\rho_2}{\rho_1}\right) = y_R \frac{\rho_1 - \rho_2}{\rho_1}$$

SOLVE

Inserting the values of the densities of water (ρ_1) and oil (ρ_2) given above and using the given value of y_R, we find

$$d = y_R \frac{\rho_1 - \rho_2}{\rho_1} = (0.10 \text{ m})\frac{(1000 - 700)(\text{kg/m}^3)}{1000 \text{ kg/m}^3} = \boxed{0.030 \text{ m}}$$

▸ ***What does it mean?***

When two fluids are in contact with each other and are at rest, the pressure must be the same on the two sides of the interface.

Barometers, Vacuums, and Measuring Pressure

U-tubes and similar types of containers can be used in many different ways. Example 10.3 shows that when a U-tube is filled with two different (and immiscible) liquids, the open surfaces of the two liquids will sit at different heights. This result can be used to measure the density of a liquid. If the density ρ_1 of liquid 1 in Figure 10.10 is known, a measurement of d as calculated in Example 10.3 then allows us to find the value of ρ_2.

A U-tube can also be used to measure pressure differences. For example, consider the situation in Figure 10.11, which shows a U-tube whose ends are connected to two different containers of gas at pressures P_L and P_R. When partially filled with liquid, the U-tube can be used to measure the pressure difference $P_L - P_R$ by applying Equation 10.5. Taking our reference point to be the top surface of the liquid on the right side of the U-tube, we find

$$P_L = P_R + \rho g h$$

where h is the difference in the heights of the two surfaces of the liquid. Hence, the difference in pressure is

$$\Delta P = P_L - P_R = \rho g h \quad (10.6)$$

Figure 10.12 shows another way to measure pressure. Here we have a straight tube, partially filled with liquid and closed at the top end, with the open end at the bottom inserted into a dish containing the same liquid. The pressure at the top of the tube is zero (this can be done with a vacuum pump), while the open surface of the

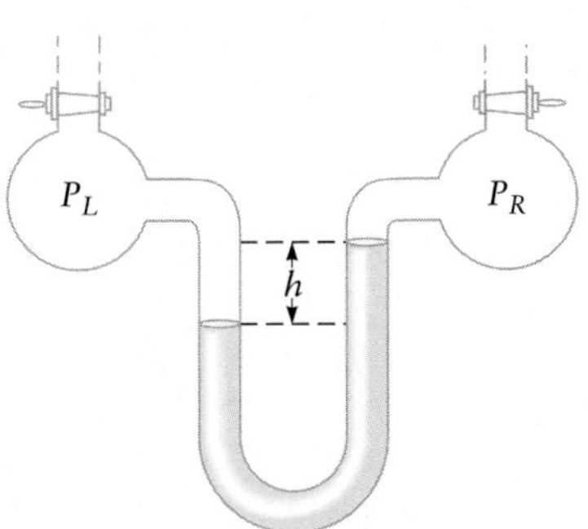

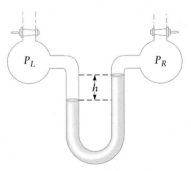

▲ **Figure 10.11** This U-tube can be used to measure the difference in pressure, $\Delta P = P_L - P_R$.

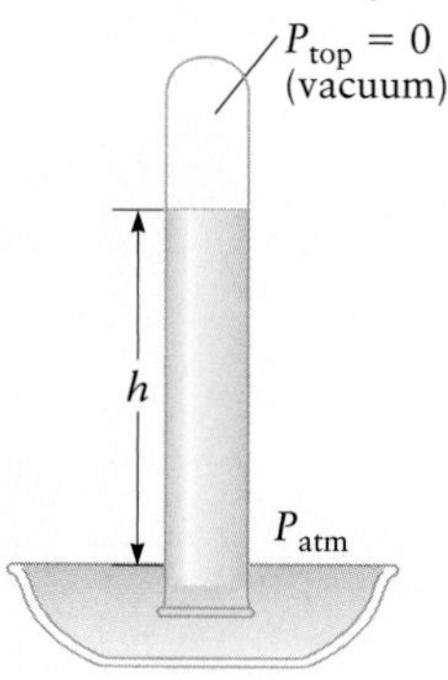

▲ **Figure 10.12** Design of a simple barometer.

liquid in the dish is exposed to the atmosphere. Applying Equation 10.5 and using the top surface of the liquid in the tube as our reference point, we have

$$P_{atm} = P_{top} + \rho gh$$

where h is the height of the liquid column in the tube. Since $P_{top} = 0$, we find

$$P_{atm} = \rho gh \tag{10.7}$$

Hence, if we measure h, we can use the device in Figure 10.12 as a ***barometer*** and measure the absolute pressure of the atmosphere.

Let's now design such a barometer and assume we are going to use water as the liquid. We want to calculate how tall the tube must be to measure the pressure in the atmosphere. We can rearrange Equation 10.7 to solve for the height of the water column h; inserting the values of atmospheric pressure and the density of water from Table 10.1 gives

$$h = \frac{P_{atm}}{\rho g} \tag{10.8}$$

$$h = \frac{1.01 \times 10^5 \text{ Pa}}{(1000 \text{ kg/m}^3)(9.8 \text{ m/s}^2)} = 10 \text{ m} \tag{10.9}$$

This barometer would require quite a long tube to measure atmospheric pressure; it would certainly not fit into a normal-sized room. That is why most barometers do not employ water. Instead, they use mercury, which has a much higher density than water, giving a smaller value of h in Figure 10.12 and Equation 10.8.

EXAMPLE 10.4 Designing a Mercury Barometer

Consider again the barometer in Figure 10.12, but now assume the liquid in the tube and dish is mercury. If the pressure to be measured is atmospheric pressure, what is h?

RECOGNIZE THE PRINCIPLE

We can use the results for pressure in an incompressible fluid and the reasoning that led to Equation 10.8, but now we replace water with mercury, so we have a different value for the density.

SKETCH THE PROBLEM

Figure 10.12 describes the problem.

IDENTIFY THE RELATIONSHIPS AND SOLVE

Using the density of mercury (which is a liquid at room temperature) from Table 10.1 along with Equation 10.8 leads to

$$h = \frac{P_{atm}}{\rho g} = \frac{1.01 \times 10^5 \text{ Pa}}{(13{,}600 \text{ kg/m}^3)(9.8 \text{ m/s}^2)} = \boxed{0.76 \text{ m}}$$

► ***What does it mean?***

This size is much more manageable than the water barometer we considered above (Eq. 10.9). Two old-fashioned units of pressure are "inches" and "millimeters of mercury." Both units refer to the value of h measured with a mercury barometer.

Units for Measuring Pressure

For most scientific work, pressure is measured in pascals, and we have seen that 1 Pa = 1 N/m^2. In the United States, we often encounter the unit pounds per square inch (lb/in.^{2}), especially when dealing with automobile tires. The pressure unit "inches" is often heard on a TV weather report. This unit originates with a barometer like the one in Example 10.4, where the pressure in the atmosphere can be determined by giving the value of h (measured in inches) for a mercury-filled barometer. Likewise, the barometer height h can be measured in millimeters, which leads to the unit "millimeters of mercury," (sometimes referred to as simply mm). Another common unit is the atmosphere (atm), which is the pressure in the Earth's atmosphere at sea level under normal weather conditions. These units are summarized in Table 10.2, which also gives the value of normal atmospheric pressure in the different units.

TABLE 10.2 Units Used for Measuring Pressure

Pressure Unit	Value of Atmospheric Pressure
Pascal	1.01×10^5 Pa
Pounds per square inch	14.7 lb/in.^{2}
Inches	29.9 in.
Millimeters of mercury	760 mm
Atmosphere	1 atm

Pumping a Liquid

Our discussion of barometers led us to Equation 10.9, showing that a water-filled barometer is not very practical. This result also has an important implication for pumping water from place to place. Figure 10.13 shows a well that brings water to the surface from deep underground. If we are going to use a pump to move the water, where is the best place to put the pump? Two possibilities are shown in Figure 10.13; in one case, the pump is located at the top of the well, and in the other, it is located underground at the bottom of the well. Placing it at the top of the well has several advantages; for example, it is simpler to fix the pump if it breaks down. There is a limitation with this location, however. If the pump is placed at the top, it must draw water up from the ground by establishing a very low pressure in the top of the pipe that carries water to the surface. This arrangement is just like the barometer in Figure 10.12, where a very low pressure (a vacuum) in the top of the tube and the atmospheric pressure at the dish act together to push water up the column. Suppose the water pump in Figure 10.13A generates a perfect vacuum at the top ($P_{top} = 0$). Since soil is porous to air, the pressure at the underground surface of the water is approximately atmospheric pressure. Following the approach we used in deriving Equation 10.8, we have, from Figure 10.13A,

$$P_{bot} = P_{top} + \rho g h$$
$$P_{atm} = 0 + \rho g h$$

which leads to

$$h = \frac{P_{atm}}{\rho g}$$

For water, then, $h = 10$ m as we found in Equation 10.9. This is the *maximum* height to which a vacuum pump can "draw" water. The design in Figure 10.13A can thus only be used if the well depth is less than 10 m.

Most water wells, however, are much deeper than 10 m. For example, the well at the author's home is approximately 50 m deep. These cases are handled by placing the pump at the bottom of the well as in Figure 10.13B. Now the pump is not a vacuum pump, but instead generates a large pressure, much greater than atmospheric pressure. We can again apply the relation for pressure in an incompressible fluid (Eq. 10.5) and get

$$P_{bot} = P_{top} + \rho g h$$

where h is again the depth, and the pressure at the top, P_{top}, is now equal to atmospheric pressure. To satisfy this relation, the pressure at the bottom of the well pipe must be larger than P_{atm}. There is no theoretical limit on the magnitude of P_{bot}; a realistic pump can generate pressures of many times atmospheric pressure, so the design in Figure 10.13B can easily pump water from typical well depths. Similar considerations arise when water must be moved or pumped to different vertical levels, such as in the water system of a tall building. (Many buildings are much taller than 10 m!)

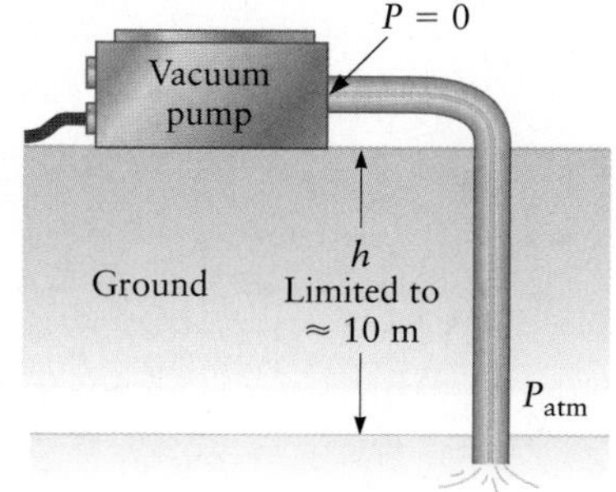

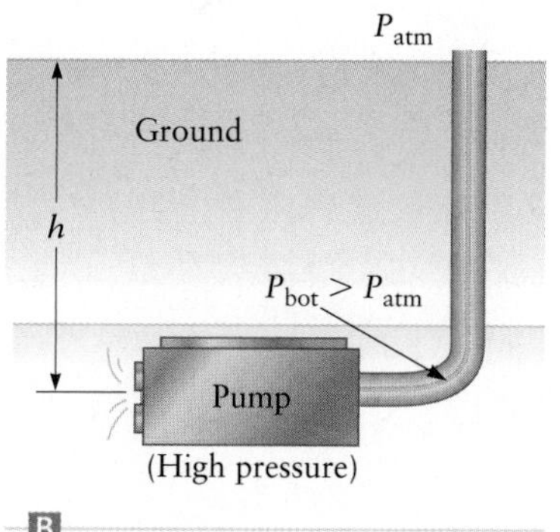

▲ **Figure 10.13** A A vacuum pump is used to pump water from a well (underground) to the surface. Since soil is porous, the pressure at the underground surface of the water is approximately P_{atm}. B Most wells use a pump placed at the bottom of the well. This pump produces a high pressure that pushes the water to the surface.

Pressure in a Compressible Fluid

Equation 10.5 tells how pressure varies with depth, but it applies only to an incompressible fluid, that is, a liquid such as water or mercury. Gases (such as air in the atmosphere) are compressible fluids and follow a different relation between pressure and depth because the density of a gas depends greatly on pressure, whereas the density of a liquid is approximately constant. The pressure in a compressible fluid still

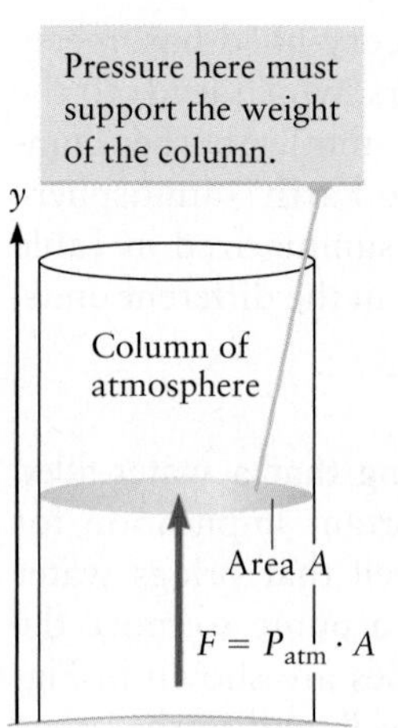

▶ **Figure 10.14** For a fluid at rest in a gravitational field, the pressure at a particular point in the fluid must support the weight of the fluid above that point. For this reason, the pressure in a fluid always increases with depth within the fluid. This result applies to both liquids and gases.

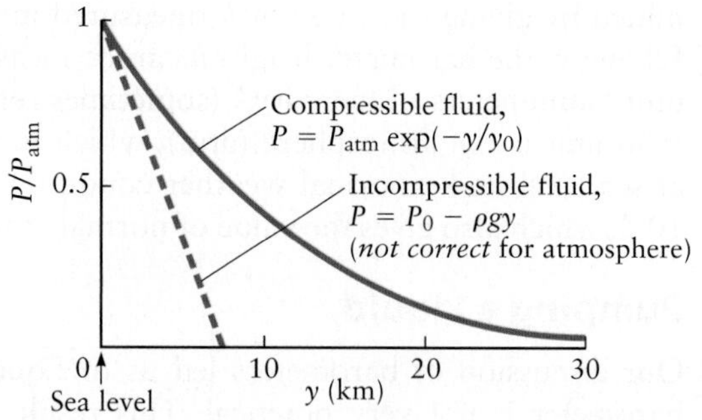

▲ **Figure 10.15** Variation of pressure in the Earth's atmosphere with height y above sea level.

increases with depth as explained in Figure 10.14. As an example, the pressure at any point in the Earth's atmosphere must support the weight of all the fluid (air) above it.

If we consider a column of fluid of area A, the upward force supporting this column is $F = P_{atm}A$. For a liquid with constant density, this analysis leads to Equation 10.5, which gives $P = P_{atm} + \rho gh$, where h is the *depth* in the fluid measured from the surface. In Figure 10.14, we denote the *height* above the surface as y, which is related to the depth by $y = -h$; we thus have for a liquid

$$P = P_0 - \rho gy \tag{10.10}$$

which is plotted as the straight dashed line in Figure 10.15.

This result does *not* apply to a gas because as y increases and the pressure drops, the density decreases. The weight of the fluid column is therefore smaller for a compressible fluid, and the pressure decreases more slowly as y increases than for an incompressible fluid. The pressure as a function of height y for a compressible fluid is also plotted in Figure 10.15 as the solid curve. This result can be derived using Newton's laws and is

$$P = P_0 \exp\left(\frac{-y}{y_0}\right) \tag{10.11}$$

where P_0 is the pressure at a reference point, y is the height as measured from the reference point, and y_0 is a constant that depends on the gas. For air with a reference point at sea level, $P_0 = P_{atm}$ and $y_0 \approx 1.0 \times 10^4$ m. The height y_0 is the altitude at which the pressure falls to $e^{-1} = 0.37 = 37\%$ of its value at the surface. This result applies for low regions of the atmosphere, where the temperature is approximately constant. Dealing with pressure variations over a wider altitude range is more complicated since the air temperature can vary substantially and the air density also depends on temperature.

Whether we are dealing with a liquid or a gas, the pressure in a fluid must vary with depth and P must increase as one goes deeper into the fluid. In both cases, the pressure difference between the top and bottom surfaces of an arbitrary volume of fluid (Fig. 10.7) must support the weight of the fluid within the volume, so the pressure increases as depth increases.

EXAMPLE 10.5 Pressure at the Top of Mount Everest

The peak of Mount Everest is 8850 m above sea level. Estimate the pressure in the atmosphere at this location.

RECOGNIZE THE PRINCIPLE

This example is a direct application of the result in Equation 10.11 for the pressure as a function of altitude in a compressible fluid (air).

SKETCH THE PROBLEM

Figure 10.15 shows the variation of pressure with altitude y in the atmosphere as described mathematically by Equation 10.11.

IDENTIFY THE RELATIONSHIPS AND SOLVE

The altitude at the top of Mount Everest is $y = 8850$ m. Using this value in Equation 10.11 with $y_0 = 1.0 \times 10^4$ m as given above leads to

$$P = P_{atm} \exp\left(\frac{-y}{y_0}\right) = (1.01 \times 10^5 \text{ Pa})\exp\left(\frac{-8850 \text{ m}}{1.0 \times 10^4 \text{ m}}\right) = \boxed{4.2 \times 10^4 \text{ Pa}}$$

► ***What does it mean?***

This result is about 40% of atmospheric pressure. According to the kinetic theory of gases (Chapter 15), the density of a gas is proportional to the pressure, assuming its temperature is constant. The temperature at the top of Mount Everest is usually a bit below (!) the temperature at sea level, which reduces the pressure on Mount Everest by about another 15%. Hence, the air density at the top of Mount Everest is less than 40% of the density at sea level. That is why mountain climbers usually bring their own oxygen when scaling this mountain.

10.3 Hydraulics and Pascal's Principle

For a liquid at rest, the pressure varies with depth according to

$$P = P_0 + \rho g h \tag{10.12}$$

Now imagine that something is done to change the pressure at a particular point in the liquid; to be specific, suppose the pressure P_0 is changed by an amount ΔP. According to Equation 10.12, the pressure everywhere in the liquid must change by an equal amount. This result is known as ***Pascal's principle.***

Hydraulics: Pascal's principle

> ***Pascal's principle:*** **If the pressure at one location in a closed container of fluid changes, this change is transmitted equally to *all* locations in the fluid.**

While Equation 10.12 (which is the same as Eq. 10.5) was derived for a liquid (an incompressible fluid), Pascal's principle is also true for a gas (a compressible fluid) as long as the gas is held in a closed container and is in static equilibrium. Pascal discovered this principle before the work of Newton. However, looking back at our derivation of Equation 10.5, we see that this principle is a direct consequence of Newton's laws.

Designing a Hydraulic Lift: Amplifying Forces

Pascal's principle is the basis of ***hydraulics.*** A simple version of a hydraulic lift is shown in Figure 10.16A. It is shaped like a U-tube and is typically filled with a liquid such as oil, with pistons at both ends of the U-tube. These pistons are movable, but they fit tightly to keep the oil from leaking out. The key point is that the two ends of the tube have different areas. Here the area A_L on the left is much smaller than the area A_R on the right.

Now imagine that a very large object such as a car sits on the piston on the right. To keep the car and its piston from moving, or to lift the car very slowly off the ground, we apply a force F_L to the left as sketched in Figure 10.16A. What force F_L is required to keep the system in static equilibrium? If the car has mass m, the car's weight produces a downward force on the right-hand piston of magnitude $F_R = mg$.

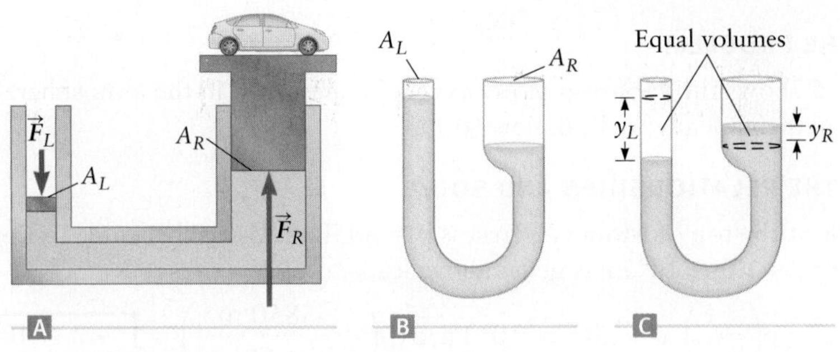

▶ **Figure 10.16** A A hydraulic jack uses a liquid (usually oil) to amplify an applied force. Here the force on the right-hand piston is greater than the force applied to the piston on the left. B and C show how the liquid moves as the left-hand piston is depressed. The volume of liquid that leaves the left-hand tube must equal the volume that appears in the right-hand tube.

From the general relation between force and pressure (Eq. 10.1), this force will lead to an increase in the pressure on the right, with

$$\Delta P = \frac{F_R}{A_R}$$

According to Pascal's principle, this pressure change is transmitted throughout the fluid, giving an increased force of magnitude $(\Delta P)A_L$ on the left-hand piston. To keep this piston from moving, there must be an external force of equal magnitude F_L applied downward on the piston. Hence,

$$F_L = (\Delta P)A_L$$

or

$$F_L = F_R \frac{A_L}{A_R} \tag{10.13}$$

The piston on the right is much larger than the one on the left ($A_R > A_L$), so

$$F_L < F_R$$

The force on the left-hand piston is thus smaller than the force on the right-hand piston. A relatively small force applied on the left can therefore lift a very massive object (the car) on the right. In other words, the hydraulic device in Figure 10.16 *amplifies forces*. This principle is used by auto mechanics every day.

The simple system in Figure 10.16 illustrates the principles of hydraulics, but practical systems are usually made as sketched in Figure 10.17, where a pump is used to produce a high pressure in the liquid in a tube connected to an elevator. The tube leading from the pump plays the role of the left-hand tube in Figure 10.16 and is smaller than the elevator tube. Here the large object to be lifted is the elevator compartment.

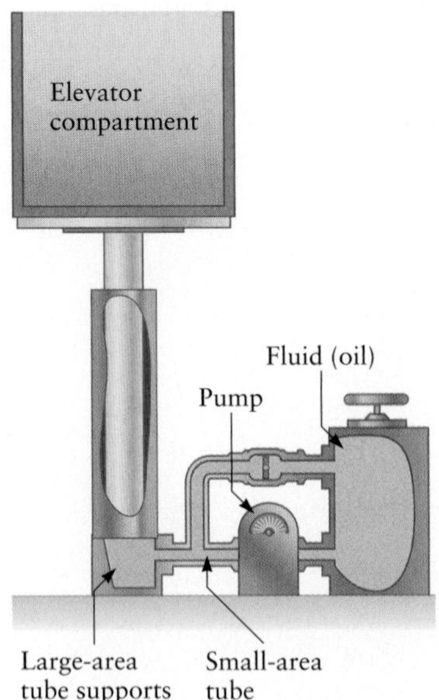

▲ **Figure 10.17** Schematic of a hydraulic system used for an elevator.

Work–Energy Analysis of a Hydraulic System

Now let's analyze the hydraulic lift in terms of work and energy. We start with the configuration in Figure 10.16B and suppose the piston on the right moves upward through a small distance y_R. The work done by the right-hand piston on the automobile is equal to the product of the force and the distance traveled:

$$W_R = F_R y_R \tag{10.14}$$

We can also consider the work done by the force F_L on the left-hand piston. That piston moves downward a distance y_L, so this work is

$$W_L = F_L y_L \tag{10.15}$$

Since we expect that the hydraulic lift will conserve mechanical energy, you should suspect that the work done *on* the left-hand piston should equal the work done *by* the right-hand piston; that is, $W_L = W_R$. Let's now show that this is indeed true.

When the two pistons move, liquid (oil) is displaced from one side of the U-tube to the other as shown in parts B and C of Figure 10.16. Because the liquid is incompressible, the volume of oil that travels to the right must equal the volume that leaves from the left. These volumes are equal to the area of the tube times the height of the oil, so

$$A_L y_L = A_R y_R \tag{10.16}$$

and

$$y_L = y_R \frac{A_R}{A_L} \tag{10.17}$$

The piston with the *smaller* area (on the left) thus moves through a *larger* distance. Combining this result with our result for the forces from Equation 10.13 leads to

$$F_R y_R = \left(F_L \frac{A_R}{A_L}\right)\left(y_L \frac{A_L}{A_R}\right) = F_L y_L$$

The term on the far left, $F_R y_R$, equals W_R (Eq. 10.14), while the term on the far right, $F_L y_L$, is W_L (Eq. 10.15). We thus have

$$W_R = W_L$$

Hence, the work done *on the piston on the left* equals the work done *by the piston on the right*, and mechanical energy is indeed conserved. A hydraulic device can amplify forces, but in doing so, it must "de-amplify" the displacement (Eq. 10.17).

EXAMPLE 10.6 Designing a Hydraulic Lift

You are given the job of designing a hydraulic device for lifting large objects as in Figure 10.16. You want this device to be able to lift an SUV of mass $m_{SUV} = 2000$ kg, when operated with an applied force of just 750 N (the approximate weight of the author). The two pistons of the hydraulic lift are cylindrical, and suppose the piston lifting the SUV has a radius of 0.50 m. Find the radius of the other piston.

RECOGNIZE THE PRINCIPLE

From Pascal's principle, the pressure increases on the two sides of a hydraulic tube are equal. This leads to Equation 10.13, which (in words) states that the ratio of the forces equals the ratio of the piston areas. Given the ratio of the weight of the SUV and the applied force, we can thus find the ratio of the piston areas.

SKETCH THE PROBLEM

Figure 10.16 shows the problem. The piston lifting the SUV must be the one with the larger area.

IDENTIFY THE RELATIONSHIPS

To lift the SUV, we must amplify the applied force of 750 N by a factor of $m_{SUV}g/(750\text{ N}) = (2000\text{ kg})(9.8\text{ m/s}^2)/(750\text{ N}) = 26$; this is the ratio of the forces F_R/F_L in Figure 10.16A. Rearranging Equation 10.13 leads to

$$\frac{F_R}{F_L} = \frac{A_R}{A_L} = 26$$

The sides of the lift are cylindrical, so for each piston $A = \pi r^2$, and

$$\frac{A_R}{A_L} = \frac{\pi r_R^2}{\pi r_L^2} = \frac{r_R^2}{r_L^2} = 26$$

SOLVE

Solving for the radius of the left-hand piston gives

$$r_L = \frac{r_R}{\sqrt{26}} = \frac{0.50\text{ m}}{\sqrt{26}} = \boxed{0.098\text{ m}}$$

(continued) ▶

▶ ***What does it mean?***
This hydraulic system will amplify force by a factor of 26 and will de-amplify displacement by the same factor. The operator will have to displace 26 m of oil through the left-hand piston to lift the car through 1 m.

10.4 Buoyancy and Archimedes's Principle

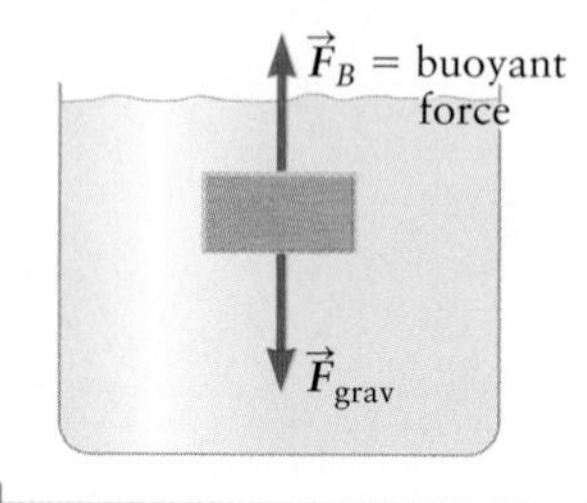

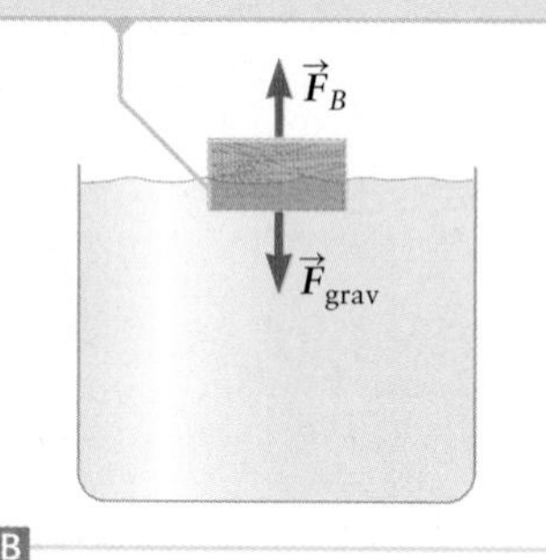

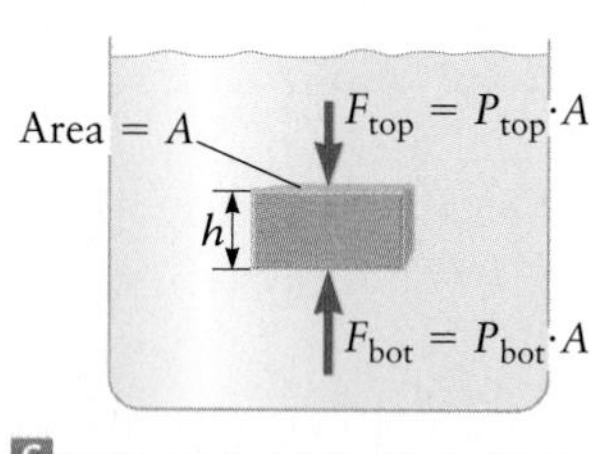

▲ **Figure 10.18** According to Archimedes's principle, an object that is immersed in a fluid will experience a buoyant force that is directed upward, whether A the object is fully immersed in the fluid or B the object floats at the surface and is partially exposed. In both cases, the buoyant force on the object equals the weight of the fluid displaced by the object. C The buoyant force is due to the increase in pressure with depth.

Archimedes lived in the third century BC and is known for several contributions to science and mathematics. While his work on fluids was done long before the time of Newton, Archimedes made some very important discoveries involving forces in fluids that anticipated Newton's laws. Consider an object that is either fully immersed or is floating with a portion exposed above the surface of the fluid as sketched in parts A and B of Figure 10.18. In either case, ***Archimedes's principle*** applies.

***Archimedes's principle:* When an object is immersed in a fluid, the fluid exerts an upward force on the object that is equal to the weight of the fluid displaced by the object.**

This upward force on the object is called the ***buoyant force***, and it is due to the increase of pressure with depth in a fluid. The pressure at the bottom surface of an object is always greater than the pressure at the top, leading to a larger (upward) force on the bottom surface. The difference between these two forces is the buoyant force.

We can calculate the buoyant force (and thus prove Archimedes's principle) by applying our relation for the variation of pressure with depth (Eq. 10.5) to the object in Figure 10.18C. Here, a solid box is completely submerged in a liquid of density ρ_L. For simplicity, we assume this box is rectangular with height h and top or bottom area A. There are pressure forces on all faces of the box, the sides as well as the top and bottom. The horizontal forces on opposite sides of the box are equal in magnitude and opposite in direction and thus cancel. The force on the top is downward with magnitude $P_{top}A$, whereas the force on the bottom is upward and of magnitude $P_{bot}A$. The total force *exerted by the liquid* on the box is along the vertical direction, with

$$F = +P_{bot}A - P_{top}A$$

If we assume the fluid is a liquid of density ρ_L, according to Equation 10.5 the pressures at the bottom and top are related by

$$P_{bot} = P_{top} + \rho_L gh$$

where h is the height of the box. Combining these results gives

$$F = +P_{bot}A - P_{top}A = (P_{top} + \rho_L gh)A - P_{top}A = \rho_L hAg \quad (10.18)$$

Hence, the force exerted by the liquid is upward; this is the buoyant force. Since the factor (hA) in Equation 10.18 equals the volume V of the box, this result is usually written as

$$F_B = \rho_L Vg \quad (10.19)$$

The mass of the liquid displaced by the box is $\rho_L V$, so the term $\rho_L Vg$ in Equation 10.19 equals the weight of the displaced liquid. In words, Equation 10.19 says that the buoyant force F_B equals the weight of the *liquid* displaced by the box. This is another statement of Archimedes's principle.

Examples and Applications of Archimedes's Principle

Figure 10.18 shows Archimedes's principle at work in two different cases. In Figure 10.18A, a box of volume V is fully immersed in the fluid. The buoyant force equals the weight of the displaced fluid, and because the box is completely immersed, the volume of displaced fluid equals the volume V of the box. The weight of the displaced fluid is thus $F_B = \rho_L g V$, where ρ_L is the density of the fluid. If the weight of this box is exactly equal to F_B, the box will remain where it is, fully submerged in the liquid. If the box is heavier than F_B, it will sink to the bottom. It might also happen that the weight of the box is smaller than the weight of the liquid it displaces, which could happen with an empty box or with a boat. In such a case, the buoyant force for a fully submerged box would be *greater* than the weight of the box, and the box would rise to the surface and float (Fig. 10.18B). A portion of the box would then be exposed above the surface of the liquid so that the weight of the *displaced* liquid (the amount below the "waterline") equals the weight of the box. For that to occur, the average density of the box in Figure 10.18B must be less than the density of the liquid (just as the average density of a boat must be less than the density of water).

As an application of Archimedes's principle, consider the situation in Figure 10.19. This box might be very heavy and submerged in a lake, and you might have the job of lifting it to the surface. The minimum force required to lift the box is called the ***apparent weight.***

Figure 10.19 shows the forces acting on the box. They are the force of gravity (the weight of the box), the force F_{lift} you are exerting as you lift the box, and the buoyant force F_B from the water. According to Archimedes's principle, F_B is the total force exerted by the fluid on the box. Since F_{lift} is the minimum force necessary to barely lift the box, we can take the acceleration of the box to be zero. Newton's second law for the box's motion along the vertical direction is then

$$\sum F = F_{\text{grav}} + F_B + F_{\text{lift}} = -mg + \rho_L V g + F_{\text{lift}} = 0$$

where the buoyant force F_B is given by Archimedes's principle, Equation 10.19. The apparent weight of the box is equal to F_{lift}. Solving for F_{lift}, we find

$$F_{\text{lift}} = mg - \rho_L V g \tag{10.20}$$

The apparent weight is thus smaller than the true weight (mg) by an amount equal to the buoyant force. You have probably already done this type of experiment in a swimming pool. To estimate the magnitude of the buoyant force, suppose the box in Figure 10.19 has a volume of 0.10 m^3 (about the size of a small suitcase) and is filled with lead. Its weight is

$$\text{true weight} = mg = \rho_{\text{lead}} V g$$

since the mass of the box is equal to the density of lead (ρ_{lead}) times the volume (V). Using the density of lead from Table 10.1, we find that the box has a true weight of

$$\text{true weight} = \rho_{\text{lead}} V g = (11{,}300 \text{ kg/m}^3)(0.10 \text{ m}^3)(9.8 \text{ m/s}^2) = 1.1 \times 10^4 \text{ N}$$

From Equation 10.20, the apparent weight of the lead-filled box equals

$$\text{apparent weight} = \rho_{\text{lead}} V g - \rho_L V g = (\rho_{\text{lead}} - \rho_L) V g$$

where ρ_L is the density of the liquid. If we assume the liquid is water (which has a density of 1000 kg/m^3), we find

$$\text{apparent weight} = (\rho_{\text{lead}} - \rho_L) V g$$
$$= (11{,}300 - 1000)(\text{kg/m}^3)(0.10 \text{ m}^3)(9.8 \text{ m/s}^2)$$
$$\text{apparent weight} = 1.0 \times 10^4 \text{ N}$$

Hence, in this case, the buoyant force makes the apparent weight about 10% smaller than the true weight.

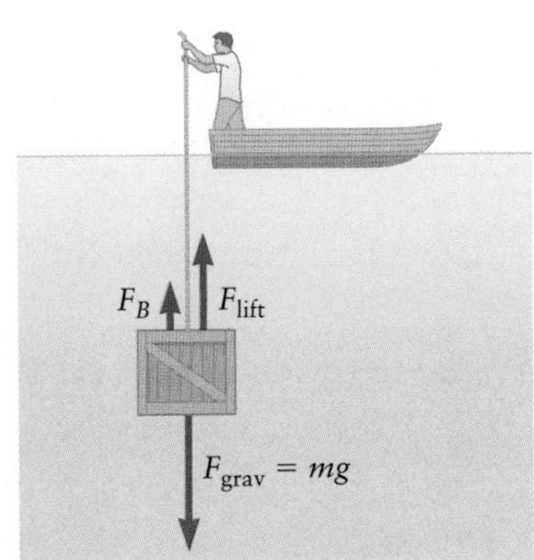

▲ **Figure 10.19** The buoyant force causes the apparent weight (F_{lift}) to be smaller than the true weight (mg).

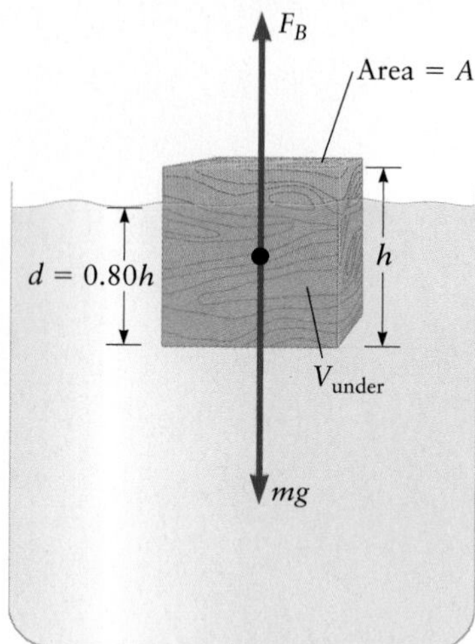

▲ **Figure 10.20** Example 10.7.

EXAMPLE 10.7 Floating in Water

A rectangular block of wood is floating at the surface of a container filled with water as shown in Figure 10.18B. You observe that 80% of the block is under water, whereas the rest is above the surface. Find the density of the wood.

RECOGNIZE THE PRINCIPLE

According to Archimedes's principle, the buoyant force equals the weight of the water displaced by the block. This upward buoyant force must equal the weight of the block in order for the block to remain in translational equilibrium.

SKETCH THE PROBLEM

Figure 10.20 shows the problem. The block of wood has a total height h. Since 80% of the block is submerged, the height of the portion under water is $d = 0.80h$.

IDENTIFY THE RELATIONSHIPS

The top and bottom of the block each have an area A, and the volume of the block that is under water is $V_{under} = dA$. The block thus displaces a volume of water equal to V_{under}, and the buoyant force equals the weight of this displaced water:

$$F_B = \rho_{water} V_{under} g = \rho_{water} dAg$$

The block is floating, so this buoyant force must equal (in magnitude) the gravitational force on the block:

$$F_{grav} = m_{wood} g = \rho_{wood} V_{total} g$$

where $V_{total} = hA$ is the total volume of the block.

SOLVE

Setting F_{grav} equal to the buoyant force F_B leads to

$$F_B = \rho_{water} dAg = \rho_{wood} V_{total} g = \rho_{wood} hAg$$

$$\rho_{water} d = \rho_{wood} h$$

$$\rho_{wood} = \rho_{water} \frac{d}{h} \tag{1}$$

Using $d = 0.80h$ together with the density of water from Table 10.1, we find

$$\rho_{wood} = \rho_{water} \frac{d}{h} = (1000 \text{ kg/m}^3)(0.80) = \boxed{800 \text{ kg/m}^3}$$

► ***What does it mean?***

We see from Equation (1) that if an object is floating in a liquid, the object must be less dense than the liquid (as we claimed in our discussion of Fig. 10.18B). Recall that the specific gravity of a substance equals the ratio of its density to the density of water. Hence, a substance will float in water if its specific gravity is less than 1.

CONCEPT CHECK 10.4 Floating in a Lake

A block of wood is initially floating in a lake at 4°C with 20% of the block exposed above the surface of the water (Fig. 10.20). The temperature of the water then increases to 20°C. Will the block move (a) up or (b) down relative to the surface of the water?

CONCEPT CHECK 10.5 Buoyant Force and a Floating Ship

A ship floats because of the buoyant force on it. A typical ship is composed of steel, and we see from Table 10.1 that the density of steel is much *greater* than the density of water. How can the buoyant force support a steel ship?

(a) Real ships are made from special low-density metals.
(b) A ship will float if its average density (including both the steel and the air inside) is less than the density of water.
(c) A ship floats because it is watertight and does not leak.

Archimedes's Principle Holds for Objects of Any Shape, in Both Incompressible and Compressible Fluids

When we derived Archimedes's result for the buoyant force, Equation 10.19, we assumed the submerged (or floating) object has the shape of a simple rectangular block (as in Figs. 10.18 and 10.20). We also used our relation for the variation of the pressure with depth in a liquid ($P = P_0 + \rho gh$). However, Archimedes's principle applies to objects of *all* shapes that are immersed in gases as well as in liquids. Archimedes's principle thus applies to a very wide variety of situations.

EXAMPLE 10.8 Operation of a Helium-Filled Balloon

NASA is experimenting with balloons for use on ultralong-duration (100-day) flights around the Earth. Consider a balloon that has the shape of a cylindrical chamber (Fig. 10.21) that is $h = 35$ m tall, has radius $r = 30$ m, and is made of extremely light plastic. (The balloon in Figure 10.22 has a more realistic shape.) If our model cylindrical balloon is filled with helium gas, what is the mass of the maximum payload it could lift from the Earth's surface?

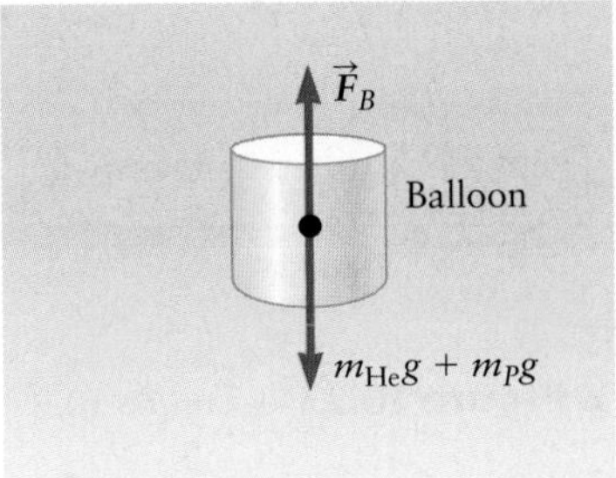

▲ **Figure 10.21** Example 10.8.

RECOGNIZE THE PRINCIPLE

The weight of the maximum payload equals the buoyant force on the balloon minus the weight of the balloon itself. We can find the buoyant force using Archimedes's principle from the volume of the balloon and the density of the air it displaces.

SKETCH THE PROBLEM

This problem is described by Figure 10.21. The balloon is fully immersed in the atmosphere.

IDENTIFY THE RELATIONSHIPS

The buoyant force on the balloon equals the weight of the air the balloon displaces. If the balloon has a volume V, this weight is

$$\text{weight of air displaced} = \rho_{\text{air}} V g = F_B$$

The "skin" of the balloon is composed of a very light plastic, so we'll assume its mass is very small compared with both the mass of the payload m_P and the mass of the helium gas inside the balloon. This helium gas has mass m_{He}, which is given by $m_{\text{He}} = \rho_{\text{He}} V$, where ρ_{He} is the density of helium gas. The buoyant force must support the weight of the helium and the weight of the payload, so

$$F_B = \rho_{\text{air}} V g = m_{\text{He}} g + m_P g = \rho_{\text{He}} V g + m_P g$$

$$m_P g = \rho_{\text{air}} V g - \rho_{\text{He}} V g$$

$$m_P = (\rho_{\text{air}} - \rho_{\text{He}}) V$$

The volume of a cylindrical balloon is equal to the area of the base multiplied by the height, $V = \pi r^2 h$.

(continued) ▶

▲ **Figure 10.22** Example 10.8. This large NASA balloon is filled with helium gas and used to lift scientific payloads to high altitudes.

SOLVE

Using the densities of air and helium from Table 10.1 along with the dimensions of the balloon, we find

$$m_P = (\rho_{\text{air}} - \rho_{\text{He}})V = (\rho_{\text{air}} - \rho_{\text{He}})(\pi r^2 h)$$

$$= (1.29 - 0.18)(\text{kg/m}^3)\pi(30 \text{ m})^2(35 \text{ m})$$

$$m_P = \boxed{1.1 \times 10^5 \text{ kg}}$$

▶ *What does it mean?*

This payload is quite large; most scientific payloads would have a much smaller mass. Why would NASA make the balloon so large for a relatively small payload? In this example, we have calculated the mass of the largest possible payload for the balloon at sea level. Air is a compressible fluid, so its density decreases as the balloon moves up in the atmosphere and the lower density at high altitudes decreases the lifting capability of the balloon. NASA designs the balloon so that it can lift the payload at the desired "cruising" altitude, so the balloon will need to have a much greater lift force at ground level.

EXAMPLE 10.9 How Can a Submarine Float?

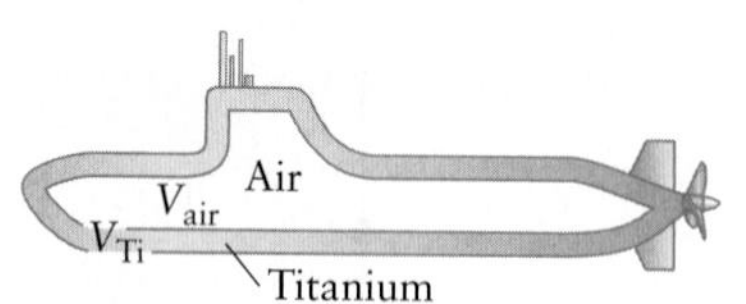

▲ Figure 10.23 Example 10.9.

Submarines are made of metal, and all metals have a density greater than that of water. So how can a submarine float? Concept Check 10.5 asks for the basic answer, but here we analyze this problem further. Figure 10.23 shows a hypothetical submarine, with an outer metal shell and an inner cavity of air. Assume the metal is titanium (a low-density metal used to make submarines) and the air is at atmospheric pressure (so that the people inside can breathe normally!). What fraction of the submarine must be occupied by the air cavity if the submarine is able to float just below the surface of the water?

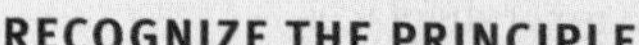

RECOGNIZE THE PRINCIPLE

According to Archimedes's principle, the buoyant force on the submarine is equal to the weight of the water it displaces. That force must be equal in magnitude to the total weight of the submarine, which includes the titanium shell and the air in the cavity. The density of titanium is 4500 kg/m³.

SKETCH THE PROBLEM

Figure 10.23 shows the problem. We denote the volume of titanium by V_{Ti} and the volume of air by V_{air}.

IDENTIFY THE RELATIONSHIPS

We denote the density of titanium and air by ρ_{Ti} and ρ_{air}, respectively. The volume of water displaced is $V_{\text{Ti}} + V_{\text{air}}$, so the weight of the displaced water is

$$\text{weight of water} = \rho_{\text{water}}(V_{\text{Ti}} + V_{\text{air}})g = F_B$$

which is equal to the buoyant force F_B. The total weight of the submarine is the weight of the titanium ($\rho_{\text{Ti}}V_{\text{Ti}}g$) plus the weight of the air ($\rho_{\text{air}}V_{\text{air}}g$). The submarine is assumed to be floating under the surface of the water, so its total weight equals the buoyant force:

$$F_B = \rho_{\text{water}}(V_{\text{Ti}} + V_{\text{air}})g = \rho_{\text{Ti}}V_{\text{Ti}}g + \rho_{\text{air}}V_{\text{air}}g \quad (1)$$

SOLVE

We can rearrange Equation (1) to solve for V_{air}. Collecting terms in V_{air} and V_{Ti}, we find

$$V_{air}(\rho_{water} - \rho_{air}) = V_{Ti}(\rho_{Ti} - \rho_{water})$$

$$\frac{V_{air}}{V_{Ti}} = \frac{\rho_{Ti} - \rho_{water}}{\rho_{water} - \rho_{air}}$$

Inserting known values of the densities (for seawater from Table 10.1 and titanium from the problem statement), we get

$$\frac{V_{air}}{V_{Ti}} = \frac{\rho_{Ti} - \rho_{water}}{\rho_{water} - \rho_{air}} = \frac{(4500 - 1025)\ \text{kg/m}^3}{(1025 - 1.3)\ \text{kg/m}^3} = \boxed{3.4}$$

keeping two significant figures in the answer.

► ***What does it mean?***

The volume of the air cavity in the hypothetical submarine in Figure 10.23 must therefore be almost 3.5 times larger than the volume of the titanium shell. Real submarines have a more complicated design, with a ballast tank that can be filled with either water or air to adjust the total weight of the submarine and thus determine if it will float at or underneath the surface.

10.5 Fluids in Motion: Continuity and Bernoulli's Equation

So far in this chapter, we have only considered fluids that are at rest. We are now ready to tackle some problems in which a fluid is in motion. To keep our discussion of fluid dynamics as simple as possible, we make the following assumptions. (1) The fluid density is constant; that is, the fluid is incompressible. (2) The flow is steady; that is, the fluid velocity is independent of time (although it can still vary from point to point in space). (3) There is no friction, that is, no *viscosity*. (We'll discuss viscosity later in this chapter.) (4) We assume there are no "smoke rings" or other kinds of complex flow patterns such as turbulence. Fluids that satisfy these assumptions are called ***ideal fluids***. Although we thus consider only a few of the simplest situations, they will still illustrate some key aspects of fluid dynamics.

Figure 10.24 shows a fluid flowing through a pipe. Because of the simple geometry of this pipe, we need only consider the velocity of the fluid along the pipe, in the x direction in Figure 10.24. We denote the fluid's speed at the pipe inlet (on the left) as v_L and the speed at the pipe outlet (on the right) as v_R. We now want to consider how v_L and v_R are related.

The amount of fluid that flows through the pipe must be *conserved*; this means that the mass of fluid flowing in from the left during a particular time interval Δt must equal the mass flowing out at the right during this interval. In other words, the rate at which the fluid flows into the pipe must equal the rate at which the fluid flows out of the pipe. This is called the ***principle of continuity***.

The basic idea behind the principle of continuity is that "what goes in must come out." In other words, the amount of fluid that leaves the pipe each second must equal the amount that flows in. So, since the outlet in Figure 10.24 is smaller than the inlet, the outlet speed (v_R) must be larger than the inlet speed. A small-diameter flow stream with a large speed can carry the same amount of fluid per unit time as a large-diameter flow stream with a small speed. That is why the speed of water coming from a garden hose increases when you put your thumb over part of the end of the hose (thus decreasing the outlet area).

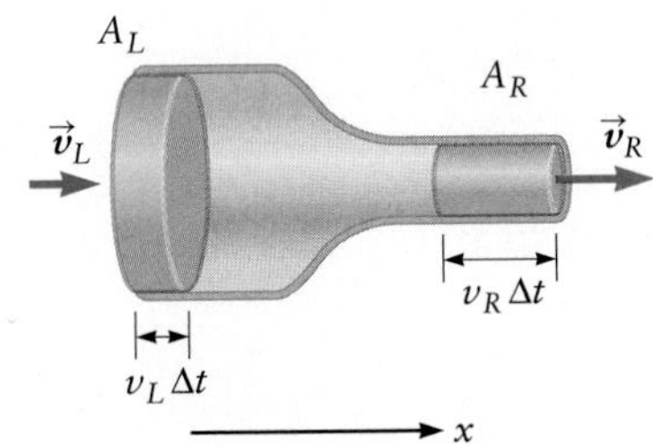

▲ **Figure 10.24** According to the principle of continuity, the rate at which liquid flows into a pipe must equal the rate at which it flows out.

Let's now give a more careful analysis of the principle of continuity. For an incompressible fluid, the *volume* of fluid flowing into a pipe must equal the volume of fluid that flows out (because mass is equal to density times volume, and the density is constant). This inflowing volume is shown in Figure 10.24; it equals the volume of fluid that flows into the pipe from the left during a time interval Δt, filling a cylinder of length $v_L \Delta t$. Since the inlet area is A_L, the volume of fluid flowing into the pipe is

$$V_{in} = (v_L \Delta t)A_L = v_L A_L \Delta t \tag{10.21}$$

Notice that here the capital V is a *volume*; the lowercase v is a *speed*. Using the same argument, the fluid flowing out of the pipe is

$$V_{out} = (v_R \Delta t)A_R = v_R A_R \Delta t \tag{10.22}$$

According to the principle of continuity, the amount that flows into the pipe must equal the amount that flows out, so

$$V_{in} = V_{out}$$

$$v_L A_L \Delta t = v_R A_R \Delta t$$

which leads to

Equation of continuity

$$v_L A_L = v_R A_R \tag{10.23}$$

This result is called the ***equation of continuity***.

The product of fluid speed and area in Equation 10.23 is also called the ***flow rate***, which we denote as Q:

$$\text{flow rate} = Q = vA \tag{10.24}$$

Hence, the equation of continuity (Eq. 10.23) says that the incoming flow rate equals the outgoing flow rate. The SI units of flow rate are cubic meters per second (m^3/s), but liters per second are also widely used. In U.S. customary units, one might use gallons per minute (gal/min). The flow rate from a garden hose is typically 2 to 5 gal/min.

CONCEPT CHECK 10.6 Flow Rate in the Aorta

The flow rate of blood through the aorta is typically 0.080 liter/s = 80 cm^3/s. If it takes 1 minute for all your blood to pass through the aorta, is the volume of blood in your body (a) 4.8 liters, (b) 10 liters, (c) 8.0 liters, or (d) 1.0 liter?

Bernoulli's Equation

Let's again consider an ideal fluid moving through a pipe as sketched in Figure 10.25 and examine the flow from the point of view of work and energy. The fluid in Figure 10.25 is flowing from left to right. We denote the pressure and speed of the fluid at the far left as P_1 and v_1, and we suppose the left end of the pipe is at a height h_1 above the reference level for gravitational potential energy. Likewise, the pressure, fluid speed, and height at the far right are P_2, v_2, and h_2. Figure 10.25 shows two snapshots of the fluid as it moves through the pipe. The first snapshot (Fig. 10.25A) is at $t = 0$, and the second snapshot (Fig. 10.25B) is a short time Δt later. During this time, the "boundaries" of this portion of the fluid have moved a certain distance; the boundary on the left moved a distance Δx_1 and the boundary on the right moved a distance Δx_2.

We now apply the work–energy theorem; our goal is to get a relationship between the pressure, speed, and height of the fluid on the two sides in Figure 10.25. There are external forces on the fluid associated with the pressure on the far left end and with the pressure on the far right end. The work done on the fluid by the pressure on the left equals the force on the boundary at the left multiplied by the distance this boundary travels. This force F_1 is related to the pressure on the left by $F_1 = P_1 A_1$, so the work done on the fluid is

$$W_1 = F_1 \Delta x_1 = P_1 A_1 \Delta x_1$$

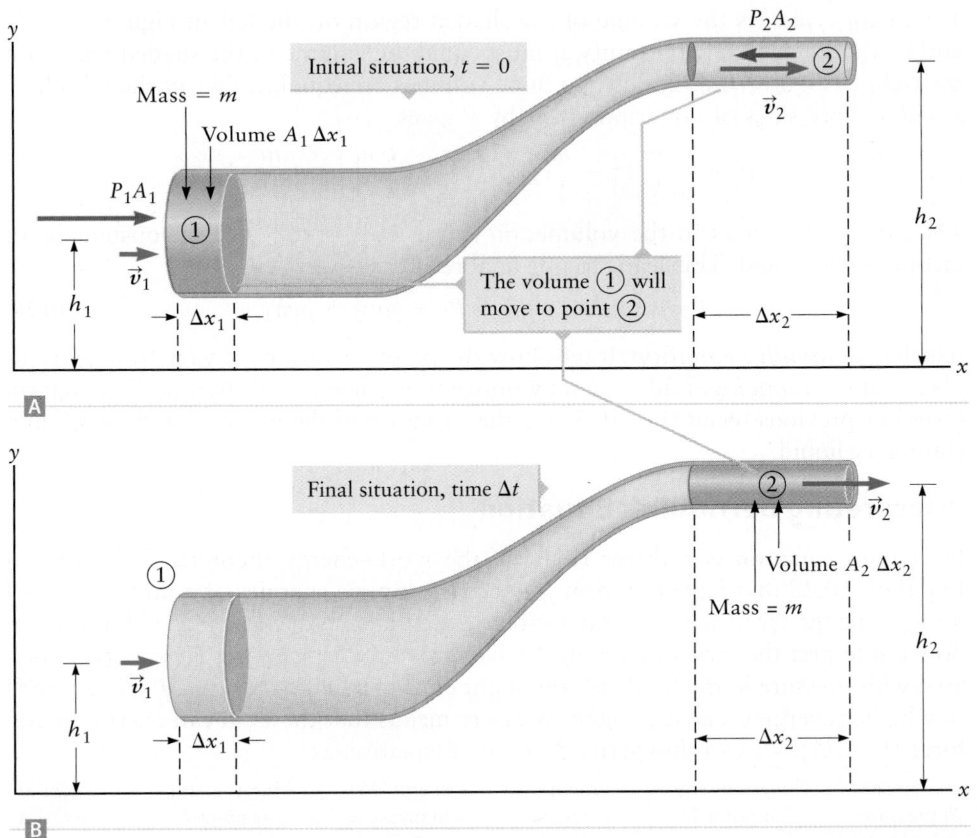

◀ **Figure 10.25** Bernoulli's equation tells how the pressure, speed, and height at one location in an ideal fluid are related to P, v, and h at any other location. As the fluid flows through this pipe, the net effect is to transport the shaded volume from the left end in A to the right end in B. The change in the energy of the entire system is then just the difference between the energies of these two shaded volumes.

To find the work done on the fluid by the pressure on the far right end, we note that because the force is a pressure force, it is directed to the left (in the $-x$ direction). The right-hand boundary of the fluid moves to the right, so the force and displacement are in opposite directions and thus the work done on the fluid is negative. We have

$$W_2 = -F_2\,\Delta x_2 = -P_2A_2\,\Delta x_2$$

The total work done on the fluid is the sum of W_1 and W_2, and according to the work–energy theorem, this sum equals the change in the mechanical energy of the fluid:

$$W_{\text{total}} = W_1 + W_2 = P_1A_1\,\Delta x_1 - P_2A_2\,\Delta x_2 = \Delta KE + \Delta PE \tag{10.25}$$

Figure 10.25 shows snapshots of the initial and final states of the system. The *only* difference between these two snapshots is that the volume of fluid shown as shaded has been moved from the left end of the pipe to the right end. So, the change in energy of the fluid in going from the snapshot in Figure 10.25A to that in Figure 10.25B is equal to the change in the energy of just this shaded region. If the shaded volume of fluid has mass m, its initial mechanical energy (kinetic plus potential) is

$$KE_i + PE_i = \tfrac{1}{2}mv_1^2 + mgh_1$$

whereas the final mechanical energy is

$$KE_f + PE_f = \tfrac{1}{2}mv_2^2 + mgh_2$$

The change in the mechanical energy is therefore

$$\Delta KE + \Delta PE = \tfrac{1}{2}mv_2^2 + mgh_2 - (\tfrac{1}{2}mv_1^2 + mgh_1) \tag{10.26}$$

Inserting this expression into Equation 10.25 leads to

$$W_{\text{total}} = P_1A_1\,\Delta x_1 - P_2A_2\,\Delta x_2 = \Delta KE + \Delta PE$$

$$P_1A_1\,\Delta x_1 - P_2A_2\,\Delta x_2 = \tfrac{1}{2}mv_2^2 + mgh_2 - (\tfrac{1}{2}mv_1^2 + mgh_1)$$

$$P_1A_1\,\Delta x_1 + \tfrac{1}{2}mv_1^2 + mgh_1 = P_2A_2\,\Delta x_2 + \tfrac{1}{2}mv_2^2 + mgh_2 \tag{10.27}$$

The factor $A_1 \Delta x_1$ is the volume of the shaded region on the left in Figure 10.25A, and by the principle of continuity, it must equal the volume of the shaded region on the right in Figure 10.25B. Because these volumes are equal, we'll call them both V. Dividing both sides of Equation 10.27 by V gives

$$P_1 + \frac{1}{2}\frac{m}{V}v_1^2 + \frac{m}{V}gh_1 = P_2 + \frac{1}{2}\frac{m}{V}v_2^2 + \frac{m}{V}gh_2$$

The ratio of the mass to the volume, m/V, is the density, which is constant in an incompressible fluid. This leads to our final result

Bernoulli's equation

$$P_1 + \tfrac{1}{2}\rho v_1^2 + \rho g h_1 = P_2 + \tfrac{1}{2}\rho v_2^2 + \rho g h_2 \qquad (10.28)$$

which is ***Bernoulli's equation***. It tells how the pressure and speed vary from place to place within a *moving* fluid. You may notice that when $v = 0$, Bernoulli's equation gives our previous result (Eq. 10.5) for the variation of the pressure with depth in a stationary liquid.

Interpreting Bernoulli's Equation

Bernoulli's equation is a direct result of the work–energy theorem. The terms in Equation 10.28 that have the form $\frac{1}{2}\rho v^2$ are the kinetic energies of a unit volume of fluid, while the terms $\rho g h$ are gravitational potential energies per unit volume. How do we interpret the terms involving P? The pressure terms represent energy associated with pressure in the fluid; as you might expect, a higher value of P corresponds to a higher energy because a higher pressure means that the fluid can exert a higher force ($F = PA$). We can thus write Bernoulli's equation as

energy due to pressure at point 1		kinetic energy at point 1		potential energy at at point 1		energy due to pressure at point 2		kinetic energy at point 2		potential energy at point 2
P_1	$+$	$\frac{1}{2}\rho v_1^2$	$+$	$\rho g h_1$	$=$	P_2	$+$	$\frac{1}{2}\rho v_2^2$	$+$	$\rho g h_2$

In words, Bernoulli's equation says that the total mechanical energy of the fluid is conserved as it travels from place to place (point 1 to point 2), but some of this energy can be "converted" from kinetic energy to potential energy and vice versa as the fluid moves.

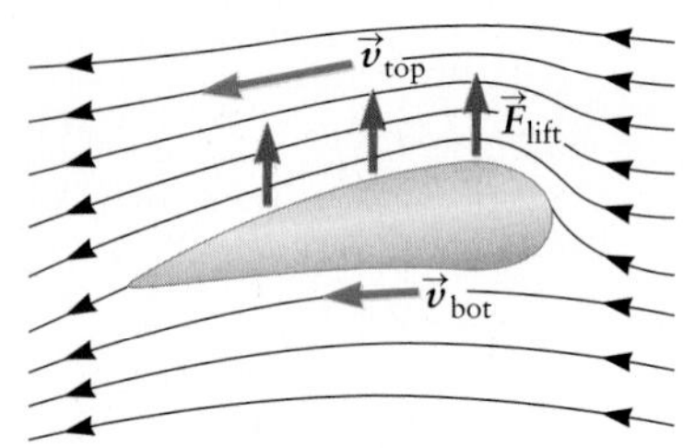

▲ **Figure 10.26** Schematic of the airflow pattern around an airplane wing. Because of the shape of the wing, the air speed is lower under the bottom of the wing than over the top. According to Bernoulli's equation, this difference leads to a higher pressure on the bottom and hence to an upward force called "lift."

Applications of Bernoulli's Equation

One of the most important applications of Bernoulli's equation involves the operation of an airplane wing. Figure 10.26 shows the pattern of airflow near a wing. Most airplane wings have an asymmetric shape, with a "bulge" on the top surface. This bulge causes the air speed to be slightly greater at the top of the wing compared with the air speed at the bottom of the wing. To see why the speed of the air on the top is greater, notice that the air flowing over the top must travel a greater distance to get around the wing (so it must travel faster) than the air that flows around the bottom. According to Bernoulli's equation (Eq. 10.28), an *increase* in the fluid speed leads to a *decrease* in pressure. Hence, the pressure on the bottom of the wing is greater than the pressure on the top. The total force due to air pressure on the wing is

$$F_{total} = P_{bot}A - P_{top}A = (P_{bot} - P_{top})A \qquad (10.29)$$

Since $P_{bot} > P_{top}$, this force is *upward*; it is the "lift" force on a wing that enables an airplane to fly.

Let's now consider the lift on an airplane wing in a little more detail. Suppose the area of the wing is 60 m^2, the air speed over the top surface of the wing is $v_{top} = 250$ m/s (about 500 mph), and the air speed under the bottom of the wing is $v_{bot} = 200$ m/s, all typical values for a commercial jet. To calculate the lift using Equation 10.29, we need to find the difference in pressure between the bottom and top of the wing. Applying Bernoulli's equation to a point at the bottom of the wing (point 1 in Eq. 10.28) and a point at the top (point 2), we have

$$P_{bot} + \tfrac{1}{2}\rho v_{bot}^2 + \rho g h_{bot} = P_{top} + \tfrac{1}{2}\rho v_{top}^2 + \rho g h_{top}$$

Insight 10.1

AIRPLANE WINGS IN AIR

In our derivation of Bernoulli's equation (Eq. 10.28), we assumed the fluid is incompressible. We then applied Bernoulli's equation to calculate the lift on an airplane wing (Eq. 10.29). The fluid in that case is air, which is compressible, but Bernoulli's equation can still be used to get an approximate value of the lift for the following reason. The pressure difference between the top and bottom of the wing (Eq. 10.30) is about 15% of atmospheric pressure, so there is a corresponding 15% variation in the air density between the bottom and top. Hence, to within this accuracy of about 15%, air under these conditions can be treated as having a constant density and our estimate for the lift force should thus be reliable to about this accuracy.

which can be rearranged to give the pressure difference across the wing:

$$P_{\text{bot}} - P_{\text{top}} = \tfrac{1}{2}\rho v_{\text{top}}^2 - \tfrac{1}{2}\rho v_{\text{bot}}^2 + \rho g h_{\text{top}} - \rho g h_{\text{bot}} = \tfrac{1}{2}\rho(v_{\text{top}}^2 - v_{\text{bot}}^2) + \rho g(h_{\text{top}} - h_{\text{bot}})$$

Here ρ is the density of air, which we know from Table 10.1. (For simplicity, we use the value of ρ at sea level and 0°C; to be more accurate, we would need to know the precise altitude and temperature.) We also know the speeds; our last step is to deal with the difference in height from top to bottom. For a typical wing of a large commercial jet, this difference is about 1 m, so we take $h_{\text{top}} - h_{\text{bot}} = 1$ m. Inserting all these values gives

$$P_{\text{bot}} - P_{\text{top}} = \tfrac{1}{2}\rho(v_{\text{top}}^2 - v_{\text{bot}}^2) + \rho g(h_{\text{top}} - h_{\text{bot}})$$

$$= \tfrac{1}{2}(1.29\ \text{kg/m}^3)[(250\ \text{m/s})^2 - (200\ \text{m/s})^2] + (1.29\ \text{kg/m}^3)(9.8\ \text{m/s}^2)(1\ \text{m})$$

$$P_{\text{bot}} - P_{\text{top}} = 1.5 \times 10^4\ \text{Pa} + 13\ \text{Pa} \qquad (10.30)$$

In the last step, we have written the pressure difference as a sum of two separate terms; the first is due to the difference in air speeds, and the second term is due to the difference in heights between the top and bottom of the wing. For this rapidly moving airplane wing, where the air speeds are large and the height difference is small, the potential energy terms (ρgh) make only a very small contribution in Bernoulli's equation. Hence, the pressure difference in Equation 10.30 and thus also the lift are due almost entirely to the speed terms in Bernoulli's equation.

To complete our calculation of the lift on the wing, we insert the pressure difference from Equation 10.30 into our expression for the lift (Eq. 10.29). We find

$$F_{\text{lift}} = (P_{\text{bot}} - P_{\text{top}})A = (1.5 \times 10^4\ \text{Pa})(60\ \text{m}^2) = 9.0 \times 10^5\ \text{N}$$

The mass of a medium-sized commercial jet is about 5×10^4 kg (weight $= mg \approx 5 \times 10^5$ N), so this lift force is indeed sufficient for the jet to fly.

The problem of lift on a real airplane wing is complicated. The Bernoulli effect (Eq. 10.28) is responsible for much of the lift, but other effects are important and must be taken into account in the design of a real airplane. To measure the lift on prototype designs, many airplane makers employ wind tunnels (Fig. 10.27).

National Museum of USAF/U.S. Air Force Photo

▲ **Figure 10.27** The aerodynamic forces on airplanes and other objects are often studied in wind tunnels. The Wright brothers used this wind tunnel to study the forces on wings they designed.

EXAMPLE 10.10 Ⓡ Speed of Water from a Faucet

Figure 10.28 shows the water system for a typical house. The source is a water tower located at some height above the houses it serves. In this particular water tower, the surface of the water in the tower is open to the atmosphere and is 30 m above the home shown in the figure. If a faucet in the home is then opened, what is the speed of the water as it comes out of the faucet?

RECOGNIZE THE PRINCIPLE

We use Bernoulli's equation (Eq. 10.28) to find the speed of the water coming out of the faucet. We take point 1 to be at the surface of the water in the tower and point 2 to be at the faucet.

SKETCH THE PROBLEM

Figure 10.28 describes the problem.

(continued) ▶

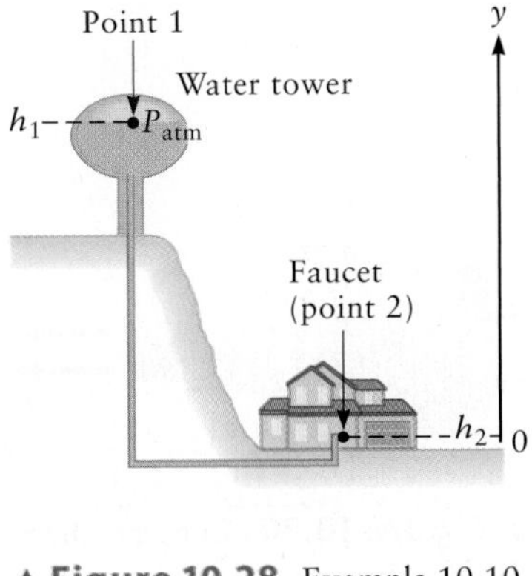

Point 1
Water tower
h_1
P_{atm}
Faucet (point 2)
y
h_2
0

▲ **Figure 10.28** Example 10.10.

▲ **Figure 10.29** The narrowing of this stream of water as it flows from a faucet can be understood using Bernoulli's equation and the principle of continuity.

IDENTIFY THE RELATIONSHIPS

Bernoulli's equation reads

$$P_1 + \tfrac{1}{2}\rho v_1^2 + \rho g h_1 = P_2 + \tfrac{1}{2}\rho v_2^2 + \rho g h_2$$

and we take locations 1 and 2 as shown in Figure 10.28. We wish to solve for the speed v_2 at the faucet, so we must find values for the pressures at the tower and faucet (P_1 and P_2, respectively), the heights of the tower and faucet (h_1 and h_2, respectively), and the speed at the tower (v_1). The water tower is open to the atmosphere, so the pressure there is atmospheric pressure and $P_1 = P_{\text{atm}}$. Likewise, the outlet of the faucet is open to the atmosphere, so it is also at atmospheric pressure and $P_2 = P_{\text{atm}}$. We can take the origin of our y coordinate axis to be at the faucet, leading to $h_2 = 0$ and $h_1 = 30$ m. We must now deal with v_1, the speed of the top surface of the water in the tower. The area of this top surface is much larger than the cross-sectional area of the faucet. As a result, this surface will fall very slowly as water flows from the faucet, and it is a good approximation to take $v_1 = 0$. This argument is equivalent to using the equation of continuity (Eq. 10.23) with the area of the water surface in the tower being much greater than the cross-sectional area of the faucet.

SOLVE

Putting all these quantities into Bernoulli's equation, we find

$$P_{\text{atm}} + 0 + \rho g h_1 = P_{\text{atm}} + \tfrac{1}{2}\rho v_2^2 + 0$$

$$\rho g h_1 = \tfrac{1}{2}\rho v_2^2$$

$$v_2 = \sqrt{2gh_1} \qquad (1)$$

Inserting our value for the height of the water tower gives

$$v_2 = \sqrt{2gh_1} = \sqrt{2(9.8\text{ m/s}^2)(30\text{ m})} = \boxed{24\text{ m/s}}$$

► ***What does it mean?***

Our result for v_2 in Equation (1) is precisely the same as the speed of an object that falls freely from height h_1, as found in Chapters 3 and 6. That is no accident. The water that flows from the water tower to the faucet in Figure 10.28 is also undergoing free fall, and we should expect it to have the same final speed we would find for a rock or a baseball falling through the same distance.

Figure 10.29 shows another example of water undergoing free fall as it flows from a faucet. Notice that the water stream becomes narrower (has a smaller cross-sectional area) the farther one gets from the faucet. This narrowing is a result of the principle of continuity together with Bernoulli's equation. According to the equation of continuity (Eq. 10.23), we have

$$v_1 A_1 = v_2 A_2$$

for any two points in a flowing stream of liquid. The speed is greater at the bottom of the stream in Figure 10.29 than at the top, so the cross-sectional area must be smaller.

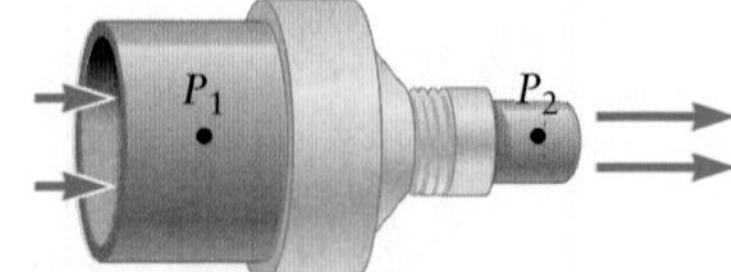

▲ **Figure 10.30** Concept Check 10.7.

CONCEPT CHECK 10.7 Pressure in a Garden Hose

Water flows through two garden hoses, one large and one small, connected as shown in Figure 10.30. How are the pressures in the two hoses related?

(a) $P_1 > P_2$ (b) $P_1 = P_2$ (c) $P_1 < P_2$

10.6 Real Fluids: A Molecular View

In all our work on moving fluids in Section 10.5, we ignored frictional forces internal to the fluid and between the fluid and the walls of the container (i.e., a pipe). These ***viscous*** forces depend on the fluid velocity and are often negligible when this velocity is small. Bernoulli's equation therefore works well when the fluid viscosity is small (e.g., when the fluid is air), when the fluid is flowing slowly, or both. However, in many situations viscosity is important. To understand the origin of viscosity and several other effects in fluids, we must consider fluids at a molecular scale.

Viscosity and Poiseuille's Law

Consider a fluid flowing through a tube as sketched in Figure 10.31. Fluid molecules near the wall of the tube experience strong forces from the molecules in the wall. In fact, these forces are so strong that in most cases the fluid molecules next to the wall are almost bound to it. You may think of the fluid in the tube as a series of cylindrical layers, a few of which are sketched in Figure 10.31A. Fluid molecules in layers near the wall move relatively slowly due to their attraction to the wall, whereas those in layers far from the wall move with a higher speed.

The motion of each fluid layer is strongly influenced by adjacent layers. Whenever the layers move at different speeds, there is a force between them very similar to the force of kinetic friction between two solid surfaces sliding past each other (e.g., a hockey puck sliding across a horizontal surface). The fluid layer denoted as layer 2 in Figure 10.31A is moving faster than the adjacent layer 1. As a result, layer 2 exerts a force on layer 1 that acts to "drag" layer 1 along with layer 2. This force is due to a property of fluids called ***viscosity***. In a "thick" fluid such as honey, drag forces are large and the fluid is said to have a large viscosity. "Thin" fluids such as water or methanol have a much smaller viscosity.

Viscous forces between adjacent fluid layers lead to the pattern of flow speeds shown in Figure 10.31B, where the fluid speed is plotted as a function of position within a pipe. Layers near the center of the tube have the highest speeds, and the flow speed approaches zero at the walls.

For a viscous fluid to move through a tube, the force due to the pressure difference across the ends of the tube must overcome the viscous force. If the pipe has a circular cross section with radius r and length L, and the pressures at the ends are P_1 and P_2, a careful application of Newton's second law shows that the *average* fluid speed in the tube is given by ***Poiseuille's law***,

$$v_{\text{ave}} = \frac{P_1 - P_2}{8\eta L} r^2 \tag{10.31}$$

Here η (the Greek letter eta) is the viscosity of the fluid. Values of the viscosity for some common fluids are given in Table 10.3. According to Poiseuille's law, increasing the pressure difference $(P_1 - P_2)$ increases v_{ave} and hence increases the flow rate. Most importantly, decreasing the tube radius r decreases the flow speed; in a smaller

TABLE 10.3 Viscosities of Some Common Fluids

Fluid	Viscosity (N · s/m²)
Water (20°C)	1.0×10^{-3}
Blood (at body temperature, 37°C)	2.7×10^{-3}
Air (20°C)	1.8×10^{-5}
Honey (20°C)	1000
Methanol (30°C)	5.1×10^{-3}
Ethanol (30°C)	1.0×10^{-3}

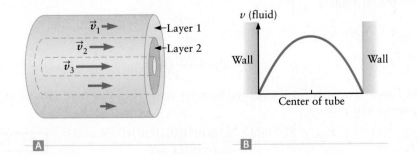

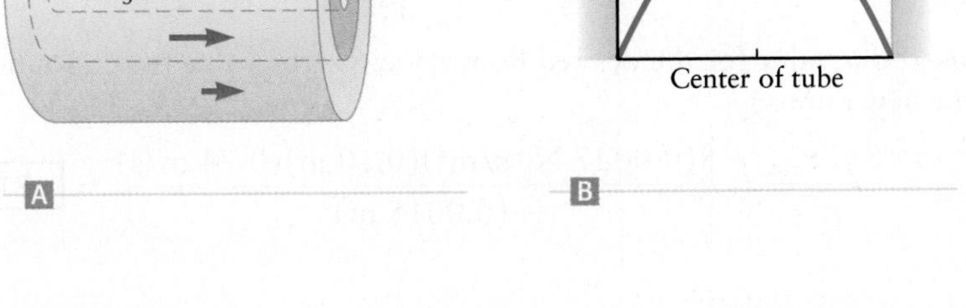

◀ **Figure 10.31** A When a fluid flows through a tube or pipe, viscosity causes fluid near the center to have a higher speed than fluid near the walls, as plotted schematically in B.

tube, a greater fraction of the fluid is close to the walls, so the viscous drag from the walls has a greater effect.

Poiseuille's law is useful for understanding fluid flow in various situations, including the flow of cooling water in an engine, the flow of crude oil through a long pipeline, and the flow of blood through the body. In many cases, one is interested in the flow rate Q (which we defined in Eq. 10.24); here Q is the rate at which fluid passes into or out of a tube or pipe. In our discussion of the principle of continuity and the volume flow into and out of a pipe, we assumed the flow velocity was constant across the radius of the pipe and found that the volume flow rate is $Q = vA$. For the case in Figure 10.31, in which the flow speed is not constant, we must use the average flow speed v_{ave}. Hence, we have

$$Q = v_{ave}A \tag{10.32}$$

where A is the area of the tube or pipe.

EXAMPLE 10.11 Blood Flow through an Artery

Consider the flow of blood through the large artery that extends from the heart to the lungs. The radius of this artery is typically 2.0 mm, and it is about 10 cm long. **(a)** If the pressure difference across the ends of the artery is 400 Pa (about 0.4% of atmospheric pressure), what is the average speed of the blood? **(b)** Now suppose this artery becomes somewhat narrower, as often happens with age. If the radius is reduced to 1.5 mm, what pressure difference is required to maintain the same average speed as in part (a)?

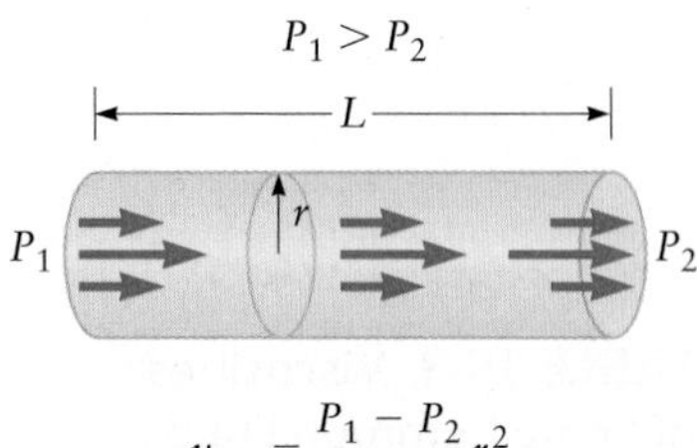

▲ **Figure 10.32** Example 10.11. Poiseuille's law tells how the average flow speed depends on the dimensions of the tube, the viscosity, and the pressure difference $P_1 - P_2$.

RECOGNIZE THE PRINCIPLE

This problem is a direct application of Poiseuille's law (Eq. 10.31), the relation between pressure difference and average speed for the flow of a viscous fluid (blood).

SKETCH THE PROBLEM

Figure 10.32 describes the problem.

IDENTIFY THE RELATIONSHIPS

According to Poiseuille's law, the average flow speed is related to the pressure across the artery $(P_1 - P_2)$ by

$$v_{ave} = \frac{P_1 - P_2}{8\eta L} r^2$$

where r is the radius of the artery and η is the viscosity.

SOLVE

(a) For blood (viscosity $\eta = 0.0027\ \text{N} \cdot \text{s/m}^2$ as listed in Table 10.3) flowing in an artery of radius $r = 2.0$ mm with a pressure difference $P_1 - P_2 = 400$ Pa, we find

$$v_{ave} = \frac{P_1 - P_2}{8\eta L} r^2 = \frac{(400\ \text{Pa})}{8(0.0027\ \text{N} \cdot \text{s/m}^2)(0.10\ \text{m})}(0.0020\ \text{m})^2 = \boxed{0.74\ \text{m/s}}$$

(b) To find the effect of changing the radius, we rearrange Poiseuille's law to solve for the pressure difference:

$$v_{ave} = \frac{P_1 - P_2}{8\eta L} r^2$$

$$P_1 - P_2 = \frac{8\eta L v_{ave}}{r^2}$$

Now we insert our value for the desired flow velocity—the value of v_{ave} found in part (a)—and the new radius:

$$P_1 - P_2 = \frac{8\eta L v_{ave}}{r^2} = \frac{8(0.0027\ \text{N} \cdot \text{s/m}^2)(0.10\ \text{m})(0.74\ \text{m/s})}{(0.0015\ \text{m})^2} = \boxed{710\ \text{Pa}}$$

▸ ***What does it mean?***
When the radius is decreased, the pressure must increase from 400 Pa to 710 Pa—quite a substantial change—to maintain the same flow speed. This example shows how what might seem to be a small change in an artery's size can lead to high blood pressure and the associated health risks.

Viscosity and Stokes's Law

Viscosity leads to a frictional force between adjacent "layers" of a moving fluid and also to a frictional force on the wall in Figure 10.31. The wall attracts molecules of the fluid, and the fluid molecules in turn exert a reaction force on the wall (Newton's third law at work again). This force on the wall is also the source of the drag force we considered in Chapter 3. According to Stokes's law (Eq. 3.23), the drag force on a spherical object of radius r that moves slowly through a fluid with a viscosity η is

$$\vec{F}_{\text{drag}} = -6\pi\eta r\vec{v} \tag{10.33}$$

where $\vec{v}$ is now the velocity of the object in the fluid. In Chapter 3, we wrote this relation as $\vec{F}_{\text{drag}} = -Cr\vec{v}$, where C is a constant; now we see that the drag coefficient C is proportional to the viscosity η. This is the *same* viscosity that enters into Poiseuille's law. The negative sign in Equation 10.33 indicates that the drag force is in the opposite direction as the object's velocity.

The Stokes's law force in Equation 10.33 is the key to understanding an important method for separating molecules according to their size. This method, based on a process called ***electrophoresis***, is shown schematically in Figure 10.33. Here we imagine a fluid containing a collection of molecules of different sizes; in a typical application, they are biological molecules such as DNA, RNA, or proteins. An electric force is then applied that pushes the molecules through the fluid such that the magnitude of the electric force is the same on each molecule.[2] Each molecule also experiences viscous drag. The Stokes drag force in Equation 10.33 applies to spherical objects, so it does not apply precisely to a real biomolecule. The basic physics of viscous drag certainly *does* apply, however, and each molecule in Figure 10.33 experiences a drag force whose magnitude depends on the molecule's size and shape. For these biomolecules, the drag force is given by the relation

$$\vec{F}_{\text{drag}} = -B\vec{v} \tag{10.34}$$

where B is a factor that depends on the size of the molecule. The value of B is a constant for any given kind of molecule—that is, for a particular type of DNA or RNA or for a particular virus—but it is different for different molecules.

If the electric force on a molecule is F_E, Newton's second law becomes

$$\sum F = F_E - Bv = ma$$

where we assume one-dimensional motion so that the forces and v are all along a single direction. This net force accelerates a molecule until it reaches a terminal speed (as discussed in Chapter 3), at which time the molecule's speed is constant and its acceleration is zero, leading to

$$\sum F = F_E - Bv_{\text{term}} = 0$$

$$v_{\text{term}} = \frac{F_E}{B}$$

Since B is different for different molecules, the terminal speed v_{term} also varies. Hence, different molecules travel through the fluid with different speeds, providing a way

[2]The origin of this force is explained in Chapter 17, where we discuss how an electric field exerts a force on an ion.

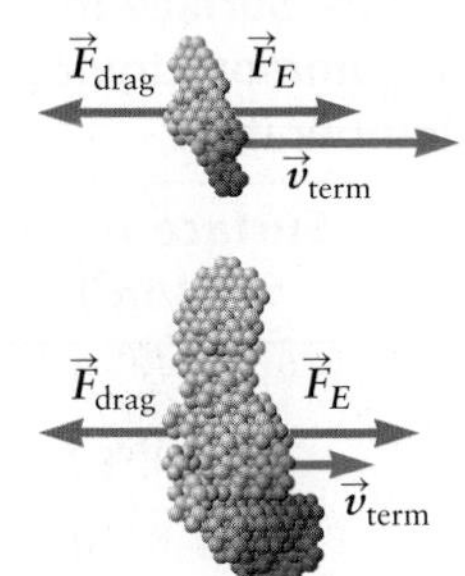

The electric forces are the same, but the drag coefficients are different, so the terminal speed is *larger* for the *smaller* molecule.

▲ **Figure 10.33** When an electric field is applied to a typical biomolecule in solution, the electric force F_E on the molecule is usually the same for all molecules because they usually have the same electric charge. The molecules also experience a drag force due to the viscosity of the fluid. This drag force depends on the size of the molecules, so different molecules move with different terminal speeds. This analysis is very similar to our calculation of the terminal speed of a skydiver in Chapter 3.

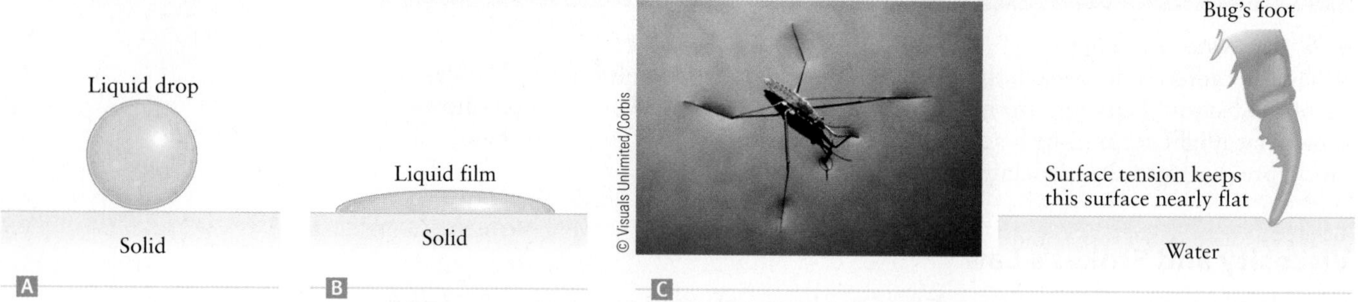

▲ **Figure 10.34** A and B Possible results for a small amount of liquid sitting on a solid surface. C Surface tension allows certain creatures to "walk" on the surface of a liquid.

to separate molecules by type. This difference is the basis for one method of DNA "fingerprinting."

Surface Tension

Consider a small amount of a liquid placed on the surface of a solid such as water sitting on a sheet of glass. Parts A and B of Figure 10.34 show two possible results; in part A, the water forms a spherical droplet, whereas in part B, the water spreads out as a flat layer. Which of these cases is found in real life? *Both* are possible, depending on the intermolecular forces. In most liquids, the liquid molecules bind more tightly to one another than to air. That is, molecules in the liquid "prefer" to be in contact with one another rather than in contact with molecules in the air. For this reason, droplets of such liquids form a shape giving the smallest possible surface area with the air, and that shape is a spherical drop.

This result can be expressed mathematically by saying that the fluid has a ***surface tension*** γ, where γ is a measure of the energy required to increase the surface area. The surface tensions of some typical fluids are given in Table 10.4. The SI unit of surface tension is joules per square meter (J/m^2).

TABLE 10.4 Surface Tensions of Some Common Liquids at Room Temperature

Liquid	Surface Tension (J/m^2)
Water	7.3×10^{-2}
Ethanol	2.3×10^{-2}
Mercury	48×10^{-2}

Surface tension is responsible for effects such as the one shown in Figure 10.34C, where a bug is "walking" on water. As with the case of a spherical drop, the surface of the water takes on a shape that keeps the surface area as small as possible, given that it also must support the bug's weight. Here, that shape is a nearly flat surface, which supports the bug's feet rather than letting them penetrate deep under the original surface. The weight of the bug is very small, so the forces that bind the water molecules together at the surface are able to overcome the added force due to the bug's weight.

Capillary Pressure

We have just stated that the molecules of a liquid such as water are attracted more strongly to other water molecules than to air molecules. When a liquid is placed in a very narrow tube, the shape of the surface also depends on the forces between the molecules in the liquid and the walls of the tube. Figure 10.35 shows water in a narrow glass tube, called a capillary tube. The surface of the liquid in a capillary tube is *not* flat; rather, the water's surface (called the meniscus) in a glass capillary tube is concave upward. Figure 10.35A also shows a molecular-scale view. Water molecules are attracted to the glass, and this attraction pulls the surface of the water up a small amount at the walls of the tube, producing the concave shape of the meniscus.

The molecules in some liquids do not bind well to the surfaces of a capillary tube. The case of mercury is shown in Figure 10.35B; here the mercury atoms are attracted more strongly to one another than to the walls, leading to a convex meniscus. Hence, the shape of the meniscus in a capillary tube depends on the molecular forces between the liquid and the walls of the tube.

The attraction of water molecules in Figure 10.35A to the walls of a glass capillary tube leads to a force that "pulls" the water upward in the tube. This force, which is associated with the ***capillary pressure***, causes the curvature of the water's surface (the meniscus), and the magnitude of this effect is proportional to the surface tension. In the simplest case, like the one shown in Figure 10.35A, the capillary pressure for a liquid in a tube of radius r is given by

$$P_{cap} = \frac{2\gamma}{r} \tag{10.35}$$

where γ is again the surface tension (Table 10.4). This result applies to water in a glass tube, where the fluid molecules are strongly attracted to the walls of the tube. For other cases, such as mercury in a glass tube (Fig. 10.35B), the capillary pressure is given by a slightly more complicated relation than Equation 10.35 and is more than we want to discuss in this book.

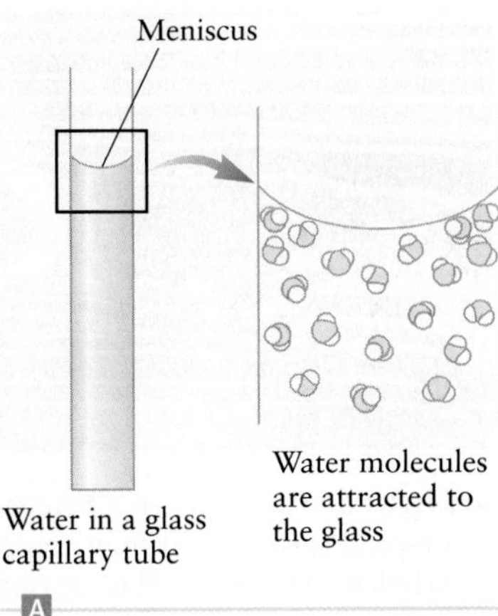

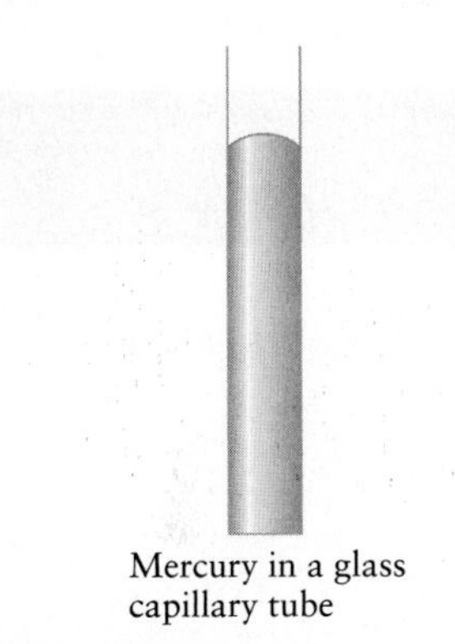

▲ **Figure 10.35** A When a thin glass capillary tube contains water, the shape of the water surface (the meniscus) is concave upward. This shape is caused by the attraction of the water molecules to the surface of the tube. B For mercury in a glass tube, the meniscus is convex.

How Plants Use Capillary Pressure

As an example of capillary pressure in action, let's consider how a tree moves water from its roots to its top branches. This problem is similar to the one we considered in Figure 10.13, where we discussed how to pump water from an underground well to the surface. There we saw that with a pressure difference of P_{atm}, one can only pump water up to a maximum height $h = (P_{atm}/\rho g) = 10$ m.

A tree is faced with a similar problem as it "pumps" water from its roots to its leaves. A tree also has a maximum pressure difference of P_{atm} to work with; hence, one might suspect a tree to have the same limitation on the maximum pumping height. So, with 10 m equal to about 30 ft, a tree cannot be taller than 30 ft and still move water from its roots to its top branches in this way. We know, though, that trees can be taller than 30 ft. How do they do it? The answer is that they use capillary pressure. The tubes that take water from tree's roots to its branches vary in size, but a radius of 1 μm ($= 10^{-6}$ m) is quite common. According to Equation 10.35, and consulting Table 10.4 for the surface tension of water, we find a capillary pressure of

$$P_{cap} = \frac{2\gamma}{r} = \frac{2(7.3 \times 10^{-2}\ \text{J/m}^2)}{1 \times 10^{-6}\ \text{m}} = 1 \times 10^5\ \text{Pa} \tag{10.36}$$

The case of a tree pumping water to its branches is then similar to the pump in Figure 10.13A, but now the capillary pressure adds to the atmospheric pressure, which enables the tree to pump water to a greater height. Recall that $P_{atm} = 1.01 \times 10^5$ Pa, so the capillary pressure in Equation 10.36 is just as great as atmospheric pressure. Narrower tubes can produce larger pressures. That is how trees do it.[3]

10.7 Turbulence

In our discussion of moving fluids and Bernoulli's equation, we explained how the pattern of airflow around an airplane wing produces lift on the wing, and we drew a hypothetical flow pattern in Figure 10.26. The flow pattern shown there is particularly simple. In many situations, fluid flow patterns can be much more complex and very difficult to predict. Indeed, in the development of the airplane, designers such as the Wright brothers used experiments in wind tunnels (Fig. 10.27) to study lift forces on wings of different shapes. Similar studies can now be done with large-scale

[3]The precise details of this process in trees are still a matter of much research. Although our discussion is certainly a simplified version, capillary pressure is the basic key to the process.

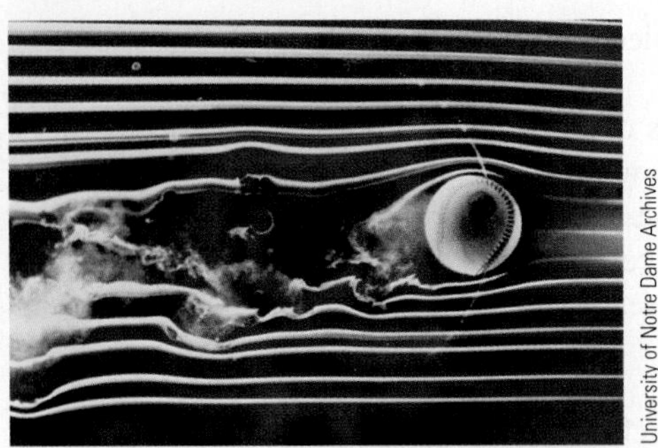

▲ **Figure 10.36** When a baseball travels at a typical speed, the airflow around the ball is turbulent. This baseball is moving from left to right.

computer simulations, but wind tunnels are still widely used to explore the aerodynamic forces on cars, bicycles, airplanes, and many other objects.

Figure 10.36 shows an example of the types of complex flow patterns that occur near even simple objects. This photo shows the flow of air around a baseball. When the speed of the incoming air is greater than about 25 m/s (about 50 mi/h), ***vortices*** are created in the baseball's wake. These vortices are circulating patterns of air, much like small tornados.

When fluid speeds are large, the flow patterns can be very complex and fluctuate erratically with time, a behavior known as ***turbulence***. Such complex flows play a role in a wide range of problems, including combustion, the weather (tornados and hurricanes), and the noise produced by rapidly moving cars and airplanes.

Summary | CHAPTER 10

Key Concepts and Principles

Pressure and density

The mechanics of fluids is described using the quantities ***pressure*** and ***density***. Pressure is equal to the force per unit area,

$$P = \frac{F}{A} \qquad (10.1)\ (\text{page } 314)$$

Pressure can be associated with the force on the surface of a container, or it can involve a surface contained within the fluid. The pressure in the atmosphere at the Earth's surface is approximately

$$P_{\text{atm}} = 1.01 \times 10^5 \text{ Pa} \qquad (10.2)\ (\text{page } 314)$$

The density of a substance is given by

$$\rho = \frac{M}{V} \qquad (10.3)\ (\text{page } 317)$$

An ***incompressible*** fluid is one for which the density is independent of pressure; most liquids are incompressible. The density of a gas varies with pressure; hence, gases are ***compressible*** fluids.

Dependence of pressure on depth

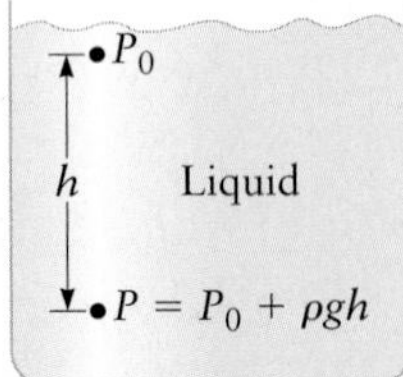

The pressure in a liquid varies with depth as

$$P = P_0 + \rho gh \qquad (10.5)\ (\text{page } 318)$$

The pressure increases as one moves deeper in a fluid because the pressure below must support the weight of the fluid above.

Real fluids

Fluids are composed of molecules, which helps explain the origin of ***viscosity*** (i.e., friction), ***surface tension***, and ***capillary pressure***. ***Poiseuille's law*** relates the average flow speed to the viscosity of the fluid and the pressure difference across the ends of a pipe:

$$v_{\text{ave}} = \frac{P_1 - P_2}{8\eta L} r^2 \qquad (10.31)\ (\text{page } 339)$$

where $P_1 - P_2$ is the pressure difference and η is the viscosity. The corresponding flow rate is

$$Q = v_{ave} A \qquad (10.32)\ (\text{page } 340)$$

Surface tension γ is the energy needed to increase the surface area of a fluid. The surface tension leads to an extra pressure when a fluid is confined to a narrow capillary. This capillary pressure is

$$P_{cap} = \frac{2\gamma}{r} \qquad (10.35)\ (\text{page } 343)$$

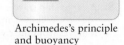

Applications

Pascal's principle

Pressure changes at one location in a closed container of fluid are transmitted equally to all locations in the fluid. This principle is the basis of ***hydraulics***. Hydraulic devices, such as a lift, are able to amplify forces. However, they de-amplify the corresponding displacement so that the work done is conserved.

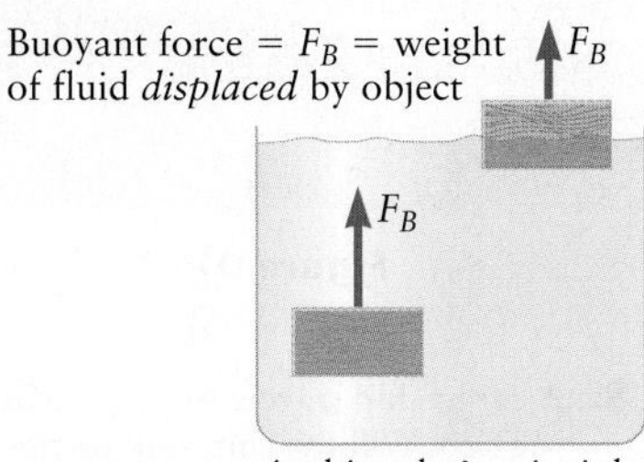

Archimedes's principle and buoyancy

Archimedes's principle

When an object is immersed in a fluid, the fluid exerts an upward force on the object that is equal to the weight of the fluid that is displaced by the object. This upward force is called the ***buoyant force***.

Principle of continuity

When a fluid flows through a pipe (or similar geometry), the ***equation of continuity*** tells how the speed v of a fluid is related to the cross-sectional area A of the pipe:

$$v_1 A_1 = v_2 A_2 \qquad (10.23)\ (\text{page } 334)$$

where the subscripts 1 and 2 denote two different sections of the pipe.

The work–energy theorem leads to ***Bernoulli's equation***, which relates the speed, pressure, and height at two different locations in an ideal fluid:

$$P_1 + \tfrac{1}{2}\rho v_1^2 + \rho g h_1 = P_2 + \tfrac{1}{2}\rho v_2^2 + \rho g h_2 \qquad (10.28)\ (\text{page } 336)$$

According to Bernoulli's equation, when the flow speed increases, the pressure must decrease, and vice versa. This leads to the lift force on an airplane wing.

Questions

SSM = answer in *Student Companion & Problem-Solving Guide* ⊗ = life science application

1. Design a hydraulic lift that could be used to move a car. Assume the input piston the person would push on has an area of $0.10\ m^2$. If you want a force of 300 N applied by the person to lift a car, estimate the area of the piston supporting the car.

2. ⊗ Explain how Archimedes's principle can be used to measure a person's percentage body fat. *Hint*: Fat has a different density than other body tissue.

3. The specific gravity of a substance is the ratio of its density to the density of water. Among other applications, measurements of the specific gravity are used by mineralogists to determine the composition of a gemstone. What are the units of specific gravity?

4. SSM Figure Q10.4 is a photograph of a graduated cylinder filled with four fluids. Starting with mercury on the bottom and going up, we have salt water, water, and vegetable oil. In addition, a solid object rests at the interface between each liquid. At the bottom is a steel ball bearing, next an egg, followed by a block of wood, and a table-tennis ball on top. (a) Rank these eight substances in terms of their densities from the highest to the lowest. (b) Which items have a specific gravity greater than 1? Less than 1?

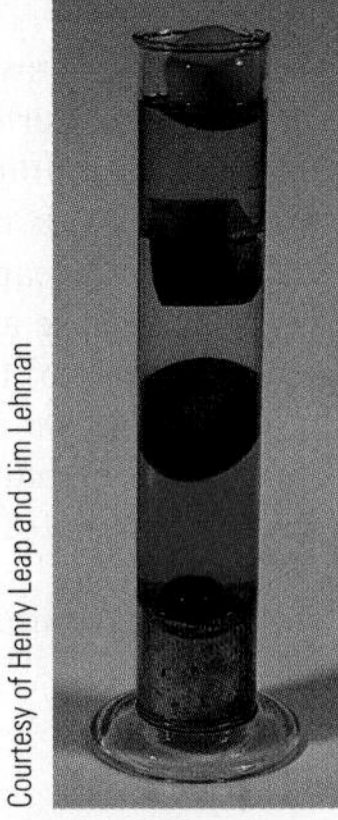

Figure Q10.4

Courtesy of Henry Leap and Jim Lehman

5. Explain why an ice cube floats in water. For ordinary water (i.e., not salt water) at room temperature, what fraction of an ice cube is above the water?

6. When water flows slowly from a faucet or pipette, the stream narrows and then eventually breaks into droplets (Fig. Q10.6). Explain why. *Hint*: Consider the effect of surface tension.

7. Explain why some people are able to float in ocean water but not in a freshwater lake (see Fig. Q10.7). (A freshwater lake contains rainwater.)

Figure Q10.6

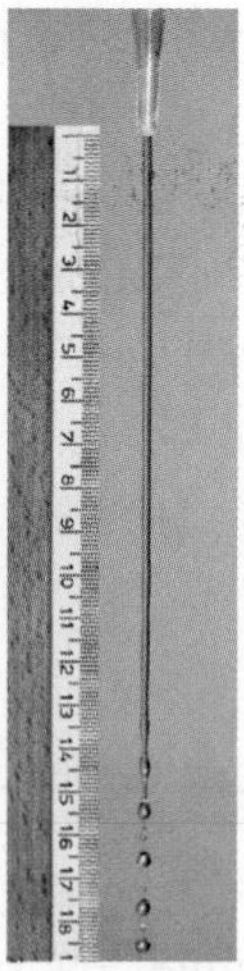

Figure Q10.7

8. A cargo ship travels from the Mississippi River into the Gulf of Mexico. Will the ship sink or rise with respect to the waterline as it moves from the river to ocean water? Why?

9. Three containers are filled with water to the same height (Fig. Q10.9). For which container is the pressure at the bottom the greatest? Or, are the pressures the same? Explain.

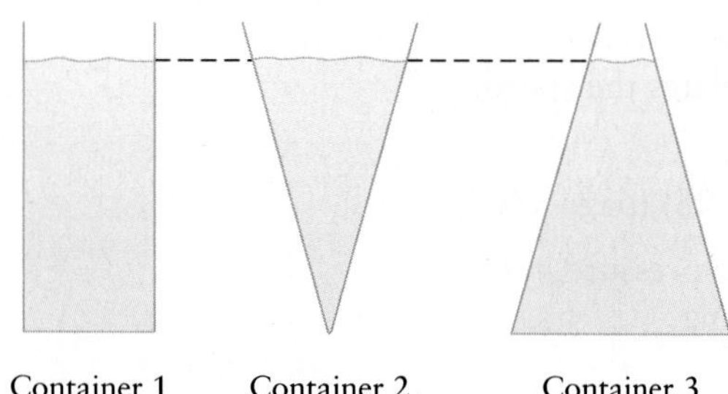

Figure Q10.9

10. Why does the "lava" in a Lavalamp (Fig. Q10.10) rise and then fall?

11. Consider a swimming pool and a tall graduated cylinder (like the one in Fig. Q10.4), both containing water and filled to the same height. (a) Is the pressure at the bottom of the pool greater than, less than, or equal to the pressure at the bottom of the graduated cylinder? (b) Is the total force exerted by the water on the bottom of the pool greater than, less than, or equal to the total force on the bottom of the graduated cylinder?

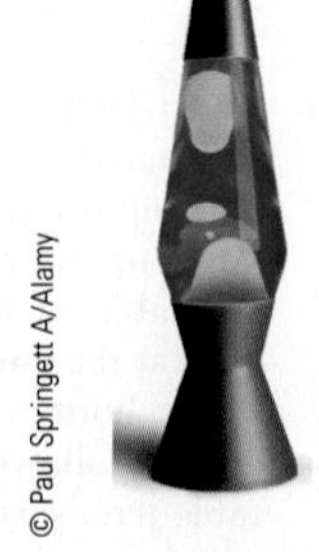

Figure Q10.10

12. Why are dams thicker at the bottom than at the top?

13. Consider a glass tube that has the shape shown in Figure Q10.13. It is filled with a fluid and is closed at the end on the left. If P_A is the pressure at point A, P_B the pressure at point B, and so forth, rank the pressures P_A, P_B, . . . from lowest to highest.

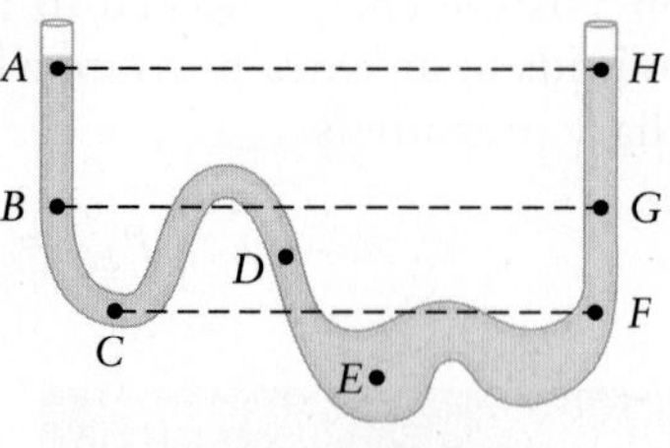

Figure Q10.13

14. Two blocks with the same volume but different densities are in a lake as shown in Figure Q10.14. Block m_1 floats at the surface with a portion above the water level, and block m_2 floats underwater. (a) Which block has the greatest mass? (b) For which block is the buoyant force greatest?

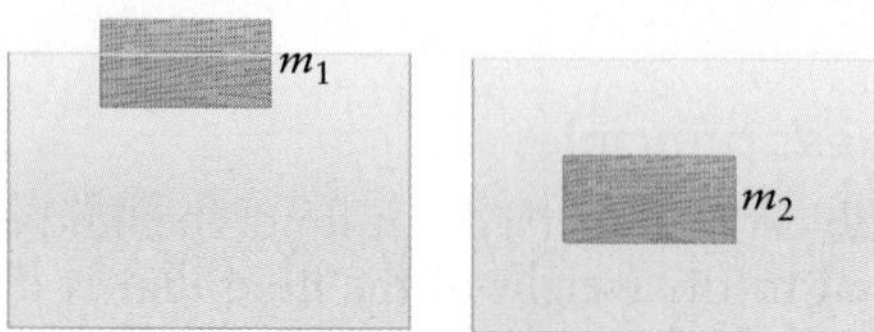

Figure Q10.14

15. When traveling in a commercial airplane, you are sometimes given snacks, such as peanuts, in a sealed bag. The bag is often "bulging" much more than when you purchase a bag of peanuts in a store. Explain why.

16. SSM Figure Q10.16 shows a popular demonstration involving a moving fluid. Here an air "jet" is aimed upward and levitates a small object such as a table-tennis ball. The ball is drawn *to the center of the jet*, where the speed is greatest. This behavior is found when the air jet is directed vertically (Fig. Q10.16 left) and also when the jet is directed at an angle (Fig. Q10.16 right). Use Bernoulli's equation to explain this behavior.

Figure Q10.16

17. Explain how there can be dust on a moving fan blade.

18. An altimeter is a device used to measure altitude. Most altimeters are based on measurements of air pressure. Explain how they work and estimate how much the pressure changes for a 100-m change in altitude. Discuss how changes in the weather can affect an altimeter reading.

19. In preparation for a bike tour of several countries, you ship your bicycle by air freight to Europe. When you arrive, you find that the tires of your bicycle have both blown out. Explain why that happened.

20. Explain how the specific gravity of a substance is connected with the buoyant force when an object is immersed in water.

21. The region near the South Pole in Antarctica has an altitude of about 3000 m, making the air pressure lower than at sea level. Explain why it is more difficult for an airplane to take off from the South Pole than from an airfield at sea level.

22. The author once had a small car with a "hatchback," a rear door that opens upward. One day, he forgot to latch it before driving on the highway, and he found that at high speeds the rear door lifted up as in Figure Q10.22B. Explain why.

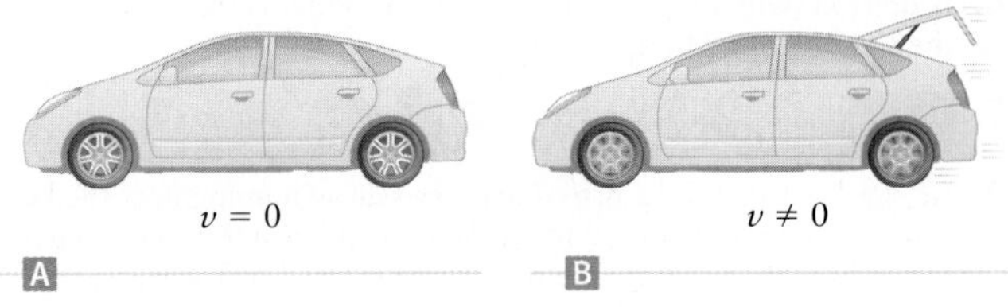

Figure Q10.22

Problems

SSM = solution in *Student Companion & Problem-Solving Guide*
★ = intermediate ✪ = challenging ⓧ = life science application
Ⓡ = reasoning and relationships problem
RT = reasoning tutorial in Enhanced WebAssign

10.1 PRESSURE AND DENSITY

1. What is the average density of the Earth?

2. In a neutron star, all the mass has "collapsed" into a relatively small volume. (a) If a particular neutron star has a radius of 1000 m and a mass of 2.0×10^{28} kg, what is its density? (b) Compare this value with the density of the Sun and with the density of steel.

3. ★ A piece of gold has a mass of 20 g. If this gold were formed into a very thin sheet of thickness 0.50 μm, find the area of the sheet. Assume it has the same density as listed in Table 10.1.

4. Consider a highway composed of concrete. If the road is 2.0 km long, 10 m wide, and 0.50 m thick, what is its mass?

5. ★ Ⓡ Estimate the mass of water in a typical bathtub. Assume the tub contains just enough water that you could be completely submerged.

6. Consider a steel sphere of mass 20 kg. Find its radius.

7. ★ ⓧ RT Knowing the density of water, estimate the volume of your body.

8. ★ Consider an airplane window (area 0.30 m^2). If the pressure inside the plane is atmospheric pressure and the pressure outside is 20% of P_{atm}, what is the force on the window? Ignore the velocity of the airplane (i.e., ignore Bernoulli's effect).

9. Consider a box with a lid 1.5 m wide and 0.70 m long. If the inside of the box is evacuated (i.e., its pressure is zero), how much force is required to open the lid? Could you open the lid?

10. ★ The pressure in the atmosphere is not constant, but fluctuates as the weather changes. If the pressure outside a window drops by 5% while the pressure inside does not change, what is the force on the window? Assume the area of the window is 1.5 m^2.

11. A block sits on a plate that in turn rests on the surface of a fluid as shown in Figure P10.11. The plate is partially submerged in the fluid, its area is 0.30 m^2, and there is air at atmospheric pressure P_{atm} above the block and liquid. If the mass of the block and plate together is 12 kg, what is the pressure P in the fluid just below the plate? Express your answer as $P - P_{atm}$.

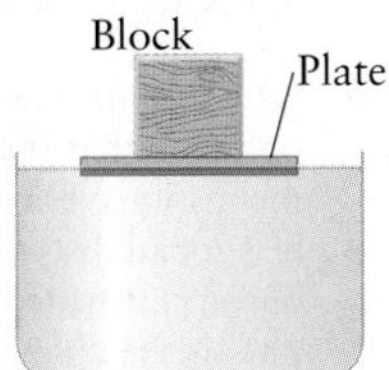

Figure P10.11

12. SSM ★ A suction cup works by virtue of a vacuum that is created within the cup. When the cup is pressed against a flat surface, most of the air is forced out, leaving a region of very low pressure. If a suction cup of area 1.0 cm^2 attached to a ceiling is able to support an object with a mass as large as 0.70 kg, what is the pressure inside the suction cup?

13. The four tires on the author's car are inflated to an absolute pressure of 2.5×10^5 Pa. If the car has a mass of 1800 kg, what is the contact area between each tire and the ground?

14. ★ RT Estimate the pressure exerted by your feet when you stand upright on the floor. Compare it to the pressure when you are wearing "stiletto" heels (use the value at the heel).

15. ★ Ⓡ When a typical 1200-page textbook is sitting on a table, it exerts a force on the table. Estimate the associated pressure.

16. The recommended pressure in your car's tires is usually specified as a gauge pressure. If a tire has a gauge pressure of 15 lb/in.^{2}, what is the absolute pressure in pascals?

17. Express a pressure of 5.0×10^6 Pa in units of pounds per square inch.

18. The water in the pipes of a typical house is at a gauge pressure of 3.0 times atmospheric pressure. What is the absolute pressure in pascals and in units of pounds per square inch?

19. ★ RT Each tire of a car contacts the road over an area of approximately 150 cm^2. Estimate the mass of the car.

20. Add a new column to Table 10.1, listing the specific gravity of each substance.

10.2 FLUIDS AND THE EFFECT OF GRAVITY

21. Find the pressure at the bottom of a swimming pool that is 2.5 m deep.

22. Consider a rectangular swimming pool of depth 2.5 m with a width and length of 15 m. What is the total force on the bottom of the pool due to the water pressure?

23. Consider again the swimming pool in Problem 22. What is the force exerted by the water on one of the sidewalls of the pool?

24. A diver is swimming at a depth of 10 m. What is the pressure in the water around him?

25. Estimate the force exerted by the water on Hoover Dam.

26. Consider a fish of average size swimming near the bottom of a lake of depth 20 m. (a) What is the water pressure next to the fish? (b) Estimate the total force exerted by the water on the outside surface of the fish.

27. SSM A U-tube contains two fluids with densities $\rho_1 = 1000$ kg/m^3 and $\rho_2 = 600$ kg/m^3 as sketched in Figure P10.27. What is the difference d in the heights of the top surfaces of the two fluids?

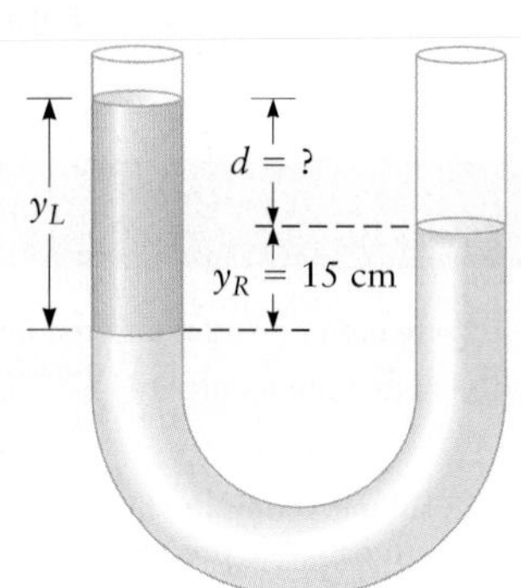

Figure P10.27

28. Figure 10.12 shows a device that can be used to measure atmospheric pressure. If $h = 6.5$ m, what is the density of the liquid?

29. An engineer is given the job of designing the water system for a tall building. He decides to locate the pumps at ground level and feed the entire building from there. If the top floor of the building is 200 m above ground level, what must the pressure be at pump level?

30. A bucket is filled with a combination of water (density 1000 kg/m^3) and oil (density 700 kg/m^3). These fluids do not mix. (The oil will float on top of the water.) If the layer of oil is 20 cm tall, what is the pressure at the interface between the oil and the water?

31. RT The blood pressure in a person's feet when he is standing is greater than the pressure in his head. Estimate this pressure difference and compare it with normal blood pressure, which is typically about 1×10^{-4} Pa.

32. When a person is given intravenous fluid, the bag containing the fluid is typically held about 1.0 m above the person's body. If the pressure in this fluid is just barely able to push the fluid into the person, what is the pressure in the person's blood?

33. The water pressure in the pipes in the basement of a tall house is 5.0×10^5 Pa. Estimate the water pressure on the second floor.

34. A swimming pool of depth 2.5 m is filled with ordinary (pure) water ($\rho = 1000$ kg/m^3). (a) What is the pressure at the bottom of the pool? (b) When the pool is filled with very salty water, the pressure changes by 5.0×10^3 Pa. What is the difference between the density of the salt water and the density of pure water?

35. The water system in a town is designed by a careless engineer. He places the new water tower at the top of an extremely tall hill, overlooking the mayor's house. The mayor then finds that some of the pipes in his house leak, which never happened with the old system. While making repairs, the mayor's plumber discovers that the pipes in the house leak when the pressure exceeds about five times atmospheric pressure. Estimate the height of the water level in the tower relative to the mayor's house.

10.3 HYDRAULICS AND PASCAL'S PRINCIPLE

36. A syringe has an area of 1.0 cm^2 in the barrel and then narrows down to an area of 0.10 mm^2 at the needle end. If a force of 5.0 N is applied to the syringe, what is the force produced at the tip of the needle?

37. Cars use a hydraulic system to transmit the force from the brake pedal to the actual brakes. If this system amplifies the force of your foot by a factor of 10, what is the ratio of the area of the hydraulic tube at the pedal to the area of the tube at a wheel?

38. SSM A block of mass 20 g sits at rest on a plate that is at the top of the fluid on one side of a U-tube as shown in Figure P10.38. The U-tube contains two different fluids with densities $\rho_1 = 1000$ kg/m^3 and $\rho_2 = 600$ kg/m^3 and has a cross-sectional area of $A = 5.0 \times 10^{-4}$ m^2. If the surfaces are offset by an amount h as shown, (a) which side of the U-tube contains the fluid with the greater density? (b) Find h. (Assume h is positive.)

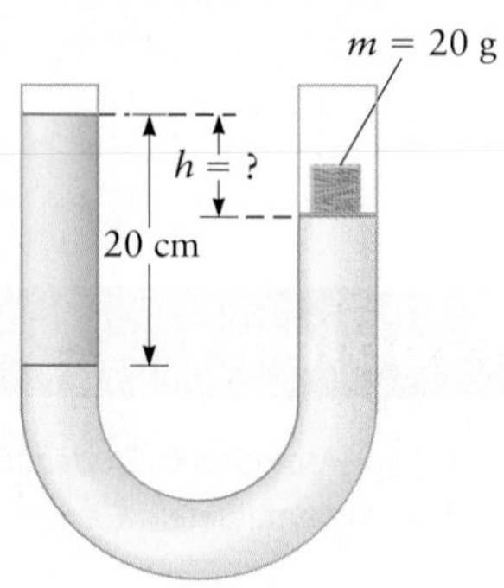

Figure P10.38

39. A car of mass 1400 kg rests on a hydraulic lift with a piston of radius 0.25 m that is connected to a second piston of radius 0.030 m. If the car is just barely lifted off the ground, what is the force at the second piston?

10.4 BUOYANCY AND ARCHIMEDES'S PRINCIPLE

40. A large suitcase full of gold sits at the bottom of a lake. A team of treasure hunters uses a crane to recover the suitcase and finds that a force of 3.0×10^5 N is required to lift the suitcase to the surface. What is the volume of the suitcase?

41. A wooden log is found to float in a freshwater lake. If 35% of the log's volume is above the surface, what is the log's density?

42. Under normal conditions in the air, you are just barely able to lift a mass of 90 kg. Your friend drops a box of volume 2.5 m^3 into a lake. If you are just able to lift it to the surface, what is the mass of the box?

43. A rectangular block of wood (density 750 kg/m^3) of height 25 cm floats in a lake. What is the height of the wood above the water level?

44. RT Estimate your apparent weight when you are under water.

45. An object has a weight of 20 N in air and an apparent weight of 5.0 N when submerged in water. What is its density? Will it float?

46. A submarine uses tanks that can be filled with either water or air so that the submarine can have "neutral buoyancy." That means its average density is equal to the density of water where it is located so that it will remain at that level under water. The density of water changes a small amount with depth in the ocean due to the changes in pressure and temperature. How much additional mass must a submarine take on to sit at the ocean bottom at a depth of 500 m? At this depth, the density of water is about 0.25% larger than at atmospheric pressure. Assume the volume of the submarine is 3.0×10^4 m^3 and it is initially floating at the surface, fully covered with water.

47. Consider a wooden block of density 700 kg/m^3 and volume 2.0 m^3. What force is required to hold it completely under water?

48. SSM RT You want to build a raft that can hold you (80 kg) plus some supplies (40 kg) for a long trip. You decide to make the raft out of logs that are each 20 cm in diameter and 3.0 m long. If the density of each log is 600 kg/m^3, how many logs do you need?

49. You want to design a helium-filled balloon that will lift a total payload of 1500 kg. What volume of helium is needed to just barely lift this payload? Assume the balloon itself has negligible mass.

50. Two metal blocks are attached by a cable as shown in Figure P10.50 and are submerged at rest near the surface of a lake. If the volumes are $V_1 = 2.0$ m^3 and $V_2 = 1.2$ m^3, what is the total buoyant force on the two blocks?

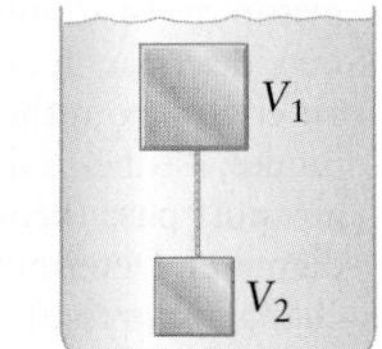

Figure P10.50 Problems 50, 51, and 52.

51. For the blocks in Figure P10.50, suppose instead of being near the surface of the lake they are just above the bottom of the lake. Find the buoyant force in this case.

52. For the blocks in Problem 50, assume they are barely submerged in water so that block 1 is just below the surface. If block 2 has a density twice that of block 1, what is the density of block 2?

53. A chest of volume 1.2 m^3 filled with gold sits at the bottom of a lake of depth 12 m. How much force is required to lift the chest to the surface?

54. A solid object is made by gluing together two pieces of equal mass. If one of the pieces is made from a material with density 500 kg/m^3, while the other has a density 1400 kg/m^3, will the object float?

55. A piece of balsa wood is placed into a tub full of ethanol. What fraction of the balsa wood is above the surface of the ethanol?

56. A piece of aluminum is attached to a string and suspended in a pool of oil with density 750 kg/m^3. If the apparent weight of the aluminum is 540 N, what is the volume of the aluminum?

10.5 FLUIDS IN MOTION: CONTINUITY AND BERNOULLI'S EQUATION

For Problems 57 through 67, assume an ideal fluid (i.e., a fluid with no viscosity).

57. Water is flowing through a pipe (area 4.0 cm^2) that connects to a faucet adjusted to have an opening of 0.50 cm^2. If the water is flowing at a speed of 5.0 m/s in the pipe, what is its speed as it leaves the faucet?

58. For the faucet in the Problem 57, how long does it take for water from the faucet to fill a bucket of volume 0.12 m^3?

59. SSM The blood flow rate through the aorta is typically 100 cm^3/s, and a typical adult has about 5.0 liters of blood. (a) How long does it take for all your blood to pass through the aorta? (b) If your aorta has a diameter of 2.0 cm, what is the speed of blood as it flows through the aorta?

60. RT It takes the author about 60 s to fill his truck with gasoline (48 liters, about 12 gal). (a) What is the flow rate in cubic meters per second? (b) Estimate the speed of the gasoline as it flows through the nozzle of the gasoline pump.

61. A tank of water sits at the edge of a table of height 1.2 m (Fig. P10.61). The tank springs a very small leak at its base, and water sprays out a distance of 1.5 m from the edge of the table. What is the water level h in the tank?

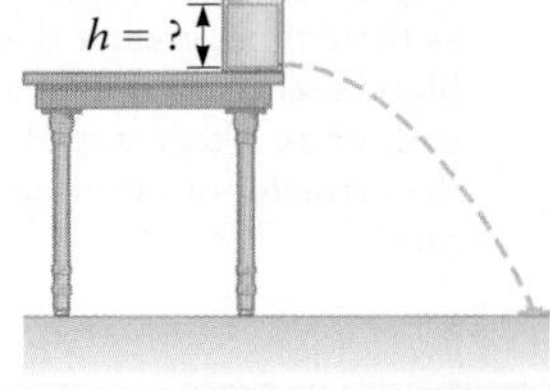

Figure P10.61

62. R A house sits in a valley and is 30 m lower than the water tower that serves it. What is the speed of water when it sprays out of a shower in the house?

63. Consider a large blood vessel that carries blood to the heart. If this blood vessel has a cross-sectional area of 1.5 cm^2 and the blood speed is 30 cm/s, what volume of blood is delivered in 1.0 s?

64. Consider an airplane wing of area 20 m^2. If the airflow speed over the top of the wing is 200 m/s, while the speed across the bottom is 150 m/s, what is the lift force on the wing at sea level at 0°C?

65. SSM R A Frisbee is observed to fly nearly horizontally, which implies that the lift force must be approximately equal to the weight of the Frisbee. If the air speed over the top of the Frisbee is 9.0 m/s, what is the flow speed across the bottom? Assume the Frisbee has a mass of 0.15 kg.

66. A fire extinguisher contains a high-pressure liquid so that it can spray the liquid out very quickly when needed. If the fluid leaves a fire extinguisher at a speed of 20 m/s, what is the pressure inside?

67. The flow rate through a blood vessel is found to be 1.5×10^{-7} m^3/s. Express this flow rate in units of cubic centimeters per second.

10.6 REAL FLUIDS: A MOLECULAR VIEW

68. A blood vessel is 20 cm long and has a radius of 2.0 mm. If blood is flowing through it with a volume flow rate of 2.0×10^{-7} m^3/s, what is the difference in the pressures at the two ends of the blood vessel?

69. An air duct in a building has a radius of 10 cm and a length of 10 m. What pressure is required to push air through the duct at a flow rate of 1.0 m^3/s?

70. A glass sphere of radius 1.0 mm is dropped into a lake. What is the terminal speed of the sphere?

71. R Estimate the terminal speed for a dust particle in air. Assume it has a radius of 1.0 μm and a density similar to that of balsa wood. Your answer will tell you why dust particles remain "suspended" in air for long periods.

72. SSM What is the capillary pressure for water in a vertical tube of diameter 0.10 mm?

73. Repeat Problem 72, but now for a tube of radius 10 nm. Express your answer in pascals and as a ratio relative to atmospheric pressure.

74. A glass capillary tube with a radius of 0.050 mm is inserted into an open container filled with an unknown liquid, and the meniscus has an appearance similar to Figure 10.35A. It is found that the surface tension causes the liquid in the tube to rise a distance of 2.5 cm relative to the surface of the liquid outside the tube. If the density of the liquid is 1000 kg/m^3, what is the surface tension?

75. The viscosity of normal (healthy) blood is about three times greater than the viscosity of water. Certain diseases such as polycythemia can cause the viscosity of blood to be as much as three times greater than normal. If the viscosity of a person's blood increases by a factor of three while the pressures at the ends of an artery stay at their normal levels, by what factor must the diameter of the artery increase to keep the same volume flow rate?

76. Estimate the terminal speed of a DNA molecule in water. Assume the only forces on the molecule are gravity and the Stokes drag force (Eq. 10.33).

77. Blood flows through an artery of diameter d and length L, and the pressure difference between the ends of the artery is ΔP. If the diameter is reduced by a factor of two and the pressure difference is kept the same, what happens to the flow rate?

Additional Problems

78. A steel ball bearing will float in mercury as shown in Figure P10.78. What fraction of the ball bearing's volume is above the surface of the mercury?

© Cengage Learning/Charles D. Winters

Figure P10.78
Problems 78 and 79.

79. What will happen if water is poured on top of the mercury and ball bearing of Problem 78 such that the ball bearing is completely submerged in the water? Will the ball bearing (a) sink farther into the mercury, (b) rise up from its former position, or (c) stay at the same position? Why? Calculate the fraction x of the ball bearing above the surface of the mercury under these new circumstances. *Hint*: If a fraction x of the volume is in the water, a fraction $(1 - x)$ of the volume will be in the mercury.

80. A small, square plank of oak floats in a beaker half full of water. The piece of oak is 6.0 cm on a side and 3.0 cm thick and floats on its side as shown in Figure P10.80. (a) Find the location of the plank's center of mass with respect to the water's surface. (b) A very light vegetable oil is poured slowly into the beaker so that the oil floats on the water without mixing and until the oak plank is a few centimeters below the surface of the oil. If the density of the oil is 700 kg/m^3, what is the new position of the center of mass of the plank with respect to the water's surface?

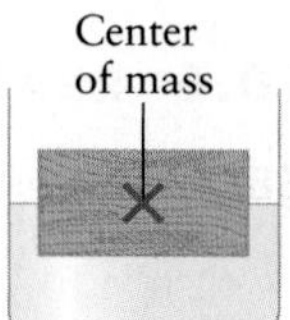

Figure P10.80

81. **Drawing (up) the longest straw.** Consider a contest where the challenge is to find the greatest height from which the contestant in Figure P10.81 is able to drink water through a straw. Find the theoretical limit on the vertical distance h between the open top surface of the liquid and the top of the straw. *Hint*: Assume (however unlikely) the contestant can create a perfect vacuum at the top of the straw.

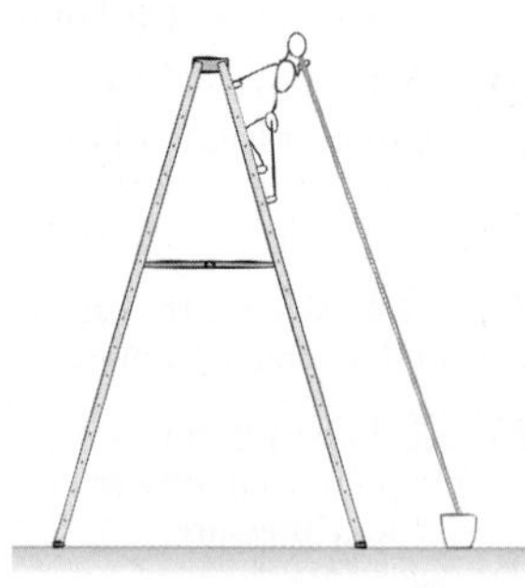

Figure P10.81

82. **Before fuel injection.** Some automobile engines (mainly older ones) use a carburetor to make the liquid fuel into vapor and mix it with air for combustion. The basic principle of carburetion is shown in Figure P10.82. A piston moves down in the cylinder, thereby drawing air from the outside through an air filter and into the carburetor. The filtered air enters from the left in Figure P10.82 and moves into the main intake, a tube of diameter 6.0 cm, with speed $v = 7.0$ m/s. The air must pass through a region of the intake that has a smaller diameter. Determine what diameter would be needed to cause a change in pressure such that fuel from the reservoir is pulled into the airflow. The surface of the fuel in the reservoir is $h = 42$ cm below the bottom of the intake, and the density of the fuel is 0.72 times that of water.

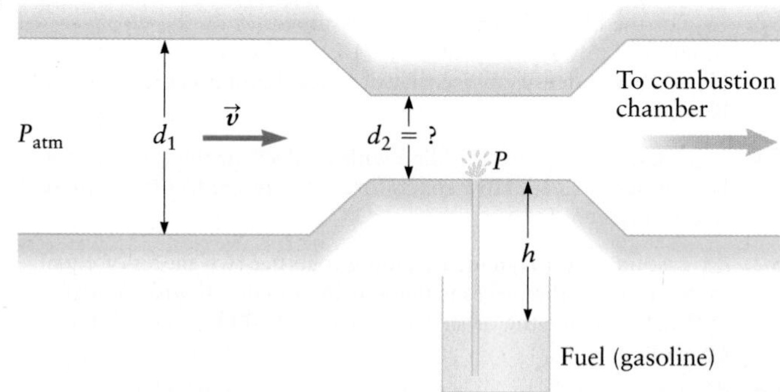

Figure P10.82

83. Consider a faucet with water running out in a smooth stream. How could one find the flow rate in cubic centimeters per second by simply taking measurements with a ruler? Following the diagram in Figure P10.83, find the rate of flow if $d_1 = 1.2$ cm, $d_2 = 0.70$ cm, and $h = 7.0$ cm.

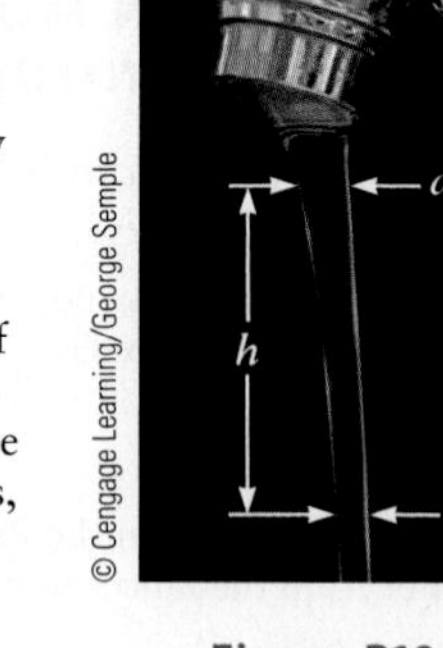

© Cengage Learning/George Semple

Figure P10.83

84. **Long snorkel.** Inhalation of a breath occurs when the muscles surrounding the human lungs move to increase the volume of the lungs, thereby reducing the air pressure there. The difference between the reduced pressure and outside atmospheric pressure causes a flow of air into the lungs. The maximum reduction in air pressure that chest muscles can produce in the lungs against the surrounding air pressure on the chest and body is about 3500 Pa. Consider the longest snorkel a person can operate. The minimum pressure difference needed to take in a breath is about 200 Pa. Find the

depth h that a person could swim to and still breathe with this snorkel (Fig. P10.84).

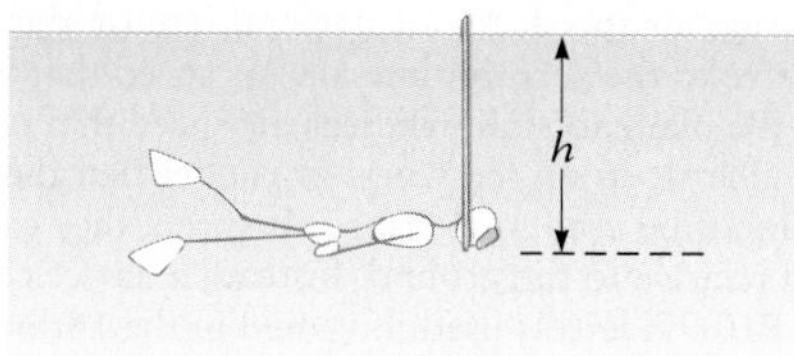

Figure P10.84

85. A scuba diver swims in a freshwater lake (Fig. P10.85). (a) If the maximum pressure in her tanks is 10 atm, to what depth h could she swim and still breathe? (b) Could she go deeper in the ocean? Calculate your answers to three significant figures.

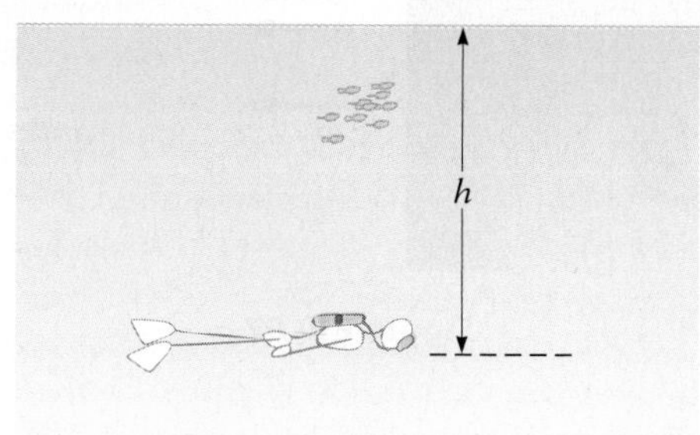

Figure P10.85

86. **Snorkeling elephants.** An elephant can use its long trunk as a snorkel while swimming or walking across the bottom of rivers (Fig. P10.86). The lining of an elephant's lung is reinforced with connective tissue not present in other mammals, allowing elephants to generate larger pressure differences when breathing. If an elephant's lungs are 2.0 m below the surface and the minimum pressure difference needed for it to inhale is 200 Pa, what is the maximum pressure difference an elephant's lungs can generate? Compare your answer to the result for human lungs discussed in Problem 84.

Figure P10.86 An elephant in the Chobe River, Botswana.

87. When an airplane is flying at a typical cruising altitude of 30,000 ft, the cabin pressure is typically about 75% of atmospheric pressure at ground level. What is the approximate force on one of the cabin windows of the airplane?

88. **Pulmonary surfactant.** Oxygen enters the blood in the hollow and roughly spherical alveoli of the lungs. The alveoli are small, approximately 170 μm in radius (Fig. P10.88), yet some 300 million of them make up 86% of the volume of each lung. The inner lining of the alveoli is, of course, wet as all living tissue must be. Thus, there is essentially a bubble of liquid just inside the lining of an alveolus having a surface tension that acts to compress the bubble, effectively generating an inward pressure that can be described by Equation 10.35. In this case, r is the radius of the sphere and P_{cap} is instead $P_{collapse}$, the inward pressure on the sphere. (a) If the liquid lining an alveolus is pure water, what is $P_{collapse}$? (b) What is the ratio of this pressure to the pressure difference generated when breathing? (c) A surfactant, a component of detergents, is a substance that greatly reduces the surface tension of a liquid. The alveoli have special cells that secrete a combination of phospholipids and proteins that give the fluid a surface tension $\gamma = 1.8 \times 10^{-2}$ J/m^2. Calculate $P_{collapse}$ for the new surface tension and compare it with the pressure difference needed for normal breathing described in Problem 84.

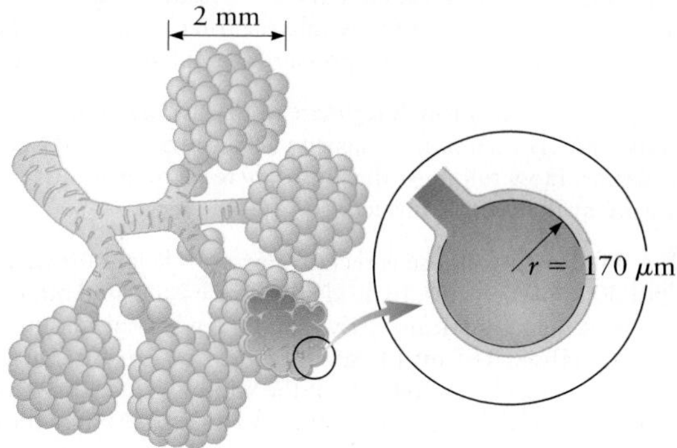

Figure P10.88

89. **Sphygmomanometer.** Blood pressure is measured at the brachial artery in the upper arm (Fig. P10.89). An inflatable strap is pumped with air until the artery is collapsed, stopping all flow. The pressure in the cuff is then slowly reduced while the operator listens with a stethoscope for a pulse downstream of the cuff. The pressure at which the pulse resumes is the ***systolic*** (maximum) pressure. The brachial artery has a radius of 0.50 cm, and the cuff is usually applied at a point 0.30 m away from the aorta at the heart. If the flow rate though the brachial artery is 1.5×10^{-5} m^3/s, what is the pressure drop between the point of measurement and the aorta? Is a measurement from the brachial artery in the arm a good representation of the pressure in the aorta at the heart? (Assume the heart and arm are at the same height.)

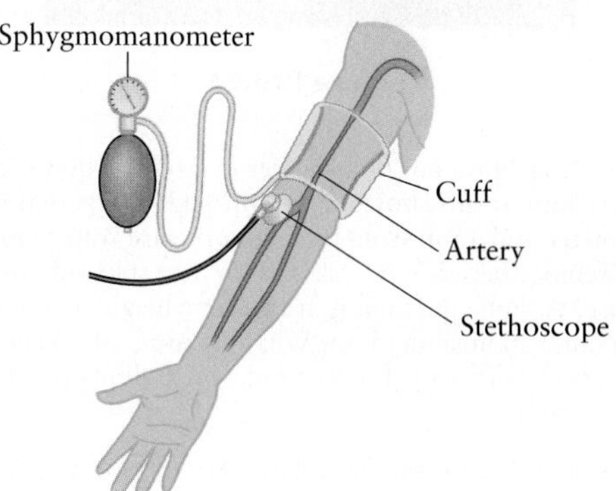

Figure P10.89

90. In Example 10.11, we considered how the pressure ΔP across the ends of a narrowed blood vessel must increase to keep the average flow speed constant. If we instead want the smaller blood vessel to have the same flow rate as the larger blood vessel, what value must ΔP have?

91. SSM The density of air decreases with height above the Earth, and when the density is too low, people get sick or even become unconscious. That is why the cabins of commercial airplanes are "pressurized" during flight. Cabin pressure, however, is lower than the pressure at sea level. Cabins typically have a pressure equal to the air pressure at an altitude of about 8000 ft above the Earth's surface. This reduced pressure is one reason some people feel fatigued after a long flight. (a) Use Equation 10.11 to calculate this cabin pressure. (b) Estimate the total net force on the walls of a Boeing 737 flying at 35,000 ft. Approximate the airplane as a cylinder of length 30 m and diameter 4 m. *Hint*: You will need to estimate the air pressure outside the plane using Figure 10.15. (c) What would the force on the walls be if the cabin pressure equals the pressure at the Earth's surface?

92. You can make a rough estimate of the thickness of the Earth's atmosphere by assuming its density is constant with the value 1.3 kg/m^3. How tall must the atmosphere be to give the observed pressure at the Earth's surface?

93. A window in a house is rectangular, with height 40 cm and width 30 cm. The pressure inside the house is normal atmospheric pressure when a hurricane arrives and the wind speed outside is 75 m/s (about 165 mi/h). (a) Does the pressure difference between the inside and outside of the window tend to push the window into the house or out of the house? (b) What is the force on the window?

94. A backyard waterfall used in landscaping (Fig. P10.94) has water cascading down at a rate of 18 liters/min. (a) If the waterfall is fed using a hose of diameter 2.5 cm, what is the speed of water in the hose? (b) If the hose is 15 m long, what is the pressure difference between the two ends of the hose?

Figure P10.94

95. (a) Water flows out of a shower in a typical house at a rate of 20 liters/min. If this shower is fed through a pipe that is 2.0 cm in diameter and 15 m from the high-pressure water tank, what is the difference between the pressure in the tank and atmospheric pressure? Assume the tank is at the same height as the shower and assume an ideal fluid. (b) What pressure difference is needed for an energy-efficient shower head, which limits the flow rate to 4.0 liters/min?

96. **Collapsing submarines.** The collapse depth of a typical military submarine—that is, the depth at which the submarine would be crushed by the force due to water pressure—is 700 m. If the pressure on the inside of the wall of such a submarine is P_{atm}, what is the force on a 1-m^2 area of the wall at the collapse depth?

97. **Measuring air speed.** An airplane pilot must always know her speed relative to the surrounding air. Air speed that is too low can cause the plane to stall, whereas air speed that is too high can cause so much stress on the wings or rudder that they fail. GPS systems cannot be used to measure air speed; they can only measure speed relative to the ground. Instead, a device called a pitot tube (Fig. P10.97, left) is used. Invented in the 18th century, a pitot tube determines the air speed v from the pressure difference $p_1 - p_2$ in Figure P10.97, right. Here the plane is flying toward the left, and the air speed is zero in the central tube (where the pressure is p_1) and v across the opening at the top where the pressure is p_2. (a) Is $p_1 - p_2$ positive or negative? (b) Use Bernoulli's equation to calculate this pressure difference for an airplane traveling with an air speed of 250 m/s (about 500 mph).

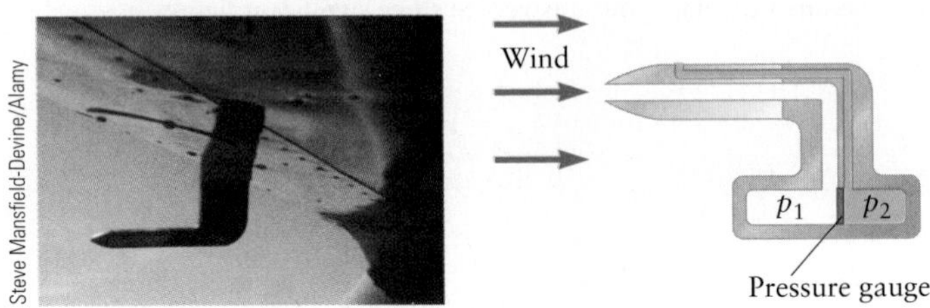

Figure P10.97

98. Ⓡ The water in the author's house is provided by a well. Water is pumped from the well to a holding tank in the basement that is pressurized so that water can flow from the tank to the upper floors of the house. The pressure in the tank is approximately 3.0×10^5 Pa. At approximately what speed does water spray out of a shower on the second floor? Ignore the viscosity of the water.

99. Ⓡ **Artesian well.** In some places, the water underground is at a pressure high enough that it will flow up and out of the ground without the need of a pump. Figure P10.99A shows an example, which is called an artesian well. Part B of the figure shows how this flow can happen. Loosely speaking, the level of the well is lower than the upper surface of the groundwater, which is also called the water table. Suppose the outlet of an artesian well is 2.5 m below the water table. Approximately what speed will the water have when it comes out of the well? Ignore the viscosity of the water.

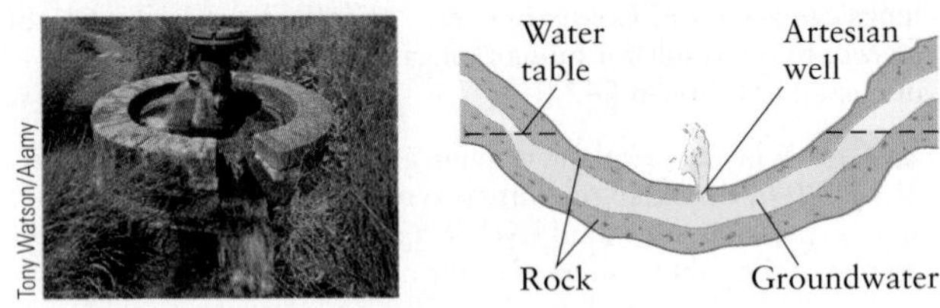

Figure P10.99

CHAPTER

11

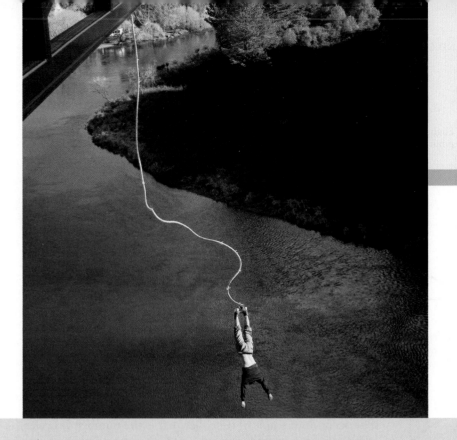

◀ *The physics of oscillations and elasticity are keys to the operation of this bungee cord (and the safety of the jumper!). (David Wall/Alamy)*

Harmonic Motion and Elasticity

In this chapter, we consider ***harmonic motion***, a type of motion in which an object moves along a repeating path over and over and over again. The universe is filled with examples of harmonic motion, such as the pendulum motion of a child on a playground swing and the motion of the Earth as it orbits the Sun. While there are many different examples of harmonic motion, certain features involving force and energy are common to all. We'll also find a close connection to many features of circular and rotational motion we have studied in previous chapters.

OUTLINE

11.1 General Features of Harmonic Motion

Figure 11.1A shows a person hanging from a spring. This spring is a little bit like the bungee cord in the photo at the start of this chapter, except that the spring in Fig. 11.1 is tightly coiled so that the force from the spring is the dominant force acting on the person (e.g., this force dominates over the force of gravity). Figure 11.1A shows several "snapshots" of the person as he "vibrates" up and down due to the force from the bungee spring.

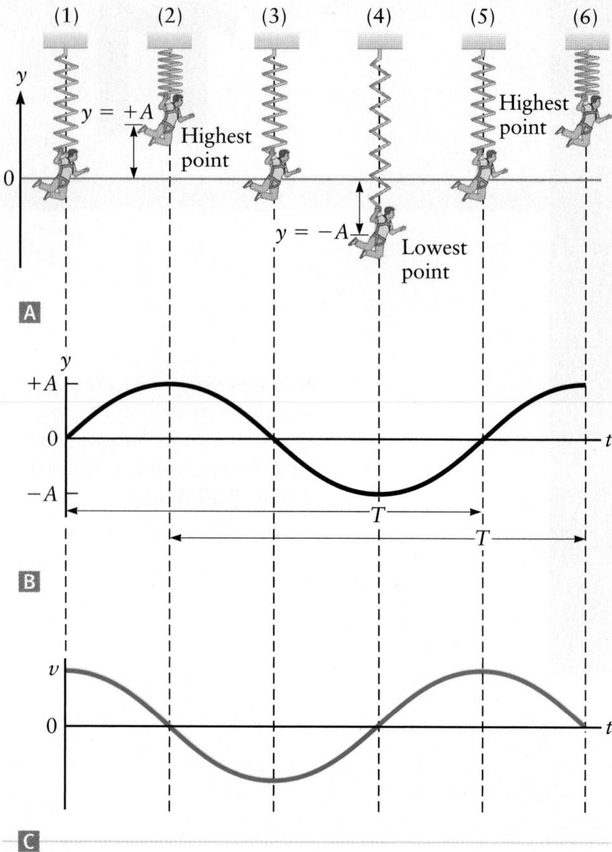

▲ **Figure 11.1** A This bungee jumper undergoes simple harmonic motion as he oscillates up and down. B Plot of the jumper's position y versus time. The period T of the motion is the time it takes to make one complete up-and-down oscillation. C Plot of the jumper's velocity versus time.

We can describe the position of the person on the bungee spring by measuring his location y on a vertical axis, and Figure 11.1B shows a plot of y as a function of time. Notice how each point along the y–t plot matches the corresponding physical location of the person in Figure 11.1A. This is an example of ***oscillatory motion***; the jumper's motion varies in a repeating fashion as he moves up and down.

To describe oscillatory motion mathematically, we use the notions of position, velocity, and acceleration, but we also need some additional quantities that relate to the repeating nature of the motion. One such variable is the ***period*** T, which is the *repeat time* of the motion. We have already encountered this quantity in our discussion of circular motion in Chapter 5. The period of our bungee jumper is the time it takes him to travel through one complete "cycle" of the motion, so it is the time interval between adjacent maxima of the position y when plotted as a function of time (Fig. 11.1B). The period T is also equal to the time interval between adjacent minima of y.

Another way to characterize harmonic motion is in terms of its ***frequency***. The frequency f is the number of oscillations that occur in one unit of time; hence, f is the number of cycles completed in 1 second. Since the period is the number of seconds required to perform one cycle, frequency and period are related by

$$f = \frac{1}{T} \tag{11.1}$$

The units of frequency are cycles per second, which are called *hertz* (Hz).

EXAMPLE 11.1 Measuring the Period

Consider a hypothetical system for which the position y oscillates as sketched in Figure 11.2. Estimate the period and frequency of the motion.

RECOGNIZE THE PRINCIPLE

The period T is the time required for one *complete* oscillation, and we can estimate T from Figure 11.2. The frequency is then given by $f = 1/T$.

SKETCH THE PROBLEM

Figure 11.2 describes the problem.

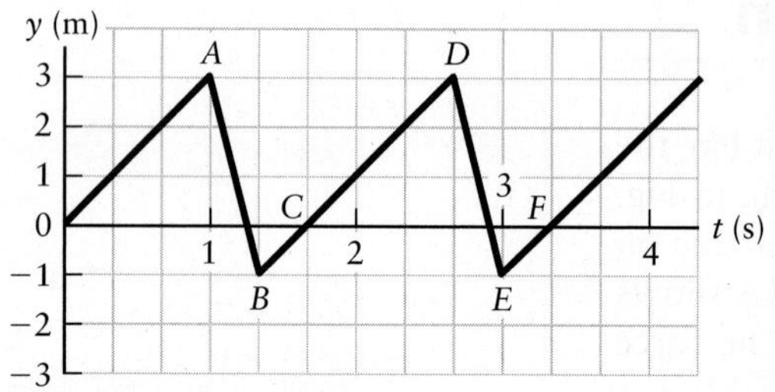

▲ **Figure 11.2** Example 11.1.

IDENTIFY THE RELATIONSHIPS

We can estimate the period from the spacing along the time axis between two corresponding (i.e., "equivalent") points of the plot in Figure 11.2. For example, we could choose two adjacent maxima, such as points A and D, or we could choose two adjacent minima (points B and E) or any other pair of adjacent "repeated" points (such as C and F).

SOLVE

From Figure 11.2, the time between points A and D is approximately

$$T \approx \boxed{1.7 \text{ s}}$$

The frequency is then

$$f = \frac{1}{T} = \frac{1}{1.7\text{ s}} = \boxed{0.59\text{ Hz}}$$

► ***What does it mean?***
Notice that the units of f are hertz.

Simple Harmonic Motion

In many cases, the motion of an oscillator is described by a simple sinusoidal variation with time,

$$y = A\sin(2\pi ft) \tag{11.2}$$

For our bungee jumper (Fig. 11.1A), y is the location along the vertical axis, and Equation 11.2 describes how y varies with time in Figure 11.1B. Systems that oscillate in a sinusoidal manner are called ***simple harmonic oscillators***, and they exhibit ***simple harmonic motion***. You might think that the behavior described by Equation 11.2 is too simple to describe any real system, but simple harmonic motion is, in fact, extremely common.

The quantity A in Equation 11.2 is called the ***amplitude*** of the motion. The bungee jumper in Figure 11.1A moves back and forth between the values $y = \pm A$ as indicated in the figure.

Figure 11.1C shows how the velocity of the bungee jumper varies with time. The velocity also "oscillates" between positive and negative values as the bungee jumper moves, and the frequency of these oscillations is the same as the frequency associated with the position y. However, the maximum values of the velocity do *not* occur when y has its maximum values. Instead, the largest values of v occur when $y = 0$. Likewise, the jumper's velocity is zero when he is at his maximum and minimum heights, at $y = \pm A$ (snapshots 2, 4, and 6 in Fig. 11.1A).

CONCEPT CHECK 11.1 Acceleration of an Object Undergoing Simple Harmonic Motion

Consider the acceleration of an object that is undergoing simple harmonic motion. Is this acceleration zero or nonzero? If it is nonzero, there must be a nonzero force acting on the object. What forces act on the bungee jumper in Figure 11.1?

The Connection between Simple Harmonic Motion and Circular Motion

A DVD spinning at a constant rate is another example of periodic, repeating motion. There is a close relationship between uniform circular motion as followed by a point on the edge of the disc and simple harmonic motion.

A particle at the edge of the DVD moves at a constant speed we call v_c around a circle of radius A (Fig. 11.3). We can specify the particle's position using a reference line drawn on the disc. If θ is the angle this reference line makes with the x axis, then θ increases with time according to

$$\theta = \omega t \tag{11.3}$$

where ω is the angular velocity of the corresponding rotational motion (Chapter 8). As the particle completes one full trip around the circle, θ varies from zero to 2π (measured in radians), corresponding to one full period of this harmonic motion. From Equation 11.3, we thus have after one full period ($t = T$):

$$\theta = 2\pi = \omega t = \omega T$$

Solving for T gives

$$T = \frac{2\pi}{\omega}$$

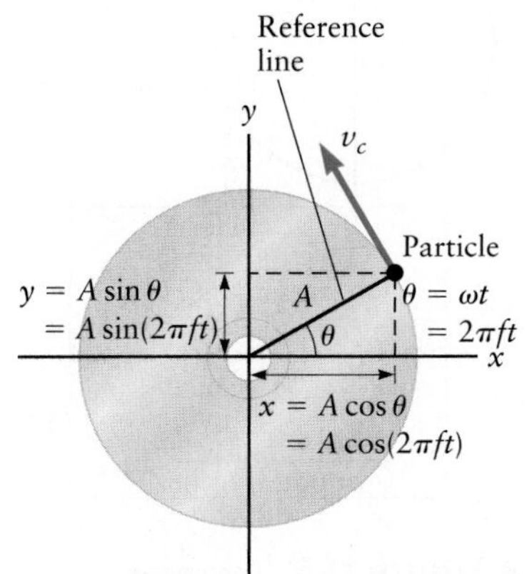

▲ **Figure 11.3** When a particle moves in a circle with constant speed v_c, both the x and y components of the particle's position undergo simple harmonic motion.

Both curves describe simple harmonic motion.

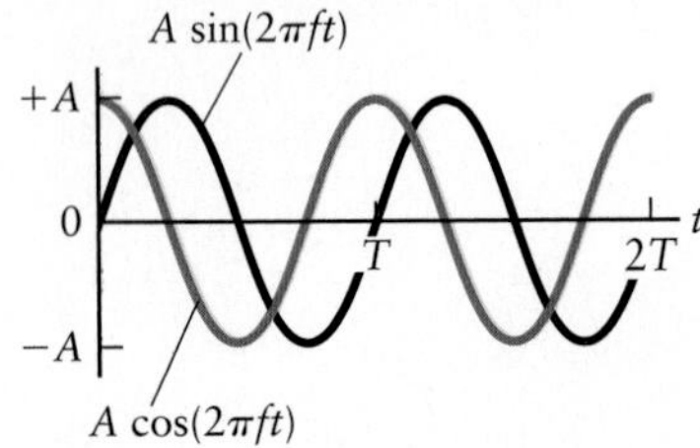

▲ **Figure 11.4** A sine-function time dependence (Eq. 11.5) and a cosine-function time dependence (Eq. 11.6) both describe simple harmonic motion. They differ only by a shift along the horizontal (time) axis.

Comparing this result with our definition of frequency in Equation 11.1, we see that harmonic motion can be described in terms of either an angular velocity ω or an equivalent frequency f, with

$$f = \frac{\omega}{2\pi} \tag{11.4}$$

The frequency f is the number of cycles a system completes in 1 second. The disc in Figure 11.3 moves through an angular displacement of 2π during each cycle, so Equation 11.4 says (in words) that the system completes an angular displacement of ω each second. For this reason, ω is also called the ***angular frequency***.

Simple Harmonic Motion: Position as a Function of Time. The connection between circular motion and simple harmonic motion becomes even clearer when we describe the motion of the particle in Figure 11.3 in terms of its coordinates in the x–y plane. For the moment, let's follow the y coordinate. According to the trigonometry in Figure 11.3, the y component of the particle's position is $y = A \sin \theta$. Since $\theta = \omega t$ (Eq. 11.3), we have

$$y = A \sin \theta = A \sin(\omega t)$$

Using $\omega = 2\pi f$ (Eq. 11.4), we also get

$$y = A \sin(2\pi f t) \tag{11.5}$$

which is precisely the same as the motion of a simple harmonic oscillator in Equation 11.2 with amplitude A.

There is nothing special about the y direction. We could just as easily make the connection between circular motion and simple harmonic motion using the x component of the particle's position. The x coordinate of the particle in Figure 11.3 is $x = A \cos \theta = A \cos(\omega t)$, which leads to

$$x = A \cos(2\pi f t) \tag{11.6}$$

Although this equation looks a bit different from our original expression for simple harmonic motion (Eqs. 11.2 and 11.5), it describes the same basic motion. This can be seen from Figure 11.4, which shows both a sine function (Eqs. 11.2 and 11.5) and a cosine function (Eq. 11.6). Both functions describe simple harmonic motion with period $T = 1/f$. The only difference is a "shift" along the time axis. One expression (the sine function) begins at $y = 0$ when $t = 0$, while the other (the cosine function) starts at a maximum when $t = 0$. In terms of the actual motion of our bungee jumper in Figure 11.1, the sine function describes a jumper who starts at the "midpoint" ($y = 0$) at $t = 0$, whereas the cosine function describes a jumper who starts at the highest point ($y = +A$). The only difference is in the way in which the system is initially set into motion.

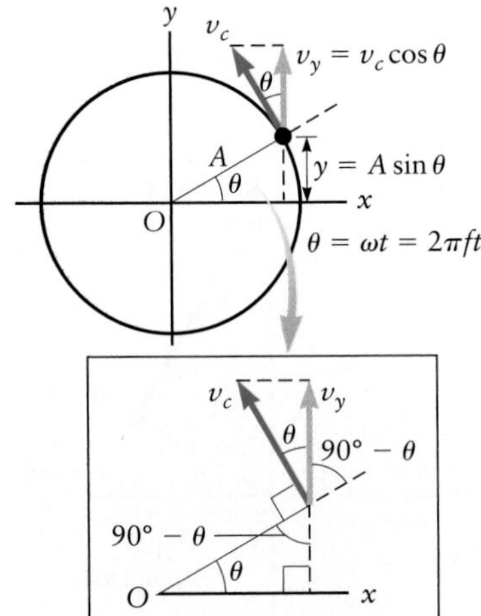

▲ **Figure 11.5** The speed v_c of the particle as it moves along the circle is constant, and the component of the velocity along y is $v_y = v_c \cos \theta$. Hence, v_y varies with time according to $v_y = v_c \cos(\omega t) = v_c \cos(2\pi f t)$.

Simple Harmonic Motion: Velocity as a Function of Time. We just showed that we can derive the position of a simple harmonic oscillator by taking either the x or y component of the position of a particle moving in a circle. We can also use the connection with circular motion to derive the velocity of a simple harmonic oscillator. Let's again follow the y coordinate of a particle moving in a circle (Fig. 11.5), so the velocity of our oscillator is v_y, the y component of the particle's velocity. Although the speed of the particle along the circular path in Figure 11.5 is constant, the y component v_y of its velocity is *not constant*. From Figure 11.5, we get

$$v_y = v_c \cos \theta = v_c \cos(\omega t)$$

$$v_y = v_c \cos(2\pi f t) \tag{11.7}$$

The particle travels once around the circle (a distance of $2\pi A$) in a time equal to one period, so its speed is $v_c = 2\pi A/T$. Using Equation 11.1 relating the period and frequency gives

$$v_c = \frac{2\pi A}{T} = 2\pi f A \tag{11.8}$$

Inserting this into Equation 11.7, we obtain

$$v_y = 2\pi fA \cos(2\pi ft)$$

Here v_y is the velocity of our simple harmonic oscillator, which we can simply call v.

We thus find that the position and velocity of a simple harmonic oscillator can be described by

$$y = A \sin(2\pi ft) \qquad v = 2\pi fA \cos(2\pi ft) \tag{11.9}$$

Position and velocity of a simple harmonic oscillator

(Compare Eq. 11.9 with the plots in Fig. 11.1.) This result for the velocity applies when we use the y component of circular motion to describe simple harmonic motion. We could also have used the x component, which leads to (see Eq. 11.6)

$$x = A \cos(2\pi ft) \qquad v = -2\pi fA \sin(2\pi ft) \tag{11.10}$$

Position and velocity of a simple harmonic oscillator

The relations in Equation 11.9 and 11.10 both describe simple harmonic motion.

CONCEPT CHECK 11.2 Analyzing a Simple Harmonic Oscillator

The position of a simple harmonic oscillator is given by $y = 25 \sin(0.22t)$, where we use SI units for position and time. What is the period of this oscillator?

(a) 4.5 s (b) 0.22 s (c) 25 s (d) 29 s

11.2 Examples of Simple Harmonic Motion

Mass on a Spring

One of the simplest examples of simple harmonic motion is sketched in Figure 11.6, which shows a block of mass m attached to a spring, with the opposite end of the spring attached to a wall. This is very similar to the person in Figure 11.1 except the block in Figure 11.6 moves on a frictionless, horizontal surface.

Let's apply Newton's second law to determine the horizontal motion of the block in Figure 11.6. Recall from Chapter 6 that the force exerted by a spring is given by Hooke's law,

$$F_{spring} = -kx \tag{11.11}$$

where x is the amount the spring is stretched or compressed away from its "natural" relaxed length. The spring constant k is a measure of the spring's strength and is different for different springs. However, the general relation in Equation 11.11 is the same for all springs, provided only that they are not stretched or compressed an excessive amount.

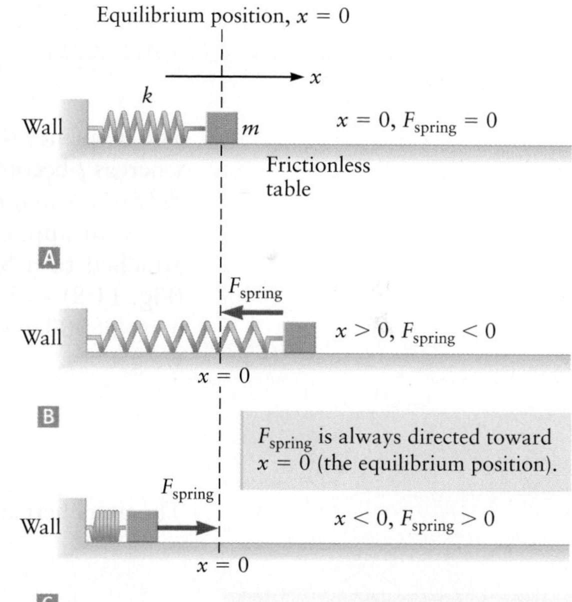

▲ **Figure 11.6** This mass on a spring is a simple harmonic oscillator. A When the block is at the equilibrium position ($x = 0$), the force exerted by the spring on the block is zero. B When the block is displaced to the right ($x > 0$), the force on the block is directed to the left. C When the block is displaced to the left ($x < 0$), the force is directed to the right. The force exerted by the spring on the block is thus always directed toward $x = 0$, the equilibrium point of the system.

The equilibrium position of the block in Figure 11.6 is at $x = 0$; when the block is at this position, the spring is in its relaxed state and the force exerted by the spring on the block is zero. If the block is displaced to the right so that $x > 0$, the force exerted by the spring on the block, according to Hooke's law (Eq. 11.11), is in the negative direction (i.e., to the left). Likewise, if the block is displaced to the left so that $x < 0$, the spring force is to the right. This force is called a ***restoring force*** because it always opposes the displacement away from the equilibrium position. Whenever there is a restoring force described by Equation 11.11, the system will exhibit simple harmonic motion.

To calculate the frequency of the oscillator, we use Newton's second law, $F = ma$, where F is the total force in the horizontal direction (parallel to x) in Figure 11.6. The force is given by Hooke's law (Eq. 11.11), so we have

$$F = -kx = ma \tag{11.12}$$

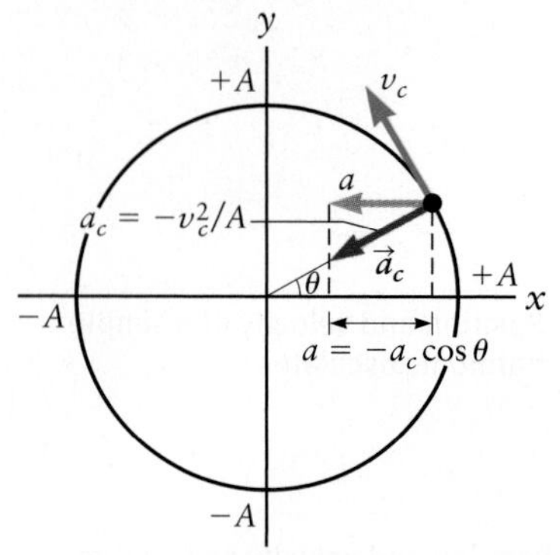

▲ **Figure 11.7** This particle undergoes uniform circular motion; its acceleration has magnitude $a_c = v_c^2/A$ and is directed toward the center of the circle. The acceleration of the corresponding simple harmonic motion is the component of $\vec{a}_c$ along one of the axes; here we use the x direction because we are modeling the oscillator in Figure 11.6.

To apply Equation 11.12, we need to find the particle's position and acceleration. We can do so by using the connection with circular motion sketched in Figure 11.7, which shows a particle moving at constant speed v_c in a circle of radius A. The x component of the particle's position is $x = A\cos\theta = A\cos(2\pi ft)$ as in Equation 11.10. The particle is moving in a circle, so it has a centripetal acceleration $a_c = v_c^2/A$ directed toward the center of the circle (since A is the radius of the circle). Here the simple harmonic motion is described by the particle's position along x, so we need to find the x component of the acceleration; from Figure 11.7, we get

$$a = -a_c\cos\theta = -\frac{v_c^2}{A}\cos\theta$$

Inserting this expression and our result for x into Equation 11.12 gives

$$-kx = -k(A\cos\theta) = ma = m\left(-\frac{v_c^2}{A}\cos\theta\right)$$

$$kA = \frac{mv_c^2}{A} \tag{11.13}$$

From Equation 11.8, we have $v_c = 2\pi fA$; using this in Equation 11.13 leads to

$$kA = \frac{mv_c^2}{A} = \frac{m(2\pi fA)^2}{A} = 4\pi^2 mf^2 A$$

We can now solve for the frequency f of this simple harmonic oscillator:

$$f = \frac{1}{2\pi}\sqrt{\frac{k}{m}} \tag{11.14}$$

The frequency thus increases as the spring constant k is increased (a "stiffer" spring), whereas f becomes smaller as the mass m is increased. Notice also that f is *independent of the amplitude*.

As an application of Equation 11.14, consider the motion of a small block attached to a Slinky, a common child's toy. A Slinky is a loosely coiled spring (Fig. 11.8) with a spring constant of about $k = 1$ N/m. A small toy might have a mass of 300 g, so we take $m = 0.3$ kg. Inserting these values into Equation 11.14, we get

$$f = \frac{1}{2\pi}\sqrt{\frac{k}{m}} = \frac{1}{2\pi}\sqrt{\frac{1\text{ N/m}}{0.3\text{ kg}}}$$

The units here are

$$\sqrt{\frac{\text{N}}{\text{m}\cdot\text{kg}}} = \sqrt{\frac{\text{kg}\cdot\text{m/s}^2}{\text{m}\cdot\text{kg}}} = \sqrt{\frac{1}{\text{s}^2}} = \frac{1}{\text{s}}$$

which is just the unit of frequency, hertz. Our result for the frequency is thus

$$f = \frac{1}{2\pi}\sqrt{\frac{1}{0.3}}\text{ Hz} = 0.3\text{ Hz}$$

with a corresponding period of

$$T = \frac{1}{f} = \frac{1}{0.3\text{ Hz}} = 3\text{ s}$$

How does this value compare with what you found the last time you played with a Slinky?

▲ **Figure 11.8** A Slinky has a spring constant of about $k = 1$ N/m.

EXAMPLE 11.2 Frequency of a Simple Harmonic Oscillator

The mass in Figure 11.9A is attached to a single spring with spring constant k, and according to Equation 11.14 it will undergo simple harmonic motion with a frequency

$$f_{\text{single}} = \frac{1}{2\pi}\sqrt{\frac{k}{m}}$$

If a second identical spring is then added as in Figure 11.9B, what is the new oscillation frequency?

RECOGNIZE THE PRINCIPLE

We need to consider how the restoring force on the mass depends on its displacement. If we can show that this restoring force has the same form as Hooke's law, we can then adapt Equation 11.14, which gives the frequency for the case of a single spring.

SKETCH THE PROBLEM

Figure 11.9 shows the problem.

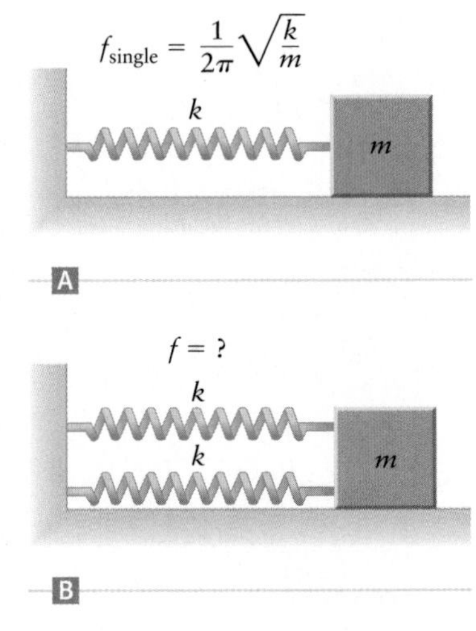

▲ **Figure 11.9** Example 11.2.

IDENTIFY THE RELATIONSHIPS

Let the position of the mass be $x = 0$ at the location where the two springs are both in their relaxed state; that is the equilibrium point. If the block in Figure 11.9B is then displaced to one side by an amount x, *each* spring will exert a force on the block according to Hooke's law. The *total* horizontal force on the block is then

$$F_{\text{both springs}} = \Sigma F = -2kx$$

This expression has precisely the same form as for the case of a single spring (compare with Eq. 11.11), so we can think of the two-spring combination as just a single "effective" spring with spring constant $k_{\text{new}} = 2k$.

SOLVE

For a single spring with spring constant k_{new}, we can calculate the oscillation frequency using Equation 11.14 and find

$$f_{\text{both springs}} = \frac{1}{2\pi}\sqrt{\frac{k_{\text{new}}}{m}} = \boxed{\frac{1}{2\pi}\sqrt{\frac{2k}{m}}}$$

▶ ***What does it mean?***

The frequency for the case of a single spring is

$$f_{\text{single}} = \frac{1}{2\pi}\sqrt{\frac{k}{m}}$$

so we can write our result for the two-spring combination as

$$f_{\text{both springs}} = \sqrt{2}\left(\frac{1}{2\pi}\sqrt{\frac{k}{m}}\right) = \boxed{\sqrt{2}f_{\text{single}}}$$

The frequency with two springs is thus $\sqrt{2}$ times higher than with a single spring.

Mass on a Vertical Spring: Bungee Jumping Revisited

The result for the frequency in Equation 11.14 was derived for a mass–spring system in which the mass moves along a frictionless, *horizontal* surface. The bungee jumper in Figure 11.1 is a lot like a mass on a spring, but in this case the mass hangs *vertically* and there are now two forces acting on the mass, the force from the spring and the force of gravity. Interestingly, the oscillation frequency for a mass on a vertical spring is precisely the *same* as the frequency for a mass on a horizontal spring. A mass hanging on a vertical spring is another example of simple harmonic motion, with an oscillation frequency given by Equation 11.14.

EXAMPLE 11.3 Ⓡ Analyzing a Bungee Oscillator

Consider the person attached to a bungee cord as shown in the photo at the start of this chapter. During the first part of the jump, the bungee cord is slack and the person undergoes (enjoys) free fall, but after a while, the person oscillates up and down in response to the forces of the bungee spring and gravity. Assume the bungee plus person can be modeled as a mass on a vertical, massless spring. For a person of average mass (e.g., your own mass), estimate **(a)** the frequency of the motion and **(b)** the spring constant of the bungee cord.

RECOGNIZE THE PRINCIPLE

The frequency of the motion is related to the period (the repeat time) by $f = 1/T$. To estimate the period, we appeal to your intuition (or experience) with bungee jumping. Once we have an approximate value for the frequency, we can use the relation for the frequency in Equation 11.14 along with the typical mass of a person to find k. This problem thus requires you to estimate two key quantities. The ability to make such estimates gives insight into how physics applies to real-world situations.

SKETCH THE PROBLEM

Figure 11.10 shows the person oscillating up and down and indicates the amplitude of the oscillation.

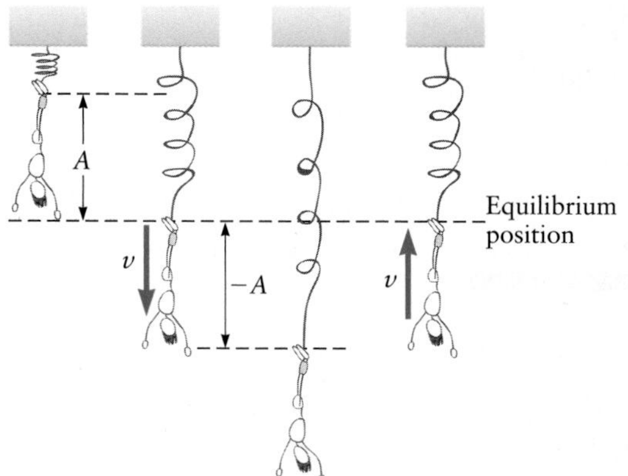

▲ **Figure 11.10** Example 11.3.

IDENTIFY THE RELATIONSHIPS

(a) The period of a bungee system depends on the length of the cord and the height of the jump. In a "typical" jump at a carnival, the person travels a distance $A = 10$ m below the equilibrium point (Fig. 11.10) and the period is about 5 s (based on the author's experience). These values are only approximate; other bungee jump experiments could be different.

(b) The mass of a typical person is about $m = 80$ kg.

SOLVE

(a) If the period of the oscillation is $T = 5$ s, the frequency is (Eq. 11.1)

$$f = \frac{1}{T} = \frac{1}{5\text{ s}} = \boxed{0.2\text{ Hz}}$$

(b) A bungee oscillator is just a mass on a spring, and its frequency is (Eq. 11.14)

$$f = \frac{1}{2\pi}\sqrt{\frac{k}{m}}$$

Rearranging to solve for k and inserting our values for f and m, we find

$$\sqrt{\frac{k}{m}} = 2\pi f$$

$$k = 4\pi^2 f^2 m = 4\pi^2(0.2\text{ Hz})^2(80\text{ kg}) = \boxed{100\text{ N/m}}$$

We include only one significant figure in our answer because the period and frequency were only known to one significant figure.

▸ What does it mean?
Other bungee jumps will have different heights, and their bungee cords will have different values of k. However, we can always use our experience (and common sense) to estimate quantities such as the period and frequency, and thus apply our results for simple harmonic motion.

The Simple Pendulum

A pendulum is another example of a simple harmonic oscillator. Pendulums can be constructed in several ways; Figure 11.11 shows one example. Perhaps the simplest pendulum is made by tying a rock of mass m to one end of a string, with the other end of the string fastened to a support as sketched in Figure 11.12. If the diameter of the rock is much smaller than the length L of the string and the string is massless, this system is a simple pendulum.

Figure 11.12A shows sketches of the rock (also called the pendulum "bob") as it moves along its path, a circular arc of radius L. It is convenient to measure the bob's displacement in terms of y, the displacement along this circular arc. To examine the pendulum motion in terms of Newton's second law, we have drawn the two forces acting on the bob: gravity and tension from the string. (We ignore the air drag force.) The force along (parallel to) the circular arc is responsible for the simple harmonic motion. Figure 11.12A shows $F_{parallel}$ at three instants in time. The bottom point on the trajectory ($y = 0$) is the equilibrium point of the pendulum, the point where $F_{parallel} = 0$. If the bob were placed at that location and given no initial velocity, it would remain there, in translational (and rotational) equilibrium (Chapters 4 and 8). As the bob moves back and forth past this equilibrium point, the magnitude and direction of $F_{parallel}$ vary with time. When the bob is on the right in Figure 11.12A, where $y > 0$, the force is directed to the left (the negative direction). Likewise, if the bob is on the left, where $y < 0$, then $F_{parallel}$ is to the right, the positive direction. Because this force always opposes the bob's displacement away from the equilibrium point, this is another example of a restoring force. (Compare with the restoring force in Fig. 11.6.)

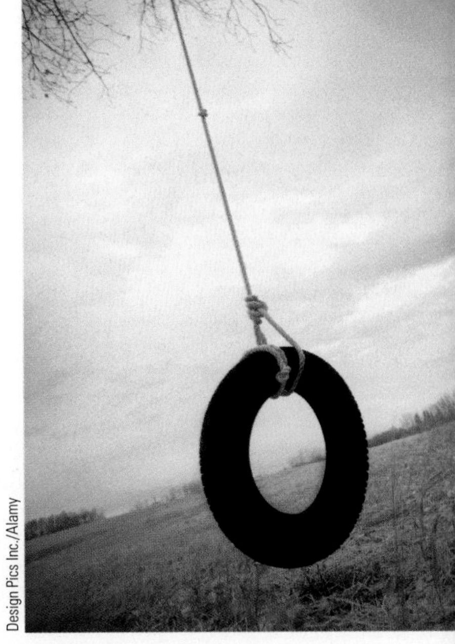

▲ **Figure 11.11** A child's tire swing is a good approximation to a simple pendulum.

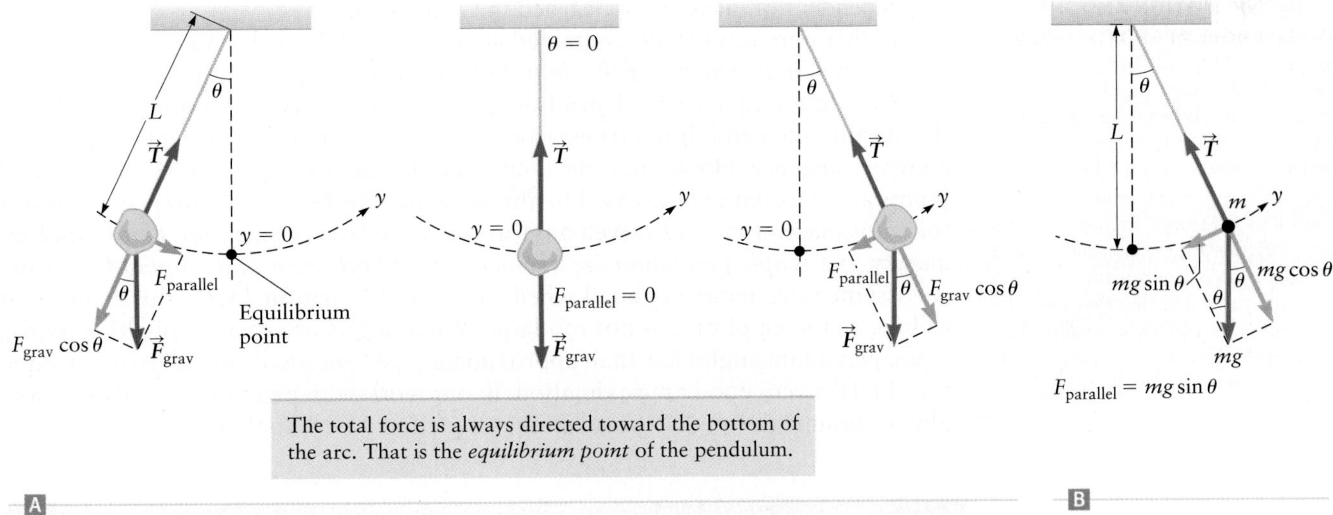

▲ **Figure 11.12** **A** This simple pendulum consists of a pendulum bob of mass m tied to the end of a very light string. There are two forces on the bob, from gravity and from the tension in the string. **B** Their vector sum is directed tangent to the circular arc along which the bob swings. This total (net) force is always directed back toward the bottom of the arc, the equilibrium point of the pendulum.

For a simple pendulum, the restoring force is due to the component of the gravitational force directed along the bob's path. Figure 11.12B shows that for a pendulum of length L with a bob of mass m, this restoring force is

$$F_{\text{restore}} = F_{\text{parallel}} = -mg \sin \theta \tag{11.15}$$

If the angle θ is small, the function $\sin \theta$ is approximately equal to θ when the angle is measured in radians. That is, $\sin \theta \approx \theta$. In addition, this angle is related to the displacement by $\theta = y/L$ according to the general definition of angle as measured in radians. Inserting all this into Equation 11.15 leads to

$$F_{\text{parallel}} = -mg\theta = -\frac{mgy}{L} \tag{11.16}$$

Newton's second law for motion along the circular arc ($F_{\text{parallel}} = ma$) leads to:

$$F_{\text{parallel}} = -\frac{mgy}{L} = ma \tag{11.17}$$

This relation has exactly the same form we found in Equation 11.12 when we applied Newton's second law to the mass on a spring. In both cases, the restoring force is proportional to the displacement from equilibrium. Comparing Equations 11.12 and 11.17, the only difference is that the spring constant k for the mass on a spring is replaced by the factor mg/L. We can therefore use the result for the oscillation frequency of a mass on a spring (Eq. 11.14) if we replace k by mg/L, which leads to

$$f = \frac{1}{2\pi}\sqrt{\frac{k}{m}} = \frac{1}{2\pi}\sqrt{\frac{mg/L}{m}}$$

Frequency of a simple pendulum

$$f = \frac{1}{2\pi}\sqrt{\frac{g}{L}} \tag{11.18}$$

Hence, given the length of the pendulum, its frequency can be calculated. This frequency is *independent* of the mass of the bob, which may at first seem surprising; you might have expected that a heavier bob will move more slowly and have a lower frequency than a lighter bob. However, the restoring force is due to gravity and is thus proportional to the bob's mass (since its weight is equal to mg). The pendulum's motion is independent of m for the same reason the velocity of a freely falling object is independent of mass (as we learned in Chapter 3).

Another important result contained in Equation 11.18 is that the frequency of the pendulum is *independent of the amplitude* of the motion. You might have expected the pendulum frequency to depend on amplitude since if the amplitude is increased (by starting the pendulum farther from $y = 0$), the pendulum bob must travel over a greater distance. However, if the amplitude A is increased, the bob's velocity at all points along its path is increased by the same amount because A enters in the results for y (displacement) and v (velocity) in the same way. Hence, *the period and frequency of a simple pendulum are independent of both mass and amplitude*. Because we assumed the angle θ is small when we derived Equation 11.18, this result holds as long as the amplitude is not too large. What angles are small enough? In typical cases, pendulum angles less than approximately 30° are small enough to make Equation 11.18 a very good approximation. In our work with pendulum oscillators, we'll always assume the angle is small so that Equation 11.18 applies.

Insight 11.1

HARMONIC MOTION AND THE SMALL ANGLE APPROXIMATION

For a simple pendulum, the restoring force is (Eq. 11.15) $F_{\text{restore}} = -mg \sin \theta$. When the pendulum angle θ is small, this force is given approximately by (Eq. 11.16) $F_{\text{restore}} = -mg \sin \theta \approx -mg\theta$, with $\sin \theta \approx \theta$ when θ is measured in radians. You can verify this approximation yourself by using a calculator to compare θ and $\sin \theta$, and you'll find that they differ by less than 10% when θ is smaller than about 45°. The approximation gets better and better as θ gets smaller.

EXAMPLE 11.4 The Frequency of a Playground Swing

If a pendulum swing on a playground has length $L = 3.0$ m, what are the frequency and period of the swing? Assume the pendulum angle is small.

RECOGNIZE THE PRINCIPLE

A pendulum swing is approximately a simple pendulum as in Figures 11.11 and 11.12, and its frequency is given by (Eq. 11.18)

$$f = \frac{1}{2\pi}\sqrt{\frac{g}{L}} \quad (1)$$

SKETCH THE PROBLEM

Figure 11.12 describes the problem.

IDENTIFY THE RELATIONSHIPS

We are given the pendulum length, so we can use Equation (1) to find the frequency. The period is related to the frequency by $T = 1/f$ (Eq. 11.1).

SOLVE

Substituting the given value of L into Equation (1) gives

$$f = \frac{1}{2\pi}\sqrt{\frac{g}{L}} = \frac{1}{2\pi}\sqrt{\frac{9.8 \text{ m/s}^2}{3.0 \text{ m}}} = 0.29\left(\frac{1}{\text{s}}\right) = \boxed{0.29 \text{ Hz}}$$

The period can be found from our general relation between frequency and period,

$$T = \frac{1}{f} = \frac{1}{0.29 \text{ Hz}} = \boxed{3.4 \text{ s}}$$

► ***What does it mean?***

These results are, as we have already noted, independent of mass, so children of different size will swing at the same frequency.

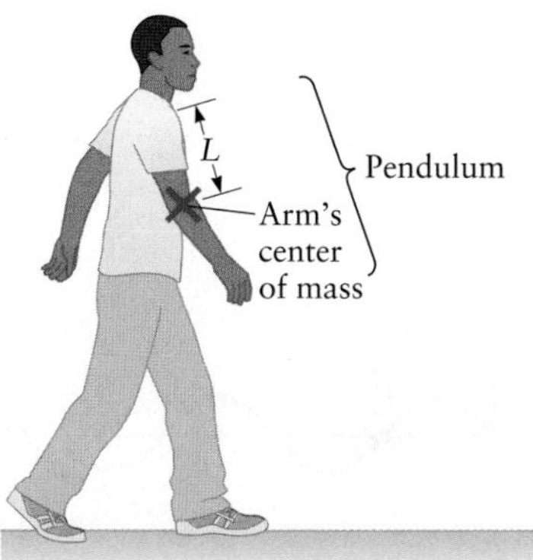

▲ **Figure 11.13** When you walk, each of your arms swings like a pendulum.

The Human Arm as a Pendulum

In Example 11.4, we calculated the period of a pendulum. The same ideas can be used to understand the way your arms swing when you walk. If your arms swing freely, they act as pendulums. Let's now estimate their period of oscillation. The arm is not quite a simple pendulum because all the mass is not at the end. We can, however, get an approximate value of the period by assuming all the mass is at the center of mass, roughly halfway along your arm (Fig. 11.13). An adult's arm is about 60 cm long, so the center of mass is about $L = 30$ cm $= 0.30$ m from the shoulder. Taking 0.30 m as the length of a pendulum, the frequency is (Eq. 11.18)

$$f = \frac{1}{2\pi}\sqrt{\frac{g}{L}} = \frac{1}{2\pi}\sqrt{\frac{9.8 \text{ m/s}^2}{0.30 \text{ m}}} = 0.91 \text{ Hz}$$

which corresponds to a period

$$T = \frac{1}{f} = 1.1 \text{ s}$$

The next time you go for a walk, you should compare this result with the way you swing your arms.

The Torsional Oscillator

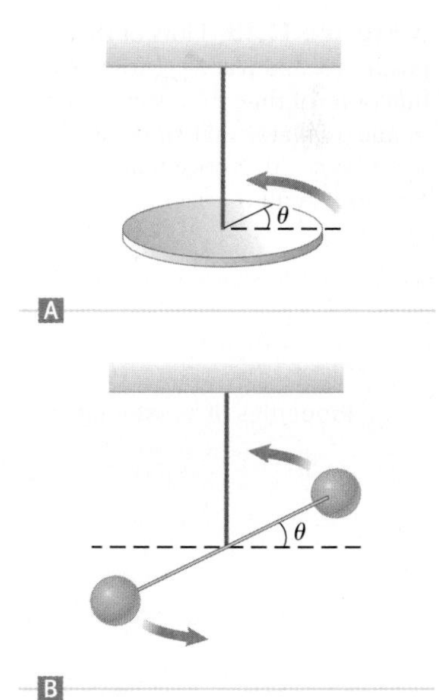

▲ **Figure 11.14** Two examples of torsional oscillators. These objects rotate about a vertical (y) axis, and the torque exerted by the fibers tends to restore the systems to their equilibrium positions.

Another system displaying simple harmonic motion is the ***torsional oscillator***. Figure 11.14 shows two examples: in part A of the figure, a circular disk is suspended by a thin wire attached to the center of the disk, and in part B, a rod is attached to masses at each end, with the rod suspended by a wire attached to its center. The wire used in both of these examples is called a ***torsion fiber***, and when it is twisted, it exerts a torque τ on whatever is connected to it, the disk in Figure 11.14A or the rod in Figure 11.14B. The torque is proportional to the twist angle θ, with

$$\tau = -\kappa\theta \tag{11.19}$$

Here, κ (the Greek letter kappa) is called the ***torsion constant*** and is a property of the particular torsion fiber. The negative sign in Equation 11.19 indicates that it is a *restoring* torque. When the disk is rotated to positive angles, the torque tends to make the disk turn back toward the "negative" angular direction, whereas if the disk is rotated to negative angles, the torque acts toward the direction of positive angles. This is just like the restoring force for the mass on a spring in Figure 11.6 and for the pendulum in Figure 11.12. Since Equation 11.19 has the same mathematical form as Hooke's law for a spring (Eq. 11.11), the torsional pendulum is also a simple harmonic oscillator, with a frequency given by

$$f = \frac{1}{2\pi}\sqrt{\frac{\kappa}{I}} \tag{11.20}$$

where I is the moment of inertia of the object attached to the fiber (the disk in Fig. 11.14A or the two masses and rod in Fig. 11.14B). This result is similar to what we found for a mass on a spring (Eq. 11.14), but the frequency of the torsional oscillator depends on the moment of inertia I of the system instead of on its mass.

Features Common to All Simple Harmonic Oscillators

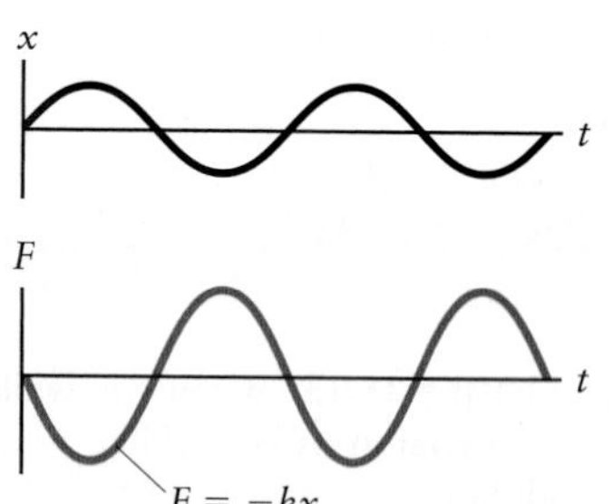

▲ **Figure 11.15** Plots of the position x and restoring force F as functions of time for a simple harmonic oscillator that starts at $x = 0$ at $t = 0$. Notice that $F = -kx$ at all times.

All simple harmonic oscillators have a number of features in common. First, their motion exhibits a sinusoidal time dependence as in Equation 11.9 or, equivalently, Equation 11.10. The variables in these equations are called "position" (x or y) and "velocity" (v). For a simple pendulum and a mass on a spring, x and v are the ordinary position and velocity, whereas for the torsional oscillator, they are the angular position (the angle θ in Fig. 11.14) and the corresponding angular velocity. The position and velocity of *all* simple harmonic oscillators have these mathematical forms.

A second feature common to all simple harmonic oscillators is that they all involve a restoring force. Figure 11.15 shows the time dependence of both the position (which we denote by x) and the restoring force F. For the horizontal mass on a spring, F is the force exerted by the spring on the mass, whereas for a simple pendulum, F is the component of the gravitational force along the path of the pendulum bob. For the torsional oscillator, we would instead plot the torque exerted by the torsion fiber. Both the displacement and the restoring force or torque vary sinusoidally with time, and F always has a sign *opposite* that of the displacement. This is the same negative sign we encountered in Hooke's law (Eq. 11.11) and for the pendulum (Eq. 11.16). This is a key feature of a restoring force and is essential for producing harmonic motion. Whenever an object experiences a *restoring* force with a magnitude proportional to the displacement from the equilibrium position, the object will undergo simple harmonic motion.

Properties of a restoring force

The Frequency of a Simple Harmonic Oscillator Is Independent of the Amplitude

A very important feature of all simple harmonic oscillators is that the frequency is *independent* of the amplitude A. The frequency depends on quantities such as the length of the pendulum string (Fig. 11.12), the mass and spring constant in Figure 11.6, or the moment of inertia and torsion constant in Figure 11.14. Hence, this frequency does depend on the properties of the oscillator, but f does *not* depend on the amplitude of the motion. The amplitude is determined entirely by how the oscillator is initially set into motion.

> **CONCEPT CHECK 11.3 Connection between Velocity and Force for a Simple Harmonic Oscillator**
>
> Figure 11.16 shows the displacement of a simple harmonic oscillator as a function of time. (a) At which point(s) is the velocity zero? (b) At which point(s) is the magnitude of the force a maximum?

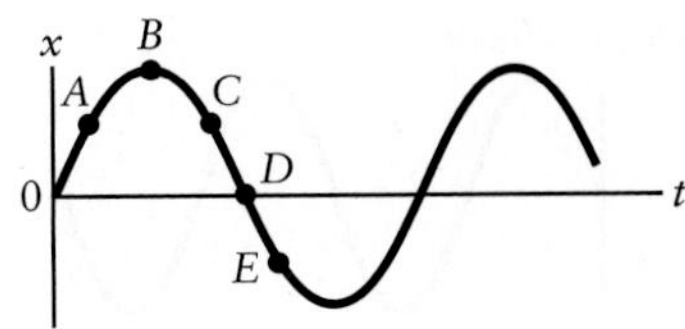

▲ **Figure 11.16** Concept Check 11.3.

11.3 Harmonic Motion and Energy

Let's next consider how the kinetic and potential energies of the simple harmonic oscillator in Figure 11.17 vary with time. The mechanical energy is the sum of the kinetic energy ($\frac{1}{2}mv^2$) and potential energy ($\frac{1}{2}kx^2$, the spring potential energy). When the mass is at $x = 0$ (parts A and C of Fig. 11.17), the potential energy is zero because the spring is relaxed, but the speed is a maximum; hence, the kinetic energy is large. When the displacement is largest, $x = \pm A$ (parts B and D of Fig. 11.17), and the potential energy is large, the speed is zero, so the kinetic energy is zero. The energy thus "oscillates" back and forth between the kinetic and potential energies as shown in Figure 11.17E.

For an ideal oscillator, there is no friction, and the total mechanical energy is conserved. Hence, the sum of the kinetic and potential energies in Figure 11.17E is constant. This also means that the maximum kinetic energy (Fig. 11.17A) must equal the maximum potential energy (Fig. 11.17B). If the maximum speed is v_{max}, the maximum kinetic energy is $\frac{1}{2}mv_{max}^2$ and the maximum potential energy is $\frac{1}{2}kA^2$ (because the maximum displacement is $x_{max} = A$). We thus have

$$\tfrac{1}{2}mv_{max}^2 = KE_{max} = PE_{max} = \tfrac{1}{2}kA^2 \qquad (11.21)$$

which gives a useful relation between the maximum speed and displacement.

These results for the time dependence of the kinetic energies and potential energies have been derived for a mass on a horizontal spring, but they apply to all simple harmonic oscillators. The *KE* and *PE* of any simple harmonic oscillator vary with

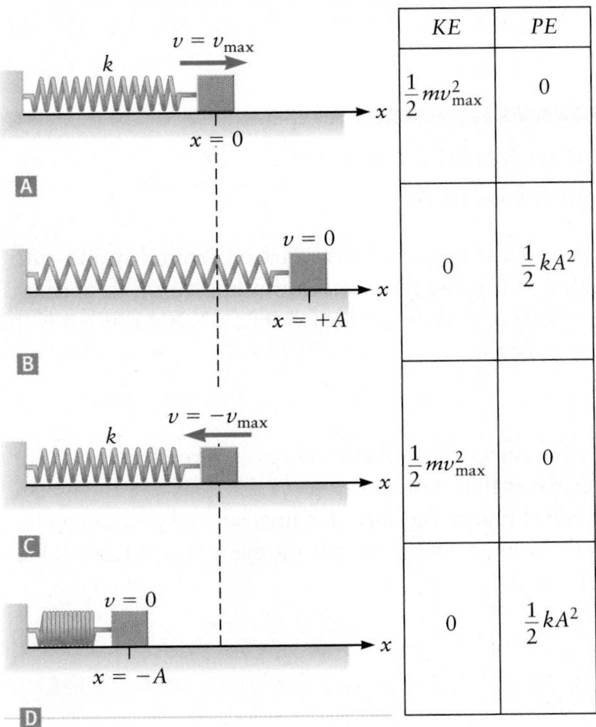

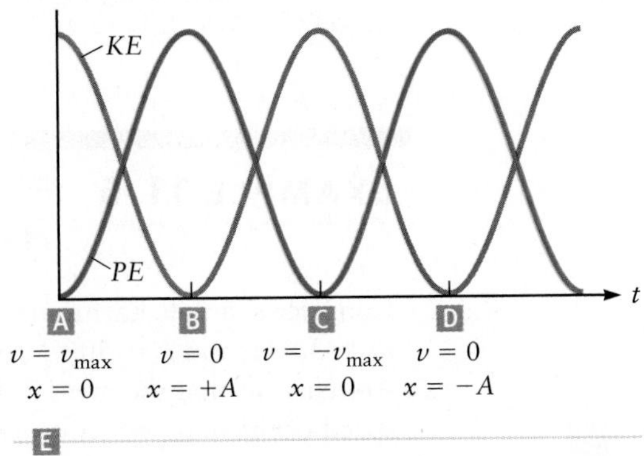

▲ **Figure 11.17** The kinetic and potential energies of a mass on a spring oscillate with time. The sum *KE* + *PE* is constant, as the mechanical energy of a simple harmonic oscillator is conserved.

time in the manner sketched in Figure 11.17, and the total mechanical energy is conserved.

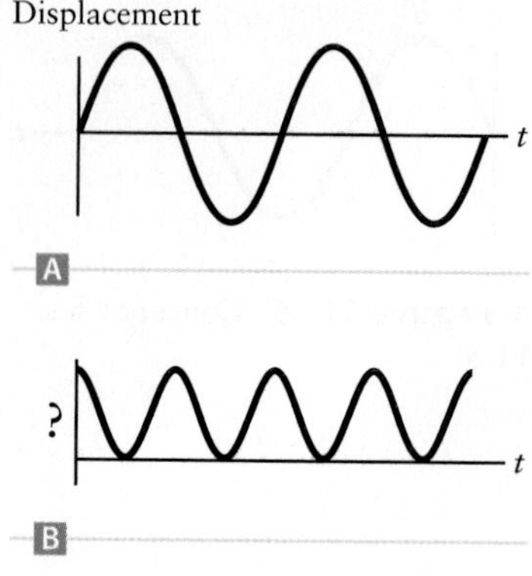

▲ **Figure 11.18** Concept Check 11.4.

CONCEPT CHECK 11.4 Energy of a Simple Harmonic Oscillator

Figure 11.18 shows the displacement of a simple harmonic oscillator as a function of time. The graph in Figure 11.18B shows either the kinetic or the potential energy. Which one?

EXAMPLE 11.5 Energy of a Simple Harmonic Oscillator

Figure 11.19 shows the velocity of a simple harmonic oscillator (like the one in Figure 11.17) as a function of time. At which of the labeled points *A*, *B*, *C*, *D*, and *E* on the curve is the kinetic energy greatest? At what point is it smallest?

RECOGNIZE THE PRINCIPLE

The kinetic energy of the oscillator is just the kinetic energy of the mass, which is $KE = \frac{1}{2}mv^2$.

SKETCH THE PROBLEM

Figure 11.19 shows the problem.

IDENTIFY THE RELATIONSHIPS AND SOLVE

From Figure 11.19, we see that the magnitude of the velocity is greatest at point *D* (even though v is negative). The kinetic energy is thus greatest at point *D*. Likewise, v is zero at point *B*, so the kinetic energy is smallest at this point.

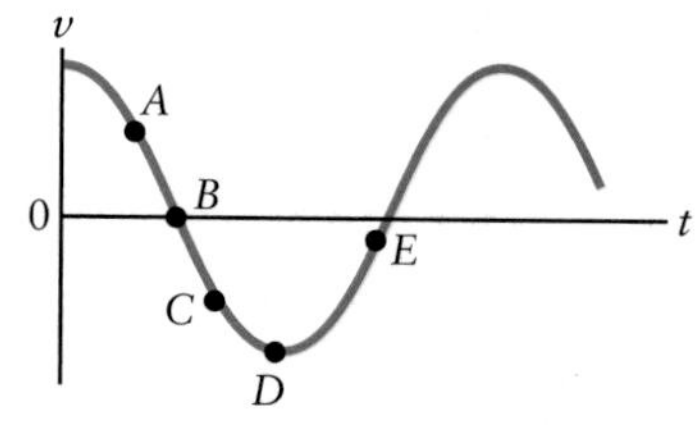

▲ **Figure 11.19** Example 11.5.

► ***What does it mean?***

The total energy of the oscillator must be conserved (i.e., constant), so the potential energy will be smallest when the speed is greatest. For the points marked in Figure 11.19, the speed is greatest at point *D*. Likewise, the potential energy is largest when the speed is zero, which occurs at point *B*.

EXAMPLE 11.6 Maximum Speed of a Simple Harmonic Oscillator

Consider a simple harmonic oscillator consisting of a mass on a horizontal spring with $m = 3.0$ kg and $k = 500$ N/m, with $x = 0$ being the equilibrium position. The spring is stretched so that the mass is at $x_i = 0.10$ m and then released (Fig. 11.20A). What is the speed of the mass when it reaches $x = 0$?

RECOGNIZE THE PRINCIPLE

Just as the mass is released, its speed is zero, so its kinetic energy is zero. The total energy thus equals the potential energy at that moment (Fig. 11.20A). When the mass is at $x = 0$, the potential energy is zero because the spring is unstretched and uncompressed, so the total energy equals the kinetic energy at that moment (Fig. 11.20B). We thus have (compare with Eq. 11.21)

$$\frac{1}{2}mv^2 = KE_{\text{max}} = PE_{\text{max}} = \frac{1}{2}kx_i^2 \quad (1)$$

We can use this relation to solve for v.

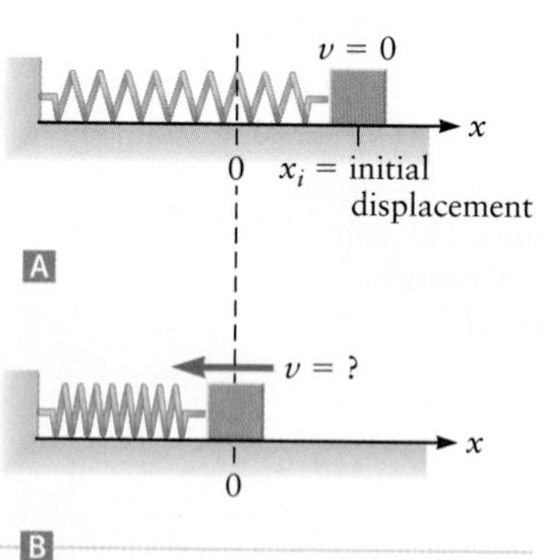

▲ **Figure 11.20** Example 11.6.

SKETCH THE PROBLEM

Figure 11.20 shows the system when the block is released and when it passes through $x = 0$.

IDENTIFY THE RELATIONSHIPS

We can rearrange Equation (1) to get

$$mv^2 = kx_i^2$$

$$v^2 = \frac{k}{m}x_i^2$$

$$v = \sqrt{\frac{k}{m}}x_i$$

SOLVE

The initial displacement is $x_i = 0.10$ m, and the values of m and k are given; we find

$$v = \sqrt{\frac{k}{m}}x_i = \sqrt{\frac{500 \text{ N/m}}{3.0 \text{ kg}}}(0.10 \text{ m}) = \boxed{1.3 \text{ m/s}}$$

► ***What does it mean?***

The maximum potential energy occurs when the displacement is greatest, whereas the maximum kinetic energy is found when the displacement is zero. This result applies to all simple harmonic oscillators, including pendulums and torsional oscillators.

CONCEPT CHECK 11.5 Amplitude and Energy of a Simple Harmonic Oscillator

If the amplitude of a simple harmonic oscillator is doubled, by what factor does the *total* energy increase?

(a) no change
(b) by a factor of two
(c) by a factor of four
(d) by a factor of eight

11.4 Stress, Strain, and Hooke's Law

According to Hooke's law, the force exerted by a spring is directly proportional to the amount that the spring is stretched or compressed, with

$$F = -kx \tag{11.22}$$

Hooke's law applies not only to the "ideal" springs found in physics books, but to many other situations as well. Consider a solid bar of metal as sketched in Figure 11.21. If you "squeeze" the bar an amount x, it will "push back." The magnitude of this "push-back" force depends on how much you squeeze and is given by Hooke's law (Eq. 11.22). The value of the constant k depends on the size of the bar and the material it is made of. This push-back force also means that virtually all objects act as springs when squeezed, leading to many applications of our results for simple harmonic motion.

In Figure 11.21, a force of magnitude F is applied to each end of the metal bar, and the bar changes length by an amount ΔL. The force in Figure 11.21 is a ***compressive*** force, making ΔL negative. If the force directions are reversed so that the bar is stretched (ΔL positive), we call F a ***tensile*** force. The value of ΔL depends on the material; a soft material compresses or stretches more than a stiff material. "Stiffness" is characterized by ***Young's modulus***, Y. In Figure 11.21, the area of the end of

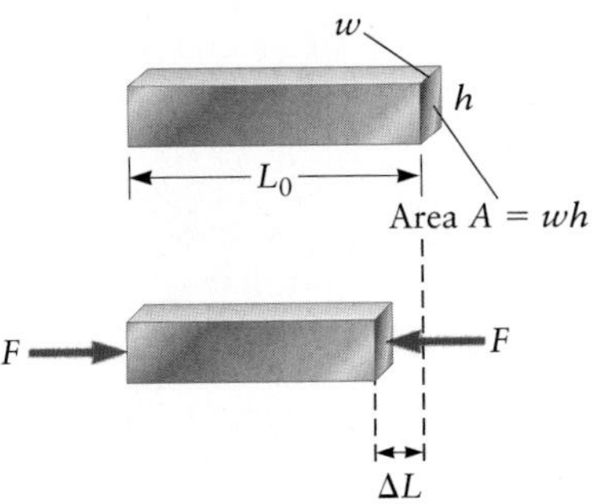

▲ **Figure 11.21** The forces F applied to the ends of this metal bar compress the bar by an amount ΔL. Tensile forces would stretch the bar instead.

Insight 11.2
ELASTIC FORCES AND HOOKE'S LAW
Hooke's law for a spring (Eq. 11.22) contains a negative sign, whereas Equation 11.23 does not, but we have still called them both "Hooke's law." The difference is due to how F is defined in the two cases. For the spring described by Equation 11.22, F is the force exerted *by* the spring, whereas in the definition of Young's modulus, Equation 11.23, F is the force *on* the bar. According to Newton's third law, the force exerted *by* the bar equals $-F$, so Equation 11.23 is indeed equivalent to Hooke's law for a spring.

TABLE 11.1 Values of Young's Modulus, Shear Modulus, and Bulk Modulus for Some Common Materials

Material	Young's Modulus, Y (Pa)	Shear Modulus, S (Pa)	Bulk Modulus, B (Pa)
Aluminum	7.0×10^{10}	2.4×10^{10}	7.1×10^{10}
Steel	2.0×10^{11}	8.1×10^{10}	1.4×10^{11}
Copper	1.3×10^{11}	4.8×10^{10}	1.4×10^{11}
Glass (Pyrex)	6.2×10^{10}	2.7×10^{9}	2.6×10^{10}
Human bone	9×10^{9} (compression) 2×10^{10} (when stretched)	8×10^{10}	
Diamond	1.2×10^{12}	4.8×10^{10}	6.2×10^{11}
Quartz	1.0×10^{11}	3.1×10^{10}	3.6×10^{10}
Teflon	4×10^{8}		6×10^{9}
Ice			9×10^{9}
Air (at atmospheric pressure)			1.0×10^{5}
Water			2.2×10^{9}
Mercury (liquid)			2.5×10^{10}

Note: These values generally vary slowly with temperature. The numbers given here are for room temperature.

the bar is $A = wh$, and L_0 is the bar's original length. The magnitude of the force F and the change in length ΔL are related by

Stress, strain, and Hooke's law

$$\frac{F}{A} = Y\frac{\Delta L}{L_0} \tag{11.23}$$

In words, Equation 11.23 says that the change in length is directly proportional to the applied force. What's more, if the material is pulled so that it is elongated by an amount ΔL (i.e., if the directions of the forces in Fig. 11.21 are reversed), Equation 11.23 still holds. Because the amount the bar is stretched or compressed is directly proportional to the magnitude of the applied force, this relation is equivalent to Hooke's law (Eq. 11.22).

The ratio F/A that appears in Equation 11.23 is called the ***stress***. By considering the force applied per unit of cross-sectional area, we can more easily compare the properties of bars of different size. For example, if we consider a second bar of the same material and length whose area is larger by a factor of two, a force twice as large is required to produce the same change in length ΔL. The stress F/A would be the same in the two cases, though. Likewise, the ratio $\Delta L/L_0$ is called the ***strain***. This ratio is the fractional change in the length. The value of Young's modulus depends on the material; values for some common materials are listed in Table 11.1. Since the ratio $\Delta L/L$ on the right-hand side of Equation 11.23 is unitless, Y has units of force/area = N/m^2 = Pa.

Elastic versus Plastic Deformations

Young's modulus describes how a material stretches or compresses in response to an applied force. We call Y an ***elastic constant*** because, according to Equation 11.23, the material (such as the bar in Fig. 11.21) will return to its original length when the force F is removed. That is what is meant by "elastic" behavior. Such behavior is found for most materials provided the strain is not too large. If the strain exceeds a certain value, however, the simple linear relation between stress and strain in Equation 11.23 will no longer be found. The value of the strain at which that happens is called the ***elastic limit***, and depends on the material. When the strain is pushed beyond this point, the material will no longer return to its original length even when F is completely removed. A typical plot of stress versus strain is shown in Figure 11.22. When the elastic limit is exceeded, the material will be permanently deformed.

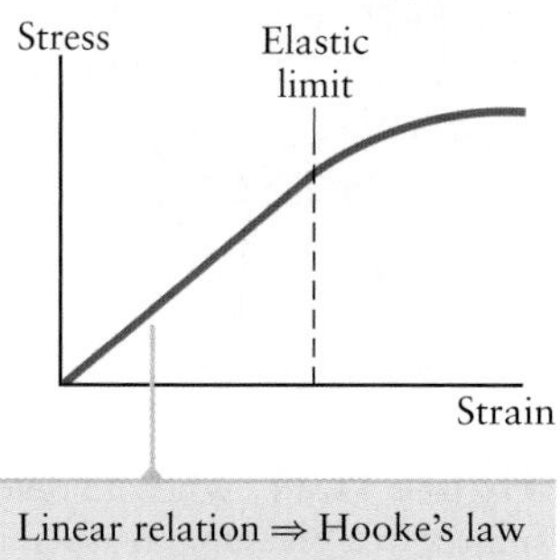
When the elastic limit is exceeded, the stress–strain curve is no longer linear.
Stress
Elastic limit
Strain
Linear relation ⇒ Hooke's law

▲ **Figure 11.22** When a material is strained (i.e., compressed, stretched, or sheared), it will initially follow a linear stress–strain relation; the magnitude of the applied force (or stress) is proportional to the amount the material is strained. When the strain exceeds a certain value, called the elastic limit, this linear relation is no longer followed. Such large stresses usually cause the object to be permanently deformed.

EXAMPLE 11.7 Ⓡ Stretching a Guitar String

When expressed in SI units, the numerical values of elastic constants are typically very large (see Table 11.1), making it difficult to get an intuitive feeling for how much an object will stretch or compress in response to typical forces. So, consider the following "everyday" problem. A guitar string is composed of steel and is under a tension of 200 N when mounted on the instrument. How much does the string stretch in response to this tension?

RECOGNIZE THE PRINCIPLE

When a force F is applied to the string, it becomes longer by an amount ΔL with (Eq. 11.23)

$$\frac{F}{A} = Y\frac{\Delta L}{L_0} \qquad (1)$$

To use Equation (1) to find ΔL, we must have values for the length and area of the string. The value of Young's modulus is given in Table 11.1, and F is just the tension, which is given.

SKETCH THE PROBLEM

We can use Figure 11.21, with the bar replaced by a guitar string.

IDENTIFY THE RELATIONSHIPS

Based on experience, the length of a typical guitar string is about $L_0 = 0.6$ m and the radius is about $r = 0.5$ mm.

SOLVE

Rearranging Equation (1) to find ΔL, we get

$$\Delta L = \frac{FL_0}{YA} = \frac{FL_0}{Y\pi r^2}$$

where $A = \pi r^2$ is the cross-sectional area of the string and F is the tensile force, given as 200 N. Inserting the value of Young's modulus for steel from Table 11.1 along with the other values estimated above, we get

$$\Delta L = \frac{FL_0}{Y\pi r^2} = \frac{(200\text{ N})(0.6\text{ m})}{(2.0 \times 10^{11}\text{ N/m}^2)\pi(0.0005\text{ m})^2} = 8 \times 10^{-4}\text{ m} = \boxed{0.8\text{ mm}}$$

▸ ***What does it mean?***

Although 0.8 mm is not a large value, it is certainly noticeable to a guitarist turning the tuning peg of a guitar.

The Shear Modulus

An applied force may deform a material in ways besides compressing or stretching, leading to other elastic constants besides Young's modulus. Consider a shearing force applied to a metal bar as shown in Figure 11.23; this force tends to make two parallel faces of the bar shift laterally by an amount Δx. The shear stress is defined as F/A, where A is the cross-sectional area of the end of the bar; and the shear strain is

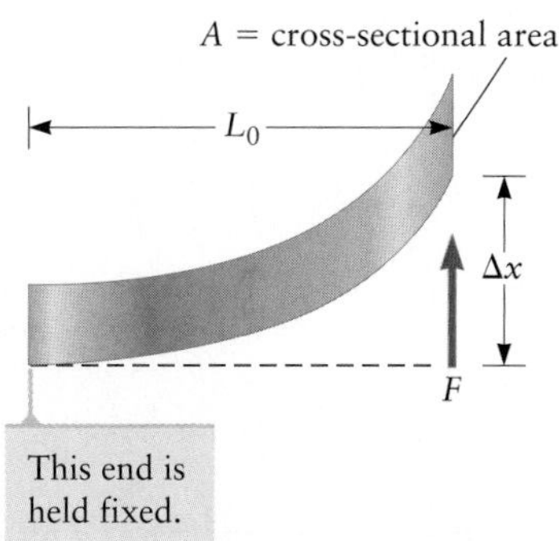
A = cross-sectional area
L0
Δx
F
This end is held fixed.

▲ **Figure 11.23** A shear force (related to the shear stress) acts to bend this bar.

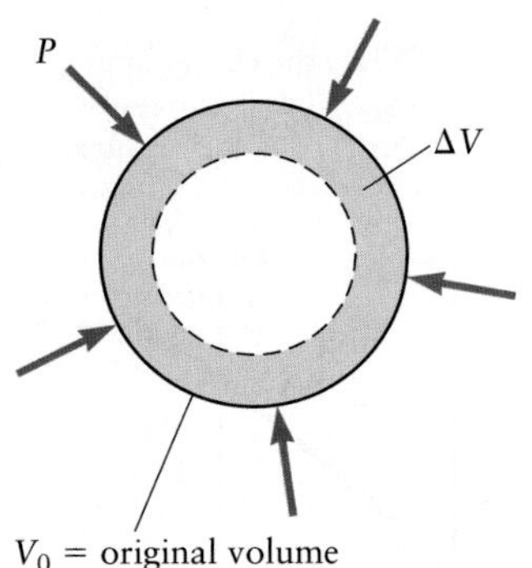

▲ **Figure 11.24** When an object is subject to a pressure (as when it is immersed in a fluid), the pressure force will compress the object, making the volume smaller by an amount ΔV.

defined as $\Delta x/L_0$, where L_0 is the length of the bar (Fig. 11.22). The shear stress and shear strain are related by

$$\frac{F}{A} = S\frac{\Delta x}{L_0} \tag{11.24}$$

where S is the ***shear modulus***. Table 11.1 lists the shear modulus for several materials.

This relation is very similar to the expression that involves a compressive stress and Young's modulus (Eq. 11.23). In both cases, the deformation of the bar is directly proportional to the stress, so both relations are equivalent to Hooke's law. These relations between elastic stress and strain describe many materials, so many material objects will act as simple Hooke's law springs when acted on by a compressive or tensile force or by a shear force. We can therefore apply our results for simple springs to a variety of situations.

The Bulk Modulus

Figure 11.24 shows an object immersed in a fluid. The pressure P in the fluid leads to a force on the surface of the object. This force is directed normal to the object's surface and causes the object's volume to change by an amount ΔV. If V_0 is the original volume of the object, the pressure and volume change are related by

$$P = -B\frac{\Delta V}{V_0} \tag{11.25}$$

where B is an elastic constant called the ***bulk modulus***. Note that the pressure and volume change are directly proportional. The presence of the negative sign in Equation 11.25 means that a positive pressure will produce a negative value of ΔV, as we should expect.

In our discussions of Young's and shear moduli, we have assumed the object in question is a solid, not a liquid or a gas. Compressive, tensile, and shear stresses (Figs. 11.21 and 11.23) can only be applied to an object that maintains its shape without the help of a container. In contrast, the concept of a bulk modulus applies to solids, liquids, and gases because all change volume in response to changes in pressure.

Elastic Properties and Simple Harmonic Motion

We learned in Section 11.2 that when an object is subject to a restoring force proportional to the displacement, the object can undergo simple harmonic motion. The same type of motion is found with the metal bars in Figures 11.21 and 11.23. If the end of the bar is displaced—that is, deformed an amount ΔL by compressing, stretching, or shearing the bar—it experiences a restoring force proportional to ΔL. If the displacement is caused by a compressive or tensile stress, the restoring force involves Young's modulus Y; for a shear displacement, the restoring force involves the shear modulus S. Some distortions produce both compression and shear, so the total restoring force may involve a combination of Y and S. In all these cases, the restoring force is proportional to the displacement and is thus described by Hooke's law. For this reason, a vibrating metal bar is closely analogous to a mass on a spring and generally behaves as a simple harmonic oscillator.

EXAMPLE 11.8 Ⓡ Oscillation Frequency of a Diving Board

When a diving board bends, the restoring force due to the board's elastic properties can be written as

$$F_{\text{board}} = -kx \tag{1}$$

where x is the displacement of the end of the board (Fig. 11.25) and k is the "spring" constant of the diving board. Estimate k for a typical diving board and use your result to find the oscillation frequency when a diver is standing at the end of the board.

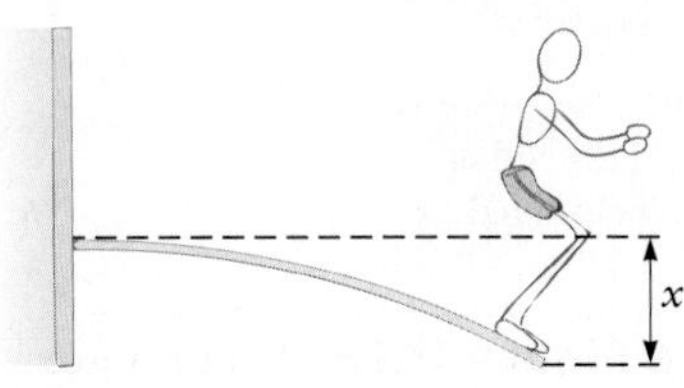

▲ **Figure 11.25** Example 11.8.

RECOGNIZE THE PRINCIPLE

We can use Equation (1) to find k if we know how much a diving board will bend in response to a known force. Assuming the mass of the board is small, when a diver of mass m stands at the end of the board, the force on the board is about mg.

SKETCH THE PROBLEM

Figure 11.25 shows the diving board bent (displaced) an amount x as the diver stands at the end of the board.

IDENTIFY THE RELATIONSHIPS

A typical person has a mass of about $m = 80$ kg. When he or she stands at rest at the end of a diving board, the displacement of the board is typically $x = 10$ cm (Fig. 11.25). Of course, this value will vary from board to board, but $x = 1$ m seems too large and $x = 1$ cm is much smaller than common experience indicates. Applying Newton's second law to the diver in Figure 11.25 leads to

$$\sum F = F_{\text{board}} - mg = -kx - mg = 0$$

$$k = -\frac{mg}{x}$$

SOLVE

Using $m = 80$ kg and taking $x = -0.1$ m (a negative value because the board is displaced downward), we find

$$k = -\frac{(80 \text{ kg})(9.8 \text{ m/s}^2)}{(-0.1 \text{ m})} = \boxed{8000 \text{ N/m}}$$

If we assume the mass of the diving board is much smaller than the mass of the person (which is only a very rough approximation), the oscillation frequency is the same as that of a simple mass-on-a-spring system (Eq. 11.14),

$$f = \frac{1}{2\pi}\sqrt{\frac{k}{m}}$$

Inserting our values for k and m gives

$$f = \boxed{2 \text{ Hz}}$$

► ***What does it mean?***

The period for the diver's oscillation is thus about $T = 1/f = 0.5$ s. You should compare this value with the value you expected based on your intuition or the last time you visited a swimming pool.

11.5 Damping and Resonance

So far, we have ignored the effect of friction on the motion of a harmonic oscillator. If there is no friction, an oscillator set into motion will oscillate forever. Of course, most oscillators eventually come to rest as their vibrations decay with time. The friction in an oscillating system is referred to as ***damping***, and we now consider the motion of a ***damped harmonic oscillator***.

Newton's second law can be used to derive the displacement y of a damped oscillator as a function of time. However, the mathematics is more complicated than we can tackle here, so let's focus on the qualitative behavior. Consider a pendulum swing moving through air. Curve 3 in Figure 11.26 shows how the displacement varies with time when the damping is weak, that is, when there is only a small amount of friction. This plot applies to any weakly damped harmonic oscillator, so

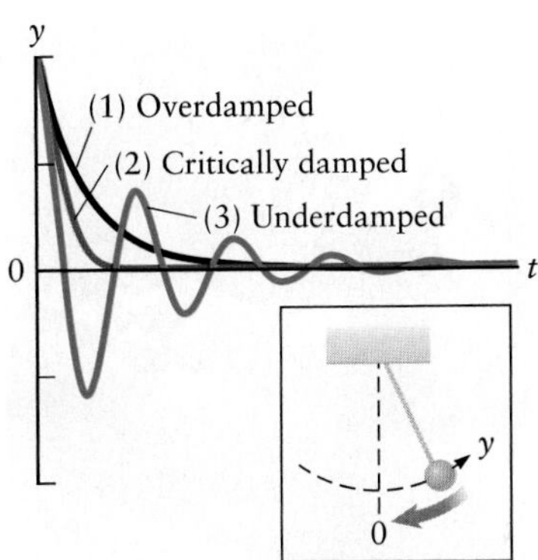

▲ **Figure 11.26** When a damped oscillator is given a nonzero displacement and then released, it can exhibit three different types of behavior: (1) overdamped, (2) critically damped, and (3) underdamped.

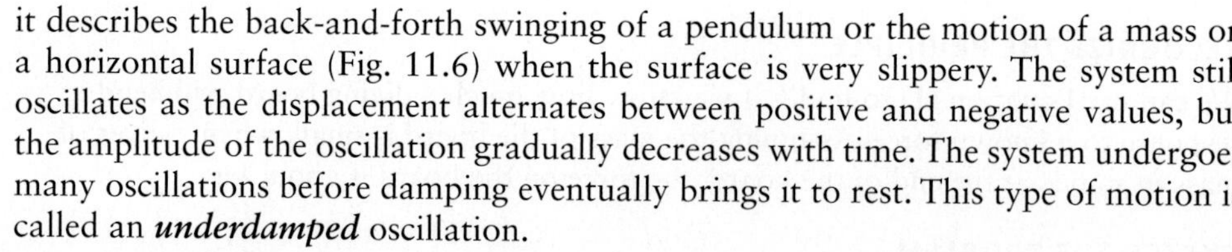

it describes the back-and-forth swinging of a pendulum or the motion of a mass on a horizontal surface (Fig. 11.6) when the surface is very slippery. The system still oscillates as the displacement alternates between positive and negative values, but the amplitude of the oscillation gradually decreases with time. The system undergoes many oscillations before damping eventually brings it to rest. This type of motion is called an ***underdamped*** oscillation.

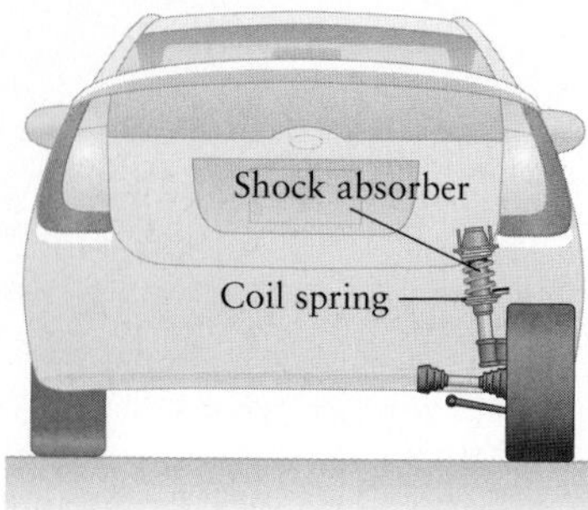

▲ **Figure 11.27** The struts on a car consist of a coil spring and a shock absorber, and act as springs with damping. The goal is usually to have a car respond to bumps in the road as a critically damped oscillator.

CONCEPT CHECK 11.6 Sources of Damping in a Real Oscillator

Consider a real pendulum or a real mass-on-a-spring system as in Figure 11.6. What are some possible sources of damping in these systems?

When friction is very large, the oscillator is said to be ***overdamped***. The resulting displacement as a function of time in this case is shown as curve 1 in Figure 11.26. This type of motion would be found if the pendulum mass were moving through a very thick fluid (such as honey) rather than air. When the bob of an overdamped pendulum is pulled to one side and then released, the bob swings *very* slowly back to the bottom ($y = 0$) as plotted in Figure 11.26 (curve 1).

The underdamped and overdamped regimes are distinguished in the following way. In the underdamped case, the displacement always passes through zero at least once, and usually many times, before the system comes to rest. In contrast, an overdamped system released from rest does not swing past the bottom, but moves to the equilibrium point without going past it. At the boundary between these two cases the system is ***critically damped***. There, the displacement falls to zero as rapidly as possible *without* moving past the equilibrium position ($y = 0$) as in curve 2 in Figure 11.26.

These different categories of damping have important implications for applications. For example, the struts in a car are essentially springs that support the car's body as shown in Figure 11.27. Struts enable the tires to move up and down according to bumps in the road without directly passing these vibrations to the car's body (and the passengers). When the car hits a bump, the strut's springs are compressed, corresponding to the initial value of the displacement in Figure 11.26. To make the ride as comfortable as possible, the struts are designed to be critically damped, which allows the body of the car to return to its original height as quickly as possible. With underdamped (worn-out!) struts, a car oscillates up and down after hitting a bump, resulting in an uncomfortable ride.

The Driven Oscillator

In many applications, an oscillator is subjected to a force at regular intervals, or even continuously. A familiar example—a child being pushed on a swing—is shown in Figure 11.28. This particular pendulum is "driven" by the force that the parent exerts on the swing. We have already seen (Eq. 11.18) that a pendulum–swing has an oscillation frequency of

$$f_{\text{nat}} = \frac{1}{2\pi}\sqrt{\frac{g}{L}}$$

which is often referred to as the ***natural frequency*** of the swing; it is not necessarily the same as the frequency associated with the driving force. Suppose the parent applies a small push to the swing at regular time intervals. If the period of this pushing is T_{drive}, we can define a ***driving frequency*** $f_{\text{drive}} = 1/T_{\text{drive}}$. The amplitude of the oscillations of the pendulum–swing depend on the driving frequency as sketched in Figure 11.29.

▲ **Figure 11.28** When the parent pushes this child, the swing acts as a driven oscillator.

If the frequency of the force from the parent does not match the pendulum's natural frequency—that is, if f_{drive} is not close to f_{nat}—the pendulum amplitude is small, even if the parent applies a large force. The amplitude is largest when the driving frequency is close to the natural frequency of a system. That happens when the parent in Figure 11.28 times her pushing to match the natural swinging frequency of the pendulum. This phenomenon is called ***resonance***. Whenever the frequency of the driving force matches the natural frequency of an oscillator, the amplitude of oscillation will be large. Figure 11.29 shows the behavior for cases of weak, medium, and strong damping. The resonance curve is narrowest and tallest when the damping is weakest, and it becomes lower and wider as the damping is increased. The peak of the resonance curve occurs at the ***resonant frequency***; for weak damping, the resonant frequency is very close to the natural frequency f_{nat} of the oscillator.

A harmonic oscillator for which damping is important is the vibrating element in a loudspeaker, the so-called speaker cone. Different musical notes have different frequencies, and when a loudspeaker is used to play music, the speaker cone is subject to a force whose frequency is determined by the frequencies present in the music. According to Figure 11.29, if the speaker cone oscillator is weakly damped, it will respond strongly to music near its natural frequency and relatively weakly to frequencies that are far from f_{nat}. A weakly damped speaker will thus produce loud sounds at some frequencies and weak sounds at others, which is *not* how one would like a loudspeaker to work. Instead, a speaker should have a frequency response that is as constant (uniform) as possible. One way to accomplish that is to make the speaker cone heavily damped. The cone of a good loudspeaker, however, should also move fairly easily so that it produces sound efficiently, hence the damping cannot be too large. This trade-off is one of the challenges faced by the speaker designer. Similar design challenges are faced in other applications of harmonic oscillators, such as the struts for a car.

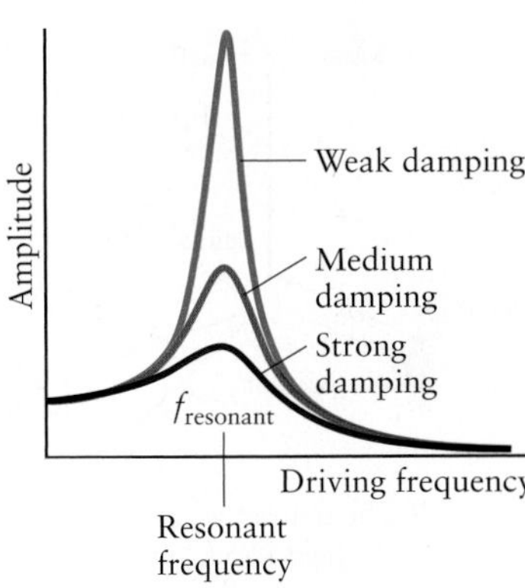

▲ **Figure 11.29** The amplitude of a driven oscillator depends on the damping. For weak damping, the oscillator will have a large amplitude when the driving frequency is near the resonant frequency, $f_{resonant}$. For a weakly damped oscillator, the resonant frequency is very close to the natural frequency.

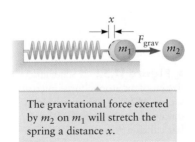

CONCEPT CHECK 11.7 Damping of a Bridge

A structure such as a bridge or a building can oscillate in response to forces from the wind or an earthquake. Do you think these structures should be designed to be (a) underdamped, (b) critically damped, or (c) overdamped?

11.6 Detecting Small Forces

The phenomena of oscillatory motion and elasticity are closely connected and come together in the problem of detecting small forces.

The Cavendish Experiment

Let's consider an experiment to measure the force of gravity between two terrestrial objects. In Chapter 5, we found that the force of gravity between two massive and closely spaced lead spheres is about $F_{grav} = 2 \times 10^{-3}$ N (Eq. 5.22). This is a small force, and we need to be clever if we want to measure it experimentally. We might try to measure this force by connecting m_1 to a spring as sketched in Figure 11.30. The gravitational force exerted by m_2 on m_1 will cause the spring to stretch an amount x, and by measuring x, we can determine the magnitude of the force. Using Hooke's law for the spring, we have (considering magnitudes only) $F_{grav} = kx$. A soft spring might have a spring constant of about $k = 1$ N/m. Using this value and $F_{grav} = 2 \times 10^{-3}$ N, we can calculate the expected value of x:

$$x = \frac{F_{grav}}{k} = \frac{2 \times 10^{-3}\ \text{N}}{1\ \text{N/m}} = 0.002\ \text{m} = 2\ \text{mm}$$

If we want to measure F_{grav} with an accuracy of 1%, we would need to measure x with a precision of 0.02 mm. This is about the diameter of a human hair, so this

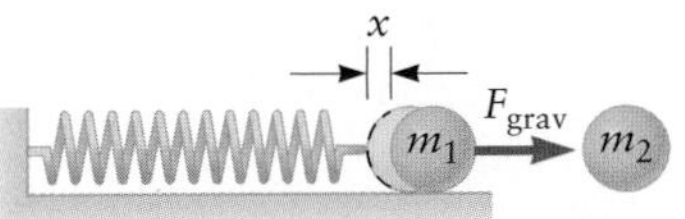

The gravitational force exerted by m_2 on m_1 will stretch the spring a distance x.

▲ **Figure 11.30** In theory, one could measure the gravitational force between two masses m_1 and m_2 by attaching m_1 to a spring and then observing how much the spring is stretched (x) when m_2 is brought nearby. This experimental design is not recommended because the displacement of the spring will be extremely small.

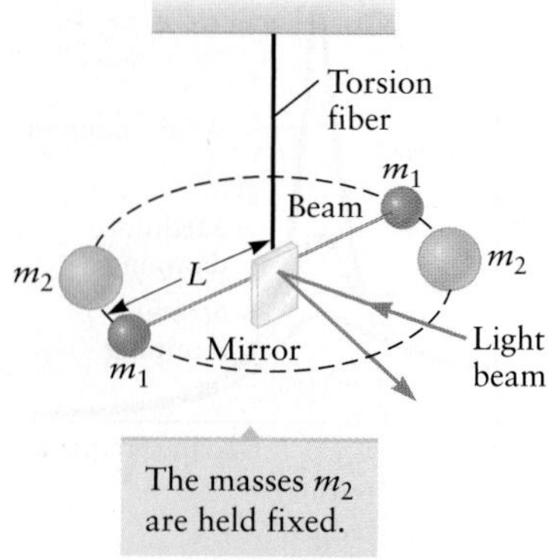

▲ **Figure 11.31** The Cavendish apparatus. The force of gravity exerted by the large masses m_2 on the small masses m_1 causes the torsion fiber to twist. The twist angle can be measured by attaching a small mirror and observing the reflection of a light beam.

method is not a very promising approach for a high precision measurement of F_{grav}. We need a different strategy.

Figure 11.31 shows how Henry Cavendish tackled this problem in an experiment he conducted about 200 years ago. He mounted two metal spheres, each of mass m_1, on a light rod and suspended the rod by a thin wire. This setup is very similar to the torsional oscillator in Figure 11.14B. Cavendish then brought two other objects, each of mass m_2, near the original spheres as shown in Figure 11.31, and the force of gravity exerted by the objects m_2 on the spheres m_1 caused the torsion fiber to twist. We saw in Section 11.2 that when a torsion fiber is twisted by an angle θ, it exerts a torque $\tau = -\kappa\theta$, where κ is the torsion constant of the fiber. Torsion fibers can have very small diameters; they can be as thin as a human hair (about 20 μm in diameter), making them *extremely* easy to twist. In fact, a value of $\kappa = 1 \times 10^{-9}$ N·m is quite common for a wire of this diameter.

With such small values of the torsion constant, Cavendish was able to measure F_{grav} accurately. However, Cavendish also had to deal with the fact that the system in Figure 11.31 acts as a torsional *oscillator*. When the two masses m_2 are brought close to the spheres on the torsion fiber, the gravitational force will set this pendulum into oscillation with a frequency (Eq. 11.20)

$$f = \frac{1}{2\pi}\sqrt{\frac{\kappa}{I}} \tag{11.26}$$

where I is the moment of inertia of the rotating part of the apparatus. Recall that the torsion constant κ is very small—that is what makes the Cavendish experiment so sensitive to small forces—but a very small value of κ in Equation 11.26 gives a very low frequency. In fact, frequencies of 0.0001 Hz are typical! That is *quite* a low frequency. The period of such an oscillator is $T = 1/f \approx 1 \times 10^4$ s, which is about 3 hours. Such a torsion oscillator is also a weakly damped system (because the friction in a torsion fiber is very small), so its oscillations will decay very slowly with time (Fig. 11.26). An experimenter will thus have to wait many periods before the oscillations stop and the twist angle can be measured. These very slow and weakly damped oscillations make the Cavendish experiment quite challenging. Legend has it that Cavendish performed his experiments very late on Sunday evenings, when the interfering vibrations from traffic on neighboring roads were smallest.

The Atomic Force Microscope

We next consider a modern invention that uses a type of spring to measure very small forces. The ***atomic force microscope*** (AFM) is illustrated in Figure 11.32A. A small bar, typically a fraction of a millimeter in length (about the diameter of the period at the end of this sentence), is fashioned so that it has an even smaller sharp tip at one

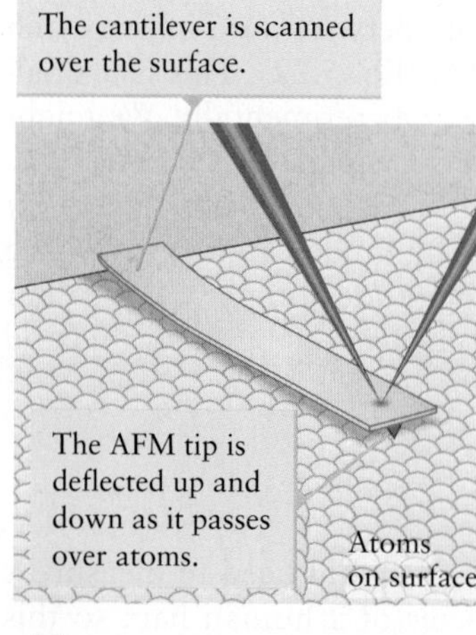

► **Figure 11.32** A Schematic design of an atomic force microscope (AFM). A very sharp tip is mounted on the end of a flexible bar called a "cantilever." The cantilever moves up and down as the atoms on the surface exert forces on the tip. The vertical deflection of the end of the cantilever is then used to construct an image of the surface profile. The red beams denote laser light reflected from the cantilever to measure its deflection. B An AFM image showing the atoms on a nickel surface. Each bump is an individual nickel atom.

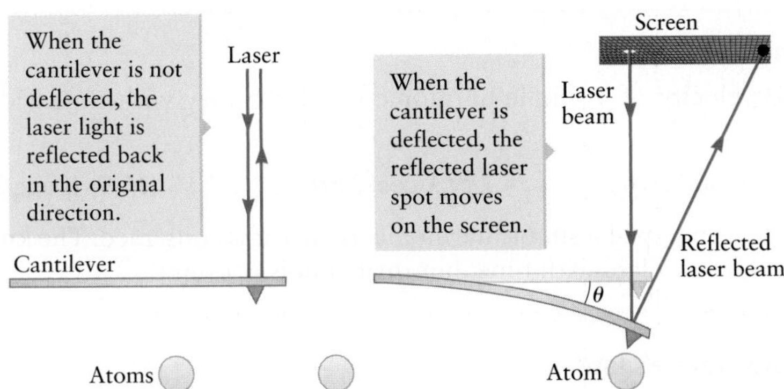

◀ **Figure 11.33** The deflection of an atomic force microscope cantilever is often measured by using the deflection of a laser beam. The location of the laser spot on a distant screen depends on the angle θ of the cantilever tip from the horizontal. By measuring the location of the laser spot, you can thus measure how much the cantilever is bent when it approaches atoms on the surface being studied.

end. This bar, called a cantilever, bends in response to any forces that act on it. These forces usually act at the tip, so the cantilever bends up and down as shown in the figure. This microscope works by scanning the tip of the cantilever over the surface of a material that is to be studied. The tip is *very* sharp, with a width at the end of only a few atoms. Although the tip usually does not directly touch the surface atoms, it is attracted to them, and the magnitude of the attractive force depends on the distance between the tip and the nearest surface atom. If an atom is "high" (i.e., sticks out and is close to the tip), the force on the tip is large, causing the cantilever to be deflected. By measuring this deflection as a function of the cantilever's position, one can make an image of the surface as shown in Figure 11.32B. An atomic force microscope is extremely sensitive to small forces. This sensitivity, along with the sharpness of the tip, makes it possible to resolve features as small as single atoms.

The cantilever behaves just like the diving board in Example 11.8 and can be treated as a simple spring obeying Hooke's law. A force of magnitude $F = kx$ is thus required to deflect the tip by a distance x. The value of k can be estimated from the shear modulus of whatever material is used to make the cantilever (using Eq. 11.24). For a typical cantilever, $k = 0.1$ N/m. The force between the tip and the surface[1] is typically 1×10^{-9} N, so the deflection of the cantilever is

$$x = \frac{F}{k} = \frac{1 \times 10^{-9}\text{ N}}{0.1\text{ N/m}} = 1 \times 10^{-8}\text{ m}$$

This very small deflection can be measured by reflecting a laser beam from the cantilever as shown in Figure 11.33. An AFM thus uses what is basically a simple spring to measure very small forces and displacements, thereby obtaining images of the atoms on the surface of a solid.

Summary | CHAPTER 11

Key Concepts and Principles

Frequency and period

Any motion that repeats with time is ***harmonic***. The repeat time is called the ***period***. The ***frequency*** (measured in hertz) is equal to the number of cycles that are completed each second. Frequency and period are related by

$$f = \frac{1}{T} \qquad \textbf{(11.1) (page 354)}$$

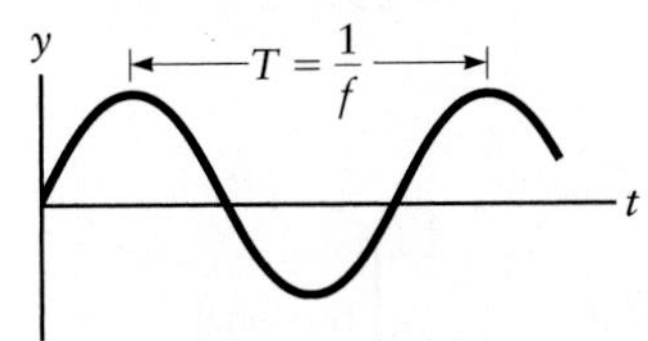

(Continued)

[1]We'll discuss this force in Chapter 29.

Simple harmonic motion

The position and velocity of a simple harmonic oscillator vary with time according to

$$y = A\sin(2\pi ft) \qquad v = 2\pi fA\cos(2\pi ft) \qquad \textbf{(11.9) (page 357)}$$

The total mechanical energy of a simple harmonic oscillator is conserved. The kinetic and potential energies oscillate with time, but their sum is constant.

Stress, strain, and Hooke's law

Elastic objects stretch or compress in response to applied forces. For a compressive or tensile force, the deformation is

$$\frac{F}{A} = Y\frac{\Delta L}{L_0} \qquad \textbf{(11.23) (page 368)}$$

where Y is an elastic constant called ***Young's modulus***. The ratio F/A is called the ***stress***, and $\Delta L/L_0$ is called the ***strain***. Other types of deformations (such as shear strains) are described by other elastic constants.

Damping and resonance

Most real harmonic oscillators are affected by friction, causing the oscillations to be damped. There are three classes of damped oscillators: ***underdamped*** (weak damping), ***overdamped*** (large damping), and ***critically damped***. When a damped harmonic oscillator is subject to a time-dependent driving force, the amplitude will be largest when the driving frequency is close to the natural frequency of the oscillator. This is called ***resonance***.

Applications

Mass connected to a spring

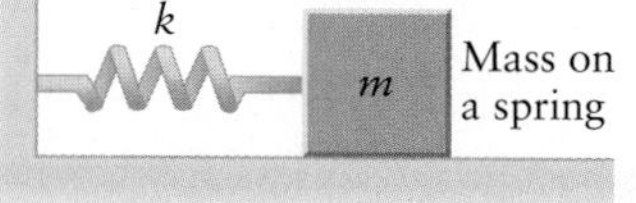

Mass on a spring

A mass connected to a spring with spring constant k oscillates with a frequency

$$f = \frac{1}{2\pi}\sqrt{\frac{k}{m}} \qquad \textbf{(11.14) (page 358)}$$

Simple pendulum

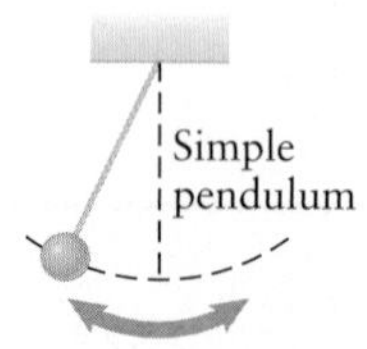
Simple pendulum

A ***simple pendulum*** consists of a mass attached to a massless string of length L. Its oscillation frequency is

$$f = \frac{1}{2\pi}\sqrt{\frac{g}{L}} \qquad \textbf{(11.18) (page 362)}$$

Torsional oscillator

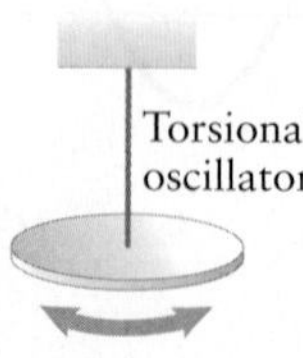
Torsional oscillator

An object with a moment of inertia I suspended by a torsion fiber is a ***torsional oscillator***. Its frequency is

$$f = \frac{1}{2\pi}\sqrt{\frac{\kappa}{I}} \qquad \textbf{(11.20) (page 364)}$$

where κ is the ***torsion constant*** of the fiber.

Properties of simple harmonic oscillators

All simple harmonic oscillators have the following properties:

- The frequency is independent of amplitude.
- The force or torque that causes the oscillatory motion is a restoring force whose magnitude is proportional to the displacement of the oscillator from its equilibrium position. Such a restoring force is described by Hooke's law.

Questions

SSM = answer in *Student Companion & Problem-Solving Guide*

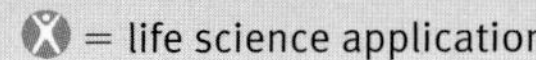 = life science application

1. A friend of yours is asked to design a swing using two ropes attached to the branch of a tree, and he shows you the design as sketched in Figure Q11.1. Explain why this design will not work very well. That is, why it is important that both of the ropes be the same length?

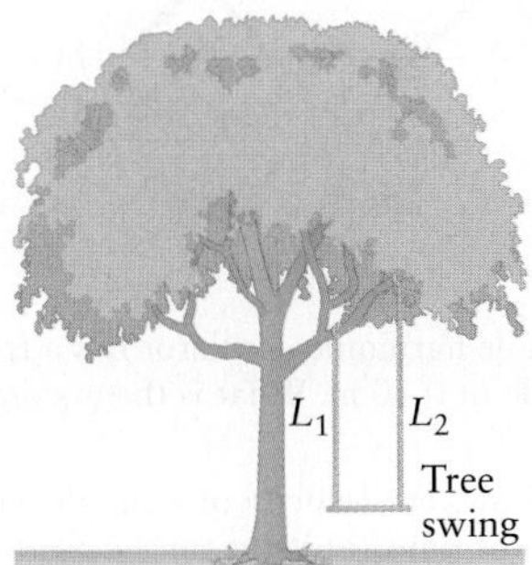

Figure Q11.1 Why is it desirable to have $L_1 = L_2$?

2. A mass hangs from a vertical spring and is initially at rest. A person then pulls down on the mass, stretching the spring. Does the total mechanical energy of this system (the mass plus the spring) increase, decrease, or stay the same? Explain your answer.
3. Design an experiment to measure the torsion constant κ of a torsion fiber. The value of κ is needed in the analysis of the Cavendish experiment (Sec. 11.6).
4. SSM Consider a simple pendulum that is used as a clock. (a) What should the length of the pendulum be to make one oscillation (one "tick" of the clock) every second when it is at sea level? (b) Will this clock "speed up" or "slow down" when it is taken to the top of Mount Everest? (c) How long will it take this clock to make 60 ticks on Mount Everest?
5. If a spring with spring constant k_0 is cut in half, what is the spring constant of one of the pieces?
6. A car mounted on struts is like a mass on a spring. If you ignore damping, how will the frequency of the oscillations change if passengers (or a heavy load) are added to the car? Will the frequency increase, decrease, or stay the same?
7. A basketball player dribbles the ball as she moves along the court. What is a typical value for the frequency of this oscillatory motion?
8. When a group of marching soldiers reach a bridge, they often "break stride" and do not walk "in step" across the bridge. Explain why.
9. Two metal rods with the same length and cross-sectional area are both subjected to a compressive force F. If the length of rod 1 changes by more than the length of rod 2, which one has the larger Young's modulus?
10. Two solid rods are made of the same material and have the same cross-sectional areas, and their lengths differ by a factor of two. If a compressive force F is applied to both, what is the ratio of their changes in length?
(a) The longer rod changes length by more, by a factor of two.
(b) The shorter rod changes length by more, by a factor of two.
(c) The lengths change by the same amount.
11. Make a sketch of how the position and kinetic energy of a harmonic oscillator vary with time. The period of the oscillations of KE can be determined from the separation in time of adjacent maxima in your KE sketch. Show that the frequency of the KE oscillations is equal to twice the frequency of the displacement oscillations (Eq. 11.14). Explain why.
12. In Section 11.3, we discussed the total mechanical energy of a mass-on-a-spring oscillator. The result in Equation 11.21 shows that the total energy is proportional to the square of the amplitude. Derive the corresponding results for the total mechanical energy of a simple pendulum and for a torsional oscillator and show that they are also proportional to the square of the amplitude.
13. SSM Use energy considerations to derive the oscillation frequency for a mass-on-a-spring oscillator. *Hint*: The maximum potential energy stored in the spring must be equal to the maximum kinetic energy of the mass.
14. Derive the frequency of oscillation of a torsional oscillator, Equation 11.20. *Hint*: Consider how the restoring force depends on the twist angle and use Newton's second law for rotational motion ($\tau = I\alpha$).
15. Pendulum clocks use a pendulum to keep time (Fig. Q11.15). This type of clock can be adjusted by varying the length of the pendulum. How should the length be adjusted in the following cases?
(a) The clock is running slow.
(b) The clock is running fast.
(c) The clock is running fine at sea level, but needs to be moved to Denver.
16. Many musicians use a metronome to help keep time when playing music. Some metronomes use a simple pendulum. If a metronome clicks 100 times per minute, what is the length of the pendulum?

Figure Q11.15

17. In Chapter 3, we learned how the apparent weight of a person in an elevator depends on the acceleration of the elevator. Consider a simple pendulum that is placed in an elevator. Does the elevator's acceleration affect the pendulum's period? If the pendulum is accelerating downward with magnitude g, what is the period of the pendulum?

18. The windshield wipers on your car are an example of periodic motion. What is the approximate period of the motion?

19. Arrange these materials according to their Young's modulus from smallest to largest: (a) Jell-O, (b) steel, (c) wood, (d) diamond.

20. Discuss the difference between the Young's modulus of a material and the strength of a material. Does a large value of Y guarantee a strong material?

Problems

SSM = solution in *Student Companion & Problem-Solving Guide*
★ = intermediate ✪ = challenging Ⓧ = life science application
Ⓡ = reasoning and relationships problem
ⓇⓉ = reasoning tutorial in Enhanced WebAssign

11.1 GENERAL FEATURES OF HARMONIC MOTION

1. A simple harmonic oscillator takes 15 seconds to undergo 25 complete oscillations. (a) What is the period of the oscillator? (b) What is the frequency of the oscillator?

2. Ⓧ A person's heart beats 70 times in 1 minute. What is the average frequency of this oscillation?

3. ★ Figure P11.3 shows the displacement of a simple harmonic oscillator as a function of time. (a) Find the period, frequency, and amplitude of the oscillator. (b) At what time(s) does the acceleration have its largest positive value? (c) When does the acceleration have its most negative value?

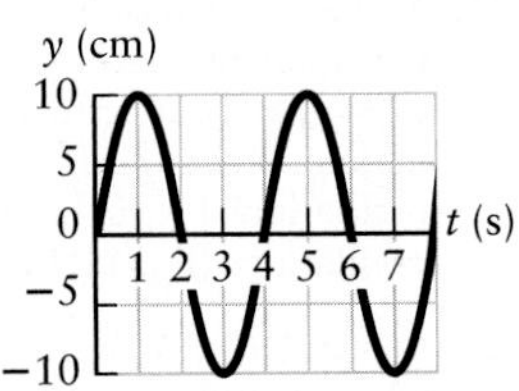

Figure P11.3 Problems 3, 5, and 15.

4. ★ Figure P11.4 shows a plot of the position as a function of time for a particle undergoing simple harmonic motion. Identify the following points on this graph: (a) Where is the speed largest? (b) Where is the velocity positive with its magnitude being largest? (c) Where is the magnitude of the acceleration largest? (d) Where is the force on the particle zero?

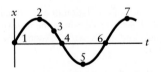

Figure P11.4 Problems 4 and 34.

5. (a) For the simple harmonic oscillator in Figure P11.3, estimate the velocity at $t = 4.0$ s and at 7.0 s. (b) At what time(s) does the velocity have its largest positive value?

6. The displacement of a harmonic oscillator is given by $y = 3.4 \sin(25t)$, where the units of y are meters and t is measured in seconds. Find (a) the amplitude and (b) the frequency.

7. ★ The displacement of a harmonic oscillator is given by $y = 9.4 \sin(15t)$, where the units of y are meters and t is measured in seconds. What is its maximum velocity?

8. The displacement of a harmonic oscillator is given by $y = 7.1 \sin(48t)$, where the units of y are meters and t is measured in seconds. Give three values of t at which the oscillator has (a) its largest (and positive) displacement and (b) its largest (and positive) velocity.

9. Figure P11.9 shows the velocity as a function of time for an oscillator. Is it a simple harmonic oscillator? Explain why or why not. What is its frequency?

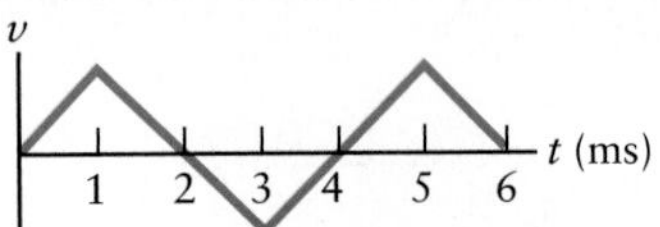

Figure P11.9

10. SSM ★ A simple harmonic oscillator has a frequency of 300 Hz and an amplitude of 0.10 m. What is the maximum velocity of the oscillator?

11. Figure P11.11 shows the velocity of a simple harmonic oscillator as a function of time. (a) Estimate the position when $v = 0$. (b) Estimate the frequency.

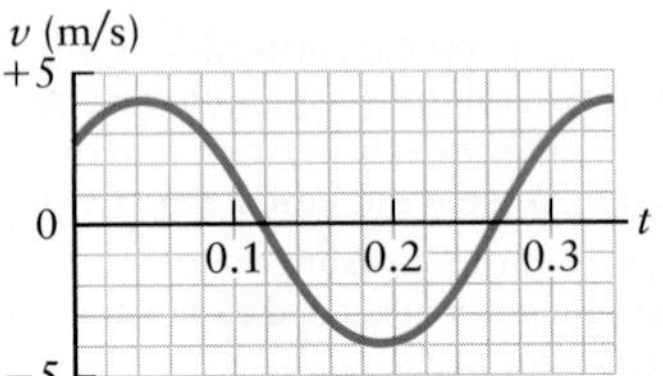

Figure P11.11

11.2 EXAMPLES OF SIMPLE HARMONIC MOTION

12. The period of oscillation for a mass-on-a-spring system is 0.22 s. If $m = 3.5$ kg, what is the spring constant of the spring?

13. Consider the mass on a spring in Figure 11.6. If the spring constant is $k = 30$ N/m and the mass is $m = 2.5$ kg, what is the period?

14. ★ A child plays on a bungee cord and oscillates with a certain frequency f. An adult with a mass that is five times greater than that of the child then uses the same bungee cord. What is the ratio of the frequency with the adult to the frequency with the child?

15. ✪ Suppose Figure P11.3 describes the displacement of a mass-on-a-spring harmonic oscillator. (a) If the mass is $m = 2.0$ kg, what is the spring constant? (b) Estimate the velocity at $t = 4.0$ s and at 5.0 s. (c) Estimate the acceleration at 1.5 s. (d) Estimate the maximum acceleration.

16. SSM ★ A mass $m = 4.5$ kg is attached to a vertical spring with $k = 200$ N/m and is set into motion. (a) What is the frequency of the oscillation? (b) If the amplitude of the oscillation is 3.5 cm, what is the maximum value of the velocity? (c) How long does it take the mass to move from $y = 1.5$ cm to $y = 2.5$ cm? (d) If the mass is oscillating with a maximum speed of 45 m/s, what is the amplitude? (e) If the spring constant is increased by a factor of two and the maximum kinetic energy of the mass is the same, by what factor does the amplitude change?

17. ✪ A mass $m = 2.4$ kg is attached to two springs as it slides along a frictionless floor, while the springs are fastened to two walls as shown in Figure P11.17. The springs both have $k = 400$ N/m and are both in their relaxed states (unstretched and uncompressed) when the mass is centered between the two walls. What is the frequency of this simple harmonic oscillator?

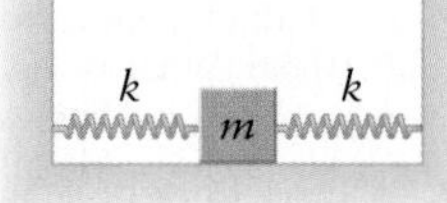

Figure P11.17

18. SSM ★ RT Estimate the spring constant for a trampoline. Assume a person is standing on the trampoline and oscillating up and down without leaving the trampoline. *Hint*: Begin by estimating the mass of the oscillator and the period of the motion.

19. ★ A mass-on-a-spring system has $m = 50$ kg and $k = 200$ N/m. The mass is pulled a distance 0.25 m from its equilibrium position and then released. (a) What is the maximum acceleration of the mass? (b) What is its maximum velocity?

20. ★ Consider the mass-on-a-spring system in Figure P11.20. Three identical springs, with the same spring constant $k = 40$ N/m, are used to connect the mass ($m = 20$ kg) to a ceiling. What is the frequency of this simple harmonic oscillator?

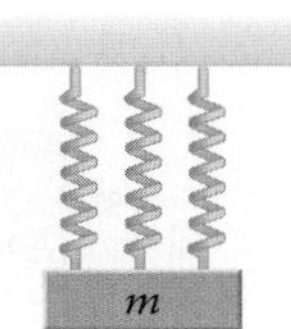

Figure P11.20

21. ✪ RT Estimate the maximum acceleration of a bungee jumper. Is it larger or smaller than the acceleration due to gravity during free fall? *Hint*: Example 11.3 gives values for several quantities that may be useful. Assume the bungee cord acts as a simple spring (which is not quite realistic) and that the mass of the person is 50 kg.

22. Consider a simple pendulum that consists of a rock of mass 3.5 kg tied to the end of a (massless) string of length 1.5 m. (a) What is the frequency of the pendulum? (b) What is the period of the pendulum?

23. A simple pendulum oscillates with a period of 3.5 s. What is its length?

24. ★ A simple pendulum has a length of 2.5 m and is pulled a distance $y = 0.25$ m to one side and then released (Fig. P11.24). (a) What is the speed of the pendulum when it passes through the lowest point on its trajectory? (b) What is its acceleration in the direction along its circular arc when it passes through the lowest point on its trajectory?

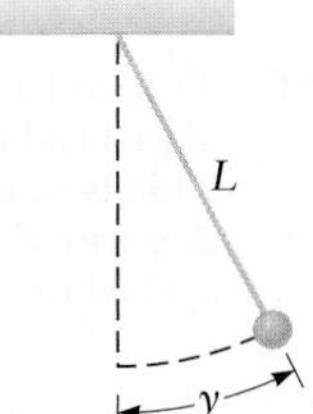

Figure P11.24

25. ★ A simple pendulum has a period of 2.5 s on the Earth. An astronaut then takes it to the surface of the Moon. What period does the astronaut measure there?

26. ★ A simple pendulum uses a steel wire as the "string." The length of this wire is 1.5 m at room temperature. If the temperature is increased by 10°C, the length of the wire increases by 0.18 mm. What is the change in the period of the pendulum?

27. ★ The length of a simple pendulum is increased by a factor of three. By what factor does the frequency change?

28. ★ RT What is the approximate period of one of the swings in Figure P11.28?

PhotoAlto/Alamy

Figure P11.28
Problems 28 and 45.

29. A torsional oscillator has a moment of inertia of 4.5 kg · m² and a torsion constant $\kappa = 0.15$ N · m. What is the frequency of the oscillator?

30. ✪ Consider the torsional oscillator in Figure P11.30. The rotating mass consists of two small spheres, each of mass $m = 2.0$ kg, that are attached to the ends of a massless rod of length $L = 1.5$ m. The frequency of this torsional oscillator is found to be 0.035 Hz. What is the value of the torsion constant κ of the torsion fiber?

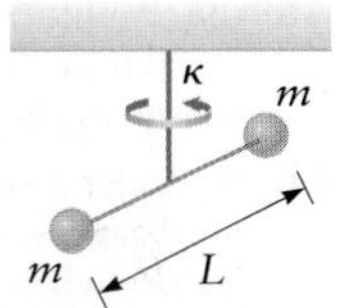

Figure P11.30
Problems 30 and 31.

31. Consider a torsional oscillator like the one in Figure P11.30 and suppose it has a frequency of 0.045 Hz. If the length of the rod is increased by a factor of three, what is the new frequency of the oscillator?

32. ✪ Figure P11.32 shows the displacement of a mass on a spring as a function of time. (a) At what point is the acceleration positive and the velocity negative? (b) Are there any times during the oscillation when the acceleration and the velocity are both positive and nonzero?

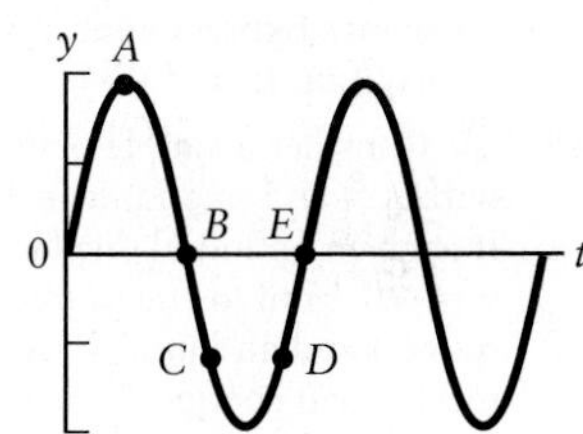

Figure P11.32

33. ✪ The cone of a loudspeaker oscillates with an amplitude of 2.0 mm. If the frequency is 2.5 kHz, what are the maximum velocity and the maximum acceleration of the cone? Assume the speaker cone moves as a simple harmonic oscillator.

11.3 HARMONIC MOTION AND ENERGY

34. ✪ Figure P11.4 shows the position as a function of time for a mass attached to a spring. At what points are (a) the magnitude of the momentum of the mass largest, (b) the kinetic energy largest, (c) the potential energy largest, and (d) the total energy largest?

35. ★ A mass-on-a-spring system has spring constant $k = 400$ N/m and mass $m = 0.50$ kg. (a) If it is given an initial displacement of 0.25 m and then released, what is the initial potential energy of the oscillator? (b) What is the maximum kinetic energy of the oscillator?

36. ✪ The position of a mass-on-a-spring oscillator is given by $y = A\sin(25t)$, where the value of t is in seconds and

$A = 0.35$ m. (a) What is the maximum kinetic energy of an oscillator of mass 1.7 kg? (b) Suppose the amplitude is increased so that the maximum kinetic energy is doubled. What is the new value of A?

37. SSM A particle attached to a spring with $k = 50$ N/m is undergoing simple harmonic motion, and its position is described by the equation $x = (5.7 \text{ m})\cos(7.5t)$, with t measured in seconds. (a) What is the mass of the particle? (b) What is the period of the motion? (c) What is the maximum speed of the particle? (d) What is the maximum potential energy? (e) What is the total energy?

38. Consider a simple harmonic oscillator whose position as a function of time is given by Figure P11.38. At what times in this plot is the kinetic energy a maximum? At what time(s) is the potential energy a maximum?

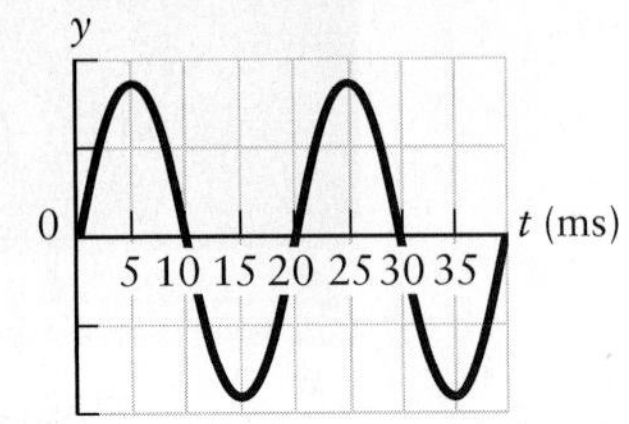

Figure P11.38

39. The position of a simple harmonic oscillator is given by $y = A\sin(2\pi ft)$, with $f = 400$ Hz. Find a value of t at which the potential energy is one quarter of its maximum value.

40. Consider a simple pendulum of length 1.2 m with $m = 2.0$ kg. If the amplitude of the oscillation is 0.15 m (as measured along the circular arc along which it moves), what is the total mechanical energy of the oscillator?

41. Consider a simple pendulum with $L = 1.2$ m and $m = 2.0$ kg. Suppose the mass is initially at rest at the lowest point on its trajectory when it is given an impulse such that it then has a velocity of 0.10 m/s. What is the total mechanical energy of the oscillator?

42. A simple pendulum has a length L and a mass m. At its highest point, the pendulum mass is $0.25L$ above its lowest point (Fig. P11.42). What is the speed of the mass when it is at its lowest point? Express your answer in terms of m, L, and g.

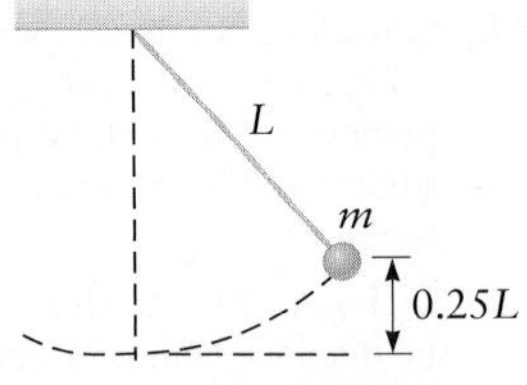

Figure P11.42

43. Consider a simple pendulum with $L = 5.2$ m and $m = 3.3$ kg. It is given an initial displacement $y = +0.15$ m (as measured along the circular arc along which it moves) and an initial velocity of $+0.20$ m/s. (a) What is the total mechanical energy of the oscillator? (b) What is the amplitude of the motion as measured along the circular arc followed by the pendulum mass? (c) What is the speed of the mass when it passes through the lowest point on its trajectory? (d) Suppose the initial velocity is instead equal to -0.20 m/s. How would the answers to parts (a), (b), and (c) change?

44. A mass-on-a-spring oscillator has $m = 7.0$ kg and $k = 300$ N/m and is oscillating with an amplitude of 12 cm. (a) If the value of the mass is increased by a factor of two but the amplitude is kept fixed, is the maximum potential energy the same, or is it different? (b) Find the original value and the new value of the maximum potential energy. (c) Explain why changing the mass does or does not change the potential energy. (d) Does changing the mass change the value of the maximum kinetic energy?

45. RT Consider one of the children in Figure P11.28. If she is swinging with a typical amplitude, what is her approximate maximum kinetic energy?

11.4 STRESS, STRAIN, AND HOOKE'S LAW

46. An aluminum bar has a length of 2.0 m and a rectangular cross setion with sides of 3.5 mm and 6.9 mm. (a) What force is required to stretch the bar 1.2 mm? (b) What is the strain for this value of the stretch?

47. Two solid rods having the same length are made of the same material with circular cross sections. Rod 1 has radius r, and rod 2 has radius $2r$. If a compressive force F is applied to both rods, their lengths are reduced by ΔL_1 and ΔL_2. Is the ratio $\Delta L_1/\Delta L_2$ (a) 4, (b) 2, (c) 1, (d) $\frac{1}{2}$, or (e) $\frac{1}{4}$?

48. Consider a steel rod of diameter 5.0 mm and length 3.5 m. If a compressive force of 5000 N is applied to each end, what is the change in the length of the rod?

49. A thick copper wire of length 10 m is observed to stretch 0.10 cm. What is the strain? If the applied force is 50,000 N, what is the diameter of the wire?

50. A shear force of 3.0×10^5 N is applied to an aluminum bar of length $L = 20$ cm, width $w = 5.0$ cm, and height $h = 2.0$ cm as shown in Figure 11.23. What is the shear deformation Δx?

51. A solid glass sphere with an initial diameter $d = 25.00$ cm is subject to a pressure of $1000 \times P_{atm}$, where atmospheric pressure $P_{atm} = 1.01 \times 10^5$ Pa. What is the new diameter of the sphere? The change in diameter is small, so keep at least four significant figures in your calculation.

52. A guitar string of diameter 0.50 mm and length 0.75 m is subject to a tension of 150 N. If the string stretches an amount 0.40 mm, what is Young's modulus of the string?

53. A steel cylinder of length 11 cm and radius $R = 3.5$ cm (Fig. P11.53A) is subjected to a compressive force of 5000 N. By what amount ΔL is the cylinder compressed?

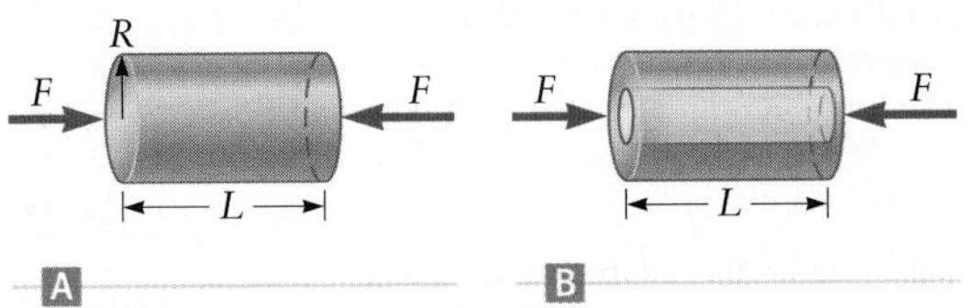

Figure P11.53 Problems 53 and 54.

54. The steel cylinder in Figure P11.53A is replaced by a steel tube (Fig. P11.53B). The outer radius of the tube is $R_1 = 3.5$ cm, and the inner radius is $R_2 = 1.5$ cm. If the compressive force is $F = 5000$ N, by what amount ΔL is the tube compressed?

55. Young's modulus for bone under compression is about 9×10^9 Pa, and the breaking stress is about 2×10^8 Pa. (a) A leg bone of length 30 cm when unstressed is stressed to its breaking point. What is the change in length of the bone? (b) If the bone has a diameter of 3 cm, how much weight can it support just before it breaks?

56. Two bars, one composed of aluminum and one of steel, are placed end to end as shown in Figure P11.56. The bars are both initially of length $L = 25$ cm, and both have a square cross section with $h = 2.0$ cm. If a compressive force $F = 10{,}000$ N is applied to each end, what is the change in length of each bar?

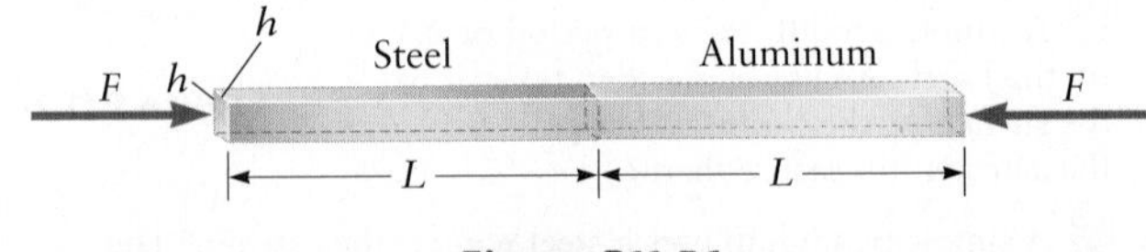

Figure P11.56

57. SSM An aluminum sphere has a radius of 45 cm on the Earth. It is then taken to a distant planet where the atmospheric pressure P is much larger than on the Earth. If the sphere has a radius of 43 cm on that planet, what is P?

58. You want to use a steel cable to tow your car ($m = 2000$ kg). The cable is solid steel with a diameter of 1.0 cm and a length of 10 m. If you are pulling the car such that its acceleration is 1.5 m/s^2, how much will the cable stretch?

59. A steel string (diameter 1.0 mm) of length 2.5 m hangs vertically. At the bottom of the string is a seat. When a child of mass m sits on the seat, she finds that the string stretches so that its new length is 1.8 mm longer than before. What is the mass of the child? Ignore the mass of the seat.

11.5 DAMPING AND RESONANCE

60. SSM A damped harmonic oscillator is displaced from equilibrium and then released. The oscillator displacement as a function of time is shown in Figure P11.60. Is this oscillator underdamped or overdamped?

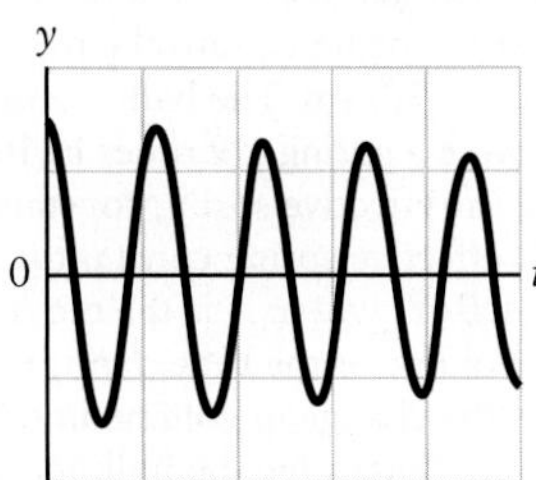

Figure P11.60

61. RT Figure P11.61 shows the displacement as a function of time for an underdamped harmonic oscillator. Estimate the fraction of the mechanical energy that is "lost" to friction during one cycle.

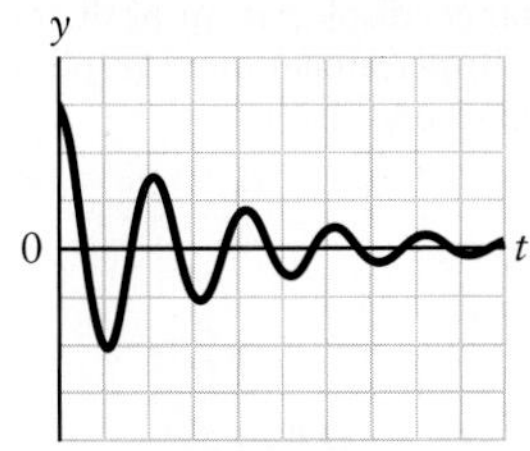

Figure P11.61

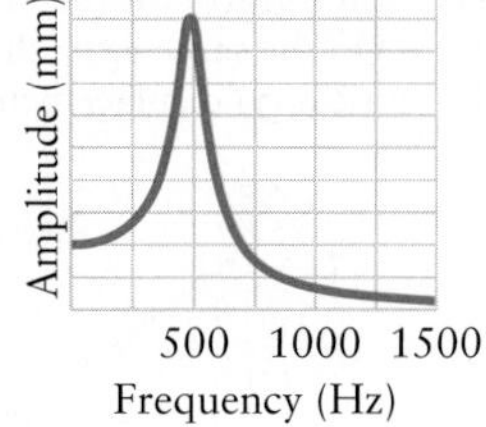

Figure P11.62

62. Figure P11.62 shows the amplitude as a function of frequency for a driven, damped oscillator. Estimate the resonant frequency.

11.6 DETECTING SMALL FORCES

63. Consider a Cavendish apparatus that employs a torsion fiber with $\kappa = 1.0 \times 10^{-8}$ N · m. Suppose the smallest twist angle that can be measured is $\theta = 2.0°$. If the distance from one of the masses m_1 to the torsion fiber is 10 cm (see Fig. 11.31), what is the smallest force F that can be measured? Assume a force F is applied to *both* of the masses m_2. Compare your result for F to the weight of a mosquito ($m \approx 10^{-3}$ g).

Additional Problems

64. SSM **Get new struts!** A car with worn-out (underdamped) struts is observed driving over a dip in the road. The car bounces up and down a total of three times in a period of 5 seconds after hitting the dip. If an average automobile strut spring has a spring constant of 6.0 kN/m, what is the approximate mass of the observed car? *Note*: The body of a car is supported on four struts.

65. A mass of 5.0 kg slides down a frictionless slope into a spring with spring constant $k = 4.9$ kN/m as depicted in Figure P11.65. (a) If the spring experiences a maximum compression of 20 cm, what is the height h of the initial release point? (b) What is the velocity of the mass when the spring has been compressed 15 cm? (c) If the mass sticks to the end of the spring, what are the period and amplitude of the oscillations that take place? (d) Determine the expressions for position as a function of time, $x(t)$, and velocity as a function of time, $v(t)$, for this oscillator. (e) What is the maximum velocity of the mass during any oscillation? (f) If the spring were to break while the mass is at the equilibrium position and moving to the right, to what maximum height would the mass rise up the slope? (g) Sketch or plot the position–time graph and velocity–time graph for the first second of the periodic motion.

Figure P11.65

66. A mass of 3.4 kg is released down a frictionless slope from a height of $h = 2.0$ m. When it reaches the bottom of the slope, it undergoes a completely inelastic collision with a mass of 1.1 kg attached to a spring of spring constant $k = 5.0$ kN/m as illustrated in Figure P11.66. The combined mass then undergoes periodic motion. Calculate (a) the maximum velocity, (b) the frequency, (c) the maximum amplitude, and (d) the period. (e) Determine expressions for position as a function of time and velocity as a function of time for the combined mass of this oscillating system.

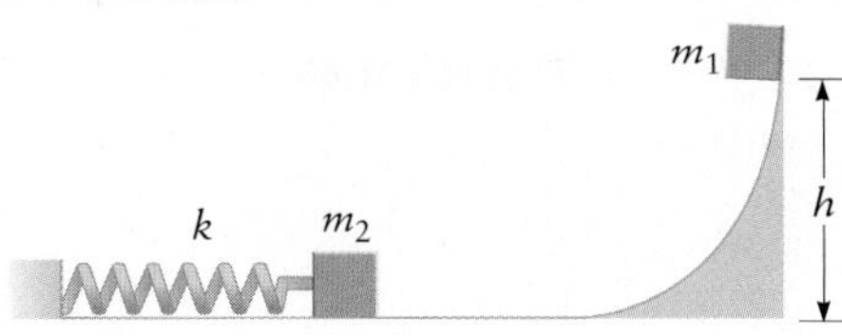

Figure P11.66

67. A mass of 3.5 kg is attached to a spring with spring constant $k = 70$ N/m. It is then compressed by 49 cm from its equilibrium position (Fig. P11.67, top) and released. At exactly the same time, an identical mass is released down a frictionless slope inclined at angle $\theta = 40°$ from the horizontal, and the two masses collide when this mass reaches the bottom of the slope (Fig. P11.67, bottom). At the moment the two masses collide, the spring is neither stretched nor compressed. The system is designed to produce periodic motion for both masses such that they always end up colliding at the same position. Assume there is no friction in the system and the collision is perfectly elastic. (a) What is the

period of this oscillating system? (b) From what height h must the second mass be released so that it will collide with the mass on the spring? *Hint*: Consider the period of motion of the first mass and the acceleration of the second mass in determining the needed release point up the slope. (c) Sketch the first 2.0 s of the displacement–time graph for the horizontal (x) displacement of the first mass. On the same axes, draw the displacement–time graph for the vertical (y) displacement of the second mass.

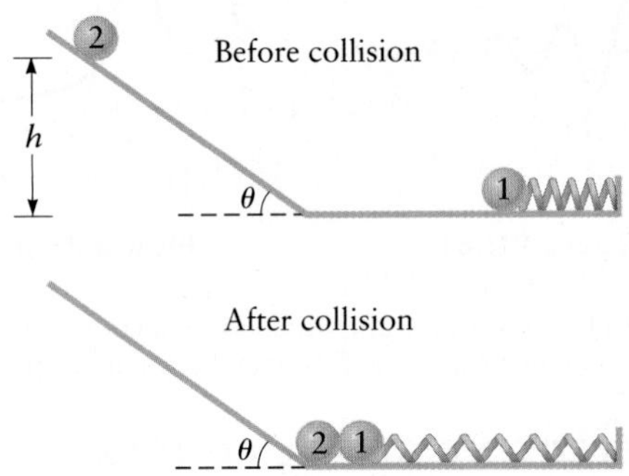

Figure P11.67

68. The bumper on a car is made of flexible plastic so that it will compress on impact and "cushion" the car's occupants during a collision. Suppose a car of mass $m = 1200$ kg is traveling slowly through a parking lot at 2.5 m/s when it strikes a brick wall. (a) What is the approximate value of the maximum compression of the bumper? (b) If the bumper can be modeled as a spring, what is approximate value of the spring constant of the bumper?

69. **Bobble-head.** A football bobble-head figure sits on your car dashboard. The 85-g head of the figure is attached to a spring with spring constant $k = 21$ N/m (Fig. P11.69). You are driving along a straight road when you notice a series of hoses lying across the road, perpendicular to your motion, that cannot be avoided, and you drive over them. If the hoses are spaced exactly 4.0 m apart, what speed should you avoid so your bobble-head figure does not sustain damage from its head vibrating too hard at maximum amplitude?

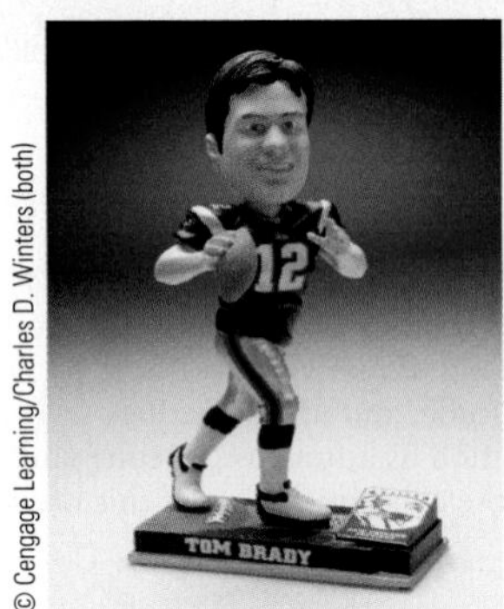

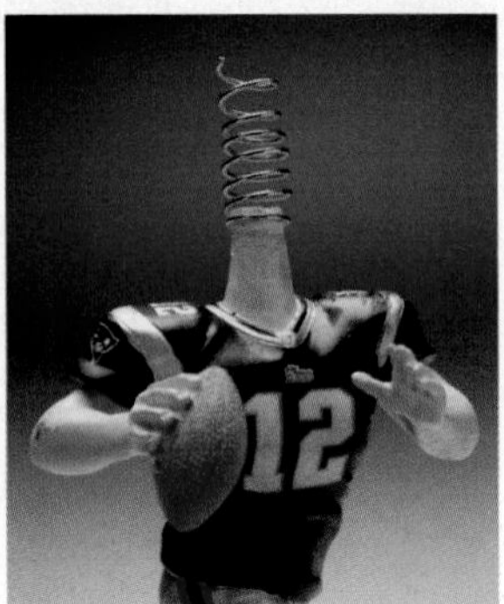

Figure P11.69

70. A solid rubber ball of radius 10 cm is submerged to 8.0 m beneath the surface of a lake. (a) If the bulk modulus of the rubber compound is $B_{\text{rubber}} = 4.0 \times 10^6$ Pa, what is the diameter of the ball at that depth? (b) A similar ball of equal radius, but made of thin rubber inflated with air, is also submerged to the same depth. If the bulk modulus of the air inside the ball under these conditions is $B_{\text{air}} = 1.4 \times 10^5$ Pa, what is the radius of this ball at that depth? (Ignore any effects of temperature.)

71. **Air spring.** The elasticity and bulk modulus of air can be measured using an apparatus consisting of a vertical glass tube and a ball bearing that just fits in the inner diameter of the tube. A 110-g ball bearing is dropped into the top of a 47-cm-high glass tube of diameter 3.0 cm. The ball is observed to oscillate as if the air column were a spring, 59 times in 10 s giving a period of 0.17 s. (a) Find the effective spring constant of the air column. (b) Show that this effective spring constant is related to the bulk modulus by $k = A^2B/V$, where A is the cross-sectional area of the tube and V is the volume of the tube. *Hint*: Start with Equation 11.25 and express the change in volume in terms of the displacement of the mass. (c) Determine the bulk modulus of air.

72. Young's modulus of granite is about 5.0×10^{10} Pa. Assuming the elastic limit of granite corresponds to a compression of 10%, estimate how tall a mountain would have to be for the granite at the bottom to just reach its elastic limit.

73. SSM Two playground swings are side by side. The children using the swings notice that one of the swings (swing 1) has a period of exactly 4.5 s. They also find that when the first swing has completed 10 oscillations, the other swing (swing 2) has completed 11 oscillations. (a) Is swing 2 longer or shorter than swing 1? (b) Find the length of swing 2.

CHAPTER 12

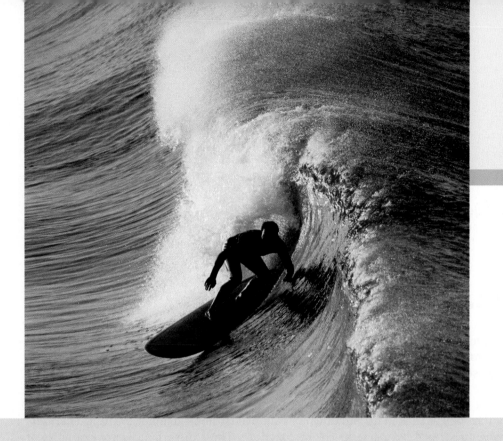

There are many types of waves, including waves on the surface of the ocean. (©Royalty-Free/Corbis/Jupiterimages)

Waves

In this chapter, we build on our work from Chapter 11 as we study ***wave motion***. A wave is *a moving disturbance that transports energy* from one place to another *without transporting matter*. This definition of a wave leaves two basic questions unanswered: (1) *what* is it that is "disturbed" by the wave, and (2) *how* is it disturbed? These questions will be answered as we proceed through this chapter. The motion associated with a wave disturbance often has a repeating form, so wave motion has much in common with simple harmonic motion.

In this chapter, we concentrate on aspects of wave motion that are common to all types of waves. The next chapter will then focus on a particularly common type of wave, *sound*. Chapters 23 through 26 are devoted to electromagnetic waves (including light), so many of the ideas that we discuss now will be revisited and developed further in those chapters.

OUTLINE

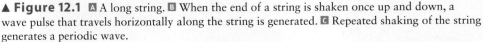

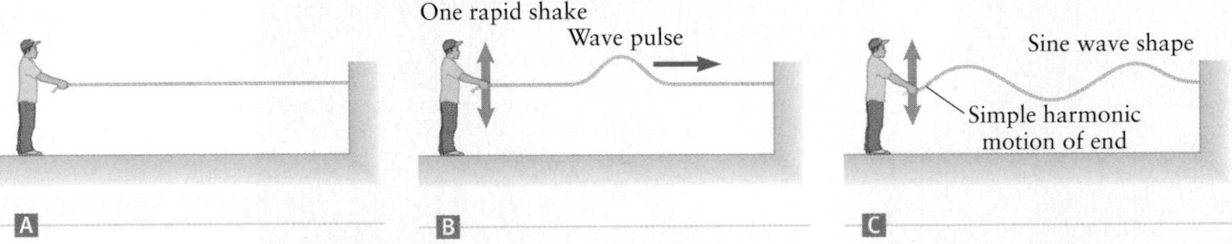

▲ **Figure 12.1** A A long string. B When the end of a string is shaken once up and down, a wave pulse that travels horizontally along the string is generated. C Repeated shaking of the string generates a periodic wave.

12.1 What Is a Wave?

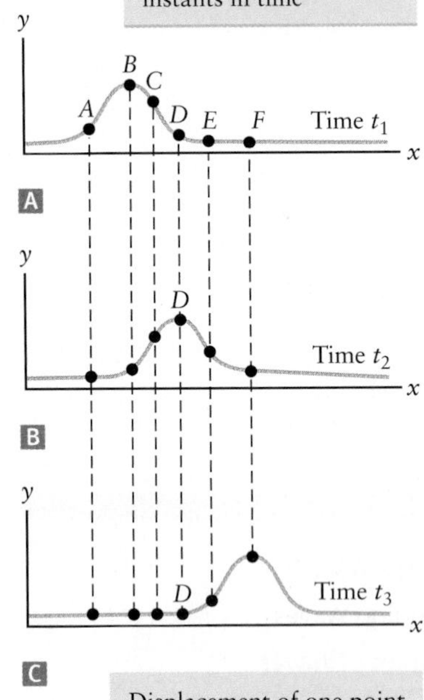

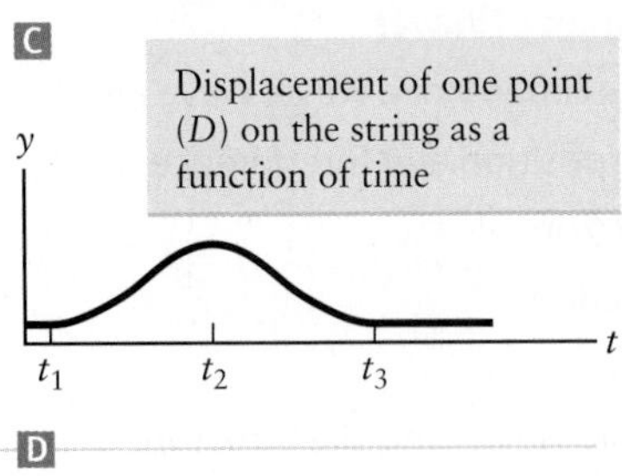

▲ **Figure 12.2** A, B, and C: Successive snapshots of a wave pulse traveling along a string with $t_1 < t_2 < t_3$. The wave pulse maintains its shape as it moves. This wave on a string is a transverse wave. Any particular point on the string (such as point D) moves in a direction perpendicular to the velocity of the wave. D Displacement of point D on the string as a function of time.

Figure 12.1A shows a person holding one end of a flexible string, with the other end tied to a wall. When the person shakes his end of the string up and down, he creates a "disturbance" that moves horizontally along the string. The shape of this disturbance depends on how the person shakes the string. In Figure 12.1B, he gives the end of the string a single shake, producing a single ***wave pulse***. In Figure 12.1C, the end of the string is shaken up and down in a simple harmonic manner, creating a ***periodic wave*** (described by a sine or cosine function). These disturbances are examples of ***waves***. Since portions of the string are moving, these portions have a nonzero velocity; hence, kinetic energy is associated with the wave. There is also elastic potential energy in the string as it stretches while the wave pulse travels by. The wave carries this energy as it travels.

Waves carry energy from one location to another. In Figure 12.1, a wave on the string gains energy from the person's hand as the force exerted by the hand does work on the string. The wave then carries this energy to the right as the wave moves along the string. Although a wave thus transmits energy from one place to another, it does *not* transport matter along with the energy. Pieces of the string do not move horizontally along with the wave disturbance. In particular, the left end of the string in Figure 12.1 does not move to the right along with the wave.

Figure 12.2 shows a wave pulse as it travels along a string. As with the wave pulse in Figure 12.1A, we suppose a person is holding the end of the string on the far left (out of the picture). We then imagine that the person gives one large up-and-down shake to the left end of the string, creating a single pulse that moves to the right as sketched in Figure 12.2. We say that this pulse ***propagates*** to the right. Parts A through C of Figure 12.2 show the string's shape at three successive times. Several points along the string are labeled to show how they move upward and then back down to the original, undisturbed configuration as the wave pulse travels by.

Let's now focus on the motion of point D on the string as the pulse moves by. The displacement of point D is plotted as a function of time in Figure 12.2D. This displacement is along the y direction and is thus *perpendicular* to the direction of propagation. At early times, when $t < t_1$, point D is at rest as the wave pulse has not yet arrived. When the pulse arrives, point D on the string first moves upward (Fig. 12.2B) and then moves back down to its original, undisturbed position (Fig. 12.2C). Other points on the string move in a similar manner. While points on the string are set into motion when the wave travels by, these points do not travel horizontally along with the wave. Hence, we again see that this wave *transports energy without transporting matter*. This is a key feature of *all* waves.

Before the wave pulse in Figure 12.2 was generated, the string was in a "relaxed" state, lying parallel to the x (horizontal) axis. When the pulse arrives, it displaces the string from this state, so we can say that it disturbs the string. All waves are a "disturbance" of some type, and the "thing" that is disturbed by the wave is called its ***medium***. In many cases, this medium is a material substance such as the string in Figure 12.1, and these waves are called ***mechanical waves***. It is also possible for the wave's medium to be a field, such as the electromagnetic field we'll learn about in later chapters. In either case, a wave must disturb something, and that something is the medium in which the wave travels. For a mechanical wave, the medium is a substance we can describe using Newton's laws of motion.

We used Figure 12.2 to compare and contrast the motion of the medium (the string in that case) with the motion of the wave. In Figure 12.2, pieces of the string move along the vertical direction (y) while the wave travels along a horizontal direction (x). Waves like this one, in which the motion of the medium is perpendicular to the direction of wave propagation, are called ***transverse waves***.

If the motion of the medium is parallel to the wave's propagation direction, the waves are called ***longitudinal***. For example, it is possible to generate longitudinal waves on a spring such as a Slinky as illustrated in Figure 12.3. Here the person holding the Slinky shakes the end back and forth along the x direction, generating a disturbance in which the loops of the Slinky alternate between being tighter (more compressed) and looser (more stretched) than normally found when the Slinky is at rest.

In nearly all cases, waves can be classified as either transverse or longitudinal, according to the motion of the medium.

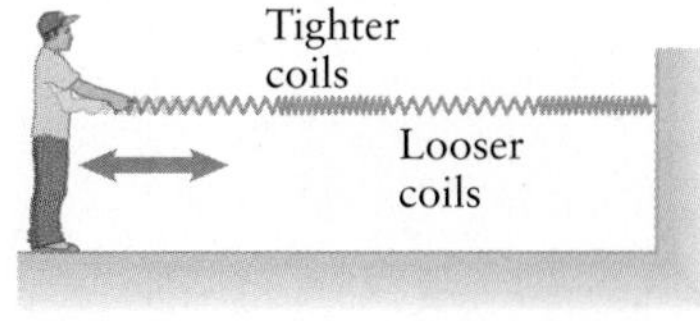

▲ **Figure 12.3** A Slinky (a loosely coiled spring) is stretched and held horizontally. Shaking one end of the Slinky back and forth in the horizontal direction generates a longitudinal wave. At some places, the Slinky coils are compressed an extra amount relative to those in an undisturbed Slinky, but at other places the coils are looser. These tight and loose places move along the Slinky, forming the wave.

EXAMPLE 12.1 Determining the Velocity of a Wave

Figure 12.4 shows three snapshots of a wave pulse traveling along a string. This pulse was generated by shaking the left end of the string, generating a disturbance that propagated to the right. (**a**) Use Figure 12.4 to estimate the speed of the wave. (**b**) Estimate the speed of point A on the string at $t = 1.5$ s as the wave pulse passed by that point.

RECOGNIZE THE PRINCIPLE

(**a**) The wave in Figure 12.4 is a single pulse. We can determine the velocity of this pulse by noting the position of its peak along the x direction as a function of time. (**b**) To estimate the velocity of a point *on the string*, we examine the motion of that point along the vertical (y) direction.

SKETCH THE PROBLEM

Figure 12.4 shows the wave pulse at three different times as well as the position y of point A.

IDENTIFY THE RELATIONSHIPS AND SOLVE

(**a**) From Figure 12.4, the peak of the wave pulse is at $x \approx 1.8$ m when $t = 1.0$ s and has moved to $x \approx 3.8$ m when $t = 3.0$ s. The average velocity during this time interval is thus along the $+x$ direction with

$$v = \frac{\Delta x}{\Delta t} = \frac{(3.8 - 1.8)\text{ m}}{(3.0 - 1.0)\text{ s}} = \boxed{1.0\text{ m/s}}$$

This is the speed *of the wave*.

(**b**) Comparing the snapshots in Figure 12.4 at $t = 1.0$ s and 2.0 s, we see that point A on the string is near its undisturbed position with $y \approx 0.03$ m at $t = 1.0$ s and that this point is at the peak of the wave when $t = 2.0$ s, where we find $y \approx 0.20$ m. The velocity of point A at $t = 1.5$ s is thus approximately

$$v_A = \frac{\Delta y}{\Delta t} = \frac{(0.20 - 0.03)\text{ m}}{(2.0 - 1.0)\text{ s}} = \boxed{0.17\text{ m/s}}$$

(continued) ▶

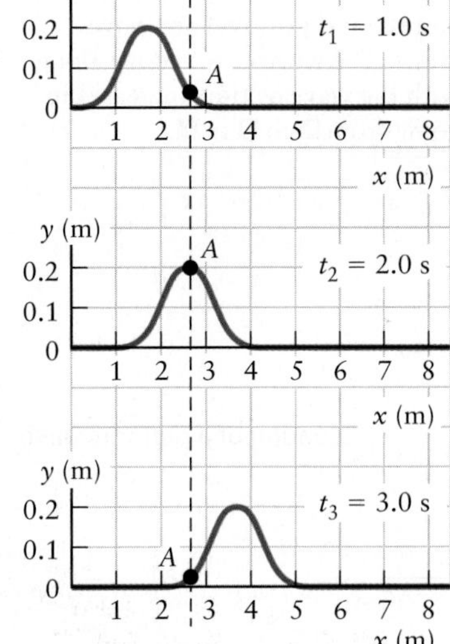

▲ **Figure 12.4** Example 12.1.

> **► What does it mean?**
> This is a transverse wave because the velocity of point A on the string is along y, which is perpendicular to the velocity of the wave. In general, the velocity of a point on the string will *not* have the same magnitude as the velocity of the wave. The velocity of a point on the string also will not be constant; here we have made an estimate of v_A during the time the wave is passing the point of interest. Before and after the wave pulse arrives, the velocity of a point on the string will be zero.

12.2 Describing Waves

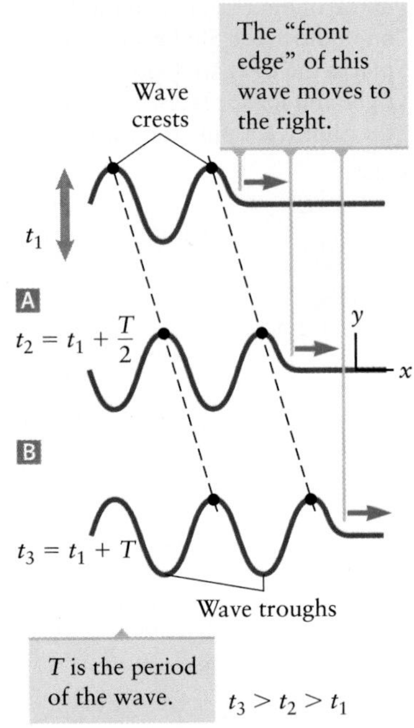

▲ **Figure 12.5** Snapshots of a periodic wave on a string. The places where the displacement is largest (and positive) are wave crests. The wave crests move along with the wave as time increases in going from A to B to C.

Wave pulses like those in Figures 12.2 and 12.4 are often found in nature, but it is also common to find periodic waves. A periodic wave can be generated by shaking the end of a string up and down repeatedly, in a periodic fashion. In the case shown in Figure 12.5, we imagine that a person (not shown in the figure) is shaking the far left end of the string in such a way that this end is undergoing simple harmonic motion along the y (vertical) direction. Figure 12.5 then shows snapshots of the resulting periodic wave as it travels along. These successive snapshots show the "front edge" of the wave, making the overall motion of the wave clear. However, if we wait until this front edge moves out of the picture to the right, all we will see is a repeating wave disturbance with a "sine-wave" form. If we examine just a single snapshot, we find places where the string displacement y is largest and positive, called wave ***crests***. Likewise, at some places called wave ***troughs***, y is equally large in magnitude but negative in sign.

Two wave crests in Figure 12.5 are marked with large dots to emphasize the motion of the wave. The wave's speed can be measured by following the position of a particular wave crest as a function of time, similar to what we did in Example 12.1.

The wave pulses in Figure 12.2 and 12.4 are examples of nonperiodic waves. In these cases, the wave disturbance is limited to a small region of space. In contrast, a periodic wave as in Figure 12.5 extends over a very wide region of space, and the displacement of the medium varies in a repeating and often sinusoidal manner as a function of both position and time. For example, if we examine a single snapshot of a periodic wave—say, in Figure 12.5C—we see that the string displacement y varies in a periodic manner as a function of *position* along the string x. We can also monitor the string displacement at a particular point on the string and find that the displacement also varies in a periodic manner as a function of *time*. Hence, *a periodic wave involves repeating motion as a function of both space and time.*

The "Equation" of a Wave

The wave in Figure 12.5 was generated by shaking the left end of the string up and down. For simplicity, let's assume the displacement at the left end vibrates as a simple harmonic oscillator with $y_{\text{end}} = A \sin(2\pi f t)$, where f is the frequency of this harmonic motion. This periodic shaking generates a periodic wave that travels along the string as sketched in Figure 12.5. The string's displacement is given by the relation

Equation of a periodic wave

$$y = A \sin\left(2\pi f t - \frac{2\pi x}{\lambda}\right) \tag{12.1}$$

where λ (the Greek letter lambda) is called the ***wavelength*** as explained below.

Equation 12.1 is a mathematical description of a periodic wave. For a wave on a string, it tells how the transverse displacement y of a point on the string varies as a function of both time t and location x along the string. It applies to many different types of waves and describes how a wave travels (propagates) away from its source. For the wave on a string in Figure 12.5, the source is the person's hand at the left end of the string. Equation 12.1 does not include what happens when the wave reaches the right end of the string. We'll learn about that in Sections 12.6 and 12.7, but first let's explore the meaning of Equation 12.1 in a little more detail.

Equation 12.1 has much in common with the displacement of a simple harmonic oscillator (Eq. 11.2). Like a simple harmonic oscillator, a periodic wave has a frequency f and an amplitude A. The frequency is related to the "repeat time" for the wave and can be understood using the snapshots in Figure 12.5. A point on the string that is at a crest in Figure 12.5A moves down to a trough in snapshot B and then returns to a crest in snapshot C, constituting one period T of the motion. The frequency of the wave is then given by $f = 1/T$. Since the sine function in Equation 12.1 repeats itself as the value of t increases, this behavior is all contained in Equation 12.1.

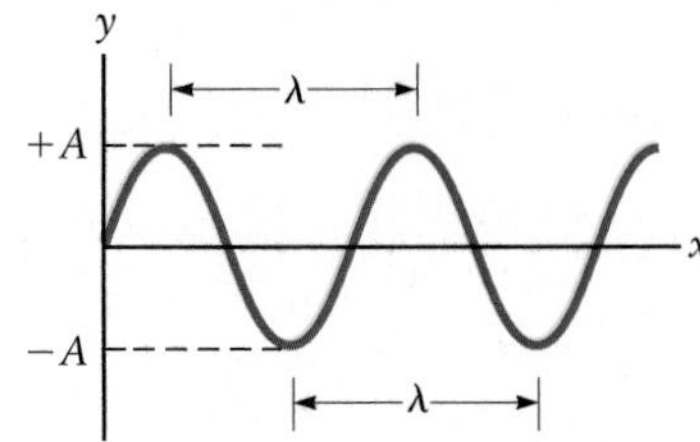

▲ **Figure 12.6** Snapshot of a periodic wave on a string. This is just a plot of Equation 12.1 showing the displacement associated with a wave as a function of x at a particular value of t. The wavelength λ is the "repeat distance." The wavelength can be measured from any point on the wave to the next "equivalent" point. The displacement oscillates between $+A$ and $-A$, where A is the amplitude.

The meaning of the wavelength λ can be understood from Figure 12.6. Here again we show a snapshot of a wave on a string. This plot corresponds to Equation 12.1, showing the transverse displacement y as a function of position x along the string at a particular value of t. Figure 12.6 shows that y is a repeating function of position. That is, if we advance x by a distance equal to λ, the value of y is repeated. One can thus think of λ as the *repeat distance* of the wave. Since the sine function in Equation 12.1 repeats itself as x varies to either increasing or decreasing values, this behavior is contained in Equation 12.1. Periodic waves thus have both a repeat *time* and a repeat *distance*. A periodic wave is a combination of two simple harmonic motions: one as a function of time and the other as a function of space.

Figure 12.6 also illustrates the meaning of the amplitude A of a wave. The wave displacement y varies between $\pm A$, so the wave crests have a height $y = +A$. At the troughs, $y = -A$.

Speed of a Wave

The mathematical description of a wave in Equation 12.1 contains three important wave properties: frequency, wavelength, and amplitude. Let's now consider how these are connected with the velocity of the wave. The snapshots in Figure 12.5 show that the wave moves a distance $\Delta x = \lambda$ during one period of the motion ($\Delta t = T$), which also follows from the definitions of wavelength and period. The speed of the wave is therefore given by

$$v = \frac{\Delta x}{\Delta t} = \frac{\lambda}{T}$$

Because frequency is related to the period by $f = 1/T$, we have

$$v = \frac{\lambda}{1/f}$$

or

$$v = f\lambda \quad (12.2)$$

The speed of a periodic wave is related to its frequency and wavelength.

Information regarding the *direction* of the wave is also contained in Equation 12.1. To see how, we focus on the motion of a wave crest. A crest occurs whenever the sine function in Equation 12.1 has its maximum value, and, because $\sin(\pi/2) = 1$, one of these maxima will occur when the argument of the sine function is equal to $\pi/2$. So,

$$2\pi ft - \frac{2\pi x}{\lambda} = \frac{\pi}{2} \quad (12.3)$$

Hence, at $t = 0$, a crest will be found at

$$2\pi f(0) - \frac{2\pi x_{\text{crest}}}{\lambda} = \frac{\pi}{2}$$

$$x_{\text{crest}} = -\frac{\lambda}{4} \quad (12.4)$$

After one period, $t = T$, and using Equation 12.3 we get

$$2\pi fT - \frac{2\pi x_{\text{crest}}}{\lambda} = \frac{\pi}{2}$$

$$x_{\text{crest}} = -\frac{\lambda}{4} + f\lambda T = -\frac{\lambda}{4} + vT$$

where we have also used the relation $v = f\lambda$. Comparing this expression with Equation 12.4, we see that x_{crest} has become larger, indicating that the wave has moved to the right. Hence, the velocity of the wave described by Equation 12.1 is positive. (The wave has moved in the $+x$ direction.)

A similar analysis shows that a wave that moves to the left is described by

$$y = A\sin\left(2\pi ft + \frac{2\pi x}{\lambda}\right) \tag{12.5}$$

Equations 12.1 and 12.5 describe periodic waves that move in opposite directions. Other similar functions can also be used to describe such traveling waves. For example,

$$y = A\cos\left(2\pi ft - \frac{2\pi x}{\lambda}\right) \tag{12.6}$$

describes a wave that moves in the $+x$ direction, whereas

$$y = A\cos\left(2\pi ft + \frac{2\pi x}{\lambda}\right) \tag{12.7}$$

is a wave that travels in the $-x$ direction. All four cases (Eqs. 12.1, 12.5, 12.6, and 12.7) describe waves with the same frequency, wavelength, speed, and amplitude. The only differences involve the direction of wave propagation (the sign of v) and the shape of the string at $t = 0$ (either a sine function or a cosine).

Interpreting the equation of a periodic wave

Many properties of a periodic wave can be read directly from the equation of the wave:

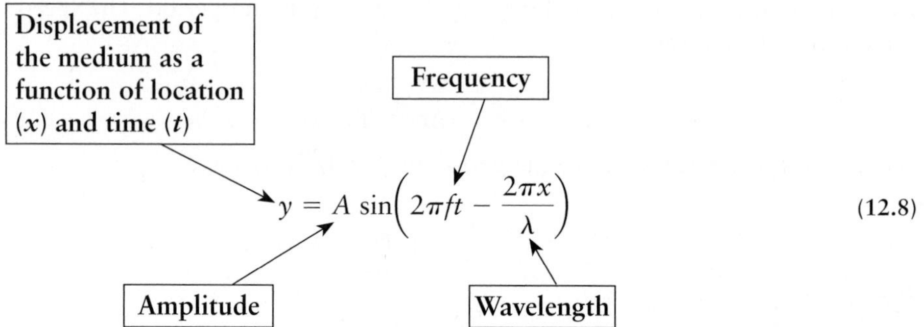

$$y = A\sin\left(2\pi ft - \frac{2\pi x}{\lambda}\right) \tag{12.8}$$

CONCEPT CHECK 12.1 Waves on a String

A certain wave on a string is found to have a frequency of 500 Hz and a wavelength of 0.60 m. If a new wave of frequency 250 Hz is then established on this string and the wave speed does not change, is the new wavelength (a) 0.60 m, (b) 0.30 m, or (c) 1.2 m? *Hint*: You do not need a calculator!

EXAMPLE 12.2 Finding the Properties of a Wave

Figure 12.7 shows snapshots of a wave on a string at two very closely spaced times, separated by $\Delta t = 0.050$ s. Assuming the wave is moving to the right, find the wavelength, amplitude, frequency, and velocity of the wave.

RECOGNIZE THE PRINCIPLE

The wavelength λ is the distance between wave crests, which we can read from Figure 12.7. We can also read the amplitude from the figure. By comparing the two snapshots, we can deduce the wave speed v and then find the frequency from the relation $f\lambda = v$.

SKETCH THE PROBLEM

Figure 12.7 describes the problem. We have also labeled two wave crests A and B, which can be used to follow how the wave moves.

IDENTIFY THE RELATIONSHIPS AND SOLVE

From Figure 12.7, the distance between the wave crests at A and B is approximately

$$\lambda = \boxed{0.65\text{ m}}$$

We can also estimate the amplitude from the height of a crest (or trough) and find

$$A = \boxed{0.40\text{ m}}$$

To find the velocity, we must compare snapshots at different times. These snapshots were taken at very closely spaced times, so we can assume the wave has only had time to move a small amount. Wave crest A in the first snapshot (at t_1) then corresponds to crest A' in the second snapshot. From Figure 12.7, this crest has moved a distance of approximately 0.25 m in a time interval $t_2 - t_1 = \Delta t = 0.050$ s, so the wave velocity is

$$v = \frac{\Delta x}{\Delta t} = \frac{0.25\text{ m}}{0.050\text{ s}} = \boxed{5.0\text{ m/s}}$$

The velocity is positive because the wave is moving along the $+x$ direction. We can find the frequency using Equation 12.2. We get

$$f = \frac{v}{\lambda} = \frac{5.0\text{ m/s}}{0.65\text{ m}} = \boxed{7.7\text{ Hz}}$$

▶ ***What does it mean?***

All the properties of a wave can be read off from snapshots like those in Figure 12.7 if you know what to look for!

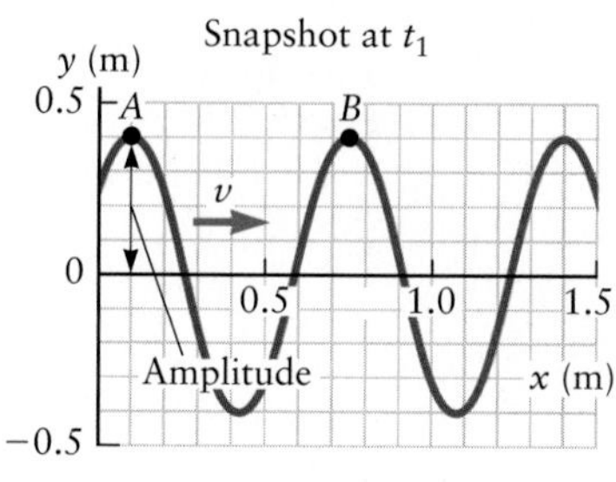

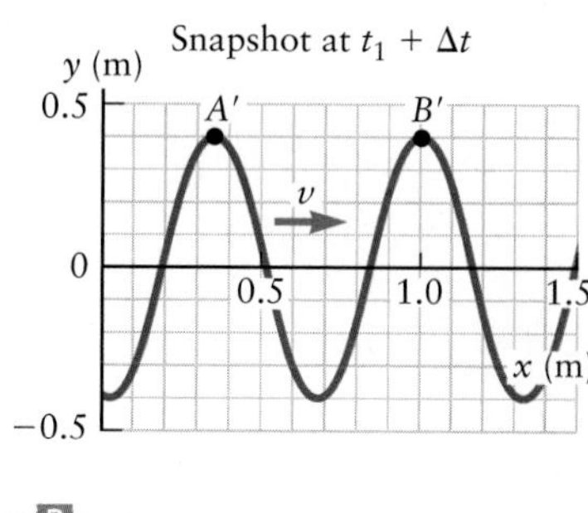

▲ **Figure 12.7** Example 12.2. Snapshots of a wave on a string at time t_1 and at a short time later, $t_1 + \Delta t$.

EXAMPLE 12.3 Analyzing the Equation of a Wave

Suppose a wave on a string is described by the equation

$$y = (0.040\text{ m})\cos(25t - 30x) \tag{1}$$

where the units of t are seconds and x is given in meters. Find the frequency, wavelength, and velocity of this wave.

RECOGNIZE THE PRINCIPLE

We can read all the properties of the wave in Equation (1) by following the outline in Equation 12.8.

SKETCH THE PROBLEM

No figure is needed.

IDENTIFY THE RELATIONSHIPS AND SOLVE

We compare Equation (1) with Equation 12.8 and read off values for the corresponding quantities. The factor that multiplies t in Equation (1) equals $2\pi f$, so we have

$$2\pi f = 25\text{ Hz}$$

(continued) ▶

The units here are hertz because that is the unit for frequency f in the SI system. Solving for f gives

$$f = \frac{25\ \text{Hz}}{2\pi} = \boxed{4.0\ \text{Hz}}$$

Likewise, the factor multiplying x in both Equation 12.8 and Equation (1) equals $2\pi/\lambda$. Setting $2\pi/\lambda = 30\ \text{m}^{-1}$ and solving for the wavelength, we find

$$\lambda = \frac{2\pi}{30\ \text{m}^{-1}} = \boxed{0.21\ \text{m}}$$

According to Equation 12.2 the speed of the wave is

$$v = f\lambda = (4.0\ \text{Hz})(0.21\ \text{m}) = \boxed{0.84\ \text{m/s}}$$

► ***What does it mean?***

Comparing the given equation again with Equation 12.1, we see that the velocity of this wave is positive; it moves in the $+x$ direction.

CONCEPT CHECK 12.2 Properties of a Wave

Figure 12.8 shows snapshots of several different waves on a string. The four strings are identical (the same tensions, etc.). Which wave has (a) the longest wavelength, (b) the highest frequency, and (c) the smallest amplitude?

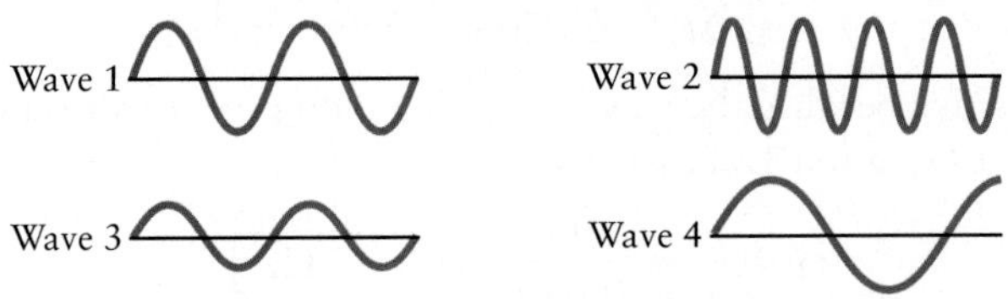

▲ **Figure 12.8** Concept Check 12.2.

12.3 Examples of Waves

We have discussed some general properties of waves and how they behave in time and space. Let's next consider some specific types of waves.

Waves on a String

Waves on a string are mechanical waves, and the medium that is disturbed is the string. For a transverse wave on a string, the velocity depends on the tension in the string and the string's mass per unit length, denoted by μ. This is not surprising; we expect that a wave will travel faster in a lighter string (one with smaller μ) since a lighter string will have a greater acceleration (and velocity) for a given value of the tension. Likewise, the wave speed is increased when the tension is increased (when the string is more "taut").

Let's denote the tension in the string by F_T, deviating from our usual symbol for tension so as not to confuse it with the period of the wave. The speed of a wave on a string can be calculated using Newton's laws of motion and is

$$v = \sqrt{\frac{F_T}{\mu}} \tag{12.9}$$

The speed of this wave does *not* depend on its frequency. The frequency is determined by how rapidly the end of the string is shaken, but the wave speed in Equation 12.9 is completely independent of f. In fact, the speed of transverse waves on a string is the *same* for periodic and nonperiodic waves (wave pulses).

EXAMPLE 12.4 Speed of a Wave on a Piano String

A typical string in the author's grand piano is composed of steel ($\rho = 7800$ kg/m^3) with radius $r = 0.50$ mm and has a tension of about 650 N. What is the speed of a wave on such a string?

RECOGNIZE THE PRINCIPLE

We can find the wave speed from $v = \sqrt{F_T/\mu}$ (Eq. 12.9). The tension F_T is given, and we can find the mass per unit length μ from the known density of steel.

SKETCH THE PROBLEM

No figure is needed.

IDENTIFY THE RELATIONSHIPS

A string is just a long cylinder; if it has length L and radius r, its volume is $\pi r^2 L$. Density ρ is mass per unit volume, so the string's mass is $m = \pi r^2 L\rho$. The density of steel is given as 7800 kg/m^3. The mass per unit length equals m/L:

$$\mu = \frac{m}{L} = \pi r^2 \rho = \pi(5.0 \times 10^{-4} \text{ m})^2(7800 \text{ kg/m}^3) = 6.1 \times 10^{-3} \text{ kg/m}$$

SOLVE

Inserting this value of μ into Equation 12.9 along with the given value of the tension, we find

$$v = \sqrt{\frac{F_T}{\mu}} = \sqrt{\frac{650 \text{ N}}{6.1 \times 10^{-3} \text{ kg/m}}} = \boxed{330 \text{ m/s}}$$

▸ ***What does it mean?***

This speed is more than 700 miles per hour! The speed is high because of the high tension. (You should compare the tension of 650 N to your own weight.)

CONCEPT CHECK 12.3 Waves on a Violin String

The tension in a violin string is 60 N, and the speed of a wave on this string is 300 m/s. If the tension is reduced to 30 N, will the new speed of a wave on this string be (a) 150 m/s, (b) 210 m/s, or (c) 600 m/s?

Sound

Sound is a mechanical wave that can travel through virtually any material substance, including gases, liquids, and solids. By far, the most familiar case is sound in a gas such as air. Sound waves can be excited in a gas by a loudspeaker as sketched in Figure 12.9. Here the surface of the speaker moves back and forth along the horizontal direction (x). As it moves, the speaker collides with nearby air molecules, and because the speaker is moving parallel to x, these collisions affect the x component of the velocity of the air molecules. The displacement of the air molecules associated with the sound wave is thus also along x. This displacement alternates between the positive and negative x directions due to the back-and-forth motion of the speaker. The displacement of the medium (the air molecules) is parallel to the wave's propagation direction, so sound

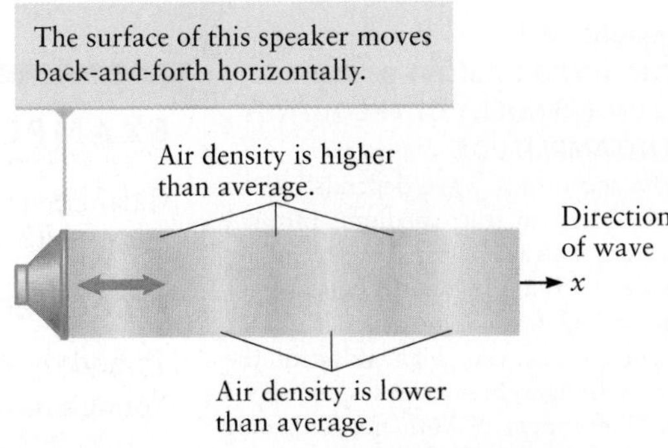
The surface of this speaker moves back-and-forth horizontally.
Air density is higher than average.
Direction of wave
x
Air density is lower than average.

▲ **Figure 12.9** Sound is a longitudinal wave. Within this wave are regions where the density of air molecules is high, separated by regions where it is lower. Compare with the compressed and loose regions of the Slinky in Figure 12.3.

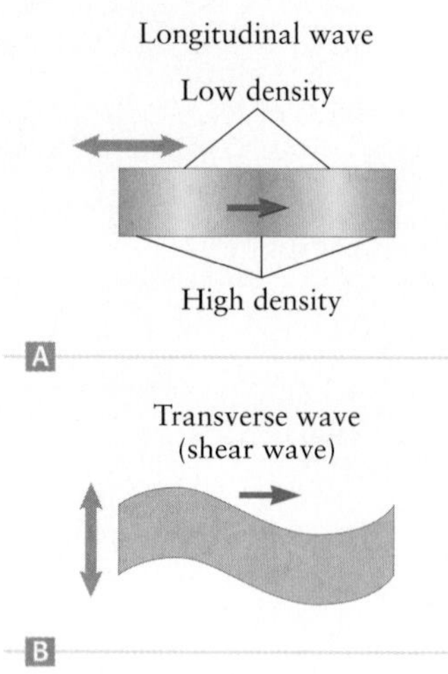

▲ **Figure 12.10** A solid, such as a metal bar, can carry A longitudinal waves or B transverse waves. In this sketch of a transverse wave, the amplitude is greatly exaggerated.

is a longitudinal wave. The speed of sound depends on the properties of the medium, such as the gas density. For air at room temperature, the speed of sound is approximately 343 m/s. This speed is independent of the frequency of the sound wave and applies to both periodic waves (steady tones) and nonperiodic waves (short "clicks" or "beeps").

Sound in a liquid or a solid also travels as longitudinal waves, with the displacements of the molecules in the liquid or solid being parallel to the propagation direction. The speed of sound is generally smallest for gases and highest for solids.

Wave Propagation in a Solid

We have just mentioned that sound (longitudinal waves) can propagate in a solid. It is also possible to have transverse waves in a solid. The motions associated with both longitudinal and transverse waves are sketched in Figure 12.10. Longitudinal waves involve motion of atoms parallel to the direction of the wave (compare Fig. 12.10A with Fig. 12.9), and such waves are therefore also called "sound." In fact, when a sound wave from air impinges on a solid, a corresponding sound wave propagates into the solid. The speed of sound waves in a solid depends on the solid's elastic properties, and sound travels faster in "stiffer" materials. For a thin bar of material, the speed of sound is given by

$$v = \sqrt{\frac{Y}{\rho}} \tag{12.10}$$

where Y is Young's modulus (Chapter 11) and ρ is the density of the material.

Transverse waves in a solid involve motion of atoms in directions perpendicular to the wave velocity (Fig. 12.10B); the velocity of these transverse waves is not given by Equation 12.10, but has a different (and slightly more complicated) dependence on the shear modulus and other elastic constants (Chapter 11). In general, the speed of these transverse waves is lower than the speed of longitudinal waves. The key points here are that *both* transverse *and* longitudinal waves can propagate in a solid and that the longitudinal waves are called *sound* because they are just like the sound waves that propagate in a gas or a liquid.

While transverse waves can travel through a solid, they *cannot* travel through liquids or gases. The displacements associated with a transverse wave involve a shearing motion (a shear stress). When a solid is subject to a shear stress, it is bent as in Figure 12.10B and then "springs back." Liquids and gases *flow* and thus do not spring back, so there is no restoring force to produce the oscillations necessary for a transverse wave. Hence, only longitudinal waves (called sound) can propagate in a liquid or a gas. This fact has some important consequences that we'll explore in Section 12.9.

Insight 12.1

THE SPEED OF MOST WAVES IS INDEPENDENT OF FREQUENCY AND AMPLITUDE

The speed of a wave depends on the properties of the medium through which it travels. This speed varies widely, from the relatively low values found for a wave on a Slinky or string to the very high value of the speed of light. In most cases, the speed is independent of both the frequency and amplitude. This speed is also the same for periodic and nonperiodic waves (wave pulses). Although the wave speed is usually independent of frequency, when we study light and optics we will see some important cases in which the speed does depend on f.

EXAMPLE 12.5 Speed of Sound in a Solid

Calculate the speed of longitudinal waves (sound) in a steel bar. Use data from Table 10.1 (p. 318) and Table 11.1 (p. 368).

RECOGNIZE THE PRINCIPLE

To find the speed of sound in steel, we apply Equation 12.10, $v = \sqrt{Y/\rho}$, where Y is Young's modulus.

SKETCH THE PROBLEM

No figure is necessary.

IDENTIFY THE RELATIONSHIPS

From Tables 10.1 and 11.1, the density of steel is $\rho = 7800 \text{ kg/m}^3$, and Young's modulus is $Y = 2.0 \times 10^{11}$ Pa.

SOLVE

Using these values in Equation 12.10 gives

$$v = \sqrt{\frac{Y}{\rho}} = \sqrt{\frac{2.0 \times 10^{11}\ \text{Pa}}{7800\ \text{kg/m}^3}} = \boxed{5100\ \text{m/s}}$$

▶ *What does it mean?*

This speed is *much* faster than that of sound in air, which is about 340 m/s.

Visible Light and Other Electromagnetic Waves

Visible light is a type of ***electromagnetic wave***. Other types include radio waves, X-rays, gamma rays, and microwaves. These are not mechanical waves, but are electrical and magnetic disturbances that can propagate even in a vacuum, requiring no mechanical medium. Because the electric and magnetic disturbances (the electric and magnetic fields) are always perpendicular to the propagation direction, electromagnetic waves are transverse.

Electromagnetic waves are classified according to their frequency, so visible light waves fall into a different range of frequencies than radio waves, X-rays, and so forth. The speed of an electromagnetic wave in a vacuum c is independent of the frequency of the wave and has the value $c = 3.00 \times 10^8$ m/s. We'll discuss the properties of electromagnetic waves in detail in Chapter 23.

Water Waves

A wave can be generated on the surface of water by simply dropping a rock into a still body of water. After the rock strikes the water (Fig. 12.11A), waves propagate outward as shown in Figure 12.11B. The up-and-down motion of the water's surface is probably familiar to you. If you were to carefully watch a bug floating on the surface, though, you would find that the motion of the water's surface is actually *both* transverse *and* longitudinal as illustrated in Figure 12.12. A water surface wave is one the few types of waves that cannot be classified as purely transverse or longitudinal. However, most waves fall into one of these two categories, making it a useful classification scheme.

CONCEPT CHECK 12.4 **What Determines the Properties of a Wave?**

What determines the (1) frequency, (2) amplitude, and (3) wavelength of a wave, (a) the properties of the wave's medium or (b) the way the wave is generated (i.e., the wave's source)?

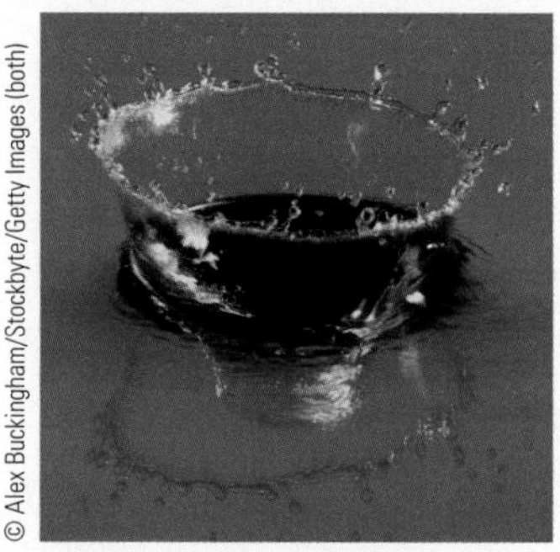

▲ **Figure 12.11** A Dropping a rock into a pool of water generates a water wave. B Waves generated by a disturbance on the surface of a pool of water. The wave crests and troughs move outward as expanding circles.

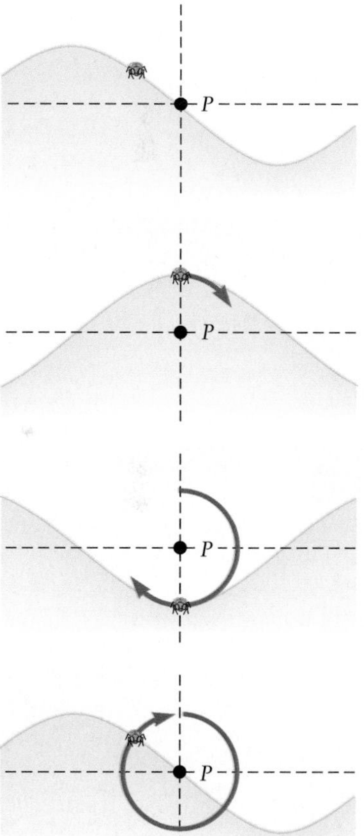

▲ **Figure 12.12** The motion associated with a wave on the surface of water has both transverse and longitudinal components. A bug on the surface of the water moves up and down—and also forward and backward—as the wave passes by, which can be seen by comparing the bug's position with that of point P.

12.4 The Geometry of a Wave: Wave Fronts

Spherical Waves

Consider a sound wave generated at the center of a large room by a small explosion. The resulting sound wave travels away from the source in a three-dimensional fashion as sketched in Figure 12.13A. The wave crests form concentric spheres centered at the source, so this is called a ***spherical wave***. These crests are also called ***wave fronts***. The direction of wave propagation is always perpendicular to the surface of a wave front. This direction is indicated by arrows called ***rays*** in Figure 12.13.

Each wave crest carries a certain amount of energy as it travels away from the source. Several quantities provide a measure of the amount of this energy, one of which is the ***power***. The total power carried by a wave equals the energy emitted by the source (and carried away by the wave) per unit time. The units of power are thus the units of energy divided by time, or joules/second = watts, just as we found for mechanical power in Chapter 6. A closely related quantity is the ***intensity*** of the wave, which is defined as the power carried by the wave over a unit area of wave front. The units of intensity are the units of power/area = watts/m^2.

Let's now follow one spherical wave front as it travels away from the source in Figure 12.13A. The amount of energy carried by a particular wave front is determined when it is created at the source, so once the wave front is emitted, the total energy it contains is fixed. As the wave front travels away from the origin and its spherical surface expands, this fixed amount of energy (and also power) is spread over a larger and larger surface. The power divided by the area thus becomes smaller, and the intensity of the wave falls as one moves away from the source. At a distance r from the source, the area of a spherical wave front is equal to $4\pi r^2$ (the surface area of a sphere of radius r), so we have

$$I = \frac{\text{power}}{\text{area}} = \frac{P}{4\pi r^2}$$

The intensity of a spherical wave thus falls with distance as

$$I \propto \frac{1}{r^2} \quad \text{(spherical wave)} \tag{12.11}$$

This dependence of the wave intensity on r is found whenever the wave front expands spherically in three-dimensional space, and it arises in many cases involving sound (as in this example) and light (such as the light waves that emanate from the Sun).

Plane Waves

Wave fronts are not always spherical. Another important case is a ***plane wave***, in which the wave fronts are (flat) planes as sketched in Figure 12.13B. For a "perfect" plane wave, each wave crest extends over an infinite plane in space, and these planes

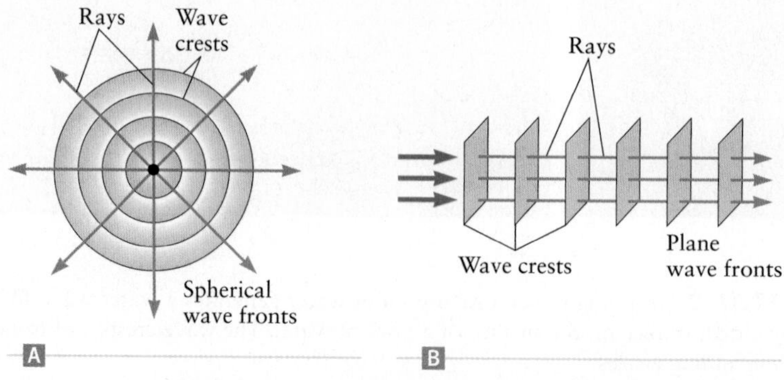

► **Figure 12.13** The geometry of a wave can be described by the shape of its wave fronts. A A spherical wave has spherical wave fronts that are concentric with the source of the wave. B The wave fronts of an ideal plane wave are parallel, flat, and infinite in extent. This idealization is not realized in practice, but is approximated by the light from a laser.

are perpendicular to the direction of propagation. Of course, such a perfect plane wave is an idealization; a real wave will not be spread over an infinite region of space. The light wave produced by a laser, however, is a very good approximation to an ideal plane wave. (We'll see other ways to generate approximate plane waves when we study geometrical optics in Chapter 24.) The wave fronts produced by a laser are very nearly flat and extend over the width of the laser beam. These wave fronts spread very little as they travel, so the wave intensity is approximately constant over long propagation distances. In contrast to the case of a spherical wave, where the intensity falls off as the square of the distance from the source (Eq. 12.11), the intensity of an *ideal* plane wave is constant, independent of distance from the source.

Intensity and Amplitude of a Wave

For all types of waves (and all types of wave fronts), the intensity of a wave is related to its amplitude A by

$$I \propto A^2$$

The reason I is proportional to A^2 can be seen by considering waves on a Slinky. The Slinky is just a spring, and the potential energy associated with a spring equals $\frac{1}{2}kx^2$, where x is the amount the spring is stretched or compressed. For a wave on a Slinky, the magnitude of this stretching and compressing is equal to the displacement of the Slinky from its undisturbed state and is thus proportional to the amplitude A of the wave. Hence, the energy carried by a wave and its intensity are both proportional to A^2.

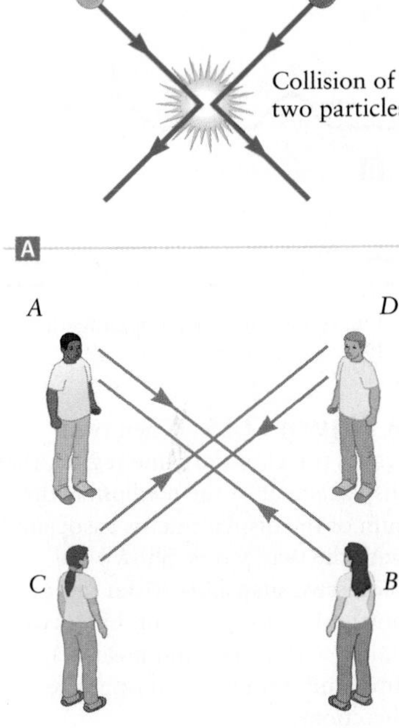

▲ **Figure 12.14** A When two particles collide, their momenta change. B If we replace the particles by waves (such as sound or light), the two waves will pass through each other. The sound waves involved in two conversations (between persons A and B and between persons C and D) do not affect each other; they propagate independently of each other.

12.5 Superposition and Interference

All our discussions so far have involved the case of a single wave propagating by itself. We now want to consider what happens when two or more waves travel at the same time through the same medium. That is, what happens when two waves collide?

When two particles collide, the momentum of each particle changes (although the total momentum of the two particles is conserved; see Chapter 7). Hence, a collision has a big effect on the motion of the particles as illustrated in Figure 12.14A, but the behavior of two colliding waves is quite different. Waves almost always propagate *independently* of one another. A wave can travel through a particular region of space without affecting the motion of another wave traveling through the same region. This behavior is illustrated in Figure 12.14B, which shows sound waves crossing in space (as in a typical classroom). Persons A and B are having a conversation, and their sound waves travel independently of the sound waves of persons C and D. The same sort of behavior is found if persons A and C shine flashlights toward their partners. The light from the flashlight at A travels independently of the light from the flashlight at C. This independence is a result of the ***principle of superposition***. When applied to wave motion, this principle can be stated as follows.

Principle of superposition

Principle of superposition

When two (or more) waves are present, the displacement of the medium is equal to the sum of the displacements of the individual waves.

For example, if we have two periodic waves with displacements y_1 and y_2 that are each described by a relation such as Equation 12.1, the total displacement of the medium is the sum $y_1 + y_2$. These two displacements each describe traveling waves, and the presence of one of the waves does not affect the frequency, amplitude, or velocity of the other. Hence, the two waves travel independently.

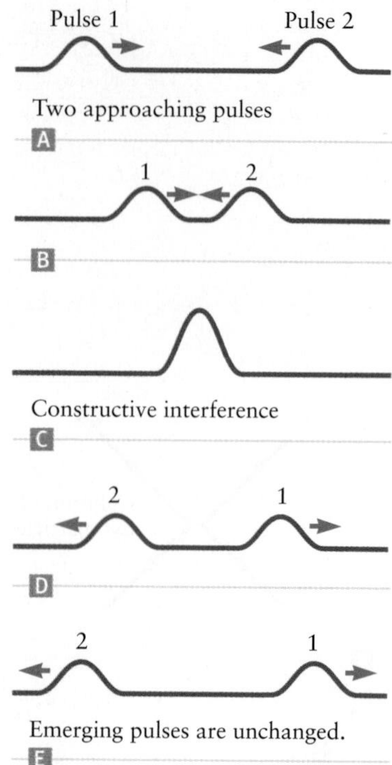

▲ **Figure 12.15** When two waves travel in the same region, the displacement of the medium is the sum of the displacements associated with the two waves. Shown are successive snapshots (from top to bottom) of a string containing two pulses with equal (and positive) amplitudes traveling in opposite directions.

Constructive and Destructive Interference

Figure 12.15 shows two wave pulses traveling in opposite directions. These pulses have equal and positive amplitudes, and the successive snapshots of the string show how the two pulses pass through each other. By the time of the final snapshot, the pulses are again isolated. Notice that the shapes of the final pulses in snapshot E are the same as the initial pulse shapes in snapshot A. According to the superposition principle, the displacement of the medium is always equal to the sum of the displacements produced by the two individual pulses, so when the two pulses overlap completely as in Figure 12.15C, the maximum displacement of the string is twice the amplitude of a single pulse. This is an example of ***constructive interference.*** The two waves add together to produce a displacement that is larger than the displacement of either of the individual waves.

Figure 12.16 shows snapshots of another experiment with two wave pulses on a string, but this time the displacement of one string is downward and that of the other is upward. Here again the two waves pass through each other, and the pulse shapes at the end in snapshot E are the same as the initial pulse shapes in snapshot A. The displacement of the string is again (according to the superposition principle) equal to the sum of the displacements of the individual waves. However, since the waves have opposite amplitudes when the pulses overlap in snapshot C, the total displacement of the string is *zero*! This is an example of ***destructive interference.*** The total displacement is smaller than the individual displacements of the two waves.

It is interesting to consider the effect of interference on the energy carried by the two waves in Figure 12.16. If you were to examine the string at the moment of perfect destructive interference in Figure 12.16C, you might think that there are no waves present and hence no propagation of energy, but even though there is a moment when the displacement of the string is zero, its velocity is *not* zero. At the instant of snapshot C, all the energy of each wave is contained in the kinetic energy of the medium (the string). In this way, two waves can interfere destructively and still carry energy independently. The term *interference* is thus a bit misleading because the presence of one wave does not "interfere" with the propagation of the other in any way.

Interference of Periodic Waves

Figure 12.17 shows an example of superposition with two periodic waves. Here two water waves are produced by dropping two rocks a short distance apart onto the surface of a still pool. Each rock produces a pattern of circular wave crests and troughs that travel away from the initial disturbance. Figure 12.17A shows the wave

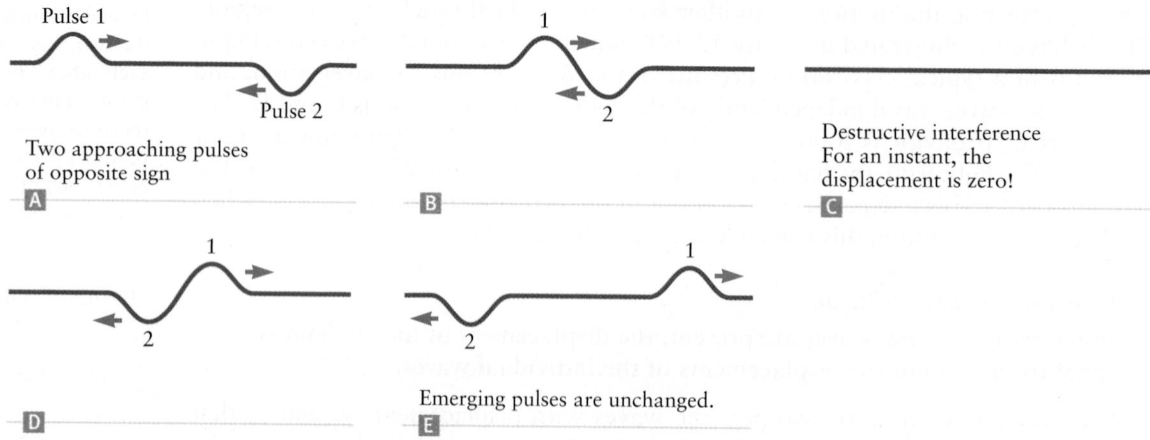

▲ **Figure 12.16** When the amplitudes of two interfering wave pulses are equal in magnitude but opposite in sign, the total displacement will be zero for an instant (snapshot C). This illustration is an example of destructive interference.

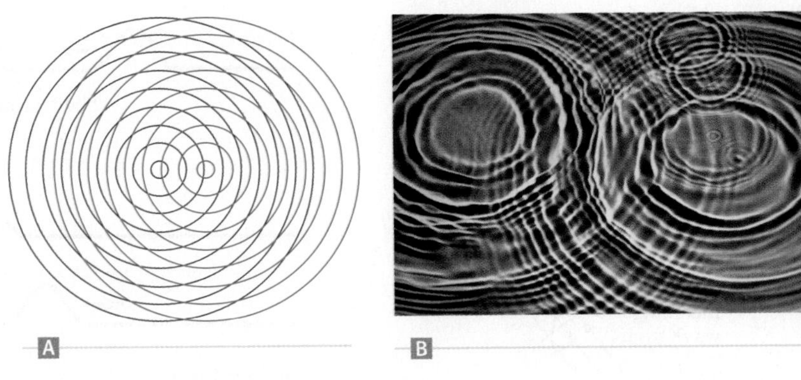
A
B
© Brand-X Pictures/Jupiterimages

◀ **Figure 12.17** A Interference of two overlapping water surface waves. Each source emits a circular pattern of wave crests and troughs. B Photo of interfering water waves on the surface of still water.

crests of these two waves separately. According to the principle of superposition, the total displacement of the water surface is the sum of the displacements of the individual waves. There is constructive interference at all points where two wave crests (or two wave troughs) in Figure 12.17A overlap, and the displacement of the surface is largest at these points. Likewise, there is destructive interference at points where a crest of one wave overlaps with a trough of the other, and at these points the displacement of the surface is small or zero. The resulting surface pattern is shown in Figure 12.17B. This ***interference pattern*** has an interesting geometry.

Insight 12.2

TWO WAVES WILL TRAVEL INDEPENDENTLY THROUGH THE SAME REGION OF SPACE

The terms *constructive interference* and *destructive interference* suggest the presence of one wave can affect the propagation of another, but quite the opposite is true. According to the principle of superposition, two waves propagate through the same region of space independently of each other. One wave does not destroy the other!

12.6 Reflection

Consider a wave pulse traveling on a string as sketched in Figure 12.18. Initially, this pulse is traveling to the right with a positive amplitude. It then ***reflects*** from the right-hand end of the string, and in the final snapshots it is traveling to the left. Another example of reflection is given in Figure 12.19, which shows a light wave from a laser reflecting from a mirror. The incoming wave approaches from the lower left; reflection from the mirror causes the wave to travel away toward the upper left edge of the photo.

In both examples, the wave on a string in Figure 12.18 and the light wave in Figure 12.19, reflection changes the propagation direction of the wave. It is useful to show this change using rays to indicate the direction of energy flow (compare with Fig. 12.13). Rays for the light wave in Figure 12.19 are shown in more detail in Figure 12.20. The rays change direction when a wave reflects from the boundary of the medium.

The rays in Figure 12.20 make an incoming (initial) angle θ_i and an outgoing (reflected) angle θ_r with a line drawn perpendicular to the surface. To understand how a wave changes direction when it is reflected, consider the component of a wave's velocity perpendicular to the surface, along with the component parallel to the surface. The perpendicular component of the wave velocity reverses direction; that is, the component of the wave that was initially traveling to the right in Figure 12.20 travels to the left after being reflected. This is just like the case with the string in Figure 12.18, where the wave is reflected back in the opposite direction. On the other hand, the parallel component of the wave in Figure 12.20 is not affected by the reflection. The final result is then as sketched in Figure 12.20, with the angle of incidence equal to the angle of reflection:

$$\theta_i = \theta_r \tag{12.12}$$

Waves are usually reflected when they reach the boundaries of the medium, and this reflection process can sometimes (but not always) *invert* the wave's displacement. For example, in Figure 12.18, the wave reflected from the end of a string that is fastened to the wall is inverted; the reflected wave pulse has a negative amplitude, whereas the incoming wave had a positive amplitude.

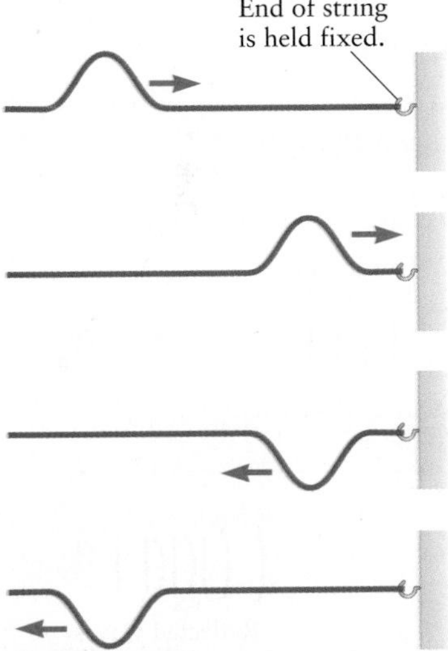
End of string is held fixed.
Wave pulse is inverted after reflection.

▲ **Figure 12.18** When a wave reaches a boundary (such as the end of a string), it can be reflected. When the end of the string is held fixed by tying it to a wall, the displacement of the reflected wave is reversed; here the displacement is changed from positive to negative.

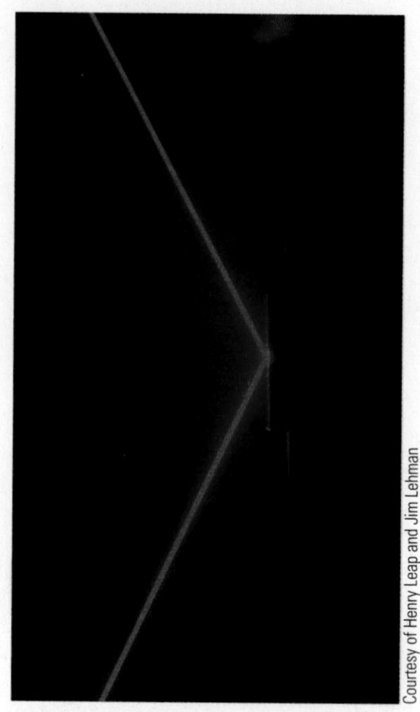
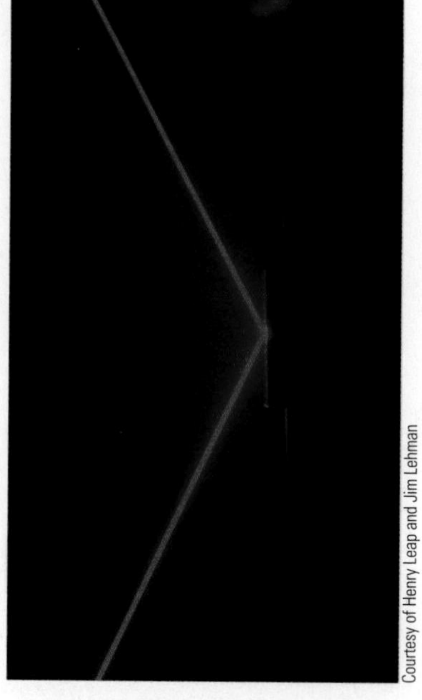

▲ **Figure 12.19** A light wave can be reflected when it strikes a boundary such as a mirror.

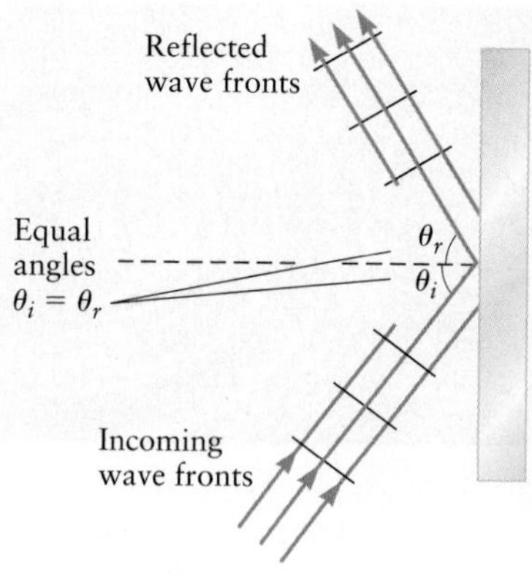

▲ **Figure 12.20** When a plane wave such as light, a water wave, or a sound wave reflects from a flat boundary, the angle of incidence θ_i is equal to the angle of reflection θ_r.

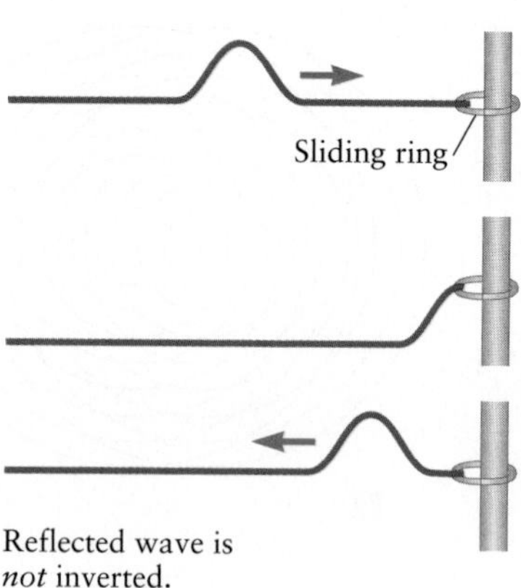

▲ **Figure 12.21** When the end of a string is attached to a ring that can slide up and down (e.g., along a pole), the reflected wave is not inverted.

While waves are often inverted on reflection, that is not always the case. Figure 12.21 shows a wave on a string where the end of the string is not fastened rigidly to the wall, but is able to slide up and down. Now the reflected wave is not inverted. In this example and for other cases of wave reflection, the properties of the medium at the boundary determine if the reflected wave will be inverted or not.

Radar

One interesting application of wave reflection is ***radar***, a technique that uses radio waves to detect objects remotely and determine their distance (Fig. 12.22). A radio wave pulse is sent from a transmitting antenna and reflects from some distant object such as an airplane. If the reflecting surface of the airplane is large and flat, the wave will be reflected at a single angle as in Figure 12.20. For radar applications, however, the surfaces on the object are usually smaller than (or comparable to) the wavelength of the radio wave, and in this case, reflected waves travel outward in all directions. A portion of the reflected wave will thus arrive back at the original transmitter, where it is detected. The distance to an object is determined by measuring the time delay Δt between the original outgoing wave pulse and the returning one. If the speed of the wave is v, the total round-trip distance traveled by the wave is $L_{\text{tot}} = v\,\Delta t$, which is twice the distance from the transmitter to the object. So, by measuring Δt, a radar system uses wave reflection to measure the distance to a remote object. By using a rotating antenna, the direction of the object can also be determined. The amplitude of the reflected wave gives information about the object's size; a large object reflects more of the wave energy than a small object and thus gives a larger signal at the detecting antenna.

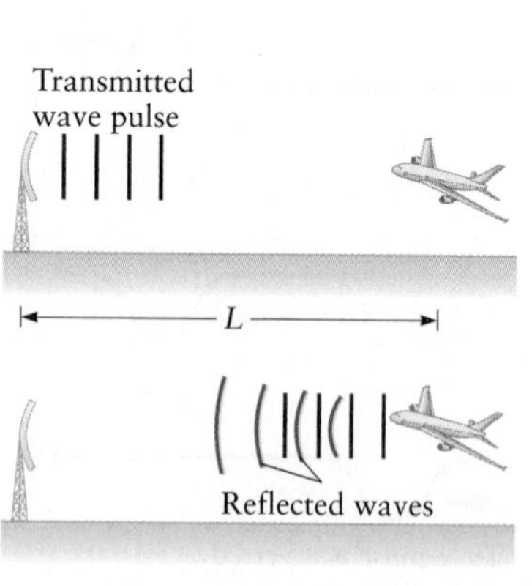

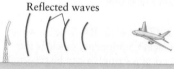

▲ **Figure 12.22** Radar uses reflected waves to measure the location of a distant object.

12.7 Refraction

In our discussions of wave fronts (Section 12.4), we assumed the speed of the wave is the same at all places in the medium. Figure 12.23A shows such an example for a plane sound wave traveling through a solid. Here the wave speed is the same everywhere, and this plane wave propagates along a straight line from left to right. In some cases, however, the wave speed may be different at different locations. A hypothetical

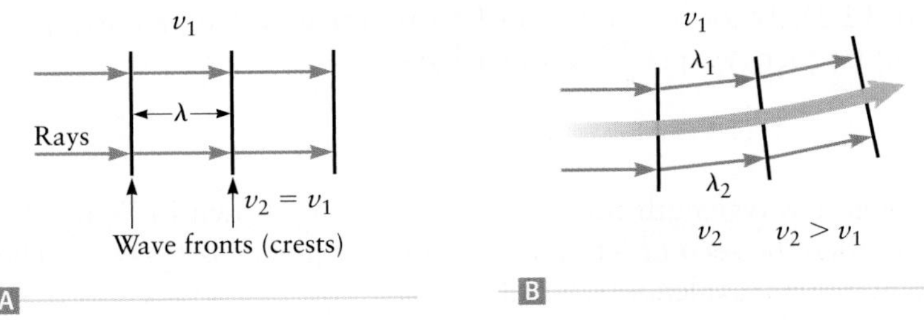

◀ **Figure 12.23** A When the wave speed is the same throughout the medium, the rays follow straight lines. B When the wave speed is different in different regions (here $v_2 > v_1$), the rays are "bent" and the waves are refracted.

case is sketched in Figure 12.23B, which shows a wave traveling through a material in which the wave speed is slightly greater at the bottom than at the top region of the figure ($v_2 > v_1$). The solid black lines show wave crests, and by definition they are separated by a wavelength. The wave frequency is the same everywhere (its value is determined by the frequency of the source), and because $\lambda = v/f$, the wavelength—and hence the distance between crests—is slightly longer in the bottom region. As a result, the wave crests tilt progressively as the wave propagates to the right, and the wave direction (the direction of energy flow) curves upward, as indicated by the blue rays, an effect called ***refraction.*** Refraction is the *change in direction of a wave due to a change in its speed*. We'll see examples of refraction when we study seismic waves in the Earth (Section 12.9) and again when we study optics.

12.8 Standing Waves

Consider waves traveling back and forth along a string of length L with both ends held fixed (i.e., tied to walls as sketched in Figure 12.24A). What will the string displacement look like as a function of both time t and position x along the string? For a very long string carrying a single periodic wave traveling to the right, the string displacement y is given by a relation such as Equation 12.1. Because the ends of the string in Figure 12.24A are fastened to rigid supports, however, the displacement must be zero at both ends. That is, the string displacement y must be zero at both $x = 0$ and $x = L$. These conditions can be satisfied by a periodic wave such as Equation 12.1 only for certain wavelengths. For example, suppose the wavelength is equal to precisely twice the length of the string so that $\lambda = 2L$. The wave will then "fit" onto the string as shown in Figure 12.24B, which shows several superimposed snapshots of the vibrating string. While the two ends are at rest with $y = 0$, the middle portion of the string oscillates up and down with a large amplitude.

The vibration in Figure 12.24B is called a ***standing wave*** because the outline of the wave appears to stand still. Such a wave pattern can be obtained by the interference of two waves, both with wavelength $\lambda = 2L$, traveling in *opposite* directions. These waves travel back and forth along the string and are reflected when they reach the ends of the string. According to the principle of superposition, the total string displacement is the sum of the displacements of these two waves, and that sum gives the pattern in Figure 12.24B. Points where the string displacement is always zero are called ***nodes*** of the standing wave. The points where the displacement is largest are called ***antinodes.*** In Figure 12.24B, there is an antinode at the center of the string, while there are nodes at both ends.

Other possible standing waves that fit on the string are shown in parts C and D of Figure 12.24. For the wave in Figure 12.24C, we have $\lambda = L$, whereas in Figure 12.24D, we find $\lambda = 2L/3$. We can generate many other standing waves by employing smaller and smaller wavelengths. The longest possible wavelength for a standing wave is the one shown in Figure 12.24B, corresponding to the *smallest* possible standing wave frequency. This is called the ***fundamental frequency*** of the string, denoted by f_1. The frequency of a standing wave is related to its wavelength by

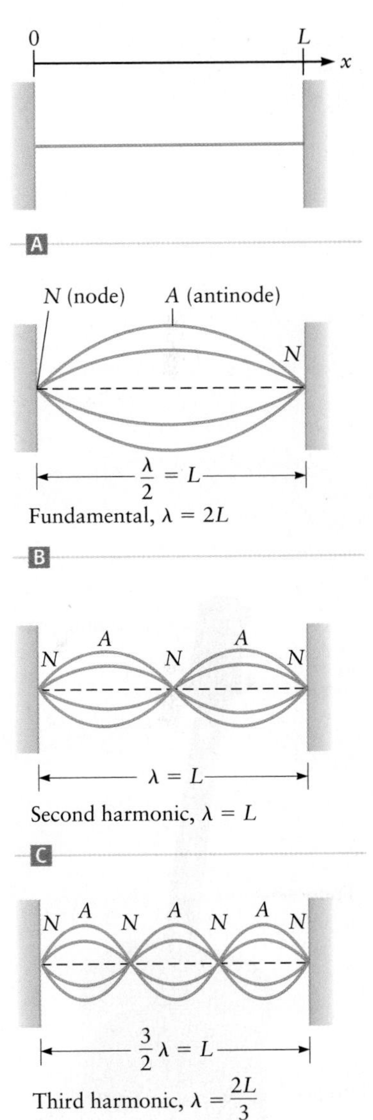

▲ **Figure 12.24** Standing waves on a string, such as on a guitar or a piano. A The string at rest. B through D Standing waves with different wavelengths. Here, N and A denote nodes and antinodes of the standing waves.

$f = v/\lambda$ (Eq. 12.2). According to Figure 12.24B, the wavelength corresponding to the fundamental frequency equals $2L$, so we have

$$f_1 = \frac{v}{2L} \tag{12.13}$$

The next longest wavelength for a standing wave is shown in Figure 12.24C. This vibration is called the ***second harmonic***, and its frequency is $f_2 = v/L$. There is a pattern to the possible wavelengths, which is

$$\lambda_n = \frac{2L}{n} \quad \text{with } n = 1, 2, 3, \ldots \quad \text{(standing waves on a string with fixed ends)} \tag{12.14}$$

In Equation 12.14, $n = 1$ gives the wavelength of the fundamental, $n = 2$ gives the second harmonic, and so forth. The corresponding pattern for the frequencies is

$$f_n = \frac{v}{\lambda_n} \tag{12.15}$$

which leads to $f_n = nv/2L$ and

$$f_n = nf_1 \quad n = 1, 2, 3, \ldots \quad \text{(standing waves on a string with fixed ends)} \tag{12.16}$$

The allowed standing wave frequencies are thus *integer multiples* of the fundamental frequency f_1.

CONCEPT CHECK 12.5 Nodes and Antinodes

Consider a standing wave on a string with a frequency equal to the fourth harmonic (Eq. 12.16 with $n = 4$). How many nodes and antinodes does this standing wave possess?

Musical Tones

Standing waves on strings are closely connected to musical tones. Many musical instruments, including guitars, violins, and pianos, employ strings as a vibrating element. A guitar string is stretched under tension from the tuning peg at the top of the neck to a point where it is fastened to the bridge and top plate of the instrument as shown in Figure 12.25. The player uses his or her finger to press the string down against a metal strip called a "fret" so that a length L of the string extending from the bridge to the fret is free to vibrate. The section of the guitar string between the fret and the bridge is thus similar to the string in Figure 12.24.

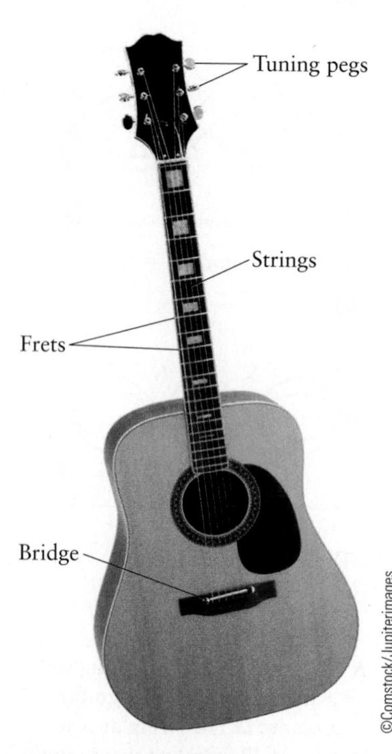

▲ **Figure 12.25** A guitar uses standing waves on a string to generate particular frequencies.

A plucked guitar string vibrates according to all the possible standing wave patterns with frequencies $f_1, f_2, f_3, \ldots$ (Eq. 12.16) all at the same time. The total string motion is, according to the superposition principle, a sum of the vibrations at these individual frequencies, although the vibrations at the lowest frequencies are usually largest. These string vibrations cause the body of the guitar to vibrate, producing the sound we hear. The frequencies of these sound waves are the standing wave frequencies of the guitar string. The sound of a single note is thus composed of many individual frequencies, which are related according to the harmonic sequence in Equation 12.16.

The ***pitch*** of a note is determined by its fundamental frequency. A guitar player can change the pitch by changing the length L of the vibrating section of the string, which can be done by holding the string against a different fret. According to our relation between the fundamental frequency and L, Equation 12.13, the pitch is inversely proportional to the length of the string. Two notes whose fundamental frequencies differ by a factor of two are said to be separated by an ***octave***. According to Equation 12.13, reducing L by a factor of two increases the frequency by one octave. This relation between musical pitch, the notes of a musical scale, and the length of a vibrating string was first discovered by Pythagoras, the mathematician famous for his work with right triangles.

EXAMPLE 12.6 Playing a Guitar

Guitarists often play several notes simultaneously by plucking or strumming two or more strings at the same time. Two notes whose fundamental frequencies are in the ratio 1.5/1 = 3/2 are known to sound pleasing when played together. This musical interval is known as a "perfect fifth." Suppose you wish to play these notes on two similar guitar strings, that is, strings that have the same value for the wave speed, v. What is the ratio of the vibrating lengths of the two strings?

RECOGNIZE THE PRINCIPLE

The fundamental frequency of a guitar string is $f_1 = v/2L$. We can apply this relation to both strings and then calculate the ratio of the lengths needed to get a frequency ratio of 3/2.

SKETCH THE PROBLEM

The problem is described by Figures 12.24 and 12.25.

IDENTIFY THE RELATIONSHIPS

Let L and L' be the vibrating lengths of the two strings. According to the relation between fundamental frequency and string length (Eq. 12.13), the fundamental frequencies are $f_1 = v/2L$ and $f_1' = v/2L'$. The ratio of the fundamental frequencies is thus

$$\frac{f_1'}{f_1} = \frac{v/2L'}{v/2L} = \frac{L}{L'}$$

SOLVE

Inserting the desired value for the ratio of the frequencies, we get

$$\frac{f_1'}{f_1} = \frac{3}{2} = \frac{L}{L'} = \boxed{1.5}$$

► ***What does it mean?***

This type of analysis can also be used to calculate how to position the frets on a guitar (see Question 8 at the end of this chapter). Changing the vibrating length of a guitar string by "one fret" changes the frequency so as to take you to the next note in the musical scale (e.g., from B to C or from C to C-sharp).

EXAMPLE 12.7 Bending a Note on a Guitar

The pitch of a guitar note is related to the length L of the vibrating portion of the string and its tension. The length is determined by the fret spacing (Fig. 12.25), whereas the tension is adjusted with the tuning pins. A guitar player can also change the pitch by "bending" a note. Bending is illustrated in Figure 12.26; the guitarist pushes the string to one side, increasing the tension and thus increasing the pitch. By doing this as a note is being played, the pitch changes continuously during the note, an effect used by many guitar players. Suppose a note is "bent" from E to F, a change in frequency of 5.9%. By what factor has the tension been increased?

RECOGNIZE THE PRINCIPLE

The pitch of a note corresponds to its fundamental frequency, with (Eq. 12.13)

$$f_1 = \frac{v}{2L} \quad (1)$$

(continued) ►

C. Betz

▲ **Figure 12.26** Example 12.7. "Bending" a note on a guitar.

To give a 5.9% change in the pitch, we therefore require a 5.9% change in the wave speed v. The wave speed depends on the tension, so we need to find the change in tension required to give this change in v.

SKETCH THE PROBLEM

Figure 12.26 shows the problem.

IDENTIFY THE RELATIONSHIPS

The wave speed varies with the tension F_T as $v \propto \sqrt{F_T}$ (Eq. 12.9). Combining Equation 12.9 with Equation (1), the frequency, wave speed, and tension for the note E are related by $f_1(\text{E}) \propto v(\text{E}) \propto \sqrt{F_T(\text{E})}$, with a similar relation for the note F. We can thus write

$$\frac{f_1(\text{F})}{f_1(\text{E})} = \sqrt{\frac{F_T(\text{F})}{F_T(\text{E})}} \tag{2}$$

SOLVE

The frequencies differ by 5.9%, so their ratio $f_1(\text{F})/f_1(\text{E})$ is 1.059. Inserting this information into Equation (2) and squaring both sides leads to

$$1.059^2 = \frac{F_T(\text{F})}{F_T(\text{E})} = \boxed{1.12}$$

► ***What does it mean?***

Here we considered a case in which the "bending" moves from one note to the next on a musical scale, but bending is often used to change the pitch by only a "fraction" of a note interval. Bending a guitar note requires strong fingers.

12.9 Seismic Waves and the Structure of the Earth

Waves can be used to probe the structure of the Earth. Any large mechanical disturbance such as an earthquake or nuclear explosion acts as a source of ***seismic waves***, another term for waves that propagate through the Earth. There are three main types of seismic waves as sketched in Figure 12.27A. One is called an S wave

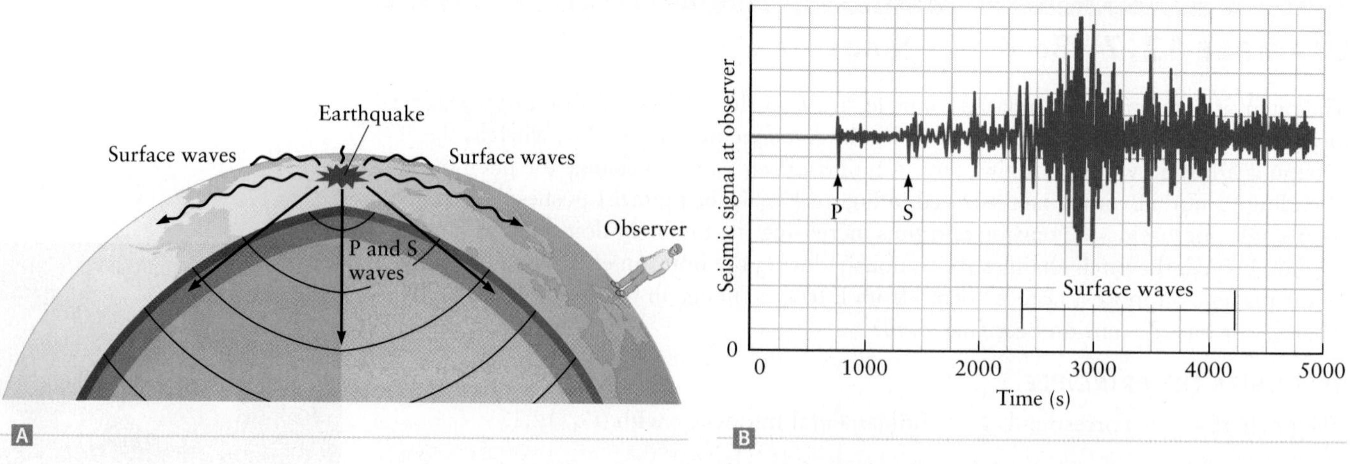

▲ **Figure 12.27** A When an earthquake occurs, several different types of seismic waves are generated, including P waves and S waves (both of which travel through the Earth) and surface waves. B These waves have different speeds, so they reach an observer at different times.

(S for "shear"). Like shear waves in a metal bar, S waves are transverse waves (Fig. 12.10B), with the displacement of the solid Earth perpendicular to the direction of propagation. Another type of seismic wave is called a P wave (P for "pressure"). P waves are longitudinal sound waves (see Fig. 12.10A and Section 12.3). Although S and P waves have different speeds, they both propagate into the body of the Earth. A third type of wave travels on the surface. These surface waves travel more slowly than S and P waves and are similar to water waves on the surface of a lake, which we described in Section 12.3.

Seismic waves can be detected by an instrument called a ***seismograph***, which measures the displacement of the Earth as a function of time. A typical seismograph signal is shown in Figure 12.27B. The P wave pulses arrive first because they travel fastest, followed by the S waves and then the surface waves. The seismograph can also measure the direction of the displacements associated with each of these wave pulses, which can be used to distinguish the different types of waves. By knowing the speeds of the three types of waves, the distance from the observer to the earthquake in Figure 12.27A can be determined.

This picture correctly describes things when the earthquake is not too far from the observer. When the earthquake and the observer are on opposite sides of the Earth, however, the observer's seismograph does not detect any S wave component. The reason is shown in Figure 12.28A. The center of the Earth contains a region called the *core*, and the area of the Earth where the S waves do not reach is in the "shadow" of the core. S waves do not propagate through the core, whereas P waves can. S waves are shear (transverse) waves, and we saw in Section 12.3 that they cannot travel in a liquid. The existence of the S wave shadow provided the first evidence that the Earth's core contains a liquid. It is now known that the core contains liquid metals, such as iron, at a very high temperature. Seismographic recordings provide the only way to measure the diameter of the core and study its properties.

Figure 12.28B shows in greater detail, in a more realistic sketch, how S and P waves propagate through the Earth. We again see the S wave shadow made by the core and also notice that the S and P waves do *not* travel in straight lines. Instead, they follow curved paths that bend toward the Earth's surface because of refraction (Section 12.7). The speeds of S and P waves depend on the composition and density of the Earth, and these speeds increase as one goes deeper into the Earth. Refraction causes the seismic waves to be deflected toward the surface, in the direction of smaller wave speed as shown in Figure 12.23B.

The amount of bending of the P and S waves depends on how the wave velocities vary from place to place within the Earth. Because it is impossible to measure the composition deep in the interior of the Earth directly, it is not possible to predict ahead of time precisely how the waves will travel. However, data taken with seismographs at the Earth's surface can be used to infer the paths actually taken by S and P waves. These inferred paths give information on how the composition of the Earth varies with depth. In this way, it was discovered that the Earth is composed of several different solid layers, two of which are called the ***crust*** and the ***mantle***, with different compositions. It has also been found that the core is not simply a spherical fluid region. Rather, the core has a solid central region (the inner core) surrounded by a fluid outer core (Fig. 12.28B). This surrounding fluid produces the S wave shadow described above.

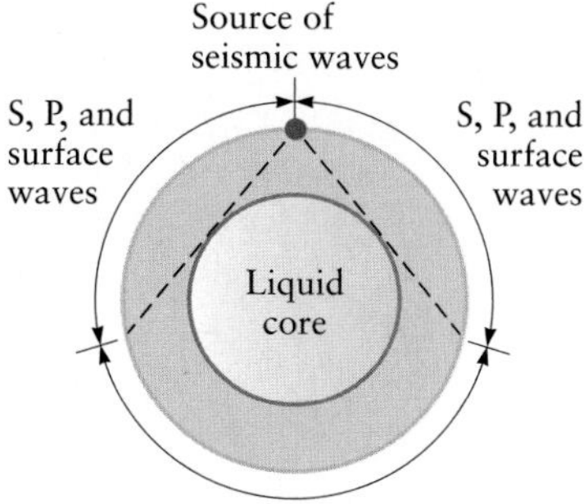

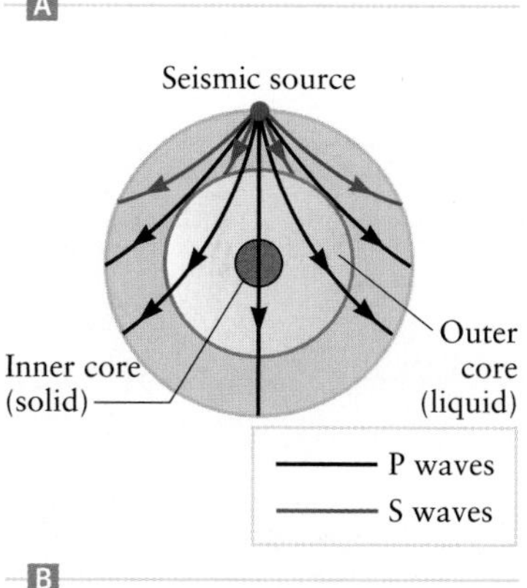

▲ **Figure 12.28** A The core of the Earth contains liquid, and transverse waves do not propagate through a liquid. Therefore, S waves do not travel through the core, and observers who are in the core's "shadow" do not record S waves from a source on the opposite side of the Earth. B Due to refraction, P and S waves do not travel in straight lines.

EXAMPLE 12.8 Ⓡ Measuring the Size of the Earth's Core

From extensive seismograph measurements, it is known that the angle θ defining the edge of the S wave shadow in Figure 12.29 is approximately 39°. Use this angle to estimate the radius of the Earth's liquid core. For simplicity, ignore the refraction (bending) of S and P waves and assume they travel in straight lines.

(continued) ▶

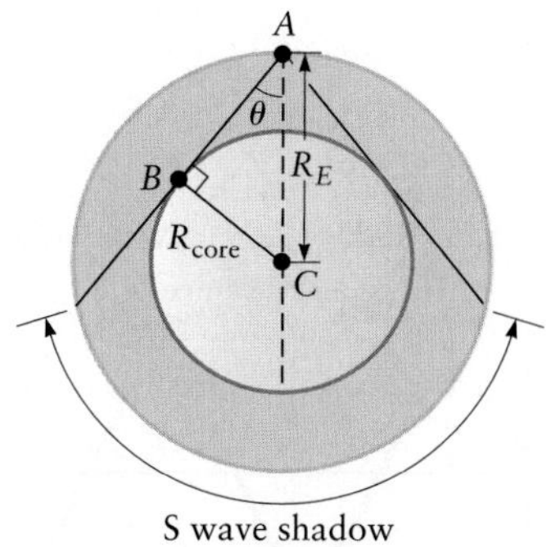

▲ **Figure 12.29** Example 12.8.

RECOGNIZE THE PRINCIPLE

We are assuming straight-line propagation, so we can apply the geometry in Figure 12.29. S waves that just graze the outer edge of the core follow a ray from A to B. The triangle ABC is a right triangle, and the side BC is the radius of the core, R_{core}, which is what we want to find.

SKETCH THE PROBLEM

Figure 12.29 shows the geometry of the problem.

IDENTIFY THE RELATIONSHIPS

The hypotenuse of triangle ABC is the radius of the Earth R_E, so we have

$$R_{core} = R_E \sin\theta$$

SOLVE

Inserting the known radius of the Earth ($R_E = 6.4 \times 10^6$ m, Table A.3) along with the given value of θ, we find

$$R_{core} = R_E \sin\theta = (6.4 \times 10^6 \text{ m})\sin(39°) = \boxed{4.0 \times 10^6 \text{ m}}$$

► ***What does it mean?***

Our result for R_{core} is about 60% of the Earth's radius. The core thus makes up a relatively large fraction of the Earth's volume.

Summary | CHAPTER 12

Key Concepts and Principles

Waves

A ***wave*** is a disturbance that transports energy from one place to another without transporting matter. Waves travel through a medium; for mechanical waves, the medium is a material substance. For ***transverse waves***, the displacement of the medium is perpendicular to the velocity of the wave. With ***longitudinal waves***, the displacement is parallel to the wave velocity. Fluids (liquids and gases) cannot support transverse waves.

Frequency and period

A wave can consist of a "single" disturbance—a wave pulse—or the disturbance can have a repetitive form. A periodic wave has a characteristic repeat time called the ***period*** T. The ***frequency*** of a periodic wave is related to the period by $f = 1/T$.

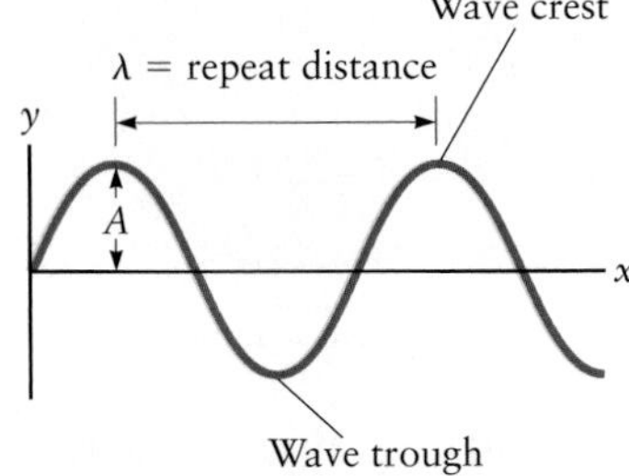

Displacement of a wave

The displacement associated with a periodic wave is often described by

$$y = A\sin\left(2\pi ft - \frac{2\pi x}{\lambda}\right) \quad \textbf{(12.1) (page 386)}$$

Here A is the amplitude of the wave, f is the frequency, and λ is the wavelength.

The frequency, wavelength, and velocity of a wave are related by

$$v = f\lambda \qquad (12.2)\ (\text{page } 387)$$

Superposition

According to the ***principle of superposition***, when two waves are present simultaneously, the total displacement of the medium is the sum of the displacements of the individual waves. One consequence of this principle is that two waves can "pass through" each other.

Wave fronts and rays

A ***wave front*** is a surface that describes how the energy carried by a wave travels through space. ***Rays*** indicate the direction of energy flow and are perpendicular to the wave fronts.

Reflection and refraction

Waves can be reflected when they reach the boundary of the medium in which they travel. ***Refraction*** is the change in direction of a wave due to a change in velocity. The wave velocity can change at a boundary or within the medium.

Applications

Examples of waves

There are many examples of waves. The velocity of a wave depends on the properties of the medium.

- ***Waves on a string*** are transverse waves and are employed in many musical instruments.
- ***Sound*** is a longitudinal wave that can exist in solids, liquids, and gases. (See also Chapter 13.)
- ***Light*** is an electromagnetic wave. Because they are not mechanical waves, light and other electromagnetic waves can travel through a vacuum as well as through a material medium such as air.

Interference and standing waves

When two waves overlap in space, they are said to ***interfere***. One example of interference is the phenomenon of ***standing waves***. A point along a standing wave where the displacement of the medium is always zero is called a ***node***, and the points where the displacement is largest are ***antinodes***. Standing waves are used to produce musical tones in stringed instruments such as guitars and pianos. The lowest possible frequency of the standing wave (corresponding to the longest possible wavelength) is called the ***fundamental frequency***. For a string held rigidly at both ends, the standing waves form a series of ***harmonics*** with frequencies

$$f_n = nf_1 \quad \text{with } n = 1, 2, 3, \ldots \quad (\text{standing waves on a string with fixed ends}) \qquad (12.16)\ (\text{page } 400)$$

where $f_1 = v/(2L)$ is the fundamental frequency.

Standing waves on a string

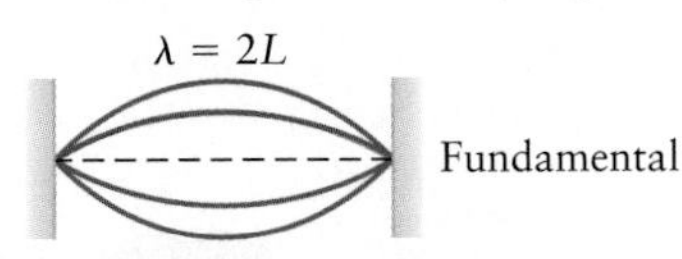

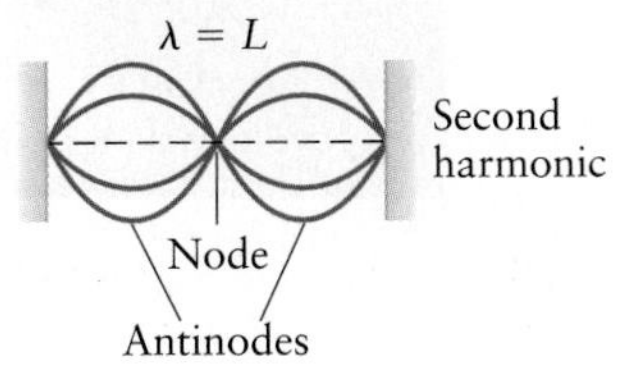

Seismic waves

Seismic waves provide information about the Earth's inner structure.

Questions

SSM = answer in *Student Companion & Problem-Solving Guide* = life science application

1. If the frequency of a wave is doubled, how does the wavelength change?
2. Identify the medium for each of the following waves:
   (a) Sound waves in a room
   (b) Waves on the surface of a pond
   (c) Radio waves
   (d) Light waves
   (e) Waves created when a guitar string is plucked
   (f) Seismic waves in the Earth
3. Identify the following waves as either longitudinal or transverse or both. Use a drawing to illustrate how the motion of the wave's medium is related to the direction of propagation of the wave.
   (a) Sound waves in a room
   (b) Waves on a guitar string
   (c) Seismic waves in the Earth
   (d) Waves on the surface of a pond
4. In a sound wave in a solid, the atoms in the solid move parallel to the propagation direction. Does that means the atoms travel along with the wave? Explain why or why not.
5. A key property of a wave is that it transports energy without transporting mass. Explain how this is possible, using work–energy ideas (i.e., $W = F\,\Delta x$). *Hint*: Consider the "leading edge" of a wave.
6. Devise or describe an experiment that uses standing waves to measure the velocity of a wave.
7. A bass guitar has longer strings than a regular guitar. Explain why these longer strings allow a bass guitarist to play "lower" notes.
8. SSM The neck of a guitar is designed with ***frets*** as shown in Figure Q12.8. A player can hold a string against one of the frets and thus shorten the vibrating length L of the string. In this way, a particular string can be used to play notes with different frequencies. The frequency of a note is determined by the fundamental standing wave mode (Fig. 12.24), and this frequency depends on L. In a musical scale, the frequencies of adjacent notes (i.e., between C and C-sharp) are in the ratio of $2^{1/12}/1 \approx 1.059$; this ratio produces the commonly used 12-tone musical scale. The artisans who make guitars position the frets according to the "rule of 18ths." According to this rule, each fret is positioned so that depressing adjacent frets shorten the vibrating length of the string by $L/18$. (See Fig. Q12.8.) Show that the rule of 18ths produces note frequencies that fit approximately on the desired musical scale.

© Tetraimages/Jupiterimages

Figure Q12.8

9. Important properties of a wave on a string include the (a) wavelength, (b) frequency, (c) amplitude, and (d) speed. Which of these properties is independent of the others? Explain.
10. Figure Q12.10 shows a child holding one end of a long Slinky. If the child wants to generate a longitudinal wave in the Slinky, in what direction should he shake the end of the Slinky, (a) in a horizontal direction or (b) in a vertical direction?

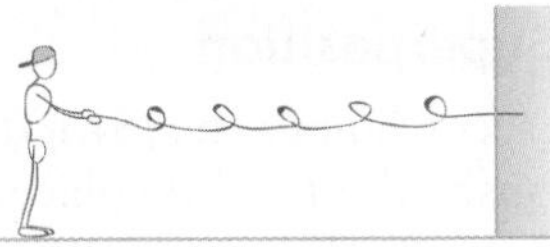

Figure Q12.10

11. A student sends a transverse wave pulse down a stretched piece of string attached to a wall. A moment later, she sends a second wave pulse. Is there any way she could alter the conditions to make the second wave overtake the first wave (i.e., before it is reflected back)? What would happen if she pulls harder on the string?
12. A rope of linear density of about 1 kg per meter is hung from the ceiling. The lower free end is then oscillated, producing waves that travel up the rope. Describe the velocity of the wave as it travels up the rope. Does it change? Why or why not? *Hint*: Is the tension the same at all points on the string?
13. SSM A woman wakes up just after 2 AM. Wondering what woke her, she goes to the window and looks around. Noting nothing out of the ordinary, she sits back down on her bed and all of a sudden a loud explosion rattles her window. The next day, she reads that an underground munitions dump exploded about 10 km from her home. She thinks that perhaps she woke up because she experienced a psychic precognition of the explosion. Can you think of a physical explanation for what might have awoken her the few moments before the sound from the blast hit her window? (See Problem 69.)
14. When a sound wave passes from air into a solid object, which of the following wave properties stays the same and which change? Explain.
   (a) The amplitude
   (b) The frequency
   (c) The wave speed
   (d) The wavelength
15. Equation 12.1 describes a wave that is traveling to the right, and Equation 12.5 corresponds to a wave of the same frequency and wavelength that is moving to the left. Show that the sum of these two waves gives a standing wave. That is, show that this sum has nodes and antinodes as sketched in the standing waves in Figure 12.24.
16. Consider a water wave as illustrated in Figure 12.11. This wave front is circular. How does the intensity of this wave vary as it moves away from its source? *Hint*: You should find that $I \propto r^p$. What is the value of p?
17. Which property of a wave is independent of its amplitude, (a) frequency, (b) wavelength, or (c) speed?
18. Most modern guitars use steel strings, whereas many historical instruments use strings made of animal gut. Suppose the strings on a guitar that was originally strung with gut are replaced with steel strings of the same diameter. Will the tension in the steel strings be greater or less than in the gut strings?
19. When you tune a guitar, what causes the frequency to change?
20. For guitars and violins, some strings are "simple" strings, and others contain a winding around a string core. Which type of string, simple or wound, is used for the low-frequency notes? Explain why.
21. **Seismic waves in the Earth.** Surface seismic waves generated by an earthquake arrive at a distant observer after the arrival of P waves and S waves (Fig. 12.27B), partly because surface waves travel more slowly than both P and S waves. Even if all these waves had the same velocities, however, the surface waves would still arrive last. Explain why.

Problems

SSM = solution in *Student Companion & Problem-Solving Guide*
★ = intermediate ✪ = challenging Ⓧ = life science application
Ⓡ = reasoning and relationships problem
RT = reasoning tutorial in Enhanced WebAssign

12.1 WHAT IS A WAVE?

1. SSM Figure P12.1 shows several snapshots of a wave pulse as it travels along a string. Estimate the speed of the wave pulse.

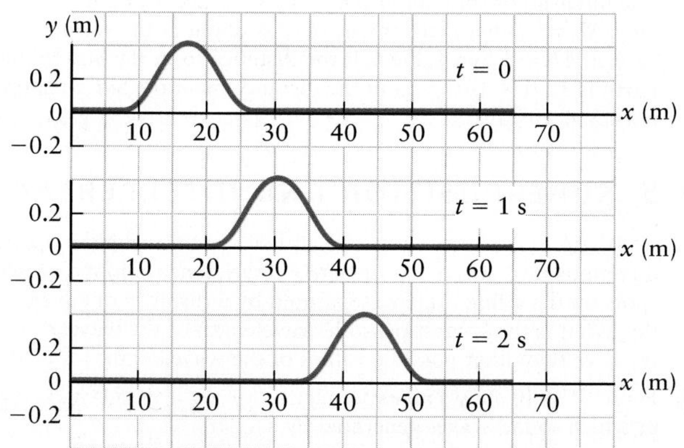

Figure P12.1 Problems 1, 2, and 3.

2. Is the wave on a string in Figure P12.1 transverse or longitudinal?

3. Ⓡ For the wave in Figure P12.1, what is the approximate magnitude of the speed *of a section of the string* as the wave pulse passes by? What is the direction of the string velocity relative to the velocity of the wave?

4. ✪ RT Consider a clothesline stretched between two supports. If one end of the line is "wiggled," a wave will be produced. Estimate the speed of this wave on a clothesline.

5. A rough rule of thumb is that light from the Sun reaches the Earth in about 8 minutes. Use this estimate along with the known speed of light to get an approximate value for the distance from the Earth to the Sun.

12.2 DESCRIBING WAVES

6. ★ Figure P12.6 shows a snapshot of a portion of a periodic wave as it travels along a string. Estimate the wavelength of the wave. Is there enough information in this figure for you to also estimate the speed of the wave? Explain.

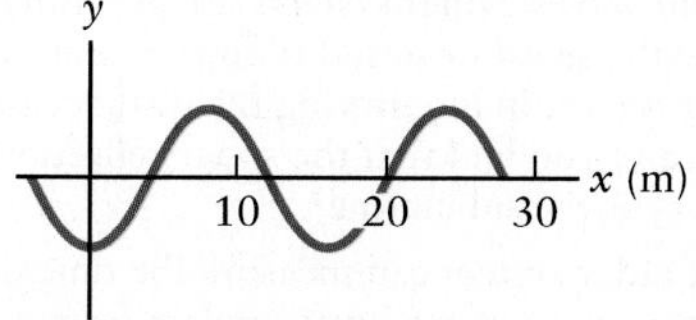

Figure P12.6

7. Figure P12.7 shows a snapshot of a wave on a string. If the frequency of this wave is 20 Hz, what is the approximate speed of the wave?

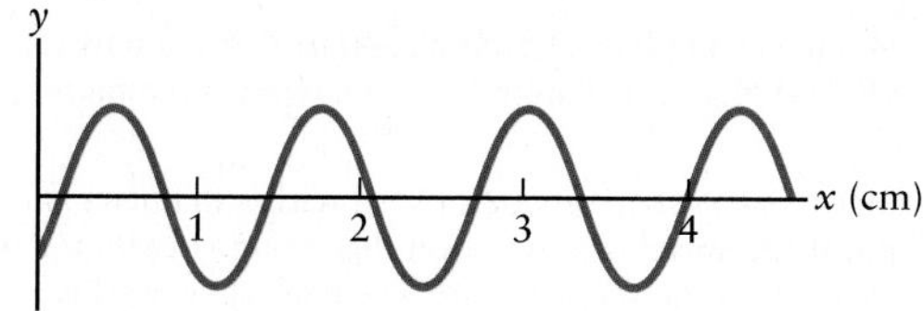

Figure P12.7

8. ★ A wave on a Slinky has a frequency of 2.3 Hz. It takes 3.5 s to travel the length of the Slinky (0.30 m). What is the wavelength?

9. Sound waves in water travel at a speed of approximately 1500 m/s. What is the wavelength of a sound wave that has a frequency of 440 Hz?

10. SSM ★ A wave on a string is described by the relation $y = A\sin(35t - 0.025x)$, where t is measured in seconds and x in meters, with $A = 0.15$ m. Find the frequency, wavelength, and speed of this wave.

11. ✪ RT A wave on a string is described by the relation $y = A\sin(72t - 0.045x)$, where t is measured in seconds and x in meters, with $A = 0.15$ m. What is the maximum speed of a point on the string?

12. ★ Using Equation 12.1, write the equation of a wave on a string that has an amplitude of 0.30 m, a frequency of 350 Hz, and a velocity of 250 m/s and travels to the right.

13. ★ Using Equation 12.1, write the equation of a wave on a string that has an amplitude of 0.50 m, a wavelength of 8.4 m, and a velocity of 330 m/s and travels to the left.

14. Consider a wave that has a frequency of 3.0 kHz and a wavelength of 0.10 m. Write an equation that might describe this wave.

15. ★ Figure P12.15 shows a snapshot of a wave on a string. This wave has a frequency of 200 Hz. Approximately how long does it take the wave crest located at point A in this snapshot to move to point B?

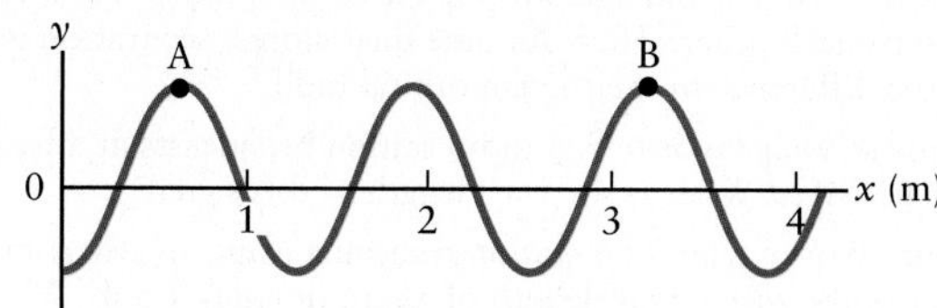

Figure P12.15

16. Draw a snapshot of a wave on a string that obeys the equation $y = (4.2\text{ mm})\cos(3000t - 0.45x)$. Assume the units of length are millimeters and the units of time are seconds. Make your drawing at $t = 1.5$ s.

17. ★ A wave on a string is described by the equation $y = (4.2\text{ mm}\cos(3000t - 0.45x)$ where t is measured in s and x is measured in mm. What are the frequency and wavelength of the wave?

18. ★ A wave on a string is described by the equation $y = (4.2\text{ mm}\cos(3000t - 0.45x)$, where t is measured in s and x is measured in mm. Make a graph of the displacement of the string at $x = 0.35$ mm as a function of time. Show two complete periods of the wave.

19. Consider a wave that has a frequency of 3.0 kHz and a wavelength of 0.10 m. What is the speed of this wave?

20. Two sound waves in air have frequencies that differ by a factor of five. What is the ratio of their wavelengths?

12.3 EXAMPLES OF WAVES

21. A wave on a string has a speed of 340 m/s. If the tension in the string is 150 N and its length is 2.0 m, what is the mass of the string?

22. Consider a piano string made of steel with a diameter of 1.0 mm. If the speed of a wave on this string is 280 m/s, what is the tension in the string? *Hint*: The density of steel is 7800 kg/m^3.

23. Waves on a violin string travel at 900 m/s. If the string has a tension of 50 N and is 0.33 m long, what is its mass?

24. A wave on a string has a speed of 200 m/s. If the diameter of the string is increased by a factor of two and the tension is held fixed, what is the speed of the wave on this new string?

25. SSM RT Consider a violin string with a vibrating length of 30 cm. If the speed of a wave on this string is 250 m/s, what is the tension in the string? *Hint*: You will have to estimate the diameter of the string. Assume the density is the same as that of steel, 7800 kg/m^3.

26. The speed of a wave on a guitar string is $v = 120$ m/s. If the tension is increased from 70 N to 110 N, what is the new value of v?

27. The speed of sound in water is 1500 m/s. Find the wavelength of a sound wave whose frequency is 300 Hz.

28. Spruce is a type of wood that is widely used in making musical instruments (such as violins and guitars) because its Young's modulus is high and its density is low. If $Y = 1.5 \times 10^{10}$ Pa and $\rho = 420$ kg/m^3, what is the speed of sound in spruce? How does that compare with the speed of sound in steel?

29. Consider the speed of sound in a hypothetical metal. If the density is reduced by a factor of three and all other properties are unchanged, by what factor is the speed of sound changed?

30. The speed of longitudinal waves in a metal is 3000 m/s. If the density is 12,000 kg/m^3, what is the Young's modulus of the metal?

31. Radio waves are electromagnetic waves and travel at $c = 3.00 \times 10^8$ m/s in a vacuum. How long does it take a radio signal to travel from Mars to the Earth when the two planets are nearest each other? *Hint*: Assume their closest separation is equal to the difference in their mean orbital radii.

32. Suppose your favorite FM radio station broadcasts at a frequency of 101 MHz. What is the wavelength of this signal?

33. Your favorite television station transmits using an electromagnetic wave with a wavelength of approximately 1.5 m. What is the frequency?

34. RT Electromagnetic waves are used to transmit television signals from a satellite that is in orbit around the Earth to the author's home. Estimate how long such a signal takes to travel from the satellite to the ground. *Hint*: This satellite completes one orbit each day and is thus in a geosynchronous orbit (Chapter 5).

12.4 THE GEOMETRY OF A WAVE: WAVE FRONTS

35. SSM A lightbulb emits a spherical wave. If the intensity of the emitted light is 1.0 W/m^2 at a distance of 2.5 m from the bulb, what is the intensity at a distance of 4.0 m?

36. For the spherical wave emitted by a small lightbulb, what is the ratio of the amplitudes of the wave at $r = 4.0$ m and $r = 2.5$ m from the lightbulb?

37. You are at a rock concert and find that when you stand 10 m from a speaker, the intensity is so high that it hurts your ears. You decide to move away from the stage so that your ears are more comfortable. If you want to reduce the intensity at your ears by a factor of three, how far from the speaker should you stand? Assume the sound emitted by a speaker at a rock concert has a spherical wave front. How would your answer change if the wave front has the shape of a hemisphere (Fig. P12.37)?

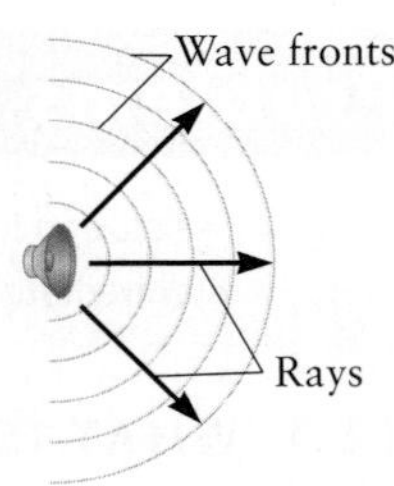

Figure P12.37

38. The intensity of sunlight at the Earth's surface is about 1000 W/m^2. What is the intensity of sunlight at the surface of Saturn? *Hint*: From Table 5.1, the distance from the Sun to the Earth is 1.50×10^{11} m, and the distance from the Sun to Saturn is 14.3×10^{11} m.

12.5 SUPERPOSITION AND INTERFERENCE

39. SSM Two waves of equal frequency are traveling in opposite directions on a string. Constructive interference is found at two spots on the string that are separated by a distance of 1.5 m. (a) What is the longest possible wavelength of the waves? (b) Give two other possible values of the wavelength.

40. Figure P12.40 shows an experiment in which sound waves generated by a singer and having a single frequency are emitted by two speakers. These sound waves interfere at a listener located at point P. If the interference at P is constructive, what is a possible value of the frequency?

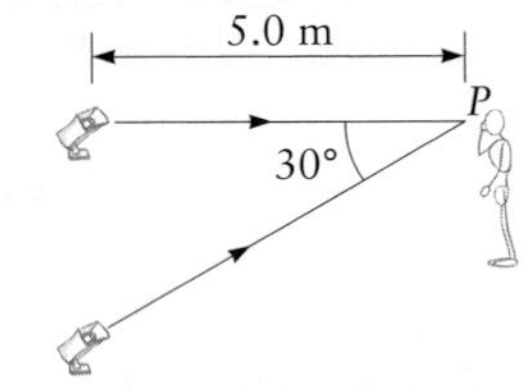

Figure P12.40

41. RT Identical sound waves of frequency f are emitted from speakers at points A and B (Fig. P12.41), which are a distance 8.5 m apart. Listeners are located at points X, Y, and Z, which are equally spaced on the line between A and B. One listener finds constructive interference of the sound waves, and the other two find destructive interference. (a) Which listener experiences constructive interference? (b) What is the lowest possible frequency at which the destructive interference in part (a) can occur? The speed of sound in air is 343 m/s.

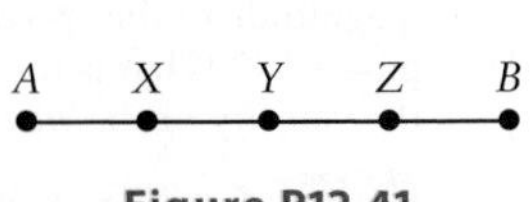

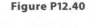

Figure P12.41

12.6 REFLECTION

42. Consider a radar system in an air traffic control center that is monitoring the distance to an approaching airplane. If the airplane is 4.0 km from the control center, what is the time delay between a transmitted wave pulse and the return of the reflected pulse?

43. Sonar is a system similar to radar except that sound is used in place of radio waves. Sonar systems are generally operated under water, where the speed of sound is approximately 1500 m/s. Suppose you are on a ship monitoring the distance to a submarine using sonar, and you find that the sonar reflections return in 24 s. How far away is the submarine?

44. SSM A radar system can measure the time delay between the transmitted and reflected wave pulses from a distant airplane with an accuracy of 1.0 picosecond. What is the accuracy in the calculated distance to the plane?

12.7 REFRACTION

45. SSM When sound waves are studied deep within the ocean, it is found that a plane wave that is initially parallel to the ocean floor will gradually bend upward and have a propagation direction that is tilted slightly toward the surface (Fig. P12.45). Does the speed of sound at these depths increase or decrease as one goes deeper within the ocean?

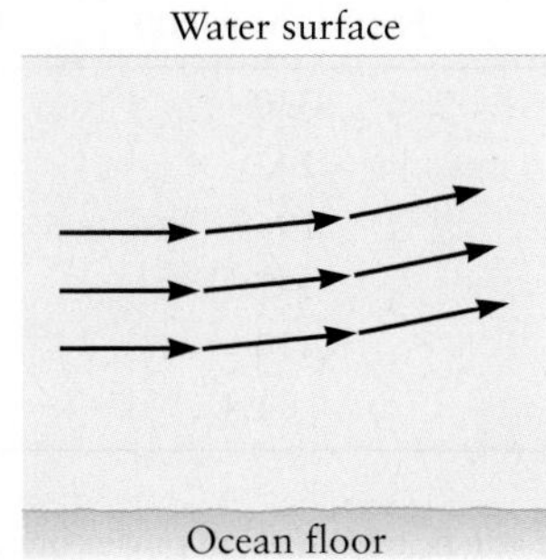

Figure P12.45

12.8 STANDING WAVES

46. What is the distance between two nodes of a standing wave?
(a) $\lambda/4$ (b) $\lambda/2$ (c) λ (d) 2λ (e) $3\lambda/2$ (f) 3λ (g) choices (c), (d), and (f) (h) choices (b) and (e)

47. What is the distance between a node and an antinode of a standing wave?
(a) $\lambda/4$ (b) $\lambda/2$ (c) λ (d) 2λ (e) $3\lambda/2$ (f) 3λ (g) choices (c), (d), and (f) (h) choices (b) and (e)

48. The antinode of a standing wave is 0.057 m away from the nearest node. What is the wavelength of the waves that combine to make the standing wave?

49. SSM ✷ Consider a guitar string that has a tension of 90 N, a length of 65 cm, and a mass per unit length of 7.9×10^{-4} kg/m. What is the fundamental frequency of the string? That is, what is the frequency of the standing wave with the lowest possible frequency? What note does this string emit when played open (i.e., without being held against a fret)? *Hint*: You may want to consult a table that gives musical notes and their corresponding frequencies for standard tuning.

50. ✪ You are studying guitar strings and are able to measure a few (but not all) of the standing wave frequencies for a particular string. If you measure frequencies of 450 Hz, 750 Hz, 900 Hz, and 1200 Hz, what is the highest possible value for the fundamental frequency?

51. ✷ A violin string is 35 cm long and has a fundamental standing wave frequency of 440 Hz. If the tension is 75 N, what is the mass of the string?

52. ✷ A vibrating string has a fundamental frequency of 250 Hz. The tension is then increased by a factor of 1.9. What is the new value of the fundamental frequency?

53. ✷ A violin string has an initial tension of 45.0 N. When playing the note A on the violin, the frequency is found to be 435 Hz. This note should actually have the frequency 440 Hz, so the player increases the tension to get the correct frequency. What is the new value of the tension? Give your answer to three significant figures.

54. ✷ A steel piano string for the note middle C is about 0.60 m long and is under a tension of 600 N. If the fundamental frequency is 262 Hz, what is the string's diameter?

55. ✷ The vibrating length of a guitar string is reduced by a factor of 2.5 when a player holds the string against a fret. What is the change in the frequency of the fundamental standing wave on the string?

56. ✷ Two nearby nodes of a standing wave are separated by 3.5 m. If the frequency of the wave is 150 Hz, what is the speed of the wave?

57. ✷ Consider a standing wave that has a frequency of 2000 Hz and a speed of 300 m/s. What is the distance between a node and the nearest antinode?

58. ✷ Figure P12.58 shows a standing wave on a string. If the frequency of the wave is 440 Hz, what is the speed of a wave on this string?

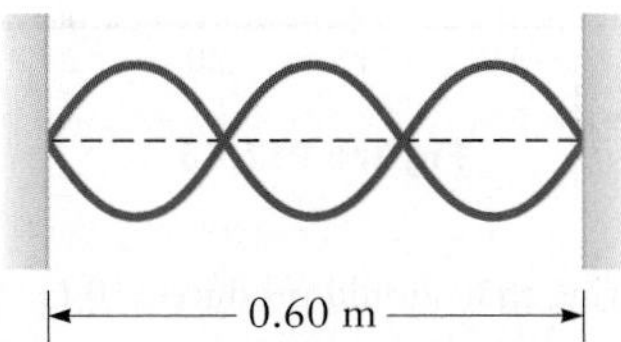

Figure P12.58

59. ✷ You want to set up a standing wave on a string that has a length of 3.5 m. You find that the lowest frequency that will work is 20 Hz. What is the speed of a wave on this string?

60. Two notes on a guitar that differ by an octave have fundamental frequencies that differ by a factor of exactly two. If these notes are played on strings that have the same diameter and tension, what is the ratio of the lengths of the strings?

61. ✷ Consider two notes on a guitar whose fundamental frequencies are in the ratio 5/4. These notes differ by a musical interval called a "major third." If these notes are played on strings that are made of the same material and have the same diameter and tension, what is the ratio of the lengths of the strings?

12.9 SEISMIC WAVES AND THE STRUCTURE OF THE EARTH

62. SSM You are operating a seismic laboratory when an earthquake occurs. Your seismographs indicate that the P wave generated by the earthquake arrives at precisely 1:55 PM, whereas the S wave arrives 5 minutes and 35 seconds later. If the speed of the P wave is 6.0 km/s and the speed of the S wave is 3.5 km/s, how far away is the epicenter of the earthquake?

63. ✪ An earthquake occurs at the South Pole. If the speed of the P waves is 6.0 km/s and the speed of the surface waves is 2.2 km/s, how long will it take these waves to reach New York City? Assume the waves travel in a straight line to New York City.

Additional Problems

64. ✷ A lift mechanic taps on an end of the main cable of a ski lift 2000 ft (610 m) long, sending a transverse pulse down the cable. He finds it takes 7.7 s for the pulse to reflect off the opposite end and return. If the steel cable is 3.8 cm in diameter, what is the tension in the cable?

65. ✪ Two cables of equal diameter and length, one made of aluminum and the other made of steel, are welded together and strung horizontally under tension. When the aluminum end is struck, a wave pulse propagates from the end at a velocity of 15 m/s. (a) The wave pulse crosses over the weld and onto the steel part of the cable. What is the velocity of the wave as it travels in the part made of steel? (b) If the total length of the cable is 10 m, how long does it take the wave pulse to reflect off the far end and return? *Hint*: Note that these are not sound waves, but transverse waves like those on a string. Also, you may want to consult Table 10.1 for the densities of aluminum and steel.

66. A wave pulse moves along a strip of rubber with a velocity of 10 cm/s as depicted in Figure P12.66. Five equally spaced points are marked on the rubber strip labeled *A* though *E* as shown. (a) Rank the points, from highest to lowest, according to the average vertical speed of each point during the next 0.50 s of the wave pulse's travel. (b) Calculate the average velocity of point *C* over the same 0.50-s time interval.

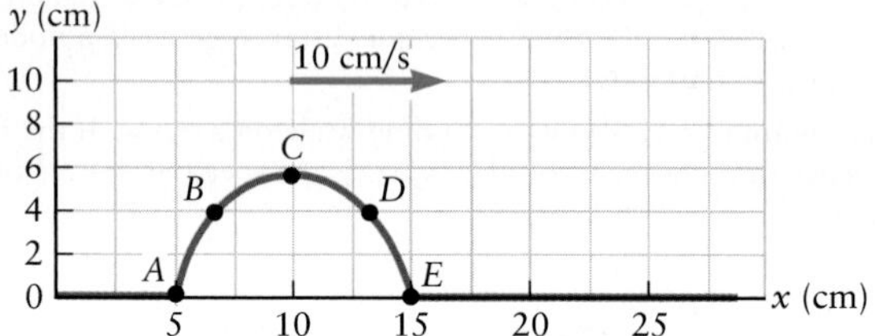

Figure P12.66

67. A guitar string that should produce 110 Hz (the note A) instead produces 105 Hz. (a) What change in tension is needed to bring the string into tune? (b) Sometimes a guitar is tuned to "drop-D" tuning, where the low E string is tuned so that it plays the note D instead. What decrease in tension is needed to bring the string from E (82.4 Hz) to D (73.4 Hz)? Express your answers as the ratio of the new tension to the old tension, and give your answers to three significant figures.

68. Two neighbors communicate with a homemade string phone (Fig. P12.68). The simple phone consists of two ends of a string (total mass of 18 g) attached to the bottom of two paper cups. The string is stretched between the two homes at a tension of 10 N over the distance of 35 m. (a) How long does it take communication to travel on the string from one home to the other? (b) How long would it take to communicate by yelling out the window? (c) How long would it take to communicate over cell phones if the nearest cell tower is 5.5 km away? (Consider only the propagation time of the radio waves and assume that the call is already connected.)

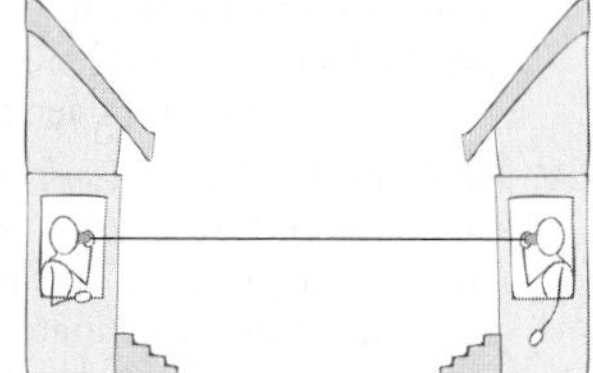

Figure P12.68

69. An explosion takes place in an underground mine 15 km away from a home located in a valley where the soil consists of a clay material with Young's modulus $Y = 2.9 \times 10^9$ Pa. The average density of the soil in that area is 3.1 g/cm^3. How long after the P waves have shaken the home does the sound of the blast finally arrive? (See also Question 13.)

70. A standard guitar is strung with six strings of increasing diameter and all of length 65 cm. Each string is to have the fundamental frequency of the notes E, B, G, D, A, amd E for standard tuning. (a) If the strings are made of a steel alloy of density $\rho = 6.3$ g/cm^3, what is the total force on the guitar neck from the combined tension in the strings? The string diameters and frequencies are listed in Table P12.70. (b) Rank the strings from highest to lowest according to the tension in each.

TABLE P12.70 Standard Tuning and Medium-Gauge Steel String Diameters for a Guitar

String	Note	Frequency (Hz)	Diameter (mm)
First	E	330	0.28
Second	B	247	0.36
Third	G	196	0.46
Fourth	D	147	0.71
Fifth	A	110	0.97
Sixth	E	82.4	1.24

71. In moderate winds, high-tension power lines sometimes "sing" or hum as a result of transverse waves on the lines that winds can generate. The cables that carry high-voltage electricity are made of aluminum with a steel core added for strength. The cross-sectional areas of the steel and aluminum in the line are 39 mm^2 and 220 mm^2, respectively. The cable is 1.8 cm in diameter and is under a tension of 4.9×10^4 N. Typically, the towers are placed about 0.20 km apart. (See Figure P12.71.) (a) What is the fundamental frequency of such a power line? (The densities of steel and aluminum are given in Table 10.1.) (b) The wind can induce vibrations when flowing around a cylinder. The frequency of the vibrations induced this way is modeled well by the equation

$$f_{\text{wind}} = \frac{1}{5}\left(\frac{v_{\text{wind}}}{d}\right)$$

where v_{wind} is the speed of the wind and d is the diameter of the cylinder. The wind will induce vibration this way only if the frequency of the induced oscillation matches a natural frequency of the power line. It is found that the power line hums when the wind blows at a steady 6.7 m/s. What harmonic of the line is this induced vibration exciting? (c) If the line is humming at the 152nd harmonic, how fast is the wind blowing?

Figure P12.71

72. SSM Figure P12.72A shows two wave pulses at time $t = 0$ s. The pulses travel at 5 cm/s in opposite directions. (a) Sketch the resulting waveform for the times $t = 1$ s, 2 s, and 3 s. (b) Figure P12.72B shows a different set of two pulses. Sketch the resulting waveform for these pulses at the times $t = 1$ s, 2 s, and 3 s.

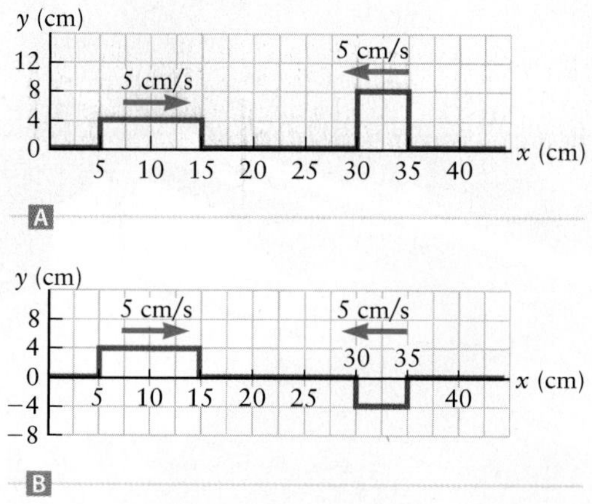

Figure P12.72

73. Wolfgang Mozart was not only an outstanding composer, but also an excellent violinist. His violin teacher was his father, Leopold, and historical records show that Leopold advised his students to tune all four strings of the violin to the same tension. A typical violin string tension is about 45 N, and the four strings of the violin (Fig. P12.73) are all 0.33 m long. Their lowest standing wave frequencies are 196 Hz (the note G), 294 Hz (D), 440 Hz (A), and 659 Hz (E). If the E string has a mass per unit length μ_E, what are the ratios μ_G/μ_E, μ_D/μ_E, and μ_A/μ_E? (Here, μ_G denotes the mass per unit length of the G string and so forth.)

Figure P12.73

74. ★ Ⓡ A mirage is created by refraction of light near the surface of the Earth. On a still day (no wind), absorption of sunlight by the Earth makes the air temperature decrease as one moves upward, away from the surface. The speed of light in air depends on the air temperature, and this variation of the speed of light causes light to refract near the surface. As a result, an observer looking along the Earth's surface (Fig. P12.74A) will see light that comes from the sky (Fig. P12.74B). Using the ray diagram in Figure P12.74A, does the existence of a mirage mean that the speed of light increases or decreases as the temperature increases?

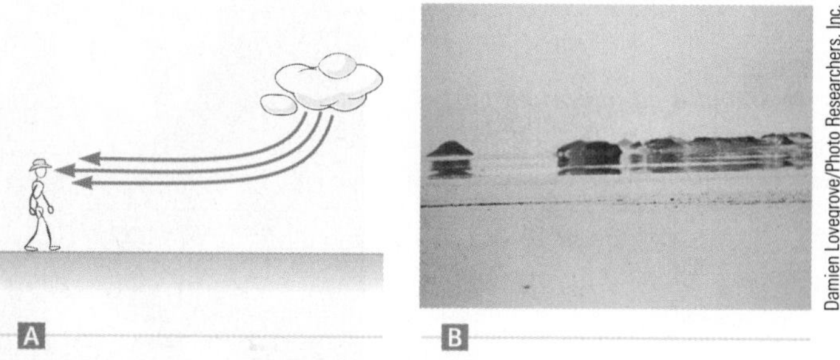

Figure P12.74

75. ★ RT Table P12.75 lists the speed of sound and the density of diamond (a crystal composed of carbon) and several metals. Some authors use the Young's modulus of a material as a measure of the material's "stiffness." According to this measure, order the materials in the table from least stiff to most stiff.

TABLE P12.75 Density and Speed of Sound in Several Materials

Material	Speed of sound (m/s)	Density (kg/m³)
Al	5,100	2,700
Cu	3,900	9,000
Diamond	12,000	10,000
Au	3,200	19,300
Fe	5,100	7,900

CHAPTER

13

▶ *Sound is an important part of everyday life. (© John Elk III/Alamy)*

Sound

OUTLINE

In Chapter 12, we discussed aspects of wave motion that are common to all types of waves. In this chapter, we focus on one particular type of wave, ***sound***. Sound is an especially important example of waves; it is central to hearing, speech, and music. We'll consider how sound waves are generated, how they propagate, and how they are detected by the ear. We also describe some medical applications of sound.

13.1 Sound Is a Longitudinal Wave

Sound waves can travel through gases, liquids, and solids. Our everyday experience is mainly with sound that travels through air, so let's begin with a molecular-scale view of sound in a gas. Air is composed mainly of oxygen and nitrogen molecules that are in constant random motion, even in "still" air. A typical air molecule at room temperature has a speed of about 500 m/s. We refer to this motion as the "background velocity" and denote this velocity by $\vec{v}_0$; the direction of this velocity is different for every molecule.

Figure 13.1 shows a tube filled with air, with a loudspeaker at one end of the tube. When the loudspeaker is turned off, the density of air molecules is the same throughout the tube (Fig. 13.1A). When the loudspeaker is turned on, a surface of

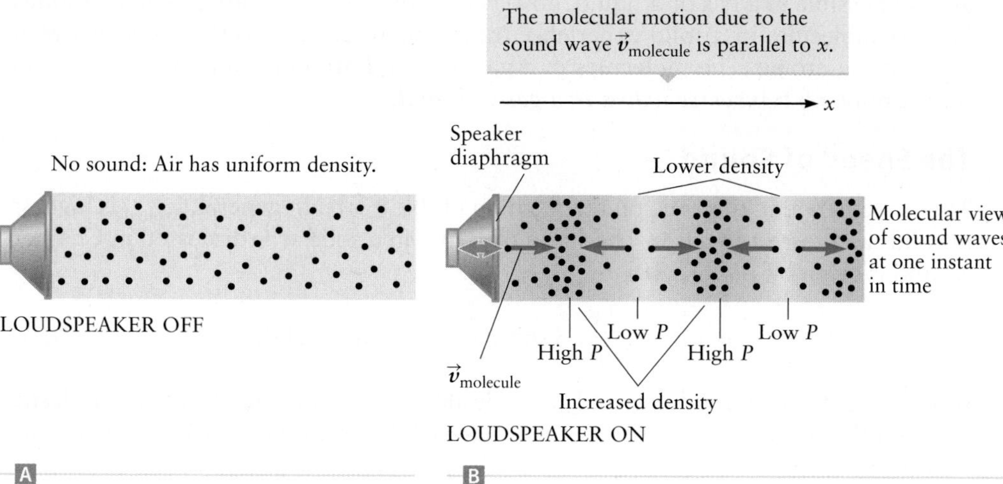

◀ **Figure 13.1** Molecular-scale view of sound in a gas. **A** An air column with no sound present. The molecules are in motion, but their velocities are in random directions and the air molecules are distributed uniformly throughout the container. **B** A sound wave generated by a loudspeaker. Each molecule now has an additional contribution to its total velocity. This contribution is denoted by $\vec{v}_{molecule}$ and is parallel to x, the direction in which the sound wave propagates.

the loudspeaker (called the diaphragm) moves back and forth along the x direction, generating a sound wave (Fig. 13.1B). As the diaphragm moves, it collides with air molecules, imparting a small extra velocity $\vec{v}_{molecule}$ to the molecules. Because the diaphragm moves back and forth along the x direction, $\vec{v}_{molecule}$ alternates along $\pm x$. At some instants, $\vec{v}_{molecule}$ at a particular location is directed to the right; at other instants, $\vec{v}_{molecule}$ at this same location is toward the left. The total velocity of a molecule is the sum of its background velocity $\vec{v}_0$ and the extra velocity $\vec{v}_{molecule}$ associated with the sound wave. This extra velocity is *not* the velocity of the sound wave; $\vec{v}_{molecule}$ is the *velocity of the medium* (the air) associated with the sound wave.

The green arrows in Figure 13.1B show $\vec{v}_{molecule}$ at several points along the wave. This snapshot only applies at a particular instant in time. Since $\vec{v}_{molecule}$ is parallel to x, the "extra" motion of the molecules—that is, the displacement of the medium—is parallel to the direction of propagation of the wave. For this reason, sound is a *longitudinal wave*.

Because the displacement alternates between the $+x$ and $-x$ directions as we move along the sound wave (along the x axis in Fig. 13.1B), the density of molecules is increased in some regions and decreased in others. Oscillations in the density cause oscillations in the pressure, so we can also view sound as a *pressure wave*.

A plot of the pressure associated with a sound wave at a particular instant in time is shown in Figure 13.2. The regions of high density and high pressure are called regions of ***condensation***. Places of low density and low pressure are called regions of ***rarefaction***. Adjacent regions of high pressure are separated by a distance equal to the wavelength λ, and adjacent regions of low pressure are separated by the same distance. The speed of the sound wave is related to its frequency and wavelength by

$$v_{sound} = f\lambda \tag{13.1}$$

It is crucial to distinguish the velocity of the wave (v_{sound} in Eq. 13.1) from the velocity $\vec{v}_{molecule}$ of particles within the medium. They are *not* the same.

Figures 13.1 and 13.2 describe a sound wave in air, and the picture is similar for sound traveling through a solid or a liquid. The sound wave causes a displacement of molecules in the solid or liquid parallel to the propagation direction, so sound is a longitudinal wave in solids and liquids as well as in gases.

In a solid, there can also be transverse waves that propagate through the solid, in which the molecular motion is perpendicular to the direction of wave propagation (such as some of the seismic waves described in Chapter 12). Such transverse waves

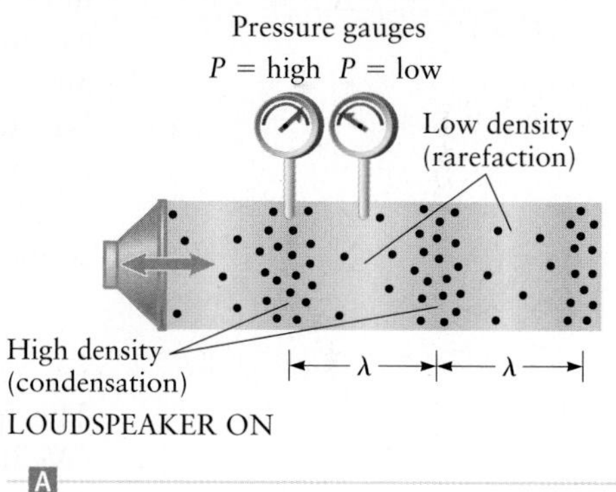

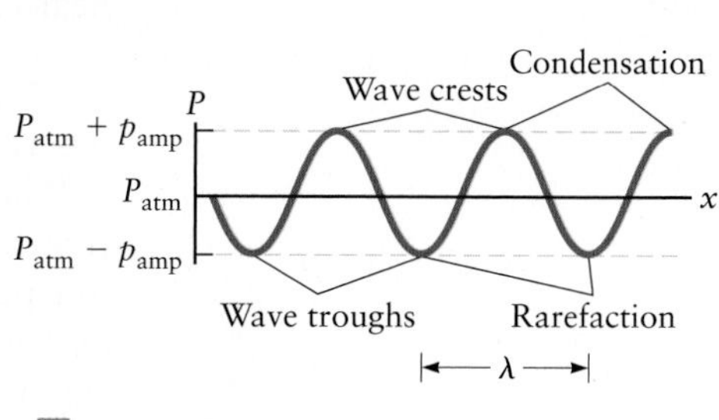

▲ **Figure 13.2** **A** The pressure in a gas is highest in the regions with the highest density. **B** For a sound wave in air, the total pressure is the sum of atmospheric pressure P_{atm} and the oscillating pressure associated with the sound wave.

TABLE 13.1 Speed of Sound in Some Common Materials

Material	Speed of Sound (m/s)
Air	343
Helium gas	1000
Hydrogen gas	1330
Oxygen gas	330
Carbon dioxide gas	269
Water	1480
Seawater	1520
Ethyl alcohol	1160
Aluminum	5100
Steel	5790
Lead	1960
Concrete	3100
Glass	5600

Note: All values are for 20°C and a pressure of 1 atm.

are *not* possible in a gas or a liquid, however. Transverse waves are possible in solids because molecules in a solid experience transverse forces due to the bonds between neighboring atoms. The molecules in a gas or liquid are not bound in this way, so there can be *no transverse waves in a gas or liquid.*

The Speed of Sound

The speed of sound depends on the medium (Table 13.1). In general, v_{sound} is highest in solids and lowest in gases. The speed of sound in a solid was discussed in Chapter 12 and is given by (Eq. 12.10)

$$v_{sound} = \sqrt{\frac{Y}{\rho}} \quad \text{(speed of sound in a solid)} \tag{13.2}$$

where Y is Young's modulus and ρ is the density. Young's modulus is an elastic constant that measures the stiffness of a material. Hence, Equation 13.2 can also be written as

$$v_{sound} = \sqrt{\frac{\text{stiffness}}{\text{density}}} \tag{13.3}$$

Equation 13.3 explains qualitatively how the speed of sound varies when we compare solids, liquids, and gases. A stiffer material (larger Young's modulus) generates a larger restoring force for a given displacement, leading to a larger particle acceleration and ultimately a greater speed of sound. Solids are stiffer than liquids, and liquids are stiffer than gases, which explains why the speed of sound is highest in solids and lowest in gases.

The density factor in the denominator in Equation 13.3 means that the speed of sound increases as the medium becomes "lighter." For example, lead has a very high density, and the speed of sound in lead is much less than that found for aluminum, which has a low density (for a metal). A similar effect is found with gases; for example, the speed of sound in helium gas is much greater than the speed of sound in air at the same pressure and temperature. This result can again be understood from Newton's second law: for a given force, a smaller mass has a larger acceleration and hence a higher speed.

The stiffness of a liquid is measured by its bulk modulus (Chapter 11), and for sound in a liquid one finds

$$v_{sound} = \sqrt{\frac{B}{\rho}} \quad \text{(speed of sound in a liquid)} \tag{13.4}$$

The speed of sound in a gas depends on its composition and also on temperature. (We'll discuss the properties of gases in Chapter 15.) The most important case is air near room temperature and pressure, for which

$$v_{sound} \approx 343 + 0.6(T - 20°\text{C})\ \text{m/s} \quad \text{(speed of sound in air)} \tag{13.5}$$

where T is the temperature in Celsius.

The general relation $v_{sound} = f\lambda$ (Eq. 13.1) connects the speed of sound with the frequency and wavelength. The speed of sound is a property of the medium and is independent of both the frequency and the amplitude of the wave. The frequency and amplitude of a sound wave depend on how the sound is generated. For example, the frequency of the sound wave in Figure 13.1 equals the oscillation frequency of the loudspeaker diaphragm. This frequency can vary over an extremely wide range, so there is no absolute lowest or highest frequency for a sound wave. The range of normal human hearing extends from approximately 20 Hz to about 20,000 Hz. These limits are only approximate and vary from person to person. The upper limit also tends to decrease with age due to a gradual stiffening of the eardrum and the other flexible connections within the ear. Sound waves can have both higher and lower frequencies than this 20–20,000 Hz range. Frequencies below 20 Hz are called ***infrasonic*** and can be generated and detected by large animals such as elephants. People can sometimes sense such very low frequencies as vibrations rather than as sound detected by the ears. Frequencies above 20,000 Hz are called ***ultrasonic*** and can be heard by many species, such as dogs and bats.

Musical Tones and Pitch

A sound wave described by a single frequency is called a ***pure tone***. Although it is possible to generate a pure tone, most sounds are a combination of many pure tones. For example, the sounds associated with speech or with the banging of a drum are complex combinations of pure tones.

In Section 12.8, we discussed standing waves on a guitar string and saw that they have frequencies given by $f_1, 2f_1, 3f_1, \ldots$. This pattern is called a ***harmonic sequence***, and f_1 is called the ***fundamental frequency***. Many sounds are a combination of frequencies that are harmonically related. For example, with an acoustic guitar, the standing wave vibrations of a string are transferred to the body of the guitar, and the vibrating surfaces of the guitar act like the loudspeaker diaphragm in Figure 13.1. The result is a combination tone containing frequencies $f_1, 2f_1, 3f_1, \ldots$.

One way to characterize such a combination tone is by a property called ***pitch***. Pitch is a subjective quality that depends on how the human brain processes sound. This processing is complicated and is still not completely understood. Under most conditions, the pitch of a pure tone is equal to its frequency, whereas the pitch of a combination of harmonically related tones (as would be produced by a plucked guitar string) is associated with the fundamental frequency. More complex sounds, as might be produced by an orchestra, are usually interpreted by the brain as a collection of tones with different pitches.

EXAMPLE 13.1 The Wavelength of Sound in Air

The frequency range that can normally be detected by a human ear is 20–20,000 Hz. What is the corresponding range of wavelengths for sound in air?

RECOGNIZE THE PRINCIPLE

The wavelength is related to the speed of sound and the frequency through $v_{\text{sound}} = f\lambda$ (Eq. 13.1). Given v_{sound} and f, we can thus find λ.

SKETCH THE PROBLEM

No figure is necessary.

IDENTIFY THE RELATIONSHIPS

Solving for the wavelength in terms of the speed of sound and the frequency gives

$$\lambda = \frac{v_{\text{sound}}}{f}$$

Here v_{sound} is the speed of sound in air at room temperature; according to Table 13.1 and Equation 13.5, $v_{\text{sound}} = 343$ m/s.

SOLVE

For the lower end of the given frequency range, we have

$$\lambda = \frac{v_{\text{sound}}}{f} = \frac{343 \text{ m/s}}{20 \text{ Hz}}$$

$$\lambda = 17 \frac{\text{m/s}}{\text{Hz}} = 17 \frac{\text{m/s}}{1/\text{s}} = \boxed{17 \text{ m}}$$

(continued) ▶

For the high-frequency end of the range of human hearing,

$$\lambda = \frac{v_{\text{sound}}}{f} = \frac{343 \text{ m/s}}{20{,}000 \text{ Hz}} = 0.017 \text{ m} = \boxed{1.7 \text{ cm}}$$

► ***What does it mean?***
The ear canal is a few centimeters in length, which is comparable to the wavelength of sound at the high-frequency end of the range that humans can hear.

13.2 Amplitude and Intensity of a Sound Wave

The oscillating pressure associated with a sound wave is sketched in Figure 13.2B, where the amplitude of the pressure oscillation is denoted p_{amp}. When a sound wave reaches your ear, the pressure on the outside of your eardrum is the sum of atmospheric pressure plus the oscillating sound pressure, $P_{\text{atm}} + p$, whereas the pressure inside your eardrum is P_{atm} (Fig. 13.3). The total force on your eardrum equals the difference in these two pressures multiplied by the area of the eardrum A. Hence, the total force equals pA, and your ear detects the sound pressure p while being insensitive to P_{atm}.

The ***intensity*** I of the sound wave equals the power carried by the wave per unit area of wave front. We showed in Chapter 12 that the intensity of a wave is proportional to the square of the amplitude, so for sound we have

Intensity is proportional to the square of the amplitude.

$$I \propto p_{\text{amp}}^2 \tag{13.6}$$

In many cases, sound waves propagate outward in all directions away from their source. In Section 12.4, we discussed the spherical wave fronts of such waves and found that the intensity at a distance r from the source varies as

$$I \propto \frac{1}{r^2} \tag{13.7}$$

Combining this expression with Equation 13.6, we see that the pressure amplitude of a spherical sound wave falls with distance as $p_{\text{amp}}^2 \propto 1/r^2$, which is equivalent to

$$p_{\text{amp}} \propto \frac{1}{r} \tag{13.8}$$

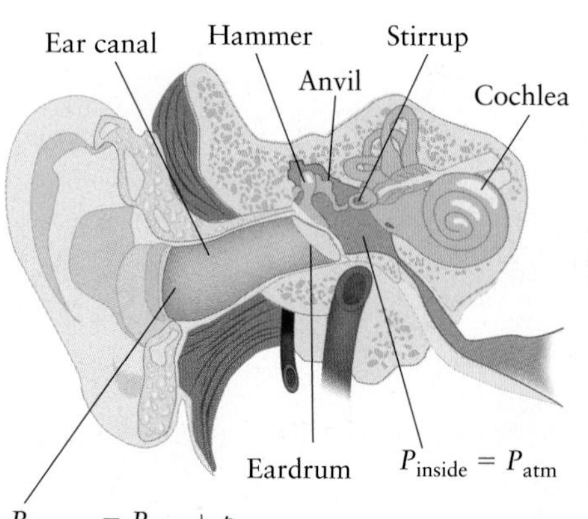

▲ **Figure 13.3** Your ear detects sound through the pressure on the eardrum. The oscillating pressure of the sound wave produces an oscillating force on the eardrum.

CONCEPT CHECK 13.1 The Intensity of a Sound Wave Falls Off with Increasing Distance from the Source

A chirping bird sits at the top of a tree, and the chirping sound has an intensity I_0 at a certain distance from the bird. If the distance to the bird is increased by a factor of three, is the new intensity (a) $I_0/3$, (b) $I_0/9$, or (c) I_0?

Decibels

The sensitivity of the ear depends on frequency. In the most favorable range of frequencies, the ear can detect sound intensities as small as about 10^{-12} W/m^2. Because values this small are rather cumbersome to write, another unit, the ***decibel*** (abbreviated dB), is widely used to measure sound intensity. This unit is computed in the following way. Suppose the intensity of a sound wave as measured in watts per square meter is I. We now introduce a constant $I_0 \equiv 1.00 \times 10^{-12}$ W/m^2, which is approximately the lowest intensity detectable by the human ear. We then define a quantity called the ***sound intensity level*** β (the Greek letter beta) by

Threshold for human hearing and the decibel scale

$$\beta = 10\log\left(\frac{I}{I_0}\right) \quad \text{(sound intensity level in decibels)} \tag{13.9}$$

The intensity level β is given in decibels, but you can see from Equation 13.9 that since β is defined through a ratio of intensities, it is, in a strict mathematical sense, unitless! When working with β and Equation 13.9, it is helpful to recall the properties of common (base 10) logarithms as reviewed in Appendix B. In particular, if $a = 10^b$, then $b = \log(a)$. We also have $\log(1) = 0$, $\log(10) = 1$, $\log(100) = 2$, and so forth.

To see how the definition of sound intensity level in Equation 13.9 is applied, consider a sound that is just at the threshold of human hearing, that is, a sound with intensity $I = 1.0 \times 10^{-12}$ W/m². The intensity level in decibels is then

$$\beta = 10\log\left(\frac{I}{I_0}\right) = 10\log\left(\frac{1.0 \times 10^{-12}\ \text{W/m}^2}{1.00 \times 10^{-12}\ \text{W/m}^2}\right) = 10\log(1)$$

Because $\log(1) = 0$, we find

$$\beta = 0\ \text{dB}$$

The intensity of this sound is thus 0 dB = "zero decibels." An intensity level of $\beta = 0$ dB does *not* mean that there is an absence of sound. It merely indicates that the intensity I is equal to the reference intensity I_0, which corresponds to a very faint sound, a sound at the limit of human hearing.

Now suppose we have another sound of intensity $I = 1.0 \times 10^{-11}\ \text{W/m}^2 = 10 \times I_0$, which is 10 times the nominal threshold for human hearing. The intensity level of this sound as measured in decibels is

$$\beta = 10\log\left(\frac{I}{I_0}\right) = 10\log\left(\frac{10 \times I_0}{I_0}\right) = 10\log(10)$$

Because $\log(10) = \log(10^1) = 1$, we get

$$\beta = 10\ \text{dB}$$

which is the result for a sound that has an intensity 10 times the threshold for hearing. If a sound has an intensity 100 times this threshold ($I = 1.0 \times 10^{-10}$ W/m²), we have

$$\beta = 10\log\left(\frac{I}{I_0}\right) = 10\log\left(\frac{100 \times I_0}{I_0}\right) = 10\log(100)$$

Because $\log(100) = 2$, we get

$$\beta = 20\ \text{dB}$$

Thus, for every increase in intensity by a factor of 10, the intensity level as measured in decibels increases by 10 dB. The intensity levels of some familiar sounds are listed in Table 13.2.

CONCEPT CHECK 13.2 Sound Intensity Level and Decibels

What is the intensity in watts per square meter of a sound with a sound intensity level of 50 dB?

(a) 1.0×10^{-12} W/m²
(b) 1.0×10^{-10} W/m²
(c) 1.0×10^{-7} W/m²
(d) 5.0×10^{-12} W/m²

TABLE 13.2 Intensity Level of Some Common Sounds

Sound	Intensity Level (dB)	Sound	Intensity Level (dB)
Threshold of pain for people	130	City traffic	70
Jet takeoff at a distance of 100 m	90–120	Normal conversation	60–70
Rock concert	110–120	Whisper	20
Thunder (nearby)	110	Pin dropping	10
Factory noise	80	Threshold of human hearing	0

CONCEPT CHECK 13.3 How Loud Is It?

The sound intensity level at a rock concert can be as high as 120 dB, whereas normal speech corresponds to about 70 dB. What is the ratio of the intensities of these two sounds?

(a) 190 (b) 50 (c) 5 (d) 10^5

Human Perception of Sound

Decibels are based on a ratio involving I_0 and logarithms, and you might now wonder why this way of measuring sound intensity was invented. The motivation comes from a feature of human hearing illustrated in Figure 13.4, which shows several contour lines obtained from extensive studies of human hearing. The lowest line plots the intensity of the minimum detectable sound as a function of the sound frequency. In the range of 500 Hz to 5000 Hz, this curve is roughly constant (horizontal) with a value close to $I_0 = 1.00 \times 10^{-12}$ W/m^2. (See the right-hand scale in Fig. 13.4.) When expressed in terms of intensity level in decibels (the left-hand scale), this curve corresponds to approximately 0 dB. The other contour curves show how the intensity varies for a fixed value of the *perceived* loudness, with each successive contour curve corresponding to an increase by a factor of two in the loudness. These results were obtained by asking many people to judge sounds of different loudness, and it was found that doubling the perceived loudness corresponds to moving from one contour to the next. Figure 13.4 shows averages obtained from many different individuals, and these contour curves represent normal human hearing. Loudness is a subjective quality that depends on how the ear functions at different frequencies, so it is fascinating that different people can agree on a loudness scale!

Over the range of 250 Hz to 8000 Hz, the contours in Figure 13.4 have intensity levels β that differ by about 10 dB. Hence, a doubling of the perceived loudness corresponds to about a 10-dB increase in the intensity level, regardless of the initial loudness. That is why the sound intensity level measured in decibels with its logarithmic definition is a very useful way to describe the intensity of a sound.

The highest contour curve in Figure 13.4 has a loudness at which the ear is damaged. Such damage occurs at an intensity level $\beta \approx 120$ dB, which is approximately 120 dB larger than the threshold for hearing. Each change in the intensity level by 10 dB corresponds to a factor of 10 change in the intensity, so an increase of β by

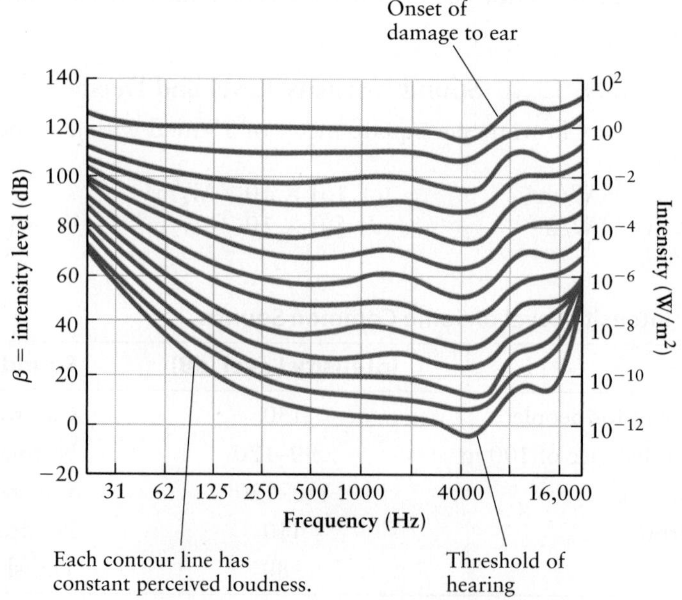

► **Figure 13.4** Contours of constant perceived loudness as a function of frequency for normal human hearing.

120 dB corresponds to an increase in the intensity by a factor of 10^{12}! Hence, as a mechanical device, the ear is able to function over a very wide range of forces.

Another important feature of Figure 13.4 is that the ear's sensitivity to sound pressure decreases at low frequencies. For example, to have the same perceived loudness (and thus fall on the same contour curve), a sound at 60 Hz must have a larger intensity than a sound at 1000 Hz.

The Ear as a Pressure Detector

Quantitative measurements of sound often involve values of the intensity, but it is also interesting to consider the corresponding values of the sound pressure. Intensity is proportional to the square of the pressure amplitude (Eq. 13.6), so there is a direct relation between the threshold intensity I_0 for human hearing and the smallest pressure oscillation that can be detected by the ear. A sound intensity $I_0 = 1.00 \times 10^{-12}$ W/m^2 corresponds to a pressure amplitude $p_0 = 2 \times 10^{-5}$ Pa. (This result can be obtained by applying Newton's laws to calculate the molecular motion of a sound wave, but the derivation is more than we can include here.) Hence, the smallest pressure amplitude that can be detected by the human ear is about $p_{amp} = p_0 = 2 \times 10^{-5}$ Pa. Since atmospheric pressure is $P_{atm} = 1.01 \times 10^5$ Pa, the ear is therefore able to detect variations in pressure that are $p_0/P_{atm} = (2 \times 10^{-5} \text{ Pa})/(1.01 \times 10^5 \text{ Pa}) = 2 \times 10^{-10}$ times smaller than atmospheric pressure! The ear is thus an extremely impressive mechanical device.

EXAMPLE 13.2 The Pressure Amplitude at a Rock Concert

According to Table 13.2, the sound intensity level at a rock concert can reach values as high as 120 dB. What is the approximate value of the corresponding pressure amplitude? The pressure amplitude at $\beta = 0$ dB is $p_0 = 2 \times 10^{-5}$ Pa.

RECOGNIZE THE PRINCIPLE

We first use the definition of decibels (Eq. 13.9) to convert the given sound intensity level β to an intensity value I. We can then use the relationship between I and p_{amp} to find the corresponding pressure amplitude.

SKETCH THE PROBLEM

No figure is necessary.

IDENTIFY THE RELATIONSHIPS

Using Equation 13.9, we can convert the intensity level at a rock concert ($\beta = 120$ dB) into intensity I in units of W/m^2. We have

$$\beta = 120 = 10 \log\left(\frac{I_{rc}}{I_0}\right)$$

where I_{rc} is the sound intensity at the rock concert (in watts per square meter) and I_0 is known to be 1.00×10^{-12} W/m^2. This leads to

$$\log\left(\frac{I_{rc}}{I_0}\right) = 12 \qquad (1)$$

According to the properties of logarithms, if $\log(a) = b$, then $a = 10^b$. Applying this fact to Equation (1), we get

$$\frac{I_{rc}}{I_0} = 10^{12}$$

$$I_{rc} = 10^{12} I_0 \qquad (2)$$

(continued) ▶

Now use the relation between the intensity I and pressure amplitude p_{amp}. At an intensity I_0 (corresponding to 0 dB), the pressure amplitude is $p_0 = 2 \times 10^{-5}$ Pa. We also know that intensity is proportional to the *square* of the pressure amplitude (Eq. 13.6). We can therefore replace the intensities in Equation (2) by the squares of the corresponding pressure amplitudes and write

$$p_{rc}^2 = 10^{12} p_0^2$$

where p_{rc} is the pressure amplitude at the rock concert.

SOLVE

Inserting the value of p_0 leads to

$$p_{rc} = \sqrt{10^{12}}\, p_0 = (10^6)(2 \times 10^{-5}\ \text{Pa}) = \boxed{20\ \text{Pa}}$$

▸ *What does it mean?*

Although a rock concert produces quite a loud sound, the resulting pressure amplitude is still only a very tiny fraction of atmospheric pressure ($P_{atm} = 1.01 \times 10^5$ Pa).

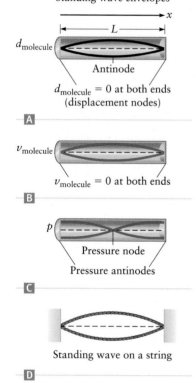

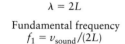

▲ **Figure 13.5** Standing sound waves in a pipe that is closed at both ends. **A** Molecular displacement, **B** velocity, and **C** sound pressure for the lowest-frequency standing wave. This is the fundamental frequency of the pipe. **D** Displacement for a standing wave on a string; it is identical to the molecular displacement in part A.

13.3 Standing Sound Waves

In Figures 13.1 and 13.2, we considered sound waves that travel in one direction inside a tube. We next want to consider what happens when these waves reach the end of the tube. This is the key to understanding how many musical instruments work, including flutes, trombones, and organs, and it is very similar to the case of standing waves on a guitar string studied in Chapter 12.

Standing Waves in a Pipe Closed at Both Ends

A standing wave is created when two waves travel in opposite directions in the same medium. As a first example, consider sound waves propagating back and forth in a pipe that is closed at both ends. A standing wave is possible only for certain wavelengths; some number of sound waves must "fit into" or match the length of the pipe. To understand this matching process, we must consider how a sound wave behaves at the ends of the pipe. Because sound is a longitudinal wave, the motion of the molecules would normally be directed along the pipe (the x direction in Fig. 13.5). The wall prevents this motion, however; for example, the right-hand wall in Figure 13.5 prevents the adjacent molecules from moving to the right. These molecules will also not move to the left (away from the wall) because that would produce a low-pressure region that pulls them back, so the velocity of the air molecules ($\vec{v}_{molecule}$ in Fig. 13.1) is zero at a wall, that is, at the closed end of a pipe. Because $\vec{v}_{molecule} = 0$ at a wall, the displacement of the molecules $d_{molecule}$ is also zero at the ends of the pipe.

A standing sound wave must therefore have zero displacement at both ends of the pipe, leading to the results in Figure 13.5. Figure 13.5A shows the displacement pattern ($d_{molecule}$) for a possible standing wave. These curves are the standing wave envelopes; that is, they show how the amplitude of the oscillation varies with position along the pipe. Places where the displacement is zero are called displacement ***nodes***, while the displacement is largest at the displacement ***antinodes.*** The wavelength of this standing wave is $2L$, where L is the length of the pipe. The standing wave pattern in Figure 13.5A is identical to the pattern found for a standing wave on a string (Fig. 13.5D; compare with Fig. 12.24B). Both have nodes at the ends of a pipe or string and an antinode in the center. The molecular velocity forms a standing wave described by the same pattern (Fig. 13.5B). The amplitude of the molecular velocity is zero at the ends of the pipe and large at the center.

A sketch of the sound pressure p is shown in Figure 13.5C. The behavior of the pressure at the ends is different from that of the molecular velocity and displacement. We have already mentioned that the air molecules cannot move past the ends of the pipe (because of the walls). During a portion of the wave's cycle, air molecules that are initially in the interior of the pipe move toward a wall, which increases the pressure at the wall to a large value because the molecules are "trapped" there. During the opposite portion of the wave's cycle, these air molecules move away from the wall, producing a low pressure at the wall. As a result, pressure oscillations are large at a closed end, forming a *pressure antinode* in the standing wave (Fig. 13.5C). Hence, both the displacement and pressure form standing waves, with the nodes of displacement located at the antinodes of p and vice versa.

Other possible standing wave patterns for the displacement and pressure are shown in Figure 13.6 for standing waves with $\lambda = L$ and $\lambda = 2L/3$. These standing wave patterns are identical to those found for a wave on a string (Fig. 12.24).

There is a mathematical pattern to the standing waves. Figures 13.5 and 13.6 show the three standing waves with the longest possible wavelengths: $\lambda = 2L$, L, and $2L/3$. There are many more possible values of the wavelength (an infinite number), and they follow the pattern

$$\lambda_n = 2L,\ L,\ \frac{2L}{3},\ \frac{2L}{4},\ \frac{2L}{5},\ldots$$

or

$$\lambda_n = \frac{2L}{n} \quad \text{with } n = 1, 2, 3, \ldots \text{ (standing waves in a pipe closed at both ends)} \tag{13.10}$$

Since frequency and wavelength are connected by $v_{\text{sound}} = f\lambda$, the frequencies of these standing waves are

$$f_n = n\frac{v_{\text{sound}}}{2L} \tag{13.11}$$

$$f_n = nf_1 \quad \text{with } n = 1, 2, 3, \ldots \text{ (standing waves in a pipe closed at both ends)} \tag{13.12}$$

where $f_1 = v_{\text{sound}}/(2L)$ is called the ***fundamental frequency*** (also known as the first harmonic). The standing wave with a frequency $f_2 = 2f_1$ is called the second harmonic, the standing wave with frequency f_3 is called the third harmonic, and so on.

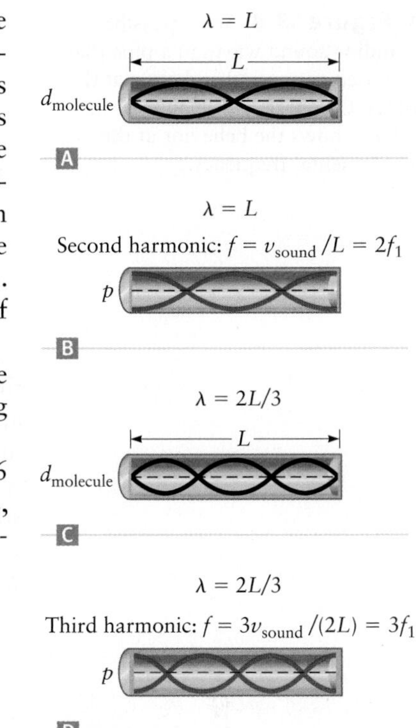

▲ **Figure 13.6** A and B: Displacement and pressure for a standing wave in a pipe closed at both ends, at the frequency of the second harmonic. C and D: Displacement and pressure at the frequency of the third harmonic.

Standing Waves in a Pipe Open at One End and Closed at the Other

Many musical instruments, including trumpets and clarinets, have an opening at one end. Let's therefore consider standing waves in a pipe that is *open* at one end and closed at the other. To analyze this situation, we must consider what happens to a sound wave at the open end of the pipe. We have already seen that there is a displacement node and a pressure antinode at the closed end of a pipe (Fig. 13.5). The behavior at an open end is quite different. At an open end, a pipe connects to the rest of the room, where the pressure is equal to atmospheric pressure P_{atm}. This outside volume is much larger than the volume of air in the pipe, so the flow of air molecules out of or into the pipe due to a sound wave has a negligible effect on the outside pressure. The pressure at the open end of the pipe is therefore held fixed at P_{atm} by the large body of air in the room, and the pressure amplitude of a standing sound wave at an open end must be $p_{\text{amp}} = 0$. Hence, there is a pressure *node* at the open end of the pipe.

Figures 13.7 and 13.8 show some of the possible standing waves for a pipe open at one end and closed at the other. There is a pressure node at the open end and (as we have seen in Fig. 13.5C) a pressure antinode at the closed end. The behaviors of

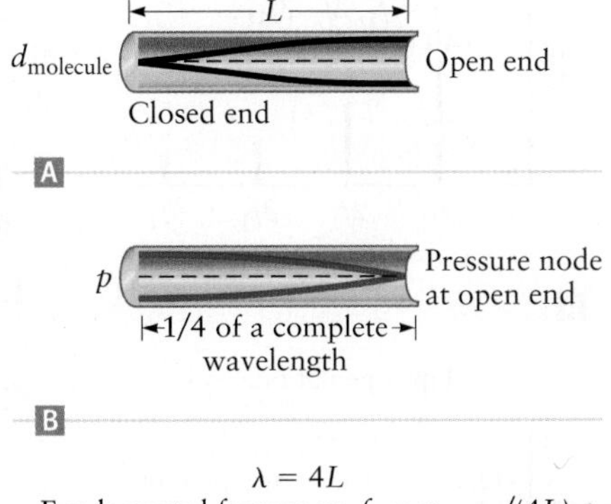

▲ **Figure 13.7** Standing sound waves in a pipe that is closed at one end and open at the other. A Molecular displacement and B pressure at the fundamental frequency of the pipe. There is a pressure node at the open end, which makes the standing wave frequencies different from those found in a pipe that is closed at both ends (Figs. 13.5 and 13.6).

▶ **Figure 13.8** Some possible standing sound waves in a pipe that is closed at one end and open at the other. Compare with Figure 13.7, which shows the behavior at the fundamental frequency.

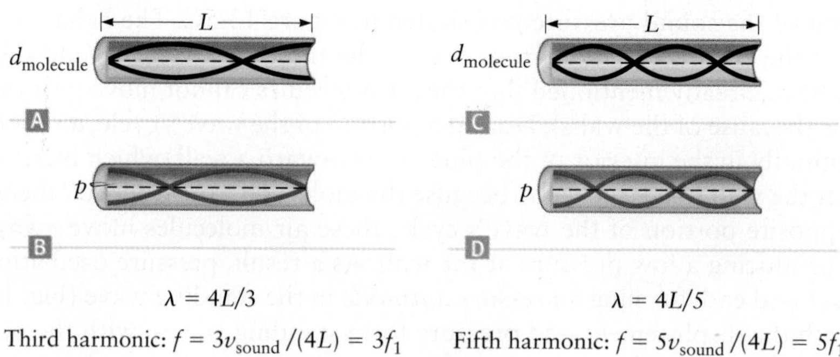

the sound pressure and displacement are linked, so there is a displacement antinode at the open end and a displacement node at the closed end. Standing sound waves must always satisfy these conditions at the ends of such a pipe. The standing wave with the longest possible wavelength is shown in Figure 13.7; it has $\lambda = 4L$. The next standing waves (in order of decreasing wavelength) are shown in Figure 13.8; they have wavelengths $\lambda = 4L/3$ and $\lambda = 4L/5$. These and the other possible standing waves for this tube follow the pattern

$$\lambda_n = 4L, \frac{4L}{3}, \frac{4L}{5}, \frac{4L}{7}, \ldots$$

$$\lambda_n = \frac{4L}{n} \quad \text{with } n = 1, 3, 5, \ldots \quad \text{(standing waves in a pipe open at one end)} \tag{13.13}$$

The frequencies of these standing waves are $f_n = v_{sound}/\lambda_n$, so

$$f_n = n\frac{v_{sound}}{4L} \tag{13.14}$$

$$f_n = nf_1 \quad \text{with } n = 1, 3, 5, \ldots$$
$$\text{(standing waves in a pipe open at one end)} \tag{13.15}$$

and the fundamental frequency is $f_1 = v_{sound}/(4L)$. Comparing with Equations 13.11 and 13.12, the fundamental frequency is lower for the pipe with one end open (if we compare two pipes of the same length L), and the pattern of harmonics is also different. For the pipe with both ends closed, the allowed frequencies form the sequence f_1, $2f_1$, $3f_1$, . . . , whereas with one end open, the sequence is f_1, $3f_1$, $5f_1$, ... Hence, the even-numbered harmonics are missing for a pipe that is open at one end. Figure 13.9 shows the pattern of standing wave frequencies in the two cases.

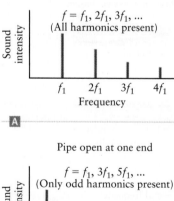

Pipe closed at both ends

$f = f_1, 2f_1, 3f_1, \ldots$
(All harmonics present)

Sound intensity

f_1 $2f_1$ $3f_1$ $4f_1$
Frequency

A

Pipe open at one end

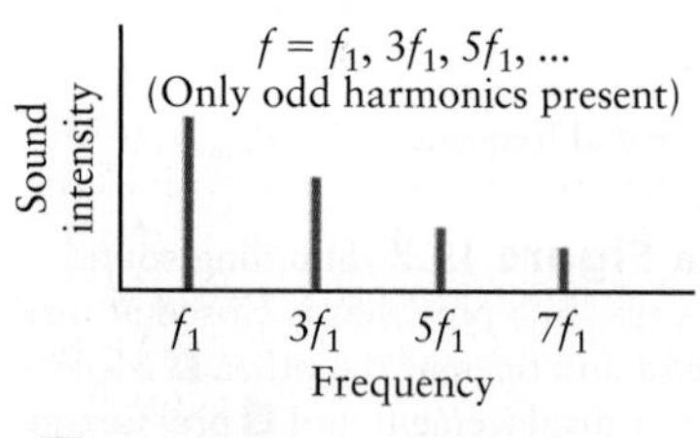

B

▲ **Figure 13.9** The pattern of standing wave frequencies in a pipe depends on whether the ends of the pipe are open or closed.

The allowed frequencies of standing sound waves in pipes can be predicted using three main facts:

1. The *closed* end of a pipe is a pressure *antinode* for the standing wave.
2. The *open* end of a pipe is a pressure *node* for the standing wave.
3. A pressure node is always a displacement antinode, and a pressure antinode is always a displacement node.

Composition of a Real Musical Tone

Pipes like those shown schematically in Figures 13.5 through 13.8 are the basis of many musical instruments, including flutes, trumpets, and organs. The sound we hear originates as a standing sound wave inside the instrument. This standing wave does not have a single frequency, but is a combination (a *superposition*) of standing waves with different frequencies. The allowed standing wave frequencies follow a pattern such as Equation 13.11 or 13.14, depending on whether the ends of the instrument are open or closed. Hence, these instruments produce combination tones as described above in our discussion of pitch. These combination tones are often said to have a property called ***timbre***, or ***tone color***. The musical timbre of a tone depends on the mix of frequencies (as in Fig. 13.9) and on the relative amplitudes of these different frequency components. The timbre is different for different instruments, and it also depends on how the instrument is played. (A "wailing" saxophone will have a different tone color than one played less forcefully.) These variations in timbre help make musical tones interesting to a listener.

EXAMPLE 13.3 The Lowest Note of a Clarinet

A clarinet is a musical instrument we can model as a pipe open at one end and closed at the other (Fig. 13.10A). A standard clarinet is approximately $L = 66$ cm long. What is the frequency of the lowest tone such a clarinet can produce?

RECOGNIZE THE PRINCIPLE

According to our work with pipes, there must be a pressure antinode at the closed end of the clarinet and a pressure node at the open end. Frequency and wavelength are related by $v_{sound} = f\lambda$, so the standing wave with the lowest frequency will be the one with the longest wavelength.

SKETCH THE PROBLEM

Figure 13.10B shows the clarinet modeled as a pipe open at one end. The allowed standing waves will be just like those found in Figures 13.7 and 13.8.

IDENTIFY THE RELATIONSHIPS

The standing wave with the longest wavelength is sketched in Figure 13.7. Its frequency is given by Equation 13.14, with $n = 1$:

$$f_1 = \frac{v_{sound}}{4L}$$

This is the fundamental frequency for the clarinet with all tone holes closed.

SOLVE

The speed of sound in air (at room temperature) is $v_{sound} = 343$ m/s and the length of the tube is $L = 0.66$ m, so we have

$$f_1 = \frac{v_{sound}}{4L} = \frac{343 \text{ m/s}}{4(0.66 \text{ m})} = \boxed{130 \text{ Hz}}$$

► *What does it mean?*

This result is the lowest frequency—that is, the lowest "note"—a clarinet can produce according to our simple model. To play other notes, the musician uncovers one of the many different tone holes along the body of the clarinet. If an open hole is sufficiently large, the pressure node moves from the end of the clarinet to a spot very near the tone hole (Fig. 13.10C), thereby shortening the wavelength of the fundamental standing wave and increasing f_1. In this way, a musician can use the many tone holes to produce many different notes.

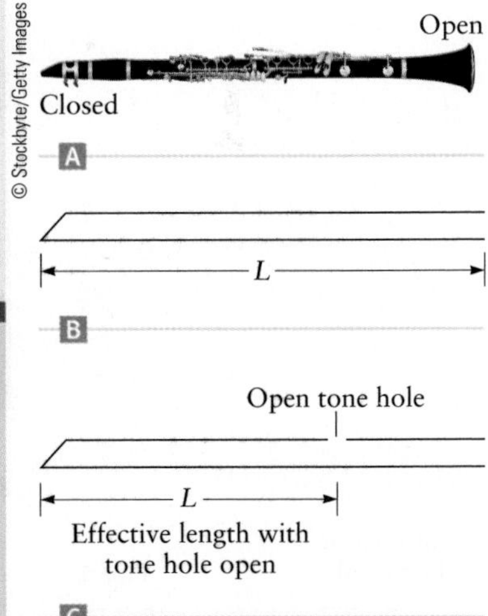

▲ **Figure 13.10** Example 13.3. A A clarinet. B Physicist's model of a clarinet, as a pipe closed at one end and open at the other. The standing waves will be the same as shown in Figures 13.7 and 13.8. C If a tone hole is opened, it will act as the open end of the pipe as far as the standing wave is concerned, changing the frequency of the standing waves.

CONCEPT CHECK 13.4 Designing a Clarinet

Suppose you design a new type of clarinet that is half the length of the ordinary clarinet in Example 13.3. How will this change affect the frequency of the lowest note on this new instrument?

(a) The frequency of the lowest note will decrease by a factor of two.
(b) The frequency of the lowest note will increase by a factor of two.
(c) The frequency of the lowest note will not change.

Pipe of length L open at both ends

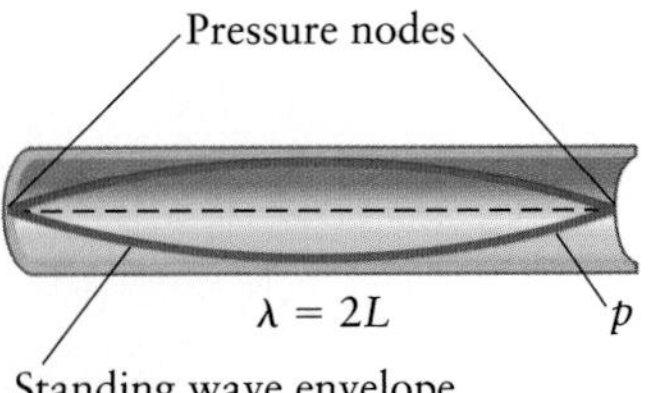

▲ **Figure 13.11** Example 13.4.

EXAMPLE 13.4 Standing Waves in a Pipe Open at Both Ends

Consider a hypothetical musical instrument that uses a pipe open at both ends. If the pipe is $L = 1.0$ m long, what is the longest wavelength that can produce a standing wave? Sketch how the pressure of this standing wave varies along the tube.

RECOGNIZE THE PRINCIPLE

A pipe open at both ends must have pressure nodes at both ends as explained on page 421.

SKETCH THE PROBLEM

Figure 13.11 shows the problem.

IDENTIFY THE RELATIONSHIPS AND SOLVE

The standing wave with the longest wavelength is sketched in Figure 13.11. For a pipe of length L, the wavelength is $\lambda = 2L$. We thus have

$$\lambda = 2L = 2(1.0 \text{ m}) = \boxed{2.0 \text{ m}}$$

▶ *What does it mean?*

Comparing Figure 13.11 with the standing wave pattern for a pipe with both ends closed (Fig. 13.5), we find that the fundamental frequency is the same for these two cases.

CONCEPT CHECK 13.5 Playing a Funky Organ

An organ uses standing sound waves in pipes of different lengths to produce different notes. Your friend is an organist, and you play a trick on him by filling his organ pipes with helium gas. The speed of sound in helium is higher than the speed of sound in air by a factor of $v_{\text{helium}}/v_{\text{air}} = 2.9$. (See Table 13.1.) How will your prank affect the tones produced by the organ?

(a) The notes will all go up in frequency by this factor.
(b) The notes will all go down in frequency by this factor.
(c) There will be no change.

An experiment similar to this one is often performed at carnivals, when a performer fills his vocal tract with helium gas and then demonstrates his "new" voice. The pitch of the human voice is determined by the frequencies of standing waves in the vocal tract.

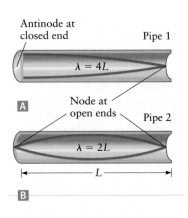

Antinode at closed end

Pipe 1

λ = 4L

A

Node at open ends

Pipe 2

λ = 2L

L

B

▲ **Figure 13.12** Example 13.5.

EXAMPLE 13.5 Two Organ Pipes

Two organ pipes have the same length and are both filled with air. Pipe 1 is open at one end and closed at the other, while pipe 2 is open at both ends. Which has the lowest fundamental frequency?

RECOGNIZE THE PRINCIPLE

For standing sound waves in a pipe, there must be a pressure antinode at a closed end and a pressure node at an open end. Using these facts, we can determine the fundamental frequencies for each pipe.

SKETCH THE PROBLEM

Figure 13.12 shows the standing wave envelopes for the pressure for each pipe.

IDENTIFY THE RELATIONSHIPS AND SOLVE

Figure 13.12A shows the standing wave envelope with the longest wavelength for pipe 1, and Figure 13.12B shows the envelope for pipe 2. Notice the pressure antinode at the closed end and pressure nodes at the open ends. The wavelength is longer for pipe 1, so pipe 1 has the lowest fundamental frequency.

▸ ***What does it mean?***

We could determine the pipe having the standing wave with the longest wavelength, and hence the lowest frequency, without any mathematics. Plots of the standing wave envelopes in Figure 13.12 were all we needed.

13.4 Beats

When two tones are played at the same time, the principle of superposition tells us that the total sound pressure is the sum of the sound pressures of the individual tones. Figure 13.13A plots the variation of pressure as a function of time for two hypothetical pure tones, one at 500 Hz and the other at 600 Hz. Figure 13.13B shows the sum of these two pressure waves, which would be the total sound pressure when the two tones are played simultaneously. This figure also shows how the amplitude of this combination pressure wave varies back and forth (oscillates) from a large value when the two waves interfere constructively to near zero when there is nearly destructive interference. These amplitude oscillations are called ***beats***. The frequency of these oscillations in this case is 100 Hz, which is called the ***beat frequency***. In general, whenever two tones of frequencies f_1 and f_2 are played simultaneously, they "beat against" each other with a beat frequency of

$$f_{\text{beat}} = |f_1 - f_2| \tag{13.16}$$

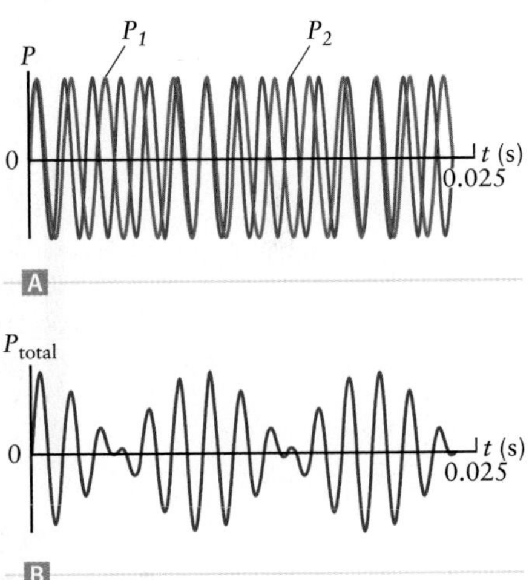

▲ **Figure 13.13** Beating of two sound waves. A Pressure as a function of time for two separate sound waves of frequency 500 Hz and 600 Hz. B The sum of these two pressure waves produces a pattern of "beats" in which the amplitude of the resulting wave oscillates with time.

The human ear is very sensitive to these beats, which are heard as variations in loudness with a frequency of f_{beat}. Most people are able to detect beat frequencies as low as about 0.2 Hz, even when the two simultaneous frequencies are near 1000 Hz or even higher. Hence, beats can be used to detect very small frequency differences, provided the tones are played simultaneously.

Beats are familiar to most musicians and can be used to tune an instrument. When the different instruments in an orchestra are "in tune," they each produce precisely the same fundamental frequency f_1 when the musicians are playing notes with the same pitch. Musicians adjust their instruments (prior to the performance!) by listening to the beat frequency between their own tone and a reference tone, and making that beat frequency as small as possible.

EXAMPLE 13.6 Beat Frequency of Two Guitar Tones

A standard guitar has six strings, so it is possible to play six tones at any particular moment. The fundamental frequency of a particular string is determined by the standing wave with the lowest possible frequency. This lowest standing wave frequency depends on the vibrating length of the string as discussed in Chapter 12 and illustrated in Figure 13.5D. By holding a string against a particular fret (Fig. 12.25), it is possible to vary the vibrating length of the string so that two different strings have the same fundamental frequency and thus play the same note. Now suppose two strings that *should* produce the same note are actually found to produce 4 beats per second. If one note has a frequency of 247 Hz, what is the frequency of the other note?

(continued) ▸

RECOGNIZE THE PRINCIPLE

If the two fundamental frequencies are f_1 and f_2, the beat frequency is $f_{beat} = |f_1 - f_2|$ (Eq. 13.16). Given the value of f_{beat} and that one of the frequencies is $f_1 = 247$ Hz, we can solve for f_2.

SKETCH THE PROBLEM

No figure is necessary.

IDENTIFY THE RELATIONSHIPS AND SOLVE

If there are 4 beats per second, the beat frequency is $f_{beat} = 4$ Hz. We know that $f_1 = 247$ Hz and $f_{beat} = |f_1 - f_2|$, so the other tone must have a frequency of either $247 + 4 = \boxed{251 \text{ Hz}}$ or $247 - 4 = \boxed{243 \text{ Hz}}$. If we are only allowed to listen to the beats, there is no way to tell which of these answers applies to a particular guitar.

► ***What does it mean?***

In practice, a guitarist will listen to the beats and adjust the tension of one of the strings to make the beat frequency as low as possible. (Recall from Chapter 12 that the fundamental frequency of waves on a string depends on the string tension.)

CONCEPT CHECK 13.6 Using Beats to Tune a Guitar

In Example 13.6, we considered the beats produced by two guitar strings and found that given only the beat frequency, there are two possible solutions for the frequency produced by the second string: $f_2 = 243$ Hz and $f_2 = 251$ Hz. Suppose the beat frequency increases when the tension in the first string is increased. Which of these solutions for f_2 is now the correct one?

(a) 243 Hz (b) 251 Hz (c) We still can't tell.

13.5 Reflection and Scattering of Sound

A standing wave can be viewed as a combination of two waves traveling in opposite directions. For the standing sound wave in a pipe, these waves travel back and forth as they are reflected from the ends of the pipe. Let's now examine the phenomenon of reflection for other geometries and also consider a related process called ***scattering***.

Figure 13.14 shows a sound wave with a plane wave front reflecting from a flat surface. It is very similar to the reflecting wave fronts we sketched in Figure 12.20. Such a ***plane wave*** reflects in a mirrorlike manner from a flat surface. The incoming angle θ_i, called the ***angle of incidence*** (see Fig. 13.14), equals the outgoing angle θ_r, called the ***angle of reflection***. We call this reflection mirrorlike because this situation is also encountered when light reflects from a mirror.

A mirrorlike reflection will occur whenever a plane wave encounters a flat surface, as long as the size of the reflecting surface is large compared to the wavelength of sound.[1] A typical sound might have a frequency of 200 Hz, in which case the wavelength is $\lambda = v_{sound}/f = (343 \text{ m/s})/(200 \text{ Hz}) = 1.7$ m. For sound in a typical room, most of the room surfaces are not large compared to this wavelength, so most of the reflections are not like the one shown in Figure 13.14. Instead, the reflected sound waves behave as in Figure 13.15, which shows an incoming plane wave reflecting from a small object. Here, there is at least a small amount of reflected sound in all directions. It is often called scattered sound because, loosely speaking, the original wave is "scattered" in many different directions. This result is found whenever a

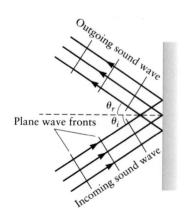

▲ **Figure 13.14** Reflection of a sound wave with plane wave fronts from a flat wall. The outgoing (reflected) angle θ_r is equal to the incoming (incident) angle θ_i.

[1]This requirement does *not* apply to a sound wave traveling in a very narrow pipe (as in Fig. 13.5 or with the clarinet in Fig. 13.10). When the diameter of the pipe is small compared to wavelength, the sound wave can only be reflected backward because it literally has nowhere else to go.

sound wave encounters an object whose size is approximately equal to, or smaller than, the wavelength of the sound. In such cases, reflected sound waves propagate out in virtually all directions from the scattering source.

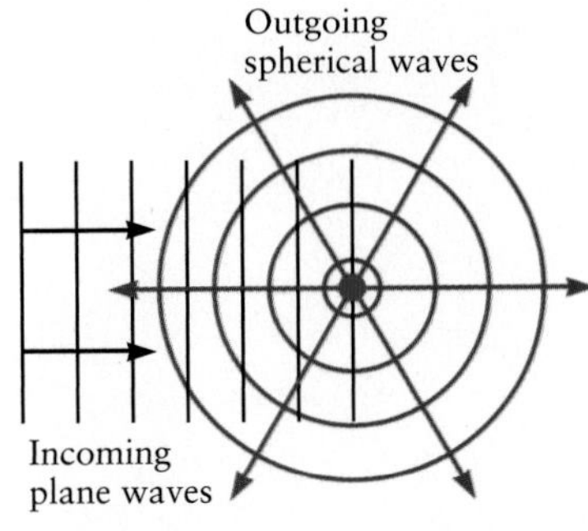

▲ **Figure 13.15** When a wave reflects from a small object (denoted by the red dot), the reflected wave travels outward in all directions. When the object is very small compared to the wavelength, the outgoing wave fronts are approximately spherical.

13.6 The Doppler Effect

Suppose the siren of an ambulance emits a pure tone of frequency f_{source}. If the ambulance is parked so that it is not moving relative to a listener standing nearby, the sound waves emitted by the siren will be as sketched in Figure 13.16A. Here we show two wave crests as they move from the source to the observer (listener) at speed v_{sound}. The source emits one wave crest during each period of the wave (one period $= 1/f_{source}$), so consecutive crests are separated by a distance equal to the wavelength $\lambda_{source} = v_{sound}/f_{source}$. The crests arrive at the observer spaced in time by $\lambda_{source}/v_{sound} = 1/f_{source}$, so one wave crest arrives at the observer every $1/f_{source}$ seconds. In other words, the frequency *as heard by the observer* is equal to f_{source}. The frequency the observer measures thus equals the frequency emitted by the siren, as you should have expected.

Let's now use the same approach to analyze the case of a siren moving toward the observer. The siren emits sound waves whose crests again travel toward the observer. These wave crests move at a speed v_{sound} *relative to the medium*, which in this case is the air. Once a sound wave leaves the source, the wave's motion is determined by the

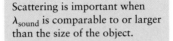

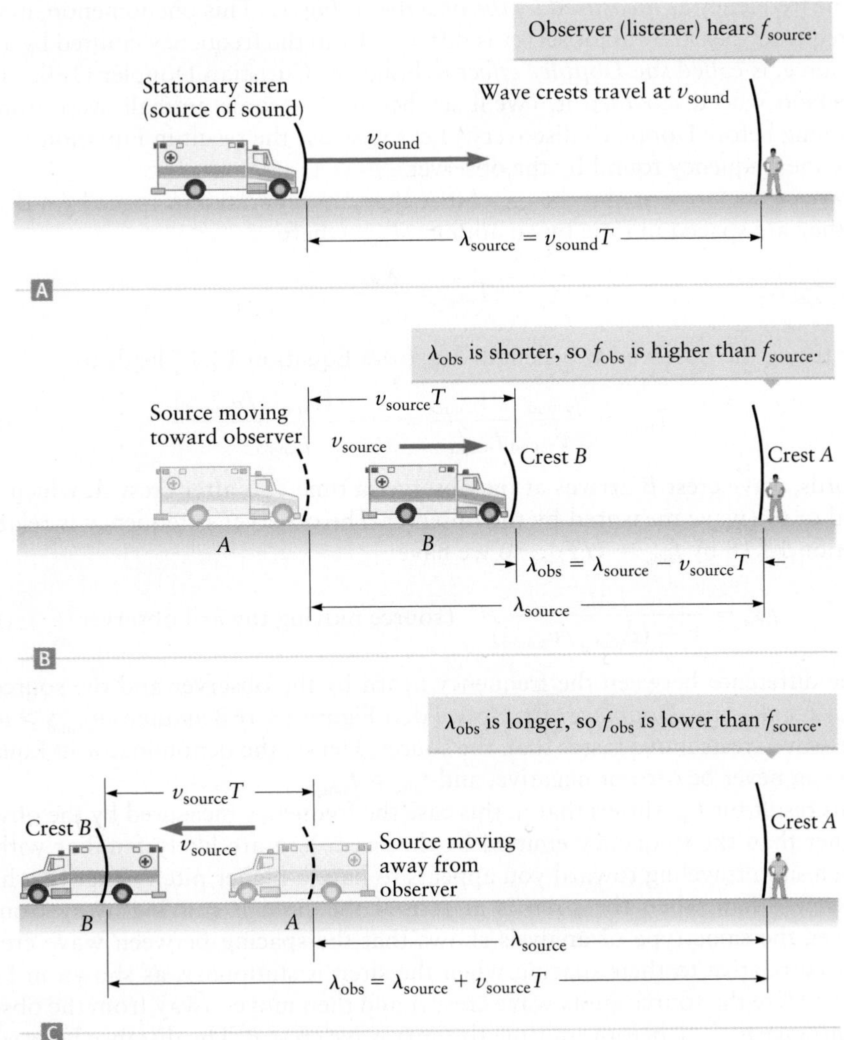

◀ **Figure 13.16** The Doppler effect. **A** When a source of sound is at rest relative to a stationary observer, the frequency heard by the observer is equal to the frequency emitted by the source. **B** When the source is moving toward the observer, the observed frequency is higher than the emitted frequency. **C** When the source is moving away from the observer, the observed frequency is lower than the emitted frequency.

medium; the observer in Figure 13.16B is stationary relative to the air, and he will find that the wave crests move at v_{sound} no matter what speed the source may have.

We must now account for the motion of the siren between the emission of each wave crest. Figure 13.16B shows two successive wave crests along with the position of the siren when it emitted these crests. This snapshot shows the siren at point *B*, having just emitted the wave crest also labeled *B*. The previous wave crest labeled *A* was emitted when the siren was at location *A*, one period earlier. The frequency of the siren, as would be measured by someone sitting in the ambulance, is still f_{source}, so the siren emits wave crests with the same period $T = 1/f_{source}$ as when the ambulance is not moving. Wave crest *A* has traveled for this amount of time at speed v_{sound}, so it has traveled a distance $v_{sound}T = v_{sound}/f_{source} = \lambda_{source}$ from the point where it was emitted.

Let's call the speed of the ambulance (and attached siren) v_{source}. By the time the siren emits wave crest *B*, it has traveled a distance $v_{source}T$ since emitting crest *A*. The spacing between crest *A* and crest *B* is thus smaller than was the case with a stationary siren. From Figure 13.16B, we see that the new distance between crests is

$$\lambda_{obs} = v_{sound}T - v_{source}T$$

The period *T* is equal to $1/f_{source}$, so

$$\lambda_{obs} = \frac{v_{sound} - v_{source}}{f_{source}} \tag{13.17}$$

Because the distance between crests is smaller when the ambulance moves toward the observer than when it is stationary, the crests arrive more frequently at the observer and the frequency as *measured by the observer* is *higher*. This phenomenon, in which the frequency heard by an observer is different from the frequency emitted by a moving source, is called the ***Doppler effect*** in honor of Christian Doppler (1803–1853), the person who discovered it. (We'll see below that some animals were using the effect long before Doppler's discovery.) Let's now use the result in Equation 13.17 to derive the frequency found by the observer.

Doppler effect

Wave crests move at speed v_{sound}. Since they are separated in space by a distance λ_{obs}, they are spaced in time by an amount T_{obs}, where

$$T_{obs} = \frac{\lambda_{obs}}{v_{sound}}$$

Substituting the observed wavelength λ_{obs} from Equation 13.17 leads to

$$T_{obs} = \frac{v_{sound} - v_{source}}{v_{sound}\, f_{source}} = \frac{1 - (v_{source}/v_{sound})}{f_{source}} \tag{13.18}$$

In words, wave crest *B* arrives at the observer a time T_{obs} after crest *A*, which is the period of the wave measured by the observer. The observer's frequency is related to Equation 13.18 by $f_{obs} = 1/T_{obs}$, so we have

$$f_{obs} = \frac{f_{source}}{1 - (v_{source}/v_{sound})} \quad \text{(source moving toward observer)} \tag{13.19}$$

The difference between the frequency heard by the observer and the source frequency is called the ***Doppler shift***. Notice that Figure 13.16B assumes $v_{sound} > v_{source}$, so the wave crests move faster than the source. Hence, the denominator in Equation 13.19 can never be zero or negative, and $f_{obs} > f_{source}$.

This result for f_{obs} shows that in this case the frequency measured by the observer is higher than the frequency emitted by the siren. You are likely familiar with this result; a siren traveling toward you appears to have a higher pitch—that is, a higher frequency—than when the siren is at rest. If the siren is moving away from the observer, the same type of analysis shows that the spacing between wave crests is increased relative to their spacing when the siren is stationary, as shown in Figure 13.16C. Here the source emits wave crest *A* and then moves away from the observer by a distance $v_{source}T$ before emitting the next wave crest *B*. The distance between the adjacent wave crests is now increased, causing the frequency heard by the observer to be lower. Taking the same approach that led to Equation 13.19, we find that the result for f_{obs} in this case is

$$f_{obs} = \frac{f_{source}}{1 + (v_{source}/v_{sound})} \quad \text{(source moving away from observer)} \qquad (13.20)$$

Let's estimate the size of the Doppler shift for a typical case. We assume the source is a siren whose frequency when at rest is $f_{source} = 500$ Hz. Suppose the siren is on a fire truck racing to an emergency at a speed of 27 m/s (about 60 mi/h). For a listener standing on a nearby sidewalk, what frequency does the siren have? If the truck is moving toward the observer, the Doppler-shifted frequency is given by Equation 13.19, where $v_{source} = 27$ m/s. Inserting these values along with $v_{sound} = 343$ m/s into Equation 13.19 gives

$$f_{obs} = \frac{f_{source}}{1 - (v_{source}/v_{sound})} = \frac{500 \text{ Hz}}{1 - [(27 \text{ m/s})/(343 \text{ m/s})]} \qquad (13.21)$$

$$f_{obs} = 540 \text{ Hz}$$

Hence, the Doppler shift is $f_{obs} - f_{source} = 40$ Hz, which is nearly 10% of the source frequency. In terms of notes on a musical scale, this value is approximately the separation between two consecutive whole notes (i.e., "do" and "re").

So far, we have considered the Doppler effect for a stationary observer. Motion of the observer also gives a Doppler shift. In the most general case, in which the observer has a speed v_{obs} and the source a speed v_{source}, the observed frequency is

$$f_{obs} = f_{source}\left[\frac{1 \pm (v_{obs}/v_{sound})}{1 \pm (v_{source}/v_{sound})}\right] \qquad (13.22)$$

Here again the speeds v_{obs} and v_{source} are measured relative to the medium, and both these speeds are assumed to be less than v_{sound}. The choices of plus and minus signs in Equation 13.22 depend on whether the source and observer are moving toward or away from each other. The signs can always be remembered by noting that the frequency measured by the observer f_{obs} is *increased when the observer and source are moving toward each other*, whereas f_{obs} is *decreased when they move away from each other*.

EXAMPLE 13.7 Doppler Shift, Coming and Going

A fire truck approaches you from one direction, passes by, and then continues on away from you, with the siren emitting sound the entire time. You have a friend (a music student) with a very keen sense of pitch who tells you that the frequency of the siren drops by precisely 50 Hz when the truck passes your location on the street. If the truck has a speed of 20 m/s, what is the frequency of the siren at rest? Take the speed of sound to be $v_{sound} = 343$ m/s.

RECOGNIZE THE PRINCIPLE

This problem is similar to our analysis of the Doppler shift for a moving truck (Eq. 13.21), but now we have to take into account *two* Doppler shifts, one as the truck moves toward you and another as it moves away from you. The observed frequency will be shifted upward to a higher frequency when the truck is moving toward you and downward to a lower frequency after the truck passes. Your friend has used his sense of pitch to estimate the difference between the "coming" and "going" frequencies.

SKETCH THE PROBLEM

The problem is described by parts B and C of Figure 13.16.

(continued) ▶

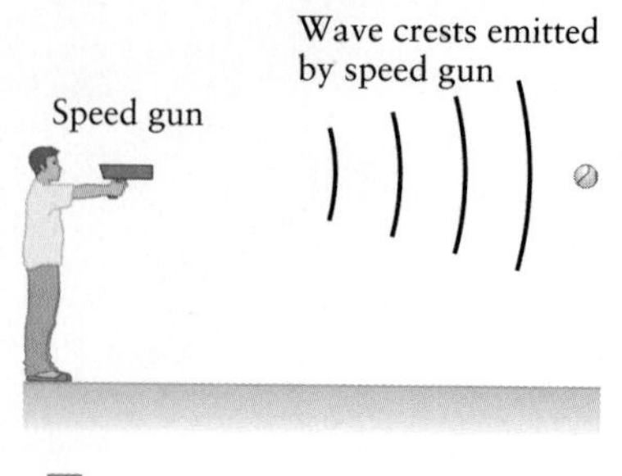

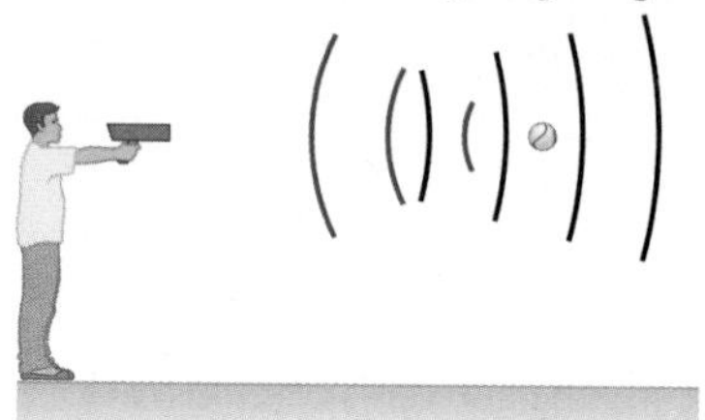

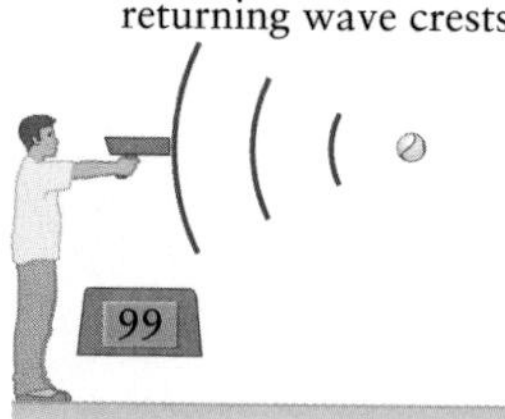

▲ **Figure 13.17** A "speed gun" (sometimes also called a "radar gun") uses the Doppler effect to measure the speed of a moving object. A Waves are emitted by the speed gun and strike the object. B These waves scatter from the object, and some of the scattered waves return to the speed gun. C The speed gun measures the difference between the frequencies of the emitted and scattered waves.

IDENTIFY THE RELATIONSHIPS

When the truck is approaching, the observed frequency is (from Eq. 13.19)

$$f_{\text{coming}} = \frac{f_{\text{source}}}{1 - (v_{\text{source}}/v_{\text{sound}})}$$

and as it moves away the frequency is (from Eq. 13.20)

$$f_{\text{going}} = \frac{f_{\text{source}}}{1 + (v_{\text{source}}/v_{\text{sound}})}$$

Your friend senses the difference $\Delta f = f_{\text{coming}} - f_{\text{going}}$, which is given by

$$\Delta f = \frac{f_{\text{source}}}{1 - (v_{\text{source}}/v_{\text{sound}})} - \frac{f_{\text{source}}}{1 + (v_{\text{source}}/v_{\text{sound}})} \tag{1}$$

SOLVE

We now want to solve for f_{source}. Rearranging Equation (1) leads to

$$\Delta f = f_{\text{source}}\left[\frac{1}{(1 - v_{\text{source}}/v_{\text{sound}})} - \frac{1}{1 + (v_{\text{source}}/v_{\text{sound}})}\right]$$

$$f_{\text{source}} = \frac{\Delta f}{\left[\frac{1}{(1 - v_{\text{source}}/v_{\text{sound}})} - \frac{1}{1 + (v_{\text{source}}/v_{\text{sound}})}\right]}$$

Inserting the given values of Δf, v_{source}, and v_{sound}, we get

$$f_{\text{source}} = \frac{50 \text{ Hz}}{\left[\frac{1}{1 - [(20 \text{ m/s})/(343 \text{ m/s})]} - \frac{1}{1 + [(20 \text{ m/s})/(343 \text{ m/s})]}\right]} = \boxed{430 \text{ Hz}}$$

► ***What does it mean?***

The Doppler shift can also be used to measure the speed of the source, as we explore in the next subsection.

Speed Guns, Bats, and the Doppler Effect

So far, we have discussed the Doppler effect using the example of sound waves from a moving siren, but the Doppler effect applies to other types of waves as well. For example, a "speed gun" used in sporting events such as baseball (Fig. 13.17) uses the Doppler effect with electromagnetic waves to measure the speed of a moving object. Electromagnetic waves from the speed gun are scattered by the ball, and some of the scattered waves then return to the speed gun. (Compare with the scattered waves in Fig. 13.15.) As the waves strike it, the ball acts like a moving "observer," whereas for the scattered waves, the ball acts like a moving source. For the case shown in Figure 13.17, with the ball moving toward the observer, the ball's motion causes the scattered wave crests to be more closely spaced than if the ball were at rest, just as we found with the moving siren in Figure 13.16B, and there is again a Doppler shift. The speed gun uses the Doppler-shifted frequency of the scattered waves to deduce v_{source}, the speed of the ball.

▲ **Figure 13.18** Some bats use reflected sound to judge the location and motion of nearby objects, including their prey.

Bats employ an approach similar to that used by the speed gun to help them navigate as they fly (Fig. 13.18). Bats generate pulses of sound waves at frequencies well above the range of human hearing (in the ultrasonic range). These ultrasonic sound pulses are reflected or scattered by objects in the bat's environment, and the bat uses the reflected sound to judge the location and motion of nearby objects. Bats have a very keen sense of timing and are able to determine accurately the time delay between an emitted pulse and its reflection. A long reflection time means the object is far away, and a small reflection time means the object is nearby (similar to the way radar works as discussed in Chapter 12). In addition to the reflection time, some bats also use the Doppler effect.[2] Suppose the bat's ultrasonic sound wave is reflected by a moth moving through the air.[3] The reflected sound is Doppler-shifted, and the bat uses this frequency shift to sense the moth's speed and obtain a meal.

Moving Sources and Shock Waves

In all our examples involving moving sources of sound, the speed of the source has (so far) been much less than the speed of sound, but it is also possible for the speed of a wave's source to be greater than the speed of the wave. For example, a jet can fly faster than the speed of sound. Figure 13.19 shows the wave crests of sound emitted by a jet in three cases. Figure 13.19A shows what happens when the jet's speed is less than the speed of sound; this arrangement of wave crests is the one we considered in our analysis of the Doppler effect in Figure 13.16. Figure 13.19B shows what happens when the jet's speed is equal to the speed of sound. The jet now travels at the same speed as the wave crests, so the crests in front are not able to move away from it. The wave crests at the front of the jet thus "pile up" at this point. According to the principle of superposition, the total sound pressure is the sum of the pressures due to the individual sound waves, so there is a very large sound intensity at the front of the jet. Figure 13.19C shows the situation when the jet moves faster than the speed of sound. Sound waves emitted at past times pile up along a conical envelope (which appears as two sides of a triangle in the projection in Fig. 13.19C). At any point on this envelope, many different sound waves, emitted at many different previous times, arrive simultaneously, producing a very large sound intensity and forming a ***shock wave***. Shock waves are the source of the ***sonic booms*** that can be heard when a supersonic jet passes overhead.

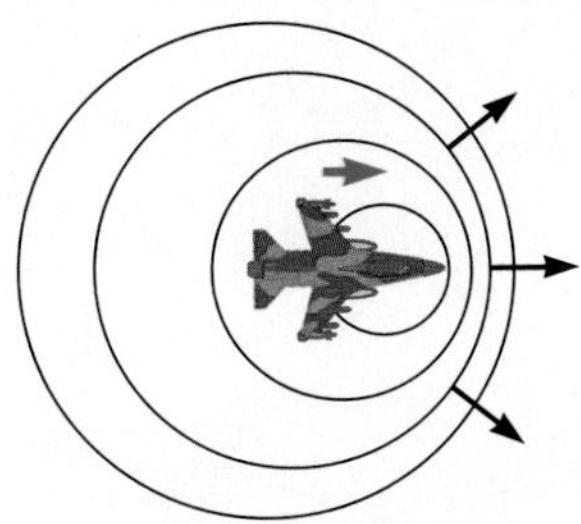

The jet does not keep up with the wave crests.

A

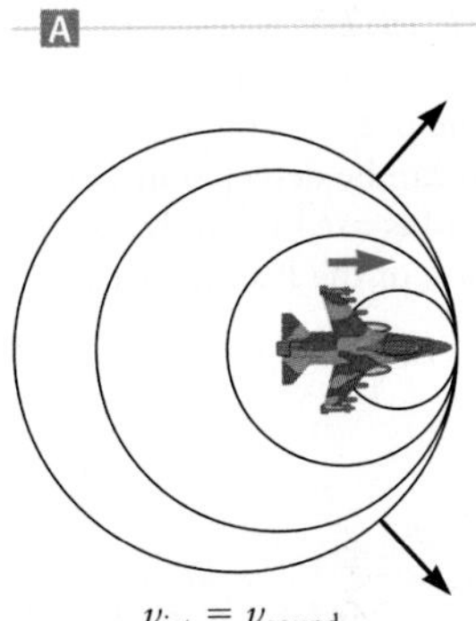

The jet keeps up with the wave crests, so the wave crests "pile" up.

B

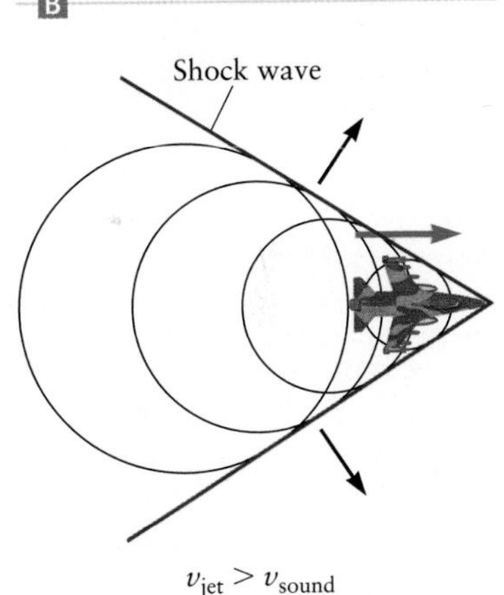

C

▲ **Figure 13.19** Wave fronts emitted by a moving source. A The speed of the source is less than the speed of sound, leading to the Doppler effect (Fig. 13.16). B When the speed of the source is equal to the speed of sound, the leading part of the wave front never "escapes" from the source. C When the speed of the source is greater than the speed of sound, shock waves are created. The direction of the shock wave depends on the ratio v_{source}/v_{sound}.

13.7 Applications

Applications involving sound are found in many places, including the home (such as the earphones of your MP3 player), in scientific research, and in medicine.

Using Sound to Study Global Warming

The speed of sound depends on a material's temperature. Table 13.1 lists v_{sound} in some typical materials at room temperature. For example, sound travels through salt water with a typical speed $v_{sound} \approx 1500$ m/s, and this value increases by approximately 4.0 m/s when the temperature is increased by 1°C. This effect is now being used in studies of the oceans. Many scientists believe that the Earth's average surface temperature is increasing, an effect known as global warming. Because changes in the Earth's average temperature as small as 1°C can greatly affect life on this planet, it is very important to confirm that such climate change is actually occurring and, if so, to measure the rate of warming. Studies of global warming are very difficult due to natural fluctuations in the Earth's temperature from place to place and from year to year. For example, the fluctuations of the average temperature over the surface of the entire Earth are typically 0.2°C from one year to the next.

A clever approach called "acoustic thermometry" has been developed to study global warming. This approach uses the temperature dependence of the speed of sound to measure changes in the temperature of the oceans. Special sound generators and detectors have been developed to send sound waves through the oceans across very large distances. For example, sound generated in the Pacific Ocean near Hawaii

[2]You can learn more about bats and the Doppler effect in C. F. Moss and S. R. Sinha, "Neurobiology of echolocation in bats," *Current Opinion in Neurobiology* 13:751–758 (2003).

[3]Because the bat is also moving, there is also a Doppler effect due to the bat's motion. The bat is able to compensate for Doppler shifts caused by its motion and isolate the shift due solely to the moth's motion.

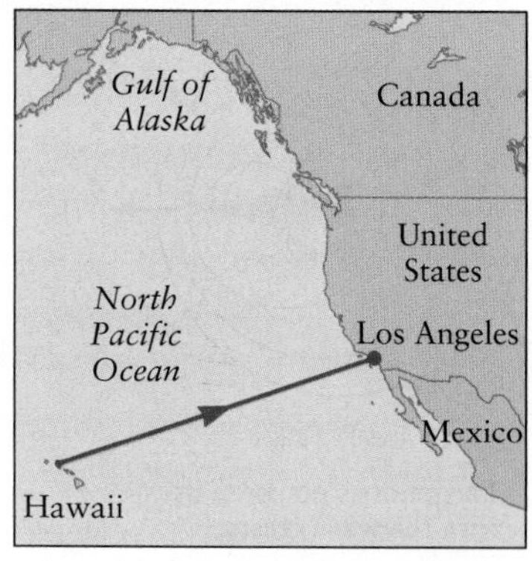

▲ **Figure 13.20** Sound can travel for very large distances through the oceans. Sound emitted in Hawaii can be detected in Los Angeles and is used to study the temperature of the Pacific Ocean.

can be detected when it reaches the coast of California (Fig. 13.20). The speed of sound in salt water is approximately 1500 m/s and the distance from Honolulu to Los Angeles is about 4100 km, so it takes about

$$t = \frac{L}{v_{\text{sound}}} = \frac{4.1 \times 10^6 \text{ m}}{1500 \text{ m/s}} = 2700 \text{ s} \approx 45 \text{ min}$$

for sound to travel this distance. The actual travel time will depend on the average temperature of seawater over this distance. Global warming would produce changes in this average temperature, so acoustic thermometry is ideal for studies of global warming. The travel time for sound can be measured very accurately; current precision approaches 20 ms (0.02 s), making it possible to measure very accurately how the average ocean temperature varies over the course of days and years. These experiments are ongoing and should yield important clues for understanding climate change.

EXAMPLE 13.8 Temperature Changes Measured by Acoustic Thermometry

What is the smallest temperature change that can be measured using acoustic thermometry? Assume the sound generator and detector are located near Honolulu and Los Angeles and are thus separated by $L = 4100$ km, and assume the sound travel time can be measured with an accuracy of 0.02 s. The speed of sound in seawater varies by 4.0 m/s for each 1.0°C change in temperature.

RECOGNIZE THE PRINCIPLE

The travel time t is related to the travel distance and the speed of sound by $t = L/v_{\text{sound}}$. To change the travel time t by 0.02 s, there must be a corresponding change in v_{sound}. This change in v_{sound} can then be used to find the change in temperature.

SKETCH THE PROBLEM

No figure is needed.

IDENTIFY THE RELATIONSHIPS

On an "average" day, the travel time for this example is approximately

$$t = \frac{L}{v_{\text{sound}}} = \frac{4.1 \times 10^6 \text{ m}}{1500 \text{ m/s}} = 2700 \text{ s}$$

Now suppose the experiment is performed on two consecutive days and the travel times are measured to be $t_1 = 2700.00$ s and $t_2 = 2700.02$ s. We can rearrange the relation between t and v_{sound} to solve for the value of v_{sound} on each day. The results are

$$v_1 = \frac{L}{t_1} = \frac{4.1 \times 10^6 \text{ m}}{2700.00 \text{ s}} = 1518.519 \text{ m/s}$$

and

$$v_2 = \frac{L}{t_2} = \frac{4.1 \times 10^6 \text{ m}}{2700.02 \text{ s}} = 1518.507 \text{ m/s}$$

Here we have kept extra significant figures because we now need to compare v_1 and v_2. The difference in these velocities is

$$v_1 - v_2 = 0.012 \text{ m/s}$$

SOLVE

This velocity difference corresponds to a temperature change ΔT of

$$\Delta T = (0.012 \text{ m/s}) \times \frac{1.0°\text{C}}{4.0 \text{ m/s}} = \boxed{0.003°\text{C}}$$

► ***What does it mean?***
Acoustic thermometry can thus be used to measure *extremely* small changes in the average temperature of the ocean. By placing the transmitters and receivers in different locations, the temperature in different ocean regions can be measured.

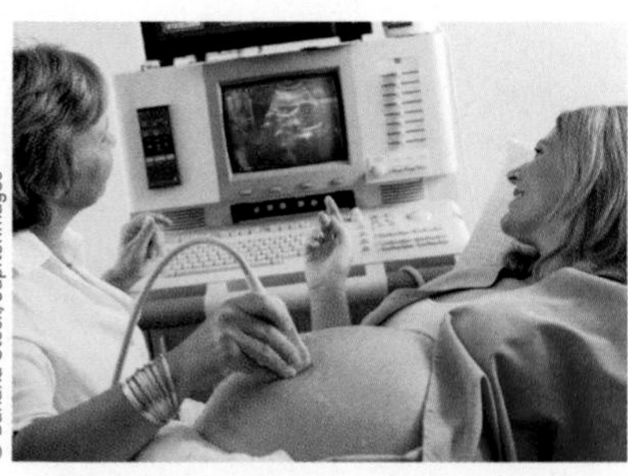

© Banana Stock/Jupiterimages

A

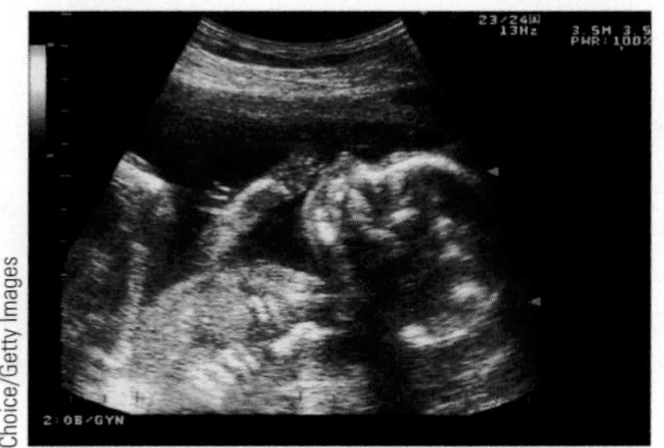

© Vincenzo Lomardo/Photographer's Choice/Getty Images

B

▲ **Figure 13.21** Ultrasonic imaging (ultrasound) uses reflected sound waves to reveal structures inside living tissue. A A transducer (a device that can both emit and detect sound) is placed on the abdomen of a pregnant woman. B Reflected waves are detected by the transducer and used to construct an image of the fetus.

Imaging with Ultrasound

Another important application of sound is the technique of ***ultrasonic imaging***. This technique uses sound waves to obtain images inside a material. The material could be virtually anything (such as an airplane wing), but the most familiar use of ultrasonic imaging produces views inside the human body. An instrument that generates sound is placed on the skin (Fig. 13.21A), and sound waves propagate into the body. A portion of the sound energy is reflected by bones and other tissue, and the reflected waves are used to construct an image (Fig. 13.21B). The depth of the bone (or whatever causes the reflection) is determined by the time it takes the sound to travel from the generator to the bone and then return to the detector.

Some medical ultrasonic imaging devices also use the Doppler effect. When ultrasound is used to construct an image of a blood vessel, sound waves are reflected by cells moving in the blood. The frequency of these reflected sound waves is Doppler-shifted as we found with the speed gun in Figure 13.17. By measuring this Doppler shift, physicians can measure the blood velocity (see Problem 75 at the end of this chapter). This information can be used to determine if blood is flowing properly and can reveal the presence of blockages or constrictions.

CONCEPT CHECK 13.7 Ultrasonic Imaging

A typical ultrasonic imaging device employs sound with a frequency of 10 MHz. If the speed of sound in tissue is approximately 1500 m/s, is the wavelength of these sound waves (a) 1.5 mm, (b) 0.15 mm, or (c) 1.5 cm?

Summary | CHAPTER 13

Key Concepts and Principles

Sound is a longitudinal wave

The molecular motion associated with a sound wave is parallel to the wave velocity, so sound is a longitudinal wave. Sound can also be viewed as a ***pressure wave***. Regions of high pressure are regions of condensation, whereas low-pressure regions are regions of rarefaction. The frequency and wavelength of a periodic sound wave are related by

$$v_{\text{sound}} = f\lambda \quad \textbf{(13.1) (page 413)}$$

The speed of sound depends on the properties of the medium and is generally high in solids and low in gases. The speed of sound in air at atmospheric pressure and room temperature is approximately 343 m/s.

Human hearing

The normal range of human hearing is approximately 20–20,000 Hz. Sounds with lower frequencies are called ***infrasonic***, and sounds with higher frequencies are ***ultrasonic***.

(Continued)

Sound intensity

The ***intensity*** of a sound wave is proportional to the square of the amplitude. Typically, the human ear can hear sounds with intensities as small as $1 \times 10^{-12}\ \text{W/m}^2$, which corresponds to a pressure amplitude of about 2×10^{-5} Pa. Sound intensity is often measured in ***decibels*** (dB). The intensity level of a sound in decibels is given by

$$\beta = 10 \log\left(\frac{I}{I_0}\right) \qquad \textbf{(13.9) (page 417)}$$

where $I_0 = 1.00 \times 10^{-12}\ \text{W/m}^2$. The human ear can function (without damage) at intensity levels up to about 120 dB.

Applications

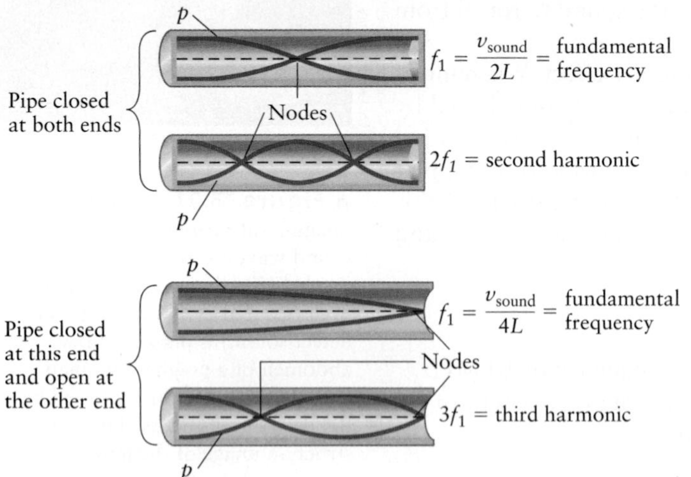

Sound in pipes

Sound propagating back and forth within a pipe forms ***standing waves*** when the wavelength of the sound (or a multiple of the wavelength) matches the length of the pipe. Standing waves have pressure ***nodes*** where the pressure amplitude is zero and pressure ***antinodes*** where the pressure amplitude is largest. The open end of a pipe is always a pressure node, while the closed end is always a pressure antinode.

For a pipe closed at both ends, the possible standing waves have frequencies

$$f_n = n\frac{v_{sound}}{2L} = nf_1 \quad \text{with } n = 1, 2, 3, \ldots$$

(13.11 and 13.12) (page 421)

where $f_1 = v_{sound}/(2L)$ is called the ***fundamental frequency***, the lowest possible standing wave frequency. For a pipe that is closed at one end and open at the other (as in many musical instruments), the standing wave frequencies are

$$f_n = n\frac{v_{sound}}{4L} = nf_1, \quad \text{with } n = 1, 3, 5, \ldots \qquad \textbf{(13.14 and 13.15) (page 422)}$$

with $f_1 = v_{sound}/(4L)$.

Beats

When two sound waves that have nearly the same frequency are combined, a listener will hear ***beats***. Beats are an oscillation of the overall sound intensity at a beat frequency equal to $|f_1 - f_2|$, where f_1 and f_2 are the frequencies of the two sound waves.

Doppler shift

When the source of a sound is moving, the frequency observed by a listener is ***Doppler-shifted***. If the frequency of the source when stationary is f_{source}, then when the source moves at speed v_{source} and the listener moves at speed v_{obs}, the frequency heard by the listener is

$$f_{obs} = f_{source}\left[\frac{1 \pm (v_{obs}/v_{sound})}{1 \pm (v_{source}/v_{sound})}\right] \qquad \textbf{(13.22) (page 429)}$$

The choices of plus and minus signs in Equation 13.22 depend on whether the source and observer are moving toward or away from each other. When using Equation 13.22, choose the signs so that the frequency measured by the observer f_{obs} is increased when the observer and source are moving toward each other, and f_{obs} is decreased when they move away from each other.

Questions

SSM = answer in *Student Companion & Problem-Solving Guide*   ⓧ = life science application

1. Rank each substance according to how fast sound travels in that medium, from that with the largest velocity to that with the slowest.
(a) Air
(b) Water
(c) Steel
(d) Aluminum

2. ⓧ A typical dog can hear sounds with frequencies as high as 45 kHz. Assuming they can hear sounds as low as 40 Hz, make a qualitative sketch of the sensitivity curve for a dog (i.e., the canine version of the dashed curve in Fig. 13.4).

3. Musicians who play the guitar and other stringed instruments use beats to tune their instruments. The value of the beat frequency tells them the difference in the frequencies of two strings, but it does not tell them which one is higher (or lower). Describe an experiment musicians could do to resolve this question.

4. Two sound waves of equal frequency interfere constructively at a listener. Which of the following statements is true?
(a) The frequency is increased by the interference.
(b) The sound amplitude is increased by the interference.
(c) The listener will hear beats.

5. How is it possible for a singer to shatter a wine glass by singing a note? Which is more important in this process, the frequency or intensity of the singing, or are they both important?

6. In this chapter, we discussed the Doppler effect that occurs when a source of sound moves directly toward or away from a listener. Do you think there will be a Doppler shift if the velocity of the source is perpendicular to a line that runs to the listener (Fig. Q13.6)? Explain why or why not.

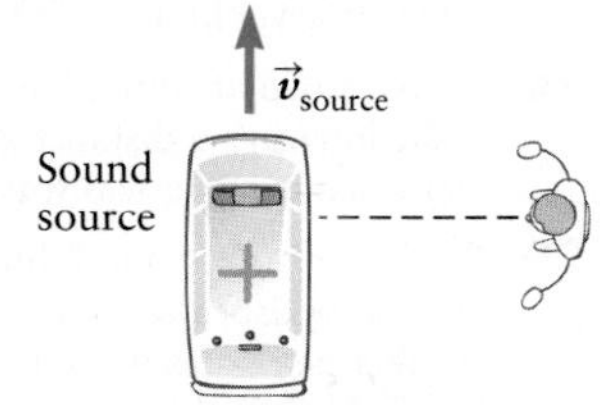

Figure Q13.6

7. ⓧ Explain why the sounds emitted by large animals usually have a lower frequency than those from small animals. *Hint*: Consider the standing waves in the vocal tract.

8. ⓧ **Ocean in a shell.** What is it that you hear when you hold a seashell (or any closed chamber like a cup or jar) to your ear? We are almost always surrounded by sound that spans every frequency (white noise). What happens if you cup your hands around an ear and vary the size of the chamber? The pitch should change as you vary the chamber. What are you hearing and why?

9. Consider the method of acoustic thermometry using the ocean. In Example 13.8, we saw that very small changes in the travel time of sound can be detected and can be used to observe very small changes in the ocean temperature. There are, however, complications that we did not mention in that example. It turns out that the velocity of sound in water also depends on the salinity of the water. Ocean water contains about 3.5% salt by volume, and the speed of sound changes by about 1.4 m/s for each 0.1% change in the salt concentration. Investigate (using the Internet or some other reference) how much the salinity of ocean water fluctuates over time and discuss how much this fluctuation will affect the measurement of temperature changes with this technique.

10. Use the general relation for the speed of sound in Equation 13.3 together with the ideal gas law (see Chapter 15 if you need a refresher on the ideal gas law) to show that the speed of sound in a gas can be written as $v_{sound} = \sqrt{P/\rho}$ where P is the pressure. Can you give an intuitive explanation of why v_{sound} increases as the pressure is increased?

11. The first television remote controllers (invented in the 1950s) used ultrasonic waves. Some people claimed they could change channels by simply jiggling their keys. Is that possible? Explain why or why not.

12. SSM ⓧ A buzzer generates sound by vibrating with a frequency of 440 Hz in air. If this buzzer is placed underwater, what frequency would a fish hear? How does placing the buzzer underwater affect the wavelength of the sound?

13. One day, while hiking in a particular spot in your favorite cave, you notice that it takes a certain time t for an echo to return to you. The next day, you find that the echo time at the same spot is shorter. How might that happen?

14. When a police car is at rest, its siren has a frequency of 750 Hz. You are a thief, and just after finishing a night's work, you hear a siren with a frequency of 700 Hz. Your fellow thief tells you to relax and assures you that there is nothing to worry about. Is she right?

15. A radio produces sound waves in air, but they can also travel from air into water so that they can be heard by a swimmer. Is the frequency of the sound in air the same or different from the frequency in water? Is the wavelength the same or different? Explain.

16. ⓧ The manufacturer of a set of earplugs advertises that they will reduce the sound intensity level by 40 dB. By what factor do they reduce the intensity?

17. SSM When you blow across the top of a cola bottle, you can generate sound (Figure Q13.17). As you drink more and more of the cola and lower the liquid level, how does the frequency of the sound change?

Figure Q13.17 Question 17 and Problem 43.

18. When you tune a flute or trombone, what causes the frequency to change?

19. Noise-canceling headphones make use of destructive interference of sound waves to eliminate "noise," while still playing music. Explain how they work.

20. If the note middle C is played on trumpet and on a piano, you can identify them as the same note but can certainly tell the difference. What feature of these sounds tells you that they are the same note? What feature of these sounds tells you that they come from different instruments?

Problems

SSM = solution in *Student Companion & Problem-Solving Guide*
★ = intermediate ✪ = challenging Ⓧ = life science application
Ⓡ = reasoning and relationships problem
RT = reasoning tutorial in Enhanced WebAssign

13.1 SOUND IS A LONGITUDINAL WAVE

1. A sound wave in air has a frequency of 220 Hz. What is its wavelength?

2. The distance along a sound wave between a point of condensation and the nearest rarefaction is 0.45 m. What is the wavelength of this sound wave?

3. Consider a sound wave in air with a frequency of 440 Hz. What is the distance between a region of condensation and an adjacent region of rarefaction?

4. Suppose a sound wave in air at room temperature has a wavelength of 400 m. What is its frequency?

5. Two ships are traveling on a very dark and foggy night and cannot see each other, even though they have their lights on. They therefore use their foghorns to broadcast their presence. If the ships are 50 m apart, how long does it take the sound from one ship to reach the other?

6. A battleship is using sonar (reflected underwater sound signals) to detect the presence of nearby submarines. It is found that a sonar reflection has a round-trip travel time (from the battleship to the submarine and back) of 17 s. How far away from the battleship is the submarine?

7. ★ A sound wave in an unknown gas has a frequency of 440 Hz and a wavelength of 2.2 m. What type of gas might it be? *Hint*: Use Table 13.1.

8. SSM ★ Consider two sounds of frequency 300 Hz, one traveling in carbon dioxide gas and the other in hydrogen gas. What is the ratio of the wavelengths in the two cases?

9. The lowest note on a piano has a fundamental frequency of about 27 Hz. What is the wavelength of this sound in air?

10. Ⓧ The ultrasonic sound waves used by owls for echolocation typically have a frequency of approximately 100 kHz. What is the wavelength of a sound with this frequency?

11. ★ Consider a sound wave of frequency f generated in air by a loudspeaker. When this sound wave reaches a nearby lake, the sound wave propagates into the water, where it has the same frequency f. What is the ratio of the wavelength of the wave in the water to the wavelength in air?

12. ✪ An automobile wheel of diameter 0.50 m is found to emit sound with a frequency of 10 Hz. What is the automobile's speed? *Hint*: Assume the frequency of the emitted sound is equal to the rotation frequency of the wheel.

13. ★ In western movies, the hero sometimes detects an approaching train by "listening" to the railroad tracks. If the tracks are made of steel, how many times faster does sound travel in the tracks than in air?

13.2 AMPLITUDE AND INTENSITY OF A SOUND WAVE

14. Ⓧ Use the data in Figure 13.4 to determine the approximate frequency at which the human ear is most sensitive to sound.

15. Ⓧ Use Figure 13.4 to find the sensitivity of the human ear at 50 Hz. That is, what is the lowest sound intensity the ear can detect at this frequency?

16. ★ A sound wave has an intensity level of 80 dB. What is the pressure amplitude of this wave?

17. SSM ✪ Ⓧ Ⓡ You are at a rock concert, and the sound intensity reaches levels as high as 130 dB. The sound pressure produces an oscillating force on your eardrum. Estimate the amplitude of this oscillating force.

18. ✪ Ⓧ The human auditory system is able to detect changes in the sound intensity level as small as 1 dB. (a) What is the corresponding percentage change in the intensity I when the sound intensity level changes from 10 dB to 11 dB? (b) What is the percentage change when the intensity level changes from 80 dB to 81 dB? (c) What are the absolute changes in the value of I in parts (a) and (b)? Why are they different, even though the percentage changes are the same?

19. ★ To decrease the intensity of a wave by a factor of four, what must you do?
(a) Decrease the pressure amplitude by a factor of two.
(b) Decrease the pressure amplitude by a factor of four.
(c) Decrease the pressure amplitude by a factor of eight.
(d) Decrease the pressure amplitude by a factor of $\sqrt{2}$.

20. ★ Two sounds differ in intensity level by 6.0 dB. What is the ratio of the intensities of these two sounds?

21. ★ If a sound intensity level increases by 16 dB, by what factor does the intensity change?

22. ✪ Ⓧ RT Estimate the amplitude of the force on your eardrum from a sound that has an intensity level of 90 dB.

23. A sound wave has an intensity of 2.5×10^{-3} W/m^2. What is the intensity level in decibels?

24. ★ A source of sound has a total emitted power of 300 W. What is the intensity a distance of 10 m from the source? Assume the source emits spherical waves.

25. ✪ Suppose the sound intensity level 0.10 m from a loudspeaker is 110 dB. Assuming the speaker generates wave fronts that have a hemispherical shape (Fig. P13.25), what is the intensity 5.5 m from the speaker?

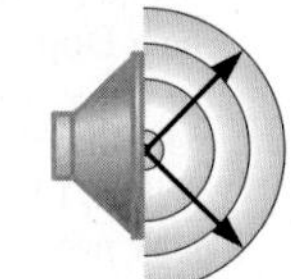

Figure P13.25

26. ★ The sound intensity level 20 m from a particular loudspeaker is 70 dB. Estimate the total power emitted by the speaker. Assume it emits spherical waves.

27. ★ What is the pressure amplitude of a sound wave if the intensity level is 40 dB?

28. ★ What is the intensity level of a sound wave if the pressure amplitude is 7.9 Pa?

29. ★ A point source emits sound with a power of 250 W. What is the intensity level in decibels at a distance of 35 m?

30. SSM ★ Ⓧ A person has severe hearing loss that reduces the intensity in his inner ear by 40 dB. To compensate, a hearing aid can amplify the sound pressure amplitude by a certain factor to bring the intensity level back to the value for a healthy ear. What amplification factor is needed for this person?

13.3 STANDING SOUND WAVES

31. For a standing sound wave, the closed end of an organ pipe forms which of the following? (More than one answer may be correct.)
(a) A displacement node
(b) A displacement antinode
(c) A pressure node
(d) A pressure antinode

32. For a standing sound wave, the open end of a flute forms which of the following? (More than one answer may be correct.)
(a) A displacement node
(b) A displacement antinode
(c) A pressure node
(d) A pressure antinode

33. Consider an organ pipe 1.2 m long that is closed at one end and open at the other. What is the fundamental frequency of the pipe?

34. The distance between a node and the nearest antinode of a standing sound wave in air is 0.24 m. What is the frequency of this wave?

35. If an organ pipe is filled with oxygen gas instead of air, will the frequency of the note go up or go down, and by what factor?

36. SSM ★ An organ pipe is open at one end and closed at the other. The pipe is designed to produce the note middle C (262 Hz). (a) How long is the pipe? (b) What is the frequency of the second harmonic? (c) How many pressure nodes of the second harmonic are found inside the pipe? Do not count nodes that may be at the ends of the pipe.

37. ★ A piccolo can be approximated as a tube that is open at both ends. If a piccolo is 20 cm long, what is the frequency of the lowest note it can play?

38. ✪ Consider an organ pipe that is closed at one end and open at the other. This pipe has resonant frequencies of 1200 Hz and 1500 Hz (among others). Find two possible values for the length of the pipe.

39. ✪ A pipe is open at both ends. The pipe has resonant frequencies of 528 Hz and 660 Hz (among others). Find two possible values for the length of the pipe.

40. ✪ RT An organ pipe that is closed at one end has a fundamental frequency of 175 Hz. There is a leak in the church roof, and some water gets into the bottom of the pipe as shown in Figure P13.40. The organist then finds that this organ pipe has a frequency of 230 Hz. What is the depth of the water in the pipe?

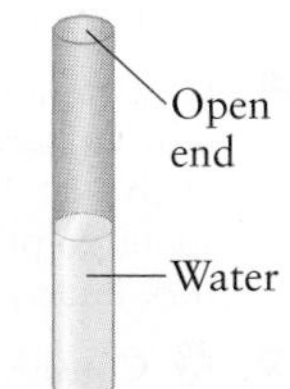

Figure P13.40

41. ★ Consider a steel rod that vibrates longitudinally. If the rod is 1.5 m long, what is the fundamental frequency? Assume the ends of the rod are displacement antinodes (i.e., it behaves as an organ pipe with both ends open).

42. ✪ R A pipe has a pattern of standing wave frequencies given in Figure P13.42. Is this pipe (a) open at both ends, (b) closed at both ends, or (c) open at one end and closed at the other? What is the length of the pipe?

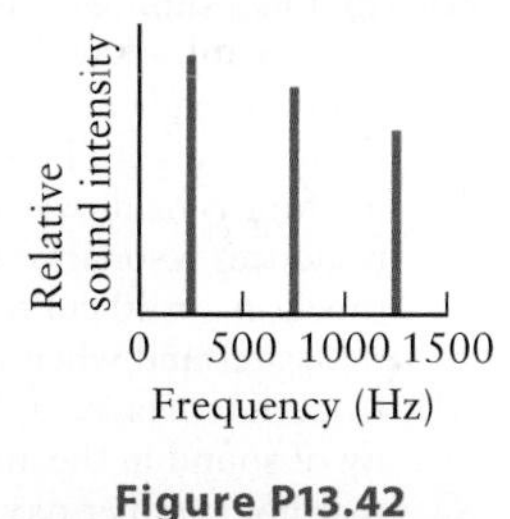

Figure P13.42

43. ★ RT Consider the soft-drink bottle on the far right in Figure Q13.17. What is the approximate value of the fundamental frequency produced when a person blows across the top of the bottle?

44. ✪ RT A trombone (Fig. P13.44) acts as a pipe that is closed at one end (the mouthpiece, where the player blows) and open at the other end (the bell of the instrument). (a) Is the lowest note played with the slide pulled as far as possible toward the player or when the slide is pushed out as far as possible? (b) What is the approximate frequency of the lowest note that can be played with the trombone in Figure P13.44?

Figure P13.44

13.4 BEATS

45. Two sound waves of frequency 80 Hz and 83 Hz are played. How many beats per second are heard?

46. A guitarist plucks two strings simultaneously. One string is tuned to a (fundamental) frequency of 340.0 Hz. It is found that beats are audible, with a beat frequency of 0.5 Hz. What are possible values for the frequency of the other string?

47. SSM ★ Two similar guitar strings have the same length and mass, but slightly different tensions. If the beat frequency is 1.5 Hz, what is the ratio of the tensions of the two strings? Assume they are both tuned to approximately 330 Hz. Give your answer to three significant figures.

48. ★ A violin string tuned to 485.0 Hz is found to exhibit 2.0 beats per second with a guitar string. The guitar string is then tightened, and the beat frequency is found to increase. What was the original frequency of the guitar string?

49. ★ A person can detect a beat rate as slow as 0.3 beat per second at a frequency of 1500 Hz. What is the percentage difference in frequency of the two notes?

50. ✪ RT One tuning fork vibrates at 440 Hz, and a second tuning fork vibrates at an unknown frequency. When the two forks are struck at the same time, beats with a frequency of 3 Hz are heard. (a) What is the frequency of the second tuning fork? Why are there two possible answers? (b) Describe an experiment you could perform with a small piece of wax to determine which of the possible answers in part (a) is correct.

13.5 REFLECTION AND SCATTERING OF SOUND

51. RT The sound from a church organ "reverberates" for many seconds after the organ stops producing sound due to reflections inside the church. That is, the sound from the organ bounces back and forth off the church walls many times before the sound intensity decays away completely. If the reverberation time is 10 s and the church has the (approximate) shape of a square box of length 25 m, approximately how many times does the sound reflect back and forth (that is, make one round trip through the room, reflecting from both walls) during the reverberation time?

52. SSM X An owl is chasing a squirrel and is using echolocation (the reflection of sound) to aid in the hunt. If the squirrel is 25 m from the owl, how long does it take sound to travel from the owl to the squirrel and then back to the owl?

13.6 THE DOPPLER EFFECT

53. A siren has a frequency of 750 Hz when it is at rest, whereas it has an apparent frequency of 800 Hz when it is moving toward a stationary observer. What is the speed of the siren in the latter case?

54. SSM ★ A siren has a frequency of 950 Hz when it and an observer are both at rest. The observer then starts to move and finds that the frequency he hears is 1000 Hz. (a) Is the observer moving toward or away from the siren? (b) What is the speed of the observer?

55. A guitar is tuned to play a note at 440 Hz. If the guitar is in a moving car and the note has a perceived frequency of 410 Hz, what is the speed of the car? Assume the car is moving directly away from a listener on the sidewalk.

56. A siren is approaching you at 35 m/s and is perceived to have a frequency of 1100 Hz. If the siren stops moving, what frequency do you now hear?

57. ✪ The siren in an ambulance has a frequency (according to you) of 1100 Hz as it moves toward you at speed v. It then makes a U-turn and travels away from you with speed v, and you hear a frequency of 950 Hz. What is the final speed v of the siren?

58. ★ X A bat is pursuing an insect and using echolocation of 60-kHz sound waves to track the insect. The insect is initially 30 cm from the bat, but then moves to a distance of 40 cm as it tries to escape. What is the difference in the two echo times the bat measures?

59. ✪ X Consider again the insect in Problem 58. If the insect moves the 10-cm distance in 0.50 s, what is the magnitude of the

Doppler shift in the bat's reflected sound wave? Assume the insect is moving directly away from the bat.

60. Two identical police cars are chasing a robber. When at rest, their sirens have a frequency of 500 Hz. A stationary observer watches as the two cars approach. The siren of one car (car 1) has a frequency of 600 Hz, and the other (car 2) has a frequency of 700 Hz. (a) Which car is moving faster? (b) What is the speed of each car? (c) If the robber hears a frequency of 650 Hz from car 2, is car 2 catching up to the robber? (d) What is the speed of the robber?

61. An ambulance with a siren emitting a whine at 1200 Hz overtakes and passes a cyclist pedaling a bike at 2.5 m/s. After being passed, the cyclist hears a frequency of 980 Hz. How fast is the ambulance moving?

13.7 APPLICATIONS

62. Ⓡ Some climate data suggest that during the period 1950 to 2000, the Earth's surface warmed an average of about 1°C. If the oceans have warmed this amount, how much will this change the time it takes a sound wave to travel from San Francisco to Honolulu?

63. SSM A child drops a rock into a vertical mineshaft that is precisely 406 m deep. The sound of the rock hitting the bottom of the shaft is heard 10.3 s after the child drops the rock. What is the temperature of the air in the shaft? Assume the temperature is the same throughout the mine. Ignore the force of air drag on the falling rock.

Additional Problems

64. The human ear canal (Fig. 13.3) is typically about 2.4 cm long, is roughly cylindrical, and is closed at one end where it is capped by the eardrum. (a) What is the fundamental frequency of the ear canal? (b) Determine the range of frequencies where the ear is most sensitive (see Fig. 13.4). How does the ear canal's fundamental frequency compare?

65. While taking a walk around your neighborhood, you see a flash of lightning hit the Earth in the distance. You immediately look at your watch and then hear thunder approximately 9 s later. (a) How far away did the lightning strike the ground? (b) Find a conversion factor for a thunder delay (measured in seconds counted) to the distance away of a corresponding lightning strike (in kilometers).

66. SSM The lowest frequency an average human ear can perceive is about 15 Hz, and the highest about 18 kHz. How tall would an organ pipe have to be to produce the lowest note? How short a tube would be needed to play the highest note? What harmonic (i.e., what value of n in Eq. 13.15) would the smaller tube excite in the larger one?

67. **Fundamental speech.** The sound of the talking or singing voice is produced by the vocal cords together with the vocal tract. When air is expelled through your throat, the vocal cords vibrate. For an adult man, this vibration has a fundamental frequency of about 110 Hz, with harmonics at 220 Hz, 330 Hz, 440 Hz, ... Sound with components at all these frequencies then passes through the vocal tract (Fig. P13.67A), which can be modeled as two pipes, one representing the larynx and another representing the oral cavity. When making the sound "ah" as in the word *father*, the larynx acts approximately as a pipe closed at both ends (with just small openings at the vocal cords and the entry to the oral cavity), while the oral cavity acts as a pipe that is open at one end (at the lips). The resonant frequencies of these two "pipes" then affects which of the harmonics of the vocal cords are strongest in the final sound. Figure P13.67B shows the spectrum of the vowel sound "ah"; the strongest harmonics are near 750 Hz, 1100 Hz, and 2500 Hz. These peaks in the spectrum are called *formants*. If the larynx has a length of 16 cm and the oral cavity a length of 11 cm, which harmonic of which section of the vocal tract gives rise to each of these formants?

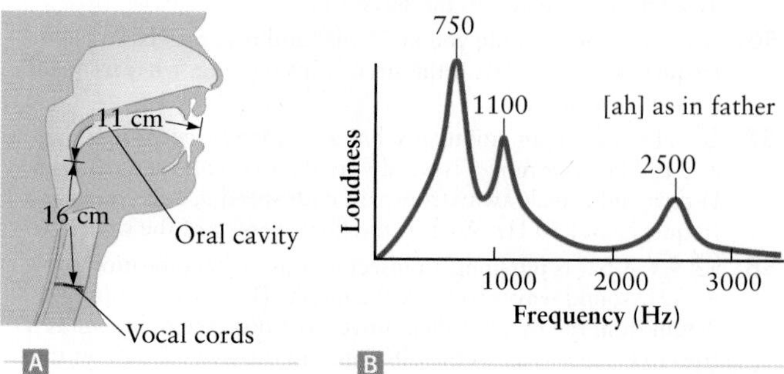

Figure P13.67 A The human vocal tract. B Loudness versus frequency for the vowel "ah" as in *father*.

68. **Drop in pitch.** A tuning fork that rings at 440 Hz is dropped down a deep well. How far down the well is the fork when the pitch is at 425 Hz? Don't forget to include the time it takes the sound to propagate to the top of the well and assume the tuning fork undergoes free fall with negligible air drag.

69. Consider the falling tuning fork in Problem 68 with frequency 440 Hz. (a) You hear it splash into the water at the bottom of the well after 1.8 s. How deep is the well? (b) What frequency is heard just before the splash? Assume the tuning fork undergoes free fall with negligible air drag.

70. On a summer afternoon, a pipe that is open at both ends is held vertically with one end submerged in water as shown in Figure P13.70. If a 220-Hz tuning fork is held over the end of the tube, it induces resonance when the top of the tube is $h_1 = 40$ cm from the surface of the water and when it is $h_2 = 120$ cm from the surface. (a) What was the velocity of sound in the air at that time on that hot summer day? (b) What was the air temperature? Is this method a good way to measure temperature? Why or why not?

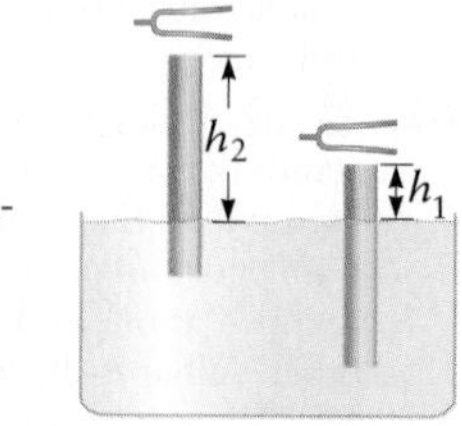

Figure P13.70

71. You buy a dog whistle that operates at 21 kHz, but when you blow into it, your dog does not show the slightest reaction. You decide to see if you can hear the whistle yourself. Two friends help you. One drives her car past you while the other blows the whistle from the car's window. (a) How fast must your friend drive the car past you to allow you to hear the whistle, assuming you can hear a maximum frequency of 19 kHz? Give your answer in meters per second and miles per hour. (b) If the whistle instead operates at 24 kHz, what speed would the car then need? Would this experiment be feasible? How could you do the experiment with two cars?

72. Dolphins use sonar as a sixth sense, giving them a three-dimensional view of their surroundings via sound. A dolphin's hearing spans the range from 15 Hz to 150 kHz, making use of the 75–110 kHz range for its echolocation sonar. Sound is emitted in a narrow beam using an organ called the melon (Fig. P13.72), which has a dense core (high sound velocity) compared with the surrounding tissue. The melon acts as an acoustic lens that focuses the sound wave pulse, and the dolphin's jaw acts as a receiver that channels the appropriate frequencies of the reflected pulse to the ear. (a) Consider a dolphin chasing a fish as in Figure P13.72. If the fish is 50 cm away, how long does it take a wave pulse to travel from the dolphin to the fish and back? (b) What is the wavelength of the sonar pulse if the frequency of the emitted sound is 90 kHz? How does this wavelength compare to the size of the fish? Why is that important? (c) Suppose the dolphin is in pursuit of the fish, with the fish swimming at a speed of 10 m/s and the dolphin swimming at 25 m/s. What frequency for the return pulse does the dolphin hear? (d) The dolphin can sense the speed of the fish, not by the small shift in frequency away from 90 kHz, but by emitting a series of pulses that sound like clicks. Each click is a pulse of 90-kHz sound with a duration of about 50 μs. Such clicks are emitted at a rate of 900 per second. What click rate would a dolphin detect reflected from the fish in part (c)?

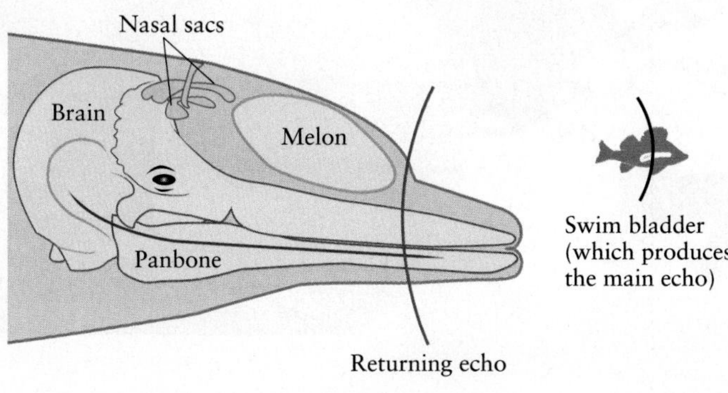

Figure P13.72

73. ★ RT A trumpet at room temperature (25°C) is tuned to play the note middle C, which has a fundamental frequency of 262 Hz. The trumpet is then taken outside, and after a few minutes, it is found that the same note has a frequency of 245 Hz. What is the outside temperature?

74. ★ Two adjacent notes in the musical scale used in Western music have fundamental frequencies whose ratio is approximately 1.059. For example, the fundamental frequencies of the notes C-sharp and C have this ratio. The frequency of a note produced by a flute, or any other woodwind instrument, depends on the temperature. How much must the temperature change to change the note C into C-sharp? Assume the instrument is initially at 20°C and that you can use Equation 13.5 for the temperature dependence of the speed of sound in air.

75. A blood flowmeter uses the Doppler effect to measure the speed of blood. Suppose a sound wave at 80,000 Hz is emitted into a vein where the speed of sound is about 1500 m/s. The wave is Doppler-shifted as it is absorbed by the red blood cells moving toward a stationary receiver. The wave is then Doppler-shifted again as it is reflected back toward the detector. Hence, there are two Doppler shifts, one as the red blood cells act as the "observer" and another as they act as a new source of sound. If the frequency of the returning sound is 80,020 Hz, what is the speed of the blood flow?

76. ★ RT The pressure amplitude associated with a very faint sound is about 2×10^{-5} Pa. (a) If this is the difference in pressure between the two sides of your eardrum, what is the approximate force on your eardrum? (b) How does this force compare to the weight of a mosquito (approximate mass 10^{-6} kg)?

CHAPTER

14

► *Although she may not be thinking about it, temperature and heat are important for this camper! (© Extreme Sports Photo/Alamy)*

Temperature and Heat

OUTLINE

All our work on mechanics in Chapters 2 through 13 has been based on Newton's laws of motion. The central thread of that work was *force*. We discussed various types of forces (such as gravity) and how these forces cause a particle to move in various situations. A careful consideration of Newton's laws led us to other important concepts such as work and energy, but the main theme was the concept of force. In this chapter and Chapters 15 and 16, we move to a different area of physics known as ***thermodynamics***, and our focus now shifts to *energy*. The key pillar of thermodynamics is the principle of conservation of energy. Although conservation of energy certainly has an important place in mechanics, it will now play an even more central role. In a nutshell, thermodynamics is about the transfer of energy between *systems* of particles and about the way changes in the energy of a system affect its properties.

For thermodynamics, we need several new quantities to describe the properties of systems and their interactions. Two of these quantities—temperature and heat—are the main subjects of this chapter.

14.1 Thermodynamics: Applying Physics to a "System"

A thermodynamic ***system*** contains multiple particles, usually a very large number of particles. Examples are a balloon full of gas molecules, a drop of water, and a sugar cube, each of which contains many atoms and molecules. The particles in a system are able to exchange energy with one another (via collisions), and systems are able to exchange energy with other systems.

Figure 14.1A shows a balloon full of oxygen molecules. What quantities are useful for describing the physics of this system? We could describe its properties by specifying the position and velocity of every oxygen molecule. We would then want to determine the forces exerted by each molecule in the balloon on all the other molecules, along with all the external forces (e.g., the forces from outside the system such as gravity). With all this information, we could use Newton's laws to calculate the motion of each individual oxygen molecule. Although we could take this approach in principle, it is not feasible in practice because of the extremely large number of atoms or molecules in a typical sample of gas. Moreover, even if it were feasible, we are really not interested in this level of detail. We can describe all the important properties of this system in a much simpler way.

Figure 14.1B shows a second balloon that also contains oxygen gas. The two balloons in Figure 14.1 are "identical" in that they are the same size and shape and contain the same number of molecules. The positions and velocities of the molecules in the two balloons will certainly not be the same, however, so these two systems will not be identical on the molecular scale. Fortunately, we are usually not interested in knowing precisely where all the molecules are located. It is much more useful to know the values of various properties of the system *as a whole*.

To describe the physics of a system fully, only a small number of systemwide properties such as pressure and temperature are needed. Also called ***macroscopic*** properties, they describe the behavior on a scale that is much larger than the scale of an individual particle. Macroscopic properties contrast with ***microscopic*** variables such as the position and velocity of a particular molecule within the system. Newton's laws provide a description at the level of these microscopic variables. Our task in this chapter is to describe a system of many particles, such as a gas-filled balloon (Fig. 14.1) or a piece of steel, at the macroscopic level. We also explore the connection between microscopic and macroscopic descriptions and use Newton's laws to gain insight into the meanings of the macroscopic properties.

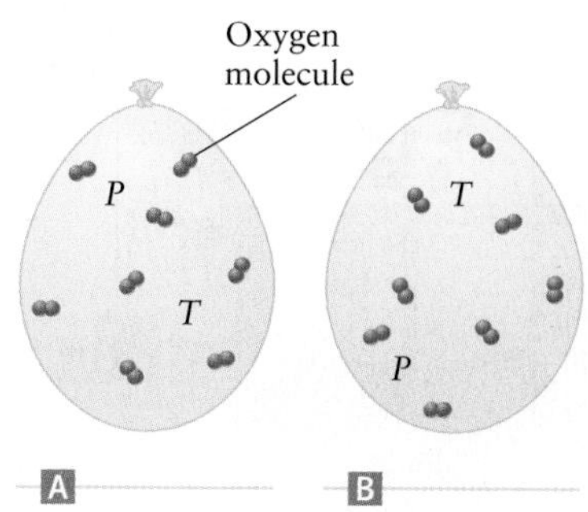

▲ **Figure 14.1** These two hypothetical balloons have the same volume and gas density. Two such systems having equal density, pressure, and temperature are said to be in the same state, even though the precise positions and velocities of the gas molecules are different.

14.2 Temperature and Heat

Temperature is not contained in or derivable from Newton's laws, which is one reason it is not easy to give a "simple" definition of temperature in the same way we gave definitions of velocity and acceleration in our work on mechanics. We can approach the notion of temperature in several different ways, each yielding useful insights.

Temperature is connected with the "hotness" or "coldness" of a system, so let's first consider a macroscopic definition. Figure 14.2A shows two hypothetical systems, which might be balloons filled with gas. If each system is initially isolated from its surroundings for a long period of time, each will have its own temperature. In general, these two initial temperatures will not be the same, and we denote them by T_1 and T_2. Suppose $T_1 > T_2$; in ordinary language, we would say that system 1 is initially hotter than system 2. If we then bring the systems into contact (Fig. 14.2B), energy is transferred spontaneously from system 1 (the hotter system) to system 2 (the colder system), and their temperatures will change. If we wait for a long time (Fig. 14.2C), the two systems will reach ***thermal equilibrium***. They then have the

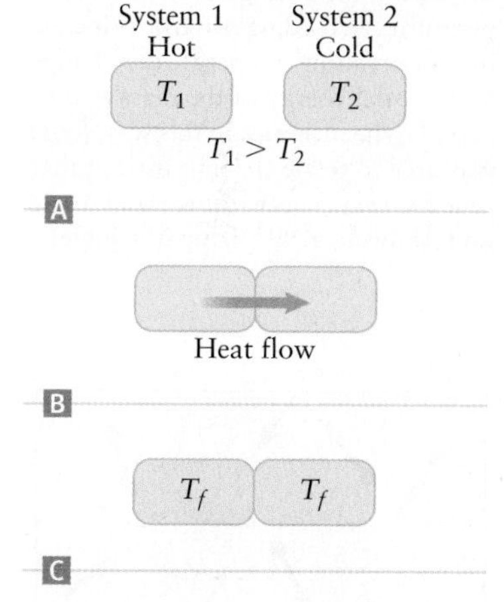

▲ **Figure 14.2** Heat (also referred to as heat energy) is energy transferred from one system to another due to a temperature difference. **A** Two systems at different temperatures. **B** Energy flows from the hotter system at temperature T_1 to the cooler system at temperature T_2 when they are brought into contact. **C** If the systems are left in contact for a long time, they will eventually come into thermal equilibrium and have the same final temperature.

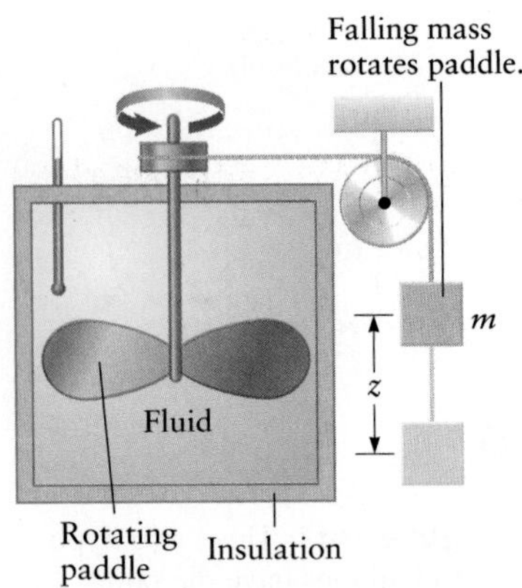

▲ **Figure 14.3** Apparatus used to measure the relation between heat energy and mechanical energy. See Insight 14.1.

Insight 14.1
THE EQUIVALENCE OF MECHANICAL ENERGY AND HEAT ENERGY
In the 1840s, James Joule measured the "mechanical equivalent" of heat energy with an apparatus like the one shown in Figure 14.3. As a mass falls through a height z, its potential energy is used to rotate a propeller in a liquid. The propeller's motion heats the liquid, increasing the liquid's temperature. According to the principle of conservation of energy, the change in potential energy of the mass is converted to heat energy. In this way, Joule was able to relate the unit used at that time for heat energy (the calorie) to the unit of mechanical energy (the joule).

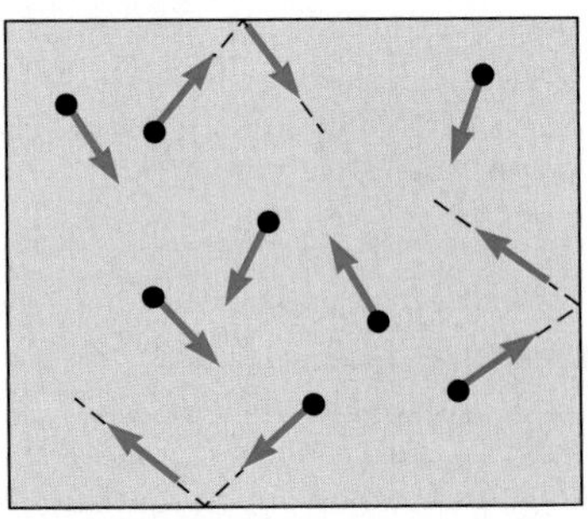

▲ **Figure 14.4** The pressure in a gas comes from collisions of gas molecules with the surface of the container. Each collision produces a force on the container walls. If the temperature is increased, the molecular speed increases, causing the collision forces and hence the pressure to increase.

same final temperature T_f, which lies somewhere between T_1 and T_2. The energy that flows between the two systems is called ***heat*** or ***heat energy***.

> ***Heat*** **is energy that passes from one system to another by virtue of a temperature difference.**

The terms *heat* and *heat energy* are often used interchangeably, especially in everyday conversation. When used in physics, these terms always refer to the transfer of energy between systems. Hence, you may also hear the term *heat flow* used to describe the process by which heat energy is transferred from one object to another. According to the principle of conservation of energy, the amount of heat energy that "leaves" system 1 in Figure 14.2B must equal the amount of heat energy that "enters" system 2. This transfer of energy can take place in different ways, but the direction of transfer—the direction of heat flow—depends only on the temperature difference.

Units of Heat

Heat is measured with the SI unit of energy, the joule (J). A unit called the ***calorie*** is also widely used in measurements involving heat. The term *calorie* comes from the 1700s when some scientists thought that heat was a sort of fluid (called a "caloric fluid") that flowed between objects. Although it is not immediately obvious that the energy that flows into your finger when you touch a hot object is the same type of physical quantity as the kinetic energy of a falling brick, experiments in the mid-1800s using an apparatus like the one in Figure 14.3 showed that heat is, in fact, a form of energy.

While most physicists measure heat in units of joules, the calorie is widely used, especially in chemistry. These units are related by

$$1 \text{ cal} = 4.186 \text{ J} \tag{14.1}$$

A unit named the "Calorie" (uppercase "C") is used to measure the energy content of foods. This "food calorie" is actually equal to 1000 cal (lowercase "c").

Temperature: A Microscopic Picture

Our definition of temperature tells how to *compare* the temperatures of two different systems. We simply place the two systems in contact and then observe if energy is transferred from one to the other and in which direction (Fig. 14.2). This procedure, however, does not give us a way to measure the *value* of the temperature of a particular system. One way to measure the temperature is to place the system of interest in contact with a gas-filled container held at constant volume and density. The pressure in the gas is caused by collisions of gas atoms and molecules with the walls of the container (Fig. 14.4). If the temperature is increased, the average speed of the gas atoms increases, causing the pressure to increase also. We'll see in Chapter 15 that for a dilute gas (one with a low density), the temperature is linearly proportional to the gas pressure, so measurement of the pressure gives a direct way to find the temperature (Fig. 14.5A). Such a device is called a ***gas thermometer***.

The picture in Figure 14.4 gives insight into the microscopic meaning of temperature. The temperature of a system of particles (such as a gas) is related to the average particle speed. A high speed corresponds to a high temperature, and a low speed corresponds to a low temperature. Because high speeds also mean high kinetic energies, this relation suggests that the temperature of a system can be increased by adding energy to it. This is another example of the connection between temperature and energy. We'll derive the relation between temperature and energy for a gas in Chapter 15.

Temperature Scales

Three different ***temperature scales*** are in common use: the ***Fahrenheit*** scale, the ***Celsius*** scale, and the ***Kelvin*** scale. The Kelvin scale is the one employed in the SI system of units, but all three scales are widely used, so you should be able to con-

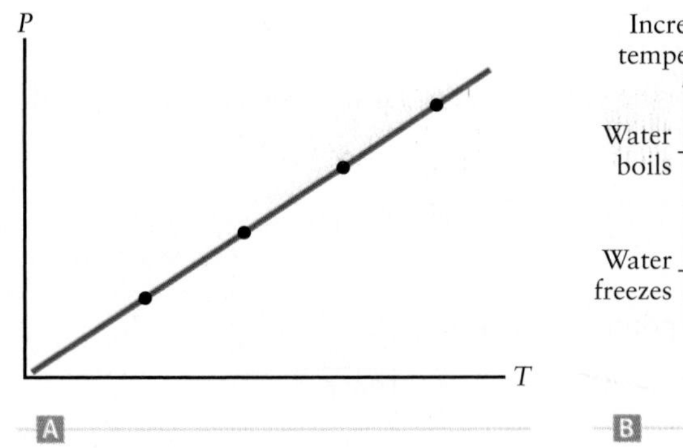

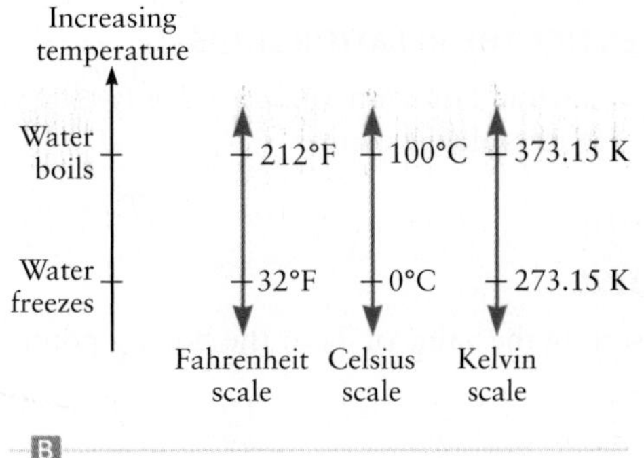

◀ **Figure 14.5** A The pressure in a dilute gas (such as air) is proportional to the temperature. B Temperature can be measured with the Fahrenheit, Celsius, or Kelvin scale.

vert temperature values from one scale to any other. This conversion process can be understood from Figure 14.5B.

The Celsius scale is defined so that at atmospheric pressure, water freezes at exactly 0°C and boils at precisely 100°C. In words, these temperatures are "zero degrees Celsius" and "100 degrees Celsius." On the Fahrenheit scale, these processes occur at 32°F and 212°F, respectively, referred to as "32 degrees Fahrenheit" and so on. On the Kelvin scale, these temperatures are 273.15 K and 373.15 K, respectively. These values are referred to as "273.15 kelvin" and so on; by convention, the term *degrees* is not used in connection with the Kelvin scale.

The relations between the different temperature scales are illustrated in Figure 14.5B. Because all three scales are linear, the values of the temperatures at which water freezes and boils on each scale define how to convert a temperature value from one scale to another. For example, to convert from a value on the Celsius scale T_C to a value on the Kelvin scale T_K, we have (according to Fig. 14.5B)

$$T_K = T_C + 273.15 \quad \text{(conversion from Celsius to Kelvin)} \tag{14.2}$$

Notice that the sizes of a temperature unit are equal in the Celsius and Kelvin scales, so a change of 1 "degree" on the Celsius scale is equal to a change of 1 kelvin on the Kelvin scale.

Conversions involving the Fahrenheit scale are not quite as simple because a 1° change on the Fahrenheit scale is not equal to a 1° change on the Celsius scale. From Figure 14.5B, we see that a change of 100°C (going from freezing water to boiling water) corresponds to a change of 212 − 32 = 180°F. A temperature change of 1°C is thus equivalent to a change of 180/100 = 9/5 = 1.8°F. We also know that 0°C is equivalent to 32°F (the freezing point of water), which leads to the relation

$$T_F = \tfrac{9}{5}T_C + 32 \quad \text{(conversion from Celsius to Fahrenheit)} \tag{14.3}$$

EXAMPLE 14.1 The Boiling Temperature of Liquid Nitrogen

At atmospheric pressure, liquid nitrogen boils at a temperature of 77 K (Fig. 14.6). Convert this value to the Celsius and Fahrenheit scales.

RECOGNIZE THE PRINCIPLE

Given the temperature of liquid nitrogen, we can use Figure 14.5B and Equations 14.2 and 14.3 to find the corresponding temperatures on the Celsius and Fahrenheit scales.

SKETCH THE PROBLEM

Figure 14.5B shows the relationship between the Kelvin, Celsius, and Fahrenheit scales.

(continued) ▶

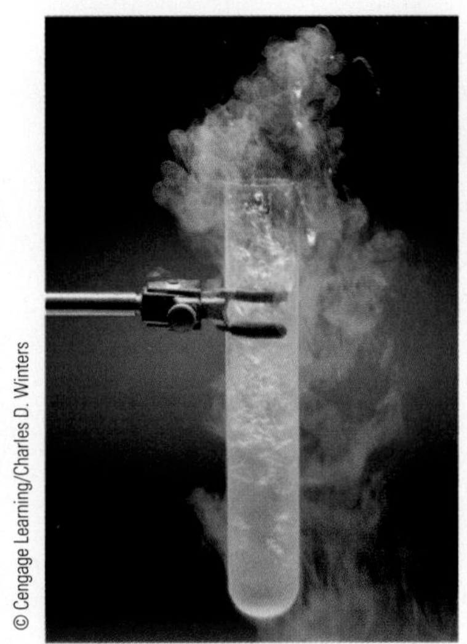

▲ **Figure 14.6** Example 14.1. At atmospheric pressure, nitrogen boils at 77 K.

IDENTIFY THE RELATIONSHIPS

Rearranging Equation 14.2 to solve for the value on the Celsius scale (T_C) in terms of the Kelvin value, we have

$$T_C = T_K - 273$$

SOLVE

Inserting the value of T_K at the boiling point of nitrogen gives

$$T_C = T_K - 273 = 77 - 273 = \boxed{-196°\text{C}}$$

To get the value on the Fahrenheit scale, we use Equation 14.3:

$$T_F = \tfrac{9}{5}T_C + 32 = \tfrac{9}{5}(-196) + 32 = \boxed{-321°\text{F}}$$

▸ *What does it mean?*

Equations 14.2 and 14.3 can be used to convert between any two temperature scales. If you forget the equations, you can always work out the conversions if you remember the temperatures at which water freezes and boils.

Very High and Very Low Temperatures: What Are the Limits?

The temperature axes in Figure 14.5B have arrows at both ends to indicate that they extend to both higher and lower temperatures. Many important temperature values lie well above the boiling point of water or well below the freezing point of water (Table 14.1). Several of these temperatures have negative values on either the Fahrenheit or Celsius scales (or both). Indeed, the boiling temperature of nitrogen, which was considered in Example 14.1, falls at such negative temperatures. However, there is a limit at the bottom ends of these scales; there are no negative values of temperature on the Kelvin scale listed in Table 14.1. The value $T_K = 0$ K is called ***absolute zero***; in Chapter 16, we'll see why 0 K is a very special temperature.

TABLE 14.1 Values of Some Important Temperatures on the Kelvin, Celsius, and Fahrenheit Scales

	Temperature Scale		
	K	**°C**	**°F**
Interior of the Sun	$\approx 1 \times 10^7$	$\approx 1 \times 10^7$	$\approx 2 \times 10^7$
Center of the Earth	≈7000	≈7000	≈13,000
Surface of the Sun	≈6000	≈6000	≈10,000
Lead melts	600	327	621
Water boils	373.15	100	212
Human body temperature	310	37	98.6
Water freezes	273.15	0	32
Nitrogen boils	77	−196	−321
Intergalactic space	3	−270	−454
Absolute zero	0	−273.15	−459.67

14.3 Thermal Equilibrium and the Zeroth Law of Thermodynamics

T_A T_B

A B

T_C

C

ZEROTH LAW: If A and B are in thermal equilibrium and if B and C are in thermal equilibrium, then A and C are also in thermal equilibrium and $T_A = T_B = T_C$.

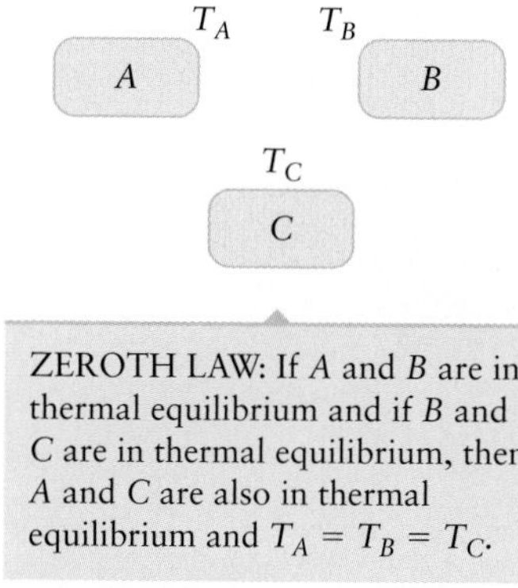

▲ **Figure 14.7** The zeroth law of thermodynamics involves the relationships between thermal equilibrium and temperature for three different systems.

Let's consider the transfer of energy between the three systems sketched in Figure 14.7. We assume these three systems are initially isolated, with temperatures T_A, T_B, and T_C. Suppose we bring systems A and B into contact and find that there is no heat flow; that is, there is no transfer of energy between them. Using our definitions of temperature and heat, we would conclude that $T_A = T_B$. We then repeat this exercise with systems B and C, and again we find that there is no heat flow between these two systems; hence, $T_B = T_C$. What, then, would we say about the temperatures of A and C? The obvious conclusion is that $T_A = T_C$, which is called the ***zeroth law of thermodynamics***.[1]

> ***Zeroth law of thermodynamics:*** **If two systems A and B are in thermal equilibrium (so that $T_A = T_B$) and systems B and C are in thermal equilibrium ($T_B = T_C$), systems A and C are also in thermal equilibrium ($T_A = T_C$).**

Although this statement may seem trivial, it has some very important implications. In words, the zeroth law of thermodynamics tells us that the basic concept of temperature is meaningful and that temperature is a unique property of a system that is allowed to come into thermal equilibrium. Indeed, we took these facts for granted in our discussion of temperature and heat in Section 14.2. It is actually an amazing fact that *only one property* of a system, its temperature, determines when heat energy will flow between it and another system.

We can illustrate one implication of the zeroth law pictorially. Figure 14.8A shows an endless waterfall in which water circulates around a closed loop, flowing downhill during each part of the loop. Of course, such a waterfall is impossible because it would violate what we know about gravitational potential energy and the principle of conservation of energy. Figure 14.8B shows an "equivalent" (and still impossible) situation, this time involving heat flow. Here heat flows from system 1 to system 2, then flows from system 2 to system 3, and finally completes the loop by

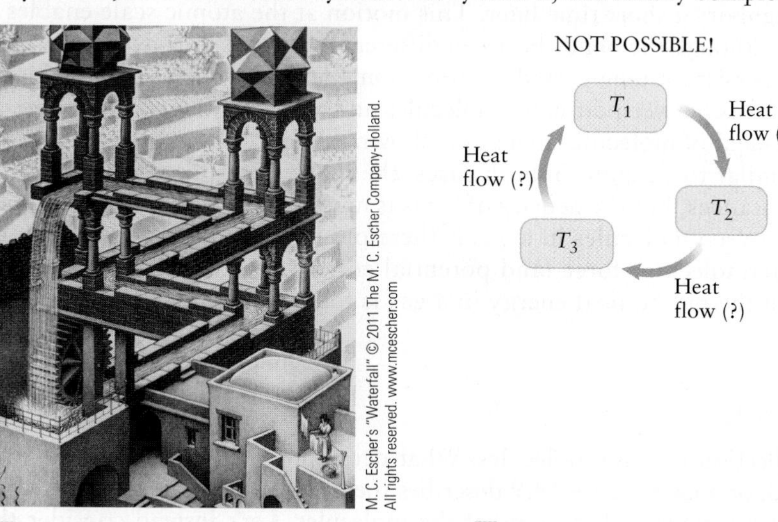

◀ **Figure 14.8** A In this famous lithograph *Waterfall* by M. C. Escher, water flows downhill endlessly around a closed loop. That is not possible in the real world because it is not consistent with the principle of conservation of energy. (It would violate the properties of gravitational potential energy.) B The zeroth law of thermodynamics states that this "circular" flow of heat energy between three different systems is *not possible*.

[1]You might wonder why physicists would start counting such laws at the number zero. The logical need for the zeroth law of thermodynamics was realized only after the first, second, and third laws of thermodynamics were already discovered and named.

flowing from system 3 to system 1. The zeroth law and our understanding of heat flow tell us that such a circular flow of heat is impossible because heat must always flow from high temperature to low temperature.

14.4 Phases of Matter and Phase Changes

The three phases of matter—***solid***, ***liquid***, and ***gas***—are illustrated in Figure 14.9. The atoms in many solids are arranged in an orderly and repeating pattern called a crystalline lattice (Fig. 14.9A), in which each atom is held in place by the forces exerted by neighboring atoms. These forces are a result of chemical bonds within the solid. For example, ionic bonds hold Na^+ and Cl^- within a lattice to form sodium chloride (NaCl), and covalent bonds hold silicon (Si) atoms in a crystalline lattice. The bonds between neighboring atoms in a solid behave essentially as "springs," giving a force similar to the Hooke's law force discussed in Chapter 11. If a particular bond is stretched or compressed, the atom experiences a restoring force proportional to its displacement. This force is directed back toward the "ideal" location of the atom within the lattice.

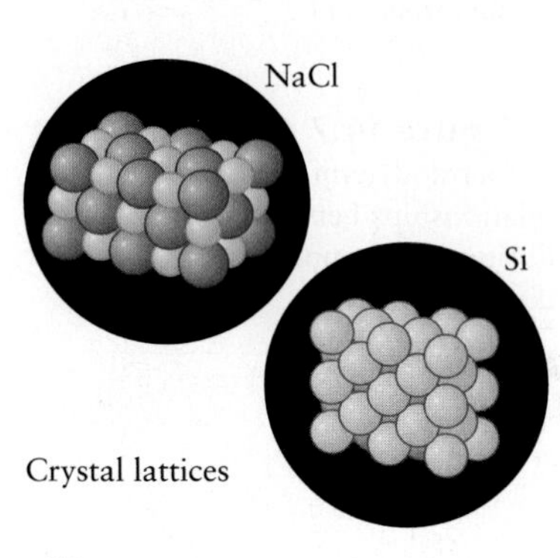

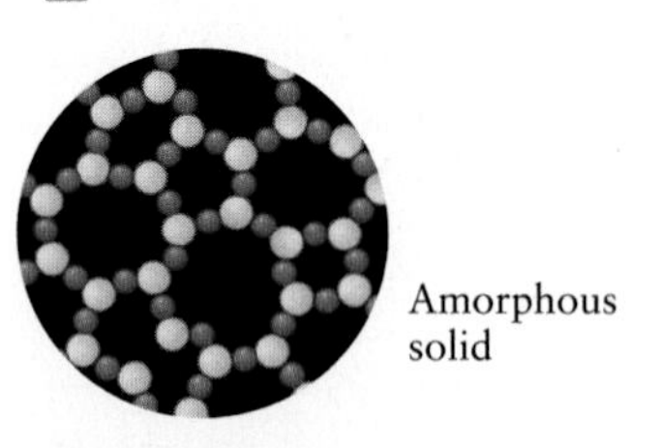

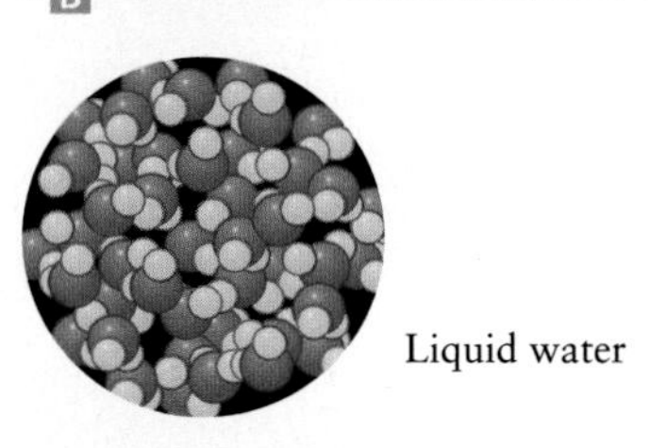

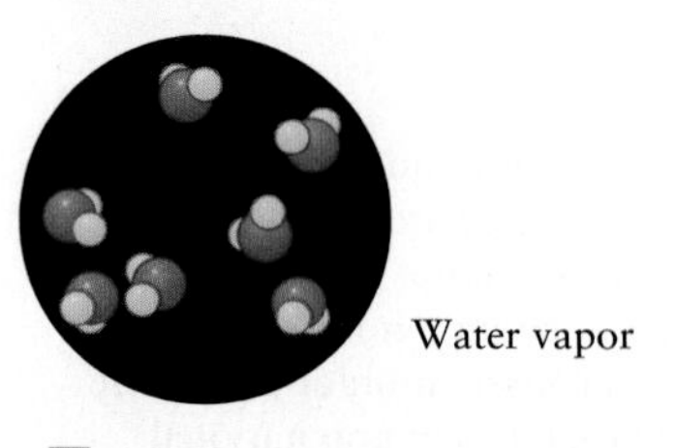

▲ **Figure 14.9** There are three phases of matter: solid, liquid, and gas. The atoms in a solid can be arranged in A a crystal lattice or B a somewhat random spatial arrangement called an amorphous solid. C In a liquid and D in a gas, the atoms do not have a regular arrangement and are able to move about.

Figure 14.9A is a simplified picture, showing all the atoms in their ideal lattice positions. More realistically, the atoms vibrate about these positions as simple harmonic oscillators. The kinetic and potential energies associated with these oscillations can be understood using the ideas we encountered in Chapter 11.

Figure 14.9B shows another example of a solid phase called an amorphous solid. A familiar example is common window glass, which is composed mainly of silicon and oxygen. The atomic arrangement in an amorphous solid does not have the regular repeating structure found in a crystal. Even so, the atoms are still held in place by chemical bonds that behave as springs obeying Hooke's law.

Figure 14.9C shows the atomic arrangement found in a liquid. The difference between liquids and amorphous solids is evident when the atomic arrangement is followed for a period of time. The atoms in a liquid are not held in fixed locations by the forces from neighboring atoms. Instead, atoms in a liquid are able to move about, and particular atoms that happen to be close neighbors at one moment are usually not neighbors a short time later. This motion at the atomic scale enables a liquid to flow. Although the bonds between different molecules in water (or in other liquids) do not persist for long periods of time, some potential energy is still associated with the forces between different molecules and atoms in a liquid.

The arrangement of molecules in a gas is shown in Figure 14.9D. In some ways, a gas is very similar to a liquid. In both cases, the molecules are able to move easily over long distances, but the density of a gas is much lower than that of a liquid. The spacing between molecules in a gas is therefore larger, and the magnitude of the average intermolecular force (and potential energy) is much smaller. For this reason, most of the mechanical energy in a gas is found in the kinetic energies of the molecules.

Internal Energy

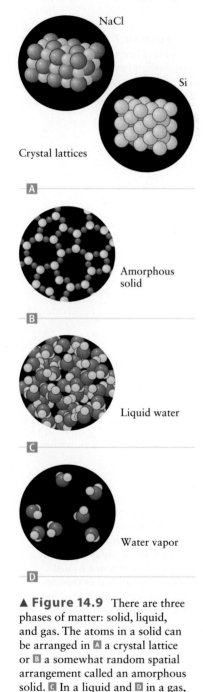

Consider a collection of H_2O molecules. What determines if these molecules form a solid, a liquid, or a gas? Figure 14.9 describes the difference between these phases in terms of the positions and motions of the molecules. Let's instead consider the energies of molecules when the system is in these different phases. We have already mentioned that the kinetic energy of a typical molecule is largest in the gas and smallest in the solid, so we can expect the mechanical energy of the molecules to be different in the different phases. This energy is called the ***internal energy*** of the system and is denoted by U. The internal energy of a system of molecules is the sum of the potential energies associated with all the intermolecular bonds plus the kinetic energies of all the molecules. The value of U increases as we go from the solid to the liquid to the gas phase. In general, the internal energy of all systems increases as the temperature is increased.

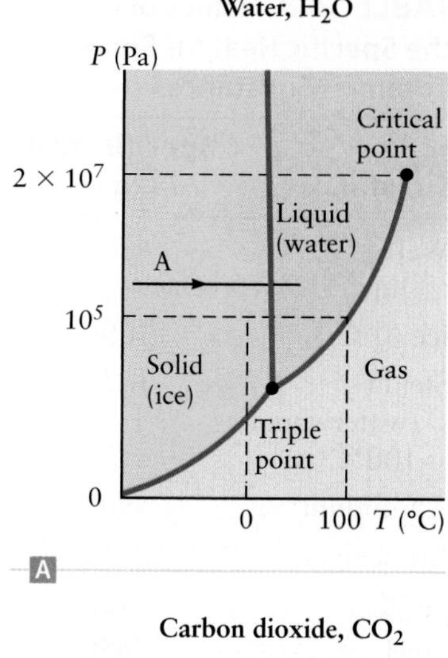

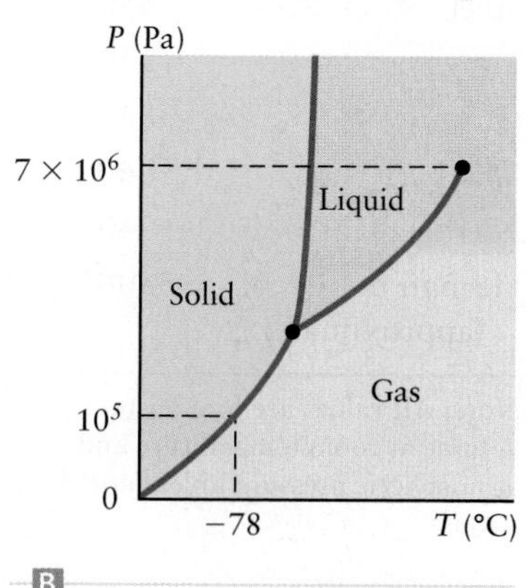

▲ **Figure 14.10** A phase diagram shows the temperatures and pressures at which phase changes occur (solid brown lines). Phase diagram for A H_2O and B CO_2. Note that 10^5 Pa is approximately atmospheric pressure.

Phase Changes

The transformation of a solid to a liquid, and of a liquid to a gas, are examples of ***phase changes***. Phase changes can be produced by changing the temperature or by changing the system's pressure as shown in the ***phase diagram*** of a substance. The phase diagram of H_2O (Fig. 14.10A) shows the phases found at different pressures and temperatures. For example, H_2O is a gas at low pressures and moderate to high temperatures, whereas it is a solid at low temperatures and moderate to high pressures. Phase changes occur at each of the solid brown lines in Figure 14.10A. For example, if we start with H_2O in the solid phase (ice) above a pressure of about 600 Pa and then increase the temperature (following, for example, path *A* in Fig. 14.10A), we eventually cross into the liquid phase. This phase change is called ***melting***. If instead we were to start in the liquid phase and then reduce the temperature, we would observe the phase change called ***freezing***. The phase diagram of H_2O includes other phase changes between liquid and gas (evaporation) and between solid and gas (sublimation). The line that separates liquid and gas ends at the ***critical point***. There is also a ***triple point***, where the solid, liquid, and gas phase regions all meet. The temperature of the triple point for water is 273.16 K.

Other substances have similar phase diagrams. For example, Figure 14.10B shows the phase diagram for CO_2. The overall arrangement of the phases is similar to the case of H_2O, although the values of temperature and pressure at which the phase changes occur vary from one substance to another. Table 14.2 lists the melting and evaporation temperatures of some common substances at atmospheric pressure.

Specific Heat and Heat Capacity

The phase diagram in Figure 14.10A shows a path labeled *A* along which the system can be taken from a low to a high temperature. In practice, this change could be accomplished by adding heat energy to the system by placing the system in contact with a second system at a higher temperature so that heat energy can flow between them. How much energy is required? Let's first consider what happens to a system when heat is added, but the system does not undergo a phase change. For example, our system might be a liquid (water) whose temperature increases but does not reach the evaporation (boiling) temperature. Let Q denote the heat energy added to the system and ΔT be the change in temperature. The ratio of these two quantities is called the ***heat capacity***:

$$\text{heat capacity} = \frac{Q}{\Delta T} \tag{14.4}$$

TABLE 14.2 Melting and Evaporation Temperatures for Some Common Substances at Atmospheric Pressure

Substance	Melting Temperature (K)	Evaporation Temperature (K)
Water	273	373
Aluminum	933	2720
Gold	1340	2930
Ethyl alcohol	159	351
Carbon dioxide	195	217
Nitrogen	63	77

TABLE 14.3 Values of the Specific Heat for Some Common Substances

Substance	Specific Heat J/(kg · K)
Water (liquid, 15°C)	4186
Ice (0°C)	2090
Steam (water vapor, 100°C)	1520
Aluminum	900
Ethyl alcohol	2450
Gold	129
Steel	450
Tungsten	130
Granite	840
Glass	840
Nitrogen gas	740
Human body (approximate)	3500

Note: All values are for constant volume at room temperature and atmospheric pressure unless noted.

This ratio does not take account of the mass of the system; a large system has a large heat capacity because a large amount of energy Q is required to change the temperature by a given amount. (Likewise, a small system has a small heat capacity.) It is therefore useful to define a closely related quantity called the ***specific heat*** c by

$$\text{specific heat} = c = \frac{Q}{m\,\Delta T} \tag{14.5}$$

The specific heat takes account of the size of the system, so c is the same for any sample of a certain substance (at a particular temperature and pressure), regardless of its mass.

We can rearrange Equation 14.5 as

$$Q = cm\,\Delta T \tag{14.6}$$

Knowing the value of c thus allows us to calculate how much heat energy is needed to increase the temperature of a system of mass m by an amount ΔT. By convention, a positive value of Q corresponds to heat energy *added to* the system, and a negative value of Q indicates that energy flows *out of* the system. In our mechanical picture involving internal energy, adding or subtracting energy changes the internal energy U of the system (the "internal" kinetic and potential energies of the atoms and molecules).

Values of the specific heat for a few common materials are given in Table 14.3. The values in Table 14.3 were (unless noted) determined at room temperature and pressure, with the volume of the system kept fixed; such values are called the specific heat at *constant volume.* The specific heat can vary with temperature, but these changes are usually small. Notice that c is always positive. Adding energy to a system (a positive value of Q in Eq. 14.6) always increases its temperature (a positive ΔT).

Values of the specific heat vary widely among different substances and tend to be larger for liquids than for solids or gases. In Table 14.3, the values of c are listed in SI units, J/(kg · K). You will often find values of specific heat quoted in terms of calories (instead of joules), especially in chemistry books, so you need to be able to convert between these two units. From Table 14.3, we have $c = 4186$ J/(kg · K) for water. Converting the joules to calories (according to Eq. 14.1), we have

$$c = 4186\,\frac{\text{J}}{\text{kg} \cdot \text{K}} \times \frac{1\text{ cal}}{4.186\text{ J}} \times \frac{1\text{ kg}}{1000\text{ g}} = 1.000\,\frac{\text{cal}}{\text{g} \cdot \text{K}} \quad \text{(specific heat of water)}$$

The value of the specific heat of water thus has a particularly "simple" (and easy to remember) value when expressed using calories. Indeed, the calorie was originally defined through the specific heat of water.

CONCEPT CHECK 14.1 Specific Heat of the Human Body

Even though the human body is not a pure substance, we can still define its specific heat through Equation 14.5. According to Table 14.3, the human body's specific heat is only a little smaller than the specific heat of water and is much larger than the specific heat of the other substances listed. Why?

Why Is Specific Heat Important?

With knowledge of the specific heat, we can calculate how the temperature of an object will change when a certain amount of heat energy is added or removed. One practical application is in understanding how oceans and lakes affect the weather. The specific heat of water is much greater than the specific heat of most other substances. (Compare the values in Table 14.3.) A large body of water thus has a very large heat capacity relative to the surrounding air and soil. As a result, the temperature of an ocean or lake changes relatively little even when relatively large amounts of heat energy are added in the summer or taken away in winter, and thus it moderates ("smooths out") temperature fluctuations.

The value of the specific heat also gives insight into the internal energy U of a substance. According to the principle of conservation of energy, if an amount of heat Q is added to a system, then U must increase by that amount. The specific heat tells how the added heat changes the temperature; hence, c tells how changes in internal energy and temperature are related. This concept is important for understanding how the microscopic atomic motions in a substance change with temperature; we'll discuss this subject more in Chapter 15.

Calorimetry

We can find the specific heat of a system by adding a known amount of heat energy Q and measuring the resulting temperature change ΔT (Eq. 14.5). A measurement of this kind is called ***calorimetry*** and is shown schematically in Figure 14.11. The system of interest is placed in contact with a reference system; the reference system is often very large, in which case it is called a ***thermal reservoir***. For example, the system of interest might be a hot piece of metal, and the thermal reservoir could be a large bucket of water into which the metal is dropped (to cool it). When defining the specific heat, Q is the total heat energy added to the system (the metal); in this example, Q would come entirely from the thermal reservoir (the water). Most calorimetry experiments are designed to control the sources of heat flow into the system. In words, the system of interest is "thermally isolated" from all but the thermal reservoir.

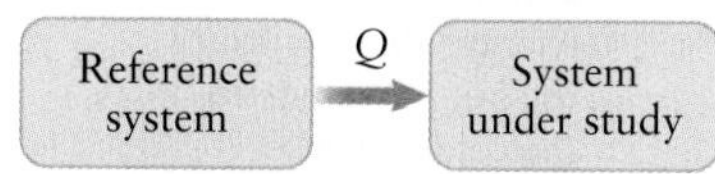

▲ **Figure 14.11** In a typical calorimetry experiment, the system of interest is arranged so that the only heat flow into or out of the system is from a reference system.

The general principles of calorimetry can also be used to analyze processes that do not involve a large thermal reservoir. For example, when a cold ice cube is dropped into a glass of warm water, heat flows between the water and the ice, changing the temperature of both. These temperature changes depend on the initial temperatures of both systems and their specific heats. In terms of conservation of energy, all the energy that leaves one system enters the other system, and that is the basis of a general way to attack all problems in calorimetry.

PROBLEM SOLVING Dealing with Specific Heat in Calorimetry

1. **RECOGNIZE THE PRINCIPLE.** All calorimetry problems are based on the principle of conservation of energy.
2. **SKETCH THE PROBLEM.** Your sketch should show the system or systems of interest and indicate how heat flows between them.
3. **IDENTIFY THE RELATIONSHIPS.** Determine (if possible) the initial and final temperatures of the system(s) and the heat energy Q added to each system. If energy flows into a system, Q is positive; if energy flows out of a system, Q is negative.
4. **SOLVE.** Apply $Q = cm\,\Delta T$ (Eq. 14.6) to relate Q and ΔT for each system and solve for the quantities of interest.

   Important note: Step 4 (and Eq. 14.6) assumes there are no phase changes; for example, the systems do not melt or evaporate. We'll see how to analyze such cases when we discuss the latent heat associated with phase changes later in this section.
5. Always *consider what your answer means* and check that it makes sense.

EXAMPLE 14.2 Specific Heat in the Blacksmith's Shop: Part 1

A blacksmith uses the heat energy from an oven to heat steel and other metals. The blacksmith also uses cold water to cool a hot horseshoe to room temperature. A steel horseshoe of mass $m_{\text{shoe}} = 0.50$ kg at a temperature of 800 K ("red hot") is dropped into a bucket of water ($m_{\text{water}} = 1.0$ kg) at 300 K (about room temperature). What is the final temperature of the horseshoe and the water? For simplicity, assume none of the water evaporates, so none of the water is converted to steam.

(continued) ▶

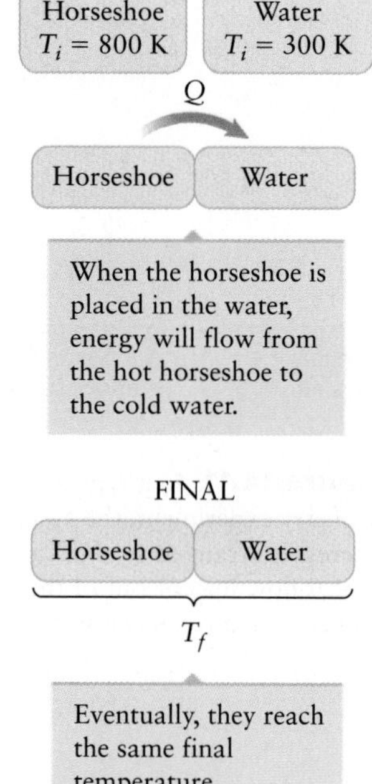
INITIAL
Horseshoe $T_i = 800$ K
Water $T_i = 300$ K
Q
Horseshoe
Water
When the horseshoe is placed in the water, energy will flow from the hot horseshoe to the cold water.
FINAL
Horseshoe
Water
T_f
Eventually, they reach the same final temperature.

▲ **Figure 14.12** Example 14.2.

RECOGNIZE THE PRINCIPLE

In this example, our two systems are the horseshoe and the water. When the horseshoe is dropped into the bucket, heat energy flows from the horseshoe to the water (Fig. 14.12). Because the horseshoe is completely immersed in the water, we can assume there is no other inflow or outflow of energy involving either system. We denote the energy that flows from the horseshoe into the water as Q. From conservation of energy principles, the energy "added" to the horseshoe is $-Q$; the negative sign here indicates that the final energy of the horseshoe is less than its initial energy. The initial temperatures of the horseshoe and the water are given, and the common final temperature T_f is the quantity we wish to find. We can calculate T_f by applying the specific heat relation between Q and ΔT for both systems.

SKETCH THE PROBLEM

Figure 14.12 shows the two systems along with their initial temperatures and the direction of heat flow.

IDENTIFY THE RELATIONSHIPS

We denote the specific heats of the horseshoe and the water by c_{steel} and c_{water}, respectively, and obtain their values from Table 14.3. Applying Equation 14.6 to the horseshoe, we get

$$Q_{\text{shoe}} = c_{\text{steel}} m_{\text{shoe}} \,\Delta T_{\text{shoe}} = c_{\text{steel}} m_{\text{shoe}} (T_f - T_{\text{shoe},\, i}) \tag{1}$$

Because the final temperature is less than the initial temperature ($T_{\text{shoe},\, i}$) of the steel horseshoe, Q_{shoe} will be negative as anticipated above. Applying the same relation to the water, we get

$$Q_{\text{water}} = c_{\text{water}} m_{\text{water}} (T_f - T_{\text{water},\, i}) \tag{2}$$

Because the water temperature increases ($T_f > T_{\text{water},\, i}$), Q_{water} will be positive. Energy is conserved, so

$$Q_{\text{shoe}} + Q_{\text{water}} = 0$$

and

$$Q_{\text{shoe}} = -Q_{\text{water}}$$

SOLVE

Using our results from Equations (1) and (2) and solving for T_f leads to

$$c_{\text{steel}} m_{\text{shoe}} (T_f - T_{\text{shoe},\, i}) = -c_{\text{water}} m_{\text{water}} (T_f - T_{\text{water},\, i})$$

$$T_f (c_{\text{steel}} m_{\text{shoe}} + c_{\text{water}} m_{\text{water}}) = c_{\text{steel}} m_{\text{shoe}} T_{\text{shoe},\, i} + c_{\text{water}} m_{\text{water}} T_{\text{water},\, i}$$

$$T_f = \frac{c_{\text{steel}} m_{\text{shoe}} T_{\text{shoe},\, i} + c_{\text{water}} m_{\text{water}} T_{\text{water},\, i}}{c_{\text{steel}} m_{\text{shoe}} + c_{\text{water}} m_{\text{water}}}$$

Inserting the given values of the various quantities, including the specific heats from Table 14.3, gives

$$T_f = \frac{[450 \text{ J/(kg}\cdot\text{K)}](0.50 \text{ kg})(800 \text{ K}) + [4186 \text{ J/(kg}\cdot\text{K)}](1.0 \text{ kg})(300 \text{ K})}{[450 \text{ J/(kg}\cdot\text{K)}](0.50 \text{ kg}) + [4186 \text{ J/(kg}\cdot\text{K)}](1.0 \text{ kg})}$$

$$T_f = \boxed{330 \text{ K}}$$

▶ ***What does it mean?***

The final temperature is quite close to the initial temperature of the water and far from the initial temperature of the horseshoe. That is because the specific heat of water is much greater than the specific heat of steel, so the water acts as a kind of thermal reservoir in this case.

EXAMPLE 14.3 Maintaining a Safe Body Temperature

A bicycle rider of mass $m = 60$ kg generates a power of approximately 400 W when riding at a moderate pace. Of this energy, about 25% is used to propel the bicycle, and the remaining 75% goes into heat generated within his body. If none of this heat were lost to the surroundings, how much would the cyclist's body temperature increase after 1 hour of riding?

RECOGNIZE THE PRINCIPLE

In this calorimetry problem, the system of interest is the cyclist along with an "internal" source (the cyclist's metabolism) that adds energy Q to this system. We must first find Q and then use it to compute the temperature rise of the system with Equation 14.6.

SKETCH THE PROBLEM

Figure 14.13 shows the system (the cyclist). An amount of heat Q is generated within, increasing his temperature.

▲ **Figure 14.13** Example 14.3. A When exercising (or even when just sitting on a couch), a person generates heat. B This heat will raise the body's temperature unless it escapes through the skin.

IDENTIFY THE RELATIONSHIPS

Power (P) is energy per unit time, so the energy Q added to the cyclist in 1 h is $Q = P\,\Delta t$. Here P is the fraction of the power that goes into heat, which is $P = 0.75(400\text{ W}) = 300$ W, with $\Delta t = 1\text{ h} = 3600$ s. We thus get

$$Q = P\,\Delta t = (300\text{ W})(3600\text{ s}) = 1.08 \times 10^6\text{ J} \quad (1)$$

where we have kept three significant figures to avoid roundoff errors below. Equation (1) is related to the change in temperature of the cyclist by (using Eq. 14.6)

$$\Delta T = \frac{Q}{cm}$$

SOLVE

Inserting Q from Equation (1) along with the given value of m and the specific heat of the human body (Table 14.3) gives

$$\Delta T = \frac{Q}{cm} = \frac{1.08 \times 10^6\text{ J}}{[3500\text{ J/(kg}\cdot\text{K)}](60\text{ kg})} = 5.1\text{ K} = \boxed{5.1°\text{C}}$$

▶ ***What does it mean?***

When a person's body temperature increases more than about 4°C (corresponding to a change ΔT of about $4°\text{C} \times \frac{9}{5} \approx 7°\text{F}$), thus giving a body temperature above about 106°F, there is danger of heatstroke. It is therefore important that our cyclist, or anyone else engaged in strenuous exercise, have some way to pass this generated heat to his surroundings. We'll consider that problem in further examples below.

Latent Heat

The specific heat tells how much the temperature will increase if a certain amount of heat is added to a system, but the definition of specific heat assumes the system does not change from one phase to another as the heat is added. When a system undergoes a phase change such as melting or evaporation, we must also account for the ***latent heat***. For example, for a block of ice at 0°C to undergo a phase change to liquid water at 0°C, an amount of heat equal to the ***latent heat of fusion*** must be added. Likewise, when water at a temperature of 0°C freezes to become ice at 0°C, this latent heat must be removed.

The latent heat associated with a phase change is connected with the energies of the atoms within the substance. For a solid, substantial energy is stored in the atomic bonds holding the atoms in place. These bonds store potential energy in

TABLE 14.4 Latent Heats for Various Phase Changes

Substance	Latent Heat of Fusion (kJ/kg)	Latent Heat of Vaporization (kJ/kg)
Water	334	2,200
Aluminum	397	11,000
Gold	67	870
Ethyl alcohol	104	850
Copper	205	5,070
Lead	23	870

a manner analogous to springs that obey Hooke's law (Chapter 6). For a solid to melt, energy is required to overcome this potential energy and break these bonds, enabling the atoms and molecules to move about freely as in a liquid. That energy is provided by the latent heat.

By convention, the latent heat of fusion L_{fusion} is a positive quantity; for a system of mass m to melt, an amount of heat

$$Q_{\text{melt}} = +mL_{\text{fusion}} \tag{14.7}$$

must be added. Likewise, when a system freezes (also called fusion), we have

$$Q_{\text{freeze}} = -mL_{\text{fusion}}$$

The negative sign indicates that energy must be removed from a system to make it freeze.

The ***latent heat of vaporization*** is associated with the phase change from a liquid to a gas (called vaporization, evaporation, or boiling) and from a gas to a liquid (condensation). To vaporize an amount of liquid of mass m, we must add an energy

$$Q_{\text{vaporization}} = mL_{\text{vaporization}} \tag{14.8}$$

and for condensation, we have

$$Q_{\text{condensation}} = -mL_{\text{vaporization}}$$

Here the negative sign indicates that energy must be removed from a system to make it condense. On an atomic scale, the latent heat of vaporization can be understood as follows. When a liquid evaporates to become a gas, the weak atom–atom interactions in the liquid become even weaker in the gas as the atoms are moved farther apart. This process requires the addition of energy, now through the latent heat of vaporization.

Table 14.4 lists the latent heats of fusion and vaporization for some common substances. Notice that for a given substance, the latent heat of fusion is *not* equal to the latent heat of vaporization.

Calorimetry: Including the Latent Heat

Figure 14.14A shows the phase diagram for a hypothetical substance. Let's consider how much heat we must add to move along the path indicated by the horizontal dashed line. Starting from a point in the solid region of the phase diagram (point A), we add heat. This increases the temperature as shown in region I in Figure 14.14B, where we plot the temperature as a function of the amount of added heat Q_{added}. The temperature increase ΔT in the solid phase is described by the specific heat through $\Delta T = Q/(cm)$ (Eq. 14.6), with c equal to the specific heat of the solid.

Point B in Figure 14.14A is at the melting transition. If we continue to add heat, the system undergoes a phase change from solid to liquid and the amount of heat needed is given by the latent heat of fusion (Eq. 14.7). As this latent heat is added and the system is melting, the temperature remains constant (region II in Fig.

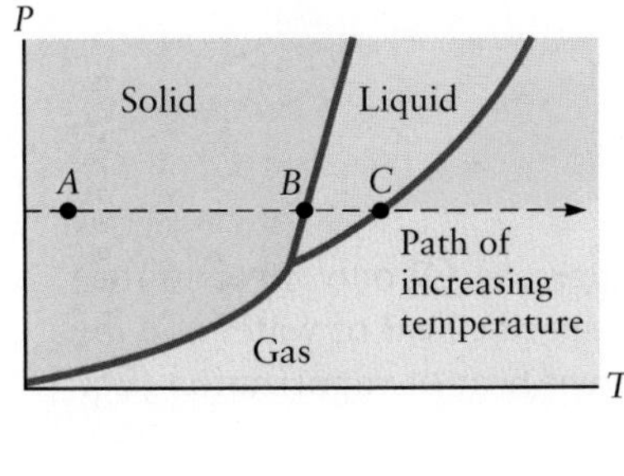

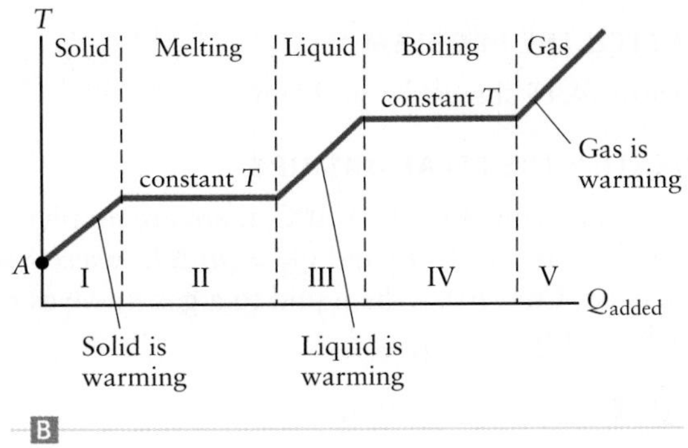

◀ **Figure 14.14** A Phase diagram of a typical system. B Results of a calorimetry experiment in which T is plotted as a function of the energy Q_{added} added to the system. The slopes in regions I, III, and V depend on the specific heat of the solid, liquid, and gas respectively. The slope is zero in regions II and IV.

14.14B). If this substance is water, we would be converting ice at 0°C into water at 0°C. When the system is completely melted, adding further heat increases the temperature according the specific heat of the liquid (region III). Eventually, the phase change to a gas is reached at point C. Now an amount of energy corresponding to the latent heat of vaporization must be added, and during this time, the temperature is again constant (region IV). If this substance is water, this latent heat converts water at 100°C to gas (water vapor) also at 100°C. Adding more heat would then increase the temperature of the gas.

Calorimetry calculations involving phase changes are similar to the problems we considered in Examples 14.2 and 14.3. The only difference is that we must now allow for the energy associated with the latent heat of any phase changes that occur.

PROBLEM SOLVING Dealing with Both Specific Heat and Latent Heat in Calorimetry

1. **RECOGNIZE THE PRINCIPLE.** All calorimetry problems are based on the principle of conservation of energy.
2. **SKETCH THE PROBLEM.** Your sketch should show the system or systems of interest and indicate how heat flows between them.
3. **IDENTIFY THE RELATIONSHIPS.** Determine (if possible) the initial and final temperatures of the system(s) and the heat energy Q added to each system. If energy flows into a system, Q is positive; if energy flows out of a system, Q is negative.
4. **SOLVE.**
   - Use $Q = cm\,\Delta T$ (Eq. 14.6) to relate Q and ΔT for temperature changes that do not involve a phase change.
   - If there is a phase change, the latent heat ($Q = mL$, Eqs. 14.7 and 14.8) must be added to or subtracted from the system, depending on the phase change.
5. Always *consider what your answer means* and check that it makes sense.

EXAMPLE 14.4 Heating a Bucket of Water

A bucket contains 2.0 kg of water (about 2 L) at 30°C (about room temperature). How much energy must be added to convert all the water to vapor at 100°C?

RECOGNIZE THE PRINCIPLE

Our system is the water, initially a liquid at 30°C (Fig. 14.15). An amount of heat energy Q_1 is added to take the liquid to 100°C, and then an additional heat energy Q_2 is needed to evaporate the liquid.

(continued) ▶

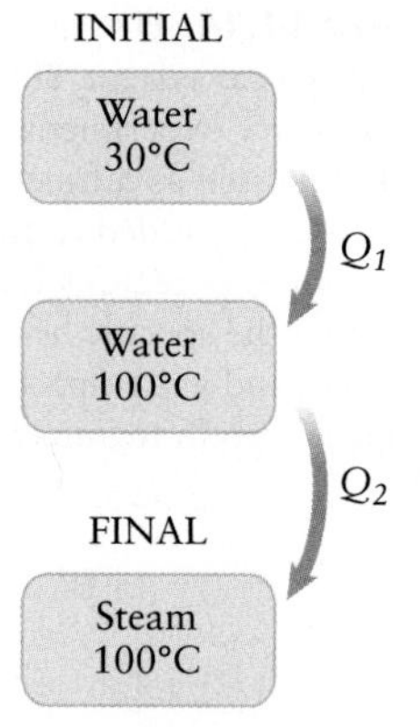

▲ **Figure 14.15** Example 14.4.

SKETCH THE PROBLEM

Figure 14.15 shows the problem.

IDENTIFY THE RELATIONSHIPS

In heating from 30°C to 100°C, there are no phase changes, so Q_1 only involves the specific heat (Eq. 14.6) and $Q_1 = cm\,\Delta T$, where c is the specific heat of water and m is the mass. Converting the liquid to a gas involves the latent heat of vaporization (Eq. 14.8), so $Q_2 = mL_{\text{vaporization}}$.

SOLVE

Using the value of c for water from Table 14.3, the given value of m, and $\Delta T = 100°\text{C} - 30°\text{C} = 70\text{ K}$, we find

$$Q_1 = cm\Delta T = [4186\text{ J/(kg}\cdot\text{K)}](2.0\text{ kg})(70\text{ K}) = 5.9 \times 10^5\text{ J}$$

Using the latent heat of water from Table 14.4 leads to

$$Q_2 = mL_{\text{vaporization}} = (2.0\text{ kg})(2200\text{ kJ/kg}) = 4.4 \times 10^6\text{ J}$$

The total heat energy required to convert all the water to steam is thus

$$Q_{\text{total}} = Q_1 + Q_2 = [(5.9 \times 10^5) + (4.4 \times 10^6)]\text{ J} = \boxed{5.0 \times 10^6\text{ J}}$$

► ***What does it mean?***

Whenever there is a phase change, we must account for the latent heat. In this case, the latent heat is the largest contribution to Q_{total}.

EXAMPLE 14.5 Specific Heat in the Blacksmith's Shop: Part 2

Consider again the blacksmith in Example 14.2, who uses a bucket of water ($m_{\text{water}} = 1.0\text{ kg}$, $T_{\text{water},\,i} = 300\text{ K}$) to cool a hot steel horseshoe. This time, he has a much larger horseshoe ($m_{\text{shoe}} = 2.0\text{ kg}$, $T_{\text{shoe},\,i} = 800\text{ K}$) and notices that when he drops the horseshoe in the water, some of the water evaporates as steam. Assuming the final temperature of the horseshoe plus water plus steam is 100°C (= 373 K), how much of the water is converted into steam?

RECOGNIZE THE PRINCIPLE

When the horseshoe is dropped into the water, an amount of heat energy Q flows from the horseshoe to the water (Fig. 14.16). From conservation of energy principles, the energy "added" to the horseshoe is $-Q$; the negative sign indicates that the final energy of the horseshoe is less than its initial energy. The initial temperatures of the horseshoe and the water are given. In the end, an unknown amount of the water undergoes a phase change to steam, and the horseshoe, water, and steam have a common final temperature $T_f = 100°\text{C} = 373\text{ K}$. We can find how much water is converted to steam using the principle of conservation of energy.

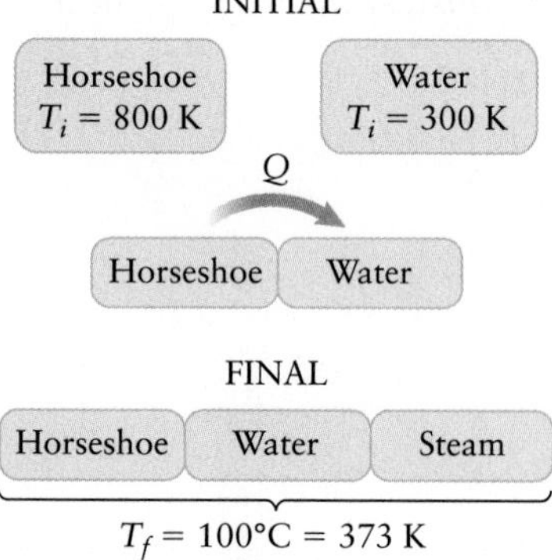

▲ **Figure 14.16** Example 14.5.

SKETCH THE PROBLEM

Figure 14.16 shows the horseshoe and the water along with their initial temperatures and the direction of heat flow.

IDENTIFY THE RELATIONSHIPS

The only unknown is the mass of water converted to steam. The heat that flows out of the horseshoe can be found from the specific heat of the horseshoe and its temperature change. Applying Equation 14.6, we get (compare with Example 14.2)

$$Q_{\text{shoe}} = c_{\text{steel}} m_{\text{shoe}}(T_f - T_{\text{shoe},\,i})$$

Consulting Table 14.3 for c_{steel} and inserting the given values, we find

$$Q_{shoe} = [450 \text{ J/(kg} \cdot \text{K)}](2.0 \text{ kg})(373 \text{ K} - 800 \text{ K}) = -3.84 \times 10^5 \text{ J} \quad (1)$$

where we have kept three significant figures to avoid roundoff errors below. The water begins at $T_{water,\,i}$, and it all reaches a final temperature of 373 K. The heat required to bring all the liquid to this temperature is

$$Q_{water} = c_{water} m_{water}(T_{water,\,f} - T_{water,\,i})$$

Using c_{water} from Table 14.3 and inserting the given values, we find

$$Q_{water} = [4186 \text{ J/(kg} \cdot \text{K)}](1.0 \text{ kg})(373 \text{ K} - 300 \text{ K}) = 3.06 \times 10^5 \text{ J} \quad (2)$$

Next we write the conservation of energy condition. The sum of the energy Q_{shoe} plus the total energy that flows into the water must equal zero. The energy that flows into the water has two contributions, one from the energy needed to heat all the water to 373 K as done in Equation (2) and another from the latent heat of the water converted to steam. We thus have (compare again with Example 14.2)

$$aQ_{shoe} + Q_{water} + Q_{steam} = 0 \quad (3)$$

We can use the results from Equations (1) and (2) to find Q_{steam}:

$$Q_{steam} = -Q_{shoe} - Q_{water} = (3.84 \times 10^5 \text{ J}) - (3.06 \times 10^5 \text{ J}) = 8 \times 10^4 \text{ J} \quad (4)$$

We want to find the mass of water m_{steam} converted to steam; it is related to Q_{steam} by (Eq. 14.8)

$$Q_{steam} = m_{steam} L_{vaporization}$$

SOLVE

Solving for m_{steam} and inserting the value of Q_{steam} from Equation (4) and the latent heat from Table 14.4, we find

$$m_{steam} = \frac{Q_{steam}}{L_{vaporization}} = \frac{8 \times 10^4 \text{ J}}{2.2 \times 10^6 \text{ J/kg}} = \boxed{0.04 \text{ kg}}$$

▸ ***What does it mean?***

The key to this problem, and to all calorimetry, is conservation of energy. In this example, it is expressed in Equation (3).

CONCEPT CHECK 14.2 Comparing Heat Energy with a Typical Kinetic Energy

The total heat energy required to cool the horseshoe in Example 14.5 is $|Q_{shoe}| = 3.8 \times 10^5$ J. Suppose this energy could be converted to the kinetic energy of a baseball ($m \approx 0.2$ kg). Would the speed of the baseball be (a) 50 m/s (about 100 mi/h), (b) 5 m/s, or (c) 2000 m/s?

14.5 Thermal Expansion

Changing the temperature of a system can affect many of its properties. In particular, the size of a system will usually change with temperature, an effect called ***thermal expansion***. Consider an object such as a piece of metal as in Figure 14.17A, with length L_0. If the temperature of the object increases by ΔT, the length changes by an amount ΔL. This change in length is proportional to the change in temperature ΔT and to the ***coefficient of linear thermal expansion*** α:

$$\frac{\Delta L}{L_0} = \alpha \, \Delta T \quad (14.9)$$

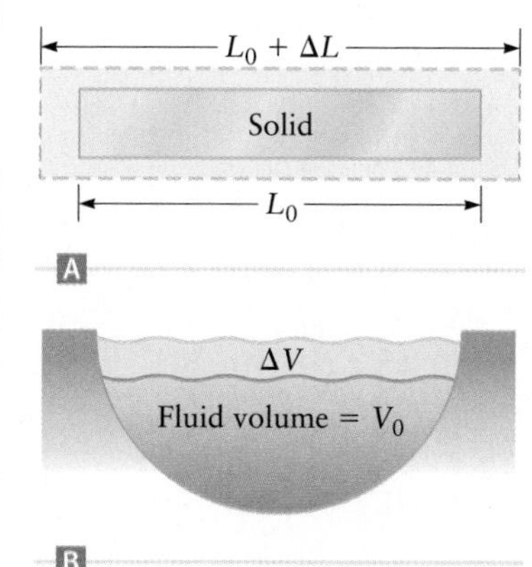

▲ **Figure 14.17** A If a solid object is heated, its length increases by an amount ΔL proportional to the coefficient of thermal expansion α. B Any substance (solid, liquid, or gas) may change volume when it is heated. This change in volume ΔV is proportional to the coefficient of volume expansion β.

TABLE 14.5 Coefficients of Linear and Volume Expansion

Substance	α (10^{-6} K^{-1})	β (10^{-6} K^{-1})
Ethyl alcohol	—	1120
Mercury	—	182
Water	—	210
Ice (0°C)	51	153
Pyrex	3.3	9.8
Glass (ordinary)	9	27
Aluminum	22.5	69
Gold	14	42
Cement	12	36
Iron	12	36
Steel	12	36
Platinum	8.8	26
Invar[a]	0.9	2.7
Wood (typical)	5.4	16
Brass	18.7	56
Copper	17	51

Note: Values are for room temperature and pressure unless noted.

[a]Invar is an alloy of iron and nickel that has very low values of the coefficients of thermal expansion, which makes it useful in applications.

As indicated in Figure 14.17A, the object's width and height also change. These changes are given by expressions similar to Equation 14.9, so both change by amounts proportional to α and ΔT, which leads to a ***volume expansion*** ΔV given by

$$\frac{\Delta V}{V_0} = \beta\,\Delta T \tag{14.10}$$

where V_0 is the initial volume and β is the ***coefficient of volume expansion.*** Values of α and β for a variety of substances are given in Table 14.5. Values of α are only given for solids because liquids and gases cannot (on their own) maintain a fixed shape as in Figure 14.17A. We can still define a coefficient of volume expansion (see Fig. 14.17B) for liquids and gases, however, and a few values are listed in Table 14.5.

CONCEPT CHECK 14.3 Linear Expansion (α) and Volume Expansion (β) Are Related

You can see from Table 14.5 that a substance with a large value of α also has a large value of β. Why?

EXAMPLE 14.6 Thermal Expansion of a Metal Bar

Consider a meterstick composed of platinum. It might be a copy of the standard meterstick as formerly employed to maintain the SI unit of length (Chapter 1). By what amount does the length of this meterstick change if the temperature increases by 1.0 K?

RECOGNIZE THE PRINCIPLE

This problem is a direct application of the definition of the thermal expansion coefficient. Given the change in temperature and the initial length, the change in length is $\Delta L = L_0\alpha\,\Delta T$ (Eq. 14.9).

SKETCH THE PROBLEM

Figure 14.17A describes the problem.

IDENTIFY THE RELATIONSHIPS AND SOLVE

From Table 14.5, we have $\alpha = 8.8 \times 10^{-6}\ \text{K}^{-1}$ for platinum. Inserting into Equation 14.9 gives

$$\Delta L = L_0 \alpha\, \Delta T = (1.0\ \text{m})(8.8 \times 10^{-6}\ \text{K}^{-1})(1.0\ \text{K}) = \boxed{8.8 \times 10^{-6}\ \text{m}}$$

► ***What does it mean?***

This increase in the bar's length is rather small (a bit smaller than the diameter of a typical human hair). For some applications, however, changes of this magnitude cannot be tolerated, especially when high precision is required. That is why standards of length that use the wavelength of light (instead of the length of a metal bar) have been developed. We'll explain in Chapter 25 how the wavelength of light can be measured very accurately.

EXAMPLE 14.7 Area Expansion

Consider a square hole in an aluminum plate (Fig. 14.18). If the aluminum is heated, will this hole become smaller or larger?

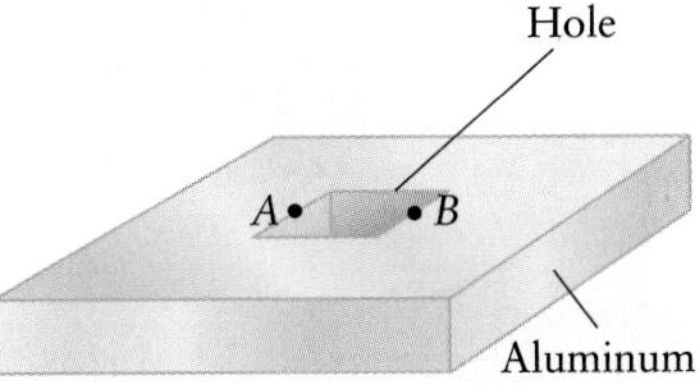

▲ **Figure 14.18** Example 14.7. What happens to the area of a hole when this plate is heated?

RECOGNIZE THE PRINCIPLE

A metal bar becomes longer when it is heated; hence, the ends of the bar are farther apart. Therefore, any two points on the bar will also be farther apart after heating.

SKETCH THE PROBLEM

Figure 14.19 shows how two points on a metal bar move farther apart when the bar is heated.

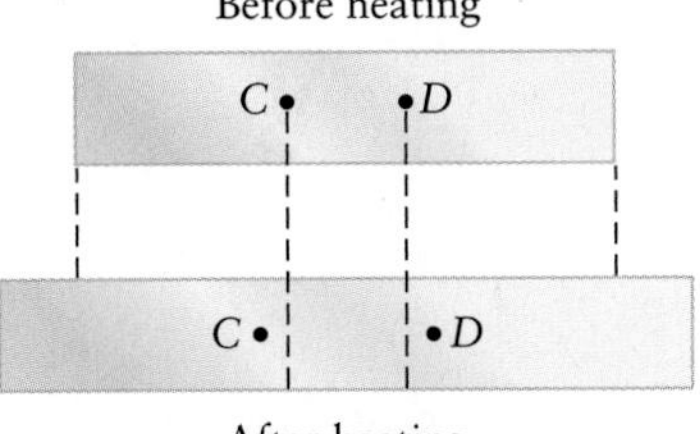

▲ **Figure 14.19** Example 14.7. When a bar expands, points C and *D* move farther apart.

IDENTIFY THE RELATIONSHIPS AND SOLVE

Figure 14.18 shows two points *A* and *B* at the edges of a square hole. These points will be farther apart after the plate is heated, so the hole will get $\boxed{\textit{larger}}$.

► ***What does it mean?***

The same reasoning applies to a circular hole and explains why heating the metal lid on a glass jar makes it easier to remove a "stuck" lid. The metal lid expands when it is heated. The glass jar also expands, but because the coefficient of linear thermal expansion of most metals is greater than the coefficient for glass (Table 14.5), the opening in the metal lid expands more than the glass jar, making it easier to remove the lid.

Effects of Thermal Expansion

The thermal expansion of the metal bar in Example 14.6 is not large, but the effect of thermal expansion can be quite significant in certain situations. For example, consider the problem of designing a large bridge. One of the first bridges built to span the Mississippi River near St. Louis, the Eads Bridge, was built with steel in the 1870s. This bridge was one of the first large-scale applications of steel. The bridge used several spans, each with a length of about 500 ft (about 150 m). Let's calculate

© Richard Choy/Peter Arnold/Photolibrary

▲ **Figure 14.20** Many bridges contain expansion joints so that they can expand and contract without breaking away from their supports or forming unsafe gaps.

the change in length of one span from a cold winter night (−10°C) to a hot summer day (40°C). From Equation 14.9, we have

$$\Delta L = L_0 \alpha \, \Delta T = L_0 \alpha (T_f - T_i)$$

The temperature change is 40°C − (−10°C) = 50°C, which corresponds to a temperature change of 50 K. Using the coefficient of linear thermal expansion for steel from Table 14.5 gives

$$\Delta L = (150 \text{ m})(12 \times 10^{-6} \text{ K}^{-1})(50 \text{ K}) = 0.090 \text{ m} \quad (14.11)$$

This change in length is about 3.5 in. and is thus quite significant; it would certainly not be desirable for a bridge that carries trains or cars to have a gap of this size. For this reason, bridges, buildings, and many other structures are designed with special expansion joints that allow for expansion and contraction without forming unsafe gaps or breaking (Fig. 14.20).

Another example of thermal expansion is sketched in Figure 14.21. A mercury thermometer uses the thermal expansion of mercury to measure changes in temperature. When T changes, the mercury (which is a liquid) expands or contracts, whereas the glass holding the mercury expands or contracts relatively little because the coefficients of thermal expansion of glass are small. The result is a change ΔL in the level of the meniscus in the thermometer tube (Fig. 14.21). Changes in the meniscus position are proportional to changes in temperature, just as are changes in the length of the bar in Example 14.6. The magnitudes of the changes, though, are much larger for the thermometer because there is a (relatively) large pool of mercury in the bulb of the thermometer, and all this mercury expands into a very narrow tube when the temperature increases.

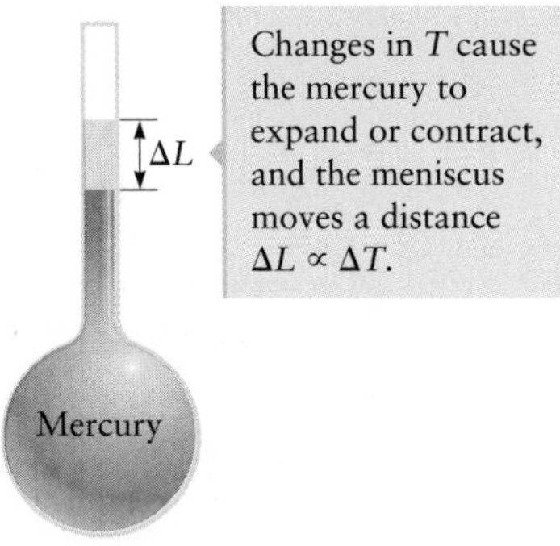
Changes in T cause the mercury to expand or contract, and the meniscus moves a distance $\Delta L \propto \Delta T$.
ΔL
Mercury

▲ **Figure 14.21** A mercury thermometer uses the thermal expansion of liquid mercury to measure temperature.

CONCEPT CHECK 14.4 Heating Your Piano

A piano string is held at a particular tension T so that its standing wave frequencies (Chapter 13) correspond to the note middle C (fundamental frequency approximately 262 Hz). Suppose the string is heated while the temperature of the rest of the piano does not change. Will the fundamental frequency of the string go up or down?

EXAMPLE 14.8 Thermal Expansion and Forces on a Bridge

Figure 14.22 shows a steel bridge, similar to the one considered in connection with Equation 14.11. In that calculation, we found that a steel beam with initial length $L_0 = 150$ m will expand an amount $\Delta L = 0.090$ m when heated by 50 K. Suppose this steel beam has cross-sectional area $A = 1.0 \text{ m} \times 0.10 \text{ m} = 0.10 \text{ m}^2$ and is part of a bridge. Also suppose the supports of the bridge hold the ends of the steel beam fixed in place so that the ends cannot move. What force must the supports exert on the ends of the beam to keep it from expanding?

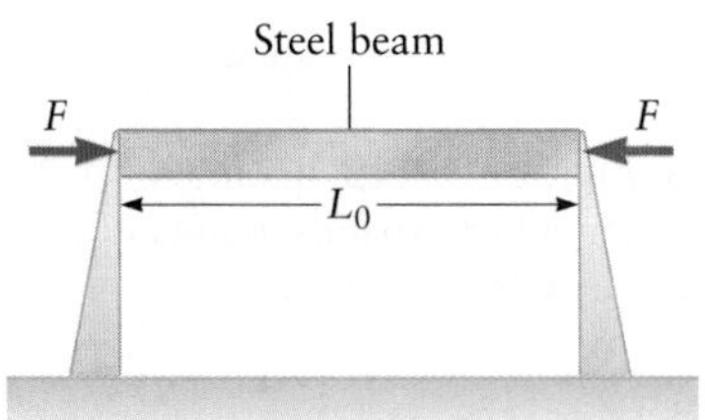
Steel beam
F
F
L_0

▲ **Figure 14.22** Example 14.8.

RECOGNIZE THE PRINCIPLE

If there are no forces on the ends of the beam, it will expand by an amount ΔL when it is heated. To prevent this expansion, the supports must exert a force that compresses the beam by an amount ΔL. This compression is described by the elastic properties of the beam and is proportional to Young's modulus (Chapter 11).

SKETCH THE PROBLEM

Figure 14.22 shows the problem along with the forces exerted by the supports on the ends of the beam.

IDENTIFY THE RELATIONSHIPS

If a force F is applied to each end of the bar, it will compress an amount ΔL. From Equation 11.23,

$$\frac{F}{A} = Y\frac{|\Delta L|}{L_0} \tag{1}$$

where Y is Young's modulus and A is the cross-sectional area of the bar. We have used the absolute value of ΔL because the applied forces compress the bar, reducing its length (making ΔL negative). The bridge supports hold the ends fixed, so this change in length must compensate for the change in length due to thermal expansion in Equation 14.11. Hence, $|\Delta L| = 0.090$ m in Equation (1).

SOLVE

Young's modulus for steel is 2.0×10^{11} Pa (Table 11.1). Solving for the force, we get

$$F = AY\frac{|\Delta L|}{L_0} = (0.10 \text{ m}^2)(2.0 \times 10^{11} \text{ Pa})\frac{(0.090 \text{ m})}{(150 \text{ m})} = \boxed{1.2 \times 10^7 \text{ N}}$$

► ***What does it mean?***

This force is quite substantial and is why metal beams and bridges sometimes "buckle." Architects and mechanical engineers are paid to prevent such buckling.

Thermal Expansion of Water

Table 14.5 lists the coefficients of thermal expansion for a number of different substances near room temperature. Strictly speaking, α and β both vary with temperature, but this variation is often small and can usually be ignored. Therefore, the length and volume of an object both change linearly with temperature, and we can usually use Equations 14.9 and 14.10 to calculate changes in length and volume even when ΔT is large (i.e., many tens or even hundreds of kelvins). There is, however, one important substance whose expansion is not a simple linear function. Figure 14.23 shows how the density of H_2O varies over a wide temperature range, and Figure 14.23B shows an expanded view near the freezing temperature ($T = 0°C$). At temperatures well above freezing, the density decreases as T increases. That is just the usual thermal expansion; an expanding sample has a lower density. At temperatures near and above room temperature (around 20°C), this expansion can be described by Equation 14.10 with $\beta \approx 2.1 \times 10^{-4}$ K^{-1}. Just above freezing, however, the density of water exhibits a maximum and cannot be described by Equation 14.10 with a constant value of β. What's more, the density of the solid (ice) near freezing is less than the density of the liquid, and H_2O thus contracts when it melts. That is why ice floats in water, as described by Archimedes's principle (Chapter 10). Most substances do not behave in this way; a solid generally does not float in its own liquid. This unusual behavior of water has many important consequences. For example, in winter a lake freezes first at the surface, which is good for ice skaters. It also enables fish to live comfortably in the water underneath and still have access to food at the lake bottom.

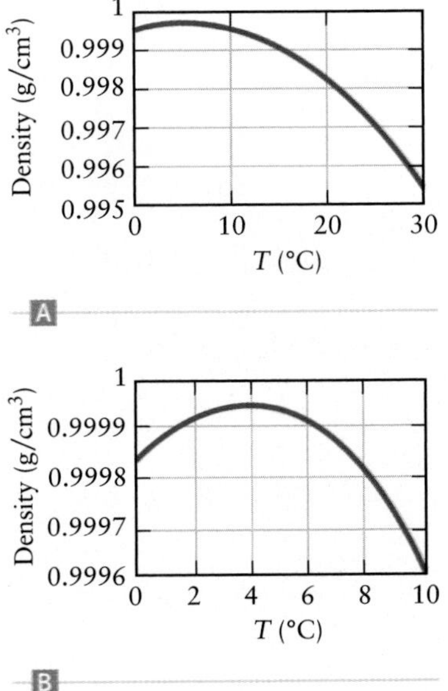

▲ **Figure 14.23** A Density of water as a function of temperature. B Expanded view of part A near the freezing temperature (0°C).

EXAMPLE 14.9 Effect of Freezing on a Cell

The fluid inside a cell is composed mainly of water, so when a cell freezes, the fluid inside expands. This expansion causes a large ("outward") pressure on the cell wall and usually causes the cell to rupture. Consider a spherical cell with initial radius $r_i = 5.0$ μm.

(continued) ►

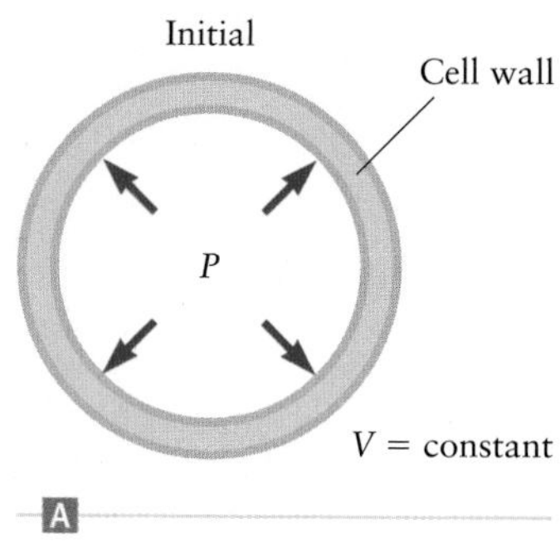

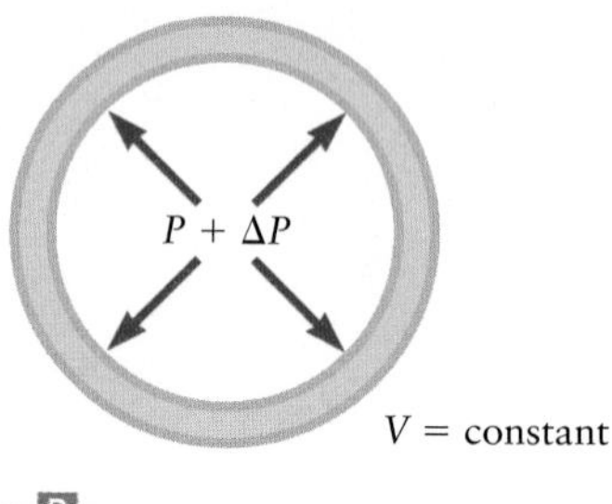

▲ **Figure 14.24** Example 14.9. If water in a cell freezes, it produces an increased pressure on the cell wall.

Calculate the increase in pressure inside the cell when it freezes. Assume the fluid inside the cell is mainly water and use the data for water from Figure 14.23; the density of water at 0°C is 999.8 kg/m³ (= 0.9998 g/cm³), and the density of ice is 917.0 kg/m³.

RECOGNIZE THE PRINCIPLE

When the water in the cell freezes, its volume expands by an amount ΔV_{ice}, which we can calculate from the given values of the densities of water and ice. We then need to calculate what pressure must be applied to the cell to compress it back to its original volume; this pressure is the extra pressure that would be exerted on the cell wall (assuming the cell does not rupture). We can find it using the bulk modulus of ice (Chapter 11).

SKETCH THE PROBLEM

Figure 14.24 describes the problem. The volume expansion that would occur due to freezing must be compensated for by an increase in the pressure.

IDENTIFY THE RELATIONSHIPS

If ΔV_{ice} is the amount the water would expand when it freezes, the pressure P from the cell wall must produce a volume change $\Delta V = -\Delta V_{\text{ice}}$. In Chapter 11, we learned that the pressure P required to compress a material to give a volume change ΔV is (Eq. 11.25)

$$P = -B\frac{\Delta V}{V_i} \tag{1}$$

where B is an elastic constant called the bulk modulus. In this example, we need to compress the ice, so ΔV is negative and P in Equation (1) will be positive. From Table 11.1, the bulk modulus of ice is approximately $B = 9 \times 10^9$ Pa.

SOLVE

Mass is equal to the product of density and volume, so $m = \rho V$. The mass of the cell is the same at the start (i) and end (f), giving

$$\rho_i V_i = \rho_f V_f \tag{2}$$

The cell is (for simplicity) spherical; hence,

$$V_i = \tfrac{4}{3}\pi r_i^3 = \tfrac{4}{3}\pi(5.0 \times 10^{-6}\ \text{m})^3 = 5.2 \times 10^{-16}\ \text{m}^3$$

Inserting this result into Equation (2) along with the given initial and final densities, we find

$$V_f = \frac{\rho_i}{\rho_f}V_i = \frac{999.8\ \text{kg/m}^3}{917.0\ \text{kg/m}^3}(5.2 \times 10^{-16}\ \text{m}^3) = 5.7 \times 10^{-16}\ \text{m}^3$$

The change of volume of the cell is thus

$$\Delta V_{\text{ice}} = V_f - V_i = 5 \times 10^{-17}\ \text{m}^3$$

Using the relation in Equation (1), we then get

$$P = -B\frac{(-\Delta V_{\text{ice}})}{V} = (9 \times 10^9\ \text{Pa})\frac{5 \times 10^{-17}\ \text{m}^3}{5.2 \times 10^{-16}\ \text{m}^3} = \boxed{9 \times 10^8\ \text{Pa}}$$

▸ ***What does it mean?***

Atmospheric pressure is only about 1×10^5 Pa, so this pressure is nearly 10,000 times atmospheric pressure! That is why cells usually rupture when they freeze.

14.6 Heat Conduction

Heat is the energy that flows between systems at different temperatures. This energy transfer can take place in three ways: ***conduction***, ***convection***, and ***radiation***. We now consider heat conduction; we'll discuss the other two methods of heat transfer in the following sections.

Figure 14.25A shows a solid bar connecting two separate systems. These systems might be large pieces of metal, and we assume they are at different temperatures T_1 and T_2 so that heat energy flows through the bar from one end to the other. If $T_1 > T_2$, energy flows from the system on the left (the hotter one) to the one on the right, and we say that the bar ***conducts*** heat. The bar itself cannot be described by a single temperature value. Instead, the bar's temperature varies smoothly from T_1 on the left to T_2 on the right (Fig. 14.25B).

We can understand this heat flow in terms of the vibrations of atoms in the bar. (If the bar is made of metal, electrons outside the ion cores in the bar will also contribute to heat flow.) In any particular region within the bar, atoms closer to the hot end (at the left) are at a slightly higher temperature than the neighboring atoms that are nearer the cold end (the right). The atoms at a higher temperature have a larger vibration amplitude, corresponding to larger potential and kinetic energies. As the atoms vibrate, some of the extra vibrational energy of "hotter" atoms is transferred to nearby "colder" atoms, similar to the way a wave transfers energy from place to place without transferring matter; energy thus "flows" from the hot end of the bar to the cold end.

The amount of energy that flows through the bar in Figure 14.25 depends on the bar's area A, length L, and the temperature difference $\Delta T = T_1 - T_2$. Heat flow also depends on a property of the bar called the ***thermal conductivity***, κ. Heat flow is measured by the energy Q that arrives per unit time t at the cold end. This heat flow is the rate at which energy leaves the hot end of the bar and arrives at the cold end and is given by

$$\frac{Q}{t} = \kappa A \frac{\Delta T}{L} \tag{14.12}$$

Values of the thermal conductivity for a variety of materials are listed in Table 14.6. Values of κ vary widely, with small values for gases (such as air) and high values for metals.

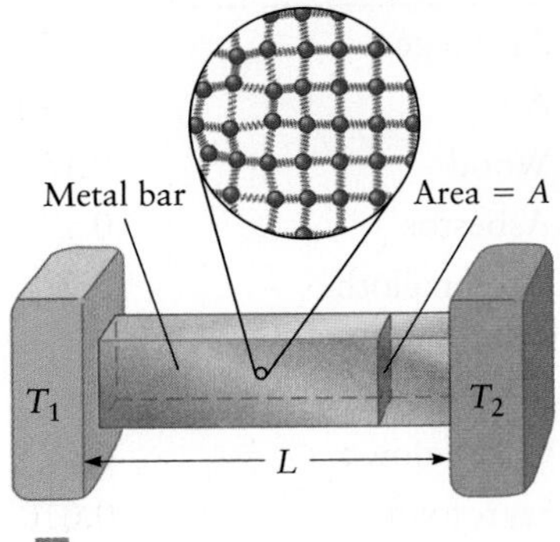

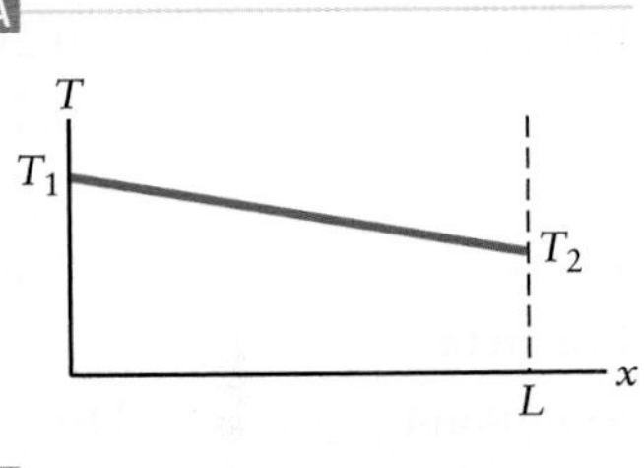

▲ **Figure 14.25** A When the ends of a solid bar are placed in contact with systems at different temperatures, the bar conducts energy from the hot end to the cold end. B Plot of the temperature as a function of position along the bar.

CONCEPT CHECK 14.5 Thermal Conductivity of Gases

The values in Table 14.6 show that the thermal conductivities of gases are generally much smaller than the thermal conductivities of solids. Use the microscopic picture of heat conduction involving atomic vibrations (Fig. 14.25) to explain why.

CONCEPT CHECK 14.6 Designing a Better Building

The walls of buildings are made from many different materials, including concrete, wood, glass, and steel. If solid walls with the same area and thickness are made from these materials, which would allow the most heat conduction through the wall for a given temperature difference, (a) concrete, (b) wood, (c) glass, or (d) steel?

TABLE 14.6 Thermal Conductivities of Some Common Substances

Material	κ W/(m · K)
Argon gas	0.016
Air	0.024
Wood	0.15
Asbestos	0.17
Cotton cloth (approximate)	0.2
Wool	2.2
Goose down	0.025
Styrofoam	0.010
Body fat	0.20
Snow	0.25
Glass	0.80
Water	0.56
Concrete	1.7
Aluminum	240
Iron	79
Steel	50
Copper	400
Silver	430
Ice (0°C)	2.2

Note: All substances are at atmospheric pressure and room temperature unless noted.

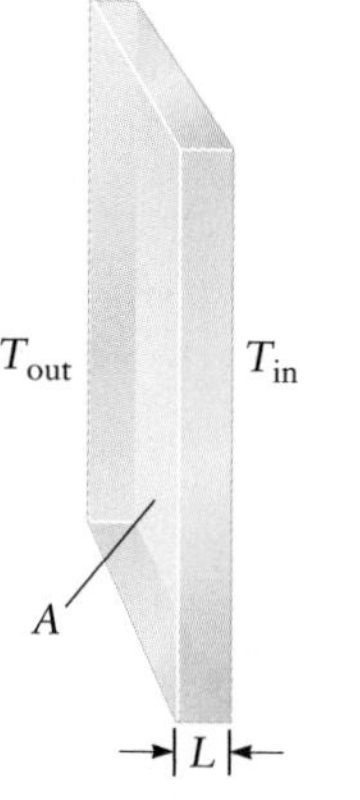

▲ **Figure 14.26** Conduction of heat through a windowpane.

EXAMPLE 14.10 Conduction of Heat through a Windowpane

Consider a glass windowpane (Fig. 14.26) with area $A = 0.50 \text{ m}^2$, thickness $L = 4.0$ mm, and temperature difference $\Delta T = 30°\text{C}$ between the two sides (i.e., between the inside of the house and the outside). What is the total amount of heat energy that flows through the windowpane in one day?

RECOGNIZE THE PRINCIPLE

We are given the dimensions of the windowpane, and we know the thermal conductivity of glass from Table 14.6. We can therefore use Equation 14.12 to calculate the rate at which heat flows through the windowpane, and from that we can calculate the total energy that flows through in one day.

SKETCH THE PROBLEM

Figure 14.26 describes the problem.

IDENTIFY THE RELATIONSHIPS AND SOLVE

The value of a temperature difference on the Celsius and Kelvin scales is the same, so $\Delta T = 30$ K. Inserting these values into Equation 14.12 with the value of the thermal conductivity κ for glass from Table 14.6, we find

$$\frac{Q}{t} = \kappa A \frac{\Delta T}{L} = \left(0.80 \frac{\text{W}}{\text{m} \cdot \text{K}}\right)(0.50 \text{ m}^2)\frac{30 \text{ K}}{0.0040 \text{ m}} = 3000 \text{ W} \quad (1)$$

The total energy that flows out through the window in one day can be calculated from Equation (1) using $t = 1$ day $= 86{,}400$ s. We get

$$Q = (3000 \text{ W})t = (3000 \text{ J/s})(86{,}400 \text{ s}) = \boxed{2.6 \times 10^8 \text{ J}} \quad (2)$$

► ***What does it mean?***

This energy comes (ultimately) from the house's furnace. In the case of an electric furnace, the cost of electrical energy is typically about \$0.02 for 10^6 J. The total cost of the energy Q in Equation (2) is thus approximately

$$\text{cost} = (2.6 \times 10^8 \text{ J})(\$0.02/10^6 \text{ J}) = \$5$$

A cost of \$5 per window per day is quite substantial because most houses have many windows; that is why energy-efficient windows (see Problem 38 at the end of this chapter) are popular.

CONCEPT CHECK 14.7 Dependence of Heat Conduction on Size

Suppose the height and width of the windowpane in Figure 14.26 are both increased by a factor of two. By what factor does the heat flow through the window change?

(a) There is no change.
(b) It increases by a factor of two.
(c) It increases by a factor of four.

EXAMPLE 14.11 Heat Flow and Cooling Your Coffee

The author likes to bring a thermos bottle of coffee to his office each morning. The walls and lid of a thermos bottle have low thermal conductivities, so the rate of heat flow out of the bottle is small (Fig. 14.27A). Suppose the coffee inside has mass $m = 0.30$ kg and an initial temperature of 65°C. Coffee is mostly water, so its specific heat is about the same as that of water. If the average rate of heat flow through the walls of the thermos bottle is $Q/t = 3.0$ W, approximately how long will it take the coffee to cool to a final temperature of 55°C? Take room temperature to be 25°C.

RECOGNIZE THE PRINCIPLE

We use the general approach outlined in the problem-solving strategies on calorimetry. We identify two systems, the coffee and the room around it. These systems are initially at temperatures $T_{\text{coffee}, i} = 65°\text{C}$ and $T_{\text{room}} = 25°\text{C}$, and the final temperature of the coffee is $T_{\text{coffee}, f} = 55°\text{C}$. Because the room is much larger than the coffee, the temperature of the room does not change significantly. (The room acts as a thermal reservoir.) The heat energy that must flow from the coffee into the room to produce this change in temperature can be found from the specific heat of the coffee and Equation 14.6. This energy equals the rate of heat flow (given to be $Q/t = 3.0$ W) multiplied by the total time.

SKETCH THE PROBLEM

Figure 14.27 describes the problem and shows the direction of heat flow.

IDENTIFY THE RELATIONSHIPS

Using the definition of specific heat (Eq. 14.6) gives

$$Q = cm\,\Delta T = cm(T_{\text{coffee}, f} - T_{\text{coffee}, i})$$

Inserting the given values for the temperatures, mass, and specific heat of water (Table 14.3), we find

$$Q = [4186\ \text{J/(kg}\cdot\text{K)}](0.30\ \text{kg})(55 - 65)\ \text{K} = -1.26 \times 10^4\ \text{J} \qquad (1)$$

where we have again kept three significant figures to avoid roundoff errors below. The negative sign indicates that energy flows out of the coffee. Here, Q equals the rate at which heat energy travels through the walls of the thermos bottle multiplied by the total time. This rate is related to Q by $rate = |Q|/t = 3.0$ W, so we have

$$|Q| = (rate)(t_{\text{total}})$$

SOLVE

Solving for the time and using the result for Q from Equation (1) leads to

$$t_{\text{total}} = \frac{|Q|}{rate} = \frac{1.26 \times 10^4\ \text{J}}{3.0\ \text{W}} = \boxed{4200\ \text{s}}$$

► ***What does it mean?***

This result is more than 1 hour, which gives plenty of time to enjoy the coffee. A better thermos bottle would have a smaller heat flow rate and would thus keep the coffee warm longer.

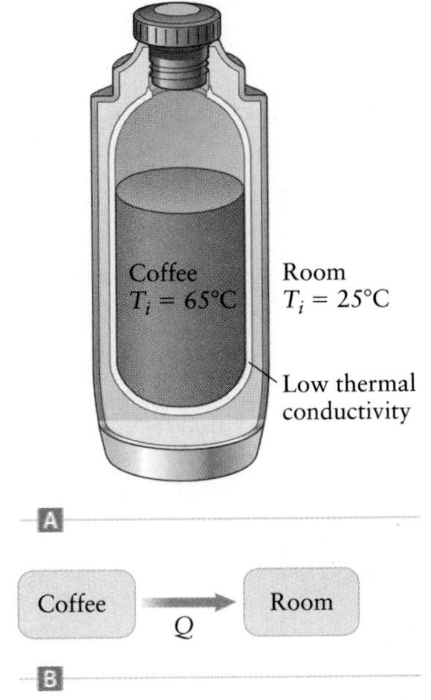

▲ **Figure 14.27** Example 14.11. A The coffee in this thermos bottle is initially much warmer than room temperature. B Energy flows via conduction from the coffee to the room, thus cooling the coffee.

Why Do Metals "Feel" Cold?

The concepts of thermal conductivity and thermal resistance can clarify a well-known aspect of heat conduction. Experience tells us that on a wintry day, an outdoor metal flagpole feels colder to the touch than a piece of Styrofoam. If the flagpole and the Styrofoam are at the same outdoor temperature, why do they feel different?

The temperature-sensitive nerves in your skin sense the difference between your inside body temperature and your skin temperature. We thus have a heat-flow problem involving a series of materials (Fig. 14.28). The thermal conductivity κ of a metal is high (Table 14.6), whereas κ for Styrofoam is low. So, according to the definition of thermal conductivity (Eq. 14.12), when your finger is in contact with a metal, the rate of heat flow Q/t from your finger to the metal is much larger than

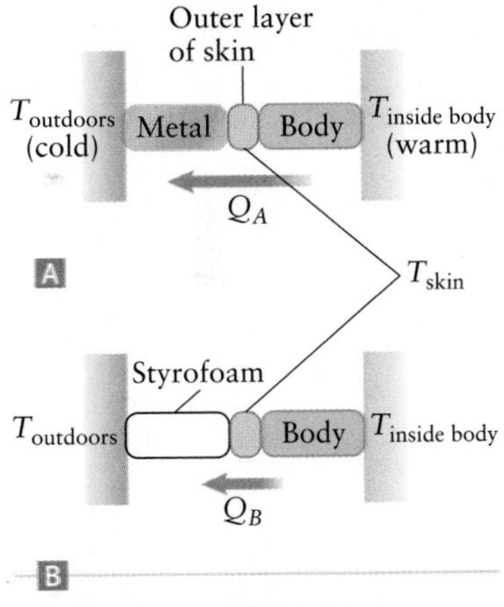

$Q_A > Q_B$ because metal is a better conductor of heat than Styrofoam. This makes T_{skin} lower in A.

▲ **Figure 14.28** Heat flowst from inside your body into your surroundings. A If the outer layer of skin is in contact with a metal (a good thermal conductor), the rate of heat flow is greater than in B, where the skin is in contact with Stryofoam (a poor thermal conductor). The higher heat-flow rate in part A makes the outer layer of skin colder than in part B.

if you were touching Styrofoam instead. This increased heat flow into the metal causes your skin to have a lower temperature than with Styrofoam. That is why a metal feels colder to the touch than Styrofoam.

14.7 Convection

Convective heat flow is based on the phenomenon of thermal expansion. Consider a pot of water being heated on a stove (Fig. 14.29A). If the pot is heated gently so that no gas bubbles are formed, heat energy is carried through the water in two ways. One way is by conduction as discussed in Section 14.6, but much more energy can transported via the motion of water sketched in the figure. Water near the bottom becomes hotter due to the conduction of heat from the burner. Thermal expansion then causes the density of this warmer water to be less than the density of the colder water that is above it in the pot. This warm low-density water moves upward due to the buoyant force associated with Archimedes's principle. As the warm water moves upward, it cools through conduction of heat to the cooler part of the pot and the air above; eventually, this water returns to the bottom of the pot, following the circular flow pattern in Figure 14.29A. A similar convective heat-flow pattern involving the air in a house causes the upper floors to be warmer than the lower floors. Convective flow also plays an important role in transporting heat energy within the oceans and the atmosphere.

Wind Chill

A process similar to convection leads to the "wind chill" effect familiar on a cold winter day. Your skin is moist, and water from it evaporates slowly, placing a small amount of water vapor into the layer of air next to your skin. Evaporation is a phase change, requiring an amount of energy equal to the latent heat. This energy comes from your skin and flows to the nearby air, causing your skin to cool (Fig. 14.30A). This process is called ***evaporative cooling***. In our previous discussion of the latent heat of vaporization of water, we assumed water was at the boiling temperature (100°C), but there is a latent heat of vaporization at all points along the liquid–gas phase change line in Figure 14.10. Evaporative cooling can thus occur over a wide temperature range.

On a windy day, the layer of water-vapor-filled air near your skin is quickly carried away, leaving a "dry" layer in its place. The rate of evaporation is faster into this dry layer than it would be into a moist layer of air. As a result, more latent heat is removed from your skin than if the air layer were filled with water vapor (Fig. 14.30B). Your skin thus loses energy faster on a windy day than on a calm day.

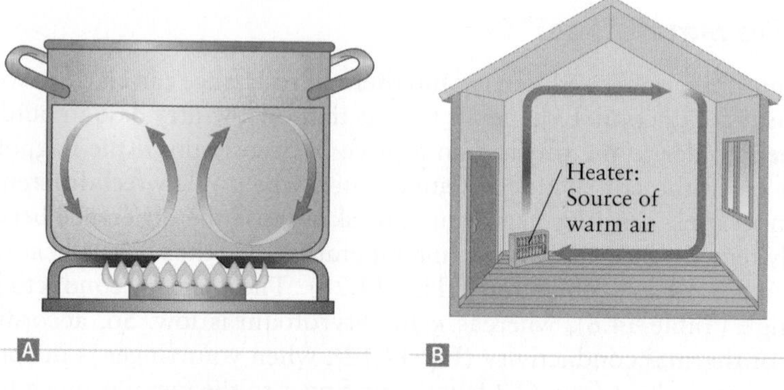

▶ Figure 14.29 Convection transports energy through the movement of matter. This movement is caused by thermal expansion.

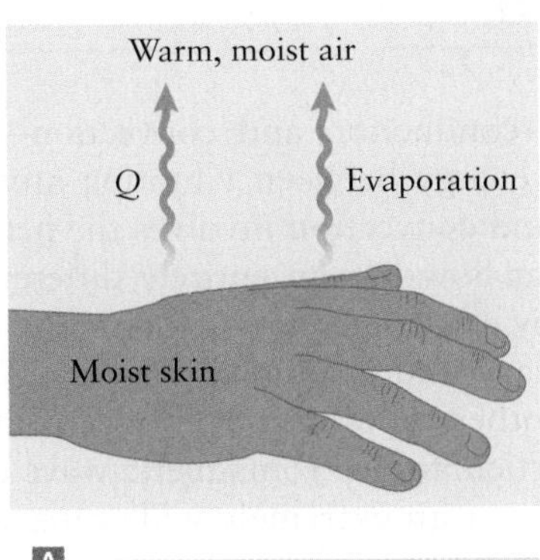

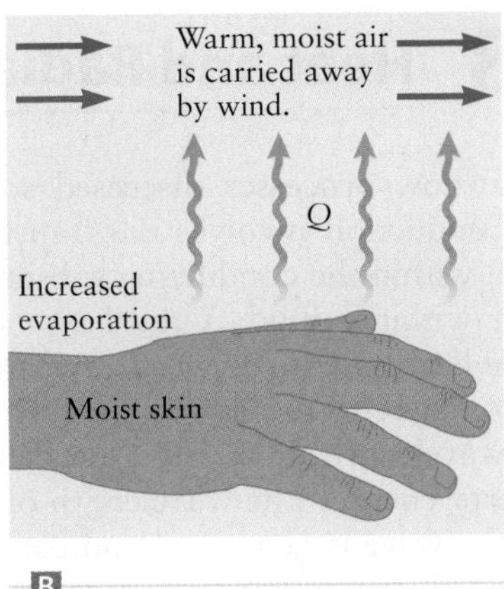

◀ **Figure 14.30** A When moisture evaporates from the body, the latent heat of evaporation leaves your skin, cooling your body and forming a layer of moist and relatively warm air near your skin. B On a windy day, this moist layer of air is rapidly blown away, increasing the evaporation rate. This evaporation increases the amount of latent heat removed from your body and lowers the skin temperature relative to part A.

EXAMPLE 14.12 Cooling Your Body by Sweating

The bicycle rider in Example 14.3 generated 300 J of heat each second. If all this energy is removed from his body by sweating, how much water would he lose each hour by sweating?

RECOGNIZE THE PRINCIPLE

When sweat evaporates from a cyclist's skin, it carries away an amount of energy equal to the latent heat of vaporization. The motion of the bicycle provides the "wind" that carries away the water vapor from a cyclist's skin.

SKETCH THE PROBLEM

Heat energy flows from the cyclist into the surrounding air as sketched in Figure 14.31.

▲ **Figure 14.31** Example 14.12. Convection or just wind carries moist air away from this cyclist, increasing the rate at which sweat evaporates.

IDENTIFY THE RELATIONSHIPS

Let m be the mass of water evaporated as sweat in 1 h. The corresponding latent heat is

$$Q = mL_{\text{vaporization}} \tag{1}$$

which must equal the amount of heat generated by the cyclist. The cyclist generates 300 J each second, so the total heat generated in 1 h (= 3600 s) is

$$\text{total heat generated} = (300\ \text{J})(3600\ \text{s}) = 1.1 \times 10^6\ \text{J}$$

SOLVE

Setting the heat generated equal to the latent heat in Equation (1) gives

$$Q = mL_{\text{vaporization}} = 1.1 \times 10^6\ \text{J}$$

Solving for m and using the latent heat of vaporization of water from Table 14.4, we get

$$m = \frac{1.1 \times 10^6\ \text{J}}{2.2 \times 10^6\ \text{J/kg}} = \boxed{0.50\ \text{kg}}$$

▶ ***What does it mean?***

This mass corresponds to about half a liter of water, which is a typical amount of water a person should consume during 1 h of strenuous exercise.

14.8 Heat and Radiation

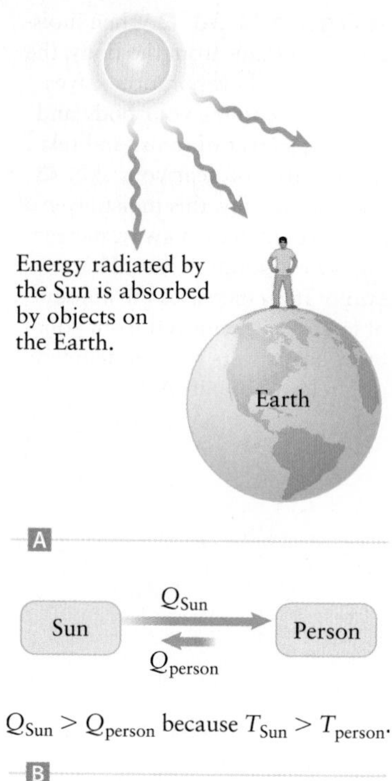

▲ **Figure 14.32** A Electromagnetic radiation from the Sun is absorbed by a person on the Earth. B The person also emits radiation, so there is a flow of energy in both directions, from the Sun to the person and from the person to the Sun.

Both heat-flow processes discussed so far—conduction and convection—involve matter. Conduction involves the transfer of energy between vibrating atoms (and electrons) within the conducting substance, and convection involves the net motion of matter within a fluid. A third type of heat flow has an entirely different basis. ***Radiative*** heat flow involves energy carried by electromagnetic radiation.

Electromagnetic radiation is a type of wave and is thus characterized by frequency, wavelength, and wave speed. Like other types of waves, electromagnetic waves carry energy. The wavelength of a particular electromagnetic wave depends on how the wave is generated and can vary over an extremely wide range. Visible light is the most familiar type of electromagnetic radiation; it has a wavelength in the range of about 400 nm to 700 nm. Other types of electromagnetic radiation correspond to different wavelengths, including ultraviolet and infrared radiation which are also important in heat flow.

Radiative heat flow involves two objects, which might be the Sun and a person on the Earth as sketched in Figure 14.32A. If these two objects are at different temperatures, there is a net transfer of energy—heat energy—from the hotter object (in this case the Sun) to the cooler one (the person).

To understand radiative heat transfer, we must consider several questions. First, why does an object, such as the Sun in Figure 14.32, emit electromagnetic radiation, and how is this radiation absorbed by another object? Second, what are the frequencies of the radiated electromagnetic waves, and how do they depend on the temperature of the emitting object? Finally, how does the amount of radiation absorbed by the cooler object depend on the temperature of that object?

Electromagnetic radiation is generated whenever a particle with an electric charge, such as an electron, proton, or ion, vibrates or undergoes an acceleration in some way. Hence, the atomic vibrations sketched in Figure 14.25A generate electromagnetic radiation that carries away energy. The vibration amplitude of the atoms in an object depends on temperature, so the radiated energy depends on temperature. This radiation—that is, the energy in these electromagnetic waves—is absorbed by another object when that radiation produces a force on the electric charges in the object.

Radiation and the Notion of a "Blackbody"

When electromagnetic radiation, including visible light, bombards an object, some of the radiation energy is absorbed and some is reflected. When you look at an object such as a wall or this book, your eye detects light reflected by the object, which determines the color of the object. An object that completely absorbs all visible light appears black. Physicists have extended this notion and defined a ***blackbody*** as an object that absorbs all electromagnetic radiation at all frequencies. A perfect blackbody does not exist, but the concept of an ideal blackbody is very useful.

The total amount of energy an object radiates depends on its temperature. An object with surface area A and temperature T radiates energy at a rate Q/t given by the ***Stefan–Boltzmann law***,

Stefan–Boltzmann law

$$\frac{Q}{t} = \sigma e A T^4 \tag{14.13}$$

where $\sigma = 5.67 \times 10^{-8}\ \text{W}/(\text{m}^2 \cdot \text{K}^4)$ is called Stefan's constant. The factor e in Equation 14.13 is the ***emissivity*** of the object, a measure of how efficiently it radiates energy. The value of the emissivity must be less than or equal to 1. Objects for which $e = 1$ are perfect blackbodies. Hence, a blackbody is not only a perfect absorber of radiation, it is also the best possible radiator of energy.

The Stefan–Boltzmann law gives the rate at which energy is emitted as electromagnetic radiation—that is, the radiated power—at all wavelengths. This power is

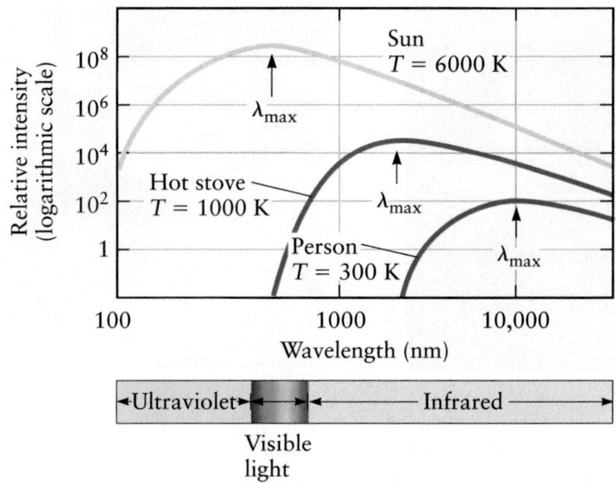

▲ **Figure 14.33** The wavelength at which an object emits radiation most strongly (λ_{max}) depends on its temperature, and that determines the "color" at which it glows. Notice how the total amount of radiation emitted increases very rapidly as T increases.

distributed as a function of wavelength as shown in Figure 14.33. For an object at temperature T, the radiated power is greatest at the wavelength λ_{max} given by ***Wien's law***:

$$\lambda_{max} = (2.90 \times 10^{-3}\ \text{m} \cdot \text{K})/T \qquad (14.14)$$

Wien's law

Figure 14.33 shows a plot of the radiated power as a function of wavelength for blackbody objects at different temperatures. According to Wien's law, the wavelength λ_{max} at which the radiated power is greatest depends on the temperature of the object. The Sun has a surface temperature of approximately 6000 K, and λ_{max} falls in the range of visible light. Some stars are much hotter than our Sun, with temperatures as high as 20,000 K, and for these stars λ_{max} is shorter than the visible range of wavelengths. These stars appear blue to the human eye because they emit more radiation at the blue end of the visible range than at the red end. For a somewhat cooler object, such as the burner of an electric stove at a typical temperature of 1000 K, λ_{max} lies in the infrared. The stove burner appears red because it emits some light at the red end of the visible spectrum in addition to its infrared radiation.

EXAMPLE 14.13 Radiation Emitted by the Human Body

The human body emits electromagnetic radiation according to the body's temperature. Estimate λ_{max} for the human body.

RECOGNIZE THE PRINCIPLE

The peak intensity of the blackbody radiation from an object occurs at a wavelength that depends on its temperature according to Wien's law. Knowing the temperature of the body, we can find λ_{max}.

SKETCH THE PROBLEM

No sketch is needed.

IDENTIFY THE RELATIONSHIPS

The temperature of the human body is approximately 37°C, which must be converted to the Kelvin scale before insertion into Wien's law. We find $T = 37 + 273.15 = 310$ K.

(continued) ▶

A

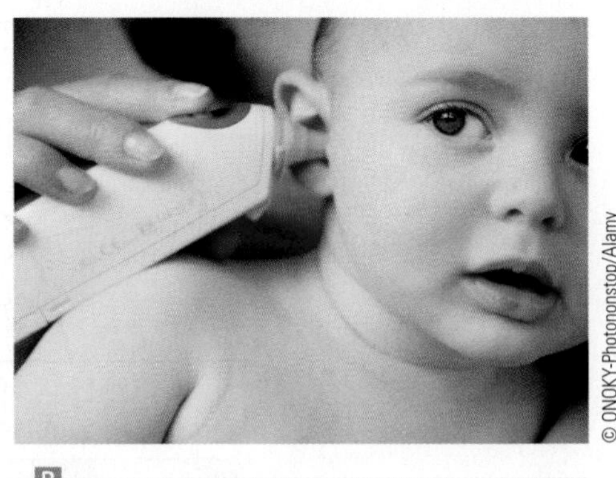

B

▶ **Figure 14.34** Example 14.13. A Night-vision goggles convert the infrared radiation emitted by an object to visible light, making it possible to see objects in the dark. B An "ear thermometer" uses the blackbody radiation from the eardrum to find the body's temperature.

SOLVE

Using this value of the temperature in Wien's law (Eq. 14.14), we find

$$\lambda_{max} = (2.90 \times 10^{-3}\ \text{m} \cdot \text{K})/T = (2.90 \times 10^{-3}\ \text{m} \cdot \text{K})/(310\ \text{K}) = \boxed{9.4 \times 10^{-6}\ \text{m}}$$

▶ ***What does it mean?***

The wavelength λ_{max} lies far into the infrared region of the spectrum (9400 nm; see Fig. 14.33). It is well outside the visible range, so your eye cannot detect this emitted radiation. Night-vision equipment is able to detect this radiation and convert it into wavelengths that can be detected by the human eye (Fig. 14.34A). Also, the "ear thermometers" used by medical professionals (Fig. 14.34B) measure your body temperature by detecting the infrared radiation emitted by your eardrum.

The Stefan–Boltzmann Law and Heat Flow

The Stefan–Boltzmann law (Eq. 14.13), which gives the total power radiated by an object at temperature T, has several noteworthy features. First, the total power varies as the fourth power of the temperature, so the radiated energy increases very rapidly as T is increased. Second, the power is proportional to the emissivity e; this factor is equal to unity for an ideal blackbody, but is smaller than 1 for any real object. In general, the value of e depends on the detailed properties of the material and will even be a function of frequency. Many objects are not far from the blackbody limit, however, so the Stefan–Boltzmann law with an emissivity factor of $e = 1$ provides an approximate description of most radiating objects. Third, the Stefan–Boltzmann law applies to all objects, at all temperatures. The Sun radiates energy that reaches a person on the Earth, but the person also radiates energy that reaches the Sun (Fig. 14.32B). Because the hotter object has a larger emitted power than the cooler one, the net transfer of energy is from the hotter to the cooler object, as we expect for heat flow. Fourth, the Stefan–Boltzmann law also describes how radiation is *absorbed* by an object. When an object of surface area A is located in an environment that has a temperature T, the heat absorbed by the object is

$$\left(\frac{Q}{t}\right)_{\text{absorbed}} = \sigma e A T^4 \qquad (14.15)$$

Insight 14.2
THE RELATION BETWEEN RADIATION AND ABSORPTION

The same relation gives the amount of radiation emitted by an object at a particular temperature (Eq. 14.13) and the amount of radiation absorbed by an object from an environment at that same temperature (Eq. 14.15). This follows from the notion of thermal equilibrium. If an object is in thermal equilibrium with its environment, both are at the same temperature. The total energy of the object is then constant, and the rate at which the object emits energy (through radiation) must equal the rate at which it absorbs energy (also through radiation) from its environment. Hence,

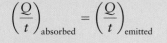

$$\left(\frac{Q}{t}\right)_{\text{absorbed}} = \left(\frac{Q}{t}\right)_{\text{emitted}}$$

Radiation from the Sun and the Temperature of the Earth

Let's apply the Stefan–Boltzmann law to the energy radiated by the Sun. The Sun has a surface temperature of approximately 6000 K, and its emissivity $e \approx 1$. The radius of the Sun is $r = 7.0 \times 10^8$ m, and the surface area through which the radiation is emitted is the area of a sphere with this radius, $A = 4\pi r^2 = 6.2 \times 10^{18}$ m^2. Inserting these values into the Stefan–Boltzmann law (Eq. 14.13), we have

$$\frac{Q}{t} = \sigma e A T^4 = [5.67 \times 10^{-8}\ \text{W}/(\text{m}^2 \cdot \text{K}^4)](1)(6.2 \times 10^{18}\ \text{m}^2)(6000\ \text{K})^4$$

$$\frac{Q}{t} = 4.6 \times 10^{26}\ \text{W} \tag{14.16}$$

which (not surprisingly) is an enormous amount of power.

We can use our result for the heat radiated by the Sun to compute the Earth's temperature. This calculation is based on the principle of conservation of energy. According to Figure 14.35, energy arrives at the Earth via radiation from the Sun, and the Earth itself then radiates energy back into space.[2] The amount of energy radiated by the Earth depends on its temperature T_E, and the value of T_E must be such that the total energy *radiated by* the Earth is equal to the energy it *absorbs* from the Sun. We'll leave the details of this calculation to the next example.

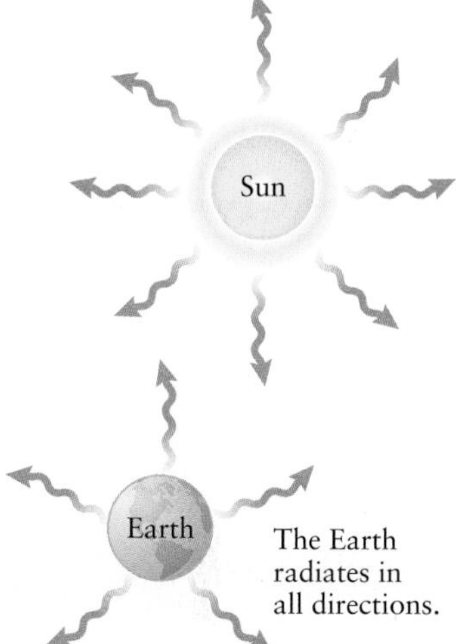

▲ **Figure 14.35** The Earth absorbs energy through radiation from the Sun and emits radiation according to its own temperature. The balance of these energies determines the temperature of the Earth.

EXAMPLE 14.14 Ⓡ Temperature of the Earth

Estimate the temperature of the Earth according to the picture of heat flow in Figure 14.35. Assume the Earth is a perfect blackbody so that $e = 1$.

RECOGNIZE THE PRINCIPLE

We need to compute the amount of energy radiated by the Sun that is absorbed by the Earth, Q_{Sun}, and the energy radiated by the Earth, Q_{Earth}. We have already calculated the total power radiated by the Sun in Equation 14.16, but only a fraction of this power reaches the Earth. That fraction is given by the area of the Earth's *disk* (not the spherical surface area of the Earth) divided by the surface area of a sphere that contains the Earth's orbit (Fig. 14.36).

SKETCH THE PROBLEM

Figure 14.36 shows the problem.

IDENTIFY THE RELATIONSHIPS

Denoting the radius of the Earth by r_E, the area of the Earth's disk is $A_E = \pi r_E^2$. If r_S is the Earth–Sun distance, the surface area of the sphere that contains the Earth's orbit (Fig. 14.36) is $A_S = 4\pi r_S^2$, and the fraction of the Sun's radiated power that strikes the Earth is the ratio A_E/A_S. The radiated power from the Sun that actually intercepts the Earth is thus

$$\frac{Q_{\text{Sun}}}{t} = \frac{A_E}{A_S}\frac{Q_{\text{total}}}{t} = \frac{\pi r_E^2}{4\pi r_S^2}\frac{Q_{\text{total}}}{t}$$

where Q_{total}/t is given in Equation 14.16. Inserting values for r_E and r_S from Appendix A, Table A.3, and our previous result for Q_{total}/t, we obtain

$$\frac{Q_{\text{Sun}}}{t} = \frac{\pi r_E^2}{4\pi r_S^2}\frac{Q_{\text{total}}}{t} = \frac{\pi(6.4 \times 10^6\ \text{m})^2}{4\pi(1.5 \times 10^{11}\ \text{m})^2}(4.6 \times 10^{26}\ \text{W}) = 2.1 \times 10^{17}\ \text{W} \quad (1)$$

The energy radiated by the Earth is also given by the Stefan–Boltzmann law, Equation 14.13. The factor A in this application of the Stefan–Boltzmann law is the area of the

(continued) ▶

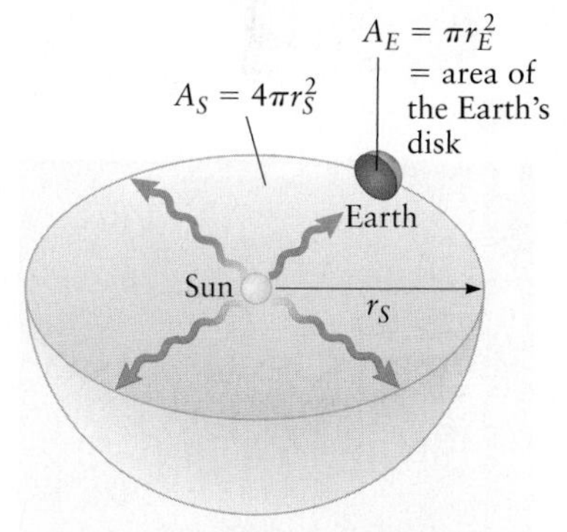

▲ **Figure 14.36** Example 14.14. The Sun emits radiation that propagates outward with circular wave fronts. The spherical wave front that intercepts the Earth has an area $A_S = 4\pi r_S^2$, where r_S is the Earth–Sun distance. The area of the Earth exposed to the Sun is $A_E = \pi r_E^2$, where r_E is the radius of the Earth. Notice that A_E is *not* the (spherical) surface area of the Earth.

[2]Energy also reaches the Earth from "deep space." In Chapter 31, we'll see that regions of space that appear dark (i.e., without stars) radiate energy as a blackbody at a temperature of 3 K. The energy that reaches the Earth in this way is negligible compared to the radiation from the Sun.

Earth's spherical surface, which is $4\pi r_E^2$. (It is *not* the same as the area of the Earth's disk in Fig. 14.36.) The temperature T_E of the Earth is the quantity that we wish to find. Inserting all these values into the Stefan–Boltzmann law gives

$$\frac{Q_{\text{Earth}}}{t} = \sigma e A T_E^4$$

SOLVE

The rate at which the Earth absorbs energy must equal the rate at which it radiates; otherwise, the Earth would be heating up or cooling down over time. Hence,

$$\frac{Q_{\text{Sun}}}{t} = \frac{Q_{\text{Earth}}}{t} = \sigma e A T_E^4$$

Solving for T_E leads to

$$T_E^4 = \frac{Q_{\text{Sun}}/t}{\sigma e A}$$

$$T_E = \left(\frac{Q_{\text{Sun}}/t}{\sigma e A}\right)^{1/4}$$

Taking the Earth's emissivity as $e = 1$, inserting Q_{Sun}/t from Equation (1), and using $r_E = 6.4 \times 10^6$ m gives

$$T_E = \left[\frac{2.1 \times 10^{17}\text{ W}}{[5.67 \times 10^{-8}\text{ W/(m}^2 \cdot \text{K}^4)](1)(4\pi r_E^2)}\right]^{1/4} = \boxed{290\text{ K}}$$

► What does it mean?

Our result for T_E is approximately 20°C. In obtaining this result, we assumed the Earth is a perfect blackbody, so $e = 1$. In fact, the Earth absorbs only about 70% of the radiation from the Sun and reflects the rest. In addition, our calculation actually gives the temperature of the atmosphere, which is somewhat cooler than the Earth's surface. Even with these approximations, our result is quite close to the actual value, confirming that the Sun is responsible for maintaining the Earth's temperature.

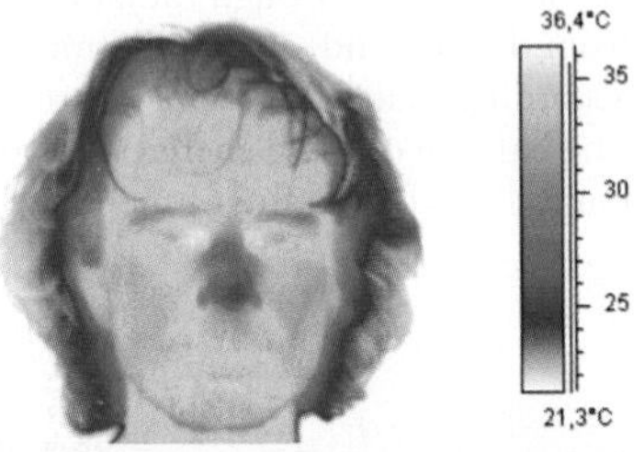

A

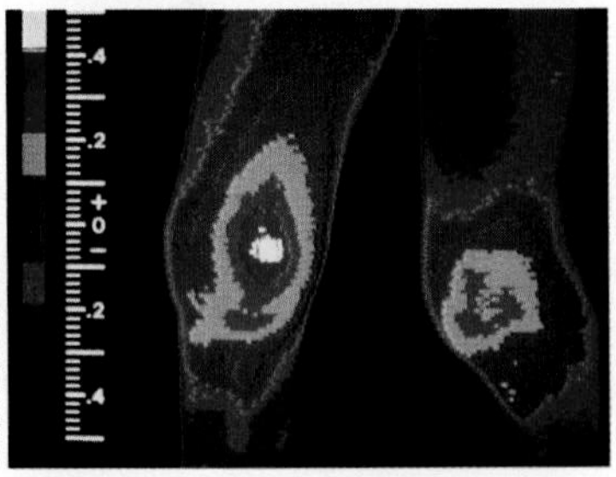

B

▲ **Figure 14.37** All objects, including people, emit radiation according to their temperature. With special detection equipment, this radiation can be used to deduce how the temperature of the skin varies from place to place on the body. A The skin temperature is lowest at the nose. B An arthritic knee (left) is warmer than a healthy one.

Medical Uses of Heat Radiation

An important application of heat radiation is medical thermography. This technique measures the heat radiation emitted by the body and forms an image based on the wavelength at which the radiated power is largest. According to Wien's law (Eq. 14.14), this wavelength depends on temperature, so these images show where a body is warmest and where it is coolest. Some examples are given in Figure 14.37; in these images, whiter regions are warmer than redder areas. Figure 14.37A shows that the nose is the coolest place on a person's face (not a surprising result!). There can be significant variations in temperature near joints; the image in Figure 14.37B shows arthritis associated with a knee joint. Other medical problems can be detected in this way; for example, cancerous tumors are often warmer than normal tissue. Other medical devices, such as ear thermometers (Example 14.13 and Fig. 14.34B), also use the human body's blackbody radiation.

The Greenhouse Effect

As a first approximation, we can treat most real objects as perfect blackbodies. Such was the approach that we took in Example 14.14, when we estimated the temperature of the Earth. A more realistic model of the Earth's temperature must go beyond this approximation, which leads us to consider the ***greenhouse effect***.

Although a complete discussion of the greenhouse effect is too complex to include here, we can understand the qualitative behavior using Figure 14.38, which shows how radiated energy is absorbed and reemitted by the Earth. This sketch is similar to Figure 14.35, but now we have included the effect of the atmosphere. We also indicate separately the infrared and visible radiation coming from the Sun; the atmosphere is not an ideal blackbody, and it absorbs these radiations differently. The Earth's atmosphere allows most of the radiation in the visible part of the spectrum through to the surface, whereas it absorbs much of the infrared radiation, allowing only a little of it to reach the Earth. Because much of the Sun's radiation is in the visible range, most of the energy from the Sun still passes through to the Earth's surface.

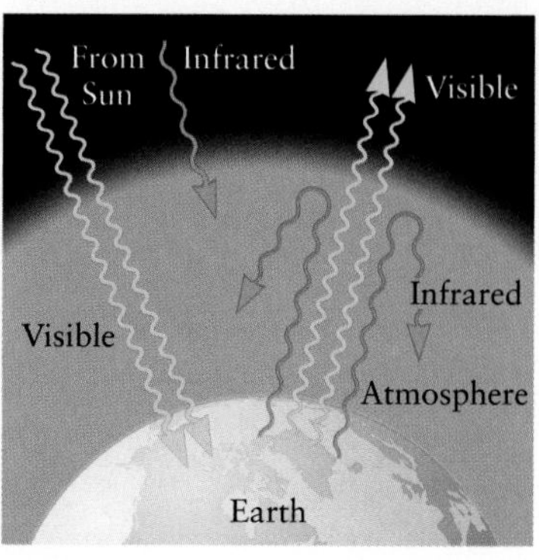

▲ **Figure 14.38** Greenhouse model of radiative heat flow between the Sun, the Earth, and the Earth's atmosphere.

When we estimated the temperature of the Earth in Example 14.14, we had to account for radiation emitted by the Earth. Because its temperature is much lower than that of the Sun, most of the power radiated by the Earth is in the infrared (Fig. 14.33). The atmosphere absorbs some of this radiation, effectively trapping it and not letting this energy escape back into space. This pattern of energy flow is similar to what takes place in a greenhouse, where the glass roof plays the role of the atmosphere. (The analogy with a true greenhouse is not perfect, but the effect is still widely referred to as "the" greenhouse effect.) The infrared radiation trapped by the atmosphere makes the Earth's surface warmer than would be found if the atmosphere were not present.

The absorption of infrared radiation by the atmosphere is caused by certain molecules such as CO_2, which are called ***greenhouse gases***. If the concentration of these gases in the atmosphere increases, the atmosphere absorbs infrared radiation more strongly and traps even more of the Earth's emitted radiation. The result is an increase in the Earth's temperature. It appears that the burning of fossil fuels is now causing an increase in the concentration of greenhouse gases, and many scientists believe that it will cause a significant increase in the Earth's temperature in the coming decades.

The details of radiative heat flow can be quite involved; even Figure 14.38 is a highly simplified picture of the heat-flow processes that determine the Earth's temperature. Our simple blackbody approximation is a very useful first step, but more accurate models are much more complex.

Summary | CHAPTER 14

Key Concepts and Principles

Zeroth law of thermodynamics

If two systems A and B are in thermal equilibrium ($T_A = T_B$) and systems B and C are in thermal equilibrium ($T_B = T_C$), systems A and C are also in thermal equilibrium ($T_A = T_C$).

In words, the zeroth law of thermodynamics asserts that temperature is a unique property of a system that determines when two systems are in thermal equilibrium. Three different ***temperature scales*** are in common use: ***Fahrenheit***, ***Celsius***, and ***Kelvin***.

(Continued)

Heat

Heat flow is the transfer of energy from one object to another by virtue of a temperature difference. The energy transferred is also called heat energy or simply heat. Heat flow can occur in three ways: ***conduction***, ***convection***, and ***radiation***.

- ***Heat conduction*** is determined by the ***thermal conductivity*** κ, with

$$\frac{Q}{t} = \kappa A \frac{\Delta T}{L} \qquad \textbf{(14.12) (page 461)}$$

  where Q is the heat energy transferred in time t, L is the length of the system, and A is the cross-sectional area of the system.
- ***Convective heat flow*** involves the movement of mass and is caused by thermal expansion and contraction.
- ***Radiative heat transfer*** involves electromagnetic radiation. The total amount of energy an object radiates depends on its temperature. An object with a surface area A and temperature T radiates energy at a rate Q/t given by the ***Stefan–Boltzmann law***,

$$\frac{Q}{t} = \sigma e A T^4 \qquad \textbf{(14.13) (page 466)}$$

  The wavelength at which this radiation is strongest depends on temperature through ***Wien's law***,

$$\lambda_{\text{max}} = (2.90 \times 10^{-3}\ \text{m} \cdot \text{K})/T \qquad \textbf{(14.14) (page 467)}$$

  Wien's law explains why the color of a glowing object depends on its temperature.

Applications

Specific heat and latent heat

If an amount of heat energy Q is added to a substance of mass m, the temperature of the system increases by an amount ΔT, where

$$Q = cm\,\Delta T \qquad \textbf{(14.6) (page 448)}$$

and c is the ***specific heat*** of the substance. Equation 14.6 applies provided there is not a phase change. If there is a phase change such as melting or evaporation, a certain amount of energy called the ***latent heat*** must be added to or taken from the system.

Thermal expansion

When the temperature of a substance changes, its physical dimensions usually change. Changes in length are proportional to the ***coefficient of thermal expansion*** α, with

$$\frac{\Delta L}{L_0} = \alpha\,\Delta T \qquad \textbf{(14.9) (page 455)}$$

and changes in volume are proportional to the ***coefficient of volume expansion*** β

$$\frac{\Delta V}{V_0} = \beta\,\Delta T \qquad \textbf{(14.10) (page 456)}$$

Questions

SSM = answer in *Student Companion & Problem-Solving Guide* = life science application

1. When the metal cap on a glass jar is stuck (i.e., when the cap is too tight to open), it sometimes helps to hold the jar and cap under hot water. Explain why.

2. Consider a hole in a block of steel such as the cylinder in a steel engine block. Will the diameter of this hole get larger or smaller when the steel is heated? By how much will it change?

3. Common bulb thermometers use liquid mercury (the familiar silver liquid) or liquid methanol (with red dye for coloring) as the working substance enclosed in glass. What are the limitations in using these devices? Could a mercury or methanol thermometer measure the temperature of dry ice? (CO_2 is a solid below $-78°C$ at atmospheric pressure.) Why or why not? Could either type of thermometer measure the temperature of molten lead?

4. A bar of steel is attached to a bar of brass side by side to form one thicker bar of the same length (Fig. Q14.4). You find that when you heat this bar of two metals, it bends. Why? Will it bend toward the brass or toward the steel? Which way would it bend if the bar were cooled instead of heated?

Figure Q14.4

5. The pressure–temperature phase diagram for a typical substance is shown in Figure Q14.5. The line that separates the solid and liquid phases has a positive slope; that is, the melting temperature increases as the pressure is increased. Give a qualitative explanation of this observation, assuming that a material expands when it is heated.

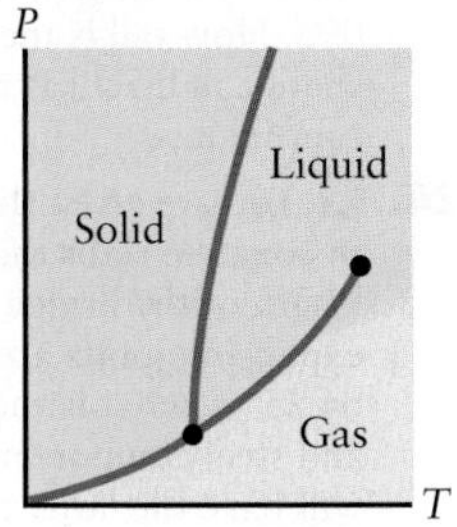

Figure Q14.5

6. When you step out of a swimming pool on a windy day, you usually feel chilly. Explain why.

7. On a hot summer day, an electric fan can make a difference in comfort. The fan does not cool the air, however. In fact, it slightly raises the temperature of the air that it circulates. How is it that a fan can make you cooler?

8. Most internal combustion engines in modern automobiles have a cooling system that uses a circulating fluid. In designing such a cooling system, should the liquid coolant have a small or a large heat capacity? Which would be best, a large or a small thermal conductivity?

9. Consider a winter mountain survival scenario. Why is it a bad idea to eat snow, even if you have no way of melting it for drinking? *Hint*: See Problem 62.

10. Given the value of α (e.g., for steel), can you calculate the value of β? Derive a general relation between α and β.

11. Explain why getting burned with a certain mass of steam at 100°C is more painful and harmful than getting burned with the same mass of water (liquid) at 100°C.

12. When you take a cookie sheet out of the oven, you use a pot holder because touching the hot cookie sheet would result in injury. The air in the oven, however, is at exactly the same temperature. Why doesn't the air burn your hand the way the cookie sheet would?

13. You pour a cup of coffee, intending to drink it a few minutes later. If you want your coffee to be as hot as possible at this later time, should you add the cream when you pour the coffee or just before you drink it? Why?

14. SSM Fruit trees bloom in the spring, and their blossoms eventually turn into oranges, apples, and so forth. Blossoms are susceptible to cold weather, so when there is the possibility of a frost, orchard owners often spray their trees with water. Explain how this practice can help prevent the blossoms from freezing.

15. Metal cookware made from steel or aluminum often has a copper bottom. Why?

16. A potato will bake faster if a nail is stuck through it. Why? Which would make the potato cook faster, an aluminum nail or a steel nail?

17. Explain why a layer of fur (Fig. Q14.17) is a very poor conductor of heat.

Figure Q14.17

18. There are many different types of glass; all are composed mainly of SiO_2, but the addition of other elements can change the color and other properties substantially. One type of glass, called Pyrex, incorporates small amounts of the element boron, which gives it a very small coefficient of thermal expansion. Most oven-safe dishes are made from Pyrex because it tends not to crack when heated or cooled. Explain why.

19. Explain why it takes longer to cook a hard-boiled egg in Denver than in San Francisco. *Hint*: Consider the phase diagram of water and how the boiling temperature depends on pressure.

20. SSM A good-quality thermos bottle is double-walled and evacuated between these walls, and the internal surfaces are like mirrors with a silver coating. This configuration combats heat loss from all three transfer methods and keeps the bottle's contents—your coffee—hot. Which design feature of a thermos bottle minimizes heat transfer by radiation, by conduction, and by convection?

21. Table 14.6 lists the thermal conductivities of several substances. Give qualitative arguments to explain the following questions. (a) Why is the thermal conductivity of goose down about the same as the thermal conductivity of air? (b) Goose down is very effective for making winter coats. Explain why goose down must be kept dry to be effective. Can you predict the value of the thermal conductivity when it is soaked with water?

22. Heat energy is "lost" from the body through the skin, so Q_{lost} is proportional to the area of the skin. Heat energy is produced throughout the body as food is metabolized, muscles flex, and so on, so $Q_{produced}$ is proportional to the body volume. Explain why these two facts tend to make it easier for a large animal to stay warm compared with a small animal.

23. An ice cube can exist for many minutes while floating in water, even when the temperature of the water is greater than 0°C. Explain why.

24. On a hot summer day, why is it easier to keep cool by wearing white clothing instead of black clothing?

25. When a wire is strung tightly between two supports, it is found that transverse waves on the wire have speed v. If the wire is heated, does the wave speed increase or decrease? Explain.

26. Why does a lake freeze first at its surface?

27. Climates near an ocean tend to be milder than climates far inland. Explain why.

Problems

SSM = solution in *Student Companion & Problem-Solving Guide*
★ = intermediate ✪ = challenging ⓧ = life science application
Ⓡ = reasoning and relationships problem
RT = reasoning tutorial in ENHANCED WebAssign

14.2 TEMPERATURE AND HEAT

1. The melting point of tin is 232°C. What is this temperature on the Fahrenheit and Kelvin scales?

2. Liquid oxygen boils at 90 K at atmospheric pressure. What is this temperature in degrees Celsius?

3. What is the value of the "absolute zero" of temperature on the Fahrenheit scale?

4. What is the melting temperature of gold on the Celsius scale?

5. The temperature inside a typical house is 70°F. What is this temperature on the Celsius and Kelvin scales?

6. ★ You and a friend each have a thermometer; yours reads in Celsius, whereas your friend's is an old-fashioned Fahrenheit model. You go outside and notice that the two thermometers read the same value. Is it hot or cold outside? What is the outside temperature?

7. The lowest temperature ever recorded outdoors is approximately −130°F. What is this temperature on the Celsius scale?

8. SSM ★ Consider Joule's experiment described in Insight 14.1. Assume the mass $m = 2.0$ kg and the fluid is water with a volume of 1.0 L. What is the increase in the temperature of the water when the mass moves vertically a distance $z = 1.0$ m?

14.4 PHASES OF MATTER AND PHASE CHANGES

9. How much heat is required to raise the temperature of a 3.0-kg block of steel by 50 K?

10. ★ You have 5.0 g of an unknown substance. To identify the substance, you decide to measure its specific heat and find that it requires 16 J of heat to increase the temperature of your sample by 25 K. If this substance is one of those listed in Table 14.3, what might it be?

11. While working in a research lab, you find that it takes 400 J of heat to increase the temperature of 0.12 kg of a material by 30 K. What is the specific heat of the material?

12. How much heat is required to increase the temperature of a glass sphere of mass 400 g by 45 K?

13. ✪ An ice cube of volume 10 cm^3 is initially at a temperature of −15°C. How much heat is required to convert this ice cube into steam?

14. ✪ You drop 500 g of solid ice at −35°C into a bucket of water (1200 g) at 45°C. Eventually, the ice all melts, and the entire system comes into thermal equilibrium. Find the final temperature of the system.

15. SSM ★ If you combine 35 kg of ice at 0°C with 75 kg of steam at 100°C, what is the final temperature of the system?

16. ★ Ⓡ A cup of coffee is initially at a temperature of 70°C and then cools to room temperature. (a) Estimate how much heat flows out of the cup during this time. (b) If this energy were converted to the kinetic energy of a baseball, what would be the speed of the ball?

17. ✪ You are a blacksmith and have been working with 12 kg of steel. When you are finished shaping it, the steel is at a temperature of 400°C. To cool it off, you drop it into a bucket containing 5.0 kg of water at 50°C. How much of this water is converted to steam? Assume the steel, the water, and the steam all have the same final temperature.

18. ★ Ⓡ (a) How much energy is required to evaporate all the water in a swimming pool of area 100 m^2 and depth 2.0 m on a typical summer day? (b) The intensity of sunlight is about 1000 W/m^2. If all this energy is absorbed by the water in the pool, how long will it take the water to evaporate?

14.5 THERMAL EXPANSION

19. One section of a steel railroad track is 25 m long. If its temperature increases by 25 K during the day, how much does the track expand?

20. ★ RT A new cement bridge will be 50 m long. The bridge contains expansion joints, so it will not crack when the temperature changes. Suppose you are the designer of this bridge, and your job is to design the expansion joints. Estimate how much change in length the joints will have to accommodate.

21. Ⓡ The Golden Gate Bridge is a famous suspension bridge in San Francisco. The main segment is approximately 1400 m long and is composed of steel. Estimate how much the length of the bridge increases as time passes from morning to afternoon on a typical day.

22. ★ Your swimming pool is square and 5.0 m on a side. It is 3.0 m deep in the morning. If the temperature changes by 20°C during the afternoon, how much does the depth of the water increase?

23. A steel building is 120 m tall when the outside temperature is 0°C. How tall is the building on a hot summer day (30°C)? The change in height is small, so express your answer with five significant figures.

24. ★ You are given the job of designing a bridge that is to be made of concrete slabs that rest on a steel support frame. The total length of the bridge is 300 m, and you want to be sure that the expansion joints are sufficiently large. If the temperature increases by 25°C, what is the difference in the new lengths of the concrete and steel components of the bridge? Which of these components will have the largest increase in length?

25. A material of unknown composition with an initial length of 2.0 m is found to expand by 0.33 mm when it is heated by 15 K. What is the coefficient of thermal expansion of this material?

26. SSM ★ You are a high-precision carpenter and use a high-precision steel ruler. If you want the dimensions of your work to be accurate to 1.0 mm over distances of 15 m, you will have to keep the temperature of your ruler constant to within an uncertainty of ΔT. Find ΔT.

27. ★ A rectangular aluminum plate has an area of 0.40 m^2 at 15°C. If it is heated until its area has increased by 4.0×10^{-6} m^2, what is the final temperature of the plate?

28. ★ The author's gold wedding ring is stuck on his finger, and he must remove it. The ring has an inside diameter of 2.00 cm, but the knuckle of the finger in question has a diameter of 2.05 cm. The author decides to get the ring off by heating it. How much must the temperature of the ring be increased for it to slide past the knuckle? Assume the author's finger does not change size when it is heated.

29. A carpenter constructs a house of length 15 m out of wood. If the temperature changes by 5.0°C, how much does the length of the house change?

30. ★ Consider a steel rod of diameter 1.0 cm and length 1.0 m. If the temperature of this rod is increased by 25°C, its length will change by a certain amount ΔL. The length can also be changed

by applying a tensile force (Chapter 11). How much force must be applied to change the length of the rod by the same amount ΔL?

31. Design a mercury thermometer. Choose dimensions for the tube and the reservoir and then estimate how much the meniscus will move in response to changes in temperature; that is, calculate how far the meniscus moves when the temperature changes by 1°C.

32. A rock has a narrow crack (width 1.0 mm) into which water can penetrate. When this water freezes, it expands. (a) Estimate the pressure on the rock when the water freezes. (b) What is the corresponding force on a rock surface that has an area of 1.0 cm^2? (c) Compare the force to your weight.

33. RT The Eiffel Tower in Paris (Fig. P14.33) is approximately 300 m tall and is composed of iron. How much does its height increase from midnight to noon on a typical summer day?

© ICP/Alamy

Figure P14.33

34. A pendulum clock uses a rock fastened to a very thin steel wire. This clock is designed so that one "tick" of the clock (one complete period of the pendulum) is equal to precisely 1 s. The clock is then heated by 10°C. Will it now run "fast" or "slow"? Find the new value of the period of the clock. Assume the thermal expansion coefficients of the rock are both zero.

35. An aluminum beam 80 m long and with a cross-sectional area of 0.10 m^2 is used as part of a bridge. The beam is clamped rigidly at both ends. (a) If the temperature of the beam is increased by 30°C, what is the extra force exerted by the beam on one of its supports? (b) What is the force if the temperature of the beam is decreased by 40°C?

14.6 HEAT CONDUCTION

36. A person has a core (inner) body temperature of 37°C, and the air temperature is 20°C. Estimate the rate at which heat flows via conduction through the person's skin. Assume a surface area of 1.5 m^2.

37. A cold-weather sleeping bag uses goose down to keep the inside warm. The rate at which a sleeping camper generates heat energy is about 70 W. Estimate the temperature of the air inside the bag and around the camper when the outside temperature of the bag is 0°C.

38. **Energy-efficient windows.** Energy-efficient windows are constructed with two panes of glass separated by an air gap (Fig. P14.38). In a typical window of this type, the air gap makes a huge difference in the rate of heat flow. The heat-flow rate is much smaller for an energy-efficient window than for a single pane of glass because the thermal conductivity of the air gap is much less than that of a glass pane. (a) To compare heat flow for the different types of windows, first calculate the heat-flow rate for a single-pane window. Assume the glass is 3.2 mm thick, it has an area of 0.50 m^2, and the temperatures on the two sides of the glass differ by 30°C. (b) Now consider an energy-efficient window consisting of two sheets of glass as in part (a), separated by an air gap that is 1.0 cm thick. Assuming the same temperature difference as in part (a), what is the heat-flow rate in this case? For simplicity, assume all the temperature difference is across the air gap and none is across the two panes of glass. (c) How thick would a single pane of glass have to be to give the same rate of heat flow as through the energy-efficient window, assuming the same temperature difference?

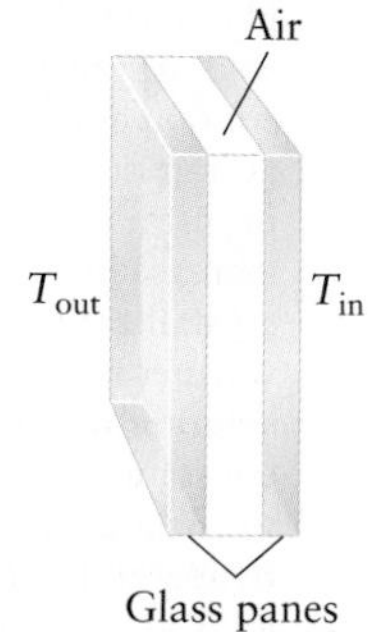

Figure P14.38

39. SSM A small dog has thick fur with thermal conductivity 0.040 W/(m · K). The dog's metabolism produces heat at a rate of 40 W, and its internal (body) temperature is 38°C. (a) If all this heat flows out through the dog's fur, what is the outside temperature? Assume the dog has a surface area of 0.50 m^2 and the length of the dog's hair is 1.0 cm. (b) Now assume it is a hot summer day with an outdoor temperature of 32°C (about 90°F). What is the body temperature of the dog now? Your answer will explain why a dog can quickly overheat on a warm day.

40. Estimate the heat-flow rate for a solid cement wall of a typical-size house and compare it with the heat flow rate for the window in Example 14.10. Assume the wall is 5.0 m wide and 2.5 m tall.

41. A winter hiker's body generates heat at a rate up to a maximum value of 120 W. She is hiking on a brisk day in which the outdoor temperature is −10°C, while wearing a jacket with goose down insulation. How thick must the jacket be so that she can maintain a normal body temperature? Assume her surface area for heat flow (the jacket) is 1.5 m^2 and that all of her body heat flows out through the jacket.

14.7 CONVECTION

42. SSM When a dog "pants," it exhausts water vapor through its mouth (Fig. P14.42). This process converts liquid water (inside the dog) into water vapor, removing heat from the dog via the evaporation of water. If 10 g of water evaporates from the dog each minute, how much heat energy does it lose during this time?

© PureStock/Jupiterimages

Figure P14.42

43. A jogger generates heat energy at a rate of 800 W. If all this energy is removed by sweating, how much water must evaporate from the jogger's skin each hour?

14.8 HEAT AND RADIATION

44. In Example 14.14, we calculated the temperature of the Earth by assuming it is a perfect blackbody with emissivity $e = 1$. (We also ignored the atmosphere.) Suppose the emissivity is $e = 0.70$. Find the temperature of the Earth (according to this simplified model) with this value of e. Explain why your answer is nearly the same as found in Example 14.14.

45. A typical incandescent lightbulb has a filament temperature of approximately 3000 K. At what wavelength is the intensity of the emitted light highest?

46. A snake is sunning itself on a rock. (a) If the snake is 0.3 m long, what is the approximate amount of radiation it absorbs from the Sun in 10 min? (b) If no heat flows out of the snake via conduction or by evaporative cooling, how much does its temperature increase during 10 min? Assume all the Sun's energy goes into heating the snake and the specific heat of the snake is equal to the specific heat of water.

47. The intensity of sunlight at the top of the Earth's atmosphere is about 1400 W/m^2. (The intensity at the surface of the Earth is somewhat smaller due to absorption and reflection by the atmosphere.) What is the total power emitted by the Sun?

48. In Problem 47, we noted that the intensity of sunlight reaching the Earth is about 1400 W/m^2. (a) What is the intensity of sunlight at Mars? *Hint*: The data in Table 5.1 may be useful. Also, the atmosphere of Mars is very "thin" and absorbs or reflects a negligible amount of the incident radiation. (b) The

Martian "rovers" (small robots that move about on wheels) sent to explore the surface of Mars use solar panels to charge their batteries. If the solar intensity on the Earth is sufficient to power such a rover, do you think the solar intensity on Mars is large enough to fully power a rover on that planet? (c) Scientists have proposed sending such rovers to investigate Europa, one of Jupiter's moons. What is the solar intensity at Europa? Do you think a solar-powered rover would work on Europa?

49. Estimate the temperature of the surface of Mars. Assume Mars acts as a perfect blackbody and the only heat input comes from the Sun. How does your result compare with the Earth's temperature?

50. SSM An incandescent lightbulb emits 100 W of radiation. If the filament is at a temperature of 3000 K, what is the surface area of the filament?

51. Estimate the temperature of the Moon's surface in two cases: (a) when it is facing toward the Sun and (b) when it is facing away from the Sun. Explain qualitatively why your answer in part (b) is probably lower than the actual temperature. *Hint*: Consider the specific heat of the surface of the Moon.

52. (a) What is the approximate energy emitted as blackbody radiation by a human being each second? (b) What is the blackbody energy absorbed by a person in 1 s in a room? Why are the answers different?

53. A typical coal-burning power plant produces energy at the rate of 1000 MW. How large in area must a solar power plant be to produce the same amount of energy during daylight hours?

Additional Problems

54. SSM An ice-cube tray containing 600 g of water at room temperature is placed into a freezer. It is found that all the water becomes solid after 60 min. What is the average rate at which heat energy flows out of the water into the freezer?

55. An incandescent lightbulb produces light using a thin tungsten filament (wire) heated to about 3000 K by an electric current that passes through it. A typical filament has a diameter of 10 μm and a length of 2.0 cm. (a) What is the rate at which heat energy is emitted by the filament? (b) When the current through the filament is turned off, the filament cools as heat is radiated away. Estimate how long it takes the filament to cool 100 K to a final temperature of 2900 K.

56. A pot of very cold water (0°C) is placed on a stove with the burner adjusted for maximum heat. The water just begins to boil after 2.0 min. How much longer will it take the water to completely boil away?

57. An aluminum block slides along a horizontal surface. The block has an initial speed of 10 m/s and an initial temperature of 10°C. The block eventually slides to rest due to friction. Assuming all the initial kinetic energy is converted to heat energy and all this energy stays in the block, what is the final temperature of the block?

58. For the aluminum block in Problem 57, it is likely that at least a little bit of the heat energy will go toward heating the horizontal surface on which the block slides. Will this effect make the final temperature of the block larger or smaller than found in Problem 57?

59. SSM **The Rankine scale.** Named after William John Macquorn Rankine (a Scottish engineer and physicist who proposed it in 1859), the Rankine scale is similar to the Kelvin scale in that the zero point is placed at absolute zero, but the size of temperature differences are the same as that of the Fahrenheit scale (e.g., a change in temperature of 1°F corresponds to a change of 1°R). (a) Determine the conversion formula to go from the Fahrenheit to the Rankine scale. (b) Find the formula to convert from Kelvin to Rankine. (c) What is the temperature of the freezing point of water on the Rankine scale? What is room temperature on this scale?

60. **Dry ice.** At standard atmospheric pressure, the solid form of carbon dioxide called "dry ice" undergoes a phase change not to a liquid, but straight to a gas. This process is called ***sublimation***, and like other phase transitions, heat energy is required. In this case, it is the latent heat of sublimation, which for carbon dioxide is 573 kJ/kg at the sublimation temperature of −78.5°C. If 150 g of dry ice is dropped into 0.50 L of water at room temperature (20°C), how much of the water will turn to ice by the time all the dry ice has sublimated? *Note*: When dry ice becomes a gas, the gas bubbles out of the system.

61. Measuring the temperature of very cold substances can be challenging. Here is one way to measure the temperature of liquid nitrogen. A 50-g flake of aluminum is submerged in liquid nitrogen and left immersed until it is in thermal equilibrium. The flake is then removed and placed in a well-insulated container with 500 ml of water at 30°C. When water and flake attain thermal equilibrium, the temperature of the water is found to be 25°C. What was the temperature of the liquid nitrogen?

62. **Winter survival.** If you are stranded in a cold and snowy environment, it is never a good idea to eat snow for hydration. Estimate the heat energy needed to convert 0.50 kg of snow to water and then heat the water to body temperature. Assume the snow is at 0°C in the survival snow cave you built as a shelter. State your answer in SI units (J) and in food calories (Cal).

63. **Phase change hand warmer.** Sodium acetate in solution has a freezing point at 58°C and a latent heat of freezing of 210 kJ/kg, but a solution of this compound can remain in liquid form at much lower temperatures. This condition, known as ***supercooling***, occurs in substances that form specific crystalline structures when in solid form. Such systems need a seed crystal to start the solidification process. Reusable hand warmers like that in Figure P14.63 are sold in sporting goods stores and contain about 60 g of sodium acetate solution. Clicking a piece of metal in the solution generates sound waves. The resulting increase in pressure causes a seed crystal to form, leading to full crystallization of the hand warmer within 4 s. Suppose a hand warmer is activated on a particularly cold morning when the solution was at a temperature of 10°C. Phase transitions occur at constant temperature, and the liquid is mostly water. (a) Calculate the heat generated by the phase transition. (b) What is the final temperature of the hand warmer?

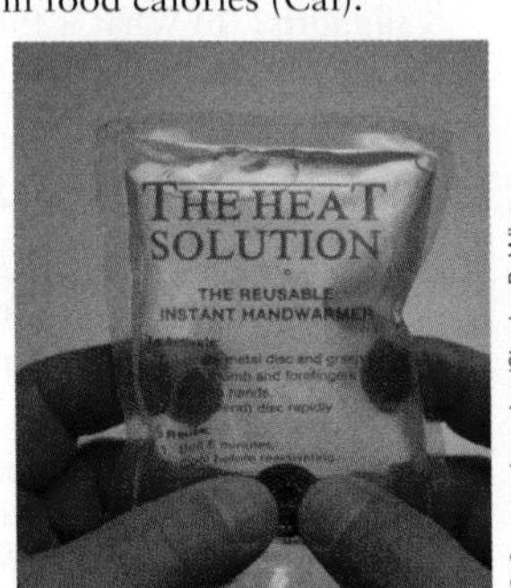

Figure P14.63 Commercial sodium acetate reusable hand warmer.

64. **Freeze plugs.** The internal combustion gasoline engines used in automobiles produce much heat, which is mostly a nuisance and must be removed from the engine to prevent deformation and melting. The heat is carried away by a cooling system, typically using a fluid that flows through the walls of the engine block and out to a radiator, where flowing air cools the fluid. The fluid is mostly water with a mix of chemicals to prevent the fluid from freezing in cold climates when the engine is not in use. Freeze plugs, a standard part of any engine block, are brass alloy plugs that are pressed into holes in the engine block (Fig. P14.64). In principle, in the event of a hard freeze of the cooling system fluid, the freeze plugs will pop out and prevent the engine block from cracking. (a) Consider a carelessly maintained engine that has only water in its cooling system. On a cold winter night in Chicago, the water becomes frozen. How much pressure is exerted on a freeze plug? (b) If the freeze plug is 5.0 cm in diameter, how much force is exerted on the plug by the frozen water?

Figure P14.64 Problems 64 and 65.

65. One of the brass freeze plugs in Problem 64 is 3.0 cm thick and just fits inside the hole in the steel of the engine block when the block is at room temperature and the brass was brought to the temperature of dry ice. (a) If the bulk modulus of brass is 6.1×10^{10} N/m^2, what pressure does the freeze plug exert on the block when both the freeze plug and the engine block are at room temperature? (b) If the coefficient of friction between brass and steel is 0.45, what force would it take to push the plug out of the block? If water were to freeze in the cooling system, would the plug pop out? See part (b) of Problem 64.

66. Figure P14.66 shows a modern electronic circuit board. It features standard miniature components such as the resistor (R516), which measures 1.6 mm long, 0.80 mm wide, and 0.20 mm high. The ceramic material of the resistor has a Young's modulus of 1.2×10^{10} Pa. When electricity flows through the circuit, the components, especially resistors, can heat up substantially. The resistor has a coefficient of thermal expansion of 2.2×10^{-6} K^{-1}, and that of the circuit board it is soldered to is 6.0×10^{-6} K^{-1}. (a) If the temperature of the circuit rises 30°C, how much longer will the circuit's "footprint" be than the resistor that is soldered to it? (b) If the resistor is stretched by this amount by the circuit board, how much stress is the resistor subjected to? (c) The resistor has a maximum stress of 25 MPa before it will break. If a power glitch raises the temperature of the circuit by 120°C, will the resistor suffer damage from the stress?

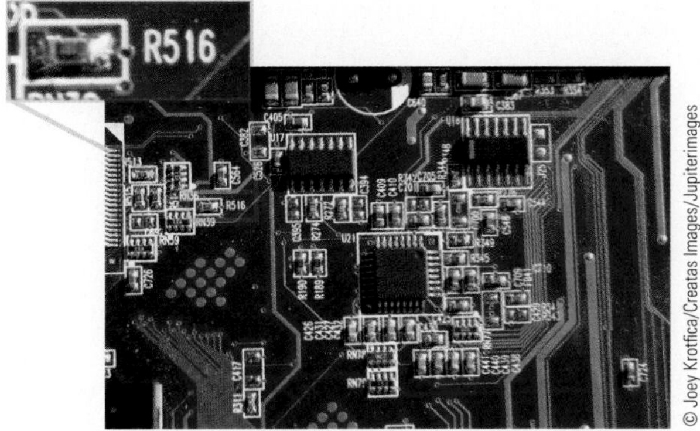

Figure P14.66

67. A bimetal strip can be used as a thermometer and thermostat, a device that converts small changes in temperature to mechanical motion (Fig. P14.67A). A thermostat can be constructed by joining two pieces of metal, brass and steel, along their length as shown in Figure P14.67B. The brass and steel are both 50 cm long and 0.10 mm thick. If this bimetal strip is heated, the brass will expand more than the steel, causing the strip to bend. What change in temperature is needed to make the bimetal strip curl into a full circle as shown in Figure P14.67C? *Hint*: The difference in the radius of the two metals making the strip is just the thickness of one metal strip.

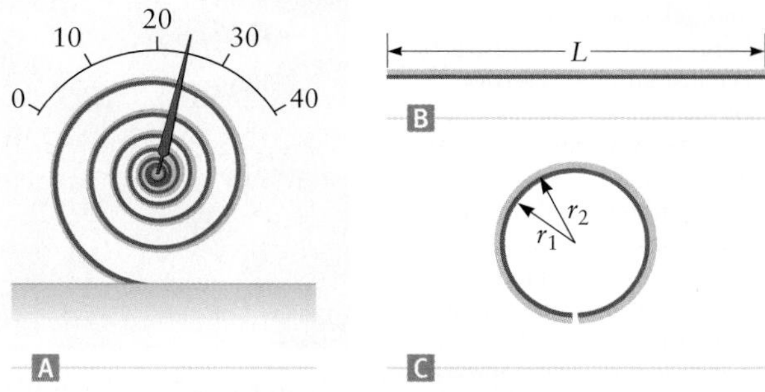

Figure P14.67

68. The owners of a cabin in the mountains forgot to turn off and purge the water out of the plumbing system before the winter season. One particularly cold night, a pipe near an outside wall freezes. Consider a length of "$\frac{1}{2}$-in." copper plumbing pipe filled with water. Although called $\frac{1}{2}$ in., the outside diameter of this standard pipe is $\frac{5}{8}$ in. and the inside diameter is 0.537 in. (a) Water freezes in a 50-cm length of this pipe capped at both ends by closed valves such that the expansion of the ice is constrained by the pipe. Calculate the pressure on the wall of the pipe. (b) Is the diameter of the pipe just before the water freezes significantly different from the diameter at room temperature? What is the percent change? (c) The pipe has a connection (Fig. P14.68). Assume the water freezes so that the separation between ice and water occurs right in the middle of the connector. Calculate the force that pulls on the connector when the water freezes.

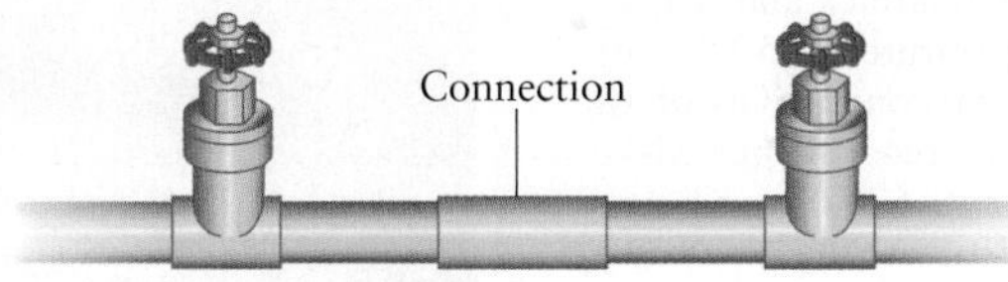

Figure P14.68

69. **Mass wasting.** Water filling a thin crack in a rock wall and then freezing can put large forces on the rock, making the cracks bigger (Fig. P14.69). (a) Calculate the force exerted by a 10-m^2 area of a thin (0.50 mm) sheet of water freezing between two slabs of rock. (b) Take the sheet of water to be oriented in a horizontal plane (unlike the case in Fig. P14.69) and assume a large slab of rock sits on top of the ice. By how much is the rock on top of the water displaced vertically? (c) If the column of rock is 15 m

Figure P14.69

tall and has a density of 5.2 g/cm^3, how much work is done by the expansion of the ice on this rock mass?

70. An extreme thermos bottle might use silica aerogel as an insulator. Figure P14.70 demonstrates the insulating properties of the aerogel compound. Aerogel can protect human skin from a blowtorch at nearly point-blank range, made possible by its very low thermal conductivity (only 0.003 W/(m · K)) and a melting point of 1200°C. Let's see if we can improve on the thermos bottle used by the author in Example 14.11, where we found that it took 1.2 h for his coffee to cool from 65°C to 55°C. (a) If the thermos bottle is a cylinder having an inner diameter of 7.0 cm, inner length of 12 cm, and wall thickness of 1.2 cm, what is the thermal conductance of the thermos material in the example? (b) If we replace the material with aerogel, how long will it take the coffee to cool the same amount? Assume the aerogel is coated with a thin protective layer of negligible thickness.

JPL/NASA

Figure P14.70 The insulating property of silica aerogel.

71. **Firewalking.** A man strides across a bed of glowing embers as seen in Figure P14.71, and due to some very specific physical circumstances, he emerges with no physical harm. Assume the glowing charcoal is covered with a thin layer of ash 0.50 mm thick and is at a temperature of 650°C. The thermal conductivity of ash and charcoal is approximately 1.5 W/(m · K). (a) Estimate the heat transferred to the feet of the firewalker as he traverses the bed of embers in six quick steps. Assume the area of the bottom of each foot comes into only 50% contact with the embers (a rough surface) and the contact lasts 0.10 s for each step. Is the quantity of heat transferred great or small? (b) Assume he steps onto the fire from slightly wet grass. If his feet are protected by the conversion of water from a liquid to a gas, how much water is needed to carry away the heat from part (a)? (c) How thick would this layer of water on the sole of each foot need to be? (d) Describe what would happen if instead he were to take the same six quick steps across a slab of iron at the same temperature.

© Sukree Sukplang/Reuters/Corbis

Figure P14.71

72. Apply the results of Example 14.14 to estimate the temperature of the dwarf planet Pluto. In your calculation, assume Pluto is at its mean distance from the Sun, 5.9×10^{12} m. Compare your answer with the measured temperature of Pluto, about 40 K. How do you think astronomers are able to measure the temperature of Pluto?

73. The rims on a regulation basketball court are exactly 10 ft above ground level. Suppose a regulation court is designed to be used at 20°C and the rims are supported by steel posts. On a very hot day, it is found that the rims are 4.0 mm too high. (a) What is the temperature? (b) Do you think it could ever be hot enough for the rims to be this high?

74. A general rule of thumb is that a person should drink about 1 L of fluid for every hour of moderate exercise. If this much fluid is lost as perspiration, what is the approximate amount of heat energy that is lost each hour?

75. **Mechanical equivalent of heat.** When you catch a fast-moving object such as a baseball with bare hands, your hands will feel warm immediately after the catch. Suppose a baseball (m = 0.22 kg) moving at 30 m/s (about 60 mi/h) is caught in one hand. How much will the temperature of the surface of your skin increase? Assume the heating is limited to a 1-mm thick layer of skin and all the initial kinetic energy of the ball is converted to heat.

76. **Sometimes even engineers make mistakes.** In 1870, the Eads Bridge was opened for traffic; it was the first bridge to span the Mississippi River at St. Louis and was one of the first large-scale applications of steel. The bridge, which is still in use, has three arches, each about 500 ft (about 150 m) long. When the last connection was being made in June 1870, it was found that the middle arch was 2.5 in. (about 10 cm) too long. Some workers then applied ice to a section of the arch, hoping to reduce its length enough so that they could make the final connections. Do you think this approach worked? Justify your answer by calculating what length of the arch would have to be cooled with ice to reduce the length by the required 10 cm.

CHAPTER

15

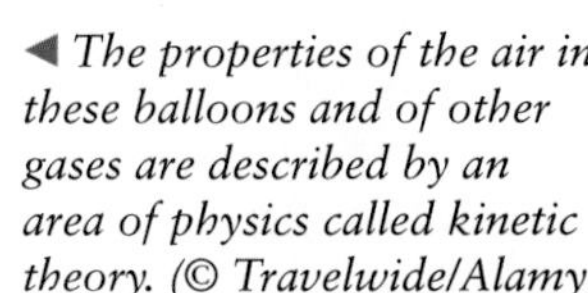

◀ *The properties of the air in these balloons and of other gases are described by an area of physics called kinetic theory. (© Travelwide/Alamy)*

Gases and Kinetic Theory

In this chapter, we apply Newton's laws to analyze the behavior of gases, the simplest phase of matter. In Chapter 14, we argued that systems containing many particles are so complicated that an approach using Newton's laws is prohibitively difficult and does not give a useful understanding of problems pertaining to heat flow and temperature. However, one special type of system—a dilute gas—is simple enough that Newton's laws *can* be applied. The atoms and molecules in a dilute gas are far apart, so they collide with one another only rarely. Hence, it is possible to apply Newton's laws in ways not possible with liquids or solids. Analysis of the properties of a dilute gas leads to important connections between temperature and motion on an atomic scale. Many properties of dilute gases are independent of composition, meaning that the same theory applies to a dilute gas of hydrogen, helium, oxygen, and so on. These ideas led to the notion of an ***ideal gas*** and an area of physics called ***kinetic theory***. This theory has a very rich and interesting history and has seen contributions from many physicists, including Newton and Einstein.

OUTLINE

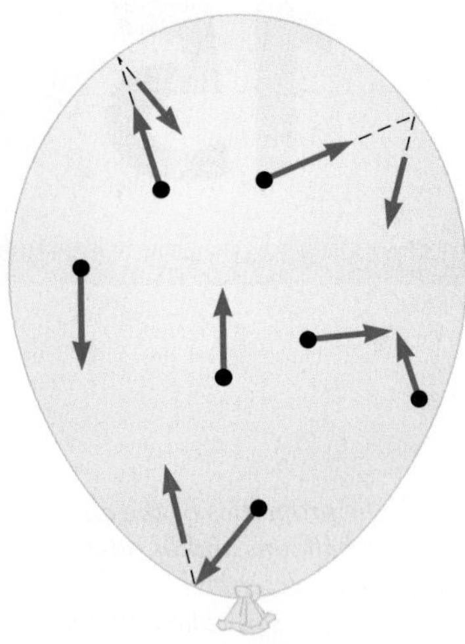

▲ **Figure 15.1** The molecules in a gas are in constant motion as they collide with one another and with the walls of their container.

15.1 Molecular Picture of a Gas

Consider a balloon containing a pure substance, so all the particles inside the balloon are identical (Fig. 15.1). These particles might be atoms, such as a balloon full of helium atoms, or molecules, such as carbon dioxide. For simplicity, we often refer to the particles as "molecules," with the understanding that all our ideas apply just as well to gases composed of single atoms, such as helium or argon.

The molecules in Figure 15.1 are all in motion, colliding with other gas molecules and with the walls of the container. The snapshot in Figure 15.1 raises a number of questions. (1) How often does a gas molecule collide with other molecules? (2) Are these collisions elastic or inelastic? (3) How are *microscopic* properties of the gas such as the molecular velocities related to *macroscopic* properties of the system such as temperature and pressure? (4) How many molecules are there in a typical sample of gas? Our job in this chapter is to answer all these questions.

A typical sample of gas contains a very large number of particles, usually on the order of 10^{23} molecules (or more). This very large number would seem to make the application of Newton's laws hopelessly complex. Actually, though, this large number of particles simplifies things by allowing us to apply statistical arguments, and because the number of particles is so large, the accuracy of a statistical analysis will be very good.

As you probably suspect, the number 10^{23} just mentioned is connected with ***Avogadro's number***, N_A. This pure number has the value

Avogadro's number

$$N_A = 6.02 \times 10^{23} \tag{15.1}$$

where N_A is the number of particles in a ***mole***.[1] Avogadro's number plays a central role in connecting the microscopic world of atoms and molecules to the macroscopic world. The ***atomic mass*** of an element is the mass (usually quoted in grams) of a natural sample of the element containing 1 mole of atoms. For example, the atomic mass of carbon is $M_C = 12.011$ g (sometimes the units are given as grams per mole). We denote atomic mass using a capital letter to distinguish it from the mass of a single carbon atom m_C; the units of m_C are grams per atom, or simply grams. From the definition of atomic mass,

$$m_C = \frac{M_C}{N_A} = \frac{12.011 \text{ g/mole}}{6.02 \times 10^{23} \text{ atoms/mole}} = 2.0 \times 10^{-23} \text{ g} = 2.0 \times 10^{-26} \text{ kg} \tag{15.2}$$

In Section 15.2, we'll show that a typical sample of gas particles contains on the order of 1 mole of particles. Thus, in discussions of kinetic theory and the properties of gases, N_A and factors like 10^{23} appear often.

EXAMPLE 15.1 Ⓡ The Number of Molecules in a Cup of Coffee

How many molecules are in a typical cup of coffee? Coffee is composed mainly of water, so for simplicity assume the sample contains only water molecules.

RECOGNIZE THE PRINCIPLE

We can calculate the number of moles of water molecules in a cup of coffee if we know the total mass of water M_{total} and the mass of 1 mole of water molecules M_{water}. We can find M_{water} from the atomic masses of hydrogen and oxygen, and although M_{total} is not given, we can use our experience to estimate a value.

[1]Avogadro's number, N_A, is defined as the number of atoms in a sample containing exactly 12 g of the carbon-12 isotope. We'll discuss isotopes in Chapter 30.

SKETCH THE PROBLEM

No figure is necessary.

IDENTIFY THE RELATIONSHIPS

A typical cup of coffee contains about 300 g (300 milliliters) of water, so we take $M_{total} = 300$ g. Water has the molecular formula H_2O (two hydrogen atoms plus one oxygen atom). The atomic mass of hydrogen is $M_H = 1.01$ g, and the atomic mass of oxygen is $M_O = 16.00$ g. (See the periodic table at the back of this book.) The mass of 1 mole of water molecules is thus

$$M_{water} = 2M_H + M_O = 2(1.01 \text{ g}) + 16.00 \text{ g} = 18.02 \text{ g}$$

which can also be written as $M_{water} = 18.02$ g/mole.

SOLVE

To find the number of moles in our sample, we divide M_{total} by the mass of 1 mole. The number of moles of water molecules in our sample of 300 g is thus

$$n_{water} = \frac{M_{total}}{M_{water}} = \frac{300 \text{ g}}{18.02 \text{ g/mole}} = 17 \text{ moles}$$

and the number of molecules N_{water} is

$$N_{water} = n_{water} \times N_A = (17 \text{ moles})(6.02 \times 10^{23} \text{ molecules/mole})$$

$$N_{water} = \boxed{1.0 \times 10^{25} \text{ molecules}}$$

► ***What does it mean?***

The number of water molecules in this sample is within two orders of magnitude of Avogadro's number. We'll find similar results for the numbers of atoms or molecules in typical macroscopic samples of gases, liquids, and solids.

CONCEPT CHECK 15.1 Mass of a Water Molecule

What is the mass of a single water molecule?

(a) 18 g (b) 3.0×10^{-23} kg (c) 3.3×10^{-27} kg (d) 3.0×10^{-26} kg

15.2 Ideal Gases: An Experimental Perspective

Experimental studies of the properties of gases began around the time of Newton and led to the discovery of a number of different "gas laws." These laws provided the clues that eventually gave rise to kinetic theory and relate various macroscopic properties of a gas such as its temperature and pressure. Most of these gas laws apply accurately only for dilute samples, which means that the spacing between molecules is much larger than the size of an individual molecule. Fortunately, that is not a serious restriction because many important cases, including air at room temperature, can be closely approximated as a dilute gas.

Avogadro's law

For a sample of gas at constant pressure and temperature, the volume is proportional to the number of molecules in the sample.

Avogadro's law

Avogadro's law tells us that the average spacing between gas particles is constant, so the density is also constant, provided the pressure and temperature are held fixed.

Boyle's law

For a sample of gas at constant temperature, the product of the pressure and volume is constant.

In mathematical terms, Boyle's law is just

Boyle's law

$$PV = \text{constant} \quad (\text{at constant } T) \tag{15.3}$$

According to Boyle's law, if we increase the pressure in a sample of gas held at constant temperature, the volume must decrease so that the product PV stays fixed. On the other hand, if we decrease the pressure, the volume must then increase, again keeping PV constant. In terms of Figure 15.1, Boyle's law says that changing the pressure changes the average spacing between particles.

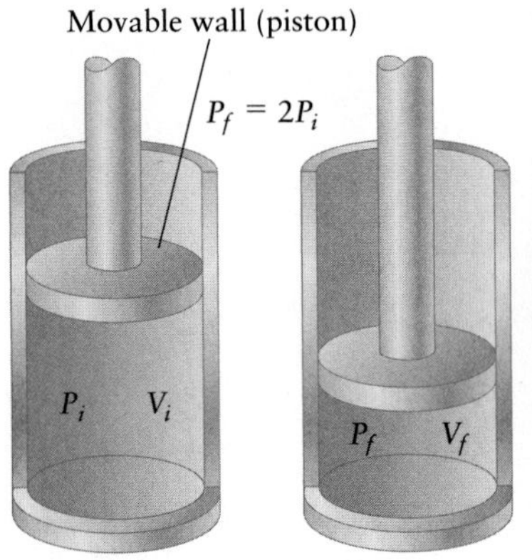

▲ **Figure 15.2** Concept Check 15.2.

CONCEPT CHECK 15.2 Density of a Dilute Gas

A cylinder (Fig. 15.2) contains a dilute gas at pressure P_i and volume V_i. One end wall of the cylinder is movable (forming a piston); moving this wall changes the volume of the gas. Suppose the gas is compressed so that the final pressure is twice the initial pressure, $P_f = 2P_i$. If the temperature is constant (so that Boyle's law applies), what is the final density?

(a) The final density is twice the initial density.
(b) The final density is half the initial density.
(c) The final density is equal to the initial density.

An important part of Boyle's law is the constant on the right-hand side of Equation 15.3. This constant must be proportional to the number of particles in the gas to be consistent with Avogadro's law. Experiments show that this constant also depends on temperature, as described by two more gas laws.

Charles's law

For a sample of gas at constant pressure, if the temperature is changed by a small amount ΔT, the volume changes by an amount ΔV, with

Charles's law

$$\Delta V \propto \Delta T \tag{15.4}$$

Gay-Lussac's law

For a sample of gas held in a container with constant volume, changes in pressure are proportional to changes in temperature.

Gay-Lussac's law

$$\Delta P \propto \Delta T \tag{15.5}$$

We have actually encountered Gay-Lussac's law in our discussion of gas thermometers in Chapter 14, where we found a linear relation between P and T in a dilute gas (Fig. 14.5A).

Absolute Temperature and the Kelvin Scale

Charles's law and Gay-Lussac's law both include temperature, and it is interesting to ask how Charles and Gay-Lussac were able to define and measure temperature in their work. Charles and Gay-Lussac used ***reference temperatures*** determined by the properties of various substances. Two convenient temperatures are those at which ice melts and at which water boils. Charles and Gay-Lussac used these reference temperatures together with a material property such as thermal expansion (Chapter 14) to interpolate between and extrapolate beyond the reference range. That is how a mercury thermometer works; the basic principle is shown in Figure 15.3.

This approach to measuring temperature was very useful, but it led to new questions. (1) What materials and properties can be used to make a thermometer (besides mercury), and can we be sure that all such thermometers will give the same results?

(2) According to Equation 15.5, changes in P for a dilute gas are proportional to changes in T. What happens to the temperature when $P = 0$, and does such a pressure (or temperature) even make sense?

Let's consider question 2 first. Careful measurements of pressure and temperature for a variety of gases give results like those sketched in Figure 15.4A; as temperature is reduced, the pressure decreases in a linear fashion (in proportion to changes in T), according to Gay-Lussac's law (Eq. 15.5). Any particular sample of gas will eventually condense and form a liquid, at which point the pressure falls below the linear P–T relation that holds at high temperature (Fig. 15.4B). This result does not conflict with Gay-Lussac's law because that relation applies only for dilute samples, and a gas is not dilute near its condensation temperature. Even though the linear behavior breaks down at low temperatures, we can still extrapolate the linear region from high temperature and pressure down to low pressures and find the temperature at which the pressure "would" be zero. If we do that with different samples of different gases, we find that they all extrapolate to zero pressure at the same temperature, $T = -273.15°C$ (Fig. 15.4A). This result applies to all gases, regardless of composition, and also holds for samples with different densities. Pressure is defined as the magnitude of the force per unit area exerted by a fluid on a surface (see Eq. 10.1), so it is not possible for P to be negative. This fact helped lead to the creation of the Kelvin temperature scale. The Kelvin scale is defined so that $T = 0$ K when the pressure of a dilute gas becomes zero, which is also called the ***absolute zero*** of temperature.

Because the linear relation between P and T is a universal property of dilute gases, it can be used as a universal way to measure temperature. One can calibrate a ***gas thermometer*** by measuring P at two reference temperatures (such as the melting point of ice and the boiling point of water) and thereby obtain an absolute calibration of the P–T relation for that particular thermometer. Most importantly, a gas thermometer can be made using any type of dilute gas.

The laws of Charles and (especially) Gay-Lussac thus lead to the notion of absolute temperature and provide an experimental method for measuring it. When we deal with gases and kinetic theory, it is most convenient to work with the Kelvin temperature scale.

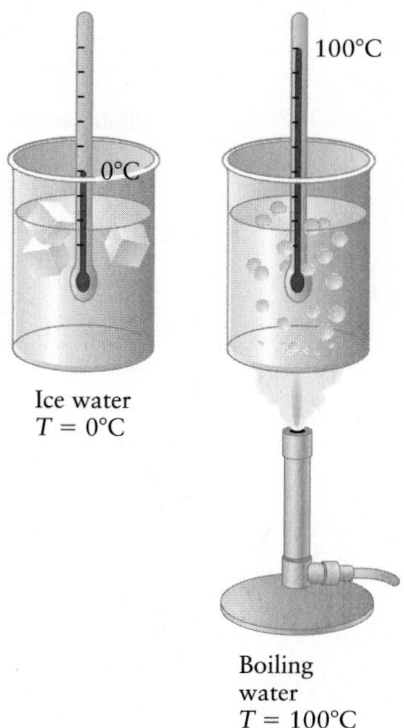

▲ **Figure 15.3** A mercury thermometer uses thermal expansion of mercury to measure changes in temperature. This thermometer is calibrated using known reference temperatures such as the melting and boiling temperatures of water.

The Ideal Gas Law

Each of the gas laws discussed above concern a particular aspect of the behavior of a dilute gas. These laws are all contained in a single relation known as the ***ideal gas law.***

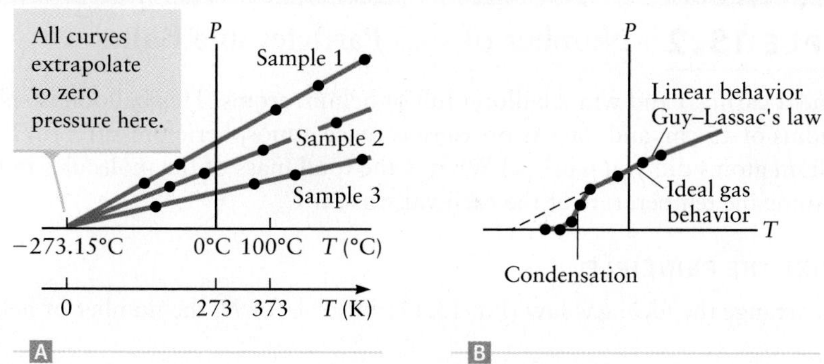

▲ **Figure 15.4** A Experiments show that the pressure of all dilute gases varies linearly with temperature when the gas is well above the condensation temperature. *All* the data at high temperatures for *all* gases extrapolate to $P = 0$ at the *same* temperature, defined to be $T = 0$ on the Kelvin scale. B At sufficiently low temperatures, all gases condense and the pressure then drops below the linear P–T dependence.

Ideal gas law
For a dilute gas composed of any substance,

Ideal gas law

$$PV = nRT \tag{15.6}$$

This relation applies to a sample of a dilute gas containing n moles of gas particles, and R is the ***universal gas constant*** with the value

$$R = 8.31 \frac{\text{J}}{\text{mole} \cdot \text{K}} \tag{15.7}$$

We can also express the ideal gas relation in terms of the number of gas particles N instead of the number of moles n. Using the definition of the mole and Avogadro's number (Eq. 15.1), the total number of particles is $N = nN_A$, so $n = N/N_A$. Substituting for n in Equation 15.6, we get

$$PV = nRT = \frac{N}{N_A} RT$$

$$PV = N\left(\frac{R}{N_A}\right)T \tag{15.8}$$

The factor R/N_A is the ratio of the gas constant to Avogadro's number; it is called ***Boltzmann's constant*** k_B and has the value

$$k_B = \frac{R}{N_A} \tag{15.9}$$

$$k_B = 1.38 \times 10^{-23} \text{ J/K} \tag{15.10}$$

Using the number of gas particles N and Boltzmann's constant, the ideal gas law (Eq. 15.8) can be written as

$$PV = Nk_BT \tag{15.11}$$

The two versions of the ideal gas law, Equations 15.6 and 15.11, are equivalent; the only difference is that one involves the number of moles n, whereas the other involves the number of particles N.

The ideal gas law contains the laws of Avogadro, Boyle, Charles, and Gay-Lussac all as special cases. For example, when the temperature and number of particles are kept fixed, the ideal gas law predicts that the product PV is a constant, which is just Boyle's law (Eq. 15.3). Most importantly, the ideal gas law applies to all dilute gases, with the same value of the gas constant R and the same value of Boltzmann's constant k_B.

EXAMPLE 15.2 Number of Gas Particles in a Balloon

You attend a carnival and win a balloon full of helium atoms. This balloon is spherical with a radius of 15 cm, and the gas pressure is twice atmospheric pressure. (**a**) How many helium atoms did you win? (**b**) What is the total mass of the molecules in the balloon? Assume the temperature at the carnival was 20°C.

RECOGNIZE THE PRINCIPLE

We can rearrange the ideal gas law (Eq. 15.11) to solve for N, the number of helium atoms:

$$N = \frac{PV}{k_BT} \tag{1}$$

We are given the pressure, volume, and temperature, so we can find N. The total mass can then be found using the atomic mass of helium.

SKETCH THE PROBLEM

No figure is needed.

IDENTIFY THE RELATIONSHIPS

The pressure is twice atmospheric pressure, and $P_{atm} = 1.01 \times 10^5$ Pa, so for the balloon we get $P = 2P_{atm} = 2.0 \times 10^5$ Pa. The sample is a sphere of radius 0.15 m; hence, the volume is

$$V = \tfrac{4}{3}\pi r^3 = \tfrac{4}{3}\pi(0.15 \text{ m})^3 = 0.014 \text{ m}^3$$

Converting the temperature to the Kelvin scale, we have $T = 273 + 20 = 293$ K.

SOLVE

(**a**) Inserting these values for the pressure, volume, and temperature into our expression for N, Equation (1), gives

$$N = \frac{PV}{k_B T} = \frac{(2.0 \times 10^5 \text{ Pa})(0.014 \text{ m}^3)}{(1.38 \times 10^{-23} \text{ J/K})(293 \text{ K})} = \boxed{6.9 \times 10^{23} \text{ atoms}}$$

which corresponds to

$$n = \frac{N}{N_A} = \frac{6.9 \times 10^{23} \text{ atoms}}{6.02 \times 10^{23} \text{ atoms/mole}} = 1.1 \text{ moles}$$

(**b**) The total mass m_{total} of the helium atoms equals the number of moles n multiplied by the mass of 1 mole. According to the periodic table, the mass of 1 mole of helium atoms is 4.00 g. Combining this with the answer for n from part (**a**) gives

$$m_{total} = n \times (4.00 \text{ g/mole}) = (1.1 \text{ moles})(0.00400 \text{ kg/mole}) = \boxed{0.0044 \text{ kg}}$$

▶ ***What does it mean?***

The total mass is equal to the mass of about two (U.S.) pennies. When using the ideal gas law in either form (Eq. 15.6 or 15.11), T must be expressed in terms of the absolute temperature, the temperature on the Kelvin scale. The answer for part (a) illustrates again that the number of molecules in a typical macroscopic sample of gas is on the order of Avogadro's number.

CONCEPT CHECK 15.3 The Ideal Gas Law and the Effect of Composition

Suppose the balloon in Example 15.2 contains nitrogen molecules instead of helium atoms. How does that affect the number of particles N in the balloon?

(a) There is no change. The number of nitrogen molecules would be equal to the number of helium atoms found in Example 15.2.

(b) The number of nitrogen molecules is twice the number of helium atoms found in Example 15.2.

(c) The number of nitrogen molecules is half the number of helium atoms found in Example 15.2.

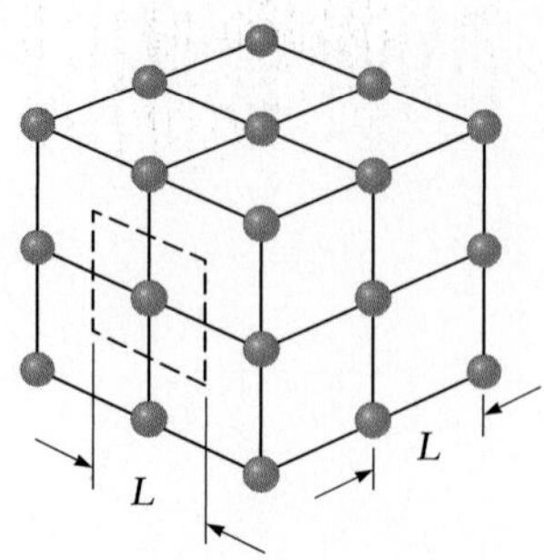

▲ **Figure 15.5** Example 15.3. To find the average molecular spacing, we can imagine the molecules are arranged in a regular array. While the molecules in any real sample of gas will be arranged in a much more random manner, each molecule will occupy, on average, a volume L^3, which is just the volume of a cube of length L.

EXAMPLE 15.3 Particle Spacing in a Gas

What is the average distance between helium atoms in Example 15.2?

RECOGNIZE THE PRINCIPLE

To find the *average* spacing, we can imagine that the atoms are arranged in a regular manner as sketched in Figure 15.5, with the distance between neighboring atoms given by L. This arrangement has atoms at the corners of cubes with edges of length L. Hence, there is a volume L^3 associated with each atom. The total gas volume is then NL^3, where the number of atoms N was found in Example 15.2.

SKETCH THE PROBLEM

Figure 15.5 shows the atoms in a regular cubic arrangement. The atoms in a real gas will not be arranged in such a simple way, but if the average distance between atoms is L, the average volume associated with each atom will still be L^3.

IDENTIFY THE RELATIONSHIPS AND SOLVE

The total gas volume from Example 15.2 is $V = 0.014\ \text{m}^3$. Setting this volume equal to NL^3 and using the result for the number of atoms N found in that example, we get $NL^3 = V$ and

$$L = \left(\frac{V}{N}\right)^{1/3} = \left(\frac{0.014\ \text{m}^3}{6.9 \times 10^{23}}\right)^{1/3} = \boxed{2.7 \times 10^{-9}\ \text{m}}$$

► ***What does it mean?***

The radius of an atom is about 5×10^{-11} m, so the spacing between particles in a gas is *much* larger than the size of an atom. The spacing between atoms in a solid or liquid is typically 0.1 nm ($= 0.1 \times 10^{-9}$ m). The average distance between molecules in this balloon, and in a typical dilute gas at room temperature, is thus about 30 times larger than the spacing between atoms in a solid or liquid.

CONCEPT CHECK 15.4 Changing the Pressure in a Gas

The helium-filled balloon in Example 15.2 contained $n = 1.1$ moles of helium atoms at room temperature and a pressure $P = 2 \times P_{\text{atm}}$ (twice atmospheric pressure). Atoms are added to the balloon so that the pressure doubles to $4 \times P_{\text{atm}}$ while the temperature and volume are kept fixed. Is the number of moles in the balloon (a) 1.1 moles, (b) 2.2 moles, (c) 4.4 moles, or (d) 0.55 mole?

EXAMPLE 15.4 Effect of Temperature on the Pressure of an Ideal Gas

A jar containing a gas of nitrogen molecules has a pressure $P_{\text{atm}} = 1.01 \times 10^5$ Pa at a temperature of 20°C. This jar is then put in a refrigerator where the temperature is 0°C. If the lid of the jar is circular with a radius of 5.0 cm, how much force is needed to overcome the pressure that is holding the lid closed?

RECOGNIZE THE PRINCIPLE

When the gas is cooled, the pressure P_{in} inside the jar will decrease according to the ideal gas law and will then be less than the pressure outside. The force required to open the jar will equal the difference in pressures multiplied by the area of the lid.

SKETCH THE PROBLEM

Figure 15.6 describes the problem.

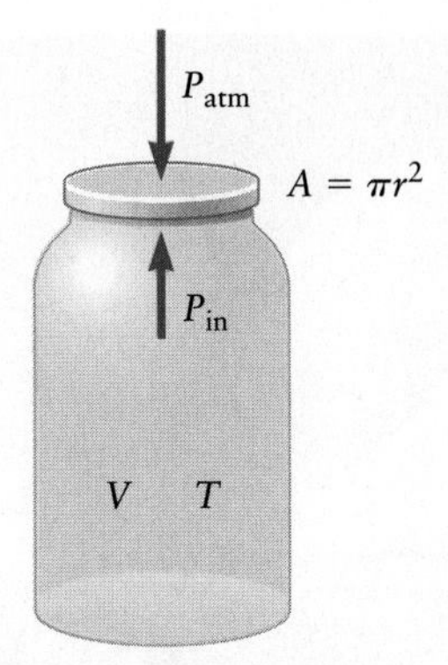

▲ **Figure 15.6** Example 15.4. The force on the lid of a jar is proportional to the difference in pressure between outside and inside the jar.

IDENTIFY THE RELATIONSHIPS

The ideal gas law reads $PV = Nk_BT$, and in this example, the volume V and the number of molecules N are held fixed. Even though we do not know the values of N and V, we can still write

$$\frac{P}{T} = \frac{Nk_B}{V} = \text{constant}$$

which is just another way of writing Gay-Lussac's law, Eq. 15.5. If we denote the initial pressure and temperature inside as $P_{in,\,i}$ and T_i and denote the final values as $P_{in,\,f}$ and T_f, we can now write

$$\frac{P_{in,\,i}}{T_i} = \frac{P_{in,\,f}}{T_f}$$

Rearranging to solve for the final pressure gives

$$P_{in,\,f} = \left(\frac{P_{in,\,i}}{T_i}\right)T_f$$

SOLVE

Inserting the given values of the initial pressure ($P_{in,\,i} = 1.01 \times 10^5$ Pa) and temperature ($T_i = 293$ K), along with $T_f = 0°\text{C} = 273$ K, we find

$$P_{in,\,f} = \left(\frac{P_{in,\,i}}{T_i}\right)T_f = \left(\frac{1.01 \times 10^5 \text{ Pa}}{293 \text{ K}}\right)(273 \text{ K}) = 9.4 \times 10^4 \text{ Pa}$$

With a pressure P_{atm} on the outside of the lid and a smaller pressure $P_{in,\,f}$ inside, the force required to open the lid is

$$F = (P_{atm} - P_{in,\,f})A$$

The area of the lid is $A = \pi r^2$, so we have

$$F = (P_{atm} - P_{in,\,f})A = (P_{atm} - P_{in,\,f})(\pi r^2)$$

$$F = (1.01 \times 10^5 \text{ Pa} - 0.94 \times 10^5 \text{ Pa})\pi(5.0 \times 10^{-2} \text{ m})^2 = \boxed{55 \text{ N}}$$

▶ ***What does it mean?***

The pressure $P_{in,\,f}$ is only slightly smaller than the initial pressure because the change in temperature is only a small fraction of the absolute temperature. Even though the resulting pressure difference across the lid is small, the force is still substantial. You could do this experiment at home.

▲ **Figure 15.7** Example 15.5. Launching a weather balloon.

EXAMPLE 15.5 Physics of a Weather Balloon

Consider a weather balloon that has a volume of 5.0 m^3 when released from a location where the temperature is 25°C and the pressure is $P_{atm} = 1.01 \times 10^5$ Pa. If the balloon eventually reaches an altitude of 16 km where the temperature is −30°C and the pressure is $0.10 \times P_{atm}$, what is the final volume of the balloon?

RECOGNIZE THE PRINCIPLE

When the balloon reaches its final altitude, its temperature and pressure are different from their initial values, so its final volume will also be different. We can use the ideal gas law to find the final volume.

SKETCH THE PROBLEM

Figure 15.7 shows a typical weather balloon just after its release.

IDENTIFY THE RELATIONSHIPS

The ideal gas law can be rearranged to read

$$n = \frac{PV}{RT}$$

This relation must hold at the initial (i) and final (f) locations, so we can write

$$\frac{P_i V_i}{RT_i} = \frac{P_f V_f}{RT_f} \qquad (1)$$

The temperatures are $T_i = 273 + 25 = 298$ K and $T_f = 273 - 30 = 243$ K.

SOLVE

We can solve Equation (1) for the final volume; inserting the given values of the pressure and temperature, we find

$$V_f = \frac{P_i T_f V_i}{P_f T_i} = \frac{P_{atm}(243 \text{ K})(5.0 \text{ m}^3)}{0.10 \times P_{atm}(298 \text{ K})} = \boxed{41 \text{ m}^3}$$

► ***What does it mean?***

When both the temperature and pressure of a gas change, we must use the ideal gas law to find the resulting change in volume. In this case, the reduction in temperature tends to make the volume smaller, whereas the reduction in pressure makes the volume larger. The latter effect dominates, so the balloon has a much larger volume at its final altitude.

15.3 Ideal Gases and Newton's Laws

We now return to some of the issues raised at the very beginning of the chapter and consider how to account for the gas laws by applying Newton's laws to microscopic molecular motions.

Pressure Comes from Collisions with Gas Molecules

The molecules in a gas move with different speeds and in all possible directions. To simplify things, we will (for the moment) consider just one typical molecule of mass

m and take its speed v to be the average molecular speed. Figure 15.8 shows this molecule as it is initially traveling along the $-x$ direction; it then collides with a container wall and travels back along the $+x$ direction. In a typical gas, such collisions are elastic, so the speed is the same before and after the collision. From Chapter 7, we know that the change in momentum $\Delta\vec{p}$ of the molecule is proportional to the force exerted by the wall on the molecule. According to the impulse theorem (Eq. 7.5), we have

$$\Delta\vec{p} = \vec{F}\,\Delta t \tag{15.12}$$

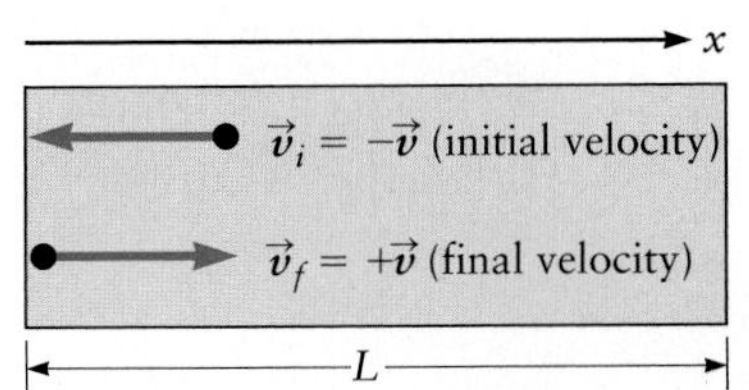

▲ **Figure 15.8** When a gas molecule collides with a wall, the force exerted by the wall leads to an impulse on the molecule, causing the molecule's momentum to change. The corresponding force exerted by the molecule on the wall contributes to the gas pressure.

The initial and final momenta in Figure 15.8 are along the x axis, and the force from the wall is parallel to the x direction. We can therefore write Equation 15.12 in terms of the components along x. The initial momentum is $-mv$, and the final momentum is $+mv$. Inserting these terms into Equation 15.12, we find

$$\Delta p = p_f - p_i = +mv - (-mv) = F\Delta t$$

$$2mv = F\Delta t$$

Solving for the force gives

$$F = \frac{2mv}{\Delta t} \tag{15.13}$$

Equation 15.13 gives the force exerted by the wall on the molecule. According to Newton's third law (the action–reaction principle), this is also the force exerted by the molecule on the wall, and it contributes to the pressure. To find the total pressure, we must add up the collision forces from all the molecules that strike the wall. Figure 15.8 shows one hypothetical molecule initially moving in the $-x$ direction; there will also be molecules moving in the $+x$, $+y$, $-y$, $+z$, and $-z$ directions. We thus have six different possible directions, so for a system with a total of N molecules, we can expect that one-sixth of them are initially moving along $-x$ and are thus candidates for striking the left-hand wall in Figure 15.8. Of these $N/6$ molecules, only those within a distance $v\,\Delta t$ strike the wall during a time interval Δt because only those molecules reach the wall during this time. The container in Figure 15.8 has a length L, so the number of molecules that collide with the wall during this time interval is

$$\frac{N}{6} \times (\text{fraction that reaches wall}) = \frac{N}{6}\frac{v\Delta t}{L} \tag{15.14}$$

The total force exerted by all these molecules on the wall is the product of the force exerted by one molecule (Eq. 15.13) and the number of such molecules (Eq. 15.14):

$$F_{\text{total}} = \left(\frac{2mv}{\Delta t}\right)\left(\frac{N}{6}\frac{v\Delta t}{L}\right) = \frac{Nmv^2}{3L}$$

Pressure is defined as the total force on the wall divided by the area A, so

$$P = \frac{F_{\text{total}}}{A} = \frac{Nmv^2}{3LA}$$

The volume of the container is $V = LA$, so the pressure can also be written as

$$P = \frac{Nmv^2}{3V}$$

or

$$PV = N\frac{mv^2}{3} \tag{15.15}$$

The Microscopic Basis of Temperature

Equation 15.15 relates a *microscopic* property of the molecules, their speed v, to a *macroscopic* property, the pressure in the gas. We also notice that the mathematical form of this result is *very* similar to the ideal gas law. According to Equation 15.11,

$$PV = Nk_BT$$

Comparing this expression with Equation 15.15, we see that they agree provided that

$$\frac{mv^2}{3} = k_BT \tag{15.16}$$

The kinetic energy of a molecule is $KE = \frac{1}{2}mv^2$, so Equation 15.16 can be written as $\frac{2}{3}KE = k_BT$ or equivalently,

Average kinetic energy of a gas atom

$$KE = \frac{3}{2}k_BT \tag{15.17}$$

While we derived this result for a dilute gas, it also holds for the average kinetic energy of atoms and molecules in a liquid or solid.

In our derivation of Equation 15.15, we assumed all the molecules move with the same speed v. An exact analysis allows for different molecules moving at different speeds and shows that v in Equations 15.15 and 15.16 is nearly equal to the average molecular speed (the difference is only a few percent). We can therefore still think of v as being the speed of a typical molecule.

In words, Equation 15.17 says that the kinetic energy of a typical molecule is proportional to the temperature of the gas. This result is remarkable because it tells how temperature is related to the microscopic motion of the gas molecules. We have thus derived the ideal gas law from Newton's laws of mechanics, provided that temperature is related to the molecular kinetic energy by Equation 15.17. This result answers one of the big questions raised in Section 15.1 because we are now able to relate the macroscopic quantities in the ideal gas law to the microscopic quantities that appear in Newton's laws.

It is interesting to use Equation 15.16 to find the speed of a typical molecule in the atmosphere. Rearranging to solve for v, we find

$$v = \sqrt{\frac{3k_BT}{m}} \tag{15.18}$$

The Earth's atmosphere is composed mainly of nitrogen, so let's calculate v for N_2 in the atmosphere. The mass of a nitrogen molecule is $m = 2(14.01\text{ g})/N_A = 4.7 \times 10^{-23}\text{ g} = 4.7 \times 10^{-26}\text{ kg}$. At room temperature, we have $T \approx 293$ K (corresponding to 20°C). Inserting these values into Equation 15.18 gives

$$v_{N_2} = \sqrt{\frac{3k_BT}{m}} = \sqrt{\frac{3(1.38 \times 10^{-23}\text{ J/K})(293\text{ K})}{4.7 \times 10^{-26}\text{ kg}}} = 510\text{ m/s} \tag{15.19}$$

which is approximately 1100 mi/h! Molecules having this speed are striking your body right now as you read this book.

15.4 Kinetic Theory

The calculations in Section 15.3 take us to the heart of the mechanics of a dilute gas, a topic called ***kinetic theory***. This theory is based on a small number of assumptions that were used either explicitly or implicitly in Section 15.3:

1. Gas atoms and molecules spend most of their time moving freely; that is, they move with a constant speed in a straight-line path. Collisions with other atoms and molecules in the gas are very infrequent.
2. Newton's laws can be used to describe the motion of individual gas particles.
3. The collisions between gas atoms and molecules, and with the walls of the container, are elastic. The average distance between collisions is called the ***mean free path***, ℓ. The value of ℓ depends on the density of gas particles, their size, and the temperature. The mean free path is not the same as the average spacing between particles (Fig. 15.9). In Example 15.3, we found that the average spacing between molecules in a typical dilute gas is approximately 3×10^{-9} m. The mean free path for a molecule in air is about 100×10^{-9} m, so in this case, ℓ is about 30 times the average spacing.
4. Even a dilute gas contains a very large number of atoms or molecules, so we can use statistical analyses to calculate its properties. We have already discussed the large number of particles present in a typical gas and have seen that it is on the order of Avogadro's number.

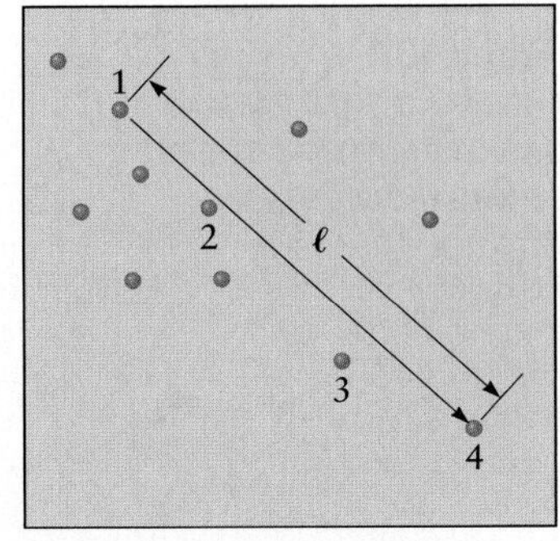

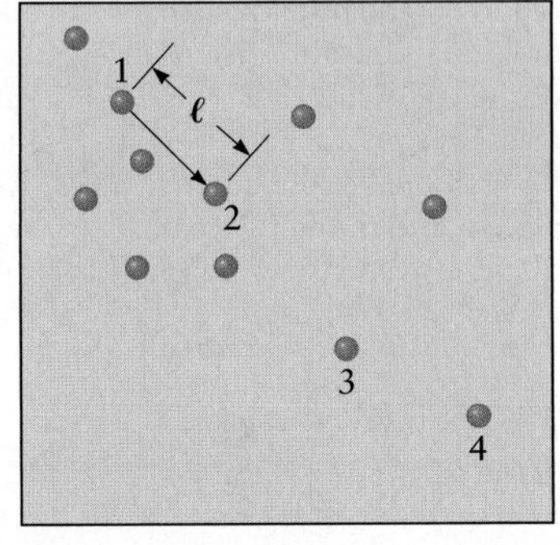

▲ **Figure 15.9** The average distance a gas molecule travels between collisions is called the mean free path. A If the molecules are very small, the mean free path is much longer than the average spacing between them. B As molecular size increases, the mean free path becomes shorter. See Insight 15.1.

These assumptions about the nature of an ideal gas are all satisfied very accurately by a dilute gas and lead to the ideal gas law. We next consider some other properties of ideal gases.

Internal Energy of an Ideal Gas

The particles in an ideal gas spend most of their time moving freely, during which time the force on a particular molecule from all the other molecules is essentially zero. The mechanical energy then equals the kinetic energy, which (as we learned in Chapters 6 and 8) has contributions from both the translational motion ($\frac{1}{2}mv^2$) and the rotational motion ($\frac{1}{2}I\omega^2$). For a monatomic gas such as helium, the rotational kinetic energy doesn't contribute to the gas properties and can be ignored. (For polyatomic gases such as N_2, one must include the rotational contribution; see below.) The translational kinetic energy is $KE_{\text{trans}} = \frac{1}{2}mv^2$. We calculated this kinetic energy in Equation 15.17:

$$KE_{\text{trans}} = \tfrac{3}{2}k_BT \quad (15.20)$$

The total energy of a system of N gas atoms is thus

$$KE_{\text{total}} = N(KE_{\text{trans}}) = \tfrac{3}{2}Nk_BT$$

which is the total energy of a monatomic ideal gas. In a sense, this energy is held "internally" by gas atoms and is not evident if we are simply given a container of the gas. For that reason, it is called the ***internal energy***, denoted by U. Hence, for a monatomic ideal gas, we have

$$U = \tfrac{3}{2}Nk_BT \quad (15.21)$$

According to Equation 15.21, the internal energy of an ideal gas depends *only* on temperature and the number of gas atoms N. It is independent of the molecular mass and pressure.

We can also express the internal energy in terms of the number of moles n in the system instead of the number of atoms N. The number of atoms is $N = n \times N_A$, so Equation 15.21 becomes

$$U = \tfrac{3}{2}(nN_A)k_BT = \tfrac{3}{2}n(N_Ak_B)T$$

From the definition of Boltzmann's constant (Eq. 15.9), we have $N_Ak_B = R$, so

$$U = \tfrac{3}{2}nRT \quad (15.22)$$

Insight 15.1

THE MEAN FREE PATH IN A GAS DEPENDS ON THE SIZE OF THE MOLECULES

Given our atomic-scale picture of a gas, you might have expected the mean free path to be approximately the same as the average spacing between molecules, but that is not the case. The mean free path is the average distance traveled between collisions, which depends on the molecular size. In Figure 15.9A, a hypothetical molecule (molecule 1) travels near molecules 2 and 3 before colliding with molecule 4. If the molecules begin at the same places but are larger (Fig. 15.9B), molecule 1 would initially collide with molecule 2, resulting in a shorter value of the collision distance and a smaller mean free path. The size of a molecule is much smaller than the average spacing, so the mean free path is longer than the average spacing.

The results in Equations 15.21 and 15.22 for the internal energy apply only for a monatomic ideal gas. For other systems such as liquids, solids, nondilute gases, and polyatomic ideal gases, the contribution from the potential energy is important and makes the internal energy a more complicated function than the simple result in Equation 15.21. We'll discuss a few such cases later in this chapter and also in Chapter 16, where the internal energy plays an important role in our work on thermodynamics.

CONCEPT CHECK 15.5 Internal Energy of Two Gases

One balloon contains exactly 3 moles of helium atoms, and a second balloon contains exactly 3 moles of argon atoms. The mass of an argon atom is about 9 times the mass of a helium atom. If the internal energies of these two systems are equal, what can we conclude?

(a) They must have the same volume.
(b) They must have the same pressure.
(c) They must have the same temperature.
(d) They must have the same pressure, volume, and temperature.

EXAMPLE 15.6 Ⓡ Internal Energy of a Gas: How Big Is It?

Helium-filled balloons are often sold at carnivals. What is the approximate internal energy of the gas in such a balloon?

RECOGNIZE THE PRINCIPLE

To calculate the internal energy of a dilute gas, we must know its temperature and the number of moles n. Neither value is given, but we can use our experience to find typical values. The temperature is about room temperature, so $T \approx 290$ K. A typical balloon has a volume of about $V = 20$ L $= 0.02$ m^3 (Fig. 15.10), and the pressure is near atmospheric pressure ($P_{atm} = 1.01 \times 10^5$ Pa); using this information, we can use the ideal gas law to find n.

SKETCH THE PROBLEM

We can estimate the balloon's volume from the photo in Figure 15.10.

IDENTIFY THE RELATIONSHIPS

We first rearrange the ideal gas law (Eq. 15.6) to solve for n:

$$n = \frac{PV}{RT}$$

Inserting our estimates for P, V, and T leads to

$$n = \frac{P_{atm}V}{RT} = \frac{(1.01 \times 10^5 \text{ Pa})(0.02 \text{ m}^3)}{[8.31 \text{ J/(mole}\cdot\text{K)}](290 \text{ K})} = 0.8 \text{ mole}$$

SOLVE

Using this value for n in our relation for the internal energy (Eq. 15.22) gives

$$U = \tfrac{3}{2}nRT = \tfrac{3}{2}(0.8 \text{ mole})[8.31 \text{ J/(mole}\cdot\text{K)}](290 \text{ K}) = \boxed{3000 \text{ J}}$$

▲ **Figure 15.10** Example 15.6.

▸ ***What does it mean?***
This result is much larger than the kinetic energy of a baseball traveling at 100 mi/h, so it is a very significant amount of energy! It is quite surprising that a simple balloon of helium gas contains such a large amount of energy "internally."

Specific Heat of an Ideal Gas

In Chapter 14, we learned that when heat is added to a system, the temperature changes according to the value of the system's specific heat. Let's apply kinetic theory to find the specific heat of an ideal gas. Heat energy added to an ideal gas causes the molecular kinetic energy to increase, thus increasing the internal energy. So, if an amount of heat energy ΔQ is added to an ideal gas, the internal energy must increase by that amount and $\Delta Q = \Delta U$. Using Equation 15.21, we can find the corresponding change in temperature ΔT. We have

$$\Delta Q = \Delta U = \tfrac{3}{2}Nk_B\Delta T$$

The heat capacity of the system (Chapter 14) is equal to $\Delta Q/\Delta T$, so

$$\text{heat capacity} = \frac{\Delta Q}{\Delta T} = \tfrac{3}{2}Nk_B$$

This is the heat capacity at *constant volume* because our derivation depended on Figure 15.8, where we assumed the walls of the container were fixed in place so that the system's volume is constant. For a system containing 1 mole of atoms, N is equal to Avogadro's number ($N = N_A$); since $N_Ak_B = R$ (Eq. 15.9), we thus get

$$\text{heat capacity of 1 mole} = \frac{\Delta Q}{\Delta T} = \tfrac{3}{2}N_Ak_B = \tfrac{3}{2}R$$

which is also called the ***specific heat*** per mole at constant volume and is denoted C_V:

$$C_V = \tfrac{3}{2}R \tag{15.23}$$

Specific heat per mole of a monatomic ideal gas

In Chapter 14, we discussed the specific heat *per unit mass*, denoted by c. Here we denote the specific heat *per mole* by C_V (an uppercase C).

Polyatomic Gases

Our results for the internal energy (Eqs. 15.21 and 15.22) and the specific heat (Eq. 15.23) apply only for monatomic ideal gases. The derivations of those results assumed all the internal energy is in the form of translational kinetic energy. For gases composed of polyatomic molecules such as O_2 or N_2 or NH_3, we must also include the energy associated with the rotational and vibrational motion of the molecules. The internal energy of these gases can also be found using kinetic theory, but the analysis is more than we can include here. The internal energy and the specific heat per mole of such gases are both greater than the values for monatomic gases.

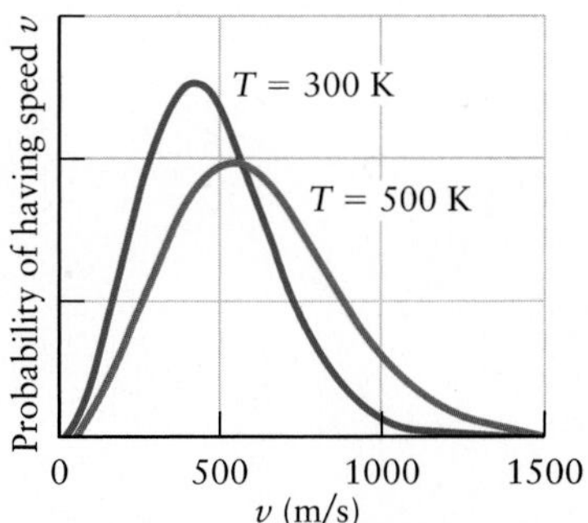

▲ **Figure 15.11** Distribution of speeds for a dilute N_2 gas. The vertical scale is the relative probability of finding a molecule moving with a particular speed v.

Distribution of Speeds in a Gas

So far, we have dealt with the speed of a typical (average) molecule. The speeds of individual molecules in a given sample are not all the same, however. Some molecules have speeds greater than the typical speed v in Equation 15.18, whereas others have lower speeds. The molecular speed distribution in a gas is an important part of kinetic theory. We will not calculate this distribution here, but it is instructive to consider it qualitatively because it gives rise to some important effects.

Figure 15.11 shows the distribution of molecular speeds in a typical case. Called the ***Maxwell–Boltzmann*** distribution, it gives the probability that a molecule will have a particular speed. The typical speed v found in Equation 15.18 is near the average speed, but a significant number of molecules have speeds that are greater or lower by a factor of two or even more. The distribution varies with temperature, and if the temperature is increased, the entire distribution curve shifts to higher speeds. This temperature dependence makes intuitive sense (because "hotter" molecules have greater kinetic energy) and is responsible for the temperature dependence of the average molecular speed in Equation 15.18.

EXAMPLE 15.7 (Absence of) Helium in the Atmosphere

Experiments show that a helium atom released into the atmosphere escapes to outer space, never to return. To understand how a helium atom is able to escape from the Earth's gravitational attraction, begin by calculating the typical speed v of a helium atom in a balloon held at atmospheric pressure and temperature. Comparing v with the escape speed for an object near the Earth's surface as calculated in Chapter 6 shows that v is smaller than the escape speed, so it would seem that a "typical" helium will not be able to escape. Given that fact, how can the distribution of molecular speeds in Figure 15.11 explain why all helium atoms do, eventually, escape from the atmosphere?

RECOGNIZE THE PRINCIPLE

We can find the speed v of a typical helium atom using the same approach we applied to a nitrogen molecule (which led to the result in Eq. 15.19). If v is greater than the escape speed, a typical helium atom will be able to escape the Earth's gravitation attraction, just as would any projectile fired from the surface of the Earth. We must also recognize that, according to the Maxwell–Boltzmann speed distribution (Fig. 15.11), some helium atoms will have a speed much greater than v, and we also need to consider the behavior of these atoms.

SKETCH THE PROBLEM

The Maxwell–Boltzmann distribution (Fig. 15.11) is a key to this problem.

IDENTIFY THE RELATIONSHIPS

According to Equation 15.18, the speed of an atom or molecule in a gas depends on the particle's mass. The mass of a helium atom (see the periodic table) is $m_{\text{He}} = 4.0\text{ g}/N_A = 6.6 \times 10^{-27}$ kg.

SOLVE

Using Equation 15.18 at room temperature ($T = 293$ K), we get

$$v_{\text{He}} = \sqrt{\frac{3k_BT}{m_{\text{He}}}} = \sqrt{\frac{3(1.38 \times 10^{-23}\text{ J/K})(293\text{ K})}{6.6 \times 10^{-27}\text{ kg}}} = \boxed{1400\text{ m/s}}$$

▶ *What does it mean?*

This speed is greater than the typical speed of N_2 molecules in the atmosphere found in Equation 15.19 because helium has a much smaller mass. In Chapter 6, we found that a projectile at the Earth's surface with a speed greater than about 11,000 m/s has enough kinetic energy to "escape" from the Earth's gravitational attraction, so the speed v of a typical helium atom is *less* than the escape speed. However, according to the distribution of molecular speeds in Figure 15.11, some helium atoms will have speeds much greater than the typical speed. A small fraction of helium atoms will be moving faster than the escape speed and hence will be able to escape from the atmosphere. All helium atoms released into the atmosphere have some probability of achieving such a high speed, and when they do, they will escape into space. That is why our atmosphere does not contain any helium gas. The helium used in carnival balloons and other applications was obtained from trapped underground deposits. What about atoms and molecules heavier than helium? A small percentage of such heavier molecules, including O_2 and N_2, will also escape from the Earth, but their rate of escape is extremely small; hence, these gases remain in the atmosphere indefinitely. (They are also replenished by natural processes.)

15.5 Diffusion

A molecule moving through a gas follows a zigzag path as it collides successively with other molecules (Fig. 15.12). This type of motion is called ***diffusion***. Similar motion is found for molecules moving through a liquid and through certain types of solids, including cell membranes.

Each particle in a gas follows a different random zigzag path, but these different trajectories can be described in an average way in terms of the typical molecular speed v between collisions and the average distance ℓ between collisions. The speed of a typical molecule is given by Equation 15.18, so v depends on both the temperature and the mass of the molecule. The distance ℓ is the mean free path we introduced earlier in this chapter. The value of the mean free path depends on the gas density; if the density is high, the gas molecules are close together, the probability for a collision is high, and ℓ is short. If the density is low, the collision probability is also low and ℓ is long. An analysis of the collision mechanics shows that the mean free path varies inversely with the density. Because the density is proportional to the number of molecules per unit volume N/V, the average distance between collisions varies as

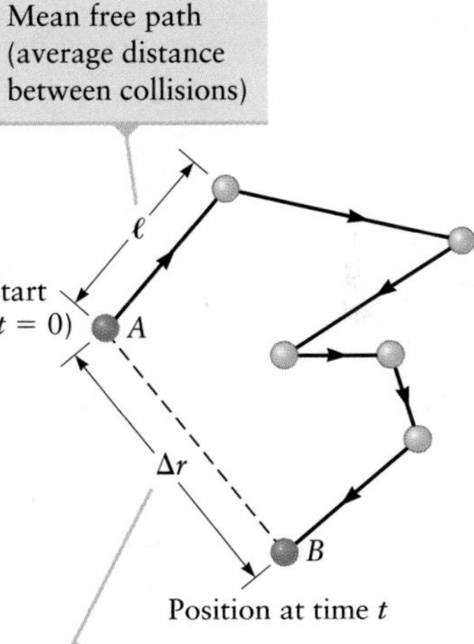

▲ **Figure 15.12** The zigzag motion of a molecule in a gas or liquid is called diffusion. (Similar motion is found for an impurity molecule moving through a solid.) The average length of each step is the mean free path. Each change of direction of a molecule is caused by a collision. The steps are in random directions, so after many steps (i.e., many collisions), the magnitude of the net displacement (Δr) is much smaller than if there had been no collisions.

$$\ell \propto \frac{V}{N} \tag{15.24}$$

for a gas in which N particles occupy a total volume V. For air at room temperature and pressure, the mean free path is approximately $\ell \approx 1 \times 10^{-7}$ m $= 100$ nm, which is about 30 times greater than the average spacing between gas molecules.

In most situations, we are interested in how the molecule travels over a relatively long distance such as when a gas molecule travels from one side of a room to the other. Figure 15.12 shows a typical diffusive path taken as a molecule travels from an initial point A to a final location B. The total distance traveled by a molecule is much longer than the length of the direct path that connects A and B. For this reason, diffusive motion between two points is much slower than motion without all the collisions.

Although we can never predict the precise path followed by a particular molecule, diffusion usually involves a large number of molecules making a large number of zigzag steps, so statistical arguments give an accurate description of an average random walk path. The magnitude of the average displacement Δr of a molecule after many steps taken over time t is

Distance traveled by a diffusing particle

$$\Delta r = \sqrt{Dt} \tag{15.25}$$

Insight 15.2

THE GRAVITATIONAL FORCE IS NOT IMPORTANT IN KINETIC THEORY

In all our discussions of the molecular motions in kinetic theory (e.g., Fig. 15.12), we have ignored the effect of the gravitational force on the molecular trajectories. This approximation is very good because the molecular speeds are so large. From mechanics (Chapter 6), when a particle falls through a distance L, the change of gravitational potential energy is mgL. If the particle is initially at rest, the final speed is $v = \sqrt{2gL}$. For $L = 1$ m, this speed is about 4 m/s, which is much less than the speed of a typical molecule in the atmosphere (we found $v \approx 500$ m/s for N_2; see Eq. 15.19). The effect of gravity on molecular trajectories in a typical gas is thus very small.

TABLE 15.1 Values of Some Typical Diffusion Constants

Diffusing Molecule or Particle	Medium	D (m²/s)
Oxygen (O_2)	Water	1×10^{-9}
Oxygen (O_2)	Air	2×10^{-5}
Nitrogen (N_2)	Air	2×10^{-5}
Hydrogen (H_2)	Air	6×10^{-5}
Oxygen (O_2)	Tissue	2×10^{-11}
Water (H_2O)	Water	3×10^{-9}
Sucrose	Water	5×10^{-10}
DNA	Water	1×10^{-12}
Hemoglobin	Water	7×10^{-11}
Pollen grain (5×10^{-7} m diameter)	Water	1×10^{-9}

Notes: All substances are at room temperature and pressure. Except for the pollen grain, the diffusing objects are all molecules.

The quantity D is called the ***diffusion constant.*** The result above can also be written as

$$t = \frac{(\Delta r)^2}{D} \qquad (15.26)$$

giving the time t required for an average molecule to diffuse a distance Δr.

The value of the diffusion constant depends on both v and ℓ, so it depends on temperature and on the mass of the diffusing particle as well as on the properties of the medium. The values of some typical diffusion constants are listed in Table 15.1. Diffusive motion is found in many situations, including molecules moving through a gas or liquid, and through a porous membrane such as a cell membrane.

A key feature of the result for Δr (Eq. 15.25) is that the net displacement is proportional to $\sqrt{t}$, the *square root* of the time. That is in contrast to motion with a constant velocity, for which the displacement increases *linearly* with time. This square root dependence means that a diffusing particle moves relatively slowly through the surrounding medium, as we explore in Example 15.8.

EXAMPLE 15.8 Diffusion in a Gas

You are at work in a chemical laboratory when a dangerous gas accidentally begins to leak from a container. Approximately how long do you have to evacuate the room safely, assuming the gas travels by diffusion alone?

RECOGNIZE THE PRINCIPLE

If the gas leaks from the center of the room, you will want to be gone long before a typical molecule has diffused to the door. This diffusion time is $t = (\Delta r)^2/D$ (Eq. 15.26). The diffusion distance Δr is not given, so to get the approximate time, we must estimate the size of a typical room.

SKETCH THE PROBLEM

Figure 15.12 describes the problem, with the distance from A to B equal to the size of the room.

IDENTIFY THE RELATIONSHIPS AND SOLVE

The diffusion constant for a typical molecule in air is $D \approx 2 \times 10^{-5}\ \text{m}^2/\text{s}$ (Table 15.1), and the size of a typical room is about $\Delta r = 3$ m. These values lead to

$$t = \frac{(\Delta r)^2}{D} = \frac{(3\ \text{m})^2}{2 \times 10^{-5}\ \text{m}^2/\text{s}} = \boxed{5 \times 10^5\ \text{s}}$$

▸ ***What does it mean?***

This result is more than 100 h, so diffusion is a very slow process. It would thus seem that you have plenty of time to escape. In most cases, however, convection currents in the room will disperse the molecules much more quickly. In addition, some molecules will have speeds much greater than the typical speed (Fig. 15.11) and these will also arrive much sooner than the typical molecule considered here.

Using Diffusion in Medicine

The slow process of diffusion is employed in a medical application called ***transdermal drug delivery***. Sometimes a drug should be delivered into the body in a slow and steady manner, which can be accomplished by placing a "patch" containing the drug in direct contact with the skin (Fig. 15.13A). The drug diffuses through the patch membrane, through the skin, and into the body. By careful choice of the membrane thickness, the drug can be made to enter the body slowly, over the course of many days or weeks. Such a slow delivery is preferred in certain cases such as the use of nicotine to help a person stop smoking.

CONCEPT CHECK 15.6 Designing a Transdermal Patch

The membrane in a transdermal patch (Fig. 15.13B) is typically about 200 μm thick. Suppose the thickness is doubled. How does that affect the rate with which the drug diffuses through the membrane?

(a) The rate is doubled.
(b) The rate is cut in half.
(c) The rate is reduced by a factor of four.

Isotope Separation

Another important application of diffusion is in the separation of different ***isotopes*** of a given element. As an example, consider the nucleus of a uranium atom. The two most common types of uranium nuclei are denoted $^{235}_{92}\text{U}$ and $^{238}_{92}\text{U}$; here the lower number is the number of protons in the nucleus, and the upper number is the number of neutrons plus protons. These two different nuclei contain different numbers of neutrons and are called isotopes of uranium. (We'll say much more about the properties of different nuclei in Chapter 30.) Naturally occurring uranium contains a mixture of $^{235}_{92}\text{U}$ and $^{238}_{92}\text{U}$, but only $^{235}_{92}\text{U}$ is able to undergo the nuclear reactions required in power plants (and nuclear bombs). To make nuclear fuel, the $^{235}_{92}\text{U}$ atoms must be removed from a uranium sample without having the other isotope come along with them. This process is called ***isotope separation*** and is used to make what is called "enriched" uranium. One way to accomplish isotope separation is with diffusion.

A diffusing molecule follows a zigzag path as sketched in Figure 15.12. On each straight portion, the molecule moves with a constant speed; for diffusion through a gas, the average speed on these different zigs and zags is just v in Equation 15.18. Since v increases as the molecular mass decreases, a light atom or molecule diffuses faster than a heavy one.

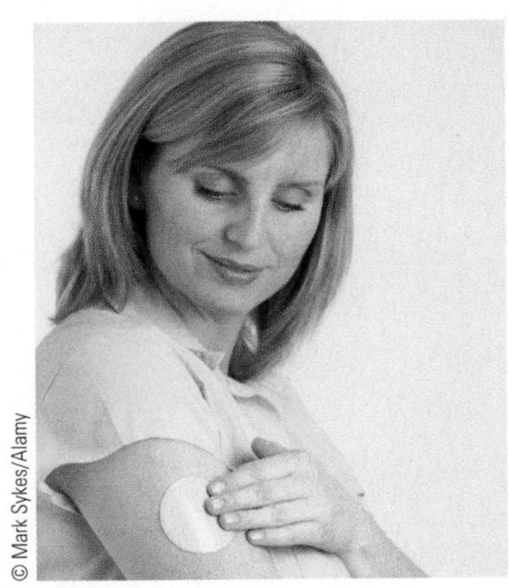

A

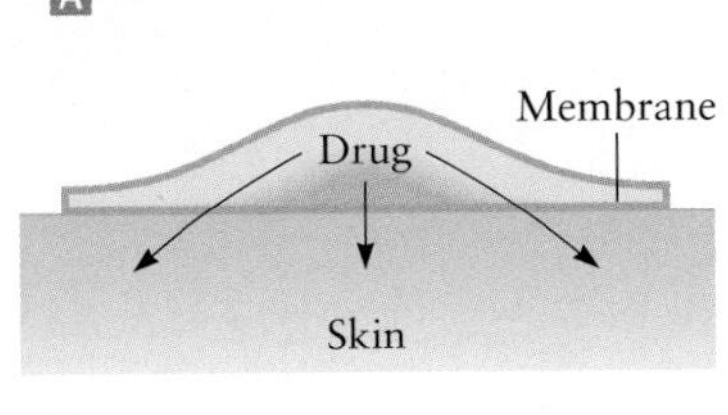

B

▲ **Figure 15.13** A A transdermal patch delivers drugs into the body at a slow, steady rate. B Structure of a transdermal patch.

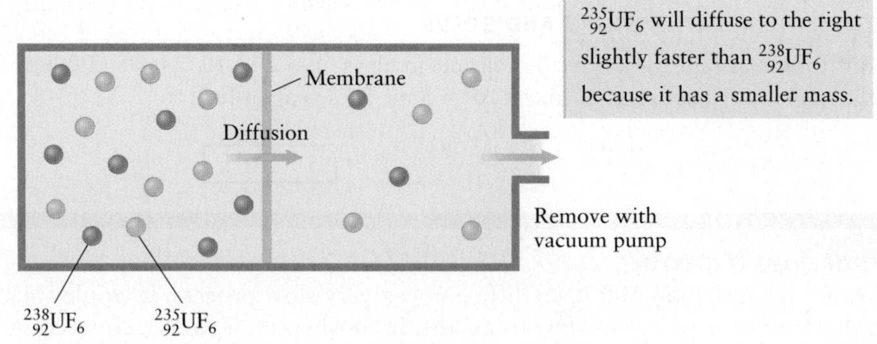
$^{235}_{92}UF_6$ will diffuse to the right slightly faster than $^{238}_{92}UF_6$ because it has a smaller mass.
Membrane
Diffusion
Remove with vacuum pump
$^{238}_{92}UF_6$
$^{235}_{92}UF_6$

▲ **Figure 15.14** The process of diffusion can be used to separate different isotopes of uranium. For simplicity, we show roughly equal numbers of $^{235}_{92}UF_6$ and $^{238}_{92}UF_6$ molecules on the left. Usually, the initial concentration of $^{235}_{92}UF_6$ is much smaller than the concentration of $^{238}_{92}UF_6$.

Diffusion experiments with uranium usually employ molecules of uranium hexafluoride containing either ^{235}U or ^{238}U and denoted as $^{235}_{92}UF_6$ and $^{238}_{92}UF_6$ respectively. Since $^{235}_{92}UF_6$ has a slightly smaller mass, it diffuses slightly faster than a molecule of $^{238}_{92}UF_6$. This difference can be used to separate isotopes as shown in Figure 15.14. The left side of this system contains an equal mixture of $^{235}_{92}UF_6$ and $^{238}_{92}UF_6$, and in the middle there is a medium (perhaps a thin membrane like the one used in a transdermal patch; Fig. 15.13) through which the molecules can diffuse. The lighter $^{235}_{92}UF_6$ molecules (with the larger diffusion constant) move through the membrane slightly faster than the $^{238}_{92}UF_6$ molecules, so there is a slightly higher concentration of $^{235}_{92}UF_6$ on the right compared with the left. Hence, we have separated (at least partially) the different isotopes; additional diffusion steps can be used to make the separation more and more complete. A similar separation process takes place in plants when they process two isotopes of carbon, $^{12}_{6}C$ and $^{13}_{6}C$, via photosynthesis.

CONCEPT CHECK 15.7 Relation of the Diffusion Distance to Time

Suppose we double the size of the room in Example 15.8. How does that affect the time required for an average molecule to diffuse across the room?

(a) The time is doubled.
(b) The time does not change.
(c) The time increases by a factor of four.

Diffusion, Brownian Motion, and the Discovery of Atoms

Figure 15.12 shows the characteristic random-walk path followed by a diffusing particle. This figure is only schematic; there is no way to observe the motion of a molecule in a gas or liquid in such fine detail. It is possible, however, to observe the diffusive motion of larger objects. This experiment was actually carried out nearly 200 years ago by botanist Robert Brown in his studies of the motion of pollen grains in water. The grains were about 1 μm in size, much larger than a molecule but still small enough to make diffusion observable with a microscope. Brown found that pollen grains do indeed follow random zigzag paths like those sketched in Figure 15.12, a motion now called ***Brownian motion*** in his honor. Just as for the molecule in Figure 15.12, the zigzag motion of Brown's pollen grains was caused by collisions with molecules in the fluid. Brown made his observations before the discovery of atoms and molecules, and there was no real explanation of this motion during his lifetime. In hindsight, his experiments provided some of the very first evidence for the existence of atoms and molecules, and detailed studies of Brownian motion led to the first accurate measurements of Avogadro's number.

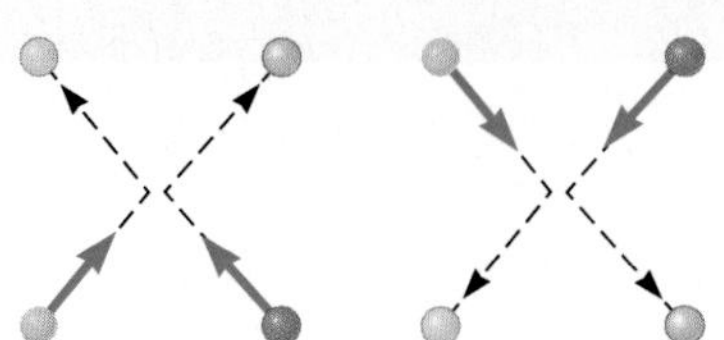

▲ **Figure 15.15** An elastic collision between two molecules in a gas is time reversible. Both forward and backward collision processes satisfy all the laws of physics (including conservation of energy and momentum).

The Arrow of Time

Consider a video recording of an elastic collision of two particles; they might be two molecules in a dilute gas. We could play this video normally or in reverse (Fig. 15.15), and both cases yield motion consistent with our intuition. In fact, both collision trajectories are consistent with Newton's laws.

Now consider the mixing of two gases A and B in Figure 15.16. Here we begin at $t = 0$ with all the molecules of type A on the left side of a container and all the molecules of type B on the right side. After a period of time, the gases mix, and there is an equal chance of finding either type of molecule anywhere in the container. If we were to videotape this mixing process, the first few frames of the video would look like Figure 15.16A and the final frames would look like Figure 15.16B. If we were to play this video backward, it would show molecules in a fully mixed gas (Fig. 15.16B) spontaneously unmix so as to end up with the two separate regions in Figure 15.16A. Your intuition should tell you that such behavior is not possible, and indeed, such unmixing of two gases does not occur in real life. In this process, the direction of time does matter.

The hypothetical unmixing process involves many collisions between molecules, and we could use Newton's laws to analyze these collisions one at a time. Each collision is time reversible, just as with the collisions in Figure 15.15, so each individual unmixing collision would appear to be perfectly reasonable. Only when we string them all together do we run into problems.

This discussion reveals a deep aspect of physics. Newton's laws are time reversible, but many processes in nature are not! Processes that are not time reversible are possible in systems that contain a very large number of particles (such as a gas). The statistical improbability of different systemwide configurations such as the mixed and unmixed gas pictures in Figure 15.16 makes certain processes such as the unmixing of a gas time irreversible. We'll see in Chapter 16 that time-irreversible behavior of systems with many particles also plays a role in the area of physics called ***thermodynamics***.

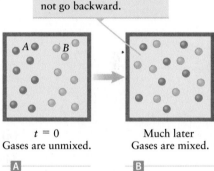

The mixing process would not go backward.

A B

$t = 0$
Gases are unmixed.

Much later
Gases are mixed.

A B

▲ **Figure 15.16** The mixing of two gases is not time reversible. A Two different types of molecules are initially separated (placed on different sides of a container). B As time passes, the molecules mix, with both types of molecules spread evenly (with equal probability) throughout the container.

15.6 Deep Puzzles in Kinetic Theory

In this chapter, we focused on the successes of classical kinetic theory. That theory is based on an application of classical mechanics—Newton's laws of motion—to the properties of gases. Classical kinetic theory is able to explain the ideal gas law, the meaning of temperature, and the universal values of the specific heats of dilute gases. However, certain properties of simple gases are *not* explained by classical kinetic theory. These failures were important in the history of physics and helped lead to the development of quantum theory. A kinetic theory based on quantum rather than classical mechanics successfully explains all the properties of dilute gases.

Another puzzle concerns the notion of an absolute temperature scale. What happens in a dilute gas, or any other phase of matter, when it is cooled to absolute zero? Do the atoms in such a gas or solid stop all motion? These important conceptual issues are not addressed by classical kinetic theory. We'll explore these issues later in this book, after we learn about thermodynamics and quantum theory.

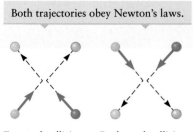

Summary | CHAPTER 15

Key Concepts and Principles

Kinetic theory

Kinetic theory is the application of Newton's laws to the mechanics of a gas of particles. This theory leads to the ***ideal gas law***; for a dilute gas composed of any substance,

$$PV = nRT \qquad \text{(15.6) (page 484)}$$

where P is the pressure, V is the volume, n is the number of moles of particles, and R is the ***universal gas constant*** with the value

$$R = 8.31 \frac{\text{J}}{\text{mole} \cdot \text{K}} \qquad \text{(15.7) (page 484)}$$

The ideal gas law can also be written in the form

$$PV = Nk_BT \qquad \text{(15.11) (page 484)}$$

where N equals the number of particles (i.e., molecules) and k_B is ***Boltzmann's constant***,

$$k_B = 1.38 \times 10^{-23} \text{ J/K} \qquad \text{(15.10) (page 484)}$$

Internal energy

The temperature of a dilute gas is related to the average kinetic energy of a molecule by

$$KE = \tfrac{3}{2}k_BT \qquad \text{(15.17) (page 490)}$$

The ***internal energy*** of a monatomic ideal gas is equal to the total kinetic energy of the molecules and is given by

$$U = \tfrac{3}{2}Nk_BT \qquad \text{(15.21) (page 491)}$$

For an ideal gas, U is independent of the molecular mass and pressure.

Applications

Gas laws

The ideal gas law contains several other gas laws as special cases.

Avogadro's law:
For a sample of gas at constant pressure and temperature, the volume is proportional to the number of molecules in the sample.

Boyle's law:
For a sample of gas at constant temperature, the product of the pressure and the volume is constant:

$$PV = \text{constant} \qquad \text{(15.3) (page 482)}$$

Charles's law:
For a sample of gas at constant pressure, if the temperature is changed by a small amount ΔT, the volume changes by an amount ΔV, with

$$\Delta V \propto \Delta T \qquad (15.4)\ (\text{page } 482)$$

Gay-Lussac's law:
For a sample of gas held in a container with constant volume, changes in pressure are proportional to changes in temperature:

$$\Delta P \propto \Delta T \qquad (15.5)\ (\text{page } 482)$$

Specific heat of a gas

The specific heat per mole of a monatomic ideal gas at constant volume is given by

$$C_V = \tfrac{3}{2}R \qquad (15.23)\ (\text{page } 493)$$

Diffusion

The random motion of individual gas molecules is described by ***diffusion***. The average net distance that a molecule diffuses in time t is given by

$$\Delta r = \sqrt{Dt} \qquad (15.25)\ (\text{page } 495)$$

where D is the ***diffusion constant.***

Questions

SSM = answer in *Student Companion & Problem-Solving Guide* — = life science application

1. Show that Avogadro's law, Charles's law, and Gay-Lussac's law all follow from the ideal gas law (Eq. 15.6).

2. In Example 15.6, we claimed that the total kinetic energy of the gas is greater than the kinetic energy of a baseball. Show that this claim is in fact true by working out the value of $\frac{1}{2}mv^2$ with realistic values of m and v for a baseball.

3. If you were to hammer on a piece of metal, the temperature of the metal would rise. Why? Describe from a microscopic point of view.

4. Is it possible to heat a can of soup by continuously lifting and dropping it on a hard surface? (Assume the can will take the abuse and not rupture.) Would it make a difference if the can were encased in an excellent thermal insulator?

5. Consider two identical containers, one full of helium and the other full of oxygen. If the molecules of both gases have the same average speed, what is the ratio of their temperatures? (Approximate both as ideal gases.)

6. SSM In Chapter 11, we learned about the bulk modulus B of a substance. The bulk modulus is related to changes in volume by

$$\frac{\Delta V}{V} = -\frac{\Delta P}{B}$$

Use the ideal gas law to calculate the bulk modulus for an ideal gas. Assume the temperature is held constant.

7. An automobile cooling system circulates a liquid called antifreeze (which is mostly water) and typically operates at a pressure higher than atmospheric pressure. Why is that desirable?

8. A family drives from a low valley to a high-altitude mountain meadow for a picnic. When they arrive at the meadow, they find that their bag of potato chips has burst open and their bag of pretzels looks like an inflated balloon. Explain the behavior of the packaging.

9. SSM A container of gas under pressure has a pinhole leak. If the content of the container is oxygen, the pressure decreases at a certain rate as the gas escapes. If the content is hydrogen gas, however, we find that the rate of pressure drop is greater under the same conditions. Why?

10. The diffusion constant D of a particle in a gas or liquid depends on temperature. If the temperature is increased, does D increase or decrease? Explain why.

11. In Example 15.8 we saw that diffusion is a very slow process. Calculate the *total distance* (not the net displacement Δr) traveled by a diffusing molecule in Example 15.8.

12. Compare the act of making tea with a tea bag in cold water versus steeping it in hot water. Which method will make the tea stronger faster? Why?

13. Consider an ideal gas composed of helium atoms. The collisions between atoms are similar to the collisions between billiard balls.

That is, whenever the distance between two atoms is less than $2r$ (where r is the radius of a helium atom), the two atoms undergo a collision. Show that the mean free path between collisions ℓ is proportional to $1/\rho$, where ρ is the density of the gas. *Hint*: Consider the motion of just one atom and assume the others are all "frozen" in place.

14. The mass of 1 mole of carbon atoms of isotope number 12 (so-called carbon 12) is defined to be exactly 12 g. The *atomic mass unit* is defined as one-twelfth the mass of a single carbon-12 atom. Use this fact to calculate the conversion factor that relates atomic mass units to kilograms.

15. In Concept Check 15.6, we considered how to slow down the release of a drug in a transdermal patch by increasing the thickness of the membrane. Which of the following strategies will decrease the rate at which drug molecules enter the body?
(a) Lower the concentration of molecules in the patch.
(b) Increase the temperature.
(c) Use larger drug molecules.

16. The air in a room consists mainly of N_2 and O_2 molecules. Which has the greater average speed? *Hint*: You do not have to calculate the average speed of either molecule to answer this question.

17. Why does the pressure in a car tire increase when the car is driven?

Problems

SSM = solution in *Student Companion & Problem-Solving Guide*
★ = intermediate ✪ = challenging Ⓧ = life science application
Ⓡ = reasoning and relationships problem
RT = reasoning tutorial in ENHANCED WebAssign

15.1 MOLECULAR PICTURE OF A GAS

1. What is the mass of a single carbon atom?

2. How many oxygen atoms are in 1.0 g of water?

3. Find the mass of a molecule of carbon dioxide.

4. ★ RT Estimate the number of atoms in your body.

5. SSM ★ You have 6.0 moles of particles of an unknown (pure) substance with a mass of 240 g. What might the substance be?

15.2 IDEAL GASES: AN EXPERIMENTAL PERSPECTIVE

6. Consider a balloon containing nitrogen molecules at room temperature (20°C) and atmospheric pressure. If the balloon has a spherical shape with radius of 20 cm, what is the number of molecules in the balloon?

7. ★ A balloon having an initial temperature of 20°C is heated so that the volume doubles while the pressure is kept fixed. What is the new value of the temperature?

8. ★ A spherical balloon containing nitrogen gas has a radius of 20 cm. If the balloon is at atmospheric pressure and room temperature, what is the mass of nitrogen in the balloon?

9. ★ Find the average spacing between molecules in the balloon in Problem 8. How does the answer change when the balloon is heated, assuming the pressure does not change?

10. ★ RT Estimate the number of air molecules in a bicycle tire.

11. Imagine you are sitting in a room full of oxygen molecules at 300 K. If the room has a volume of 1.0×10^4 m^3, what is the number of oxygen molecules in the room?

12. You are given the job of calibrating a gas thermometer. You do some experiments and find that the gas pressure in the thermometer is 40,000 Pa at $T = 280$ K. If the pressure later reads 60,000 Pa, what is the new temperature?

13. For the gas thermometer in Problem 12, what is the pressure when $T = 230$ K?

14. SSM ★ The gas in a cylindrical container has a pressure of 1.0×10^5 Pa and is at room temperature (293 K). A piston (a movable wall) at one end of the container is then adjusted so that the volume is reduced by a factor of three, and it is found that the pressure increases by a factor of five. What is the final temperature of the gas?

15. ★ A balloon contains 3.0 moles of helium gas at normal atmospheric pressure ($P_{atm} = 1.01 \times 10^5$ Pa) and room temperature (293 K). (a) What is the volume of the balloon? (b) The weather becomes very stormy (a hurricane is approaching), and the volume of the balloon increases by 15%. What is the new pressure?

16. How many molecules are there in 1.0 cm^3 of air? *Hint*: The average mass of a mole of molecules in the air is approximately 29 g.

17. Ⓧ The lungs of a typical adult hold about 2 L of air (0.002 m^3). How many moles of air molecules are in a typical adult's lungs?

18. ★ A weather balloon filled with helium gas has a volume of 300 m^3 and pressure 1.4×10^5 Pa at the Earth's surface (where the temperature is 293 K). The balloon is then released and moves to a high altitude, where the pressure in the balloon is 25,000 Pa and the temperature is 250 K (−23°C). What is the new volume of the balloon?

19. Ⓧ A tank of compressed oxygen gas at a doctor's office has a pressure of 100 times atmospheric pressure. If the volume of the tank is 500 cm^3, how many oxygen molecules does it contain?

20. ★ Ⓡ Approximately how many moles of gas molecules are in your classroom?

21. An ultrahigh vacuum system can reach pressures in the laboratory that are less than 1.0×10^{-7} Pa. If that is the pressure in a system of volume 1000 cm^3, how many molecules are in the system?

22. ✪ RT Estimate the number of helium molecules in a blimp filled with helium gas. *Hint*: Assume the pressure is a factor of two larger than atmospheric pressure (so that it will not collapse).

23. SSM ★ The pressure in a gas thermometer is 5000 Pa at the freezing point of water. (a) What is the pressure in this thermometer at the boiling point of water? (b) If the pressure is 6000 Pa, what is the temperature?

24. Some endurance athletes (such as cyclists and long-distance runners) use a hypobaric chamber to enhance their ability to use oxygen. A hypobaric chamber mimics the effect of being at a high altitude, where the amount of oxygen available to the lungs is less than at sea level. A typical hypobaric chamber simulates the environment at an altitude of 9000 ft above sea level, where the air pressure is approximately 7×10^4 Pa. What is the amount of oxygen available at this altitude compared with that at sea level? Express your answer as a fraction (without units).

15.3 IDEAL GASES AND NEWTON'S LAWS

15.4 KINETIC THEORY

25. What is the average speed of a molecule of H_2O in the atmosphere? Assume a nice spring day with $T = 20°C$ (= 293 K).

26. SSM You place 80 moles of hydrogen gas in a balloon of volume 2.5 m³, and find the pressure to be 1.5 times atmospheric pressure. What is the typical speed of a hydrogen molecule?

27. For the balloon in Problem 26, what is the total kinetic energy of all the hydrogen molecules in the balloon?

28. You have a container of an ideal gas at room temperature (20°C) and pressure, and you want to use it to store energy in the form of the kinetic energy of the gas. If you want to double the energy in the system, what must be the temperature's new value?

29. The surface of the Sun has a temperature of about 6000 K. What is the speed of a typical hydrogen molecule in the Sun's atmosphere?

30. What is the ratio of the speed of a typical hydrogen molecule at the Sun's surface to the speed of a typical hydrogen molecule in the Earth's atmosphere? The temperature at the surface of the Sun is approximately 6000 K.

31. What is the total internal energy of 15 moles of helium gas at room temperature? How does the answer change if the helium is replaced by argon?

32. SSM A balloon of volume 1.5 m³ contains argon gas at a pressure of 1.5×10^5 Pa and room temperature (20°C). What is the total internal energy of the gas?

33. Calculate the speed v of a typical hydrogen molecule in the Earth's atmosphere. Is v less than or greater than the escape speed? Experiments show that a hydrogen molecule released into the atmosphere escapes into outer space. If v is less than the escape speed, why is this so?

15.5 DIFFUSION

34. Approximately how long does it take a molecule of nitrogen to diffuse a distance of 1 cm through the atmosphere?

35. SSM How long would it take a nitrogen molecule in the atmosphere to travel a distance of 1 cm if there were no collisions with other molecules?

36. Approximately how long does it take a water molecule to diffuse a distance of 3 cm through a glass of water?

37. About how far does a typical water molecule diffuse (in water) in 1 minute?

38. Approximately how long does it take an oxygen molecule to diffuse across the interior of a cell that is 100 μm in diameter?

39. SSM What is the approximate time required for a hemoglobin molecule to diffuse through a cell membrane? *Hint*: The diffusion constant D is approximately proportional to $1/r$, where r is the diameter of the diffusing particle or atom. Estimate the diffusion constant of a hemoglobin molecule from the value of D for an oxygen molecule in Table 15.1.

40. One way for a protein to move from one side of a cell to another is through diffusion. If the diameter of a cell is increased by a factor of four, by what factor does the time required to diffuse across the cell increase?

41. Consider an oxygen molecule as it diffuses in your blood. How far will a typical molecule diffuse in 1 second? Assume the value of D for oxygen in blood is the same as for oxygen in water.

42. Approximately how long does it take a hydrogen molecule to diffuse across a small room?

43. In our discussion of diffusion, we mentioned that between collisions, a molecule in air travels in a straight line with a constant speed. For a hydrogen molecule in air, this speed is approximately 3000 m/s. If the hydrogen molecule in Problem 42 were moving with this speed and did not collide with any air molecules, how long would it take to travel across the room?

44. A cell membrane is typically 10 nm (1×10^{-8} m) thick. How long does it take an oxygen molecule to diffuse this distance through the membrane?

45. Calculate the net distance a pollen grain of radius 5×10^{-7} m will diffuse in water in 1 minute.

Additional Problems

46. You place 3.5 g of water at room temperature in a pressure cooker (a sealed pot) of volume 2.5 L and heat the water to 200°C. What is the pressure in the cooker?

47. **Caesar's last breath.** (a) How many molecules from Julius Caesar's dying breath are currently in your lungs? The total mass of the Earth's atmosphere is approximately 5×10^{18} kg, it is composed mainly of N_2 (78%) and O_2 (21%), and an average breath (called the tidal volume) is 1.5 L. (b) Is it possible that Caesar's last breath would be evenly diffused around the globe by now? Many processes would distribute the molecules of his last breath, including convection, absorption and release in water, and diffusion. Calculate how far a nitrogen molecule will travel by the process of diffusion alone in the approximately 2000 years since Caesar's demise. How much of a role does diffusion play in distributing and diluting emissions into the atmosphere?

48. Calculate the volume occupied by 1 mole of each of the following substances (in both cubic centimeters and liters): (a) water at room temperature, (b) ice at 0°C, (c) air at room temperature

and atmospheric pressure, (d) the element aluminum at room temperature, and (e) helium at 0°C and atmospheric pressure. (f) Rank these substances from largest volume to smallest. (g) Rank these substances according to density from smallest to largest.

49. The continent of North America has an area of approximately 2.4×10^7 km^2. If the entire continent were evenly covered with Avogadro's number of BBs (spherical projectiles 4.5 mm in diameter), how deep would the layer of BBs be? *Hint*: If spheres are randomly poured into a jar, the fraction of the jar's volume occupied by spheres when the jar is "full" is about 65% if the diameter of the jar is large compared to the size of the spheres.

50. An air bubble ascends from the bottom of a lake 15 m deep. The temperature at the bottom of the lake is 4°C, and near the surface it is 20°C. How many times larger in volume is the bubble at the top of the lake compared with its volume at the bottom?

51. A scuba diver descends to 8.2 m below the surface of the ocean. (a) What pressure will her regulator need to supply so that her lungs can fill as normal? (b) When the diver is at 8.2 m, how many times more molecules of air are in her lungs than at the surface? Treat air as an ideal gas and assume the temperature is the same in both instances, which will be somewhat less than body temperature.

52. A diver uses an air cylinder with a volume of 11 L and filled to a pressure of 200 atm (Fig. P15.52). He dives at a leisurely pace, taking about 15 breaths per minute, and each breath is 1.2 L on average. (a) If he must leave 17% of the air for reserve, how long can he dive at 20 m deep? *Hint*: The pressure must be considered as in Problem 51. (b) How long will his air supply last if he dives at 30 m? Approximate air as an ideal gas.

Figure P15.52 A standard 80 diving tank: capacity 88 ft^3 at 200 atm.

53. A steel cylinder is filled with an ideal gas at 20°C and pressure of 8.0 times atmospheric pressure. (a) The cylinder is submerged in a bath of boiling water and allowed to reach thermal equilibrium. Calculate the pressure inside the cylinder. (b) Some gas is then allowed to escape so that the pressure returns to its original value, at which time it is resealed. Find what fraction of the original gas remains. (c) If the cylinder is then allowed to cool to its original temperature, what would the pressure inside be?

54. The human nose is very sensitive to certain molecules. For example, it can sense the presence of the chemical CH_3SH (methyl mercaptan) at levels as small as 2 parts per billion. Another especially smelly substance is ammonia, NH_3. A student in a lab with very still air has two vials placed 20 cm from his nose; one contains methyl mercaptan, the other ammonia. He begins his experiment and removes the caps from both vials, and after 30 minutes, he notices the smell of ammonia. How soon after does his nose detect the methyl mercaptan? Is diffusion a good model for how fast odors travel?

55. **Bug breath.** Insects do not have lungs or a blood circulatory system. Instead, a system of openings in the exoskeleton (spiracles) lead to branching tubes of decreasing diameter called trachea, the smallest of which are about 1 μm in diameter and connect to every cell in the insect's body (Fig. P15.55). For many insects, each cell's steady supply of oxygen is provided by diffusion of air along the trachea. The average distance to a cell along a trachea for a small beetle is 0.25 cm. (a) Starting from one of the insect's spiracle openings, how long does it take an oxygen molecule on average to reach a cell? (b) Repeat the calculation for a human being, assuming people obtain their needed oxygen by these means. Assume the average distance to a cell along a trachea is now 10 cm (since human beings are much larger than beetles). (c) These calculated diffusion times are proportional to the rate at which oxygen can be delivered to a cell. Why don't we ever see beetles the size of cows, or mosquitoes the size of birds?

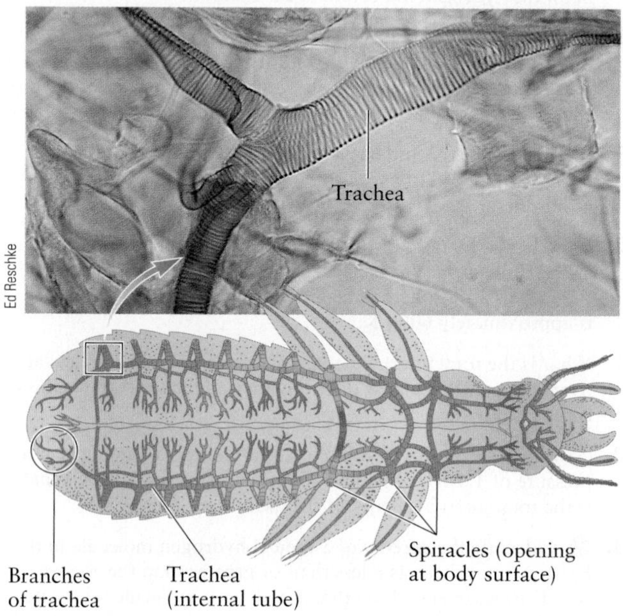

Figure P15.55 Tracheal system of an insect.

56. (a) Find the speed for typical Ne, Ar, and Kr atoms and for typical H_2, LiF, and Cl_2 molecules in the atmosphere at room temperature. (b) Compare these typical speeds to the escape speed for an object in the Earth's atmosphere. (c) Explain why the typical speed can be less than the escape speed, yet the atom or molecule would still not be found in the atmosphere.

57. SSM In our discussions of kinetic theory, we have focused on the motion of the molecules in a gas. The result for the typical speed in Equation 15.18, however, also applies to molecules in a liquid. Use this result to calculate the speed of the amino acid molecule glutamine ($C_5H_{10}N_2O_3$) in solution at room temperature.

58. Deuterium is an isotope of the element hydrogen that has a nucleus consisting of one proton and one neutron. In Chapter 30, we show how two deuterium nuclei can undergo a reaction called fusion in which 6.4×10^{-13} J is given off. Seawater contains about 154 ppm of deuterium (the ratio of deuterium to hydrogen atoms in the water.) If all the deuterium atoms in a cup of seawater undergo fusion, how much energy would be released? Your answer will explain why fusion is a potential solution to the world's energy needs.

59. ★ Ⓡ Which number is closest to Avogadro's number? (a) The number of people on the Earth. (b) The number of stars in our galaxy. (c) The number of O_2 molecules in your lungs. (d) The number of hairs on your head.

60. ✪ Ⓧ Suppose a particular molecule is diffusing a certain distance through a particular substance. By what factor will the diffusion time change if (a) the distance is doubled, (b) the absolute temperature is increased by a factor of two, and (c) the mass of the molecule is doubled?

CHAPTER

16

▶ *Heat engines have a long and interesting history. This is a reconstruction of a steam engine invented by Denis Papin of France, circa 1690. (CNAM, Conservatoire Nationale des Arts et Metiers, Paris/Archives Charmet/The Bridgeman Art Library)*

Thermodynamics

OUTLINE

In Chapter 14, we discussed heat energy and how the flow of heat into or out of a system can affect its properties. In Chapter 15, we developed the kinetic theory of a dilute gas and saw how the mechanics of molecular motion is connected to the temperature of a gas. The central thread of Chapters 14 and 15 was the principle of conservation of energy. In this chapter, we continue along that thread, exploring the area of physics called ***thermodynamics***. Our goal now is more ambitious: we wish to understand the rules that govern the flow of heat and the exchange of other forms of energy between all types of systems, including gases, liquids, and solids. We are especially interested in *fundamental limits* on how heat can flow from one system to another. This understanding will lead us to some very surprising and powerful results, including the impossibility of building a perpetual motion machine and the best possible efficiency of an engine.

16.1 Thermodynamics Is About the Way a System Exchanges Energy with Its Environment

Two systems are in the same state if their macroscopic properties are the same.

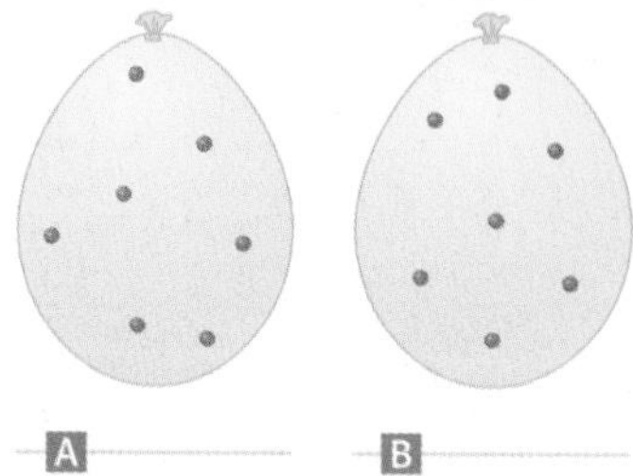

▲ **Figure 16.1** A system of many particles such as the gas in a balloon can be described by macroscopic quantities such as temperature and pressure.

Thermodynamics is concerned with the properties of systems composed of many particles such as a gas, liquid, or solid. We need variables for describing the state of a system and for describing the interactions of one system with another. In mechanics, we describe the state of motion of a particle in terms of its position, velocity, and acceleration, and the interactions between particles are described by forces. However, a typical system such as 1 mole of a dilute gas contains far too many values of position, velocity, and acceleration to keep track of, even with any computers that are likely to be available in your lifetime. In addition, even if we had such a computer, it would be virtually impossible to make sense of all that information.

Fortunately, we can use a small number of macroscopic quantities such as temperature, pressure, and volume to describe systems and their interactions. For example, we might have two balloons containing helium gas as sketched in Figure 16.1. These balloons have the same volume V and contain the same number N of helium atoms, with the same temperature T and pressure P. Although P, V, N, and T are each the same in the two balloons, the precise locations of the helium atoms and their velocities are certainly *not* the same, so these balloons are not microscopically identical. They are nevertheless macroscopically identical, and their properties as systems will be the same. For example, they will have the same values for the velocity of sound and will exhibit the same condensation temperature. Thermodynamics is based on a description involving such macroscopic quantities; the extra detail of the microscopic variables is not needed.

Our gas balloon may interact with the air around it as sketched in Figure 16.2. In such cases, when one of the systems (the air around the balloon) is very much larger than the other (the balloon), the larger system is often referred to as the ***environment***. The interactions between two systems or between a system and its environment generally involve forces and the transfer of energy. These interactions can be described in terms of the work done by one system on another or in terms of the heat that flows between them. One goal of thermodynamics is to understand how energy is exchanged between systems.

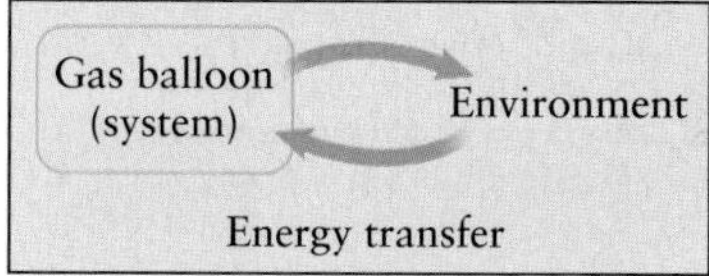

▲ **Figure 16.2** Thermodynamics is concerned with the interactions of a system with its environment. These interactions transfer energy between the two.

As with Newton's laws of mechanics, thermodynamics is based on a small set of physical laws. The laws of thermodynamics each seem quite "innocent" or obvious. One of them (the first law) is a restatement of the principle of conservation of energy, and another (the second law) is a simple statement about the relation between heat flow and temperature. When the laws of thermodynamics are put together, however, they yield remarkable results, including fundamental limits on what is and is not possible.

16.2 The Zeroth Law of Thermodynamics and the Meaning of Temperature

There are four laws of thermodynamics, beginning with the "zeroth" law, which is a statement about the concept of temperature. We encountered the zeroth law in our study of heat in Chapter 14, but it is important to repeat it here.

Zeroth law of thermodynamics

If two systems A and B are both in thermal equilibrium with a third system, then A and B are in thermal equilibrium with each other.

Zeroth law of thermodynamics

This statement of the zeroth law is based on the notion of ***thermal equilibrium*** and hence involves the properties of heat. You should recall from Chapter 14 that heat is the energy that flows from one system to another as a result of a temperature difference. We can use this fact as the basis of an experiment to tell if two systems, which we can call A and B, are in thermal equilibrium with each other. The experiment begins with the two systems separated from each other so that no heat can flow between them. We then allow the systems to interact by putting them into direct contact so that heat can flow between them via conduction (Chapter 14). If our experiment finds that heat flows between A and B, the two systems were not initially in thermal equilibrium; that is, they had different temperatures prior to the experiment. The notions of thermal equilibrium and temperature are thus intimately connected. Two systems in thermal equilibrium with each other have the same temperature. So, the zeroth law of thermodynamics can be restated as follows.

Zeroth law of thermodynamics: alternative form

Zeroth law of thermodynamics (restatement)

Suppose the temperature of system A is equal to the temperature of system C; that is, $T_A = T_C$ (so that A and C are in thermal equilibrium). In addition, the temperature of system B is equal to the temperature of system C; that is, $T_B = T_C$. Then $T_A = T_B$.

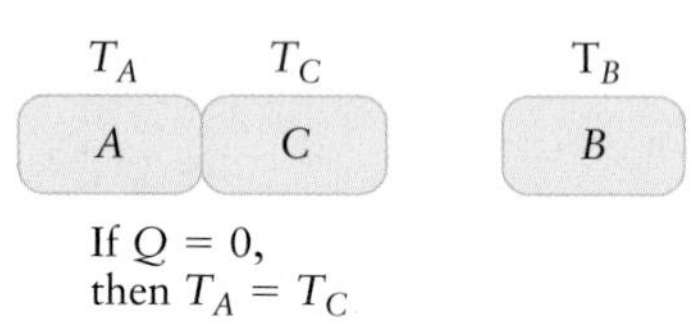

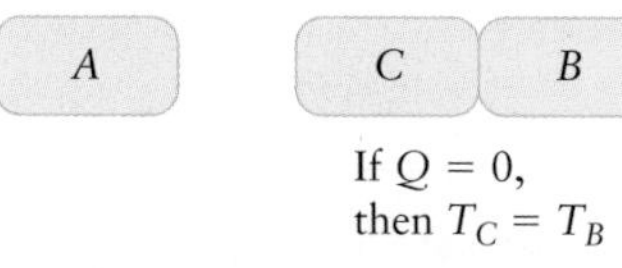

ZEROTH LAW:
If $T_A = T_C$ and $T_C = T_B$,
then $T_A = T_B$.

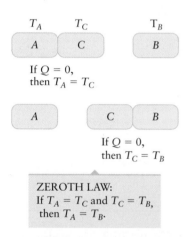

▲ **Figure 16.3** The zeroth law of thermodynamics involves the notion of thermal equilibrium as applied to three (or more) systems. See also Figure 14.7.

The zeroth law is illustrated schematically in Figure 16.3.

From the point of view of arithmetic, the statements of the zeroth law regarding the temperatures of three systems seem quite obvious. So why does it deserve to be called a law of physics? The answer is that the zeroth law declares that a property called *temperature* does indeed exist (!) and that this quantity determines the way heat flows between systems. This profound statement cannot be derived from Newton's laws or any other laws of physics.

16.3 The First Law of Thermodynamics and the Conservation of Energy

One property central to thermodynamics is ***internal energy***, denoted by U. We first introduced internal energy in Chapter 14; it is the total energy associated with all the particles in a system. For an ideal monatomic gas, all this energy resides in the kinetic energy of the particles, whereas in other systems such as liquids and solids, the chemical bonds or interactions between the particles also contribute to U. Physicists would say that these bonds and interactions give rise to potential energy because they can be modeled as springlike forces obeying Hooke's law. (See spring potential energy, Equation 6.24.) A chemist would say that the bonds store chemical energy because this energy can be released (or gained) in chemical reactions. Both views are correct. We'll adopt the physicist's point of view of the internal energy of a system as the sum of the molecular kinetic energies plus the potential energies due to the interactions between particles within the system.[1]

The internal energy of a system is a function of macroscopic variables such as temperature, pressure, and volume. For example, in Chapter 15 we showed that the internal energy of a monatomic ideal gas containing N atoms at temperature T is

$$U = \tfrac{3}{2}Nk_BT \qquad (16.1)$$

where k_B is Boltzmann's constant. It is useful to write Equation 16.1 in terms of the number of moles n in the gas instead of the number of atoms. The result is

$$U = \tfrac{3}{2}nRT \qquad (16.2)$$

[1]By definition, internal energy does not include the kinetic energy of the system "as a whole," that is, the kinetic energy associated with either the motion of the center of mass or the rotational motion of the system.

where R is the gas constant. The internal energy of an ideal gas thus depends only on temperature and the number of particles and is independent of the pressure and volume. Because this internal energy function is so mathematically simple, we'll use the monatomic ideal gas in many of our thermodynamic examples; however, the same basic ideas apply to all systems, although with more complicated expressions for U.

Consider a system in some initial state described by temperature T_i, pressure P_i, and volume V_i as shown in Figure 16.4A. We now allow the system to interact with its environment as shown schematically in Figure 16.4B. These interactions with the environment might result in some heat flow into or out of the system. Forces might also be exerted between our system and the environment, causing the system to do work on the environment or the environment to do work on the system (Fig. 16.4B). As a result of the interactions, the system ends up in some final state (Fig. 16.4C), with a temperature T_f, pressure P_f, and volume V_f. This change in our system is called a ***thermodynamic process***. The first law of thermodynamics is an application of the principle of conservation of energy to such processes.

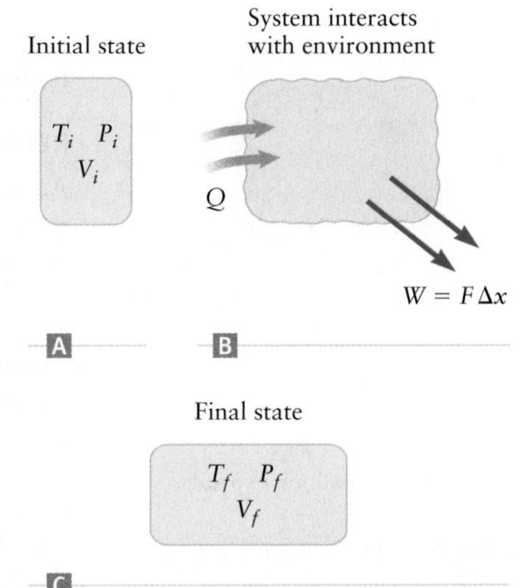

▲ **Figure 16.4** Schematic description of a thermodynamic process. **A** A system begins in some initial state. **B** The system then interacts with its environment. This interaction involves the flow of heat (Q) and mechanical work (W). **C** The system is then left in some final state.

First law of thermodynamics

If an amount of heat Q flows *into* a system from its environment and an amount of work W is done *by* the system on its environment, the internal energy of the system changes by an amount

$$\Delta U = U_f - U_i = Q - W \tag{16.3}$$

The First Law of Thermodynamics: The Meaning of Q and W

Definitions of Q and W in the first law of thermodynamics

There are two ways to change the energy of a system. One way is to arrange for heat to flow into (or out of) the system. By convention, a positive value for Q means that heat flows into the system (and energy is added); hence, the internal energy increases (ΔU is positive) as indicated by Equation 16.3. Another way to change the internal energy is through the action of an external force, that is, a force exerted on the system by the environment. We saw in Chapter 6 that the work done *on* a system is equal to the change in the energy of the system. By convention, a positive value of W in Equation 16.3 means that a positive amount of work is done *by* the system on its environment, so if W is positive, the internal energy of a system must decrease. To apply Equation 16.3 correctly, you must remember that Q is the heat energy that flows *into the system*, whereas W is the work done *by the system*.

Nowadays, we take it for granted that heat is a form of energy, so it is natural to include heat energy along with kinetic and potential energy whenever we discuss the principle of conservation of energy. However, when the science of thermodynamics was being developed around 200 years ago, the nature of "heat" was a matter of much debate. For example, when you place your hand over a flame, you know that "something" flows from the flame to your hand. It is not obvious that the "stuff" that flows from the flame to your hand is similar to the kinetic energy of a moving baseball or the potential energy of a compressed spring. Eventually, physicists discovered that heat is indeed a form of energy, so heat can be measured with either joules (the SI unit for energy) or calories (an alternative unit of energy defined in Chapter 14).

EXAMPLE 16.1 The First Law of Thermodynamics Applied to an Ideal Gas

Consider a flexible balloon containing $n = 3.5$ moles of an ideal monatomic gas such as helium. The balloon is placed over a flame so that $Q = 700$ J of heat flows into the

(continued) ▶

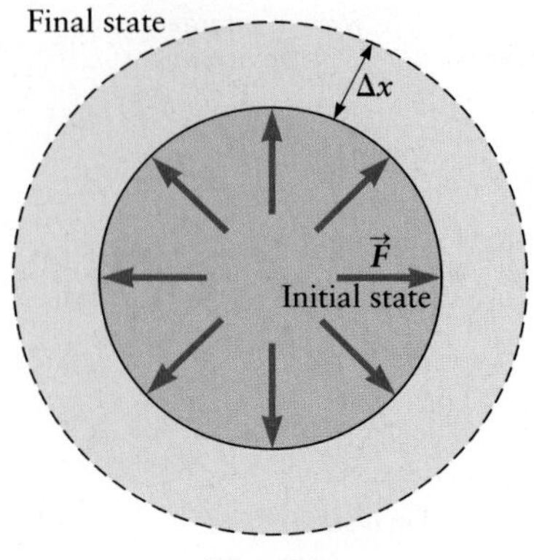

$W = F\,\Delta x$
If $W > 0$, balloon expands.

▲ **Figure 16.5** Example 16.1. If a balloon expands, the gas inside does work on the surface of the balloon. The thermodynamic quantity W is defined as the work done by the system (the gas) on its environment (the wall of the balloon), so for the balloon in this sketch, $W > 0$.

gas, and its temperature increases by 10 K. How much work is done by the gas in the balloon on its environment during this process?

RECOGNIZE THE PRINCIPLE

A central thread of thermodynamics is energy and how it is exchanged between systems. The principle of conservation of energy is thus at the heart of all problem solving in thermodynamics. From the first law of thermodynamics,

$$\Delta U = Q - W \tag{1}$$

The internal energy of a monatomic ideal gas is given by $U = \frac{3}{2}nRT$ (Eq. 16.2), so changes in U are related to changes in temperature by

$$\Delta U = \tfrac{3}{2}nR\,\Delta T \tag{2}$$

In this example, the change in temperature is known, so we can use it to calculate the corresponding change in the internal energy ΔU. Because Q is given, we can then use the first law of thermodynamics, Equation (1), to find the work W done by the gas.

SKETCH THE PROBLEM

Figure 16.5 shows the initial and final states of the balloon. An expanding balloon does work on its environment because the pressure force and displacement are parallel, corresponding to a positive value of W in the first law of thermodynamics.

IDENTIFY THE RELATIONSHIPS

We first find the change in the internal energy using Equation (2):

$$\Delta U = \frac{3}{2}\,nR\,\Delta T = \frac{3}{2}(3.5 \text{ moles})\left(8.31\,\frac{\text{J}}{\text{mole}\cdot\text{K}}\right)(10\text{ K}) = 440\text{ J}$$

Notice that this value is positive. The internal energy of an ideal gas increases when its temperature is increased.

SOLVE

Inserting this result for ΔU with the given value of Q into Equation (1) leads to

$$W = Q - \Delta U = (700 - 440)\text{ J} = \boxed{260\text{ J}}$$

► ***What does it mean?***

Recall again that W is the work done *by* the gas in the balloon *on* its environment. For W to be positive, the force exerted by the gas in the balloon must be parallel to the associated displacement. Here that is the displacement Δx of the surface of the balloon, whereas the force F is produced by the pressure of the gas on this surface (see Fig. 16.5). A positive value for W thus means that the balloon *expands*, as shown in the figure.

CONCEPT CHECK 16.1 A Thermodynamic Process with $W = Q$

Consider a thermodynamic process involving a monatomic ideal gas in which, by some coincidence, $W = Q$. Which statement best describes the process?

(a) The temperature increases.
(b) The temperature decreases.
(c) The temperature does not change.
(d) The balloon expands.
(e) The balloon contracts.
(f) The volume of the balloon does not change.

The Signs of *Q* and *W* in the First Law of Thermodynamics

A common pitfall or source of confusion involves the signs of Q and W in Equation 16.3. These signs are such an important part of thermodynamics that it is worth repeating and emphasizing a few crucial points.

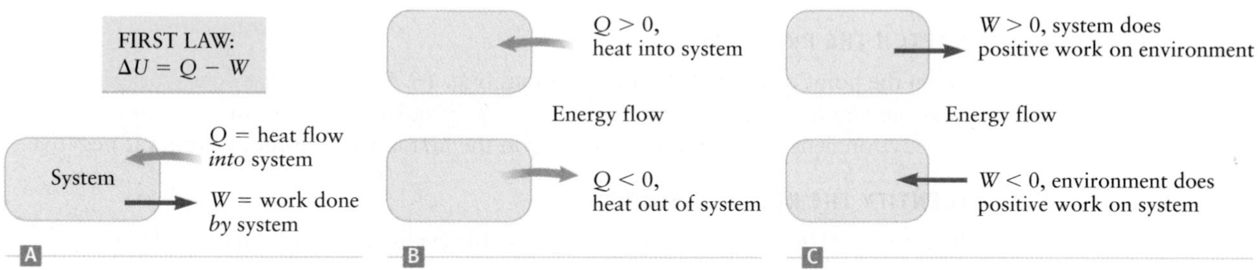

▲ **Figure 16.6** A In the first law of thermodynamics, Q is defined as the heat added to the system and W is the work done by the system. B Meaning of positive and negative Q. C Meaning of positive and negative W.

The first law of thermodynamics reads

$$\Delta U = Q - W \tag{16.4}$$

and is indicated pictorially in Figure 16.6A. By convention, a *positive* value of Q indicates that heat flows *into* the system, which means that the energy of the system *increases*; hence, the change in the internal energy is positive. In any particular case, it is possible for Q to have a *negative* value, and if so, heat energy flows *out of* the system *into* the environment and the internal energy of the system decreases. These two cases are shown schematically in Figure 16.6B.

By convention, a *positive* value of W indicates that the system does a positive amount of work *on* its environment. The internal energy of the system must therefore decrease while the energy of the environment increases (Fig. 16.6C). It is also possible for W to have a *negative* value, which leads to an increase in the internal energy of the system.

These sign conventions for Q and W are a crucial part of the first law of thermodynamics. You can always remember (or re-derive) the correct signs in Equation 16.4 by using energy conservation principles as illustrated in Figure 16.6.

EXAMPLE 16.2 Thermodynamics and Your Diet

A human being is an example of a thermodynamic system. A person riding a bicycle at a moderate speed (Fig. 16.7A) does mechanical work on the pedals at a rate of about

$$P = \frac{W}{t} \approx 100 \text{ W} \tag{1}$$

At the same time, she gives off some heat to her surroundings. Because this heat is flowing out of the cyclist and into the environment, the term Q in the first law of thermodynamics is negative, and measurements on typical cyclists give

$$\frac{Q}{t} = -400 \text{ W} \tag{2}$$

Suppose a cyclist rides for 1 h. What is the change in her internal energy?

RECOGNIZE THE PRINCIPLE

The principle of conservation of energy is at the heart of all problem solving in thermodynamics. From the first law of thermodynamics,

$$\Delta U = Q - W$$

To calculate the change in the internal energy, we need to find both Q and W. The value of Q/t is given, and W can be found from the power exerted by the cyclist.

(continued) ▶

A

B

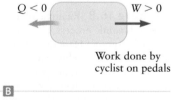

▲ **Figure 16.7** Example 16.2.

SKETCH THE PROBLEM

From the point of view of thermodynamics (Fig. 16.7B), the cyclist does work on her environment (through the force exerted on the pedals), so W is positive. Heat is lost to the environment through sweating, so Q in the first law of thermodynamics is negative.

IDENTIFY THE RELATIONSHIPS

In 1 h ($t = 3600$ s), the total work done by the cyclist is, using Equation (1),

$$W = Pt = (100\ \text{W})(3600\ \text{s}) = +3.6 \times 10^5\ \text{J}$$

where we have included the positive sign explicitly to emphasize that the cyclist does a positive amount of work on the pedals. In 1 h, the total heat Q is

$$Q = (-400\ \text{W})(3600\ \text{s}) = -1.4 \times 10^6\ \text{J}$$

Because heat flows from the cyclist to her environment, Q is negative.

SOLVE

Inserting these values of W and Q into the first law gives

$$\Delta U = Q - W = (-1.4 \times 10^6\ \text{J} - 3.6 \times 10^5\ \text{J}) = \boxed{-1.8 \times 10^6\ \text{J}} \qquad (3)$$

▶ *What does it mean?*

The negative value of ΔU means that the internal energy of the cyclist decreases as the result of her exercise, as expected. The cyclist can replenish this energy by eating. For example, according to dietary tables, the total energy available from a typical fast-food hamburger is about 2.0×10^6 J. So, our cyclist would have to eat about one hamburger to fuel her hour of exercise.

16.4 Thermodynamic Processes

Thermal Reservoirs

Energy can flow via conduction from one system to another when the two systems are in contact. For example, suppose an ice cube is placed outside on a sidewalk on a warm day (Fig. 16.8). Because the ice cube is colder than its environment, heat flows from the air and the sidewalk into the ice cube and Q is positive. Eventually, the ice cube melts and the puddle of water reaches a final temperature equal to the temperature of its surroundings. In this example, the environment around the ice cube acts as a ***thermal reservoir***. By definition, a reservoir is much larger than the system to which it is connected, so there is essentially no change in the temperature of the reservoir when heat enters or leaves it. The notion of a thermal reservoir comes up often in discussions of thermodynamic processes.

▲ Figure 16.8 Example of a thermodynamic process involving a thermal reservoir. Here the system is the H_2O (the ice plus water). The air and the surroundings act as a thermal reservoir; their temperature does not change during the thermodynamic process in which the ice cube melts.

Calculating the Work Done in a Thermodynamic Process

To understand how to calculate the work associated with a thermodynamic process, consider a container filled with gas (Fig. 16.9A). One wall of this container is a movable piston. Due to its pressure, the gas exerts a force on the piston, moving the piston to the right through a distance Δx (Fig. 16.9B). If F is the force exerted by the gas on the piston, the work done by the gas on the piston is

$$W = F\,\Delta x \qquad (16.5)$$

The piston is part of the environment of the gas (the piston might be connected to other components of an engine), so W is also the work done by the gas on its environment.

If the pressure in the gas is P and the area of the piston is A, the force on the piston is $F = PA$ and the work done by the gas on the piston is (using Eq. 16.5)

$$W = F\,\Delta x = PA\,\Delta x$$

The product $A\,\Delta x$ equals the change in volume of the gas ΔV, so we have

$$W = P\,\Delta V \tag{16.6}$$

We have derived this result for a gas in a piston-type container, but the result is very general. The only restriction on Equation 16.6 is that the force and hence also the pressure are constant. Whenever the volume of a system changes by a small amount ΔV, the work done by the system is equal to $P\,\Delta V$.

We can apply this result to cases in which the volume changes by a large amount by viewing the process as plotted in Figure 16.10. Here the initial (i) and final (f) states of the system are indicated by points in the P–V plane. A particular thermodynamic process such as the expansion of the gas in Figure 16.9B corresponds to a line or curve in the P–V plane that connects the initial state in which the system has pressure P_i and volume V_i to the final state with pressure P_f and volume V_f. The entire process produces a large change in the volume, but we can also view it as a series of small changes and apply Equation 16.6 to each small change (Fig. 16.10A). The total work done during the entire process thus equals *the area under the corresponding curve in the P–V plane.*

In the hypothetical process in Figure 16.10A, the final volume is greater than the initial volume, so the gas expands. The force exerted by the system is thus parallel to Δx (Fig. 16.9A and B), and the work done by the gas on the environment is positive ($W > 0$). The work done by a system can also be negative. For W to be negative, the force exerted by the gas on the piston must be opposite to the piston's displacement. In Figure 16.9C, the piston moves to the left, which corresponds to compression of the gas, giving a negative value of ΔV in Equation 16.6. This situation can occur if there is an external force on the system as might be exerted by a person pushing on the piston. Such a process is shown in the P–V plane in Figure 16.10B. The magnitude of the work done by the system during this process is still equal to the area under the P–V curve, but we are traversing this curve from right to left, so ΔV is negative and hence W is negative.

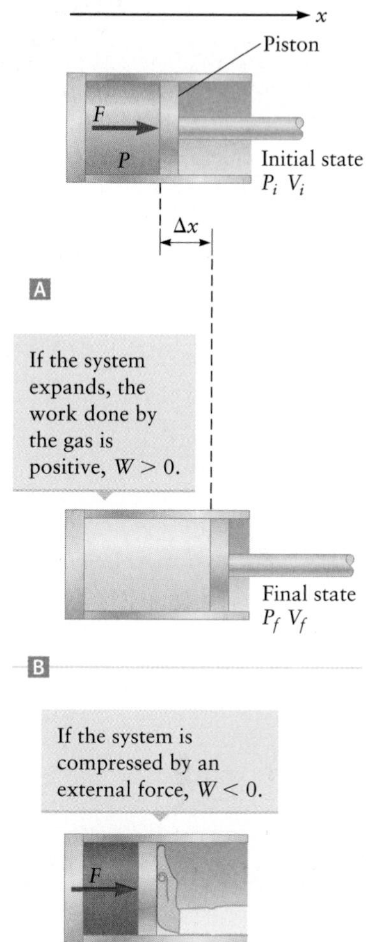

▲ **Figure 16.9** A gas expands as the piston moves to the right in going from A to B. The work done by the gas on the piston is thus positive, and $W > 0$ in Equation 16.5. C If an outside agent exerts a force on the piston so as to compress the gas, the work done *by the gas* is still $W = F\,\Delta x$ (where F is the force exerted by the gas on the piston), but now $W < 0$.

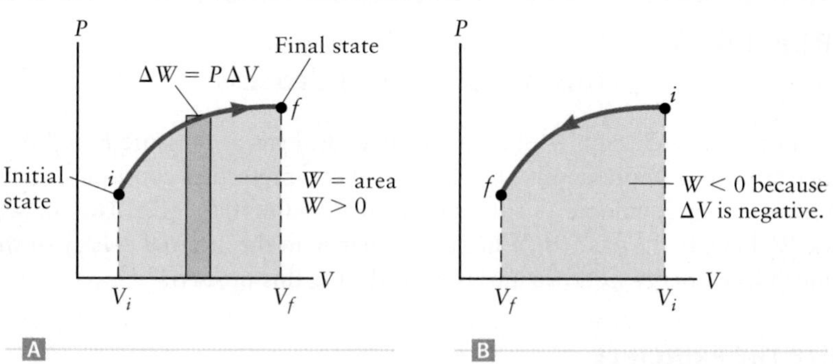

▲ **Figure 16.10** The work W is the area under the path in the P–V plane. A In this case, the gas expands ($V_f > V_i$) and does a positive amount of work on its environment ($W > 0$). B Here the gas is compressed ($V_f < V_i$), and $W < 0$.

CONCEPT CHECK 16.2 Interpreting a P–V Diagram

Consider a thermodynamic process involving an ideal gas that is described by the P–V diagram in Figure 16.11. Which of the following statements is true?

(a) The gas contracts, and W (the work done by the gas) is negative.
(b) The gas pressure drops, so W is negative.
(c) The gas expands, and W is positive.

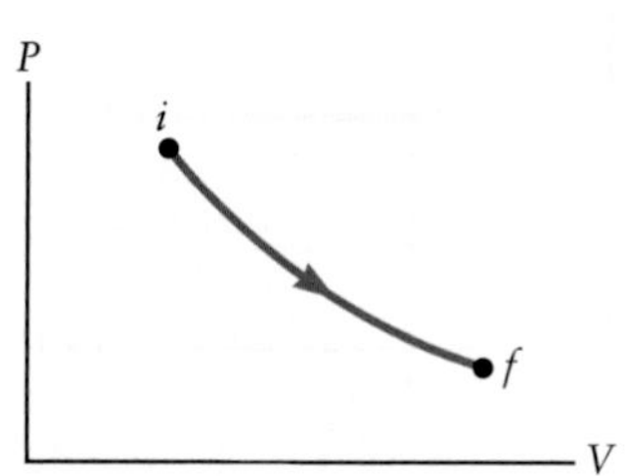

▲ **Figure 16.11** Concept Check 16.2.

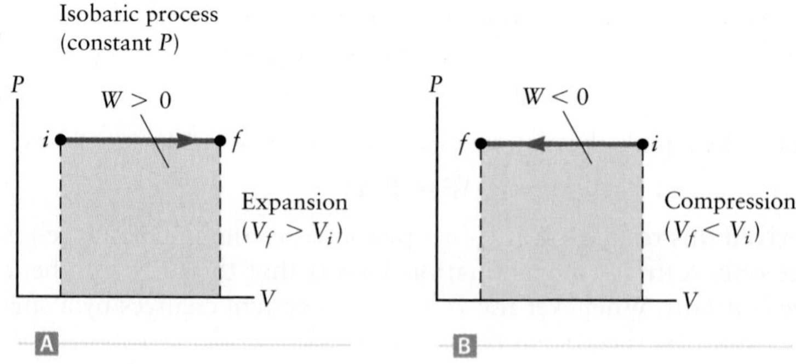

▲ **Figure 16.12** A P–V diagram for an isobaric (constant pressure) expansion. B An isobaric compression.

Examples of Thermodynamic Processes

Thermodynamic processes are classified according to how quantities such as P, V, and T change or remain constant during the course of the process.

Changes at Constant Pressure (Isobaric). If the pressure is constant during a process, the process is called ***isobaric***. When viewed on a P–V diagram, an isobaric process is a horizontal line as shown in Figure 16.12. In the hypothetical process shown in Figure 16.12A, the system expands and the work done by the gas is positive. The work W is equal to the area under the P–V curve, and because P is constant, we can write

$$W = P\,\Delta V = P(V_f - V_i) \quad \text{(isobaric process)} \tag{16.7}$$

For an expansion, $V_f > V_i$ so W is positive, whereas for a contraction, $\Delta V < 0$ and the work done is negative (Fig. 16.12B). An isobaric process may involve some heat flow, as we illustrate in Example 16.3.

EXAMPLE 16.3 Analyzing an Isobaric (Constant-Pressure) Process

A cylinder contains $n = 5.0$ moles of a monatomic ideal gas at pressure $P = 2.0 \times 10^5$ Pa (about twice atmospheric pressure). The gas is then compressed at constant pressure (isobarically) from an initial volume $V_i = 0.090$ m^3 to $V_f = 0.040$ m^3 (Fig. 16.13). (**a**) What is the work W done by the gas? (**b**) What is the change in the internal energy of the gas? (**c**) How much heat energy Q flows into the gas during this process?

RECOGNIZE THE PRINCIPLE

We can find the work done by the gas using $W = P\,\Delta V$ (Eq. 16.7) along with the given initial and final volumes. The change in the internal energy can be calculated using $U = \frac{3}{2}nRT$ for an ideal gas (Eq. 16.2) together with the initial and final temperatures, which can be found using the ideal gas law. When we know W and ΔU, we can use the first law of thermodynamics to get Q.

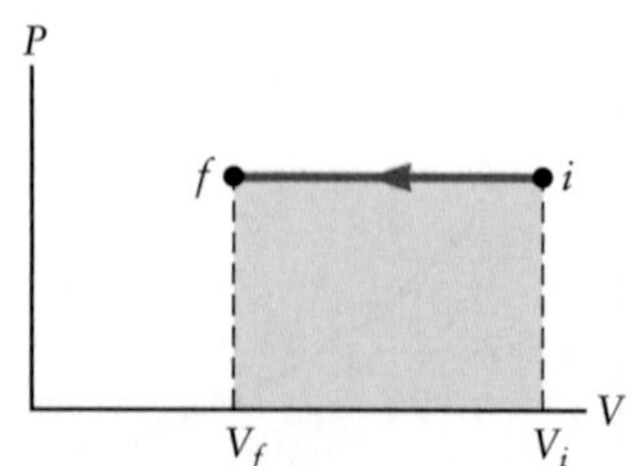

▲ **Figure 16.13** Example 16.3.

SKETCH THE PROBLEM

A P–V diagram describing the process is given in Figure 16.13. The work done by the gas is the area under the curve that connects the initial (i) and final (f) states.

IDENTIFY THE RELATIONSHIPS AND SOLVE

(a) The magnitude of W is the area under the P–V curve in Figure 16.13. We find

$$W = P\,\Delta V = P(V_f - V_i) = (2.0 \times 10^5 \text{ Pa})(0.040 \text{ m}^3 - 0.090 \text{ m}^3)$$

$$W = \boxed{-1.0 \times 10^4 \text{ J}}$$

Because $V_f < V_i$, ΔV is negative and W is negative.

(b) The internal energy of an ideal gas is

$$U = \tfrac{3}{2}nRT \tag{1}$$

so to find the change in U, we must first calculate the initial and final temperatures. Rearranging the ideal gas law $PV = nRT$ to solve for T gives

$$T_i = \frac{PV_i}{nR} = \frac{(2.0 \times 10^5 \text{ Pa})(0.090 \text{ m}^3)}{(5.0 \text{ moles})[8.31 \text{ J/(mole}\cdot\text{K)}]} = 430 \text{ K}$$

For the final temperature, we have

$$T_f = \frac{PV_f}{nR} = \frac{(2.0 \times 10^5 \text{ Pa})(0.040 \text{ m}^3)}{(5.0 \text{ mole})[8.31 \text{ J/(mole}\cdot\text{K)}]} = 190 \text{ K}$$

Using these values with Equation (1) leads to

$$\Delta U = \tfrac{3}{2}nR\,\Delta T = \tfrac{3}{2}nR(T_f - T_i)$$

$$\Delta U = \tfrac{3}{2}(5.0 \text{ moles})[8.31 \text{ J/(mole}\cdot\text{K)}](190 \text{ K} - 430 \text{ K}) = \boxed{-1.5 \times 10^4 \text{ J}}$$

(c) According to the first law of thermodynamics,

$$\Delta U = Q - W$$

Using the values of W and ΔU from parts (a) and (b), we find

$$Q = \Delta U + W = (-1.5 \times 10^4 \text{ J}) + (-1.0 \times 10^4 \text{ J}) = \boxed{-2.5 \times 10^4 \text{ J}}$$

► ***What does it mean?***

In an isobaric process, the pressure stays constant and the work can be found using $W = P\,\Delta V$. The temperature may or may not change, depending on the value of W and on how much heat Q flows into or out of the system.

Changes at Constant Temperature (Isothermal). Consider the expansion of an ideal gas in Figure 16.14A, in which the gas moves the piston to the right. Let's also assume the walls of the container are in contact with a thermal reservoir (surrounded with air at room temperature or immersed in water at a particular temperature). Contact with the thermal reservoir keeps the temperature of the gas constant during the course of the expansion. A thermodynamic process in which the temperature is constant is called an ***isothermal process.***

Even though the temperature is constant in an isothermal process, the pressure and volume can both change. Figure 16.14B shows an isothermal process for an ideal gas, drawn as a path on a P–V diagram. The ideal gas law is $PV = nRT$, so if the temperature is constant, the pressure and volume are related by

$$P = \frac{nRT}{V} \propto \frac{1}{V} \quad \text{(isothermal process)} \tag{16.8}$$

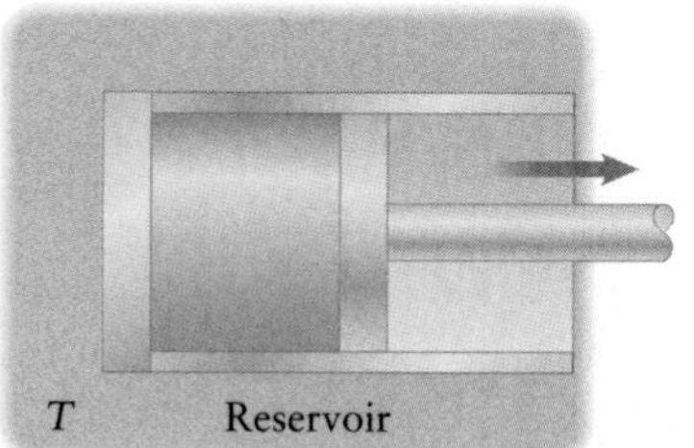

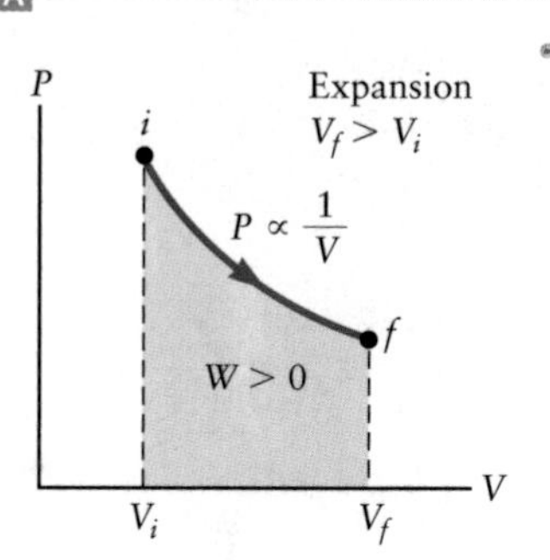

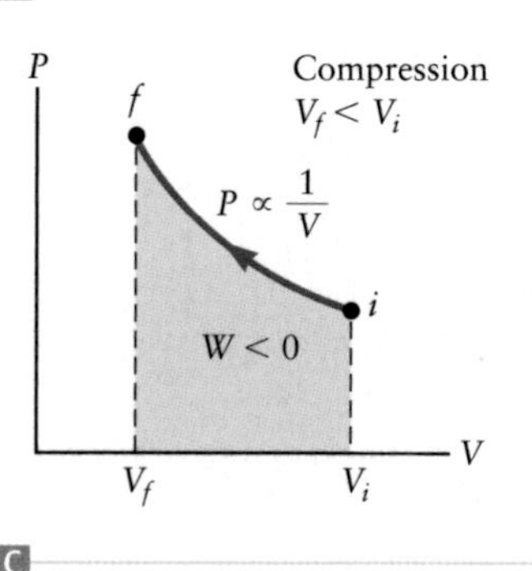

▲ **Figure 16.14** A In an isothermal process, the temperature of the system is held fixed, usually because it is in contact with a thermal reservoir. B Isothermal expansion of an ideal gas in the P–V plane. C Isothermal compression.

which is the curve plotted in Figure 16.14B. The work done by the gas is again equal to the area under the P–V curve; for an expansion described by Equation 16.8 with T held constant, the result is[2]

$$W = nRT\ln\left(\frac{V_f}{V_i}\right) \quad \text{(isothermal process)} \tag{16.9}$$

Here, "ln" is the natural logarithm function, which is reviewed in Appendix B. In an expansion the volume increases, so $V_f > V_i$ and the logarithmic term in Equation 16.9 is positive. Hence, the work done by the gas is positive in this case as we would expect from Figure 16.14B and from our previous discussions of the work done by an expanding gas. Conversely, for an isothermal compression (Fig. 16.14C), the factor $\ln(V_f/V_i)$ in Equation 16.9 is negative and the work done by the gas is negative.

EXAMPLE 16.4 Analyzing an Isothermal (Constant-Temperature) Process

Consider the isothermal expansion of an ideal gas sketched in Figure 16.14B. The system contains $n = 4.0$ moles of argon gas at room temperature. If the system then expands to double its initial volume, what are (**a**) the work done by the system and (**b**) the amount of heat that flows into the system from its surroundings?

RECOGNIZE THE PRINCIPLE

Our discussion of isothermal processes has emphasized the work done (Eq. 16.9), but there may also be heat flow either into or out of the system during the process. In this example, the system expands to fill a greater volume. For the gas to maintain a constant temperature (because we have an isothermal process), some heat Q must flow into the gas from its environment. We can calculate Q using the first law.

SKETCH THE PROBLEM

Figure 16.14B describes an isothermal expansion. The work W is the area under the P–V curve, and because the gas expands, W is positive.

(a) IDENTIFY THE RELATIONSHIPS

The work done by an ideal gas in an isothermal process is given by Equation 16.9:

$$W = nRT\ln\left(\frac{V_f}{V_i}\right)$$

The system is at room temperature ($T \approx 293$ K) and the final volume is twice the initial volume ($V_f = 2V_i$), so

$$W = (4.0 \text{ moles})[8.31 \text{ J/(mole} \cdot \text{K)}](293 \text{ K})\ln\left(\frac{2V_i}{V_i}\right)$$

SOLVE

Evaluating this expression for W gives

$$W = \boxed{6800 \text{ J}}$$

[2]This result can be derived by using calculus to compute the area under the curve specified by Equation 16.8.

(b) IDENTIFY THE RELATIONSHIPS

To find the heat that flows into the gas from its environment during this expansion, we use the first law of thermodynamics,

$$\Delta U = Q - W$$

The internal energy of a monoatomic ideal gas (such as argon) is $U = \frac{3}{2}nRT$. Because we are dealing with an isothermal process, the final temperature is equal to the initial temperature, and there is no change in the internal energy.

SOLVE

We thus have $\Delta U = 0$, and the first law of thermodynamics gives

$$\Delta U = 0 = Q - W$$

Using the result for W from part (**a**), we find

$$Q = W = \boxed{6800\ \text{J}}$$

▸ *What does it mean?*

In this example, the internal energy of the system does not change. Hence, as expected from the principle of conservation of energy, the energy added to the system (Q) must equal the energy that leaves the system (W).

Changes with $Q = 0$ (Adiabatic). Sometimes, a system is thermally isolated from its surroundings as shown schematically in Figure 16.15A. For example, the system might be the liquid in a thermos bottle. The wall of a thermos bottle contains a vacuum, making the heat flow through the wall extremely small. If there is no heat flowing into or out of a system during a thermodynamic process, then $Q = 0$ and the process is called ***adiabatic***.

To analyze an adiabatic process for a particular system, we need to know the relation between temperature, pressure, and volume. Let's again consider an example with an ideal gas (because we know its internal energy function). The path for an adiabatic process in the P–V plane can be calculated using the ideal gas law and the expression for the internal energy (Eq. 16.2). We will omit the derivation, but the result is

$$PV^{\gamma} = \text{constant}$$

where $\gamma = \frac{5}{3}$ for a dilute monatomic gas.[3] The pressure thus varies as

$$P \propto \frac{1}{V^{\gamma}} \quad \text{(adiabatic process)} \tag{16.10}$$

On a P–V diagram, an adiabatic path (Fig. 16.15B) is steeper than the path for an isothermal process that begins from the same initial state. (Recall that an isothermal path follows $P \propto 1/V$; Eq. 16.8.)

The value of W for an adiabatic process is again equal to the area under the P–V curve (Fig. 16.15B). According to the first law of thermodynamics, $\Delta U = Q - W$, and because $Q = 0$ for an adiabatic process, we have

$$\Delta U = -W \quad \text{(adiabatic process)} \tag{16.11}$$

[3]The parameter γ is the ratio of the specific heats at constant pressure and constant volume. This ratio depends on the type of molecules in the gas. For a monatomic gas, $\gamma = \frac{5}{3}$, and for a diatomic gas, $\gamma = \frac{7}{5}$.

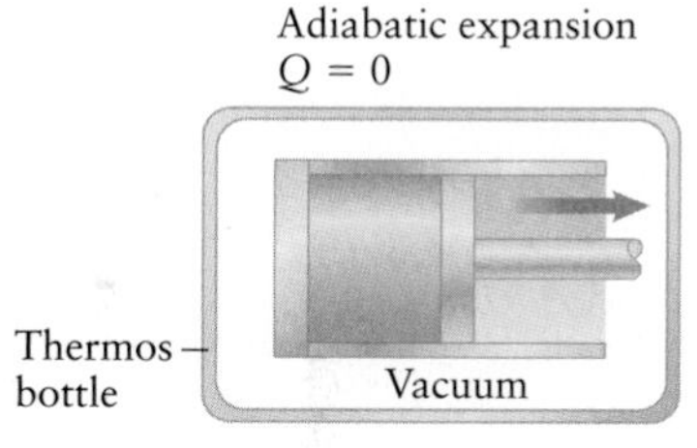

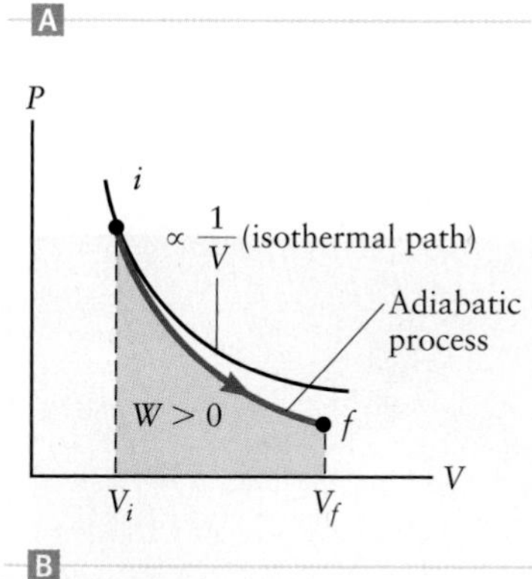

▲ **Figure 16.15** A In an adiabatic process, no heat flows into or out of the system ($Q = 0$). B Adiabatic expansion of an ideal gas in the P–V plane.

Isochoric (constant-volume) process: $V_i = V_f$

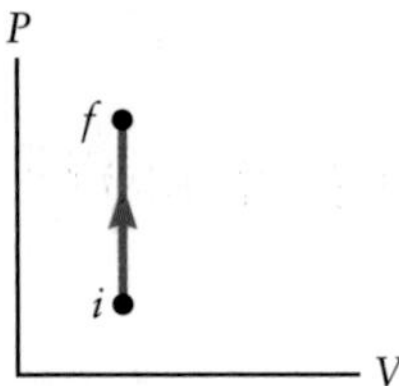

▲ **Figure 16.16** In an isochoric process, the volume does not change.

A hypothetical adiabatic process is the expansion of a gas against a piston. As the gas expands, it does a positive amount of work on the piston, and $W > 0$. According to Equation 16.11, the internal energy of the gas must decrease. For a monatomic ideal gas, $U = \frac{3}{2}Nk_BT$, meaning that the temperature decreases; that is, the gas cools. Although $Q = 0$ in an adiabatic process, other properties of the system such as the temperature can change during the process.

Changes at Constant Volume (Isochoric). An ***isochoric*** process is one for which the volume does not change; it is also called an ***isovolumetric*** process. The corresponding path on a P–V plot is shown in Figure 16.16. A key feature of all isochoric processes is discussed in Example 16.5.

CONCEPT CHECK 16.3 Heating a Gas at Constant Volume

Figure 16.17 shows P–V plots for two different isochoric (constant-volume) processes. In one case, the pressure increases, whereas in the other, the pressure decreases. In which case does the temperature increase?

▶ **Figure 16.17** Concept Check 16.3.

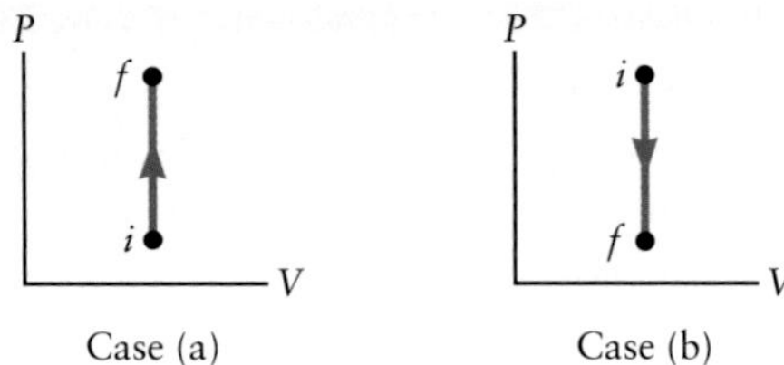

EXAMPLE 16.5 A Constant-Volume (Isochoric) Process

Consider an ideal gas that is heated under isochoric conditions. What is the work done by the gas?

RECOGNIZE THE PRINCIPLE

Because the volume is kept fixed, an isochoric process is a vertical line in the P–V plane.

SKETCH THE PROBLEM

The P–V diagram for this process is shown in Figure 16.16. In this example, the gas is heated, so the final pressure is greater than the initial pressure.

IDENTIFY THE RELATIONSHIPS AND SOLVE

The work done by the gas is the area under this curve, which is zero. So, for this process (and for *all* isochoric processes), $\boxed{W = 0}$.

▶ ***What does it mean?***

When analyzing a thermodynamic process, it is always useful to consider the P–V diagram. The work done by the system is equal to the area under the P–V curve.

CONCEPT CHECK 16.4 What Kind of Process Is It?

A bear has a very thick fur coat, providing very good thermal insulation from its environment (Fig. 16.18). When the bear hibernates, is it an example of an isothermal or an adiabatic process, or both? Explain.

▲ **Figure 16.18** Concept Check 16.4.

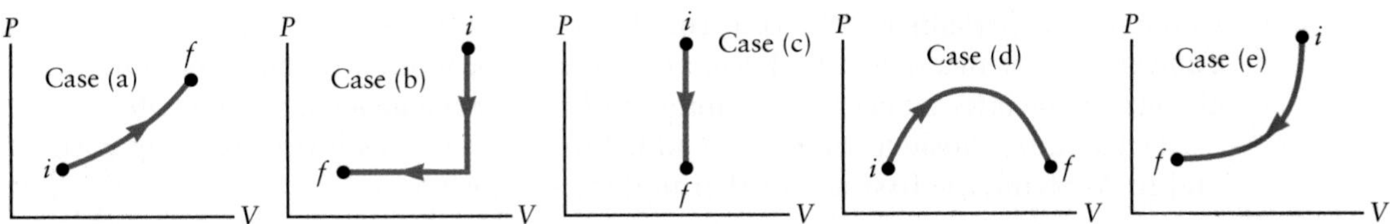

▲ **Figure 16.19** Concept Check 16.5.

CONCEPT CHECK 16.5 Work Done in a Thermodynamic Process

Figure 16.19 shows several thermodynamic processes. Without doing a calculation, determine if W is positive, negative, or zero for each process.

Some Properties of Q and W

A thermodynamic process can be described by a path on the P–V diagram that takes the system from some initial state i to a final state f. Many paths connect a particular initial state to a given final state; Figure 16.20 shows two paths that connect states i and f. One of these paths passes through point A, and we'll refer to it as "path A." The other passes through point B ("path B"). Let's consider how ΔU, W, and Q compare for these two paths, that is, for these two different thermodynamic processes.

We can write the change in internal energy as

$$\Delta U = U_f - U_i \tag{16.12}$$

In words, the change in U depends only on the internal energies of the initial and final states, so ΔU is the same for *any* path that connects states i and f. This result is reminiscent of our work in Chapter 6, where we learned that changes in the potential energy are independent of the path taken.

The work done in going from state i to state f along path A is the sum of the work done in going from i to A and the work done in going from A to f; hence,

$$W_A = W_{iA} + W_{Af}$$

Here W_{iA} denotes the work done on the path from i to A in Figure 16.20 and so forth for W_{Af}. Work is the area under the P–V curve, so

$$W_{iA} = P\,\Delta V = P_i(V_A - V_i)$$

The work along the portion of the path from A to f is zero ($W_{Af} = 0$) because there is no change in the volume. The total work done along path A is thus

$$W_A = W_{iA} + W_{Af} = P_i(V_A - V_i)$$

The volume is the same at point A and f ($V_A = V_f$); hence,

$$W_A = P_i(V_f - V_i) \tag{16.13}$$

We can find the work done along path B in a similar way:

$$W_B = W_{iB} + W_{Bf}$$

with $W_{iB} = 0$ because there is no volume change on this portion of the path, and

$$W_{Bf} = P_f(V_f - V_B)$$

The result for the work on path B is then

$$W_B = W_{iB} + W_{Bf} = P_f(V_f - V_B)$$

Using $V_B = V_i$, this expression becomes

$$W_B = P_f(V_f - V_i) \tag{16.14}$$

The change in internal energy is the same: $\Delta U_A = U_f - U_i = \Delta U_B$ **but** $W_A \neq W_B$ and $Q_A \neq Q_B$

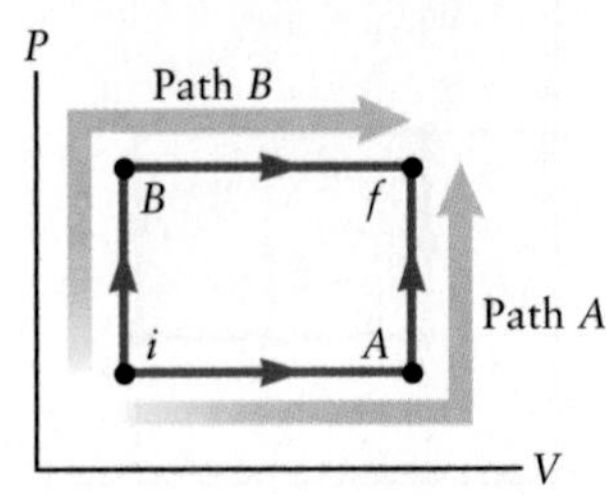

▲ **Figure 16.20** Two different thermodynamic processes (two paths in the P–V plane) can connect the same initial and final states.

Comparing the results for W_A (Eq. 16.13) and W_B (Eq. 16.14), we see that the work done by the system along path A is *not* the same as the work done along path B. Unlike changes in the internal energy, the amount of work done *depends on the path*.

We can also consider the heat Q added to the system as it travels along paths A and B. Applying the first law of thermodynamics, we have

$$\Delta U_A = Q_A - W_A \quad \text{(along path } A\text{)} \tag{16.15}$$

$$\Delta U_B = Q_B - W_B \quad \text{(along path } B\text{)} \tag{16.16}$$

We have already seen that changes in the internal energy are independent of the path, so $\Delta U_A = \Delta U_B$. We have also just shown that the work done is different along the two paths, so $W_A \neq W_B$. Using these facts with Equations 16.15 and 16.16, we see that

$$Q_A \neq Q_B$$

In words, the heat added to the system also *depends on the path* taken.

The principle of conservation of energy thus leads us to the following conclusions.

Properties of U, W, and Q for thermodynamic processes

1. **The internal energy of a system depends only on the current state of the system. Changes in the internal energy are thus independent of the path taken in a thermodynamic process. The change in internal energy ΔU depends only on the internal energies of the initial and final states.**
2. **The work done during a thermodynamic process depends on the path taken. Even with the same initial and final states, two different thermodynamic paths can have different values of W.**
3. **The heat added to a system during a thermodynamic process depends on the path taken. Even with the same initial and final states, two different thermodynamic paths can have different values of Q.**

EXAMPLE 16.6 Analyzing a Cyclic Process

Consider the thermodynamic process described in Figure 16.21. This P–V diagram is similar to the one in Figure 16.20 except that now we begin and end at the same point (the same state). The system in Figure 16.21 starts in state 1, moves as indicated to state 2, state 3, and state 4, and then moves back to state 1. This set of changes is called a ***cyclic process*** because the system returns to its initial state. Find the work done during this process.

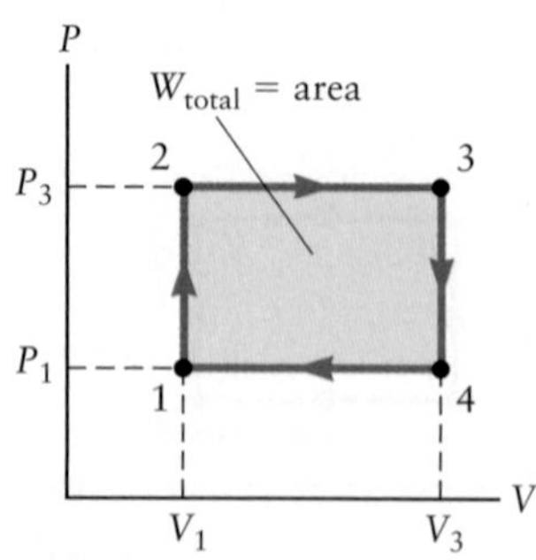

▲ **Figure 16.21** Example 16.6. A cyclic process is one that returns to its original state.

RECOGNIZE THE PRINCIPLE

We can use $W = P\,\Delta V$ (Eq. 16.6) to find the work done along each portion of the path. Adding those results gives the total work done for the entire process.

SKETCH THE PROBLEM

Figure 16.21 shows the P–V diagram.

IDENTIFY THE RELATIONSHIPS

The volume does not change along the path from state 1 to state 2 or along the path from state 3 to state 4, so $\Delta V = 0$ and the work done in these parts of the process is zero. Using $W = P\,\Delta V$ for the path from state 2 to state 3 gives

$$W_{23} = P\,\Delta V = P_3(V_3 - V_1)$$

Likewise, along the path from state 4 to state 1, we get

$$W_{41} = P_1(V_1 - V_3)$$

Notice that W_{41} is negative because $V_1 < V_3$.

SOLVE

Combining these results gives

$$W_{total} = W_{23} + W_{41} = P_3(V_3 - V_1) + P_1(V_1 - V_3) = \boxed{(P_3 - P_1)(V_3 - V_1)}$$

► *What does it mean?*

From the geometry of Figure 16.21, W_{total} is just the area enclosed by the path in the P–V diagram.

Work Done by a Cyclic Process

Example 16.6 considered the work done in a cyclic process and showed that W is equal to the area enclosed by the path on the P–V diagram. The process analyzed in that example had an especially simple path, but the general result applies to all cyclic processes. The work done in any cyclic process equals the area enclosed by the path in the corresponding P–V diagram (Fig. 16.22). We'll use this result when we analyze the performance of heat engines and other devices in Section 16.6.

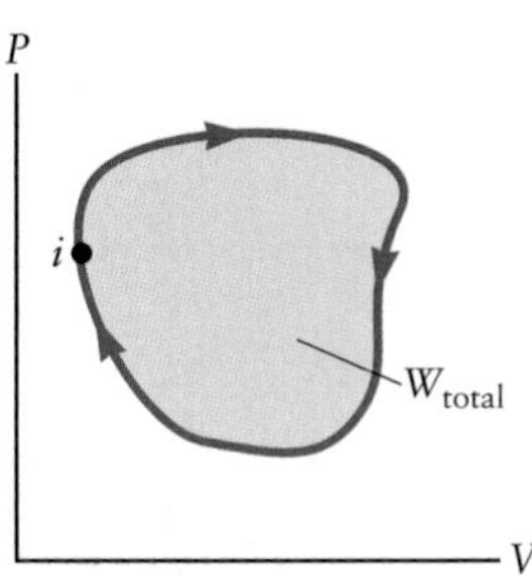

▲ **Figure 16.22** In any cyclic thermodynamic process, the work done by the system equals the area enclosed by the path on the P–V diagram.

CONCEPT CHECK 16.6 Thermodynamic Processes

Figure 16.23 shows several thermodynamic processes. Which of these processes could be (a) an isothermal expansion, (b) an adiabatic compression, (c) an isobaric expansion, and (d) an isobaric compression?

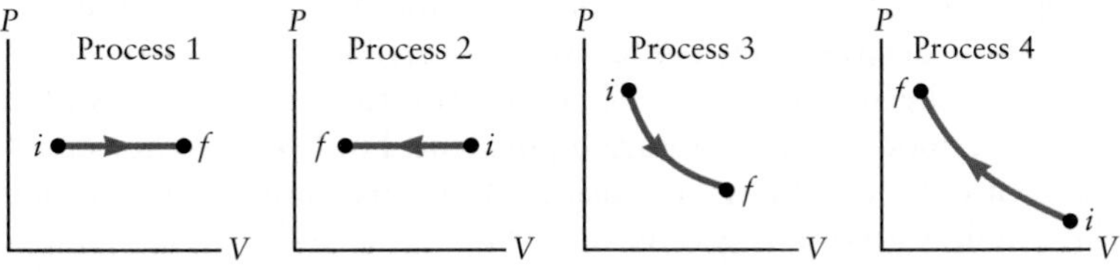

◄ **Figure 16.23** Concept Check 16.6.

16.5 The Second Law of Thermodynamics

Figure 16.24A shows an isothermal expansion of a dilute gas in which the system starts in an initial state i and moves to a final state f. The piston moves to the right as the gas does some amount of work $W > 0$, and during this time the gas absorbs some heat $Q > 0$ from the environment. This example shows a thermodynamically ***reversible*** process. We could imagine a reversed process in which the gas begins in state f and is isothermally compressed back to state i (Fig. 16.24B). During this

Reversible versus irreversible processes

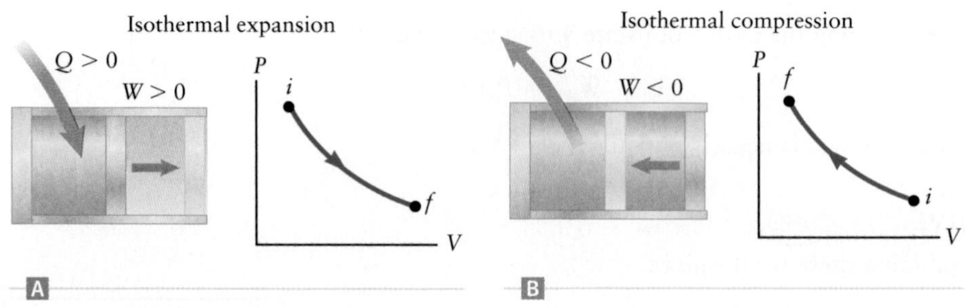

▶ **Figure 16.24** Schematic of a thermodynamically reversible process. A Isothermal expansion from state i to state f. B Isothermal compression from state f back to state i.

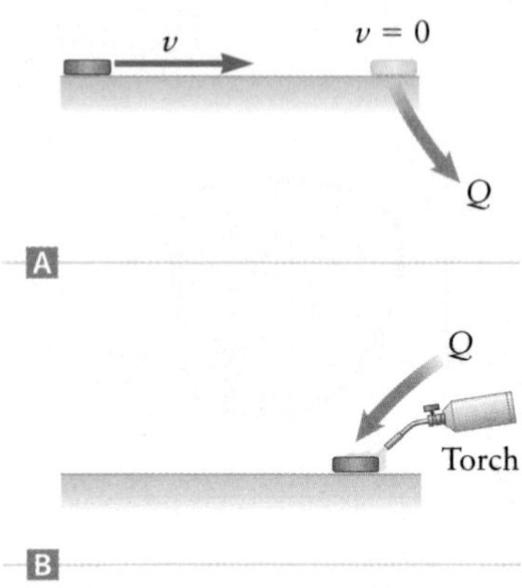

▲ **Figure 16.25** A A hockey puck sliding across a floor with friction is an example of a thermodynamically irreversible process. B Heating a hockey puck will *not* cause it to move back to its original position!

compression, the gas does a negative amount of work ($W < 0$) on the piston because an outside agent must push on the piston to compress the gas. At the same time, heat would flow out of the gas into the surroundings. The work done by the gas during the compression is equal in magnitude but opposite in sign compared with W during the expansion. Also, the amount of heat that flows out of the gas during compression is equal in magnitude to the heat that flows in during expansion. Hence, both the system and the environment are brought back to precisely their original states, which is the key feature of a thermodynamically reversible process.

An example of an ***irreversible*** process is sketched in Figure 16.25. Here a hockey puck is initially sliding to the right on a horizontal surface. There is some friction between the puck and the surface, causing the puck to come to a stop as all its initial kinetic energy is converted to heat energy. This heat energy flows from the region where the puck is in contact with the horizontal surface into its surroundings (Fig. 16.25A). Suppose we now try to reverse this process by arranging for heat to flow from the surroundings into the puck, perhaps with a torch as sketched in Figure 16.25B. Intuition tells us that this will not cause the puck to begin moving! Of course, we could start the puck in motion again by striking it with a hockey stick, but for a process to be thermodynamically reversible, it must be possible to return both the system (the puck) and its surroundings (the rest of the universe) back to their original states. There is nothing we could do to reverse the process in Figure 16.25; it is simply *not possible* to take the puck and the rest of the universe back to precisely their original states. This is the essential feature of an irreversible process. And, as in Figure 16.25, irreversible processes usually involve friction.

The distinction between reversible and irreversible processes is the subject of the second law of thermodynamics. There are several different ways of stating the second law of thermodynamics; all are equivalent, but because they emphasize different aspects of the second law, it is useful to discuss more than just one of them. Perhaps the simplest version is given as follows.

Second law of thermodynamics

Second law of thermodynamics

Heat flows spontaneously from a warm object to a colder one. It is not possible for heat to flow spontaneously from a cold object to a warmer one.

This statement certainly seems obvious. The part about heat flowing from a warm object to a colder one is implied by the zeroth law of thermodynamics. Two crucial aspects of the second law go beyond the zeroth law, however. The first is the notion of a *spontaneous* process. This is a process that takes place "on its own" without work done by any added or external forces. For example, the flow of heat from the surroundings into the ice cube in Figure 16.8 is a spontaneous process. The other crucial part of the second law of thermodynamics is the statement that heat cannot flow spontaneously from a cold object to a warmer one. This statement leads to many important results, including the impossibility of building a perpetual motion machine.

16.6 Heat Engines and Other Thermodynamic Devices

One of the discoverers of the second law of thermodynamics was an engineer named Sadi Carnot (1796–1832), who was interested in how the science of thermodynamics could be applied to practical devices such as steam engines and other marvels of the Industrial Revolution that were being developed around that time. Carnot's engineering approach leads to a rather different way of thinking about the second law of thermodynamics.

Imagine you are an engineer and you have been asked to invent the world's best heat engine. A heat engine takes heat energy, as might be generated by a coal-burning furnace, and converts this energy into work. For example, this work might rotate a shaft as would be useful in a steamboat, locomotive, or automobile. With that in mind, we can draw a heat engine in the highly schematic form shown in Figure 16.26A. An amount of heat Q_H is extracted from the furnace and fed into some sort of mechanical device, which then does an amount of work W. It is your engineering challenge to design the mechanical device. The principle of conservation of energy and the first law of thermodynamics tell us that the work done cannot be larger than the energy input Q_H, but you might hope to design and patent a device that reaches this limit. The world's best heat engine would perform an amount of work $W = Q_H$, and from the principle of conservation of energy, you would be absolutely guaranteed that no one could design an engine better than that one.

That is the problem Carnot tackled. Although he was not able to construct the heat engine in Figure 16.26A, Carnot was able to show that such an engine is *impossible*! He showed that it is simply not possible to construct a device whose sole effect is to convert heat into work, that is, to convert a given amount of heat energy completely into mechanical energy. Instead, all heat engines must be diagrammed as in Figure 16.26B. Here we again have a mechanical device that takes an amount of heat Q_H from a hot reservoir and does an amount of work W. However, Carnot showed that all real heat engines must also expel some heat energy Q_C to a colder reservoir.

Applying conservation of energy to the heat engine in Figure 16.26B, we have

$$W = Q_H - Q_C \qquad (16.17)$$

Your goal as an engine designer is get the most work (the largest W) possible out of the engine, so we define the ***efficiency*** e of the engine as

$$e = \frac{W}{Q_H} \qquad (16.18)$$

Combining Equation 16.18 with Equation 16.17 gives

$$e = \frac{W}{Q_H} = \frac{Q_H - Q_C}{Q_H}$$

$$e = 1 - \frac{Q_C}{Q_H} \qquad (16.19)$$

When comparing heat engines, a larger value of e means that the engine converts a larger fraction of the input energy Q_H into work.

Carnot's Engine

To maximize the efficiency e, one must make Q_C as small as possible, which makes W as large as possible. Carnot considered this problem for a heat engine that makes

Imaginary heat engine (NOT possible)

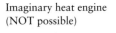

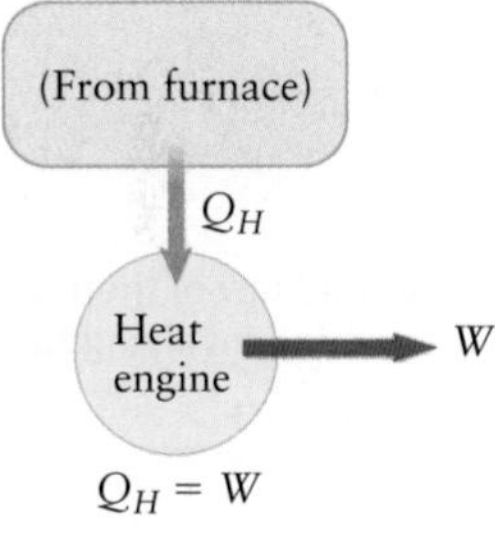

Real heat engines all expel some heat to a cold reservoir.

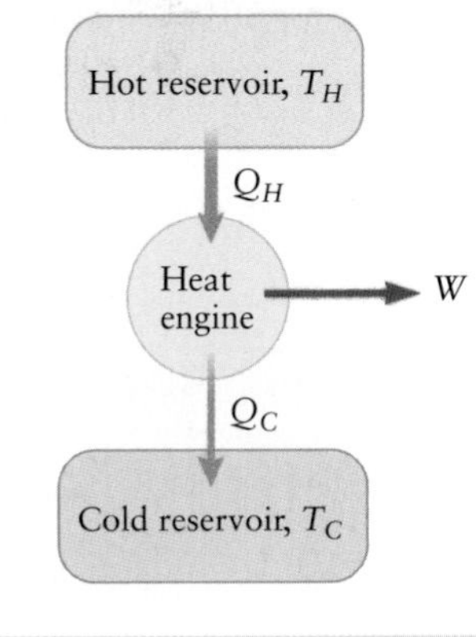

▲ **Figure 16.26** Schematic diagrams for a heat engine. A A hypothetical engine, which the laws of thermodynamics tell us is *not* possible. B All real heat engines must expel some heat to a cold reservoir.

use of the reversible compression and expansion of an ideal gas. Because Carnot's engine uses only reversible processes, it is an example of a reversible heat engine. He was able to prove that all reversible heat engines have the same efficiency as his particular design. Carnot also showed that for a hot reservoir at temperature T_H (providing the heat input Q_H) and a cold reservoir at temperature T_C, the heat expelled to the cold reservoir satisfies

$$\frac{Q_C}{Q_H} = \frac{T_C}{T_H} \tag{16.20}$$

Here the temperatures T_H and T_C must be measured in Kelvin units. Combining Equation 16.20 with the definition of efficiency (Eq. 16.19) leads to

Efficiency of a reversible heat engine

$$e = 1 - \frac{T_C}{T_H} \tag{16.21}$$

Equation 16.21 is an amazing result: it applies to *all* reversible engines, no matter how they are constructed. Carnot did not stop here, but discovered one more crucial point: no engine can have an efficiency that is better than the efficiency of a reversible engine. In practice, all heat engines will always be irreversible to some extent, so Carnot's result sets an *absolute limit* on the efficiency of all real heat engines. Carnot's results are the basis for an alternative statement of the second law of thermodynamics.

Second law of thermodynamics (alternative form)

The efficiency of a reversible heat engine is given by

Second law of thermodynamics: formulation involving heat engines

$$e = 1 - \frac{T_C}{T_H}$$

No heat engine can have a greater efficiency than this.

We have now seen two statements of the second law, this one involving the efficiency of heat engines and another (in Section 16.5) involving the direction of heat flow. Although it is not obvious, these two statements are *completely equivalent.* Given either one of them as a starting point, the other one can be derived!

The result in Equation 16.21 shows that the efficiency of an engine depends on the temperatures of the reservoirs. If the temperature of the cold reservoir is zero ($T_C = 0$ K), the efficiency $e = 1$. In this case, we also have $Q_C = 0$, and it would thus seem that we have created the hypothetical heat engine in Figure 16.26A—a heat engine that expels no heat—which we said was impossible. One way out of this apparent contradiction is to hypothesize that it is impossible to lower the temperature of a reservoir to $T = 0$ K, the temperature called absolute zero. We'll see this hypothesis again in Section 16.8 when we discuss the third law of thermodynamics.

EXAMPLE 16.7 Efficiency of a Steam Engine

Steamboats were used extensively in the 1800s to carry people and goods (Fig. 16.27). These boats used heat engines: a hot reservoir contained steam ($T_H \approx 200°C$) that was heated by burning wood or coal, and the cold reservoir was provided by the air or river water ($T_C \approx 10°C$). The engine turned a shaft that drove the boat's paddle wheels. Suppose one of these steam engines does $W = 10{,}000$ J of work each second. If it is a reversible heat engine, how much heat Q_H does it absorb from the hot reservoir in 1 s?

RECOGNIZE THE PRINCIPLE

The efficiency of any heat engine is defined as (Eq. 16.18)

$$e = \frac{W}{Q_H} \qquad (1)$$

For a reversible heat engine, Carnot showed that e is also equal to

$$e = 1 - \frac{T_C}{T_H} \qquad (2)$$

We can use these two relations to find the heat absorbed from the hot reservoir Q_H in terms of the given quantities W, T_H, and T_C.

▲ **Figure 16.27** Example 16.7. Steamboats in the 1800s used heat engines to drive their paddle wheels.

SKETCH THE PROBLEM

Figure 16.27 shows a picture of a real steamboat, but the diagram of a heat engine in Figure 16.26B is more useful for analyzing the operation of the engine.

IDENTIFY THE RELATIONSHIPS

Combining Equations (1) and (2) gives

$$\frac{W}{Q_H} = 1 - \frac{T_C}{T_H}$$

Rearranging to solve for Q_H, we have

$$Q_H = \frac{W}{1 - (T_C/T_H)} \qquad (3)$$

SOLVE

When using Equation (3), we must be careful to use values of the temperature on the Kelvin scale; Carnot's result in Equation (2) assumes the temperatures are measured on this scale. Converting the given values of T_H and T_C gives $T_H = (200 + 273)\text{ K} = 473\text{ K}$ and $T_C = (10 + 273)\text{ K} = 283\text{ K}$. Inserting these values of the reservoir temperatures and the given value of W into Equation (3), we find

$$Q_H = \frac{W}{1 - (T_C/T_H)} = \frac{10{,}000\text{ J}}{1 - (283\text{ K}/473\text{ K})} = \boxed{25{,}000\text{ J}}$$

► ***What does it mean?***

Because the engine did 10,000 J of work each second, the energy expelled to the cold reservoir in that time was $Q_C = Q_H - W = 15{,}000$ J. Most of the energy extracted from the hot reservoir was expelled to the cold reservoir and was thus "wasted." When analyzing the operation of a heat engine, temperature values should always be expressed in Kelvin units.

EXAMPLE 16.8 Ⓡ Efficiency of a Modern Automobile Engine

The internal combustion engine found in most cars is a type of heat engine (Fig. 16.28). A hot reservoir created by exploding gasoline provides heat energy, causing the gas in a piston to expand and do work as it rotates the driveshaft of the engine. Some heat is then expelled to a cold reservoir (the air) via the radiator and the exhaust. What is the maximum possible efficiency of such an engine?

(continued) ►

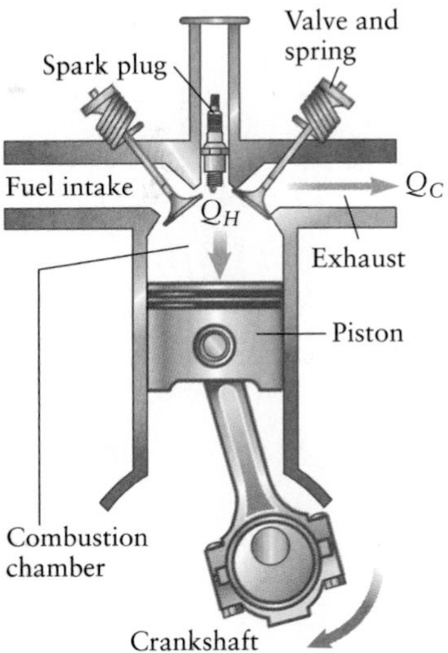

▲ **Figure 16.28** Example 16.8. A real automobile engine is a type of heat engine following the general schematic form in Figure 16.26B. The laws of thermodynamics place rigorous limits on the efficiency of all real heat engines.

Insight 16.1

EFFICIENCY OF A DIESEL ENGINE

A diesel engine is similar to a gasoline internal combustion engine (Example 16.8). One difference is that a gasoline engine ignites the fuel mixture with a spark from a spark plug, whereas a diesel engine ignites the fuel mixture purely "by compression" (without a spark). The compression of the fuel mixture is therefore much greater in a diesel engine, which leads to a higher temperature in the hot reservoir. According to Equation 16.21 and Example 16.8, this higher temperature gives a higher theoretical limit on the efficiency of a diesel engine.

RECOGNIZE THE PRINCIPLE

A real internal combustion engine is irreversible, so its efficiency will always be less than that of a reversible engine. This irreversibility is caused by friction between the moving parts inside the engine and by other factors. Nevertheless, an upper limit on the efficiency is set by the reversible case (Eq. 16.21)

$$e = 1 - \frac{T_C}{T_H} \tag{1}$$

To apply Equation (1), we need to know the temperatures of the two reservoirs T_C and T_H; these values are not given, so we need to use some common sense and knowledge about this type of engine to find approximate values.

SKETCH THE PROBLEM

The schematic heat engine in Figure 16.26B describes the problem. The corresponding hot and cold reservoirs for a real automobile engine are shown in Figure 16.28.

IDENTIFY THE RELATIONSHIPS

The cold reservoir (the radiator and exhaust) is at air temperature, so we estimate $T_C \approx 300$ K. For the hot reservoir, we can use reference data (from the Internet or elsewhere) to find that the temperature of exploding gas in a piston is about $T_H \approx 600$ K.

SOLVE

Inserting our estimates for the reservoir temperatures into Equation (1) gives

$$e = 1 - \frac{T_C}{T_H} = 1 - \frac{300\text{ K}}{600\text{ K}} = \boxed{0.5} \tag{2}$$

► ***What does it mean?***

Even in the best possible case, an automobile engine can only convert about half the chemical energy in the gasoline into useful work. This limit on the efficiency could be improved by increasing the temperature of the hot reservoir (the temperature in the piston), which can be accomplished by using a suitable blend of gasoline with a higher octane rating. Such high-performance engines and fuels are more expensive than typical engines and fuels, however.

In a real engine, a substantial amount of energy is lost to friction. Only about 25% of the chemical energy in the gasoline is typically converted into work.

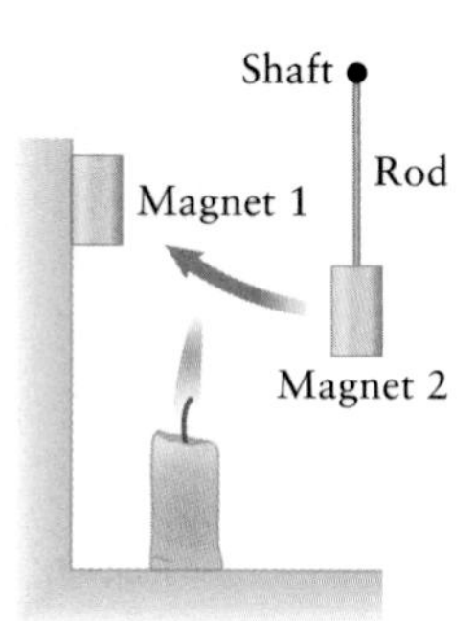

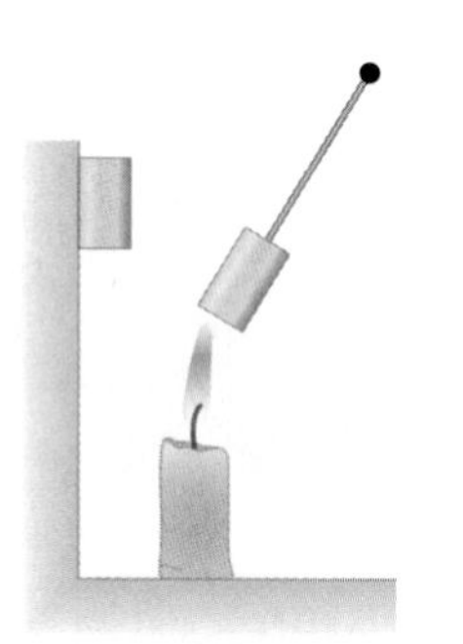

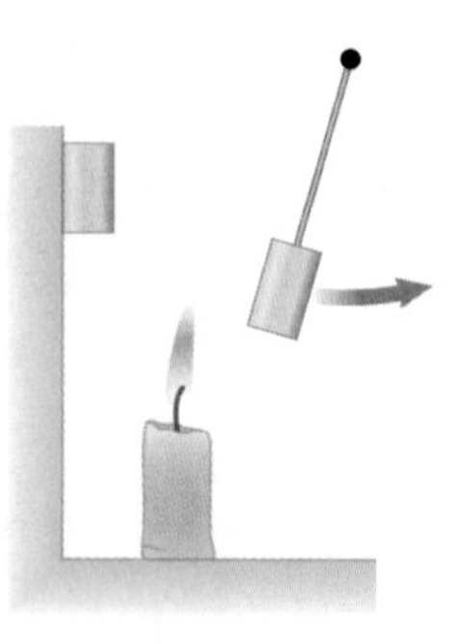

▲ **Figure 16.29** Example 16.9. A simple heat engine.

EXAMPLE 16.9 A Magnetic Heat Engine

Figure 16.29 shows a heat engine that uses two permanent magnets and a candle. Magnet 1 is attached to a wall, while magnet 2 swings on a rod that can take it over the candle's flame. Although we have not yet discussed permanent magnets (see Chapter 20), we can still analyze this engine; we only need to know that these two permanent magnets attract each other except when one of them is very hot, in which case the hot one ceases to behave as a permanent magnet. This heat engine works as follows. (1) Magnet 2 is initially far from the flame of the candle (Fig. 16.29A). Magnet 2 is attracted to magnet 1, causing the rod to rotate clockwise. (2) Magnet 2 eventually passes over the flame and is heated so that it is no longer attracted to magnet 1 (Fig. 16.29B), and the torque from gravity causes the rod to rotate back counterclockwise (Fig. 16.29C). (3) Magnet 2 then moves away from the flame, it cools to room temperature, and its permanent magnetism is restored. It is then attracted to magnet 1, and the cycle repeats. This system thus behaves as an engine that can do work via the motion of the rod. What are the hot and cold reservoirs of this heat engine?

RECOGNIZE THE PRINCIPLE

Without the heat from the candle, magnet 2 would always be attracted to magnet 1 and the rod would never swing back counterclockwise. At the same time, this engine would not work if magnet 2 were not able to cool back to room temperature, which it does through its interaction with the surrounding air.

SKETCH THE PROBLEM

Figure 16.29 shows the problem.

IDENTIFY THE RELATIONSHIPS AND SOLVE

The heat from the candle and the cool air are both necessary for this engine to work. The candle acts as the hot reservoir and the air acts as the cold reservoir.

► ***What does it mean?***

Many mechanical devices can be viewed as heat engines and analyzed using the principles discovered by Carnot. Room temperature is about 300 K. Although the temperature of a candle's flame depends on the type of wax and other factors, it is typically around 1500 K. The maximum efficiency of this heat engine is thus about $e = 1 - T_C/T_H = 1 - (300\text{ K})/(1500\text{ K}) = 0.8$. The actual efficiency will be less than this value because the magnet will probably not reach 1500 K when it is nearest the flame.

Perpetual Motion and the Second Law of Thermodynamics

The construction of a perpetual motion machine has been the quest of many inventors. Various laws of physics imply that such machines are impossible. One popular perpetual motion machine design sketched in Figure 16.30 makes use of the hypothetical (and impossible) heat engine in Figure 16.26A. The machine in Figure 16.30 takes some heat energy Q from a hot reservoir (the atmosphere) and converts it all to an amount of work W. This work is then used to power a car. As the car moves, it loses energy to friction due to air drag, the wheels, and so on. This loss causes all the energy W to be converted to heat energy, which returns to the atmosphere. The net effect is that the car has moved, whereas the energy in the atmosphere is back to its original value; hence, the process can be repeated, and the car can move on forever.

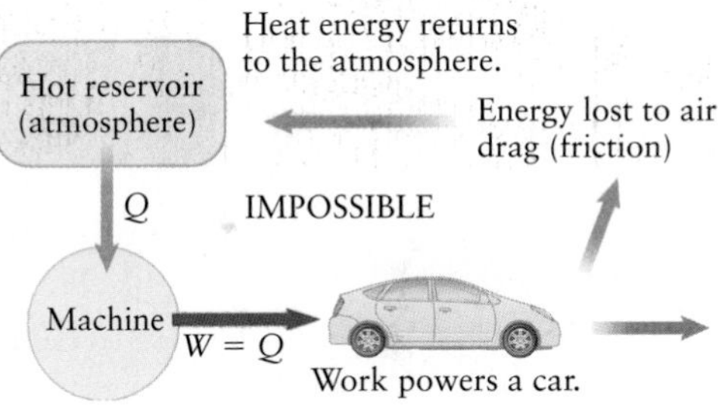

▲ **Figure 16.30** Hypothetical (and *impossible*) design of a perpetual motion machine. This machine is not possible because it violates the second law of thermodynamics.

Carnot's work and the second law of thermodynamics tell us that the perpetual motion machine in Figure 16.30 is *impossible* because a real heat engine must always expel some heat energy to a cold reservoir. Hence, all the heat energy Q extracted from the atmosphere cannot later be returned to the atmosphere. In this way, thermodynamics rules out many hypothetical perpetual motion designs. In fact, the second law along with conservation of energy principles rule out *all* perpetual motion machines.

The Carnot Cycle

We have so far represented heat engines in a highly schematic form as in the block diagram in Figure 16.26B and have ignored how one might construct a real engine. Carnot showed how to use an ideal gas and piston to make a reversible heat engine. Figure 16.31A shows the operation of Carnot's engine as a path in the P–V plane. The corresponding changes of the gas and piston are shown in Figure 16.31B, where

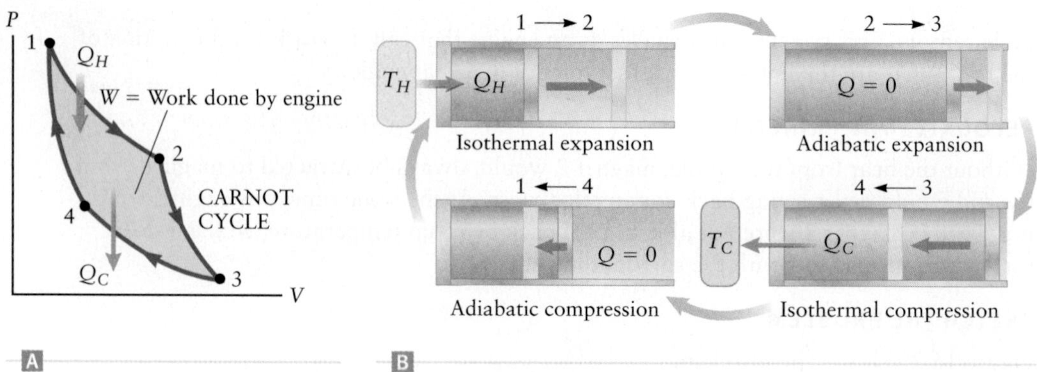

▲ **Figure 16.31** An example of a thermodynamically reversible heat engine. This engine uses a cyclic process called a Carnot cycle. A The work done during a Carnot cycle is equal to the area enclosed on the P–V diagram. B The four steps in the Carnot cycle.

we show only the mechanical part of the heat engine; the thermal reservoirs could be air or water held at temperatures T_H and T_C.

The system begins in state 1 on the P–V diagram. It then moves progressively to states 2, 3, and 4 and finally back to state 1.

Change from state 1 to state 2: The system is placed in contact with the hot reservoir, during which time the gas absorbs an amount of heat Q_H and expands to state 2. Because the system is in contact with the hot reservoir, this process is an *isothermal expansion* at temperature T_H.

Change from state 2 to state 3: The system is isolated from its surroundings (detached from the hot reservoir) and allowed to expand to state 3. Because no heat is absorbed from or expelled to the environment, this process is an *adiabatic expansion.*

Change from state 3 to state 4: We place the system in contact with the cold reservoir, and an amount of heat Q_C flows out of the gas and into that reservoir. This change causes the gas volume to decrease and is an *isothermal compression* because it occurs at the fixed temperature T_C of the cold reservoir.

Change from state 4 to state 1: The system is again isolated from its surroundings and *compressed adiabatically* back to its initial state 1.

This cyclic path on the P–V diagram is called a ***Carnot cycle.*** Each step in the cycle involves a reversible process, and we saw in Section 16.4 how each of these processes (isothermal expansions and compressions, and adiabatic expansions and compressions) can be analyzed. The total work done by the gas in one cycle is the area enclosed by the path on the P–V diagram. This area is the total work W done by the heat engine during one complete cycle and is indicated in Figure 16.31A for the Carnot cycle.

CONCEPT CHECK 16.7 Work Done in a Carnot Cycle

Consider the steps in the Carnot cycle in Figure 16.31B. During which steps is the work W done by the system on the piston positive and during which steps is W negative?

Thermodynamic Devices: The Refrigerator

The Carnot heat engine in Figure 16.31 is reversible because each step in the cycle involves a reversible process. It is therefore possible to run this engine in reverse, that is, to run "backward" along the P–V path through states $1 \rightarrow 4 \rightarrow 3 \rightarrow 2 \rightarrow 1$. An amount of heat Q_C is then extracted from the cold reservoir, an amount of heat Q_H is expelled to the hot reservoir, and a certain amount of work W is done on the gas as indicated schematically in Figure 16.32A. This diagram is very similar to the heat engine in Figure 16.26B except that all the arrows showing the direction of energy flow are reversed.

Running a heat engine in reverse may seem like a crazy thing to do, but it is actually of great practical importance. Such a device is called a ***refrigerator.*** In a typical case (Fig. 16.32B), an electric motor does an amount of work W on a refrigeration "unit," which extracts an amount of heat Q_C from inside the refrigerator, keeping your food cold. The thermodynamic details of a refrigeration unit are more than we can discuss here, but they can be based on (you guessed it) cyclic processes involving the compression and expansion of gases. A refrigerator expels some heat Q_H to a hot reservoir, which in this case is the air around the refrigerator. Hence, refrigerators warm the air around them, usually underneath or behind the appliance. Applying conservation of energy to the refrigerator in Figure 16.32, we have

$$W + Q_C = Q_H \tag{16.22}$$

Carnot's results apply in this case, just as they apply to a heat engine, so we again have (from Eq. 16.20)

$$\frac{Q_C}{Q_H} = \frac{T_C}{T_H} \tag{16.23}$$

A good refrigerator extracts a great deal of heat Q_C from the cold reservoir while using as little work W as possible. The performance of a refrigerator is therefore measured in terms of the ***refrigeration efficiency*** e_{refrig} (also called the coefficient of performance), where

$$e_{\text{refrig}} = \frac{Q_C}{W} \tag{16.24}$$

The goal of a refrigerator designer is to make e_{refrig} as large as possible. Thermodynamics, however, places a limit on this efficiency. As with heat engines, the highest efficiency is found with a reversible refrigerator, and we can use Equations 16.22 and 16.23 to calculate e_{refrig}. Combining those relations, we find

$$W + Q_C = Q_H = Q_C\frac{T_H}{T_C}$$

$$W = Q_C\left(\frac{T_H}{T_C} - 1\right) \tag{16.25}$$

Inserting this result for W into the definition of refrigeration efficiency (Eq. 16.24) leads to

$$e_{\text{refrig}} = \frac{Q_C}{W} = \frac{Q_C}{Q_C[(T_H/T_C) - 1]} = \frac{1}{(T_H/T_C) - 1}$$

$$e_{\text{refrig}} = \frac{T_C/T_H}{1 - (T_C/T_H)} = \frac{T_C}{T_H - T_C} \tag{16.26}$$

A refrigerator is used to keep things cold, so T_C is usually low, but a low value of T_C leads to a small value for e_{refrig}. We can see that by inserting $T_C \approx 0$ into Equation 16.26, which gives

$$e_{\text{refrig}} = \frac{T_C}{T_H - T_C} = \frac{0}{T_H - 0} = 0$$

So, when the cold reservoir is at a very low temperature, the efficiency is also very low. That is an unavoidable consequence of the second law of thermodynamics.

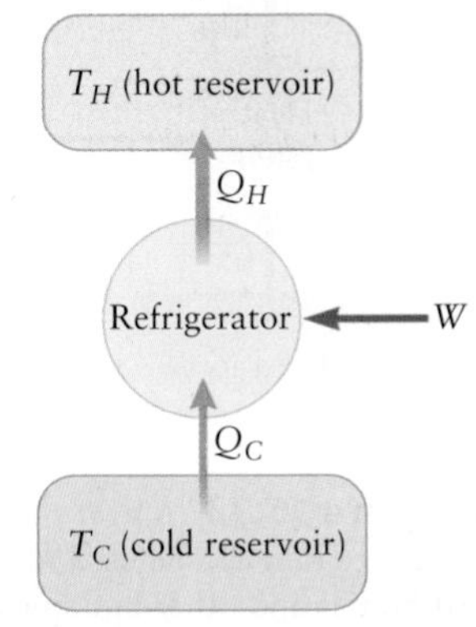

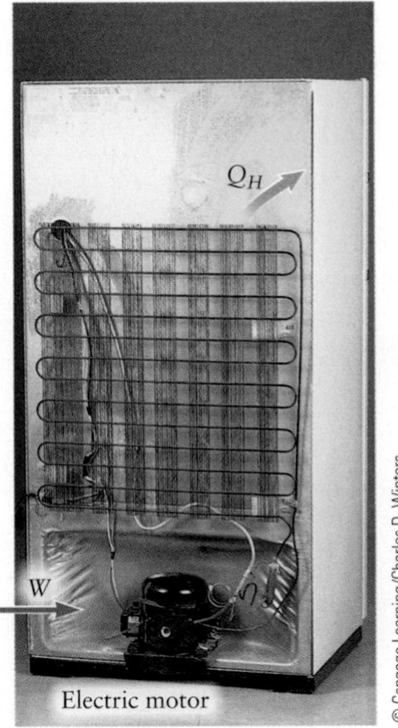

▲ **Figure 16.32** A When a heat engine is operated in "reverse," it functions as a refrigerator. It extracts heat energy from a cold reservoir that could then be used to store food. B In a real refrigerator, the cold reservoir is the air inside the refrigerator, and the hot reservoir is the room air.

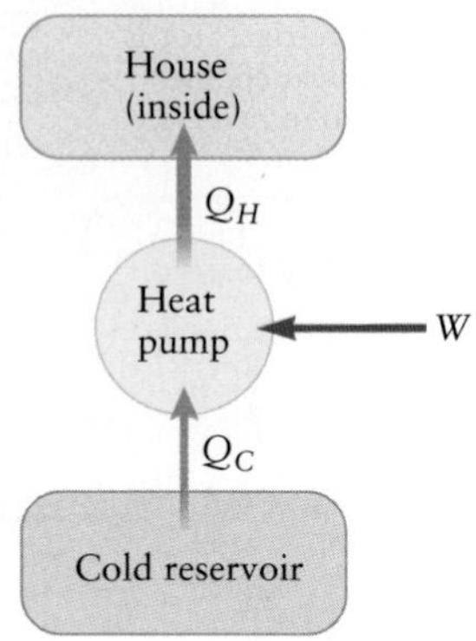

▲ **Figure 16.33** A heat pump is a device that can be used to heat homes and other buildings. It "pumps" heat from a cold reservoir into the building (the hot reservoir).

The schematic diagram of a refrigerator in Figure 16.32A shows that heat is removed from a cold reservoir and expelled to a hot reservoir. Hence, a refrigerator causes heat energy to flow from a cold object to a warmer one. Doesn't that violate the second law of thermodynamics? The answer is no; it does not violate the second law because the second law applies to the *spontaneous* flow of heat energy. The heat flow in a refrigerator is possible only because of the work W done, so the heat flow in Figure 16.32A is not spontaneous.

Heat Pumps

We have seen that a refrigerator must emit some heat; this heat is Q_H in Figure 16.32. A device called a ***heat pump*** uses this principle as part of a furnace for heating a house (Fig. 16.33). An amount of heat Q_C is extracted from a cold reservoir, which might be the outside air or perhaps water from underground. This device literally "pumps heat" from the cold reservoir into your house.

For a heat pump to be effective, you want the amount of heat that goes into your house to be as large as possible while at the same time keeping W as small as possible. So, the efficiency of a heat pump is defined as

$$e_{\text{pump}} = \frac{Q_H}{W} \tag{16.27}$$

and a good heat pump should have a large value of e_{pump}. Equations 16.22 and 16.23 apply to a heat pump as well as to a refrigerator (because a heat pump is thermodynamically identical to a refrigerator; compare Figs. 16.32A and 16.33), and we can use them to calculate e_{pump} for a reversible heat pump. They lead to

$$W = Q_H - Q_C = Q_H - Q_H \frac{T_C}{T_H}$$

$$W = Q_H[1 - (T_C/T_H)]$$

Inserting this result into the definition of e_{pump} (Eq. 16.27), we get

$$e_{\text{pump}} = \frac{Q_H}{W} = \frac{Q_H}{Q_H[1 - (T_C/T_H)]}$$

$$e_{\text{pump}} = \frac{1}{1 - (T_C/T_H)} \tag{16.28}$$

Efficiency values much greater than 1 are mathematically possible and achievable in practice, as we explore in Example 16.10.

EXAMPLE 16.10 Ⓡ Heating the Author's House

The author uses a heat pump to keep his house warm in the winter. The cold reservoir is water that comes from an underground well. What is the approximate efficiency of such a heat pump, assuming it is reversible?

RECOGNIZE THE PRINCIPLE

To find the efficiency of the heat pump (Eq. 16.28), we must know the reservoir temperatures. These temperatures are not given, so we need to estimate values from the description of the heat pump.

SKETCH THE PROBLEM

No sketch is needed.

IDENTIFY THE RELATIONSHIPS

The "hot" reservoir is the temperature inside the house, which is about 20°C = 293 K = T_H. The temperature of the cold reservoir is the temperature of the water in the well; for cold water from underground, it is about 10°C = 283 K = T_C.

SOLVE

Using these values for the reservoir temperatures, the efficiency is

$$e_{\text{pump}} = \frac{1}{1 - T_C/T_H} = \frac{1}{1 - (283\text{ K})/(293\text{ K})} \approx \boxed{29}$$

▶ *What does it mean?*

The work W that acts as the "input" to the heat pump (Fig. 16.33) is provided by an electric motor. With the value of e_{pump} for this heat pump, each 1 J of energy used to run the electric motor leads to about 29 J of heat energy "pumped" into the house, which is quite a good deal! An alternative scheme, called "electric resistive heating," produces a maximum of 1 J of heat for each 1 J of electrical energy consumed. Hence, a heat pump can be a very efficient way to heat a house. Notice, however, that we have calculated the efficiency of a *reversible* heat pump. The efficiency of a real heat pump will be lower than this value. (See Problem 67.)

CONCEPT CHECK 16.8 Heat Engine or Refrigerator?

Figure 16.34 shows the P–V diagrams for two cyclic thermodynamic processes. Which one describes a heat engine, and which one describes a refrigerator?

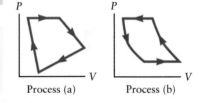

P V Process (a)

P V Process (b)

▲ **Figure 16.34** Concept Check 16.8.

16.7 Entropy

Before the work of Carnot and other scientists who developed the theory of thermodynamics, the nature of heat energy was not completely understood. Some physicists proposed (incorrectly) that heat was a sort of "fluid" that can flow from one object to another, much like water can flow from one container to another. This picture implied that an object "contains" only a certain amount of heat; if that amount of heat were somehow removed, it would then be impossible to extract any more heat from the object. We now know that this picture is not correct. However, there is a quantity called ***entropy*** that does have some of these properties.

Entropy, denoted by S, is a macroscopic property of a system much like pressure and temperature. If a small amount of heat Q flows into a system, its entropy changes by

$$\Delta S = \frac{Q}{T} \tag{16.29}$$

where T is the temperature of the system. The units of entropy are thus joules per kelvin. Let's apply Equation 16.29 to compute the total entropy change when heat flows between the two systems in Figure 16.35. The quantity Q in Equation 16.29 is the heat that flows *into* a system; in Figure 16.35, an amount of heat Q flows into system 2, so the entropy of this system changes by

$$\Delta S_2 = \frac{+Q}{T_2} \tag{16.30}$$

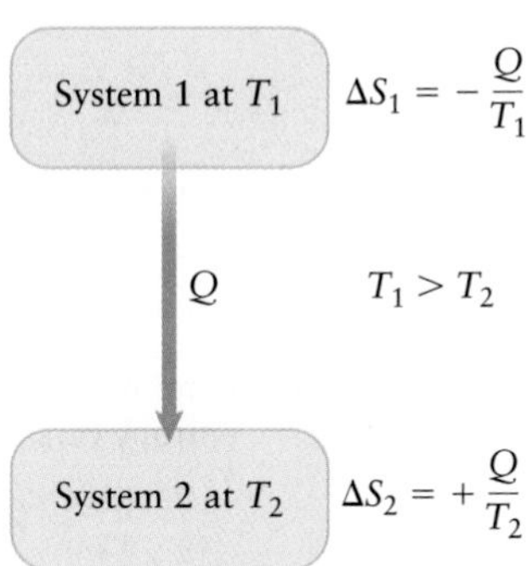

▲ **Figure 16.35** Whenever heat flows into or out of a system, the entropy of the system changes by an amount $\Delta S = Q/T$. The entropy change of a system can be positive or negative, depending on the direction of heat flow (the sign of Q).

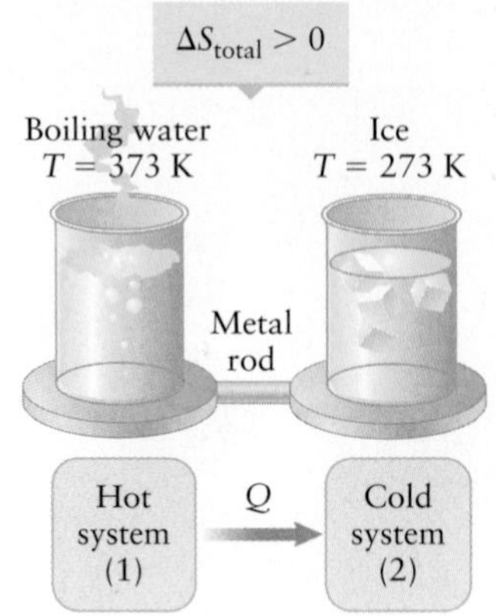

▲ **Figure 16.36** Heat flow from a hot system (1) to a cold system (2). The entropy change for the entire process, including both systems, is ΔS_{total}.

Likewise, an amount of heat Q flows *out* of system 1, so we have

$$\Delta S_1 = \frac{-Q}{T_1} \tag{16.31}$$

Hence, if heat flows from system 1 to system 2 as shown in Figure 16.35, the entropy of system 2 increases and the entropy of system 1 decreases.

A typical heat-flow experiment might involve a system at 373 K (the boiling temperature of water) in contact with a system at 273 K (the freezing temperature of water) as sketched in Figure 16.36. In this example, heat flows via conduction through a metal rod connecting the two systems. If a total amount of heat Q flows from the hot system (system 1) to the cold system (system 2), the total change in entropy is (using Eqs. 16.30 and 16.31)

$$\Delta S_{total} = \Delta S_1 + \Delta S_2 = \frac{-Q}{T_1} + \frac{Q}{T_2}$$

$$\Delta S_{total} = Q\left(-\frac{1}{T_1} + \frac{1}{T_2}\right) \tag{16.32}$$

Inserting $T_1 = 373$ K and $T_2 = 273$ K gives

$$\Delta S_{total} = \frac{-Q}{T_1} + \frac{Q}{T_2} = Q\left(-\frac{1}{373\text{ K}} + \frac{1}{273\text{ K}}\right) \tag{16.33}$$

The quantity in parentheses is positive, so ΔS_{total} is also positive.

Now suppose the temperatures of the two systems are nearly equal; in fact, let's assume we have an isothermal process so that $T_1 = T_2$. According to Equation 16.32, the total entropy change of both systems combined is zero. Such an isothermal process is reversible and gives an example of a process in which the entropy does not change. In fact, the total entropy change is zero for *all* reversible processes:

$$\Delta S_{total} = 0 \quad \text{(for a reversible process)} \tag{16.34}$$

We could also say that the total entropy change of the universe is zero for any reversible process.

The heat-flow process in Figure 16.36 involves systems at different temperatures and is irreversible because heat flows spontaneously from the hot to the cold system but not in the reverse direction (according to the second law of thermodynamics). According to Equation 16.32, ΔS_{total} in this case is always positive, which leads to the result

$$\Delta S_{total} > 0 \quad \text{(for an irreversible process)} \tag{16.35}$$

Equation 16.35 applies to *all* irreversible processes.

The results in Equations 16.34 and 16.35 form the basis for another statement of the second law of thermodynamics.

Second law of thermodynamics: entropy version

Second law of thermodynamics (entropy version)

In any thermodynamic process, $\Delta S_{universe} \geq 0$.

We now have three separate versions of the second law of thermodynamics. All are equivalent; any one version can be used to prove the other two. Each version gives different insights into the meaning and consequences of the second law.

Entropy: A Microscopic View

Let's now describe entropy in terms of a microscopic picture. Consider an ice cube placed in a warm environment. The ice cube is initially at a lower temperature than its environment, so heat will flow from the surroundings into the ice cube, causing it to melt. Because heat flows into the ice cube, its entropy will increase (according to Eq. 16.29). A molecular-scale view of this process is shown in Figure 16.37. How can we tell from this picture that the entropy has changed?

Before the ice cube melts, the water molecules are in regular positions characteristic of a solid, whereas after melting, the molecules are in the disordered arrangement characteristic of a liquid state. The system thus becomes much more *random* after the heat has been added. In microscopic terms, *entropy is a measure of the amount of disorder* or randomness in a system. Increasing the temperature of a gas or liquid increases the kinetic energy of the molecules and increases the amount of disorder. Adding heat to a system always increases its entropy.

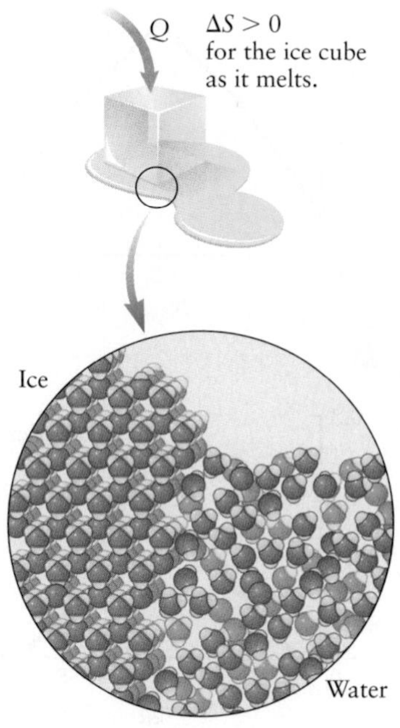

▲ **Figure 16.37** When the entropy of a system increases, the system becomes more disordered.

Entropy and Probability

Let's consider a system of gas molecules in a box and ask how the molecules are distributed. Are they all on one side of the box, are they distributed randomly throughout the box, or are they distributed in some other way? For simplicity, we can divide our box into two halves and ask how many molecules are in each half at any given "snapshot" in time. By symmetry, a particular molecule is equally likely to be on one side as on the other, so the probability of finding any given molecule on the left side is $p = \frac{1}{2}$. For a system containing only a single molecule, we thus have two possibilities—we call these two possible "states" of the system—as shown in Figure 16.38A.

We can do the same analysis with 2 molecules. Because each molecule can be on either the right or left side of the box, we now have $2 \times 2 = 4$ possible states for the system—four different possible arrangements of the molecules—as shown in Figure 16.38B. Figure 16.38C shows all the different possible states with 4 molecules. Notice how the number of possible states grows very rapidly as we add molecules to the system. With 4 molecules, we already have 16 possible states, and by the time

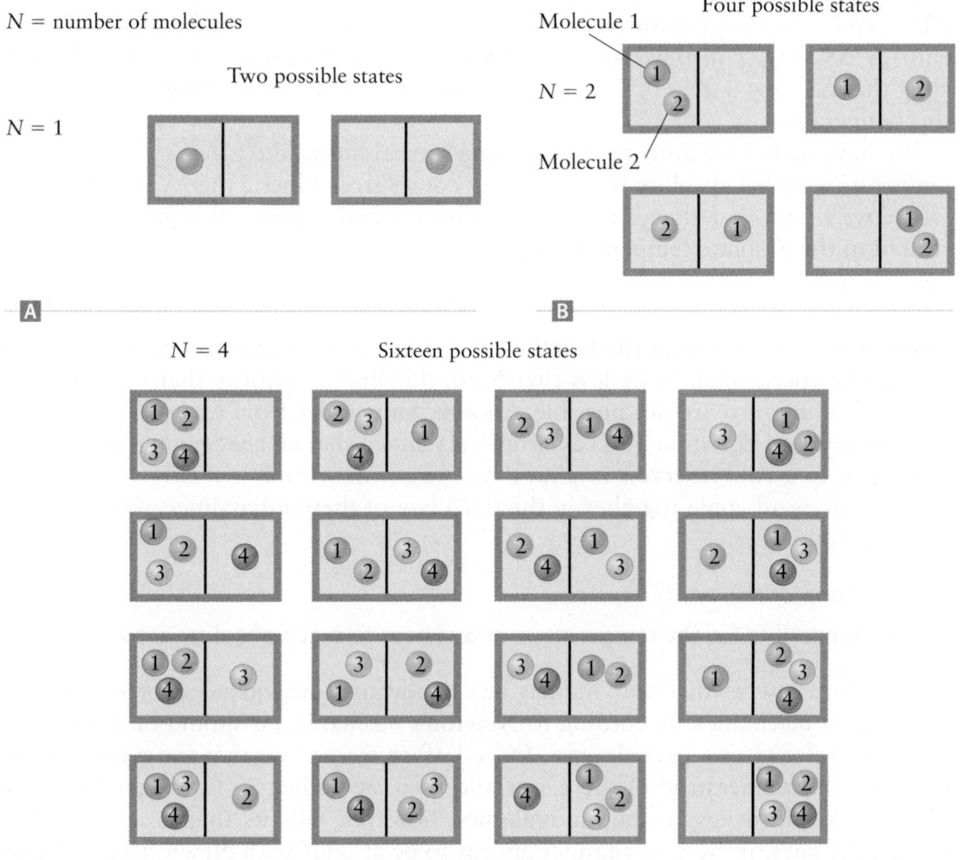

◀ **Figure 16.38** A A box containing a single gas molecule. We divide the box into two equal parts; the molecule might be found in either of these two possible "states." B If the box contains 2 molecules, they can be arranged in the two parts of the box in four different ways. C With 4 molecules, there are 16 different ways to arrange the molecules. There are thus 16 different possible states of the system, and all are equally likely to occur.

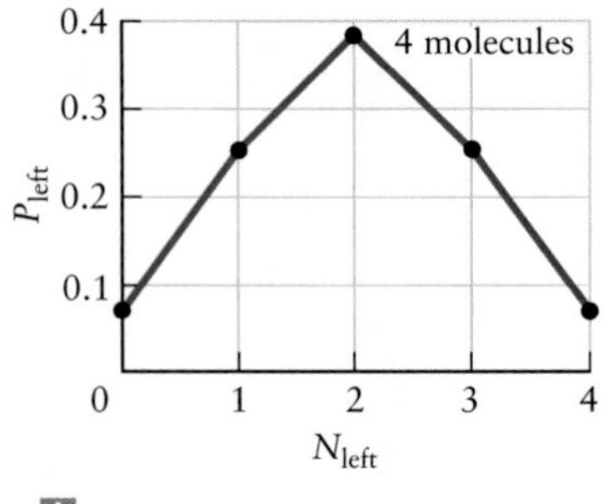

A

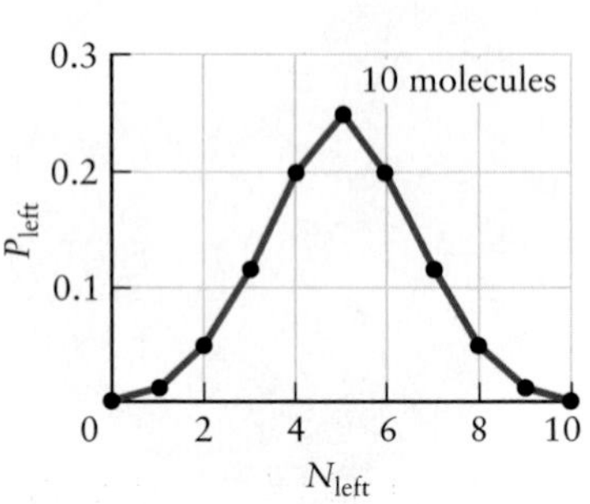

B

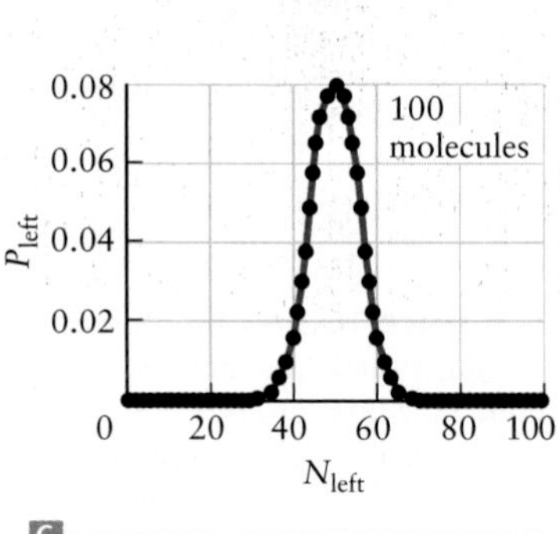

C

▲ **Figure 16.39** Probability for finding different numbers of molecules (N_{left}) on the left side of the box. A For 4 molecules. When $N_{left} = 2$, there are equal numbers of molecules on both sides. B For 10 molecules. C For 100 molecules. The greatest probability is found when $N_{left} = 50$, which means that the number of molecules on the left equals the number of molecules on the right.

we reach 10 molecules, the number of states is more than 1000. A key assumption of thermodynamics is that each of these states is equally likely. With this assumption, we can calculate P_{left}, the probability of finding a molecule on the left side of the box; when $P_{left} = 0.5$, the molecules are equally distributed on the two sides. Hence, the expected random distribution of molecules corresponds to values of P_{left} near 0.5. Figure 16.39 shows results for 4 molecules, 10 molecules, and 100 molecules. As the number of molecules grows, the probability peak becomes narrower, meaning that the probability of finding the molecules fairly equally distributed becomes very high.

What do the probability curves in Figure 16.39 have to do with entropy? Large values of the entropy correspond to a greater amount of randomness in a system. The results in Figures 16.38 and 16.39 show that there are many more ways of distributing the gas molecules so that there are equal numbers on the two sides of the box—that is, so that $P_{left} = 0.5$—than if all the molecules are on one side or on the other. As the number of molecules becomes very large, the most likely state of the system (the state with the highest probability) will be one in which the number of molecules on the left and on the right are approximately equal. Since there are many molecular arrangements that achieve this condition, this state has the greatest randomness and hence the largest entropy.

16.8 The Third Law of Thermodynamics and Absolute Zero

We defined the three common temperature scales (Fahrenheit, Celsius, and Kelvin) in Chapter 14. The Kelvin scale is most closely connected to kinetic theory and thermodynamics. For example, the ideal gas law has the mathematical form $PV = nRT$ when temperature is measured on the Kelvin scale, and the definition of entropy $\Delta S = Q/T$ also assumes T is given on the Kelvin scale. It is certainly puzzling that $\Delta S \rightarrow \infty$ if $T = 0$. The implication is that there is something special about this temperature.

We have had other hints that something special must take place at $T = 0$ K, the temperature called absolute zero. One hint comes from kinetic theory (Chapter 15), where we found that the average translational kinetic energy of a gas molecule is related to the absolute temperature by

$$KE = \tfrac{3}{2}k_B T \tag{16.36}$$

Thus, at $T = 0$, we expect the kinetic energy of a gas molecule to be zero! Because the kinetic energy cannot be less than zero, this finding implies that temperatures below absolute zero are not possible. Another hint comes from Carnot's work with heat engines and Equation 16.20. Carnot's statement that all heat engines must expel some heat to a cold reservoir implies that a reservoir with $T = 0$ K is not possible. These notions all come together in the third law of thermodynamics, which can be stated as follows.

Third law of thermodynamics

Third law of thermodynamics

It is impossible for the temperature of a system to reach absolute zero.

The third law of thermodynamics has important consequences for our understanding of mechanics. According to Newton's mechanics, it should be possible to make the velocity of a particle zero. Indeed, if we were to do that for every particle in a gas, the temperature of the gas would then, according to Equation 16.36, be $T = 0$. The third law of thermodynamics, however, tells us that is not possible. Hence, the laws of thermodynamics appear to be at odds with Newton's laws! This conflict was not resolved until the development of quantum mechanics in the early twentieth century.

16.9 Thermodynamics and Photosynthesis

In our work on thermodynamics, we made extensive use of the ideal gas because its properties, such as the internal energy and the ideal gas law, have simple mathematical forms. Thermodynamics, however, can be applied to many different physical systems. In this section, we show how thermodynamic arguments can be applied to photosynthesis.

Figure 16.40 represents the process of photosynthesis in a way that shows its similarity to a heat engine. An amount of heat Q_H from the Sun enters a plant, and (in the spirit of a heat engine) the plant "does an amount of work" W. At the same time, the plant expels some heat Q_C to its surroundings (the cold reservoir). In physics and engineering, the heat engine is a mechanical device that does work (e.g., by turning a shaft). In photosynthesis, a plant stores chemical energy in certain molecules, so the product of the photosynthetic heat engine in Figure 16.40 is actually potential energy stored in chemical bonds. This chemical energy can be used later to produce mechanical work when, for instance, an animal eats the plant, so the analogy with a heat engine still holds.

Sun
Q_H
Plant
W
Q_C
Atmosphere

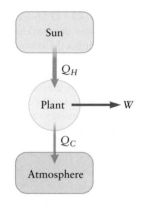

▲ **Figure 16.40** Photosynthesis is a thermodynamic process and can be viewed as a type of heat engine. Compare this diagram with Figure 16.26B.

Let's now analyze the efficiency of a photosynthetic heat engine. Assuming it is a reversible engine, this efficiency depends on the temperatures of the reservoirs and is given by $e = 1 - (T_C/T_H)$ (Eq. 16.21). The cold reservoir is the air or soil around the plant; hence, $T_C \approx 300$ K. The Sun serves as the hot reservoir, so our first estimate for the temperature of the hot reservoir is the temperature of the surface of the Sun, which is $T_H \approx 6000$ K as mentioned in Chapter 14. (The temperature of the hot reservoir is not the temperature of the plant. The leaves are *not* at 6000 K!) Inserting these values into Equation 16.21 gives

$$e_{\text{photosyn}} = 1 - \frac{T_C}{T_H} = 1 - \frac{300 \text{ K}}{6000 \text{ K}} = 0.95$$

A real photosynthetic engine does not approach this efficiency for several reasons, one of which can be seen from Figure 16.41. Radiation emitted by the Sun is distributed over a range of wavelengths throughout the infrared, visible, and ultraviolet regions (the blackbody distribution in Fig. 14.33). Plants absorb solar radiation mainly in the range of wavelengths $\lambda \approx 400$–700 nm and do not absorb much of the radiation outside this range. A mathematical analysis shows that plants absorb only about 50% of the Sun's total radiation, thus reducing the maximum efficiency of photosynthesis to approximately $0.95 \times 0.5 \approx 48\%$. There are other inefficiencies in photosynthesis, and the actual efficiency of the entire process is somewhat lower than 48%.[4] The important point is that thermodynamics and the physics of blackbody radiation place absolute physical limits on the efficiency of a real process.

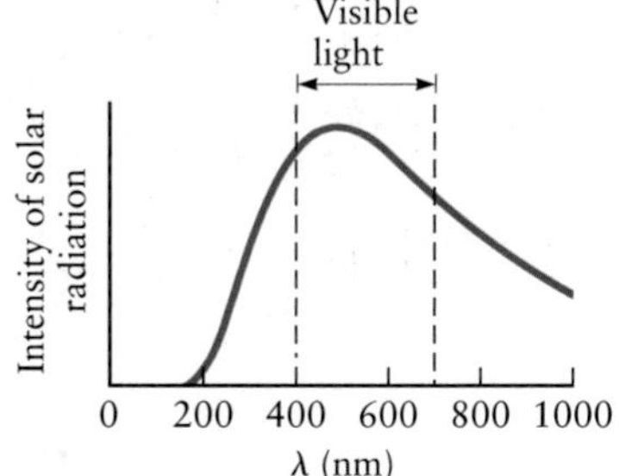

▲ **Figure 16.41** Plants absorb light only in the visible portion of the spectrum. Visible radiation corresponds to only about half the total radiated energy from the Sun. Shown is the blackbody distribution from Figure 14.33, but here it is plotted on a linear scale.

16.10 Converting Heat Energy to Mechanical Energy and the Origin of the Second Law of Thermodynamics

One key tenet of thermodynamics is that heat is a form of energy. The second law of thermodynamics, however, also tells us that heat energy differs from kinetic and potential energy in a subtle way. If you are given a certain amount of energy in the

[4]For comparison, solar cells convert sunlight into electrical energy with an efficiency of typically 15% to 20%. Photosynthesis is thus more efficient than manufactured solar cells.

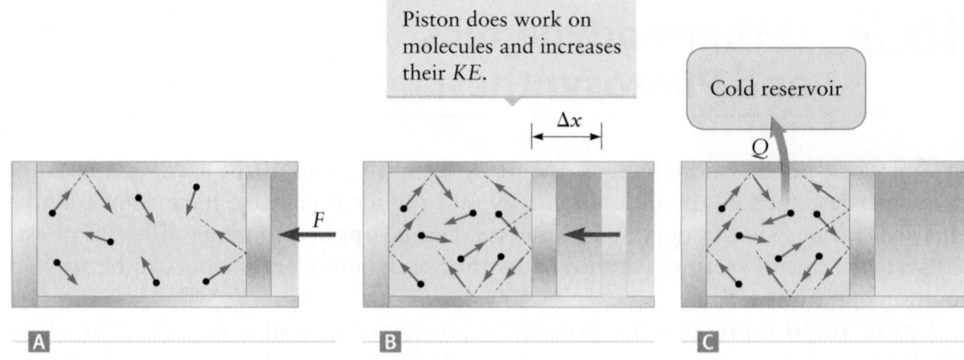

▶ **Figure 16.42** A Molecular-scale view of how the movement of a piston changes the energy of a gas. B When the gas is compressed, the piston does a net positive amount of work on all the gas molecules with which it collides. C This increased energy can later be transferred (via conduction) to a colder reservoir.

form of the kinetic energy of a baseball or the potential energy of a compressed spring, it is possible to convert all this energy to heat energy. Suppose, however, you are given the same amount of heat energy and asked to convert it to mechanical energy. The second law of thermodynamics as applied to a heat engine tells us that *only a fraction* of this energy can be converted to mechanical energy; this fraction is equal to the efficiency of a reversible heat engine (Eq. 16.21). A conversion from mechanical energy to heat can be done with 100% efficiency; why can't we convert heat energy to work with the same 100% efficiency?

To gain some insight into this difference, we first consider an example in which mechanical energy is converted to heat: the compression of an ideal gas. A molecular-scale view of this process is given in Figure 16.42, showing gas compressed by moving a piston. As the piston moves to the left (Figs. 16.42A and B), it does work on the gas molecules with which it collides. The work done by an external force acting on the piston is $F\,\Delta x$; this amount of energy is converted to additional kinetic energy of the molecules and shows up thermodynamically as an increase in the internal energy of the gas. If we now place this warm gas in contact with a colder thermal reservoir, heat flows from the gas into the reservoir (Fig. 16.42C). In this way, we could convert all the mechanical energy used to move the piston to heat energy.

Now consider the possibility (actually, the *impossibility*) of running the process in Figure 16.42 backward. To run in reverse, all the gas molecules in Figure 16.42C would have to collide with the piston in synchrony so as to exert a force on the piston to the right. The randomness of the gas makes such perfect synchrony impossible, however; some of the molecules may collide with the piston in this way, but the inherent disorder associated with heat energy prevents all the molecules from simultaneously "giving energy back" to the piston. For this reason, it is not possible to extract all the molecular kinetic energy from the gas. The disordered motion of the gas molecules makes it impossible to convert heat energy completely to mechanical energy.

Summary | CHAPTER 16

Key Concepts and Principles

Thermodynamics is about the properties of systems

Thermodynamics is concerned with the behavior of systems containing many particles such as a liquid, a gas, or a solid. Such systems interact with their environment, and the laws of thermodynamics describe these interactions.

Zeroth law of thermodynamics

If two systems A and B are both in thermal equilibrium with a third system, then A and B are in thermal equilibrium with each other.

First law of thermodynamics

If an amount of heat Q flows *into* a system from its environment and an amount of work W is done *by* the system on its environment, the internal energy of the system changes by an amount

$$\Delta U = U_f - U_i = Q - W \qquad \textbf{(16.3) (page 509)}$$

Second law of thermodynamics

The second law of thermodynamics can be stated in several different ways.

- ***Second law of thermodynamics (formulation involving heat flow):*** Heat flows spontaneously from a warm object to a colder one. It is not possible for heat to flow spontaneously from a cold object to a warmer one.
- ***Second law of thermodynamics (formulation involving heat engines):*** The efficiency of a reversible heat engine operating between reservoirs at temperatures T_H and T_C is

$$e = 1 - \frac{T_C}{T_H} \qquad \textbf{(16.21) (page 524)}$$

  No heat engine can have a greater efficiency.
- ***Second law of thermodynamics (entropy version):*** In any thermodynamic process, $\Delta S_{\text{universe}} \geq 0$.

Third law of thermodynamics

It is impossible for the temperature of a system to reach absolute zero.

Applications

Thermodynamic processes

A ***thermodynamic process*** takes a system from some initial state with an initial temperature T_i, pressure P_i, and volume V_i to a final state described by T_f, P_f, and V_f. Such a process can be described as a path on a ***P–V diagram***. The work done by the system on its environment is the area under this P–V curve. An ***isothermal process*** is one in which the temperature is constant ($T_i = T_f$). In an ***adiabatic process***, there is no heat flow into or out of the system ($Q = 0$).

Heat engines

A ***heat engine*** takes heat energy from a hot reservoir, expels heat energy to a cold reservoir, and does an amount of work on its environment. The ***efficiency of a heat engine*** is defined as

$$e = \frac{W}{Q_H} = \frac{Q_H - Q_C}{Q_H}$$

(Continued)

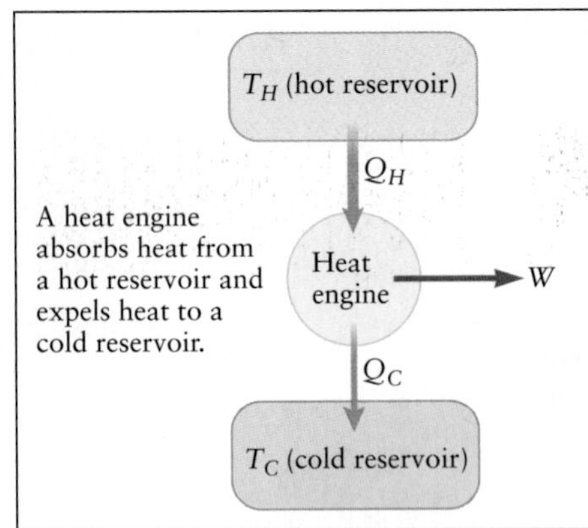

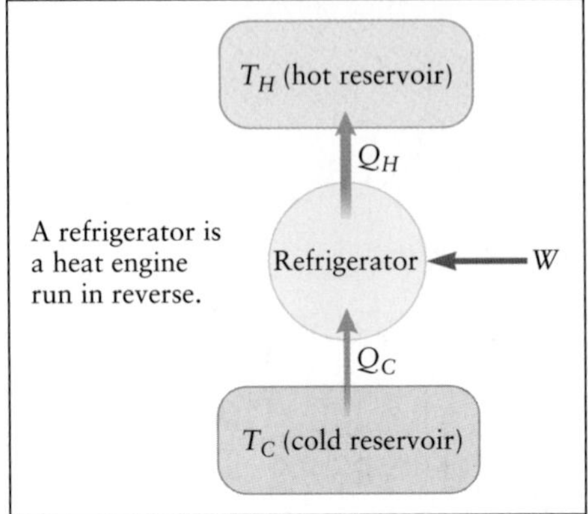

The efficiency of a thermodynamically reversible heat engine is given by the laws of thermodynamics as

$$e = 1 - \frac{Q_C}{Q_H} = 1 - \frac{T_C}{T_H} \qquad \text{(16.19 and 16.21) (pages 523 and 524)}$$

No heat engine can have a greater efficiency.

Refrigerators and heat pumps

In a ***refrigerator***, work is done by the environment on the system, which then extracts heat from a cold reservoir and expels heat to a hot reservoir. The ***efficiency of a refrigerator*** is $e_{\text{refrig}} = Q_C/W$. A thermodynamically reversible refrigerator has an efficiency of

$$e_{\text{refrig}} = \frac{T_C/T_H}{1 - (T_C/T_H)} = \frac{T_C}{T_H - T_C} \qquad \text{(16.26) (page 529)}$$

No refrigerator can have a greater efficiency.

Heat pumps use the same thermodynamic arrangement as in a refrigerator to heat your house. The hot reservoir in the case of a heat pump is the interior of your house.

Questions

SSM = answer in *Student Companion & Problem-Solving Guide*   (X) = life science application

1. (X) Two blocks, one of wood and the other aluminum, are at room temperature. When you touch them, the aluminum feels cold and the wood feels warm. Why? Is there a temperature at which both blocks would feel the same temperature to you? Explain.

2. It is a very hot day, and your friend decides to cool his apartment by opening the door of his refrigerator. Use thermodynamics to answer the following question: Will this method cool his apartment?

3. Can heat flow either into or out of a substance without the substance changing temperature? If not, why? If so, explain and give an example of such a process.

4. Consider an ideal gas that is expanding in a cylinder with a piston. As the gas expands, the piston moves such that a constant pressure is maintained. Does the temperature rise or fall? Does heat flow into or out of the gas in this process?

5. What quantity is represented by an area enclosed by a curve on a P–V diagram? What device is described by following such a curve in a clockwise direction? If the process follows the curve counterclockwise, what kind of device is being represented?

6. Consider a window air conditioner in terms of the thermodynamic block diagram for a refrigerator in Figure 16.32A. Identify the hot reservoir, the cold reservoir, and the source of mechanical work.

7. Explain how a heat pump can deliver more heat energy into a house than the pump consumes as work.

8. SSM Consider a thermodynamic process in which an ice cube melts to become a liquid (water). Is the change in internal energy of the H_2O positive or negative? Is the change in entropy of the H_2O positive or negative?

9. (a) Many heat engines need both a working fluid and a source of heat, often from a chemical process like combustion. Identify the working fluid and the source of heat for a steam engine like the one in Figure 16.27. (b) Many refrigerators require a working fluid and some device that does work. Identify the working fluid and the device that does work for a refrigerator like the one in Figure 16.32B. (c) Do *all* heat engines have a working fluid? Explain.

10. Figure Q16.10 shows a thermodynamic process in which a system first expands and is then compressed. Indicate the area on the P–V diagram that corresponds to the work done by the system.

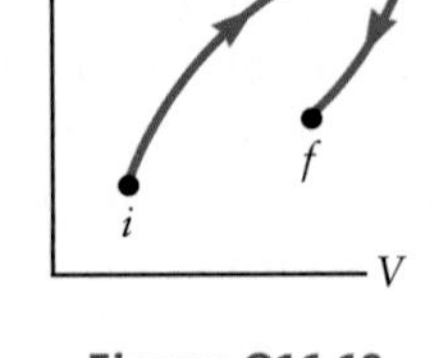

Figure Q16.10

11. In Section 16.4, we asserted that the work done in any cyclic process is equal to the area enclosed by the path on a P–V plot. In Example 16.6, we proved this assertion for a cyclic process described by a rectangular path. Prove this result for a path of any shape. *Hint*: Use a graphical argument involving area.

12. Give an example of a mechanical process that is reversible.

13. Give an example of a mechanical process that is irreversible. *Hint*: Include friction.

14. Suppose the second law of thermodynamics is *not* true and it is possible for heat to flow spontaneously from a cold object to a warmer one. Show that this process would allow one to construct a perpetual motion machine. *Hint*: Consider a machine that functions in two or more separate steps, with one step being the flow of heat from a cold reservoir to a hot reservoir. Then, in a separate step, energy from the hot reservoir is used in some way.

15. Consider the entropy changes that occur during the Carnot cycle sketched in Figure 16.31. For each stage of the cycle, is the entropy change of the engine positive, negative, or zero? What is the total entropy change of the engine, and of the environment, during one complete cycle? Is there anything that would prevent a real engine from having the same entropy changes for each stage of its cycle?

16. Under what conditions will the efficiency of a Carnot engine approach 100%?

17. If an ideal gas is compressed adiabatically, does the internal energy of the gas increase or decrease?

18. SSM Explain why heat pumps that use air as the cold reservoir are most efficient in mild climates. (They are called "air-to-air" heat pumps.) Why does a heat pump that uses water from deep underground as the cold reservoir have a thermodynamic advantage over air-to-air heat pumps when the weather is very cold?

19. In Example 16.10, we found that for typical temperatures of the hot and cold reservoirs, the theoretical efficiency of a home heat pump can be quite large. Investigate the efficiency obtained with real heat pumps (i.e., do some research on the Internet). How does the efficiency of a real heat pump compare with the theoretical limit in Equation 16.28? What factors make the efficiency of a real heat pump smaller than this theoretical limit?

20. Two ideal heat engines A and B operate using the same hot reservoir. It is found that engine B is more efficient than engine A. Explain how that can be.

21. Consider the common expressions (a) "There's no such thing as a free lunch" and (b) "You can't even break even." Which of the laws of thermodynamics might these expressions refer to? Explain.

22. A hydroelectric power plant converts potential energy of water stored behind a dam to electrical energy with an efficiency close to 100%. A coal-burning (or nuclear) power plant only approaches 40% for conversion of the chemical (or fission) energy to electricity. Why?

Problems

SSM = solution in *Student Companion & Problem-Solving Guide*
★ = intermediate ✪ = challenging ⊗ = life science application
R = reasoning and relationships problem
RT = reasoning tutorial in Enhanced WebAssign

16.3 THE FIRST LAW OF THERMODYNAMICS AND THE CONSERVATION OF ENERGY

1. What is the total internal energy of a monatomic ideal gas that contains 4.0 moles at a temperature of 400 K?

2. A system takes in 5000 J of heat and does 4000 J of work on its surroundings. What is the change in the internal energy of the system?

3. SSM ★ The temperature of an ideal monatomic gas increases by 100 K while 3000 J of heat are added to the gas. If the gas contains 4.0 moles, how much work does it do on its surroundings?

4. ★ The temperature of a sample of dilute argon gas with $n = 7.5$ moles decreases by 200 K during a thermodynamic process. If 35,000 J of heat are extracted from the gas, what is the work done by the gas on its surroundings?

5. One gallon of gasoline releases approximately 1.2×10^8 J when it is "burned" (i.e., undergoes combustion) in a car engine. If the engine releases 4.0×10^7 J of heat during this time, what is the mechanical work done by the engine?

6. ★ ⊗ A doughnut contains about 300 Calories. (*Note*: These are "food" Calories.) After eating this doughnut, you decide to compensate by taking the stairs instead of using the elevator. If each stair has a height of 20 cm, how many stairs must you climb to compensate for the doughnut? Assume all the energy in the doughnut can be converted to mechanical energy.

7. ✪ ⊗ Table 16.1 lists the energy that can be extracted from various foods, and Table 16.2 gives the rate at which work is done and heat is generated by a person engaged in various activities. Use these data to analyze the following cases.
   (a) How much heat energy is generated during a normal night (8 h) of sleep? How many apples must a person consume to compensate for this energy loss? (Round to the next highest whole number.)
   (b) What is the total energy expended when a person runs for 30 min at a moderate pace? How many soft drinks does this energy correspond to? (Round to the next highest whole number.)
   (c) What is the total energy expended when an elite cyclist races for 6 h at a rapid pace (i.e., at the Tour de France bicycle race)? How many hamburgers does this energy correspond to? (Round to the next highest whole number.)

TABLE 16.1 Energy Obtained by Eating Various Foods

Food	Approximate Energy Obtained (J)
Apple	2.2×10^5
Banana	4.6×10^5
Egg	2.7×10^5
Fast-food hamburger	2.0×10^6
Soft drink (12-oz size)	5.7×10^5

TABLE 16.2 Approximate Values of the Work W and Heat Q Associated with Various Human Activities

Activity	W/t (W)	Q/t (W)
Walking	50	200
Bicycle riding (moderate speed)	100	400
Bicycle racing	400	1600
Sleeping	0	80
Running (moderate pace)	200	800
Shoveling snow	15	500

Note: Work W and heat Q are both expressed in terms of the rate at which they are done.

8. An ideal gas absorbs 400 J of heat from its environment and does 800 J of work on the environment. Find the change in the internal energy of the gas and the change in the internal energy of the environment.

16.4 THERMODYNAMIC PROCESSES

9. Match the following processes (at left) with the descriptions (at right).

(a) Isobaric	(i) The volume remains constant.
(b) Isothermal	(ii) The pressure remains constant.
(c) Adiabatic	(iii) No heat energy is added to or subtracted from the system.
(d) Isochoric	(iv) The temperature remains constant.

10. In which of these processes does the internal energy of an ideal gas not change?
(a) An isothermal process
(b) An adiabatic process
(c) An isobaric process
(d) A process at constant volume

11. A gas expands in an isobaric (constant pressure) process, with $P = 4.5 \times 10^5$ Pa. If the volume changes from 4.5 m^3 to 7.8 m^3, how much work is done by the gas?

12. Consider a system that undergoes an expansion from an initial state (i) to a final state (f) in Figure P16.12. Approximately how much work is done by the system in this process?

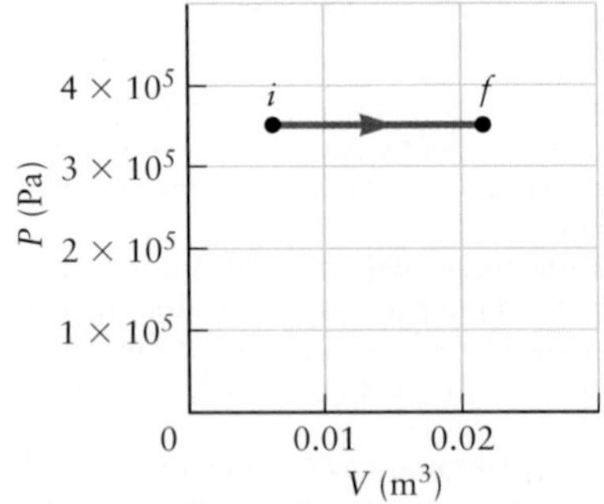

Figure P16.12 Problems 12, 14, and 15.

13. Approximately how much work is done by the system in the thermodynamic processes in the two parts of Figure P16.13? Are they expansions or contractions?

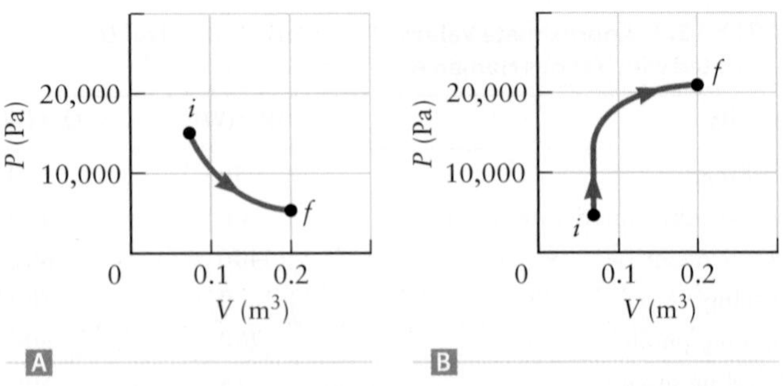

Figure P16.13 Problems 13 and 14.

14. For the thermodynamic processes in Figures P16.12 and P16.13, what is the approximate value of the work done *on* the system?

15. For the expansion in Figure P16.12, if the internal energy of the system decreases by 4500 J, what is the heat added to the system?

16. A system is compressed adiabatically, during which time the environment does 3500 J of work on the system. What is the change in the internal energy of the system?

17. SSM Consider a balloon of volume 2.0 m^3 that contains 3.5 moles of helium gas at a temperature of 300 K. If the gas is compressed adiabatically so as to have a final temperature of 400 K, what is the work done by the gas?

18. RT A container holds 1.5 kg of water that is initially at 320 K and is converted to steam at 400 K. Assuming the volume of the system does not change, what is the change in the internal energy of the system?

19. Water vapor (n = 1.0 mole) is heated in a pressure cooker so that the volume is held constant. If the temperature increases from 500 K to 600 K, what is the change in the internal energy of the water vapor?

20. You have 3.2 moles of a monatomic ideal gas that are compressed isothermally. (a) If the pressure changes from 4.5×10^5 Pa to 8.5×10^5 Pa and the initial volume was 0.50 m^3, what is the final volume of the system? (b) For the compression in this problem, what are the initial and final temperatures?

21. A system undergoes a two-step process as sketched in Figure P16.21. The system first expands at constant pressure from state i to state A and then it is compressed with the pressure again held fixed, ending with the system in state f. What is the approximate amount of work done by the system during each step?

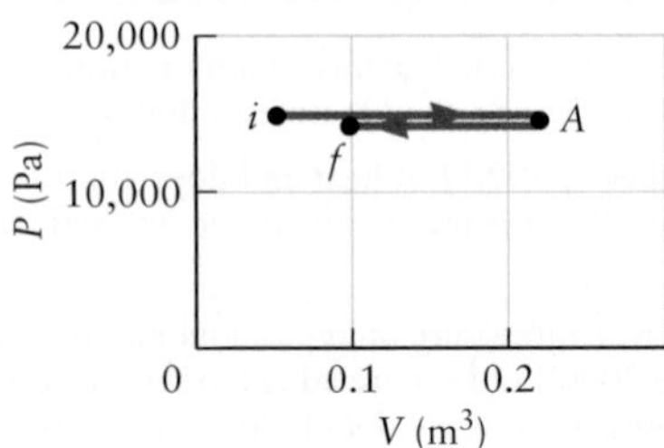

Figure P16.21 Problems 21 and 22.

22. Consider the process described by Figure P16.21. If 500 J of heat are added to the system during the second step, what is the change in the internal energy of the system during this step?

23. SSM Figure P16.23 shows a system that is compressed from an initial state i to a final state f. If the internal energy of the system decreases by 5000 J, approximately how much heat was added to the system?

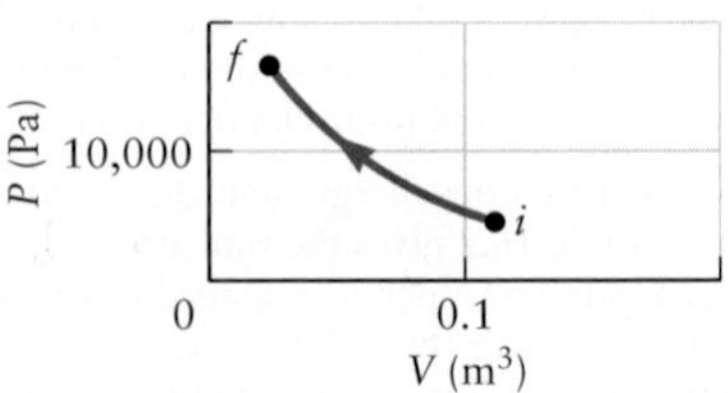

Figure P16.23

24. An ideal monatomic gas expands adiabatically from an initial temperature of 400 K and volume of 4.5 m^3 to a final volume of 12 m^3. If the initial pressure is $1.5 \times P_{atm}$, how much work is done on the gas?

25. You have 2.0 moles of helium gas initially at a temperature of 340 K and a pressure of $0.20 \times P_{atm}$. The gas is compressed isothermally to a pressure of $0.50 \times P_{atm}$. (a) Find the final volume of the gas. Assume it behaves as an ideal gas. (b) Find the work done by the gas. (c) How much heat flows into the gas?

16.6 HEAT ENGINES AND OTHER THERMODYNAMIC DEVICES

26. A reversible heat engine operates between thermal reservoirs at 900 K and 200 K. If 9000 J are extracted from the hot reservoir, how much work is done by the heat engine?

27. An engine absorbs 3500 J of heat energy each cycle and expels 1200 J. (a) How much work does it do in one cycle? (b) If it is a reversible heat engine and the hot reservoir is at 750 K, what is the temperature of the cold reservoir?

28. A gasoline engine absorbs 4000 J of heat and performs 3000 J of mechanical work for each turn of the engine shaft. How much heat is expelled from the engine after 500 turns of the shaft?

29. Consider a reversible heat engine that operates between a hot reservoir at 550 K and a cold reservoir at 150 K. If this engine has a power rating of 750 W, how much heat is expelled to the cold reservoir in 10 min?

30. A reversible heat engine operates between a hot reservoir at 580 K and a cold reservoir of unknown temperature. After 5 min, the engine has absorbed 1000 J of energy from the hot reservoir and expelled 500 J of energy to the cold reservoir. What is the temperature of the cold reservoir?

31. SSM Assume the gasoline engine in your car is a heat engine operating between a hot reservoir at 800 K and a cold reservoir at 280 K and that your engine produces a peak power output of 250 hp (horsepower). If the temperature of the hot reservoir is increased to 900 K, what is the theoretical peak power output?

32. Consider two heat engines: Engine 1 operates between thermal reservoirs at 450 K and 170 K, and engine 2 operates between reservoirs at 50 K and 10 K. Which engine has a greater maximum theoretical efficiency?

33. A reversible heat engine consumes five times more heat Q_H from its hot reservoir than it expels to its cold reservoir, Q_C. What is the ratio T_H/T_C?

34. Consider a heat engine that operates between thermal reservoirs at 500 K and 250 K. Suppose the engine does 5000 J of work while it expels 1000 J of energy to the cold reservoir. Is this engine allowed by the second law of thermodynamics?

35. A container filled with 1.0 L of water that is initially at 0°C is placed in a reversible refrigerator. Suppose the hot reservoir for this refrigerator is the room air at 30°C. How much work must be done on the refrigerator to turn all this water into ice at 0°C? Assume the only thing that is cooled by the refrigerator is the water.

36. RT It is a cold winter day in the author's neighborhood. One of the houses is heated with a reversible heat pump that uses the outside air as the cold reservoir. Suppose 40,000 J of energy are lost from the house due to heat conduction through the walls, windows, and one broken window. Estimate how much energy must be used by the heat pump (as input through mechanical work) to compensate for this loss of energy.

37. A reversible heat pump operating between thermal reservoirs at 25°C and 10°C consumes 500 J of energy each second. How much heat energy is "pumped" into the house each second?

38. A heat pump operates between reservoirs at 293 K (inside the house) and 265 K (outside). It is not a reversible heat pump; it only pumps 75% of the maximum possible heat into the house. How much work must be done to "pump" 6000 J of heat energy into the house?

39. A reversible heat engine operates between thermal reservoirs at 600 K and 150 K. What is the efficiency of this engine?

40. Suppose the heat engine in Problem 39 is operated "in reverse" as a refrigerator. What is its refrigeration efficiency?

41. A Carnot engine extracts 2000 J of heat from a hot reservoir and does 1200 J of mechanical work. Suppose the cold reservoir is a very large block of ice at 0°C. What mass of ice would melt during this time?

42. A heat engine operates using the P–V cycle shown in Figure P16.42. Approximately how much work is done by the system during each cycle?

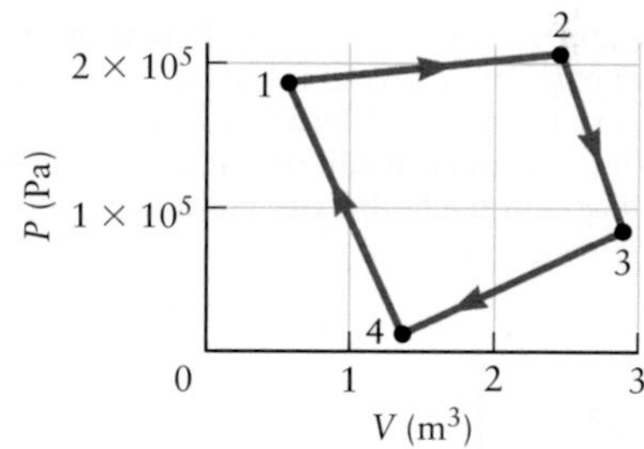

Figure P16.42 Problems 42 and 43.

43. For the heat engine in Figure P16.42, during which portion(s) of the cycle does the engine do positive work on its environment and during which portion(s) does the environment do work on the system?

44. SSM A heat engine absorbs 8000 J of heat from a hot reservoir and produces 5500 J of work. What is the efficiency of the engine?

45. SSM RT Suppose the refrigerator in your house is a reversible refrigerator. Estimate how much heat is extracted from the fresh-food compartment when 1000 J of work are done. *Hint:* You will first need to identify the hot and cold reservoirs and estimate their temperatures. Ignore the freezer compartment.

16.7 ENTROPY

46. Consider a reversible heat engine that employs a hot reservoir at a temperature of 750 K and a cold reservoir at 230 K. (a) What is the entropy change of the hot reservoir during a period in which 5000 J are extracted from the hot reservoir? (b) What is the change in the entropy of the cold reservoir? (c) Find the change in the entropy of the engine itself during this time.

47. SSM ★ Suppose 7000 J of heat flow out of a house that is at 293 K to the outside environment at 273 K. Find the change in entropy of the house, the change in the entropy of the environment, and the total change in entropy of the universe. Is this process reversible or irreversible?

48. ★ Suppose 5000 J of heat flow from a reservoir at 400 K to a colder reservoir. If the total entropy change of the universe is +200 J/K, what is the temperature of the cold reservoir? *Note*: Because the reservoirs are assumed to be very large, their temperatures do not change during this process.

49. ★ A large piece of steel (initial temperature 400 K) is placed into a large bucket of ice water. If 4.0 J of heat flow from the steel into the ice water, what are the entropy changes of the steel, the ice water, and the universe? Assume this small amount of heat flow does not change the temperatures of the steel or the ice water by any significant amount.

50. ★ A pot contains 250 g of water at 100°C. If this water is heated and all evaporates to form steam at 100°C, what is the change in the entropy of the H_2O?

51. ★ A 1.0-kg block of ice is initially at 0°C. Heat is added to this system, and the entropy is increased by 300 J/K. How much of the ice melts?

52. ★ Ⓑ The Sun emits heat via radiation. Estimate ΔS for the Sun during one day. The surface of the Sun is approximately 6000 K.

Additional Problems

53. ★ Consider the following systems and how closely each approximates a reversible process. Rank them from approximately reversible to irreversible with greatest change in entropy. *Hint*: Consider a video of each in reverse.
(a) A child swinging on a swing
(b) Mars orbiting the Sun
(c) Making a milkshake in a blender
(d) A yo-yo climbing up its string

54. SSM ★ Figure P16.54 shows the P–V diagrams for a variety of different processes involving an ideal gas. (a) Which diagrams describe an expansion? (b) Which diagrams describe a compression? (c) Which diagrams might describe an adiabatic process? (d) Which diagrams might describe an isothermal process?

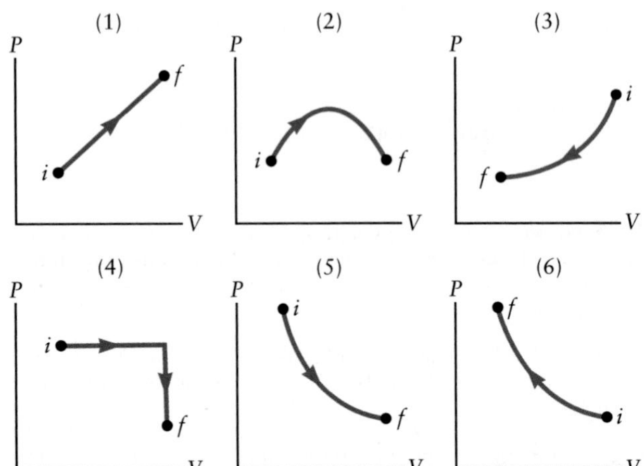

Figure P16.54 Problems 54, 55, and 56.

55. ★ Suppose process (6) in Figure P16.54 is an isothermal process involving an ideal gas. What is the value of the ratio $(P_f/P_i)/(V_i/V_f)$?

56. ★ RT (a) For which processes in Figure P16.54 is W positive? (b) For which processes is it negative?

57. ★ Figure P16.57 shows several cyclic processes involving an ideal gas. (a) Which processes might describe a heat engine? (b) Which processes might describe a refrigerator? (c) Which process might be a Carnot cycle?

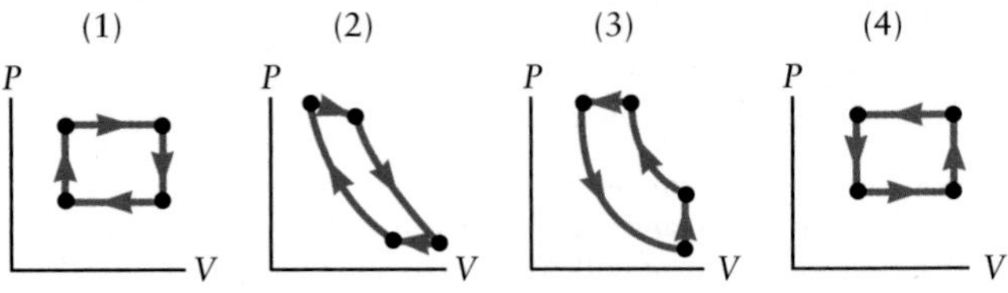

Figure P16.57

58. ★ (a) Calculate the efficiency of a Carnot engine running between a low temperature of 27.0°C and a high temperature of 387°C. Determine the change in efficiency if you were to (b) raise the high temperature by 10.0°C or (c) lower the low temperature by 10.0°C. Which improves the efficiency more? *Note*: Keep three significant figures in your calculations for this problem.

59. ★ **One engine versus two.** A Carnot engine produces 1000 J of work when running between reservoirs of 470°C and 100°C. (a) How much heat flows from the hot reservoir? *Hint*: Consider the efficiency. (b) Find the heat flow to the cold reservoir. (c) Consider a second engine that uses the heat flow out of the first engine as the hot reservoir and has a cold reservoir at room temperature, 20°C. How much work can the second engine produce? (d) Finally, consider a third engine that runs between the hot reservoir of the first engine and the cold reservoir of the second engine. Calculate the work done by this engine and compare it with the sum of the work of the first two engines.

60. ★ **The drinking bird.** A popular novelty bird toy consists of two semievacuated glass chambers (the head and the body), which are connected by a glass tube as shown in Figure P16.60 (left). The head is covered with an absorbent coating, and the bottom chamber is partially filled with a liquid that has a boiling point near room temperature. The bird goes through a cycle illustrated in Figure P16.60 (right). The bird starts with all the liquid in the bottom chamber and is suspended at an off-center pivot point so that it leans slightly forward. Next the fluid rises slowly up the tube, which is open at the bottom but submerged in the fluid. The fluid reaches the top chamber, and the bird tips forward. Finally, at full tilt, the bottom of the tube breaks the surface and the liquid pours back into the bottom chamber. The bird then returns to the upright position, and the cycle repeats. (a) Is work done by this cyclic process? Is the bird a heat engine? Describe. (b) The bird dips its head in the water during the cycle. Why is that important for the toy's operation?

Figure P16.60

61. Figure P16.61 shows the P–V diagrams for idealizations of three important real engines: the Brayton (gas turbine) engine, the Otto (four-stroke) automobile engine, and the diesel engine. For each of these cycles, there is no heat flow for the processes along the curves $2 \rightarrow 3$ and $4 \rightarrow 1$. (a) Using the diagram for each engine, describe the sequence and name each thermodynamic process that constitutes the cycle, starting at point 1. (b) Identify the points of highest and lowest temperature for each engine.

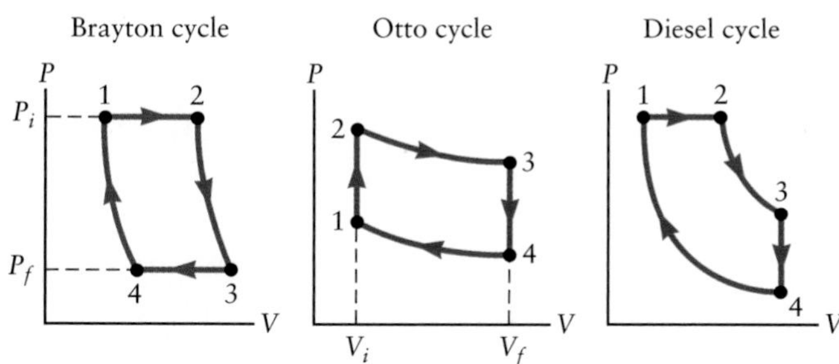

Figure P16.61 The Brayton cycle, Otto cycle, and diesel cycle for Problems 61 through 64.

62. The Brayton cycle shown in Figure P16.61 is the idealized cycle for a gas turbine or jet engine. Consider an engine using the Brayton cycle that operates on 0.070 mol of ideal gas between the pressures $P_i = P_1 = P_2 = 4.9$ atm and $P_f = P_3 = P_4 =$ 3.5 atm with volumes $V_1 = 0.50$ L, $V_2 = 1.00$ L, $V_3 = 1.22$ L, and $V_4 = 0.61$ L. *Note*: Retain three significant figures in all calculations. (a) Determine the work done for the displacements $1 \rightarrow 2$ and $3 \rightarrow 4$. (b) Determine the temperature of the ideal gas at each point. (c) Calculate the work done for the adiabatic expansion $2 \rightarrow 3$ and contraction $4 \rightarrow 1$. *Hint*: Consider the change in internal energy. (d) What is the total work done by one cycle? (e) Determine the amount of heat flowing in, $Q_H = Q_{1 \rightarrow 2}$ during the isobaric expansion and the heat flowing out of the gas during the isobaric contraction, $Q_C = Q_{3 \rightarrow 4}$. *Hint*: The specific heat per mole for an ideal gas under constant pressure is given by $C_P = C_V + R$. (f) Determine the efficiency for this Brayton engine (Eq. 16.18). *Note*: The heat loss and friction in a real Brayton engine will reduce this ideal efficiency. (e) Calculate and compare to the efficiency of a Carnot engine running between the same high and low temperatures.

63. The Otto cycle shown in Figure P16.61 is the idealized cycle for an internal combustion automobile engine. Consider an engine using the Otto cycle that operates on 0.070 mol of ideal gas between the volumes $V_i = V_1 = V_2 = 0.2$ L and $V_f = V_3 = V_4 = 1.6$ L. (a) In which parts of the cycle ($1 \rightarrow 2$, $2 \rightarrow 3$, $3 \rightarrow 4$, $4 \rightarrow 1$) is work done on or by the system? In which parts does heat flow into or out of the system? (b) The lowest temperature is 323 K at point 4, where the exhaust valve is open and the pressure is 1.0 atm. Find the temperature at point 1. *Hint*: Use Equation 16.10 to first find the pressure of the ideal gas. Assume that $\gamma = \frac{5}{3}$. (c) The combustion of fuel in stroke $1 \rightarrow 2$ supplies 1000 J of energy to the gas. Calculate the change in temperature of the gas and find the temperature at point 2. (d) What is the pressure at point 2? (e) Find the pressure and temperature at point 3, the end of the "power stroke." (f) Determine the heat loss during stroke $3 \rightarrow 4$. (g) Determine the efficiency for this Otto engine (Eq. 16.18). *Note*: The heat loss and friction in a real Otto engine will reduce this ideal efficiency. (h) Calculate and compare to the efficiency of a Carnot engine running between the same high and low temperatures.

64. **Compression ratio and engine efficiency.** The Otto cycle shown in Figure P16.61 is the idealized cycle for an internal combustion automobile engine. Corresponding to positions of the piston, the cycle is in four parts (two isochoric and two adiabatic). (a) Show that Equation 16.19 describing the efficiency can be expressed as

$$e = 1 - \left(\frac{T_3 - T_4}{T_2 - T_1}\right)$$

for the Otto cycle, where the temperatures are taken from the four points on the cycle. (b) If an ideal gas is used as the working fluid, it turns out that the efficiency for the Otto cycle is independent of the temperature and depends only on the ratio V_f/V_i, called the ***compression ratio***. Show that the efficiency can also be written in terms of the initial and final volumes as

$$e = 1 - \left(\frac{V_i}{V_f}\right)^{\gamma - 1}$$

where $\gamma = \frac{5}{3}$ for an ideal gas. *Hint*: Use the ideal gas law and Equation 16.10 to write each temperature in terms of volume and pressure.

65. A cylinder and piston arrangement is made of an excellent conductor in contact with a thermal reservoir at 38°C and is filled with an ideal gas. The gas is slowly compressed in an isothermal process in which 80 J of work are done on the gas. (a) What is the change in entropy of the gas? (b) What is the change in entropy of the thermal reservoir? (c) What is the net change in entropy of the universe?

66. RT A particular engine runs for a period of time during which it is found that $W = 2{,}000$ J, $Q_H = 12{,}000$ J, $T_C = 450$ K, and $T_H = 600$ K. (a) Is this engine (i) a reversible heat engine, (ii) an irreversible heat engine, or (iii) not possible according to the laws of thermodynamics? (b) If your answer to part (a) is option (ii) or option (iii), give a new value of one of the quantities (W, Q_H, T_C, or T_H) so that we are now describing a reversible heat engine. (c) If your answer to part (a) is option (i) or (ii), give a new value of one of the quantities (W, Q_H, T_C, or T_H) that would make this heat engine impossible.

67. The efficiency of a reversible heat pump is given by Equation 16.28. Real heat pumps, however, may achieve an efficiency of only about 20% of this ideal limit. A heat pump with this reduced efficiency is used to heat a house for which $T_H = 300$ K. What is the lowest value of the temperature of the cold reservoir for which it makes sense to use this heat pump instead of just using electric resistive heating, which converts the input energy W directly to the same amount of heat energy Q_H?

68. Table 16.2 (after Problem 7) lists the work W associated with several human activities. For walking, the table lists $W/t = 50$ W, where t is the time spent walking. Use Newton's laws of motion to estimate the value of W/t for a typical person walking. *Hint*: Consider the work done in lifting the body during each footstep and assume that potential energy is not "returned" when a foot comes back down to the ground.

APPENDIX

Reference Tables

TABLE A.1 Some Useful Physical Constants and Data

Quantity	Symbol	Value
Universal gravitational constant	G	$6.674 \times 10^{-11}\ \text{N} \cdot \text{m}^2/\text{kg}^2$
Acceleration due to gravity near the Earth's surface	g	$9.8\ \text{m/s}^2$
Speed of light in vacuum	c	$2.997\ 924\ 58 \times 10^8\ \text{m/s}$
Mass of the electron	m_e	$9.109\ 382 \times 10^{-31}\ \text{kg}$
Mass of the proton	m_p	$1.672\ 621\ 6 \times 10^{-27}\ \text{kg}$
Mass of the neutron	m_n	$1.674\ 927\ 2 \times 10^{-27}\ \text{kg}$
Magnitude of the electron charge	e	$1.602\ 176\ 5 \times 10^{-19}\ \text{C}$
Permittivity of free space	ε_0	$8.854\ 187\ 817 \times 10^{-12}\ \text{C}^2/(\text{N} \cdot \text{m}^2)$
Coulomb constant	$k = \dfrac{1}{4\pi\varepsilon_0}$	$8.987\ 551\ 788 \times 10^9\ \text{N} \cdot \text{m}^2/\text{C}^2$
Permeability of free space	μ_0	$4\pi \times 10^{-7}\ \text{T} \cdot \text{m/A}$
Boltzmann's constant	k_B	$1.380\ 650 \times 10^{-23}\ \text{J/K}$
Planck's constant	h	$6.626\ 069 \times 10^{-34}\ \text{J} \cdot \text{s}$
Avogadro's number	N_A	$6.022\ 142 \times 10^{23}\ \text{mol}^{-1}$
Universal gas constant	R	$8.314\ 472\ \text{J/(mol} \cdot \text{K)}$
Stefan–Boltzmann constant	σ	$5.670\ 40 \times 10^{-8}\ \text{W/(m}^2 \cdot \text{K}^4)$
Atomic mass unit	u	$1.660\ 538\ 8 \times 10^{-27}\ \text{kg}$

Note: 2006 CODATA recommended values.

TABLE A.2 SI Base Units

Quantity	Unit	Symbol	Quantity	Unit	Symbol
Length	meter	m	Electric current	ampere	A
Mass	kilogram	kg	Temperature	kelvin	K
Time	second	s	Luminous intensity	candela	cd
			Amount of substance	mole	mol

TABLE A.3 Solar System Data

Body	Radius at Equator (km)	Mass (kg)	Average Density (kg/m³)	Orbital Semimajor Axis (km)	Orbital Period (years)	Orbital Eccentricity
Mercury	2,440	3.30×10^{23}	5430	5.79×10^{7}	0.241	0.205
Venus	6,050	4.87×10^{24}	5240	1.08×10^{8}	0.615	0.007
Earth	6,370	5.97×10^{24}	5510	1.50×10^{8}	1.000	0.017
Mars	3,390	6.42×10^{23}	3930	2.28×10^{8}	1.881	0.094
Jupiter	71,500	1.90×10^{27}	1330	7.78×10^{8}	11.86	0.049
Saturn	60,300	5.68×10^{26}	690	1.43×10^{9}	29.46	0.057
Uranus	25,600	8.68×10^{25}	1270	2.87×10^{9}	84.01	0.046
Neptune	24,800	1.02×10^{26}	1640	4.50×10^{9}	164.8	0.011
Pluto	1,200	1.3×10^{22}	1750	5.91×10^{9}	247.7	0.244
Sun	6.95×10^{5}	1.99×10^{30}	1410			
Moon	1,740	7.35×10^{22}	3340	3.84×10^{5}	27.3 days	0.055

Note: From NASA Planetary Fact Sheet (2006).

TABLE A.4 Properties of Selected Isotopes

Atomic Number Z	Element	Chemical Symbol	Chemical Atomic Mass (u)*	Mass Number A	Atomic Mass (u)	Percent Abundance	Half-Life (if radio-active) $T_{1/2}$	Primary Decay Modes†
—	(Proton)	p		1	1.007276			
0	(Neutron)	n		1	1.008 665		10.4 min	β^-
1	Hydrogen	H	1.007 94	1	1.007 825	99.988 5		
	Deuterium	D		2	2.014 102	0.011 5		
	Tritium	T		3	3.016 049		12.33 years	β^-
2	Helium	He	4.002 602	3	3.016 029	0.000 137		
				4	4.002 603	99.999 863		
3	Lithium	Li	6.941	6	6.015 122	7.5		
				7	7.016 004	92.5		
4	Beryllium	Be	9.012 182	7	7.016 929		53.1 days	EC
				9	9.012 182	100		
5	Boron	B	10.811	10	10.012 937	19.9		
				11	11.009 306	80.1		
6	Carbon	C	12.010 7	10	10.016 853		19.3 s	EC, β^+
				11	11.011 434		20.4 min	EC, β^+
				12	12.000 000	98.93		
				13	13.003 355	1.07		
				14	14.003 242		5 730 years	β^-
7	Nitrogen	N	14.006 7	13	13.005 739		9.96 min	EC, β^+, β^-
				14	14.003 074	99.632		
				15	15.000 109	0.368		
8	Oxygen	O	15.999 4	15	15.003 065		122 s	EC, β^+
				16	15.994 915	99.757		
				18	17.999 160	0.205		

*Atomic mass values given are averaged over isotopes in the percentages in which they exist in nature.

†EC denotes decay by electron capture, SF denotes spontaneous fission.

Atomic Number Z	Element	Chemical Symbol	Chemical Atomic Mass (u)*	Mass Number A	Atomic Mass (u)	Percent Abundance	Half-Life (if radio-active) $T_{1/2}$	Primary Decay Modes†
9	Fluorine	F	18.998 403 2	19	18.998 403	100		
10	Neon	Ne	20.179 7	20	19.992 440	90.48		
				22	21.991 385	9.25		
11	Sodium	Na	22.989 77	22	21.994 437		2.60 years	EC, β^+
				23	22.989 770	100		
				24	23.990 963		14.96 h	β^-
12	Magnesium	Mg	24.305 0	24	23.985 042	78.99		
				25	24.985 837	10.00		
				26	25.982 593	11.01		
13	Aluminum	Al	26.981 538	27	26.981 539	100		
14	Silicon	Si	28.085 5	28	27.976 926	92.229 7		
15	Phosphorus	P	30.973 761	31	30.973 762	100		
				32	31.973 907		14.26 days	β^-
16	Sulfur	S	32.066	32	31.972 071	94.93		
				35	34.969 032		87.3 days	β^-
17	Chlorine	Cl	35.452 7	35	34.968 853	75.78		
				37	36.965 903	24.22		
18	Argon	Ar	39.948	40	39.962 383	99.600 3		
19	Potassium	K	39.098 3	39	38.963 707	93.258 1		
				40	39.963 999	0.011 7	1.28×10^9 years	EC, β^+, β^-
20	Calcium	Ca	40.078	40	39.962 591	96.941		
21	Scandium	Sc	44.955 910	45	44.955 910	100		
22	Titanium	Ti	47.867	48	47.947 947	73.72		
23	Vanadium	V	50.941 5	51	50.943 964	99.750		
24	Chromium	Cr	51.996 1	52	51.940 512	83.789		
25	Manganese	Mn	54.938 049	55	54.938 050	100		
26	Iron	Fe	55.845	56	55.934 942	91.754		
27	Cobalt	Co	58.933 200	59	58.933 200	100		
				60	59.933 822		5.27 years	β^-
28	Nickel	Ni	58.693 4	58	57.935 348	68.076 9		
				60	59.930 790	26.223 1		
29	Copper	Cu	63.546	63	62.929 601	69.17		
				65	64.927 794	30.83		
30	Zinc	Zn	65.39	64	63.929 147	48.63		
				66	65.926 037	27.90		
				68	67.924 848	18.75		
31	Gallium	Ga	69.723	69	68.925 581	60.108		
				71	70.924 705	39.892		
32	Germanium	Ge	72.61	70	69.924 250	20.84		
				72	71.922 076	27.54		
				74	73.921 178	36.28		
33	Arsenic	As	74.921 60	75	74.921 596	100		

*Atomic mass values given are averaged over isotopes in the percentages in which they exist in nature.

†EC denotes decay by electron capture, SF denotes spontaneous fission.

Atomic Number Z	Element	Chemical Symbol	Chemical Atomic Mass (u)*	Mass Number A	Atomic Mass (u)	Percent Abundance	Half-Life (if radio-active) $T_{1/2}$	Primary Decay Modes†
34	Selenium	Se	78.96	78	77.917 310	23.77		
				80	79.916 522	49.61		
35	Bromine	Br	79.904	79	78.918 338	50.69		
				81	80.916 291	49.31		
36	Krypton	Kr	83.80	82	81.913 485	11.58		
				83	82.914 136	11.49		
				84	83.911 507	57.00		
				86	85.910 610	17.30		
37	Rubidium	Rb	85.467 8	85	84.911 789	72.17		
				87	86.909 184	27.83	4.75×10^{10} years	β^-
38	Strontium	Sr	87.62	86	85.909 262	9.86		
				88	87.905 614	82.58		
				90	89.907 738		158 s	β^-
39	Yttrium	Y	88.905 85	89	88.905 848	100		
40	Zirconium	Zr	91.224	90	89.904 704	51.45		
				91	90.905 645	11.22		
				92	91.905 040	17.15		
				94	93.906 316	17.38		
41	Niobium	Nb	92.906 38	93	92.906 378	100		
42	Molybdenum	Mo	95.94	92	91.906 810	14.84		
				95	94.905 842	15.92		
				96	95.904 679	16.68		
				98	97.905 408	24.13		
43	Technetium	Tc		98	97.907 216		4.2×10^6 years	β^-
				99	98.906 255		2.1×10^5 years	β^-
44	Ruthenium	Ru	101.07	99	98.905 939	12.76		
				100	99.904 220	12.60		
				101	100.905 582	17.06		
				102	101.904 350	31.55		
				104	103.905 430	18.62		
45	Rhodium	Rh	102.905 50	103	102.905 504	100		
46	Palladium	Pd	106.42	104	103.904 035	11.14		
				105	104.905 084	22.33		
				106	105.903 483	27.33		
				108	107.903 894	26.46		
				110	109.905 152	11.72		
47	Silver	Ag	107.868 2	107	106.905 093	51.839		
				109	108.904 756	48.161		
48	Cadmium	Cd	112.411	110	109.903 006	12.49		
				111	110.904 182	12.80		
				112	111.902 757	24.13		
				113	112.904 401	12.22	7.7×10^{15} years	β^-
				114	113.903 358	28.73		

*Atomic mass values given are averaged over isotopes in the percentages in which they exist in nature.

†EC denotes decay by electron capture, SF denotes spontaneous fission.

Atomic Number Z	Element	Chemical Symbol	Chemical Atomic Mass (u)*	Mass Number A	Atomic Mass (u)	Percent Abundance	Half-Life (if radio-active) $T_{1/2}$	Primary Decay Modes†
49	Indium	In	114.818	115	114.903 878	95.71	4.4×10^{14} years	β^-
50	Tin	Sn	118.710	116	115.901 744	14.54		
				118	117.901 606	24.22		
				120	119.902 197	32.58		
51	Antimony	Sb	121.760	121	120.903 818	57.21		
				123	122.904 216	42.79		
52	Tellurium	Te	127.60	126	125.903 306	18.84		
				128	127.904 461	31.74	2.2×10^{24} years	β^-
				130	129.906 223	34.08	7.9×10^{20} years	β^-
53	Iodine	I	126.904 47	127	126.904 468	100		
				129	128.904 988		1.6×10^{7} years	β^-
54	Xenon	Xe	131.29	129	128.904 780	26.44		
				131	130.905 082	21.18		
				132	131.904 145	26.89		
				134	133.905 394	10.44		
				136	135.907 220	8.87	$\geq 2.36 \times 10^{21}$ years	β^-
55	Cesium	Cs	132.905 45	133	132.905 447	100		
56	Barium	Ba	137.327	137	136.905 821	11.232		
				138	137.905 241	71.698		
57	Lanthanum	La	138.905 5	139	138.906 349	99.910		
58	Cerium	Ce	140.116	140	139.905 434	88.450		
				142	141.909 240	11.114	$> 2.6 \times 10^{17}$ years	β^-
59	Praseodymium	Pr	140.907 65	141	140.907 648	100		
60	Neodymium	Nd	144.24	142	141.907 719	27.2		
				144	143.910 083	23.8	2.3×10^{15} years	α
				146	145.913 112	17.2		
61	Promethium	Pm		145	144.912 744		17.7 years	EC, α
62	Samarium	Sm	150.36	147	146.914 893	14.99	1.06×10^{11} years	α
				149	148.917 180	13.82		
				152	151.919 728	26.75		
				154	153.922 205	22.75		
63	Europium	Eu	151.964	151	150.919 846	47.81		
				153	152.921 226	52.19		
64	Gadolinium	Gd	157.25	156	155.922 120	20.47		
				158	157.924 100	24.84		
				160	159.927 051	21.86		
65	Terbium	Tb	158.925 34	159	158.925 343	100		
66	Dysprosium	Dy	162.50	162	161.926 796	25.51		
				163	162.928 728	24.90		
				164	163.929 171	28.18		
67	Holmium	Ho	164.930 32	165	164.930 320	100		

*Atomic mass values given are averaged over isotopes in the percentages in which they exist in nature.

†EC denotes decay by electron capture, SF denotes spontaneous fission.

Atomic Number Z	Element	Chemical Symbol	Chemical Atomic Mass (u)*	Mass Number A	Atomic Mass (u)	Percent Abundance	Half-Life (if radio-active) $T_{1/2}$	Primary Decay Modes†
68	Erbium	Er	167.6	166	165.930 290	33.61		
				167	166.932 045	22.93		
				168	167.932 368	26.78		
69	Thulium	Tm	168.934 21	169	168.934 211	100		
70	Ytterbium	Yb	173.04	172	171.936 378	21.83		
				173	172.938 207	16.13		
				174	173.938 858	31.83		
71	Lutecium	Lu	174.967	175	174.940 768	97.41		
72	Hafnium	Hf	178.49	177	176.943 220	18.60		
				178	177.943 698	27.28		
				179	178.945 815	13.62		
				180	179.946 549	35.08		
73	Tantalum	Ta	180.947 9	181	180.947 996	99.988		
74	Tungsten	W	183.84	182	181.948 206	26.50		
				183	182.950 224	14.31		
				184	183.950 933	30.64	$> 3 \times 10^{19}$ years	α
				186	185.954 362	28.43		
75	Rhenium	Re	186.207	185	184.952 956	37.40		
				187	186.955 751	62.60	4.1×10^{10} years	β^-
76	Osmium	Os	190.23	188	187.955 836	13.24		
				189	188.958 145	16.15		
				190	189.958 445	26.26		
				192	191.961 479	40.78		
77	Iridium	Ir	192.217	191	190.960 591	37.3		
				193	192.962 924	62.7		
78	Platinum	Pt	195.078	194	193.962 664	32.967		
				195	194.964 774	33.832		
				196	195.964 935	25.242		
79	Gold	Au	196.966 55	197	196.966 552	100		
80	Mercury	Hg	200.59	199	198.968 262	16.87		
				200	199.968 309	23.10		
				201	200.970 285	13.18		
				202	201.970 626	29.86		
81	Thallium	Tl	204.383 3	203	202.972 329	29.524		
				205	204.974 412	70.476		
				208	207.982 005		3.053 min	β^-
				210	209.990 066		1.30 min	β^-

*Atomic mass values given are averaged over isotopes in the percentages in which they exist in nature.

†EC denotes decay by electron capture, SF denotes spontaneous fission.

Atomic Number Z	Element	Chemical Symbol	Chemical Atomic Mass (u)*	Mass Number A	Atomic Mass (u)	Percent Abundance	Half-Life (if radio-active) $T_{1/2}$	Primary Decay Modes†
82	Lead	Pb	207.2	204	203.973 029	1.4	$\geq 1.4 \times 10^{17}$ years	α
				206	205.974 449	24.1		
				207	206.975 881	22.1		
				208	207.976 636	52.4		
				210	209.984 173		22.2 years	α, β^-
				211	210.988 732		36.1 min	β^-
				212	211.991 888		10.64 h	β^-
				214	213.999 798		26.8 min	β^-
83	Bismuth	Bi	208.980 38	209	208.980 383	100		
				211	210.987 258		2.14 min	α, β^-
84	Polonium	Po		210	209.982 857		138.38 days	α
				214	213.995 186		164 μs	α
85	Astatine	At		218	218.008 682		1.5 s	α, β^-
86	Radon	Rn		222	222.017 570		3.823 days	α, β^-
87	Francium	Fr		223	223.019 731		22 min	α, β^-
88	Radium	Ra		226	226.025 403		1 600 years	α
				228	228.031 064		5.75 years	β^-
89	Actinium	Ac		227	227.027 747		21.77 years	α, β^-
90	Thorium	Th	232.038 1					
				228	228.028 731		1.912 years	α
				232	232.038 050	100	1.40×10^{10} years	α
91	Protactinium	Pa	231.035 88	231	231.035 879		3.28×10^{4} years	α
92	Uranium	U	238.028 9	232	232.037 146		69 years	α
				233	233.039 628		1.59×10^{5} years	α
				235	235.043 923	0.720 0	7.04×10^{8} years	α, SF
				236	236.045 562		2.34×10^{7} years	α, SF
				238	238.050 783	99.274 5	4.47×10^{9} years	α
93	Neptunium	Np		237	237.048 167		2.14×10^{6} years	α
94	Plutonium	Pu		239	239.052 156		2.412×10^{4} years	α
				242	242.058 737		3.75×10^{5} years	α
				244	244.064 198		8.1×10^{7} years	α, SF

*Atomic mass values given are averaged over isotopes in the percentages in which they exist in nature.

†EC denotes decay by electron capture, SF denotes spontaneous fission.

Sources: Chemical atomic masses are from T. B. Coplen, "Atomic weights of the elements 1999," a technical report to the International Union of Pure and Applied Chemistry and published in *Pure and Applied Chemistry* 73(4):667–683 (2001). Atomic masses of the isotopes are from G. Audi and A. H. Wapstra, "The 1995 update to the atomic mass evaluation," *Nuclear Physics* A595(4):409–480 (December 25, 1995). Percent abundance values are from K. J. R. Rosman and P. D. P. Taylor, "Isotopic compositions of the elements 1999," a technical report to the International Union of Pure and Applied Chemistry and published in *Pure and Applied Chemistry* 70(1):217–236 (1998). Data on decay modes and half-lives are from Lawrence Berkeley Laboratory Isotopes Project Listing and the National Nuclear Data Center, Brookhaven National Laboratory.

APPENDIX

B Mathematical Review

B.1 Mathematical Symbols

The following symbols are used to express mathematical relationships.

Symbol	Meaning	Example	Explanation
$+$	Addition	$a + b = c$	a plus b equals c
$-$	Subtraction	$a - b = c$	a minus b equals c
$\times$ or $\cdot$	Multiplication	$a \times b = c$ $a \cdot b = c$	a times b equals c
$/$ or $\div$	Division	$a/b = c$ $a \div b = c$	a divided by b equals c
$>$	Greater than	$a > b$	a is greater than b
$<$	Less than	$a < b$	a is less than b
$\geq$	Greater than or equal to	$a \geq b$	a is greater than or equal to b
$\leq$	Less than or equal to	$a \leq b$	a is less than or equal to b
$\approx$	Approximately equal to	$a \approx b$	a is approximately equal to b
$\lim\limits_{x \to 0}$	Limit as x approaches 0	$y = \lim\limits_{x \to 0} f$	y equals the value of f as the variable x approaches zero

B.2 Algebra and Working with Equations

The basic rules of algebra are reviewed in Section 1.6. When solving an equation to find the value of a single variable, you should rearrange the equation to put the *unknown* variable (the variable you wish to solve for) on one side of the equation and the *known* quantities on the other side. For example, suppose the variable a in the equation

$$4a + 7 = T - 9$$

is the unknown. The goal is to isolate a on one side of the equal sign (usually the left). In this example, we can subtract 7 from both sides to get

$$4a = T - 16$$

We next divide both sides by the constant factor 4 to get the solution

$$a = \frac{T}{4} - 4$$

When dealing with two equations containing two unknowns, the usual approach is to use one equation to eliminate one of the unknowns from the other equation. For example, suppose a and T are unknown in the equations

$$3a = T - 9 \quad \textbf{(B2.1)}$$

$$2a = -T + 24 \quad \textbf{(B2.2)}$$

We can rearrange Equation B2.2 to get

$$T = -2a + 24$$

We then substitute this expression for T in Equation B2.1 to find

$$3a = T - 9 = (-2a + 24) - 9 = -2a + 15 \tag{B2.3}$$

We can now solve for a:

$$5a = 15$$
$$a = 3$$

This value can then be used in Equation B2.1 to find the value of T; we get $T = 18$.

Another approach when working with two equations and two unknowns is to simply add the two equations. For example, if we add Equations B2.1 and B2.2, we will eliminate T:

$$\begin{aligned} 3a &= T - 9 \\ 2a &= -T + 24 \\ \hline 5a &= -9 + 24 \end{aligned}$$

which is another way to get Equation B2.3.

The equations considered in Equations B2.1 through B2.3 all involve the unknowns (e.g., a and T) raised to the first power. In a ***quadratic equation***, the unknown quantity is raised to the second power. If x is an unknown, an equation of this type is

$$ax^2 + bx + c = 0$$

The solution is given by the ***quadratic formula***,

$$x = \frac{-b \pm \sqrt{b^2 - 4ac}}{2a} \tag{B2.4}$$

There are two solutions as indicated by the appearance of the $\pm$ sign.

B.3 Scientific Notation

Scientific notation is a convenient way to express extremely large or extremely small numbers. For example, the number 640,000 is written as 6.4×10^5 in scientific notation. To express a number in scientific notation, move the decimal point in the original number to obtain a new number between 1 and 10. Count the number of places the decimal point has been moved; this number will become the exponent of 10 in scientific notation. If you started with a number greater than 10 (such as 150,000,000,000), the exponent of 10 is positive (1.5×10^{11}). If you started with a number less than 1 (such as 0.000055), the exponent is negative (5.5×10^{-5}).

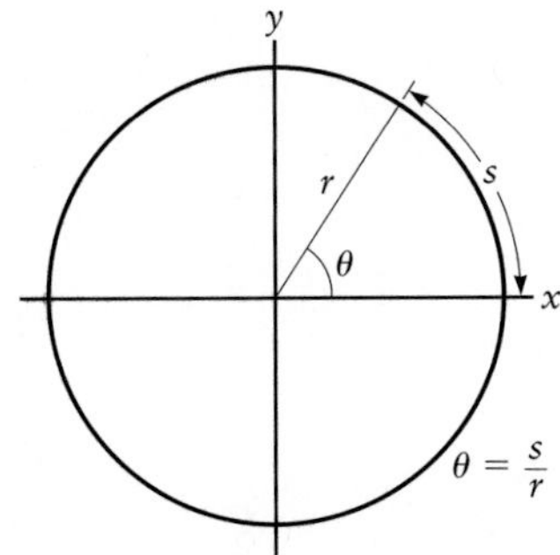

▲ **Figure B.1** The angle θ measured in radians equals the ratio of the length s measured along the circular arc divided by the radius r of the circle.

B.4 Geometry and Trigonometry

One way to define and measure angles is shown in Figure B.1. The angle θ equals the ratio of the arc length s measured along the circle divided by the radius r of the circle:

$$\theta = \frac{s}{r} \tag{B4.1}$$

This relation gives the value of θ in ***radians***. To convert to degrees, we use the fact that one complete "trip" around the circle in Figure B.1 corresponds to both 2π radians and to 360°. We thus have

$$\theta \text{ (in radians)} \times \frac{360°}{2\pi \text{ radians}} = \theta \text{ (in degrees)} \tag{B4.2}$$

Figure B.2 shows a right triangle with sides x, y, and r, where r is the hypotenuse. According to the ***Pythagorean theorem***,

$$x^2 + y^2 = r^2 \tag{B4.3}$$

The trigonometric functions ***sine***, ***cosine***, and ***tangent*** are defined as

$$\sin\theta = y/r \tag{B4.4}$$
$$\cos\theta = x/r \tag{B4.5}$$
$$\tan\theta = y/x \tag{B4.6}$$

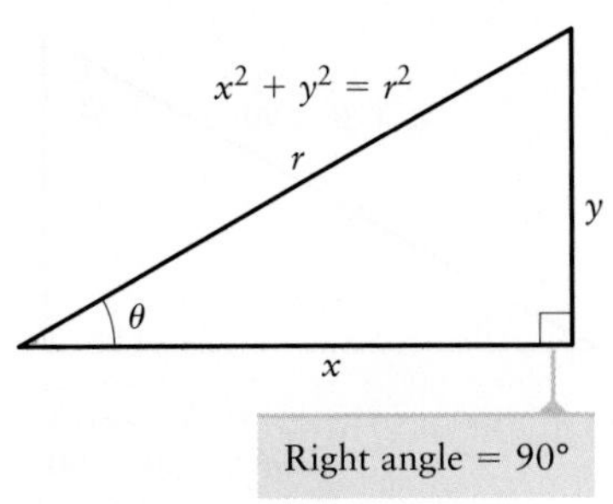

▲ **Figure B.2** The Pythagorean theorem is a relation between the lengths of the sides of a right triangle.

The Pythagorean theorem (Eq. B4.3) implies that

$$\sin^2\theta + \cos^2\theta = 1 \tag{B4.7}$$

for any value of the angle θ. Equations B4.3 through B4.6 also lead to a number of other trigonometric relations (called ***identities***), some of which are the following:

$$\sin(-\theta) = -\sin\theta \qquad \cos(-\theta) = \cos\theta \qquad \tan(-\theta) = -\tan\theta$$
$$\sin(\theta + 90°) = \sin(\theta + \pi/2) = \cos\theta$$
$$\cos(\theta + 90°) = \cos(\theta + \pi/2) = -\sin\theta$$
$$\sin(\alpha \pm \beta) = \sin\alpha\cos\beta \pm \cos\alpha\sin\beta$$
$$\cos(\alpha \pm \beta) = \cos\alpha\cos\beta \mp \sin\alpha\sin\beta \tag{B4.8}$$

The relations in Equation B4.8 hold for any values of the angles θ, α, and β. If $\alpha = \beta$, the relation for $\sin(\alpha + \beta)$ becomes

$$\sin(\alpha + \alpha) = \sin(2\alpha) = \sin\alpha\cos\alpha + \cos\alpha\sin\alpha$$

We can also replace the angle α with θ; rearranging then leads to

$$\sin\theta\cos\theta = \tfrac{1}{2}\sin(2\theta) \tag{B4.9}$$

Equations B4.4 through B4.6 tell how to calculate the sine, cosine, and tangent of an angle from the lengths of the sides of the associated right triangle. We can also use this approach to find the value of the angle. For example, if we know the values of y and r, we can solve for the angle as

$$\sin\theta = y/r$$
$$\theta = \sin^{-1}(y/r) \tag{B4.10}$$

The function $\sin^{-1}$ is the ***inverse*** sine function, also known as the ***arcsine*** ($=$ arcsin $= \sin^{-1}$). In words, Equation B4.10 says that θ equals the angle whose sine is y/r. Scientific calculators all contain this function (as a single "button"). If the values of y and r are known, you must only take the ratio y/r and then compute the arcsine of this value to find θ in Equation B4.10. Likewise, there are also functions for the inverse cosine ($\cos^{-1} =$ arccos) and inverse tangent ($\tan^{-1} =$ arctan):

$$\theta = \cos^{-1}(x/r) \quad \text{and} \quad \theta = \tan^{-1}(y/x) \tag{B4.11}$$

We often need to deal with triangles like the one in Figure B.3, which shows a right triangle with interior angles α and β. The sum of the three interior angles of a triangle is always 180°. For the triangle in Figure B.3, one of the angles is 90° (it is a right triangle); hence,

$$\alpha + \beta = 90° \tag{B4.12}$$

Two angles whose sum is 90° are called complementary angles. From the definitions of the sine and cosine functions, for any pair of complementary angles α and β we have

$$\sin\alpha = \cos\beta$$
$$\cos\alpha = \sin\beta$$

Two approximate relations involving trigonometric functions are also useful. When the angle θ is small,

$$\sin\theta \approx \theta \quad \text{and} \quad \tan\theta \approx \theta \tag{B4.13}$$

which are good approximations (accurate to a few percent or better) when θ is smaller than about 30°. Notice that the relations in Equation B4.13 apply only when the angle is measured in radians.

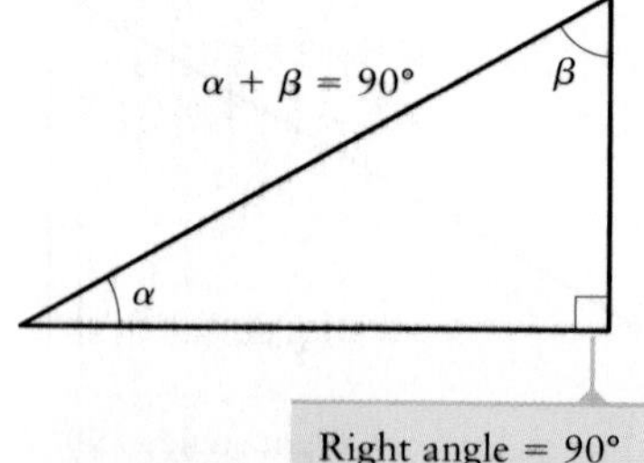

▲ **Figure B.3** The sum of the three interior angles of a right triangle is 180°. Hence, $\alpha + \beta = 90°$.

B.5 Vectors

A vector quantity has both a magnitude and a direction. There are two ways to do calculations with vectors; one is a graphical approach, which is useful for approximate or qualitative calculations, and the other involves the components of the vectors.

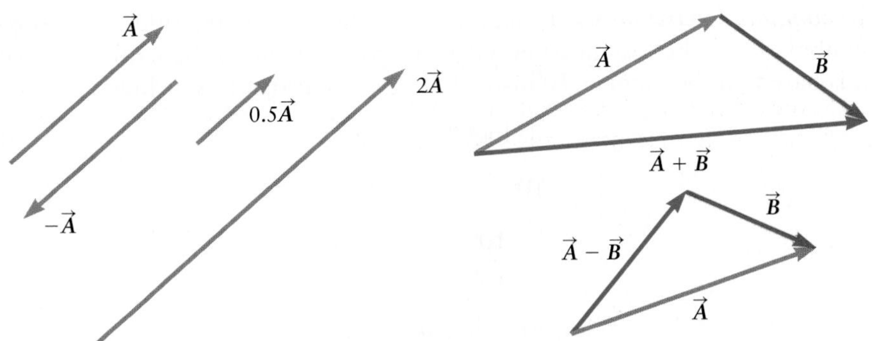

◀ **Figure B.4** Examples showing the effect of multiplying a vector by a scalar, adding two vectors, and subtracting one vector from another.

The graphical approach is illustrated in Figure B.4, which shows two vectors $\vec{A}$ and $\vec{B}$ along with their sum $(\vec{A} + \vec{B})$ and difference $(\vec{A} - \vec{B})$. Multiplying a vector by a scalar (e.g., 0.5 or 2) changes the length but does not change the direction. Multiplying by a negative number (e.g., -1) reverses the direction. The vectors $-\vec{A}$, $0.5\vec{A}$, and $2\vec{A}$ are also shown in Figure B.4.

The components of a vector can be obtained using trigonometry. The length of a vector $\vec{A}$, also called its ***magnitude***, is denoted by either A or $|\vec{A}|$. If this vector makes an angle θ with the x axis, the ***components*** are

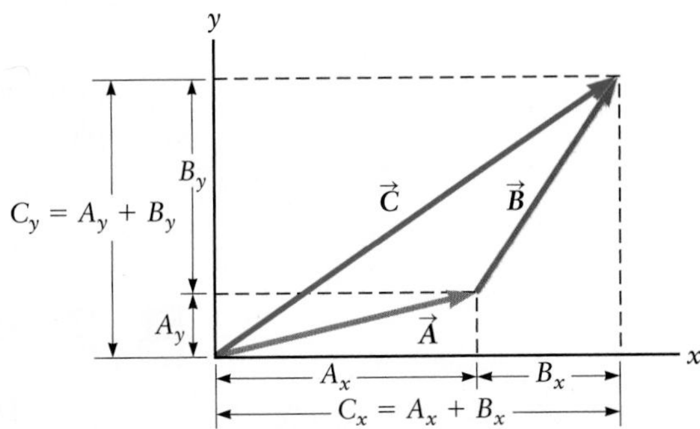

▲ **Figure B.5** To compute the sum of two vectors, we add their components. Here we add the x components of $\vec{A}$ and $\vec{B}$ to get the x component of $\vec{C}$. We follow the same procedure to find the y component of $\vec{C}$.

$$A_x = A\cos\theta = |\vec{A}|\cos\theta \text{ and } A_y = A\sin\theta = |\vec{A}|\sin\theta \tag{B5.1}$$

The sum of two vectors can be found by adding components, as illustrated in Figure B.5. If

$$\vec{C} = \vec{A} + \vec{B} \tag{B5.2}$$

in terms of the components of these vectors we then have

$$C_x = A_x + B_x \quad \text{and} \quad C_y = A_y + B_y \tag{B5.3}$$

B.6 Exponential Functions and Logarithms

Exponential functions and logarithms play a role in many areas of physics, including the behavior of fluids (Chapter 10), sound (Chapter 13), electric circuits (Chapters 19 and 22), and radioactive decay (Chapter 30).

The number $e = 2.71828\ldots$ plays an important role. Useful numerical values involving the number e are $e^{-1} \approx 0.37$ and $1 - e^{-1} \approx 0.73$.

There are two versions of the logarithm function. The ***natural logarithm*** function is denoted by the symbol "ln" and is related to the exponential function by

$$\ln(e^a) = a \tag{B6.1}$$

where a is a number. Other useful properties of these functions are

$$\begin{gathered} e^a \times e^b = e^{a+b} \\ \frac{e^a}{e^b} = e^{a-b} \\ \ln(a) + \ln(b) = \ln(ab) \\ \ln(a) - \ln(b) = \ln(a/b) \\ \ln(a^b) = b\ln a \end{gathered} \tag{B6.2}$$

The *common logarithm* function (also called the base 10 logarithm) is denoted by the symbol "log"; the log function has properties similar to the natural logarithm, but it is based on the number 10 instead of e. These properties include

$$\log(10^a) = a \tag{B6.3}$$

$$\begin{gathered} 10^a \times 10^b = 10^{a+b} \\ \frac{10^a}{10^b} = 10^{a-b} \\ \log(a) + \log(b) = \log(ab) \\ \log(a) - \log(b) = \log(a/b) \\ \log(a^b) = b \log a \end{gathered} \tag{B6.4}$$

Exponential functions may also involve base numbers besides e and 10. For example, for any number y, one has

$$y^a \times y^b = y^{a+b} \quad \text{and} \quad \frac{y^a}{y^b} = y^{a-b} \tag{B6.5}$$

B.7 Some Relations Useful in Special Relativity

In special relativity, factors such as v/c and v^2/c^2, where c is the speed of light and v is the speed of an object, appear often. The speed of light is very large, so the ratio v/c is usually very small. When that is the case, certain algebraic expressions can be simplified. For example, when A is a very small number, the approximation

$$\frac{1}{1 + A} \approx 1 - A \tag{B7.1}$$

is very accurate. If we replace A by v/c, we find

$$\frac{1}{1 + v/c} \approx 1 - v/c \tag{B7.2}$$

and in a similar way, we get

$$\frac{1}{1 + (v/c)^2} \approx 1 - (v/c)^2 \tag{B7.3}$$

Another expression that arises in special relativity, especially in problems involving time dilation and length contraction, is $\sqrt{1 - (v/c)^2}$. When v/c is very small, to a good approximation we have

$$\sqrt{1 - v^2/c^2} \approx 1 - \frac{v^2}{2c^2} \tag{B7.4}$$

and

$$\frac{1}{\sqrt{1 - v^2/c^2}} \approx 1 + \frac{v^2}{2c^2} \tag{B7.5}$$

The accuracy of these approximations depends on the value of v/c. If $v/c = 0.1$, the approximation in Equation B7.2 is good to about 1%, whereas Equations B7.4 and B7.5 are accurate to about 0.005%. The accuracy improves for smaller values of v/c.

Answers to Concept Checks and Odd-Numbered Problems

CHAPTER 1

Concept Checks

1.1 Significant Figures

Quantity	Length or Distance in Meters (m)	Number of Significant Figures
Diameter of a proton	1×10^{-15}	1
Diameter of a red blood cell	8×10^{-6}	1
Diameter of a human hair	5.5×10^{-5}	2
Thickness of a piece of paper	6.4×10^{-5}	2
Diameter of a compact disc	0.12	2
Height of the author	1.80	3
Height of the Empire State Building	443.2	4
Distance from New York City to Chicago	1,268,000	4
Circumference of the Earth	4.00×10^{7}	3
Distance from Earth to the Sun	1.5×10^{11}	2

1.2 Using Prefixes and Powers of 10

Quantity	Length or Distance in Meters (m)	Same Length or Distance Using Prefix
Diameter of a proton	1×10^{-15}	1 femtometer = 1 fm
Diameter of a red blood cell	8×10^{-6}	8 micrometers = 8 μm
Diameter of a human hair	5.5×10^{-5}	55 micrometers = 55 μm
Thickness of a sheet of paper	6.4×10^{-5}	64 micrometers = 64 μm
Diameter of a compact disc	0.12	1.2 decimeters = 1.2 dm = 12 centimeters = 12 cm
Height of the author	1.80	1.80 meters = 1.80 m
Height of the Empire State Building	443.2	443.2 meters = 443.2 m
Distance from New York City to Chicago	1,268,000	1.268 megameters = 1.268 Mm
Circumference of the Earth	4.00×10^{7}	40.0 megameters = 40.0 Mm
Distance from Earth to the Sun	1.5×10^{11}	150 gigameters = 150 Gm

1.3 (a), (c), and (d)

Problems

1. (R) 7×10^6 pieces of paper
3. (RT) Approximately 2×10^6 steps
5. TABLE 1.2

Quantity	Number of Significant Figures for Each Time Listed
Time for light to travel 1 meter	3
Time between heartbeats (approximate)	1
Time for light to travel from the Sun to Earth	1
1 day	exact
1 month (30 days)	exact
Human lifespan (approximate)	1
Age of the universe	1

TABLE 1.3

Object	Number of Significant Figures for Each Mass Listed
Electron	2
Proton	2
Red blood cell	1
Mosquito	1
Typical person	1
Automobile	2
Earth	2
Sun	2

7. 8.9 km
9. 5×10^3 kg/m^3
11. (a) 2.99792458×10^8 m/s (b) 2.998×10^8 m/s
13. (RT) About 1×10^8 m; about the size of Uranus
15. (a) 6.37×10^6 m (b) 3.84×10^5 km (c) 5.97×10^{24} kg
17. (R) About 2500 cm^3
19. (a) 1.10×10^2 m (b) 1.10×10^5 mm (c) 3.60×10^2 ft (d) 4.32×10^3 in
21. (a) 5.0×10^2 cm (b) 16 ft (c) 2.0×10^2 in (d) 3.1×10^{-3} mi
23. 0.273 kg
25. (a) 1.2×10^3 cm^3 (b) 1.2×10^{-3} m^3
27. (a) 1×10^6 cm^3 (b) 1×10^9 mm^3
29. The Corvette C4 engine has 7.44 L of combustion volume, larger than that of the Corvette Z06 with just 7.0 L. Yes, the newer car has more power despite having a smaller engine.
31. (a) ML/T (b) L^2/T^2 (c) L^2/T
33. 380 m/s
35. 1.008 g/mol (hydrogen); 12.01 g/mol (carbon)
37. (a) 5.8 m (b) 40° = 0.70 rad; 50° = 0.87 rad (c) 0.64; 0.77; 0.82 (d) 0.77; 0.64; 1.2
39. 0.76 m; 0.64 m; 1.0 m or any set of lengths with the same ratios
41. 71.6°
43. 95 m
45. 0.52 km
47. 59 m; 23°
49. Vectors (a), (b), (c), and (f) have the same magnitude; (e) and (h) have the same magnitude; and (d) and (g) have the same magnitude.
51. (c)
53. 47 km; 250°
55.

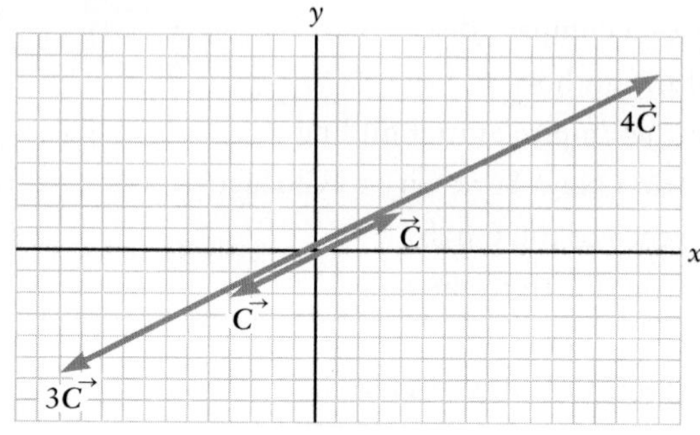

57. (a) $C_x = 22$; $C_y = 30$ (b) $C_x = 5.0$; $C_y = -17$ (c) $C_x = 48$; $C_y = 100$ (d) $C_x = -73$; $C_y = -171$
59. 2.3 km; 34° north of east
61. (RT) 1.3×10^8 beats
63. (a) 640 acres/mi^2 (exact) (b) 4050 m^2/acre (three significant figures) (c) 43,560 ft^2/acre (exact); 27,878,400 ft^2/mi^2 (exact)
65. 97
67. (R) 2×10^{15} cells
69. (R) 3.3
71. 49 m
73. Earth: 5500 kg/m^3; Mars: 3900 kg/m^3; Jupiter: 1200 kg/m^3; Saturn: 620 kg/m^3. Saturn would float in water.
75. (R) 30 miles ≈ 50 km

CHAPTER 2

Concept Checks

2.1 (a) is inconsistent, (b) is inconsistent, (c) may be consistent.

2.2 Yes, at $t \approx 1.6$ s in Figure 2.9.

2.3 (a) graph 3; (b) graph 2; (c) graph 1.

2.4 (b) and (d)

2.5 (b)

2.6 (a) and (c) are not action–reaction pairs.

Problems

1. (a), (d), (e), and (f)
3. (b), (c), (d), and (f)
5. 7.0 m/s

7. (a) 2.3 m/s (b) v_{max} = 3.5 m/s; v_{min} = 0. The average is closer to the maximum velocity.

9.

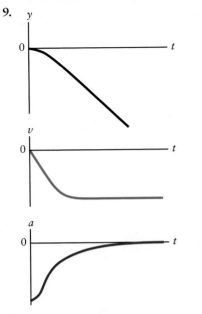

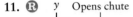

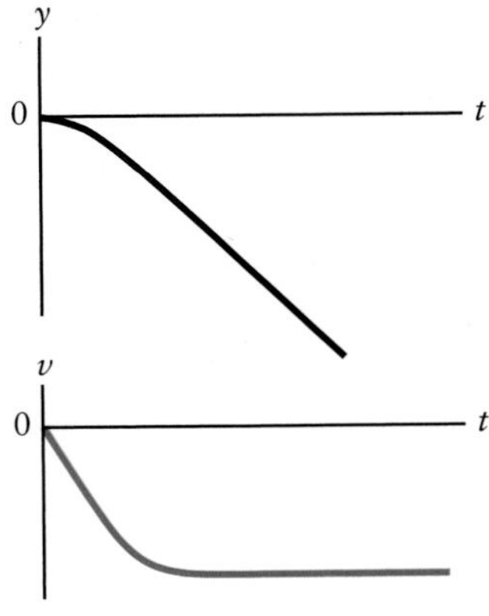

11. Ⓡ

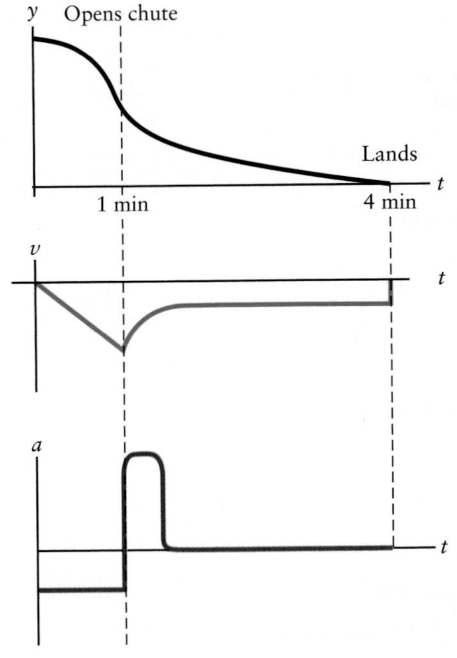

13.

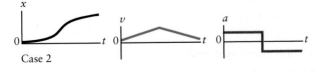

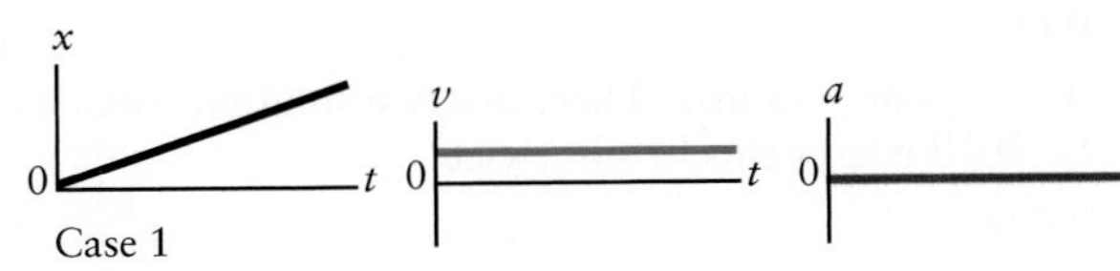

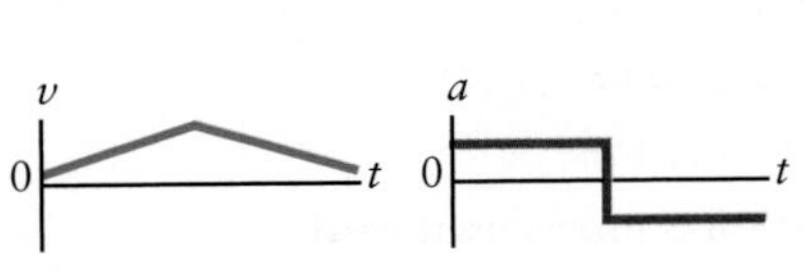

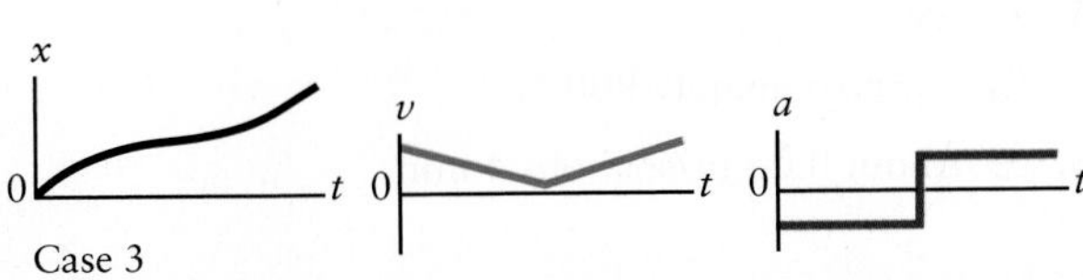

15.

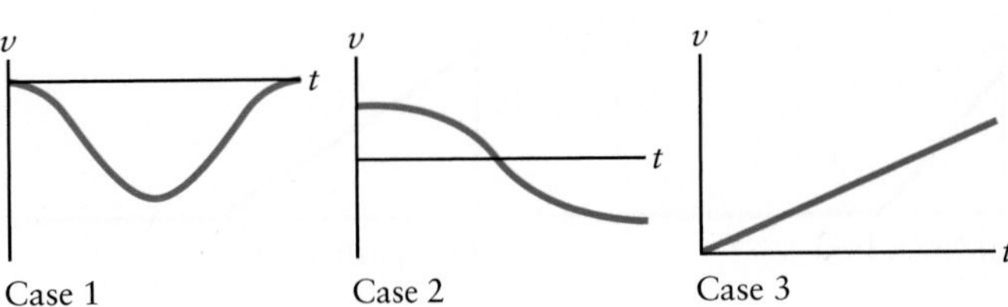

17.

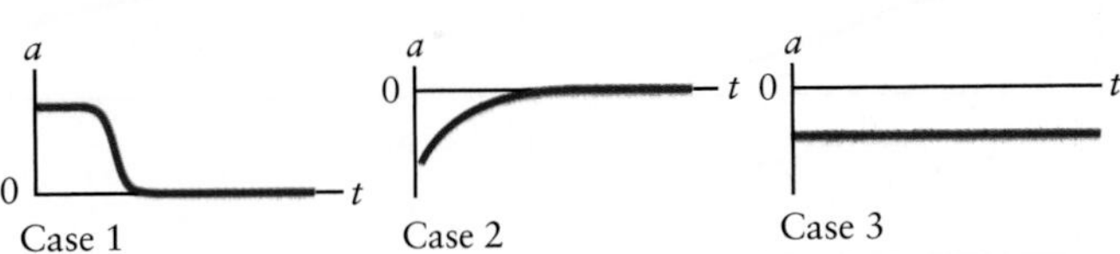

19. Case 1: A car starting from a traffic light (at rest), accelerating to the speed limit, and traveling at that speed; Case 2: A car slowing to a stop at a traffic light; Case 3: A ball thrown upward from the top of a building. The ball starts with some initial speed that decreases linearly due to gravity.

21. Case 1: A driver keeps his foot steady on the gas pedal to keep the car moving at constant acceleration. The driver then shifts gears and the car is still accelerating, but the value of *a* is then smaller than at the start. Case 2: A motorist applies his brakes so as to slow down to match the speed of a car in front of him. Case 3: A car is moving initially at constant velocity in a lane on the freeway; the driver then decides to momentarily accelerate and change to a faster-moving lane.

23. (a) Matches Case 2 in Figure P2.23
(b) Matches Case 4 in Figure P2.23
(c) Matches Case 3 in Figure P2.23
(d) Matches Case 1 in Figure P2.23

25. (a) About 0.2 m/s (b) About 0.4 m/s

27. -10 m/s^2

29. 0.3 m/s^2, 0.2 m/s^2

31. (a) (b) (c)

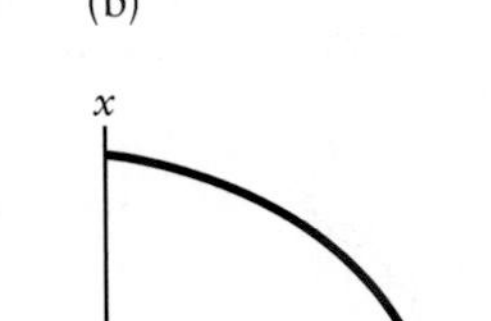

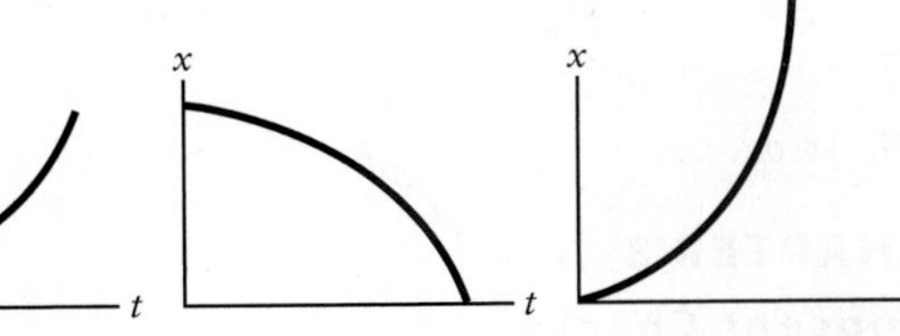

33. -5.0 m/s, -15 m/s, -10 m/s^2

35. (a) 1.5 m/s^2 (b) About 1.0 m/s^2 (c) About 6 m/s^2 (d) 0

37. (a) 340 s (b) 1200 s

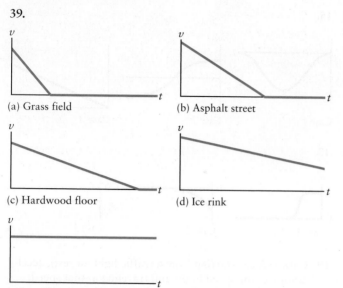

39.

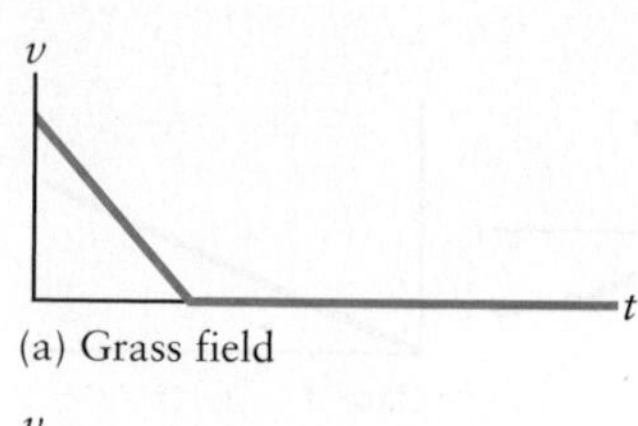

(a) Grass field

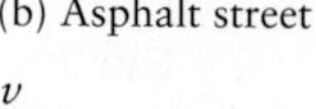

(b) Asphalt street

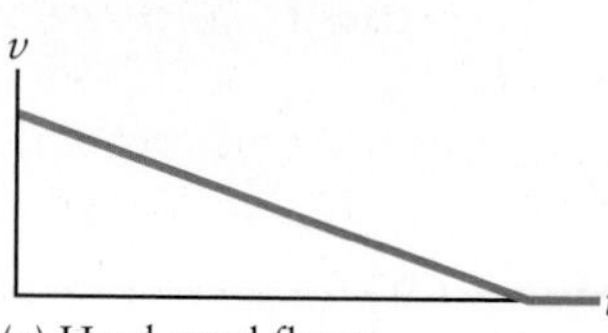

(c) Hardwood floor

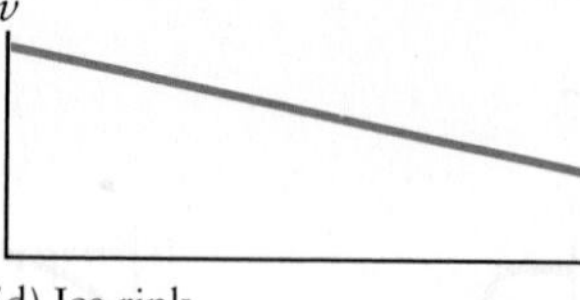

(d) Ice rink

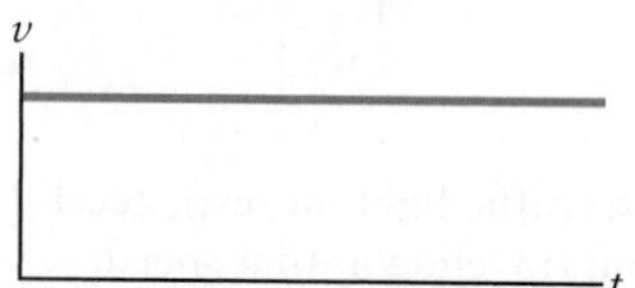

(e) Ideal frictionless surface

41. (a), directly below the drop point. From the point of view of the man on the railcar, the path of the ball is identical to what it would look like were he to drop the ball while standing on a sidewalk.

43. 2.0 m/s^2

45. (c), (d), and (e)

47. 670 N

49. 11,000 kg

51. (a) Highway 99; 4 min (b) 5.5 min

53. (a) 3.0 m/s (b) 4.5 m/s (c) 1.5 m/s (d) The average of the answers found for parts (b) and (c) must be equal to the answer to part (a).

55. (a) 2.4 s

(b)

(c)

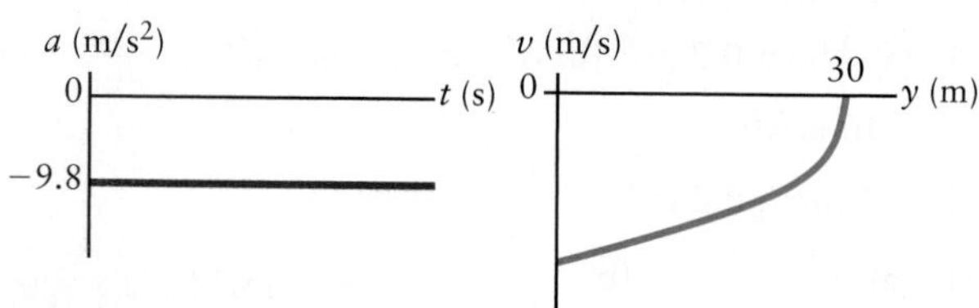

57. 5.8×10^9 m

59. 18 m

CHAPTER 3

Concept Checks

3.1 In Figure 3.7, $\vec{F}_{grav}$ and $\vec{N}$ are *not* an action–reaction pair. The reaction force to $\vec{N}$ is the contact force of the person on the floor.

3.2 The driver can drive with an acceleration smaller than $a_{crate,\,max}$ for a very long time and thereby reach a very large speed, and the crate will not slip.

3.3 (a) Opposite; the velocity is to the right, while the acceleration is directed to the left.
(b) Parallel; $\vec{v}$ and $\vec{a}$ are both directed to the right.
(c) Parallel; both $\vec{v}$ and $\vec{a}$ are directed downward.

3.4 Only statement (a) is true.

3.5 (b)

3.6 The bumper experiences a force equal to three times the weight of the car!

3.7 (d)

Problems

1. 0.015 m/s^2

3. 140 m

5. 6.4 m/s^2

7. +12,000 N

9. 260 N

11. (a) 1.3 m/s^2 (b) 710 m

13. 1.4 m

15. (a) 21 m/s^2 (b) 92 m/s

17.

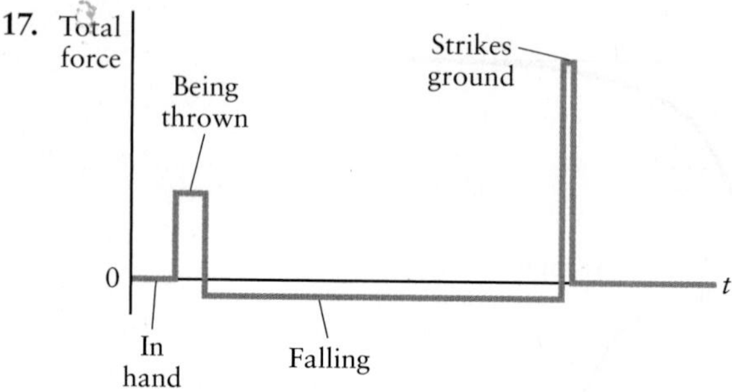

19. 16 m

21. Yes

23. (a) 35 m/s (b) 39 m/s

25. 2300 m

27. 51 kg

29. 1600 N

31. −4.0 m/s^2

33. (a) 50 N (b) 830 s

35. (a) 500 N (b) 530 N

37. 0.51

39. 75 N

41. 0.49 m/s^2

43. (a) 0.18 (b) 0.14

45. (a) 0.42 (b) Static friction

47. 0.66

49. +35 m/s or −35 m/s. There are two solutions because the baseball passes this height twice.

51. 9.2 m

53. 3.3 s

55. 7.3 m

57. 260 m

59. 6.2 m/s

61. 8.3 m/s^2 downward

63. 26 N, 90 N

65. 250 N

67. RT Approximately 900 N

69. R About 0.04 m/s; about 4 min

71. RT Approximately 2000 N
73. R Approximately 14 m/s
75. R Rank (b), (f), (e), (d), (a), (c)
77. 22 m/s or 49 mi/h. You would not likely survive impact with the ground, but would have a good chance in the case of a water impact.
79. RT Approximately 20 μm/s
81. RT (a) 4 m/s (b) v_f increases by a factor of 1.4
83. (a) 0.30 s
(b)

Δt (s)	Δx (cm)
0.14	10
0.16	13
0.18	16
0.20	20
0.22	24

85. 0.76
87. RT About 0.25 s
89. (a)

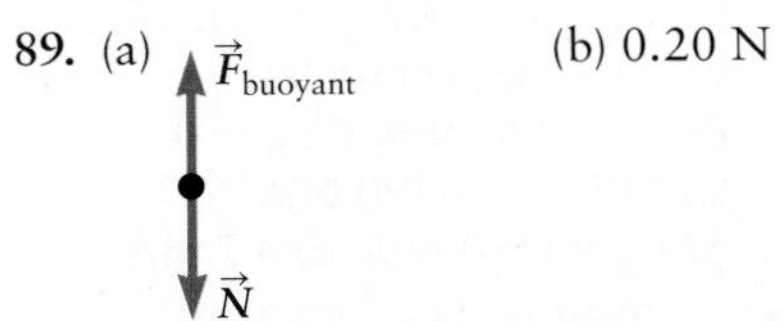

(b) 0.20 N

91. (a) 11 m/s² (b) 4.6 s (c) Changing the mass of the block does not alter the needed acceleration.
93. (a) 9.9 N (b) 17 N
95. (a) 5.6×10^5 m/s² (b) 5.4×10^{-4} s (c) 5.6×10^3 N
97. (a) 22 m/s (b) 61 m/s²
99. R About 6 N

CHAPTER 4

Concept Checks

4.1 (b)

4.2 (b)

4.3 (a) The acceleration is *never* zero. (b) $v_y = 0$ at the instant the ball is at the highest point on its trajectory. (c) The force is *never* zero. (d) The speed is *never* zero.

4.4 The acceleration of m_1 is *not* the same in the two cases.

4.5 (b)

4.6 Suppose a spacecraft is sitting at rest in deep space. Now consider the spacecraft as viewed from a noninertial reference frame; that is, it is viewed by an observer who has an acceleration $+a$ along a particular direction. As viewed by this observer, the spacecraft will appear to have an acceleration $-a$. Since there are no forces acting on the spacecraft, Newton's first law is violated in this noninertial reference frame.

Problems

1. 54 N, −17°
3. 46 N, −22°
5.

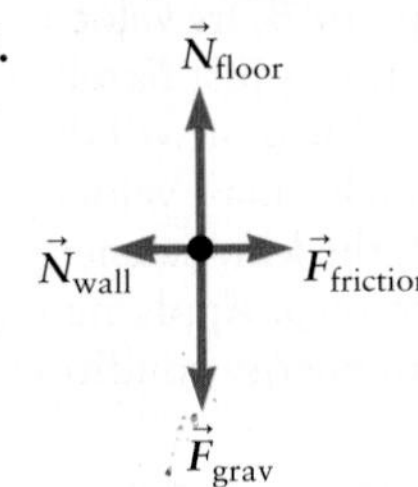

7. $T_1 = 260$ N, $T_2 = 440$ N, $T_3 = 510$ N
9. 27 N, 64°
11. 17°
13. 20 kg
15. (a) 2.3 s (b) 45 m (c) 30 m/s, −48°
17. 19 m/s
19. (a) 8.2 m (b) 27 m/s (c) 70 m
21. Yes
23. (a) 320 m/s (b) 320 m/s
25. 0.83 m, 2.7 ft
27. 37 m/s (b) 8.6° below the x axis (c) 2.6 s
29. R 20 m/s, 2 s
31. 48°
33. (a) 5.1 m (b) It is about double the record. A high jumper is not able to "convert" all his initial horizontal velocity into motion in the vertical direction.
35. (a) 1700 s (b) 1100 s (c) 3800 s (d) The time "lost" swimming against the current is greater than the time "gained" swimming with the current.
37. 90 m
39. 2100 s
41. (a) 240 m/s (b) 240 m/s (c) 240 m/s
43. (a) 1.3 m/s², −0.47° (b) 7.7 s
45. 25 m
47. 0.12
49. 0.27
51. 34°
53. 82 m
55. R 6×10^5 N
57. $\theta = \tan^{-1}\left(\frac{a_x}{g}\right)$
59. R 20 m/s
61. R 20 m/s
63. RT 1×10^7 N
65. 1.2×10^3 N, 32°
67. (a) 670 N (b) 15 kg
69. (a) 250 ft (b) 86 mi/h; goggles are recommended.
71. (a) Perpendicular to the shore (b) 2.0 m/s (c) 81° from a line perpendicular to the shore (d) 50 s (e) 100 m
73. It is better to have Joe pull and Paul push ($a_x = 5.0$ m/s²). If Joe pushes and Paul pulls, then $a_x = 3.3$ m/s².
75. 3.0 m/s²
77. (a) 25 m/s, 72°
79. (a) 0 (b) 0 (c) 4.5 N
81. RT 20 m/s
83. (a) 9.4 m/s (b) 11 m/s
85. (a) 58 m (b) It increases by 39 m, to a new distance of 97 m. (c) 43 m
87. The acceleration will be larger in scenario 1.
89. RT (a) 110 N (b) $T_{left} = 100$ N, $T_{right} = 170$ N
91. R (a) $-x$ direction (b) 150 mi/h (c) 1.7 (d) A 70% difference would certainly be significant.

CHAPTER 5

Concept Checks

5.1 Statements (a), (b), and (c) are all correct.

5.2 (c)

5.3 (c)

5.4 The observers in (a) and (b) are moving at constant velocities, so their accelerations are zero. Hence, they are both inertial frames. In (c) and (d), the frames are accelerating, so they are noninertial frames.

5.5 (d)

5.6 The answer is shown in the accompanying figure.

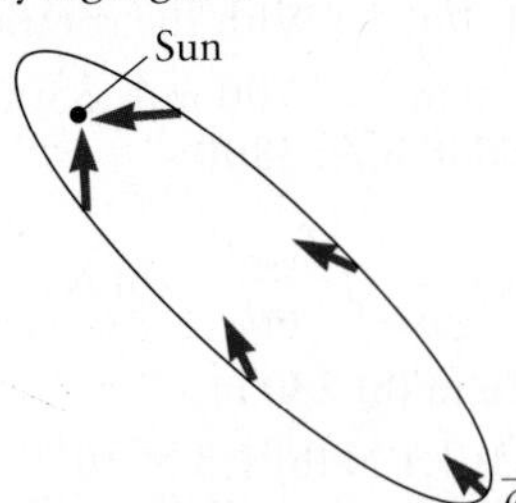

5.7 (b)

Problems

1. (a) 12 m/s^2 (b) 1.0 s
3. RT (a) 470 m/s (b) 350 m/s
5. 0.17 m/s^2
7. (a) 130 s (b) Cutting r in half would double the centripetal acceleration.
9. (a) 3.4×10^{-2} m/s^2 (b) 2.9×10^{-2} m/s^2
11. (a) 0.75 (b) Static friction
13. R 30 m/s^2, about $3 \times g$
15. 6.6 m/s
17. 2.2 m/s
19. 8.1 m/s
21. (a) 4200 N (b) 20 m/s
23. 20 m
25. 0.054
27. 8.6 m/s
29. 7.0×10^3 N
31. 29 N
33. R 4×10^4 s
35. 1.8×10^{-8} N
37. 4.0×10^{-9} N, 34° below the negative x axis
39. 3.5×10^{22} N
41. RT 8×10^{-8} N
43. 3.5×10^8 m
45. 0.9998
47. 470 N
49. 1.4×10^{12} m
51. 4.4×10^9 s or about 140 years
53. 2.8×10^4 s or 7.7 h
55. Satellite 3100 m/s, Earth 3.0×10^4 m/s
57. R 0.94
59. 6.7×10^{-4} percent
61. Less, 0.997
63. RT Doctor's head 1.6×10^{-8} N, Mars 2.4×10^{-8} N
65. (a) 0.30 (b) 0.59 loops/s (c) Top 3700 N, bottom 6800 N
67. R 1200 N
69. (a) 20 s (b) 31 m/s (c) $0.75mg$ (d) $1.3mg$ (e) With the rotation
71. (a) 8.8 m/s^2, 7700 m/s, 5500 s (b) 7600 m/s, 5800 s
73. (a) $v_{\max} = \sqrt{\dfrac{T_{\max} r^2}{mL}}$ (b) $N = 9$
75. (a) 26 m (b) 230 m
77. R (a) 1.1 N (b) 1.3×10^{-10} m/s^2 (c) 6.5×10^4 m (d) The asteroid will hit the Earth.

CHAPTER 6

Concept Checks

6.1 (1) corresponds to case (a); (2) corresponds to case (c); (3) corresponds to case (b).

6.2 (c)

6.3 As the roller coaster in Figure 6.13 moves higher on the track, approaching point B, its velocity will decrease. If it is just barely able to make it to point B, it will do so with an extremely small velocity. In such a case, the kinetic energy at point B will be zero. Applying the conservation of energy condition then gives

$$PE_A + KE_A = PE_B + KE_B = PE_B$$
$$KE_A = \tfrac{1}{2}mv_A^2 = PE_B - PE_A$$
$$= mgh_B - mgh_A$$

which can then be solved for h_B.

6.4 (1) and (2)

6.5 No reasonable bicycling hill would require 1000 Calories to climb it.

6.6 (3)

Problems

1. 370 N
3. 250 J
5. 2400 J
7. −2400 J
9. −140,000 J
11. (a) 23,000 J (b) −23,000 J (c) 45,000 J
13. (a) 5500 J (b) 0 J
15. 16 J
17. 390,000 J
19. 9.9 m/s
21. 94 m/s
23. RT 40 m/s
25. 64 m/s
27. −8200 J
29. 11 kg
31. (a) 220,000 J (b) 63 m/s
33. (a) 2.2×10^5 J (b) -1.7×10^5 J
35. -7.3×10^6 J
37. $h = 5.7$ m
39. (a) 8.9 m/s (b) 1600 N (c) 0 (d) 2.2 m above the top edge of the half-pipe
41. (a) Yes. The ramp is frictionless, so the only force acting on the puck is the force of gravity, which is conservative. (b) In principle, the origin of the y axis can be placed anywhere. For this problem, it is convenient to place $y = 0$ at ground level.

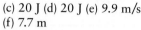

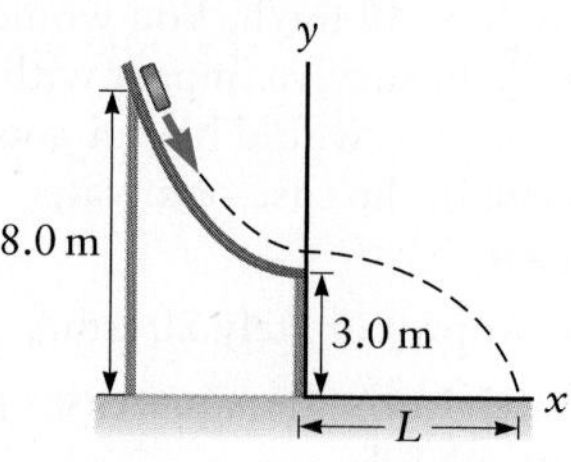

(c) 20 J (d) 20 J (e) 9.9 m/s (f) 7.7 m
43. 8.6 m/s
45. (a)

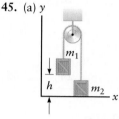

(b) $KE_{1i} = 0$, $KE_{2i} = 0$; taking $y = 0$ to be ground level, we have $PE_{1i} = (390\text{ N})h$, $PE_{2i} = 0$ (c) $\Delta PE_1 = -(390\text{ N})h$; $\Delta PE_2 = (150\text{ N})h$ (d) 4.7 m/s
47.

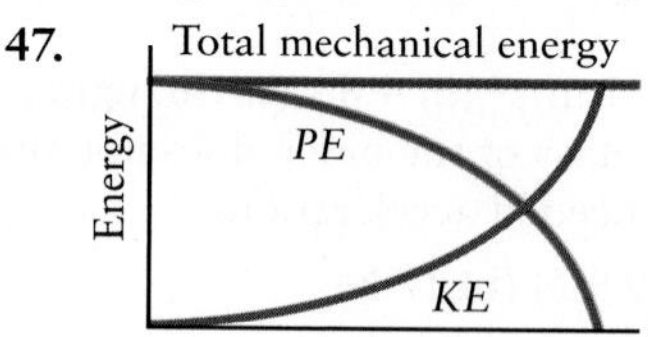

49. (a) 1.5 m/s (b) 39 N
51. (a) $v_f = \sqrt{v_0^2 + 2gh}$ (b) Since energy is a scalar quantity, the final magnitude of the bullet's velocity will depend only on the magnitude of the initial velocity.
53. (a) 1.1×10^4 m/s (b) The change in escape speed is very small. This would not be a good way to save fuel.
55.

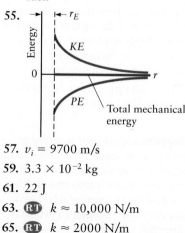

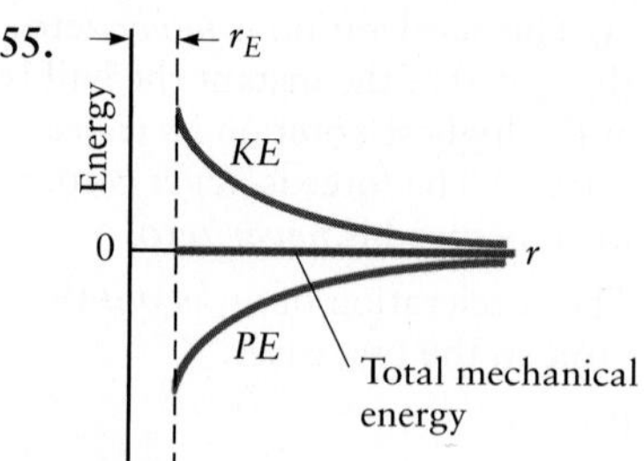

57. $v_i = 9700$ m/s
59. 3.3×10^{-2} kg
61. 22 J
63. RT $k \approx 10{,}000$ N/m
65. RT $k \approx 2000$ N/m

67. (a) 0.14 J (b) 0.3 m

69. 0.2 J. The lump of clay dissipates all its *KE* (2.0 J) as heat and in deformation. The Super Ball dissipates only one-tenth its initial *KE* by similar processes.

71. 37 m/s

73. 0.29 m/s

75. (a) 10 m/s (b) -3×10^5 J (c) -500 W

77. RT 60 m/s

79. 20 pN

81. 5×10^{-18} W

83. R The weight of a typical amino acid is about 3×10^{-24} N. The myosin molecular motor can apply a force about 12 orders of magnitude greater than that.

85. RT About half a kilometer

87. R ≈ 1

89. (a) 1.1 kN (b) Yes, 21°

91. (a) 4300 J (b) 3500 J (c) Two factors enable the vaulter to go higher than expected from the answer to part (a): (1) The vaulter keeps running after planting the pole, giving extra *PE* to the pole; and (2) the vaulter is not a point particle—his center of mass (Chapter 7) can actually pass under the bar while his body passes over it!

93. (a) 7.0 cm (b) 80 cm/s (c) 16 cm

95. 9.4×10^4 W = 94 kW

97. 280 m/s; this is very close to the reported value, suggesting that air drag at these altitudes is indeed quite small.

99. (a) 20 kJ (b) 67 W (c) 270 W (d) 0.24 (e) Her energy was dissipated by controlled frictional resistance in the device.

101. $2^{1/3} \approx 1.3$

103. *KE* of the 500-kg mass = gravitational *PE* $\approx$ 15,000 J; *KE* of goose $\approx$ 25,000 J; the goose has more energy!

CHAPTER 7

Concept Checks

7.1 (a)

7.2 (b)

7.3 (b)

7.4 (b)

7.5 (f)

7.6 (a)

7.7 (c)

7.8 (c)

7.9 (a)

Problems

1. 6.3 kg·m/s

3. 18,000 kg·m/s

5. (a) 5.0 kg·m/s (b) 42 m/s

7. (a) 2.4 kg·m/s, directed upward (b) 0.13 s

9. (a) -0.85 kg·m/s (b) -0.85 kg·m/s

11. (a) 12 kg·m/s, $\theta = 26°$ (b) $F = 1500$ N, more than 1000 times the ball's weight (1.4 N) (c) 12 kg·m/s, $\theta = 206°$

13. (a) The impulses are equal. (b) The average force on the egg must be greater than the force on the tomato because the collision time for the egg is smaller.

15. RT ≈ 360 N

17. R ≈ 3 N·s

19. RT ≈ 7 N·s

21. (a) $v_f = -1.7$ m/s (b) 2.0×10^2 J

23. $v_{1f} = -15.3$ m/s; $v_{2f} = 4.2$ m/s

25. 8.9 m/s, 34° above the x axis

27. (a) 8.7 m/s (b) $-28°$

29. 140 kg

31. 0.50 m/s southward

33. -0.088 m/s

35. 680 kg

37. 6.9 m/s

39. 180 kg

41. $x = -0.75$ m

43. (a) Object 1 (b) $\frac{m_1}{m_2} = \frac{7}{5}$

45. RT Case 1: $x_{CM} = 0$, $y_{CM} = 1.5$ m; case 2: $x_{CM} = 1$ m, $y_{CM} = 1$ m; case 3: $x_{CM} = 1$ m, $y_{CM} = 1$ m; case 4: $x_{CM} = 0.5$ m, $y_{CM} = 0.5$ m

47. 4.5×10^5 m; the Sun's radius is 7.0×10^8 m

49. R $\approx 5 \times 10^{-25}$ m

51. 53 m/s; 101° counterclockwise from the x axis

53. RT $\approx 20{,}000$ N

55. R About 100 N

57. 4.7×10^6 m; the radius of the Earth is about 6.4×10^6 m, so the center of mass of the Earth–Moon system lies just inside the Earth.

59. (a) 0.013 m (b) 0.22 m

61. (a) $v_{1f} = 0$, $v_{2f} = v_0$ (b) $v_{1f} = -0.96\, v_0$, $v_{2f} = 0.038\, v_0$ (c) $v_{1f} = 0.96\, v_0$, $v_{2f} = 1.96\, v_0$ (d) $v_{1f} = -v_0$, $v_{2f} = 0$ (e) $v_{1f} = v_0$, $v_{2f} = 2v_0$

63. RT (a) 2.0×10^4 N (b) 2.0×10^3 N (c) 11 ms

65. 0.97 km

67. (a) 5.6 m (b) 1.4 m

69. 3.4×10^2 N

71. 400 m/s

73. R 0.4

CHAPTER 8

Concept Checks

8.1 (h)

8.2 (c)

8.3 (b)

8.4 The torques are (a) negative, (b) positive, (c) positive, and (d) positive.

8.5 (a)

8.6 (b)

8.7 (d)

8.8 (a)

Problems

1. 1.5 rad

3. (a) 3.5 rad/s (b) 0.62 m/s

5. 750 m

7. -67 rad/s^2

9. (a) 7.3×10^{-5} rad/s (b) 7.3×10^{-5} rad/s

11. 8.3×10^{-7} rad/s

13. -1.0×10^{-1} rad/s, 60 s; 5.0 cm

15. 2.2×10^{-2} rad/s

17. -5300 N·m

19. (a) $\phi = 28°$ (b) Positive

21. By a factor of six

23. R $\approx 50{,}000$ N·m

25. RT $T_{left} \approx 12$ N; $T_{right} \approx 32$ N

27. 130 N; 60 N

29. 231 N, 25° above x axis

31. Positive

33. 40°

35. 290 N

37. RT ≈ 80 kg

39. R 45°

41. (a) 14 kg·m^2 (b) 36 kg·m^2 (c) 22 kg·m^2 (d) 14 kg·m^2

43. 2800 kg·m^2

45. RT 0.07 kg·m^2
47. 5.1/1
49. $\frac{1}{3}mL^2$
51. 13 rad/s
53. 15 km
55. 3.3×10^{-2} rad/s^2; 60 rad
57. (a)

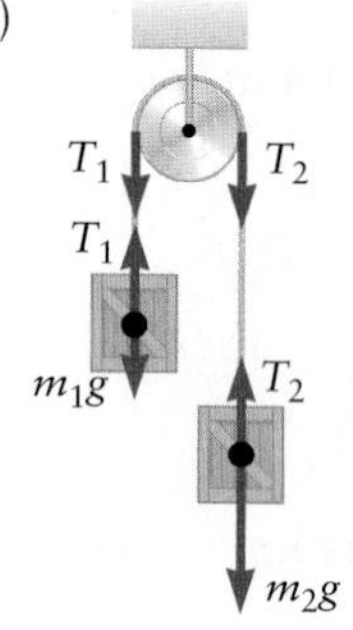

(b) Left crate: $\Sigma F = -m_1g + T_1 = m_1a$; right crate: $\Sigma F = -T_2 + m_2g = m_2a$; pulley: $\Sigma\tau = T_2R_{\text{pulley}} - T_1R_{\text{pulley}} = I_{\text{pulley}}\alpha$ (c) $a = \alpha R_{\text{pulley}}$ (d) $a = 2.5$ m/s^2 (e) $T_1 = 62$ N; $T_2 = 66$ N
59. 4.5 m/s^2
61. R $\approx 1.2 \times 10^4$ m
63. -1.2×10^{-3} rad/s^2; 3.0 rev
65. (a) 0.38 m/s^2 (b) 1.5 rad/s^2 (c) 1200 rad (d) 300 m
67. 300 rad/s
69. 10 m/s^2
71. R $\approx$ 1600 revolutions
73. (a) 1.1 m/s^2 (b) $R_b = 45$ cm, $R_f = 15$ cm
75. (a) $\frac{11}{12}L$ (b) $\frac{25}{24}L$ (c) 31
77. (a) 1.2 kN (b) 26 N·m
79. 890 N
81. 96 N
83. 5.3 m from the left end of the board

CHAPTER 9

Concept Checks

9.1 (b)
9.2 (d)
9.3 (c)
9.4 (c)
9.5 The larger ball has a greater angular momentum.
9.6 In both cases, the system is able to rotate even though the angular momentum is zero and there are no external torques.

Problems

1. 0.14 J
3. 1/3
5. 29 N
7. RT 5.3 J
9. R $\approx$ 91,000 J
11. RT 0.13
13. 8.1×10^5
15. 1.7 m/s
17. 2.3 m/s^2
19. $v_B = 0.71$ m/s, $v_C = 1.4$ m/s
21. $h = 14$ m
23. 13 m/s
25. 12 m/s
27. 1
29. (a) 7.9 J (b) 3.9 J
31. 5.3 m/s
33. 1
35. 4.8 kg·m^2/s
37. R $\approx$ 31 kg·m^2/s
39. (a) 2.4 m/s (b) 0.88 J
41. 6.0×10^{-4} kg·m^2/s
43. R (a) 11 rad/s (b) $KE_{\text{arms out}} = 50$ J, $KE_{\text{arms in}} = 110$ J
45. RT 870 kg·m^2/s
47. (a) 120 kg·m^2/s (b) 120 kg·m^2/s (c) The mass has the same tangential speed at a distance of 2.0 m from any point located along the x axis.
49. 0.44
51. RT $\approx 3 \times 10^{34}$ kg·m^2/s
53. Into the page
55. $\approx 7.1 \times 10^{33}$ kg·m^2/s along the rotation axis and toward the North Pole
57. $\vec{L}$ $\vec{L}$
59. 11.8 m
61. RT $\approx$ 1.9 rad/s
63. (a) 4.6 J (b) 9.1×10^{-3} kg·m^2/s
65. (a) $\approx$ 1 day (b) 31 days (c) The orbital radii of Jupiter and Saturn are so large that although their masses are much less than that of the Sun, their angular momenta are still a significant fraction of the total angular momentum of the solar system.
67. (a) $I = 6.9$ kg·m^2, $M = 3.4 \times 10^2$ kg (b) 8.0 J (c) 0.18 m/s (d) 0.70 m
69. (a) 1.5×10^{-11} rad/s (b) 0.02 s
71. (a) 22 J (b) 0.042 kg·m^2/s (c) This angular momentum is too small to notice.
73. R $\approx 6 \times 10^{-3}$ kg·m^2
75. 90 rad/s
77. 2.5
79. (a) 4/5 (b) 14/15
81. RT There will be no measurable change in the rotation rate of the Earth.

CHAPTER 10

Concept Checks

10.1 (a)
10.2 (c)
10.3 (a)
10.4 (b)
10.5 (b)
10.6 (a)
10.7 (a)

Problems

1. 5500 kg/m^3
3. 2.1 m^2
5. R 200 kg
7. RT 7×10^{-2} m^3
9. 1.1×10^5 N; you could not lift the lid.
11. 390 Pa
13. 300 cm^2
15. R 1000 Pa
17. 720 lb/in.^{2}
19. RT About 1000 kg
21. 1.3×10^5 Pa
23. 4.3×10^6 N
25. R About 5×10^{10} N
27. 10 cm
29. 2.1×10^6 Pa
31. RT 2×10^4 Pa, equal to the maximum systolic blood pressure
33. R 4×10^5 Pa
35. ≥ 41 m
37. 1:10
39. 200 N
41. 650 kg/m^3
43. 6.3 cm
45. 1.3×10^3 kg/m^3; the object will sink.
47. 5.9×10^3 N
49. 1.4×10^3 m^3

51. 3.1×10^4 N

53. 2.2×10^5 N

55. 85%

57. 40 m/s

59. (a) 50 s (b) 32 cm/s

61. 47 cm

63. 45 cm^3

65. Ⓡ About 6 m/s

67. 1.5×10^{-1} cm^3/s

69. 4.6 Pa

71. Ⓡ 2×10^{-5} m/s

73. 1.5×10^7 Pa, 150:1

75. Ⓡ The diameter must increase by a factor of 1.3.

77. The flow rate is reduced by a factor of 16.

79. Rises up, 0.46

81. 10 m

83. The water undergoes free fall after it leaves the faucet, so we can use the relations for free fall to relate the speeds at points 1 and 2. The speeds and areas at these two points are also related by the equation of continuity. This gives enough information to find the flow rate, 47 cm^3/s.

85. (a) 92.8 m (b) No; the maximum depth is 90.5 m.

87. RT About 3000 N

89. 50 Pa. Yes; the measurement is good to at least three significant figures.

91. (a) 7.9×10^4 Pa (b) 1.8×10^7 N (c) 2.6×10^7 N

93. (a) It tends to push the window out of the house. (b) 410 N

95. (a) The pressure in the tank exceeds atmospheric pressure by 550 Pa. (b) 23 Pa

97. (a) $p_1 - p_2 > 0$ (b) 4.0×10^4 Pa

99. Ⓡ 7 m/s

CHAPTER 11

Concept Checks

11.1 The acceleration of an object undergoing simple harmonic motion is nonzero unless the object is at the equilibrium position. There are forces on a bungee jumper due to the bungee cord and gravity. In general these forces do not cancel, and the acceleration of the jumper is also nonzero.

11.2 (d)

11.3 (a) Point *B* (b) Point *B*

11.4 Kinetic energy

11.5 (c)

11.6 Air drag and friction between the mass and the table

11.7 (c)

Problems

1. (a) 0.60 s (b) 1.7 Hz

3. (a) Period ≈ 4.0 s, frequency ≈ 0.25 Hz, amplitude ≈ 10 cm (b) At about $t \approx 3.0$ s and $t \approx 7.0$ s (c) At about $t \approx 1.0$ s and $t \approx 5.0$ s

5. (a) $v(t = 4.0\text{ s}) \approx 0.15$ m/s, $v(t = 7.0\text{ s}) \approx 0.025$ m/s (b) At about $t = 0$ s and $t = 4$ s

7. 140 m/s

9. It is not a simple harmonic oscillator because the velocity does not vary sinusoidally with time; $f = 250$ Hz.

11. (a) ≈ ±0.18 m (b) 3.3 Hz

13. 1.8 s

15. (a) 4.9 N/m (b) $v(t = 4.0\text{ s}) \approx 15$ cm/s, $v(t = 5.0\text{ s}) \approx 0$ cm/s (c) -17 cm/s^2 (d) ≈ 24 cm/s^2

17. 2.9 Hz

19. (a) 1.0 m/s^2 (b) 0.50 m/s

21. RT 16 m/s^2; larger

23. 3.0 m

25. 6.2 s

27. $f_{new} = 0.58 f_{old}$

29. 0.029 Hz

31. 0.015 Hz

33. $v_{max} = 3.1 \times 10^4$ mm/s, $a_{max} = 4.9 \times 10^8$ mm/s^2

35. (a) 13 J (b) 13 J

37. (a) 0.89 kg (b) 0.84 s (c) 43 m/s (d) 810 J (e) 810 J

39. 2.1×10^{-4} s

41. 0.010 J

43. (a) 0.14 J (b) 21 cm (c) 0.29 m/s (d) The answers would remain the same.

45. RT Approximately 100 J

47. (a)

49. Strain $= 1.0 \times 10^{-4}$; diameter = 7.0 cm

51. 0.2497 m

53. $\Delta L = 7.1 \times 10^{-7}$ m

55. (a) 6.7×10^{-4} m (b) 140,000 N

57. 9.1×10^9 Pa

59. 12 kg

61. RT About 75% of the energy is lost with each cycle.

63. 1.8×10^{-9} N; it is about 5000 times smaller than the weight of a mosquito.

65. (a) 2.0 m (b) 4.1 m/s (c) Period = 0.20 s, amplitude = 0.20 m (d) $x(t) = (0.20\text{ m})\sin[(31\text{ rad/s})t]$, $v(t) = (6.3\text{ m/s})\cos[(31\text{ rad/s})t]$ (e) 6.3 m/s (f) 2.0 m

(g)

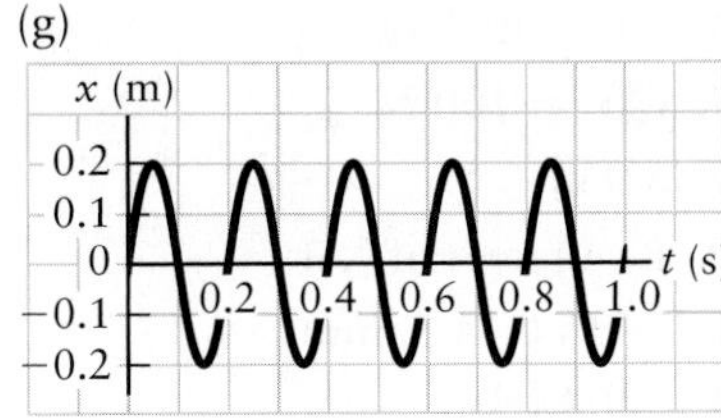

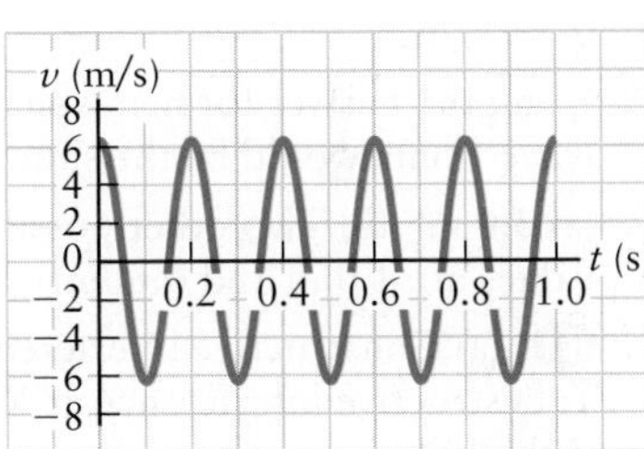

67. (a) 0.70 s (b) 25 cm

(c)

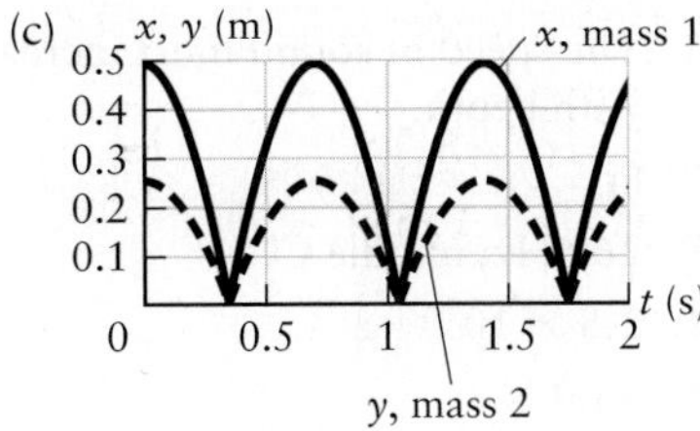

69. 10 m/s = 36 km/h

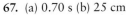

71. (a) 150 N/m (c) 1.0×10^5 Pa

73. (a) Swing 1 is longer. (b) 4.2 m

CHAPTER 12

Concept Checks

12.1 (c)

12.2 (a) Wave 4 (b) Wave 2 (c) Wave 3

12.3 (b)

12.4 The frequency and amplitude depend on the way the wave is generated. The wavelength depends on both the medium and the way the wave is generated because it depends on both the frequency and the speed.

12.5 Five nodes and four antinodes.

Problems

1. 13 m/s
3. (R) ≈ 0.40 m/s; perpendicular to the wave velocity
5. 1.4×10^{11} m
7. 26 cm/s
9. 3.4 m
11. (RT) 11 m/s
13. $y = 0.50 \sin(250t + 0.75x)$ m
15. 0.010 s
17. $f = 480$ Hz, $\lambda = 14$ mm
19. 300 m/s
21. 0.0026 kg
23. 2.0×10^{-5} kg
25. (RT) ≈ 100 N
27. 5.0 m
29. By a factor of $\sqrt{3}$
31. 260 s (= 4.3 min)
33. 200 MHz
35. 0.39 W/m^2
37. 17 m; the answer for hemispherical wave fronts would be the same.
39. (a) 3.0 m (b) Two other possible values are 1.0 m and 0.60 m
41. (RT) (a) Constructive interference at *Y*; destructive interference at *X* and *Z* (b) 40 Hz
43. 18,000 m
45. The speed of sound must increase with depth.
47. (a)
49. 260 Hz; middle C
51. 2.8×10^{-4} kg
53. 46.0 N
55. The frequency will increase by a factor of 2.5.
57. 0.038 m
59. 140 m/s
61. 4/5
63. Surface waves = 1.8 h, P waves = 0.54 h
65. (a) 8.8 m/s (b) 1.8 s
67. (a) 1.10 (b) 0.793
69. 28 s
71. (a) 0.58 Hz (b) 128th harmonic (c) 7.9 m/s
73. $\mu_G/\mu_E = 11$, $\mu_D/\mu_E = 5.0$, $\mu_A/\mu_E = 2.2$
75. (RT) Al, Cu, Au, Fe, diamond

CHAPTER 13

Concept Checks

13.1 (b)
13.2 (c)
13.3 (d)
13.4 (b)
13.5 (a)
13.6 (a)
13.7 (b)

Problems

1. 1.6 m
3. 0.39 m
5. 0.15 s
7. Helium
9. 13 m
11. $\lambda_{water}/\lambda_{air} = 4.3$
13. 17
15. About 1×10^{-8} W/m^2
17. (R) About 0.006 N
19. (a)
21. It increases by a factor of 40.
23. 94 dB
25. 3.3×10^{-5} W/m^2
27. 0.002 Pa
29. 100 dB
31. (a) and (d)
33. 71 Hz
35. The frequency will go down to 96% of the frequency in air.
37. 860 Hz
39. 1.3 m and any multiple thereof, such as 2.6 m
41. 1.9×10^3 Hz
43. (RT) About 400 Hz
45. 3
47. 0.991 or 1.01
49. 0.02%
51. (RT) The sound makes about 140 trips across the church.
53. 21 m/s
55. 25 m/s
57. 25 m/s
59. 70 Hz
61. 37 m/s
63. 12°C
65. (a) 3 km (b) 0.34 km/s
67. Formant at 750 Hz is due to fundamental frequency of the oral cavity, formant at 1100 Hz is due to fundamental frequency of the larynx, and formant at 2500 Hz is due to third harmonic of oral cavity.
69. (a) 15 m (b) 420 Hz
71. (a) 35 m/s = 81 mi/h (b) 90 m/s = 200 mi/h; few automobiles can go this speed. Using two cars driven in opposite directions, each with half of this speed (100 mi/h) on a two-lane road, would still be very dangerous.
73. (RT) −12°C
75. 0.19 m/s

CHAPTER 14

Concept Checks

14.1 The body is composed mainly of water, so its specific heat is close to that of water.
14.2 (c)
14.3 The coefficient of thermal expansion α determines how the length of an object changes when it is heated. It applies to the length, width, and height, and if all three change a large amount, the corresponding changes in the volume and hence the value of β are also large.
14.4 The frequency will go down.
14.5 Figure 14.25A shows how heat conduction is due to the transfer of energy between vibrating atoms. Solids have a higher density of atoms than gases, so (if all else is similar) solids can conduct a greater amount of heat.
14.6 (d)
14.7 (c)

Problems

1. $T_F = 450$°F and $T_K = 505$ K
3. −460°F
5. $T_C = 21$°C and $T_K = 294$ K
7. −90°C
9. 6.8×10^4 J
11. 110 J/(kg·K)
13. 2.7×10^4 J
15. 100°C
17. 0.26 kg

19. 7.5×10^{-3} m
21. Ⓡ 0.30 m
23. 120.04 m
25. 1.1×10^{-5} K^{-1}
27. 155°C
29. 4.1×10^{-4} m
31. Ⓡ 3×10^{-4} m
33. RT 4×10^{-2} m
35. (a) -4.7×10^{6} N (compression) (b) 6.3×10^{6} N (tension)
37. Ⓡ 20°C
39. (a) 18°C (b) 52°C
41. Ⓡ 1.5 cm
43. 1.3 kg/h
45. 9.7×10^{-7} m
47. 4.0×10^{26} W
49. Ⓡ 240 K
51. Ⓡ (a) Approximately 330 K (b) Approximately 3 K; the calculation for the dark side of the Moon is low because we assumed that all the heat input is coming from the 3-K background in space. The surface of the Moon facing the Sun gets very warm by Earth standards during the lunar "day." (Our 330-K estimate is actually low in equatorial regions, which have an equilibrium temperature more like 107°C!) The rock will retain some of that heat at "night," especially because there is no atmosphere and cooling takes place by radiation only.
53. Ⓡ About 1×10^{6} m^2
55. Ⓡ (a) 2.9 W (b) 1.3×10^{-4} s
57. 10.06°C
59. (a) $T_R = T_F + 459$ (b) $T_R = \frac{9}{5}T_K$ (c) Freezing point of water = 491°R; room temperature ≈ 530°R
61. −208°C
63. (a) 1.3×10^{4} J (b) 70°C
65. (a) 2.2×10^{8} Pa, with the engine block pushing on the plug (b) 4.7×10^{5} N; yes
67. 190 K
69. (a) 8.1×10^{9} N (b) 0.00055 m (c) 4.2×10^{3} J
71. Ⓡ (a) 4 kJ, which is a small amount of energy that will raise the temperature of 1 kg of water by only about 1°C. (b) 2 g (c) 0.1 mm (d) The walker would almost certainly receive severe burns from a slab of iron.
73. Ⓡ (a) 130°C (b) This is *not* likely to happen!
75. RT About 10 K

CHAPTER 15

Concept Checks

15.1 (d)
15.2 (a)
15.3 (a)
15.4 (b)
15.5 (c)
15.6 (c)
15.7 (c)

Problems

1. 2.0×10^{-26} kg
3. 7.3×10^{-26} kg
5. Two possibilities are elemental calcium (Ca) and magnesium oxide (MgO).
7. 313°C
9. 3.4×10^{-9} m; the average spacing will increase with heating.
11. 2.5×10^{29} molecules
13. 33,000 Pa
15. (a) 0.072 m^3 (b) 8.8×10^{4} Pa
17. About 0.08 mole
19. 1.2×10^{24} molecules
21. 2.5×10^{10} molecules
23. (a) 6830 Pa (b) 55°C
25. 640 m/s
27. 5.7×10^{5} J
29. 8.6×10^{3} m/s
31. 55 kJ; changing the gas to argon does not change the internal energy.
33. 1900 m/s, well below the escape speed for the Earth. Hydrogen escapes because it has a distribution of speeds as explained in Example 15.7.
35. 2.0×10^{-5} s
37. 4.2×10^{-4} m
39. Ⓡ 5×10^{-7} s
41. 3.2×10^{-5} m
43. Ⓡ About 8×10^{-4} s
45. 2.4×10^{-4} m
47. Ⓡ (a) ≈ 10 molecules for each breath (b) Distributed but not diffused; N_2 diffuses ≈ 1 km in 2000 years. Gaseous emissions disperse much more quickly than typical diffusion times; convection and other meteorological phenomena play a much larger role than does diffusion in mixing atmospheric constituents.
49. RT ≈ 2000 m
51. (a) 1.8 atm (b) 1.8
53. (a) 10 atm (b) 0.80 (c) 6.3 atm
55. (a) 0.31 s (b) 8.3 min (c) Such animals would be too large to allow sufficient oxygen to diffuse throughout their bodies at a rate fast enough to sustain metabolism.
57. 220 m/s
59. (c)

CHAPTER 16

Concept Checks

16.1 (c)
16.2 (c)
16.3 (a)
16.4 Both. Because bear fur is very thick, very little heat flows between the bear and its environment, so this situation is approximately adiabatic. During hibernation, the bear produces very little heat because its metabolic rate is very low. A balance between the heat it creates and the heat flow out through its fur keeps the bear approximately isothermal.
16.5 (a) *W* is positive. (b) *W* is negative. (c) *W* = 0. (d) *W* is positive. (e) *W* is negative.
16.6 (a) Process 3 could be an isothermal expansion. (b) Process 4 could be an adiabatic compression. (c) Process 1 is an isobaric expansion. (d) Process 2 is an isobaric compression.
16.7 *W* is positive during steps 1 → 2 and 2 → 3. *W* is negative during steps 3 → 4 and 4 → 1.
16.8 Process (a) is a heat engine; process (b) is a refrigerator.

Problems

1. 20 kJ

3. -2000 J
5. 8.0×10^7 J
7. (a) 2.3×10^6 J; 11 apples (b) 1.8×10^6 J; 3 soft drinks (c) 4.3×10^7 J; 22 hamburgers
9. (a) Description ii (b) Description iv (c) Description iii (d) Description i
11. 1.5×10^6 J
13. Process A: expansion, about +1300 J; Process B: expansion, about +2300 J
15. About +1100 J
17. -4400 J
19. 1200 J
21. $W_1 \approx 2600$ J; $W_2 \approx -1800$ J
23. -6000 J
25. (a) 0.11 m^3 (b) -5200 J (c) -5200 J
27. (a) 2300 J (b) 257 K
29. 1.7×10^5 J
31. 270 hp
33. 5:1
35. (R) 37,000 J
37. 9900 J
39. 0.75
41. 2.4 g
43. During transitions $1 \to 2$ and $2 \to 3$, the system does positive work on the environment. During transitions $3 \to 4$ and $4 \to 1$, the environment does positive work on the system.
45. (RT) 20,000 J
47. $\Delta S_{house} = -24$ J/K; $\Delta S_{environment} = +26$ J/K; $\Delta S_{universe} = +1.8$ J/K; irreversible process
49. $\Delta S_{steel} = -0.010$ J/K; $\Delta S_{ice\ water} = +0.015$ J/K; $\Delta S_{universe} = +0.005$ J/K
51. 0.25 kg
53. (b) $<$ (a) $<$ (d) $<$ (c)
55. 1
57. (a) Processes 1 and 2 (b) Processes 3 and 4 (c) Process 2
59. (a) 2000 J (b) 1000 J (c) 220 J (d) 1200 J. The work done by the first two engines combined is 1000 J + 220 J = 1220 J $\approx$ 1200 J, which is the same (to within rounding errors) as the work done by the engine operating between the highest and lowest reservoir temperatures.
61. (a) **Brayton cycle:** From states $1 \to 2$: isobaric expansion; from states $2 \to 3$: adiabatic expansion; from states $3 \to 4$: isobaric compression; from states $4 \to 1$: adiabatic compression. **Otto cycle:** From states $1 \to 2$: isochoric heating; from states $2 \to 3$: adiabatic expansion; from states $3 \to 4$: isochoric cooling; from state $4 \to 1$: adiabatic compression. **Diesel cycle:** From states $1 \to 2$: isobaric expansion; from states $2 \to 3$: adiabatic expansion; from states $3 \to 4$: isochoric cooling; from states $4 \to 1$: adiabatic compression. (b) **Brayton cycle:** Gas reaches its highest temperature at state 2 and its lowest temperature at state 4. **Otto cycle:** Gas reaches its highest temperature at point 2 and its lowest temperature at point 4. **Diesel cycle:** Diesel fuel reaches its highest temperature at state 2 and its lowest temperature at state 4.
63. (a) Work done on system: $4 \to 1$; work done by system: $2 \to 3$; heat flow into system: $1 \to 2$; heat flow out of system: $3 \to 4$ (b) 1100 K (c) $\Delta T = 1100$ K; $T_2 = 2300$ K (d) 6.6×10^6 Pa (e) 2.1×10^5 Pa; 560 K (f) 210 J leaves the system (g) 0.79 (h) 0.86. The Carnot engine is an idealization that exceeds what can be done in any real engine.
65. (a) $\Delta S_{gas} = -0.26$ J/K (b) $\Delta S_{reservoir} = 0.26$ J/K (c) $\Delta S_{universe} = 0$ J
67. 240 K ≈ -27°F

Index

Note: Information presented in tables and figures is denoted by t and f respectively.

D

E

N

O

S

T

Trigonometric Relations

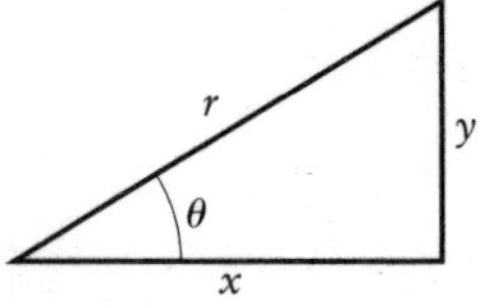

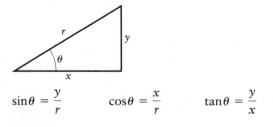

$\sin\theta = \frac{y}{r}$ $\cos\theta = \frac{x}{r}$ $\tan\theta = \frac{y}{x}$

Pythagorean Theorem: $x^2 + y^2 = r^2$

$\sin^2\theta + \cos^2\theta = 1$

$\sin(\theta + 90°) = \cos\theta$

$\cos(\theta + 90°) = -\sin\theta$

$\sin(0) = 0$	$\sin(\pi/4) = 1/\sqrt{2}$	$\sin(\pi/2) = 1$	$\sin(\pi) = 0$
$\cos(0) = 1$	$\cos(\pi/4) = 1/\sqrt{2}$	$\cos(\pi/2) = 0$	$\cos(\pi) = -1$

Degrees and Radians

2π rad $= 360°$

1 revolution = 1 rev = $360° = 2\pi$ rad

$1° = 60$ arcmin $= 60'$

1 arcmin = 60 arcsec = $60''$

$45° = \pi/4$ rad

$90° = \pi/2$ rad

Powers of 10

Prefix	Factor
atto (a)	10^{-18}
femto (f)	10^{-15}
pico (p)	10^{-12}
nano (n)	10^{-9}
micro (μ)	10^{-6}
milli (m)	10^{-3}
centi (c)	10^{-2}
deci (d)	10^{-1}
kilo (k)	10^{3}
mega (M)	10^{6}
giga (G)	10^{9}
tera (T)	10^{12}
peta (P)	10^{15}
exa (E)	10^{18}

Useful Mathematical Formulas

Circumference of a circle = $2\pi r$

Area of a circle = πr^2

Area of a triangle = $\frac{1}{2}$(base × height)

Surface area of a sphere = $4\pi r^2$

Volume of a sphere = $\frac{4}{3}\pi r^3$

Volume of a cylinder = $\pi r^2 L$

$$ax^2 + bx + c = 0 \Rightarrow x = \frac{-b \pm \sqrt{b^2 - 4ac}}{2a}$$

Exponential Functions and Logarithms

$x = e^y \Leftrightarrow \ln x = y$

$e^{xy} = e^x e^y$

$\ln(xy) = \ln x + \ln y$

$\ln\left(\frac{x}{y}\right) = \ln x - \ln y$

$x = 10^y \Leftrightarrow \log x = y$

$e^0 = 1$

$\ln(1) = 0$

$10^0 = 1$

$\log(1) = 0$

$\log(10) = 1$